The Encapsulation Phenomenon

Yan Voloshin • Irina Belaya
Roland Krämer

The Encapsulation Phenomenon

Synthesis, Reactivity and Applications of Caged Ions and Molecules

Springer

Yan Voloshin
Nesmeyanov Institute of Organoelement
Compounds, Russian Academy
of Sciences and Gubkin Russian State
University of Oil and Gas
Moscow
Russia

Irina Belaya
Nesmeyanov Institute of Organoelement
Compounds, Russian Academy
of Sciences
Moscow
Russia

Roland Krämer
Anorganisch Chemisches Institut
Ruprecht-Karls-Univers. Heidelberg
Heidelberg
Germany

ISBN 978-3-319-27737-0 ISBN 978-3-319-27738-7 (eBook)
DOI 10.1007/978-3-319-27738-7

Library of Congress Control Number: 2016938266

Printed on acid-free paper

This Springer imprint is published by Springer Nature
The registered company is Springer International Publishing AG Switzerland

Y.Z.V. dedicates this book to his beloved son Sasha and daughter Lisa

Preface

Nowadays, a separate realm of modern chemistry is populated by compounds with encapsulated species. Those may be organic and inorganic, anions, cations or neutral molecules that are almost completely isolated from external factors (other ligands, solvents, etc.) by 3D caging (encapsulating) ligands to give molecular and polymeric capsules. The caged species have unique physicochemical properties responsible for specific reactivity and selectivity of their binding and release. The resulting cage complexes may be used for recognition, separation, carrying, storage, and detection of various encapsulated species in the field of molecular and supramolecular devices and modern materials, such as catalysts, carriers of therapeutically active compounds, and many others.

Depending on the nature of chemical or supramolecular bonds between precursors (ligand syntones, some of them are shown in *italic*) of the ligands (shown in **bold**), the cage complexes can be divided into three main classes: covalent, supramolecular, and coordination capsules.

Covalent capsules (Chap. 2) include imine cages and their reduced derivatives; amine, amide, and pyrrole encapsulating ligands; cryptophanes and cryptands; and arene- and cavitand-based capsules. These subclasses reflect the nature of their caging ligands. The pioneering works of J.-M. Lehn, D.J. Cram, and J. Nelson should be emphasized; further development in this field has been achieved by M. Mastalerz, A.P. Davis, A.I. Cooper, V. Amendola, J.-P. Dutasta, F. Huang, K. Severin, and their coworkers.

Supramolecular capsules (Chap. 3) can be divided into two main subclasses: arene- and cavitand-based assemblies. J. Rebek Jr. and D.N. Reinhoudt are pioneers in the field of these "soft" encapsulating ligands and their cage complexes. General experimental and theoretical (such as so-called "Rebek's rule") approaches to the design, synthesis, and study of supramolecular capsule have been also introduced by J. Rebek Jr., with a valuable input from B.C. Gibb, D. Ajami, J. L. Atwood, D. Rudkevich, and A. Shivanyuk.

Coordination capsules (Chap. 4), formed by coordination-driven self-assembly of complimentary organic syntones containing donor groups with appropriate metal ions, can be divided into three main subclasses depending on the nature of their cage frameworks (4n- and 6n-capped capsules) and coordination modes (bridging or cross-linking) of the metal ions. Pioneering efforts in synthetic and physical chemistry of these capsules have been made by M. Fujita and K.N. Raymond. Several new types of coordination capsules have been also designed and prepared by J.R. Nitschke, R.W. Saalfrank,

M.J. Hardie, M.D. Ward, K. Severin, G.H. Clever, P.D. Beer, M. Shionoya, and their coworkers.

Continuous progress in the chemistry of these classes of compounds with encapsulated species has been summarized in several reviews by J. Rebek Jr. and coworkers [1–17] and, very recently, by M. Yoshizawa [18]. The coordination capsules have been nicely covered by M. Fujita [19–25], K.N. Raymond [26–31], and others [32–40] and, more recently, by J.R. Nitschke [41–45], G.H. Clever [46], and M. Shionoya [47]. In 2002, we published the book *Clathrochelates: Synthesis, Structure and Properties* [48] that summarized general concepts and features of the complexes with an encapsulated metal ion.

Main developments in the design and chemistry of covalent capsules and their practical applications as they appeared to date are reported in many reviews [21, 34, 35, 48–65] and books [66–69] by J.-M. Lehn, K. Bowman-James, V. Amendola, A.P. Davis, and others in 1988–2013 and, very recently, by M. Mastalerz [70].

The aim of this book is to generalize the main progress in the design and synthesis of capsules with various caged species (some of them also shown in *italic*) that are important for their classification, reactivity, and practical applications recognized in 2014.

The authors would like to thank Dr. Ekaterina Lebed and Dr. Yulia Nelubina (INEOS RAS) for their valuable contribution in writing the book. We are also much indebted to Prof. M. Fujita for fruitful suggestions and comments.

Moscow, Russia — Yan Z. Voloshin
Moscow, Russia — Irina G. Belaya
Heidelberg, Germany — Roland Krämer

References

1. Rebek J Jr (1996) Assembly and encapsulation with self-complementary molecules. Chem Soc Rev 25(4):255–264
2. Rebek J Jr (1996) Molecular recognition and assembly. Acta Chem Scand 50(8):707–716
3. Rebek J Jr (1996) Molecular assembly and encapsulation. Pure Appl Chem 68(6):1261–1266
4. Conn MM, Rebek J Jr (1997) Self-assembling capsules. Chem Rev 97(5):1647–1668
5. Rebek J Jr (2000) Host-guest chemistry of calixarene capsules. Chem Commun 8:637–643
6. Hof F, Craig SL, Nuckolls C, Rebek J Jr (2002) Molecular encapsulation. Angew Chem Int Ed 41:1488–1508
7. Palmer LC, Rebek J Jr (2004) The ins and outs of molecular encapsulation. Org Biomol Chem 2(21):3051–3059
8. Rebek J Jr (2004) Some got away, but others didn't. J Org Chem 69(8):2651–2660
9. Rebek J Jr (2005) Simultaneous encapsulation: molecules held at close range. Angew Chem Int Ed 44(14):2068–2078
10. Schramm MP, Rebek J Jr (2006) Moving targets: recognition of alkyl groups. Chem Eur J 12(23):5924–5933
11. Biros SM, Rebek J Jr (2007) Structure and binding properties of water-soluble cavitands and capsules. Chem Soc Rev 36(1):93–104
12. Rebek J Jr (2007) Contortions of encapsulated alkyl groups. Chem Commun 27:2777–2789

13. Rebek J (2009) Molecular behavior in small spaces. Acc Chem Res 42(10): 1660–1668
14. Berryman OB, Dube H, Rebek J Jr (2011) Photophysics applied to cavitands and capsules. Isr J Chem 51(7):700–709
15. Durola F, Dube H, Ajami D, Rebek J Jr (2011) Control of nanospaces with molecular devices. Supramol Chem 23(1 & 2):37–41
16. Avram L, Cohen Y, Rebek J Jr (2011) Recent advances in hydrogen-bonded hexameric encapsulation complexes. Chem Commun 47(19):5368–5375
17. Ajami D, Rebek J Jr (2013) More chemistry in small spaces. Acc Chem Res 46(4): 990–999
18. Yoshizawa M, Klosterman JK (2014) Molecular architectures of multi-anthracene assemblies. Chem Soc Rev 43(6):1885–1898
19. Fujita M (1998) Metal-directed self-assembly of two- and three-dimensional synthetic receptors. Chem Soc Rev 27:417–425
20. Maurizot V, Yoshizawa M, Kawano M, Fujita M (2006) Control of molecular interactions by the hollow of coordination cages. Dalton Trans 2750–2756
21. Fujita N, Shinkai S, James TD (2008) Boronic acids in molecular self-assembly. Chem Asian J 3:1076–1091
22. Yoshizawa M, Fujita M (2005) Self-assembled coordination cage as a molecular flask. Pure Appl Chem 77:1107–1112
23. Yoshizawa M, Klosterman JK, Fujita M (2009) Functional molecular flasks: new properties and reactions within discrete, self-assembled hosts. Angew Chem Int Ed 48:3418–3438
24. Fujita M, Umemoto K, Yoshizawa M, Fujita N, Kusukawa T, Biradha K (2001) Molecular paneling via coordination. Chem Commun 6:509–518
25. Yoshizawa M, Fujita M (2010) Development of unique chemical phenomena within nanometer-sized, self-assembled coordination hosts. Bull Chem Soc Jpn 83(6): 609–618
26. Caulder DL, Raymond KN (1999) Supermolecules by design. Acc Chem Res 32:975–982
27. Caulder DL, Raymond KN (1999) The rational design of high symmetry coordination clusters. J Chem Soc Dalton Trans 8:1185–1200
28. Johnson DW, Raymond KN (2001) The role of guest molecules in the self-assembly of metal-ligand clusters. Supramol Chem 13(6):639–659
29. Yeh RM, Davis AV, Raymond KN (2003) Supramolecular systems: self-assembly. In: Fujita M, Powell A, Creutz A (eds) Comprehensive coordination chemistry II, vol 7. Elsevier, Amsterdam, pp 327–355
30. Fiedler D, Leung DH, Bergman RG, Raymond KN (2005) Selective molecular recognition, C–H bond activation, and catalysis in nanoscale reaction vessels. Acc Chem Res 38:349–358
31. Pluth MD, Raymond KN (2007) Reversible guest exchange mechanisms in supramolecular host-guest assemblies. Chem Soc Rev 36(2):161–171
32. Amijs CHM, Van Klink GPM, Van Koten G (2006) Metallasupramolecular architectures, an overview of functional properties and applications. Dalton Trans 308–327
33. Cookson J, Beer PD (2007) Exploiting the dithiocarbamate ligand in metal-directed self-assembly. Dalton Trans 1459–1472
34. Koblenz TS, Wassenaar J, Reek JNH (2008) Reactivity within a confined self-assembled nanospace. Chem Soc Rev 37:247–262
35. Ballester P (2010) Anion binding in covalent and self-assembled molecular capsules. Chem Soc Rev 39:3810–3830
36. Tranchemontagne DJ, Ni Z, O'Keeffe M, Yaghi OM (2008) Reticular chemistry of metal-organic polyhedral. Angew Chem Int Ed 47:5136–5147
37. Severin K (2006) Supramolecular chemistry with organometallic half-sandwich complexes. Chem Commun 3859–3867
38. Han Y-F, Jia W-G, Yu W-B, Jin G-X (2009) Stepwise formation of organometallic macrocycles, prisms and boxes from Ir, Rh and Ru-based half-sandwich units. Chem Soc Rev 38:3419–3434
39. Ward MD (2009) Polynuclear coordination cages. Chem Commun 4487–4499
40. Seidel SR, Stang PJ (2002) High-symmetry coordination cages via self-assembly. Acc Chem Res 35:972–983

41. Breiner B, Clegg JK, Nitschke JR (2011) Reactivity modulation in container molecules. Chem Sci 2:51–56
42. Takezawa H, Murase T, Fujita M (2012) Twisting of a tetrasubstituted olefin via inclusion in a Td-symmetric cage. Paper presented at the 40 international conference on coordination chemistry, Valencia, 9–13 Sept 2012
43. Smulders MMJ, Riddell IA, Browne C, Nitschke JR (2013) Building on architectural principles for three-dimensional metallosupramolecular construction. Chem Soc Rev 42(4):1728–1754
44. Castilla AM, Ramsay WJ, Nitschke JR (2014) Stereochemical communication within tetrahedral capsules. Chem Lett 43(3):256–263
45. Nitschke JR (2014) Supramolecular and dynamic covalent reactivity. Chem Soc Rev 43(6):1798–1799
46. Han M, Engelhard DM, Clever GH (2014) Self-assembled coordination cages based on banana-shaped ligands. Chem Soc Rev 43(6):1848–1860
47. Tashiro S, Shionoya M (2014) Cavity-assembled porous solids (CAPSs) for nanospace-specific functions. Bull Chem Soc Jpn 87(6):643–654
48. Voloshin YZ, Kostromina NA, Krämer R (2002) Clathrochelates: synthesis, structure and properties. Elsevier, Amsterdam
49. Meyer CD, Joiner CS, Stoddart JF (2007) Template-directed synthesis employing reversible imine bond formation. Chem Soc Rev 36:1705–1723
50. Bernhardt PV, Moore EG (2003) Functionalized macrocyclic compounds: potential sensors of small molecules and ions. Aust J Chem 56:239–258
51. Lehn J-M (1988) Supramolecular chemistry – scope and perspectives molecules, supermolecules, and molecular devices (Nobel Lecture). Angew Chem Int Ed 27:90–112
52. Wichmann K, Antonioli B, Söhnel T, Wenzel M, Gloe K, Gloe K, Price JR, Lindoy LF, Blake AJ, Schröder M (2006) Polyamine-based anion receptors: extraction and structural studies. Coord Chem Rev 250:2987–3003
53. Gimeno N, Vilar R (2006) Anions as templates in coordination and supramolecular chemistry, novel triptycene-derived hosts: synthesis and their applications in supramolecular chemistry. Coord Chem Rev 250:3161–3189
54. Chen C-T (2011) Novel triptycene-derived hosts: synthesis and their applications in supramolecular chemistry. Chem Commun 47:1674–1688
55. Zhang M, Zhang M, Zhu K, Huang F (2010) Improved complexation of paraquat derivatives by the formation of crown ether-based cryptands. Chem Commun 46: 8131–8141
56. Mastalerz M (2010) Shape-persistent organic cage compounds by dynamic covalent bond formation. Angew Chem Int Ed 49:5042–5053
57. Davis AP (2009) Synthetic lectins. Org Biomol Chem 7:3629–3638
58. Rue NM, Sun J, Warmuth R (2011) Polyimine container molecules and nanocapsules. Isr J Chem 51(7):743–768
59. Holst JR, Trewin A, Cooper AI (2010) Porous organic molecules. Nat Chem 2:915–920
60. Mastalerz M (2012) Permanent porous materials from discrete organic molecules – towards ultra-high surface areas. Chem Eur J 18(33):10082–10091
61. Mateus P, Delgado R, Brandão P, Carvalho S, Félix V (2009) Selective recognition of tetrahedral dianions by a hexaaza cryptand receptor. Org Biomol Chem 7:4661–4673
62. Bull SD, Davidson MG, Van den Elsen JMH, Fossey JS, Jenkins ATA, Jiang Y-B, Kubo Y, Marken F, Sakurai K, Zhao J, James TD (2013) Exploiting the reversible covalent bonding of boronic acids: recognition, sensing, and assembly. Acc Chem Res 46:312–326
63. Takemura H, Shinmyozu F, Inazu T (1996) Nitrogen-bridged macrocycles: synthesis, structures and inclusion phenomena. Coord Chem Rev 156:183–200
64. Kang SO, Llinares JM, Day VW, Bowman-James K (2010) Cryptand-like anion receptors. Chem Soc Rev 39:3980–4003
65. Amendola V, Bonizzoni M, Esteban-Gómez D, Fabbrizzi L, Licchelli M, Sancenón F, Taglietti A (2006) Some guidelines for the design of anion receptors. Coord Chem Rev 250:1451–1470
66. Gokel GW (1991) Crown ethers and cryptands. Royal Soc Chem, Cambridge

67. Lehn J-M (1995) Supramolecular chemistry: concepts and perspectives. Wiley, Weinheim
68. Gerbeleu NV, Arion VB, Burgess FJ (2000) Template synthesis of macrocyclic compounds. Wiley-VCH, Weinheim
69. Diederich F, Stang PJ, Tykwinski RR (2008) Modern supramolecular chemistry: strategies for macrocycle synthesis. Wiley-VCH, Weinheim
70. Zhang G, Mastalerz M (2014) Organic cage compounds – from shape-persistency to function. Chem Soc Rev 43(6):1934–1947

Contents

Main Abbreviations

acac	Acetylacetone
ADH	Alcohol dehydrogenase
ADMA	4-[3-(9-Anthryl)propyl]-N,N-dimethylaniline
Bn	Benzyl radical
bpy	2,2′-Bipyridine
bpz	2,2′-Bipyrazol
Bu	*n*-Butyl radical
Bz	Benzene sulfonate radical
CD	Circular dichroism
cod	Cyclooctadiene
Cp	Cyclopentadiene
Cp*	Pentamethyl cyclopentadiene
CT	Charge transfer
CTB	Charge transfer band
DBU	1,8-Diazabicyclo[5.4.0]undec-7-ene
DLS	Dynamic light scattering
DMA	Dimethylacetamide
DMF	Dimethylformamide
DMSO	Dimethyl sulfoxide
ECD	Electronic circular dichroism
EIE	Equilibrium isotope effect
en	Ethylenediamine
ESP	Electron spin polarization
ET	Electron transfer
Fc	Ferrocenyl radical
FRET	Fluorescence resonance energy transfer
GFC	Gel filtration chromatography
Hal	Halogen
Im	Imidazole
IEC	Ion-exchange chromatography
L	Ligand
LIESST	Light-induced excited spin state trapping
M	Metal ion
MD	Molecular dynamics
Me	Methyl radical
MM	Molecular modeling
Ms	Mesityl radical

NBS	*N*-bromosuccinimide
NMP	*N*-methylpyrrolidone
PEG	Polyethylene glycol
PET	Photoinduced electron transfer
PGSE	Pulse gradient spin-echo
Py	Pyridine
RTP	Rotating tube processor
r_i	Physical (Shannon) ionic radius
SCO	Spin-crossover
SDP	Spinning disk processing
TAP	Trigonal antiprism
tempo	2,2,6,6-Tetramethylpiperidine-1-oxyl
THF	Tetrahydrofuran
Tol	Tolyl radical
TP	Trigonal prism
tren	Tris(2-aminoethylamine)
trpn Ts	Tris(3-aminopropylamine) Tosyl radical
VCD	Vibrational circular dichroism

1 General Considerations

1.1 Scope and Goal

Before going into any details, we should identify molecular and supramolecular compounds that are considered cage complexes with encapsulated species (organic or inorganic anions, cations, and neutral molecules). The main criteria for the formation of such a complex are (*i*) 3D cavity in a caging ligand (capsule, cage frameworwk) and (*ii*) inclusion (encapsulation) of chemical species within this cavity via coordination (donor–acceptor) bonds or supramolecular interactions (hydrogen bonds, electrostatic (Coulombic) and dipole–dipole interactions, etc).

Typically, caging ligands contain at least three organic or metal-organic cyclic or pseudocyclic (including macrocyclic and pseudomacrocyclic) fragments formed by covalent, coordination, or supramolecular bonds and cross-linking (apical, capping, vertex) atoms, chemical groups, or metal ions belonging to at least three of these cyclic or pseudocyclic entities. Such caging ligands almost completely isolate encapsulated species from external factors (e.g., solvent effects, side reactions), resulting in unusual or even unprecedented chemical and physicochemical properties. This size- and shape-discriminating encapsulation allows stabilizing unusual oxidation and spin states of an encapsulated metal ion [1], highly reactive organic and inorganic guests [2–4] (including catalytically active intermediates and organic radicals), conformational and structural [5] isomers of caged organic guests (including their tautomeric forms [6]), etc. Cationic and anionic encapsulating (caging) ligands also demonstrate charge-discriminating properties by binding mostly oppositely charged guests or neutral molecules. Differences in the guest selectivity of cage complexes arise from their topology (spatial arrangement of donor groups and the shape of inner cavity) and rigidity of an encapsulating ligand [7]. Severe steric restrictions within the cavity determine high selectivity of the encapsulation; those, however, can be tuned. Indeed, depending on the size of the facial ligand syntones, not only the cavity size but also the port size of the cage framework could be controlled. This offers potential possibilities for developing suitable nanosized encapsulating hosts that can accommodate suitable small molecular guests or chiral catalysts [8]. If a cage molecule contains large pores, guest species in it can be quickly exchanged between the interior and the outside [9]. Such a structure could be advantageous if some reactive intermediates of a chemical reaction can be stabilized in the interior, while the final products are released quickly. This would be an entry into catalysis, with a caging ligand acting as the catalyst [9]. A general and powerful method [10] for determining the structure of highly reactive, short-lived species that can be hardly observed by conventional methods uses encapsulation of the precursor by a caging ligand followed by in situ generation of the target

Y. Voloshin et al., *The Encapsulation Phenomenon: Synthesis, Reactivity and Applications of Caged Ions and Molecules*, DOI 10.1007/978-3-319-27738-7_1

species in a crystal. Moreover, a given caging ligand can have an ability to bind a given guest with high affinity and remarkable selectively, not through simple encapsulation of a reactive precursor of the guest but by transforming it into an optimal guest that is not observed in the host's absence [11].

The modern approach uses designed reactive caging ligands as a special class of hosts: stimulus-controllable encapsulation by them allows obtaining delivery vehicles in material science [12]. These so-called molecular containers display an impressive ability to selectively encapsulate guests of appropriate size, shape, and chemical identity within their rigidly shaped, sized, and functionalized cavities [13]. Typically, their caging ligands are formed by coordination or hydrogen bonds, or dynamic covalent functionality [12]; the selective encapsulation of guests with the internal cavity enables developing of a dynamically controllable nanocapture of important guest molecules. The capsules can entrap a small neutral molecule in the inner space, but the guest is released when the cage structure is adopted; so they are prospective for molecular transport systems [14], including drug delivery systems in physiological media [15]. In particular, coordination capsules offer a strategy for delivering drugs to cancer cells to avoid additional side effects [16]. Such encapsulation has been proposed in [17] for keeping reactive molecules away from others until an appropriate signal opens a caging ligand [3] or for keeping toxic drug molecules away from sensitive tissues pending their delivery [17]. The tight, reversible, and selective encapsulation of a suitable guest by a caging ligand allows delivering a drug to the vicinity of a target area where "opener" molecules would be selectively placed [18]. The capsules are described in [19] to fit with externally directed groups permitting solubility in a given solvent system or possibly facilitating recognition and binding to a given target in a biological system. The caged guest can then diffuse out, either in response to a specific signal such as a competitive guest or simply as a part of its equilibration process. More complex derivatives of these cage compounds can serve as subsystems within larger systems that are able to provide complex responses to applied stimuli [19]. By using endohedral functionalization with functionalizing organic substituents of different nature for "endohedral molecular coating," well-defined nanospace surrounded by various types of the interior surfaces can be provided [20]. Such capsules can be regarded as structural biomimetics of functional biological giant hollow structures such as a family of spherical viruses that effectively bind substrates within their inner functionalized cavities [20]. In contrast to other molecular containers, the caging nanotubes reported in [21] have unique geometrical features and a nanoscale size. Those allow for the simultaneous entrapment of multiple guests in a 1D fashion, a novel phenomenon in molecular encapsulation [21].

Entrapment of guest molecules in self-assembled caging ligands is reported in [22] to be possible under ambient conditions through a controlled "disassembly–reassembly" process, facilitated by using SDP. This type of application of SDP with its practical convenience has the potential as a versatile method for assessing and controlling interactions at the molecular and macromolecular level; it can facilitate a guest–host exchange within nanosized containers and potentially replace RNA from capsids. The continuous flow technology can be used to load molecular cargos within cage containers on a large scale [22]. These molecular containers have the potential as drug delivery systems and as affordable biological cell models. Moreover, molecular encapsulation may provide valuable information for a better understanding of enzymes or other complex biological systems [23]. As it has been pointed in [24], extensive studies have been made on the formation of peptide secondary structures from de novo designed oligopeptides. Unlike those incorporated in a protein scaffold, short peptide fragments cannot form stable secondary structures, and the folding of secondary structures of peptides has been achieved by peptide encapsulation within a hydrophobic cavity. This strategy is reminiscent of a natural system, as the secondary structures are sustained only by weak interactions within the cavity. However, oligopeptide encapsulation by synthetic hosts is a difficult task, as most of the artificial hosts have a relatively small cavity. The most efficient way to provide cavities large enough to bind such peptides is to construct caging ligands with extraordinarily large cavities [24].

Easy access to nanosized complex-in-complex cage compounds {so-called Russian-doll-type (*Matreshka*) assemblies} with encapsulated inorganic cations, metal complexes, small hydrocarbons, and biologically active compounds (such as choline and cisplatin) is reported in [25] to demonstrate the potential for molecular encapsulation by providing a unique environment for investigations of molecular interactions. Specific conformation and orientation of guests within the cavity of the caging ligand allowed probing their geometry and dynamics through host–guest interactions [26]. Encapsulation of H-acidic and basic guests dramatically decreases the rate of proton transfer [23], and a designed caging ligand is reported in [27] to behave as a proton sponge. The suitable coordination caging ligands also allowed for both AND and OR bimolecular recognition of guests [28].

Formation of cage complexes is used in the fabrication of advanced functional supramolecular systems, such as molecular switches and molecular machines [29]; some of them can be so-called molecular gyroscopes [30]. The cage molecules can exhibit a reversible change in chirality controlled by a solvent. This establishes the feasibility of a new mode of stimulation for such molecular switches; the ratio of the rates of the clockwise and anticlockwise tilting motions of a cage molecule showed that the solvent governs the rotational motion of an encapsulating ligand [31]. The prospective advantageous use of the cage compounds is also within PET reactions [32]; their discrete molecules can undergo physically addressable electronic spin switching occurring by thermal, light, or solvent perturbation [33]. Formation of the cage complex is reported in [34] to exert strong effects on the kinetics of heterogeneous ET reactions, which can be used to effectively destroy or disrupt these complexes. Those with encapsulated stacked aromatic molecules are reported in [35] to be conductive and suitable for ET, whereas the initial empty framework is not. The excited-state chemistry and physics of guest molecules are distinctly different from those in organic solvents [36]; for example, the discrete stacking between encapsulated large aromatic guest molecules within the cavities of the caging ligands is reported in [37] to produce unique photochemical and electrochemical properties that are not common for isolated or infinitely stacked π-systems.

Capsules that are rigid, kinetically and thermodynamically stable, and metrically well defined are useful as building blocks in nanotechnological applications [38]. They and their cage complexes are suitable for the design of mechanically interlocked structures, such as rotaxanes and catenanes. For example, they can envisage polymerization of diacetylene components to generate unique π-conjugated polymer materials and mechanically interlocked polymers with π-conjugated backbones either by direct use of diacetylene-containing caging ligands or after the formation of interlocked structures [39].

Paramagnetic coordination capsules and their cage complexes have the potential for fine-tuning of magnetic interactions in response to their geometry, the nature of an encapsulating ligand, and the choice of a paramagnetic cation [40]; they are also prospective for creation of new types of single-molecule magnets [41]. The efficient noncovalent host–guest magnetic interactions through the encapsulation of open-shell metal complexes in a radical caging ligand are reported in [42]: its TP cavity ideally suited for encapsulation and control of geometry and spin interactions of planar guests.

Cage complexes and their encapsulating ligands are practically useful for separation, purification, and reaction control in organic synthesis [43]. The coordination capsules are described in [44] to be readily available and their cavities extraordinarily large. This makes it possible to create new chemical reactivity within the localized microspace of discrete molecules. Such reactions can proceed within a cavity quantitatively with remarkable rate acceleration and perfect regio- and stereocontrol [44]. The use of caging hosts (including those with a chiral interior) as molecular flasks allows for self-discrimination and preorganization of suitable guest molecules, reactants for given chemical reactions within their inner cavities. These reactions are controlled only by spatial restrictions governed by the caging ligand, being one of the most important essences of enzyme reactions [45]. The control over relative orientation of substrates by the cavity makes a specific reaction pathway allowed, while the others are forbidden, thus mimicking reaction control normally exhibited by enzymes [46]. High regio- and stereoselectivities attained in these reactions are explained in [47] as a consequence of precise preorganization of the encapsulated substrates. Upon encapsulation, the guests are isolated from the bulk solution and set in a restricted chemical environment that can reduce the activation energy of reactions and thus control the reaction pathways. In particular, polymerization within a well-defined nanocavity results in fine control over the rate of polymerization and the molecular weight of the resultant polymer as well [48]. Moreover, the enantiomerically defined capsules provide useful chiral spaces for enantioselective catalysis, and the differential molecular recognition of enantiomers and diastereomerically defined capsules ensures discrimination and transformation of diastereomeric guests [49].

In the case of photochemical reactions, encapsulation by a caging ligand allows control of their selectivity and enhances reactivity during direct excitation owing to the ability of such ligands to localize, preorient (even in the solid state), and restrict the mobility of guest reactants in the short excited singlet state lifetime; the ability of photoreactions to be conducted in water is an extra bonus [50]. Confinement within the cavity of a caging ligand is described in [51] to suppress photochemical pathways that are normally favored in solution; the pathway is modulated as a function of shape complementarity between the potential guest and the cavity of the encapsulating host. Variation in the photochemistry of homological guests within the cavity of a caging ligand allowed [52] to highlight the importance of small variations in supramolecular structure for the selectivity within a confined nanoscale reactor. This allows control of the excited behavior of guests, all of which behave in an identical manner in solution. It is a common knowledge that enzyme active sites can impart specificity on the bound substrate. The carefully designed synthetic hosts can do likewise in terms of both their ability to interact with the guest and how templation can influence the reaction course [52]. Photoexcitation of a caging ligand accommodating photochemically inert guests is reported in [53] to allow for regioselective oxidation of a guest within the cavity; fluorescent caging ligands containing a photoactive fragment are potentially useful as a new generation of catalysts for hydrogen production from water [54].

Thus, these capsules and their cage complexes provide wide opportunities of transferring solution host–guest chemistry into the solid state using new encapsulation phenomena [55]. They consist of closed hollow cage frameworks within which encapsulated guests are isolated from external factors, and in such environment otherwise reactive molecules can be stabilized [56]. Charge-, size-, and shape-selective guest encapsulation is observed, and variations in the host–guest properties can be explained by the differences in cavity shape and accessibility; cavity-directed chemical transformations represent one of the most important features in 3D host chemistry [45]. Another intriguing possibility is stabilization of labile molecules by encapsulation: in most cases, the encapsulated reagents are stable, mild, and size–shape selective, and their reactivity as well as the reaction selectivity can be controlled by the hosting cavity [57].

1.2 Caging Ligands and Cage Complexes: Main Classes and Subclasses, Types and Subtypes

The classification of caging (encapsulating) ligands (capsules) used here is based on the nature of chemical bonds between the ligand syntones and the nature of these syntones belonging to different classes of organic and coordination compounds and supramolecular systems as well.

Pre-synthesized organic caging ligands and their cage complexes belong to the first class of these compounds. Among them, the following types of encapsulating ligands are described: (*i*) Schiff-base compounds obtained by imine condensation, their polyamine derivatives, and the corresponding metal complexes with a vacant cavity for intraligand metal-assistant binding of a guest molecule; (*ii*) amine, amide, and pyrrole encapsulating ligands and their complex derivatives; (*iii*) macrobicyclic and macropolycyclic cryptophanes and cryptands and their cage compounds; and (*iv*) arene- and cavitand-based caging ligands and their complexes. In comparison to coordination capsules, purely organic cages are much rarer, probably due to the fact that most covalent bonds are formed by irreversible reactions that do not allow for a structural "self-correction" often accompanied by low overall yields in a cyclization step [58]. Use of reversible formations of covalent bonds leads to organic cage compounds obtained from small and readily accessible precursors in reasonable to very high yields.

The second class of compounds with encapsulated species includes capsules and their cage complexes formed by supramolecular (mainly hydrogen-bonded) self-assembly from (*i*) arene ligand syntones, (*ii*) cavitand building blocks, and (*iii*) cyclotriveratrylene platforms. Such supramolecular capsules are described in [59] as nanoscale structures made from multiple synthetic subunits held together by weak intermolecular forces and acting as hosts that can completely surround small guests of the appropriate size, shape, and chemical nature. In most cases, the subunits of these caging ligands are identical and form homomeric assemblies of high symmetry, but small variations in their structures are tolerated and lead to heteromeric assemblies with slightly different recognition properties. These ligands are dynamic, with lifetimes from milliseconds to hours, and allow performing direct spectroscopic observation of smaller molecules under ambient conditions at equilibrium in solution [59].

The third class includes capsules obtained by coordination-driven self-assembly through the formation of donor–acceptor bonds between the vertex or bridging metal ions cross-linking their ligand syntones with donor groups. The authors of [60] also described a ligand syntone that can play a role of blocking and bridging donor fragments. Two main approaches for the synthesis of these species have been developed. First of them uses a "blocking ligand" to protect coordination sites at the metal center, thus controlling the positions available for binding of bridging ligands and enabling the construction of the desired architecture, while the second strategy uses bridging chelating ligands to control the metal geometry. Diastereoselective formation of coordination cage complexes has been achieved in [61] using chiral bridging ligand syntones, thus resulting in substantial chiral amplification associated with adoption of a helical conformation by all these syntones upon binding. Vertex metal ions can be divided into two groups depending on their coordination polyhedra: (*i*) square planar and tetrahedral polyhedra of tetracoordinate metal ions (so-called 4n end-capping metal ions) and (*ii*) pseudooctahedral (TP, TAP, etc.) polyhedra of hexacoordinate metal ions (so-called 6n end-capping metal ions) as well as bridging and cross-linking tripodal and tetrapodal metal ions. Several coordination-inspired highly symmetrical capsules of trigonal pyramidal, square pyramidal, trigonal bipyramidal, TP, TAP, square prismatic, tetrahedral, cuboctahedral, and octahedral geometries have been reported to date [8]. Their ligand syntones can be divided depending on their binding modes (face-capping and edge-bridging syntones) that lead to different polyhedral structures. The combinations of these syntones can generate coordination capsules that are not available using one ligand's type only [62]. Self-assembly of highly symmetric capsules through coordination with a directional bond uses two different approaches. The first one is edge-directed self-assembly of the linear syntones linked at the corners, and the second one is face-directed self-assembly of the facial syntones linked either at the edges or at the corners [8–554]. Coordination self-assembly is reported in [63] to provide one of the most efficient methods for creating large cavities that are able to bind substrates in fixed orientations. The design of appropriate caging ligands is the key element in driving the self-assembly toward the formation of their target cage complexes. In the specific case of heteronuclear coordination capsules, it is crucial to differentiate the type, number, and position of the ligand syntones, to reach the correct metal-to-ligand cross-reactivity [64].

Each of the groups within these three classes can be divided into three subgroups depending on the nature of a guest caged molecule (i.e., free cages and neutral guests, anionic and cationic species).

References

1. Voloshin YZ, Varzatskii OA, Vorontsov II, Antipin MY (2005) Tuning a metal's oxidation state: the potential of clathrochelate systems. Angew Chem Int Ed 44:3400–3402
2. Riddell IA, Smulders MMJ, Clegg JK, Nitschke JR (2011) Encapsulation, storage and controlled release of sulfur hexafluoride from a metal–organic capsule. Chem Commun 47:457–459
3. Mal P, Breiner B, Rissanen K, Nitschke JR (2009) White phosphorus is air-stable within a self-assembled tetrahedral capsule. Science 324:1697–1699
4. Raymond KN (2009) Supramolecular chemistry: phosphorus caged. Nature 460(7255):585–586
5. Liu S, Gan H, Hermann AT, Gibb BC, Rick SW (2010) Kinetic resolution of constitutional isomers controlled by selective protection inside a supramolecular nanocapsule. Nat Chem 2:847–852
6. Kuberski B, Szumna A (2009) A self-assembled chiral capsule with polar interior. Chem Commun 1959–1961
7. Takemura H, Kon N, Yasutake M, Shinmyozu T (2003) Structurally isomeric two pyridino macrocycles: complexation and structures. Tetrahedron 59: 427–431
8. Moon D, Kang S, Park J, Lee K, John RP, Won H, Seong GH, Kim YS, Kim GH, Rhee H, Lah MS (2006) Face-driven corner-linked octahedral nanocages: M6L8 cages formed by C3-symmetric triangular facial ligands linked via C4-symmetric square tetratopic PdII ions at truncated octahedron corners. J Am Chem Soc 128:3530–3531
9. Albrecht M, Janser I, Burk S, Weis P (2006) Self-assembly and host–guest chemistry of big metallosupramolecular M_4L_4 tetrahedra. Dalton Trans 2875–2880
10. Kawano M, Kobayashi Y, Ozeki T, Fujita M (2006) Direct crystallographic observation of a coordinatively unsaturated transition-metal complex in situ generated within a self-assembled cage. J Am Chem Soc 128:6558–6559
11. Meng W, Clegg JK, Nitschke JR (2012) Transformative binding and release of gold guests from a self-assembled Cu_8L_4 tube. Angew Chem Int Ed 51: 1881–1884
12. Kataoka K, Okuyama S, Minami T, James TD, Kubo Y (2009) Amine-triggered molecular capsules using dynamic boronate esterification. Chem Commun 1682–1684
13. Fairchild RM, Holman KT (2005) Selective anion encapsulation by a metalated cryptophane with a π-acidic interior. J Am Chem Soc 127:16364–16365
14. Huang Y, Yan Z (2005) Probing the local environments of phosphorus in aluminophosphate-based mesostructured lamellar materials by solid-state NMR spectroscopy. Angew Chem Int Ed 44: 2727–2731
15. Corbellini F, Knegtel RMA, Grootenhuis PDJ, Crego-Calama M, Reinhoudt DN (2005) Water-soluble molecular capsules: self-assembly and binding properties. Chem Eur J 11:298–307
16. Mattsson J, Zava O, Renfrew AK, Sei Y, Yamaguchi K, Dyson PJ, Therrien B (2010) Drug delivery of lipophilic pyrenyl derivatives by encapsulation in a water soluble metalla-cage. Dalton Trans 39:8 248–8255
17. Meng W, Breiner B, Rissanen K, Thoburn JD, Clegg JK, Nitschke JR (2011) A self-assembled M_8L_6 cubic cage that selectively encapsulates large aromatic guests. Angew Chem Int Ed 50:3479–3483
18. Mal P, Schultz D, Beyeh K, Rissanen K, Nitschke JR (2008) An unlockable–relockable iron cage by subcomponent self-assembly. Angew Chem Int Ed 47:8297–8301
19. Hristova YR, Smulders MMJ, Clegg JK, Breiner B, Nitschke JR (2011) Selective anion binding by a "chameleon" capsule with a dynamically reconfigurable exterior. Chem Sci 2:638–641
20. Tominaga M, Suzuki K, Murase T, Fujita M (2005) 24-fold endohedral functionalization of a self-assembled $M_{12}L_{24}$ coordination nanoball. J Am Chem Soc 127(34):11950–11951
21. Organo VG, Sgarlata V, Firouzbakht F, Rudkevich DM (2007) Long synthetic nanotubes from calix[4] arenes. Chem Eur J 13:4014–4023
22. Swaminathan IK, Norret M, Dalgarno SJ, Atwood JL, Raston CL (2008) Loading molecular hydrogen cargo within viruslike nanocontainers. Angew Chem Int Ed 47:6362–6366
23. Dimitrov-Raytchev P, Martinez A, Gornitzka H, Dutasta J-P (2011) Encaging the verkade's superbases: thermodynamic and kinetic consequences. J Am Chem Soc 133:2157–2159
24. Tashiro S, Tominaga M, Yamaguchi Y, Kato K, Fujita M (2006) Peptide recognition: encapsulation and α-helical folding of a nine-residue peptide within a hydrophobic dimeric capsule of a bowl-shaped host. Chem Eur J 12:3211–3217

25. Shivanyuk A (2007) Nanoencapsulation of calix[4] arene inclusion complexes. J Am Chem Soc 129: 14196–14199
26. Kusukawa T, Yoshizawa M, Fujita M (2001) Probing guest geometry and dynamics through host–guest interactions. Angew Chem Int Ed 40:1879–1984
27. Bazzicalupi C, Bellusci A, Bencini A, Berni E, Bianchi A, Ciattini S, Giorgi C, Valtancoli B (2002) A new dipyridine-containing cryptand for both proton and Cu(II) encapsulation. A solution and solid state study. J Chem Soc Dalton Trans 2151–2157
28. Yoshizawa M, Tamura M, Fujita M (2004) AND/OR bimolecular recognition. J Am Chem Soc 126: 6846–6847
29. Zhang M, Zheng B, Xia B, Zhu K, Wu C, Huang F (2010) Synthesis of a bis(1,2,3-phenylene) cryptand and its dual-response binding to paraquat and diquat. Eur J Org Chem 2010:6804–6809
30. Clever GH, Tashiro S, Shionoya M (2009) Inclusion of anionic guests inside a molecular cage with palladium(II) centers as electrostatic anchors. Angew Chem Int Ed 48:7010–7012
31. Martinez A, Guy L, Dutasta J-P (2010) Reversible, solvent-induced chirality switch in atrane structure: control of the unidirectional motion of the molecular propeller. J Am Chem Soc 132:16733–16736
32. Podkoscielny D, Philip I, Gibb CLD, Gibb BC, Kaifer AE (2008) Encapsulation of ferrocene and peripheral electrostatic attachment of viologens to dimeric molecular capsules formed by an octaacid, deep-cavity cavitand. Chem Eur J 14:4704–4710
33. Duriska MB, Neville SM, Moubaraki B, Cashion JD, Halder GJ, Chapman KW, Balde C, Létard J-F, Murray KS, Kepert CJ, Batten SR (2009) A nanoscale molecular switch triggered by thermal, light, and guest perturbation. Angew Chem Int Ed 48: 2549–2552
34. Qiu Y, Yi S, Kaifer AE (2011) Encapsulation of tetrathiafulvalene inside a dimeric molecular capsule. Org Lett 13:1770–1773
35. Kiguchi M, Takahashi T, Takahashi Y, Yamauchi Y, Murase T, Fujita M, Tada T, Watanabe S (2011) Electron transport through single molecules comprising aromatic stacks enclosed in self-assembled cages. Angew Chem Int Ed 54:5708–5711
36. Yang H, McLaughlin CK, Aldaye FA, Hamblin GD, Rys AZ, Rouiller I, Sleiman HF (2009) Metal-nucleic acid cages. Nat Chem 1:390–396
37. Yoshizawa M, Nakagawa J, Kumazawa K, Nagao M, Kawano M, Ozeki T, Fujita M (2005) Discrete stacking of large aromatic molecules within organic-pillared coordination cages. Angew Chem Int Ed 44:1810–1813
38. Caulder DL, Brückner C, Powers RE, König S, Parac TN, Leary JA, Raymond KN (2001) Design, formation and properties of tetrahedral M_4L_4 and M_4L_6 supramolecular clusters. J Am Chem Soc 123:8923–8938
39. Xu Z, Huang X, Liang J, Zhang S, Zhou S, Chen M, Tang M, Jiang L (2010) Efficient syntheses of novel cryptands based on bis(m-phenylene)-26-crown-8 and their complexation with paraquat. Eur J Org Chem 2010:1904–1911
40. Escuer A, Harding CJ, Dussart Y, Nelson J, McKee V, Vicente R (1999) Constrained ferromagnetic coupling in dinuclear μ 1,3-azido nickel(II) cryptate compounds. Crystal structure and magnetic behaviour of $[Ni_2(L_1)(N_3)(H_2O)][CF_3SO_3]_3{\cdot}2H_2O{\cdot}EtOH$ $\{L_1 = N[(CH_2)_2NHCH_2(C_6H_4\text{-}m)CH_2NH(CH_2)_2]_3N\}$. J Chem Soc Dalton Trans 223–228
41. Lee IS, Long JR (2004) Synthesis and magnetic behavior of the tetrahedral cage complex $[(cyclen)_4V_4(CN)_6]^{6+}$. Dalton Trans 3434–3436
42. Ozaki Y, Kawano M, Fujita M (2009) Engineering noncovalent spin–spin interactions in an organic-pillared spin cage. Chem Commun 4245–4247
43. Sato S, Iida J, Suzuki K, Kawano M, Ozeki T, Fujita M (2006) Fluorous nanodroplets structurally confined in an organopalladium sphere. Science 313:1273–1276
44. Yoshizawa M, Takeyama Y, Kusukawa T, Fujita M (2002) Cavity-directed, highly stereoselective [2+2] photodimerization of olefins within self-assembled coordination cages. Angew Chem Int Ed 41:1347–1349
45. Yoshizawa M, Fujita M (2005) Self-assembled coordination cage as a molecular flask. Pure Appl Chem 77:1107–1112
46. Yamaguchi T, Fujita M (2008) Highly selective photomediated 1,4-radical addition to o-quinones controlled by a self-assembled cage. Angew Chem Int Ed 47:2067–2069
47. Horiuchi S, Nishioka Y, Murase T, Fujita M (2010) Both [2+2] and [2+4] additions of inert aromatics via identical ternary host–guest complexes. Chem Commun 46:3460–3462
48. Kikuchi T, Murase T, Sato S, Fujita M (2008) Polymerisation of an anionic monomer in a self-assembled $M_{12}L_{24}$ coordination sphere with cationic interior. Supramol Chem 20:81–94
49. Meng W, Clegg JK, Thoburn JD, Nitschke JR (2011) Controlling the transmission of stereochemical Information through space in terphenyl-edged Fe_4L_6 cages. J Am Chem Soc 133:13652–13660
50. Kaanumalle LS, Ramamurthy V (2007) Photodimerization of acenaphthylene within a nanocapsule: excited state lifetime dependent dimer selectivity. Chem Commun 1062–1064
51. Kaanumalle LS, Gibb CLD, Gibb BC, Ramamurthy V (2005) A hydrophobic nanocapsule controls the photophysics of aromatic molecules by suppressing their favored solution pathways. J Am Chem Soc 127:3674–3675
52. Gibb CLD, Sundaresan AK, Ramamurthy V, Gibb BC (2008) Templation of the excited-state chemistry of α-(n-alkyl) dibenzyl ketones: how guest packing within a nanoscale supramolecular capsule influences photochemistry. J Am Chem Soc 130:4069–4080
53. Yoshizawa M, Miyagi S, Kawano M, Ishiguro K, Fujita M (2004) Alkane oxidation via photochemical excitation of a self-assembled molecular cage. J Am Chem Soc 126:9172–9173

54. Hao H-G, Zheng X-D, Lu T-B (2010) Photoinduced catalytic reaction by a fluorescent active cryptand containing an anthracene fragment. Angew Chem Int Ed 49:8148–8151
55. Inokuma Y, Yoshioka S, Fujita M (2010) A molecular capsule network: guest encapsulation and control of Diels–Alder reactivity. Angew Chem Int Ed 49:8912–8914
56. Tashiro S, Fujita M (2006) Selective recognition of Trp- and Tyr-Rich oligopeptides by self-assembled coordination hosts. Bull Chem Soc Jpn 79:833–837
57. Zyryanov G, Rudkevich DM (2003) Encapsulated reagents for nitrosation. Org Lett 5(8):1253–1256
58. Zhang G, Presly O, White F, Oppel IM, Mastalerz M (2014) A permanent mesoporous organic cage with an exceptionally high surface area. Angew Chem Int Ed 53:1516–1520
59. Ajami D, Hou JL, Dale TJ, Barrett E, Rebek J Jr (2009) Disproportionation and self-sorting in molecular encapsulation. Proc Natl Acad Sci U S A 106: 10430–10434
60. Dolomanov OV, Blake AJ, Champness NR, Schröder M, Wilson C (2003) A novel synthetic strategy for hexanuclear supramolecular architectures. Chem Commun 682–683
61. Argent SP, Riis-Johannessen T, Jeffery JC, Harding LP, Ward MD (2005) Diastereoselective formation and optical activity of an M_4L_6 cage complex. Chem Commun 4647–4649
62. Argent SP, Adams H, Riis-Johannessen T, Jeffery JC, Harding LP, Ward MD (2006) High-nuclearity homoleptic and heteroleptic coordination cages based on tetra-capped truncated tetrahedral and cuboctahedral metal frameworks. J Am Chem Soc 128(1):72–73
63. Sharrad CA, Lüthi SR, Gahan LR (2003) Embracing ligands. A synthetic strategy towards new nitrogen–thioether multidentate ligands and characterization of the cobalt(III) complexes. Dalton Trans 3693–3703
64. Gruppi F, Boccini F, Elviri L, Dalcanale E (2009) Self-assembly of a cavitand-based heteronuclear coordination cage. Tetrahedron 65:7289–7295

Encapsulation by Covalent Capsules 2

Overwhelming majority of covalent capsules are neutral compounds that have been prepared using classical organic reactions and synthetic approaches from classical [1–9] and modern [10–12] macrocyclic chemistry. Those are based on "bottom-to-up" principle, such as imine condensations (in many cases, followed by reduction of the cage Schiff bases formed to their amine analogs) and condensations in high dilution conditions of diamine- or dihydroxyl-containing components (mostly the reactive macrocyclic precursors) with terminal dihalogenoalkanes or chloroanhydrides of suitable dicarboxylic acids. In contrast to supramolecular and coordination capsules (Chaps. 3 and 4), the template effect of encapsulated species is less pronounced, but in several cases, templation of these reactions by alkali or alkali-earth metal cations, neutral guests, and caged anions is reported.

2.1 Schiff-Base Imine Encapsulating Ligands and Their Polysaturated Cage Derivatives

2.1.1 Free Cages and Encapsulation of Neutral Molecules

First representatives of macrobicyclic Schiff bases **1–4**, known as prospective caging polynucleating ligands for transition metal ions and polar and nonpolar organic guests, have been synthesized by J. Nelson and coworker [13] by *one-pot* 2:3 imine condensation of tripodal amine *tren* with aromatic *meta*-dialdehydes by Scheme 2.1.

The reduction of a series of the imine macrobicycles of this type with $NaBH_4$ has been performed in [14] by Scheme 2.2 to give the corresponding amine capsules.

The Schiff-base macrobicyclic compounds **6** and **11** have been prepared in [15] by Scheme 2.3 using imine template condensation on Ba^{2+} ion as a matrix.

The thiophene-based imine capsules **4** and **12** have been synthesized in [16] using template condensation on silver(I) ion as an appropriate matrix [17, 18] by Scheme 2.4; the reduction of **4** with $NaBH_4$ gave its amine derivative **13**. This macrobicyclic ligand cannot encapsulate bulky anions but can accommodate neutral molecules of some solvents [19]. Its host–guest 1:4 cage complex contains four water molecules forming a cyclic pentamer in an "envelope" conformation with a protonated amino group (see an insert in Scheme 2.4); the N…N distance between the cross-linking nitrogen atoms of **13** (7.89 Å) is substantially higher than those in its 1:1 cage complexes with encapsulated chloride and nitrate anions and positively charged protonated tripodal amine capping fragments (6.10 and 5.53 Å, respectively [20, 21]). These experimental results have been found [19] to be in a good agreement with ab initio DFT calculations.

Y. Voloshin et al., *The Encapsulation Phenomenon: Synthesis, Reactivity and Applications of Caged Ions and Molecules*, DOI 10.1007/978-3-319-27738-7_2

Scheme 2.1

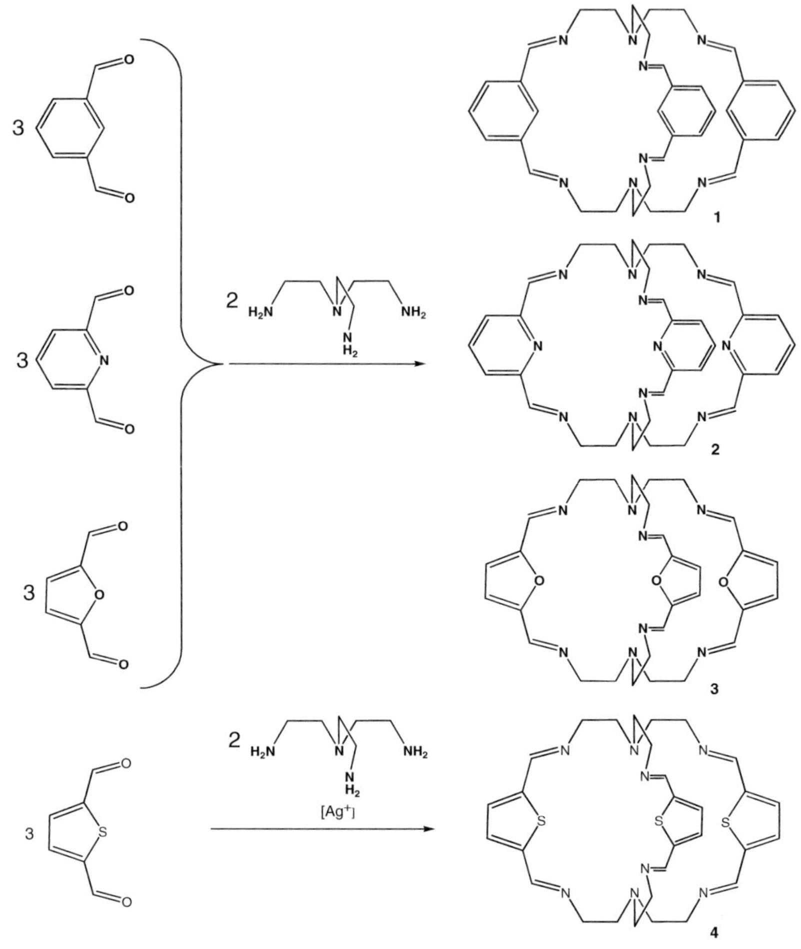

An octaamine macrobicyclic ligand **14** with N_4-donor sets forming a channel of water layers allowed isolating and characterizing by X-ray diffraction a water-based $(H_2O)_{45}$ cluster [22]. Each of the three C_3-symmetric capsules in its asymmetric unit interacts with 15 water molecules, and 45 water molecules in total are included in the cluster. Each of the three $(H_2O)_{15}$ subunits has a shape of a puckered eight-membered ring that is conjugated with the square $(H_2O)_4$ fragment and with two water dimers and one water molecule via hydrogen bonds. The latter species also form NH…O hydrogen bonds with secondary amino groups of the macrobicyclic ligand **14**. According to the TGA and IR data, this 2D layer of water molecules undergoes reversible desorbing (with temperature)–absorbing (in the presence of water) processes by Scheme 2.5 [22].

The macrobicyclic ligand **15** and its host–guest 1:2 complexes with encapsulated water molecules and with caged silver(I) cations have been prepared in [23] by Scheme 2.6. These cage complexes contain the macrobicyclic ligand in its *endo,endo*-conformation with a distance between the capping ternary nitrogen atoms of approximately 19.6 and 18.35 Å, respectively. The encapsulated Ag^+ ions have an irregular AgN_4-coordination polyhedron and coordinates three ribbed secondary and one apical ternary amine nitrogen atoms; the distance between them is approximately 14.7 Å. In the cage complex of **15** with encapsulated water molecules, the distance between the cross-linking nitrogen atoms does not exceed 12.3 Å because each of such two guests forms only one hydrogen bond, thus making the encapsulating ligand **15** rather labile. As a

Scheme 2.2

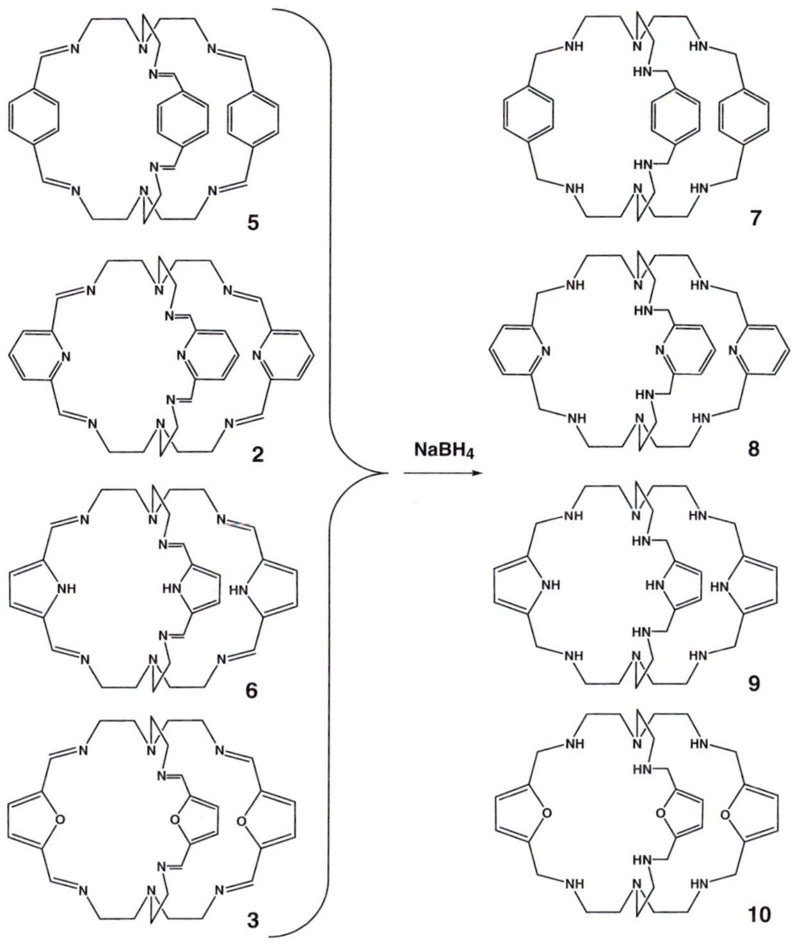

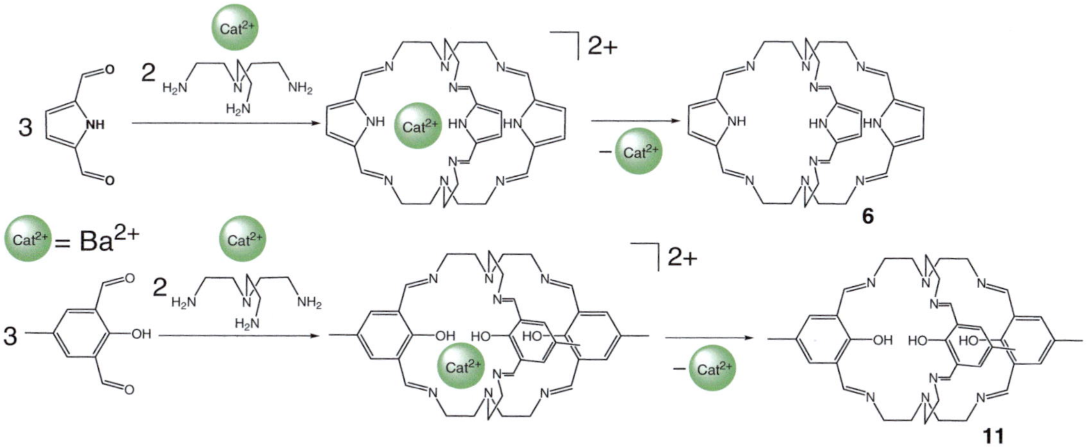

Scheme 2.3

result, its macrobicyclic framework undergoes a rotation–contraction distortion along the C_3-symmetry pseudoaxis passing through these apical nitrogen atoms, whereas in the analogous dinuclear silver(I) complex the ribbed fragments of **15** are practically parallel [23].

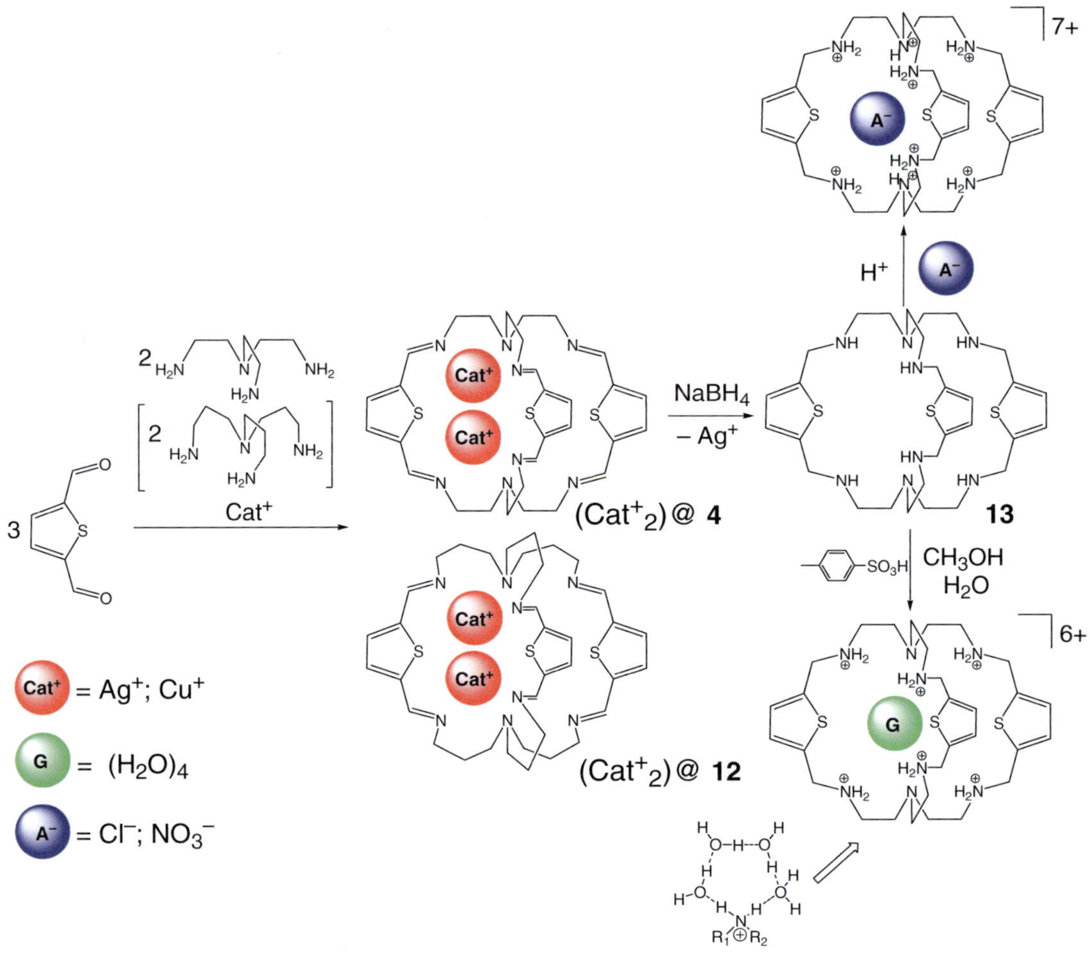

Scheme 2.4

Scheme 2.5

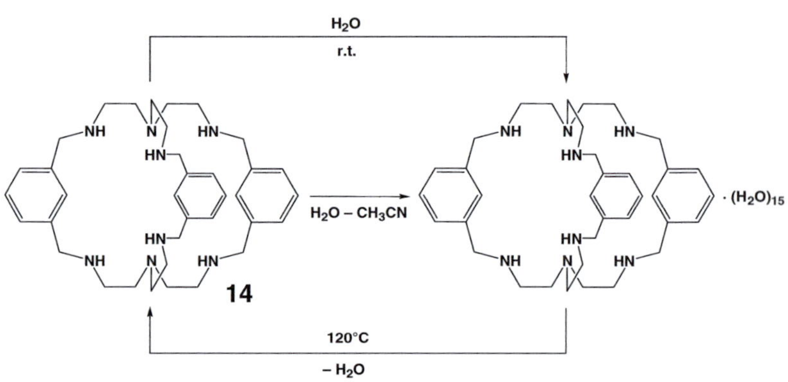

Selectivity of encapsulation of trihydroxybenzenes by the bipyridine-based covalent capsules **16–18** shown in Scheme 2.7 has been studied in [24]. The ligands **16** and **17** encapsulate phloroglucinol and 1,2,4-trihydroxybenzene to give 1:1 cage complexes, but **18** does not. This result is explained [24] by host–guest hydrogen bonding within the inner cavities of **16** and **17**, which have

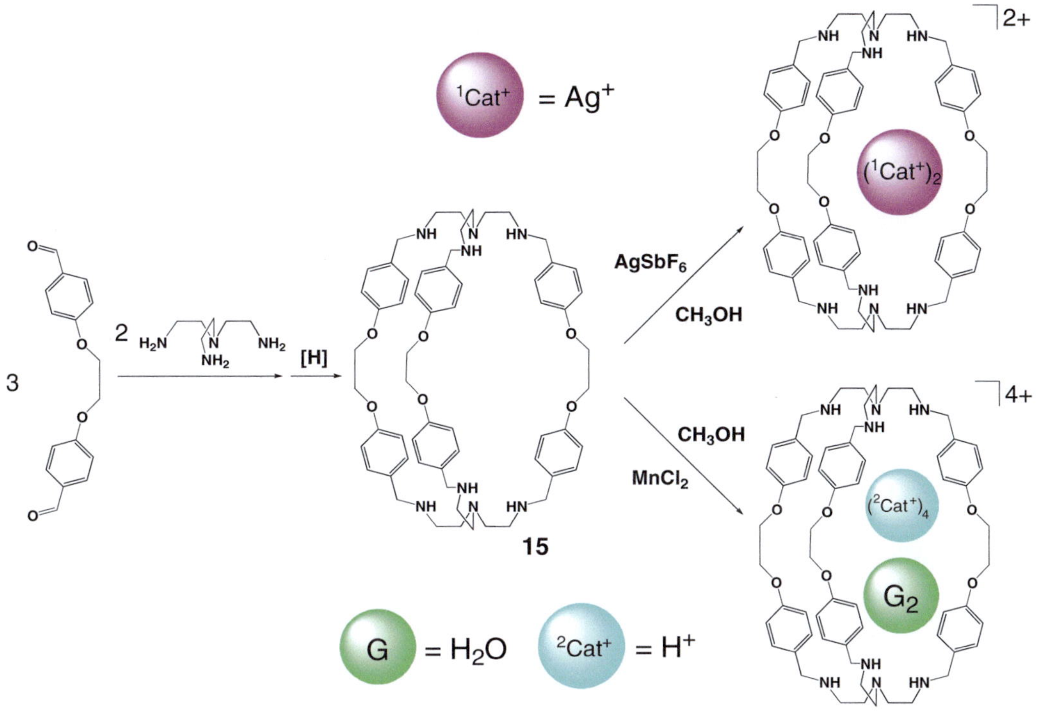

Scheme 2.6

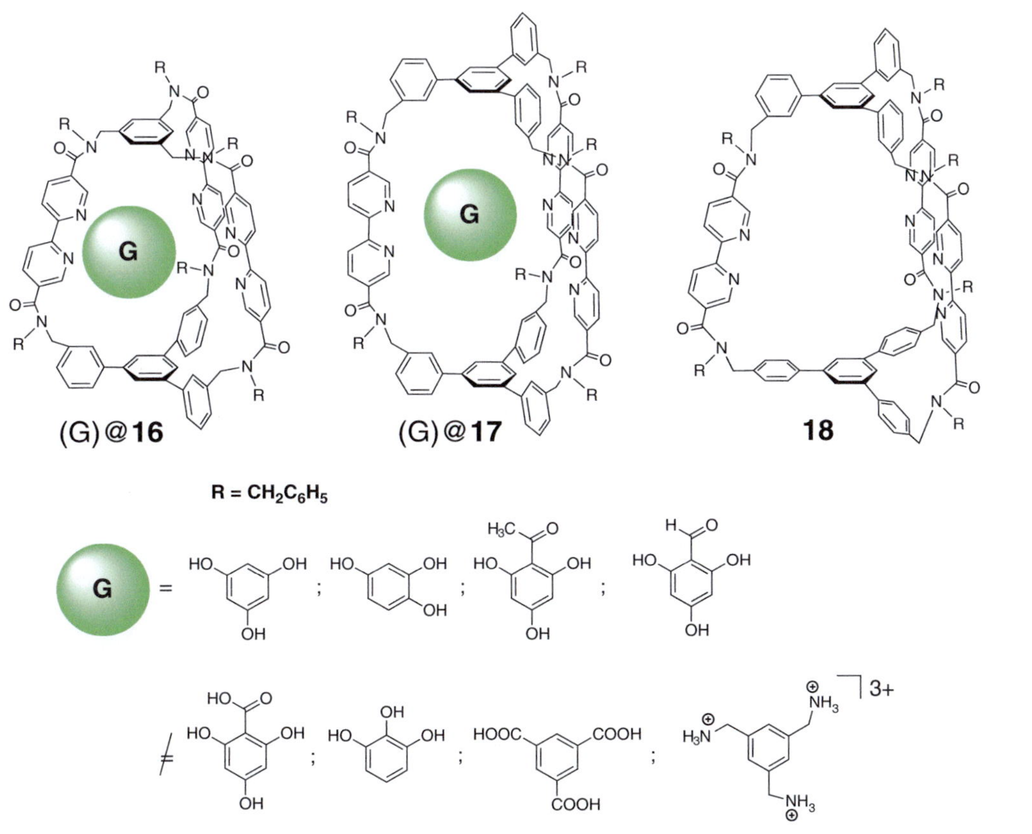

Scheme 2.7

suitable size and can discriminate between phloroglucinol derivatives. All attempts to encapsulate tricarboxylic acids such as 1,3,5-benzenetricarboxylic acid or the tris(amrnoniurn) salt of 1,3,5-tris(aminomethy1)benzene by **16–18** have failed. Thus, even a small structural change, such as the introduction of methyl or carboxyl substituents, can prevent encapsulation by these caging ligands [24].

Highly symmetrical cube-shaped polyamine caging ligand **19** has been prepared in [25] by imine condensation of tripodal amine *tren* with formaldehyde by Scheme 2.8; its T_d-symmetric hexahedral cage framework encapsulates two water molecules.

A very efficient *one-pot* synthesis of the octahedral hexacavitand caging ligand **20** with a cavity volume of approximately 1700 Å^3 has been performed in [26] by Scheme 2.9 through 18-component imine condensation of 6 aldehyde-functionalized cavitand ligand syntones and 12 terminal diamine linkers, followed by the reduction of a Schiff-base cage intermediate with $NaBH_4$.

Similar multicomponent dynamic imine condensation by Scheme 2.10 gave a Schiff-base rhombicuboctahedral cage framework **21** with a cavity volume of approximately 5000 Å^3 co-encapsulating tetraalkylammonium salts and solvent molecules [27].

A pyrrole-based caging ligand **22** has been designed in [28] for selective encapsulation of *β*-glucopyranosides by Scheme 2.11. This covalent capsule specifically recognizes *β*-anomer of *D*-glucose and the corresponding alkyl glucosides with complete *β*-/*α*-selectivity and effectively discriminates *β*-monosaccharides, the glucoderivatives, from both the *α*- and *β*-anomers of the galacto- and mannoderivatives.

Hybrid boron ester–Schiff-base hemicarcerand capsules **23–28** have been prepared in [29] by imine condensation of 3-aminophenylboronic acid with the corresponding carbonyl-containing ligand syntones by Scheme 2.12; these covalent capsules encapsulate solvent benzene molecules, thus giving the host–guest 1:2 cage complexes.

Condensation of various aromatic and heteroaromatic dialdehydes (e.g., *ortho*-, *meta*-, and *para*-isomers of diformylbenzene) with 1,3,5-trisubstituted aromatic amine by

Scheme 2.8

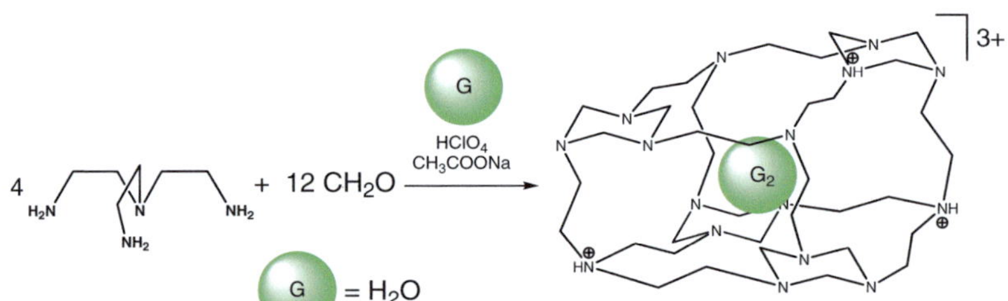

$((H_2O)_2)$@ **19**

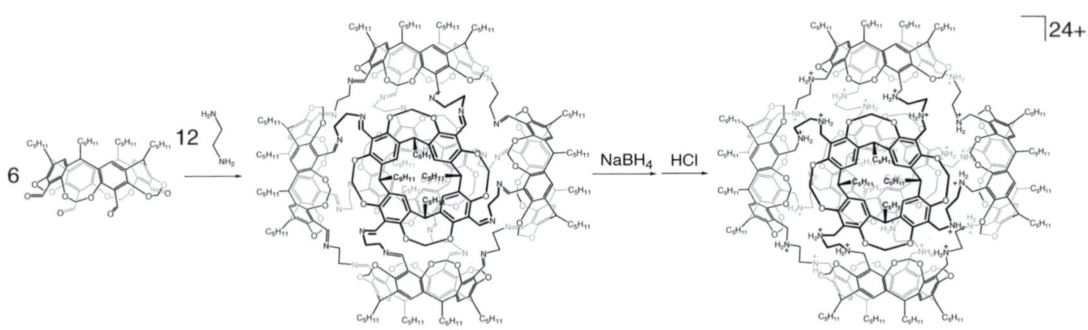

20

Scheme 2.9

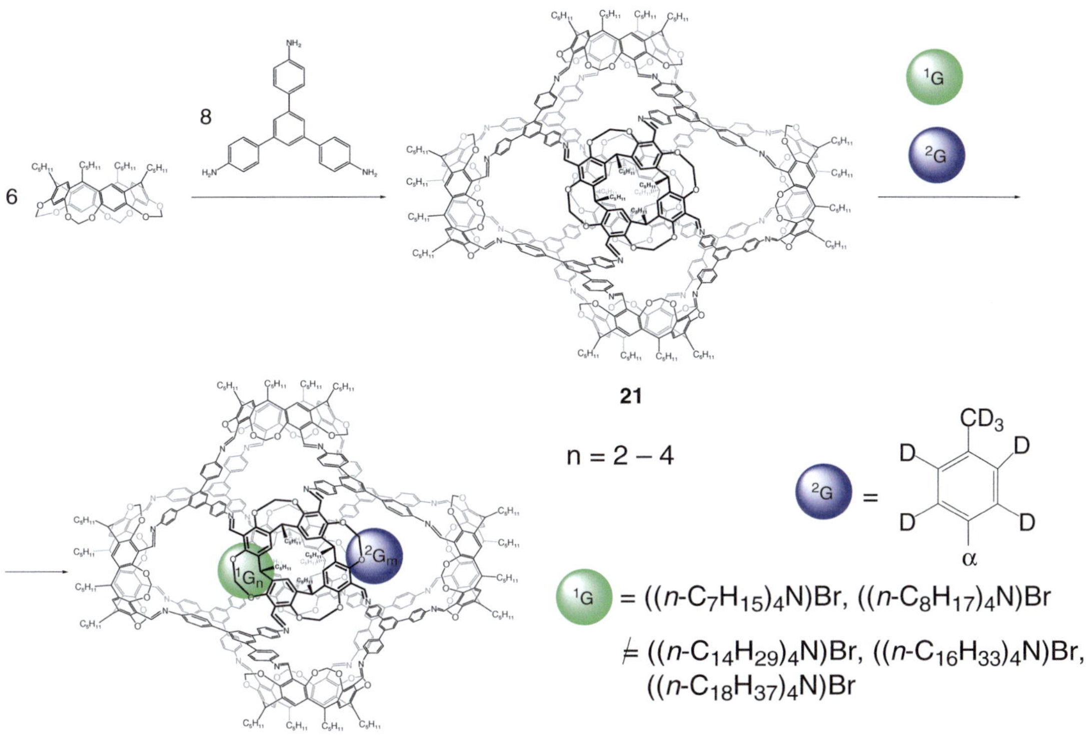

Scheme 2.10

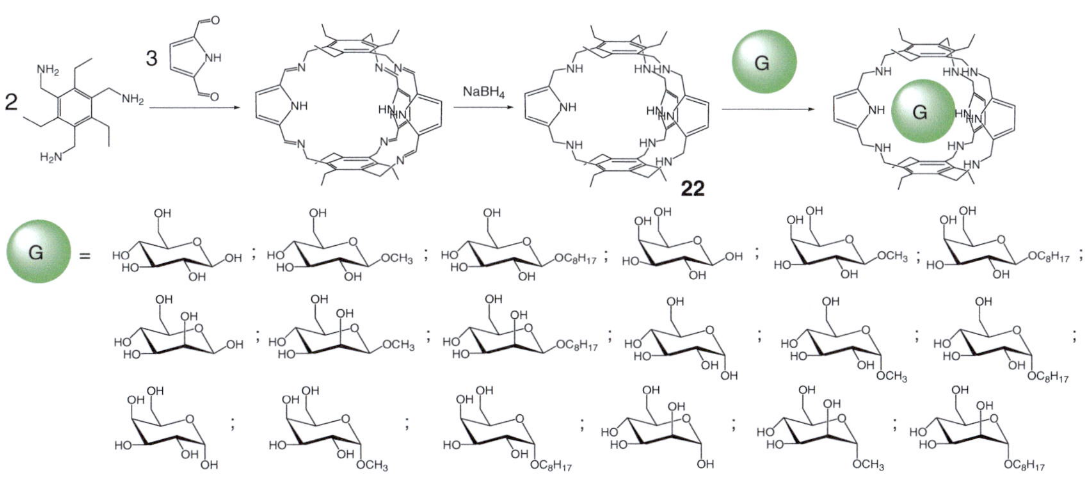

Scheme 2.11

Scheme 2.13 gave the corresponding hexaimine capsules **29–32** only in the case of 1,3-dialdehyde syntones [30]. Their reduction with $NaBH_4$ afforded the amine macrobicyclic ligands **33–36**. The early experiments used secondary amine groups as donors of hydrogen bonds to bind the encapsulated species, while the authors of [30] performed alkylation and acylation of these ligands to study the hydrogen-donor ability of ternary amino groups. Such alkylation reactions allowed attaching the functionalizing pendant arming groups; so the capsule **33** has

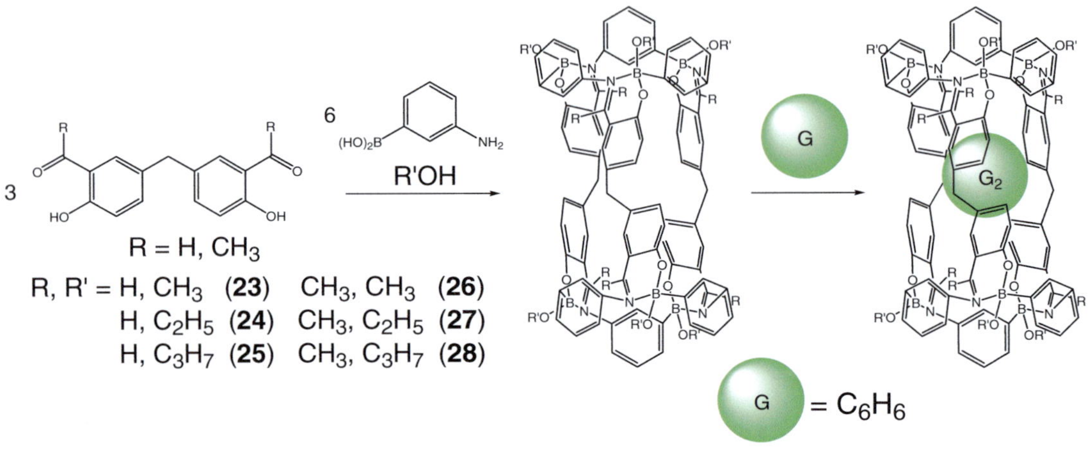

Scheme 2.12

been permethylated in [30] using the standard Eschweiler–Clarke reaction conditions, while its hexafold Michael addition with acrylates gave the hexafunctionalized cage framework **38** with six methyl ester pendants by Scheme 2.14. The prolonged tosylation of the capsules **33** and **36** as macrobicyclic precursors with tosyl chloride in the presence of triethylamine afforded their persulfonylated derivatives **39** and **40**. The X-ray diffraction data for **37**, **38**, and **40** demonstrate an influence of tertiarization of amino groups of such cage frameworks with three ribbed aromatic fragments forming the faces of a regular TP on their conformation. This alkylation caused the elongation of these frameworks: two parallel apical aromatic fragments of the C_3-symmetric alkylated macrobicycles **37** and **38** are at 10 Å from each other, whereas in **33** this distance is approximately 7.6 Å. In **40**, these nitrogen atoms are strictly sp^2-hybridized and planar, while their bulky tosyl substituents cause the distortion of a TP macrobicyclic framework. As a result, the apical aromatic fragments are nonparallel resulting in lowering of its molecular symmetry [30].

According to the X-ray diffraction data of [31], the covalent capsule **33** encapsulates acetonitrile and propionitrile solvate molecules giving 1:1 cage complexes by Scheme 2.15. The caged propionitrile molecule is located between two aromatic fragments of the encapsulating ligand, and its C≡N group occupies the center of a distorted tetrahedron formed by four nitrogen atoms of secondary amino groups belonging to two di(aminomethyl)pyridine fragments of **33**; at the same time, it did not encapsulate a relatively large molecule of benzonitrile [31].

A dynamic combinatory library of pyridine-2,6-dicarbaldehyde and tripodal triamine *tren* and ethylene glycol diamine (Scheme 2.16) has been studied in the absence and in the presence of Ca^{2+} ions [32]. The imine capsule **41**, presenting in the equilibrium reaction mixture, has been reduced with $NaBH_4$ to its octaamine analog **42**. This amine capsule is the major product of the *one-pot* subsequent imine condensation–hydrogen reduction synthetic procedure in the absence of Ca^{2+} ion [32].

The same approach (i.e., the condensation of aromatic triamine with aromatic dialdehyde followed by reduction of the Schiff-base product) has been used in [33] for the synthesis of a series of covalent capsules **43**–**46** designed for selective carbon dioxide uptake (Scheme 2.17); the analogous reaction of the corresponding aromatic trialdehyde syntone with aromatic diamine by Scheme 2.18 gave a similar macrobicycle **47**. Among a series of the oligomeric products of such reversible imine condensation, only the TP cage framework, which is enthalpy-favored due to the absence of angle strains and entropy-favored because of minimum of its building blocks, dominates in this reaction mixture [33]. Within the range of the so-called "cage-to-framework" and "bottom-to-up" strategies, a

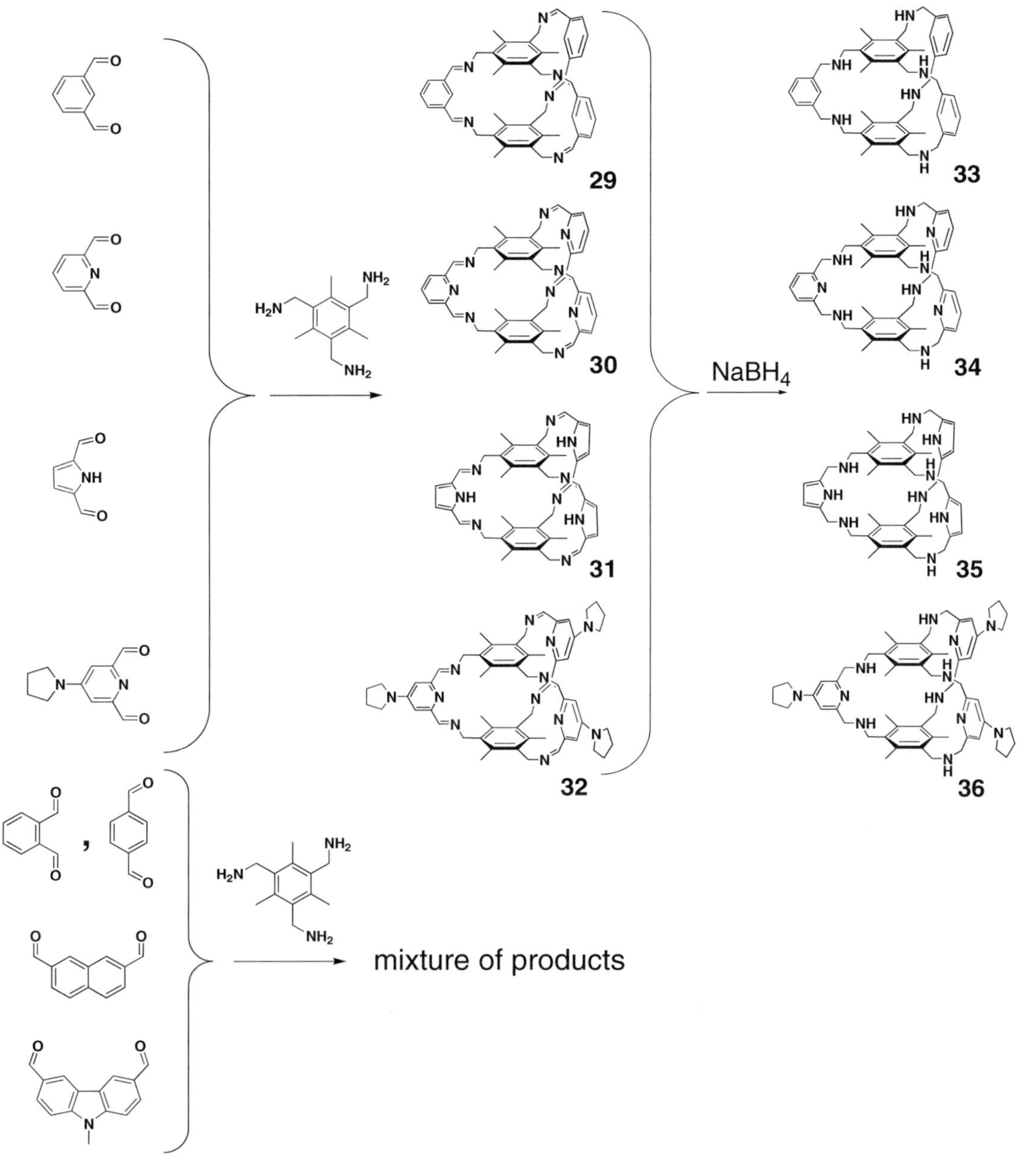

Scheme 2.13

polybromine macrobicyclic precursor **45** has been transformed in [33] by Sonogashira reaction with 1,4-diethylbenzene to a pentacage ligand **48** (Scheme 2.19). All these caging ligands have high affinity to and selectivity of CO_2 uptake over that of N_2; the largest selectivity (138:1) has been observed in the case of a covalent capsule **46**. The molar % absorption capacity of CO_2 is practically the same for all the monocage ligands **43–46**; their polycage analog **48** has four times higher absorption capacity due to the presence of five vacant cavities per one of its molecule. Such CO_2 uptake strongly depends on the density of amino groups in the caging ligand and is substantially

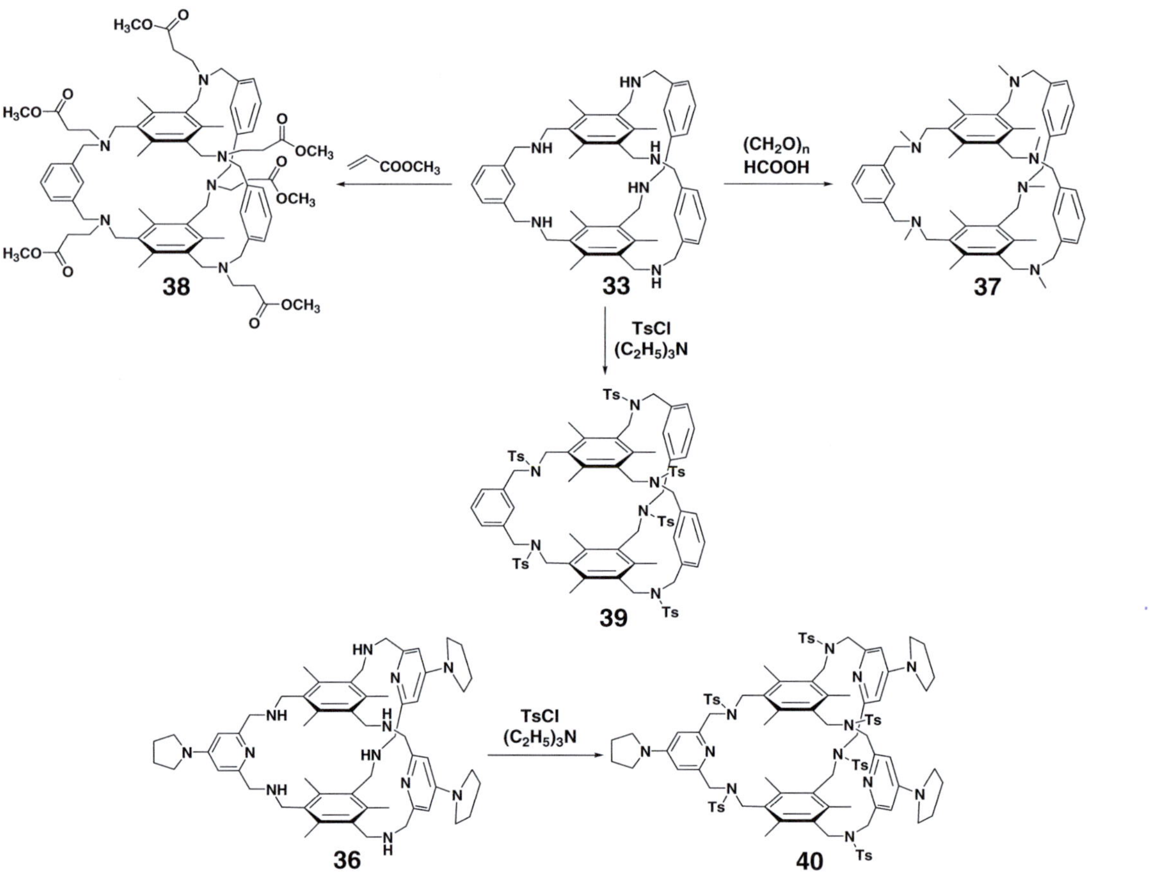

Scheme 2.14

Scheme 2.15

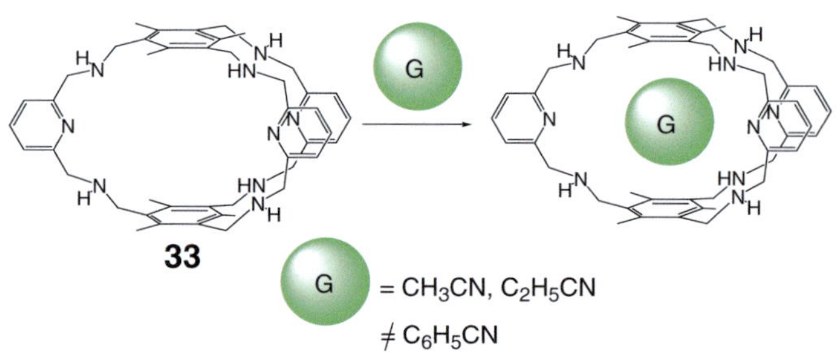

less sensitive to the size of its pores. In contrast, nitrogen uptake by these capsules substantially increases with their size. Moreover, in the case of the polycage compound **48**, N_2 uptake becomes eight times higher due to the cooperative absorption effect of its monocage fragments [33].

A [4+6] condensation by Scheme 2.20 of benzene-1,3,5-tricarbaldehyde with ethylenediamine quantitatively afforded a macropolycyclic Schiff-base imine capsule **49** with an enlarged cavity [34].

The Schiff-base chiral imine covalent capsules **50** and **51** have been prepared in [35] using [8+12] condensation of chiral diamines (*R*,*R*)-1,2-cyclohexanediamine and (*R*,*R*)-1,2-cyclohex-4-enediamine with aromatic trialdehyde syntone *2* by Scheme 2.21. As follows from X-ray diffraction data, their cage frameworks with a cavity

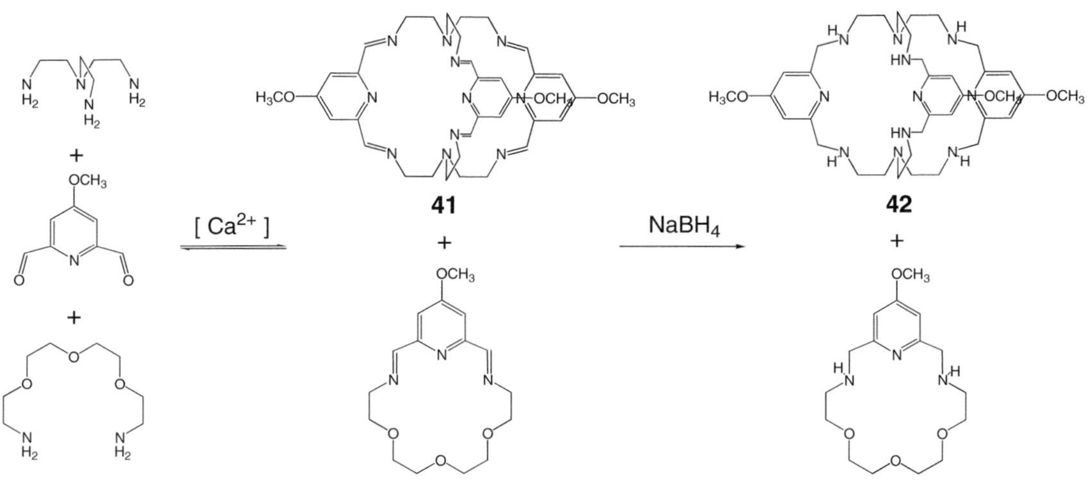

Scheme 2.16

volume of approximately 1500 Å^3 have a tetrahedral T symmetry and are reported to encapsulate solvent guest molecules.

One-pot synthesis of a shape-persistent *endo*-functionalized imine cage framework **52** using reversible imine condensation of a triamine ligand syntone *3* and salicylic dialdehyde by Scheme 2.22 has been performed by M. Mastalerz in [36]. The covalent capsule **52** has an adamantoid T_d symmetry with six endohedral directed hydroxyl groups forming a regular octahedron with O…O distance between the opposite oxygen atoms of 15.7 Å and an edge length of 11.3 Å. The outer diameter of its cage framework is 27.2 Å, and the smallest cavity volume is estimated in [36] to be approximately 680 Å^3. In its C_3-symmetric tetrahedral cage framework, three adjacent edges are twisted around their common vertex by about 32°, and two different types of imine bonds are observed in [37]: one of them forms a six-membered ring stabilized by hydrogen bond, whereas the other does not. The hydroxyl groups are directed into the cavity of this covalent capsule, thus forming a slightly distorted octahedron with an average edge length of r_i(O…O)=10.53 Å and a cavity volume of approximately 550 Å^3. This imine framework **52** is self-assembled through stacking interactions between the phenolic arene rings that are coplanar to each other at a distance of 3.65 Å. It is reported in [37] to have a very high surface area and CO_2/CH_4 selectivity of 10:1 (w/w).

A [4+6] condensation of triptycene triamine with the corresponding salicyl dialdehydes by Scheme 2.23 afforded 12-imine covalent capsules **52**–**56**; the pseudooctahedral cage frameworks of **52** and **53** are reported in [38] to be self-assembled through stacking interactions of their aromatic fragments.

Exo-functionalized imine [2+3] capsules **57** and **58** have been prepared in [39] using condensation of two bis-salicylaldehyde ligand syntones with triptycene triamine by Scheme 2.24. As follows from the X-ray diffraction data, the covalent capsule **57** with the distances of 7.81 or 7.66 Å between its two interior triptycene bridgehead hydrogen atoms has a shape-persistent cage framework with the intramolecular distance between its apical protons of 9.81 Å.

Imine synthesis of the Schiff-base [4+6] covalent capsules **53**, **54**, and **59** (Scheme 2.25), having shape-persistent frameworks and permanent porosity, has been performed in [40]. *One-pot* condensation by Scheme 2.26 gave an *exo*-functionalized imine capsule **60** of the same stoichiometry with shape-persistent cage framework that showed a very high specific surface area [41].

The Schiff-base [6+4] covalent capsules **61**–**65** can be obtained using two different synthetic approaches shown in Scheme 2.27: the direct

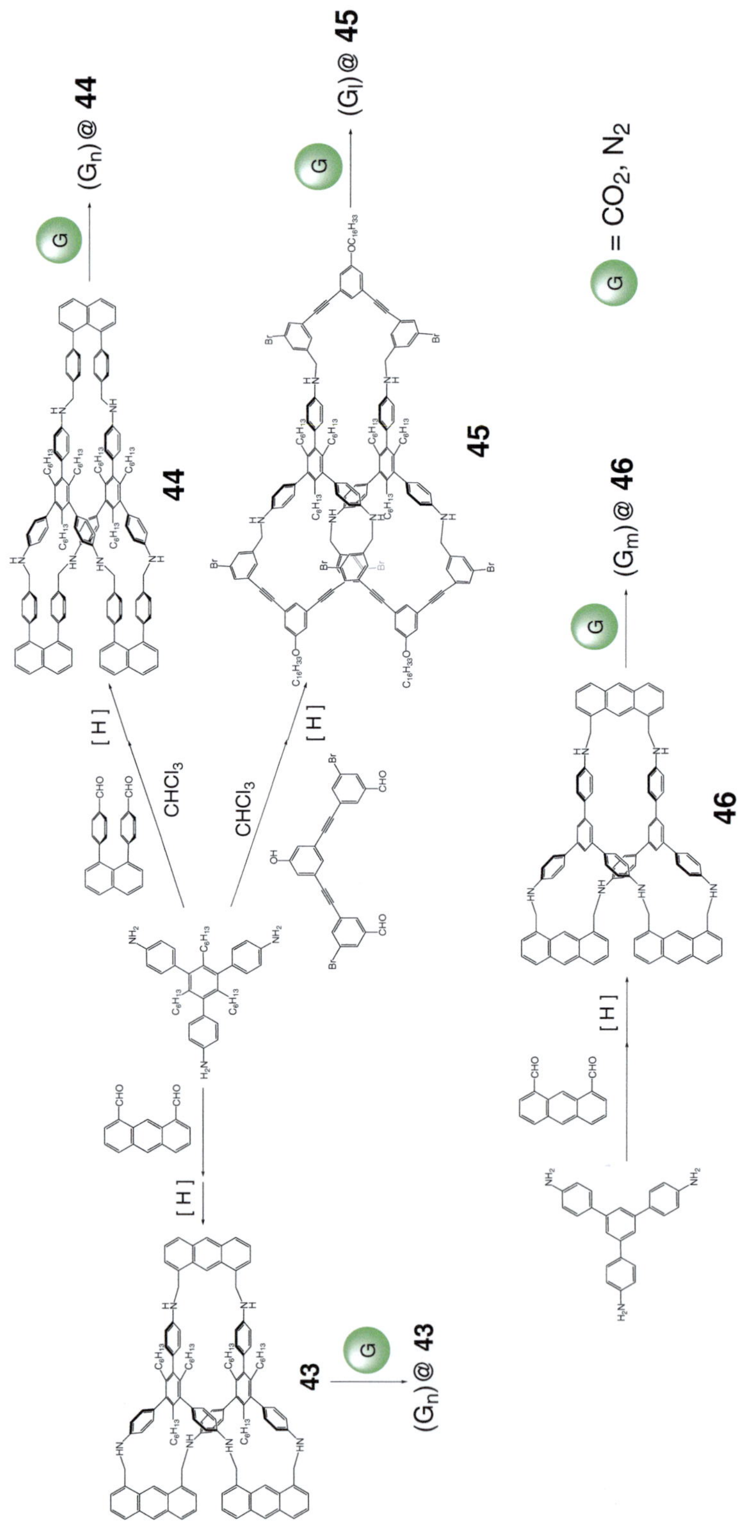

Scheme 2.17

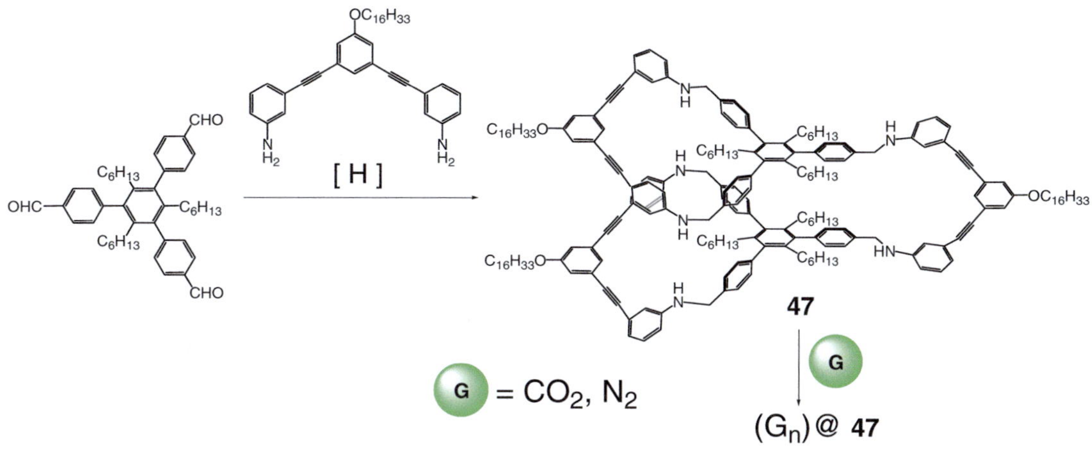

Scheme 2.18

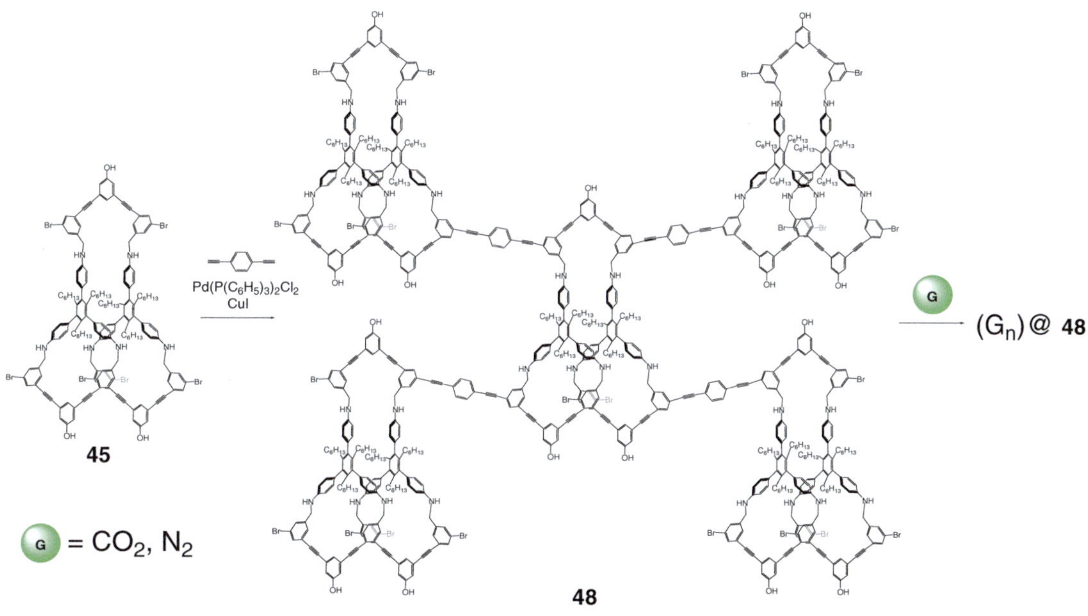

Scheme 2.19

Scheme 2.20

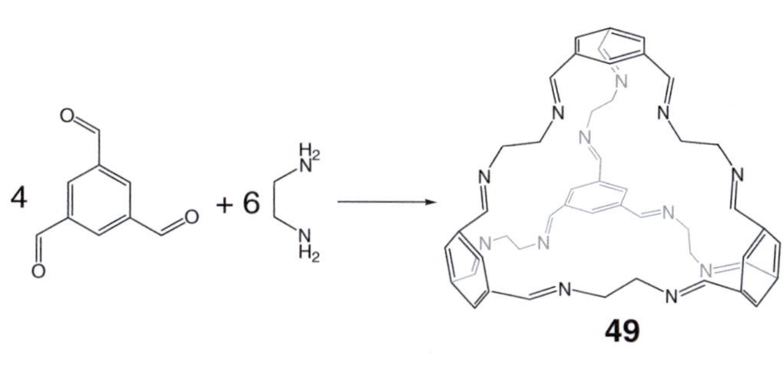

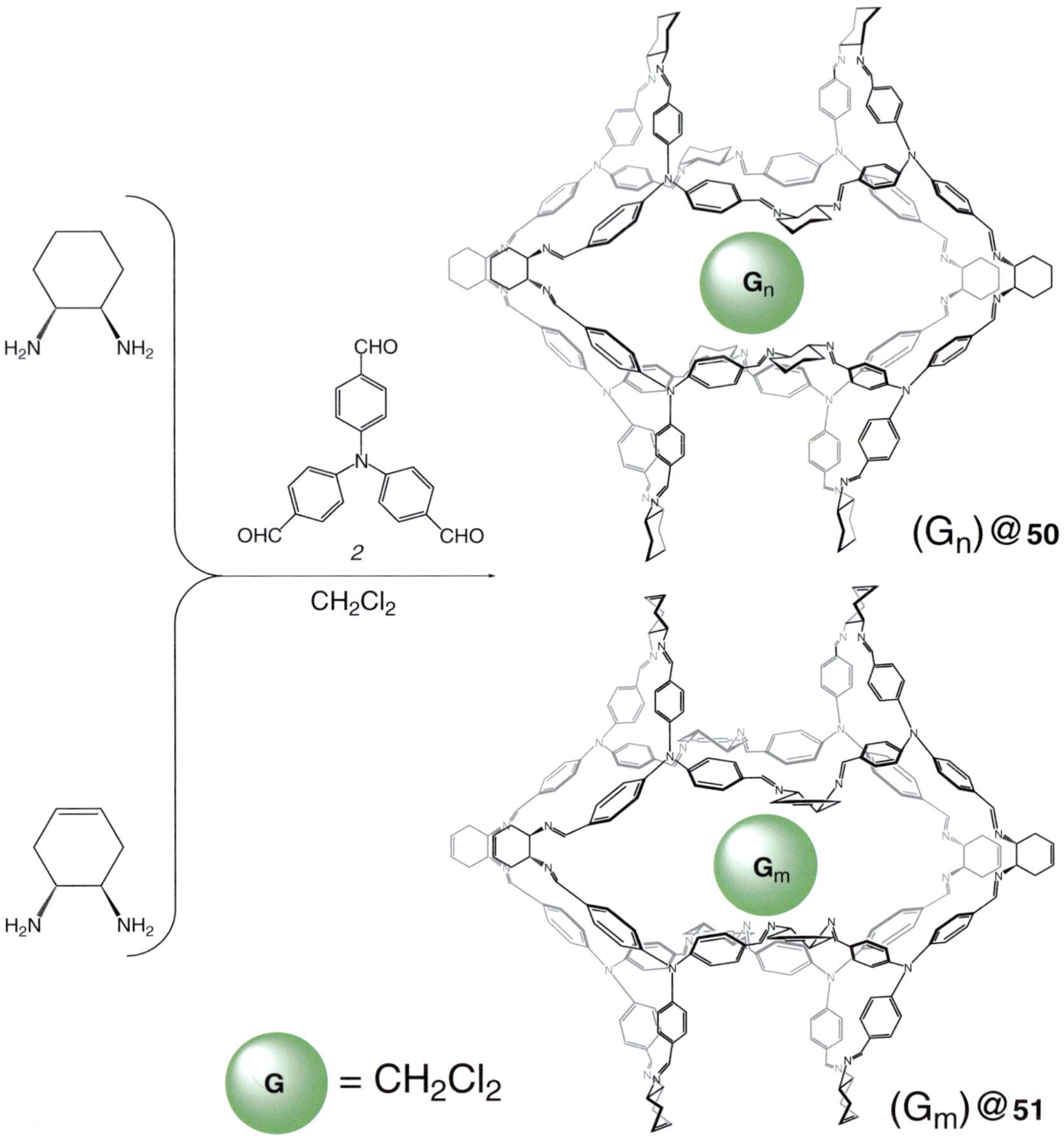

Scheme 2.21

12-fold imine condensation (Pathway I) was found [42] to be uneffective, but the postsynthetical modification of the cage precursor **50** by sixfold Williamson ether formation (Pathway II) afforded the target capsules in moderate yields. The capsule **61** has a C_3 symmetry, and its inner triptycene bridgehead protons form a slightly distorted tetrahedron with an average edge length of 11.4 Å; six methyl carbon and oxygen atoms are in the vertexes of a distorted octahedral cage framework **50** with an average edge lengths of 8.53 and 9.70 Å, respectively [42].

The [4+4] imine capsules **66** and **67** with unique pseudocubic frameworks have been synthesized in [43] by Scheme 2.28 using the corresponding tripodal triamine and aromatic trialdehyde ligand syntones.

A Schiff-base large cuboctahedral covalent capsule **68** with 24 imine bonds has been prepared in [44] using thermodynamically driven [8+12] condensation by Scheme 2.29; its O_h-symmetric

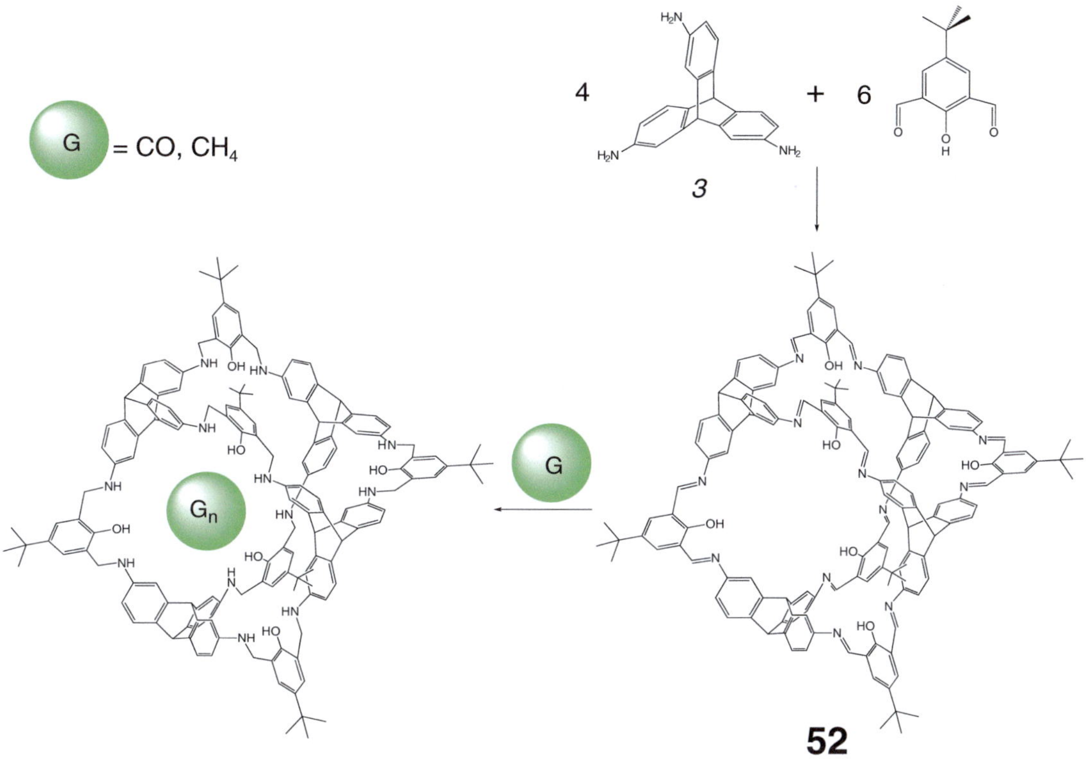

Scheme 2.22

cage framework with a cavity volume of 2680 Å^3 can be approximated into a sphere with the radius of 8.52 Å. The structural stability of this strain-free capsule is explained [44] by *E*-configuration of its imine fragments forming hydrogen bonds with vicinal hydroxyl group.

The [12+8] imine condensation of the corresponding triptane tetraols with aromatic triboronic acid by Scheme 2.30 gave the target cubooctahedral capsule **69** only in the case of a diethyl-containing triptycene precursor *4* [45]. This host having a cavity with minimum and maximum inner diameters of 26 and 31 Å, respectively, is capable of encapsulating solvent molecules forming the corresponding cage complexes. To activate their porous crystals for gas sorption, these guests have been removed giving a mesoporous organic material with a very high specific surface area [45].

The [4+4] polycondensation of C_3-symmetric triamine ligand syntones *5* and *6* with the designed metallamacrocyclic building blocks containing reactive aldehyde groups has been used in [46] for the synthesis of covalent capsules **70–73** by Scheme 2.31. The capsule **70** has an approximate tetrahedral *T* symmetry, which is in agreement with its solution NMR spectra. Trinuclear metallamacrocyclic capping fragments of **70** occupy its four vertices with the largest Ru…Ru distance of approximately 23 Å, while the triphenylmethane groups of the corresponding ligand syntone span each of the four faces of its cage framework. All four capping fragments have the same configurations with respect to their rotation around the C_3-pseudosymmetry axes connecting such vertex with an opposite face, and the X-rayed crystal of **70** is the equimolar racemic mixture of these stereoisomers. A cavity volume of this imine covalent capsule is estimated in [46] to be approximately 500 Å^3, while that of its larger analog **73** with maximal Ru…Ru distance of 29 Å is calculated to be approximately 1500 Å^3.

K. Severin and coworkers also used more complex (arene)ruthenium trinuclear

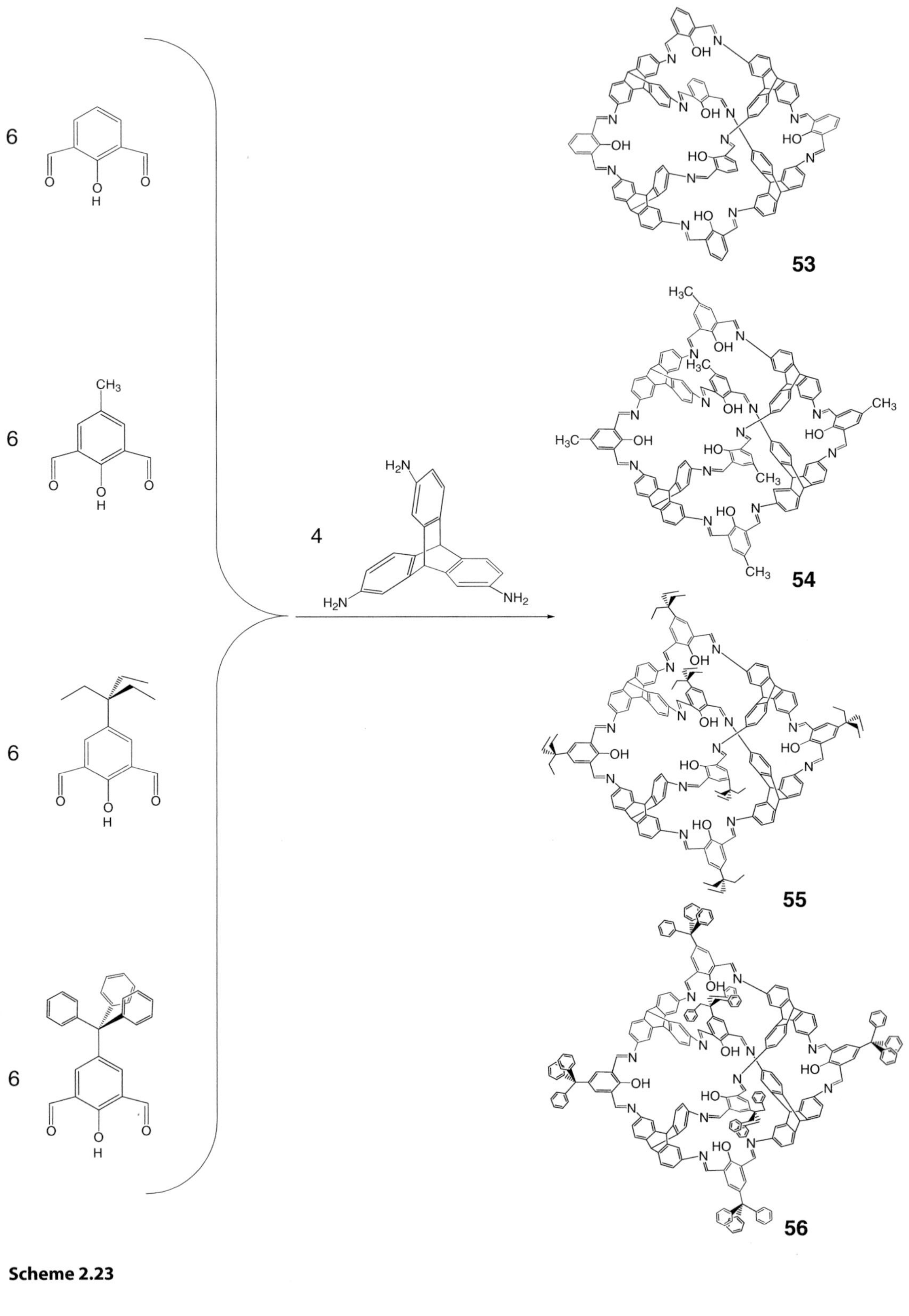

Scheme 2.23

Scheme 2.24–1394

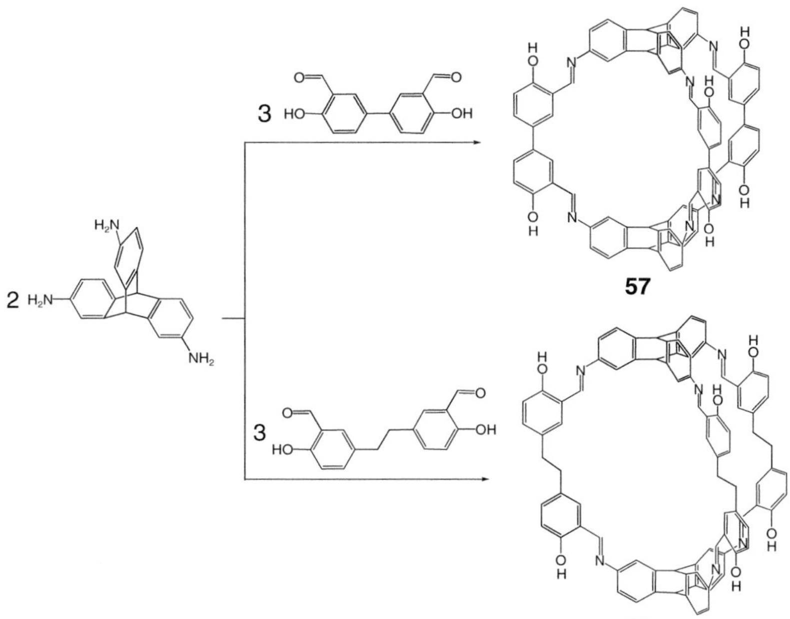

57

58

Scheme 2.25

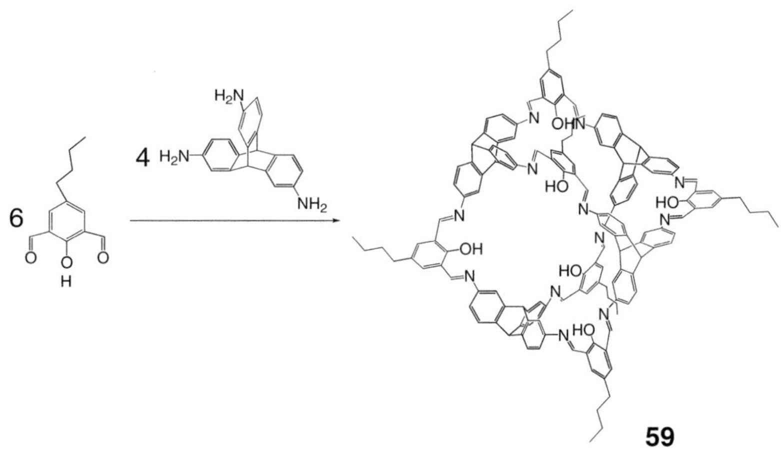

59

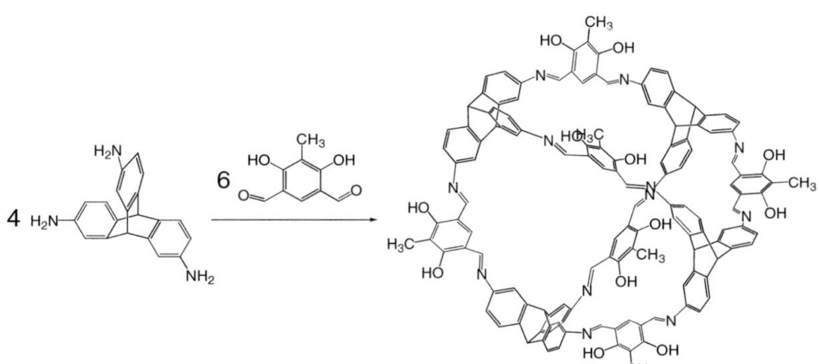

Scheme 2.26

60

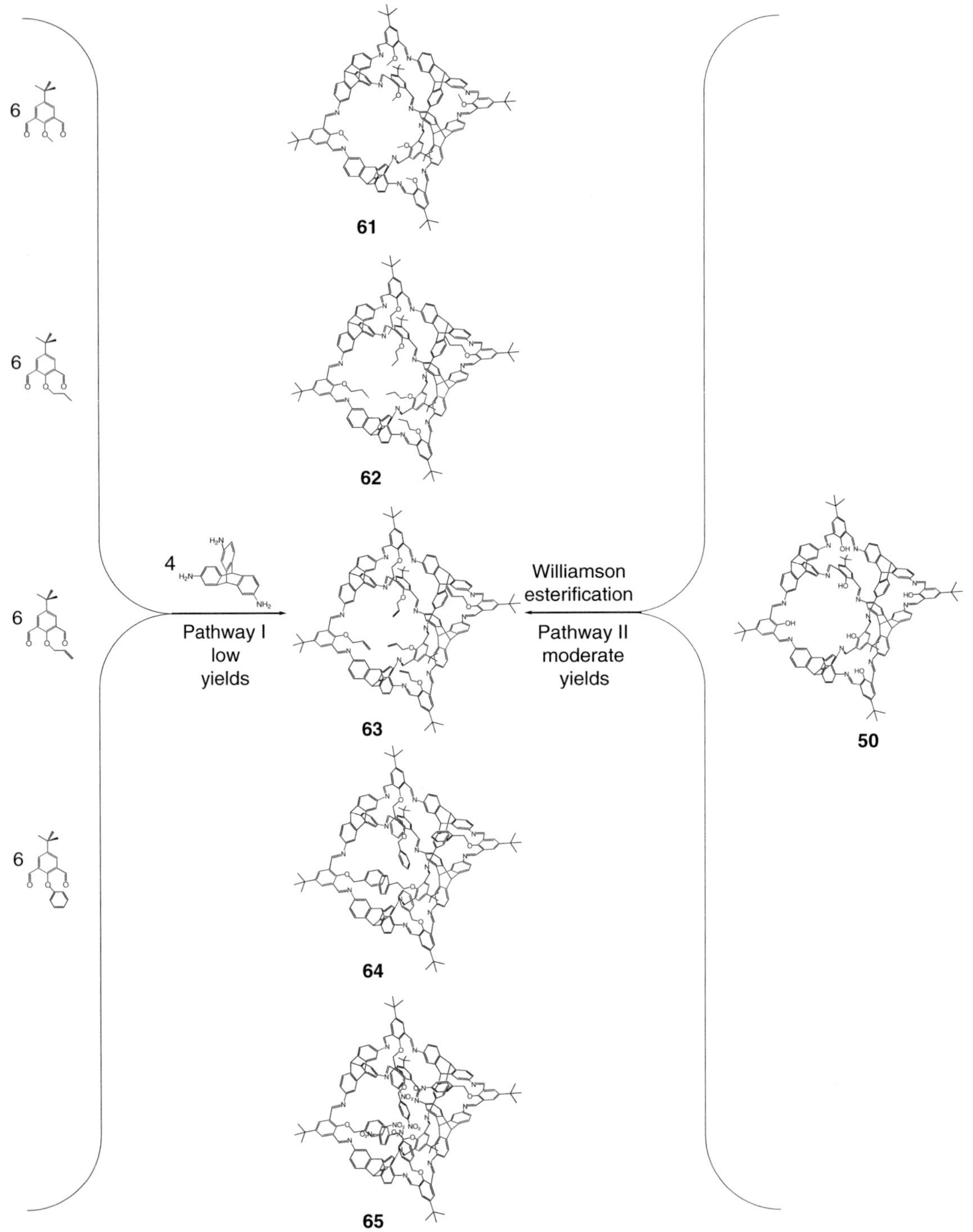
6
6
6
6
4
Pathway I
low
yields
Williamson
esterification
Pathway II
moderate
yields
61
62
63
64
65
50

Scheme 2.27

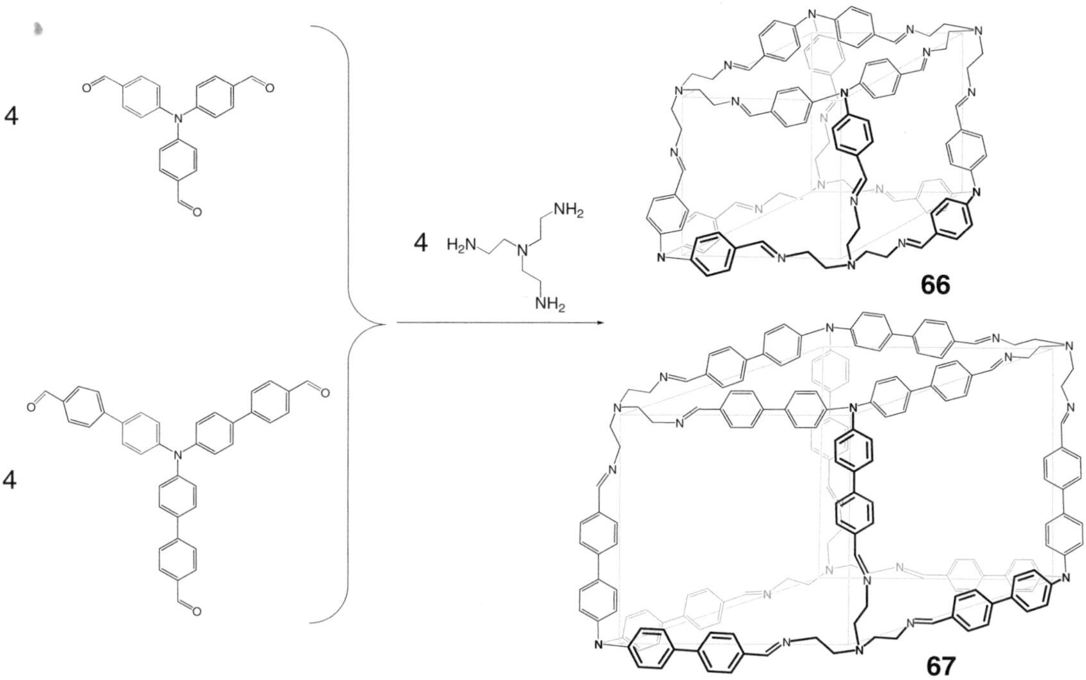

Scheme 2.28

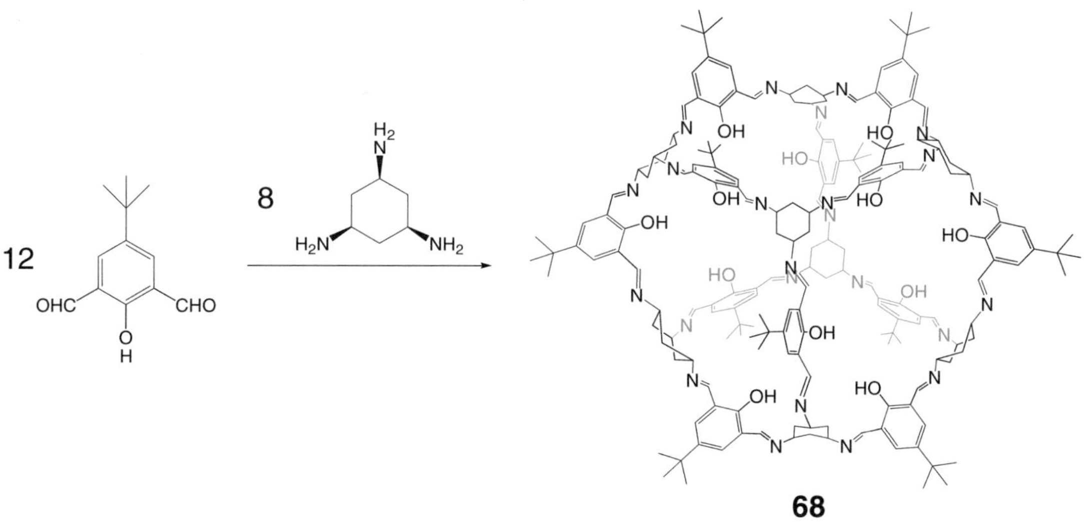

Scheme 2.29

metallamacrocycles *7–11* (Scheme 2.32) with three terminal aldehyde groups per molecule as ligand syntones. The macrocycles *7–9* are reported in [47] to be inert in self-assembly reaction conditions, whereas their *meta*-phenylene-containing homologs *10* and *11* are more labile; the C_3-symmetric triamines *12–14* have been used as amine components of these condensation reactions. Schiff-base dodecanuclear cage products **74–82** of the diastereoselective [4+4] imine condensations by Schemes 2.33, 2.34, and 2.35 have been characterized using ESI-MS and

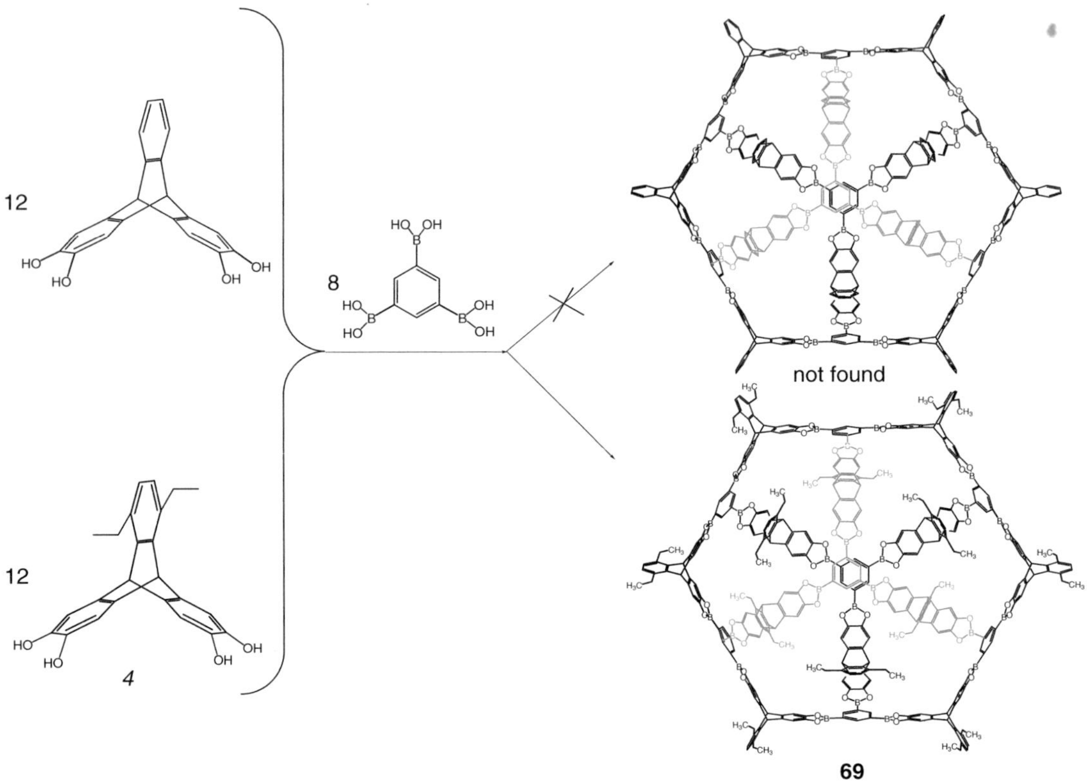

Scheme 2.30

NMR methods. The self-assembly of *12* and *13* with a tris-*iso*-propyl(arene)ruthenium precursor *9* gave no target cage products due to bulkiness of its arene π-ligand [47]. In contrast, these (arene) ruthenium trialdehydes form the covalent capsules with a labile tripodal aliphatic compound *tren* as an amine component (Schemes 2.33, 2.34, and 2.35). As follows from single-crystal X-ray diffraction data [47], all these $Ru_{12}L_4$ imine capsules have a similar structure: four apical trinuclear ruthenium metallamacrocycles with the same configuration are linked by the bridging triamine fragments to form a *T*-pseudosymmetric dodecanuclear cage framework encapsulating the disordered solvent molecules; their ruthenium metallocenters are stereogenic.

Imine condensation of *7* with 1,3,5-tris(aminomethyl)benzene in a molar ratio 2:3 gave a hexanuclear (*para*-cymene)ruthenium covalent capsule **83** as the only reaction product. Its cage framework with a length of approximately 25 Å has a cylindrical shape, and the trinuclear (arene)ruthenium cross-linking groups have opposite chirality (achiral *meso*-form). This hexanuclear framework **83** in a chloroform solution undergoes transformation into its dodecanuclear derivative **84** by Scheme 2.36.

The cylindrical covalent capsules **85–88** have been synthesized in [47] by Scheme 2.37 using tris-(*para*-cymene)ruthenium trialdehyde *7* as a metallamacrocyclic precursor. Its imine condensation with *para*- and *meta*-xylenediamines gave a mixture of two diastereomers of an imine capsule. One of these diastereomers with a D_3-symmetric cylindrical shape has the apical trinuclear metallamacrocyclic fragments of the same chirality, whereas the C_{3v}-symmetric cylindrical diastereomer has these fragments of opposite chiralities.

Condensation of the trialdehyde building block **7** with ethylenediamine and 2,7-bis-(aminomethyl)-3,6-dimethoxynaphthalene

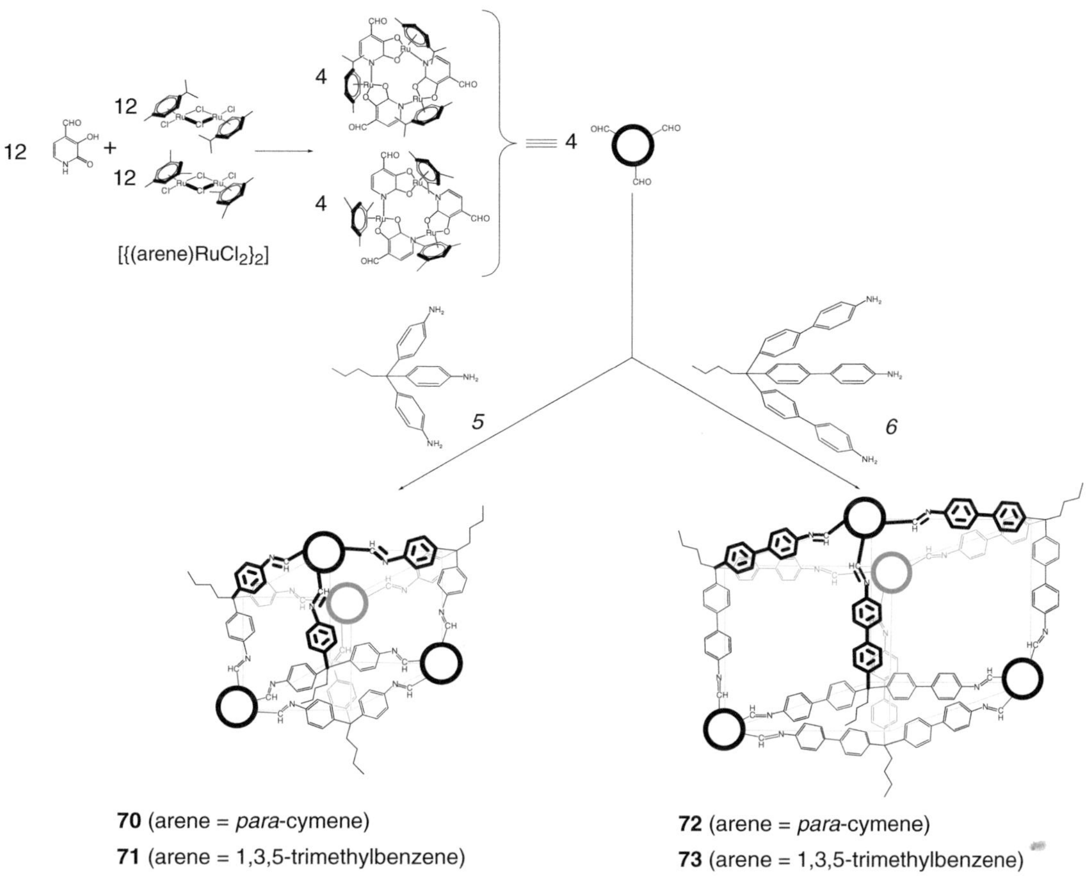

70 (arene = *para*-cymene)
71 (arene = 1,3,5-trimethylbenzene)

72 (arene = *para*-cymene)
73 (arene = 1,3,5-trimethylbenzene)

Scheme 2.31

afforded single diastereomeric forms of D_3-symmetric cylindrical covalent capsule **87** and tetrahedral polyimine cage framework **82**, respectively. These ligand syntones [47] demonstrate self-sorting behavior: only two products (i.e., **82** and **87**) of imine condensation by Scheme 2.38 have been isolated from the reaction mixture in the case of the system metallamacrocyclic trialdehyde *7* – ethylenediamine – *tren* in their molar ratio 6:3:4.

The metallamacrocyclic trialdehydes *10* and *11* undergo [1 + 1] condensation by Scheme 2.39 with tris(aminomethyl)benzene as a rigid triamine ligand syntone giving the capsules **89** and **90**; those are not formed with more nucleophilic tripodal alkylamines. A cylindrical hexanuclear capsule **91** and its trinuclear analog **92** have been prepared in [47] by *one-pot* self-assembly of [(*para*-cymene)$RuCl_2]_2$, the corresponding pyridone, and di- and triamine ligand syntones, respectively, by Scheme 2.40. At the same time, the attempts to obtain large tetrahedral cage frameworks of this type are reported in [47] to be unsuccessful. So, *one-pot* reaction is suitable for the synthesis of more simple covalent capsules of this type, whereas their more complex analogs can be obtained using a two-stage synthetic procedure.

2.1.2 Encapsulation of Anions

Hexaprotonated forms of the covalent capsules **8** and **10** [48] can encapsulate perchlorate and hexafluorosilicate anions, respectively, thus forming 1:1 cage complexes by Scheme 2.41. The latter guest has been preferentially encapsulated by $H^+_6 \cdot$ **8** even in the presence of a large excess of tetrafluoroborate ion [48].

Scheme 2.32

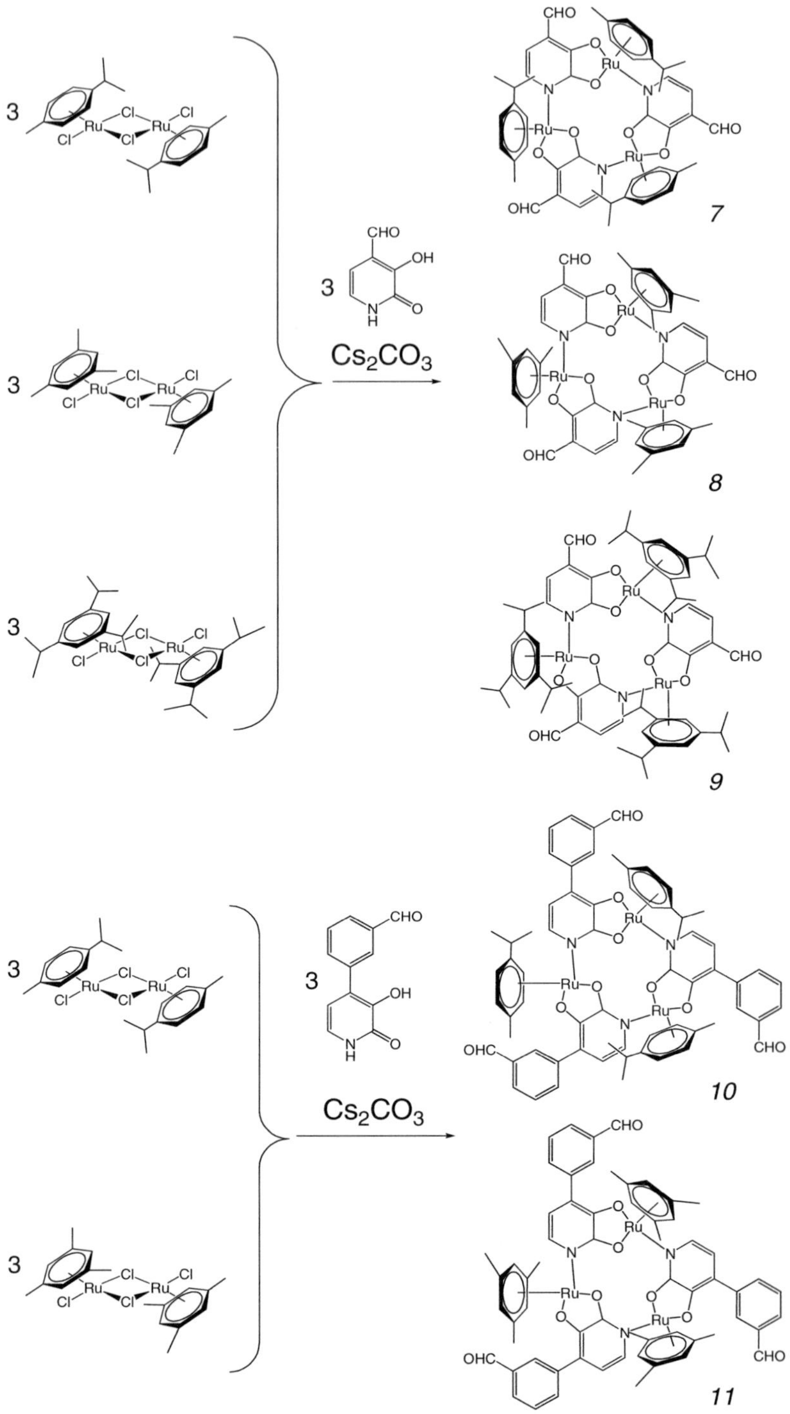

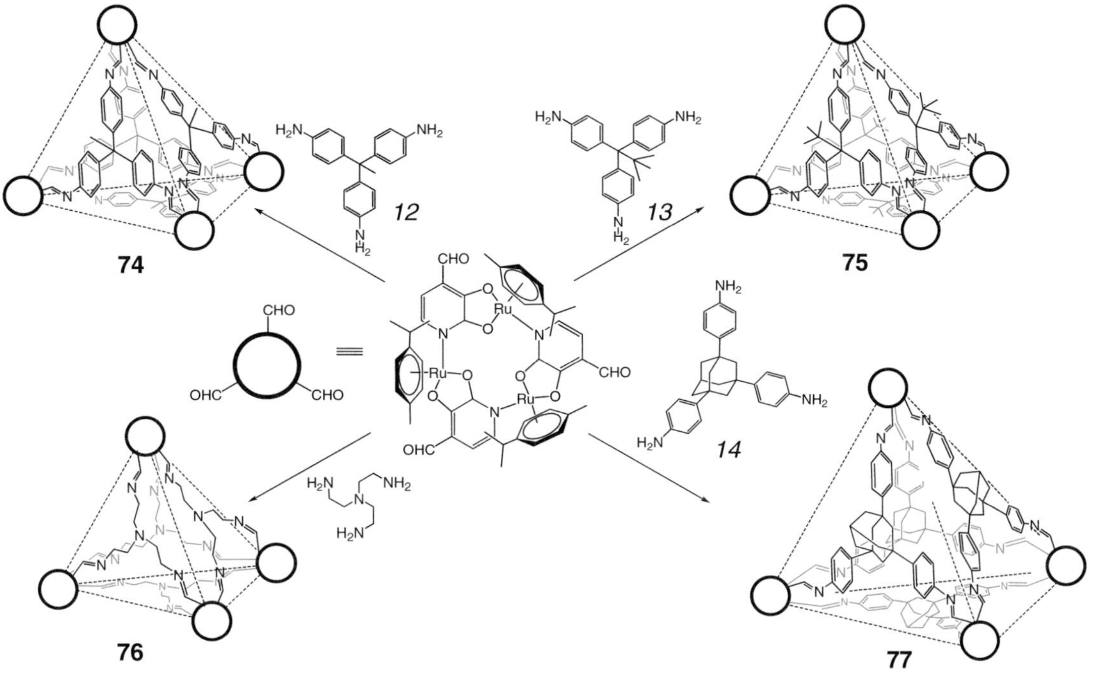

Scheme 2.33

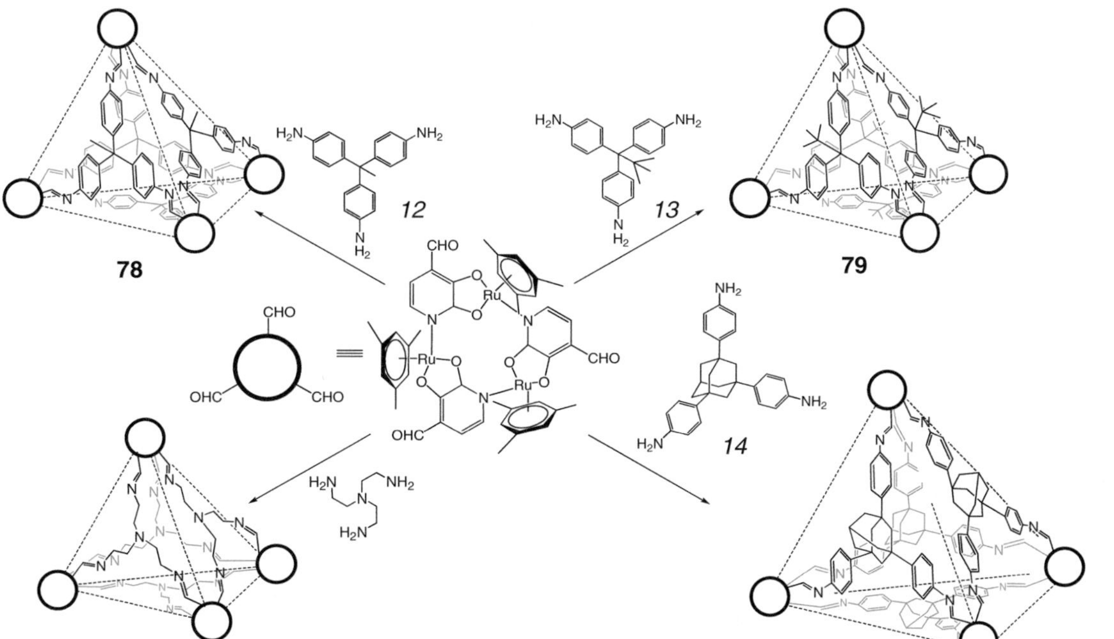

Scheme 2.34

Scheme 2.35

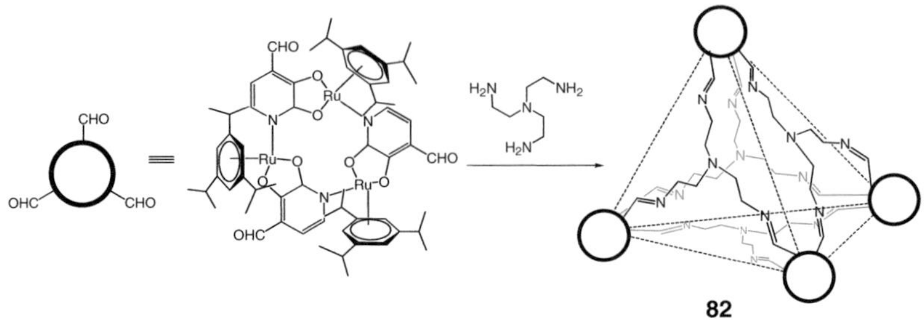

Scheme 2.36

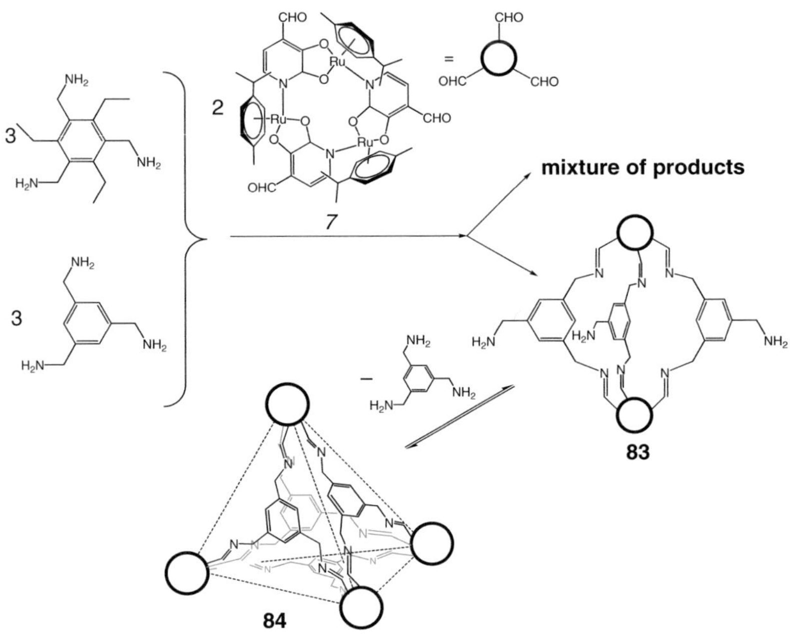

According to the data from potentiometric titration, solution ^{1}H NMR, and single-crystal X-ray diffraction [49], the hexaprotonated form of **14** encapsulates two nitrate anions, thus giving the corresponding 1:2 cage complex shown in Schemes 2.42. Encapsulation of nitrate and perchlorate anions by the hexaprotonated forms of caging ligands **10** and **14** has been studied in detail [50] using potentiometric and NMR titration methods; they showed dominant formation of 1:1 cage complexes for both of these guests (Scheme 2.43). Good geometric host–guest complementarity has been confirmed for the perchlorate ion by X-ray diffraction experiment. The nitrate anion is also reported in [50] to also form the corresponding host–guest 1:2 cage complexes in the presence of its large excess [50].

A hexaamine capsule **14** is described in [51] to bind a fluoride anion in a wide range of pH. Its hexaprotonated form encapsulates one F^- ion and a solvent water molecule, thus giving a heteroguest 1:1:1 cage complex by Scheme 2.42. As follows from X-ray diffraction data, these caged guest species are shifted from the geometrical center of this covalent capsule in the direction of one of its three ribbed fragments.

Detailed X-ray diffraction and solution phase studies of encapsulation of double negatively charged guest oxoanions $S_2O_3^{2-}$, SO_4^{2-} and CrO_4^{2-} by hexaprotonated forms of covalent capsules **8**, **10**, and **14** have been performed in [52]. The encapsulation is reported to involve a complex balance of competing factors such as anion free energy of hydration, host basicity and solvation, and complementarity of steric matching between these hosts and thiosulfate, sulfate, and chromate dianionic guests. No correlation between the sta-

Scheme 2.37

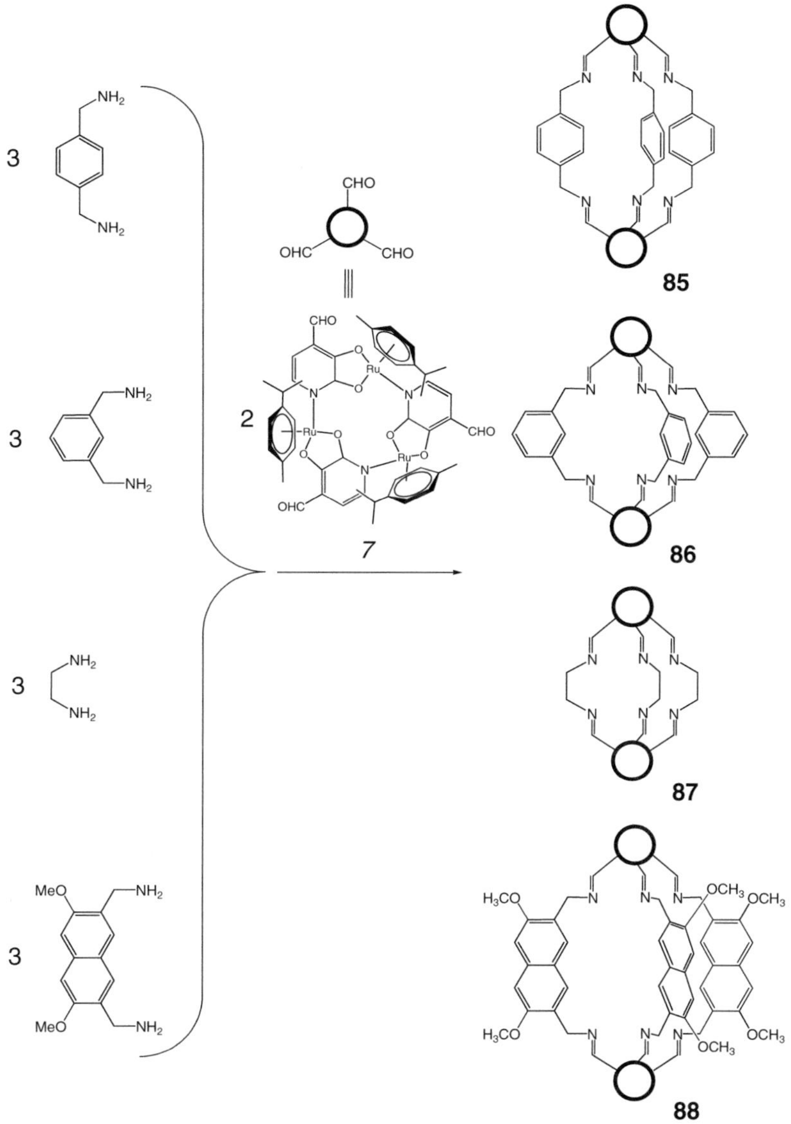

bility constants and host–guest hydrogen-bonding distances has been observed, thus suggesting that substantial entropic effects (first of all, solvation–desolvation) contributed to the free energies of such encapsulation. This deduction is supported in [52] by observation of higher stability constants in the case of least highly hydrated oxoanions.

Caging ligands of appropriate size and with suitable donor groups are reported in [53] to be very promising for selective encapsulation of perrhenate and pertechnetate oxoanions. In particular, polyamine capsules, the derivatives of tripodal amine *tren*, in their protonated forms of varying size and geometry, are good candidates for encapsulation of these anions through multiple hydrogen bonds and electrostatic interactions involving protonated amino groups [53]. The selectivity of this binding is governed by geometric complementarity between the guest oxoanion and the cavity of a capsule depending on the degree of protonation: at low pH, the cavity of its fully protonated form is expanded due to electrostatic repulsions thus representing the best condition for selectivity [54]. Polyamine caging ligands **7**, **8**, **10**, and **14** and their per-

Scheme 2.38

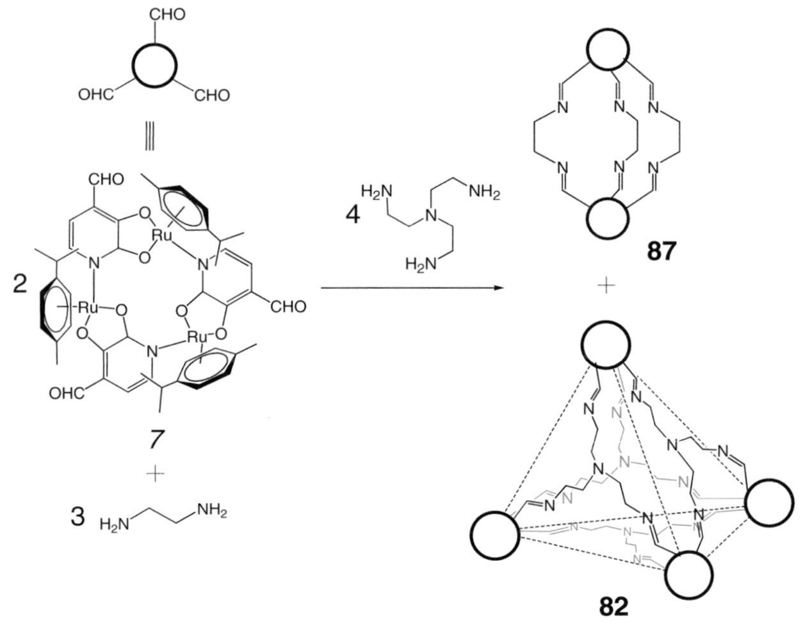

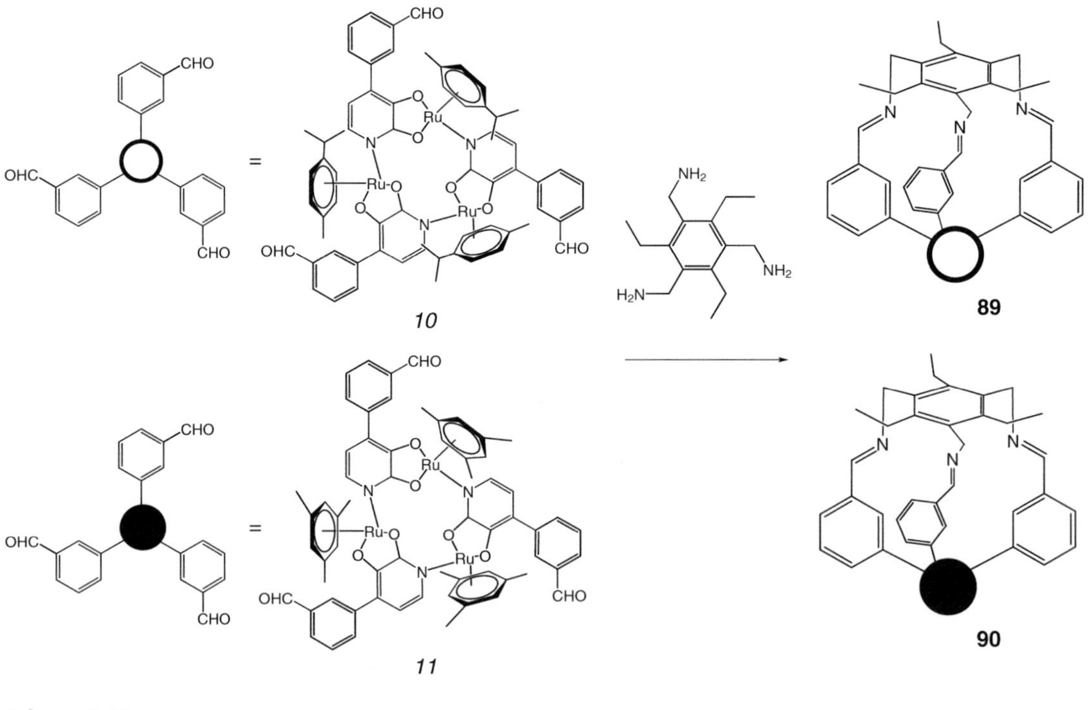

Scheme 2.39

methylated derivatives **93–96** (Scheme 2.44) have been studied in [55] as potent extractants for ReO_4^- and TcO_4^- anions. In all cases, the pertechnetate anion has been extracted better than ReO_4^- anion due to higher lipophilicity of the former. The extractive ability of these macrobicyclic ligands increases in a row **7** < **8** < **94** ≈ **93** < **95** < **96**, and *N*-methylated capsules showed better extractive abilities than their amine precursors. The *N*-methylated

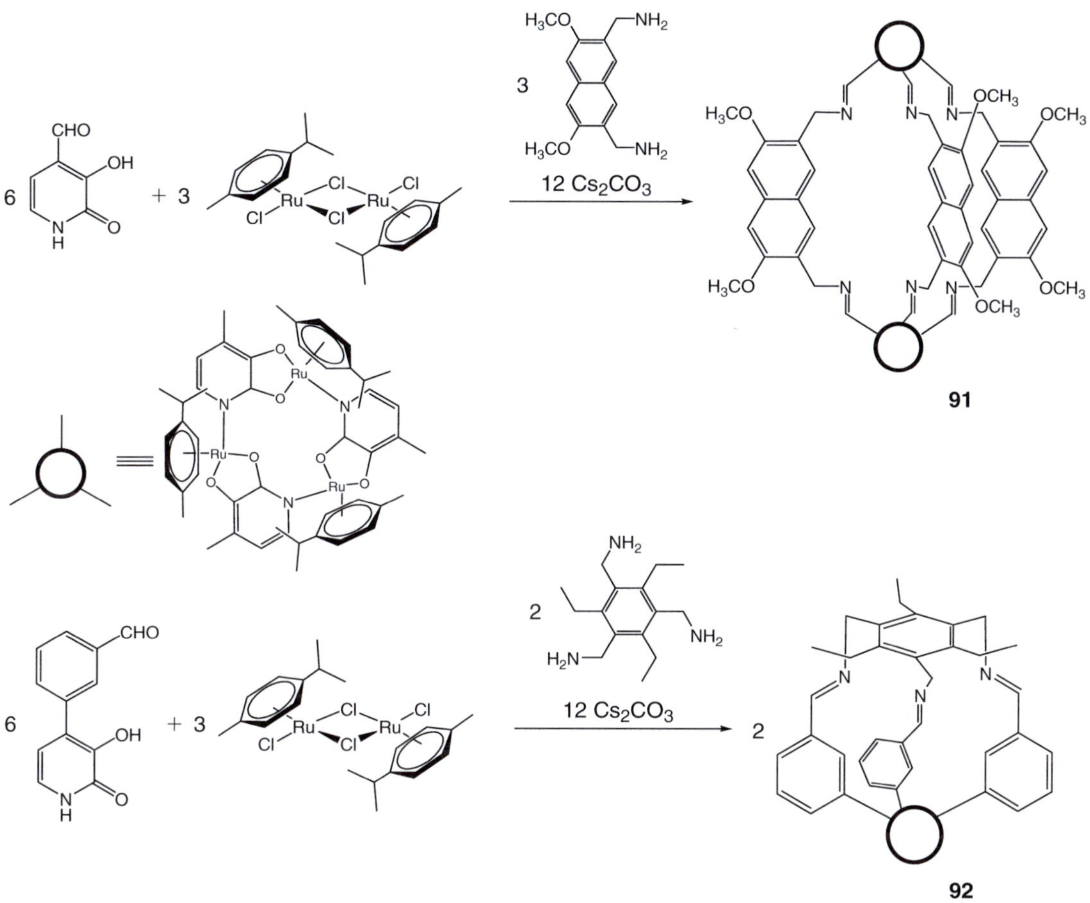

Scheme 2.40

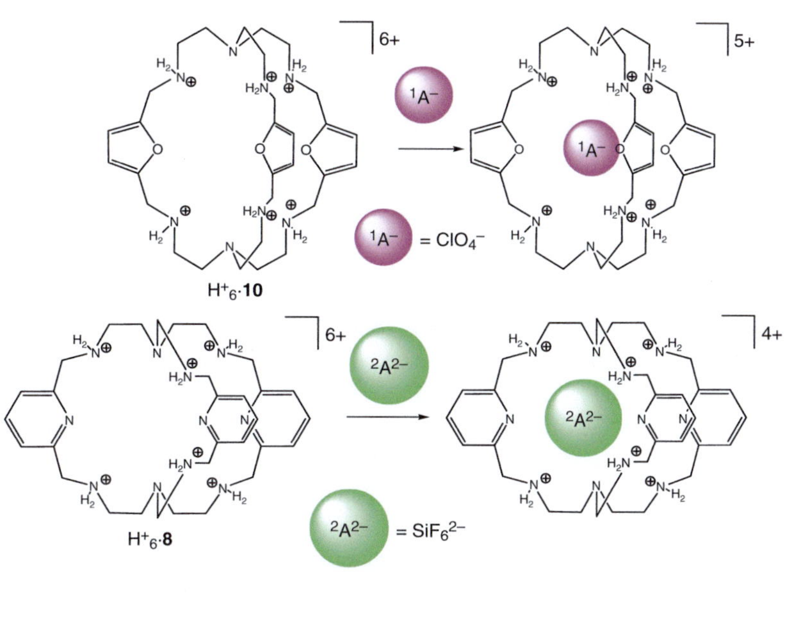

Scheme 2.41

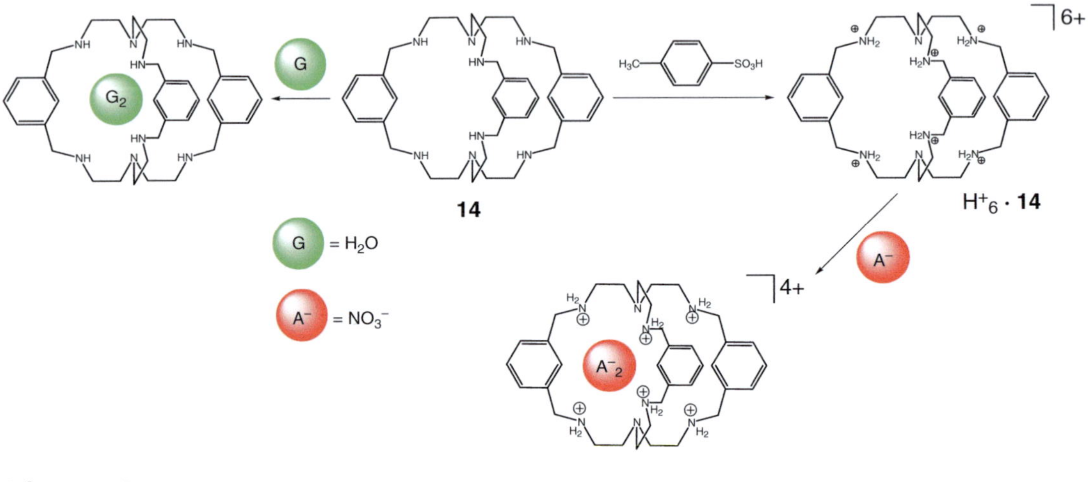

Scheme 2.42

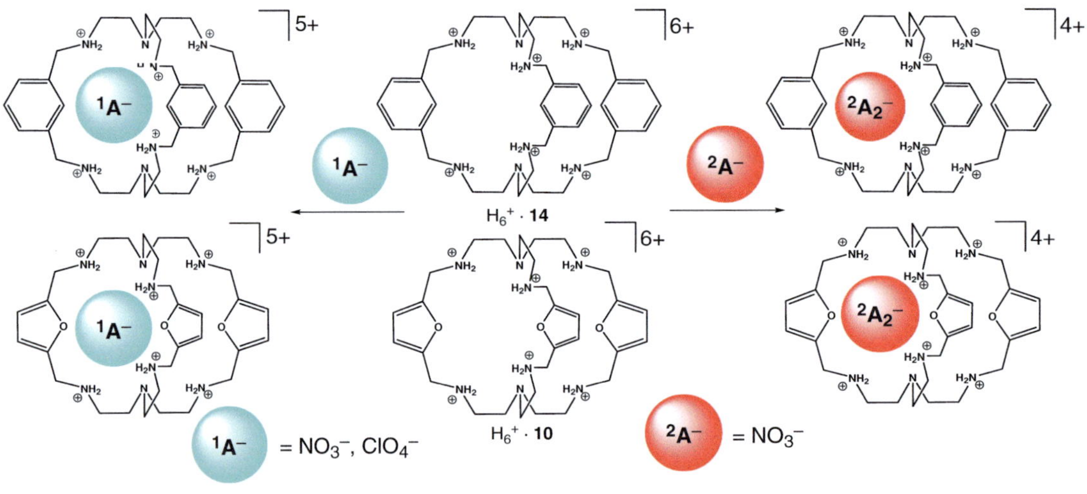

Scheme 2.43

pyridine-based ligand **96** has the highest extractive ability [55].

Macrobicyclic ligands **7**, **8**, **10**, **14**, and **97** with different ribbed fragments have been studied in [56] as extragents for perrhenate anion. The binding ability of their hexaprotonate forms decreases in a row H^+_6·**7** > H^+_6·**14** >> H^+_6·**10** ≈ H^+_6·**8**. These forms of **7** and **14** also encapsulate ReO_4^- to give 1:1 cage complexes. The highest stability of the cage complex of encapsulating ligand H^+_6·**7** is explained by complimentary of its cavity to the caged anion occupying the center of the covalent capsule (as follows from X-ray diffraction data [56]) and forming hydrogen bonds with ribbed secondary amino groups and with apical tertiary amine fragments as well. This macrobicyclic host binds $^{99}TcO_4^-$ anion in acidic aqueous solutions giving a 1:1 cage complex, which is more stable than its analog with perrhenate anion [53].

The solid phases **98**–**100** (Scheme 2.45) with immobilized polyamine covalent capsules have been designed in [57] for binding, separation, and extraction of perrhenate and pertechnetate anions from their aqueous solutions. While the Amberlite CG50 derivative **100** was unsuitable for this purpose, the silica-based sorbents **98** and **99** showed a promising performance. Their

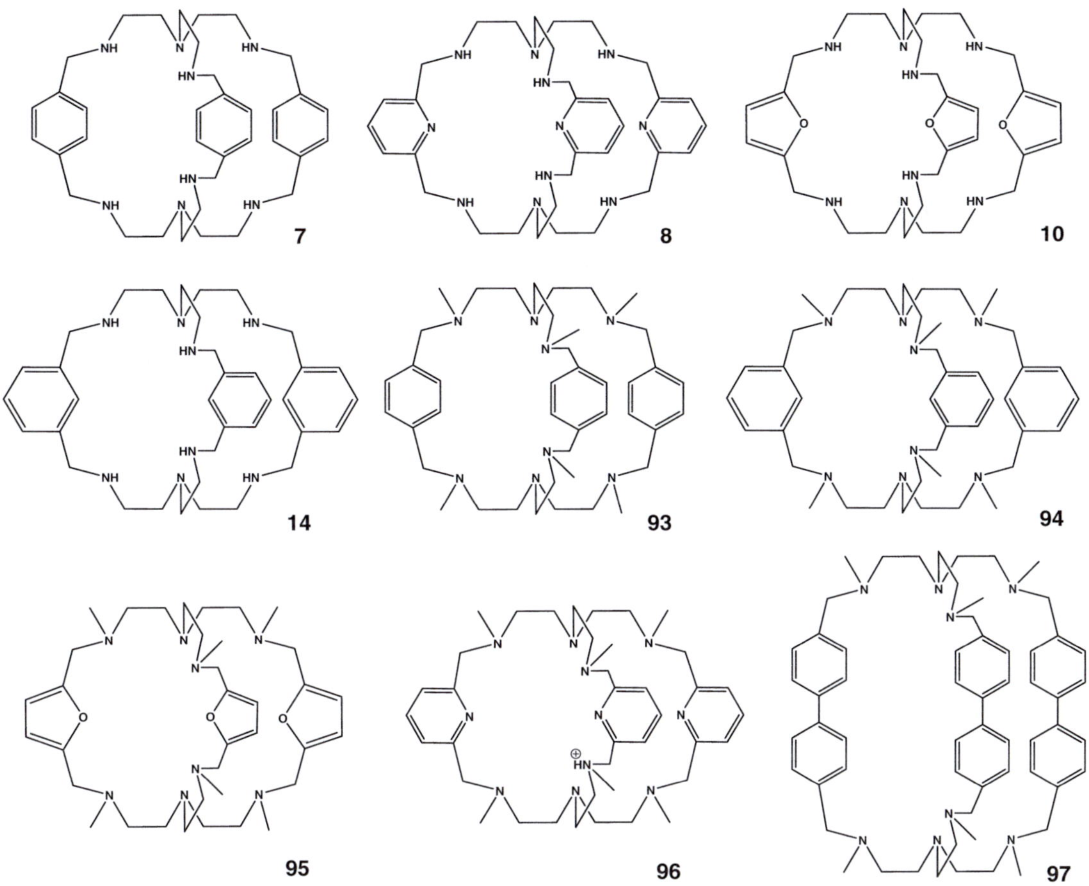

Scheme 2.44

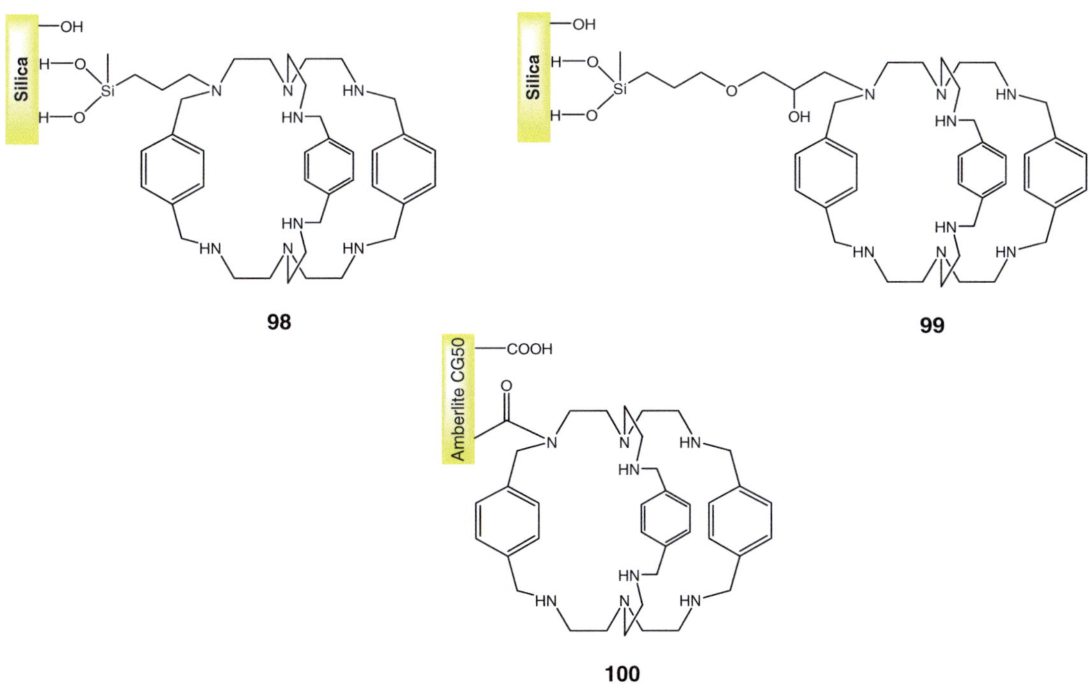

Scheme 2.45

spacer fragments are slightly different in length and flexibility; the sorbent **98** has a higher concentration of active sites, but its analog **99** displays higher affinity to these oxoanions and faster kinetics of their encapsulation [57].

Introduction of a fluorescent anthracene ribbed fragment into the polyamine cage framework **101** (Scheme 2.46) has been used in [58] for selective recognition and fluorescent detection of pertechnetate anion in its micromolar acidic aqueous solutions. The encapsulation by hexaprotonated form of **101** caused quenching of the ligand's anthracene emission due to the formation of a stable 1:1 cage complex by multiple strong hydrogen bonds between the protonated amino groups of this caging ligand and TcO_4^- anion.

The protonated form of the octaamine macrobicyclic ligand **7** encapsulates perchlorate, hydrosulfate, and hydrohexafluorosilicate monoanions by Scheme 2.47 to give 1:1 cage complexes; those have been studied in [59] by X-ray diffraction. Hexaprotonated cage framework of **7** in its supramolecular assembly with encapsulated ClO_4^- anion and octaprotonated form of **7** in its cage complexes with HSO_4^- and $HSiF_6^-$ anions have an *endo,endo*-conformation with the N…N distances between the apical nitrogen atoms of approximately 9.85, 7.76, and 7.57 Å, respectively. These caged tetrahedral and octahedral anions form NH…O and NH…F hydrogen bonds with the protonated secondary amino groups of $H^+_6 \cdot$ **7** and $H^+_8 \cdot$ **7** [59].

Condensation of tripodal amine *tren* with dialdehyde *15* followed by reduction of the hexaimine Schiff-base product afforded an amine macrobicyclic ligand **102**; its hexaprotonated form gave 1:1 complexes with a series of mono-, bis-, and tris-carboxylate anions (Scheme 2.48).

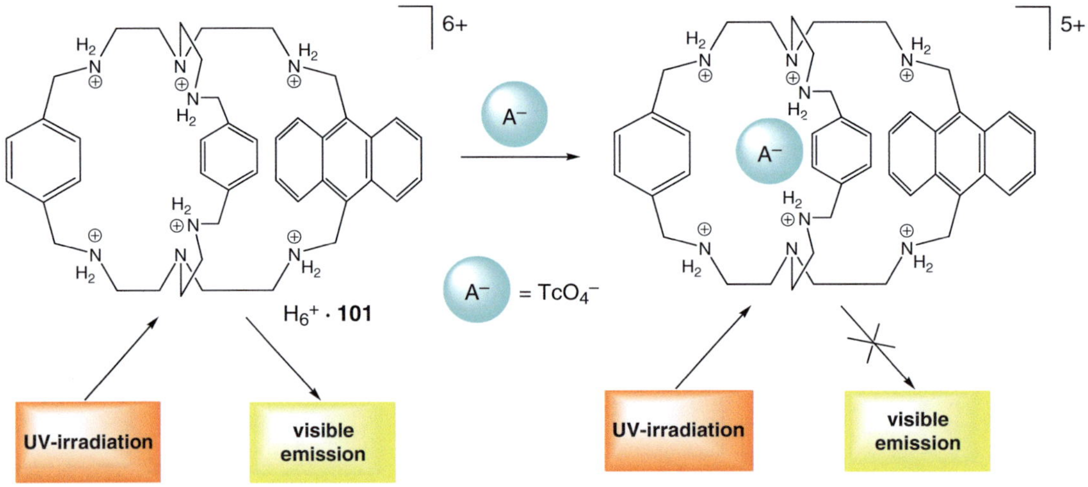

Scheme 2.46

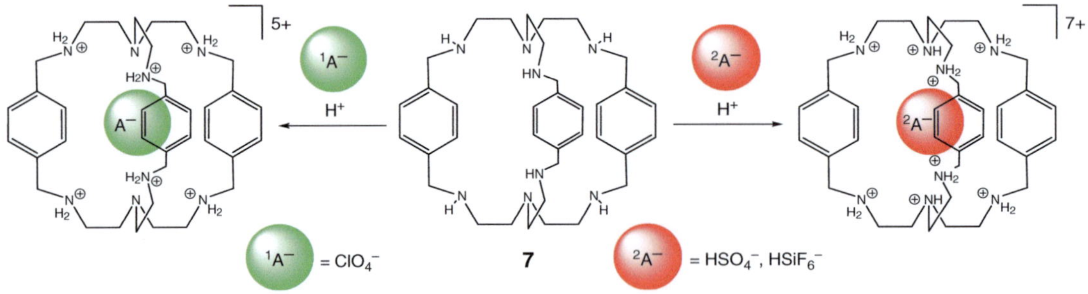

Scheme 2.47

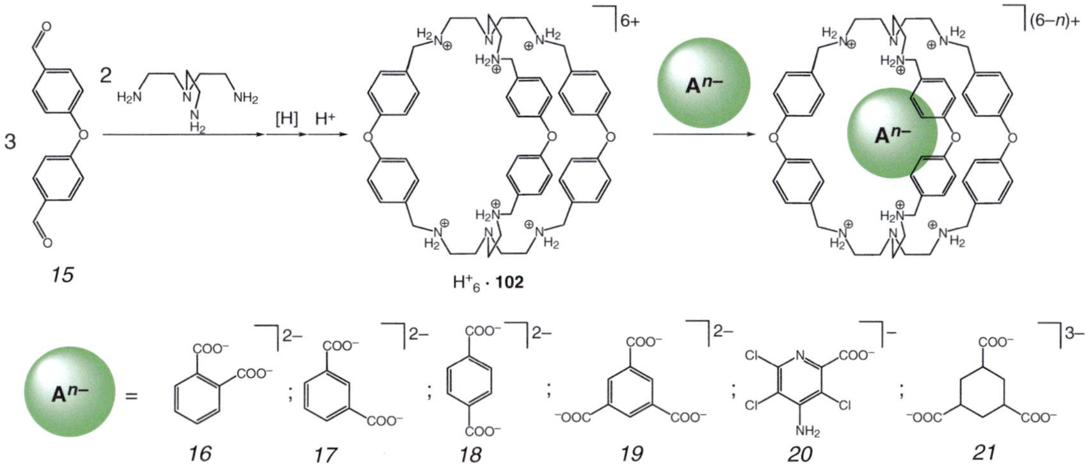

Scheme 2.48

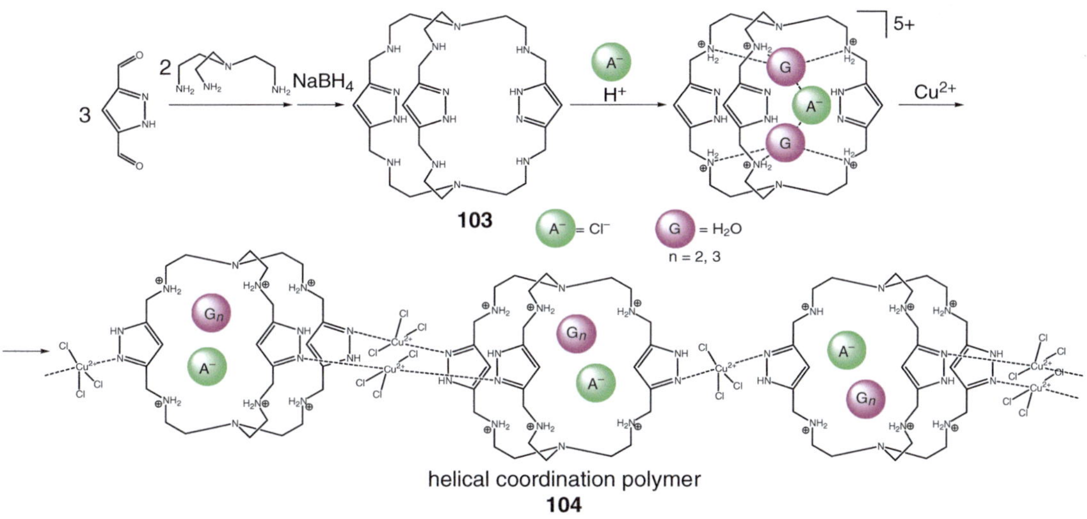

Scheme 2.49

The binding constants for this caging ligand at pH = 5.5 decrease in a guest row *19>18>16~17>21~20*. It has been shown [60] that the electrostatic (Coulombic) interactions, hydrogen bonds, and structural and size matching between the macrobicyclic host and the substrate guest species [60] played the key roles in such molecular recognition.

The hexaprotonated form of tris-1*H*-pyrazole polyamine capsule **103**, prepared in [61] by Scheme 2.49 using an imine condensation of 3,5-pyrazoldicarbaldehyde with tripodal amine *tren* followed by reduction, is described in [62] to be a macrobicyclic receptor for chloride anion. The caging ligand H^+_6·**103** encapsulates one chloride ion and two water molecules, thus forming a heteroguest 1:1:2 cascade cage complex by hydrogen bonding between the caged species. A 1D helical coordination polymer **104** (Scheme 2.49) formed by the protonated polyamine capsules, which are cross-linked by copper(II) ions, also encapsulates chloride anion behaving as a multianionic receptor. Switching from the monomeric capsule **103** to the corresponding polymeric receptor **104** is described in [62] to be performed by metal ions and pH.

An octaprotonated form of the covalent capsule **13** is described in [20] to encapsulate chloride anion by Scheme 2.50 giving a 1:1 cage complex. In this case, the guest forms hydrogen bonds only with two protonated ternary capping amino groups of the caging ligand in *in–in* conformations. As a result, this cage complex is contracted along its C_{3h}-symmetry axis (with tripodal apical groups shifted inward the macrobicyclic cavity), whereas the protonated secondary amino groups of the ribbed fragments are shifted outward due to the expansion of the macrobicyclic framework along its C_2-symmetry axis [20].

Imine macrobicyclic complexes **105** and **106** with two and four encapsulated metal(II) ions respectively, have been prepared in [63] by Scheme 2.51; their reduction with $NaBH_4$ gave tris-bipyridine capsules **107** and **108** with tripodal amine capping fragments. These covalent capsules are described in [63] to be more flexible and thus more suitable for the encapsulation of large anionic guests.

2.1.3 Encapsulation of Cations

A Schiff-base macrobicyclic ligand **109** has been designed in [64] for efficient encapsulation of complementary aromatic molecules: its 1,3,5-substituted benzene bases are able to form strong stacking interactions with aromatic guests, while the lability of ethylene glycol ribbed fragments with polar donor centers allows for fine-tuning the shape of the encapsulating ligand to match that of the guests as well as for binding them with these donor centers. Template condensation by Scheme 2.52 afforded [2]-rotaxane as the sole product of this multicomponent template reaction in the presence of dumbbell-shaped sterically hindered aromatic dication as a matrix. According to the force-field modeling calculations, [2]-rotaxane is formed by both stacking and hydrogen bonding interactions as the origin of thermodynamic selectivity [64].

2.1.4 Intraligand Metal-Assisted Binding

Tren-capped binucleating caging ligand **110** and its *N*-methylated derivative **111** (Scheme 2.53) have been synthesized for the first time by J.-M. Lehn and coworkers [65, 66] using the macrocyclic precursor *22*. The ligand **110** undergoes anion-depended stepwise protonation: the lower protonation constants are substantially higher in the presence of chloride and nitrate anions than those with 2,4,6-trinethylbenzenesulfonate and perchlorate ions (probably, due to the encapsulation of Cl^- and NO_3^- anions by the protonated capsule $H^+_6 \cdot$ **110**). This ligand also encapsulates zinc, copper, nickel, and cobalt(II) ions in a stepwise manner to give host–guest 1:1 and 1:2 cage complexes; its protonated form gave the same mono- and binuclear complexes by Scheme 2.54. Their stability constants decrease in a row Cu^{2+} > Zn^{2+}, Ni^{2+} > Co^{2+}, and those for binding of the second guest metal ion are lower by the factor of 10^4–10^5. These binuclear complexes encapsulate hydroxide ion to give cascade 1:2:1 heteroguest capsules with a bridging OH^- anion between their encapsulated metallocenters; the binding of the second hydroxide ion is less common. The binuclear cobalt(II) complex of **110** is described in [65] to reversibly bind molecular oxygen, forming doubly bridged species with both

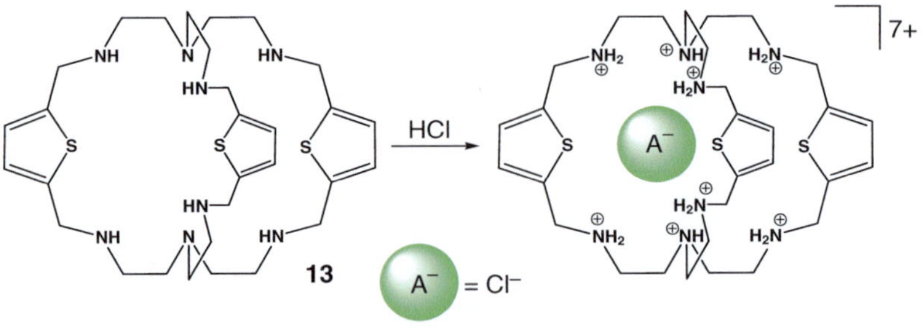

13

Scheme 2.50

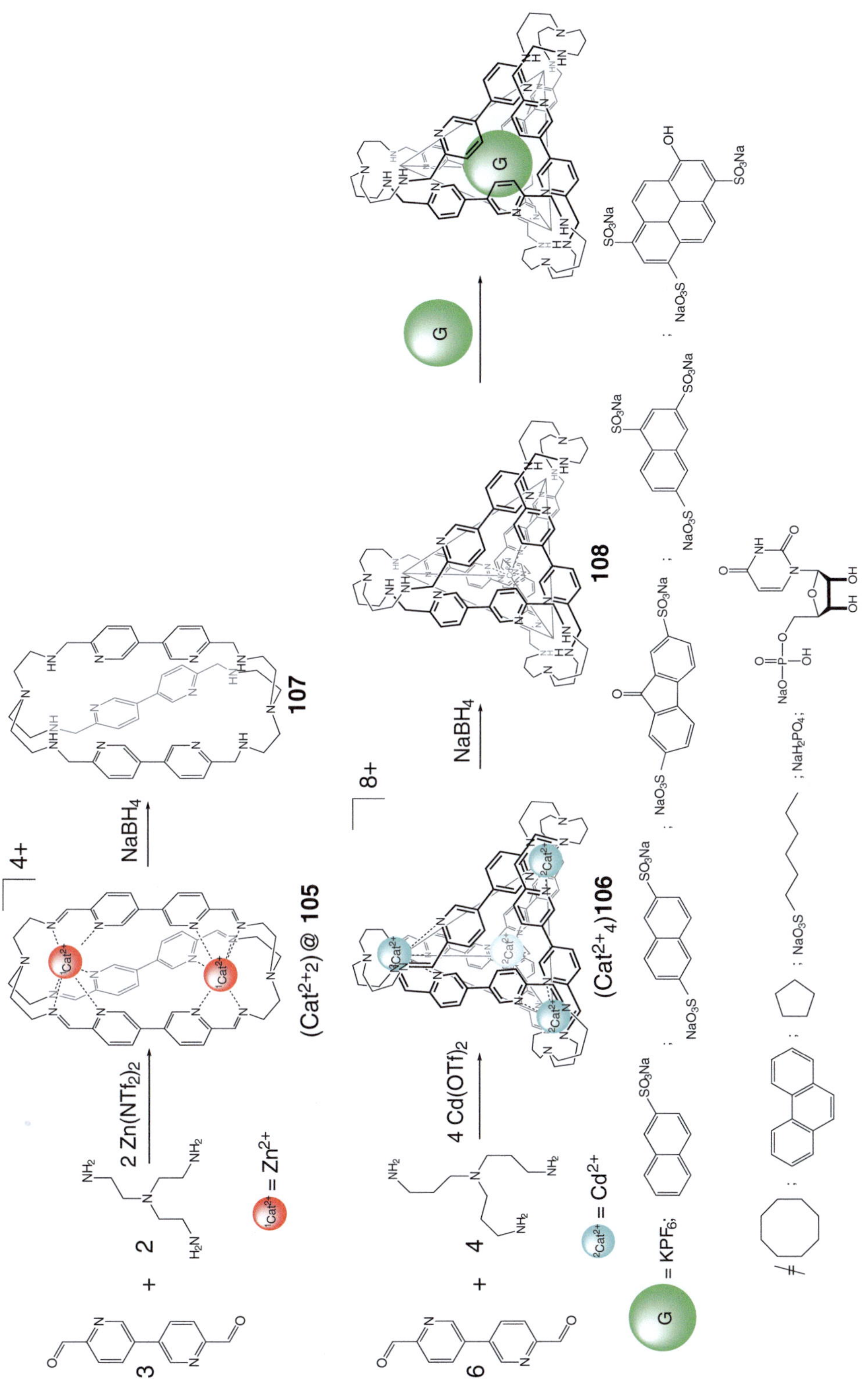

2 Zn(NTf2)2
NaBH4
(Cat2+2)@105
107
4 Cd(OTf)2
(Cat2+4)106
108
= Zn2+
= Cd2+
= KPF6;
; NaH2PO4;

Scheme 2.51

Scheme 2.52

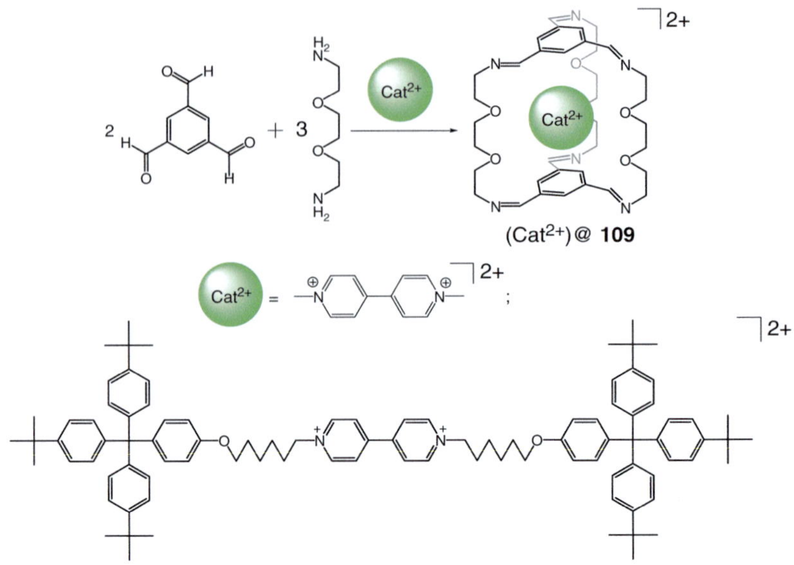

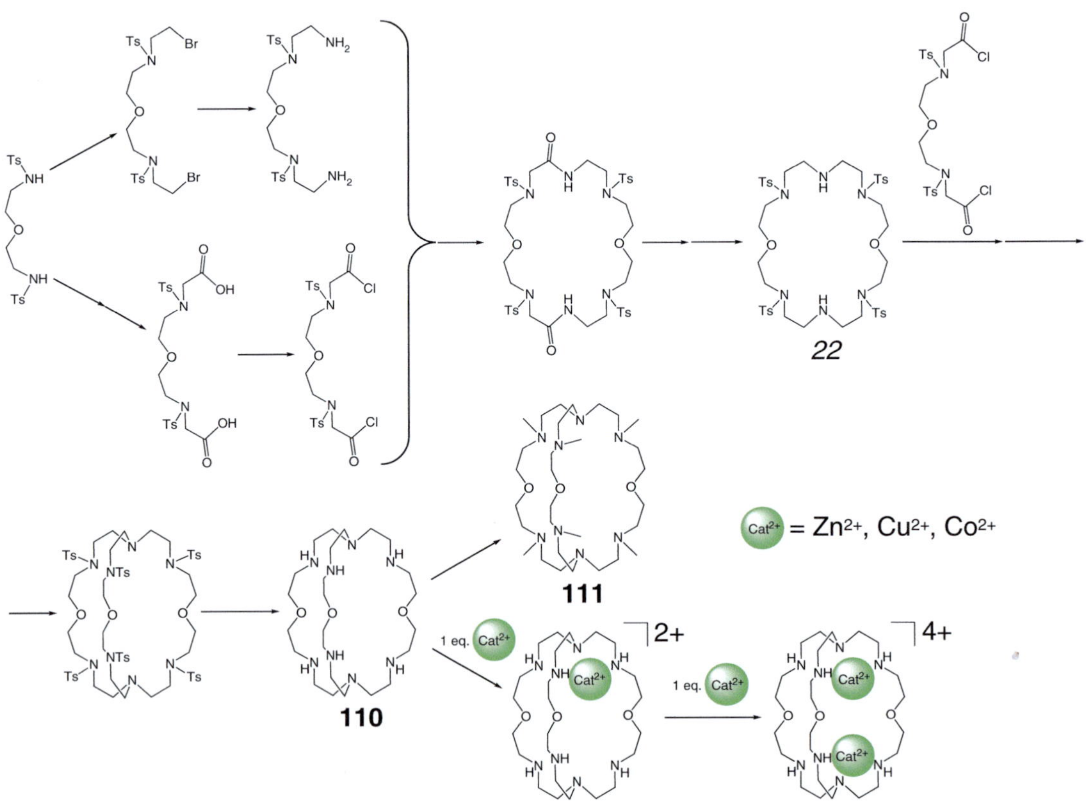

Scheme 2.53

Scheme 2.54

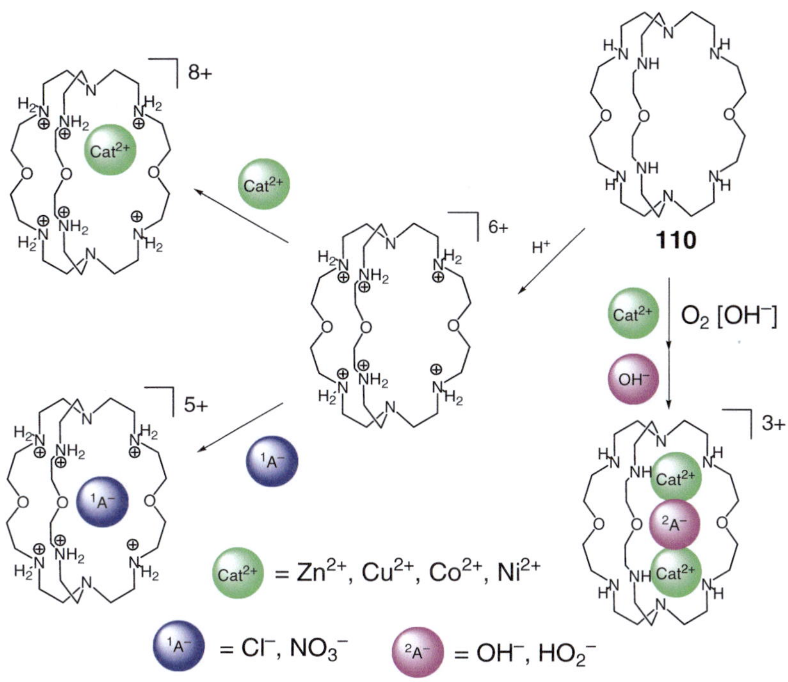

hydroxo and peroxo bridges within a cavity of the cage framework; however, its affinity to O_2 is relatively low due to the steric hindrances within this cavity. In the case of a binuclear cobalt(II) complex with encapsulated dioxygen molecule (Scheme 2.55), the oxygenation constant is three orders lower than those for regular polyamine cobalt(II) chelates [67]. UV-vis spectrum of this dioxygen-containing heteroguest 1:2:1 cage complex contains the intensive CTB at 380 nm that is characteristic of oxygenated cobalt polyamines. The band disappears after heating of an aqueous solution of this complex up to 90 °C and re-appears after its cooling back to 25 °C, suggesting reversibility of the oxygenation process during cycling of temperature. An irreversible oxidation of the 1:2:1 cobalt(II) complex to its cobalt(III)-containing oxygen-inactive analog by Scheme 2.56 is observed in [67] above 90 °C.

Stepwise protonation of a macrobicyclic ligand **110** and its complexation with copper(II) ion have been thoroughly studied in [68]. This ligand undergoes protonation to give up to heptaprotonated macrobicyclic species; this heptacation has been detected in the presence of HF_2^- anion formed in the system HF–NaF. Its hexaprotonated form H^+_6 · **110** is reported in [68] to be the most efficient caging ligand for the encapsulation of halide ions by Scheme 2.57 with high affinity toward fluoride anion. The copper(II) ion gave mono- and binuclear cage complexes with the covalent capsule **110**, followed by cascade encapsulation of a halide ion by Scheme 2.57. These 1:1 and 1:2 copper(II) cage complexes most efficiently bind fluoride anion (as compared to chloride ion) by strong hydrogen bonding within the cavity of **110**. In the case of its binuclear cage complex, iodide anion gave a more stable cascade compound than chloride ion, as I^- ion better fits into the vacant cavity between the two caged copper(II) cations. The OH^- ion competes with these bridging halide anions due to its polarity and the ability to form hydrogen bonds with oxygen atoms of the ethylene glycol ribbed fragments of the macrobicyclic ligand **110**. Thus, the encapsulation of hydroxide ion by this binuclear copper(II) cage complex at high pH values is reported in [68] to exclude its binding with halide anions.

Protonation and binding properties of a pentamethylene-containing ligand **111** as

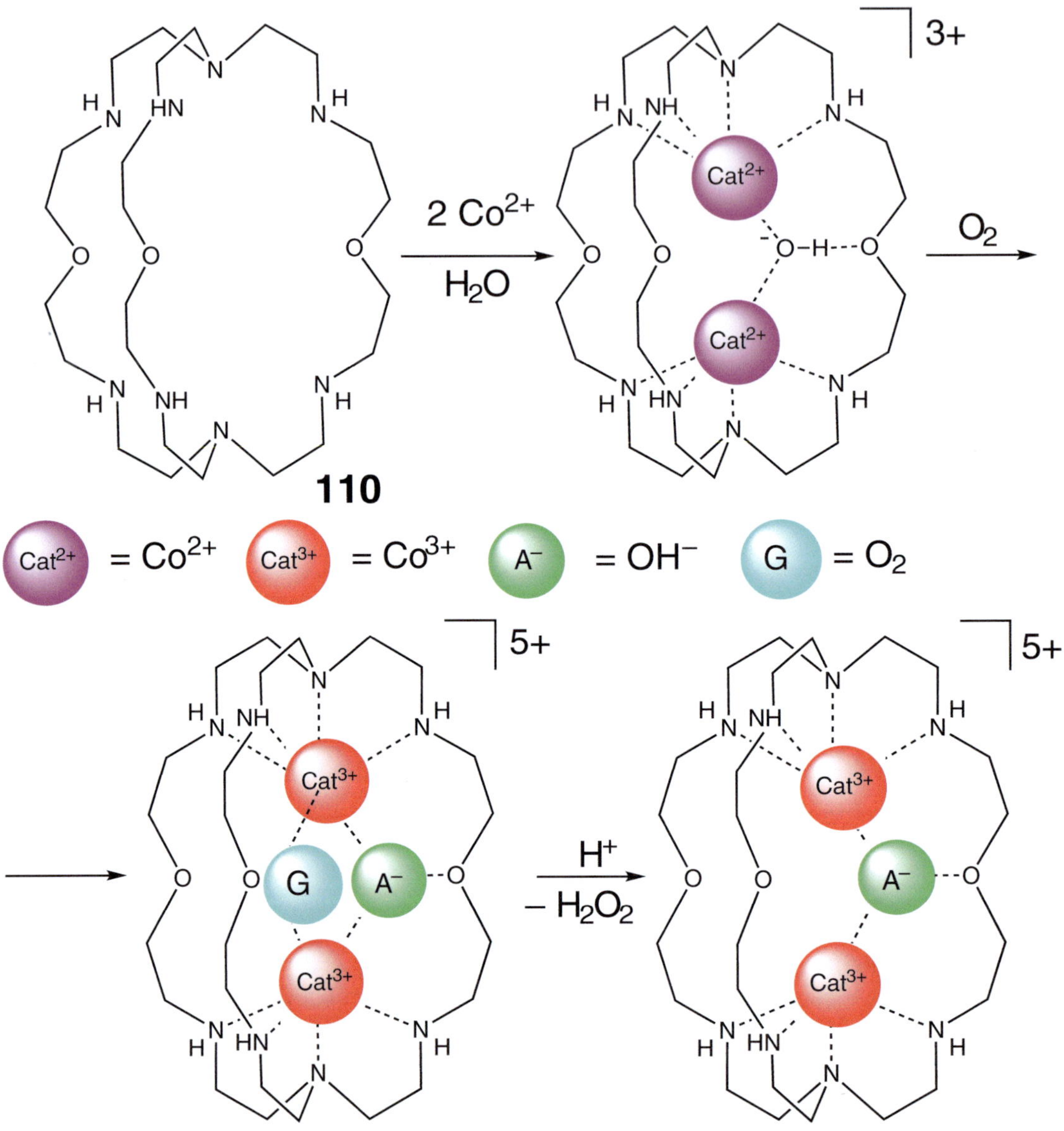

Scheme 2.55

aliphatic analog of ethylene glycol-based macrobicycle **110** have been studied in [69] by potentiometric titration. The initial polyamine capsules have six successive protonation constants and bind the seventh H^+ ion only in the presence of fluoride ion by Scheme 2.58. In both these cases, its binding causes an increase in the protonation constants, and the seventh protonation step takes place only with an encapsulation of HF_2^- anion. This process is more favorable for the trioxygen-containing capsule **110** than for its aliphatic analog **111** with fully extended aliphatic chains and causes an increase in the distance between its apical *tren* fragments. Moreover, the exterior of a cavity in this case is more hydrophobic than that of **110** [69]. For both these encapsulating ligands, their hexaprotonated forms most efficiently bind F^- and HF_2^- anions, while the stability constant for a 1:1 cage complex of an aliphatic macrobicycle **111** with one encapsulated copper(II) ion is two orders lower than that for its ethylene glycol-based analog **110** due to less efficiently preorganized donor groups in **111**. The same constant for the second guest copper ion significantly exceeds that of **110** due to the above preorganization on the first complexation stage. Substantial increase observed experimentally in the

Scheme 2.56

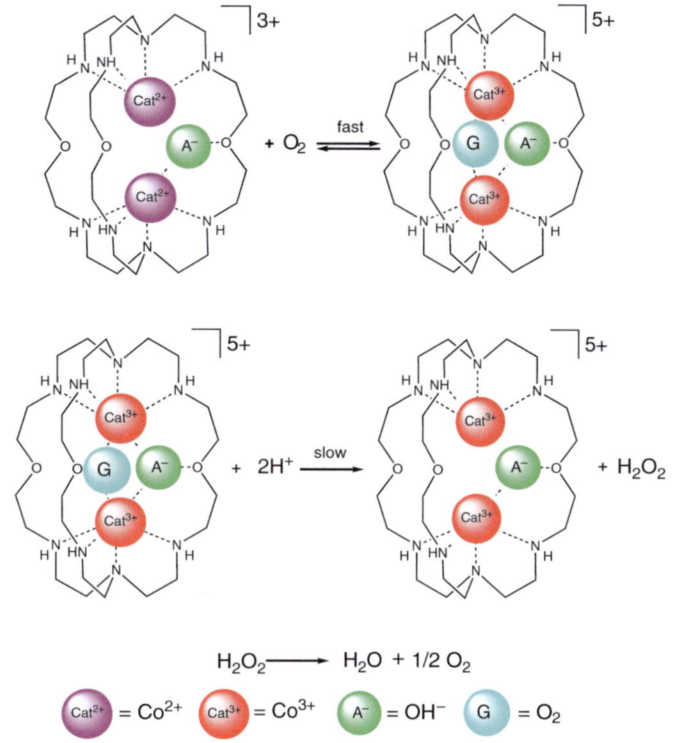

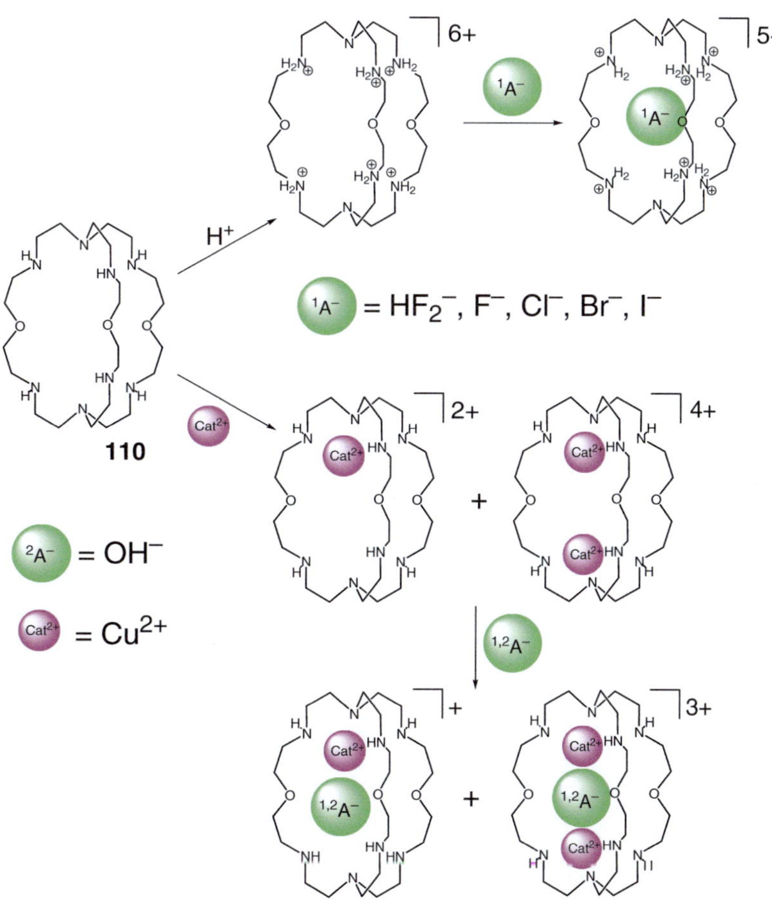

Scheme 2.57

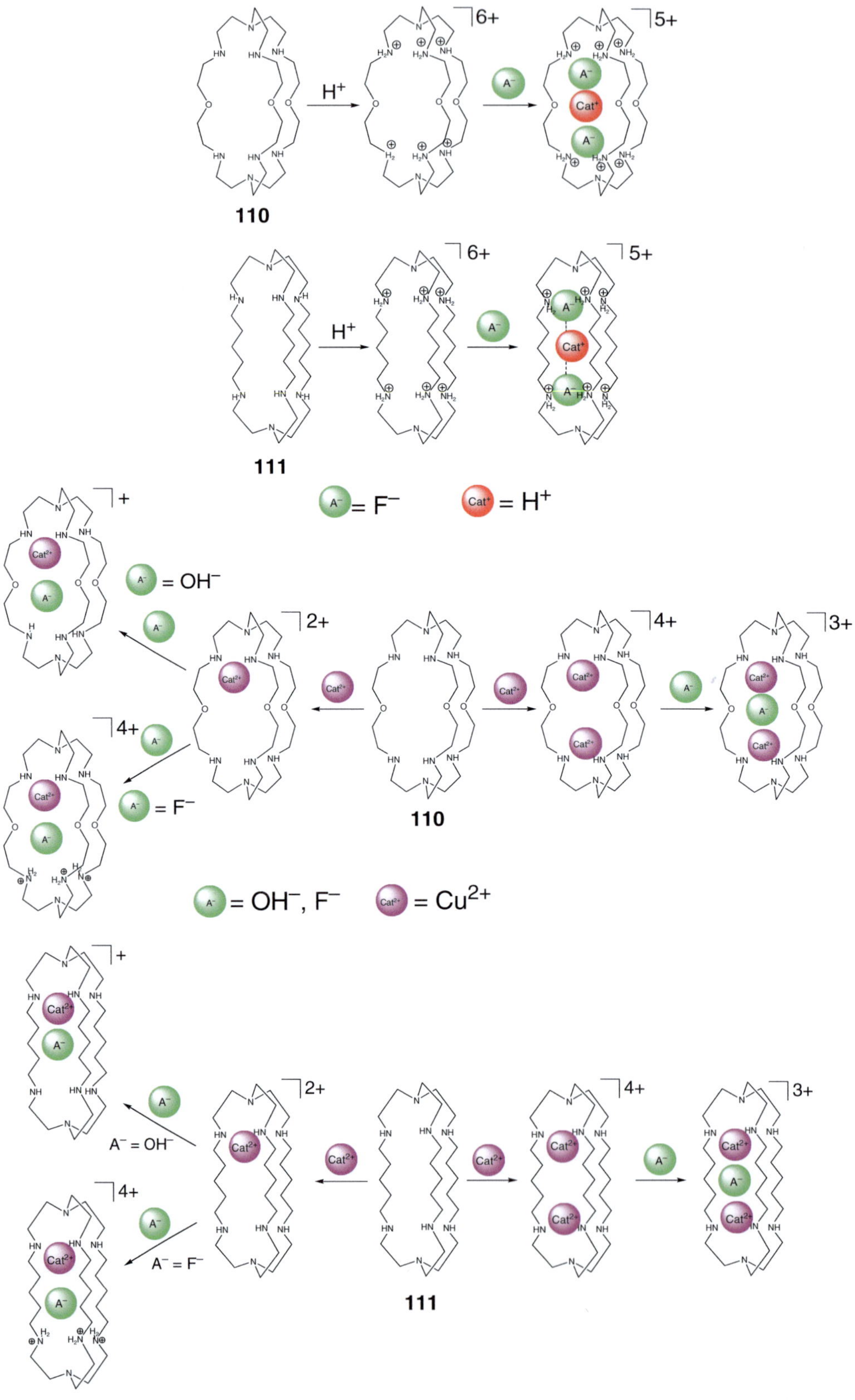

Scheme 2.58

protonation constants for the caging ligand **111** is explained in [69] by higher basicity of its amino groups due to electrometric effects of the ribbed fragments. The dramatic difference between their 1:2 cage complexes with copper(II) ion has been observed in the case of an encapsulation of hydroxide ion: the complex of the ethylene glycol-based macrobicyclic ligand **110** is stabilized by hydrogen bonds of this bridging ion with oxygen atom of one of its three ribbed fragments. Mononuclear copper(II) complexes of these capsules undergo hydrolysis only in very basic media. Both the mono- and binuclear copper(II) cage complexes encapsulate fluoride ion, but in the case of **110** this process is more efficient than for **111** due to higher stability of the initial binuclear complex [69].

An octaamine caging ligand **7** effectively encapsulates silver(I), copper, cobalt, and nickel(II) cations by Scheme 2.59 to give host–guest 1:2 cage compounds [70]. In the case of Co^{2+} ion, the cascade complex with a bridging μ-OH^{-} anion has been isolated in this work; imidazolate and azide anions are also described to form the same cascade complexes. In the case of iron and manganese(II) triflates, the tetraprotonated form of the ligand **7** has been formed: it has the intramolecular hydrogen bonds preventing the encapsulation of solvent molecules and metal cations [70].

As follows from X-ray diffraction data [71], 1:2 cage complex of the furan-based hexaamine macrobicyclic ligand **110** with two encapsulated copper(II) ions also binds μ-hydroxo guest anion, thus giving heteroguest 1:2:1 compound by Scheme 2.60. The intramolecular $Cu^{II}\ldots Cu^{II}$ distance is approximately 3.9 Å and its linear $Cu^{II}\ldots OH\ldots Cu^{II}$ moiety has a trigonal bipyramidal geometry around the caged metal ions [71].

The hexaimine binucleating caging ligand **1** forms cascade 1:2:1 cage complex with two copper(II) ions and one bridging OH^{-} anion by

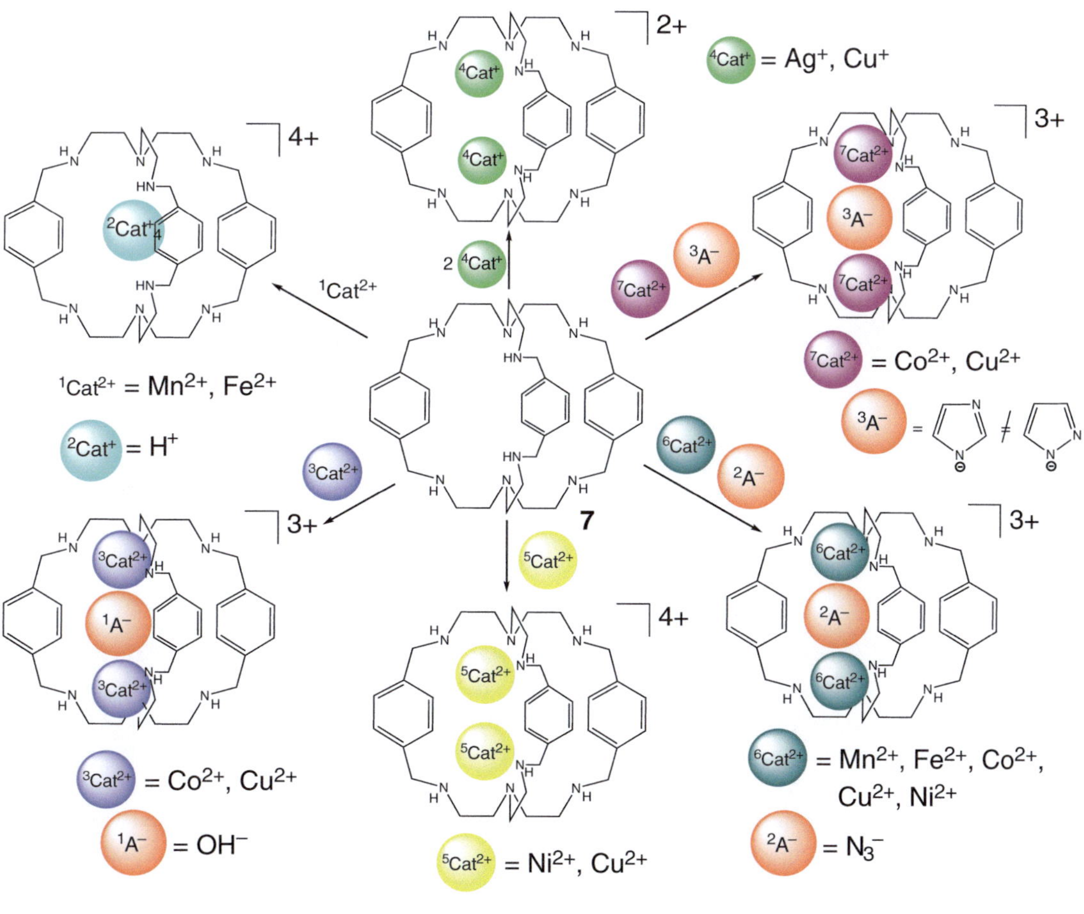

Scheme 2.59

Scheme 2.60

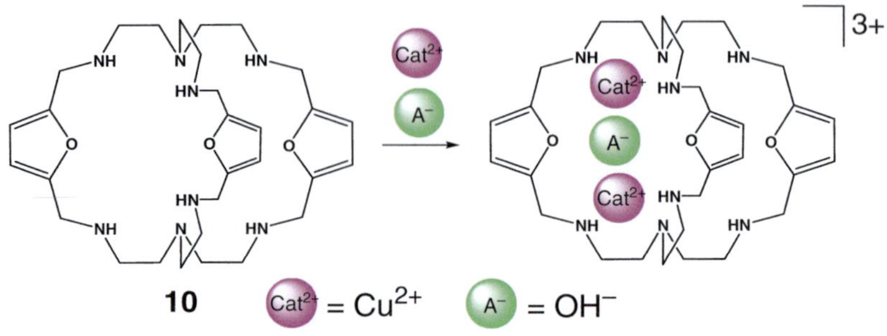

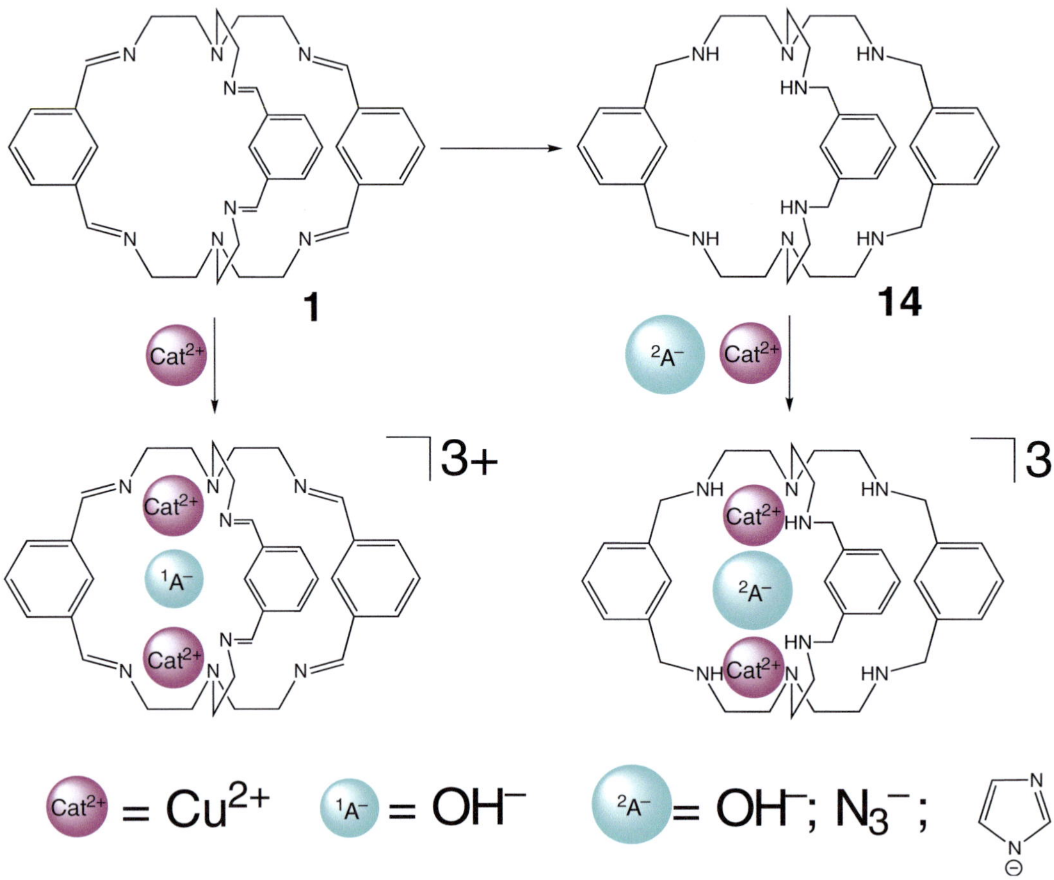

Scheme 2.61

Scheme 2.61 [72]. This compound is diamagnetic, which suggests efficient antiferromagnetic exchange between the caged metallocenters. Its octaamine derivative **14** also encapsulates copper(II) ions to give heteroguest 1:2:1 cascade complex with nonlinear bridging μ-hydroxo anion between its metallocenters. This binuclear copper(II) complex can also accommodate larger N_3^- and Im^- ions. As follows from single-crystal X-ray diffraction data, in the imidazolate-containing complex this bridging ion is sandwiched between two ribbed aromatic fragments. The encapsulated copper(II) ions have trigonal bipyramidal coordination polyhedra with the largest Cu–N distance for the third ribbed fragment that does not form stacking interactions with a caged anionic guest [72].

Cascade cage complexes of polyamine capsules **7** and **14** (Scheme 2.62) with encapsulated $\mu_{1,3}$-azide anion have been synthesized and structurally characterized in [73]. The manganese(II) ion is reported in this work to be too large to form such a cage complex with the *meta*-phenylene capsule **14**. In the cascade nickel(II) complex of this ligand, two caged nickel(II) ions are bridged by the linear μ_2-azide monoanion; each of these nickel(II) ions has a distorted trigonal bipyramidal N_5-coordination polyhedron. This anionic guest forms stacking interactions with ribbed aromatic fragments of **14**.

A fluorescent covalent capsule **112** with an extended cage framework and a 9,10-anthracene ribbed substituent has been synthesized in [74] by Scheme 2.63. Its cavity can accommodate two metal ions, thus allowing a cascade binding of suitable ambidentate anions such as N_3^- and NCO^-. As the copper(II) ion is a fluorescence quencher (so its complexes cannot be used as fluorosensors), the authors of [74] choose a 1:2 cage complex with encapsulated redox-inactive zinc(II) ions [74]. In acidic media, the fluorescence of the initial capsule **112** quenches in the range pH 4–6 because its amino groups display reducing properties and are able to ensure the ET to the neighboring excited anthracene fragment that quenches fluorescence. Addition of two equivalents of a zinc(II) salt causes a decrease in the fluorescence up to pH 7 (with plateau of pH 8.5). This result has been explained [74] by the coordination of the donor amino groups of **112** to the encapsulated metal ions that prevented the ET to the anthracene fluorophore. The fluorescence intensity goes down with pH due to the formation of a hydroxide-containing cage complex with encapsulated pentacoordinate zinc(II) ions: the bridging OH^- anion is included in this ET process and in quenching of fluorescence. The same effect has been observed in the presence of N_3^- anion: in its heteroguest 1:2:1 cage complex, the ET from electron-rich encapsulated N_3^- ion to the neighboring anthracene fragment quenches its fluorescence [74].

The potentiometric titration has been used in [75] to study coordination–chemical properties of the furan-based macrobicyclic ligand **10** with copper(II) ions at different pH in the absence and in the presence of halide ions. This ligand forms homoguest 1:2 and heteroguest 1:2:1 copper(II) cage complexes by Scheme 2.64 at pH 4–7 and 6–11, respectively. In the heteroguest complex, the encapsulated hydroxide ion is located between the copper(II) ions as a bridging moiety. The spectrophotometric titration of these cage complexes in the presence of Cl^-, Br^-, and I^- ions at pH > 3 showed the encapsulation of these anions in the manner similar to OH^- ion with an intensive $Hal^- \rightarrow Cu^{2+}$ CTB in the visible range. This suggestion is confirmed by X-ray diffraction study of 1:2:1 cage complex with encapsulated bromide anion: this bridging guest occupies the vacant cavity between the two caged metallocenters. These heteroguest halide-encapsulating capsules are stable only in the narrow range of pH and easily undergo exchange reactions to hydroxide anion. The binuclear copper(II) cage complex of **10** has an affinity to chloride ion; the corresponding binding constant is two orders higher than those for bromide and iodide ions. It is also able to encapsulate linear N_3^- and NCS^- anions owing to the lability of its 2,5-dimethylfuran-containing ribbed fragments [75].

Anion-binding properties of the thiophene-based macrobicyclic ligand **13** (Scheme 2.65) have been studied [76] by potentiometric titration. Three binuclear copper(II) cage complexes have been found to exist in the range of pH 5–12; the one with the vacant cavity was the most suitable for the encapsulation of anions. At neutral pH, monohydroxide-encapsulating heteroguest 1:2:1 cage complexes dominate over other complex forms, and binding of a wide range of monoanions proceeds through hydroxide exchange reaction. In the case of N_3^-, NCO^-, and NCS^- ions, the UV-vis spectra contained characteristic $Cu^{2+} \rightarrow A^-$ CTBs in the visible range, thus suggesting the formation of cascade 1:2:1 cage complexes with encapsulated pseudohalide ions. The binding of halide anions by **13** has not been observed. In the azide-encapsulating binuclear copper(II) cage complex, bridging anion and two encapsulated copper(II) cations form a linear $Cu^{2+}...N{=}N - N^-...Cu^{2+}$ fragment. Due to the intrinsically higher length of the cage framework **13**, it cannot form stable complexes with spherical monoanions [76].

Encapsulation of three copper(II) cations in basic conditions by the extended analog of **8** shown in Scheme 2.66 in the presence of imidazole allowed

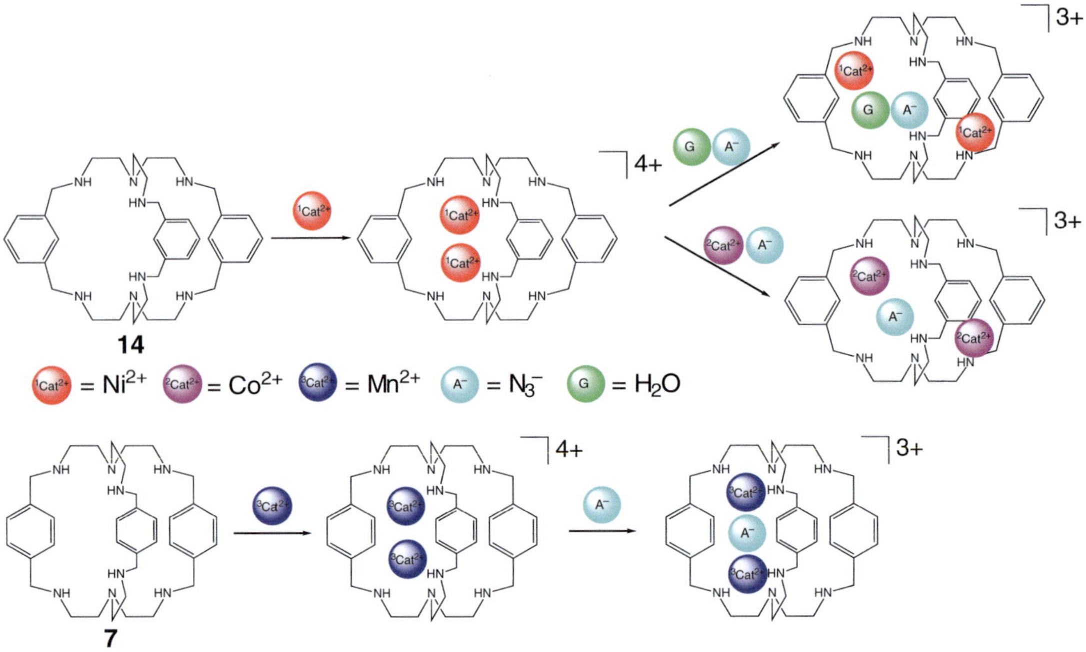

Scheme 2.62

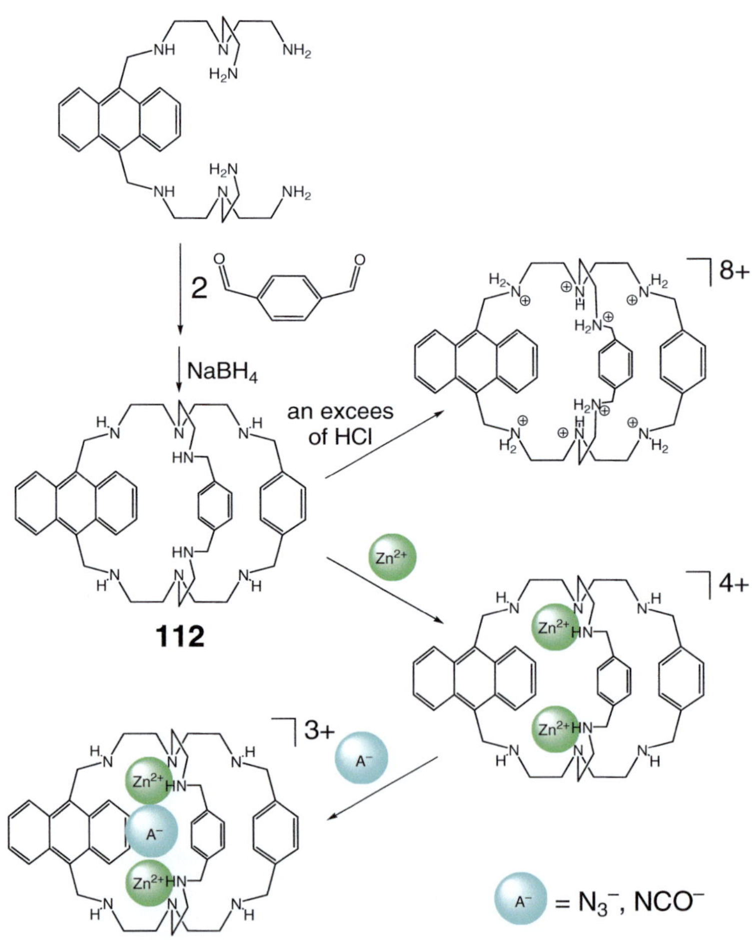

Scheme 2.63

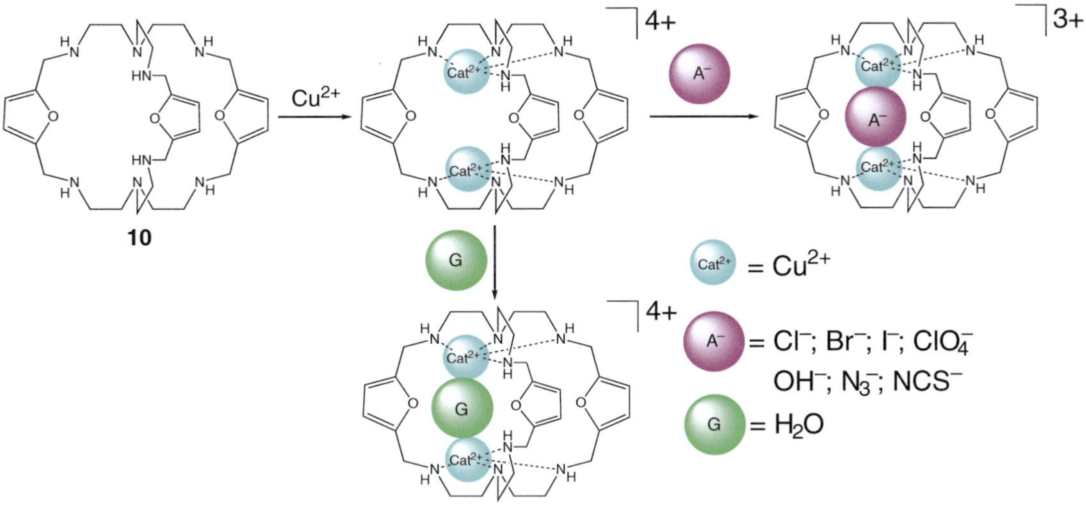

Scheme 2.64

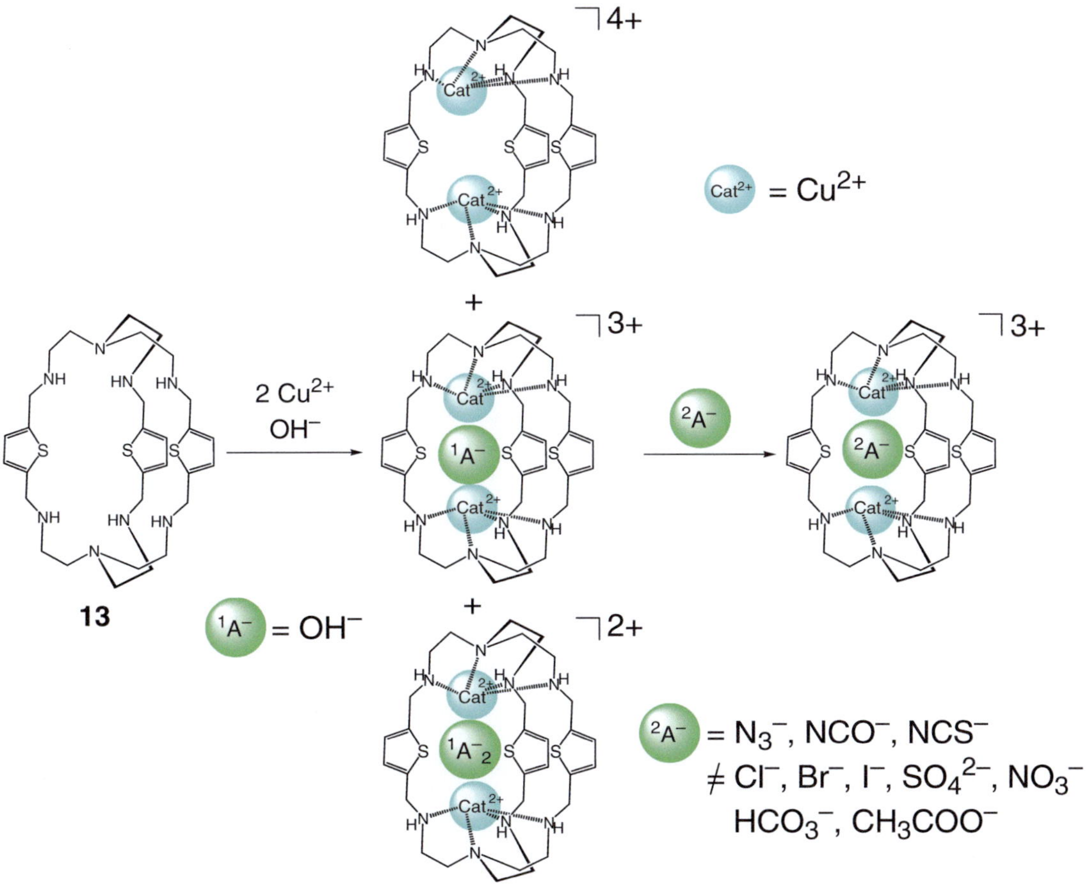

Scheme 2.65

obtaining a cascade imidazolate-encapsulating cage complex 1:3:1 with an anionic guest bridging two of these three caged Cu^{2+} ions [77].

Similar dicopper cascade complexes of the caging ligand **14** with encapsulated $\mu_{1,3}$-cyanamide and dicyanamide bridging anions (Scheme 2.67) have been isolated and structurally characterized in [78]. *Para*-phenylene analog **7** of this macrobicyclic ligand is reported in [79] to form a $H_3O_2^-$-bridged dicopper covalent capsule shown Scheme 2.68.

At the same time, the *N*-methylated macrobicyclic ligand **113** (Scheme 2.69) encapsulates only one copper(II) cation in the presence of oxygen-centered cationic co-guest H_3O^+ species that form strong hydrogen bonds with its three *N*-methylated amino groups. This hydrogen bonding successfully competes with coordination of this covalent capsule to second Cu^{2+} ion, thus suppressing the formation of the corresponding dicopper cage complex [80].

Encapsulation of cyanide ion by binuclear copper(II) complexes of the octaamine capsules **7**, **10**, and **14** [71] by Scheme 2.70 has been performed in [81]. In these complexes, a bridging CN^- anion is located between the metallocenters with a nearly

Scheme 2.66

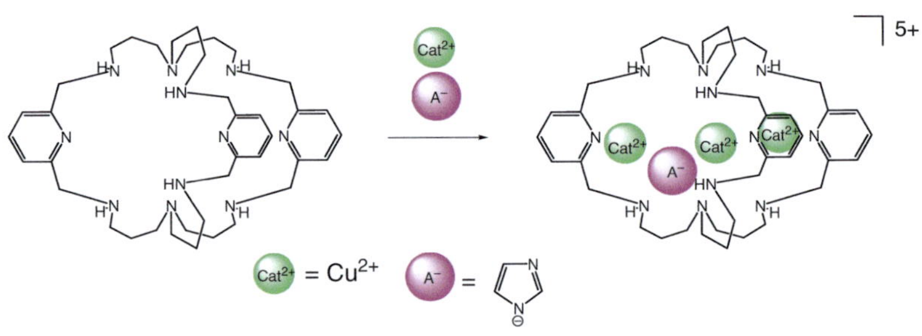

Scheme 2.67

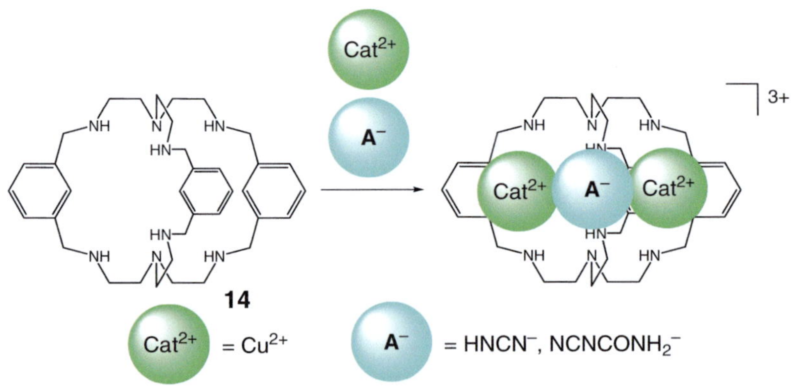

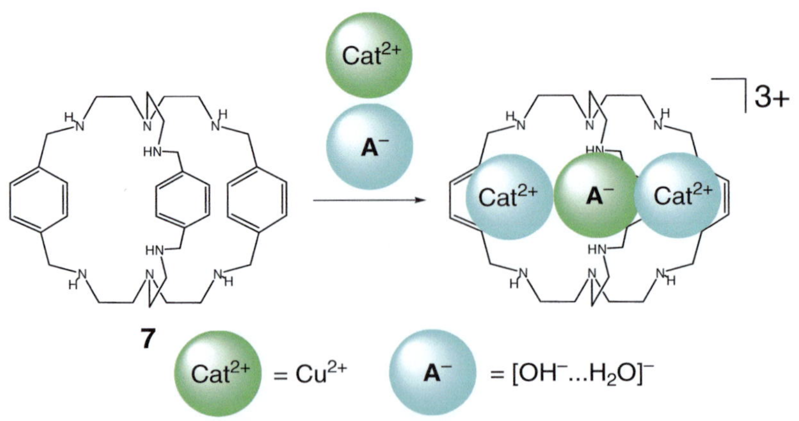

Scheme 2.68

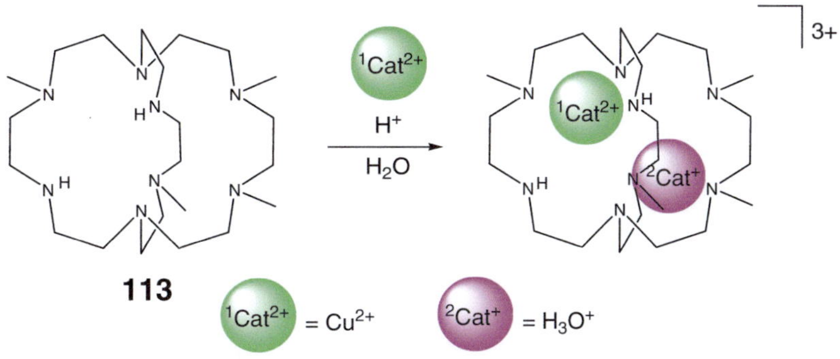

Scheme 2.69

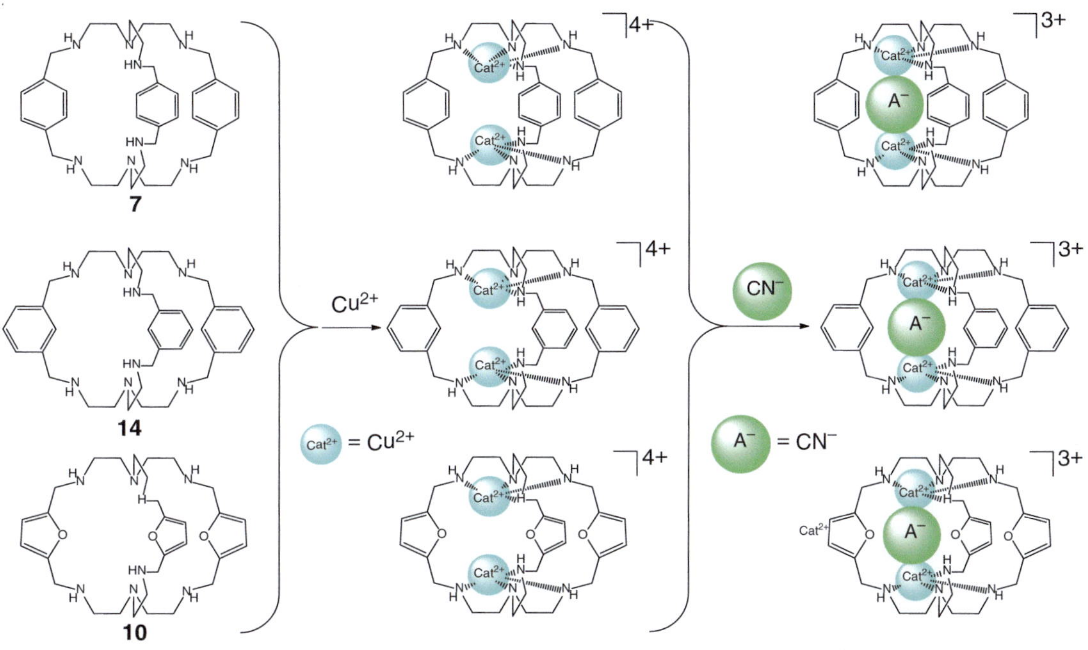

Scheme 2.70

linear geometry. These metallocenters have slightly distorted trigonal bipyramidal polyhedra and the copper(II) ions are substantially (by approximately 0.18–0.27 Å) shifted in the direction of the geometrical center of the cage framework [81].

The X-ray diffraction study [82] of the analogous cage complex of **10** with an encapsulated chloride ion (Scheme 2.71) showed the guest to be linked with two trigonal bipyramidal CuN_4Cl-coordination polyhedra and to interact with furan rings of the ribbed fragments [82]. The spectrophotometric titration of this 1:2 cage copper(II) complex with tetraalkylammonium chloride caused a change in its color from light blue to bright yellow, and the intensity of the $Cl^- \rightarrow Cu^{2+}$ CTB increased up to their equimolar concentrations and then went down to zero at one molar excess of the chloride ions. These experimental results have been explained in [82] by the processes shown in Scheme 2.71: on the initial stage, bridging chloride anion links two copper(II) ions of the different cage species, and then it undergoes encapsulation by the binuclear copper(II) macrobicyclic complex; further addition of chloride ions results in the corresponding cage complex by Scheme 2.71. The analogous coordination transformations (i.e., the formation of three types of macrobicyclic species) have been also observed in the case of bromide ion and another cage complex of this type (Scheme 2.72). The spectrophotometric titration with bulky iodide

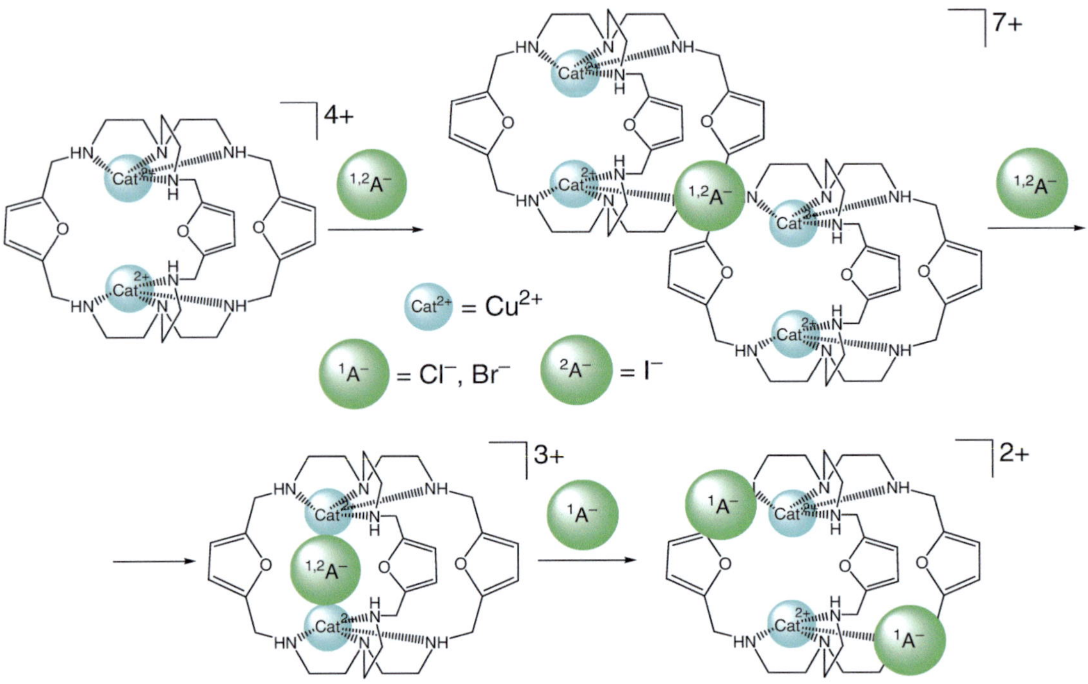

Scheme 2.71

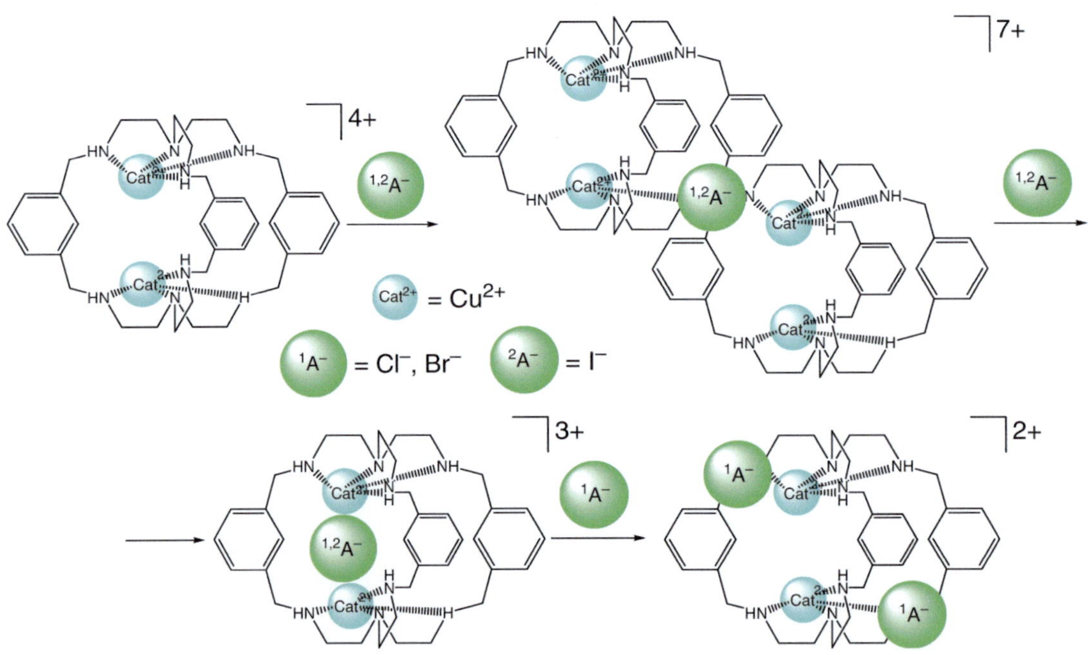

Scheme 2.72

ion performed in [82] evidences the formation of only two types of such species for both these covalent capsules (Schemes 2.71 and 2.72).

In the C_2-symmetric binuclear copper(II)-containing heteroguest 1:2:1 cage complex of **7**, the encapsulated μ_2-azide anion forms a linear $Cu^{2+}...N = N - N^{-}...Cu^{2+}$ chain, and stacking interactions between it and the aromatic fragments of the encapsulating macrobicyclic ligand at the distance of approximately 3.06 Å have been observed in [83].

An analogous *tren*-based capsule **114** has been synthesized in [84] by self-assembly of *para*-formylphenylboronic acid, pentaerythritol ligand syntone, and this tripodal amine by Scheme 2.73. The macrobicyclic ligand **114** forms a host–guest 1:2 cage complex with two encapsulated copper(I) ions, which is suitable for caging of anions, thus giving their cascade complexes.

The anion-encapsulating ability of the per-*N*-methylated *para*- and *meta*-phenylene capsules **93** and **94** have been studied in [85]. As follows from X-ray diffraction data, cage framework of **93** is doubly protonated through its two ternary apical amine fragments, and this structure is stabilized by trifurcated hydrogen bonding with all three ribbed nitrogen-containing fragments of a macrobicyclic ligand. This per-*N*-methylated capsule does not form heteroguest binuclear 1:2:1 cage complexes with copper(II) cations and isocyanide and azide counterions: those are outside the ligand cavity. The same type of outer-sphere complexes has been

Scheme 2.73

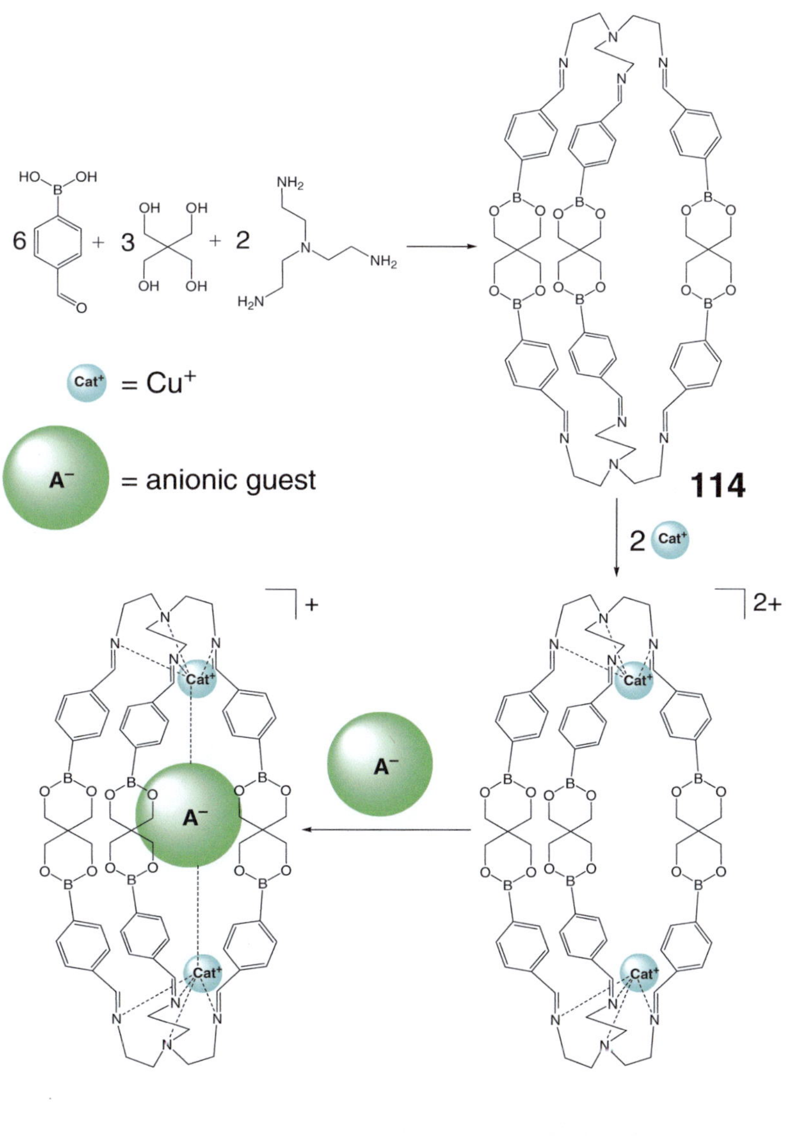

observed in [85] for its *meta*-phenylene analog **94**: it binds atmosphere carbonate dianion in the presence of copper(II) ions by Scheme 2.74, thus forming the 1:2:1 complex with an open-ended basket conformation of the ligand **94**. The *N*-methylation of the macrobicycles **93** and **94** is reported in [85] to reduce their encapsulation ability [85].

At the same time, hexaprotonated form of the parent caging ligand **7** is described in [86] to form cascade cage complexes with two encapsulated fluoride ions and one bridging water molecule by Scheme 2.75; an attempt to obtain a heteroguest F^-, Cl^--encapsulating complex is reported in this work to be unsuccessful.

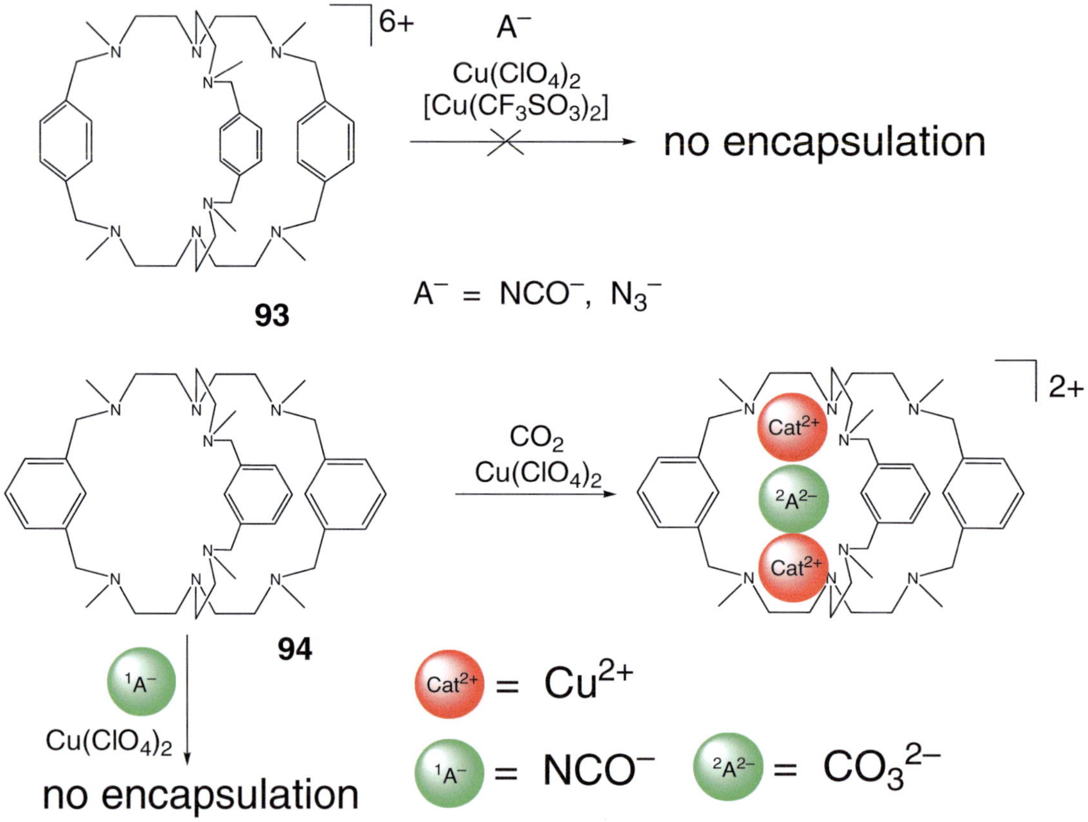

Scheme 2.74

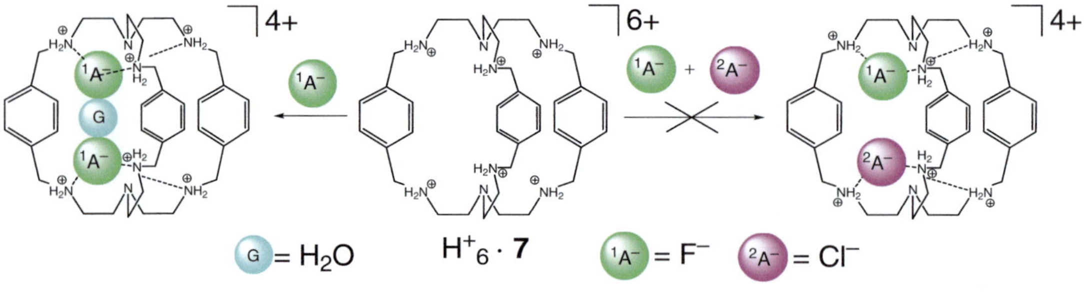

Scheme 2.75

2.2 Amine, Amide, and Pyrrole Caging Ligands

2.2.1 Free Cages and Encapsulation of Neutral Molecules

A hexaamide capsule **115** has been designed by A.P. Davis and coworkers for selective recognition of carbohydrates in their organic solutions [87]; the synthetic strategy shown in Scheme 2.76 used a biphenyl-containing unsymmetrically protected intermediate *23* that resulted from two [2+2] macrocyclizations. According to ^{1}H NMR and fluorescence data, the caging ligand **115** forms 1:1 cage complexes with octylated pyranosides in their chloroform solutions and selectively binds monosaccharide *24* [87]; there was a 50-fold difference in binding constants for the monosaccharides *24* and *25*.

A similar synthetic approach with two key [2+2] macrocyclization stages allowed obtaining a lipophilic caging ligand **116** (Scheme 2.77) for the membrane transport of monosaccharides [88]. The peripheral substituents in the phenylene fragments of a cage framework of **116** only slightly affect its binding properties. According to ^{1}H NMR data, it forms a stable 1:1 cage complex with *n*-octyl-*β*-*D*-glucopyranoside *26* in chloroform and is an efficient extractant of monosaccharides *27–33* from their aqueous solutions to chloroform media. The ligand **116** also showed

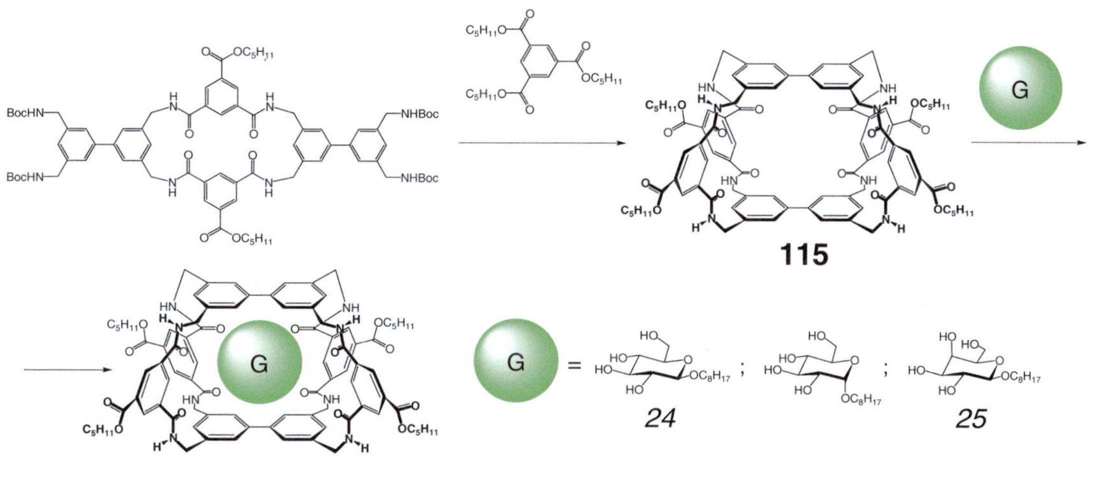

Scheme 2.76

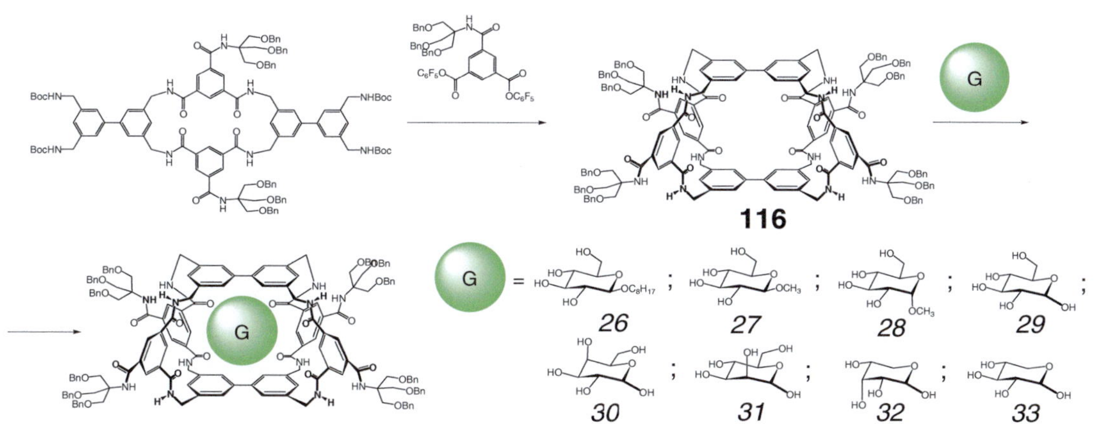

Scheme 2.77

the selective extraction of glucose relative to galactose and mannose. This suggests stereospecific interactions of a glucose guest with this encapsulating receptor: the internal groups in its cavity can bind substrates with suitable equatorial substituents. Xylose also can be efficiently transferred to the organic phases from its aqueous solutions using a capsule **116** [88].

An extended octaamide caging ligand **117** for the efficient binding of disaccharides has been prepared in [89] using two synthetic approaches: one of them (Pathway I on Scheme 2.78) gave a poorly separable mixture of the target macropolycyclic product with its regioisomer, while the second one (Pathway II) allowed isolating the capsule **117**. Its geometry is similar to that of the above macropolycyclic receptor **115**, and this covalent capsule is suitable for selective encapsulation of equatorially substituted carbohydrates. In particular, all-equatorial *n*-octyl-*β*-*D*-cellobioside *34* interacts with **117** to form a 1:1 cage complex with loss of symmetry (as follows from NMR, fluorescence emission, and CD data). Octyl glycosides *35* and *36* as well as their disaccharide analogs *37–39* caused no changes in these spectra; so this extended ligand **117** has been reported in [89] to selectively bind the disaccharide *34*.

The octaamide caging ligand **115** is described in [87] to efficiently encapsulate octylated glycosides (mainly, the *β*-glycosylated derivatives of *35*) in organic media (see above), but the main goal is the binding of saccharides in aqueous solutions. Unfortunately, this tetracarboxylate capsule underwent aggregation under these conditions; the attempts [88] to use the macrobicyclic ligand **116** with tris(hydroxybenzyl)methyl substituents were also unsuccessful. As a result, the dodecacarboxylate capsule **118**, designed for the recognition in aqueous media, has been synthesized in [90] by Scheme 2.79. As follows from NMR titration experiments, this ligand encapsulates carbohydrates *40–54* in chloroform media, and the binding constants obtained as well as the corresponding NMR chemical shifts were similar to those for the analogous amide macrobicyclic ligands. The same effects have been observed in aqueous solutions of these carbohydrates: both the NMR and fluorescent titration experiments suggest the formation of 1:1 cage complexes. The binding constants vary with history of the glucose aqueous solution from 4.6 $mol^{-1}cm^3$ for freshly dissolved substrate with anomeric *α*/*β* ratio 72:28 to 9.2 $mol^{-1}cm^3$ in the case of the ratio 60:40 due to the selective binding of its *β*-anomer by the caging ligand **118**. The binding of other carbohydrates shown in Scheme 2.79 has been also studied in [90] using NMR and fluorescent methods. Similar to the analogs **115** and **116**, this ligand selectively encapsulates *β*-glycosylated guests with higher binding constants for methyl *β*-*D*-glucoside *35* and cellobiose *52*; the selectivity to *52* as compared to lactose *53* and maltose *54* is observed. This result is in a good agreement with the experimental data [89] for **115** in organic media and those for the extended terphenyl-based macrobicyclic ligand **117**. The dodecacarboxylate capsule **118** also efficiently binds other carbohydrates with equatorial substituents (2-deoxy-*D*-glucose *47* and *D*-xylose *48*), while the termination of hydroxyl groups (the saccharide *47* and glucose) causes no increase in the binding constants; the receptor **118** poorly encapsulates rhamnose *50* and fucose *51*. The encapsulation of glucose and its interactions with the interior of **118** has been confirmed in [90] by NOESY experiments.

Functionalization of such macrobicyclic platform with various groups in different positions by Scheme 2.80 has been performed in [91] to increase the efficiency of binding of glucose and the selectivity to *β*-glycosylated carbohydrates relative to others (including *N*-acetylglucosamine, GlcNAc). For this purpose, 4,4′-substituents in its biphenyl moieties are more prospective than those in 2,2′-positions as well as the groups in the *iso*-phthalimide spacer, as 4,4′-substituents only slightly affect the conformation of a macrobicyclic framework but are fold inward between the *iso*-phthalimide to reinforce the cavity walls [91]. As follows from ^1H NMR titration data, this causes an increase in the stability constants of the corresponding 1:1 cage complexes with encapsulated saccharide guests. In a series of the capsules **119–122** (Scheme 2.80), their methoxylation (**119**) causes an increase in the affinities except that for the carbohydrate *59*. In the homological

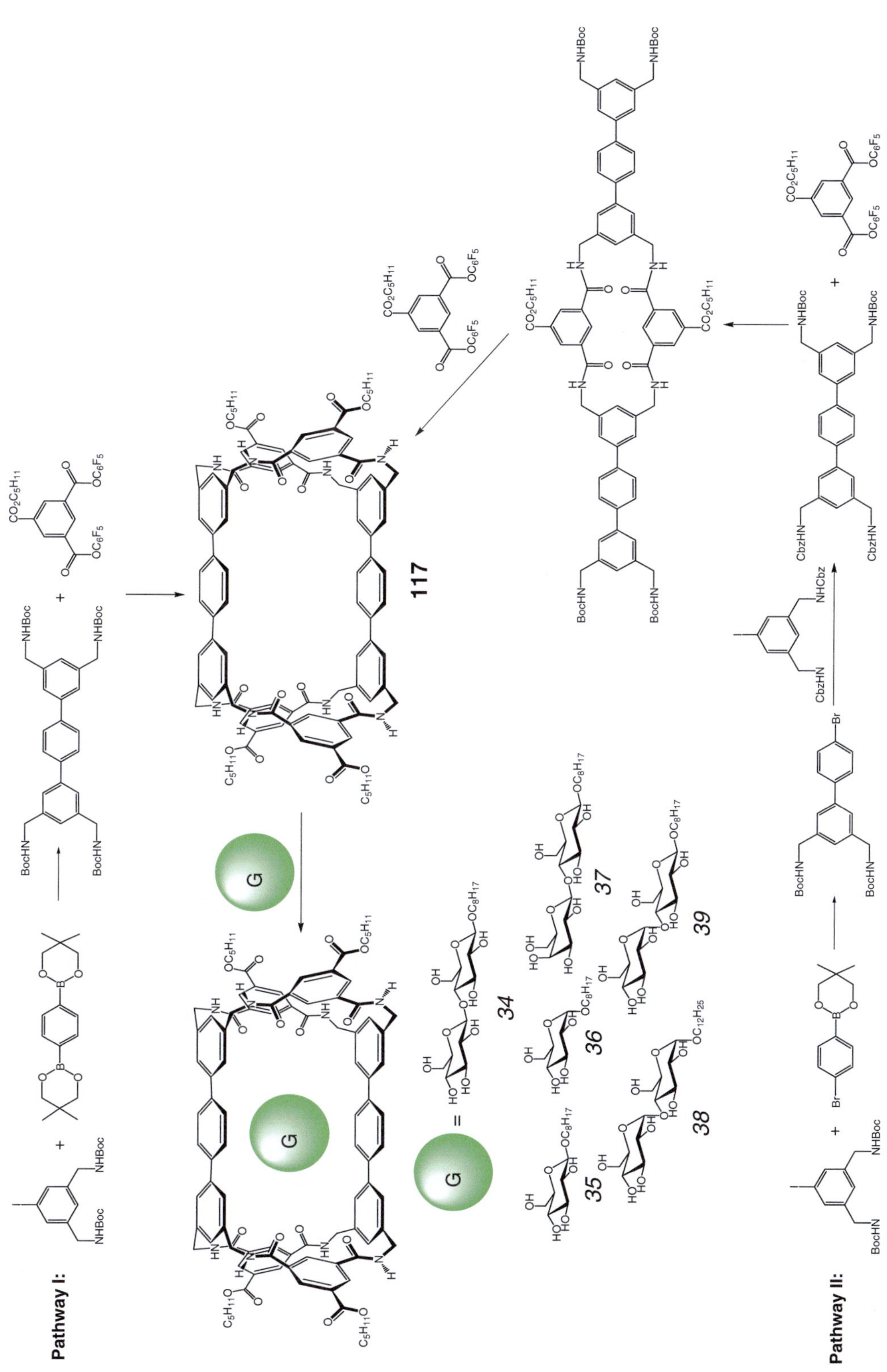

Scheme 2.78

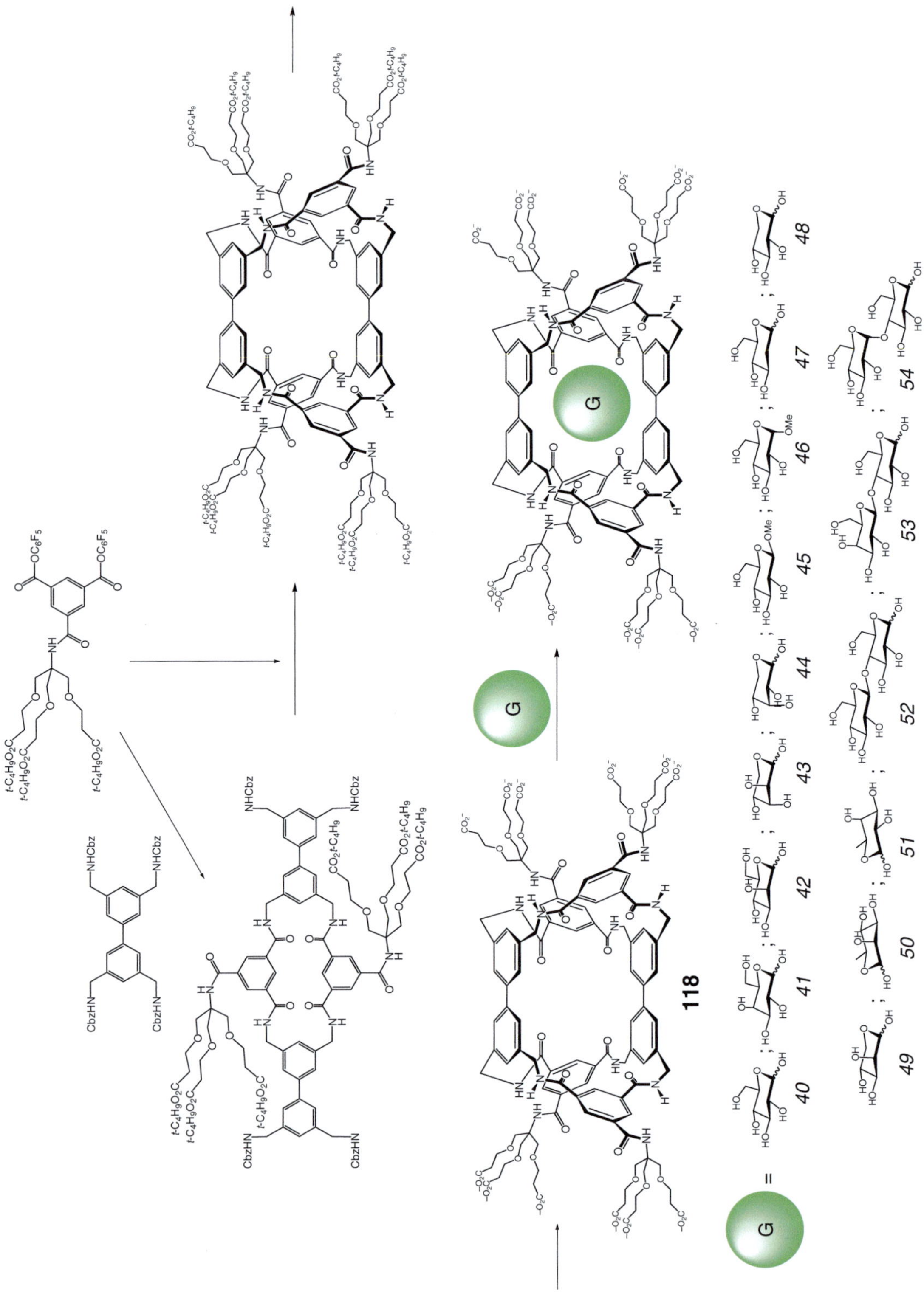

Scheme 2.79

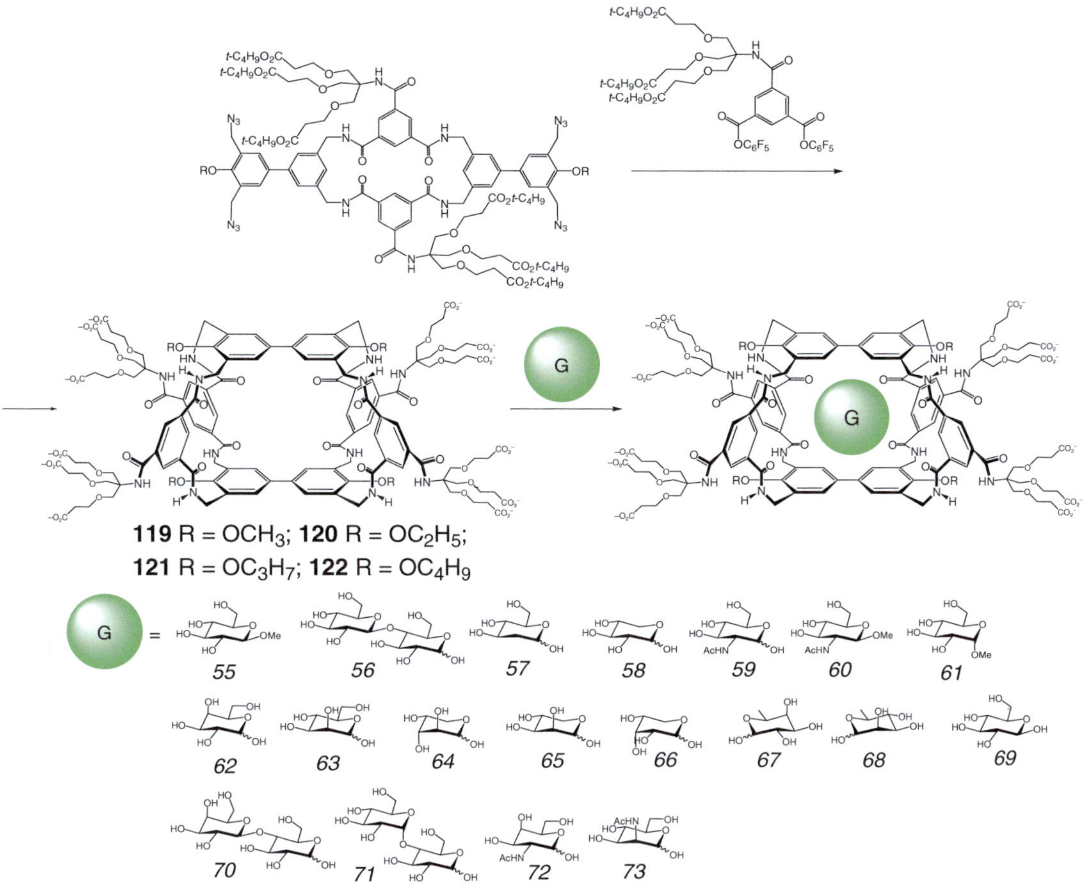

Scheme 2.80

series of the alkoxyl substituents (macrobicycles **119–121**), the binding constants increase for the *β*-glycosylated substrates such as *69*, *55*, and *56* and for their closely related all-equatorial saccharides *57* and *58*; the binding of the substrates *59* and *60* substantially decreases. Going to the macrobicyclic ligand **122** with even longer alkyl spacers causes an increase in its affinity for all the carbohydrates studied in [91]. As compared with an unfunctionalized macrobicyclic platform **118**, the caging ligand **121** encapsulates glucose much more efficiently and selectively: in particular, the selectivities of **121** for the carbohydrate pairs glucose/galactose and glucose/GlcNAc are 20:1 and 9:1, respectively, while those for **118** are 4.5:1 and 1:6. Thus, the affinity of the caging ligand **121** to glucose is much worse than those of the natural lectins but is described in [91] to be still enough for the blood glucose monitoring. The authors also mentioned that normal blood contained no soluble components that interfere with the binding of glucose. Two 1:1 cage complexes (*60*)@**119** and (*69*)@**121** have been theoretically and experimentally studied in [91] by MM calculations and NMR spectroscopy. The molecular structure of (*60*)@**119** is practically the same as that of the cage complex (*60*)@**118** [92]. The stability constant for the latter compound is equal to 730 $mol^{-1}cm^3$, while the complex (*69*)@**121** is considerably less stable (K = 60 $mol^{-1}cm^3$). Only *β*-glucose can be encapsulated by these caging receptors. The substantial difference in the binding of glucose and GlcNAc by them has been explained in [91] by steric reasons: the NHAc moieties of GlcNAc are directed to the narrow portals, and an increase in the size of alkoxyl substituents causes the hindrances disfavoring the encapsulation of this guest [91].

Synthesis of a polyamine capsule **123** by Scheme 2.81 is described in [93]; its protonation has been studied by ^1H and ^{13}C NMR and potentiometric titration methods.

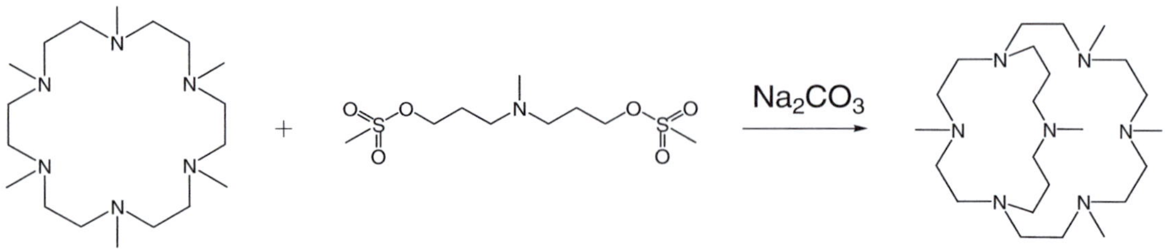

Scheme 2.81

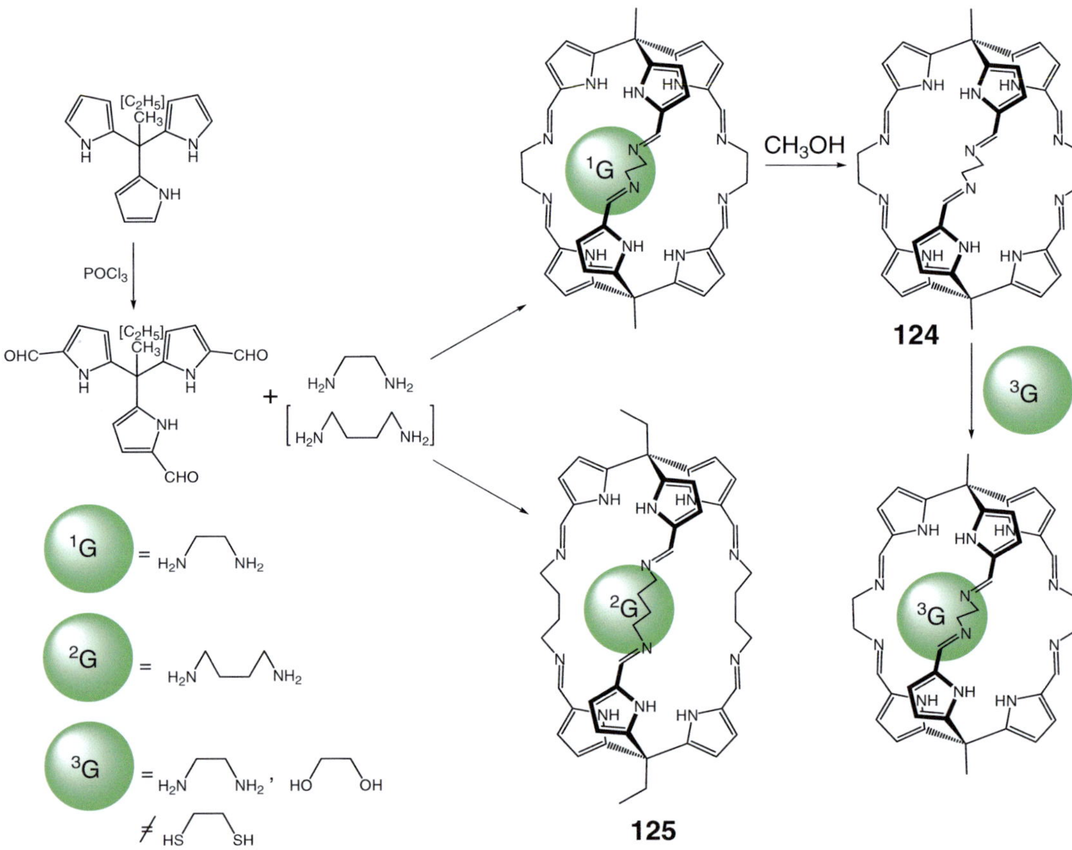

Scheme 2.82

Tripodal tris-aldehyde ligand syntones, which have been prepared in [94] using Vilsmeier–Haack-type formylation of the corresponding tris-pyrroles by Scheme 2.82, gave the covalent capsules **124** and **125**, the derivatives of ethylene-1,2- and butan-1,4-diamines, respectively. These macrobicyclic ligands encapsulate parent diamines during the template condensation with a diamine guest as a matrix to give the corresponding 1:1 cage complexes. In these complexes, the guests form intramolecular hydrogen bonds within the ligand's cavity. In particular, hydrogen bonding between two terminal amino groups of butan-1,4-diamine as guest and seven hydrogen atoms of **125** is observed in such a manner that only two of the three pyrrole groups of each of the tripodal capping fragments are oriented inward the host cavity and form hydrogen bonds with a guest molecule, while the third outward-oriented pyrrole group forms only weak C–H(diamine)…π(pyrrole) interactions. The encapsulated butan-1,4-diamine molecule has a less energetically preferable *gauche*, *gauche*, *trans*-conformation, while the caged ethylene-1,2-diamine is in a *gauche* form and can be extracted from the ligand's cavity with methanol. According to NMR titration data, the binding constants of the macrobicyclic ligand **124** with ethane-1,2-diamine and ethane-1,2-diol as guests are approximately 1500 and 1060 mol^{-1} cm^3, respectively, whereas ethane-1,2-dithiol does not form the same cage complex due to a reduced hydrogen bonding ability, larger size, and substantially lower stability of its *gauche* conformation [94].

The first urotropin-like capsule **126** has been prepared in [95] using *one-pot* synthetic procedure by Scheme 2.83 from its diamine and dibromine syntones. In this covalent capsule, all four lone pairs of the bridgehead nitrogen atoms are oriented toward the center of its cage framework, but the cavity size of **126** is described in [95] to be too small to encapsulate guest species except H^+ ions.

Covalent capsules **127–129** with donor pyridyl groups and their 1:1 cage complexes with an encapsulated potassium(I) cation have been synthesized in [96] using similar synthetic approaches (Scheme 2.84). These complexes underwent demetalation with hydrochloric and nitric acids. A reversible protonation–deprotonation cycle of the capsule **128** [96] is shown in Scheme 2.84. Treatment of tetraprotonated form of this ligand with benzyltrimethylammonium hydroxide as a strong organic base gave not the target fully deprotonated macrobicycle **128** but its monoprotonated form. The latter reacts with potassium hydroxide giving a 1:1 cage complex with an encapsulated potassium(I) cation [96]. This encapsulated cation can be also substracted from its complex by 18-crown-6 giving the covalent capsule **128** by Scheme 2.85 [97]. The caging ligands **128** and **129** are described in [97] to slowly form their host–guest 1:1 complexes with encapsulated water molecule: the X-ray diffraction study excluded the possibility of encapsulation of H_3O^+ cation. The caging ligand **129** binds chloride anion, thus giving the corresponding 1:1 cage complex [97]. The methoxyl- and chlorine-containing analogs **130** and **131** (Scheme 2.85) of the bis-pyridine capsule **127** have been synthesized in [97] starting from the corresponding 4-substituted bis-aminomethylpyridine ligand syntones.

2.2.2 Encapsulation of Anions

J.-M. Lehn and coworkers are pioneers in the use of covalent capsule (so-called "anion cryptates") for anion encapsulation, separation, and detection.

Tetraprotonated form of the oxoamine covalent capsule **132** is reported in [98] to form 1:1 cage complex with chloride anion by Scheme 2.86; this anionic guest forms four

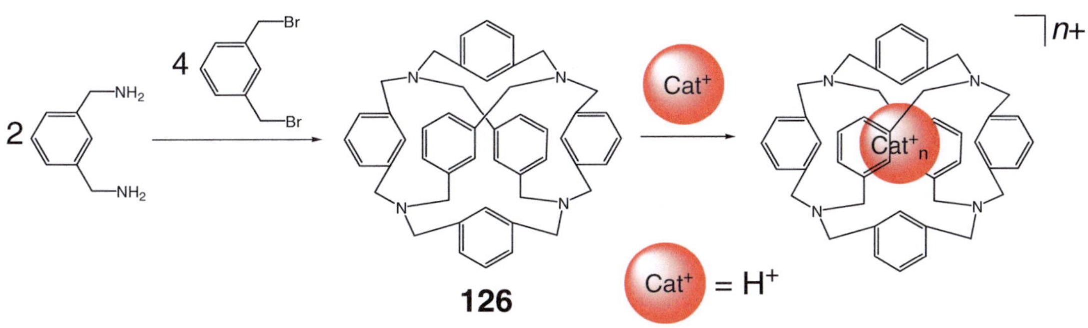

126

Scheme 2.83

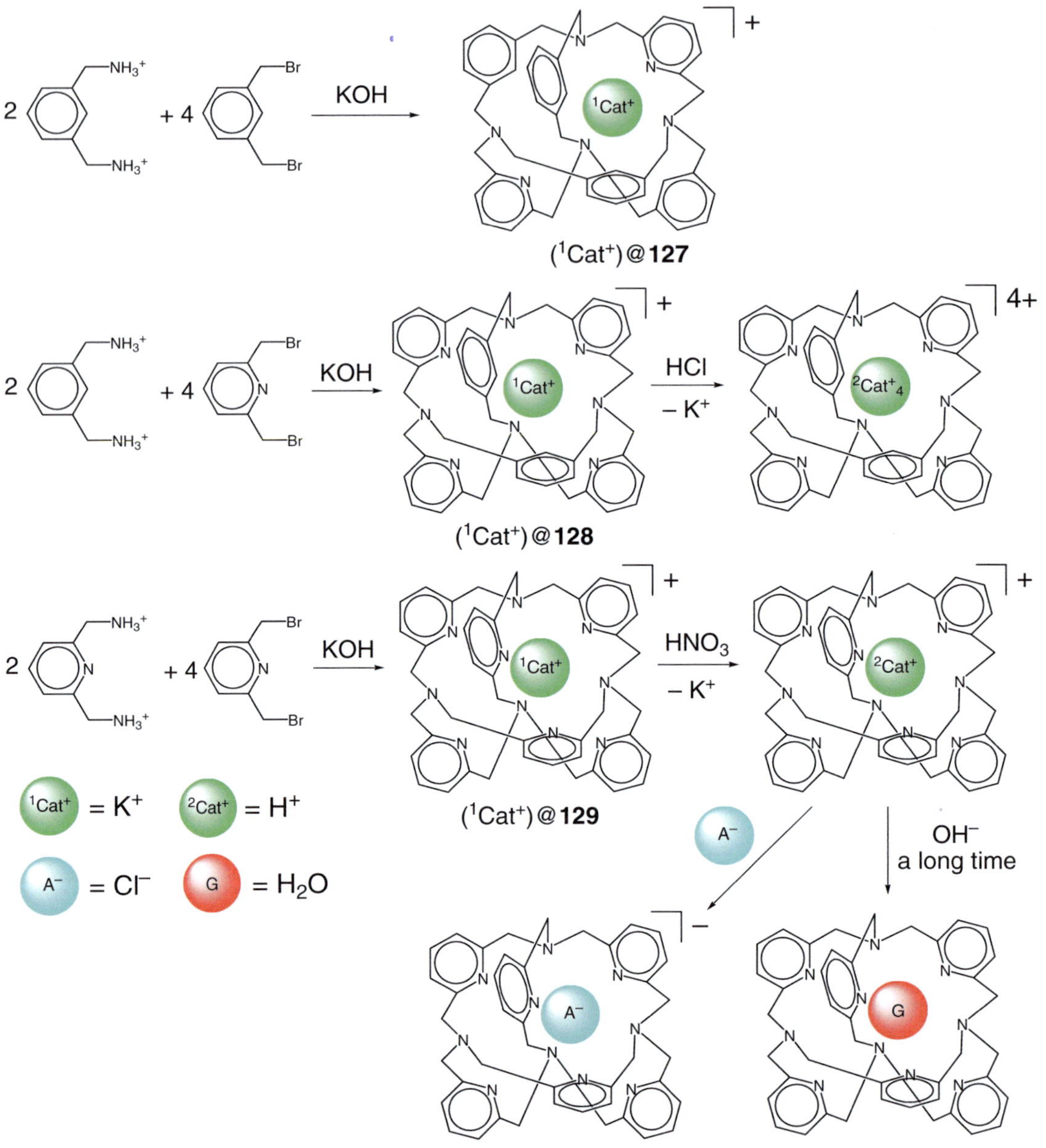

Scheme 2.84

hydrogen bonds with the protonated amino groups of a caging ligand.

The covalent capsule **134** has been prepared in [99] by Eschweiler–Clarke methylation of its amine precursor. As follows from ^{13}C NMR data, protonated forms of this macrobicyclic ligand and its analogs **132** and **133** selectively bind chloride, fluoride, and bromide ions but discriminate other monoanions shown in Scheme 2.87 [99].

Protonated form of a macrobicyclic receptor **110** has been used in [100] for encapsulation of linear triatomic azide anion, forming 1:1 cage complex by Scheme 2.88. The single-crystal X-ray diffraction structures of 1:1 cage complexes of $H^+_6 \cdot$ **110** with fluoride, chloride, bromide, and azide anions have been determined in [100]. The small fluoride ion is tetracoordinated by this macrobicyclic ligand, while chloride and bromide

Scheme 2.85

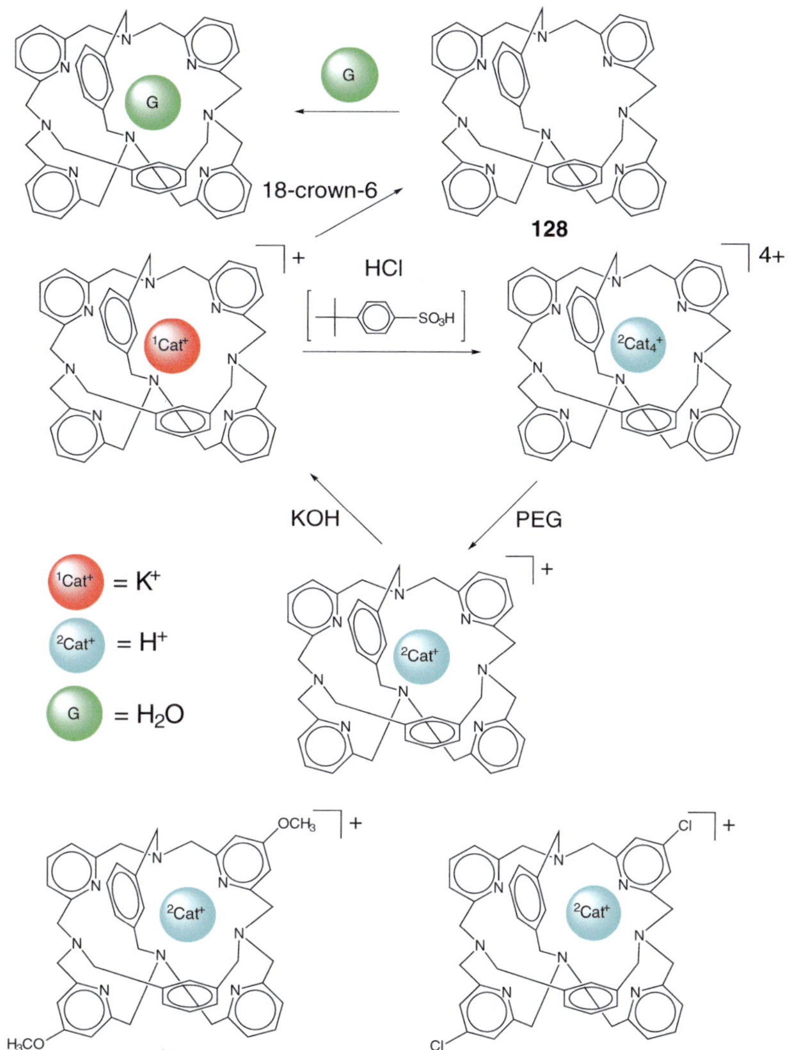

anions form six hydrogen bonds with it. The non-complementarity between these spherical anionic species and an ellipsoidal cavity of $H^{+}_{6} \cdot$ **110** caused the distortions of its cage framework. Such structural complementarity has been observed in [100] for the linear N^{-}_{3} ion, which is bound by two pyramidal arrays of three hydrogen bonds, each interacting with a terminal nitrogen atom of this anion. As follows from potentiometric titration data, the macrobicyclic ligand $H^{+}_{6} \cdot$ **110** also binds other anionic guests shown in Scheme 2.88.

^{35}Cl NMR spectroscopy has been used in [101] for the study of encapsulation of chloride anion by the protonated forms of covalent capsules **110**, **132**, **133**, **135**, and **136** (Scheme 2.89), forming its 1:1 cage complexes.

The synthesis of macrobicyclic polyamine ligands **110** and **137–140** using tripode–tripode coupling of a protected derivative of the tripodal amine *tren* and its propylene-containing homolog with corresponding protected tripodal polyether ligand syntones by Schemes 2.90 and 2.91 is reported in [102]. This synthesis sequence is described to be appreciably shorter than the stepwise construction via macrocyclic precursor and to give better yields.

The protonated form of covalent capsule **138** is described in [103] to selectively form stable

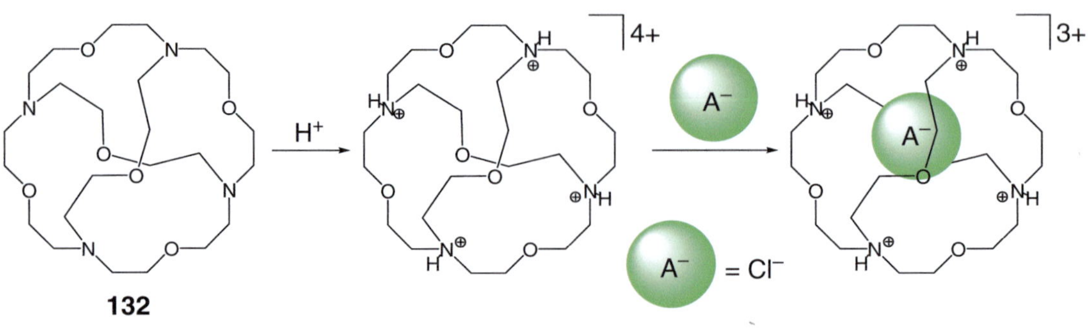

Scheme 2.86

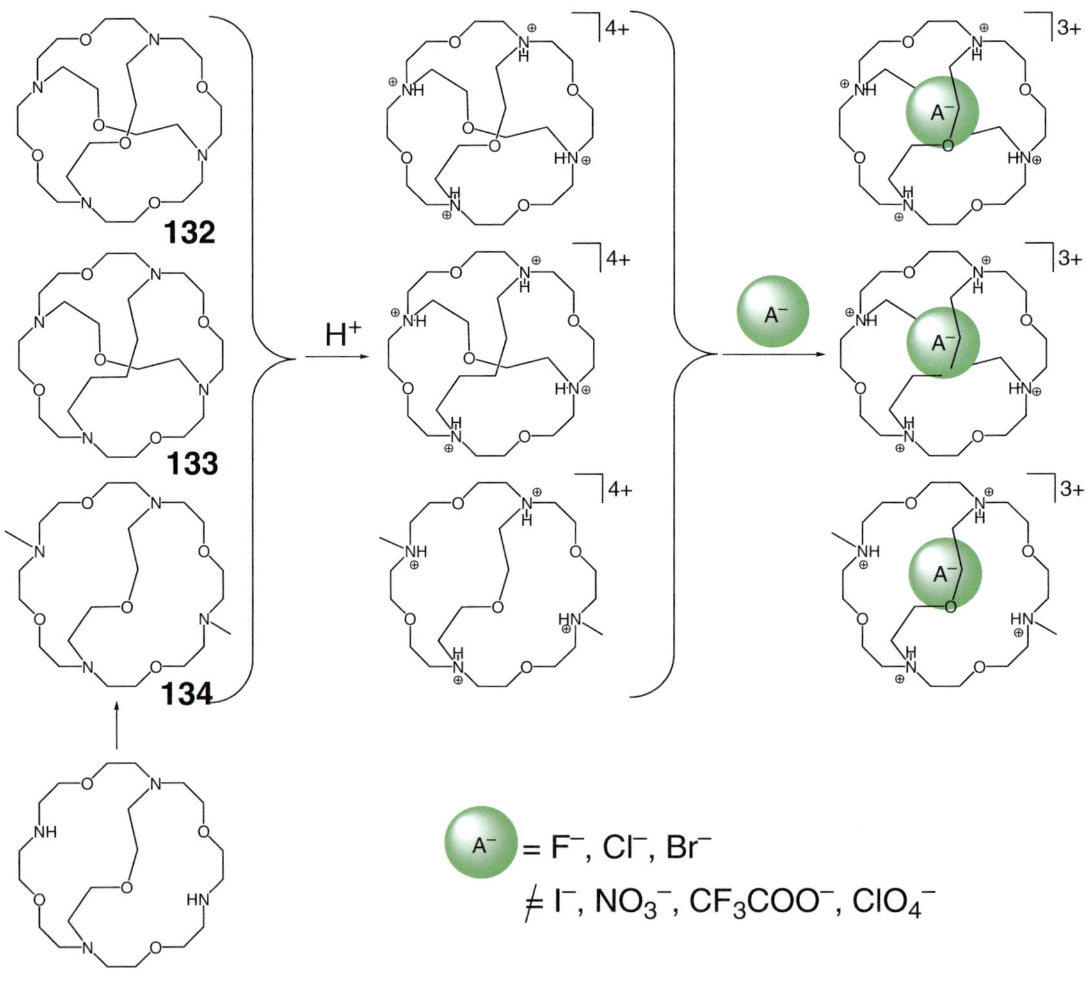

Scheme 2.87

Scheme 2.88

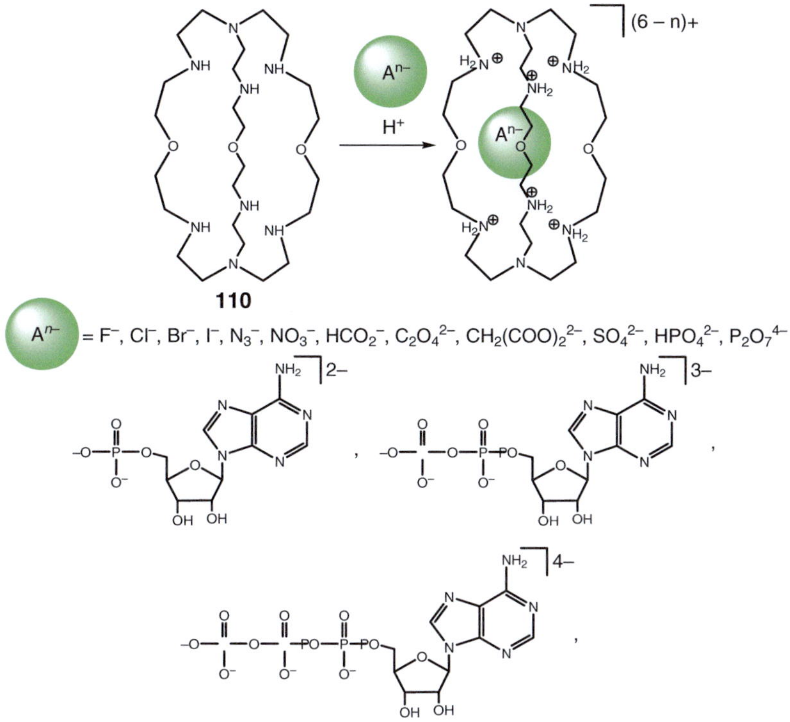

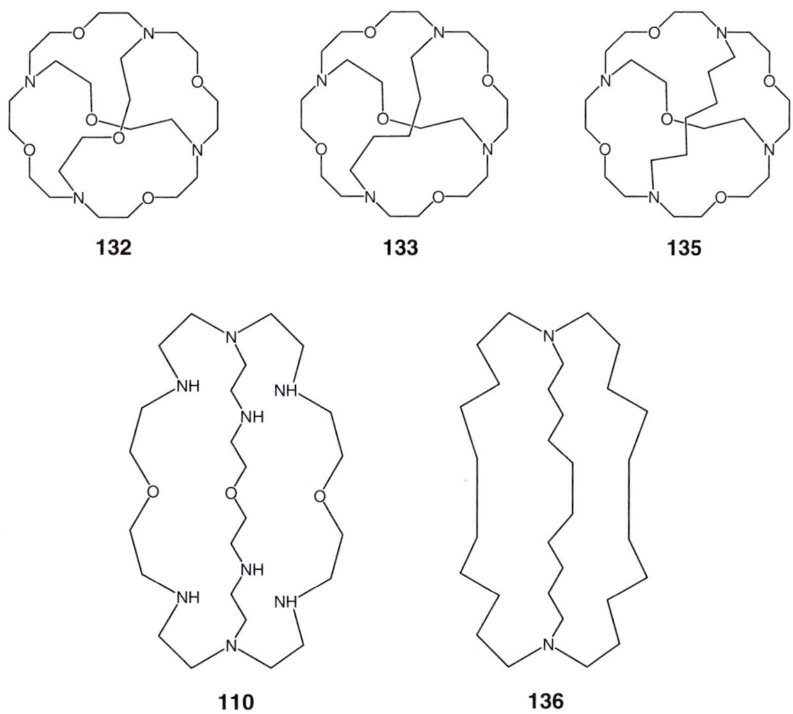

Scheme 2.89

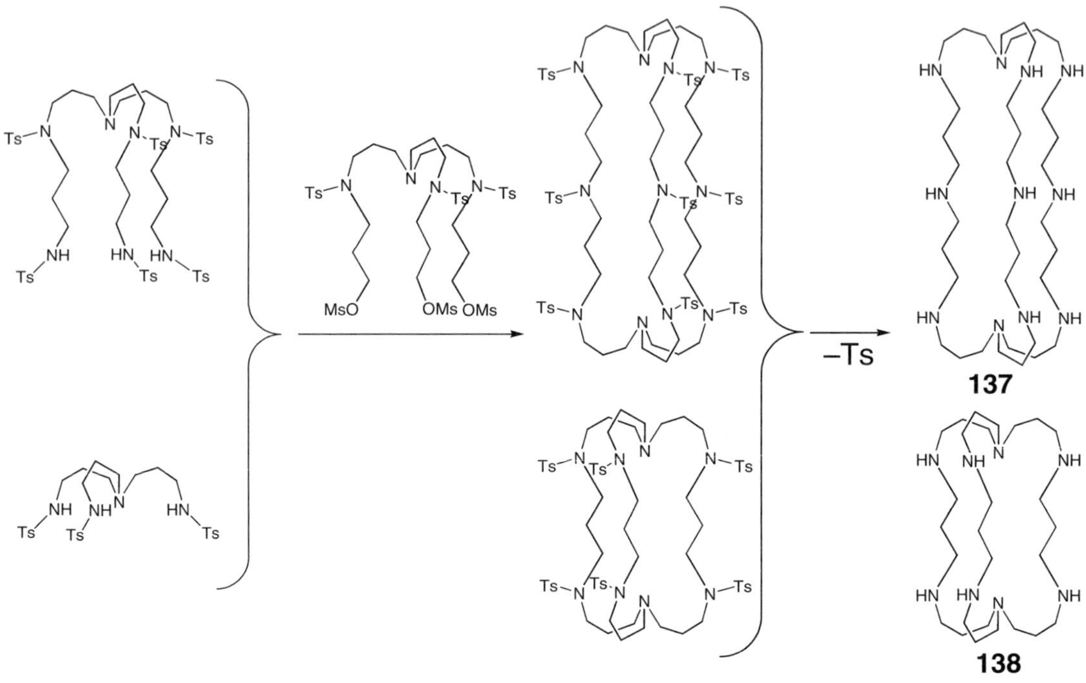

Scheme 2.90

1:1 cage complexes with encapsulated sulfate and oxalate dianions. At the same time, more bulky malonate anion cannot enter its cavity [103]. Like in the case of **110**, this protonated form also binds monoanions (first of all, halide ions) by Scheme 2.92. At pH 3–5, the resulting cage complexes are formed by their octa- and heptaprotonated forms, while the hexaprotonated form of **138** dominates in the range pH 5–7. The stability of these 1:1 complexes with halide guests increases in a row $Cl^- < Br^- < I^-$. Thus, the encapsulating ligand **138** with large and spherical cavity has affinity for a bigger iodide ion, whereas the opposite sequence of the stability constants $I^- < Br^- < Cl^-$ has been observed in [103] for the capsule **110**.

A tripodal–tripodal coupling of the corresponding protected triether and triamine ligand syntones by Scheme 2.93 gave 1,3,5-cyclophane-like capsules **141**–**143** having a cylindrical shape; the protonation of these ligands afforded their hexaprotonated forms. According to ^{1}H NMR and X-ray diffraction data, those form 1:1 cage complexes with nitrate, sulfate, chloride, tartrate, and dithionate anions both of the inclusive (with an encapsulation of an anion) and exclusive (without such caging) types [104].

An octaamine macrobicyclic ligand **144** has been first prepared by Scheme 2.94 using stepwise synthetic procedure in high-dilution reaction conditions [105]. Its hexaprotonated form with a spherical cavity binds the sterically complementary fluoride anion that is coordinated within a distorted TP N_6-coordination polyhedron formed by six secondary amino groups of this covalent capsule.

An optimized non-template synthetic pathway by Scheme 2.95 to the amine caging ligand **144** using reduction of its imine macrobicyclic precursor **145** has been elaborated in [106]. Both these cage molecules have a D_3 pseudosymmetry, and the capping tertiary amine nitrogen atoms with the lone pairs located within their cavities are 6.37 and 6.84 Å apart, respectively [106].

Protonation of the encapsulating ligand **144** and its cage complexes with fluoride anion has been studied in [107] by ^{1}H NMR spectroscopy. This strongly basic macrobicycle (log $K^H_1 = 11.18$) undergoes stepwise protonation, and the log K^H_{2-4} values decrease (and, therefore, basicity lowers) with the degree of proton-

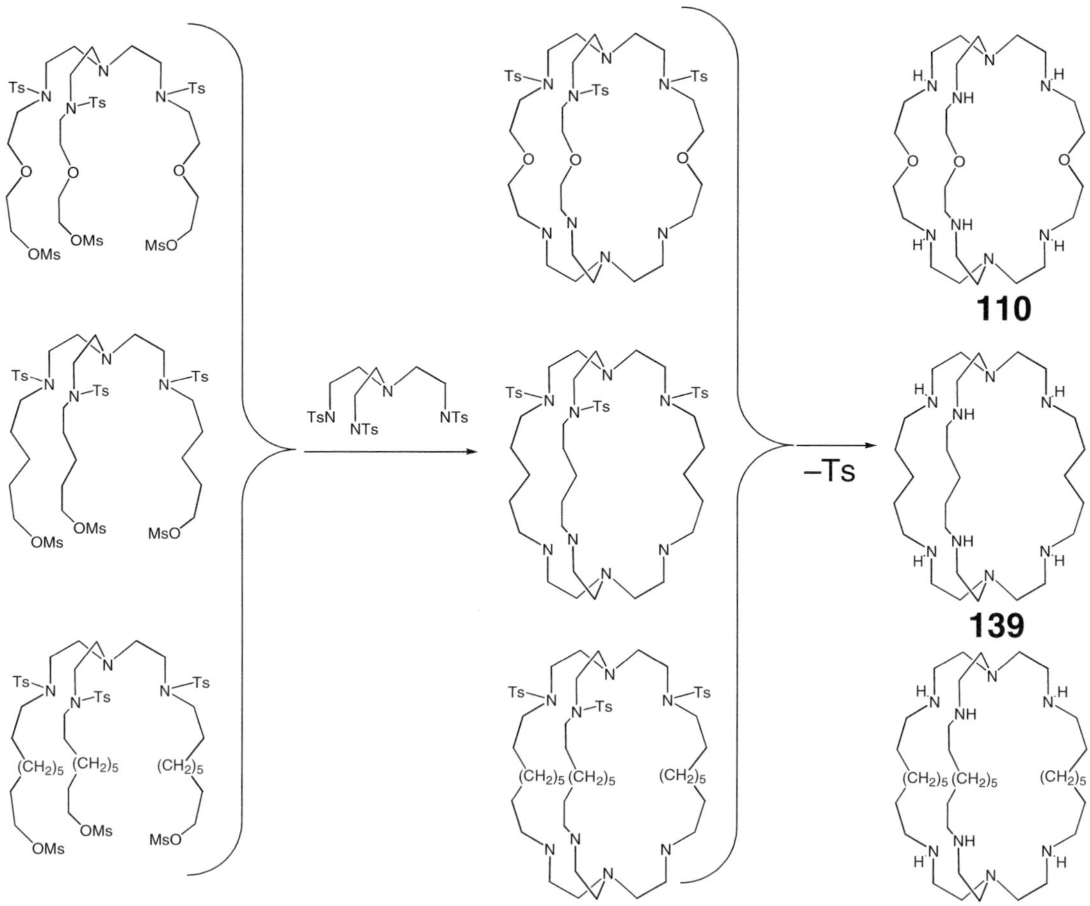

Scheme 2.91

ation. The decrease in pH from 8.8 at which the doubly protonated species mainly exist to pH = 3.6 leads to predominate formation of a tetra-protonated form of this ligand. Further lowering of pH results in three new signals that eventually replace initial ones due to the rupture of hydrogen bond network within its cavity (these intramolecular hydrogen bonds are formed by inward-oriented secondary amine groups at pH above 4) and the following conformation changes including the outward rotation of the protonated positively charged amino groups because of their electrostatic (Coulombic) repulsions [107]. The cage complexes with an encapsulated fluoride anion are substantially more basic than the initial macrobicyclic ligands [107]. Increase in a degree of their protonation leads to the increase in fluoride affinity relative to the chloride binding up to seven orders of magnitude. This F^-/Cl^-selectivity has been also confirmed by the MM calculations and is explained in [107] by the cavity size suitable for binding of small fluoride ion forming strong hydrogen bonds (especially, in the case of the hexaprotonated form of this caging ligand) and decreasing electrostatic (Coulombic) repulsions in hexapositively charged macrobicyclic framework [107].

The protonated forms of octaamine macrobicyclic ligands **138**, **144**, and **146** are reported in [108] to encapsulate chloride and bromide anions by (Scheme 2.96), thus giving 1:1 cage complexes, which were structurally characterized by single-crystal X-ray diffraction data. The cova-

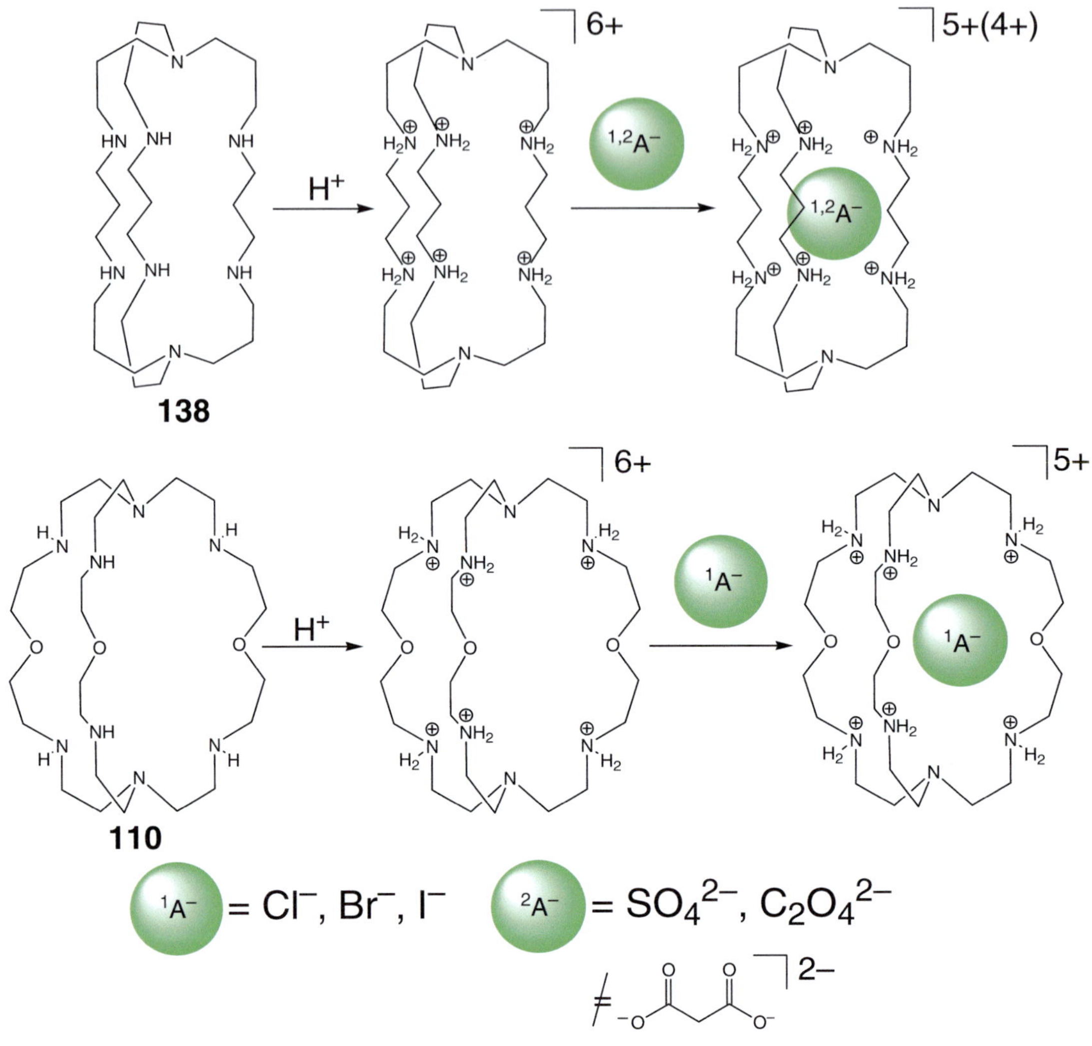

Scheme 2.92

lent capsule H^+_6 · **144** binds fluorine anion with higher selectivity than the chloride ion (selectivity coefficient $F^-/Cl^- > 10^8$). The ligand **138** binds strongly both F^- and Cl^- anions [108].

Condensation of *tren* with aromatic dichloride *74* by Scheme 2.97 gave a hexaamide caging ligand **147** for anion binding [109]. As follows from NMR titration data, this covalent capsule in polar organic media selectively encapsulates a series of monoanions; its binding constants are in a row $F^- > Cl^- > CH_3COO^- > H_2PO_4^-$, and it only slightly binds NO_3^-, HSO_4^-, and Br^- ions. The increase in such constants by two orders in a row $F^- > Cl^- > Br^-$ is explained in [109] by both the size effects and the hydrogen-bonding ability. The difference in binding constants for trigonal ions $CH_3COO^- \gg NO_3^-$ and for tetrahedral anions $H_2PO_4^- \gg HSO_4^-$ is explained in [109] by only weak hydrogen-bonding abilities of nitrate and hydrosulfate ions. The monoprotonated form of capsule **147** encapsulates a hexacoordinated chloride anion that is shifted from the geometrical center of the ligand's cavity in the direction of one of the two apical cross-linking nitrogen atoms. A caged fluoride anion in its 1:1 complex with hexaprotonated capsule H^+_6 · **147** occupies its center, thus forming hydrogen bonds with all six amide protons of this ligand [109].

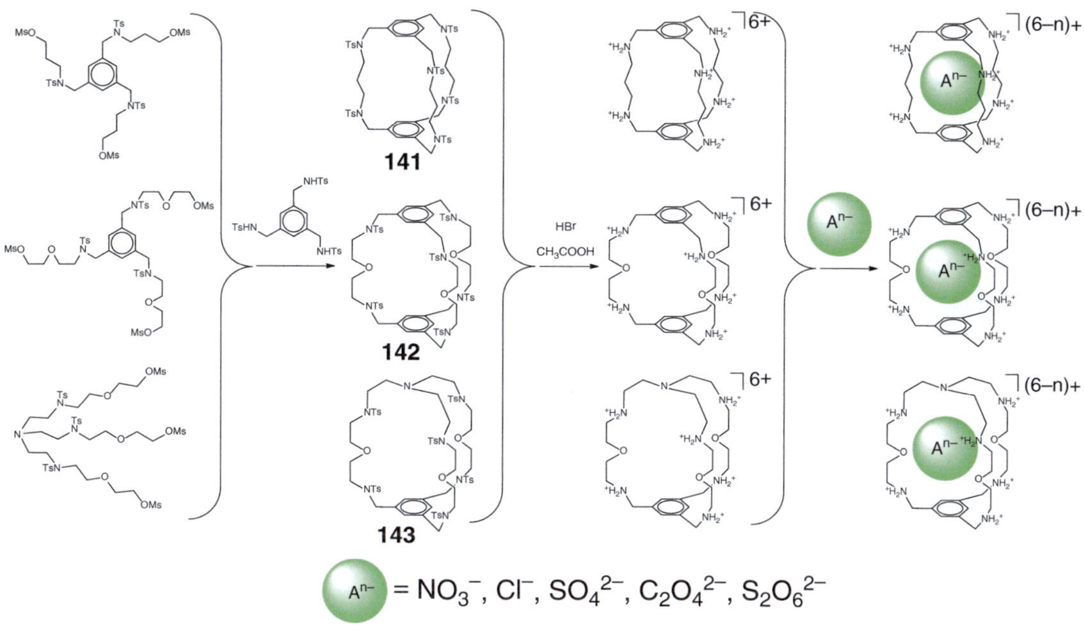

Scheme 2.93

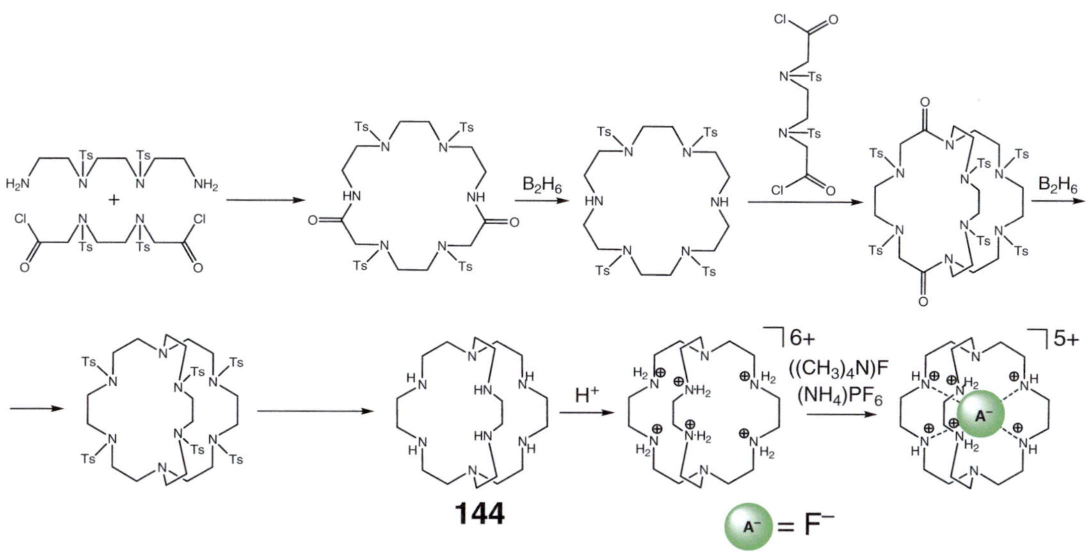

Scheme 2.94

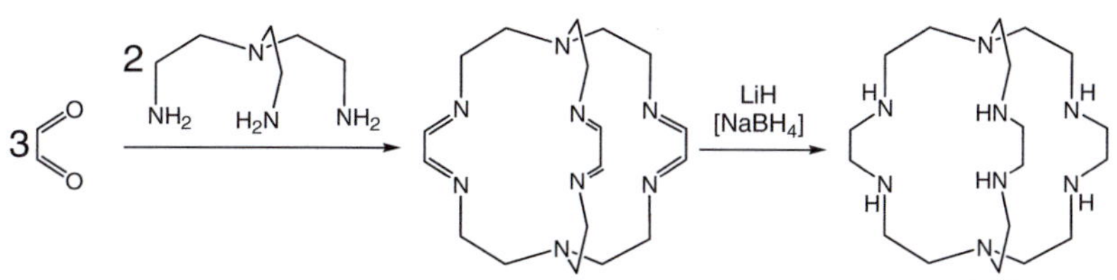

Scheme 2.95

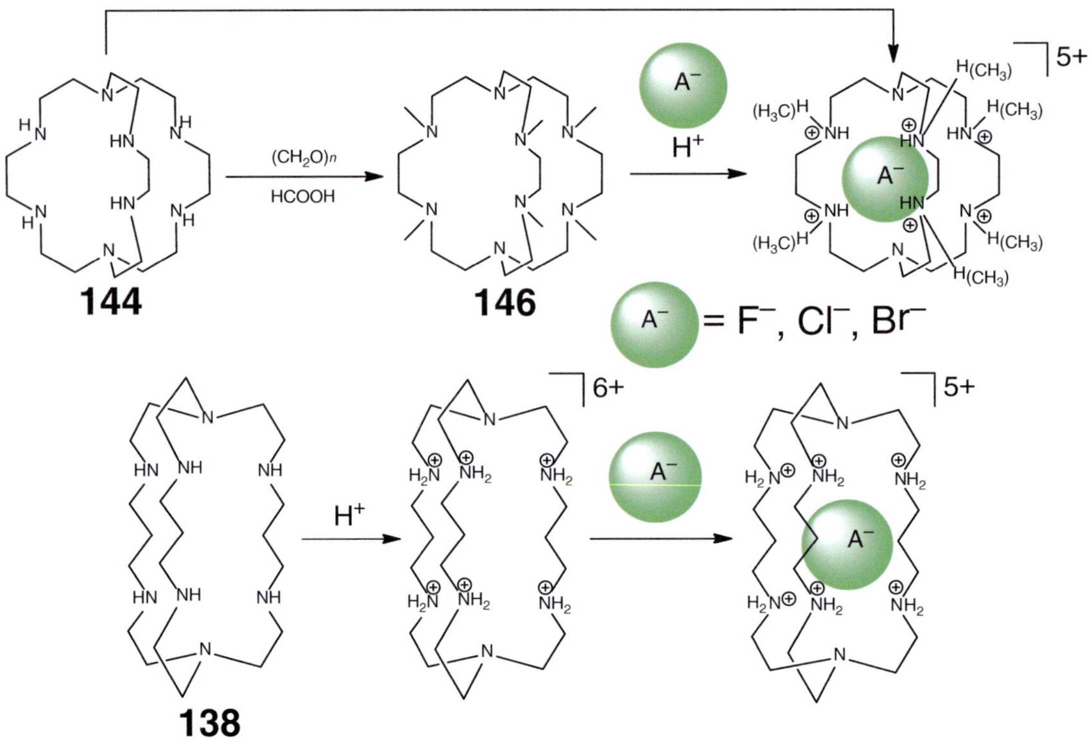

Scheme 2.96

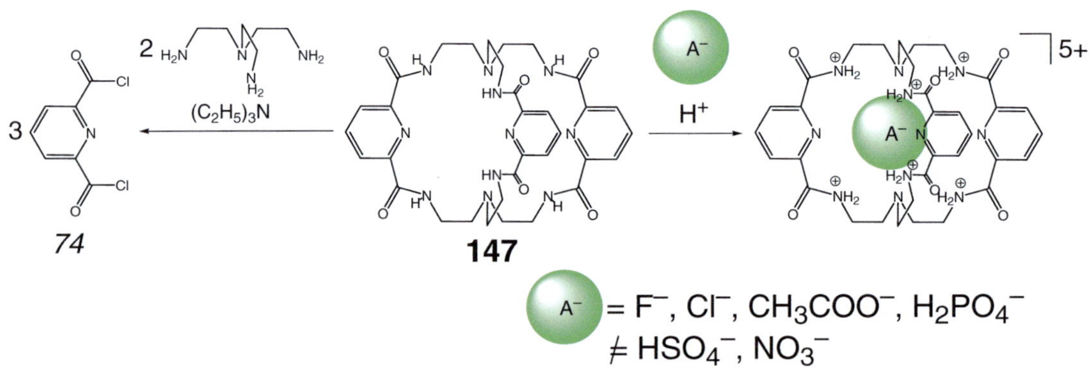

Scheme 2.97

1:1 cage complexes of the hexaamide capsule **147** and its hexaamine analog **14** have been prepared in [110] by their reactions with $(NH_4)_2SO_4$ and sulfuric acid, respectively (Scheme 2.98). In the cavity of octaprotonated ligand $H^+_8 \cdot$ **147**, an encapsulated sulfate dianion is hydrogen-bonded only with five of its eight protonated amino groups. In the intracomplex (SO_4^{2-})@ $H^+_2 \cdot$ **147**, the caged dianion forms such bonds with all amino groups of the doubly protonated capsule **147** [110].

A hexaamide covalent capsule **148** with an extended macrobicyclic framework has been described in [111] as a first polynucleating amide caging ligand with two of three pyridine-containing ribbed fragments having the same orientation and the third one looking at the

Scheme 2.98

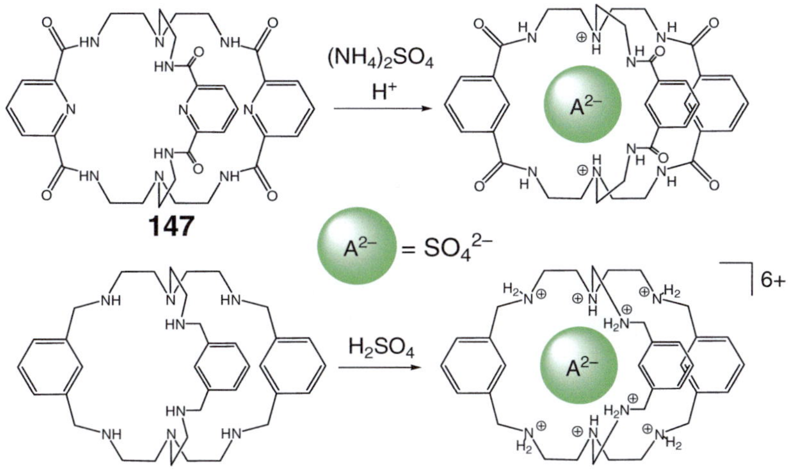

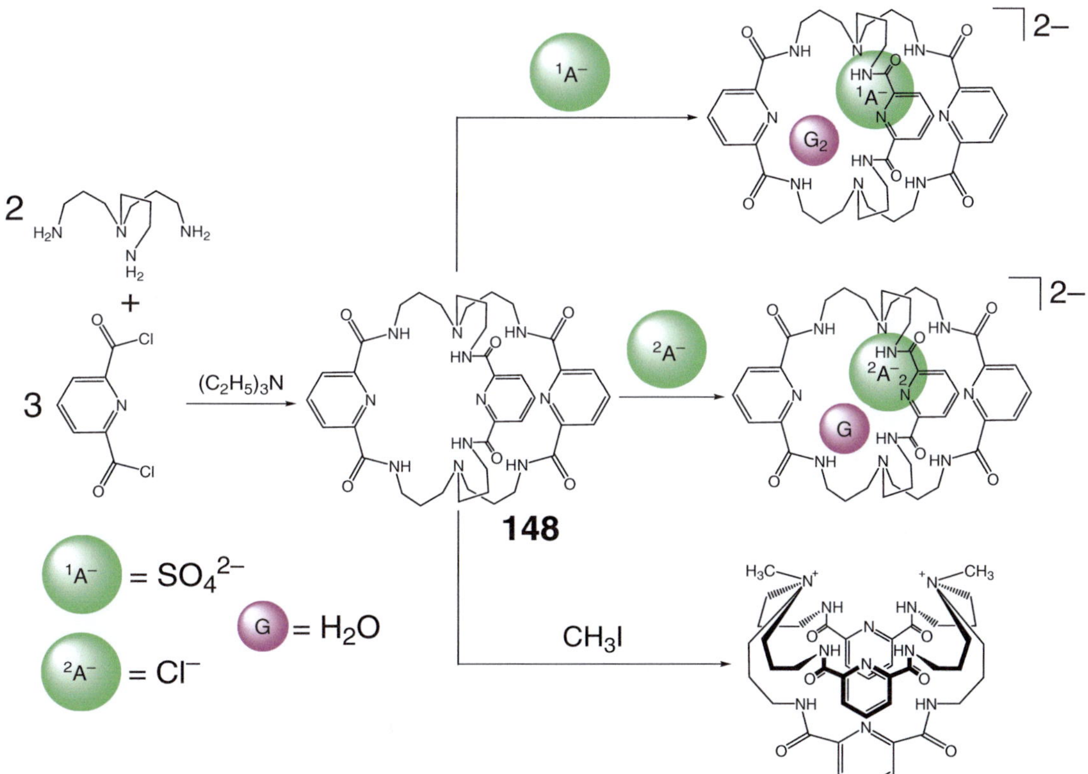

Scheme 2.99

opposite directions. This ligand encapsulates three guest species in its cavity by Scheme 2.99; those are either two chloride anions and one water molecule or one sulfate ion and two water molecules. In contrast, its *N*-methylated derivative **149** has a bowl-like shape with the same orientation of all ribbed hydrogen-donor fragments [111].

A similar synthetic approach has been used [112] to obtain a hybrid N_3O_3-macrobicyclic ligand **150** (Scheme 2.100) with an extended cavity, which gave a 1:1 cage complex with bromide anion. This complex contains a hydrogen-bonded encapsulated Br^- ion and its C_3-symmetry axis passes through this guest and the capping nitrogen atoms of a caging ligand **150**. Six adamantanoid $(H_2O)_{10}$ clusters form a 3D cage-like supramolecular network via hydrogen bonding with other water molecules and bromide ions. The cage complex (Br^-)@**150** is located within this supramolecular cage-like framework, thus forming a "Matreshka" complex-in-complex assembly [112].

Condensation of a tris-pyrrole ligand syntone *75* with its diformylated derivative *76* by Scheme 2.101 has been used in [113] for the synthesis

Scheme 2.100

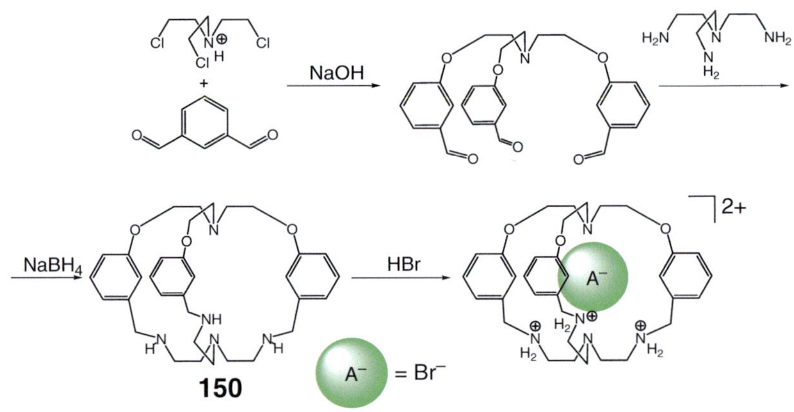

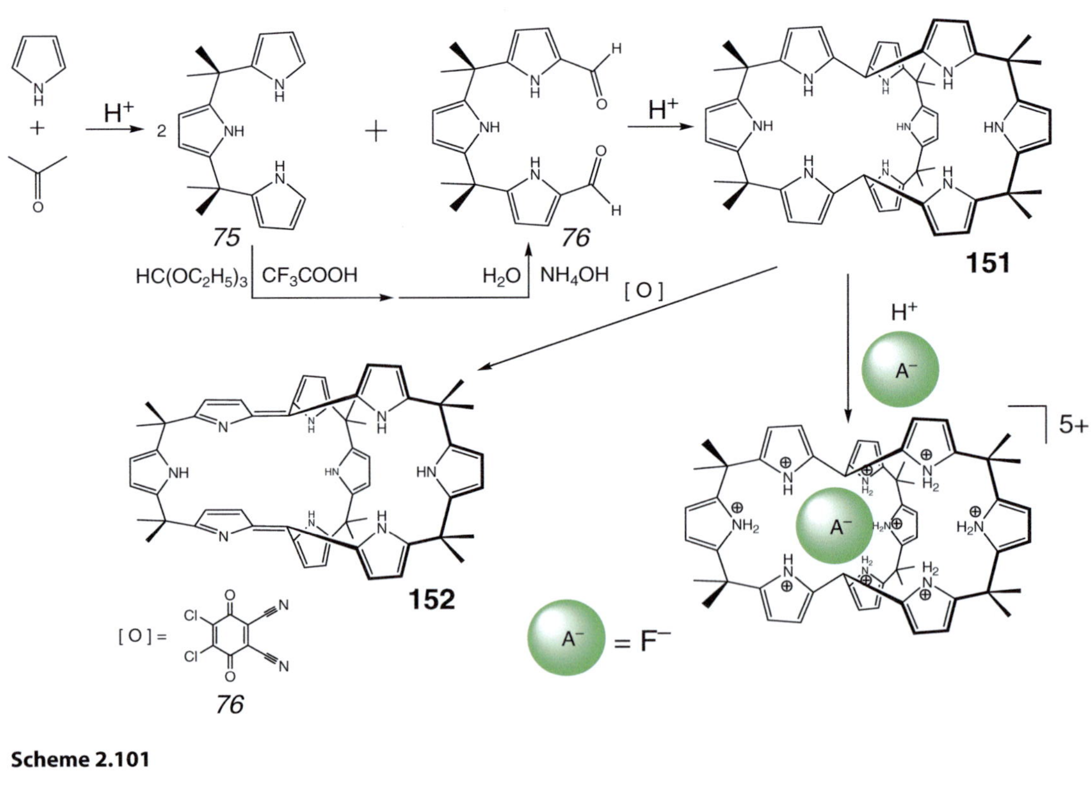

Scheme 2.101

thesis of the first macrobicyclic calixpyrrole **151** with bridgehead carbon atoms showing their *in–in* configuration with *meso*-like bridging carbon atoms, which are oriented inward of the cavity of **151** (as follows from X-ray diffraction and NMR data). This caging ligand encapsulates only small spherical fluoride anion from a series of monoanions. In their 1:1 cage complex, two of the three ribbed tris-pyrrole fragments form a large pocket in which six calixpyrrole NH groups are hydrogen-bonded with the encapsulated F^{-} ion; the third ribbed moiety does not interact with this guest. In contrast, chloride ion forms a host–guest 2:1 bis-cage complex with this ligand in which it is hydrogen-bonded with 12 NH groups of two molecules of **151**. When deposited in air, this complex is described in [114] to undergo oxidation leading to the change in its coloration from colorless to light red due to the formation of a calixphyrine analog **152** of its macrobicyclic calixpyrrole precursor **151**. The stoichiometric oxidation of **151** with oxidant *76* gave in 60 % yield a bright-red calixphyrine product of lower symmetry; six signals of non-equivalent *β*-pyrrole protons instead of three and no signals of methine protons of apical groups have been observed in its ^{1}H NMR spectrum. The covalent capsule **152** has more opened cavity and, therefore, is more suitable for encapsulation of guests than that of **151** [114].

A labile octaaminophenolic macrobicyclic ligand **153**, prepared in [115] by Scheme 2.102, and its hexaprotonated form are able to encapsulate metal cations as well as suitable anionic guests (first of all, chloride ion), thus forming 1:1 cage complexes. The covalent capsule **153** undergoes a stepwise protonation; its initial deprotonated and first two protonated forms are strong

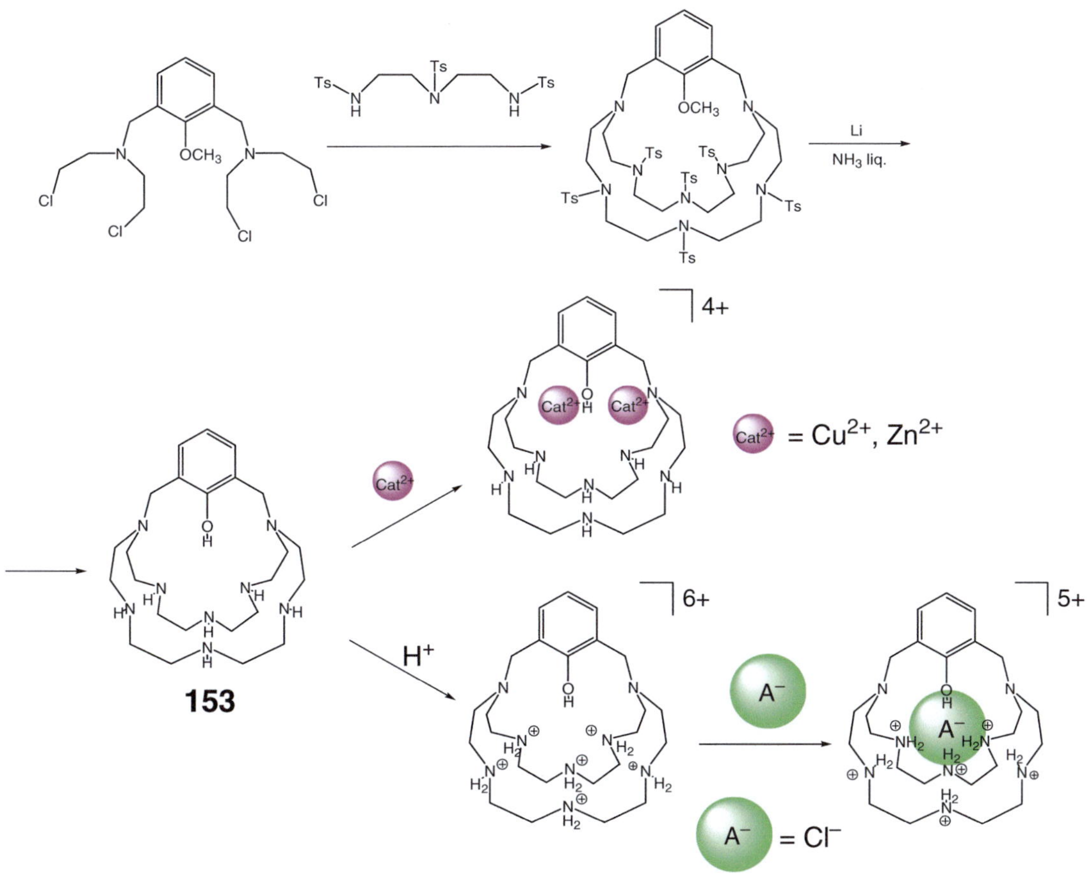

Scheme 2.102

bases. This ligand forms both 1:1 and 1:2 cage complexes with encapsulated copper and zinc(II) ions. The former complexes undergo protonation, while the latter are stable at pH >6. As follows from ^{35}Cl NMR spectra, hexa-, penta-, and tetraprotonated forms of **153** encapsulate chloride anion by Scheme 2.102 [115].

A bis-cyclopeptide covalent capsule **154** for selective binding of sulfate dianion has been prepared in [116] using copper(I)-catalyzed "click" reaction–cycloaddition of the corresponding azide to alkyne ligand syntones by Scheme 2.103 and characterized using NMR and X-ray diffraction data. Thermodynamics and kinetics of the encapsulation reaction for this guest have been studied by ITC and temperature-depended ^{1}H–^{1}H NOESY NMR methods, respectively.

Condensation of tripodal amine *tren* with a phthalimide-protected dicarboxylate ligand syntone *77* by Scheme 2.104 afforded a binucleating hexacarboxamide caging ligand **155**; its C_3-unsymmetric macrobicyclic framework easily accommodates cobalt, iron, nickel, and zinc(II) ions to give host–guest 1:2 complexes [117, 118]. The encapsulated metal ions in all these isostructural cage complexes occupy apical *tren* cavities having the distorted trigonal pyramidal N_4-coordination polyhedra with a cofacial orientation; the dipropoxyphenoxyl substituents are splayed out to the periphery of these capsules. The $M^{II}...M^{II}$ distance is the smallest for the binuclear manganese(II) cage complex (approximately 6.08 Å), and the metal ions are shifted from the mean plane of equatorial donor nitrogen atoms of a caging ligand. For the zinc(II) complex, this distance is approximately 6.42 Å, and their metallocenters are relaxed into its tripodal *tren*-based binding pockets. This fact is explained in [118] by a difference in sizes of the encapsulated metal ions. The analogous iron(II) complex underwent both the electrochemical and chemical oxidations giving its high-spin diiron(III)-containing derivative (as it follows from the ^{57}Fe Mössbauer spectrum). The re-reduction of this iron(II) compound with cobaltocene gave the initial binuclear iron(II) cage complex, which has been identified in [118] by NMR spectroscopy. These electrochemical and chemical redox reactions are reversible. The host–guest 1:2 iron- and cobalt(II)-encapsulating hexacarboxamide cage derivatives of **155** bind a cyanide anion by Scheme 2.104, thus forming cascade complexes. According to the X-ray diffraction, IR, and ^{1}H NMR data, this anion is located between two metallocenters substantially changing the geometry of their coordination polyhedra: the encapsulated cations are pulled from *tren*-based apical pockets causing elongation of Co–N bonds (and, therefore, their coordination polyhedra may be described as tetrahedral) and nonequivalence of the encapsulated high-spin iron(II) ions due to their different (C or N) coordination mode (as follows from the ^{57}Fe Mössbauer spectrum), respectively. According to magnetometry data, these 1:2 cage complexes of paramagnetic transition metal ions are in a high-spin state with antiferromag-

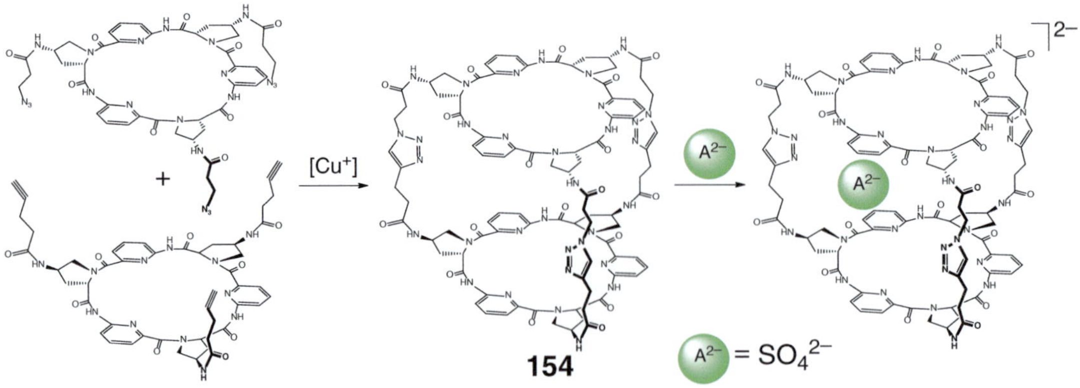

Scheme 2.103

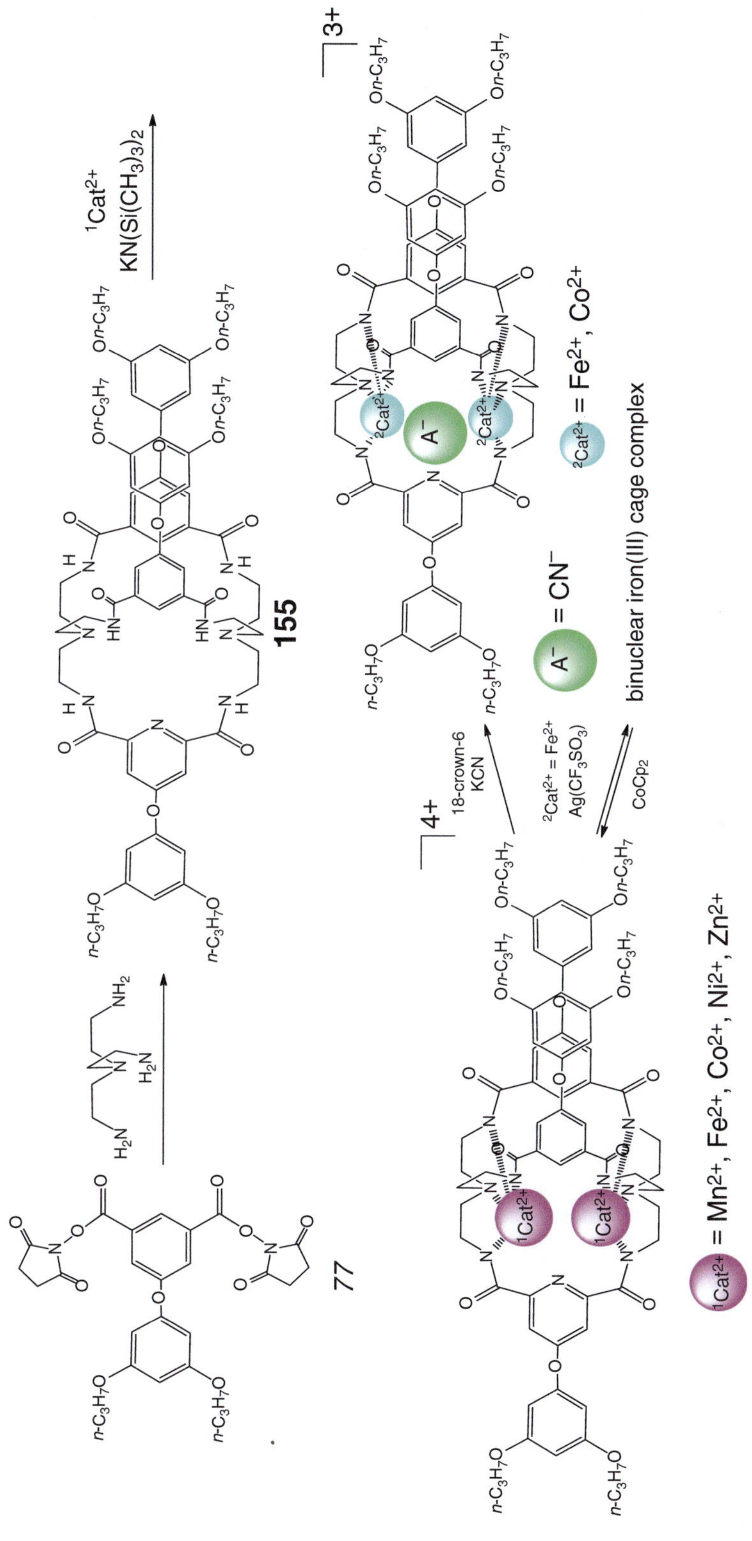
77
155
1Cat2+
KN(Si(CH3)3)2
18-crown-6
KCN
2Cat2+ = Fe2+
Ag(CF3SO3)
CoCp2
4+
3+
A− = CN−
binuclear iron(III) cage complex
2Cat2+ = Fe2+, Co2+
1Cat2+ = Mn2+, Fe2+, Co2+, Ni2+, Zn2+

Scheme 2.104

netic and extremely weak magnetic coupling between their remote caged metallocenters [118].

Two covalent capsules **156** and **157** with an azol-modified cyclic peptide platform have been designed in [119] for the selective binding of anions, as their thiourea groups in the side peptide chains are able to form strong hydrogen bonds with these anionic guests. The condensation of a cyclic tris-amine peptide syntone *78* with tris-isocyanides *79* and *80* by Scheme 2.105 afforded these macrobicyclic ligands. According to X-ray diffraction data, those encapsulate solvate molecules. In the case of **156**, the guest molecule of methanol forms strong hydrogen bonds with the capping ternary amine fragment in its *endo*-conformation, and the thiourea hydrogen-donor groups of **156** are directed inward the cavity of its cage framework. One of the two independent macrobicyclic species in the unit cell of the crystal **157** encapsulates one acetonitrile and two water molecules, while the second one contains three molecules of water. The observed difference in the caging properties of these capsules **156** and **157** is explained in [119] by larger size of the cavity of the 1,3,5-triethylbenzene-based ligand **157** as compared with that in *tren*-based analog **156**; the thiourea groups of **157** are also directed inward its cavity forming the hydrogen bonds with the encapsulated solvate molecules. The binding of monoanions by these caging ligands has been studied in [119] using their titration with *n*-tetrabutylammonium salts of F^-, Cl^-, Br^-, and CH_3COO^- monoanions. In all cases except for F^- ion, the fast exchange of these guests has been observed. The ligand **156** binds the Cl^- and Br^- anions three times stronger than its analog **157**, and both of them efficiently encapsulate small halogenide anions; their binding with I^-, NO_3^-, and HSO_4^- ions is much weaker. Both of these encapsulating ligands also form stable 1:1 cage complexes with acetate anion (in the case of **157**, the difference with other anions becomes more than three orders of magnitude). The observed lower selectivity of the macrobicyclic ligand **156** is explained in [119] by lability of its tripodal aliphatic amine cap as compared with the rigid 1,3,5-triethylbenzene-based capping group of **157**.

Encapsulation of halide anions by the tetracationic quaternary ammonium N_4-capped capsules **158–160** (Scheme 2.106) has been observed in [120]. Their 1:1 cage complexes with bromide and iodide anions are substantially more stable than those with Cl^- ion. The ligands **159** and **160** also form the stable cage complexes with other anionic guests shown in this scheme. Those are reported in [121] to discriminate between these anions according to their size, and the stability of the corresponding cage complexes depends on both the electrostatic and hydrophobic interactions, dominating for **159** and for **160**, respectively [121]. In the molecule of 1:1 cage complex of **159** with iodide anion, this spherical guest is encapsulated symmetrically within the spherical cavity of this ligand and, because of its electrostatic interactions with the four quaternary ammonium centers, the latters are arranged at the corners of an almost ideal tetrahedron; the average $N^+...I^-$ distance is approximately 4.54 Å [122].

The apically functionalized neutral analogs **161** and **162** of the above N_4-capped cationic capsules have been used in [123] as tetracationic caging hosts for encapsulation of a series of monoanions shown in Scheme 2.107.

The first borane–amine caging ligand **163** of this type has been prepared in [124] by Scheme 2.108 and characterized by single-crystal X-ray diffraction data. This covalent capsule also encapsulates anionic guests, thus giving the corresponding 1:1 cage complexes.

Both the side chain of the amino acid pendant substituents and the substituents at the aromatic tripodal fragment are reported in [125] to affect chloride binding ability of the macrobicyclic amide ligands **164–166** obtained in [125] by Scheme 2.109. The covalent capsules having the 1,3,5-benzene moieties showed low affinity to chloride anion, whereas those with triple methyl- and ethyl-substituted benzene panels displayed stronger binding by one order of magnitude [125].

2.2.3 Encapsulation of Cations

The covalent capsule **132** is reported in [98] to encapsulate one ammonium cation, thus giving a

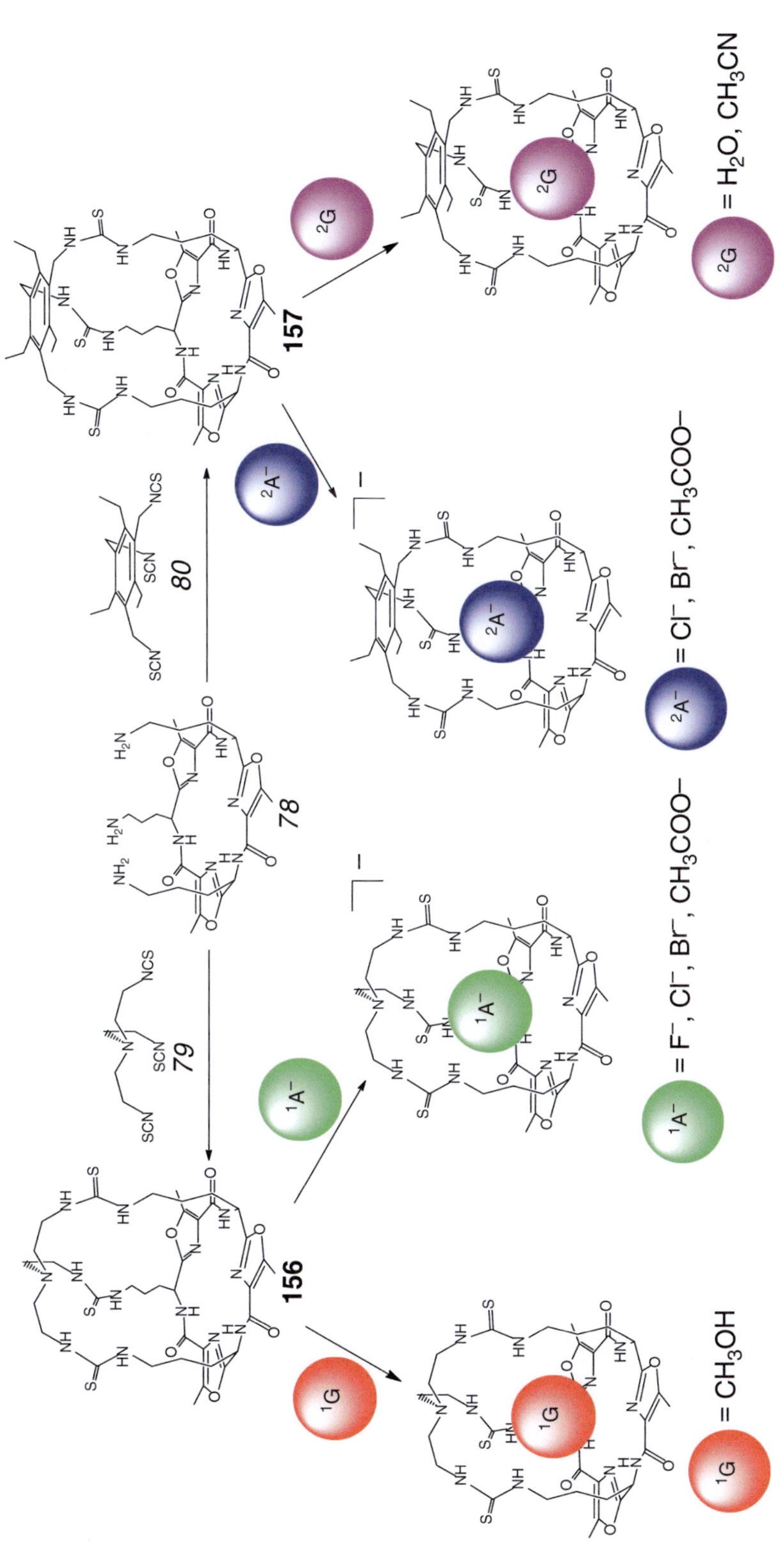

Scheme 2.105

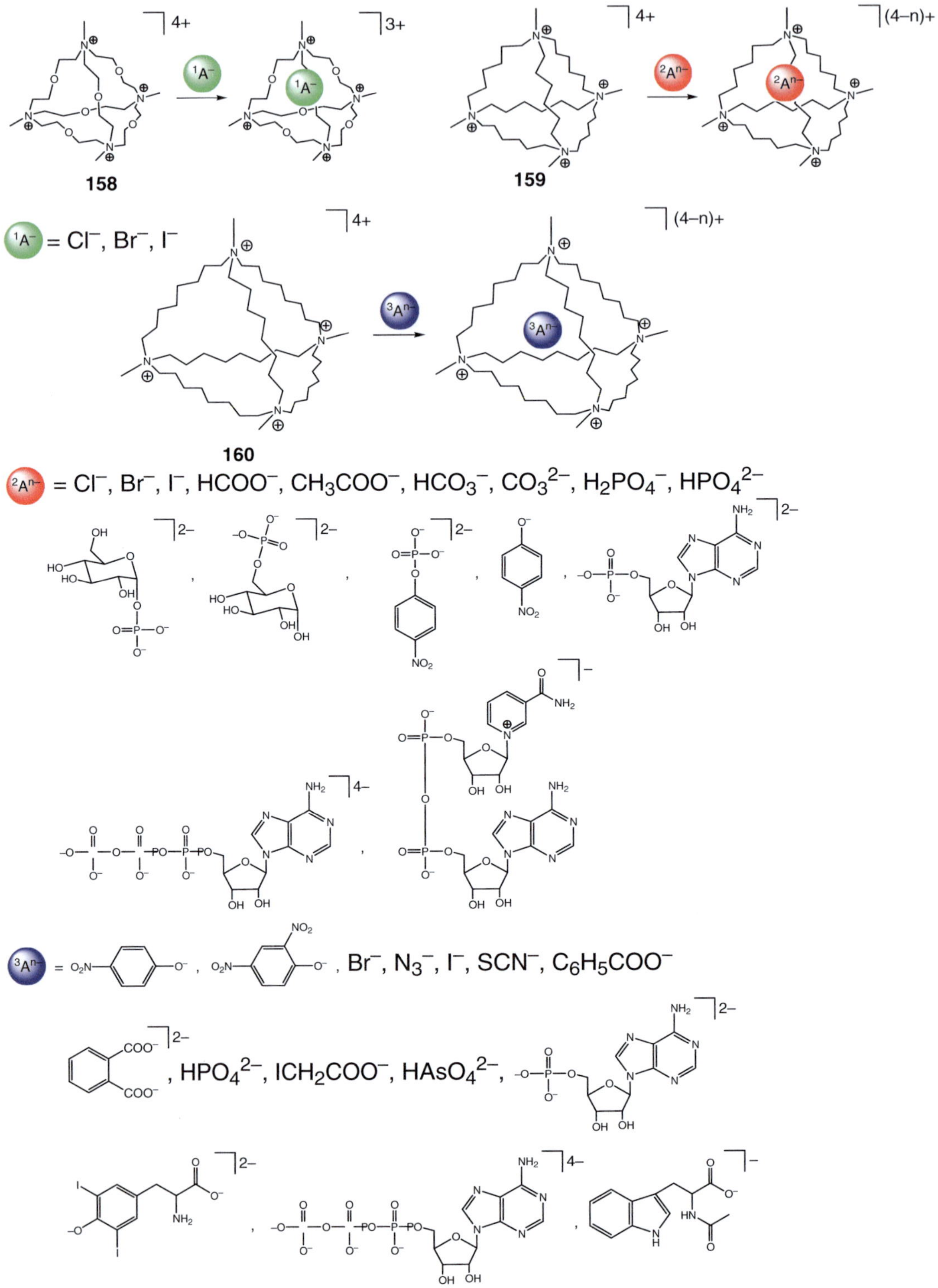
158
159
160
$^1A^-$ = Cl⁻, Br⁻, I⁻
$^2A^{n-}$ = Cl⁻, Br⁻, I⁻, HCOO⁻, CH₃COO⁻, HCO₃⁻, CO₃²⁻, H₂PO₄⁻, HPO₄²⁻
$^3A^{n-}$ = Br⁻, N₃⁻, I⁻, SCN⁻, C₆H₅COO⁻
HPO₄²⁻, ICH₂COO⁻, HAsO₄²⁻

Scheme 2.106

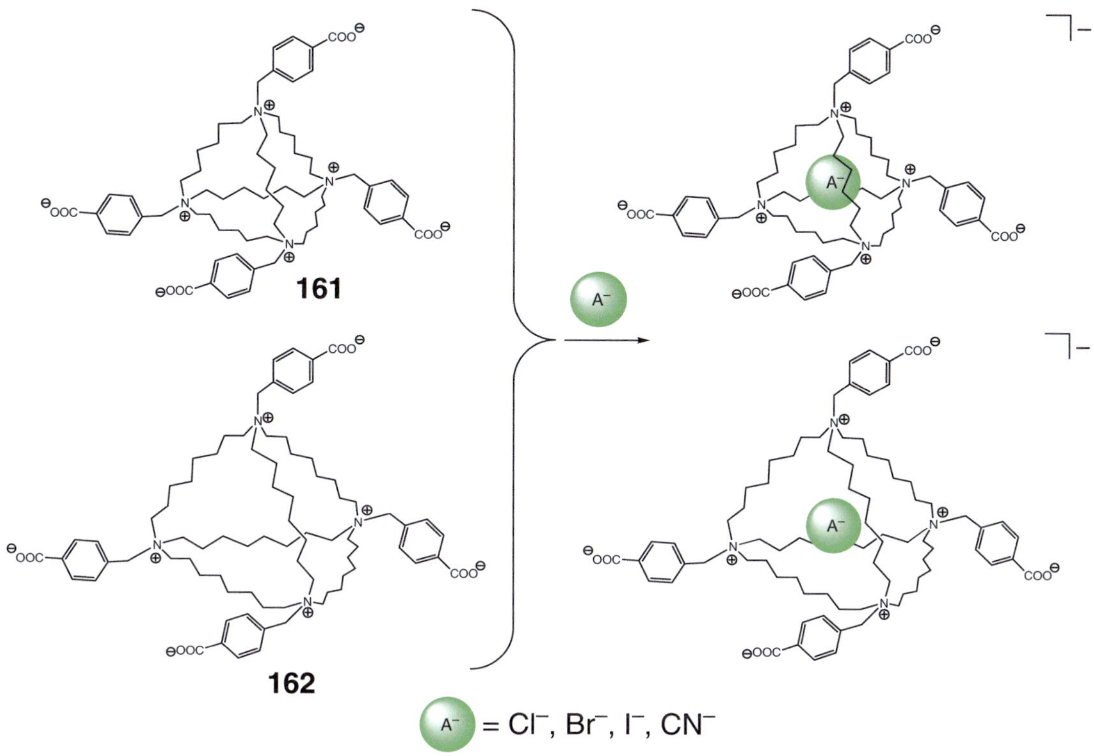

Scheme 2.107

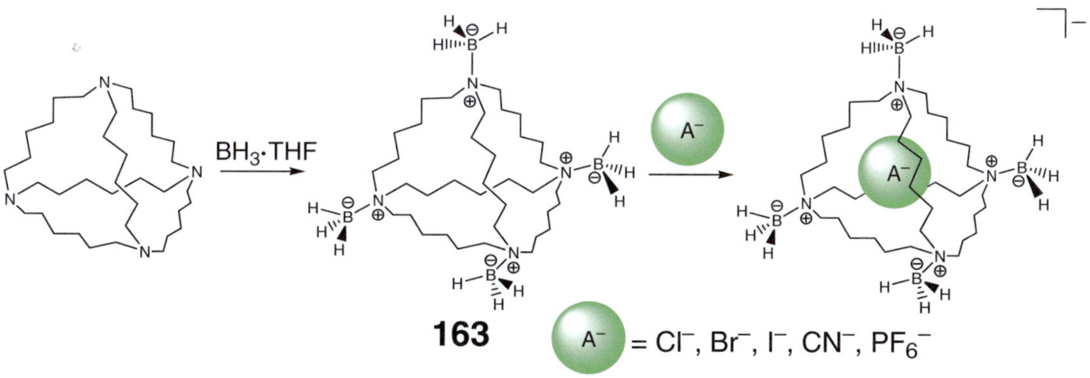

Scheme 2.108

1:1 cage complex by Scheme 2.110. Such caged cationic guest forms four NH_4^+...N hydrogen bonds stabilizing this supramolecular assembly.

A combined synthetic strategy including subsequent imine condensation–hydrogen reduction reactions and condensation of a diamine-terminated macrocyclic precursor with the corresponding aromatic dichloroanhydride by Scheme 2.111 gave a 1:1 cage complex of macrobicyclic ligand **167** [126]. It contains one encapsulated water molecule that forms N–H...O and O–H...N hydrogen bonds with four nitrogen-containing fragments. The distance between the center of its crown ether moiety with donor oxygen atoms and that of N_4-polyhedron is approximately 6.19 Å, thus allowing for the

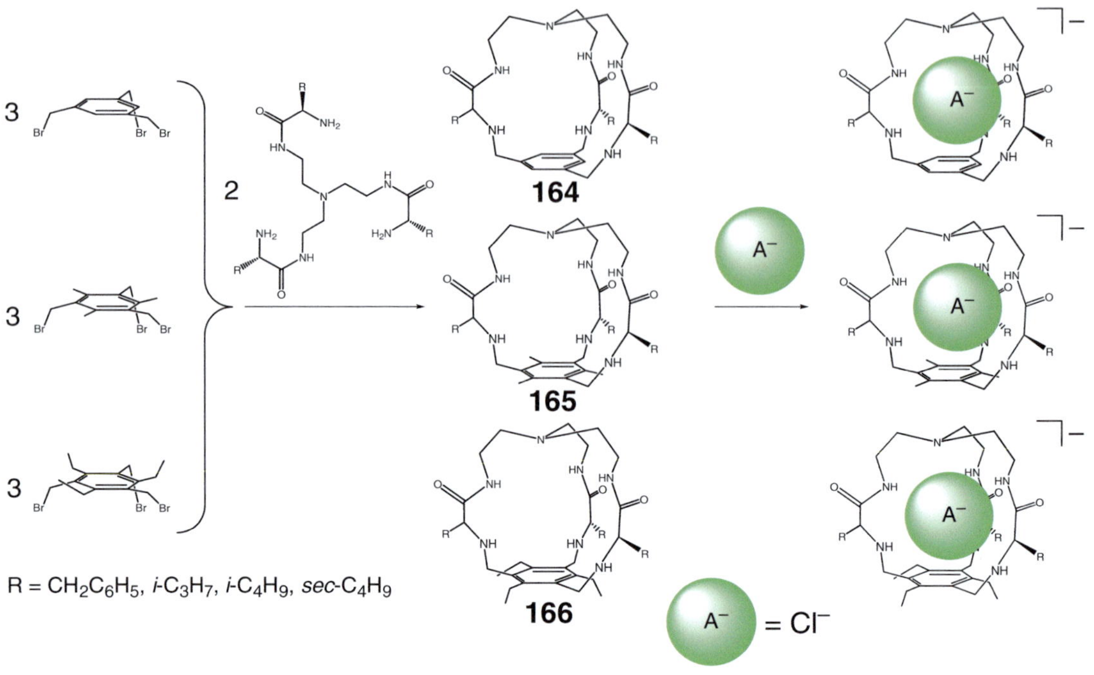

Scheme 2.109

Scheme 2.110

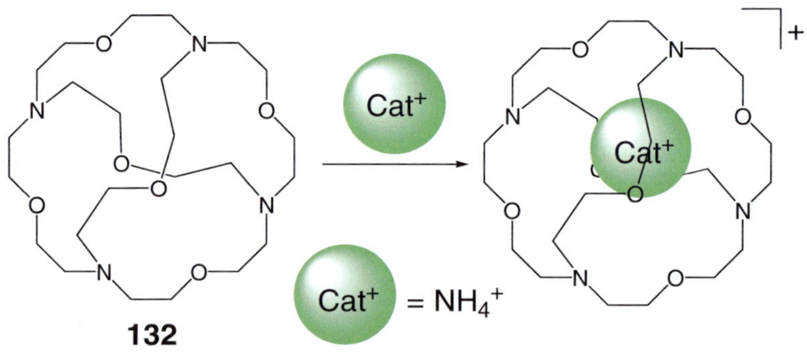

encapsulation of one cationic or one anionic substrate. The extraction capacity of **167** for encapsulation of alkylammonium salts, alkali metal halides, and amino acids has been studied in [126] using ^{1}H NMR and ESI-MS spectra. This ligand extracts potassium halides from their aqueous solutions to chloroform media, whereas neither solid–liquid nor liquid–liquid extractions are observed in the case of methylammonium chloride and several amino acids (Scheme 2.111). Thus, the presence of potassium(I) cation is described in [126] to play an important role in these extraction processes. The binding constants with halide anions strongly depend on the extractant; however, in all the cases studied they decrease in a row $I^- > Br^- > Cl^- > F^-$. Therefore, all these halide anionic substrates are bonded to the same sites of the covalent capsule **167** via the same interaction pattern [126]. This caging ligand undergoes protonation of its capping (cross-linking) ternary amine fragments; the doubly protonated form of **167** weakly binds trifluoroacetate anion that may be easily exchanged by another anionic guests. This form gave 1:1 cage complexes with halide anions, and its affinity decreased in a row $Cl^- > Br^- > I^- \approx F^-$. The observed discrimination between these spherical monoanions is explained in [126] by the size-match between the cavity of the doubly protonated

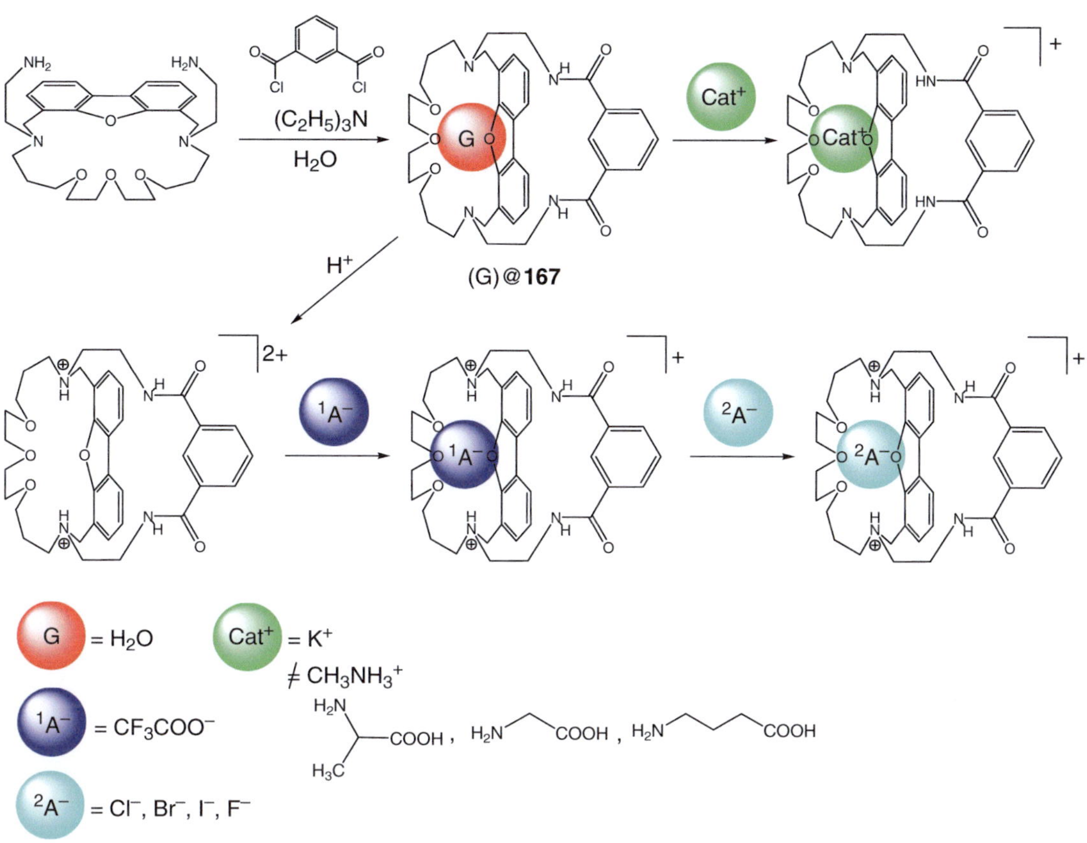

Scheme 2.111

caging ligand and that of an anionic substrate, which displays the best fit for chloride ion. Going from initial neutral caging ligand **167** to its protonated forms, the selectivity of chloride anion encapsulation and the corresponding binding constants increase, whereas other halide anions demonstrate an opposite effect [126].

Template condensation of the aromatic trialdehyde ligand syntone *81* with tripodal amine *tren* on a cesium(I) ion as a matrix by Scheme 2.112 has been used in [127] for the synthesis of a hybrid O_3N_3-macrobicyclic ligand **168**, which contains a tripodal O_3-donor cross-linking fragment as well as a Schiff-base tris-azomethine N_3-donor capping apical group. Its reduction with $NaBH_4$ gave a trioxatriamine capsule **169** in a moderate yield. This ligand forms 1:1 cage complexes with transition metal ions; in particular, it easily encapsulates Cu^{2+} ions having a pseudotetrahedral CuN_4-coordination polyhedron within a cavity of **169** (as follows from UV-vis and EPR data). The N_3O_3-macrobicyclic ligand **169** is described in [127] also to encapsulate various organic and inorganic mono- and dications shown in this Scheme.

2.3 Cryptophane and Cryptand Capsules

2.3.1 Free Cages and Encapsulation of Neutral Molecules

D_3- and C_{3h}-symmetric cryptophane capsules **170–176** have been prepared in [128] using two-step synthetic procedures by Scheme 2.113. This series of cryptophane caging ligands has been

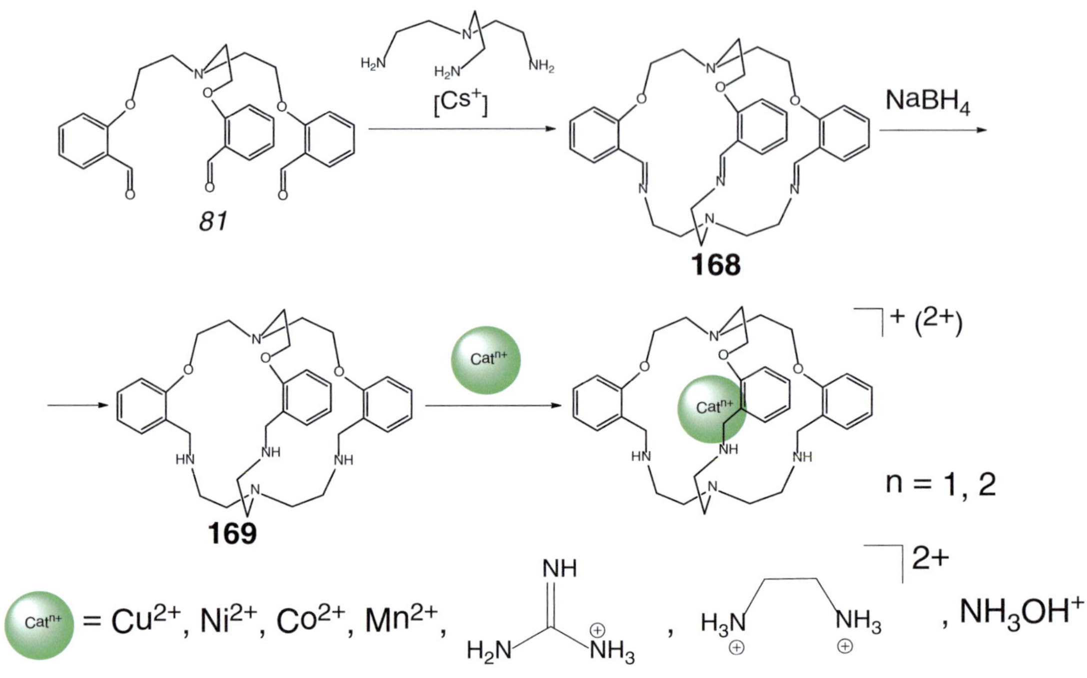

Scheme 2.112

proposed for xenon encapsulation; the capsule **177** and its congeners (Scheme 2.114) are also described in [129] as prospective encapsulating ligands for a wide range of neutral molecules. The cavity of **177** with an approximate volume of 81 Å^3 can also efficiently encapsulate xenon, as static dimensions of its portals are smaller than van der Waals diameter of this guest (approximately 4.3 Å). The binding constant for this caging ligand is 3.5 times higher than that for **170** [129].

Enantiopure cryptophane capsules **178–183** have been prepared in [130] by Scheme 2.115. Experimental ECD and VCD methods and quantum chemical calculations allowed studying their chiroptical properties and determining of their absolute configurations.

Optical resolution of bromochlorofluoromethane was performed in [131] using its encapsulation by Scheme 2.116 within the cavity of a chiral caging host **184**, cryptophane-C, designed to fit substrates of a given size. The diastereomeric 1:1 cage complex formed by the (–)-macrobicyclic ligand and the (–)-guest molecule is more stable than those formed by hosts and guests with opposite signs of their optical rotations. Absolute configuration of the enantioselectively caged molecule within the cavities of *R*- and *S*-capsules and the corresponding free energy difference have been also evaluated in [132].

Monofunctionalized cryptophane-A caging ligands **185–190** with achiral and chiral substituents such as amino acid residues (Scheme 2.117) have been designed in [133] for xenon encapsulation. Their binding properties have been theoretically and experimentally studied in this work using quantum chemical calculations and ^{129}Xe NMR spectra, respectively.

Scalable synthesis of the capsule **177** by Scheme 2.118, giving a target product in a high yield, is described in [134]. This ligand exhibits high affinity for xenon encapsulation in organic media and easily undergoes bromination and iodination, thus affording monohalogen-containing cryptophanes **191** and **192**. The catalytic hydrodehalogenation, lithiation, and palladium-catalyzed Heck reactions of these reactive precursors allowed obtaining monofunctionalized capsules **193–195**.

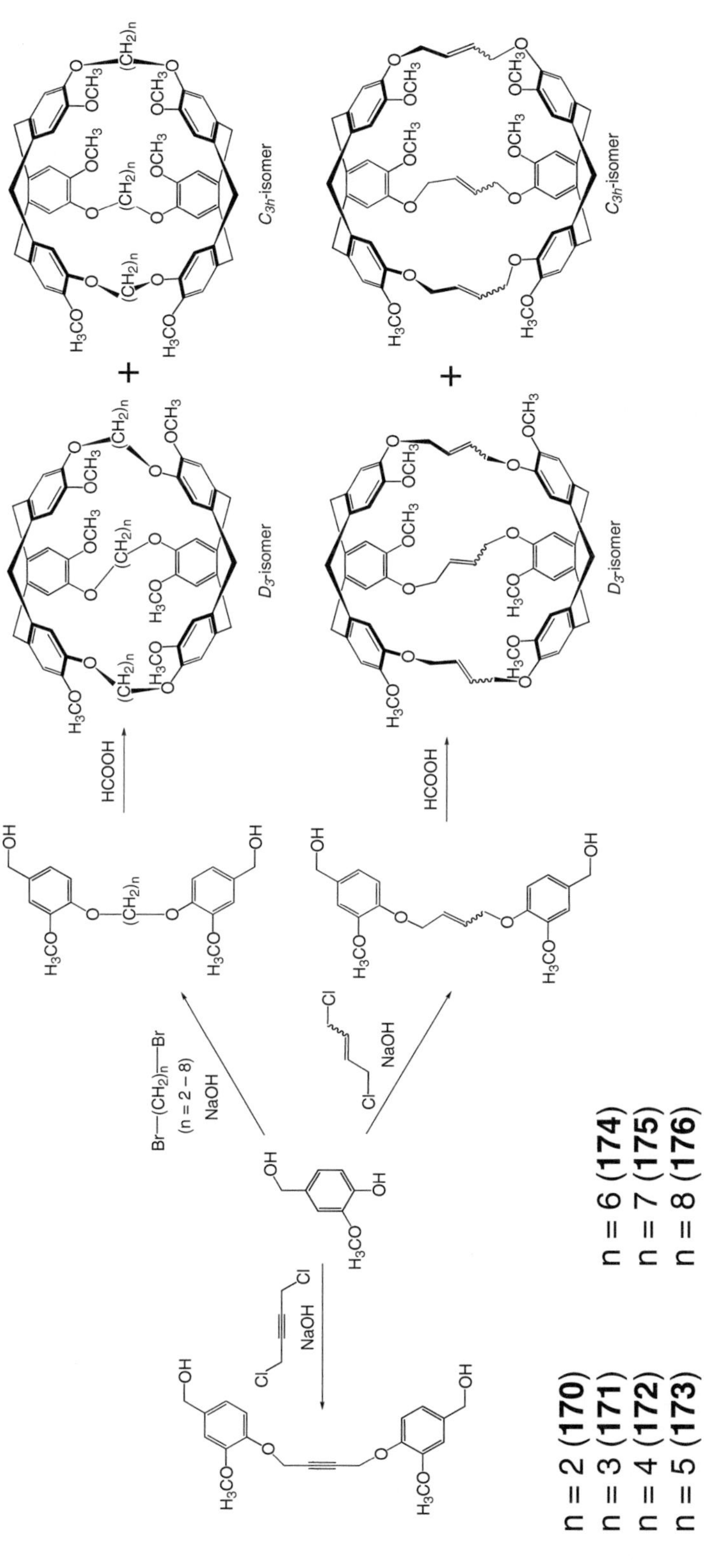

Scheme 2.113

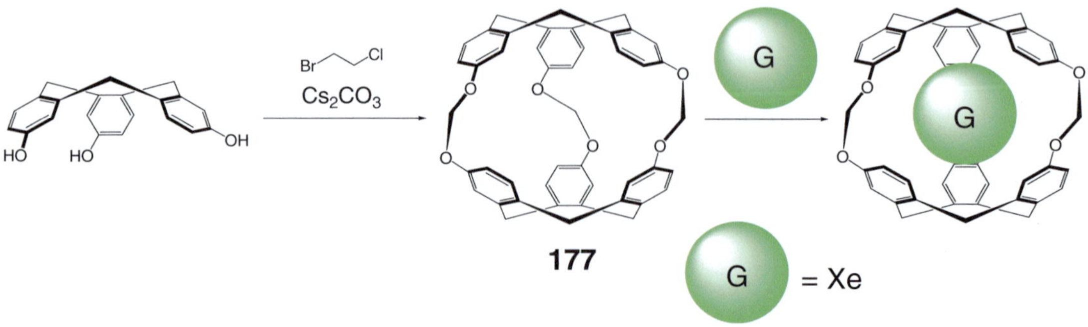

Scheme 2.114

Scheme 2.115

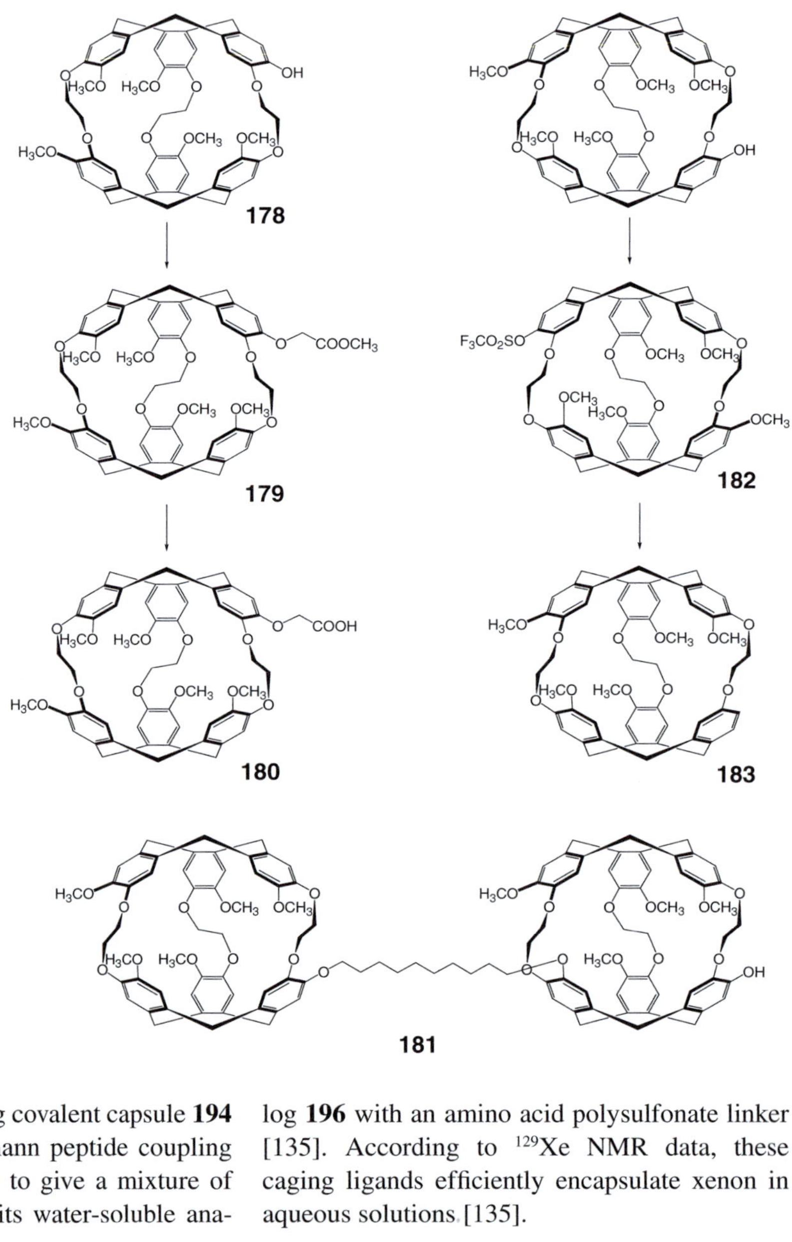

The carboxyl-containing covalent capsule **194** undergoes Schotten–Baumann peptide coupling reaction by Scheme 2.119 to give a mixture of racemic diastereomers of its water-soluble analog **196** with an amino acid polysulfonate linker [135]. According to ^{129}Xe NMR data, these caging ligands efficiently encapsulate xenon in aqueous solutions [135].

Scheme 2.116

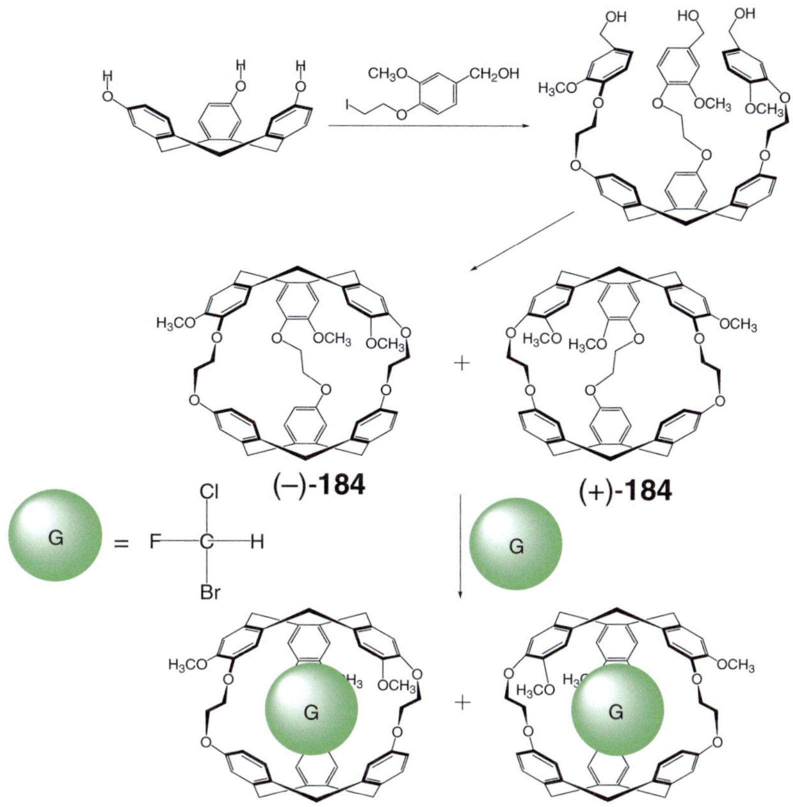

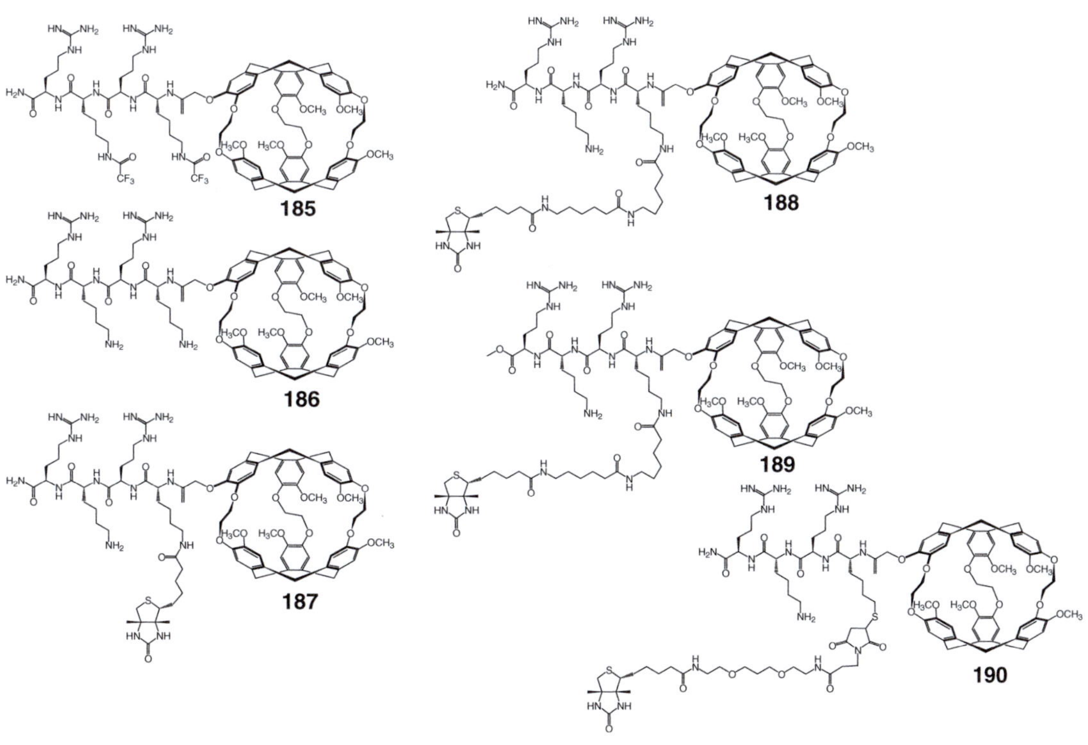

Scheme 2.117

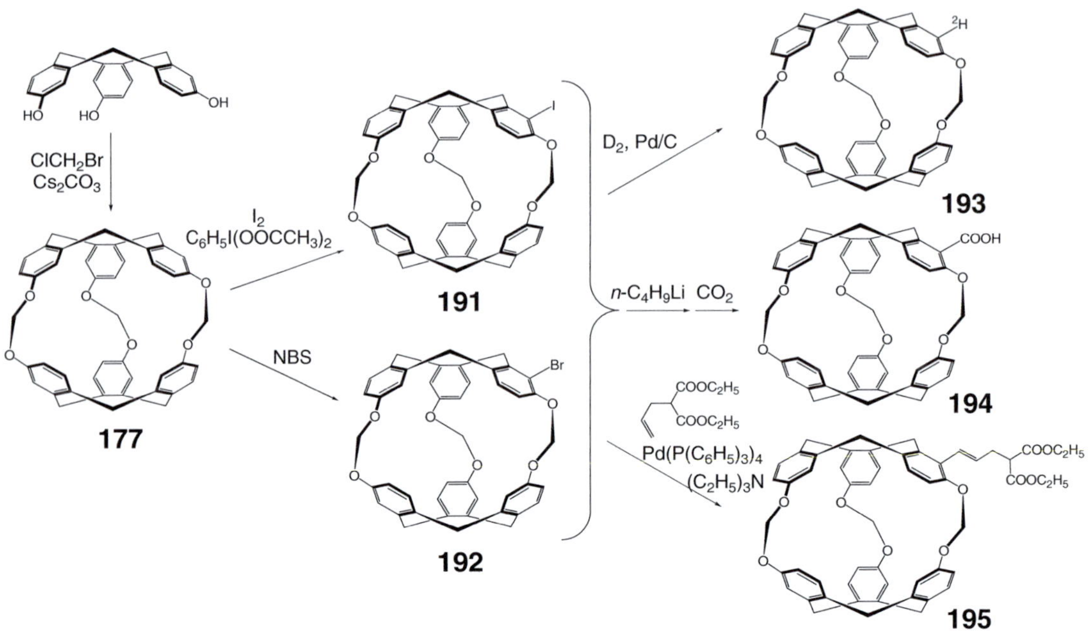

Scheme 2.118

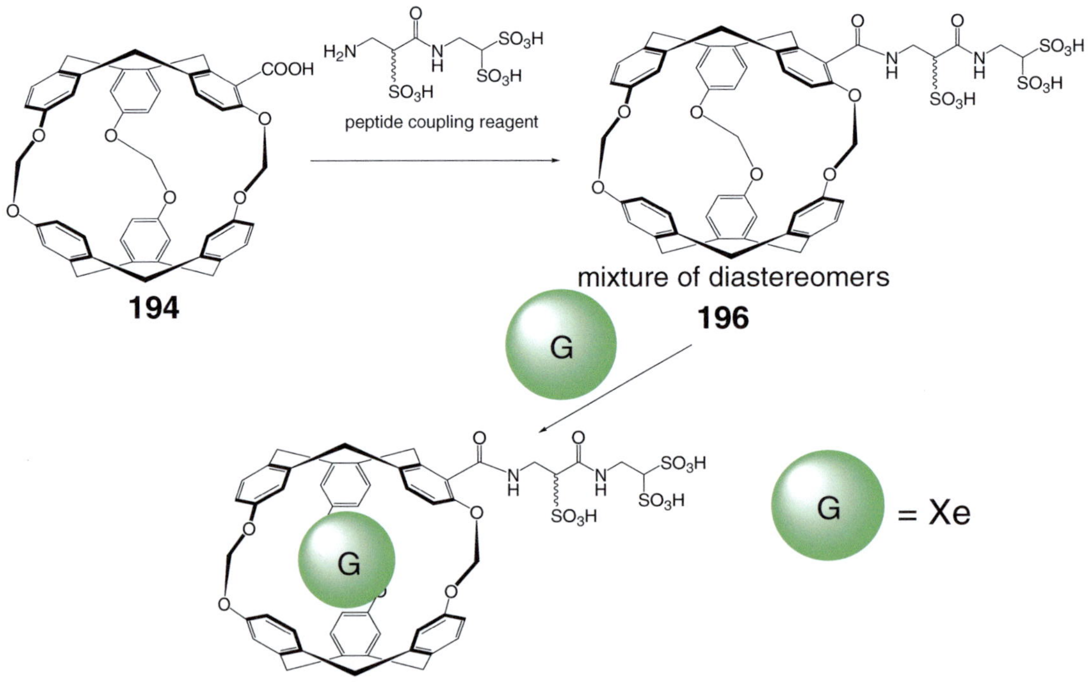

Scheme 2.119

A water-soluble pentahydroxyl-containing cryptophane capsule **198** has been prepared in [136] from its precursor **197** by Scheme 2.120. Chiroptical properties of their *MM*- and *PP*-enantiomers have been studied by polarimetry, ECD, and VCD methods. This covalent capsule **198** undergoes guest-induced conformational changes upon encapsulation of guest solvent molecules [136].

A facile synthetic procedure based on copper(I)-mediated [3 + 2] azide–alkyne Huisgen

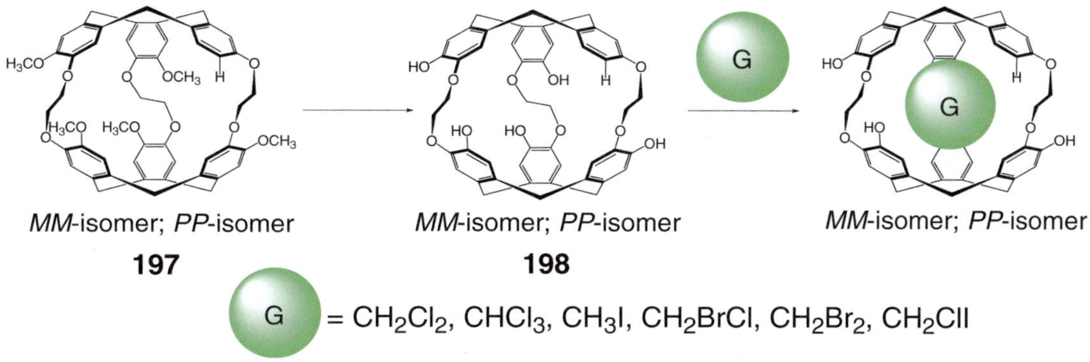

Scheme 2.120

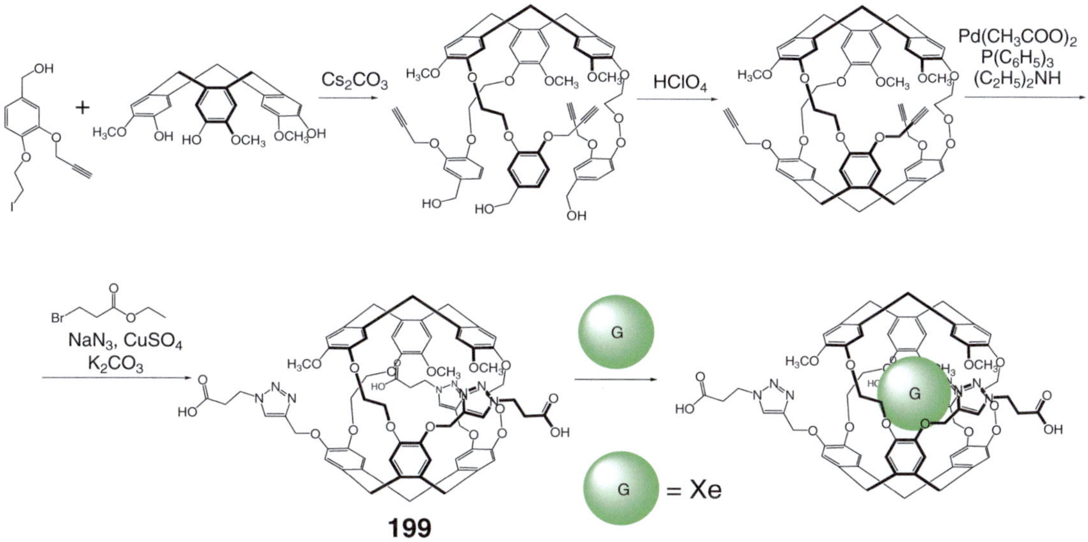

Scheme 2.121

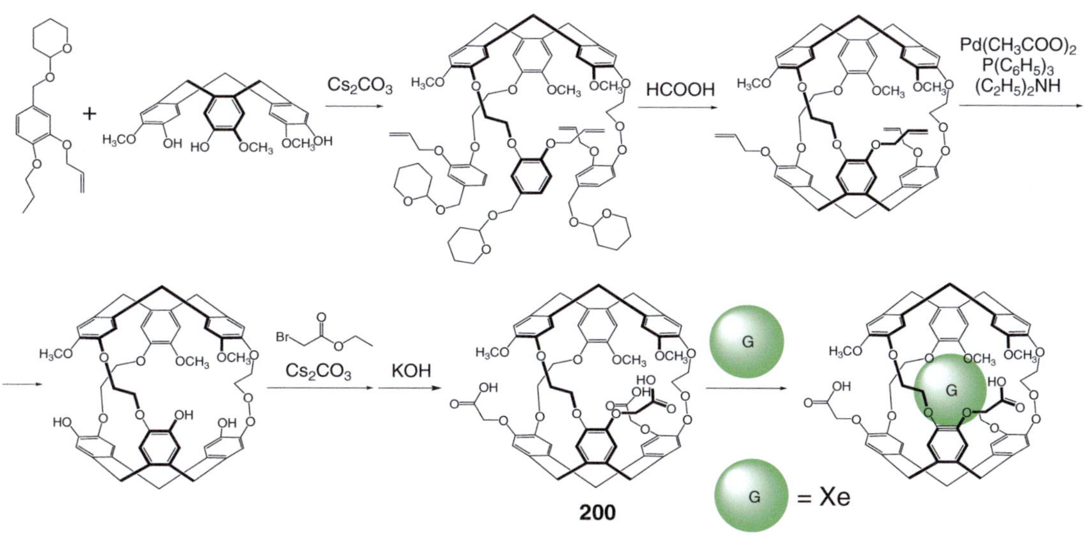

Scheme 2.122

cycloaddition has been developed in [137] to obtain a water-soluble caging ligand **199** by Scheme 2.121 [137]. According to fluorescence quenching and ITC data, this ligand shows high affinity toward xenon, especially at physiological temperature [137].

A triacetate cryptophane capsule **200** has been synthesized in [138] by Scheme 2.122. Its binding constant with xenon is much higher than that for the above tris-(triazolopropionic acid)-containing ligand **199**. The fluorescence studies showed that the encapsulated xenon guest quenched fluorescence by promoting intersystem crossing from the first singlet excited state of a chromophore (S_1) to the lowest triplet state (T_1). The lifetime experiments suggest the presence of two isomers of **200**: major C_3-symmetric crown–crown isomer (approximately 95 %), which can encapsulate this guest (as both its cyclotriveratrylene fragments have a concave form allowing to form a molecular cavity), and its minor crown–saddle conformer (approximately 5 %) [138]. The deprotonated carboxyl groups of **200** are poorly solvated due to them being close to the cryptophane cage framework that causes partial destabilization of the encapsulating crown–crown conformer. The higher fluorescence lifetime of 1:1 cage complex (Xe)@**199** than that of (Xe)@**200** suggests that xenon is a better fluorescence quencher of **200** than of **199**.

Microwave-assisted modification of the cryptophane precursor **177** with an excess of $[Cp^*Ru(\mu_3\text{-}Cl)]_4$ (*82*) by Scheme 2.123 afforded a water-soluble permetallated cryptophane **201** [139]. According to X-ray diffraction data, its empty cavity adopts a contracted conformation with a cavity volume of approximately 32 Å^3. Such behavior of the cryptophane capsule **201** is unusual, as other macrobicyclic species collapse upon emptying via partial inversion of one of their caps. As follows from ^{129}Xe NMR data, this caging ligand encapsulates xenon; the process is described [139] to be significantly affected by six cationic arene-containing electron-withdrawing substituents.

A tris-triazoloethylamine capsule **202** of this type with high xenon and radon affinity has been prepared in [140] using copper-catalyzed 1,3-cycloaddition "click" reaction of appropriate alkyne- and azide-containing syntones by Scheme 2.124.

Encapsulation of small gas guests by the cryptophane ligand **177** has been studied in [141] by variable-temperature ^{1}H NMR spectroscopy. This ligand forms a more stable 1:1 cage complex with methane than that its analog **170**. The high negative entropy of this complexation reaction in the case of **177** suggests strong reorganization of the guest and the rigid structure of a transition state of its cage complex [141]. According to ^{1}H NMR data, covalent capsule **177** also encapsulates carbon dioxide, ethane, ethylene, and molecular hydrogen, but discriminates chloro- and dichloromethanes and propane as guests. The binding constants for the cage complexes with encapsulated ethane and

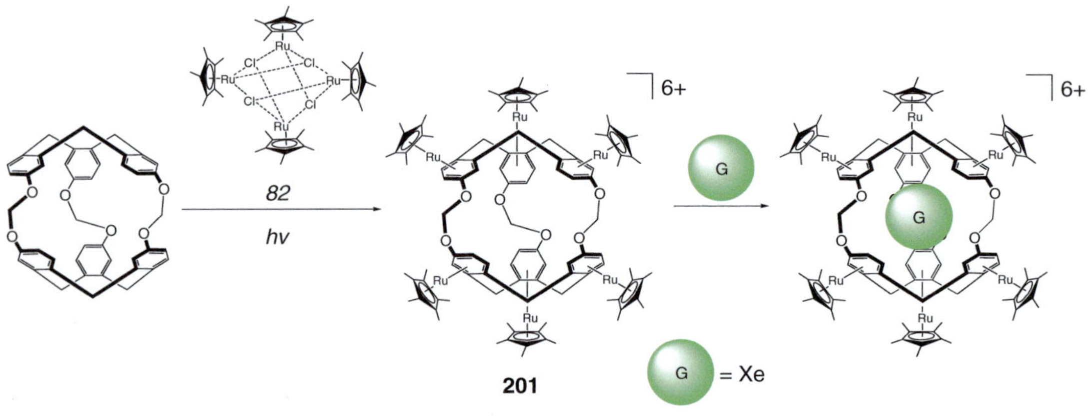

Scheme 2.123

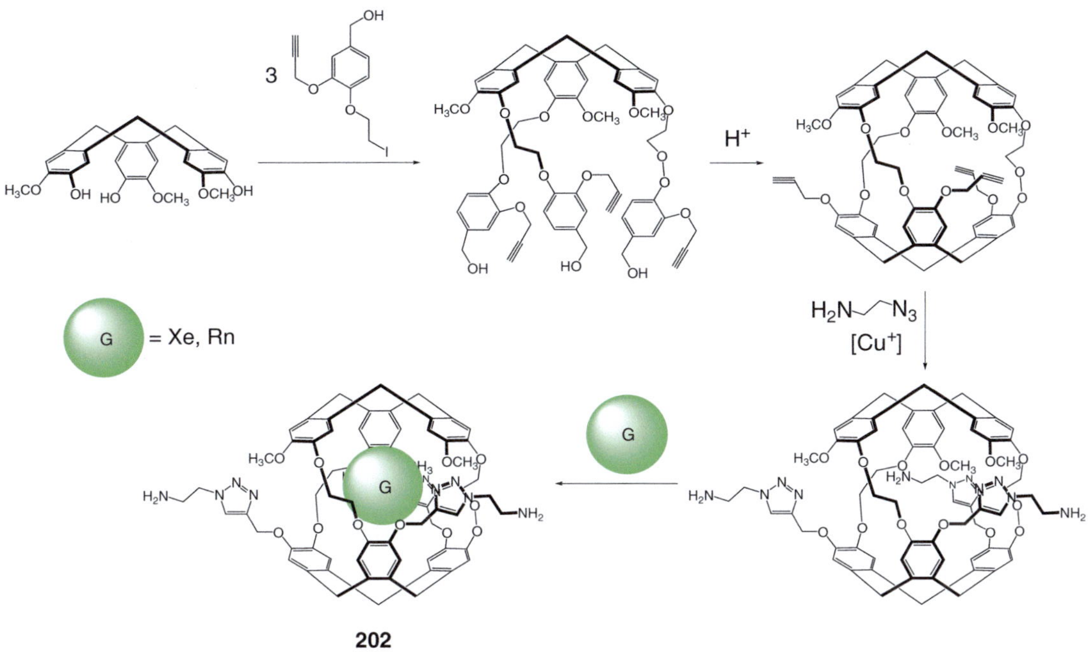

Scheme 2.124

Scheme 2.125

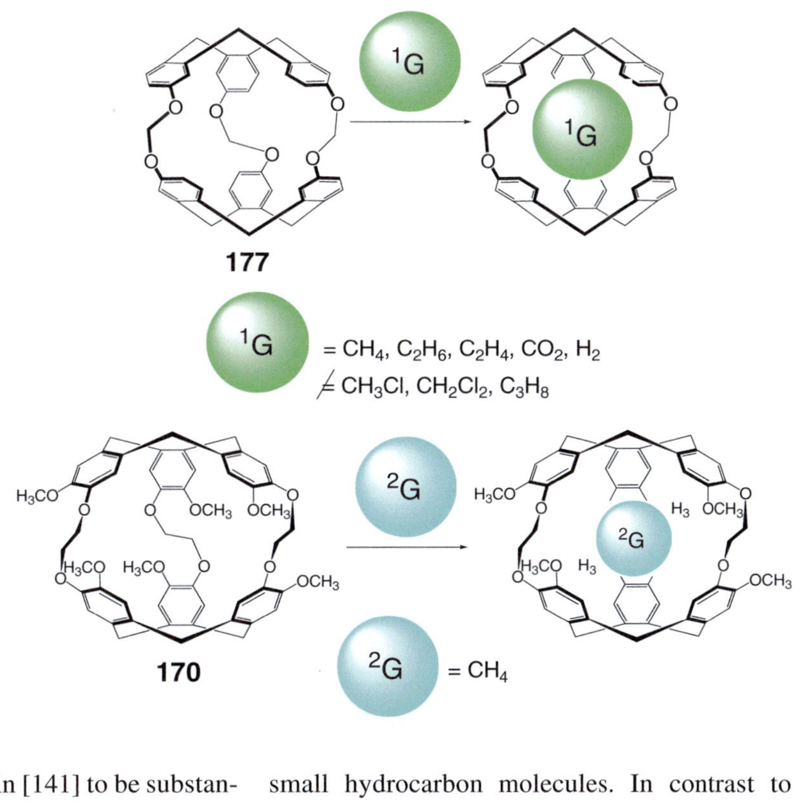

ethylene have been found in [141] to be substantially lower than that for methane. Thus, the caging ligand **177** is highly selective host for small hydrocarbon molecules. In contrast to methane, the encapsulation of ethane and ethylene is enthalpically driven and entropically dis-

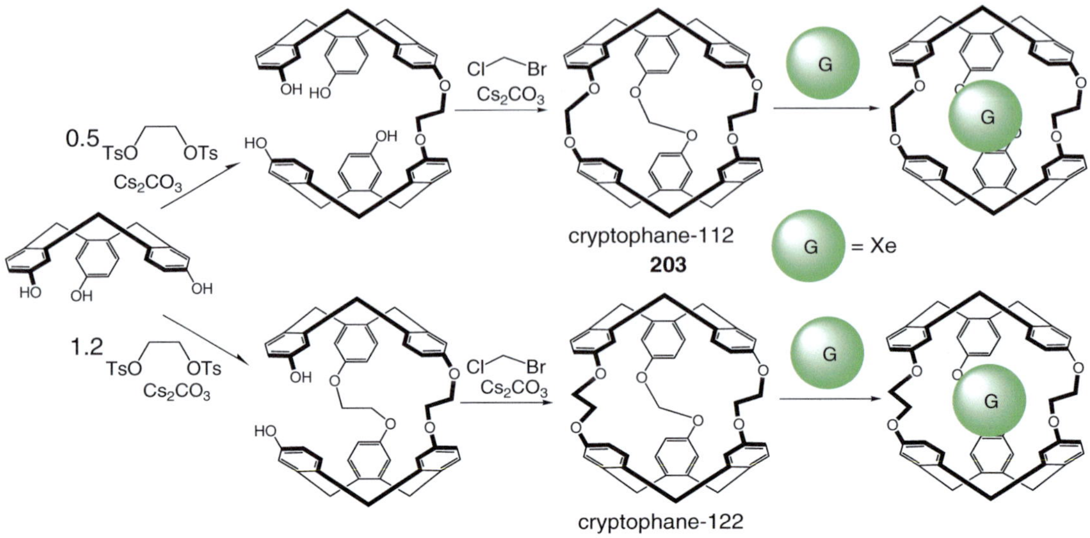

Scheme 2.126

favored; the encapsulation of ethane by **177** needs more free energy in its transition state than those of ethylene and methane [141] (Scheme 2.125).

Cryptophane capsules **203** and **204** have been obtained by template macrobicyclization of their reactive macrocyclic syntones with $ClCH_2Br$ (Scheme 2.126) in the presence of an excess of Cs_2CO_3 both as a base and as a matrix (Cs^+ ion). Chiral HPLC data suggest the presence of two enantiomers of their chiral D_3-symmetric *anti*-form; as follows from ^{129}Xe NMR spectra, both these ligands effectively encapsulate xenon [142].

Diastereomeric hemicryptophanes **205** and **206** and their enantiomers have been synthesized in [143] by intramolecular H^+-catalyzed macrobicyclization of its chiral tripodal ligand syntone by Scheme 2.127 in the presence of scandium(III) triflate. The use of (*S*)-(+)-enantiomer of glycidyl tosylate allowed obtaining *S,S,S*-diastereomeric analogs of the capsules **205** and **206**. The C_3-symmetry of all these cage species and their absolute configurations have been obtained from NMR and CD spectra [143]. The reaction of the ligand **205** with vanadium(V) oxytriisopropoxide afforded the corresponding vanadium(V)-containing C_3-symmetric capsule **207** with a cavity volume of approximately 110 Å [143].

The vanadium(V)-containing capsule **207** exist as diastereomeric mixtures with *P*- and *M*-handedness of the cyclotriveratrylene fragments and chiral ether groups in their *S*-configuration (Scheme 2.128) [144]. Diastereomerically pure chiral forms of **207** can exist as two diastereomeric conformers *Λ*-*M*-(SSS)-**207**/*Δ*-*M*-(SSS)-**207** and *Λ*-*P*-(SSS)-**207**/*Δ*-*P*-(SSS)-**207** corresponding to the *Λ* and *Δ* forms of their caged vanatrane fragments. A stereoconversion between the forms $\Lambda \leftrightarrow \Delta$ in solution is described in [144] to have unexpectedly high energetic barrier. As follows from ^1H NMR data, this process depends on the nature of a solvent that governs the direction of the rotation motion (clockwise or anticlockwise) of such vanatrane fragments. For these two diastereomers, the change of a solvent to that of an opposite nature is described in [144] to cause a preference to the clockwise rotation.

S,S,S-diastereomers of the capsule **207** have been synthesized in [145] by Scheme 2.129 starting from the corresponding precursor **205**. The solution ^1H and ^{53}V NMR spectra of each of these diastereomers contain two sets of the signals suggesting the presence of two C_3-symmetric cage species. This result is explained in [145] by a rigidity of the cyclotriveratrylene fragment and propeller arrangement around its V–N bond. This

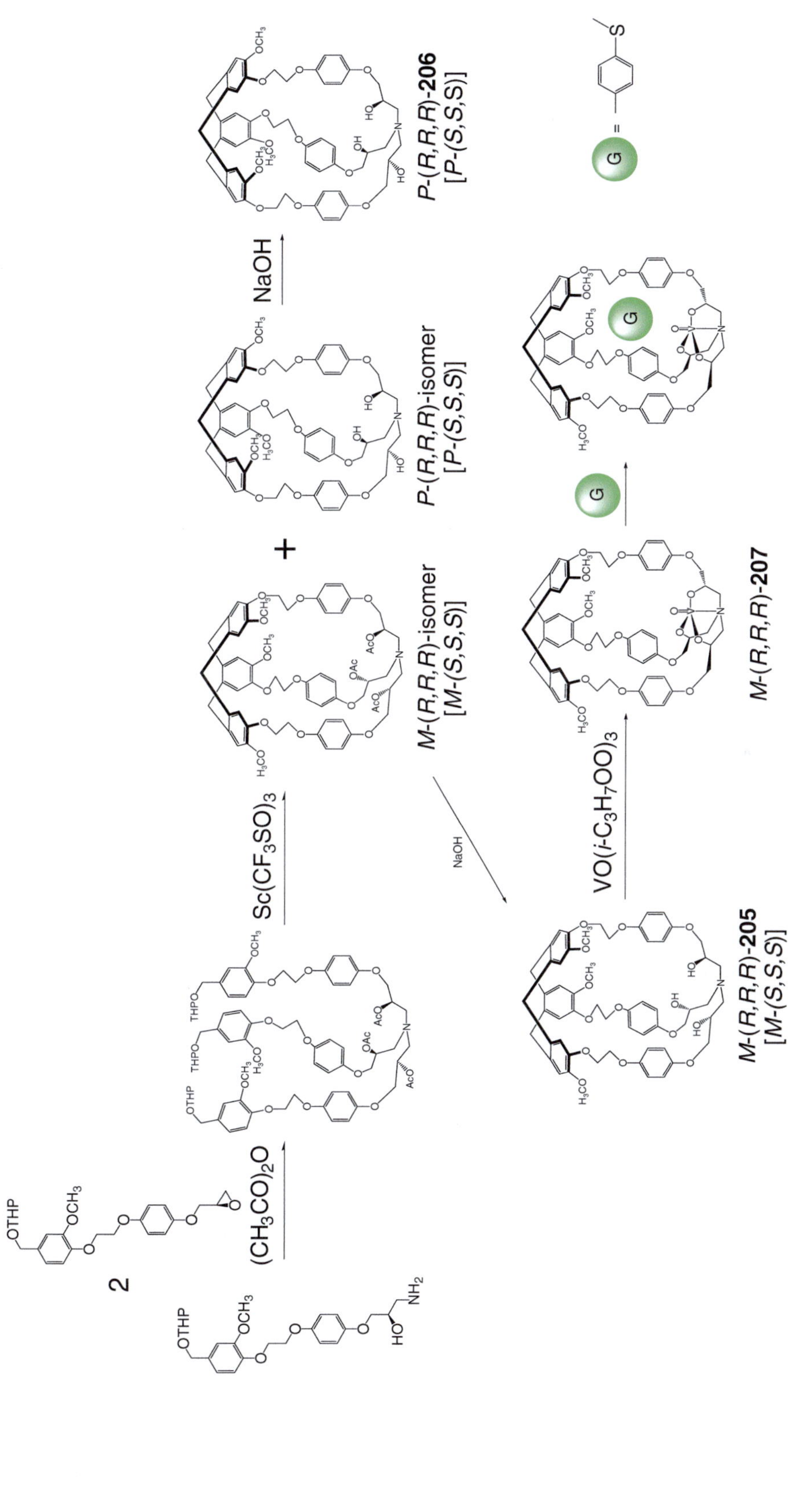
2
$(CH_3CO)_2O$
$Sc(CF_3SO)_3$
NaOH
$VO(i\text{-}C_3H_7OO)_3$
+
M-(R,R,R)-isomer
[M-(S,S,S)]
P-(R,R,R)-isomer
[P-(S,S,S)]
P-(R,R,R)-206
[P-(S,S,S)]
M-(R,R,R)-205
[M-(S,S,S)]
M-(R,R,R)-207
G =

Scheme 2.127

Scheme 2.128

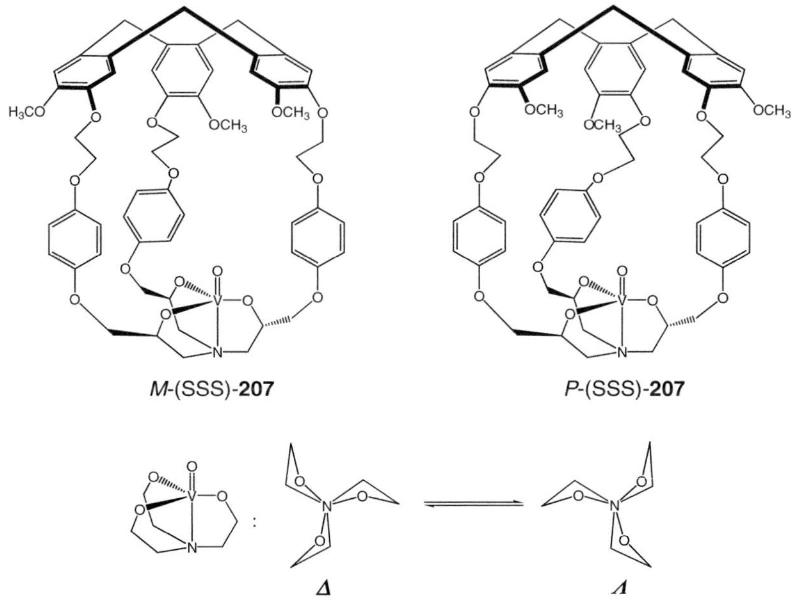

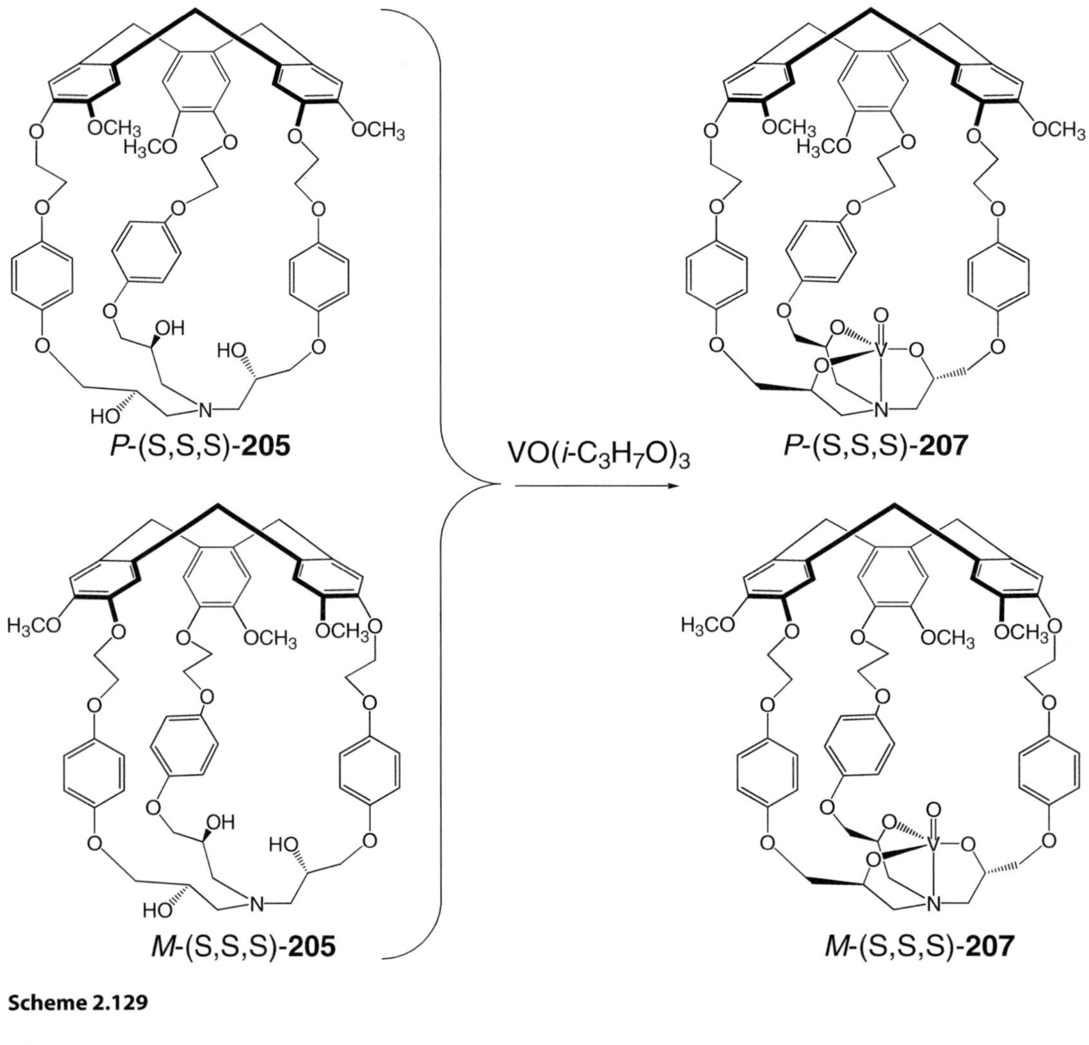

Scheme 2.129

rigidity decreases the conformational mobility of three atrane-type five-membered rings and caused the formation of the *Λ* and *Δ* diastereomers. Thus, a low rate of interconversion between them on the NMR time scale allowed detecting two diastereomers of the caging ligands (*M*-(*S*,*S*,*S*-*Λ*)-**207**/*M*-(*S*,*S*,*S*-*Δ*)-**207** and *P*-(*S*,*S*,*S*-*Λ*)-**207**/*P*-(*S*,*S*,*S*-*Δ*)-**207**) in both these cases. *P*-(*S*,*S*,*S*)-diastereomer of **207** has a compact structure with a cyclotriveratrylene cavity that is occupied by one of the linkers, so its atrane fragment is pushed outward. This molecule does not retain its C_3 symmetry in solution, as suggested from ^{1}H NMR spectrum. Single crystals of only one *P*-(*S*,*S*,*S*-*Λ*)-diastereomer have been obtained in [145]. Initially, the ^{1}H NMR spectrum of its $CDCl_3$ solution contained one set of the signals, and the second minor set of signals appear in time; so the *Λ* form dominates in crystal as well as in chloroform solution over its *Δ* form. These experimental data have been confirmed in [145] by quantum chemical DFT calculations. A conversion between these *Λ* and *Δ* forms, which are at equilibrium in solution, is described in [145] to have a concerted character.

A new class of C_3-symmetric hemicryptophane caging ligands, the triamide- and *tren*-based macropolycyclic covalent capsules **208** and **209**, have been prepared in [146] by Scheme 2.130. As follows from ^{1}H NMR data, their cage frameworks have C_3-symmetry axis passing through their cyclotriveratrylene fragment in solution. In the X-rayed crystal of **208**, its molecule is asymmetric due [146] to inward orientation of one amide C=O bond that is stabilized by hydrogen bonding with the neighboring NH groups. This ligand encapsulates one dichloromethane molecule, while **209** forms 1:1 cage complex with a caged *n*-pentane guest [146]. The mixture of enantiomeric forms of the hemicryptophane capsule **208** has been separated by HPLC. These optically resolved cage compounds have been characterized by CD spectra, and the absolute configuration assignment of these inherently chiral hemicryptophane species has been performed in [146].

The gram-scale synthetic procedure for preparation of the hemicryptophane capsule **209** by Scheme 2.131 has been elaborated in [147]; its isomers **210** and **211** have been also prepared using similar synthetic procedures. This synthetic

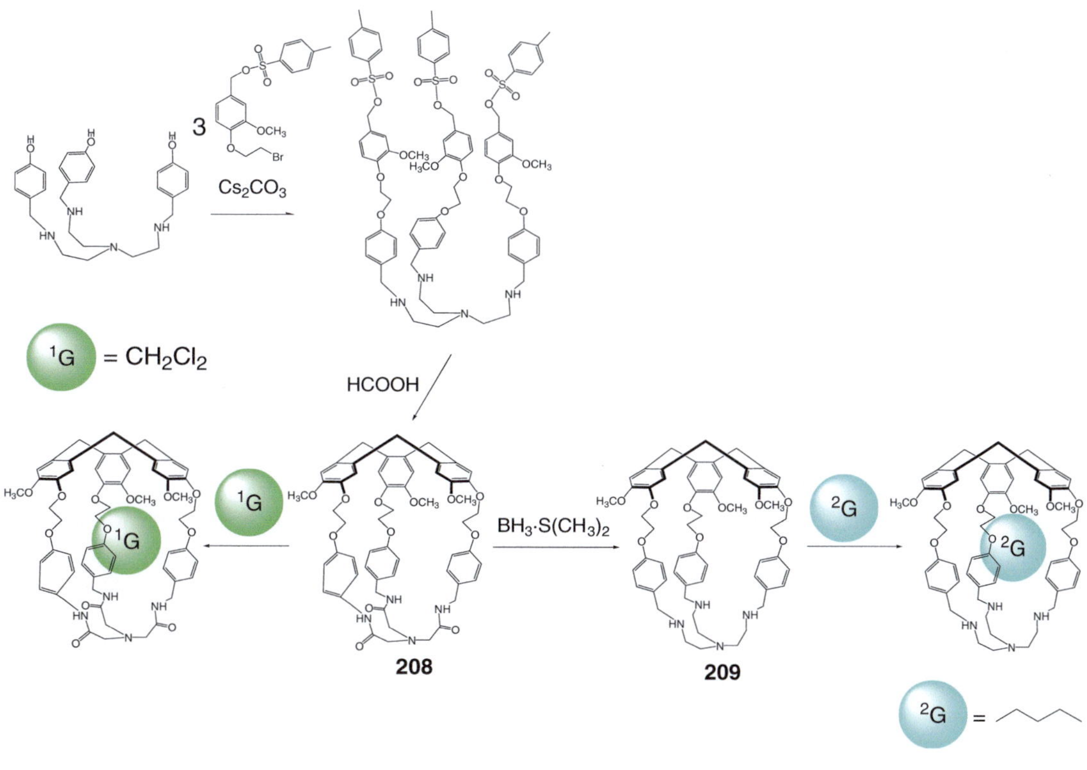

Scheme 2.130

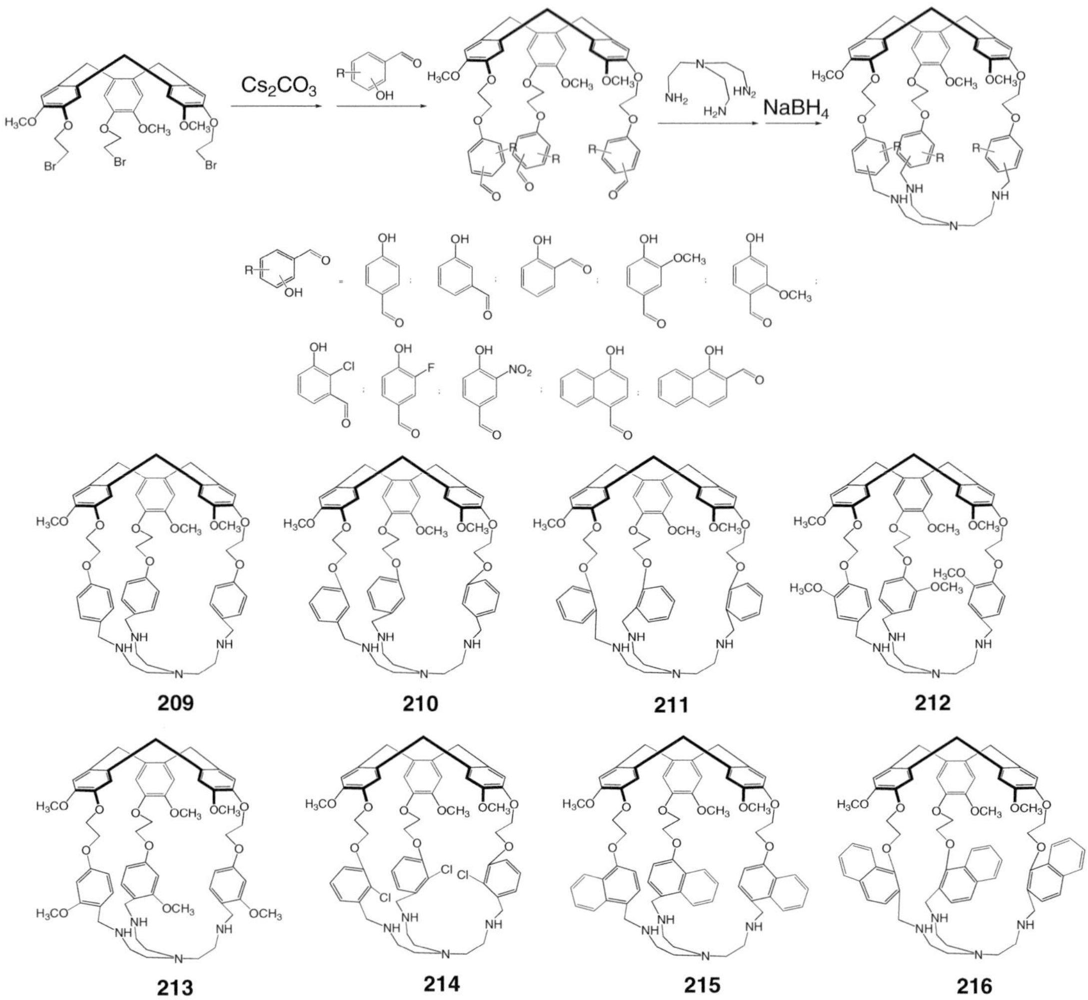

Scheme 2.131

approach allowed obtaining the hemicryptophane ligands with different shape and functionality and in various cavity sizes; the latter has been shown in [147] to be affected by the position of aldehyde and hydroxyl groups in the aromatic linkers. According to X-ray diffraction data, the ligand **209** encapsulates one of its own aromatic linkers. The methoxyl- and chlorine-substituted capsules **212**–**214** (but not those with fluorine and nitro substituents) and the caging ligands **215** and **216** with naphthalene linker moieties have been synthesized in [147]; the preparation of their isomer **217** by Scheme 2.132 is described in [148].

The hemicryptophane ligand **208** is described in [149] to give 1:1 cage complexes with inorganic anions as well as with organic cations and their ionic associates by Scheme 2.133. As follows from ^{1}H NMR titration data, the encapsulated guest anions form strong hydrogen bonds within the cavity of this ligand with its triamide fragment. Affinity of the ligand **208** to spherical halogenide anions decreases with their hydrogen bond-accepting ability in a row $F^- > Cl^- > Br^- > I^-$. A more basic tetrahedral dihydrophosphate anion forms a more stable cage complex than the hydrosulfate guest, while that of acetate ion is less stable due to its angular nature [149]. At the same time, the covalent capsule **208** encapsulates tetramethyl- and tetraethylammonium cations by Scheme 2.133 to give 1:1 cage complexes and discriminates their

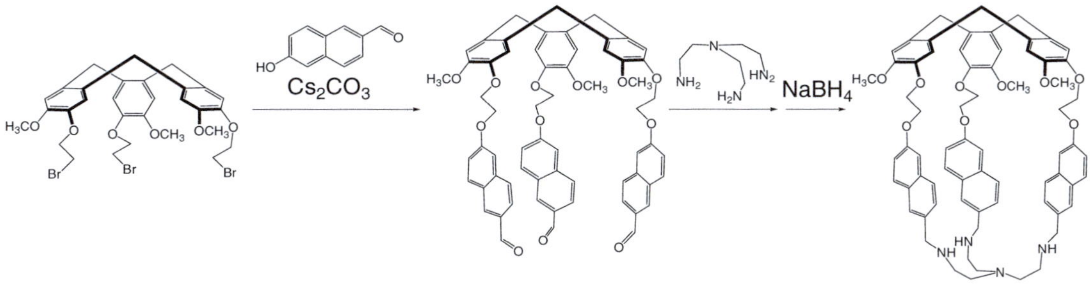

Scheme 2.132

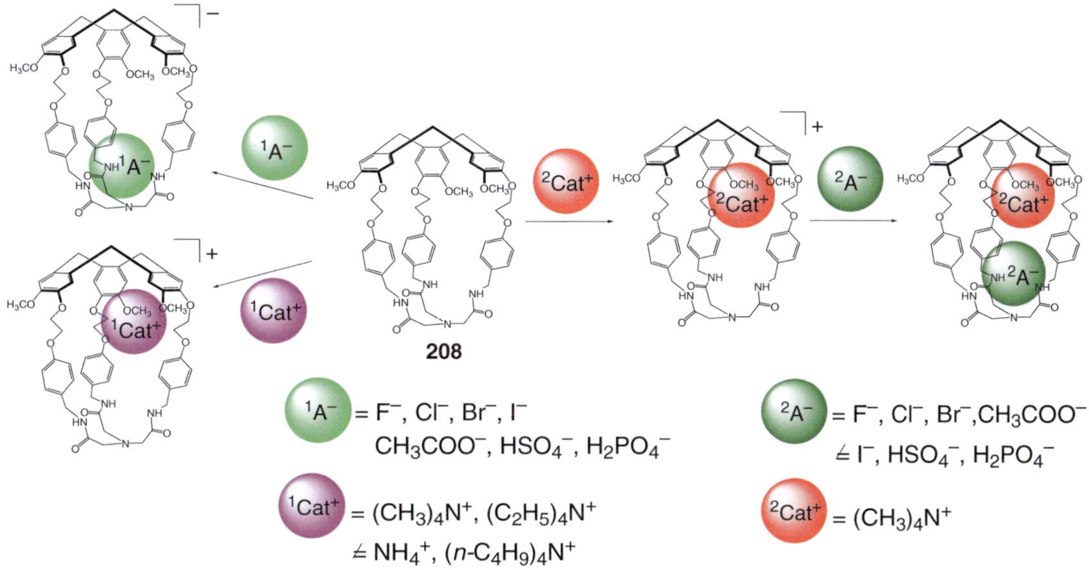

Scheme 2.133

tetra-*n*-butyl-containing analog due to steric clashes; those also cause preferential binding of tetramethylammonium cation over that of its tetraethyl-containing homolog. These encapsulated cations form C–H…π and cation…π interactions with a lipophilic cyclotriveratrylene fragment of the cage framework of **208**. As these anionic and cationic guests occupy different sites in the inner cavity of **208**, the authors of [149] have also studied cooperative binding of the ionic pairs formed by tetramethylammonium cation. In this case, F^-, Cl^-, Br^-, I^-, and CH_3COO^- counteranions form heteroguest 1:1:1 cage complexes more efficiently, and a factor of cooperativity (i.e., a ratio of the binding constants of anions in the absence and in the presence of $(CH_3)_4N^+$) increases in a row $Br^- < CH_3COO^- < F^- < Cl^-$. Thus, inversion of selectivity against chloride and fluoride anions takes place, and tetramethylammonium chloride is described in [149] to be optimal ionic associate for its cooperative encapsulation.

The ability of covalent capsule **208** to encapsulate ionic pairs has been used in [150] for recognition of taurine neurotransmitters (Scheme 2.134) in their aqueous solutions. These zwitterions very selectively form 1:1 cage complexes of the ligand **208** in the presence of other related substrates (including glycine). This result is explained in [150] by simultaneous action of three main factors: their zwitterionic character,

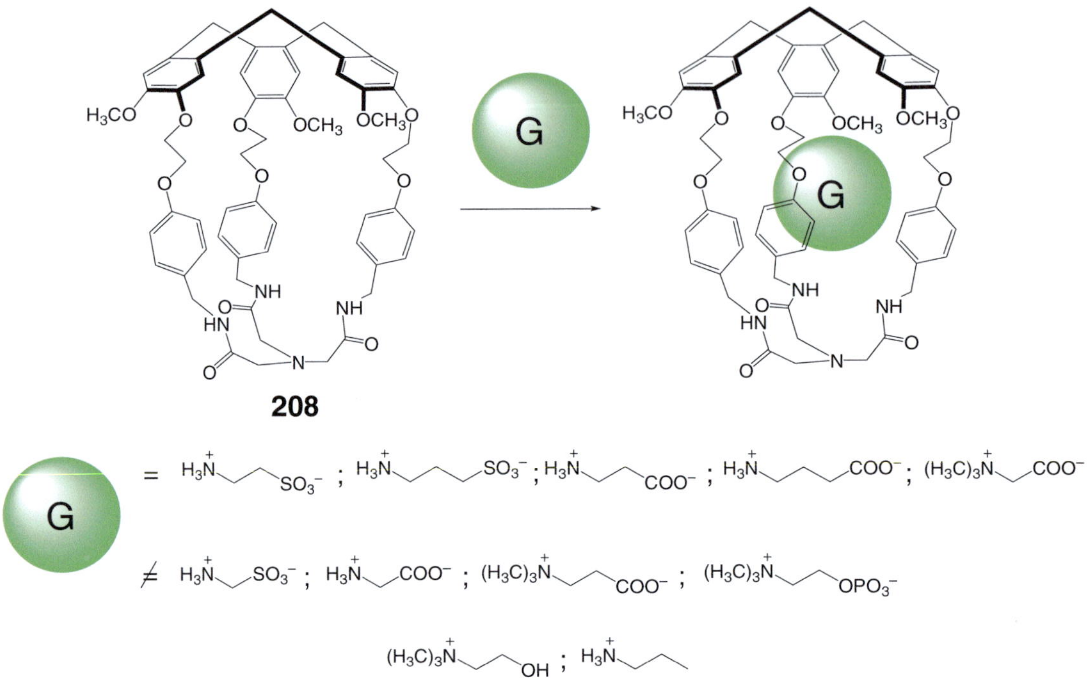

Scheme 2.134

an appropriate shape, and van der Waals volume. At the same time, according to ^{1}H NMR data, this caging ligand did not encapsulate ionic pairs of NH_4^+, $(CH_3)_4N^+$, or $(n\text{-}C_4H_9)_4N^+$ cations with $CH_3SO_3^-$, CH_3COO^-, or $H_2PO_4^-$ anions in their acetonitrile–water solutions due to an entropic factor. Going from taurine and *β*-alanine to their methyl-containing analogs as well as to homologs with other number of methylene chains in a spacer fragment also caused a substantial decrease in the corresponding binding constants; these data are in a good agreement with the calculated packing coefficients. The DFT calculations of the cage complexes of **208** with zwitterionic guests suggest only weak intermolecular cation…π interactions of a cyclotriveratrylene fragment with hydrophobic cavity as well as hydrogen bonding of three of its amide moieties with the negatively charged SO_3^- group of a guest [150].

Enantiomerically pure *M*- and *P*-forms of **208** have been used in [151] for encapsulation of the diastereomeric glucoside guests *83* and *84* by Scheme 2.135 to give 1:1 cage complexes (as follows from ^{1}H NMR data). These encapsulating hosts showed diastereomeric discrimination: the *M*-form of **208** binds the guest *83* three times better than its analog *84* and seven times better than the *P*-form; the latter does not encapsulate the *β*-anomer of the glucoside guest. In contrast, the *β*-anomer is described in [151] to form more stable cage complexes with enantiomeric pure hemicryptophane ligands *M*-(*SSS*)-**205**, *M*-(*RRR*)-**205**, *P*-(*SSS*)-**206**, and *P*-(*RRR*)-**206** [143] by Scheme 2.136 than those of the α-analog *83*. The *β*-/*α*-diastereomeric discrimination strongly depends on the stereochemistry of the encapsulating ligands **205** and **206**: its value increases from 1.2 for *M*-(*SSS*)-**205** and 1.8 for *M*-(*RRR*)-**205** to an exceptional binding of the diastereomer *84* by their enantiomer *P*-(*SSS*)-**206**. These results demonstrate high diastereoselectivity of the hemicryptophane capsules in the recognition of the glucoside guests. The hemicryptophane ligands with the *M*-cyclotriveratrylene fragment encapsulate the derivatives of *D*-glucoside more effectively than their analogs with the *P*-cyclotriveratrylene fragment: the binding constant

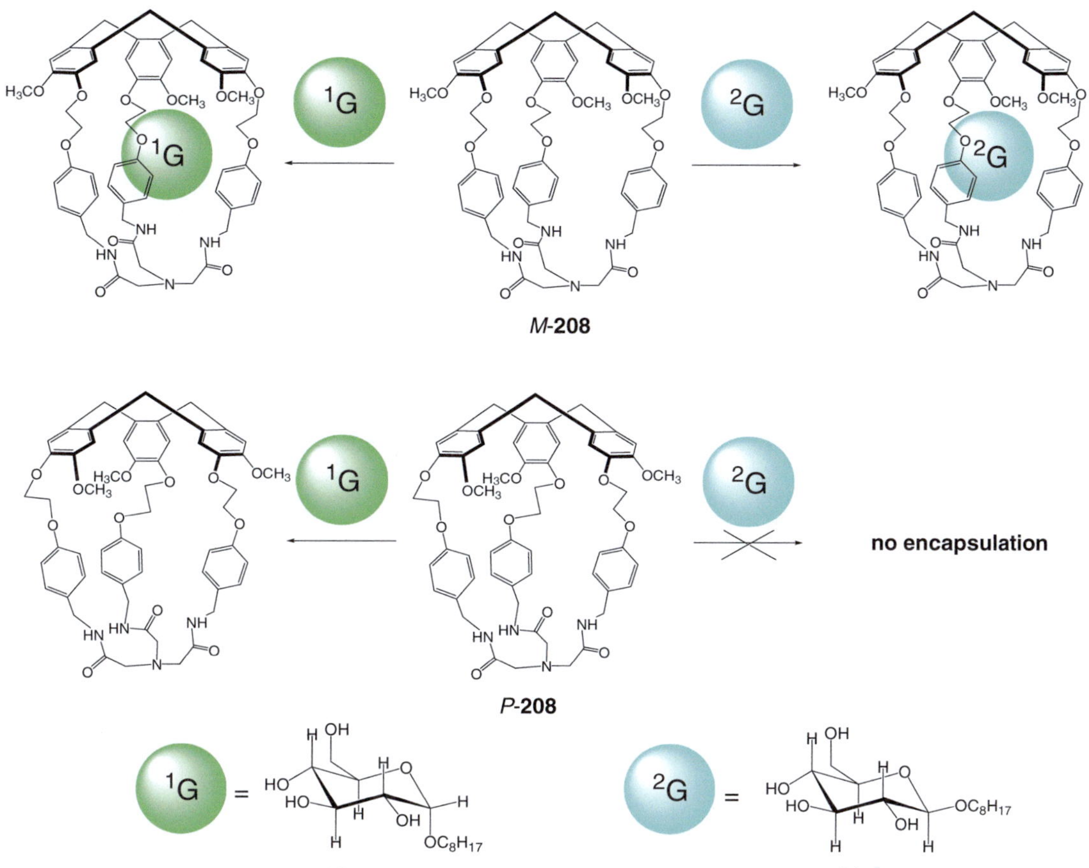

Scheme 2.135

of *84* for the caging ligand *M*-(*RRR*)-**205** is two times higher than that for its enantiomer *P*-(*SSS*)-**206**. The latter ligand showed high enantioselectivity with respect to *P*-(*RRR*)-**531** that does not encapsulate a glucoside guest. The guest *83* forms a 1:1 cage complex only with the hemicryptophane ligand **205** containing the *M*-cyclotriveratrylene fragment. At the same time, three stereogenic carbon atoms only slightly affect the stereoselective binding: the binding constants of this glucoside for covalent capsules *M*-(*SSS*)-**205** and *M*-(*RRR*)-**205** are close to each other [151].

The first hybrid cage complex of hemicryptophanophosphatrane **218** with a superbasic center (so-called "Verkade's superbase") has been synthesized in [152] by Scheme 2.137. This complex contains one encapsulated dichloromethane molecule, which forms supramolecular bonds with aromatic rings of the cyclotriveratrylene fragment of its caging ligand. The caged phosphatrane superbase is substantially more basic than its acyclic analog *85*, whereas the deprotonation rate for this analog is described in [152] to be approximately 500 times higher than that for **218**.

A hemicryptophane capsule **219** has been synthesized in [153] by Scheme 2.138. As follows from ^{1}H NMR titration experiments, it is an efficient heteroditopic receptor encapsulating zwitterionic neurotransmitters, and its binding constants with them are higher by up to three orders of magnitude than those for the caging ligand **208** [150]. This result is explained in [153] by the presence of three additional hydrogen-donor amide groups in **219** that strongly stabilize anionic species within its inner cavity. The most stable is 1:1 cage complex with

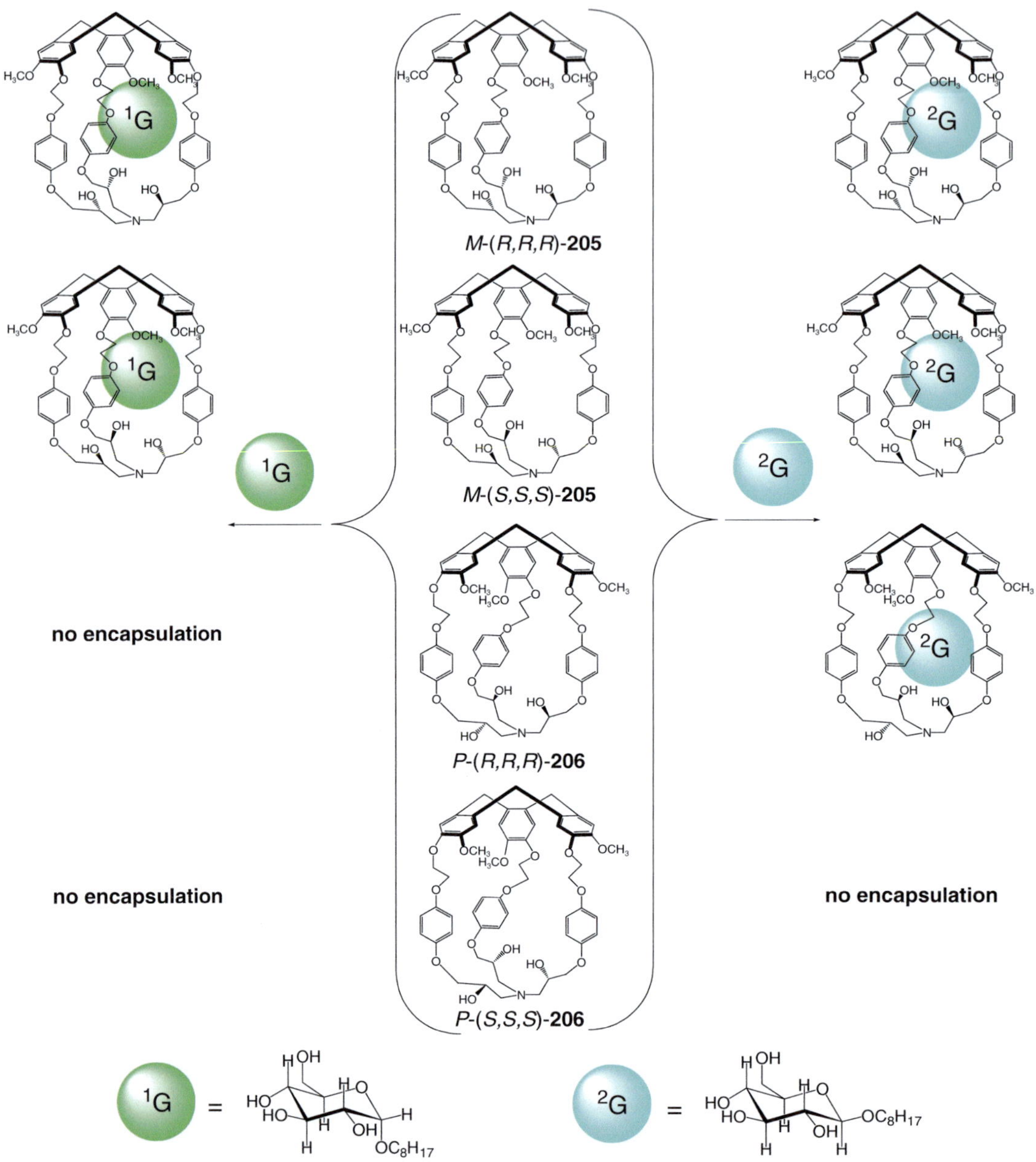

Scheme 2.136

an encapsulated taurine *87*: the size and shape of this zwitterionic guest are complementary to those of **219** [153].

Arsenic-, antimony-, and bismuth-capped M_2L_3 covalent capsules **220–222** have been prepared in [154] by Scheme 2.139 using two synthetic approaches: by direct capping (cross-linking) of their dithiolate syntone with pnictogen trihalides and by transmetalation of the initially synthesized antimony-containing macrobicyclic precursor with arsenic and bismuth trichlorides; the latter reaction has been used for the synthesis of phosphorus-capped macrobicycles **223** and **224**. These capsules have similar structures, although going from arsenic- and antimony- to bismuth-capped ones

Scheme 2.137

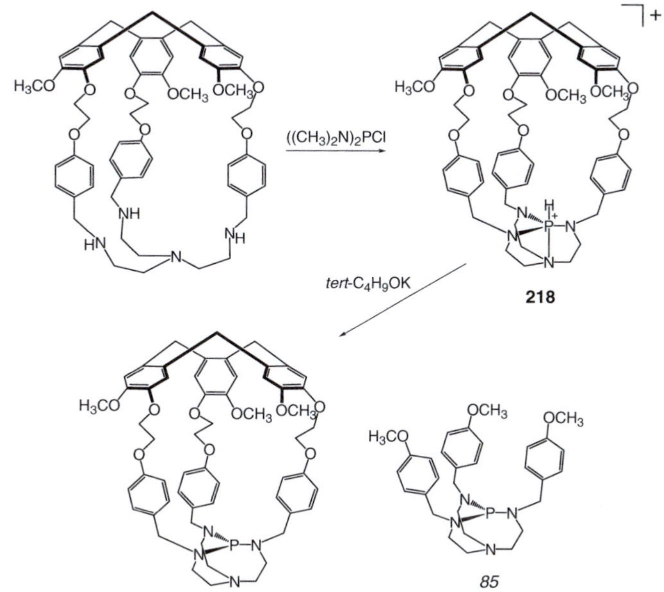

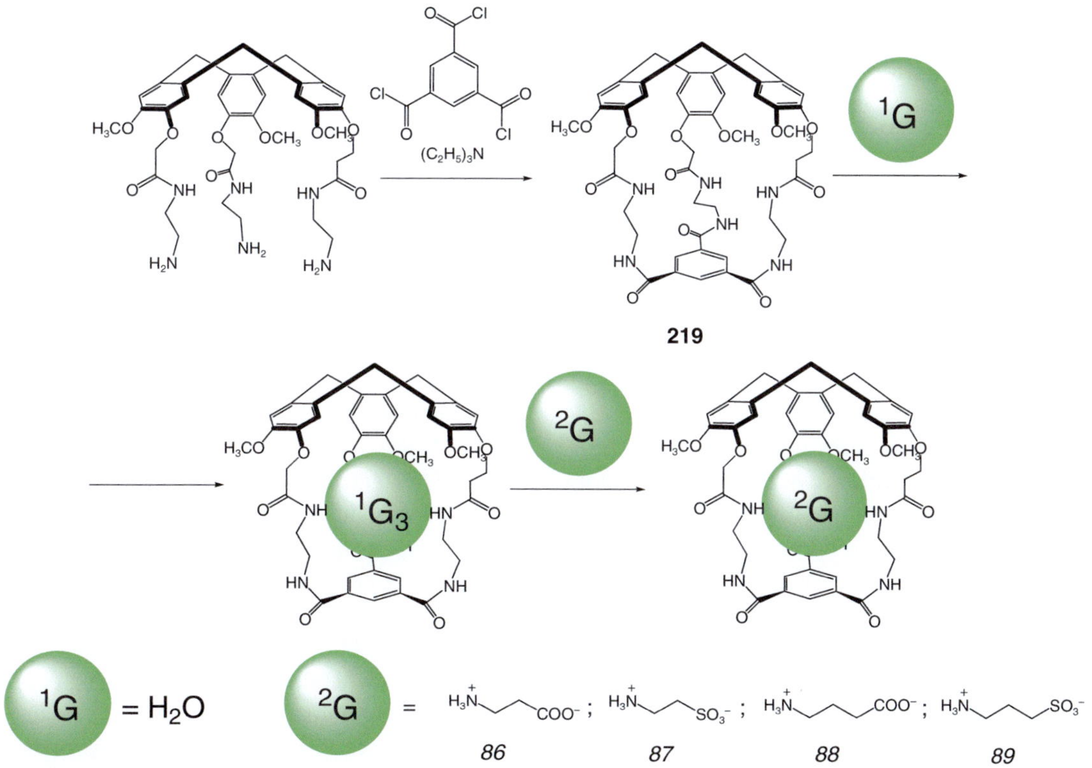

Scheme 2.138

caused elongation of E–S bond lengths and shortening of intramolecular E…E distances, accompanied by a decrease in S–E–S bond angles. This allows retaining the same geometry of a macrobicyclic framework in all these covalent capsules [154]. Each of them is stabilized

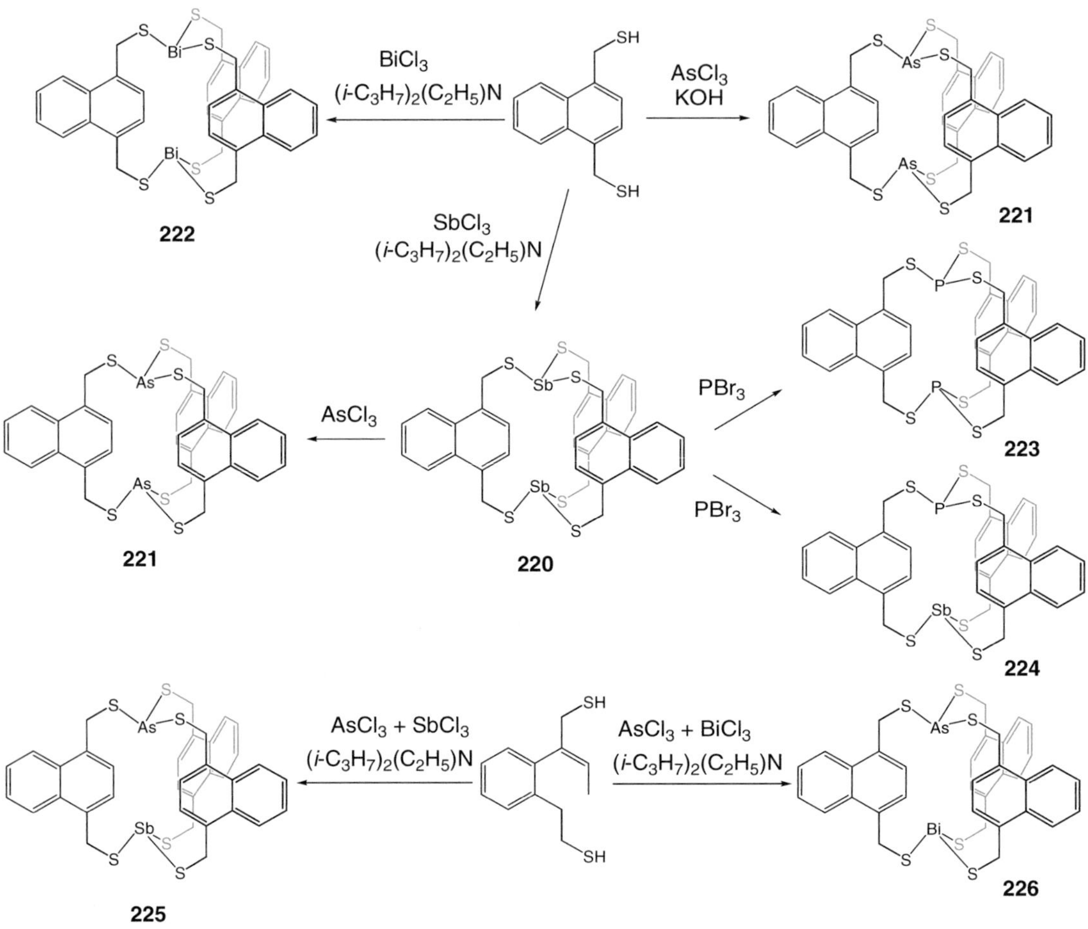

Scheme 2.139

by six intramolecular η^2 E…π interactions caused by π-donation of the ligand's aryl fragments to the σ*-orbital of the corresponding capping pnictogen atom; for their phosphorus-containing macrobicyclic analog **223**, the intramolecular P…π interactions have not been observed. Moreover, if the P_2L_3 macrobicycle has been found to adopt only one conformation in its solution, other capsules can adopt two conformations. First one with a C_{3h} symmetry is identical to that in crystal, and in the other one of its three ribbed fragments is flipped so that the macrobicyclic framework has a C_s-symmetry. This conformer dominates in the case of the Bi_2L_3 macrobicycle, while its content is substantially lower for the antimony- and arsenic-containing analogs **220** and **222** (47 and 5 %, respectively). These experimental ^{1}H NMR data [154] are in a good agreement with the theoretical DFT calculation.

The use of a mixture of pnictide trichlorides as cross-linking agents allowed obtaining heterocapped $AsSbL_3$ and $AsBiL_3$ capsules **225** and **226** by Scheme 2.139. Their phosphorus-containing $SbPL_3$ analog **224** has been synthesized by transmetalation of the labile antimony-containing complex with phosphorus(III) tribromide. The same synthetic approach based on the lability of the antimony-containing macrobicyclic precursors has been widely used for the synthesis of various hybrid and heterocapped tris-dioximate and tris-oximehydrazonate clathrochelates [155–161].

Main bond angles and bond distances in the AsSb, AsBi, and PSb-heterocapped capsules

224–226 are intermediate between those for their homocapped analogs, and they also form six intramolecular E…C_{aryl} contacts (and, therefore, E…π interactions). All these C_3-symmetric heterocapped capsules are chiral in solution. This result is explained in [154] by the rigidity of cage frameworks due to the non-equivalence of their capping atoms.

The bis(*meta*-phenylene)-32-crown-10-based covalent capsules **227–230** have been prepared in [162] by Scheme 2.140 in pseudo-high-dilution conditions. These caging ligands strongly bind paraquat derivatives as dicationic guest species, and such efficient encapsulation is explained in this work by a combination of the preorganization of these macrobicyclic hosts and the introduction of additional binding sites to their cage frameworks. In particular, the pyridine-containing caging ligand **227** has the highest binding constant with paraquat dication, and this constant is approximately four orders of magnitude greater than those of its crown ether analog.

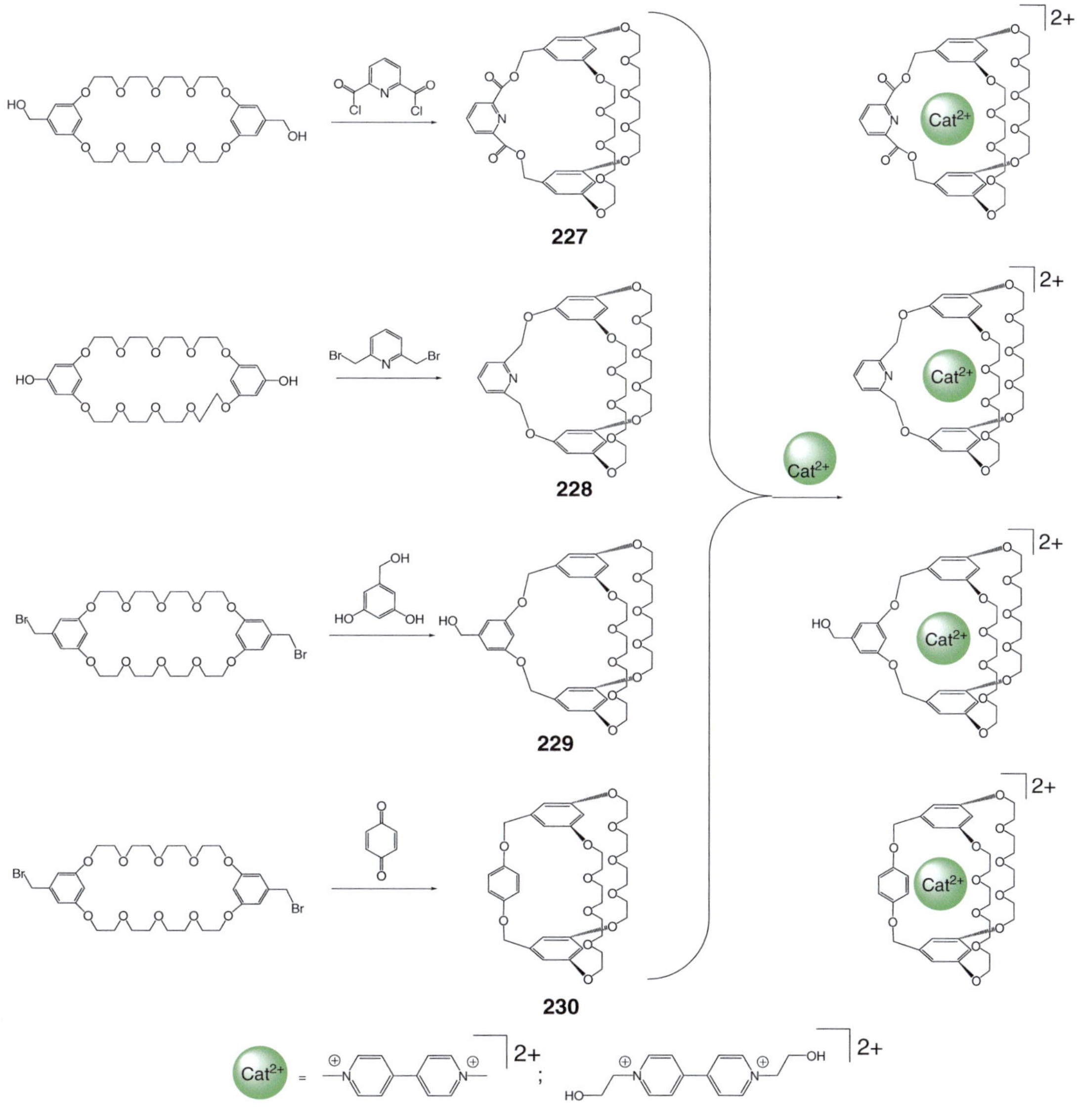

Scheme 2.140

An efficient synthetic procedure of [163] allowed obtaining the bis(*meta*-phenylene)-26-crown-8-based caging ligand **231** in a moderate yield under mild reaction conditions by Scheme 2.141.

Homologous bis(*meta*-phenylene)-32-crown-10-based covalent capsules **232**–**239** with a different size of their ligand's cage frameworks have been prepared in [164] by Scheme 2.142.

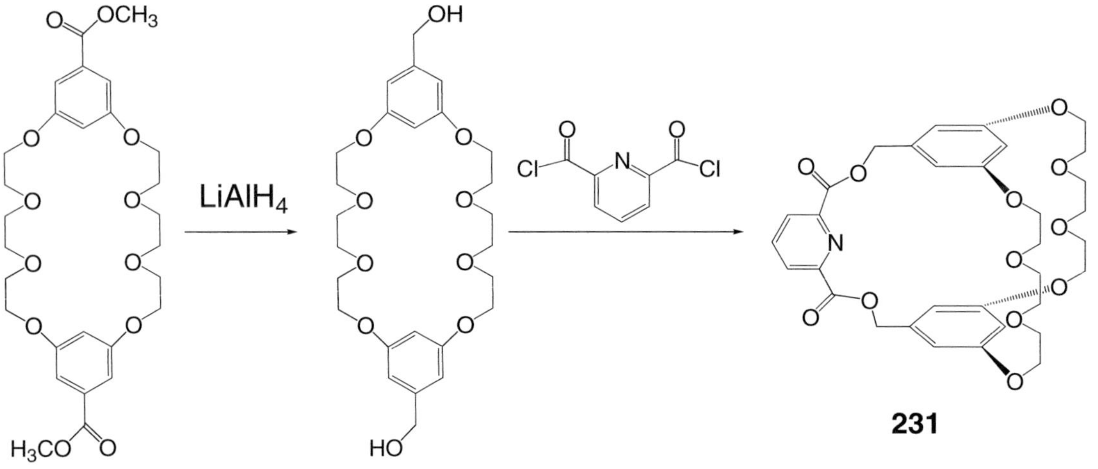

Scheme 2.141

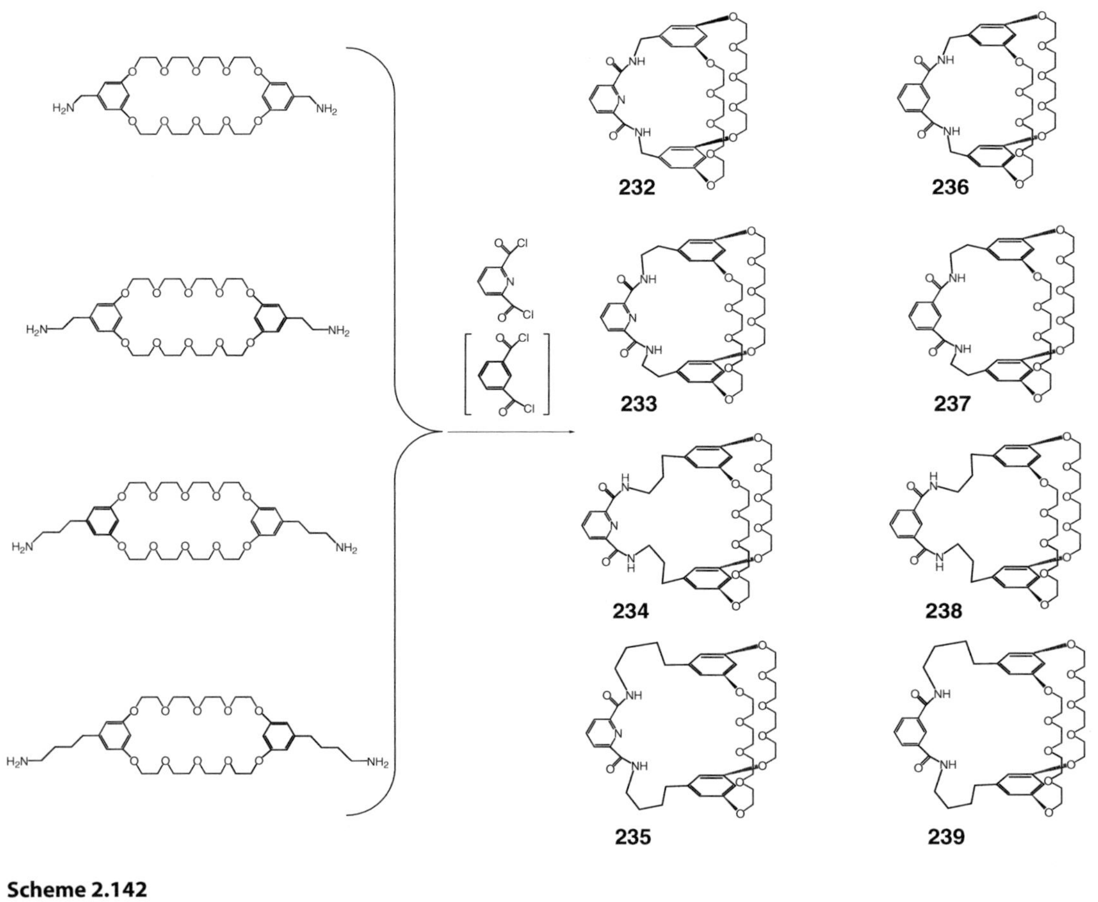

Scheme 2.142

The crown ether caging ligand **240** is reported in [165] to only slightly bind the aromatic dione *90* as guest. However, encapsulation process is more pronounced in the presence of potassium(I) cations as co-guests giving a heteroguest 1:1:2 cage complex by Scheme 2.143; the latter has been characterized by NMR spectra and X-ray diffraction data.

Tris-di-yne macrobicyclic compounds **241**–**243** have been obtained in [166] from the corresponding C_3-symmetric tripodal precursors with terminal alkyne bonds using their copper(I)-catalyzed homocoupling reactions by Scheme 2.144.

Bis(*meta*-phenylene)32-crown-10-based capsule **244** has been prepared in [167] by

Scheme 2.143

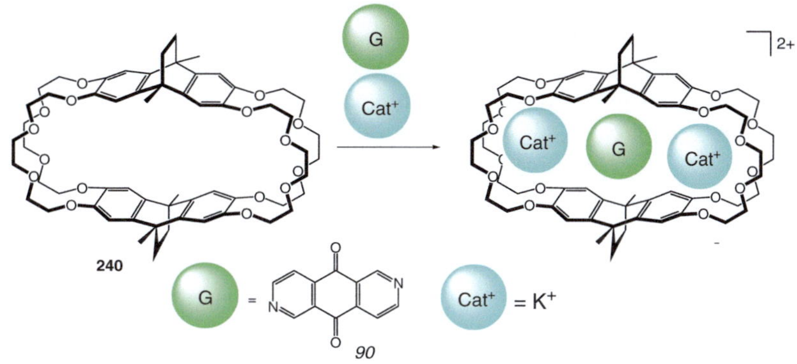

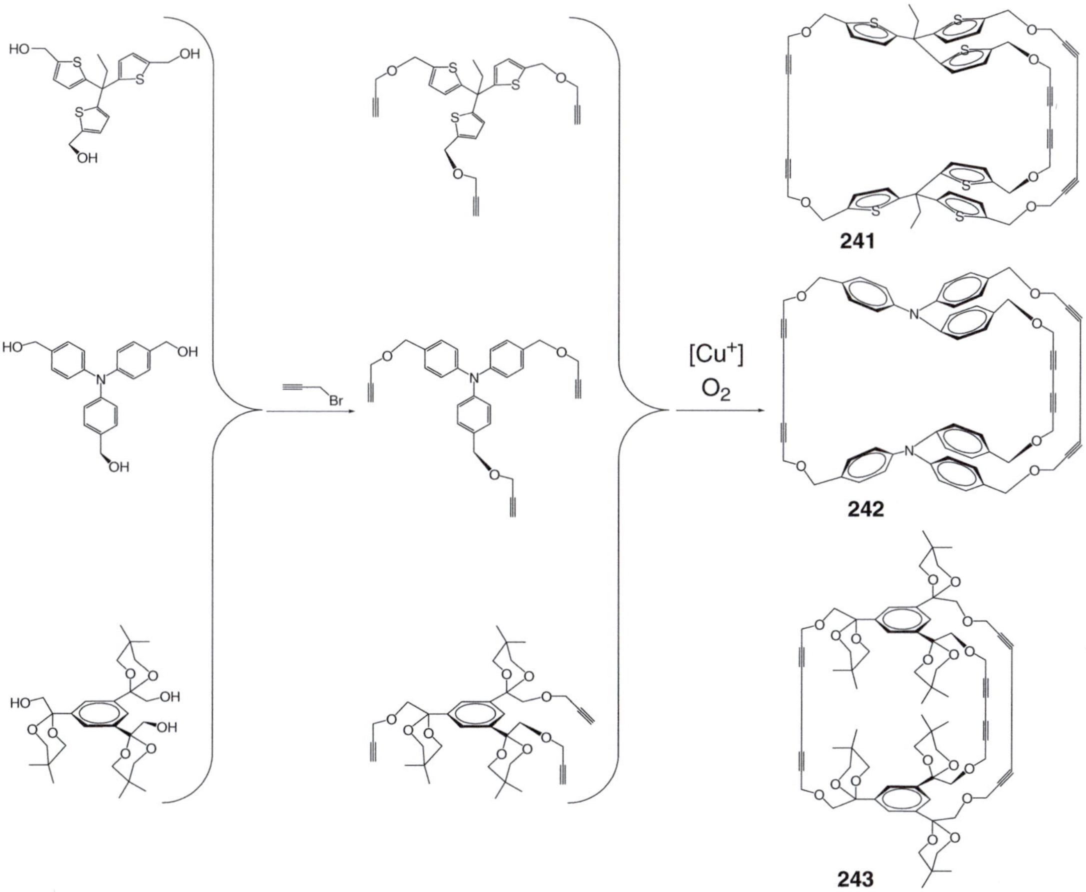

Scheme 2.144

Scheme 2.145 under pseudo-high-dilution reaction conditions.

Eglinton homocoupling of the tripodal rigid alkyne-terminated precursor *91* in high-dilution conditions by Scheme 2.146 in the presence of large excess of copper(I) and copper(II) salts has been used in [168] for the synthesis of a robust covalent capsule **245**. Its calculated energy-minimized model had a distorted TP geometry with vertical and horizontal inner sizes of 13.5 and 12 Å, respectively. According to X-ray diffraction data, this macrobicycle has a cavity volume of approximately 1300 Å^3 and a geometry that is in a good agreement with the calculated one; its crystal contains 1D channels comprised of the adjacent cage frameworks connected by windows approximately 4 Å wide [168].

2.3.2 Encapsulation of Anions

A π-basic cryptophane capsule **171** (see Sect. 2.3.1) is described in [169] as an efficient encapsulating ligand for small organic cations and neutral molecules but not for anions. The electrophilicity of its arene carbon atoms can substantially be increased by η^6-metalation with certain electron-withdrawing transition metal-containing moieties, in particular the [(arene)Ru]$^{2+}$ dications. The metalation of **171** by Scheme 2.147 has been used in [169] for the synthesis of a π-acidic [Cp*Ru]$^+$-functionalized caging ligand **246**, which is able to encapsulate different anions (e.g., triflate and hexafluorostibate ions, as follows from X-ray diffraction and NMR data). The addition of tetrafluoroborate anions does not affect the NMR spectrum of the cage complex with an encapsulated triflate anion,

Scheme 2.145

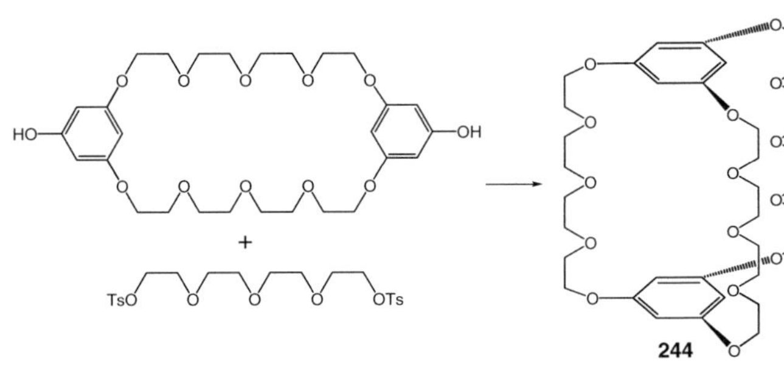

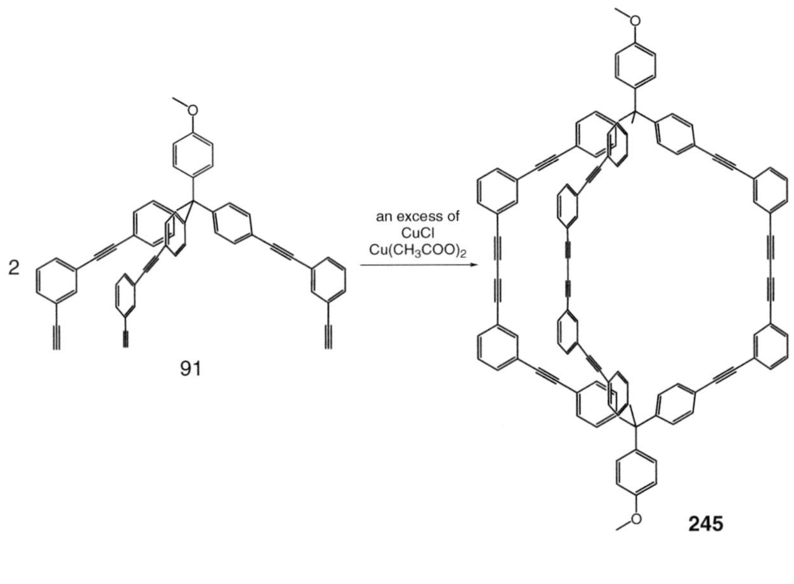

Scheme 2.146

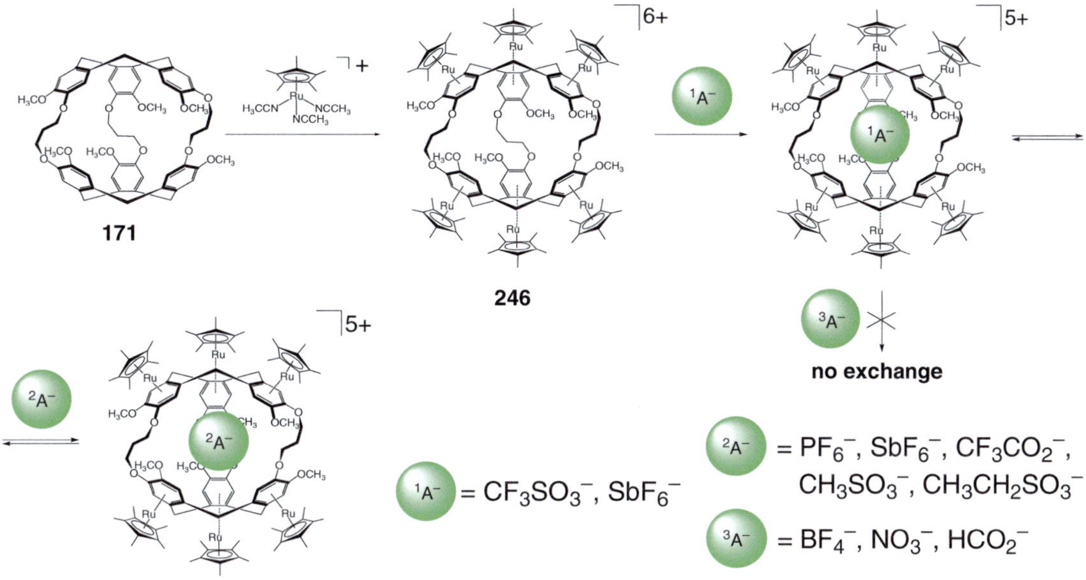

Scheme 2.147

Scheme 2.148

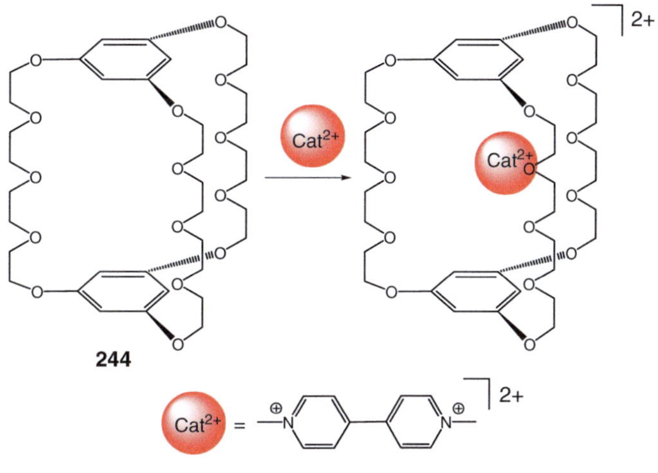

whereas hexafluorophosphate and hexafluorostibate anions very slowly exchange this guest. The $CF_3CO_2^-$, $CH_3SO_3^-$, and $CH_3CH_2SO_3^-$ anions compete with encapsulated triflate ion and decrease its overall binding, whereas the NO_3^- and HCO_2^- anions are described in [169] to turn the cavity binding off.

2.3.3 Encapsulation of Cations

The crown ether-based capsule **244** (see Sect. 2.3.1) is described in [167] to encapsulate paraquat dication by Scheme 2.148 giving a 1:1 cage complex. Such caged aromatic dication forms hydrogen bonds as well as stacking interactions with two arylene fragments of macrobicyclic ligand **244** [167].

The covalent capsules **227**–**230** are reported in [162] to form host–guest complexes with encapsulated aromatic dications *92* and *93* by Scheme 2.149. In solution and gas phase, only 1:1 cage complexes have been detected by ^1H NMR and ESI-MS methods. In the case of **228** its 2:1 cage complex with paraquat dicationic guest *92* has been characterized by X-ray diffraction

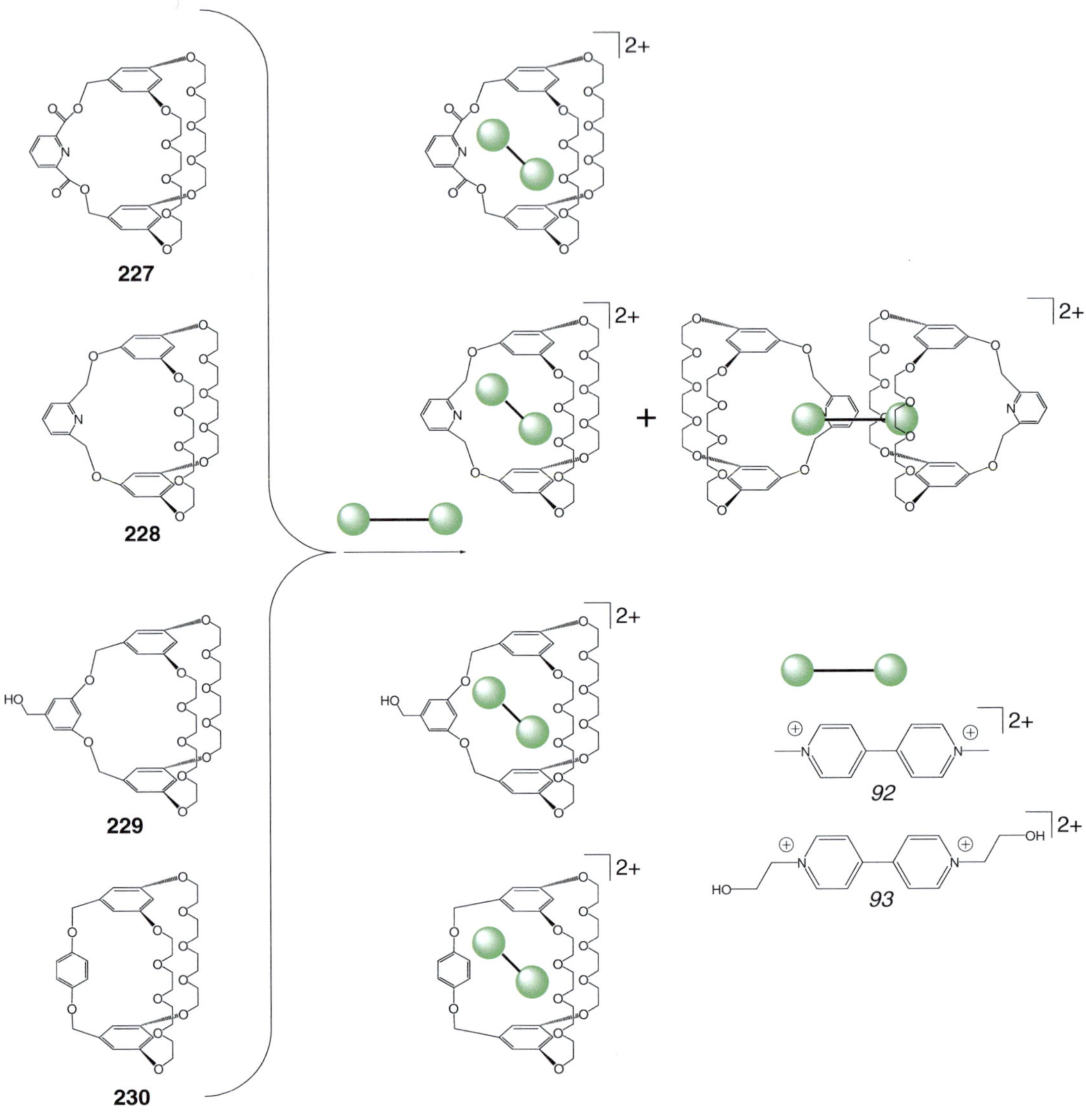

Scheme 2.149

data. The guest dication with parallel pyridinium fragments and *N*-methyl hydrogen atoms forming hydrogen bonds with interiors of cage frameworks is encapsulated by two macrobicyclic ligands [162].

A comparative study of the binding properties of 32-crown-10-based capsules **227** and **244** toward diquat dication has been performed in [170]. These ligands gave 1:1 cage complexes by Scheme 2.150; those have been characterized using ^{1}H NMR spectra and by X-ray diffraction experiments. As follows from X-ray diffraction data, this encapsulated dication forms hydrogen bonds and face-to-face stacking interactions of its two electron-deficient pyridinium fragments with two electron-rich phenylene bases of these capsules [167]. The binding constant for the pyridine-containing caging ligand **227** has been found in [167] to be substantially higher than that for its analog **244**.

Condensation of a dihydroxyl-containing macrocyclic precursor *94* with the corresponding dichloroanhydride by Scheme 2.151 has been used in [171] for the synthesis of a functionalized

Scheme 2.150

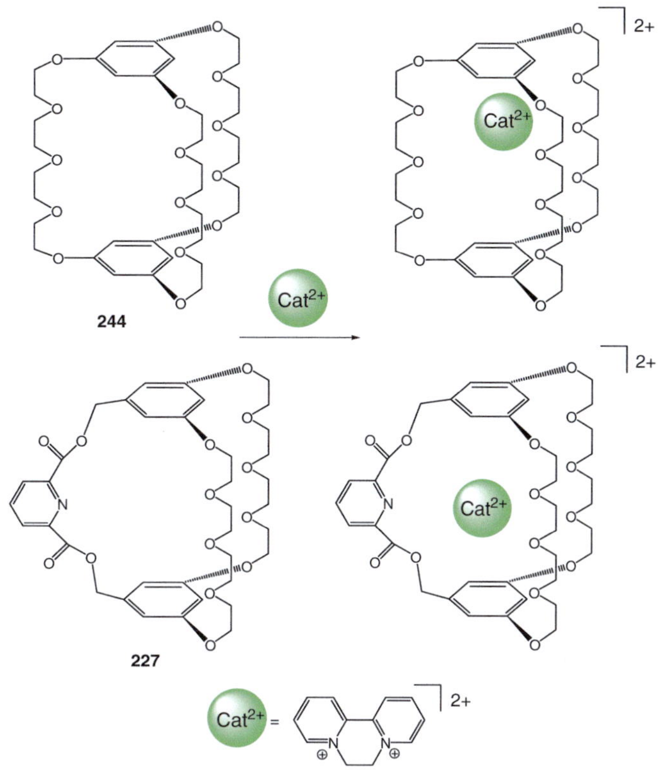

capsule **247**, which gave a 1:1 cage complex with an encapsulated paraquat dication. This complex is stabilized by hydrogen bonds and face-to-face stacking interactions with the caging macrobicyclic ligand. According to X-ray diffraction and ESI-MS data, its molecules form dimers both in crystal and in solution through intermolecular face-to-face stacking and dipole–dipole interactions. The electron-rich aromatic apical fragment of **247** forms a dipole with the pyridinium groups of the guest, and two such dipoles are arranged to give π…π interactions between the donor–acceptor pairs [171].

Effect of the guest size on its encapsulation by *meta*-phenylene-based covalent capsules **227** and **244** (Scheme 2.152) has been studied in [172]. The UV-vis and ESI-MS spectra and X-ray diffraction experiments showed that both these macrobicyclic ligands form 1:1 cage complexes with an encapsulated diazapyrene cation both in solution and in crystal. This diazapyrene guest better fits in the capsule **244** than the paraquat ion, and due to this geometrical effect and stronger face-to-face stacking interactions, it forms a substantially (by approximately 30 times) more stable cage complex. In contrast, paraquat dication forms a more stable cage complex with **227** than its diazapyrene analog. These results were explained in [172] by the effects of host–guest size fit and the absence of hydrogen bonding with pyridyl nitrogen atoms in the case of the encapsulating ligand **244**.

The covalent capsule **248** also forms cage complexes of the same stoichiometry with paraquat and diquat dications [173]. According to X-ray diffraction data, the molecule with an encapsulated paraquat ion has a mechanically interlocked geometry and is stabilized by hydrogen bonding, CT, and aromatic edge-to-face stacking interactions. As follows from NMR spectra, this complex undergoes a reversible protonation–deprotonation in solution (Scheme 2.153), and the protonation causes an extrusion of the encapsulated organic cation [173].

A new and efficient synthetic pathway (Scheme 2.154) toward the covalent capsules **249**

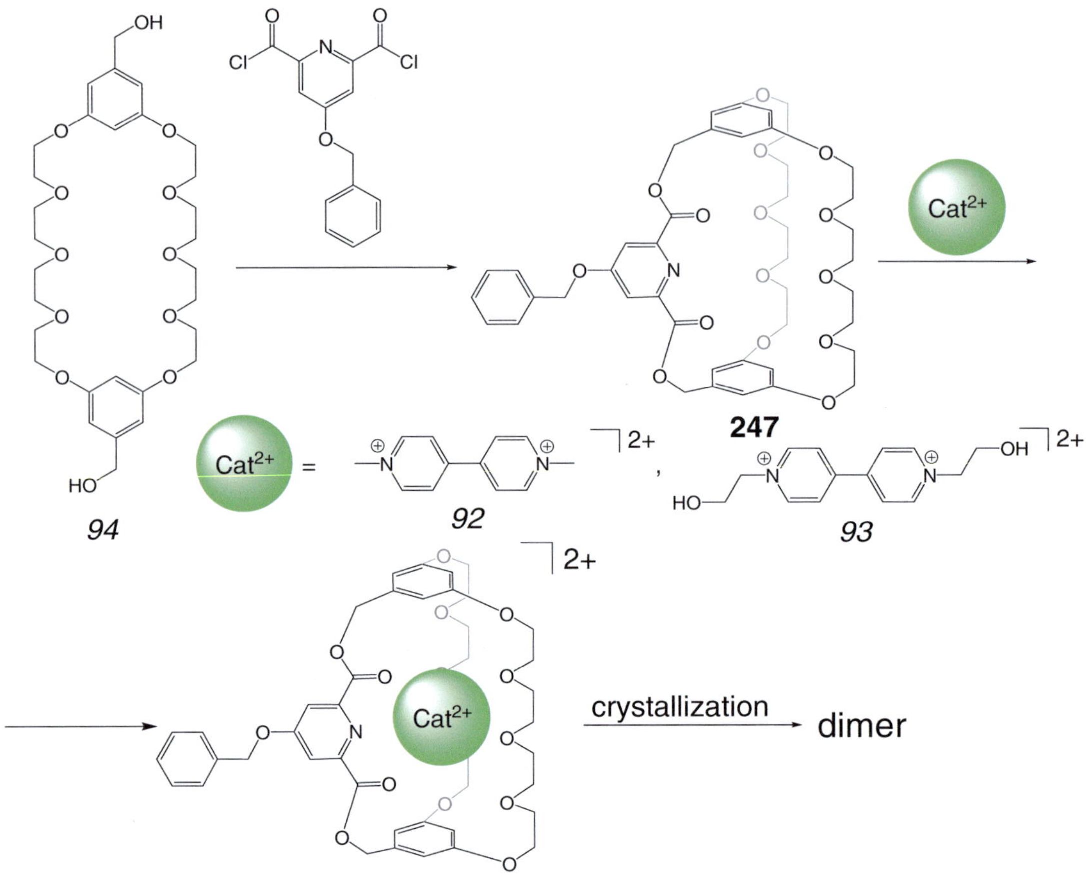

Scheme 2.151

and **250** has been proposed in [174]. As follows from X-ray diffraction data of [174], their cavities are suitable for encapsulation of guest species. According to ESI-MS and ^{1}H NMR spectra, those form stable 2:1 cage complexes with paraquat dication shown in Scheme 2.154. These complexes have [3]-pseudorotaxane structures with host–guest hydrogen bonding and face-to-face stacking interactions between the aromatic host bases and pyridinium fragments of this dication and with CT interactions between them as well [174]. These ligands **249** and **250** also form [2]-rotaxane 1:1 cage complexes with encapsulated dication *94* having the terminal stoppers; such complexes have been also obtained in [175] using a *one-pot* synthetic procedure.

Condensation of the dibromomacrocyclic precursor **251** with 4,4′-dihydroxybenzene allowed obtaining by Scheme 2.155 a capsule **251** with a photoresponsive azobenzene fragment. This ligand is described in [176] to be able to change the size of a cavity and its encapsulating properties by photoswitchable *cis–trans* isomerization of the azobenzene fragment. According to NMR data, the cage framework adopts a *trans*-conformation in the dark after heating for 1 h; the UV irradiation causes a *trans*-to-*cis* transition of this framework, whereas that with visible light resulted in the opposite process. As follows from ESI-MS and ^{1}H NMR data, *cis*-isomer of **251** forms 1:1 cage complexes with 2,7-diazapyrenium dications. The X-ray diffraction and MM data suggest that its *trans*-isomer cannot encapsulate these guests due to geometrical restrictions. In the case of the

Scheme 2.152

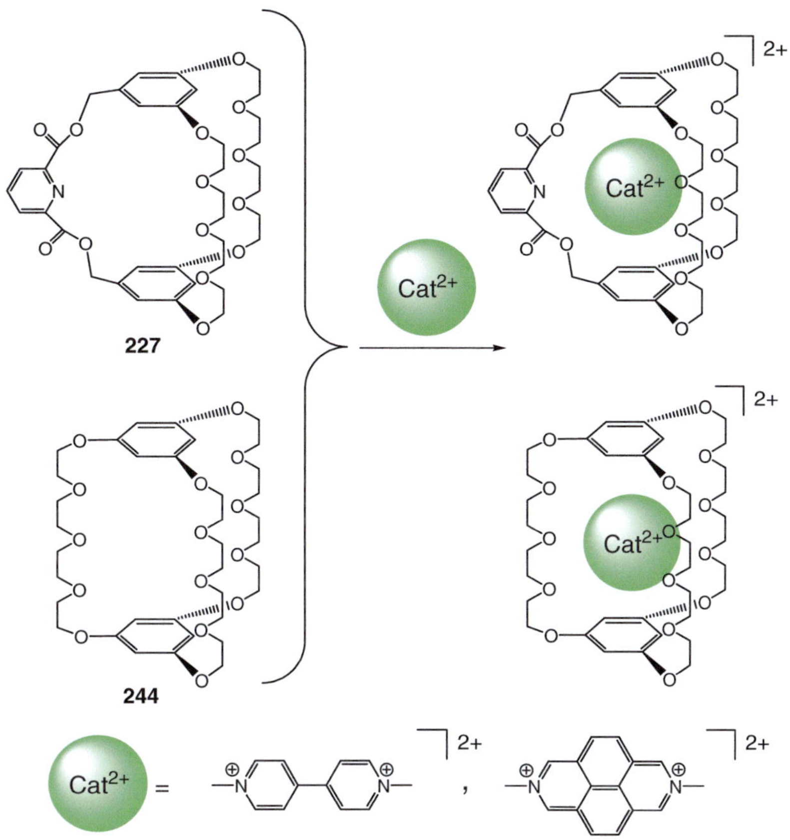

Scheme 2.153

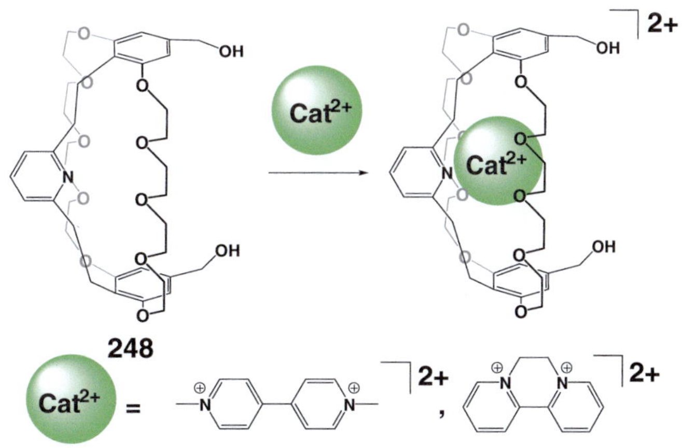

cis-form of **251**, the π-electron-rich benzene rings of both the host and guest are almost parallel that results in the efficient stacking interactions and the ET between them. The intensive fluorescence of 2,7-diazapyrenium guests allowed revealing encapsulation of these dications and *cis–trans* isomerization of **251** [176].

The *cis*-dibenzo-24-crown-8- and bis(*meta*-phenylene)-32-crown-10-based capsules **227** and **252** (see Sect. 2.3.1) are described in [177] to form host–guest 1:1 and 2:1 cage complexes (Scheme 2.156), respectively, with an encapsulated *N*-methylated 1,2-bis-(4-pyridinium)ethane dication. In them, the *N*-methyl group of this

Scheme 2.154

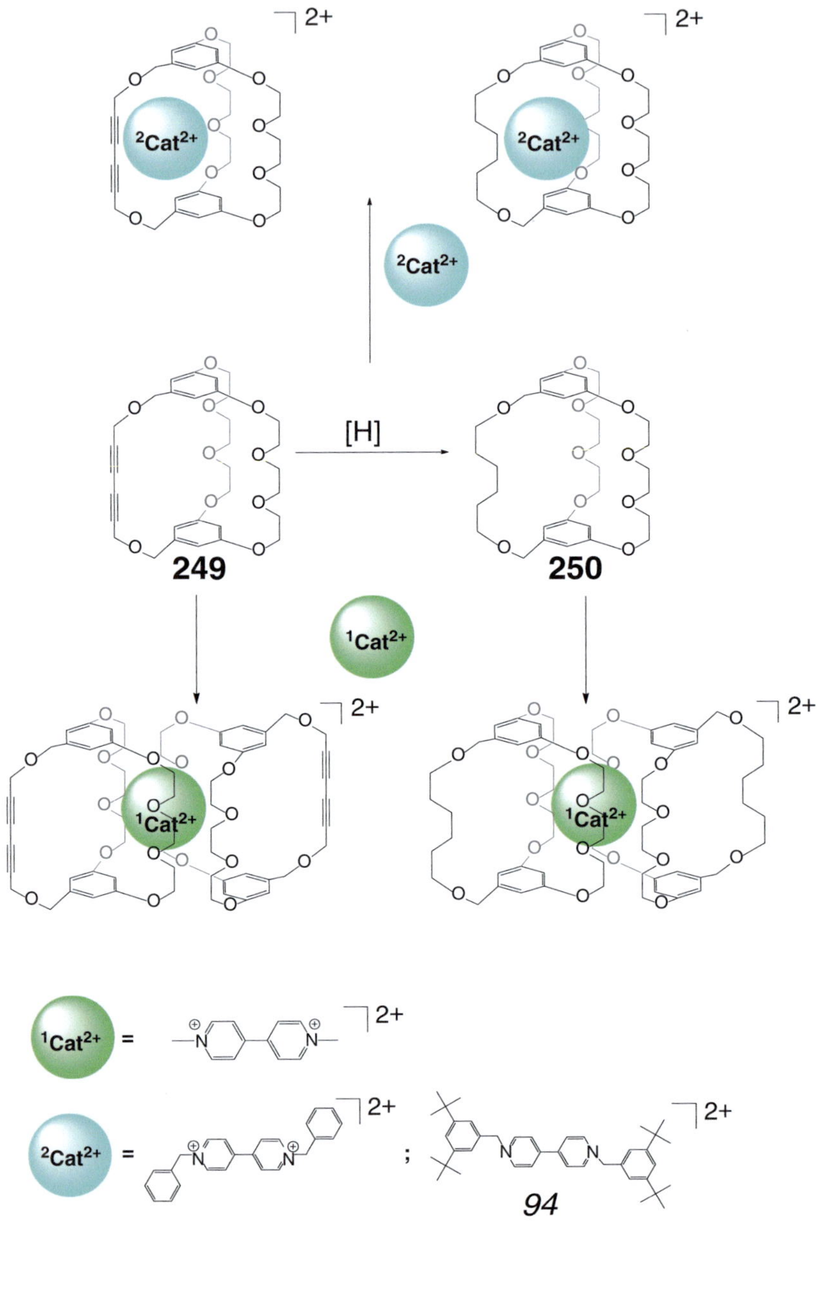

dicationic guest is oriented toward the dibenzo-24-crown-8 fragments of the macrobicyclic ligands forming hydrogen bonds and edge-to-face and face-to-face stacking interactions [177]. These ligands and their bis(*meta*-phenylene)-26-crown-8-based analog **253** encapsulate tropylium cation by Scheme 2.157, thus giving 1:1 cage complexes by the formation of C–H…O hydrogen bonds, face-to-face stacking, and CT interactions [178].

Hybrid cavitandomacrocyclic ligands **253**–**255** have been synthesized in [179] by Scheme 2.158; their 18-membered N_3O_3-macrocyclic fragments selectively bind the

Scheme 2.155

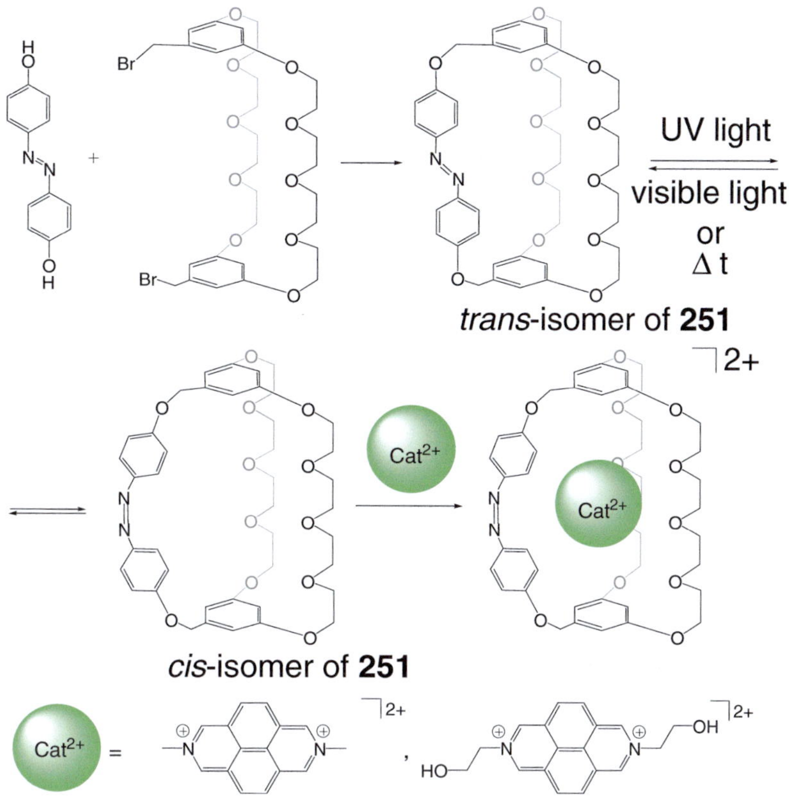

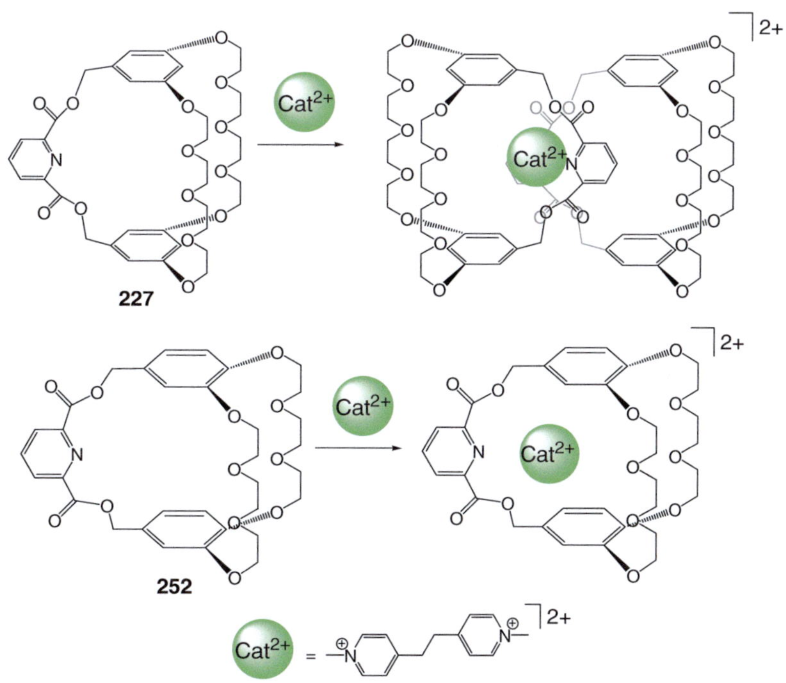

Scheme 2.156

Scheme 2.157

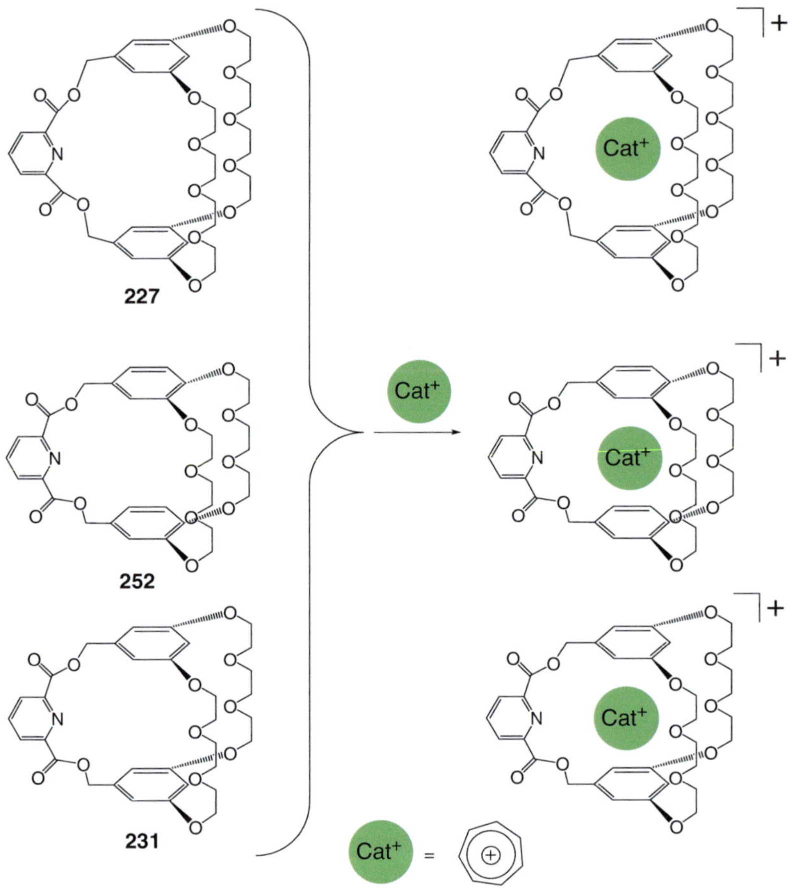

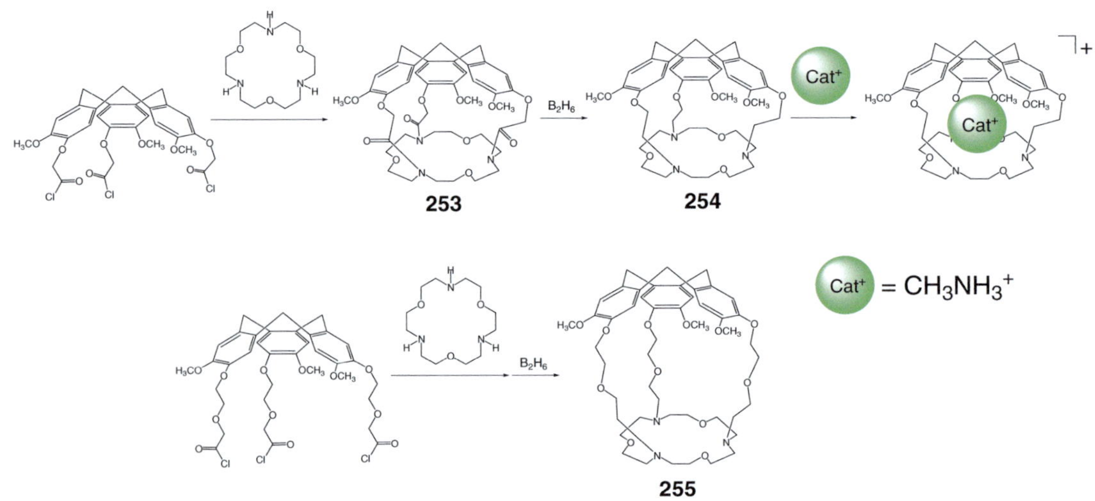

Scheme 2.158

cationic species. In particular, the capsule **254** efficiently encapsulates methylammonium cation but discriminates its homologs with longer aliphatic substituents [179].

The ditopic crown ether–amide caging ligand **256**, prepared in [180], co-encapsulates cationic and anionic species by Scheme 2.159. This ligand transports NaCl, KCl, NaBr, and KBr salts from their aqueous solutions through a liquid organic membrane with the selectivity changing in a row $K^+ > Na^+ > Li^+$. The selectivity order is reversed when this covalent capsule extracts solid alkali metal chlorides and bromides into organic phases [180].

2.4 Arene- and Cavitand-Based Caging Ligands

2.4.1 Free Cages and Encapsulation of Neutral Molecules

Condensation of stoichiometric amounts of a tetrathiol cavitand syntone *95* and its tetrachloride analog *96* in the presence of cesium carbonate by Scheme 2.160 has been used in [181, 182] to obtain a bis-cavitand covalent capsule (carcerand) **257**. The latter gives 1:1 cage complexes (carceplexes) with solvent and gas guest molecules as well as those with encapsulated inorganic cations and anions.

Its tetramethylene-cross-linked analog **258** has been synthesized in [183] by condensation of two tetraol cavitand syntones with dichloromethane in the presence of cesium carbonate (Scheme 2.161). This covalent capsule forms 1:1 cage complexes with various solvent guests.

Bis-cavitand capsules **257**, **259**, and **260** have been prepared [184, 185] in high-dilution conditions by Scheme 2.162. They encapsulate solvent guests as well as alkali metal cations and their salts to give the corresponding cage complexes.

Enantiomerically pure complex $(CHCl_3)@R_4$-**261** and its S_4-containing analog have been synthesized in [186] by Scheme 2.163 using rigid bowl-shaped cavitand syntone *97* and enantiomerically pure *R*- and *S*-cross-linking dibromide *98*; they form 1:1 carceplexes with various neutral guest molecules.

Cage complexes of the bis-cavitand capsules **262–264** (Scheme 2.164) with encapsulated solvent molecules have been isolated in [187] as the products of cross-linking of their deep-cavitand hemispherical precursors, containing phenolic hydroxyl groups, with the molecules of $BrCH_2Cl$ in the presence of potassium carbonate as a strong inorganic base. The carbonyl group of the encapsulated DMF molecule in its C_2-symmetric 1:1 cage complex with **262** is described to point toward the portal of this covalent capsule. Their cage complexes release encapsulated solvent molecules upon heating to give the corresponding capsules by Scheme 2.164. Those are described in [187] to bind other solvents (CH_3CN, CS_2, CH_2Cl_2, CH_2Br_2, and H_2O) and gases (O_2, N_2, and Xe) in their $CDCl_3$ solutions, forming host–guest 1:1 cage complexes. Among them, only one with the encapsulated xenon is kinetically stable. The tetrabutylene-containing capsule **264**, obtained by a similar synthetic procedure [188], is reported in [189] to encapsulate aromatic guests shown in Scheme 2.164, including a reactive benzocyclobutenedione molecule (see Sect. 5.2).

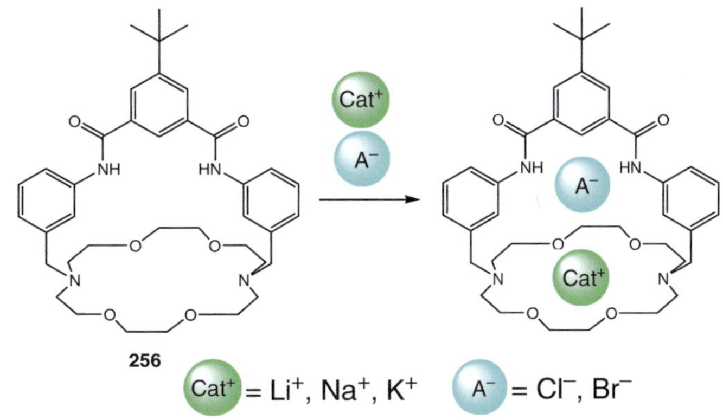

Scheme 2.159

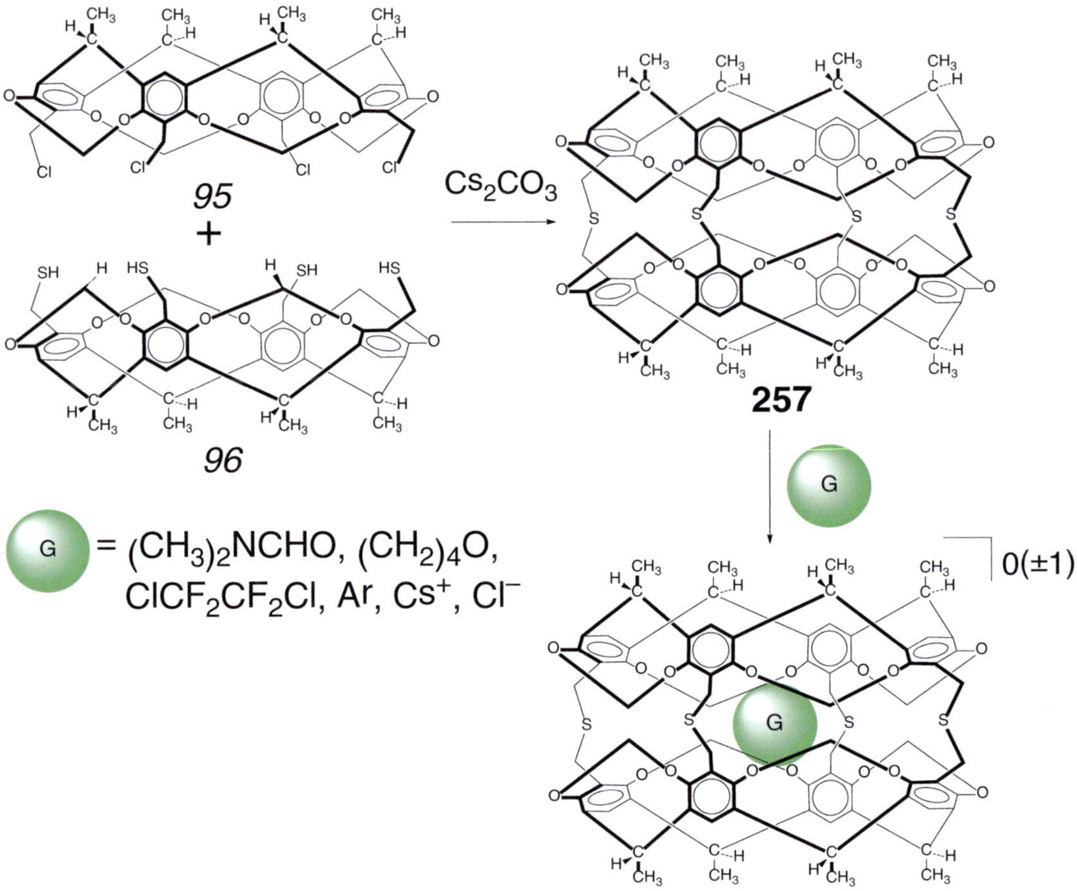

Scheme 2.160

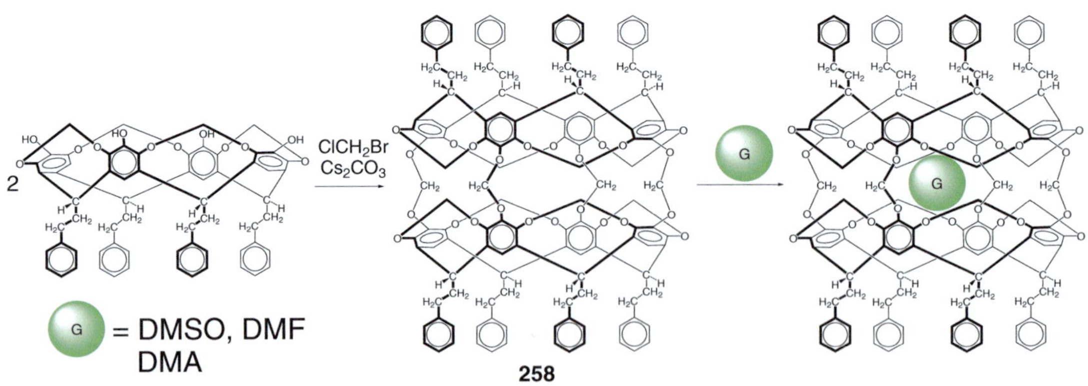

Scheme 2.161

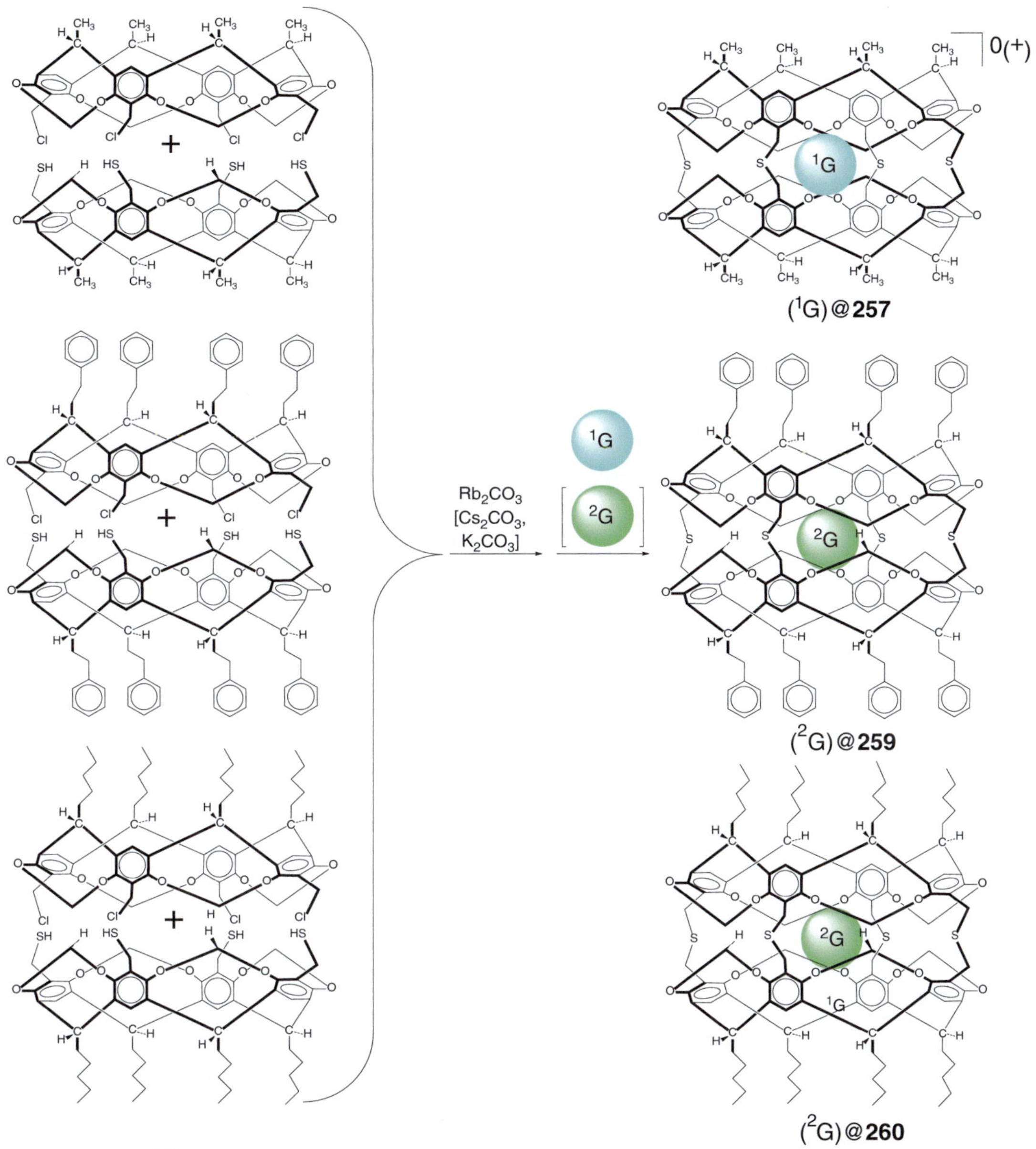

^1G = Cs^+, $Cs^+ \cdot H_2O$, $Cs^+ \cdot (CH_3)_2NCHO$, $(CH_3)_2NCHO$, $(CH_2)_4O \cdot H_2O$, $(CH_3)_2NCHO \cdot Ar$, $(CH_3)_2NCHO \cdot CsCl$, $CsCl \cdot Cs^+$

^2G = $CH_3OH \cdot HOCH_3$, $CH_3CN \cdot NCCH_3$, CH_3CN, CH_3CH_2OH, $(CH_3)_2NCHO$, $CH_3COCH_2CH_3$, $CH_3CH_2COCH_2CH_3$

Scheme 2.162

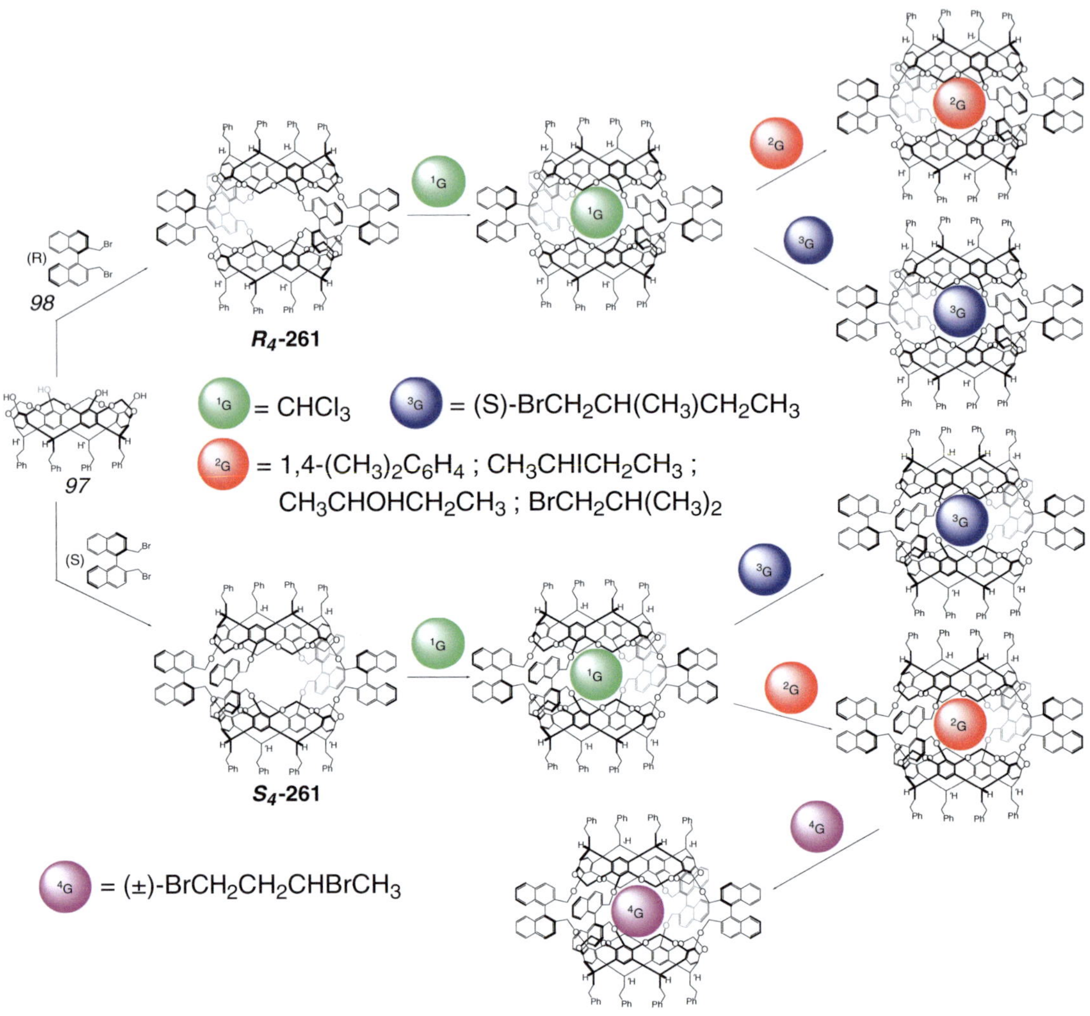

Scheme 2.163

Rigid piperazine bridging fragments have been used in [190] to obtain bis-resorcinarene covalent capsules **265**–**267** by Scheme 2.165. According to ^{1}H NMR data, they are symmetric in solution and encapsulate solvent molecules of ethanol, methanol, and dichloromethane during the Mannich reaction. These encapsulated molecules can leave the cavities of covalent capsules **265**–**267**, and the vacant cavities of these ligands are described in [190] to adopt acetonitrile and nitromethane solvent molecules. In one of the two X-rayed single crystals of **266**, upper and lower resorcinarene capping fragments are similar with the helix turn of approximately 54.3°, and they both have a *cone*-conformation stabilized by four intramolecular hydrogen bonds between the neighboring hydroxyl groups and those of four remaining groups with piperazine nitrogen atoms. Its caging ligand with a cavity volume of 166 Å^3 encapsulates two disordered methanol molecules. The second crystal of this capsule contains two independent cage species with resorcinarene fragments in a *cone*-conformation; the helix turn between them is equal to 52.6°. These species with a cavity volume of 177 Å^3 are more open, so they can encapsulate one disordered *iso*-propanol molecule [190].

Chiral hemicarcerand cage complexes of covalent capsules **268** and **269** with encapsulated chloroform molecules have been synthesized in

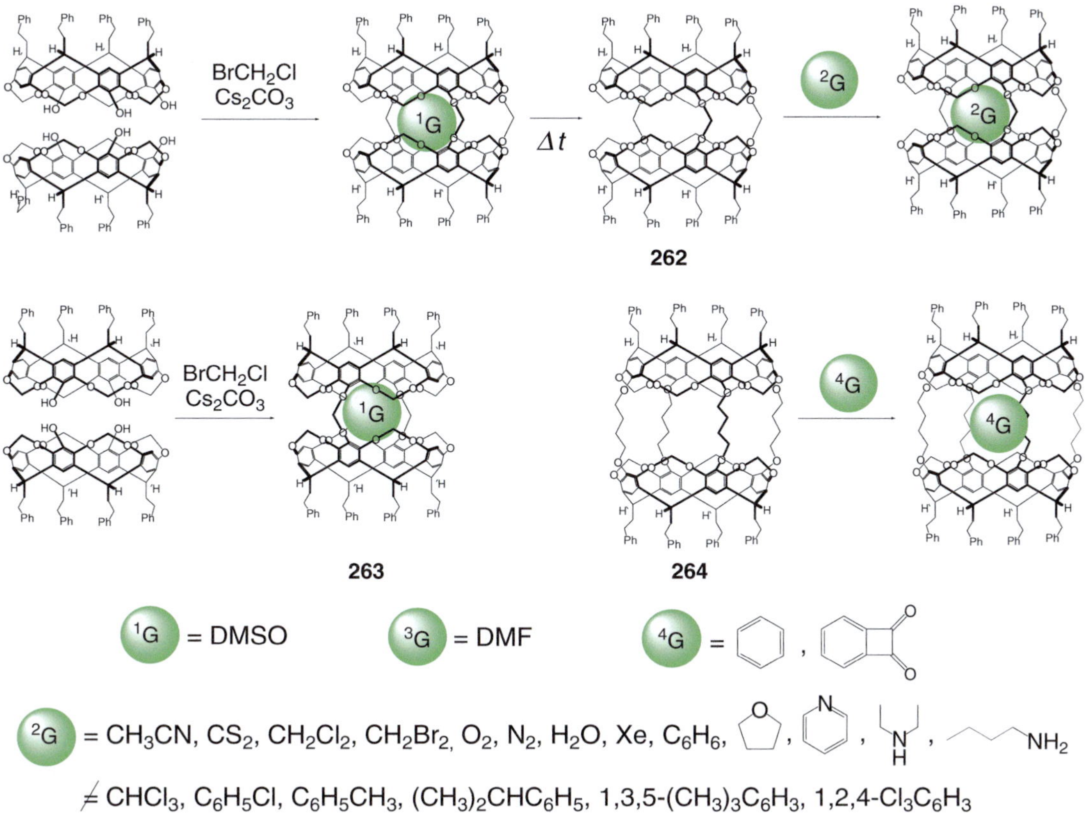

Scheme 2.164

[191] by Scheme 2.166 using the corresponding diol and enantiomerically pure (*S*)-(–)- and *S*,*S*-ligand syntones.

Two cyclotriveratrylene capsules **270** and **271** have been prepared in [192] by Scheme 2.167 using intramolecular cyclization of their hydrogen bonding–preorganized tripodal syntones with the formation of three imine bonds. This reaction gave only their *anti*-isomers with an opposite orientation of the methoxyl and imine groups relative to the linker fragments. According to one- and two-dimensional ^{1}H NMR data, the cavities of **270** and **271** can accommodate fullerenes C_{60} and C_{70} to give 1:1 cage complexes. This efficient binding is explained in [192] by strong stacking interactions between fullerene guests and two aromatic fragments of these encapsulating ligands.

The cavitand syntone *99* with opposite benzyl-protected hydroxyl groups (those prevent the formation of the corresponding monocage product) has been used in [193] to obtain a tetracavitand bis-cage capsule **272** by Scheme 2.168. This ligand efficiently encapsulates two pyrazine molecules forming host–guest 1:2 complex (one guest molecule per a host cavity). These caged guests do not leave the cavities even after boiling this complex in 1,2,4-trichlorobenzene for a long time [193]. The bis-caging ligand **272** also encapsulates methyl acetate molecules by Scheme 2.169 in the presence of DBU to give the corresponding 1:2 cage complex. Such an encapsulation is described in [194] to be highly cooperative: an analogous 1:1 complex is not found in appreciable amounts. In contrast to nitrobenzene, one chloroform molecule can be co-encapsulated by its cavity, and the corresponding heteroguest 1:1:1 complex has been detected in [194]. Addition of methyl acetate causes the exchange of the chloroform guest by methyl acetate, thus

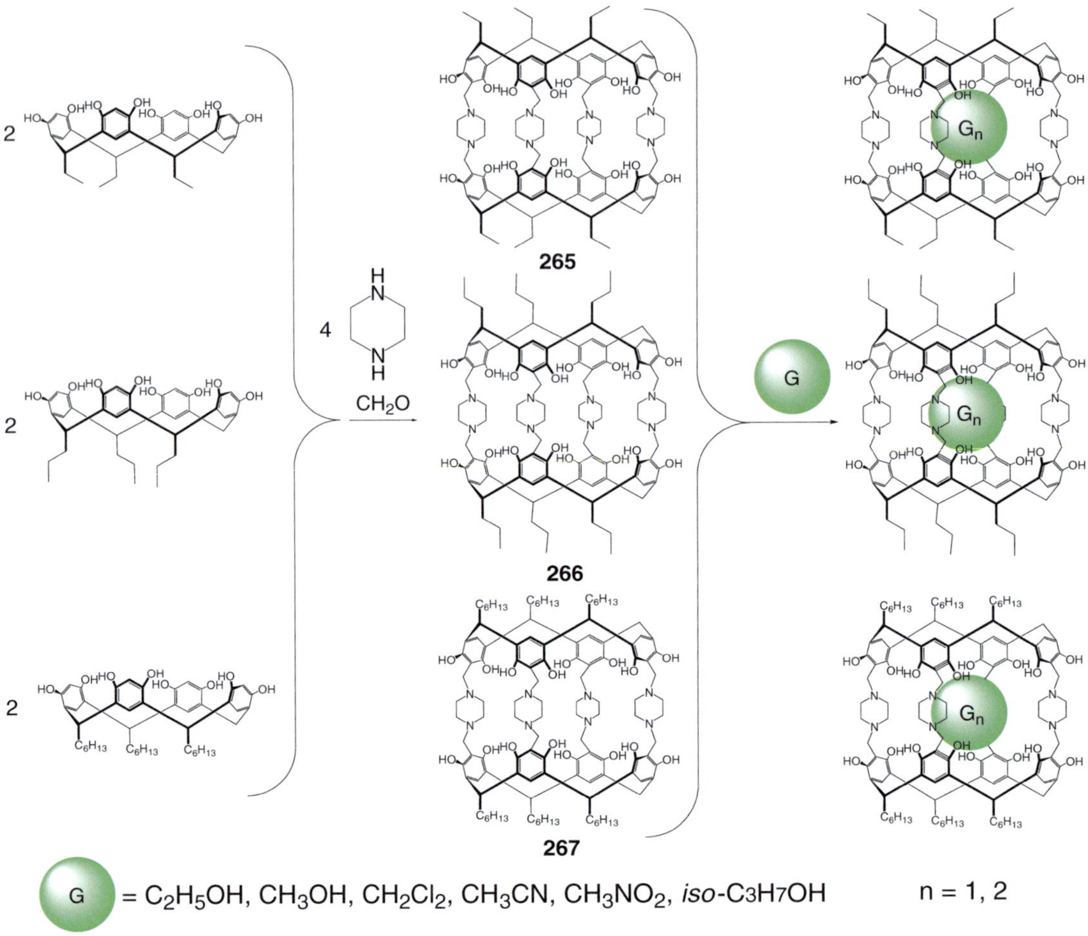

Scheme 2.165

giving a homoguest 1:2 cage complex. Such unusual communication between the neighboring guests co-encapsulated by **272** has been detected in [194] using NMR spectroscopy.

Analogous tris-caging capsules **273** and **274** have been synthesized in [195] by Scheme 2.170 using the hexacavitand ligand syntone *100*. They encapsulate three methyl acetate molecules, thus forming 1:3 cage complexes. As in the case of their bis-caging analogs, no formation of the corresponding 1:1 and 1:2 complexes has been observed in this work.

The first representative of C_3-symmetric calix[6]azacrown caging ligand, the capsule **275**, has been prepared in [196] by Scheme 2.171. According to NMR data, the functionalizing methoxyl substituents are remoted from its symmetry axis, and all the *tert*-butyl groups have the same orientations due to hydrogen bonding between its anisole fragments and amino groups of the tripodal cap [196].

Two synthetic strategies with similar efficiencies shown in Scheme 2.172 have been used in [197] for the synthesis of a tris-pyridyl calix[6]arene capsule **276**. According to ^{1}H NMR data, its cage framework has a C_{3v}-symmetric flattened *cone*-conformation and is able to co-encapsulate Na^+ cation and solvate ethanol molecule, giving a heteroguest 1:1:1 cage complex. At the same time, **276** does not form similar complexes with an encapsulated K^+ cation in the absence of ethanol, and vice versa. The X-rayed crystal of the protonated form of **276** contains three cage molecules, two of which

Scheme 2.166

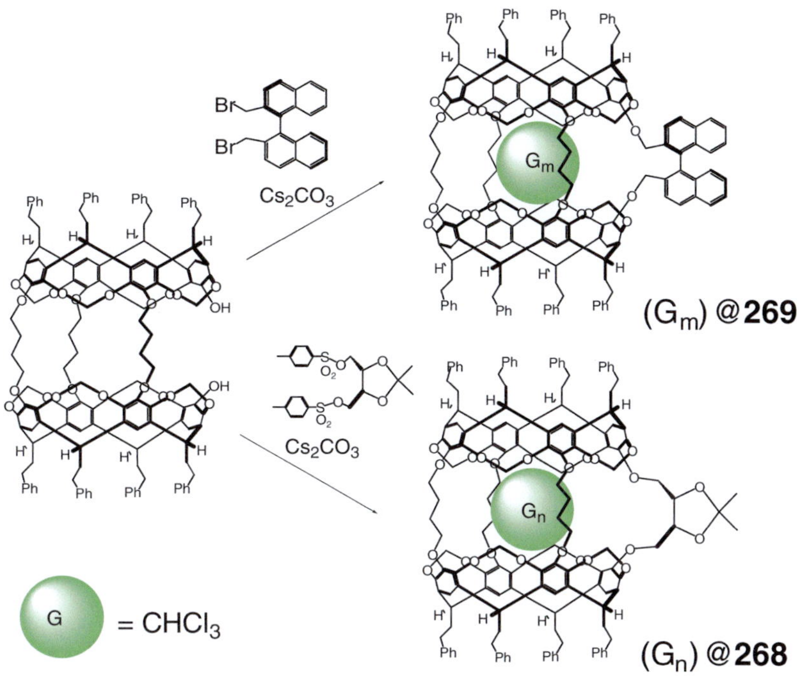

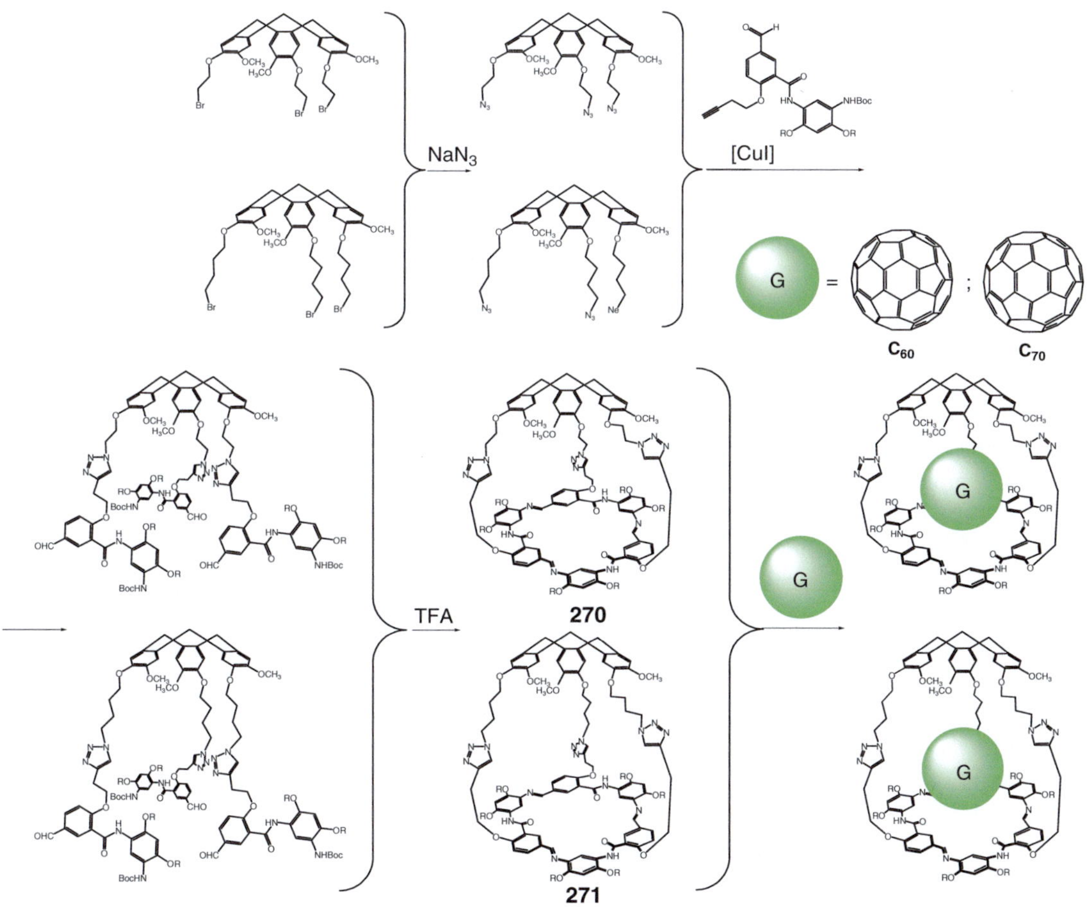

Scheme 2.167

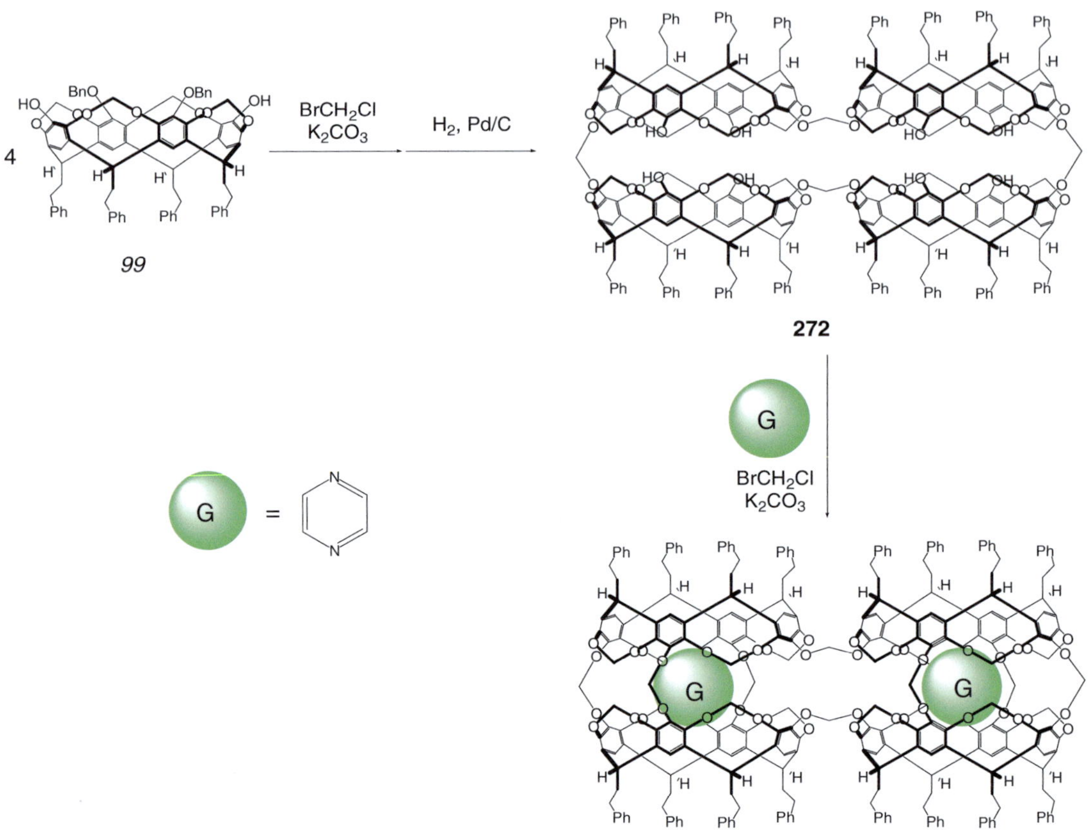

Scheme 2.168

have a flattened *cone*-conformation, and the third one is an *1,3*-alternate. According to NMR data of [197], the capsule H^+ · **276** does not encapsulate neutral solvate molecules but forms C_{3v}-symmetric 1:1 cage complexes of encapsulated ammonium cations, the derivatives' primary amines (Scheme 2.172).

A bis-hemicarcerand caging ligand **277** is reported by D. Rudkevich and coworkers in [198] to encapsulate solvent molecules and gases shown in Scheme 2.173 in solid state.

Condensation of the corresponding tetrakis(dihydroxyboryl)-containing cavitand ligand syntone and 1,2-bis(3,4-dihydroxyphenyl) ethane by dynamic formation of boronic ester gave the covalent capsule **278** [199]. This caging ligand highly selectively and solvent-dependently encapsulates polyaromatic guests shown in Scheme 2.174, thus giving the corresponding 1:1 cage complexes.

Self-assembly of analogous ligand syntone *101* with a bis-catechol *102* by Scheme 2.175 afforded a covalent capsule **279** in a quantitative yield [200]. This ligand and its analog **278** form 1:1 cage complexes with different aromatic guests. In the cage complex of **279** with an encapsulated 4,4′-diacetoxybiphenyl, two cavitand entities are linked by eight boronic ester bonds of four bis-catechol linkers, and acetoxyl groups of the guest are oriented toward the aromatic capping fragments, thus maximizing their C–H…π interactions [200]. The same orientation of the caged 4,4′-diacetoxy-2,2′-bis(methoxycarboxyl) biphenyl molecule within the cavity of **278** has been elucidated in [200] using ^1H NMR method in solution. Methyl ester groups of the guest are oriented toward the equatorial portals of the caging ligand, and solvent effect on the encapsulation has been observed. According to thermodynamic data, the encapsulation in

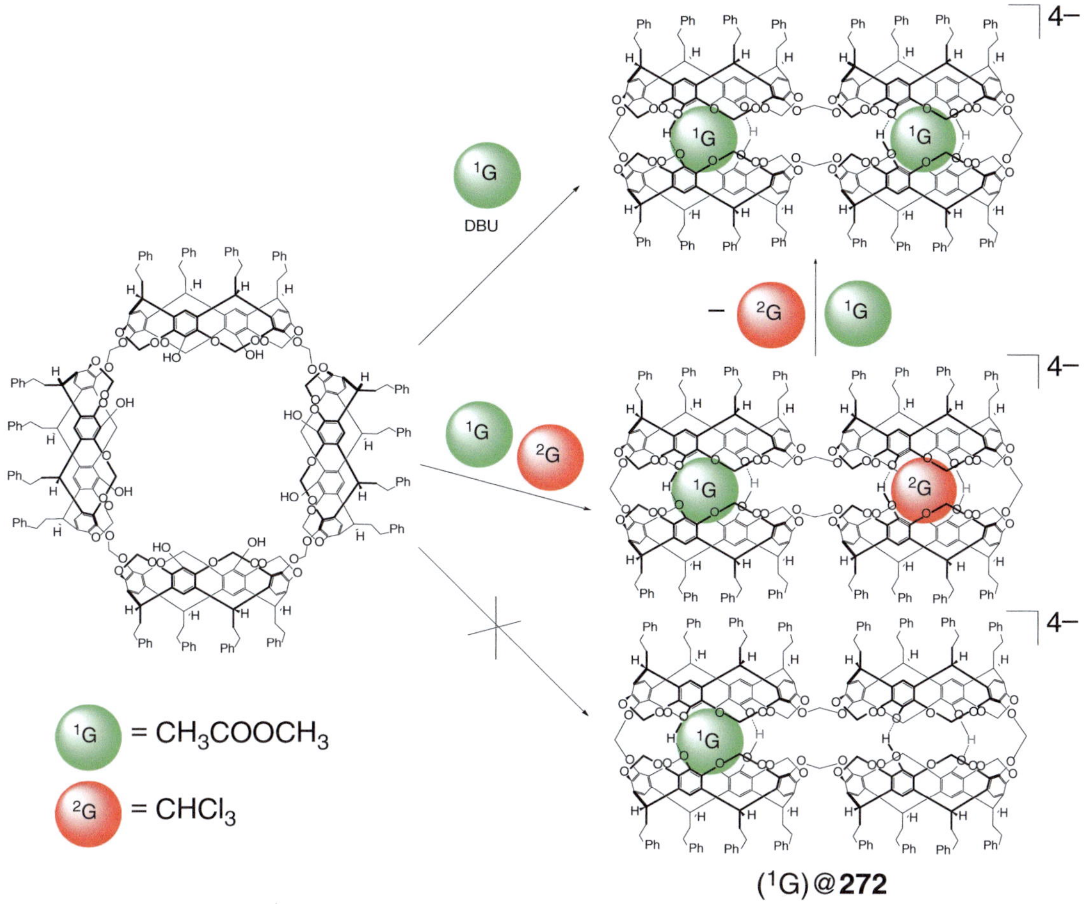

Scheme 2.169

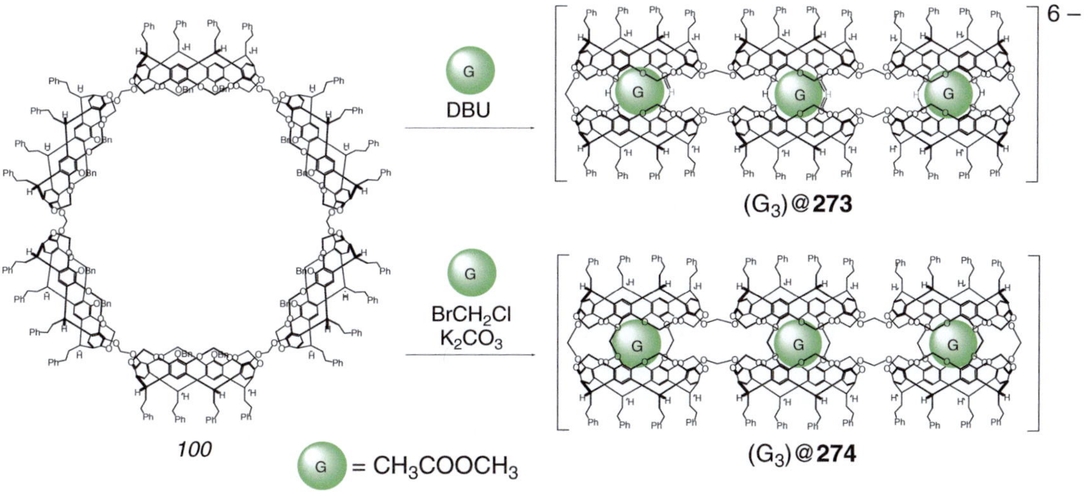

Scheme 2.170

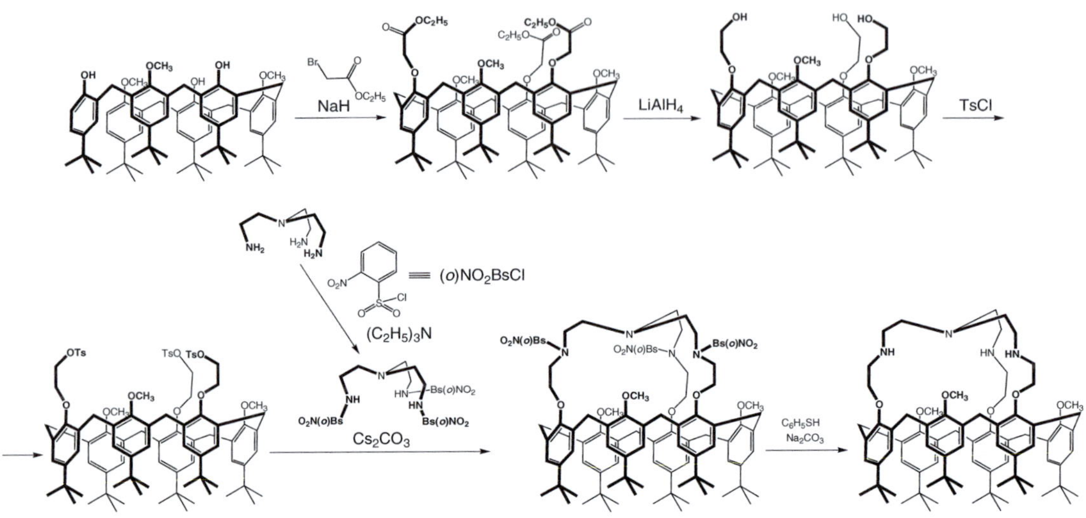

275

Scheme 2.171

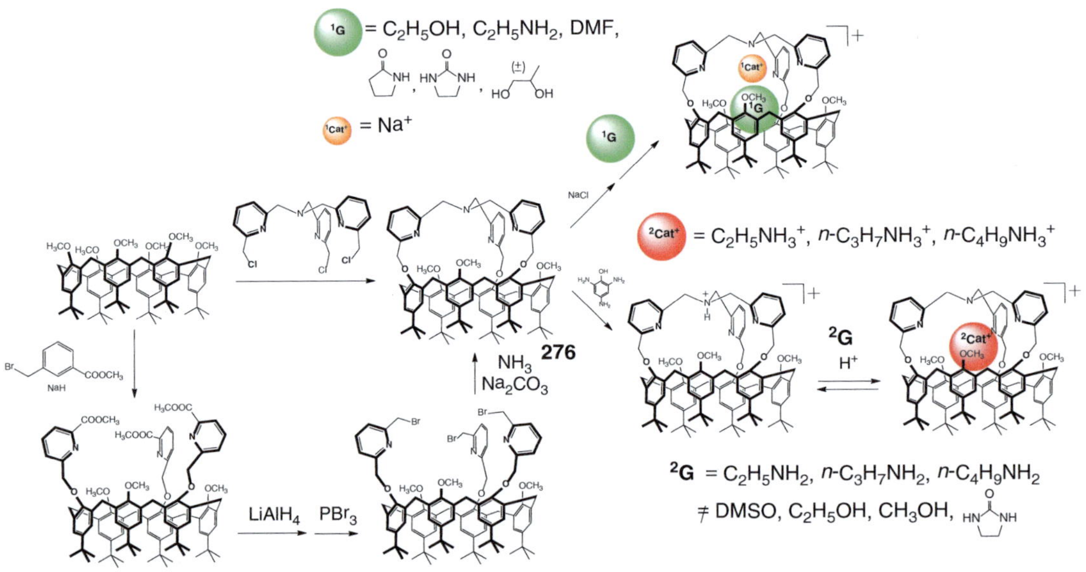

Scheme 2.172

deuterobenzene solution is enthalpically driven, whereas in deuterochloroform it is both enthalpically and entropically driven. The same effect has been observed going from a more polar guest to its less and least polar analogs [200]. The ligand **278** strictly discriminates between the terminal groups of the guest (via C–H…π and C–H…O=C interactions within the ligand's cavity) as well as between the guests of the different sizes (Scheme 2.175). Three possible pathways of this caging process are discussed in [200]. The first one is an extended portal mechanism based on conformational changes in the linker's moieties; however, this mechanism does not agree with X-ray diffraction data. The second pathway includes complete dissociation of the linker, but it was not observed experimentally on the NMR time scale. Therefore, the most probable

Scheme 2.173

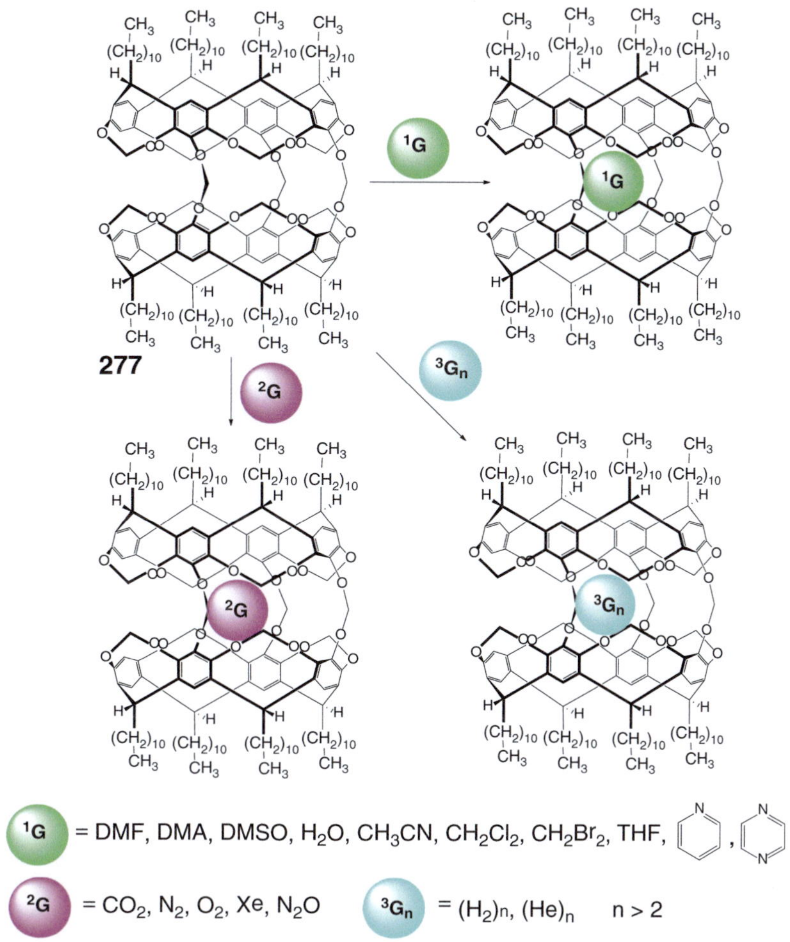

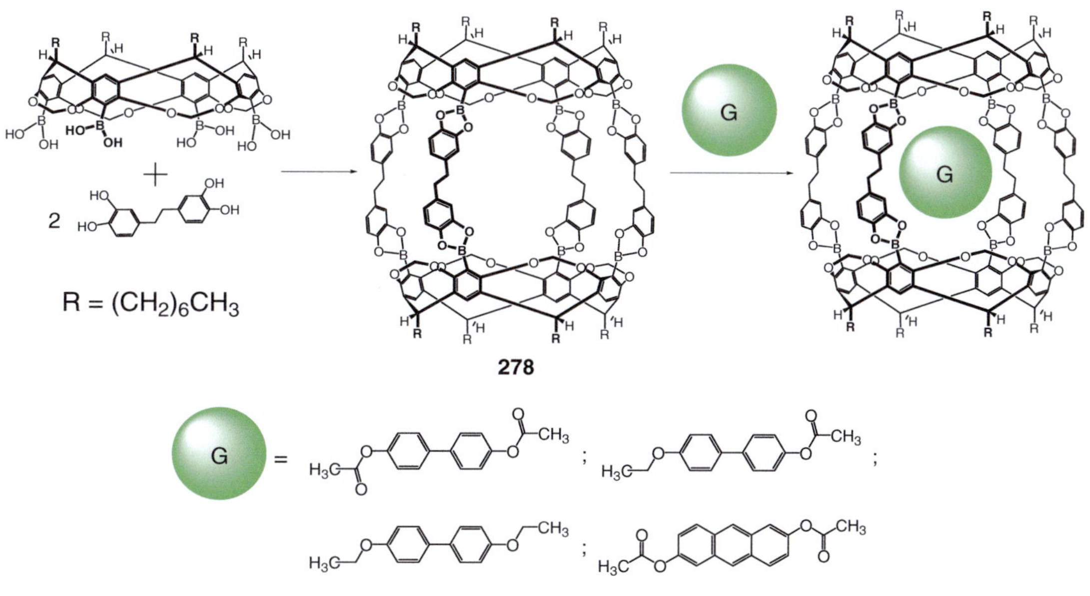

Scheme 2.174

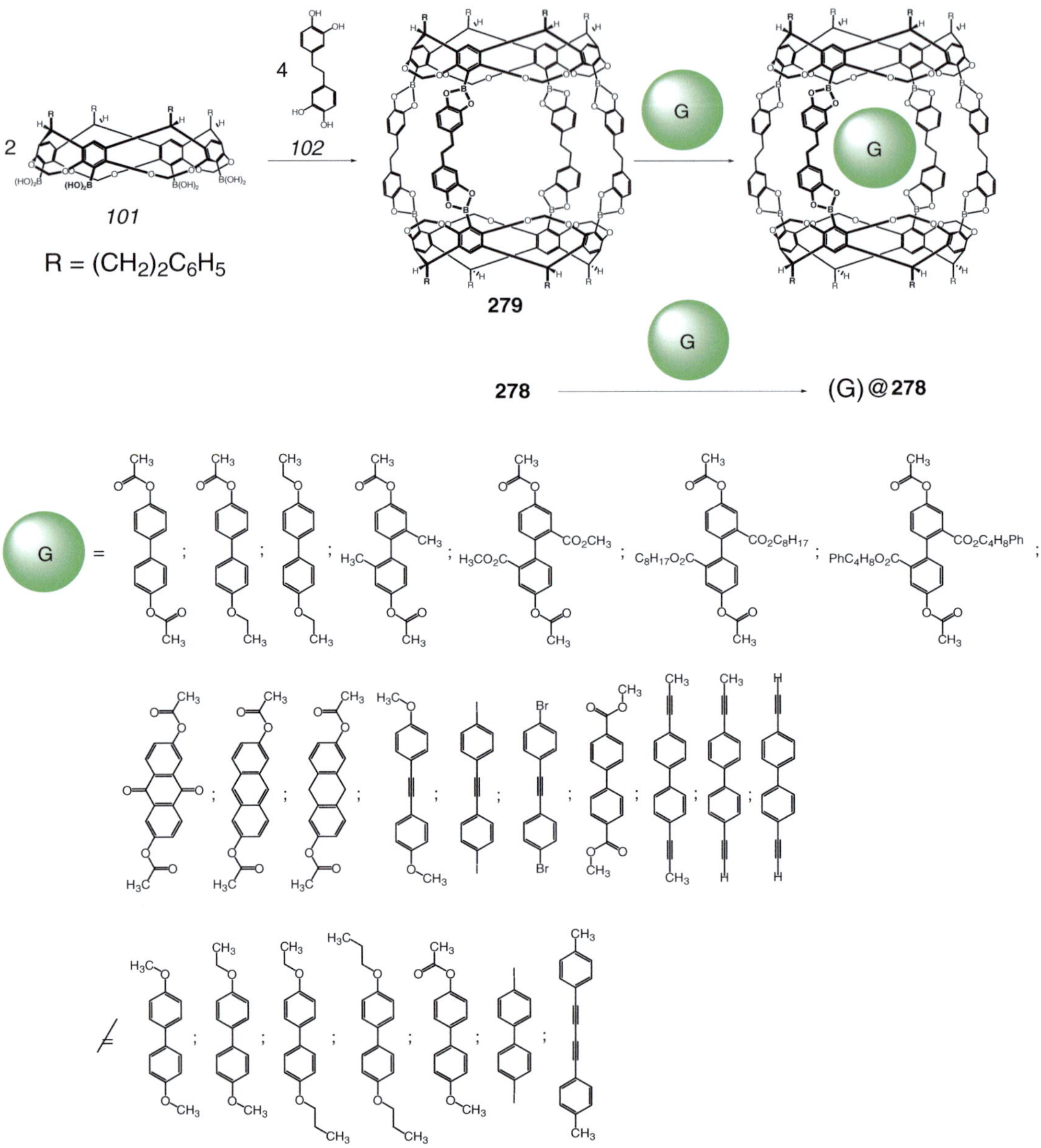

Scheme 2.175

mechanism of encapsulation and release of guest molecules includes partial dissociation of each one of the respective two boronic ester bonds of the two linkers. Such mechanism has been confirmed in [200] by kinetic studies: these encapsulation reactions follow second-order kinetics, while the release of a guest molecule is first-order reaction.

Ion pair-driven formation of the boron ester-based heterodimeric arene covalent capsule **280** by Scheme 2.176 is described in [201].

Another boronic ester-based arene covalent capsule **281** has been prepared [202] using [6+3+2] condensation by Scheme 2.177.

Dissymmetric tetraureacalix[4]pyrrole–tetraureacalix[4]arene caging ligands **282** and **283**

Scheme 2.176

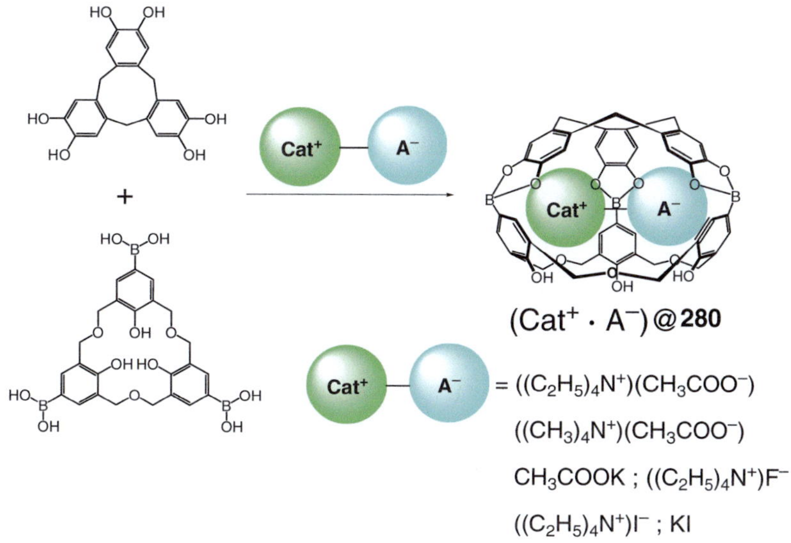

Scheme 2.177

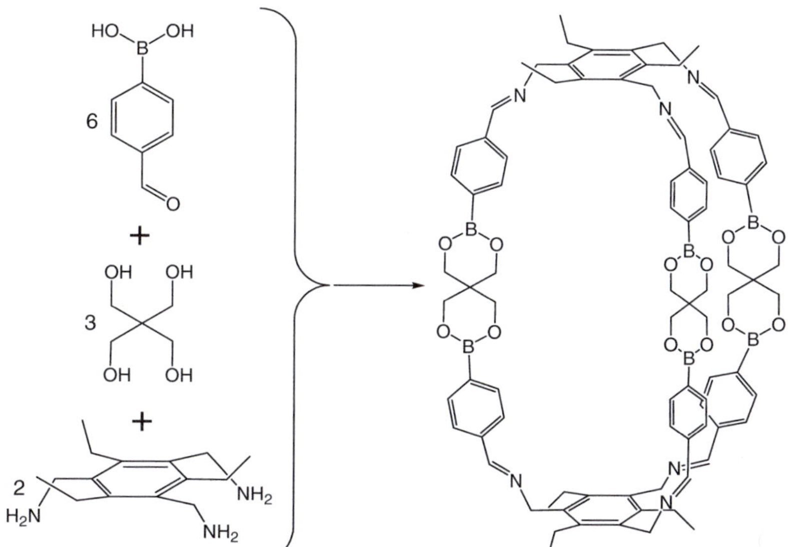

with two mechanically locked hemispheres have been synthesized in [203] by Scheme 2.178. Bismacrocycle *103* acts both as a reactive ligand syntone and as a matrix for template coupling with alkenyl calix[4]pyrrole *104* that proceeds only in the presence of *N*-oxides *105* and *106*. The resulting 1:1 cage complexes of **282** underwent further metathesis and catalytic hydrogenation reactions giving the capsule **283**. Its calix[4]pyrrole fragment can accommodate guest molecules by hydrogen bonding, while the calix[4]arene semisphere provides strong supramolecular cation…π and C–H…π interactions. According to ^{1}H NMR data, **283** has C_{2v} symmetry in solu-

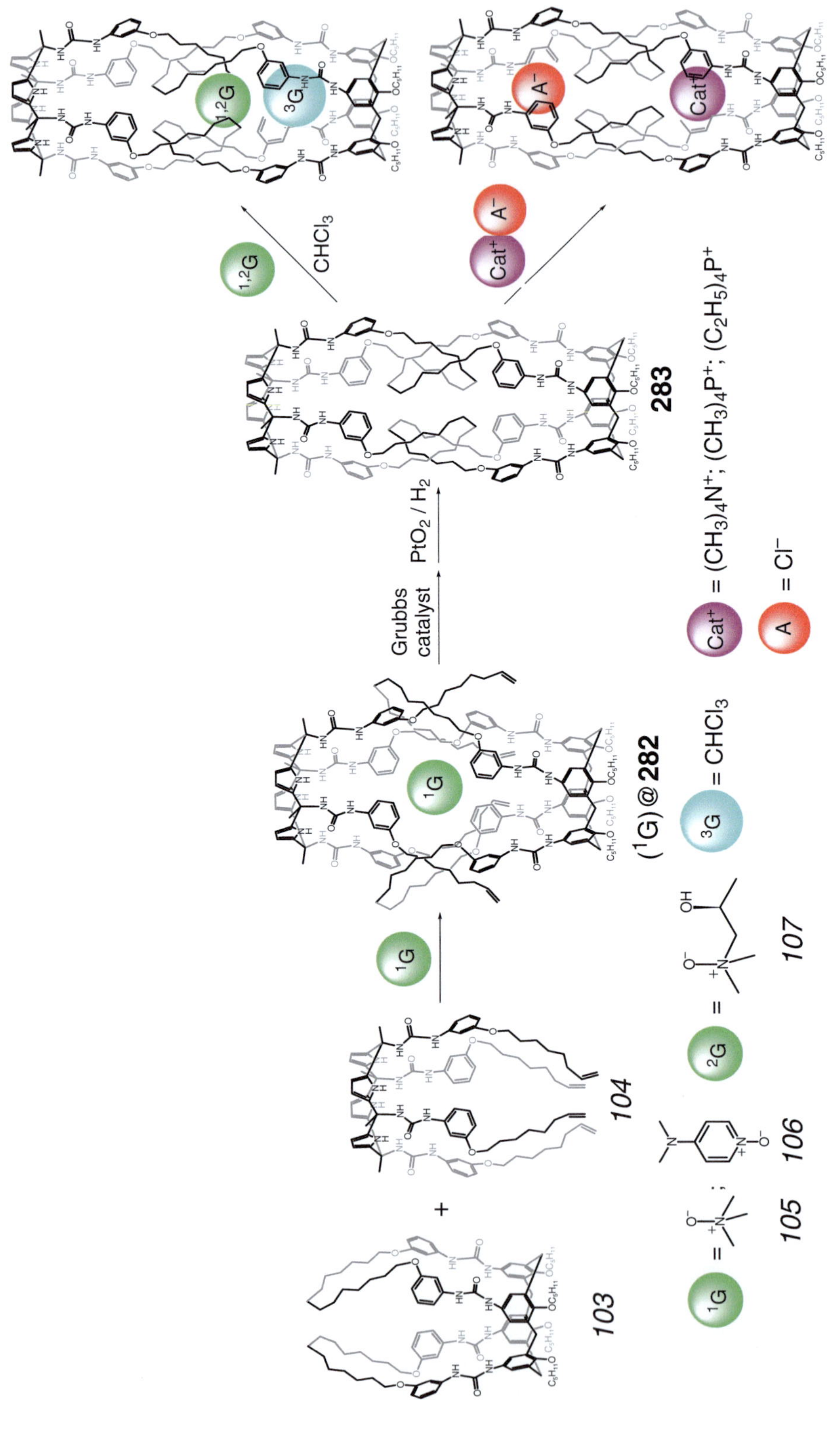
103
+
104
1G
(1G)@282
Grubbs catalyst
PtO2 / H2
283
1,2G
CHCl3
Cat+
A−
1G = 105 ; 106
2G = 107
3G = CHCl3
Cat+ = (CH3)4N+; (CH3)4P+; (C2H5)4P+
A = Cl−

Scheme 2.178

tion but is dissymmetric in solid state. Racemic bis-[2]-catenane and its supramolecular assemblies with encapsulated achiral guests are mixtures of two pairs of cyclodiastereomers that are enantiomers. In contrast, encapsulation of enantiomerically pure *N*-oxide *107* by the racemic caging ligand **283** afforded a mixture of four diastereomeric 1:1 cage complexes; ^{1}H NMR spectrum of this mixture was assigned in [203] after HPLC separation of the two enantiomers {(+)- and (−)-**283**}. The enantiomerically pure capsule **283** showed complete lack of chiral discrimination in recognition of chiral guests. It has a cavity volume of approximately 280 Å^3, and the *N*-oxide guests *105* (V ≈ 80 Å^3) and *106* (V ≈ 120 Å^3) are too small to fit in. As a result, those are co-encapsulated with one solvate chloroform molecule (as follows from ROESY NMR data), whereas the co-encapsulation is not observed in the case of their more bulky analog *107* (V ≈ 180 Å^3) [203]. The cage framework of **283** is formed by the calix[4]arene fragment with a lipophilic cavity formed by electron-rich aromatic rings (and, therefore, is suitable for binding of tetraalkylammonium cations) and by the calix[4]pyrrole hemisphere with more deep aromatic cavity that is able to accommodate anions and electron-rich guests. As a result, it encapsulates tetraalkylammonium and tetraalkylphosphonium chlorides as ionic pairs [203].

The hemicryptophane capsule **209** (see Sect. 2.3.1) is described in [204] as an efficient receptor for alkylammonium cations; the binding constants are three orders of magnitude higher than those for other ligands of this type. According to NMR titration data, these cationic guests form 1:1 cage complexes by Scheme 2.179, and the affinity of **209** to them decreases in a row $C_6H_5CH_2NH_3^+ > CH_3NH_3^+ >$ *tert*-$C_4H_9NH_3^+ \approx$ *n*-$C_3H_7NH_3^+$. It is governed by a set of hydrogen bonds between the cationic guest and the tripodal amine capping fragment of **209** as well as by the fit and C–H…π interactions between the guests and the interior of this covalent capsule. Its high affinity to benzylammonium cation is explained in [204] by the presence of additional π–π interactions between their aromatic fragments. The caging ligand **209** also efficiently binds dopamine monocation, thus being a potent receptor for neurotransmitters [204].

The largest of the two cyclotriveratrylene caging ligands **284** and **285** (Scheme 2.180), prepared in [205], showed a higher affinity to C_{70} as compared to C_{60} allowing for their separation and purification directly from fullerene extracts.

2.4.2 Encapsulation of Anions

Synthesis of calix[4]arene-based capsules **286–289** with tripodal amine and crown ether capping fragments has been performed in [206] by

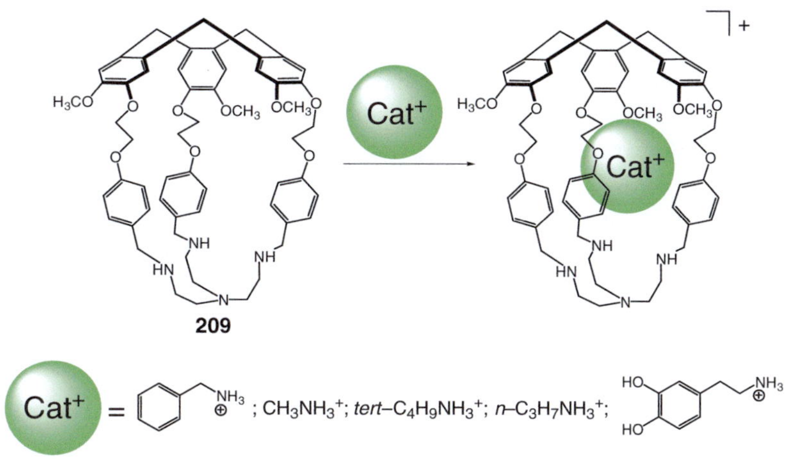

Scheme 2.179

Scheme 2.180

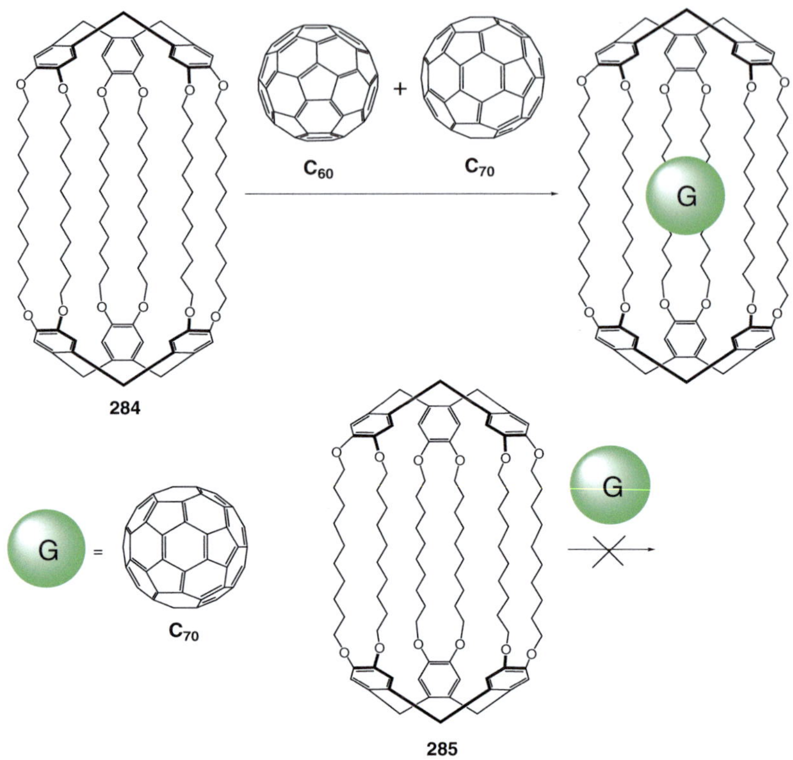

Scheme 2.181 using the *para-tert*-butylcalix[4]arene as a ligand syntone. According to X-ray diffraction data, one of the ethoxy-containing benzyl cross-linking moieties of $H^+_4 \cdot$ **287** passes through its cavity that is formed by two such chains. ^{1}H NMR spectrum of $H^+_4 \cdot$ **289** suggests higher symmetry of its glycolic chains. The capsules **287** and **289** have different structures, and the sizes of their cavities is also different. As a result, they and their tetraprotonated forms demonstrate different binding properties toward cations and anions; Scheme 2.181 summarizes the results obtained in [206]. Among the tested spherical (F^-, Br^-, and I^-), trigonal planar (CO_3^{2-}), angular (AsO_2^-), and tetrahedral ($H_2PO_4^-$, HPO_4^{2-}, SO_4^{2-}, and PO_4^{3-}) anions, fluoride and sulfide ions do not enter into the corresponding frameworks as encapsulated species, as they do not fit well in to the cavity. The tetraprotonated forms of **287** and **289** gave 1:1 cage complexes with Br^-, I^-, and NO_3^- anions; their stability increases in a row $Br^- < I^- < NO_3^-$ in the presence of Na^+ counterion. The ligand $H^+_4 \cdot$ **287** binds nitrate anion more efficiently than $H^+_4 \cdot$ **289** due to the better fit of its cavity size to that of this ion [206]. In the case of tetrabutylammonium counterion, iodide anion forms the most stable cage complex with $H^+_4 \cdot$ **289**. The presence of potassium(I) cation causes 1.5-fold increase in the binding of bromide anion by the encapsulating ligand $H^+_4 \cdot$ **287**. Therefore, the crown ether fragment of this capsule has higher affinity to K^+ than to Na^+. The binding of the alkali metal cation changes structural organization of the ligand $H^+_4 \cdot$ **287** making it more suitable for the binding of anions. The binding of Br^- and I^- anions by $H^+_4 \cdot$ **289** decreases in the presence of Na^+ and K^+ cations [206].

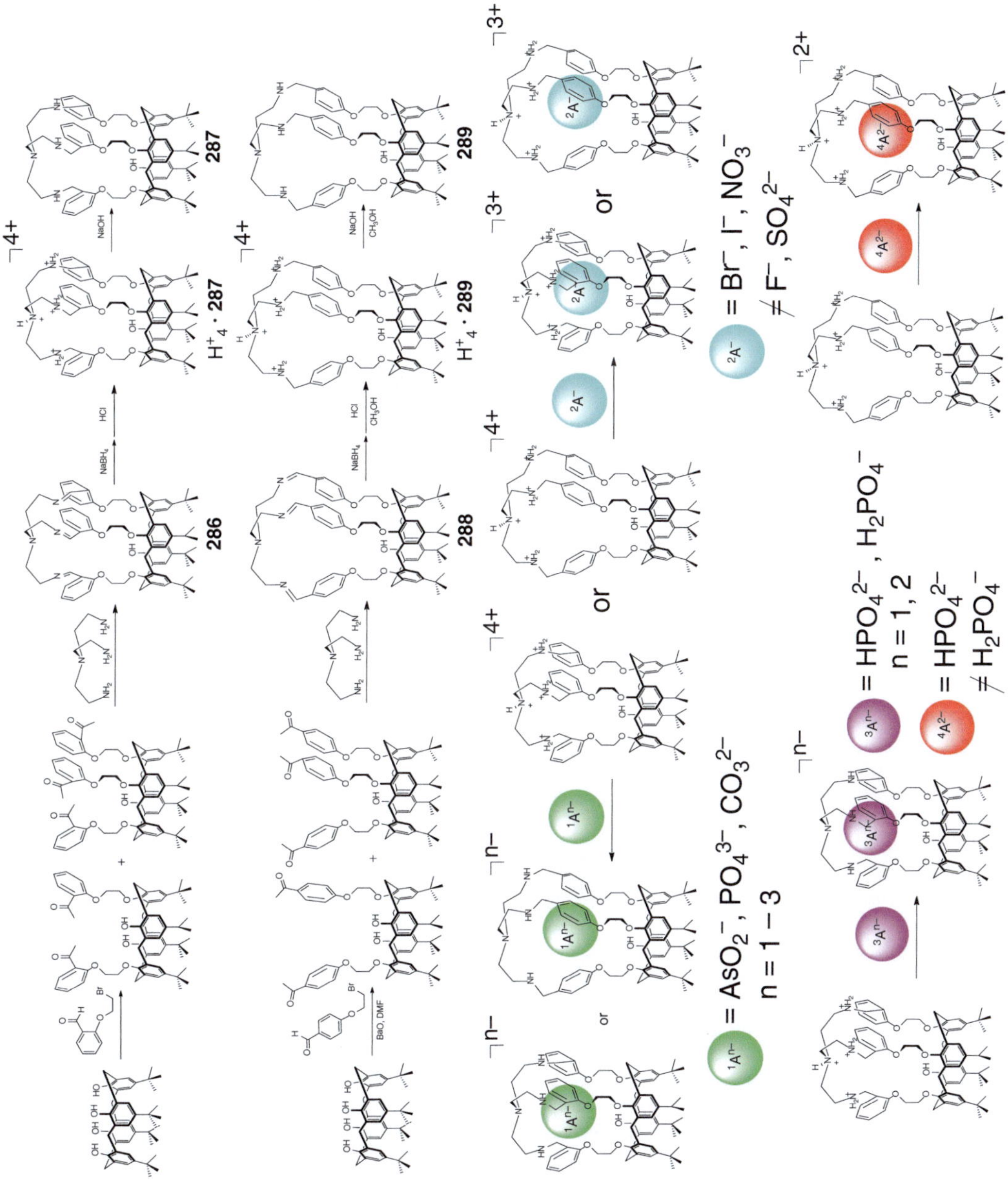
NaBH4
HCl
NaOH
CH3OH
BaO, DMF
286
287
288
289
H+4 · 287
H+4 · 289
or
= AsO2−, PO43−, CO32−
n = 1 − 3
= Br−, I−, NO3−
≠ F−, SO42−
= HPO42−, H2PO4−
n = 1, 2
= HPO42−
≠ H2PO4−

Scheme 2.181

References

1. Melson GA (ed) (1980) Coordination chemistry of macrocyclic compounds. Plenum Press, New York
2. Gokel GW, Korzeniowski SH (1982) Macrocyclic polyether chemistry. Springer, Berlin
3. Hiraoka M (1982) Crown compounds: their characteristics and applications. Elsevier, Amsterdam
4. Vögtle F, Weber E (eds) (1985) Host-guest complex chemistry: macrocycles. Synthesis, structures, application. Springer, Heidelberg
5. Izatt RM, Christensen JJ (eds) (1987) Synthesis of macrocycles. The design of selective complexing agents. Wiley-Interscience, New York
6. Yatsimirskii KB, Kolchinskii AG, Pavlischuk VV, Talanova GG (1987) Synthesis of macrocyclic compounds. Naukova Dumka, Kiev
7. Gokel GW (1991) Crown ethers and cryptands. The Royal Society of Chemistry, Cambridge
8. Parker D (ed) (1996) Macrocycle synthesis: a practical approach. Oxford University Press, Oxford
9. Gerbeleu NV, Arion VB, Burgess FJ (2000) Template synthesis of macrocyclic compounds. Wiley-VCH, Weinheim
10. Diederich F, Stang PJ, Tykwinski RR (2008) Modern supramolecular chemistry: strategies for macrocycle synthesis. Wiley-VCH, Weinheim
11. Fitzpatrick DW, Ulrich HJ (2010) Macrocyclic chemistry: new research developments. Nova Science Pub, New York
12. Davis F, Higson S (2011) Macrocycles: construction, chemistry and nanotechnology applications. Wiley, Chichester
13. MacDowell D, Nelson J (1988) Facile synthesis of a new family of cage molecules. Tetrahedron Lett 29:385–386
14. Chen D, Martell AE (1991) The synthesis of new binucleating polyaza macrocyclic and macrobicyclic ligands: dioxygen affinities of the cobalt complexes. Tetrahedron 47:6895–6902
15. McKee V, Dorrity MRJ, Malone JF, Marrs D, Nelson J (1992) Novel group 1 cation cryptates: X-ray and ^{23}Na NMR studies. J Chem Soc Chem Commun 383–386
16. Drew MGB, Harding CJ, Howarth OW, Lu Q, Marrs DJ, Morgan GG, McKee V, Nelson J (1996) Thiophene-linked azacryptand sites for dicopper and disilver; thiophene sulfur as an inert spacer J Chem Soc Dalton Trans 3021–3030
17. Drew MGB, Marrs D, Hunter J, Nelson J (1992) Divergent and convergent forms of a new Schiff-base cryptand; X-ray crystallographic and molecular mechanics investigations J Chem Soc Dalton Trans 11–18
18. Drew MG, McDowell D, Nelson J (1988) A new ditopic polyaza macrobicyclic ligand: X-ray crystallographic structure determination. Polyhedron 7:2229–2232
19. Saeed MA, Wong BM, Fronczek FR, Venkatraman R, Hossain MA (2010) Formation of an amine-water cyclic pentamer: a new type of water cluster in a polyazacryptand. Cryst Growth Design 10:1486–1488
20. Saeed MA, Fronczek FR, Hossain MA (2009) Encapsulated chloride coordinating with two in–in protons of bridgeheadamines in an octaprotonated azacryptand. Chem Commun 6409–6411
21. Saeed MA, Fronczek FR, Huang M-J, Hossain MA (2010) Unusual bridging of three nitrates with two bridgehead protons in an octaprotonated azacryptand. Chem Commun 46:404–406
22. Lakshminarayanan PS, Suresh E, Chosh P (2005) Formation of an infinite 2D-layered water of $(H_2O)_{45}$ cluster in a cryptand–water supramolecular complex: a template effect. J Am Chem Soc 127:13132–13133
23. Ma Z, Cao R (2005) An N_8O_6 cryptand: the host for Ag^+ guests and H_2O molecules. J Mol Struct 738:137–142
24. Ebmeyer F, Vögtle F (1989) Selective molecular recognition of trihydroxybenzenes. Angew Chem Int Ed 28:79–81
25. Bing-guang Z, Hong M, Chun-ying D, Cheng H, Qing-jin M, Zhe-ming W, Chun-hua Y (2001) Novel highly symmetrical cube-shaped cation with 16-nitrogen donors. Chem Commun 2652–2653
26. Liu X, Liu Y, Li G, Warmuth R (2006) One-pot, 18-component synthesis of an octahedral nanocontainer molecule. Angew Chem Int Ed 45:901–904
27. Liu Y, Liu X, Warmuth R (2007) Multicomponent dynamic covalent assembly of a rhombicuboctahedral nanocapsule. Chem Eur J 13:8953–8959
28. Francesconi O, Ienco A, Moneti G, Nativi C, Roelens S (2006) A self-assembled pyrrolic cage receptor specifically recognizes β-glucopyranosides. Angew Chem Int Ed 45:6693–6696
29. Barba V, Betanzos I (2007) Direct synthesis of polymacrocyclic boron compounds: a convenient method for the synthesis of hemicarcerands. J Organomet Chem 692:4903–4908
30. De Rycke N, Marrot J, Couty F, David ORP (2010) Synthesis and characterization of hexasubstituted azacryptands. Tetrahedron Lett 51:6521–6525
31. Higa T, Fukui M, Fukui K, Naganuma Y, Kajita Y, Inomata T, Ozawa T, Funahashi Y, Hideki M (2010) Simplified bicyclic cage-type molecule as a C_3-symmetric host: X-ray and FTIR characterization of encapsulation of a nitrile molecule. J Incl Phenom Macrocycl Chem 66:171–177
32. Schleef F, Lüning U (2011) Macrocycle or cage the presence or absence of Ca^{2+} template ions controls the equilibrium in an oligoimine dynamic combinatorial library. Eur J Org Chem 2011:2062–2065
33. Jin Y, Voss BA, Jin A, Long H, Noble RD, Zhang W (2011) Highly CO_2-selective organic molecular cages: what determines the CO_2 selectivity. J Am Chem Soc 133:6650–6658
34. Lydon DP, Campbell NL, Adams DJ, Cooper AI (2011) Scalable synthesis for porous organic cages. Synth Commun 41:2146–2151
35. Jelfs KE, Wu X, Schmidtmann M, Jones JTA, Warren JE, Adams DJ, Cooper AI (2011) Large

self-assembled chiral organic cages: synthesis, structure, and shape persistence. Angew Chem Int Ed 50:10653–10656
36. Mastalerz M (2008) One-pot synthesis of a shape-persistent endo-functionalised nano-sized adamantoid compound. Chem Commun 4756–4758
37. Mastalerz M, Schneider MW, Oppel IM, Presly O (2011) A salicylbisimine cage compound with high surface area and selective CO_2/CH_4 adsorption. Angew Chem Int Ed 50:1046–1051
38. Schneider MW, Oppel IM, Ott H, Lechner LG, Hauswald H-JS, Stoll R, Mastalerz M (2012) Periphery-substituted [4+6] salicylbisimine cage compounds with exceptionally high surface areas: influence of the molecular Structure on nitrogen sorption properties. Chem Eur J 18:836–847
39. Schneider MW, Oppel IM, Mastalerz M (2012) Exo-functionalized shape-persistent [2+3] cage compounds: influence of molecular rigidity on formation and permanent porosity. Chem Eur J 18(14):4156–4160
40. Schneider MW, Lechner LG, Mastalerz M (2012) Uniform porous nanospheres of discrete shape-persistent organic cage compounds. J Mater Chem 22(15):7113–7116
41. Schneider MW, Hauswald H-JS, Stoll R, Mastalerz M (2012) A shape-persistent exo-functionalized [4 + 6] imine cage compound with a very high specific surface area. Chem Commun 48(79):9861–9863
42. Schneider MW, Oppel IM, Griffin A, Mastalerz M (2013) Post-modification of the interior of porous shape-persistent organic cage compounds. Angew Chem Int Ed 52(13):3611–3615
43. Briggs ME, Jelfs KE, Chong SY, Lester C, Schmidtmann M, Adams DJ, Cooper AI (2013) Shape prediction for supramolecular organic nanostructures: [4 + 4] macrocyclic tetrapods. Cryst Growth Des 13:4993–5000
44. Skowronek P, Warżajtis B, Rychlewska U, Gawroński J (2013) Self-assembly of a covalent organic cage with exceptionally large and symmetrical interior cavity: the role of entropy of symmetry. Chem Commun 49:2524–2526
45. Zhang G, Presly O, White F, Oppel IM, Mastalerz M (2014) A permanent mesoporous organic cage with an exceptionally high surface area. Angew Chem Int Ed 53:1516–1520
46. Granzhan A, Riis-Johannessen T, Scopelliti R, Severin K (2010) Combining metallasupramolecular chemistry with dynamic covalent chemistry: synthesis of large molecular cages. Angew Chem Int Ed 49:5515–5518
47. Granzhan A, Schouwey C, Riis-Johannessen T, Scopelliti R, Severin K (2011) Connection of metallamacrocycles via dynamic covalent chemistry: a versatile method for the synthesis of molecular cages. J Am Chem Soc 133:7106–7115
48. Morgan G, McKee V, Nelson J (1995) Caged anions: perchlorate and perfluoroanion cryptates. J Chem Soc Chem Commun 1649–1652
49. Clifford T, Danby A, Llinares JM, Mason S, Alcock NW, Powell D, Aguilar JA, Garcia-Espana E, Bowman-James K (2001) Anion binding with two polyammonium macrocycles of different dimensionality. Inorg Chem 40(18):4710–4720
50. Hynes MJ, Maubert B, McKee V, Town RM, Nelson J (2000) Protonated azacryptate hosts for nitrate and perchlorate. J Chem Soc Dalton Trans 2853–2859
51. Aguilar JA, Clifford T, Danby A, Llinares JM, Mason S, Garcia-Espana E, Bowman-James K (2001) Fluoride ion receptors: a comparison of a polyammonium monocycle versus its bicyclic corollary. Supramol Chem 13:405–417
52. Nelson J, Nieuwenhuyzen M, Pál I, Stephan H, Town RM (2004) Dinegative tetrahedral oxoanion complexation; structural and solution phase observations. Dalton Trans 2303–2308
53. Amendola V, Bonizzoni M, Esteban-Gomez D, Fabbrizzi L, Licchelli M, Sancenon F, Taglietti A (2006) Some guidelines for the design of anion receptors. Coord Chem Rev 250:1451–1470
54. Kang SO, Llinares JM, Day VW, Bowman-James K (2010) Cryptand-like anion receptors. Chem Soc Rev 39:3980–4003
55. Farrell D, Gloe K, Gloe K, Goretzki G, McKee V, Nelson J, Nieuwenhuyzen M, Pál I, Stephan H, Towna RM, Wichmann K (2003) Towards promising oxoanion extractants: azacages and open-chain counterparts. Dalton Trans 1961–1968
56. Amendola V, Alberti G, Bergamaschi G, Biesuz R, Boiocchi M, Ferrito S, Schmidtchen F-P (2012) Cavity effect on perrhenate recognition by polyammonium cages. Eur J Inorg Chem 2012:3410–3417
57. Alberti G, Amendola V, Bergamaschi G, Colleoni R, Milanese C, Biesuz R (2013) Supramolecular receptors in solid phase: developing sensors for anionic radionuclides. Dalton Trans 42:6227–6234
58. Amendola V, Bergamaschi G, Boiocchi M, Alberto R, Braband H (2014) Fluorescent sensing of ^{99}Tc pertechnetate in water. Chem Sci 5:1820–1826
59. Ravikumar I, Lakshminarayanan PS, Suresh E, Ghosh P (2009) Structural studies on encapsulation of tetrahedral and octahedral anions by a protonated octaaminocryptand cage. Beilstein J Org Chem 5. doi:10.3762/bjoc.5.41
60. Carvalho S, Delgado R, Félix V (2010) Evaluation of the binding ability of a macrobicyclic receptor for anions by potentiometry and molecular dynamics simulations in solution. Tetrahedron 66:8714–8721
61. Lamarque L, Navarro P, Miranda C, Aran VJ, Ochoa C, Escarti F, Garcia-Espana E, Latorre J, Luis SV, Miravet JF (2001) Dopamine interaction in the absence and in the presence of Cu^{2+} ions with macrocyclic and macrobicyclic polyamines containing pyrazole units. Crystal structures of $[Cu_2(L1)(H_2O)_2](ClO_4)_4$ and $[Cu_2(H\text{-}1L3)](ClO_4)_3{\cdot}2H_2O$. J Am Chem Soc 123(43):10560–10570
62. Pitarch-Jarque J, Belda R, Garcia-Espana L, Llinares JM, Pan FF, Rissanen K, Navarro P, Garcia-Espana E (2015) From isolated ^{1}H-pyrazole cryptand

anion receptors to hybrid inorganic-organic [1]D helical polymeric anion receptors. Dalton Trans 44(17):7761–7764
63. Mosquera J, Zarra S, Nitschke JR (2014) Aqueous anion receptors through reduction of subcomponent self-assembled structures. Angew Chem Int Ed 53(6):1556–1559
64. Kivansky LM, Koshkakaryan G, Cao D, Liu Y (2009) Linear π-acceptor-templated dynamic clipping to macrobicycles and [2]rotaxanes. Angew Chem Int Ed 48:4185–4189
65. Motekaitis RJ, Martell AE, Lehn JM, Watanabe E (1982) Bis(2,2′,2″-triaminotriethylamine) cryptates of cobalt(II), nickel(II), copper(II), and zinc(II). Protonation constants, formation constants, and hydroxo bridging. Inorg Chem 21:4253–4257
66. Lehn J-M, Pine SH, Watanabe E, Willard AK (1977) Binuclear cryptates. Synthesis and binuclear cation inclusion complexes of bis-tren macrobicyclic ligands. J Am Chem Soc 99:6766–6768
67. Motekaitis R, Martell AE (1988) The dioxygen carrier properties of the dicobalt-obistren cryptate in aqueous solution. Chem Commun 1020–1022
68. Motekaitis R, Martell AE, Murase I (1986) Cascade halide binding by multiprotonated 7,19,30-trioxa-1,4,10,13,16,22,27,33-octaazabicyclo[11.11.11] pentatriacontane (BISTREN) and copper(II) BISTREN cryptates. Inorg Chem 25:938–944
69. Motekaitis R, Martell AE, Murase I, Lehn JM, Hosseini MW (1988) Comparative study of the copper(II) cryptates of C-BISTREN and O-BISTREN. Protonation constants, formation constants, and secondary anion bridging by fluoride and hydroxide. Inorg Chem 27:3630–3636
70. Drew MGB, Hunter J, Marrs DJ, Nelson J, Harding C (1992) Cascade complexes of an octaaza cryptand: co-ordinated azide with linear M-NNN-M geometry. Dalton Trans 3235–3242
71. Lu Q, Latour J-M, Harding CJ, Martin N, Marrs DJ, McKee V, Nelson J (1994) Dicopper cryptates with 1,1 and 1,3 bridging ligands: spectroscopic, magnetic and electrochemical properties. J Chem Soc Dalton Trans 1471–1478
72. Harding CJ, Jin Q, Malone JF, Marrs DJ, Martin N, McKee V, Nelson J (1995) Hydrolytically-sensitive hexaimino and hydrolytically-inert octaaminocryptand hosts for dicopper. Dalton Trans 1739–1747
73. Escuer A, Harding CJ, Dussart Y, Nelson J, McKee V, Vicente R (1999) Constrained ferromagnetic coupling in dinuclear $\mu_{1,3}$-azido nickel(II) cryptate compounds. Crystal structure and magnetic behaviour of $[Ni_2(L1)(N_3)(H_2O)][CF_3SO_3]_3{\cdot}2H_2O{\cdot}EtOH$ {L1 = $N[(CH_2)_2NHCH_2(C_6H_4\text{-}m)CH_2NH(CH_2)_2]_3N$}. Dalton Trans 223–228
74. Fabbrizzi L, Faravelli I (1998) A fluorescent cage for anion sensing in aqueous solution. Chem Commun 971–972
75. Amendola V, Bastianello E, Fabbrizzi L, Mangano C, Pallavicini P, Perotti A, Lanfredi AM, Ugozzoli F (2000) Halide-ion encapsulation by a flexible dicopper(II) bis-tren cryptate. Angew Chem Int Ed 39:2917–2920
76. Amendola V, Fabbrizzi L, Mangano C, Pallavicini P, Zema M (2002) A di-copper(II) bis-tren cage with thiophene spacers as receptor for anions in aqueous solution. Inorg Chim Acta 337:70–74
77. Chen Q-Y, Pan Z-Q, Luo Q-H, Zhen L-M, Hu X-L, Wang Z-L, Zhou Z-Y, Yeung C-H (2002) Synthesis, crystal structure and properties of the first trinuclear copper(II) cryptate bridged by an imidazole anion. Dalton Trans 1315–1318
78. Escuer A, McKee V, Nelson J, Ruiz E, Sanz N, Vicente R (2004) Ferromagnetic Interaction in $\mu_{1,3}$-cyanamido-derived copper(II) cryptates. Chem Eur J 11:398–405
79. Bond AD, Derossi S, Jensen F, Larsen FB, McKenzie CJ, Nelson J (2005) Squeezing the $[Cu{-}OH{\cdots}H_2O{-}Cu]^{3+}$ bridge by cryptate encapsulation. Inorg Chem 44:5987–5989
80. Farrell D, Harding CJ, McKee V, Nelson J (2006) Effect of methylation on the coordination of copper by small azacryptands; the role of geometrically constrained hydrogen bonding in stabilizing terminally coordinated oxygen species. Dalton Trans 3204–3211
81. Bond AD, Derossi S, Harding CJ, McInnes EJL, McKee V, McKenzie CJ, Nelson J, Wolowska J (2005) Cascade complexation: a single cyano bridge links a pair of Cu(II) cations. Dalton Trans 2403–2409
82. Amendola V, Bergamaschi G, Boiocchi M, Fabbrizzi L, Poggi A, Zema M (2008) Halide ion inclusion into a dicopper(II) bistren cryptate containing 'active' 2,5-dimethylfuran spacers: the origin of the bright yellow colour. Inorg Chim Acta 361:4038–4046
83. Ravikumar I, Suresh E, Ghosh P (2006) A perfect linear Cu–NNN–Cu unit inside the cryptand cavity and perchlorate entrapment within the channel formed by the cascade complex. Inorg Chem 45:10046–10048
84. Christinat N, Scopelliti R, Severin K (2008) Multicomponent assembly of boronic acid based macrocycles and cages. Angew Chem Int Ed 47:1848–1852
85. Derossi S, Farrell DT, Harding CJ, McKee V, Nelson J (2007) N-methylation of macrobicycles reduces encapsulation ability. Dalton Trans 1762–1772
86. Hossain MA, Llinares JM, Mason S, Morehouse P, Powell D, Bowman-James K (2002) Parallels in cation and anion coordination: a new class of cascade complexes. Angew Chem In Ed 41:2335–2338
87. Davis AP, Wareham RS (1998) A tricyclic polyamide receptor for carbohydrates in organic media. Angew Chem In Ed 37:2270–2273
88. Ryan TJ, Lecollinet G, Velasco T, Davis AP (2002) Phase transfer of monosaccharides through noncovalent interactions: selective extraction of glucose by a lipophilic cage receptor. Proc Natl Acad Sci 99:4863–4866
89. Lecollinet G, Dominey AP, Velasco T, Davis AP (2002) Highly selective disaccharide recognition

by a tricyclic octaamide cage. Angew Chem Int Ed 41:4093–4096
90. Klein E, Crump MP, Davis AP (2005) Carbohydrate recognition in water by a tricyclic polyamide receptor. Angew Chem Int Ed 44(2):298–302
91. Barwell NP, Crump MP, Davis AP (2009) A synthetic lectin for β-glucosyl. Angew Chem Int Ed 48:7673–7676
92. Ferrand Y, Klein E, Barwell NP, Crump MP, Jiménez-Barbero J, Vicent C, Boons G-J, Ingale S, Davis AP (2009) A synthetic lectin for O-linked β-N-acetylglucosamine. Angew Chem Int Ed 48:1775–1779
93. Bencini A, Bianchi A, Garcia-Espana E, Fusi V, Micheloni M, Paoletti P, Ramirez JA, Rodriguez A, Valtancoli B (1992) Synthesis and protonation behavior of the macrocyclic ligand 1,4,7,13-tetramethyl-1,4,7,10,13,16-hexaazacyclooctadecane and of its bicyclic derivative 4,7,10,17,23-pentamethyl-1,4,7,10,13,17,23-heptaazabicyclo[11.7.5]pentacosane. A potentiometric and proton and carbon 13 NMR study. J Chem Soci Perkin Trans 2: Phys Org Chem 7:1059–1065
94. Fox OD, Rolls TD, Drew MGB, Beer PD (2001) The binding of difunctional neutral guest molecules by novel bis(tripyrrolyl) cryptands. Chem Commun 1632–1633
95. Takemura H, Hirakawa T, Shinmyozu T, Inazu T (1984) Synthesis and structure of "hexametaxylylenetetramine" a urotropin-like cage compound. Tetrahedron Lett 25:5053–5056
96. Takemura H, Shinmyozu T, Inazu T (1988) Highly symmetrical cage compounds, in which all six, eight or ten lone pairs on nitrogen atoms at bridgehead or intervening pyridine rings are oriented toward the central cavity. Tetrahedron Lett 29:1789–1792
97. Takemura H, Shinmyozu T, Inazu T (1991) Syntheses and properties of highly symmetrical cage compounds: pyridine analogs of hexa-m-xylylenetetraamine. J Am Chem Soc 113:1323–1331
98. Metz B, Rosalky JM, Weiss R (1976) Cryptates: X-ray crystal structures of the chloride and ammonium ion complexes of a spheroidal macrotricyclic ligand. J Chem Soc Chem Commun 533b–534
99. Graf E, Lehn JM (1976) Anion cryptates: highly stable and selective macrotricyclic anion inclusion complexes. J Am Chem Soc 98(20):6403–6405
100. Dietrich B, Guilhem J, Lehn JM, Pascard C, Sonveaux E (1984) Molecular recognition in anion coordination chemistry. Structure, binding constants and receptor-substrate complementarity of a series of anion cryptates of a macrobicyclic receptor molecule. Helv Chim Acta 67:91–104
101. Kintzinger JP, Lehn JM, Kauffmann E, Dye JL, Popov AI (1983) Anion coordination chemistry. Chlorine-35 NMR studies of chloride anion cryptates. J Am Chem Soc 105(26):7549–7553
102. Dietrich B, Hosseini MW, Lehn JM, Sessions RB (1985) Synthesis of macrobicyclic polyamines by direct macrobicyclisation via tripode-tripode coupling. Helv Chim Acta 68:289–299
103. Hosseini MW, Lehn JM (1988) Anion-receptor molecules: macrocyclic and macrobicyclic effects on anion binding by polyammonium receptor molecules. Helv Chim Acta 71:749–756
104. Heyer D, Lehn JM (1986) Anion coordination chemistry – synthesis and anion binding features of cyclophane type macrobicyclic anion receptor molecules. Tetrahedron Lett 27:5869–5872
105. Dietrich B, Lehn JM, Guilhem J, Pascard C (1989) Anion receptor molecules: synthesis of an octaazacryptand and structure of its fluoride cryptate. Tetrahedron Lett 30:4125–4128
106. Smith PH, Barr ME, Brainard JR, Ford DK, Freiser H, Muralidharan S, Reilly SD, Ryan RR, Silks LA, Yu WH (1993) Synthesis and characterization of two nitrogen-donor cryptands. J Org Chem 58:7939–7941
107. Reilly SD, Khalsa GRK, Forf DK, Brainard JR, Hay BP, Smith PH (1995) Octaazacryptand complexation of the fluoride ion. Inorg Chem 34:569–575
108. Dietrich B, Dilworth B, Lehn JM, Souchez JP, Cesario M, Guilhem J, Pascard C (1996) Anion cryptates: synthesis, crystal structures, and complexation constants of fluoride and chloride inclusion complexes of polyammonium macrobicyclic ligands. Helv Chim Acta 79(3):569–587
109. Kang SO, Llinares JM, Powell D, VanderVelde D, Bowman-James K (2003) New polyamide cryptand for anion binding. J Am Chem Soc 125:10152–10153
110. Kang SO, Hossain MA, Powell D, Bowman-James K (2005) Encapsulated sulfates: insight to binding propensities. Chem Commun 328–330
111. Kang SO, Powell D, Bowman-James K (2005) Anion binding motifs: topicity and charge in amidocryptands. J Am Chem Soc 127:13478–13479
112. Das MC, Bharadwaj PK (2007) Molecular ice with hybrid water–bromide network around a cryptand with a bromide ion included in the cavity to form a host-within-a-host-like structure. Eur J Inorg Chem 2007:1229–1232
113. Bucher C, Zimmerman RS, Lynch V, Sessler JL (2001) First cryptand-like calixpyrrole: synthesis, X-ray structure, and anion binding properties of a bicyclic[3,3,3]nonapyrrole. J Am Chem Soc 123:9716–9717
114. Bucher C, Zimmerman RS, Lynch V, Sessler JL (2003) Synthesis and X-ray structure of a three dimensional calixphyrin. Chem Commun:1646–1647
115. Ambrosi G, Dapporto P, Formica M, Fusi V, Giorgi L, Guerrir A, Micheloni M, Paoli P, Pontellini R, Rossi P (2006) Synthesis of a large amino-phenolic cage. Synthesis, crystal structures, and acid–base and coordination behavior toward cations and anions. Inorg Chem 45:304–314
116. Fiehn T, Goddard R, Seidel RW, Kubik S (2010) A cyclopeptide-derived molecular cage for sulfate ions that closes with a click. Chem Eur J 16:7241–7255
117. Alliger GE, Müller P, Cummins CC, Nocera DG (2010) Cofacial dicobalt complex of a binucleating

hexacarboxamide cryptand ligand. Inorg Chem 49:3697–3699
118. Alliger GE, Müller P, Do LH, Cummins CC, Nocera DG (2011) Family of cofacial bimetallic complexes of a hexaanionic carboxamide cryptand. Inorg Chem 50:4107–4115
119. Young PG, Clegg JK, Bhadbhade M, Jolliffe KA (2011) Hybrid cyclic peptide-thiourea cryptands for anion recognition. Chem Commun 47:463–465
120. Schmidtchen FP (1977) Inclusion of anions in macrotricyclic quaternary ammonium salts. Angew Chem Int Ed 16:720–721
121. Schmidtchen FP (1981) Macrocyclische quartäre ammoniumsalze. Chem Ber 114(2):597–607
122. Schmidtchen FP, Müller G (1984) Anion inclusion without auxiliary hydrogen bonds: X-ray structure of the iodide cryptate of a macrotricyclic tetra-quaternary ammonium receptor. J Chem Soc Chem Commun 1115–1116
123. Worm K, Schmidtchen FP (1995) Molecular recognition of anions by zwitterionic host molecules in water. Angew Chem Int Ed 34:65–66
124. Worm K, Schmidtchen FP, Schier A, Schäfer A, Hesse M (1994) Macrotricyclic borane-amine adducts: the first uncharged synthetic host compounds without Lewis acid character, for anionic guests. Angew Chem Int Ed 33:327–329
125. Martí I, Rubio J, Bolte M, Burguete MI, Vicent C, Quesada R, Alfonso I, Luis SV (2012) Tuning chloride binding, encapsulation, and transport by peripheral substitution of pseudopeptidic tripodal small cages. Chem Eur J 18:16728–16741
126. Bernier N, Carvalho S, Li F, Delgado R, Félix V (2009) Anion recognition by a macrobicycle based on a tetraoxadiaza macrocycle and an isophthalamide head unit. J Org Chem 74:4819–4827
127. Ragunathan KG, Bharadwaj PK (1992) Template synthesis of a cryptand with hetero-ditopic receptor sites. Tetrahedron Lett 33:7581–7584
128. Canceill J, Collet A (1988) Two-step synthesis of D_3 and C_{3h} cryptophanes. J Chem Soc Chem Commun 582–584
129. Fogarty HA, Berthault P, Brotin T, Huber G, Desvaux H, Dutasta J-P (2007) A cryptophane core optimized for xenon encapsulation. J Am Chem Soc 129:10332–10333
130. Cavagnat D, Buffeteau T, Brotin T (2008) Synthesis and chiroptical properties of cryptophanes having C_1-symmetry. J Org Chem 73:66–75
131. Canceill J, Lacombe L, Collet A (1985) Analytical optical resolution of bromochlorofluoromethane by enantioselective inclusion into a Tailor-made "cryptophane" and determination of its maximum rotation. J Am Chem Soc 107:6993–6996
132. Costante-Crassous J, Marrone TJ, Briggs JM, McCammon JA, Collet A (1997) Absolute configuration of bromochlorofluoromethane from molecular dynamics simulation of its enantioselective complexation by cryptophane-C. J Am Chem Soc 119:3818–3823
133. Ruiz EJ, Sears DN, Pines A, Jameson CJ (2006) Diastereomeric Xe chemical shifts in tethered cryptophane cages. J Am Chem Soc 128:16980–16988
134. Traoré T, Delacour L, Garcia-Argote S, Berthault P, Cintrat J-C, Rousseau B (2010) Scalable synthesis of cryptophane-1.1.1 and its functionalization. Org Lett 12:960–962
135. Traore T, Clavé G, Delacour L, Kotera N, Renard P-Y, Romieu A, Berthault P, Boutin C, Tassali N, Rousseau B (2011) The first metal-free water-soluble cryptophane-111. Chem Commun 47:9702–9704
136. Bouchet A, Brotin T, Linares M, Ågren H, Cavagnat D, Buffeteau T (2011) Conformational effects induced by guest encapsulation in an enantiopure water-soluble cryptophane. J Org Chem 76:1372–1383
137. Hill PA, Wei Q, Eckenhoff RG, Dmochowski IJ (2007) Thermodynamics of xenon binding to cryptophane in water and human plasma. J Am Chem Soc 129:9262–9263
138. Hill PA, Wei Q, Troxler T, Dmochowski IJ (2009) Substituent effects on xenon binding affinity and solution behavior of water-soluble cryptophanes. J Am Chem Soc 131:3069–3077
139. Fairchild RM, Joseph AI, Holman KT, Fogarty HA, Brotin T, Dutasta J-P, Boutin C, Huber G, Berthault P (2010) A water-soluble Xe@cryptophane-111 complex exhibits very high thermodynamic stability and a peculiar ^{129}Xe NMR chemical shift. J Am Chem Soc 132:15505–15507
140. Jacobson DR, Khan NS, Collé R, Fitzgerald R, Laureano-Pérez L, Bai Y, Dmochowski IJ (2011) Measurement of radon and xenon binding to a cryptophane molecular host. Proc Natl Acad Sci 108:10969–10973
141. Chaffee KE, Fogarty HA, Brotin T, Goodson BM, Durasta J-P (2009) Encapsulation of small gas molecules by cryptophane-111 in organic solution. 1. Size- and shape-selective complexation of simple hydrocarbons. J Phys Chem A 113:13675–13684
142. Kotera N, Delacour L, Traoré T, Tassali N, Berthault P, Buisson D-A, Dognon J-P, Rousseau B (2011) Design and synthesis of new cryptophanes with intermediate cavity sizes. Org Lett 13:2153–2155
143. Gautier A, Mulatier J-C, Crassous J, Durasta J-P (2005) Chiral trialkanolamine-based hemicryptophanes: synthesis and oxovanadium complex. Org Lett 7:1207–1210
144. Martinez A, Guy L, Dutasta J-P (2010) Reversible, solvent-induced chirality switch in atrane structure: control of the unidirectional motion of the molecular propeller. J Am Chem Soc 132:16733–16736
145. Martinez A, Robert V, Gornitzka H, Dutasta J-P (2010) Controlling helical chirality in atrane structures: solvent-dependent chirality sense in hemicryptophane-oxidovanadium(V) complexes. Chem Eur J 16:520–527
146. Dimitrov-Raytchev P, Perraud O, Aronica C, Martinez A, Dutasta J-P (2010) A new class of C_3-symmetrical hemicryptophane hosts: triamide- and tren-hemicryptophanes. J Org Chem 75:2099–2102
147. Chatelet B, Payet E, Perraud O, Dimitrov-Raytchev P, Chappellet L-L, Dufaud V, Martinez A, Dutasta J-P (2011) Shorter and modular synthesis of hemicryptophane-tren derivatives. Org Lett 13:3706–3709

148. Perraud O, Tommasino J-B, Robert V, Albela B, Khrouz L, Bonneviot L, Dutasta J-P, Martinez A (2013) Hemicryptophane-assisted electron transfer: a structural and electronic study. Dalton Trans 42:1530–1535
149. Perraud O, Robert V, Martinez A, Dutasta J-P (2011) The cooperative effect in ion-pair recognition by a ditopic hemicryptophane host. Chem Eur J 17:4177–4182
150. Perraud O, Robert V, Martinez A, Dutasta J-P (2011) A designed cavity for zwitterionic species: selective recognition of taurine in aqueous media. Chem Eur J 17:13405–13408
151. Perraud O, Martinez A, Dutasta J-P (2011) Exclusive enantioselective recognition of glucopyranosides by inherently chiral hemicryptophanes. Chem Commun 47:5861–5863
152. Dimitrov-Raytchev P, Martinez A, Gornitzka H, Dutasta J-P (2011) Encaging the verkade's superbases: thermodynamic and kinetic consequences. J Am Chem Soc 133:2157–2159
153. Perraud O, Robert V, Gornitzka H, Martinez A, Dutasta J-P (2012) Combined cation–π and anion–π interactions for zwitterion recognition. Angew Chem Int Ed 51:504–508
154. Cangelosi VM, Carter TG, Crossland JL, Zakharov LN, Johnson DW (2010) Self-assembled E_2L_3 cryptands (E = P, As, Sb, Bi): transmetalation, homo- and heterometallic assemblies, and conformational isomerism. Inorg Chem 49:9985–9992
155. Voloshin YZ, Kostromina NA, Krämer R (2002) Clathrochelates: synthesis, structure and properties. Elsevier, Amsterdam
156. Voloshin YZ, Varzatskii OA, Korobko SV, Maletin YA (1998) Macrobicyclic iron (II) oximehydrazonates and α-dioximates formed by capping with antimony (V) triorganyles: the first synthesis of antimony-containing clathrochelates. Inorg Chem Commun 1:328–331
157. Voloshin YZ, Varzatskii OA, Korobko SV, Antipin MY, Vorontsov II, Lyssenko KA, Kochubey DI, Nikitenko SG, Strizhakova NG (2004) Pathways of directed synthesis of iron(II) clathrochelates and polyclathrochelates with non-equivalent capping groups starting from antimony- and germanium-containing precursors. Inorg Chim Acta 357:3187–3204
158. Voloshin YZ, Varzatskii OA, Korobko SV, Chernii VY, Volkov SV, Tomachynski LA, Pehn'o VI, Antipin MY, Starikova ZA (2005) Ditopic macropolycyclic complexes: synthesis of hybrid phthalocyaninoclathrochelates. Inorg Chem 44:822–824
159. Voloshin YZ, Varzatskii OA, Tomilova LG, Breusova MO, Magdesieva TV, Bubnov YN, Krämer R (2007) First hybrid oximehydrazonate phthalocyaninoclathrochelates: the synthesis and properties of lutetium phthalocyanine-capped, cage iron(II) complexes. Polyhedron 26:2733–2740
160. Voloshin YZ, Varzatskii OA, Belov AS, Starikova ZA, Dolganov AV, Magdesieva TV (2008) New antimony-capped iron(II) and cobalt(III) clathrochelate precursors of the polytopic hybrid cage complexes: synthesis, X-ray structures and electrochemistry. Polyhedron 27:325–334
161. Voloshin YZ, Korobko SV, Dolganov AV, Novikov VV, Vologzhanina AV, Bubnov YN (2011) New types of the germanium-capped clathrochelate iron(II) and cobalt(III) tris-dioximates: the synthesis, structure and electrochemical properties. Inorg Chem Commun 14:1043–1047
162. Huang F, Switek KA, Zakharov LN, Fronczek FR, Slebodnick C, Lam M, Golen JA, Bryant WS, Mason PE, Rheingold AL, Ashraf-Khorassani M, Gibson HW (2005) Bis(m-phenylene)-32-crown-10-based cryptands, powerful hosts for paraquat derivatives. J Org Chem 70:3231–3241
163. Wang F, Zhou Q, Zhu K, Li S, Wang C, Liu M, Li N, Fronczek FR, Huang F (2009) Efficient syntheses of bis(*m*-phenylene)-26-crown-8-based cryptand/paraquat derivative [2]rotaxanes by immediate solvent evaporation method. Tetrahedron 65:1488–1494
164. Zhu K, Wu L, Yan X, Zheng B, Zhang M, Huang F (2010) Anion-assisted complexation of paraquat by cryptands based on bis(*m*-phenylene)-[32]crown-10. Chem Eur J 16:6088–6098
165. Yen M-L, Chen N-C, Lai C-C, Liu Y-H, Peng S-M, Chiu S-H (2011) Controlling the rotation of a complexed guest within a molecular cage. Dalton Trans 40:2163–2166
166. Pascanu V, Cîrcu M, Socaci C, Terec A, Soran A, Grosu I (2013) Synthesis of cryptands with di-yne units via acetylenic homocoupling reactions of C_3 tripodands. Tetrahedron Lett 54:6133–6136
167. Bryant WS, Jones JW, Mason PE, Guzei I, Rheingold AL, Fronczek FR, Nagvekar DS, Gibson HW (1999) A new cryptand: synthesis and complexation with paraquat. Org Lett 1(7):1001–1004
168. Avellaneda A, Valente P, Burgun A, Evans JD, Markwell-Heys AW, Rankine D, Nielsen DJ, Hill MR, Sumby CJ, Doonan CJ (2013) Kinetically controlled porosity in a robust organic cage material. Angew Chem Int Ed 52:3746–3749
169. Fairchild RM, Holman KT (2005) Selective anion encapsulation by a metalated cryptophane with a π-acidic interior. J Am Chem Soc 127:16364–16365
170. Huang F, Slebodnick C, Switek KA, Gibson HW (2006) Bis(*meta*-phenylene)-32-crown-10-based cryptand/diquat inclusion [2]complexes. Chem Commun 1929–1931
171. Huang F, Zhou L, Jones JW, Gibson HW, Ashraf-Khorassani M (2004) Formation of dimers of inclusion cryptand/paraquat complexes driven by dipole-dipole and face-to-face [small pi]-stacking interactions. Chem Commun 2670–2671
172. Zhang X, Zhai C, Li N, Liu M, Li S, Zhu K, Zhang J, Huang F, Gibson HW (2007) Host size effect in the complexation of two bis(m-phenylene)-32-crown-10-based cryptands with a diazapyrenium salt. Tetrahedron Lett 48:7537–7541
173. Zhang M, Zheng B, Xia B, Zhu K, Wu C, Huang F (2010) Synthesis of a bis(1,2,3-phenylene) cryptand and its dual-response binding to paraquat and diquat. Eur J Org Chem 2010:6804–6809
174. Xu Z, Huang X, Liang J, Zhang S, Zhou S, Chen M, Tang M, Jiang L (2010) Efficient syntheses of novel cryptands based on bis(m-phenylene)-26-crown-8 and their complexation with paraquat. Eur J Org Chem 2010:1904–1911

175. Xu Z, Jiang L, Feng Y, Zhang S, Liang J, Pan S, Yang Y, Yang D, Cai Y (2011) One-pot synthesis of donor-acceptor [2]rotaxanes based on cryptand-paraquat recognition motif. Org Biomol Chem 9:1237–1243
176. Liu M, Yan X, Hu M, Chen X, Zhang M, Zheng B, Hu X, Shao S, Huang F (2010) Photoresponsive host–guest systems based on a new azobenzene-containing crytpand. Org Lett 12:2558–2561
177. Yan X, Chi X, Wei P, Zhang M, Huang F (2012) [*n*] Pseudorotaxanes ($n = 2, 3$) from self-assembly of two cryptands and a 1,2-bis(4-pyridinium)ethane derivative. Eur J Org Chem 2012:6351–6356
178. Wu X, Li J, Yan X, Zhou Q (2013) Crown ether-based cryptand/tropylium cation inclusion complexes. Tetrahedron 69:9573–9579
179. Canceill J, Collet A, Gabard J, Kotzyba-Hibert F, Lehn J-M (1982) Speleands. Macropolycyclic receptor cages based on binding and shaping sub-units. Synthesis and properties of macrocycle cyclotriveratrylene combinations. Preliminary communication. Helv Chim Acta 65:1894–1897
180. Mahoney JM, Beatty AM, Smith BD (2004) Selective solid–liquid extraction of lithium halide salts using a ditopic macrobicyclic receptor. Inorg Chem 43:7617–7621
181. Cram DJ, Karbach S, Kim YH, Baczynskyj L, Kallemeyn GW (1985) Shell closure of two cavitands forms carcerand complexes with components of the medium as permanent guests. J Am Chem Soc 107(8):2575–2576
182. Cram DJ, Karbach S, Kim YH, Baczynskyj L, Marti K, Sampson RM, Kalleymeyn GW (1988) Host-guest complexation. 47. Carcerands and carcaplexes, the first closed molecular container compounds. J Am Chem Soc 110(8):2554–2560
183. Sherman JC, Cram DJ (1989) Carcerand interiors provide a new phase of matter. J Am Chem Soc 111(12):4527–4528
184. Bryant JA, Blanda MT, Vincenti M, Cram DJ (1991) Host-guest complexation. 55. Guest capture during shell closure. J Am Chem Soc 113(6):2167–2172
185. Bryant JA, Blanda MT, Vincenti M, Cram DJ (1990) Shell closures of tetrabenzyl chloride cavitands with tetrabenzylthiol cavitands provide carceplexes in which one or two guest molecules are incarcerated. J Chem Soc Chem Commun 1403–1405
186. Judice JK, Cram DJ (1991) Stereoselectivity in guest release from constrictive binding in a hemicarceplex. J Am Chem Soc 113:2790–2791
187. Cram DJ, Tanner ME, Knobler CB (1991) Host-guest complexation. 58. Guest release and capture by hemicarcerands introduces the phenomenon of constrictive binding. J Am Chem Soc 113:7717–7727
188. Robbins TA, Knobler CB, Bellew DR, Cram DJ (1994) A highly adaptive and strongly binding hemicarcerand. J Am Chem Soc 116:111–122
189. Warmuth R (1997) o-Benzyne: strained alkyne or cumulene–NMR characterization in a molecular container. Angew Chem Int Ed 36:1347–1350
190. Beyeh NK, Valkonen A, Rissanen K (2010) Piperazine bridged resorcinarene cages. Org Lett 12:1392–1395
191. Yoon J, Cram DJ (1997) Chiral recognition properties in complexation of two asymmetric hemicarcerands. J Am Chem Soc 119:11796–11806
192. Wang L, Wang G-T, Zhao X, Jiang X-K, Li Z-T (2011) Hydrogen bonding-directed quantitative self-assembly of cyclotriveratrylene capsules and their encapsulation of C_{60} and C_{70}. J Org Chem 76:3531–3535
193. Chopra N, Sherman JC (1997) A bis(carceplex) from a cyclic tetramer of cavitands. Angew Chem Int Ed 36:1727–1729
194. Chopra N, Naumann C, Sherman JC (2000) Bis-capsules: cooperative reversible encapsulation of two molecules in adjacent separate chambers. Angew Chem Int Ed 39:194–196
195. Mungaroo R, Sherman JC (2002) Formation of a tris-capsule and a tris-carceplex from a cyclic six-bowl assembly. Chem Commun 1671–1673
196. Jabin I, Reinaud O (2003) First C_{3v}-symmetrical calix[6](aza)crown. J Org Chem 68:3416–3419
197. Zeng X, Coquière D, Alenda A, Garrier E, Prangé T, Li Y, Reinaud O, Jabin I (2006) Efficient synthesis of calix[6]tmpa: a new calix[6]azacryptand with unique conformational and host–guest properties. Chem Eur J 12:6393–6402
198. Leontiev AV, Rudkevich DM (2004) Encapsulation of gases in the solid state. Chem Commun 1468–1469
199. Nishimura N, Kobayashi K (2008) Self-assembly of a cavitand-based capsule by dynamic boronic ester formation. Angew Chem Int Ed 47:6255–6258
200. Nishimura N, Yoza K, Kobayashi K (2010) Guest-encapsulation properties of a self-assembled capsule by dynamic boronic ester bonds. J Am Chem Soc 132:777–790
201. Kataoka K, James TD, Kubo Y (2007) Ion pair-driven heterodimeric capsule based on boronate esterification: construction and the dynamic behavior. J Am Chem Soc 129:15126–15127
202. Christinat N (2008) Self-assembly of boron-based supramolecular structures. Dissertation, École Polytechnique Fédérale de Lausanne
203. Chas M, Ballester P (2012) A dissymmetric molecular capsule with polar interior and two mechanically locked hemispheres. Chem Sci 3:186–191
204. Perraud O, Lefevre S, Robert V, Martinez A, Dutasta J-P (2012) Hemicryptophane host as efficient primary alkylammonium ion receptor. Org Biomol Chem 10:1056–1059
205. Li M-J, Huang C-H, Lai C-C, Chiu S-H (2012) Hemicarceplex formation with a cyclotriveratrylene-based molecular cage allows isolation of high-purity (≥99.0%) C_{70} directly from fullerene extracts. Org Lett 14:6146–6149
206. Tuntulani T, Thavornyutikarn P, Poompradub S, Jaiboon N, Ruangpornvisuti V, Chaichit N, Asfari Z, Vicens J (2002) Synthesis of tripodal aza crown ether calix[4]arenes and their supramolecular chemistry with transition-, alkali metal ions and anions. Tetrahedron 58:10277–10285

Encapsulation by Hydrogen-Bonded and Other Supramolecular Capsules

3

Most of the supramolecular capsules are neutral compounds that are self-assembled by hydrogen bonding between complementary organic ligand syntones to form unimolecular, dimeric, tetrameric, or hexameric hydrogen-bonded cage frameworks. In several cases, electrostatic (Coulombic) and stacking interactions have been used for self-assembly of the corresponding supramolecular capsules. The template effect of encapsulated species (mostly of neutral guests or co-guests) has been observed in many systems. Moreover, co-encapsulation of various guests caused the specific types of stereoisomerism (so-called "social isomerism" and "isomeric (diastereomeric) constellation", see Sect. 3.1.1) to appear, which are characteristic of these "soft" capsules. Another important feature of these capsules is the so-called Rebek's rule postulating that the encapsulation proceeds efficiently at the packing coefficient of the guest of approximately 55 % (see below).

3.1 Arene-Based Encapsulating Supramolecular Assemblies

3.1.1 Free Cages and Encapsulation of Neutral Molecules

A novel type of stereoisomerism within the cavity of a calix[4]arene-based caging ligand **290** as a result of different orientations of solvent guests (social isomerism) has been reported for the first time by D.N. Reinhoudt and coworkers in [1]: the corresponding 1:1 cage complex switches between its two isomeric states **A** and **B** (an example for DMA-containing compound is shown in Scheme 3.1).

Reversible formation by Scheme 3.2 of a self-assembled hydrogen-bonded capsule **291** through hybridization of two nonequivalent calix[4]arene ligand syntones having carboxylic acid residues or pyridyl groups on their upper rim is described in [2].

Synthesis and self-assembly of self-complementary calix[4]arene ligand syntone shown in Scheme 3.3 are performed by J. Rebek, Jr. and coworker; this syntone has been designed in [3] to reversibly dimerize and to form an egg-shaped supramolecular capsule **292**. Hydrogen-bonding interactions in this capsule are formed by four self-associating urea fragments giving a cyclic array with 16 hydrogen bonds. The caging ligand **292** is able to encapsulate neutral aromatic guests such as *para*-xylene, ethyl-, and fluorobenzene. An asymmetry of this dimeric capsule **292** is explained [3] by slowed rotation about its aryl–urea bonds; the circular hydrogen-bonded array forces all eight urea groups to orient in the same direction.

Urea-functionalized calix[4]arene syntones shown in Scheme 3.4, fixed in their *cone*-conformation, have been synthesized in [4]. These syntones undergo dimerization and hybridization in apolar solvents, leading to hybrid dimeric capsules with two nonequivalent urea-substituted arene fragments and their homodimeric analogs as well.

Y. Voloshin et al., *The Encapsulation Phenomenon: Synthesis, Reactivity and Applications of Caged Ions and Molecules*, DOI 10.1007/978-3-319-27738-7_3

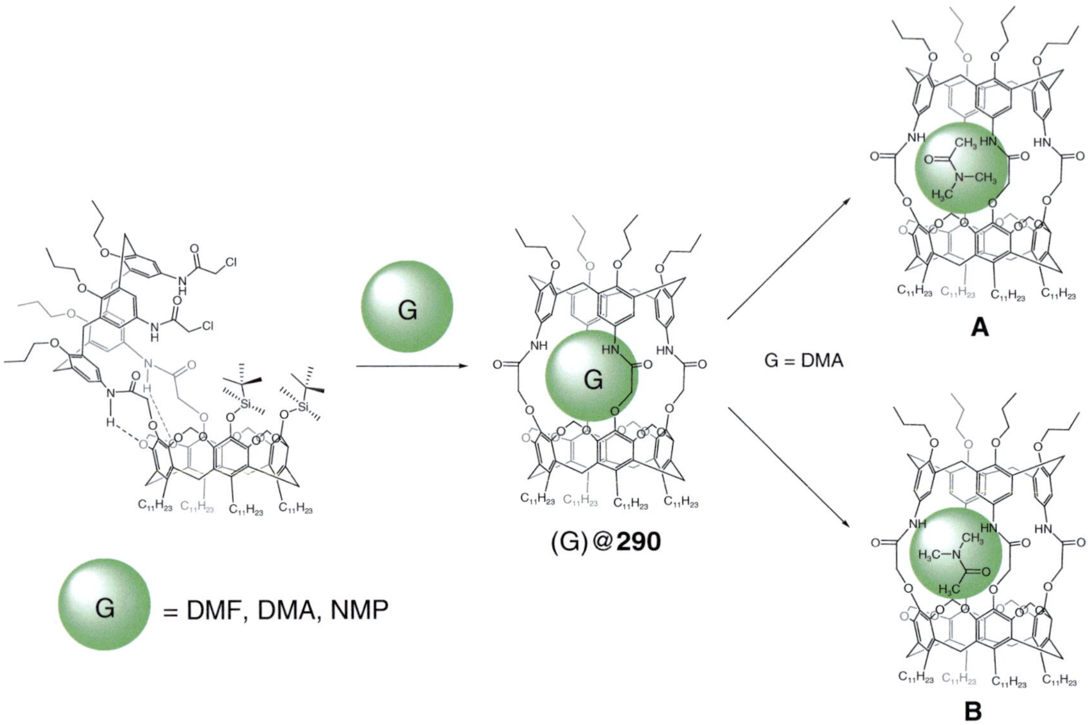

Scheme 3.1

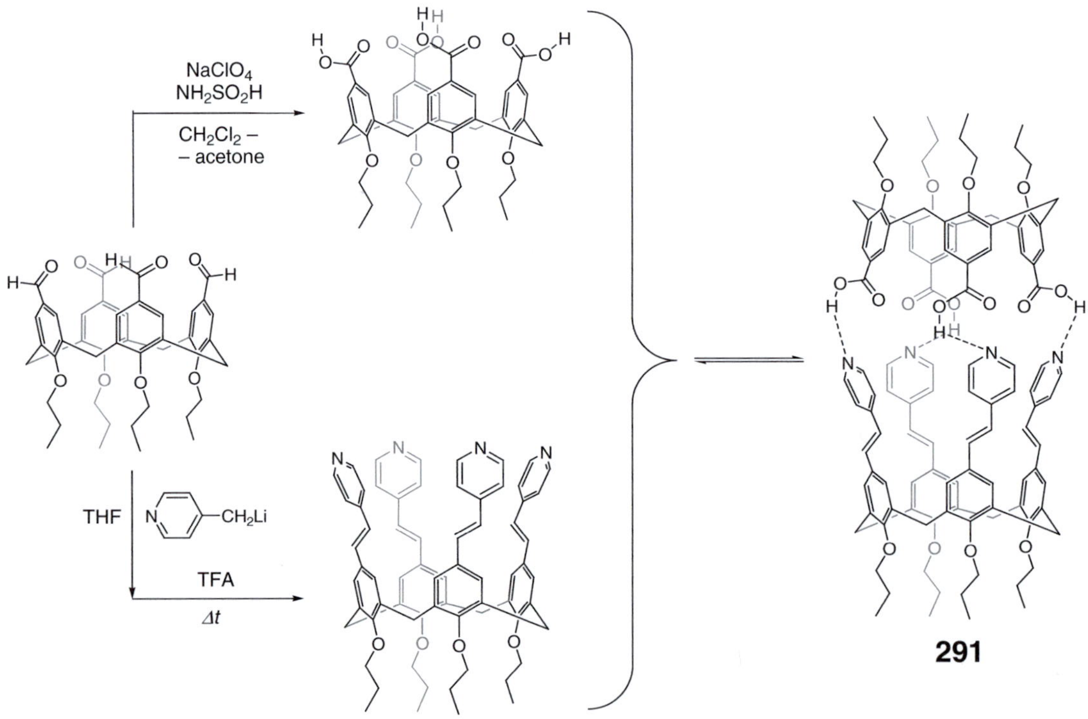

Scheme 3.2

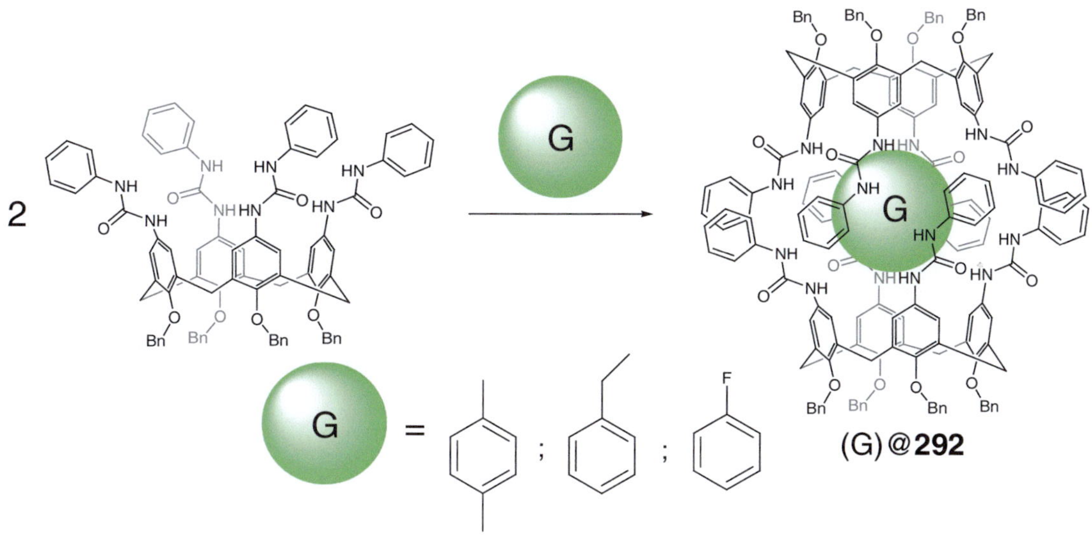

Scheme 3.3

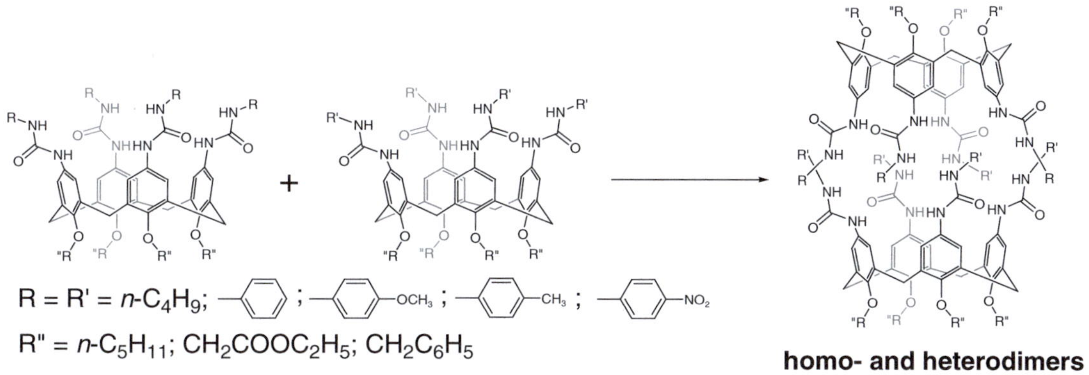

Scheme 3.4

In particular, self-assembly of two calix[4]arene ligand syntones *108* with four pendant urea groups at their upper rim by Scheme 3.5 afforded a hydrogen-bonded dimeric capsule **293**, characterized in [5] by single-crystal X-ray diffraction. Its cage framework with a cavity volume of 202 Å is formed by two crystallographically independent calix[4]arene fragments of a slightly different shape that are hydrogen-bonded through their urea residues; all analogous atoms of their four phenolic units are located at the corners of regular squares [5].

An analogous C_{2v}-symmetric calix[4]arene ligand syntone *109*, having different ether residues attached to the narrow rim, is reported in [6] to form a hydrogen-bonded dimer **294**. Its guest-templated formation by Scheme 3.6 has been induced by the presence of a suitable guest such as deuterobenzene.

Bis(ureido)calix[4]arene ligand syntones in their pinched *cone*-conformation underwent dimerization by Scheme 3.7 giving the hydrogen-bonded capsules **295** and **296** [7].

An upper-rim-functionalized calix[4]arene tetracarboxylic syntone *110* is described in [8] to form in chloroform media the hybrid hydrogen-bonded capsules **297** and **298** with lower-rim-functionalized tetra(4-pyridyl)- and tetra(3-pyridyl)-calix[4]arenes shown in Scheme 3.8.

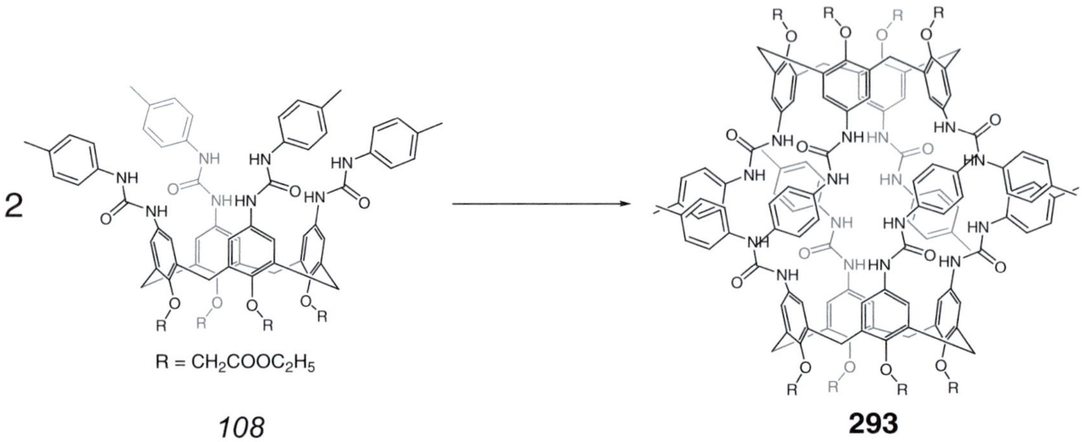

Scheme 3.5

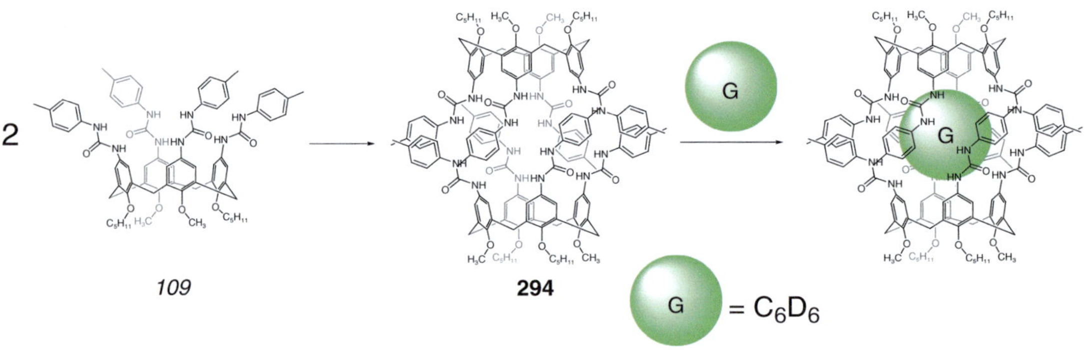

Scheme 3.6

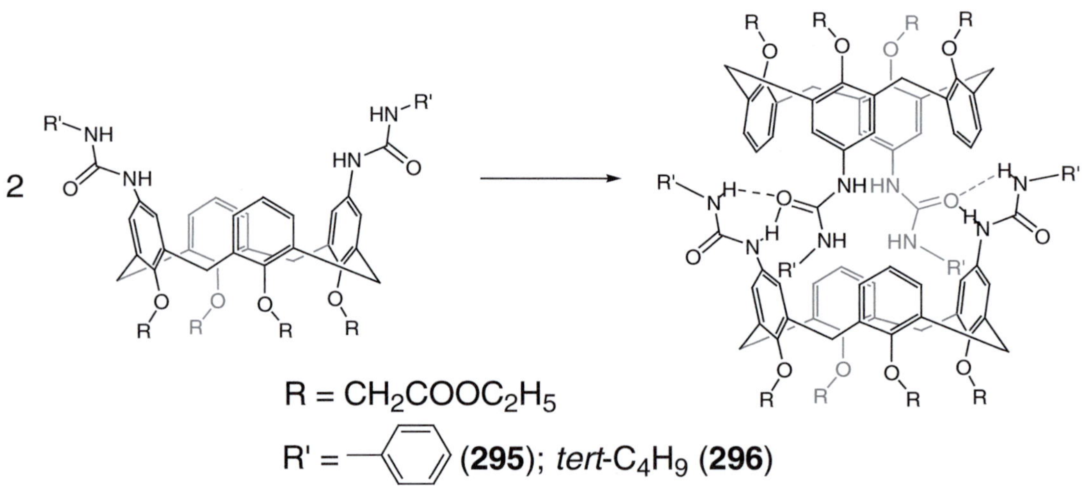

Scheme 3.7

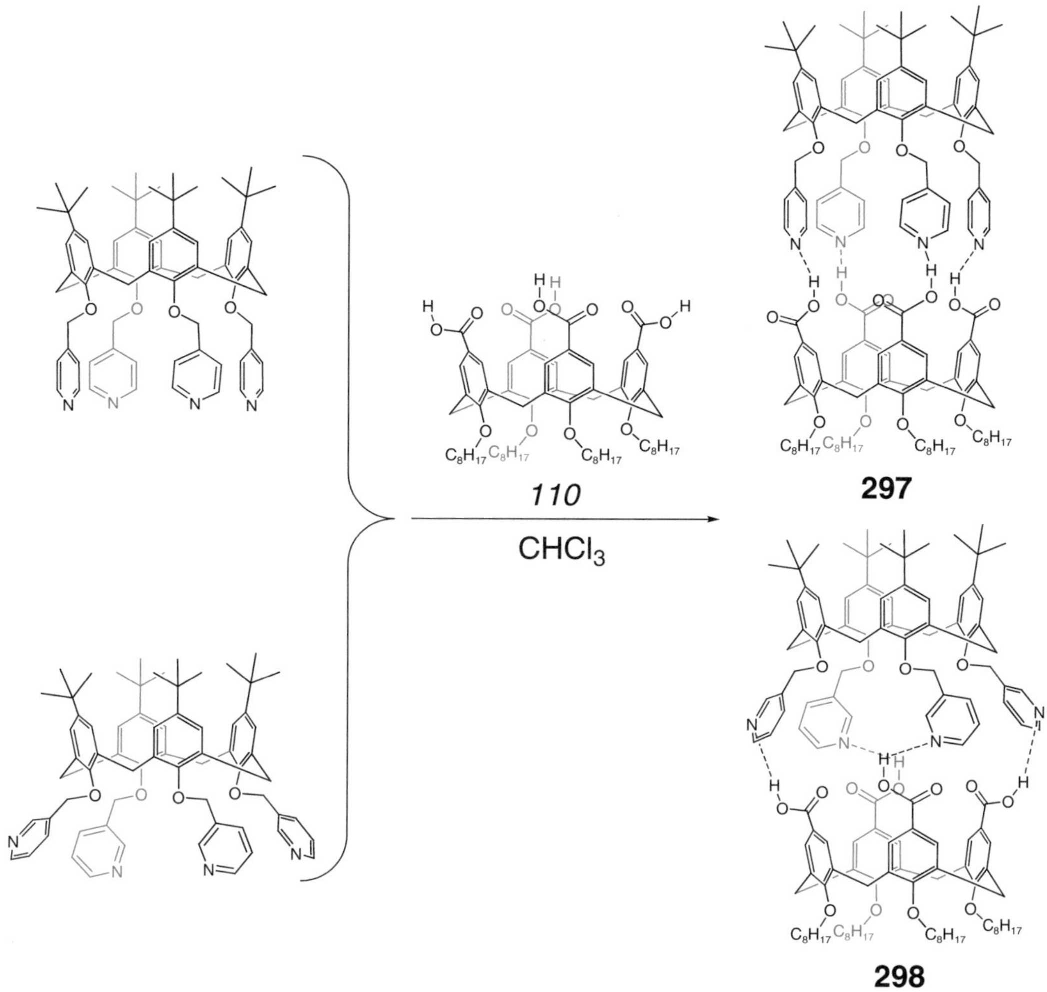

Scheme 3.8

Arene supramolecular capsules **299–302** (Scheme 3.9) have been self-assembled in [9] through multiple electrostatic (Coulombic) interactions between positively charged calix[4]arene ligand syntones and negatively charged calix[4] arene species in polar solvents.

First unimolecular capsule **303**, designed by J. Rebek, Jr and coworkers, has been self-assembled in [10] by Scheme 3.10 using the ligand syntone *111*; different options for its self-assembly are also shown in this Scheme. The hydrogen-bonded caging ligand **303** readily encapsulates ammonium salts, thus providing ion labeling for their cage complexes [10].

The dimeric capsule **304** has been prepared in [11] by Scheme 3.11 through self-assembly of a self-complementary ligand syntone *112*, which has specific features that allow it to dimerize into tennis ball-like cage framework. Cut along its seam, this capsule gives two identical pieces: the ends are complementary to the middle, and the subunits feature curvatures that dictate the overall spherical shape of the dimer with a vacant cavity. Indeed, the lactam functions of its ligand syntone *112* are reported to provide self-complementary hydrogen bond donors and acceptors. The curvature of the molecule along its length is a consequence of folding caused by seven-membered

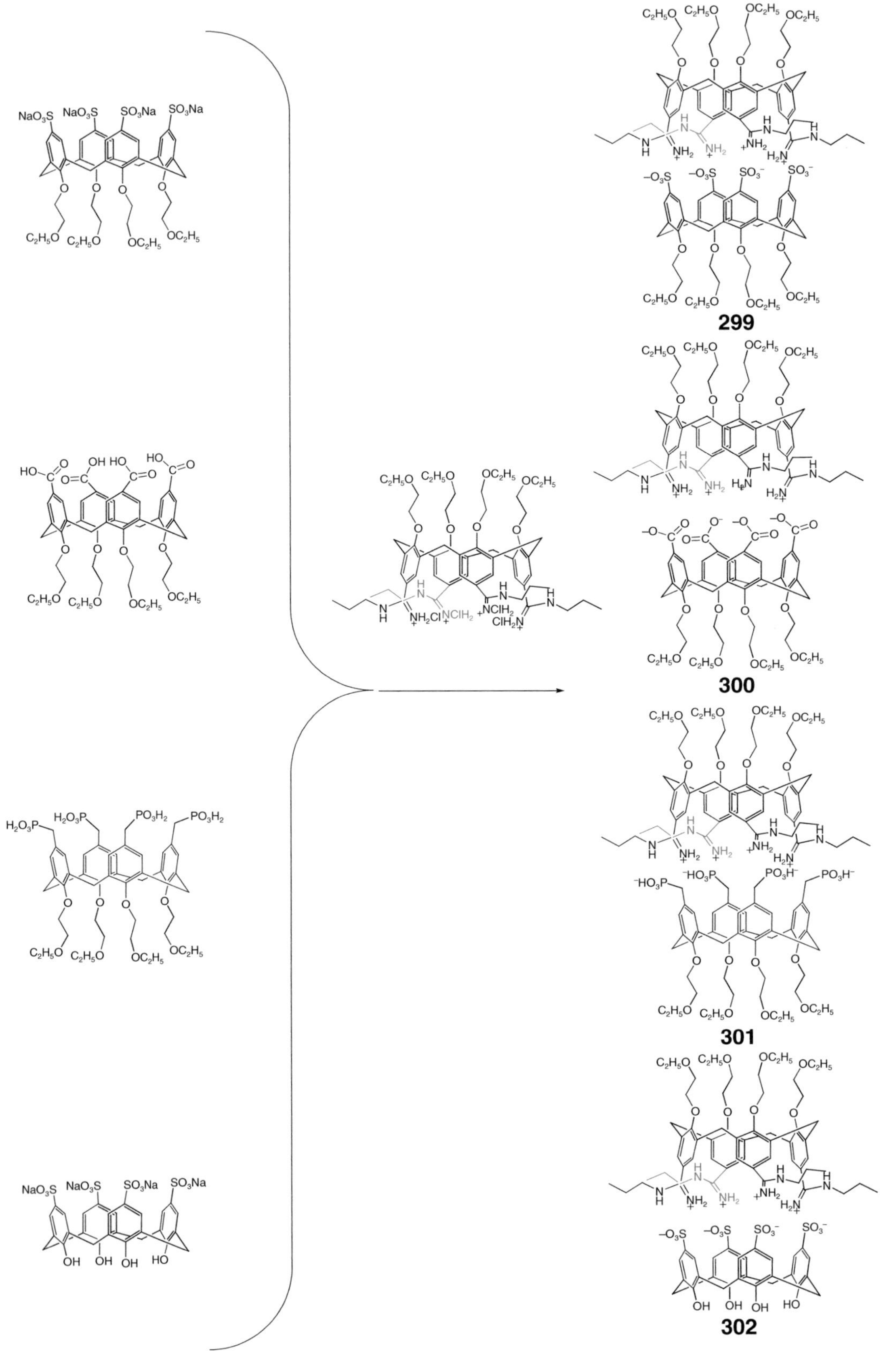

Scheme 3.9

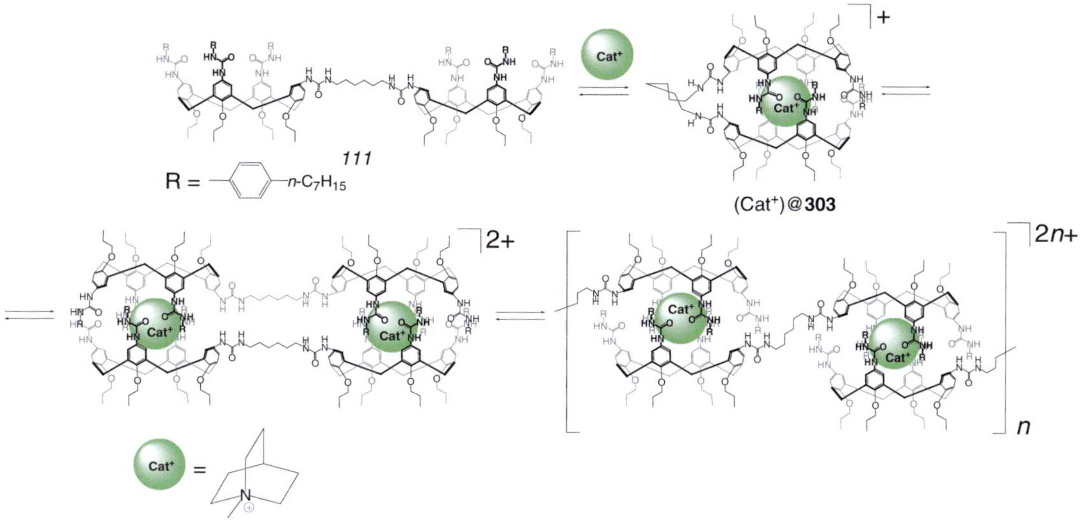

Scheme 3.10

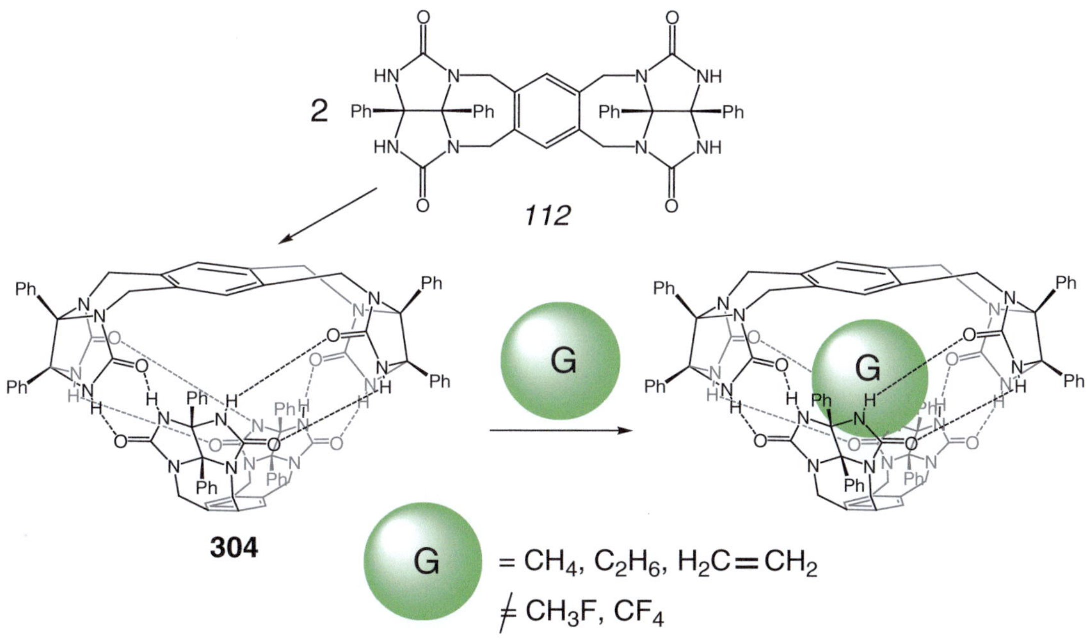

Scheme 3.11

rings when the phenyl units are all on the same face of the molecule. The curvature along its width is caused by *cis*-fusion of five-membered rings. Thus, two self-complementary syntones *112* assemble to form the supramolecular capsule **304** with a closed-shell, three-dimensional surface through a network of hydrogen bonds. As follows from ^{1}H NMR data of [11], this hydrogen-bonded capsule is able to selectively encapsulate few small molecules such as methane, ethane, and ethylene, discriminating other, more bulky gaseous guests shown in Scheme 3.11.

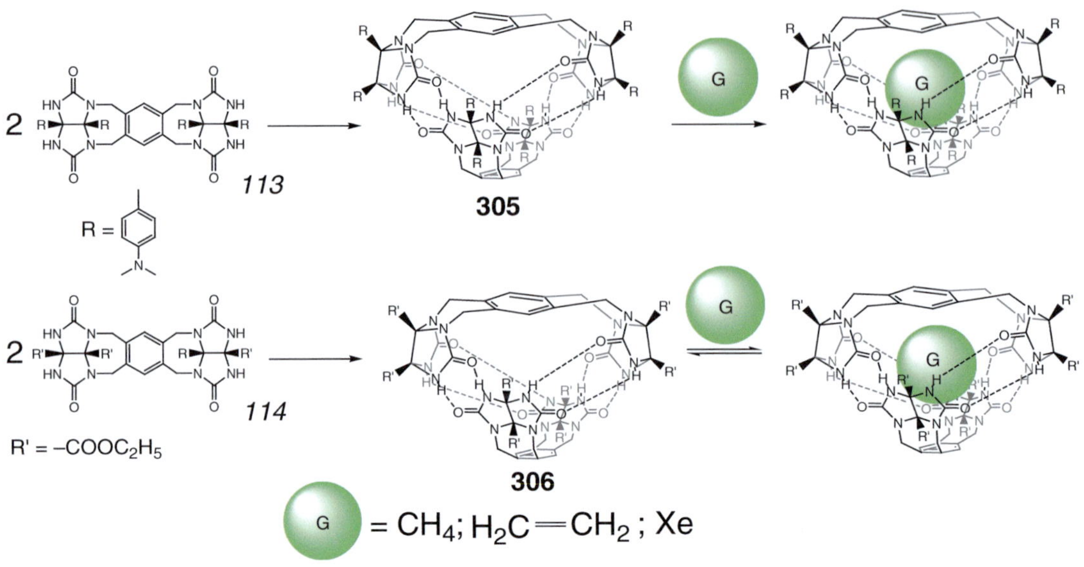

Scheme 3.12

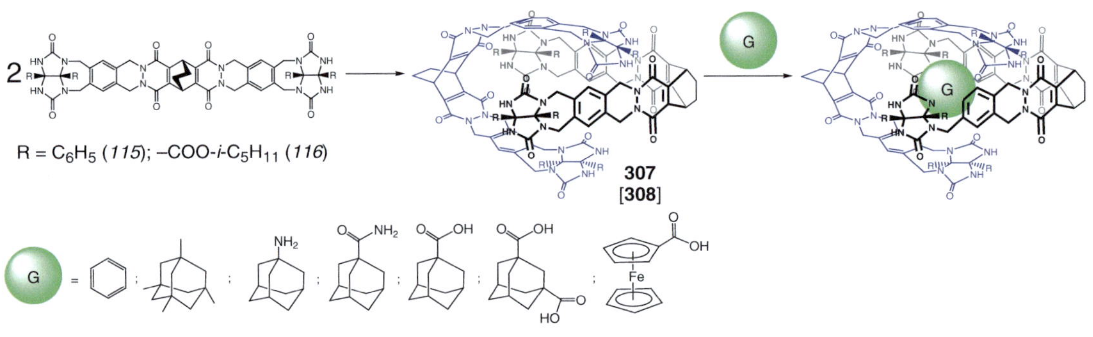

Scheme 3.13

Two self-complementary ligand syntones *113* and *114* are described in [12] to assemble into dimeric capsules **305** and **306** by Scheme 3.12. The size and shape of their cavities are similar to those of **304**; however, peripheral functional groups cause their enhanced solubility. Peripheral basic sites of one of these capsules **305** permit the control over the assembly by protonation–deprotonation reactions. These caging ligands are able to encapsulate methane, ethylene, and xenon guests, forming 1:1 cage complexes; the encapsulation of xenon has been detected in [12] for a capsule **306** using ^{1}H and ^{129}Xe NMR spectra.

Self-assembly of the tetraurea ligand syntones *115* and *116* performed in [13] gave hydrogen-bonded capsules **307** and **308** (Scheme 3.13). Guests of appropriate size and shape such as benzene, adamantanes, and ferrocenes can be encapsulated by these pseudospherical dimeric hosts; those that fit best into their cavities and offer chemical complementarity to **307** and **308** are reported to be preferentially caged by them.

As follows from ^{1}H NMR data of [14] for their cage analog **309**, this self-assembled hydrogen-bonded ligand encapsulates guests shown in Scheme 3.14 in a reversible and entropy-driven process, giving rise to unusual temperature dependence of encapsulation. The positive entropy of this process is explained [14] by caging of more than one solvent molecule into the

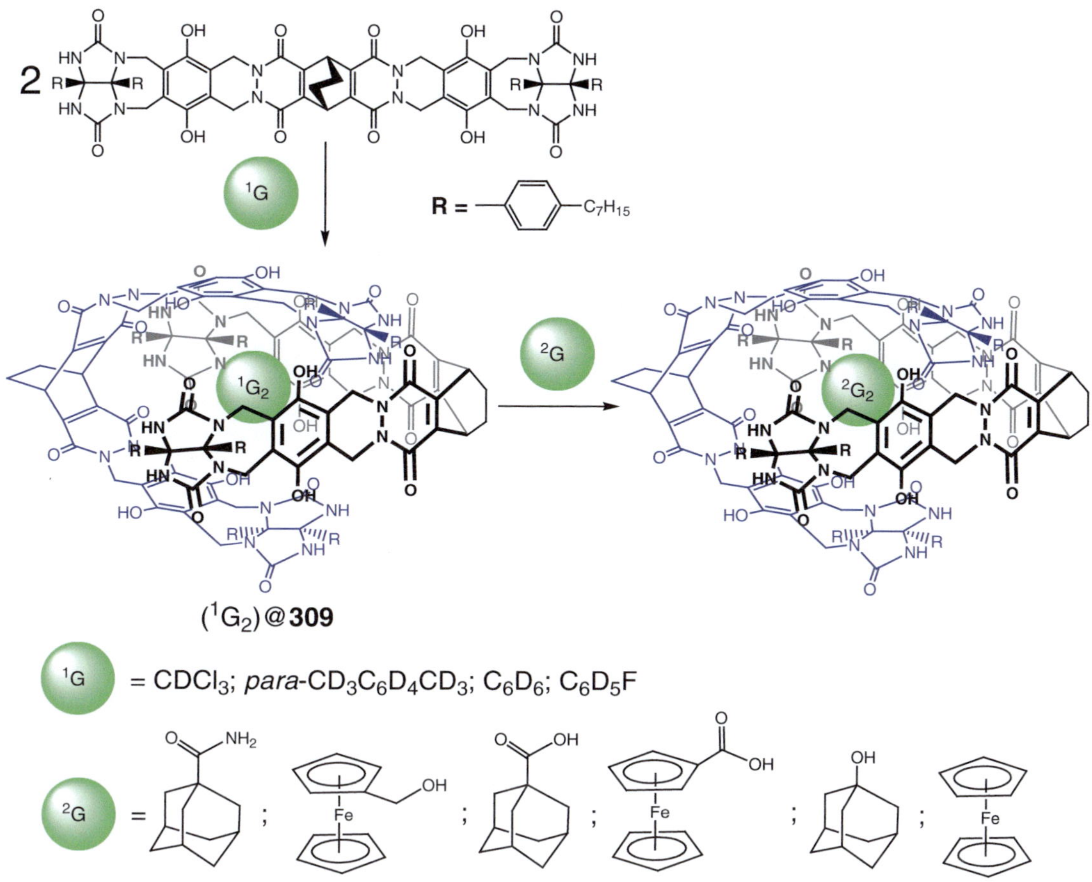

Scheme 3.14

ligand's cavity making their replacement by a single large guest molecule (in particular, adamantane or ferrocene derivatives) entropically favorable [14].

Ligand syntones *117–119*, having glycoluril subunits held apart by various rigid spacers {ethylene (*117*), naphthalene (*118*), and ethenoanthracene (*119*)}, are described in [15] and [16] to assemble through hydrogen bonds by Scheme 3.15 into pseudospherical homo- and heterodimeric supramolecular frameworks. A hydrogen-bonded capsule **306** was studied using X-ray diffraction method. This capsule contains eight almost linear (167–178°) hydrogen bonds with r_i = 2.78–2.89 Å that hold these ligand syntones together; the phenyl spacers and all four methylene groups connecting the glycoluril species are coplanar resulting in an almost tetrahedral geometry of its cage framework. These caging ligands (except of **311**) are able to encapsulate guests in a reversible manner as shown by ^{1}H NMR spectroscopy. Hydrogen-bonded homodimeric capsules undergo disproportionation to give corresponding heterodimeric frameworks, which have been characterized using various NMR techniques; those feature inner cavities of varying sizes and shapes and the formation of hybrid dimers **313–315** (Scheme 3.15) have been observed in solutions containing mixtures of two corresponding homodimeric capsules. Their disproportionation equilibria have been tuned by addition of appropriately sized solvents: even the energetically unlikely hybrid hydrogen-bonded capsules **314** and **315** were found in [15] and [16] to be the dominant species when suitable guests such as $CDBr_3$ are present in solution.

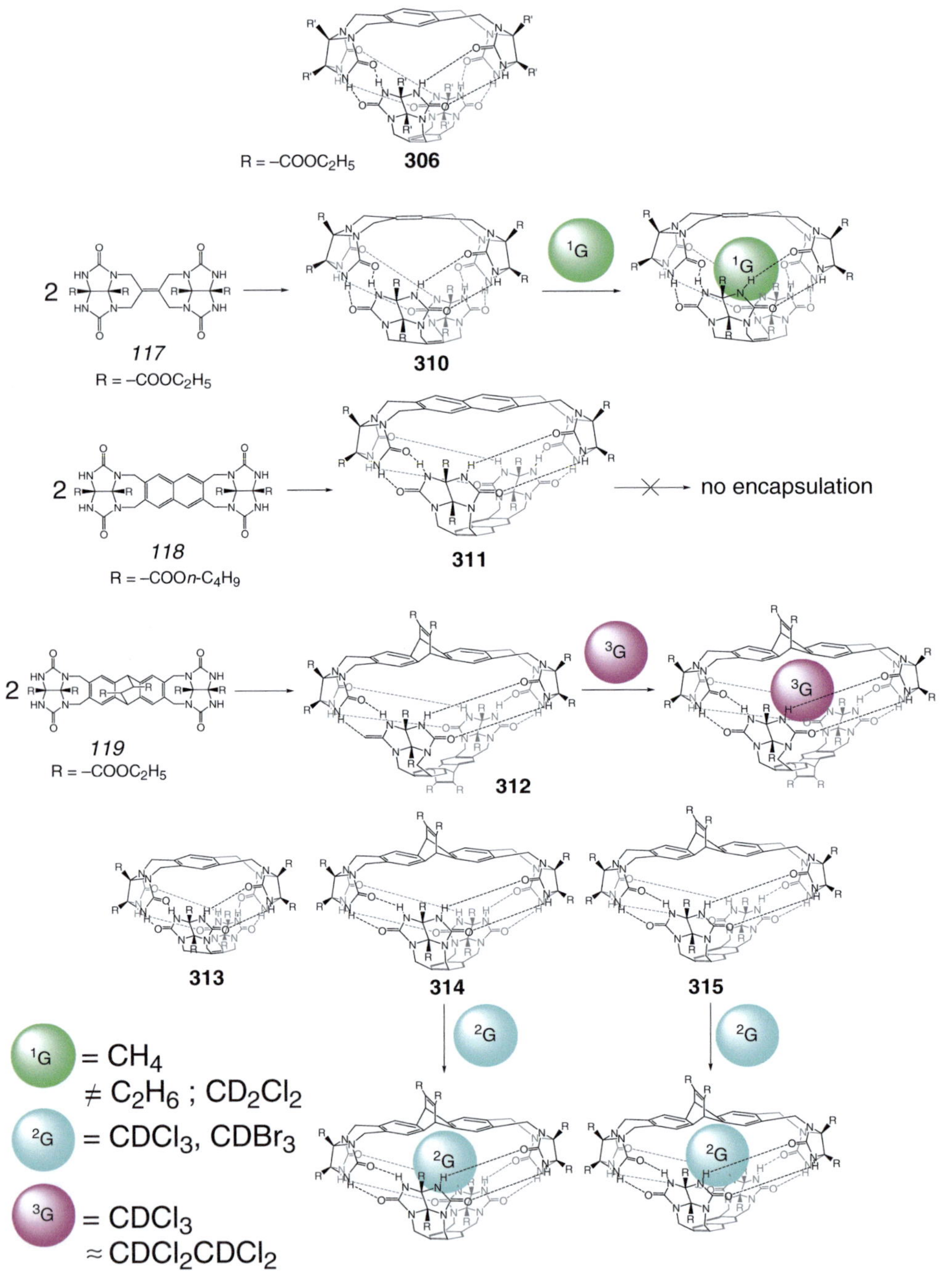

Scheme 3.15

Detailed study of binding selectivity of a dimeric hydrogen-bonded capsule **310** (Scheme 3.16) in solution is performed in [17]. This bis-glycoluril caging ligand, formed by a series of intramolecular hydrogen bonds, is able to discriminate between methane and ethane in $CDCl_3$ medium. Direct evidence for this selectivity is a consequence of host-to-guest size match (the host's cavity is sufficiently small to encapsulate the larger ethane molecule); the

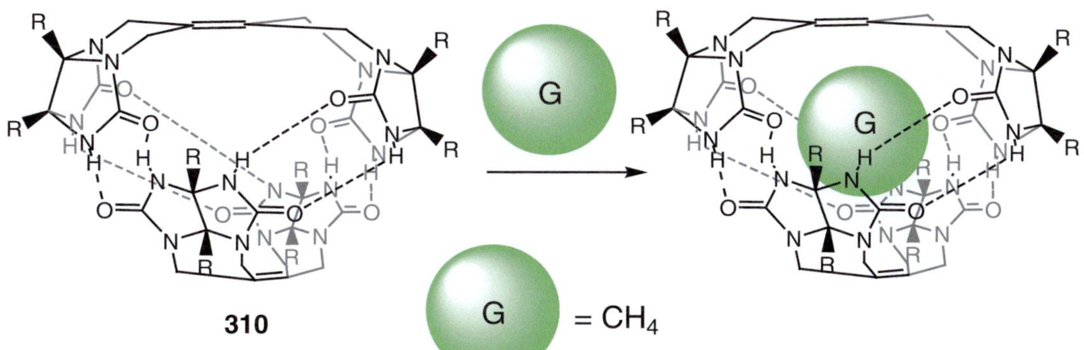

310

R = $-COOC_2H_5$

Scheme 3.16

thermodynamic parameters for the binding of methane were obtained from variable-temperature ^{1}H NMR data.

Analogs of **306** with electron-rich or electron-deficient inner cavities and enhanced solubility in organic media have been prepared in [18] by Scheme 3.17. In addition to self-assembly of the ligand syntones *120–122* into homodimeric hydrogen-bonded capsules **316** and **317**, their disproportionation into corresponding heterodimeric cage frameworks **318–320** has been observed in the mixtures of hydroquinone *121*, quinone *122*, or both with the capsule **306**. These encapsulating ligands form 1:1 cage complexes with methane, ethane, and fluoromethane, but they are unable to bind CF_4. Qualitative evidence for the encapsulation of nitric oxides within the capsule **317** has been obtained in [18].

The threefold symmetric ligand syntone shown in Scheme 3.18 has been designed in [19] for efficient encapsulation of disk-shaped guest molecules. As follows from ^{1}H and ^{13}C NMR data, its hydrogen-bonded derivative **321** binds a suitable solvent molecule discriminating other neutral guests. The driving force for this encapsulation process has an enthalpic nature arising from favorable van der Waals interactions between the convex surface of the guest and the concave surface of the host [19]. At the same time, the loss of entropy due to the decreased translational freedom of the confined guest molecule hinders their encapsulation, and the disk-shaped molecules, which form the multiple host–guest interactions and fill the cavity without a strain, are most suitable guests in this case. Small molecule of $CDCl_3$ has been replaced by benzene within the cavity of **321**, while the larger *para*-xylene guest, which is poorly accommodated by this caging ligand, exchanges readily by cyclohexane that is more complementary for this hydrogen-bonded capsule [19]. The ring inversion dynamics of a caged cyclohexane molecule within the cavity of **321** and its bis-calixarene analog **322** (Scheme 3.19) has been studied in [20] using ^{1}H NMR method; their host–guest C–H…π interactions are reported in this work to raise the ring-inversion barrier of the guest.

Threefold- and tetrafold-symmetric ligand syntones *123–128* have been used in [21] for the design of D_{3d}- and D_{4d}-symmetric hydrogen-bonded capsules and their analogs. Dimerization and hybridization of these syntones by Schemes 3.20 and 3.21 gave homo- and heterodimeric ligands **323–329**, respectively, which form 1:1 cage complexes with various encapsulated mono- and dications shown in these Schemes.

Reversible encapsulation of suitable guests by the fluorine-containing bis-calixarene capsule **330** has been studied in [22]. As follows from ^{1}H and ^{19}F NMR data, this hydrogen-bonded capsule forms 1:1 cage complexes by Scheme 3.22. If the calculated volume of the fluorine-free capsule **331** is approximately 210 Å^3, the shape of the cavity of its fluorinated analog **330** is more complex: each of the calix[4]arene ligand syntones can be represented by a square pyramid with its

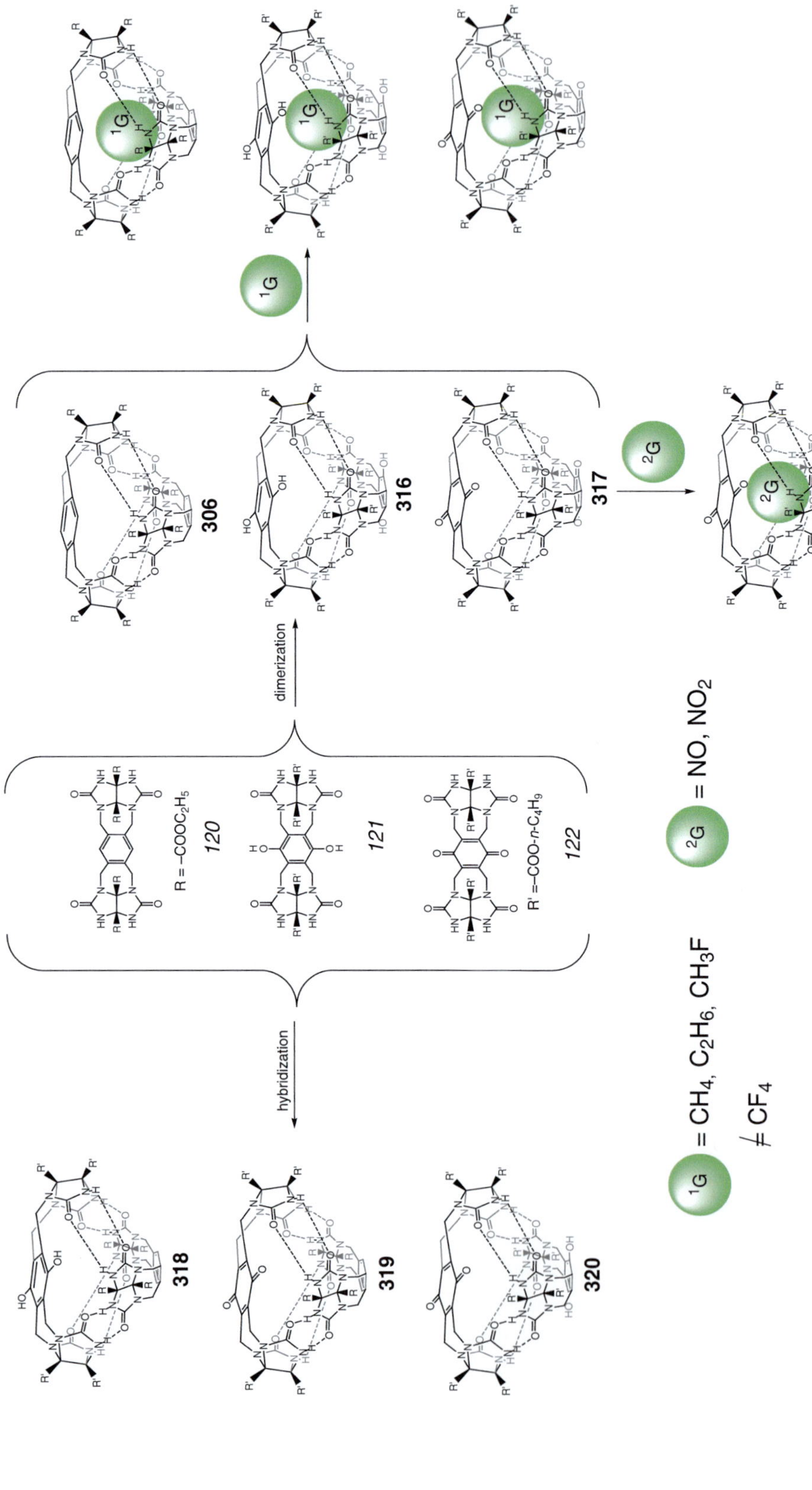

306
316
317
318
319
320
120
121
122
R = –COOC2H5
R' = –COO-n-C4H9
hybridization
dimerization
1G = CH4, C2H6, CH3F
≠ CF4
2G = NO, NO2

Scheme 3.17

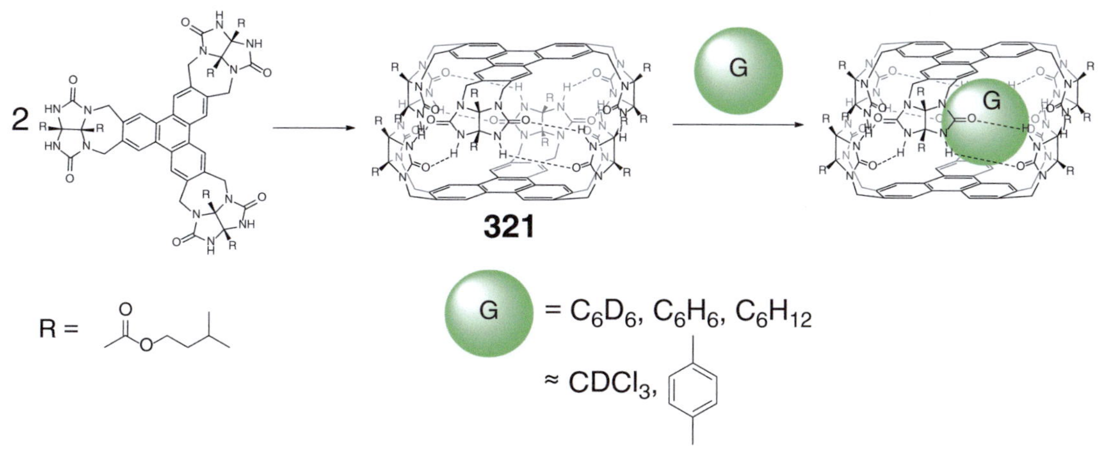

Scheme 3.18

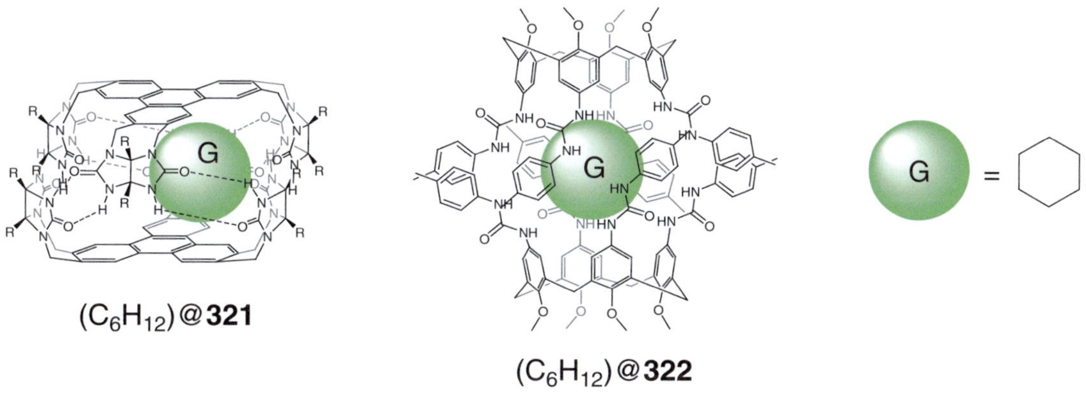

Scheme 3.19

triangular faces corresponding to the phenyl rings of the macrocycle, and after dimerization, these pyramids become offset by 45°. Their corners are cut off to simulate the eight interleaved urea groups forming the equator of the dimeric capsule, while a horizontal slice through the plane of the urea group yields its octagonal cavity. Its vertical slice gives a diamond-shaped inner cavity of **330** with the height of approximately 10 Å and width of 7–8 Å, having different angles and lengths for the top and bottom sections [22].

Asymmetric microenvironments from S_8 to racemic and optically active cavities, induced by appropriate chiral guests within the homo- and heterodimeric calix[4]arene-based capsules **332–337** shown in Scheme 3.23, are described in [23]. These dimeric caging ligands provide a set of increasingly asymmetric environments, but their peripheral asymmetric centers are not well positioned to provide the steric information to encapsulated guests. The reasons for the exclusive formation of these heterodimeric hydrogen-bonded capsules and their encapsulation abilities toward neutral guests with different shapes and sizes have been theoretically and experimentally studied in [24] using MD and ^{1}H NMR methods, respectively.

The calix[4]arene-based ligand syntones *129–133* with either aryl urea or sulfonylurea fragments on the upper rims in their *cone*-conformation gave small hydrogen-bonded capsules **338–342** by Scheme 3.24, while their heterodimeric capsules have been exclusively formed when both aryl- and sulfonylureas-containing syntones are present in solution [25]. All these homo- and heterodi- and tetrameric capsules are able to reversibly encapsulate suitable guests in organic media.

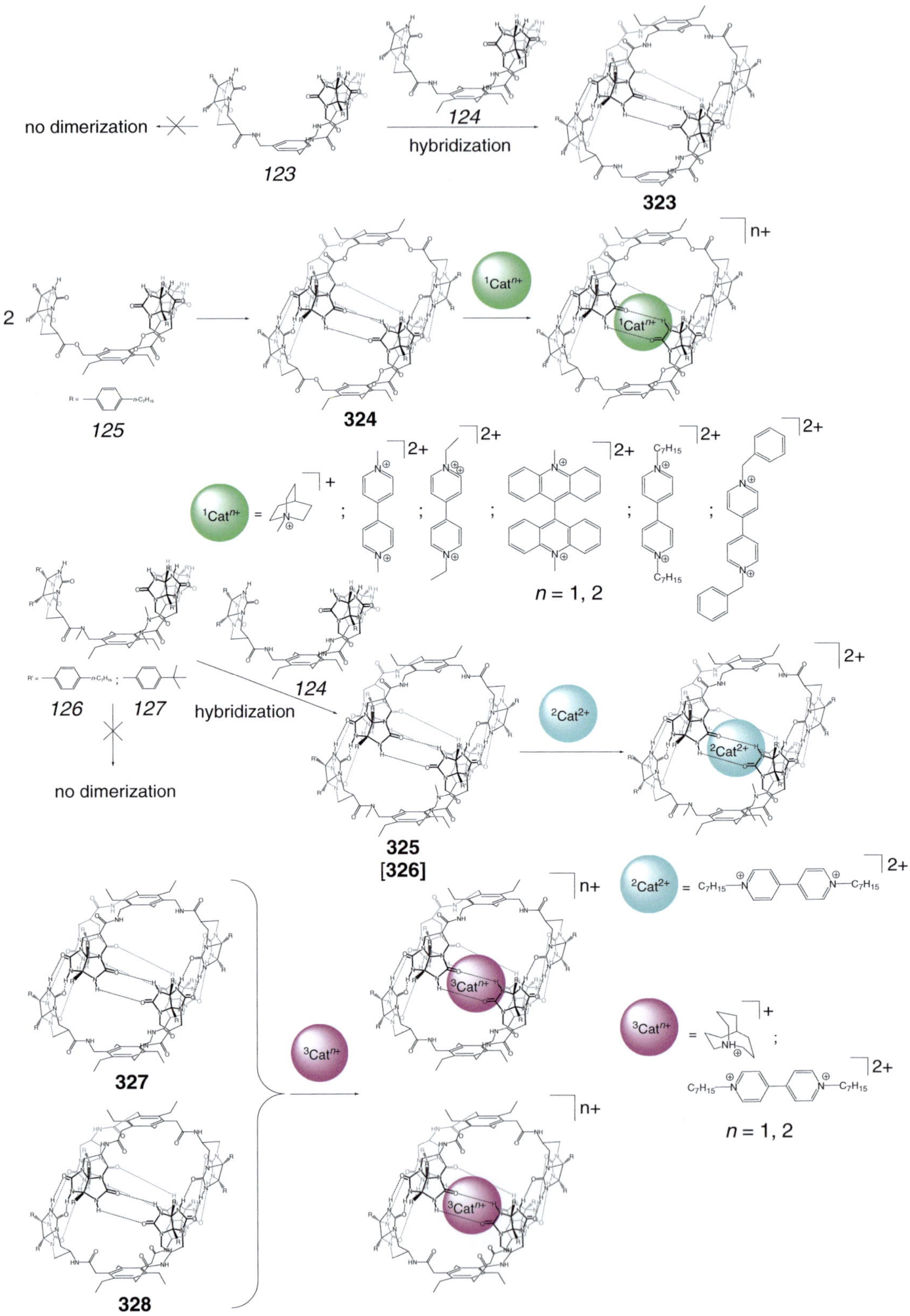

no dimerization
123
124
hybridization
323
2
125
324
$^1Cat^{n+}$
n+
$n = 1, 2$
126
127
no dimerization
124
hybridization
325
[326]
$^2Cat^{2+}$
2+
327
328
$^3Cat^{n+}$
n+
$n = 1, 2$

Scheme 3.20

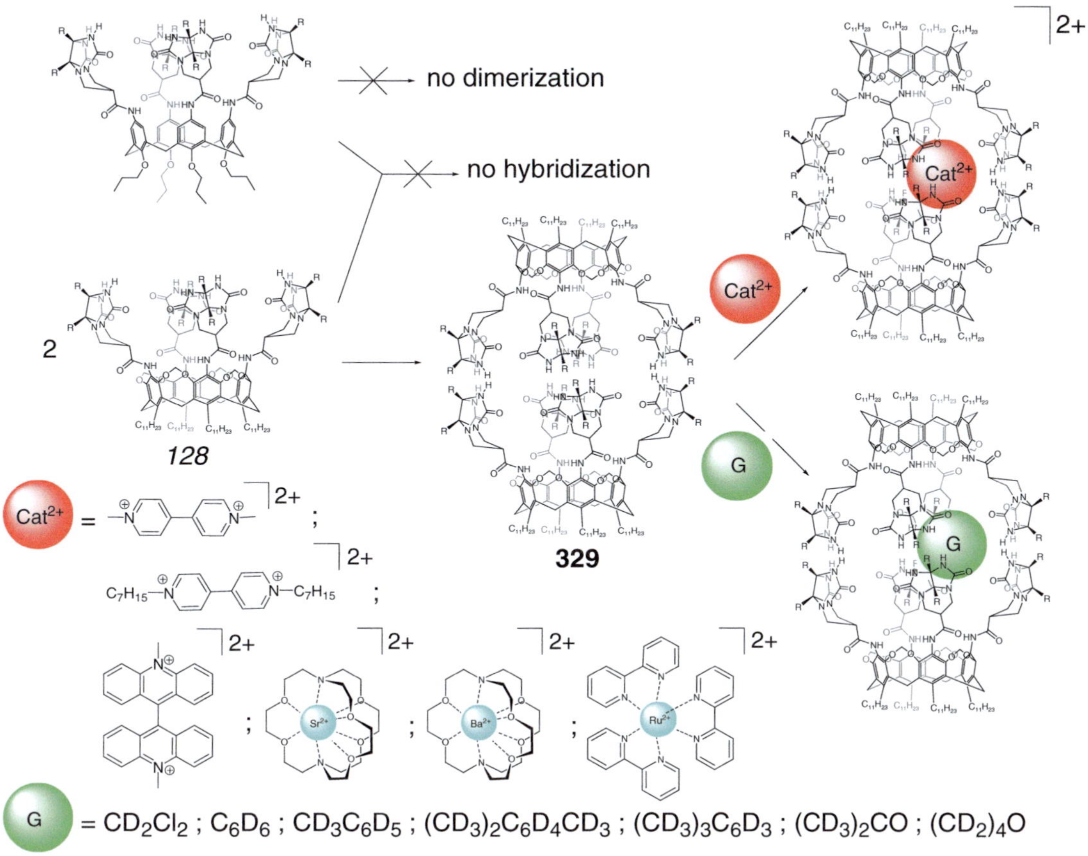

Scheme 3.21

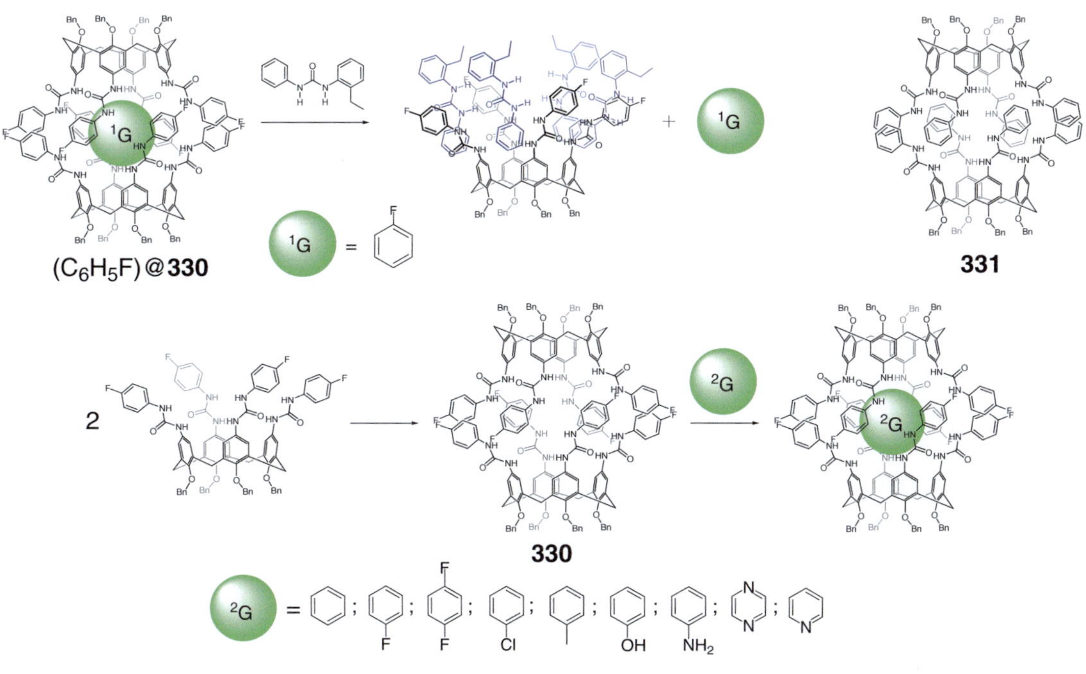

Scheme 3.22

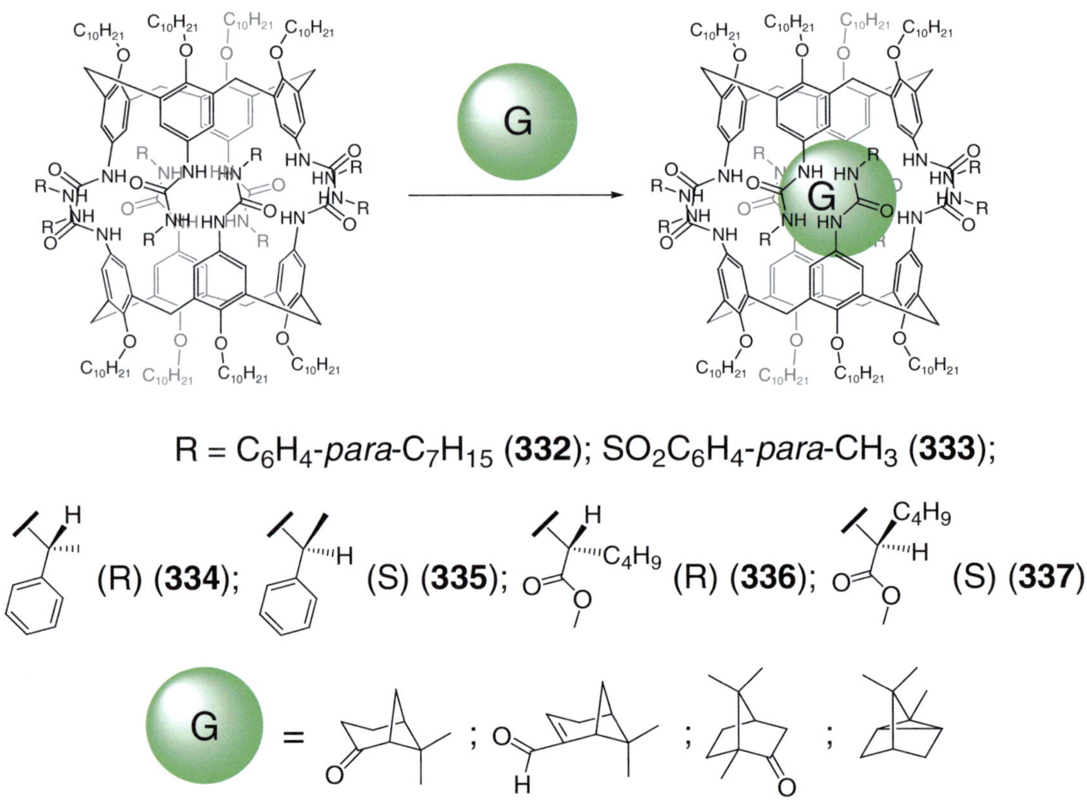

Scheme 3.23

The resorcinarene ligand syntones *134* and *135* having extended pedant substituents and self-complementary hydrogen-bonding sites are reported in [26] to undergo dimerization and hybridization by Scheme 3.25. The resulting dimeric capsules **343–345** encapsulate two heptyl or octyl chains of their ligand syntones thus giving the self-encapsulated cage complexes.

Expanded bis-calix[4]arene homo- and heterodimeric capsules **346–348** have been self-assembled in [27] from an appropriate ligand syntones *136* and *137* by Scheme 3.26. As follows from NMR, CIS, and UV-vis data, the ligand **348** forms 1:1 cage complexes with encapsulated cationic guests.

Binding affinities of the supramolecular capsule **308** (Scheme 3.27) for the reversible encapsulation of neutral guests of suitable size and shape have been studied in [28]. The variable-temperature ^{1}H NMR studies allowed calculating positive enthalpies and entropies of these encapsulation processes to show that they are entropy driven. The caging host in its resting state contains two caged solvent molecules, and the encapsulation of a single large guest liberates these molecules, resulting in an increase in a number of free species. This suggestion has been improved in [28] by experiments in solvent mixtures and MD calculations. Sizable guest such as adamantane or ferrocene derivatives (specifically, 1-adamantanecarboxylic acid, 1,3-dicarboxylic acid, 1-adamantylamine, and ferrocene acid) are reported in [28] to undergo autoencapsulation.

Hydrogen-bonded capsules **307–309**, which accelerate Diels–Alder reactions between dienes and dienofiles, have been prepared in [29–31] using dimerization of the ligand syntones *115*, *116*, and *138* by Schemes 3.28 and 3.29; kinetic and thermodynamic studies of these reactions within the above capsules were performed in [31]. Two solvent molecules occupying their cavities form the corresponding 1:2 cage complexes,

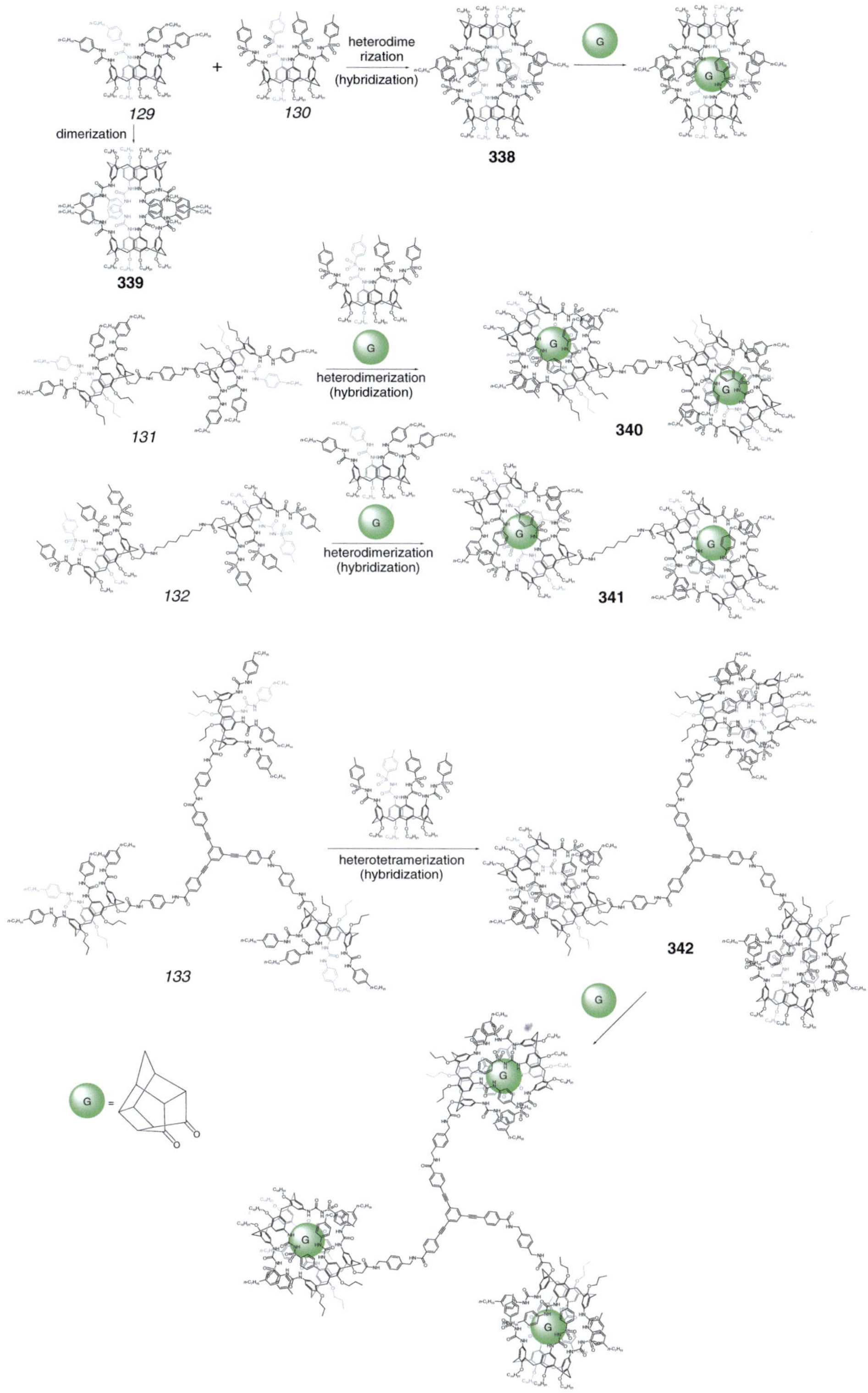
heterodime rization (hybridization)
129
130
338
dimerization
339
heterodimerization (hybridization)
131
340
heterodimerization (hybridization)
132
341
heterotetramerization (hybridization)
133
342
G
G =

Scheme 3.24

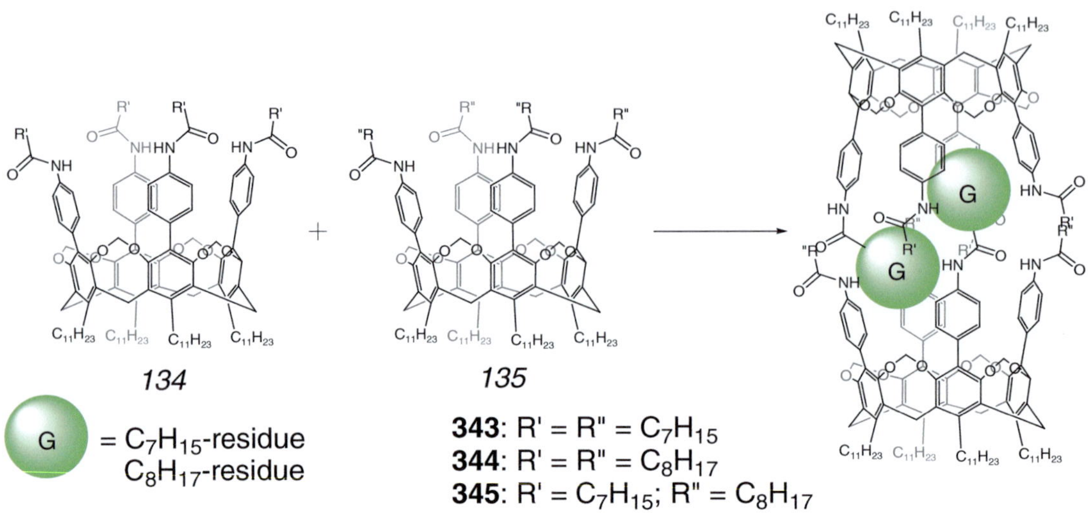

Scheme 3.25

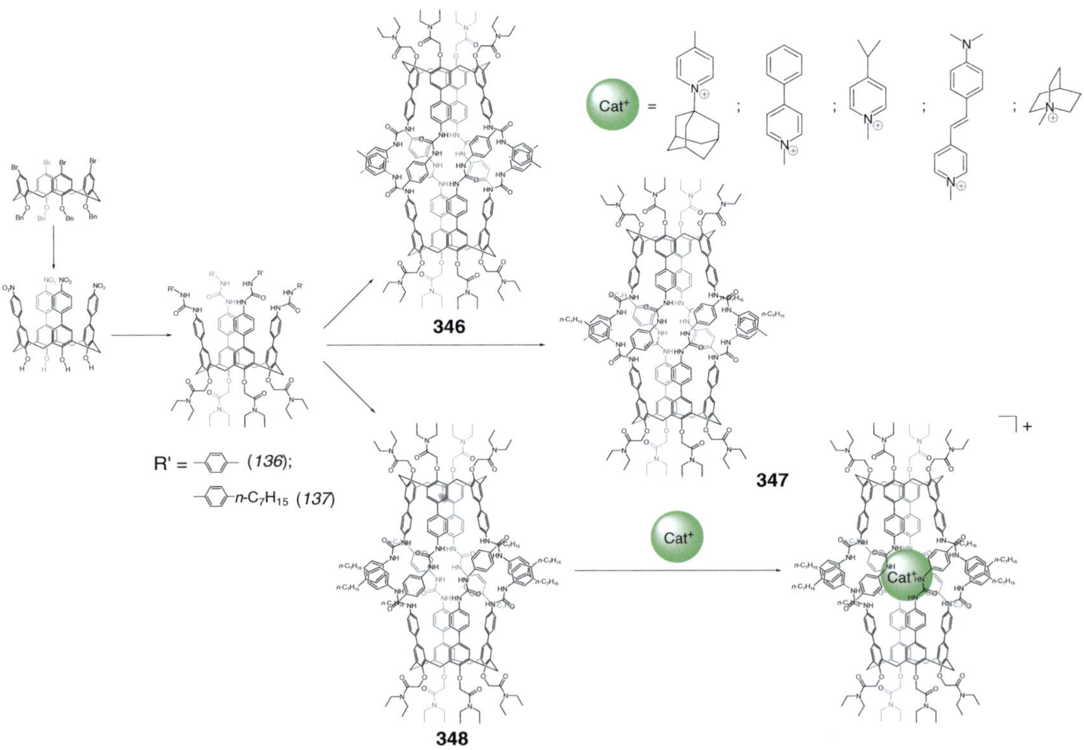

Scheme 3.26

Scheme 3.28

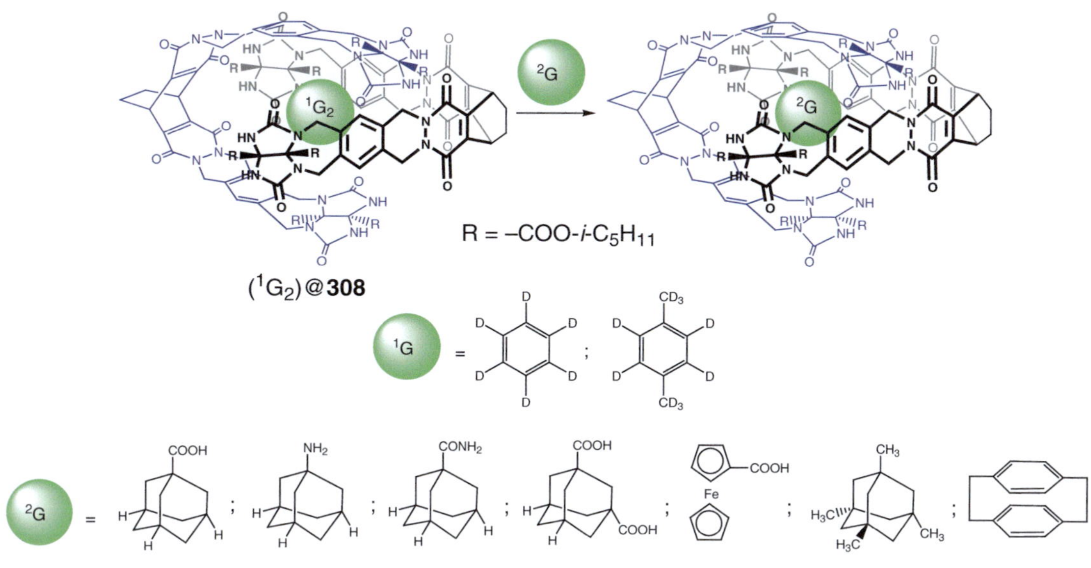

^{2}G
R = –COO-*i*-C$_5$H$_{11}$
(^{1}G$_2$)@**308**
^{1}G =
^{2}G =

Scheme 3.27

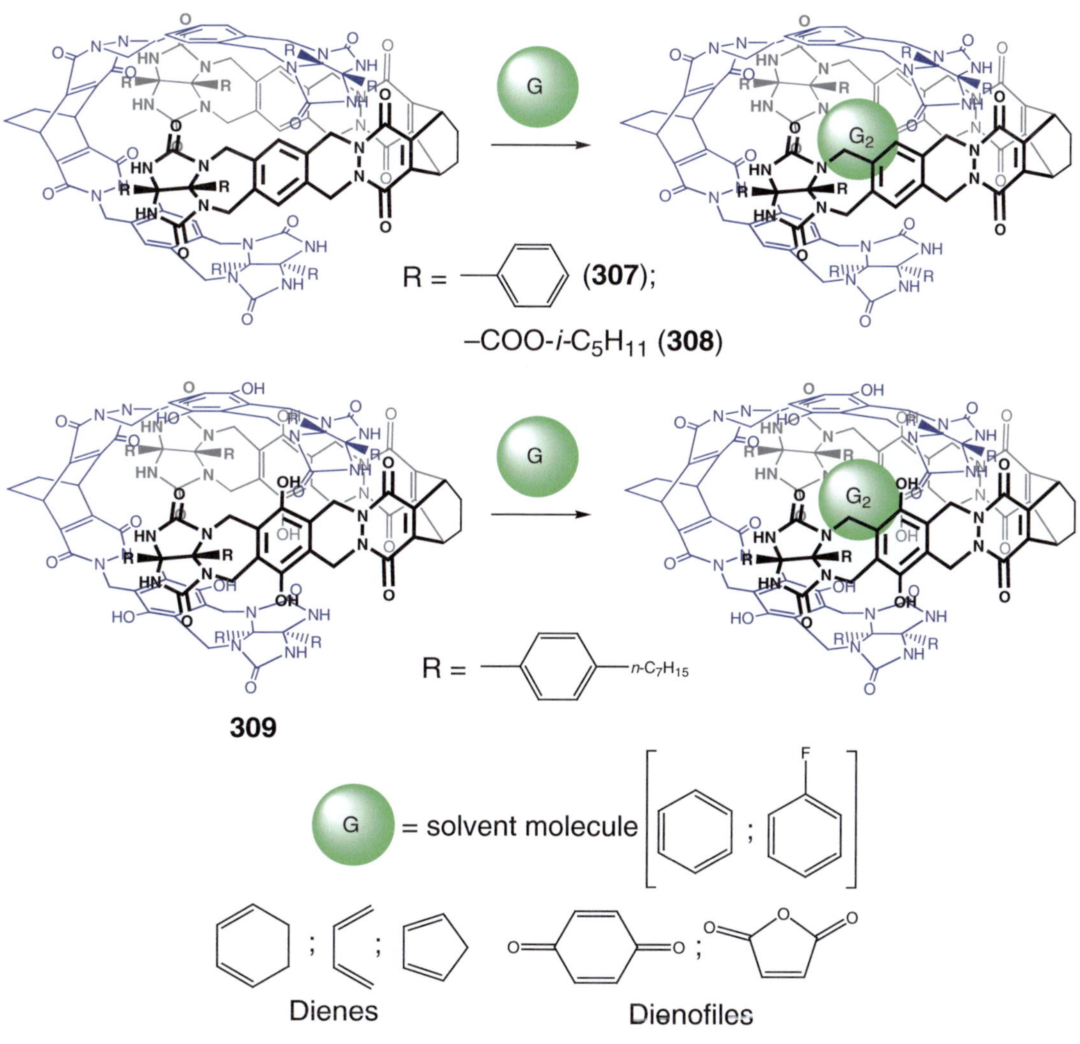

G
R = (**307**);
–COO-*i*-C$_5$H$_{11}$ (**308**)
R = *n*-C$_7$H$_{15}$
309
G = solvent molecule
Dienes
Dienofiles

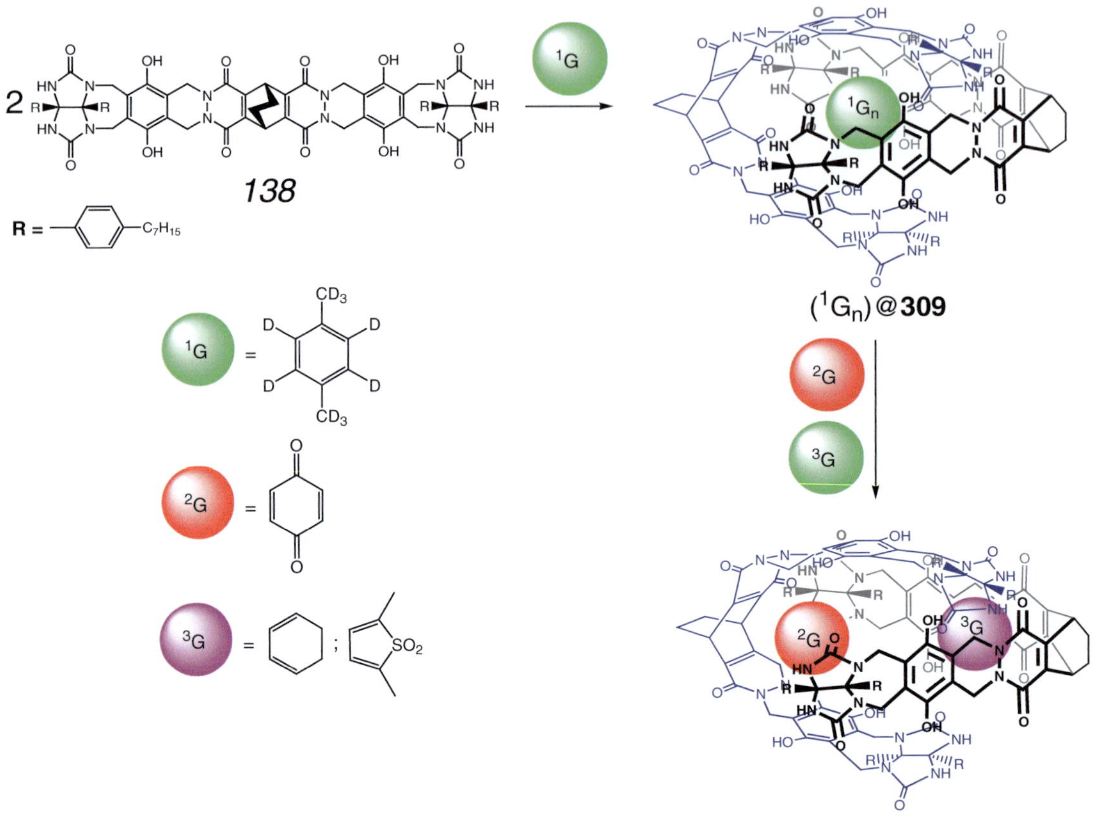

Scheme 3.29

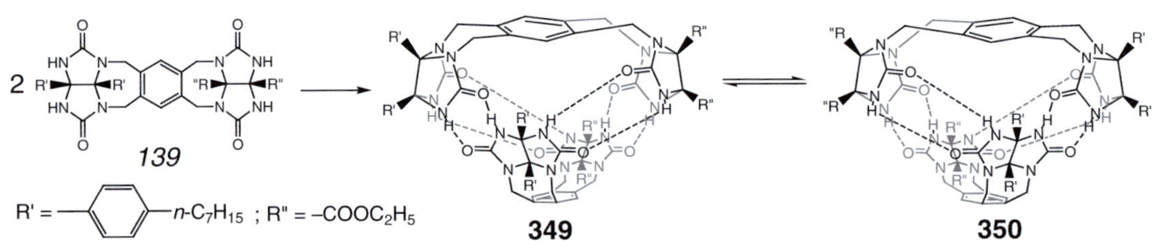

Scheme 3.30

and liberation of these guests are responsible for unexpected thermodynamic parameters of encapsulation processes observed in [31]; these supramolecular capsules also bind various dienes and dienofiles (Schemes 3.28 and 3.29).

The dimerization of nonsymmetric ligand syntone *139* by Scheme 3.30 has been used [32] for the preparation of diastereomeric hydrogen-bonded capsules **349** and **350**.

Template effect of guest solvent molecules such as benzene, toluene, and *para*-xylene has been observed in [33] for the formation of a pseudospherical hydrogen-bonded capsule **308**, resulted from dimerization of self-complementary ligand syntone *140* by Scheme 3.31. On the other hand, the amount of its *C*-shaped molecules in chloroform and dichloromethane solutions was found to be lower than expected statistically, as these solvents favored the formation of *S*-shaped isomer of *140* [33].

The supramolecular capsules **351** and **352** with dissymmetric cavities, presented in

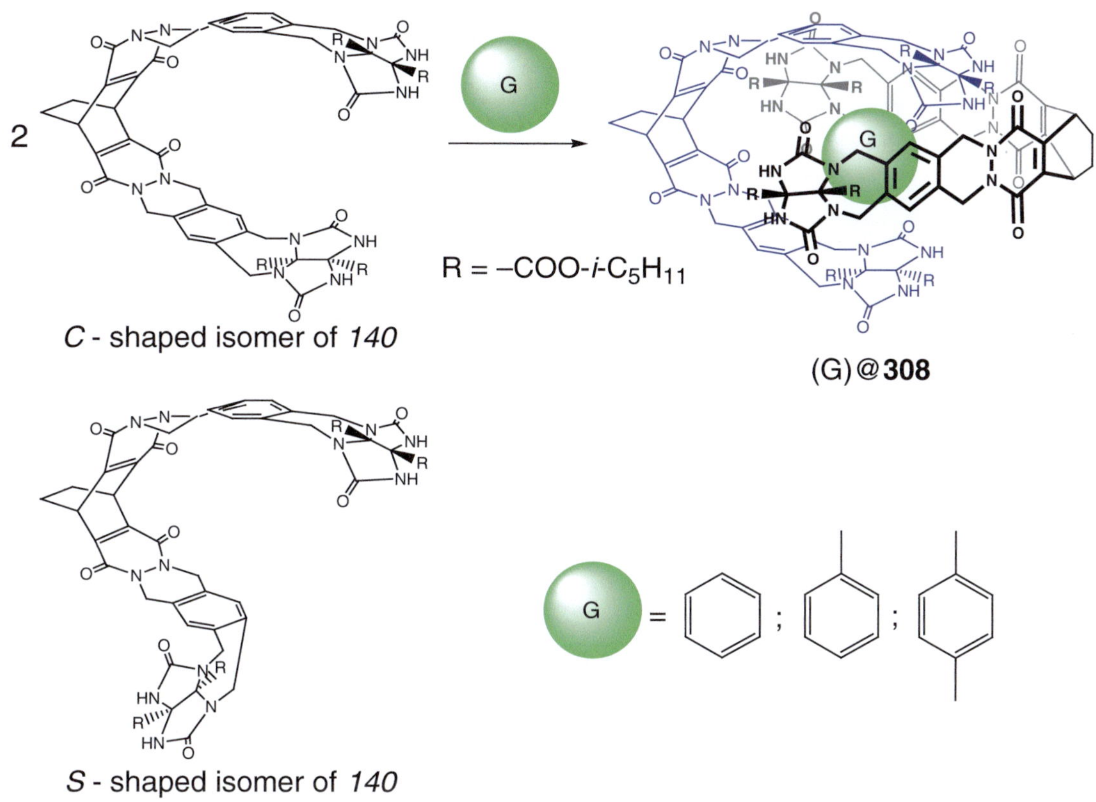

Scheme 3.31

Scheme 3.32, are able to perform molecular recognition of chiral guests, such as terpenes [34]. In particular, the cavities of **351** and **352** have calculated volumes of 241 and 231 Å^3, respectively; those may vary within ~10 % via "breathing" dynamics of their labile hydrogen-bonded cage frameworks. The guests with suitable shape have been bonded more strongly if their volumes were approximately 55 ± 9 % of the cavity volume. The hydrogen-bonded capsules **351** and **352** are reported to encapsulate deuterated benzene and toluene solvent molecules. Those also bind chiral guests *141–145* forming diastereomeric 1:1 cage complexes. As follows from the ^{1}H NMR data of [34], one of these two possible complexes is formed with some degree of preference; with a rather small guest such as *141*, no selectivity has been observed. It increases with the size of the guest, and the maximum selectivity was found for the one largest *145*. The selectivities for the same guests are higher with the smaller capsule **352** that forms more host–guest interactions. The presence of functionalizing groups in the guest molecule that can form hydrogen bonds with the caging host is also important. The greatest selectivity has been observed in [34] for guests of appropriate size that share the C_2 symmetry of the capsules, making extensive surface contacts with their inner cavities and offering a specific array of hydrogen-bonding sites to interact with the seam of these encapsulating ligands. Those are flexible enough to arrange appropriately around a caged guest while maintaining enough rigidity to be formed mostly in the presence of a given chiral guest. Thus, these dissymmetric capsules are accessible through self-assembly by weak intermolecular forces, and guest enantioselectivity is described to be possible even for labile supramolecular caging ligands [34].

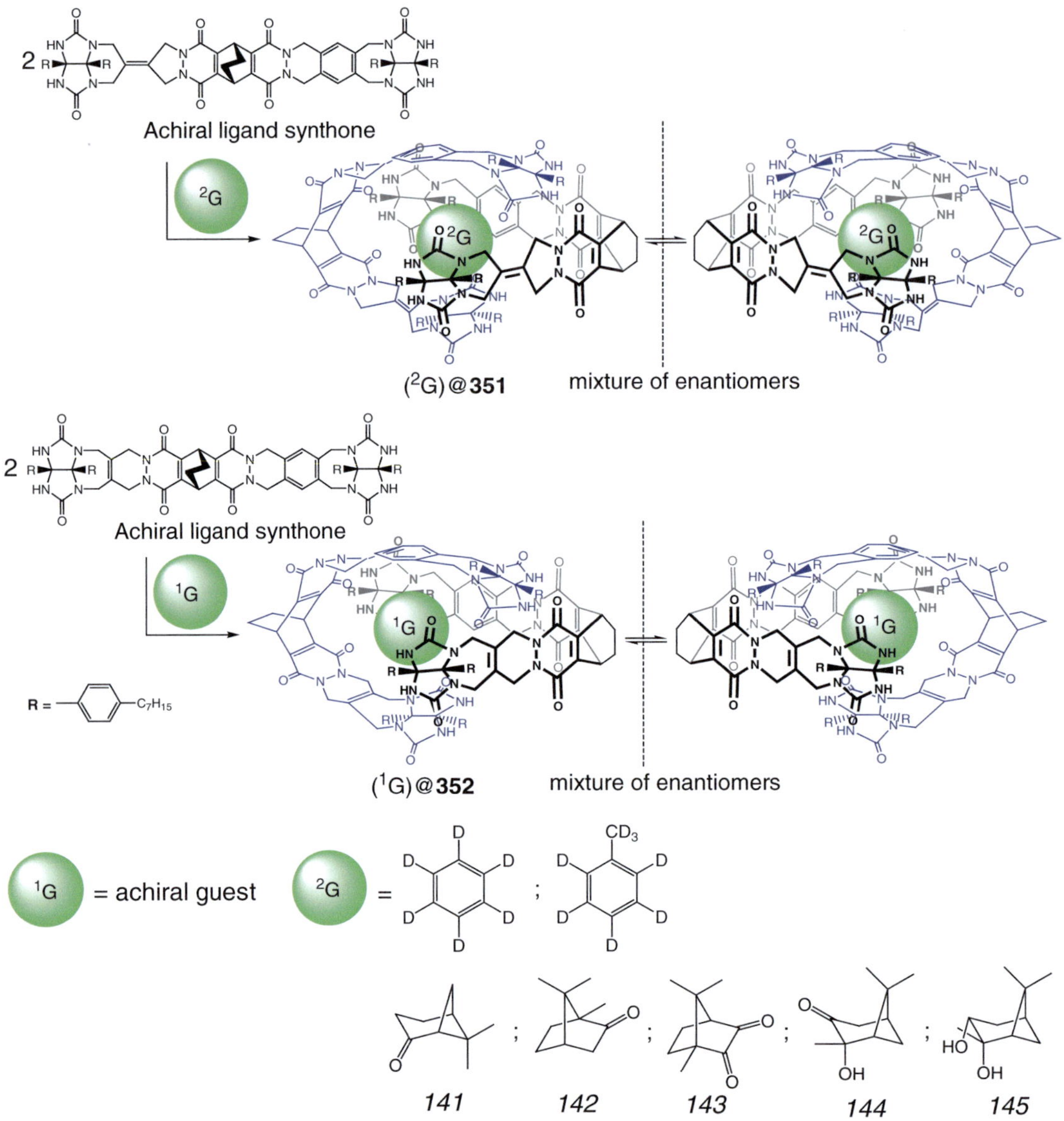

Scheme 3.32

The homodimeric capsules **353** and **355** and their heterodimeric derivative **354** are also able to encapsulate a series of bulky organic guest molecules shown in Scheme 3.33 to give the corresponding 1:1 cage complexes [35].

Template effect of appropriate solvent guests on self-assembly of dimeric supramolecular capsules **308**, **356**, and **357** by Scheme 3.34 is described in [36]. These hydrogen-bonded solvent-occupied cage complexes have been formed predominantly in aromatic solvents such as benzene, toluene, and *para*-xylene by *C*-shaped ligand syntones *116* and *146*. In the case of the solvents that do not suit for dimerization ($CHCl_3$ and CH_2Cl_2) or those competing for hydrogen bonds (DMSO and THF), only statistical amounts of *C*-, *S*-, and *W*-shaped stereoisomers of these ligand syntones have been detected.

Molecular recognition by the above "soft" supramolecular capsules through encapsulation

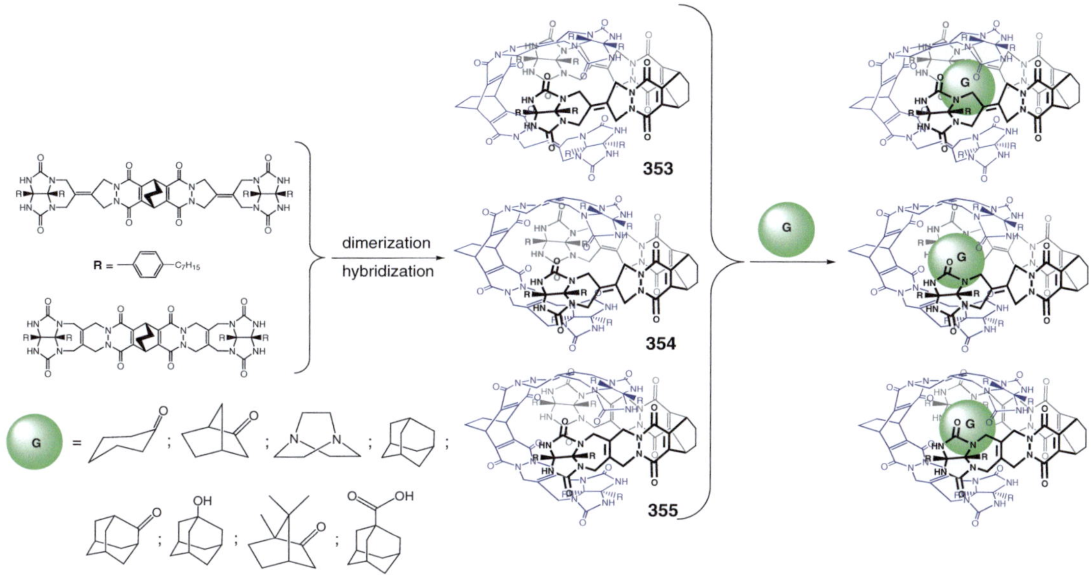

Scheme 3.33

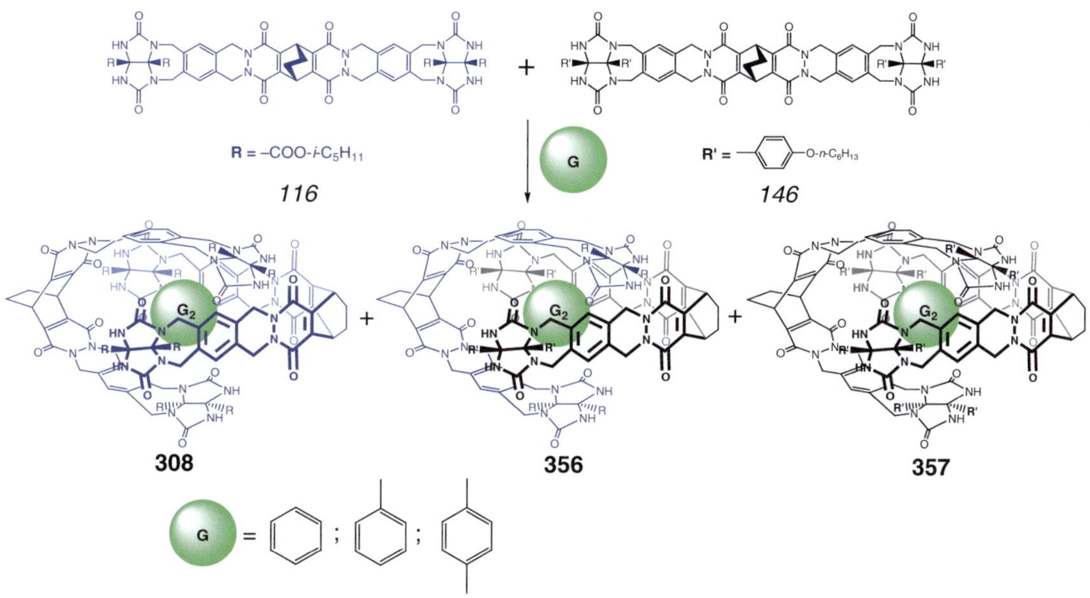

Scheme 3.34

processes is reported by J. Rebek, Jr. [37] to be largely determined by the volumes of the guest molecule and of the cavity of such caging ligand. The encapsulation of guests of suitable dimensions in solution by Scheme 3.35 takes place when the packing coefficient (the ratio of the guest's volume to the host's volume) is in the range of 0.55 ± 0.09; the larger packing coefficients (up to 0.70) have been reached if the cage complex was stabilized by strong supramolecular interactions such as hydrogen bonds. These considerations are also applicable for multiguest

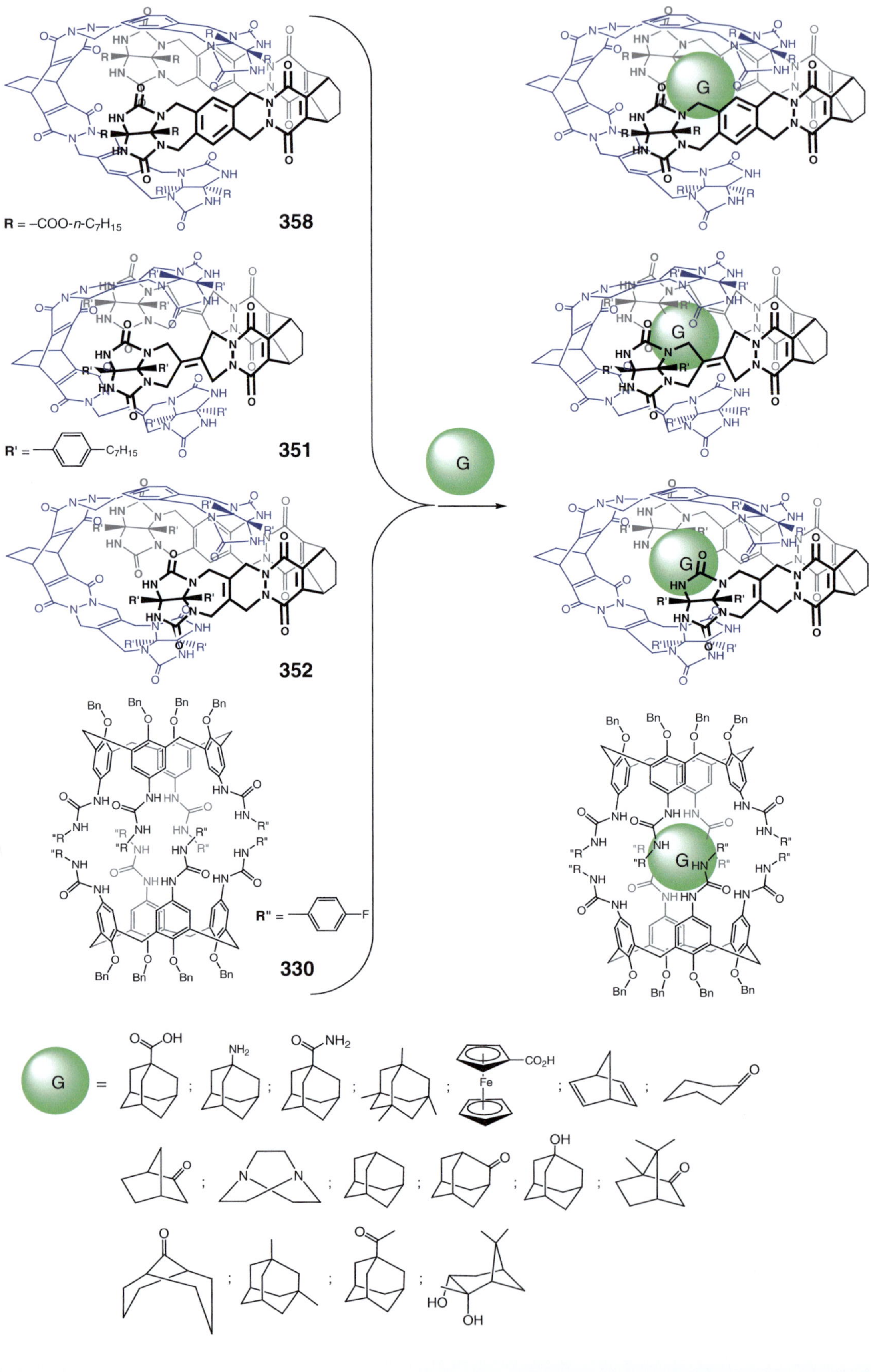
R = –COO-n-C7H15
358
R' =
C7H15
351
352
R" =
F
330
G
G =
Fe
CO2H

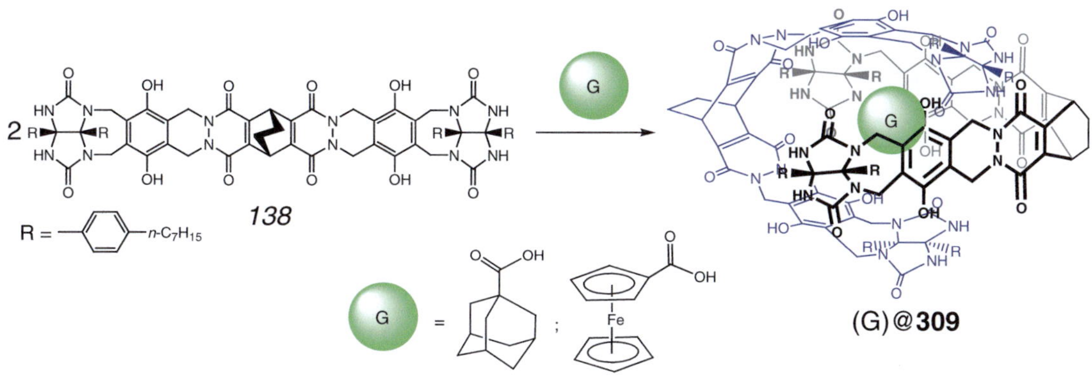

Scheme 3.36

capsules. The best binding is reached when the packing coefficient is approximately 0.55, if the guest has the shape that well suits for given caging ligand [37]. In particular, the rigid ligand syntone *138* is reported in [38] to form 1:1 cage complexes of hydrogen-bonded capsule **309** with encapsulated 1-adamantane- and 1-ferrocenecarboxylic acids by Scheme 3.36.

Guest-induced chirality within the cavity of hydrogen-bonded "softballs" **307**, **308**, **359**, and **360** (a racemic mixture of the latter is formed as a result of dimerization of nonsymmetric ligand syntone *148*) after encapsulation of chiral camphor derivatives (Scheme 3.37) has been examined [39] by ^{1}H NMR method.

Cage complexes of reversibly formed dimeric supramolecular capsules **351** and **361** (Scheme 3.38) with asymmetric microenvironment within their cavities, which hydrogen bonds were imprinted by long-departed guests as chiral templates, have been synthesized in [40]. As follows from ^{1}H NMR data [40], these chiral caging ligands remain after removal of the template-caged molecules, and their recognition abilities persist for hours in organic media. The chiral capsule **351** is formed by dimerization of two corresponding self-complementary ligand syntone *149* in organic solvents through a seam of eight hydrogen bonds. These syntones are achiral and their dimerization gives a distorted spheric supramolecular capsule with C_2-symmetry axes only, existing as a pair of enantiomers. These enantiomers undergo interconversion (racemization) by complete dissociation and recombination of such entities. As a result, the asymmetric guests recognize one enantiomer of the caging ligand over its mirror image. The dimerization of the ligand syntone *150* in a number of solvents gave a more robust hydrogen-bonded capsule **361**; its phenolic groups form four additional hydrogen bonds that slow its racemization. This template-formed host is described in [40] to have memory effects: as follows from the competition experiments data, it still prefers the template guests after multiple guest exchanges (Scheme 3.39) and physical manipulations. The pyridazinyl cyclic fragment of the ligand syntone *150* having a very low inversion barrier creates openings in its dimeric capsule **361** that are large enough to easily accommodate incoming and outgoing guests; each of them in turn experiences imprinted asymmetric cavity of this hydrogen-bonded capsule [40].

The complimentary ligand syntones *125* (Scheme 3.21) and *151* are described in [41] to undergo dimerization and hybridization by Scheme 3.40 leading to D_{3d}-symmetric supramolecular capsules **327**, **362**, and **363**. The

◄ **Scheme 3.35**

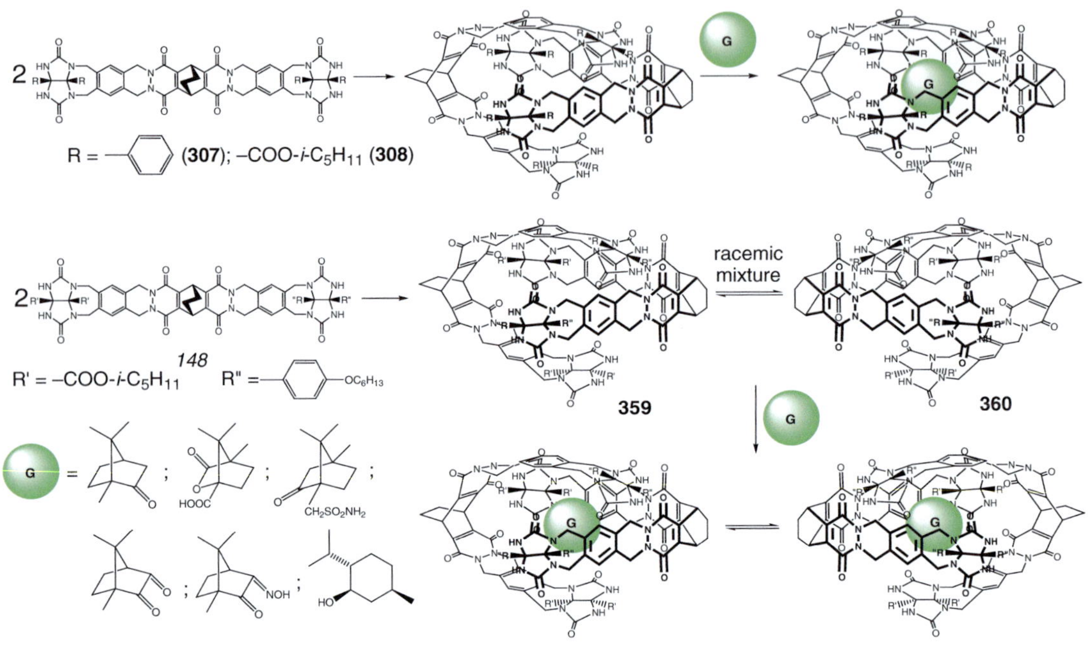

Scheme 3.37

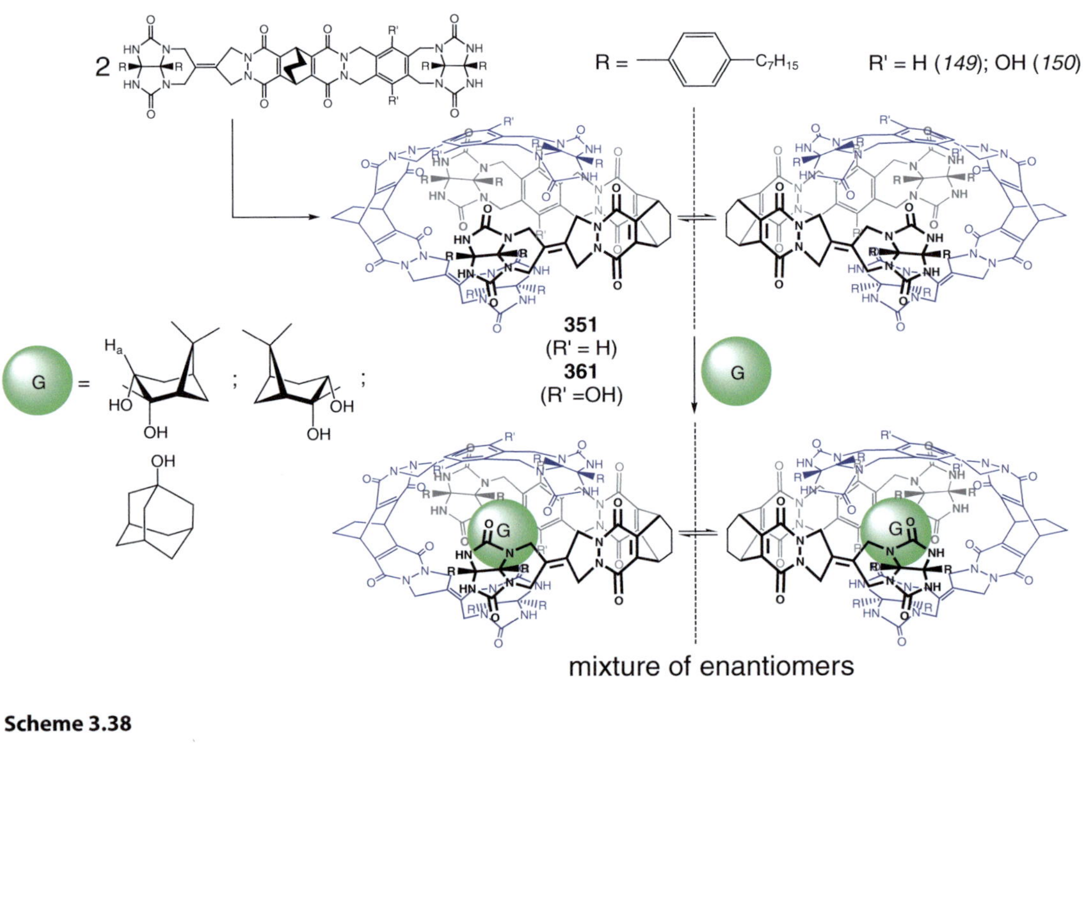

Scheme 3.38

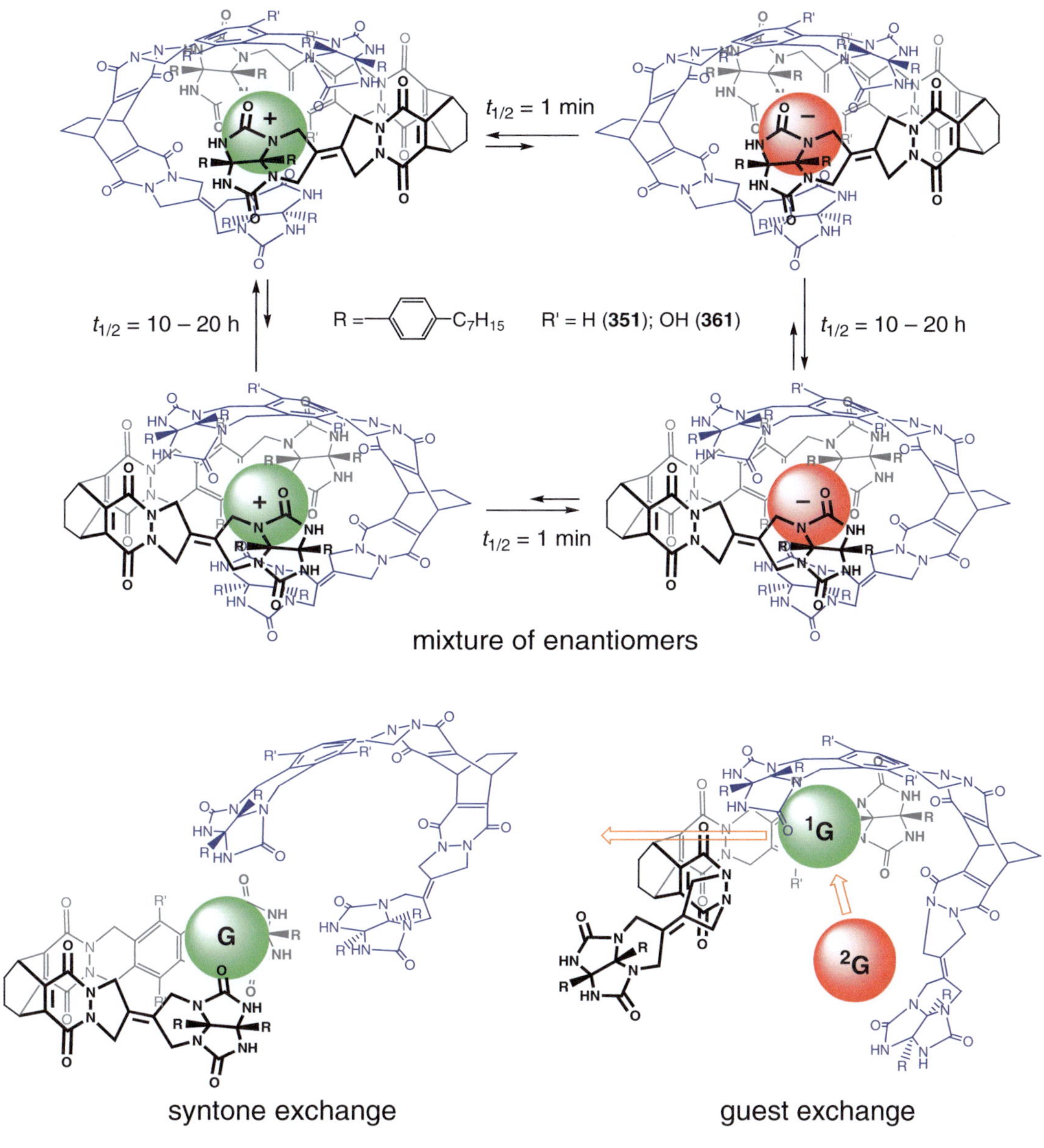

Scheme 3.39

hydrogen-bonded capsule **327** in $CDCl_3$ solution does not form cage complexes with a variety of neutral guests (in contrast with cationic species [21], see Scheme 3.21), whereas **363** encapsulates several organic molecules in $CDCl_3$ medium. For example, the latter ligand readily binds ferrocene derivatives, whereas the capsule **327** does not encapsulate them even at their large exterior concentrations. This result is explained in [41] by the difference in size and shape of the supramolecular capsules **327** and **363**. Encapsulation of large guest molecules by these porous caging ligands, which have been regarded in [41] as self-assembled sieves, does not depend on entropically favorable release of caged solvent molecules and thus is an enthalpy-driven process.

Hydrogen-bonded bis-calix[4]arene capsules **364** and **365** have been prepared in [42] by

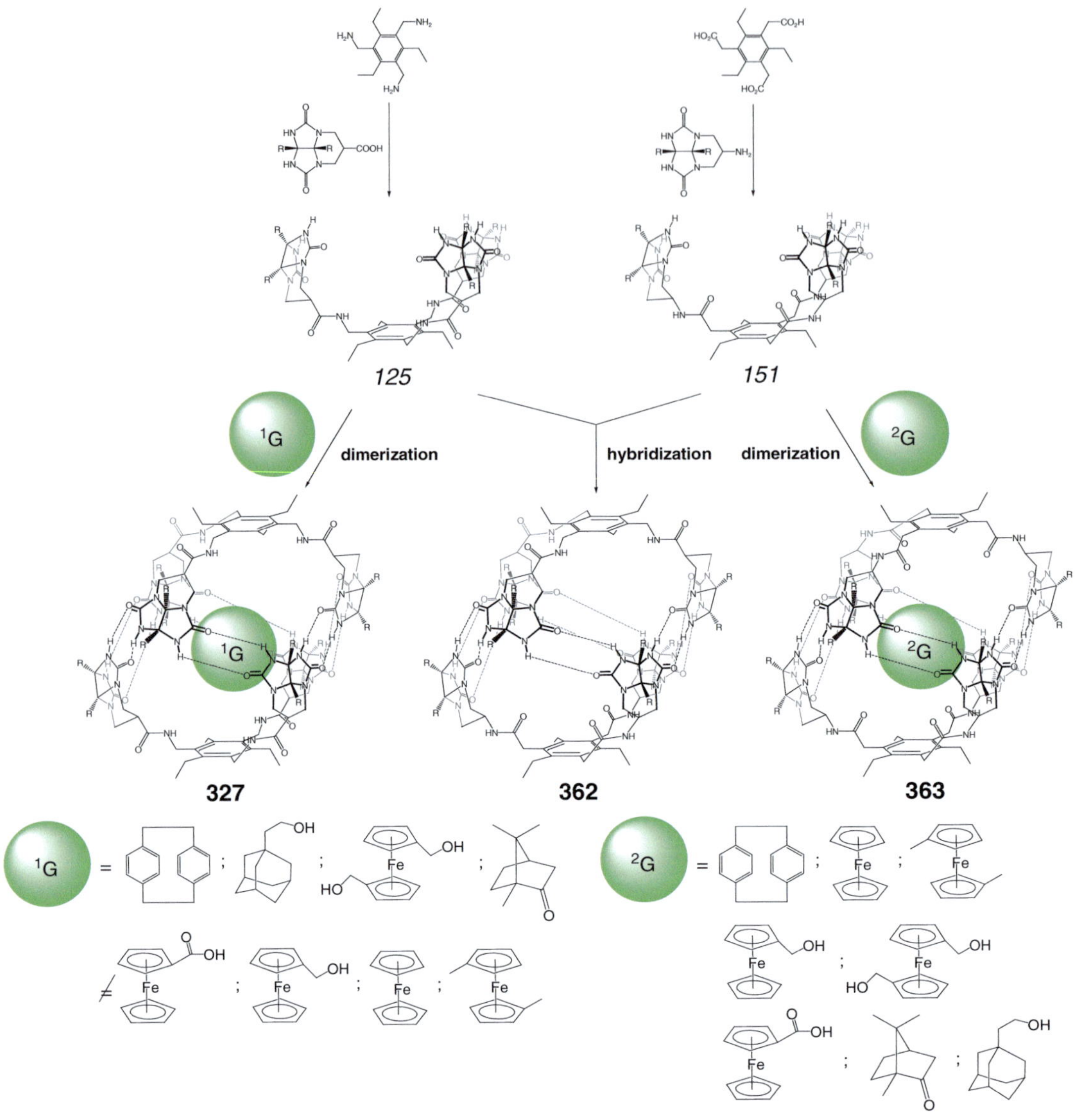

Scheme 3.40

dimerization of their ligand syntones (Scheme 3.41) with four urea substituents at the upper rims and four secondary amide fragments at the lower rims, forming inter- and intramolecular hydrogen bonds in apolar solvents. The communication between the calixarene rims of these syntones affects the size and shape of the cavity of the self-assembled ligands **364** and **365** with *cone*-conformation of their calixarene entities. These capsules are stabilized by intramolecular C=O…HN hydrogen bonds at the lower rim of these fragments, and their seam is cycloenantiomeric with either clockwise or counterclockwise arrangements of the head-to-tail amides. Encapsulation of sodium cation by these capsules breaks hydrogen bonds at the lower rim of the calixarene fragments but the cage framework remains intact. The presence of amide groups at the lower rims causes substantial difference in the binding properties of **364** and **365** as compared to those of their ester-containing analogs **366** and **367** [42].

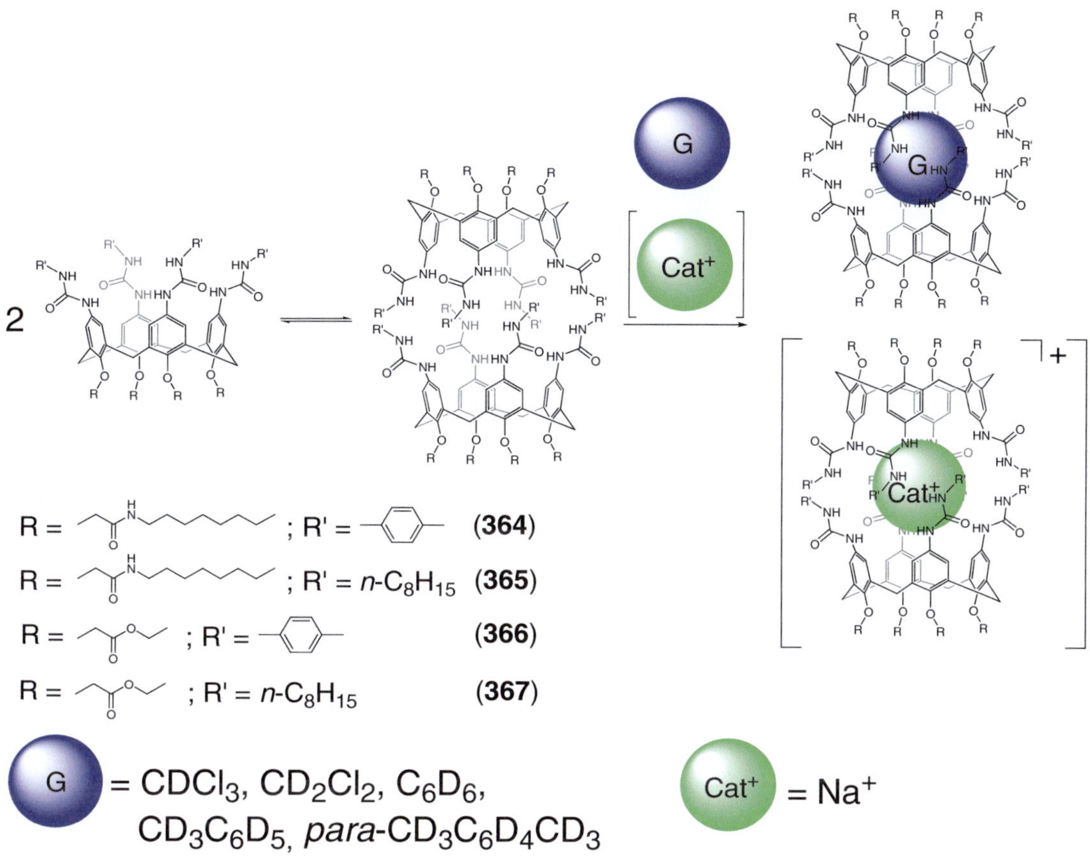

Scheme 3.41

Different congeners of the ligand syntones *150*, *152–156* shown in Scheme 3.42 are proposed in [43] for enantioselective encapsulation. Two different spacer elements in these syntones give chiral capsules, although they themselves are achiral. The resulting capsules **351** and **368–372** have dissymmetric spheric cavities with the volumes of 190–390 Å^3, and asymmetric guests such as terpenes are reported to discriminate one enantiomer of the corresponding capsule over its mirror image with moderate selectivities. The complexation studies suggested that these hydrogen-bonded capsules (except of **368**) were flexible enough to arrange themselves comfortably around the guest but still maintain enough rigidity to be affected by its occupancy. These enantiomeric cage frameworks are reported in [43] to undergo interconversion (racemization) through dissociation and recombination of their ligand syntones.

The pyrrogallolarene ligand syntones *157* and *158* are reported in [44] to undergo dimerization by Scheme 3.43 in the presence of suitable cationic guests giving the hydrogen-bonded capsules **373** and **374** that are highly stable in competing with solvents such as alcohols [44]. The solvent molecules participate in this self-assembly as bridging fragments cooperating with direct hydrogen bonds between the ligand syntones. Strong cation…π and CT interactions between the encapsulated species and the π-basic pyrogallol rings stabilize the bis-pyrrogallolarene cage complexes with encapsulated organic cations. The reversible encapsulation of tropylium cation prevents its alcoholysis, and a characteristic change of color caused by host–guest CT interactions is observed [44].

Supramolecular electrostatically driven assembly of the *L*-alanine-functionalized calix[4]arene

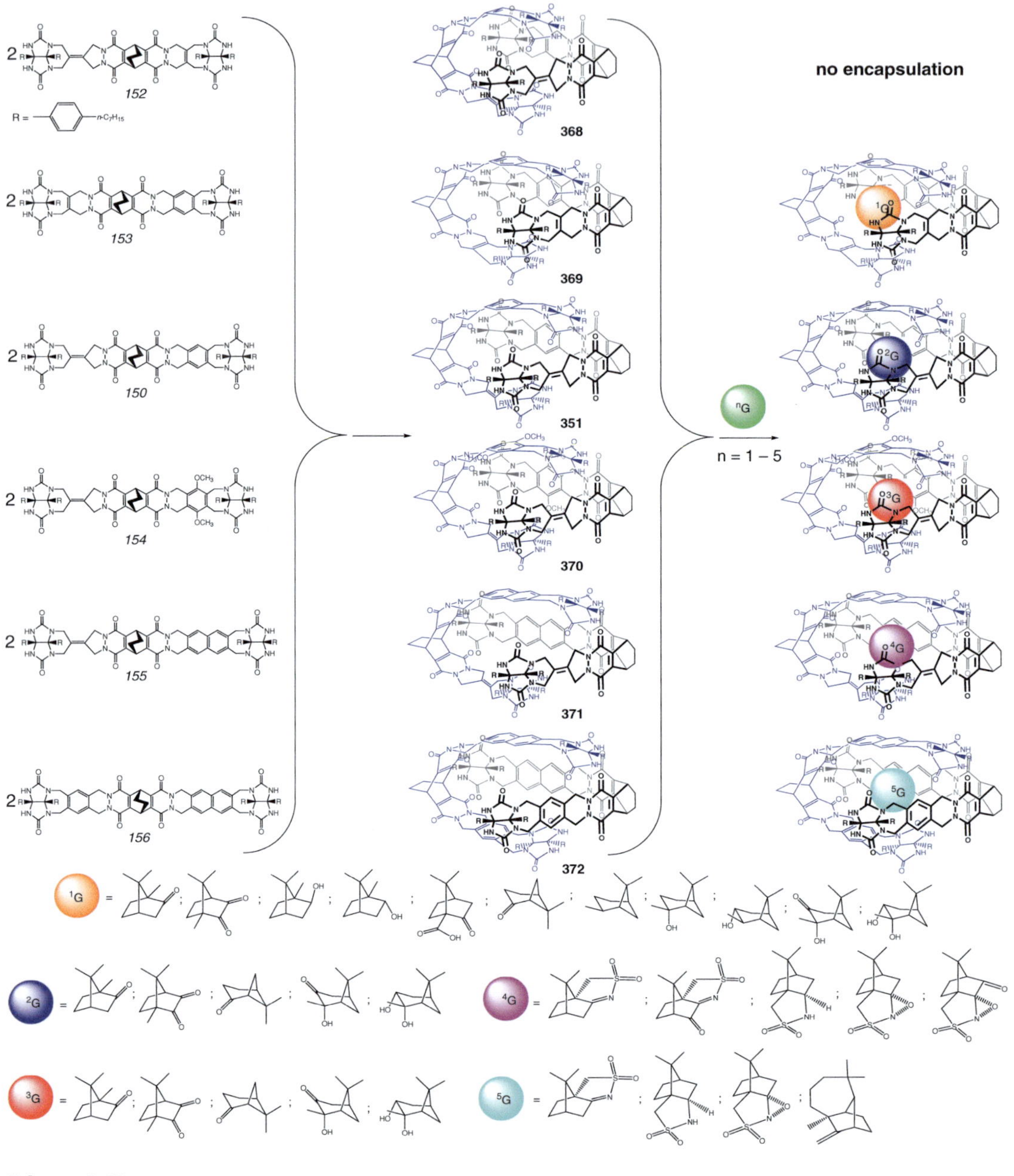

Scheme 3.42

159 with negatively charged carboxylate pendant groups and the calix[4]arene *160* with positively charged amidinium groups by Scheme 3.44 is reported in [45] to give a bis-calixarene capsule **375**. This ligand encapsulates *N*-methylquinuclidinium cation to give a 1:1 cage complex, which was characterized using ITC, ESI-MS, and NMR methods.

The pyridine[4]arene ligand syntone *161* undergoes dimerization by Scheme 3.45 giving the hydrogen-bonded capsule **376** [46]. This capsule selectively encapsulates several carboxylic acids, acetamide and trifluoroacetamide, but not more bulky molecules of butyric, 2-methyl- and 2,2-dimethyl-containing carboxylic acids [46].

Assembly of the negatively charged calix[4]arene *162* and its analogs *163* and *164* (Scheme 3.46) with *165* and the caging properties of supramolecular bis-calixarene capsules **377–379** have been studied in [47] using ESI-MS, ^{1}H NMR, and ITC meth-

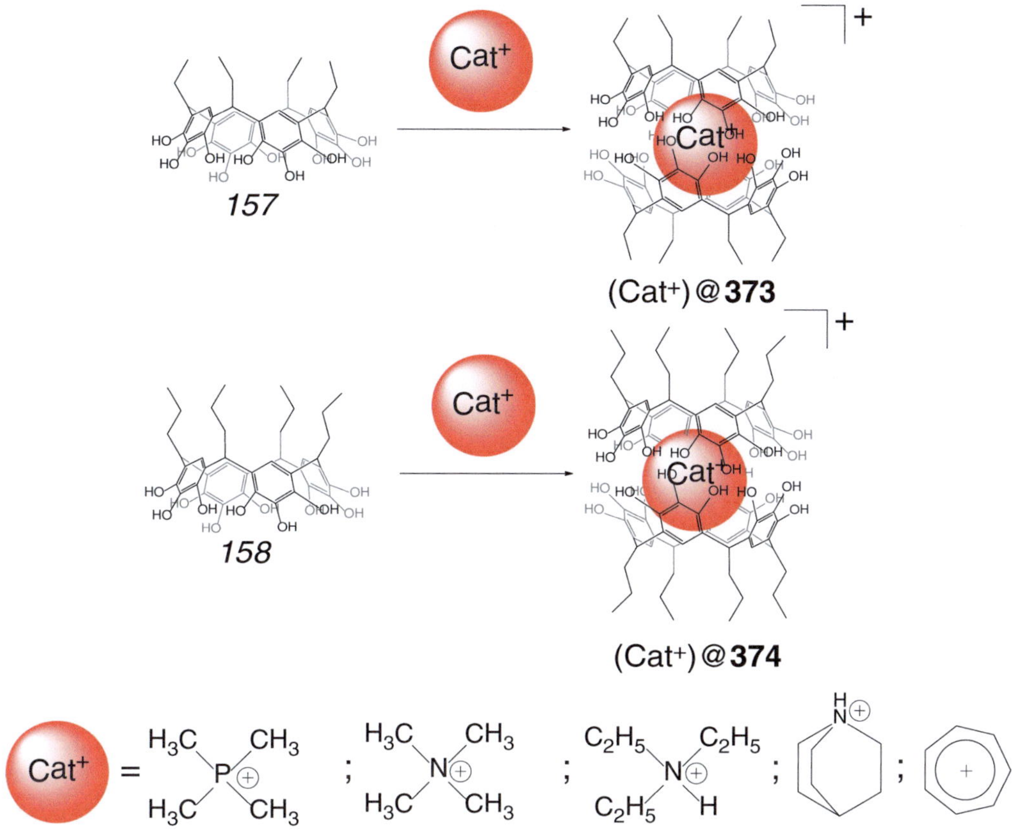

Scheme 3.43

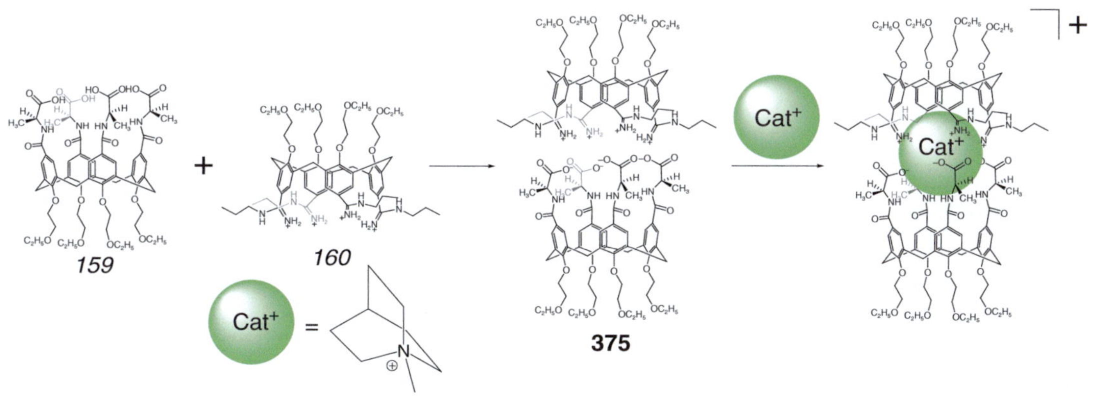

Scheme 3.44

ods. According to ^{1}H NMR data, those encapsulate *N*-methylquinuclidinium cation as well as various neutral aromatic guests. At the same time, they do not form the cage complexes with acetylcholine and tetramethylammonium cation [47].

Self-complementary *L*-phenylalanine-functionalized resorcinarene syntone *166* resulted from Mannich condensation has been used in [48] for the design and preparation by Scheme 3.47 of a hydrogen-bonded capsule **380**

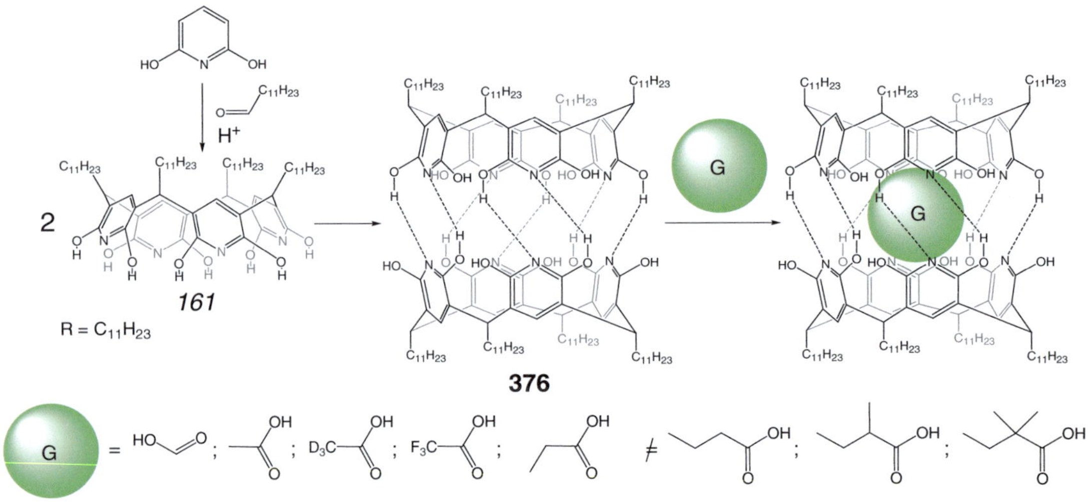

Scheme 3.45

with both polar and non-polar interior functionalities. This capsule contains complementary amine and carboxyl functionalizing groups and has an almost D_4 symmetry with the dihedral angle between the resorcinarene fragments of 44°. Besides the dipole–dipole interactions between them, these fragments also form hydrogen bonds involving their hydroxyl and carboxyl substituents. As a result, the cage framework of **380** has dense packing without holes and it encapsulates two nitromethane and four water solvent molecules. The guest water molecules form hydrogen bonds with H-bond donor amino groups of this capsule, while the latter gave hydrogen bonds with its H-bond acceptor carboxyl substituents [48]. Possibility for the formation of the heterodimeric analog of homodimeric capsule **380** has been also studied in this work. According to ^{1}H NMR data, the hybrid capsule **381** is not formed even upon heating of a mixture of *D*- and *L*-phenylalanine-functionalized resorcinarene ligand syntones for 2 weeks; however, few dissolve–evaporate cycles of this mixture in the presence of methanol afforded the target cage compound in a quantitative yield. The heterodimeric capsule **381** is more thermodynamically stable than its kinetically stable homodimeric analog, and this ligand and its cage complexes with encapsulated solvent chloroform, water, and ethanol molecules and a self-encapsulated phenylalanine chain have been characterized in [48] using NMR spectra.

The self-encapsulation with the caging ligand **380** "biting its own tail" is reported in [49] to occur in the absence of water and in appropriate solvents. According to NMR and X-ray diffraction data, water molecules can also be encapsulated by this homodimeric capsule (Scheme 3.48) causing the changes in both the structure of its cage complexes and in the guest dynamics within its cavity. In the case of malonic acid, water molecules also affect the stoichiometry of host–guest complexation: 1:2 and 1:1 cage complexes with D_4- and C_4- symmetric molecules are formed without and in the presence of water, respectively, together with encapsulation from two to four water molecules in the latter case. Such co-encapsulation is described in [49] to affect the NMR spectra due to structural and dynamical changes of the caged guest. The encapsulated water molecules lead to slight decrease in the chiral discrimination of the guest hydroxyl-containing carbonic acids and play a role of plasticizer causing the binding site to be more adaptable. This effect is, however, too small to discriminate the enantiomers [49].

The use of RTP allowed encapsulation a guest organic molecule by dimeric *para*-phosphonic acid-containing calix[5]arene capsule **382** [50]. This hydrogen-bonded capsule binds carboplatin

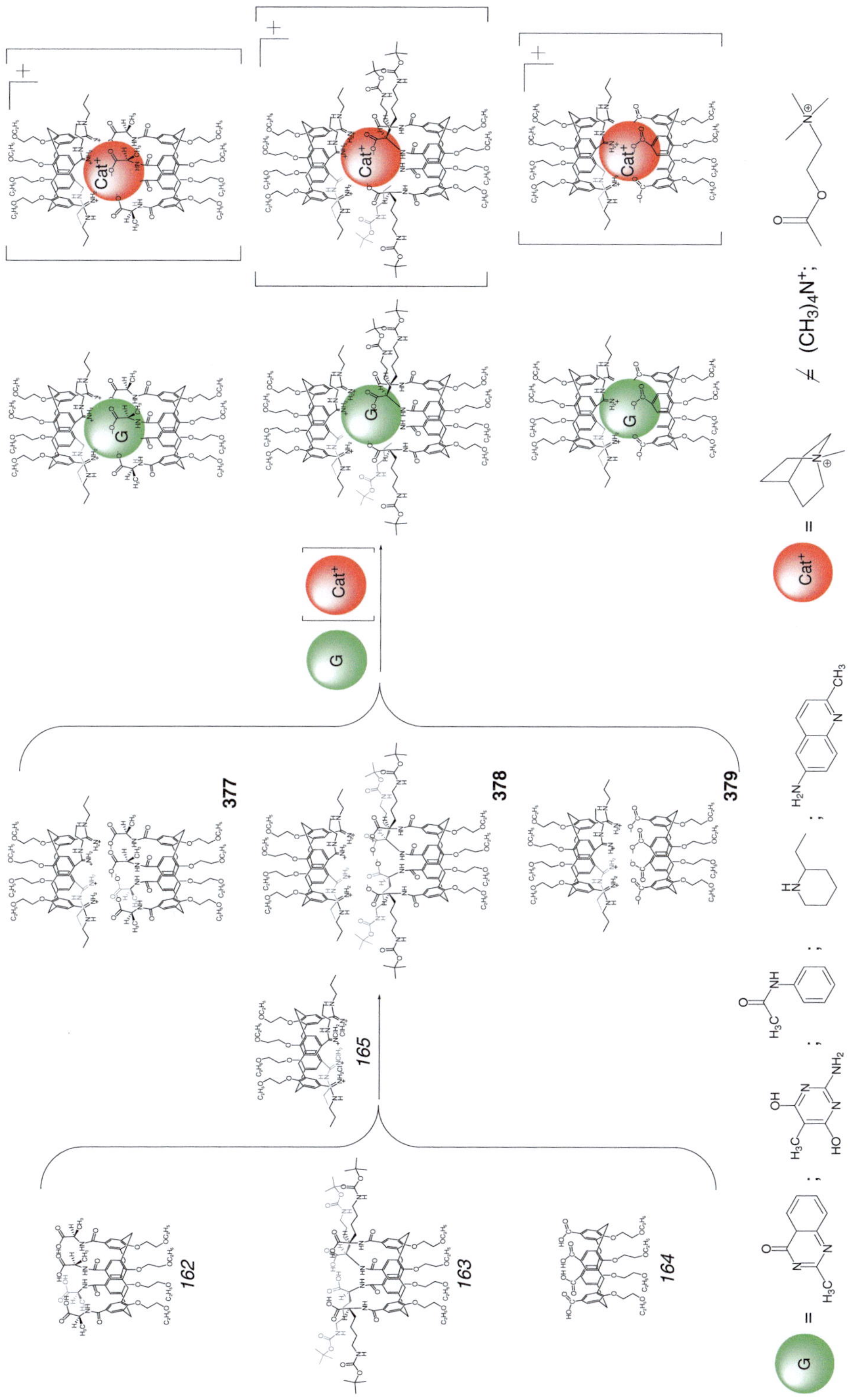

Scheme 3.46

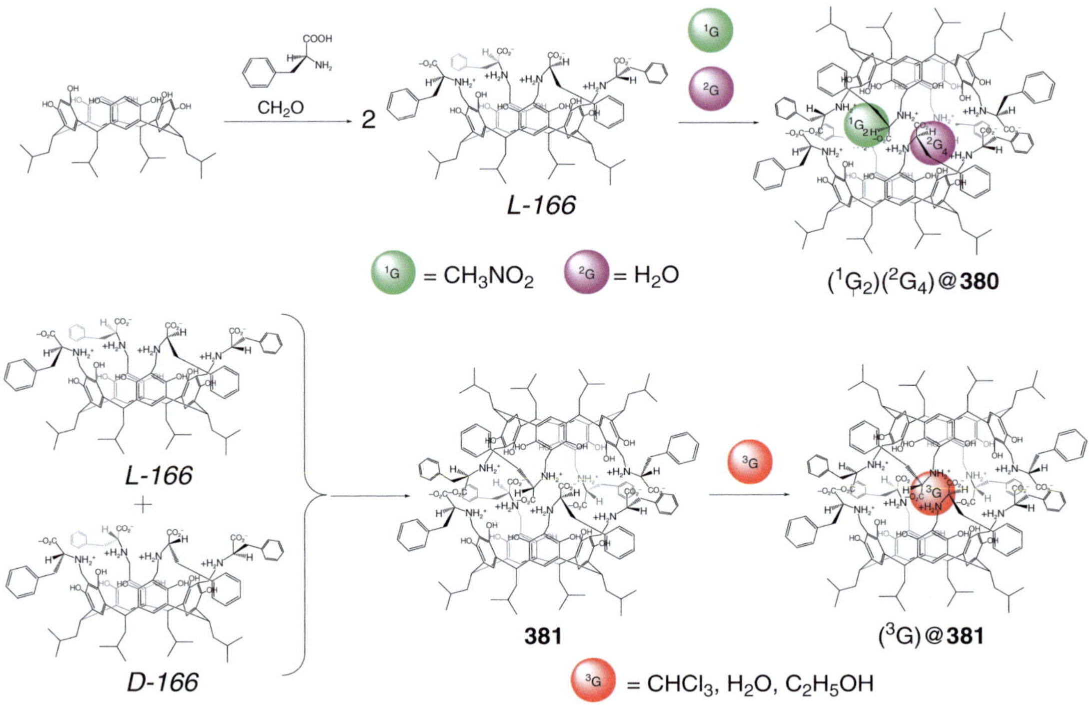

Scheme 3.47

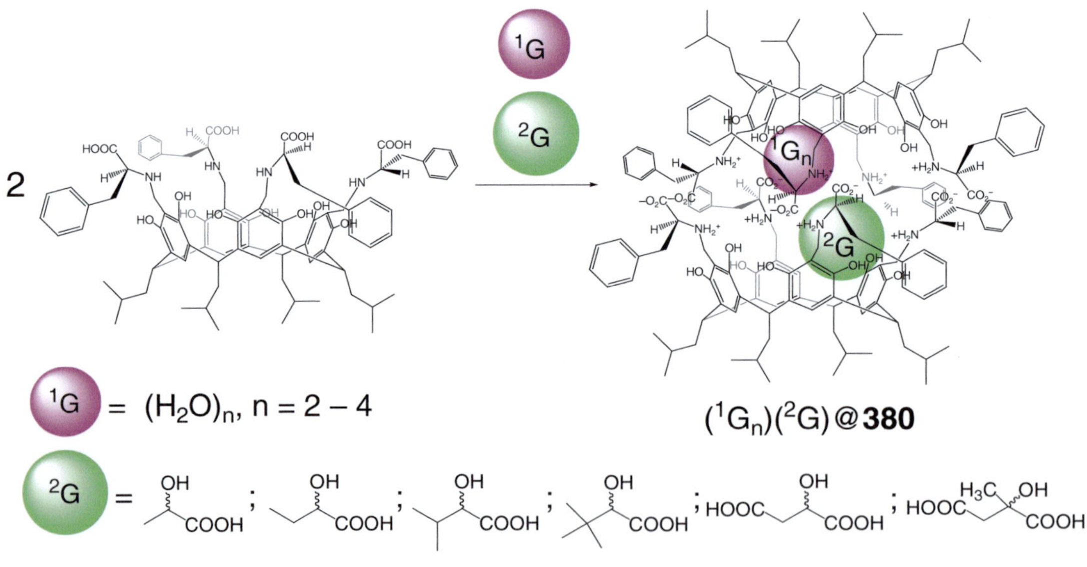

Scheme 3.48

by Scheme 3.49 giving a 1:1 cage complex in aqueous solution after stirring the reaction mixture inside inclined rotating tube more than 1500 rpm under ambient conditions for 5 min; this complex has been characterized using multinuclear NMR and ESI-MS methods. At rotational speeds less than 1500 rpm, the cage framework of **382** does not appear [50]. Cavity volume of

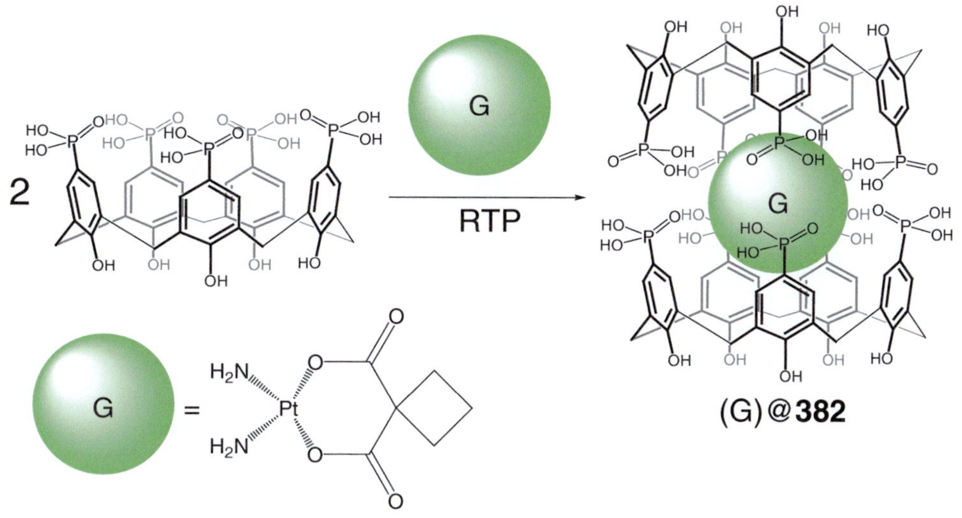

Scheme 3.49

this encapsulating ligand is approximately 550 Å^3, while the volume of the caged guest does not exceed 170 Å^3. Coordination capsule **382** gives uniform 2 nm layer on the surface of single-wall carbon nanotubes even under the excess of calix[5]arene [50].

Tetrameric hydrogen-bonded encapsulating ligands **383** and **384** shown in Scheme 3.50 are reported in [51] to give 1:1 cage complexes with adamantane derivatives *167–173* and appropriate dimers *174–176* formed by ureas and cyclic amides. Complementarity of their ligand syntones plays a key role in these supramolecular capsules being formed by weak interactions directing the self-assembly to one aggregate rather than another. The suitable hydrogen-bonding interactions and an appropriate molecular curvature favored the supramolecular assembly of four self-complementary ligand syntones into the corresponding hydrogen-bonded pseudospherical cage framework, the filling of which with complementary guest molecules provides the final instruction for the building process [51]. In particular, a highly ordered pseudospherical 1:1 cage complex has been formed by encapsulation of adamantane; an exchange of its molecule is slow on the NMR time scale. The shape and volume of this guest, which forms only weak van der Waals interactions with the concave surface of the host's interior, fits into the cavity of a caging ligand. Its interactions overcome entropic disadvantages caused by assembly of five individual species into a highly ordered cage complex. Such adamantane molecule competes successfully for its cavity with the solvent CD_2Cl_2 molecules, and this tetrameric cage self-assembly is reported in [51] to be stabilized by the presence of H-bond acceptor groups in the encapsulated guest through the formation of bifurcated hydrogen bonds with the glycoluril NH groups of its caging ligand. In particular, carbonyl-containing 2-adamantanone and adamantane-2,6-dione filling the appropriate volume of its cavity are shown to be very good guests. At the same time, 1,4-cyclohexanedione with carbonyl groups available for hydrogen-bonding fragments of the caging tetrameric host does not fill its cavity properly, and thus it does not form such a capsule. Urea *174* and sulfamide *175* have been used in [51] to test their host–guest interactions within the cavity of this hydrogen-bonded capsule. Dilution experiments with these individual guests and with their 1:1 mixture showed that heterodimerization of their molecules giving the heteroguest 1:1:1 cage complexes is preferred over homodimerization forming the homoguest 1:2 capsules.

Solid-state and solution studies of a tetrameric hydrogen-bonded capsules **385** and **386**

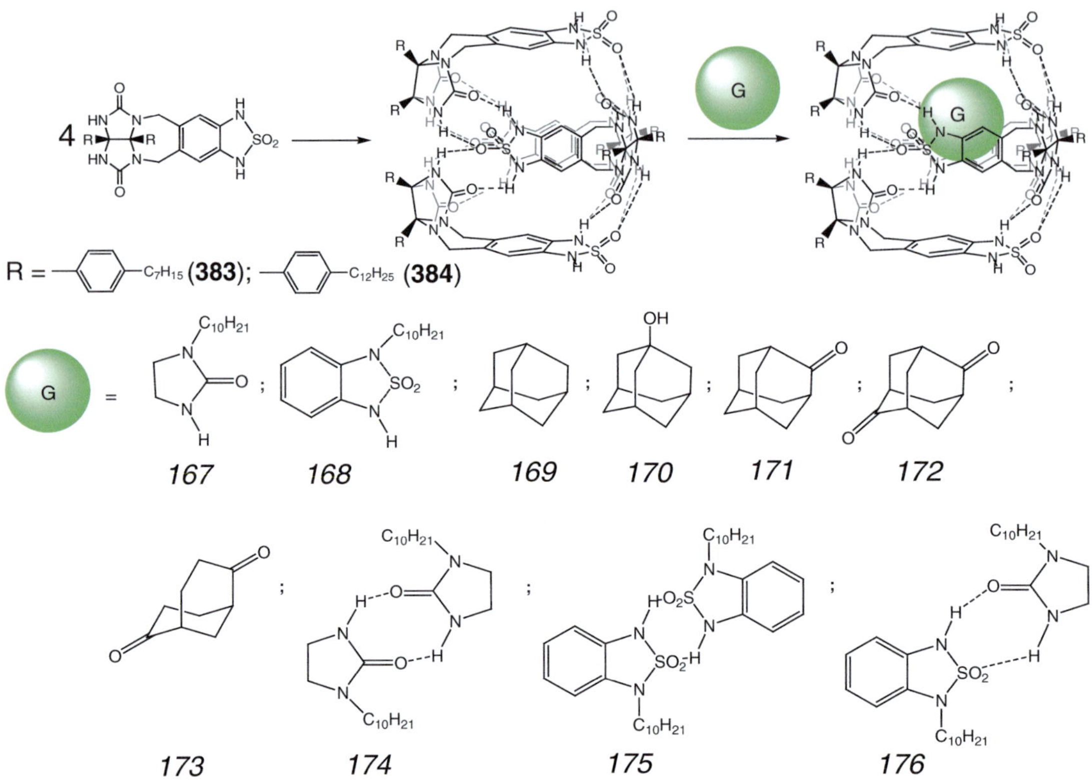

Scheme 3.50

and their 1:1 cage complexes with encapsulated adamantane derivatives (Scheme 3.51) have been performed in [52]. The nature of these ligand syntones allows 24 hydrogen bonds to form between these four entities. As follows from X-ray diffraction data for the cage complex of **385** with encapsulated adamantane-2,6-dione, those are held in a head-to-tail arrangement of their glycoluril (urea) and sulfamide functionalities. Its cage framework also contains additional equatorial eight phenolic hydrogen bonds, which stabilize this supramolecular capsule. The hydroxyl groups on opposing sides of each of these ligand syntones donate hydrogen bonds to one intermolecular and one intramolecular acceptors, thus alternating this structural motif in such a way that the mirror symmetry of each ligand syntone disappears leading to a chiral tetrameric capsule that crystallizes as a racemate [52]. The distance between the opposite aromatic walls of this capsule is approximately 10 Å, and the distance between the two glycoluril–sulfamide hydrogen-bonded seams is approximately 8 Å, resulting in a cavity volume of 184 Å^3. Adamantane-2,6-dione guest matches for this cavity, so it forms a set of CH…π interactions and van der Waals contacts with the electron-rich aromatic wall of the caging ligand **385**. The position of such encapsulated molecule is well defined: its long axis is oriented along the short axis of the host cavity through hydrogen bonds. The oxygen atoms of this guest, which form polar interactions with the glycoluril–sulfamide seam of **385**, are located in each end of its cavity. Each carbonyl group accepts hydrogen bonds from four identical donor atoms, giving eight bifurcated hydrogen bonds and four CH…π host–guest interactions with space filling of approximately 72 % [52]. These cage complexes of **386** with encapsulated adamantane derivatives have D_{2d} symmetry that results from fast exchange between its two slightly twisted D_2-symmetric enantiomers. Solution and solid-state

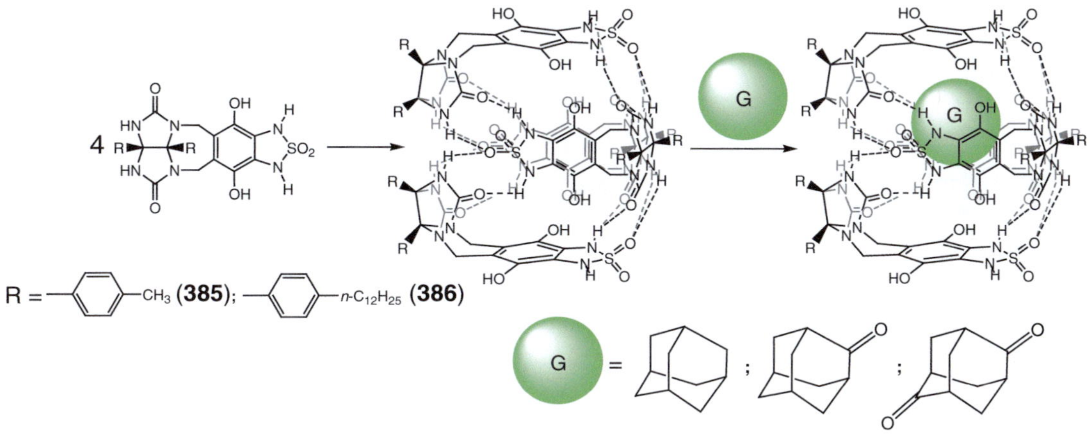

Scheme 3.51

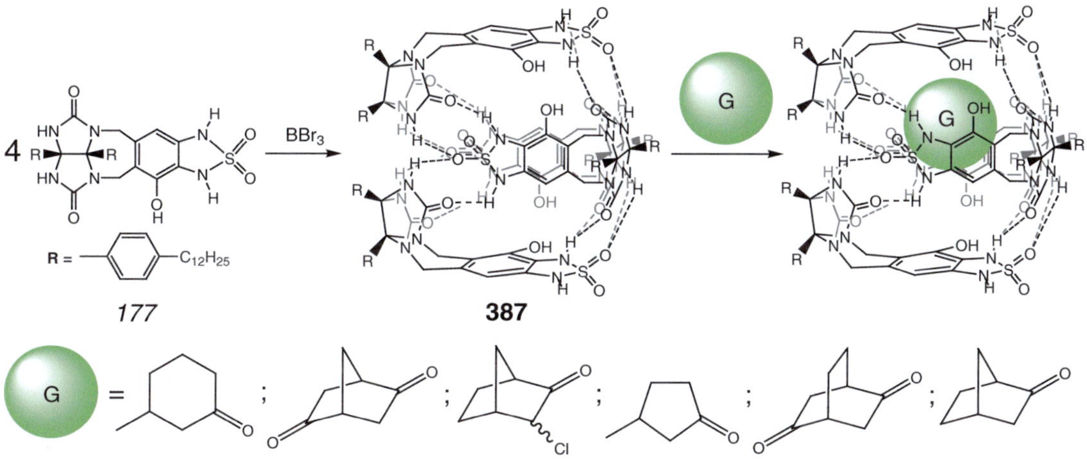

Scheme 3.52

studies performed in [52] showed that host–guest hydrogen bonding not only is an important driving force for the formation of this supramolecular capsule but also dictates its guest-encapsulating selectivity.

The chiral microenvironments within the self-assembled capsule **387** arised from tetramerization of its non-racemic ligand syntone *177* (Scheme 3.52). The self-assembly of this chiral syntone yielded multiple functional groups, which are asymmetrically arranged within the cavity of **387**. This tetrameric capsule showed a special affinity for chiral ketones, discriminating between their enantiomers in solution. The caging ligand **387** achieves enantioselective recognition through multiple interfaces between the lining of its cavity and a chiral guest, despite relative weakness and elasticity of the hydrogen bonds that are responsible for its cage framework [53].

Hydrogen-bonded tetrameric capsule **388** that encapsulates neutral polycyclic molecules has been prepared in [54] using guest-templated self-assembly of the corresponding ligand syntone *178* by Scheme 3.53. The spin polarization transfer experiments suggested that these cage complexes underwent rapid guest-exchange processes; however, they are kinetically stable even at elevated temperatures [54].

A tris-3-pyridyl hexagram-shaped ligand syntone *179* is reported in [55] to aggregate in

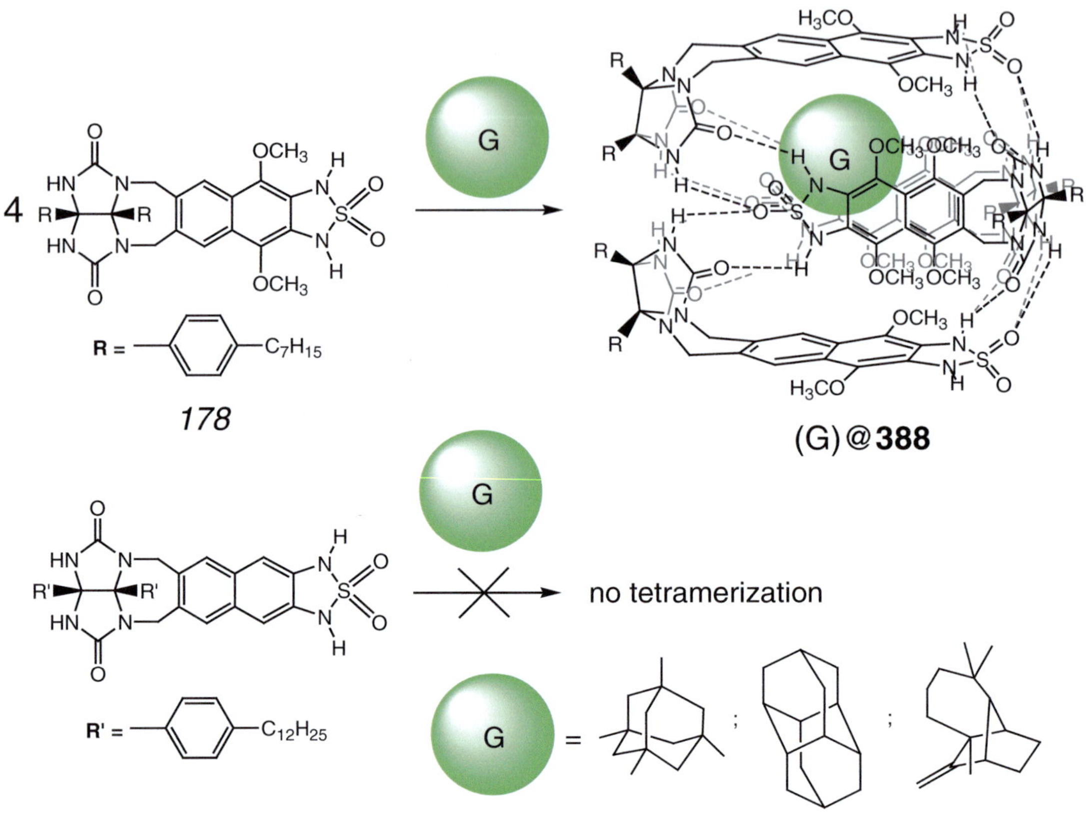

Scheme 3.53

aqueous methanol giving a box-shaped cage framework **389** that is self-assembled through hydrophobic van der Waals and CH…π interactions. Its hydrophobic cavity encapsulates two guest molecules of hexasubstituted benzene derivatives thus giving 1:2 cage complexes by Scheme 3.54. In this hexameric supramolecular capsule with two stacked face-to-face tribromomesitylene guests, the ligand syntones *179* do not have their original C_3-symmetry axes because of the pyridyl and *para*-tolyl substituents became nonequivalent [55].

Small spherical molecules, such as adamantane, are reported in [56] to be templating guests that form tetrahedral 1:1 cage complexes of the ligand **390** through rearrangement of the hexameric supramolecular capsule **389** by Scheme 3.55. This rearrangement strongly depends on both the size and shape of a template molecule, the encapsulation of which arises mainly from hydrophobic host–guest interactions. This tetrahedral supramolecular capsule is able to reversibly encapsulate and to release the adamantane molecule through acid–base controlled dissociation and reconstruction of a cage framework of **390**. The X-ray diffraction data showed that four hexagram-shaped C_3-symmetric ligand syntones form the supramolecular capsule with the hydrophobic inner cavity that encapsulates one adamantane molecule. The pyridyl groups of **390** are situated between the clefts formed by other hexagram-shaped syntones, and six pairs of these stacked groups are located on the six sides of its framework. Thus, the tetragonal tetrameric 1:1 cage complex of **390** formed by four hexagram-shaped ligand syntones has been self-assembled [56] exclusively in an induced-fit manner in the presence of a spherical template molecule. In contrast, a hexameric box-shaped capsule **389** has been formed without the

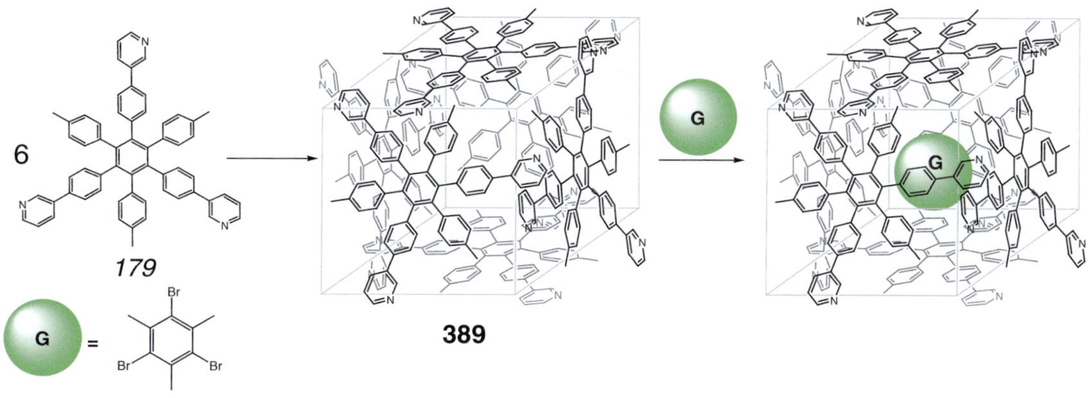

Scheme 3.54

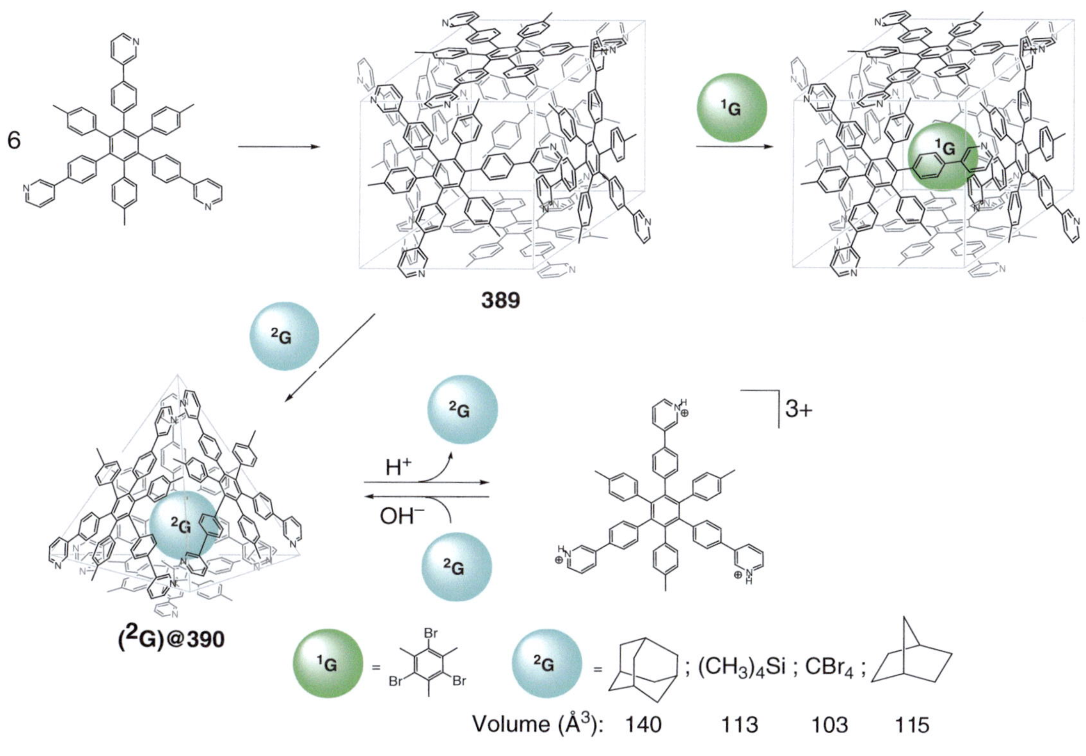

Scheme 3.55

template molecule. The reversible encapsulation and release of the guest molecule as a result of the dissociation and reconstruction of this capsule was achieved in [56] by acid–base control.

Self-assembly of six and four gear-shaped ligand syntones *180–182* and similar guest-templated reactions (Scheme 3.56) have been tested in [57] to obtain highly stable and water-soluble, box-shaped supramolecular capsule **391** and its tetrahedral analog **392** with encapsulated 2,4,6-tribromomesitylene and adamantane molecules. The design of the ligand syntone *180* has been based on the structure of the hexameric capsule **389** with hydrophilic pyridyl nitrogen atoms

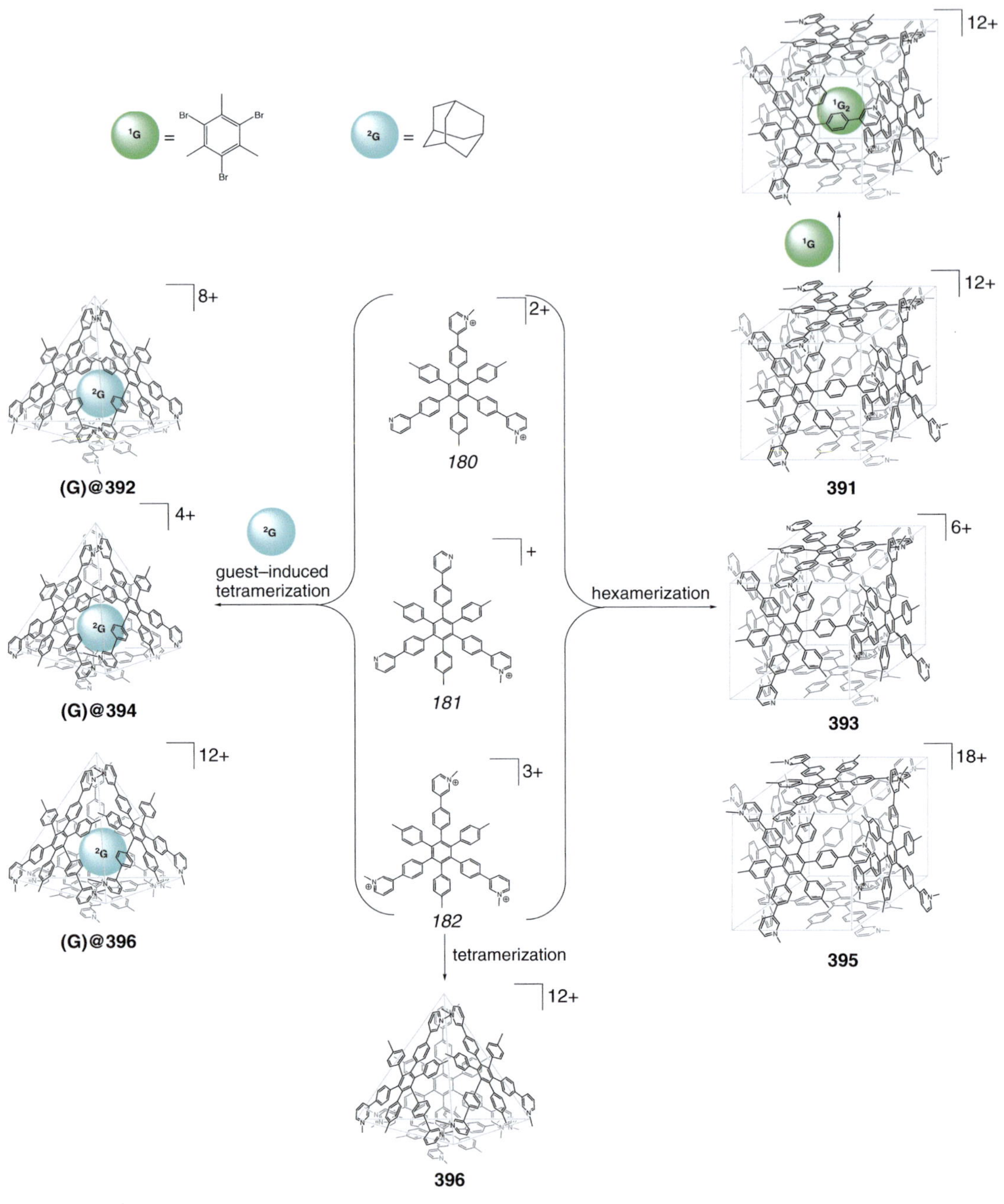

Scheme 3.56

put outward and the hydrophobic surface of the hexaphenylbenzene part buried deep. Replacement of its pyridyl substituents with *N*-methylpyridinium groups in **391** made this capsule remarkably soluble in water and more thermodynamically stable due to electrostatic interactions between the positively charged pyridinium and the electron-rich pyridyl nitrogen atoms in the triply stacked aromatic rings of the caging ligand. Indeed, this highly symmetric box-shaped capsule and its 1:2 cage complex have been exclusively formed in D_2O solution and in solid state. In the case of the ligand

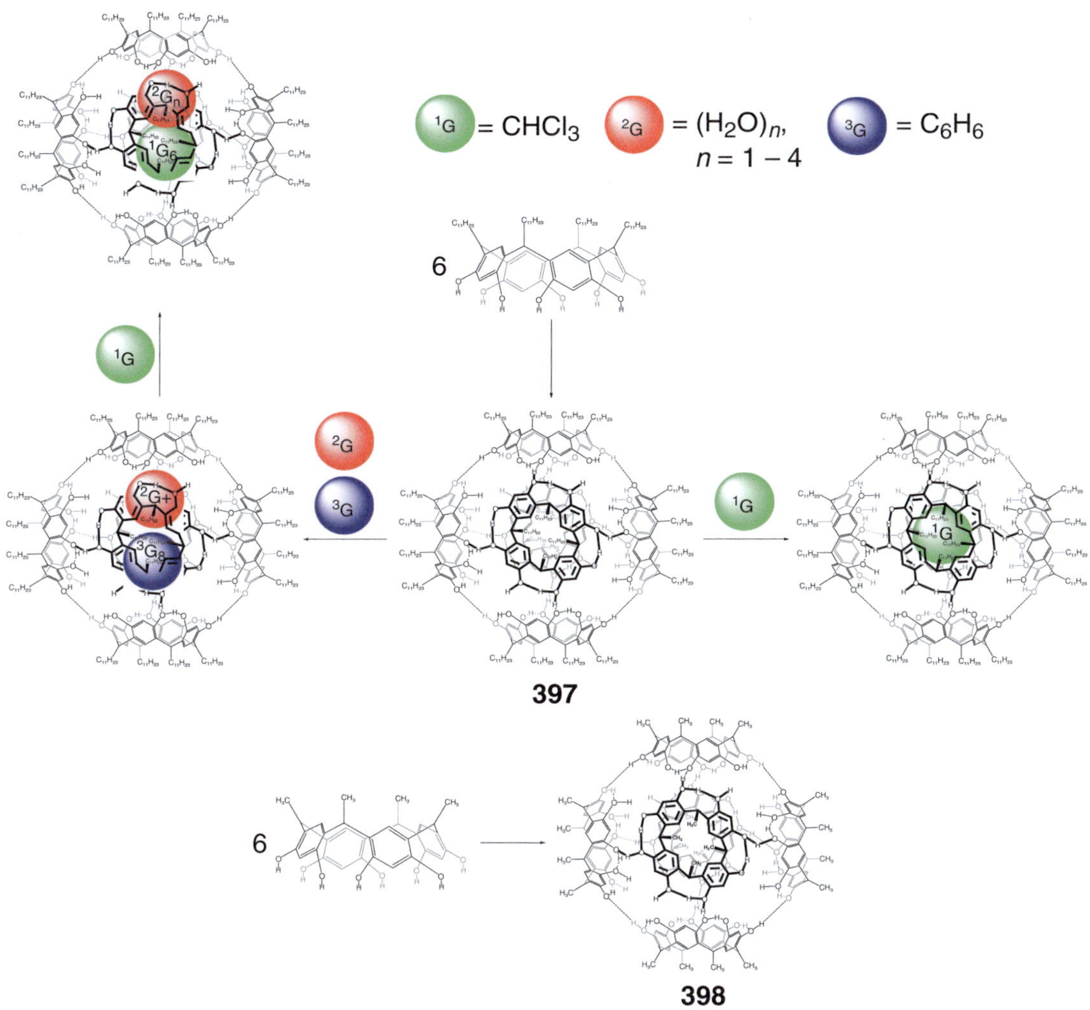

Scheme 3.57

syntone *181* with one *N*-methylpyridinium group, the corresponding hexameric capsule that exists as several structural isomers has been formed in aqueous methanol, but it was insoluble in water. Guest-templated self-assembly of four gear-shaped ligand syntones *182* with three *N*-methylpyridinium groups each and the template adamantane molecule gave a water-soluble tetrahedral 1:1 cage complex of the supramolecular capsule **396**. The relative stability of this tetrahedral capsule is explained in [57] by electrostatic repulsion between its three pyridinium groups. As a result, the cage complex of **396** is selectively formed by hydrophobic effects and template effects of a guest as well.

The possibility of the formation of hexameric hydrogen-bonded capsules has been first shown by J. L. Atwood and coworker [58] for *C*-undecyl- and *C*-methylcalix[4]resorcinarene ligand syntones giving the supramolecular assemblies **397** and **398**, respectively.

The hexaresorcinarene capsule **397** in its wet chloroform solution co-encapsulates chloroform and water molecules, forming the corresponding heteroguest cage complexes by Scheme 3.57 [59]. This caging ligand also co-encapsulates benzene and water, and the benzene co-caged molecules easily undergo exchange with chloroform as a concurrent guest giving the corresponding

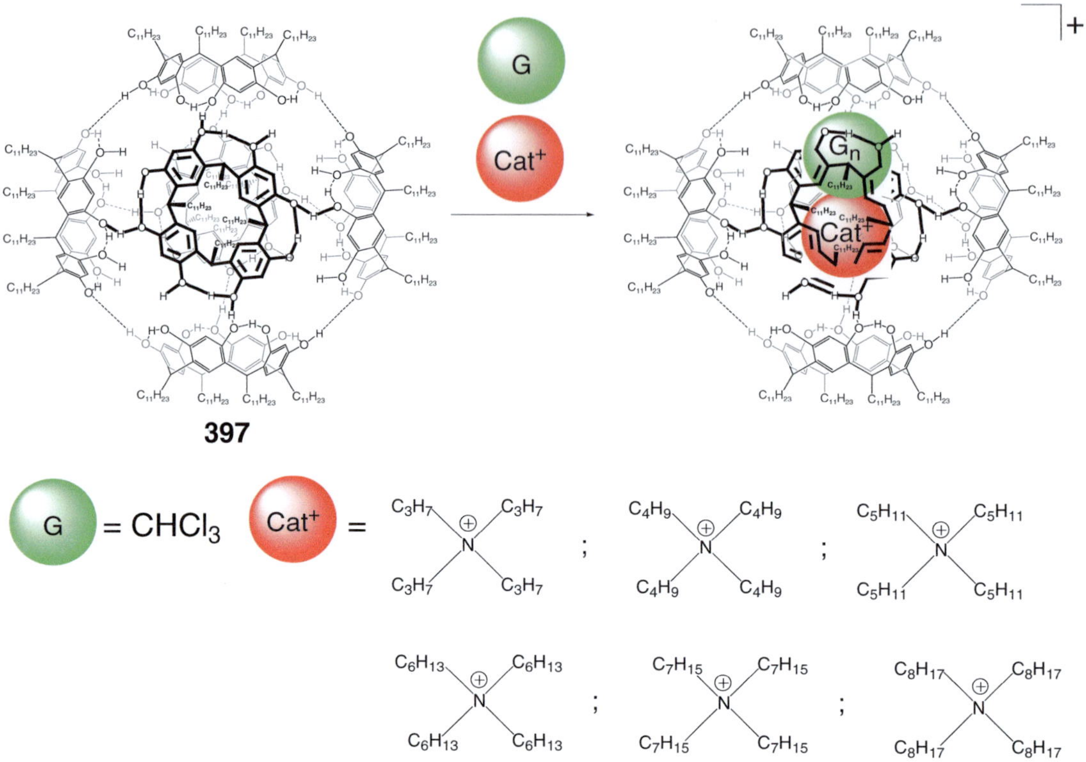

Scheme 3.58

cage complexes with co-encapsulated chloroform and water molecules [59].

Kinetics and thermodynamics of the formation of the cage complexes of the hydrogen-bonded hexameric capsule **397** with encapsulated organic monocations have been studied in [60]. The corresponding ligand syntone forms these 1:1 cage complexes by Scheme 3.58 in wet $CDCl_3$ solution in the presence of an appropriate tetraalkylammonium salt. The remaining space within the cavity of **397** is occupied by co-encapsulated solvent molecules, and a maximum of three and a minimum of one $CHCl_3$ co-guest molecule(s) were found to be co-encapsulated with tetrapropyl- and tetraheptylammonium cations, respectively. Encapsulation of these cationic guests is reported in [60] to be endothermic and entropically favored by the liberation of several caged solvent molecules. Thus, the number of co-encapsulated species decreases with the increase in the cation size. This allows controlling the reversible encapsulation of solvent by this hydrogen-bonded caging ligand. The stabilities of its cage complexes and the rates of their guest exchange decrease for larger cations as a result of [60] higher activation barriers for both *in–out* exchange by conformational restraints in the transition state. The dissociation of one resorcinarene ligand syntone from this hexameric capsule **397** is required for guest exchange.

Detailed study of self-assembly, stability, and guest affinity of the pyrogallol- and resorcin[4]arene capsules in their deuterochloroform solutions in the presence of polar solvents by diffusion NMR spectroscopy has been performed in [61]. The diffusion coefficients of 1:1 cage complexes of the hexameric encapsulating ligands **397** and **399–401** with guest solvent molecules (Scheme 3.59) in chloroform were found to be relatively low and to increase in the presence of deuteromethanol due to their disaggregation giving the corresponding monomeric arenes. A new

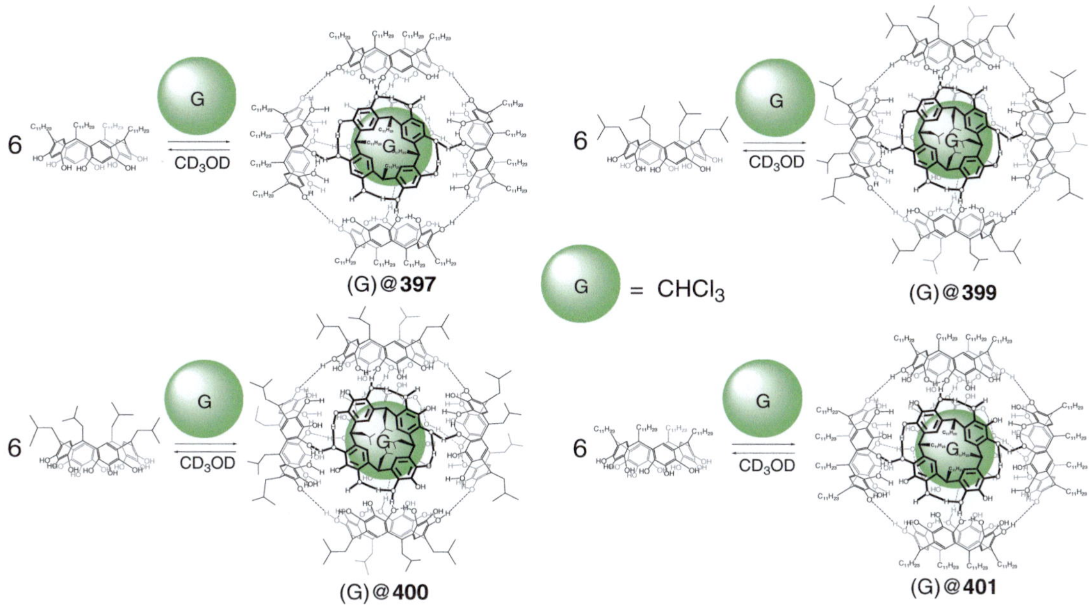

Scheme 3.59

signal of encapsulated chloroform appearing in the ^{1}H NMR spectra of these cage complexes disappeared after addition of this polar solvent [61]. The pyrogallol[4]arene capsules **400** and **401** are more stable than their resorcin[4]arene analogs **397** and **399**, and more lipophilic capsules **397** and **401** are more stable than their homologs **399** and **400**. According to diffusion NMR data and in contrast to their pyrogallol[4]arene analogs, the resorcine[4]arene capsules **397** and **399** contain eight water molecules that form hydrogen bonds with the resorcin[4]arene entities in its cage framework. Six binar mixtures of four different arene monomers have been tested in [61] to study self-recognition in such systems. According to NMR data, the self-assembly of these arenes gave only the mixtures of the corresponding homohexameric capsules. The resorcine[4]arene caging ligands **397** and **399** encapsulate both the tetraalkylammonium cations and trialkylamines by Scheme 3.60, whereas their pyrogallol[4]arene analogs **400** and **401** can accommodate only neutral guests [61].

According to ^{1}H NMR data of [62, 63], the caging ligand **400** with a cavity volume of approximately 1300 $Å^3$ encapsulates polar solvent molecules and those of initial reagents shown in Scheme 3.61.

The hydrogen-bonded hexapyrogallol[4]arene capsule **397** is reported in [64] to co-encapsulate normal, branched, and cyclic alkanes, tetraethylsilane and tetrabutylgermane by Scheme 3.62. These caged guest molecules undergo exchange processes that are slow on the NMR time scale. As follows from NMR data of [65], this ligand is also able to bind 1,2-*cis*-cyclohexanediol and *L*-phenylalanine as neutral guests to give the corresponding 1:1 cage complexes.

Encapsulation of ADMA as a fluorescent probe by Scheme 3.63 has been used in [62] to examine microenvironment of **397**; the changes in its fluorescent properties resulted from conformational restrictions and movement of the guest within the cavity of this caging ligand. Their 1:1 cage complex has been characterized both in solid state and in solution using single-crystal X-ray diffraction data and solution ^{1}H NMR and fluorescent spectra, respectively. In the former case, encapsulation of the relatively small guest molecule is reported to substantially change the crystal packing of this large and robust cage framework due to the formation of a series of supramolecular host–guest interactions. The fluorescence emission intensity of the anthracene moiety of the encapsulated ADMA molecule in

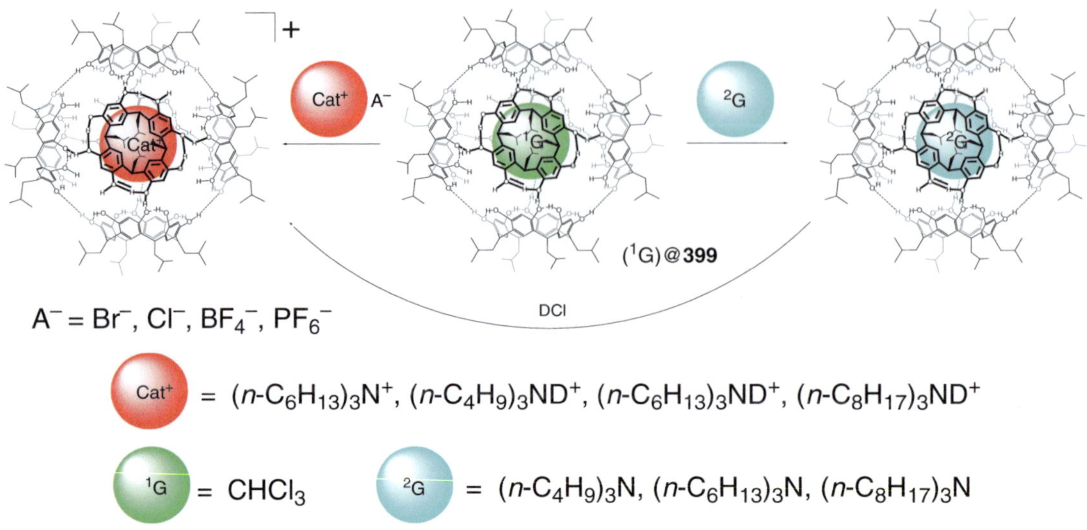

Scheme 3.60

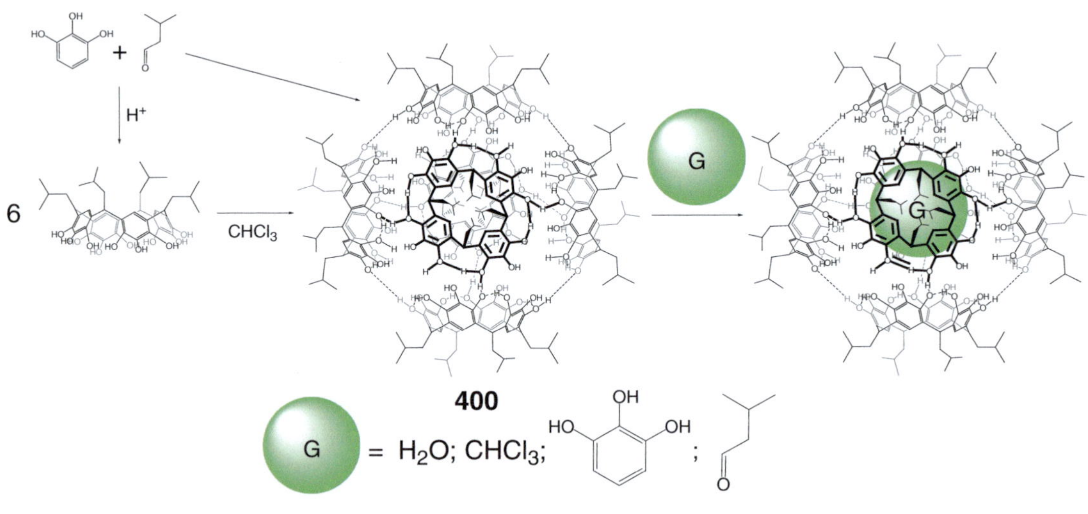

Scheme 3.61

solution is dramatically increased compared to free ADMA. This effect is attributed in [62] to structural constraints imposed by the caging ligand **397**, minimizing the possibility of nonradiative excited-state deactivation such as collisional quenching by the solvent molecules.

FRET has been used in [66] to probe dynamic behavior of the hexaresorcinarene ligands **402** and **403**, and their cage complexes. The resorcinarenes *183* and *184* with donor and acceptor fluorophore pyrene- and perylene-containing substituents, respectively, have been used as ligand syntones. Each of these syntones easily forms the corresponding hexameric caging ligand in the presence of tetrahexylammonium bromide, a well-known suitable guest for such a self-assembly (Scheme 3.64). As follows from luminescent spectra and change in FRET signal, mixing of these homohexamers **402** and **403** resulted in the formation of hybrid caging ligands by Scheme 3.65. FRET has been also observed in the case of the encapsulation

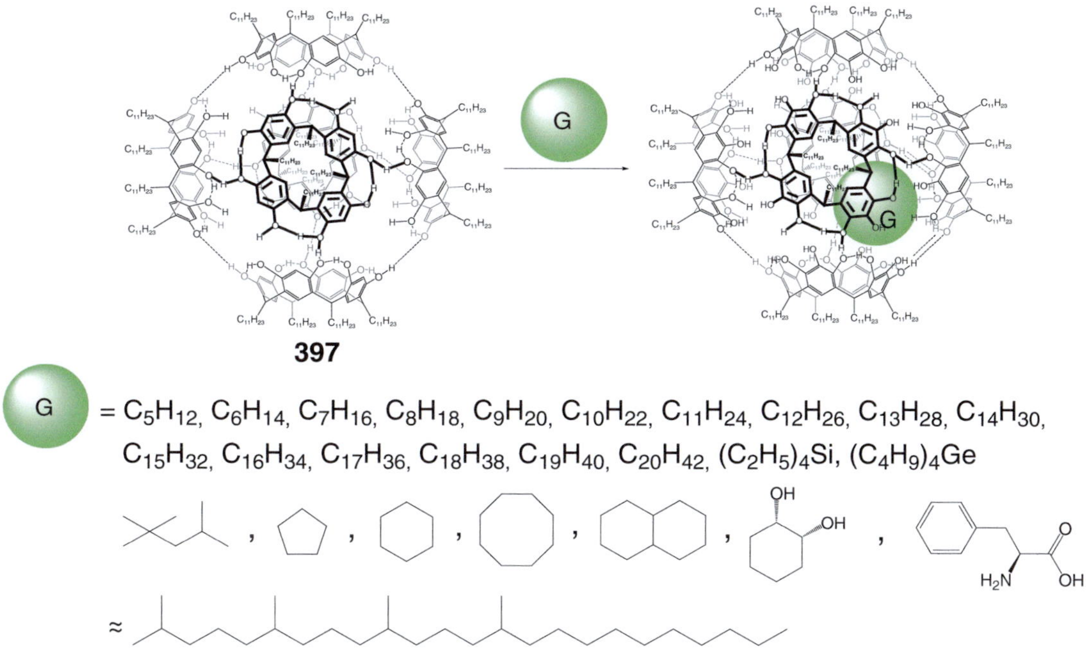

Scheme 3.62

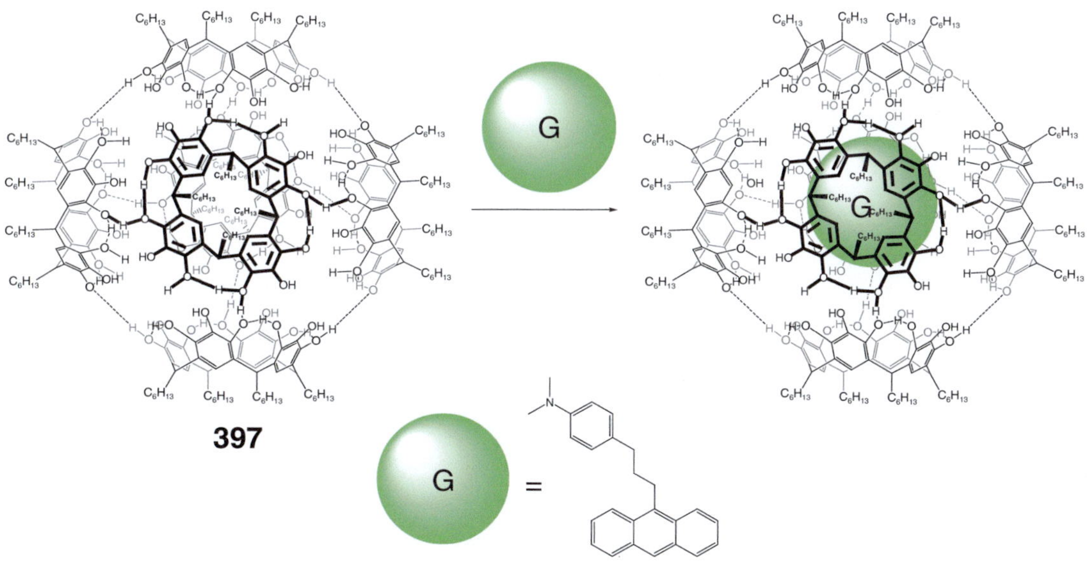

Scheme 3.63

of a pyrene-containing azaadamantane guest by the donor-containing hexaresorcinarene caging ligand **403**; in all these cases, FRET signals disappeared after addition of methanol [66].

The same approach based on FRET between the donor and acceptor pendant substituents has been used in [67] to study dynamic behavior of the hexapyrogallolarene (**404** and **405**) and hexaresorcinarene (**402** and **403**) capsules using concentration and temperature variations. The importance of water for self-assembly of their arene syntones and sensitivity of the cage

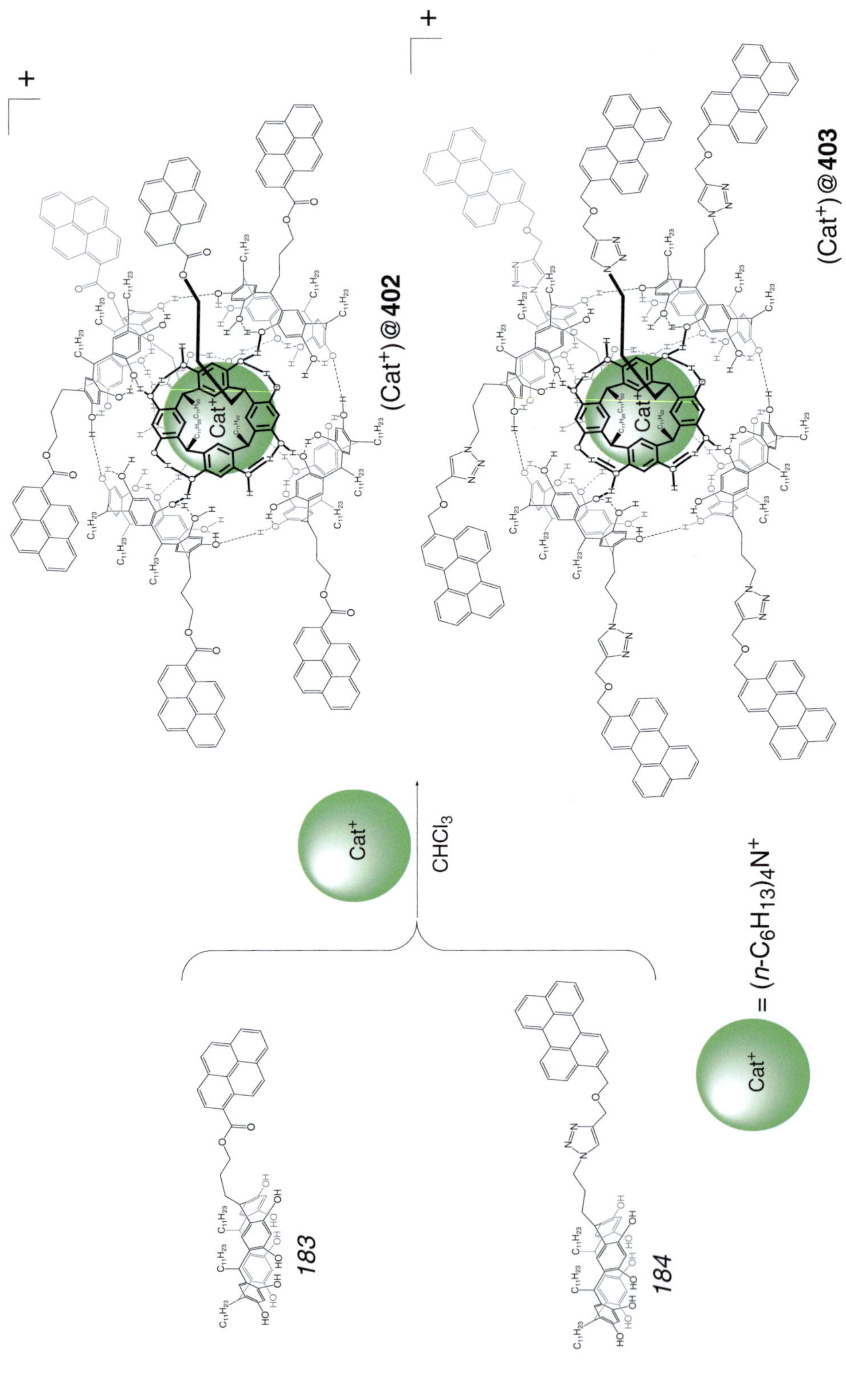

Scheme 3.64

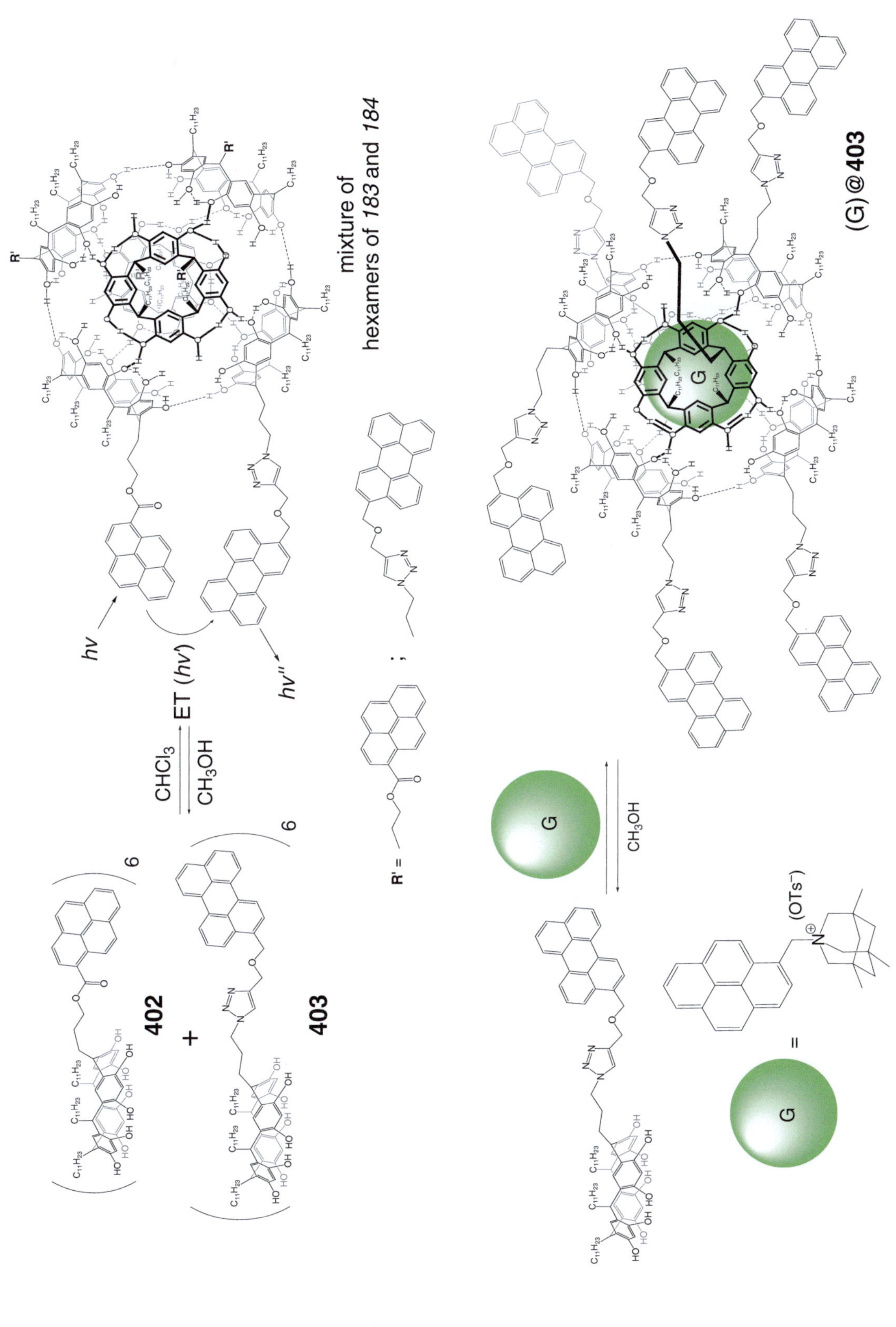
402
403
CHCl3
CH3OH
ET (hν')
hν
hν''
mixture of
hexamers of 183 and 184
R' =
(OTs−)
G
CH3OH
(G)@403

Scheme 3.65

framework stability to polar solvents have been FRET tested in [67]. Such study of the pyrene- and perylene-labeled hexapyrogallolarene syntones (Scheme 3.66) at their different concentrations showed that the exchange of the pyrogallolarene syntones was faster at their low concentrations, while higher ET was observed at high concentrations. Kinetics of the formation of these pyrogallolarene capsules **404** and **405** is temperature-dependent: the elevation of temperature by 20° C causes the decrease in a half-life time to reach equilibrium from *ca.* 96 to *ca.* 8 min. This time also decreases in the presence of water impurities, which shift the equilibrium in the direction of monomeric ligand syntones. In contrast, the formation of hexaresorcinarene capsules **402** and **403** is only slightly concentration- and temperature-dependent, and the process requires the presence of water. During the self-assembly reactions between the donor- and acceptor-containing pyrogallolarene and resorcinarene syntones by Scheme 3.67, no formation of the hybrid capsules has been detected in [67]. Thus, these pyrogallolarene and resorcinarene syntones undergo self-assembly and self-sorting exclusively into the homohexameric capsules in solution even at their low concentrations [67].

Encapsulation of *185* as a guest that contains a fluorophoric anthracene fragment and a quenching dimethylaniline group with possible CT between them by the hexapyrogallol[4]arene capsules **397** and **406** by Scheme 3.68 has been studied in [68]. According to ^{1}H NMR data, population of the cage framework of **397** in the presence of acetonitrile and ethyl acetate becomes only 50 % due to the conformational lability of an encapsulated molecule *185* and its transition from sandwich-like cage framework with stacking interactions between this guest and polyaromatic fragments of the host to fully extended conformation (without such interactions). 1:1 cage complexes of **397** and **406** with encapsulated molecule *185* in the presence of ethyl acetate have very week, if any, exciplex emission. Its intensity increases approximately two- and five-fold in the presence of ethyl acetate and acetonitrile, respectively. The same results have been obtained in [68] for the self-assembly of these hexameric capsules in the presence of *185* and for its interaction with the initially formed caging ligands; the steady-state data showed that the encapsulated guest molecule is in extended conformation. Heteroguest 1:1:1 cage complex of **397** with encapsulated acetonitrile and *185* molecules undergoes slow disproportion in time (from 10 to 40 % per 12 days) causing a substantial increase in the exciplex emission of the free molecules of *185* in their folded conformation. In contrast, the analogous 1:1 cage complex with an encapsulated ethyl acetate molecule is stable [68].

Pathways of disassembly and reassembly of the hexa-*C*-alkylpyrogallol[4]arene capsules **401** and **407–410** (Scheme 3.69) and their possibility to be used as dihydrogen molecular cargo have been discussed in [69]. The formation of the target capsules and their cage complexes has been detected by DLS and ^{1}H NMR methods. Mixing of the capsules **407** and **410** in chloroform gave only homodimeric assemblies that under SDP conditions underwent disassembly on a disk followed by reassembly to give their heterodimeric derivatives. The flash disassembly–reassembly process has been used in [69] to encapsulate molecular hydrogen by Scheme 3.70. Such encapsulation is hampered for the homohexameric caging ligands, and the caged H_2 molecule can be released from their cavities by heating or during a long-time storage [69].

Self-assembled hexaresorcinarene capsules **397** and **398** (Scheme 3.71) have been used in [70] as potent hosts for paramagnetic nitroxide spin probes (e.g., the derivatives of *tempo*). According to EPR data, they encapsulate cationic and amine nitroxide guests, which are complementary to the mostly aromatic inner cavities of these caging ligands, but discriminate oxo- and hydroxyl-functionalized nitroxides [70].

A highly efficient solvent-free synthetic procedure has been elaborated in [71] to obtain the hexapyrogallol[4]arene capsule **406** and its kinetically trapped cage complexes with encapsulated organic guests by Scheme 3.72. Using pyrene as a model guest, three solvent-free methods have been tested: dry grinding with mortar and pestle, melting of the mixtures with heat gun, and melting of the mixture

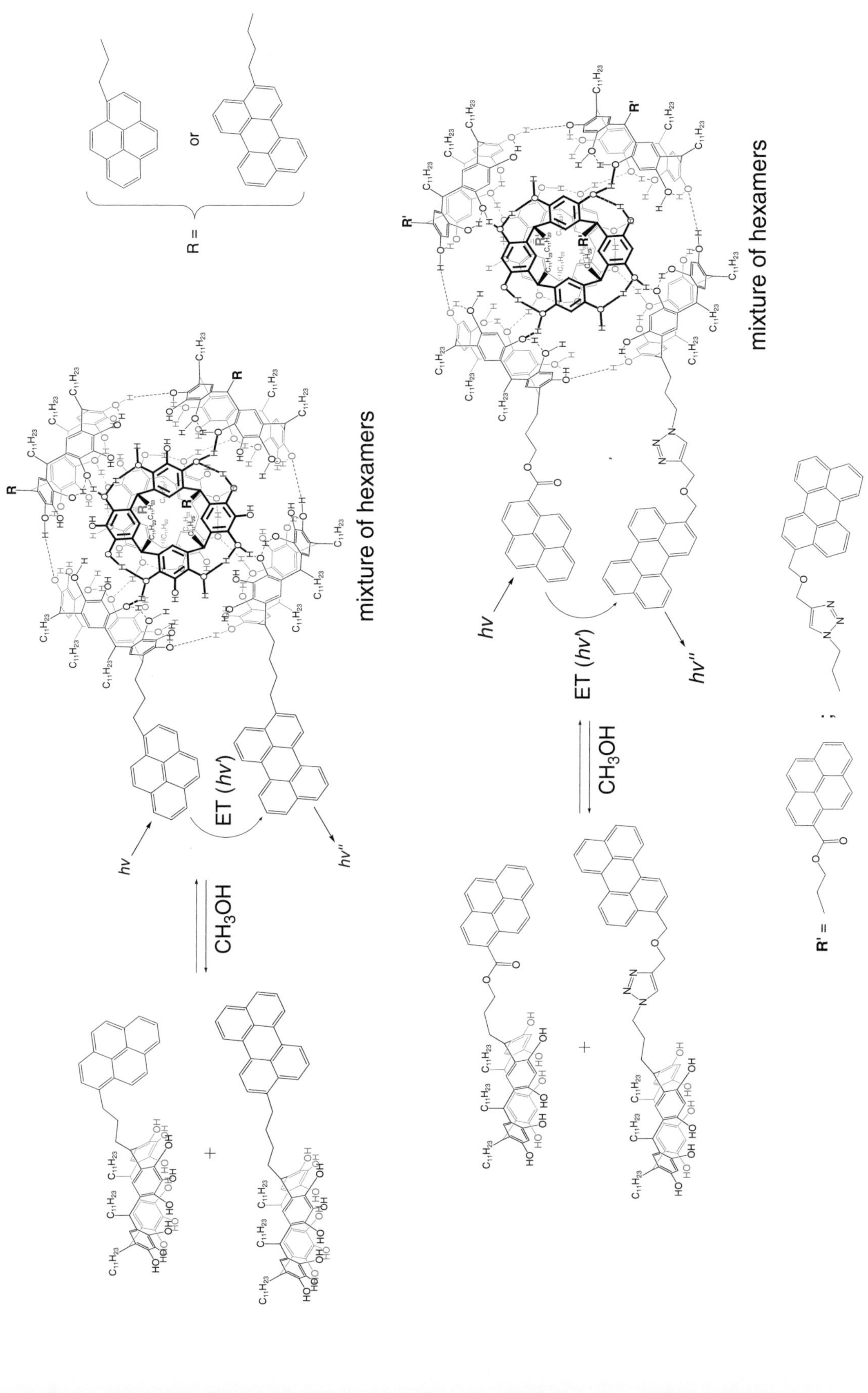

Scheme 3.66

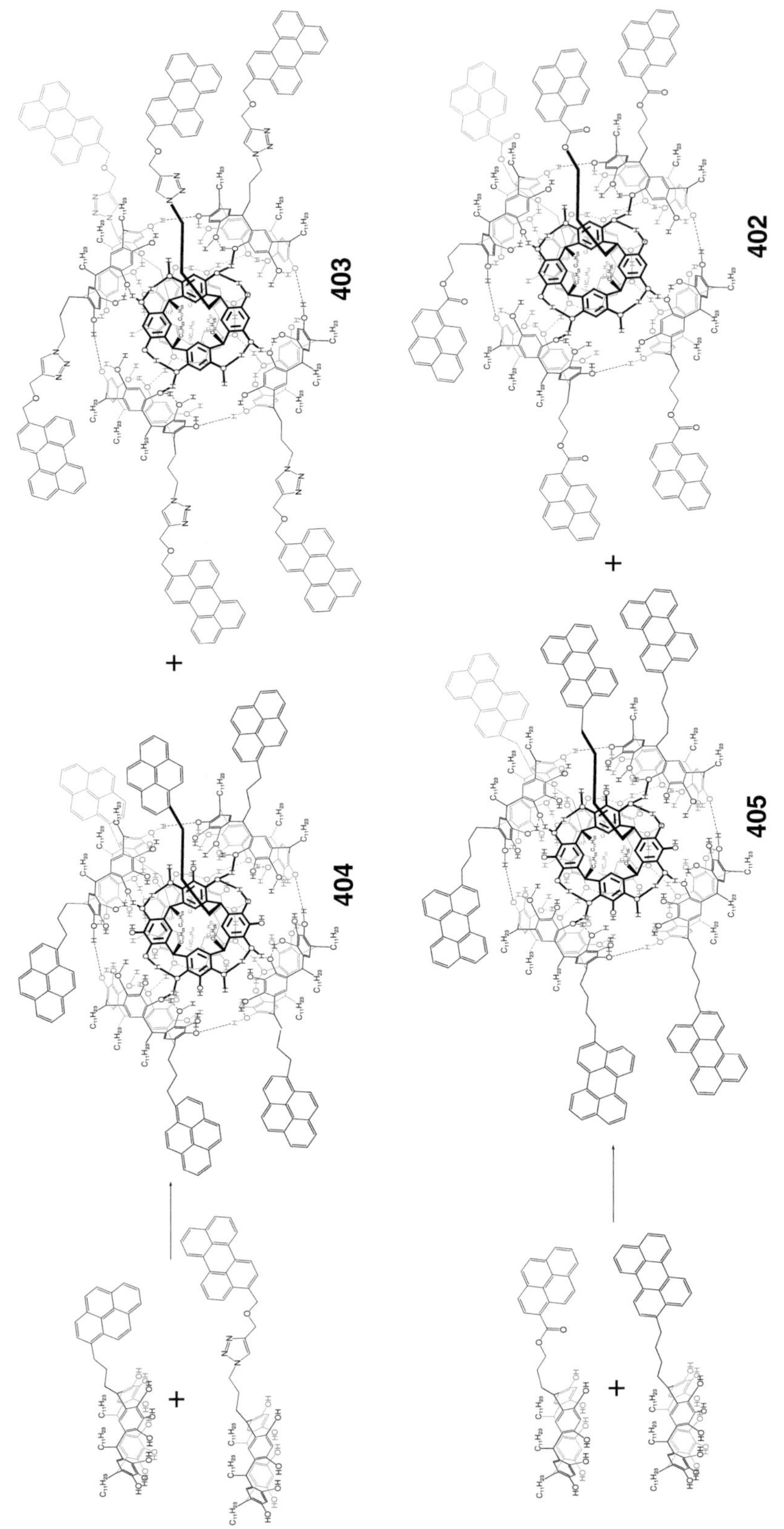

Scheme 3.67

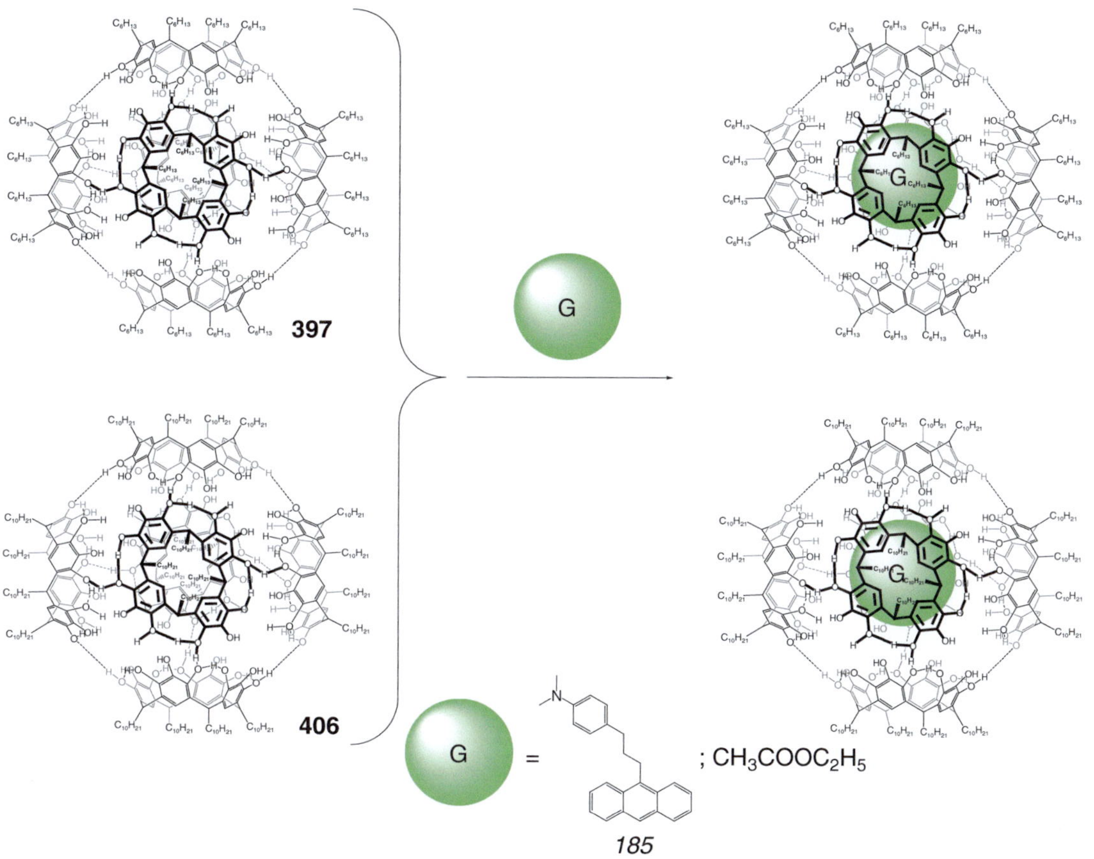

Scheme 3.68

plus paraffin wax as a "solvent." First one gave no target compound, whereas the melting by the second method resulted in the formation of the cage complex with an encapsulated pyrene molecule. This method is, however, limited by the temperature-resistant guests with a low melting point. The use of paraffin wax as a "solvent" is described in [71] to be the most convenient method of the solvent-free synthesis of the hexameric capsules of this type. According to NMR data, the capsule **406** adopts a metastable configuration that undergoes re-configuration in time giving the very stable structure. Its cage complex (pyrene)@**406** releases the caged pyrene molecule upon heating up to 70° C, and the caging ligand **406** re-encapsulates the solvent chloroform molecules. It also efficiently binds a wide range of organic guests but discriminates some bulky molecules shown in Scheme 3.72.

NMR study of binding of various alcohols shown in Scheme 3.73 by the hexametric resorcin[4]arene capsule **397** has been performed in [72]. 2-octyl-1-dodecanol and 1-octadecanol did not interact with it, whereas bulkier and branched alcohols such as 3-ethyl-3-pentanol and 2-ethyl-1-butanol underwent encapsulation. At the same time, small and unbranched alcohols such as 3-pentanol and 2-methyl-1-butanol are reported in [72] to be a part of this hydrogen-bonded capsule.

The capsule **397** is reported in [73] to be a reasonably strong Brönsted acid with high affinity toward tertiary amines, their salts, and other guests shown in Scheme 3.74. This allowed performing highly substrate-selective Witting reaction and substrate-selective diethyl acetal hydrolysis within its cavity.

Scheme 3.69

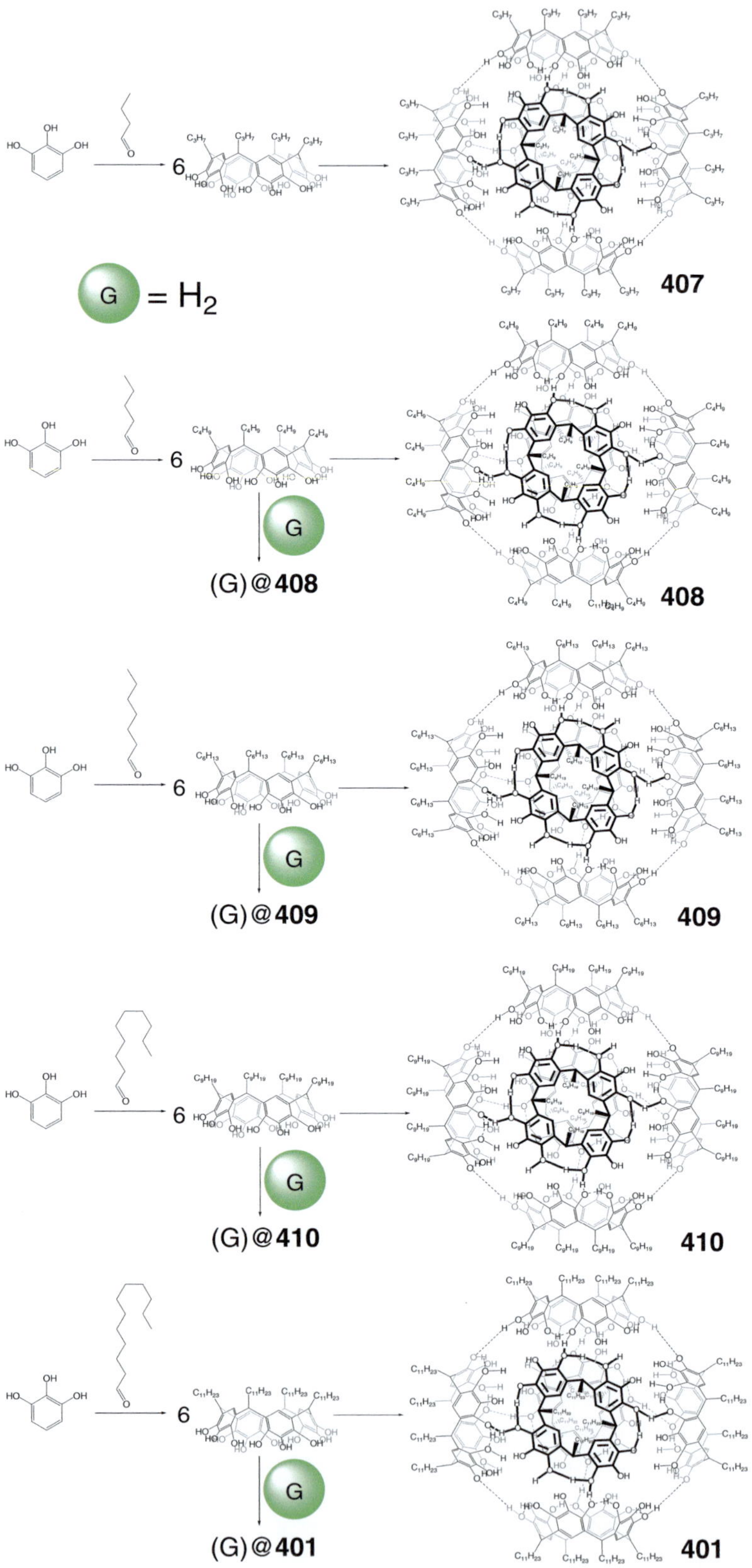

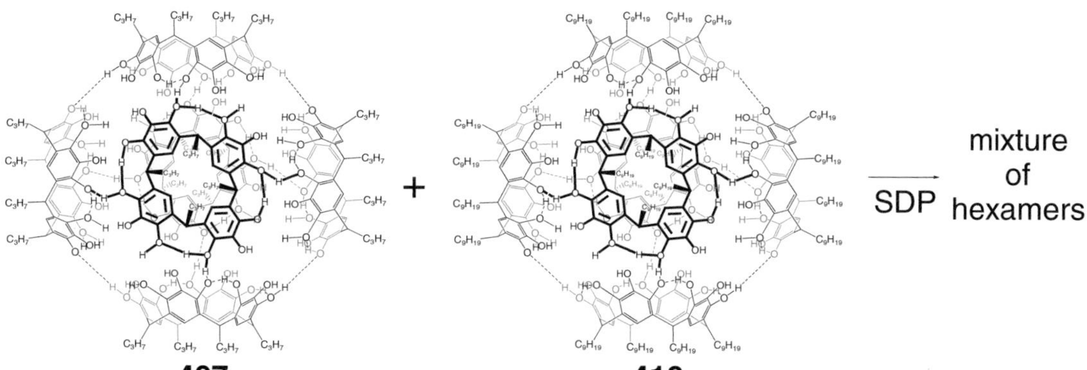

Scheme 3.70

3.1.2 Encapsulation of Anions

Formation of a 1:1 cage complex of the supramolecular capsule **412** by dimerization of a tetracationic calix[4]arene ligand syntone *186* (Scheme 3.75) in the presence of the dianionic guest *187* has been studied in [74] using NMR, ESI-MS, and ITC methods. The double-headed guest molecule with aromatic fragments is deeply included into a lipophilic "pocket" in the ligand's cavity. The polar anionic sulfonate groups are located close to the positively charged upper rim of the calix[4]arene platform due to electrostatic interactions between them, while the ethylene glycol spacer is in equatorial region between the facing calixarene entities [74].

3.1.3 Encapsulation of Cations

Hydrogen-bonded capsules **309**, **353**, **355**, and **413** formed by self-complimentarily ligand syntones are reported in [75] to encapsulate organic cations by Scheme 3.76; the formed cage complexes remain even in gas phase under the conditions of ESI-MS experiments.

Cage complexes of homo- and heterodimeric hydrogen-bonded capsules **414–416**, the derivatives of calixarene tetraurea-containing ligand syntones and their analogs, with encapsulated organic cations shown in Scheme 3.77, have been detected in gas phase using ESI-MS method [76]. The competition experiments with different guest cations revealed a clear dependence of the encapsulation process on their size and shape. The formation of dumbbell-like assemblies **417** and **418** containing two positively charged guest species by Scheme 3.78 has been also observed. The hydrogen-bonded triple-capsule **419** shown in Scheme 3.79 is reported to contain seven entities: one tripodal cross-linking fragment, three cage frameworks, and three encapsulated cations.

A D_{4d}-symmetric hydrogen-bonded capsule **420** with a cavity volume of approximately 950 Å^3 is reported in [77] to encapsulate suitable cryptand and cryptate guests (including their ionic associates with SCN^- and CN^- anions) having the volumes of 390–420 Å^3 thus forming the host–homoguest and host–heteroguest 1:1 and 1:1:1 "*Matreshka*" complex–in–complex assemblies by Scheme 3.80.

The tetrameric hydrogen-bonded capsules **383** and **384** with encapsulated cations and neutral guests, formed by self-assembly of their self-complimentary ligand syntones (Scheme 3.81), have been characterized in [78] both in solution and in gas phase using NMR and ESI-MS methods, respectively. The data of competition experiments with a series of different guest cations suggest that size selectivity is characteristic of these hydrogen-bonded capsules. MS experiments have revealed the formation of heterotetrameric cage complexes from different ligand syntones that could not be determined using NMR spectra. At the same time, the NMR method

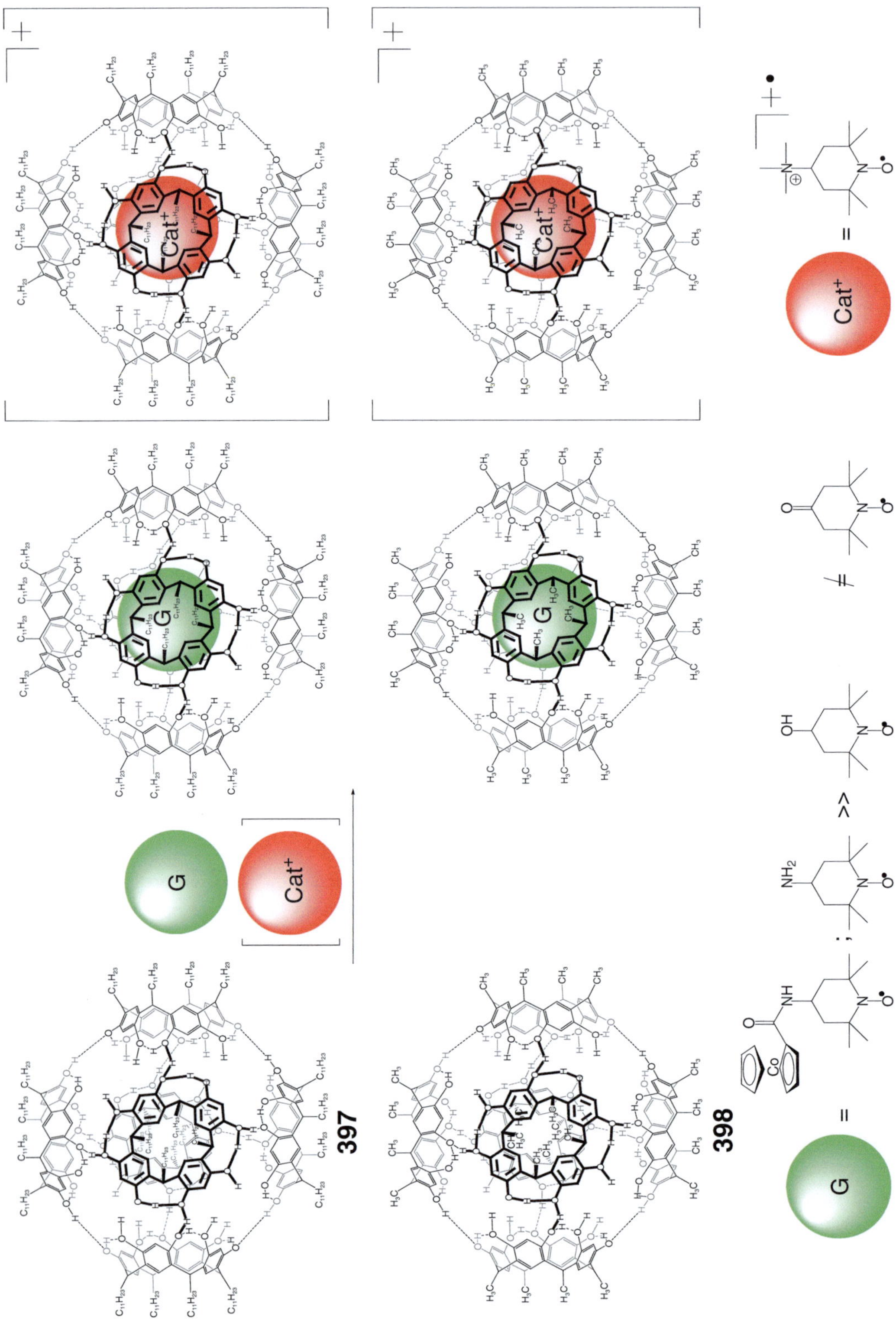

Scheme 3.71

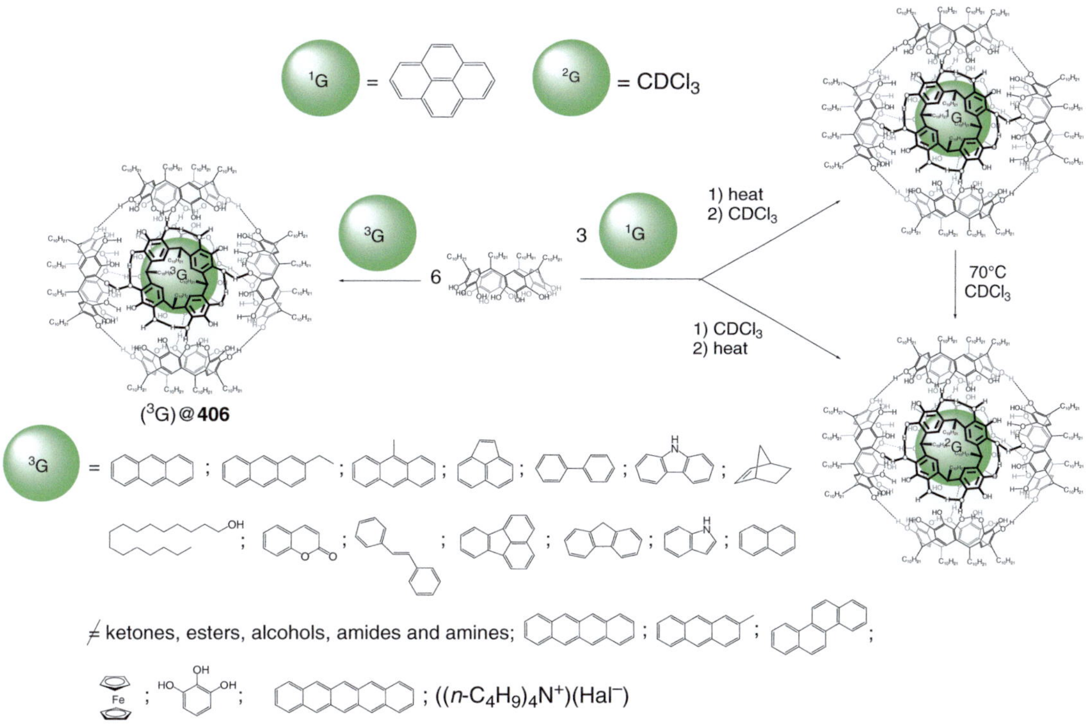

Scheme 3.72

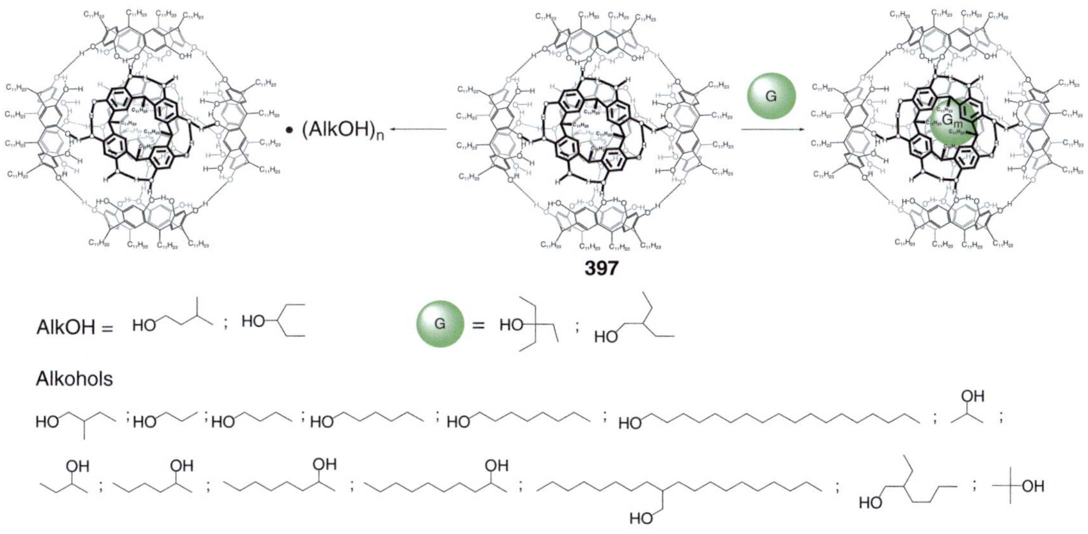

Scheme 3.73

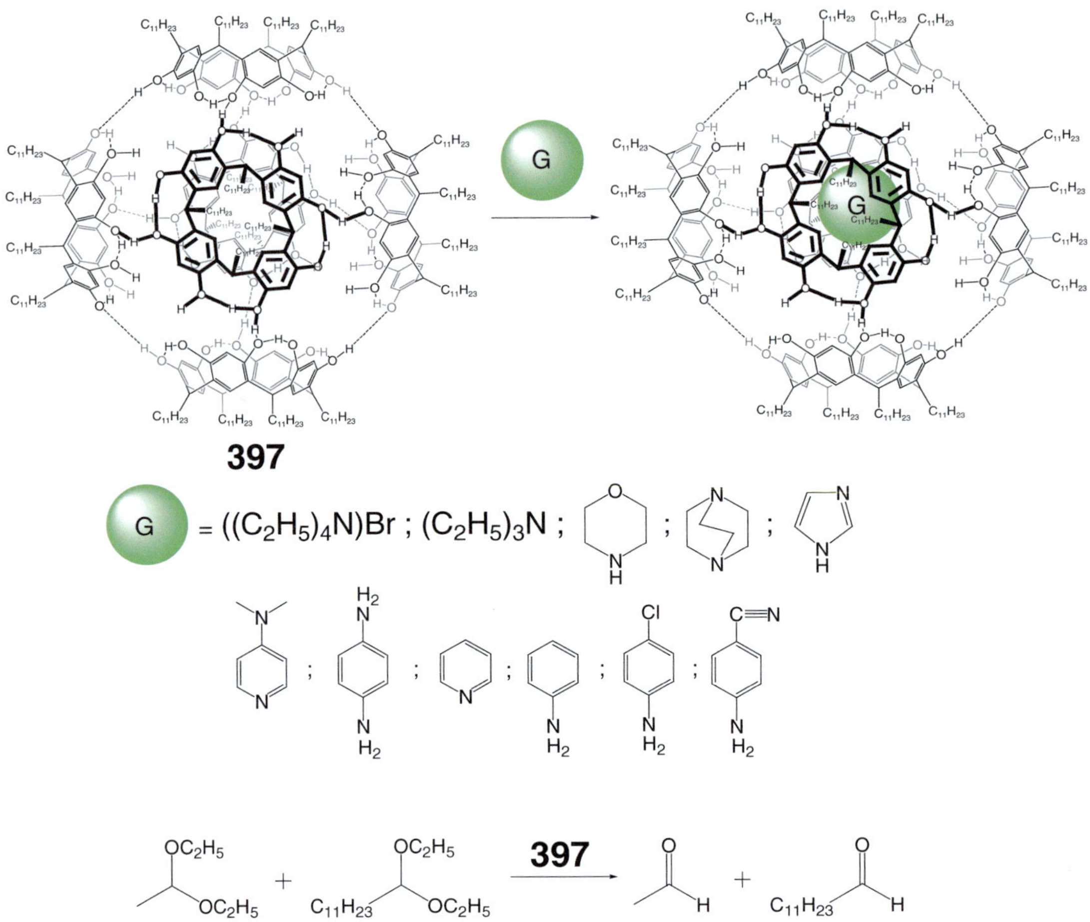

Scheme 3.74

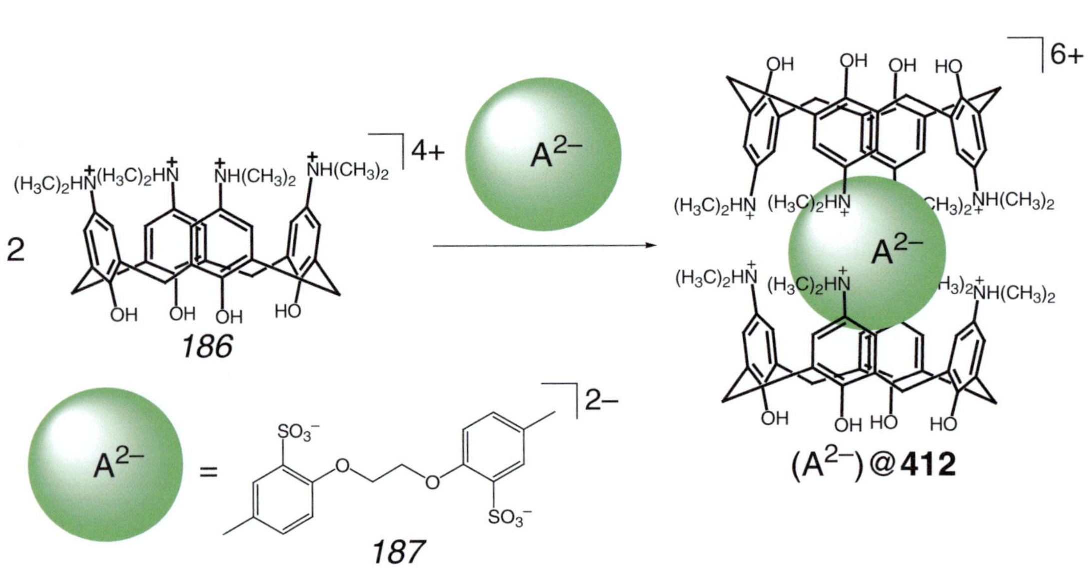

Scheme 3.75

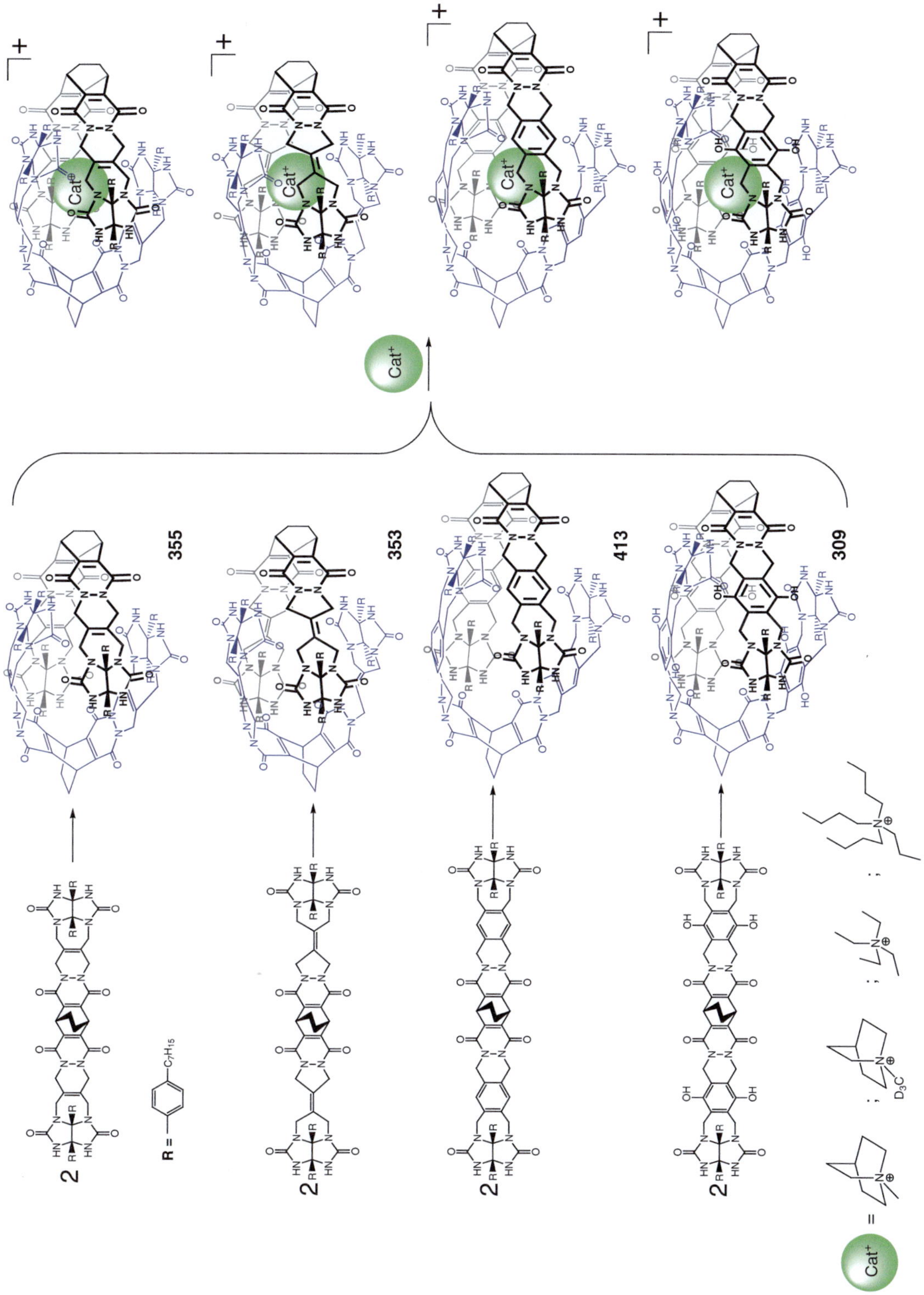

Scheme 3.76

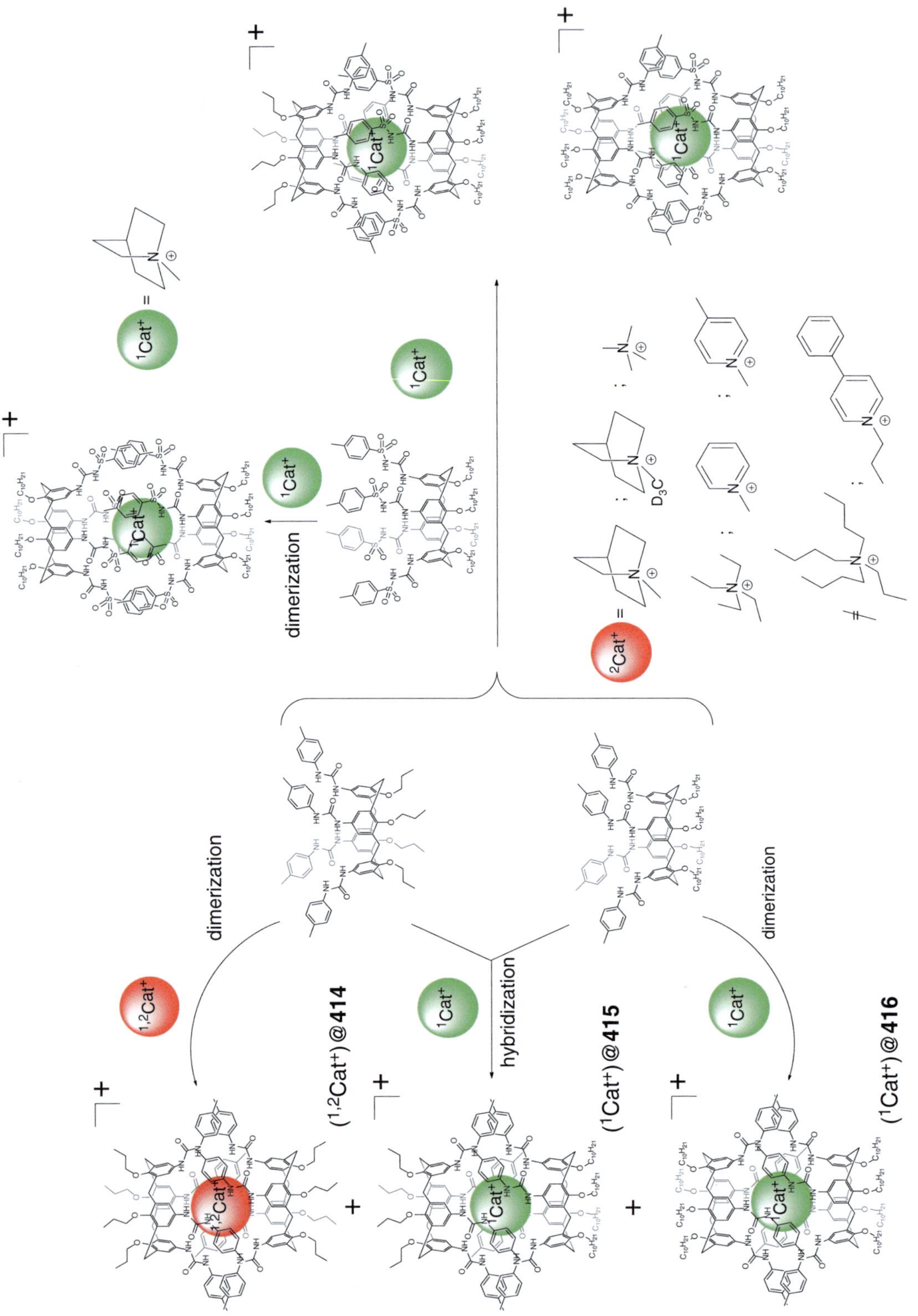

dimerization
hybridization
dimerization
$^{1,2}Cat^+$
$^{1}Cat^+$
$^{2}Cat^+$
$(^{1,2}Cat^+)@$**414**
$(^{1}Cat^+)@$**415**
$(^{1}Cat^+)@$**416**

Scheme 3.77

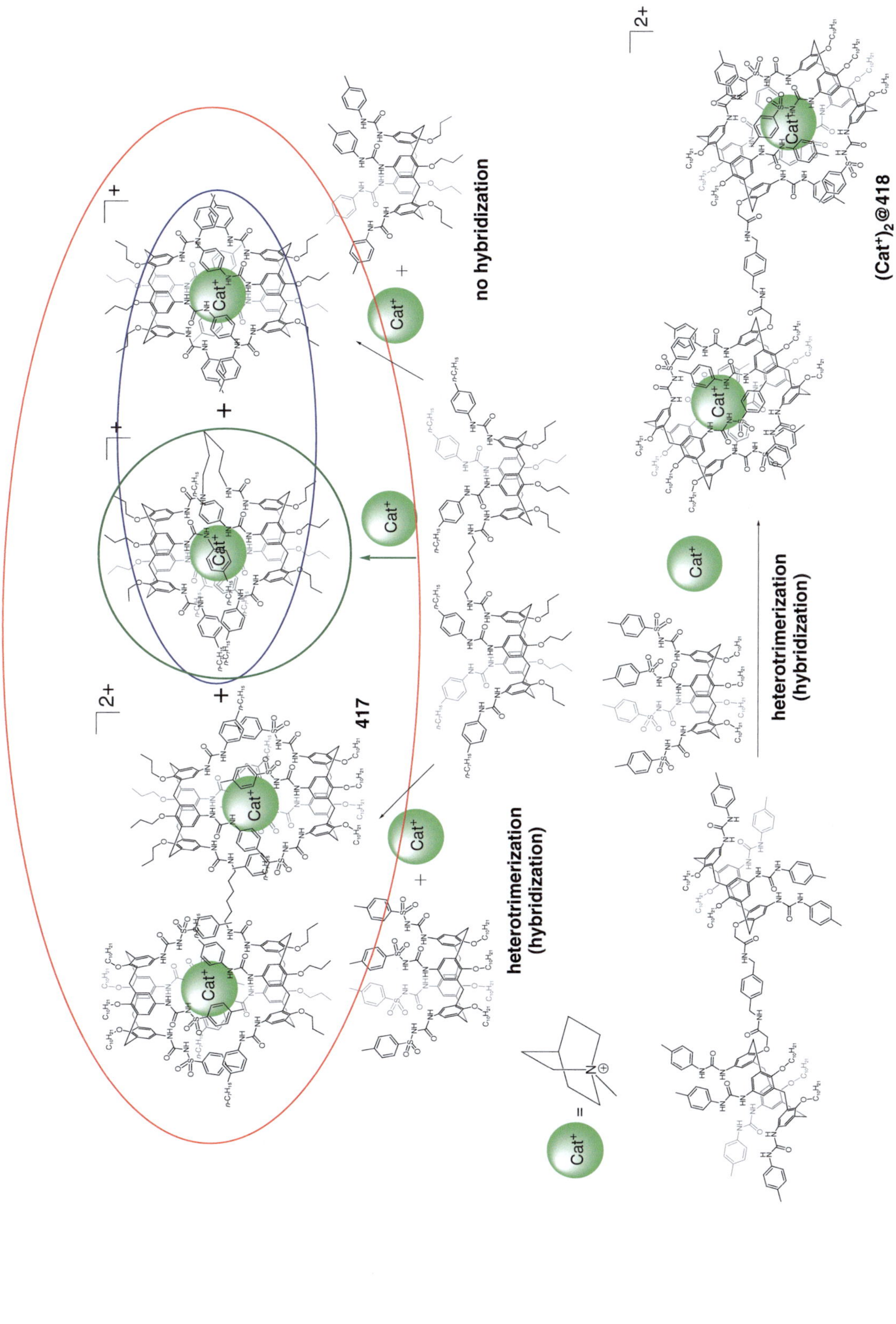
no hybridization
heterotrimerization
(hybridization)
417
$(Cat^+)_2$@418
Cat+

Scheme 3.78

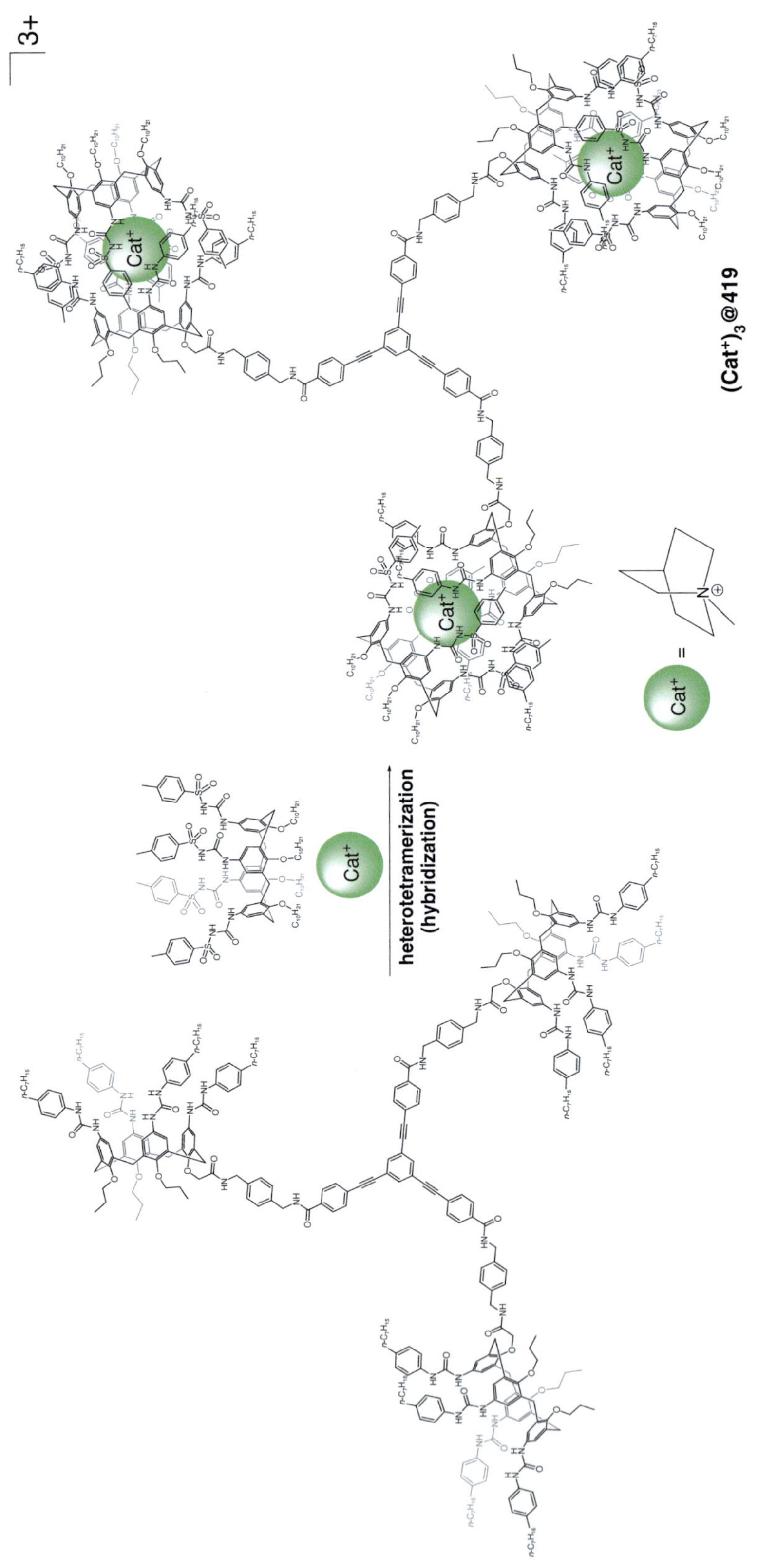

$(Cat^+)_3$@419

Scheme 3.79

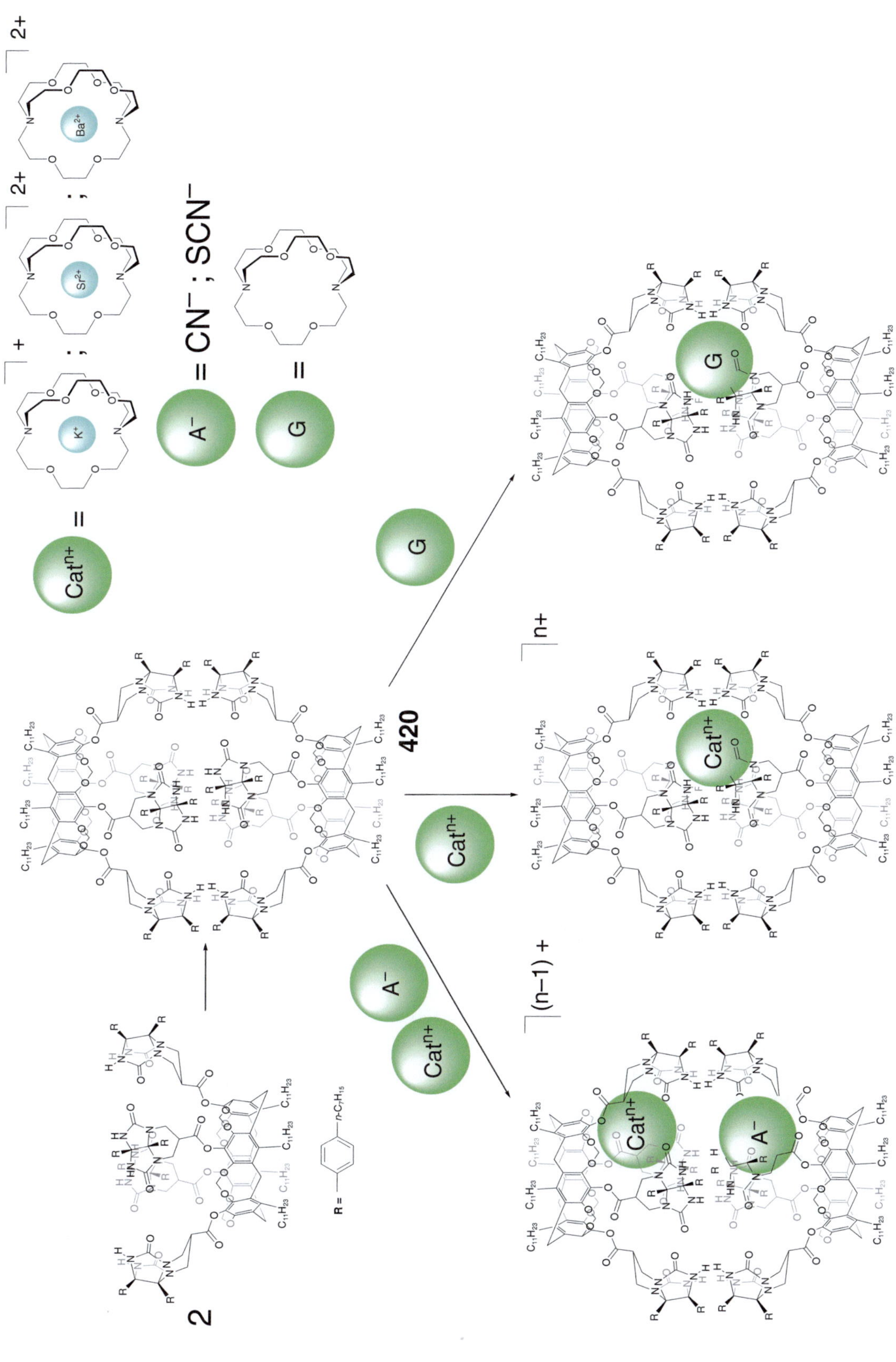

Scheme 3.80

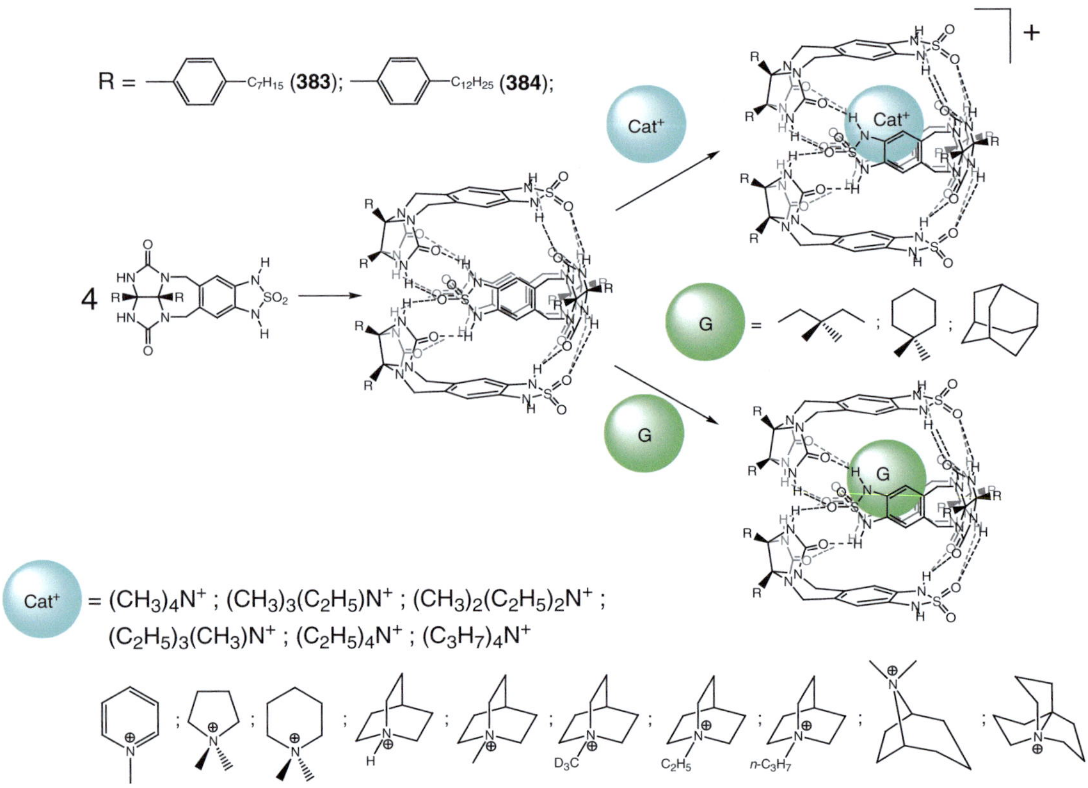

Scheme 3.81

allowed observing in [78] the competitive encapsulation of neutral and cationic guest species that could not be detected using mass spectra.

Template dimerization of the self-complementary resorcarene tetraester syntones *188–190* by Scheme 3.82 in the presence of tropylium cation as a matrix is reported in [79] to give 1:1 cage complexes of the homodimeric capsules **421–423** with an encapsulated tropylium cation. These ligands, however, do not encapsulate benzene or toluene as potent space-filling guests. In the case of a mixture of ligand syntones, a heterodimeric assembly ($C_7H_7^+$)@**424** has been detected in [79] by ^{1}H NMR method. As follows from X-ray diffraction data, a 1:1 cage complex of **423** with an encapsulated tropylium cation is stabilized by eight intermolecular C–O…H–O bonds.

A detailed study of the homo- and heterodimeric resorcinarene capsules **425–427** and their cage complexes, the derivatives of the tetraester syntone *191* and its analogs *192* and *193*, has been performed in [80]. These hydrogen-bonded capsules in solution gave a statistical mixture of the homo- and heterodimeric products. Quantum chemical calculations performed in [80] suggest that the heterodimers formed by π-donor and π-acceptor resorcarene syntones should be more stable than their homodimeric analogs. Indeed, the ^{1}H NMR study of their self-assembly in the presence of tetramethylammonium cation confirmed the formation of the corresponding 1:1 cage complexes by Scheme 3.83. The benzyl carboxylate analog of these syntones does not form the same cage complexes with the encapsulated tetramethylammonium cation due to the weak hydrogen-bonding properties and lability of its benzyl substituents. On the other hand, the caging ligands **425–427** do not encapsulate BF_4^- anion and triethyl- and tetraethylammonium cations. At the same time, their resorcarene ligand syntones form the heterodimeric cage complexes with the encapsulated tetraethylammonium cation (see an example of the assembly (Cat^+)@**428** on Scheme 3.84) [80].

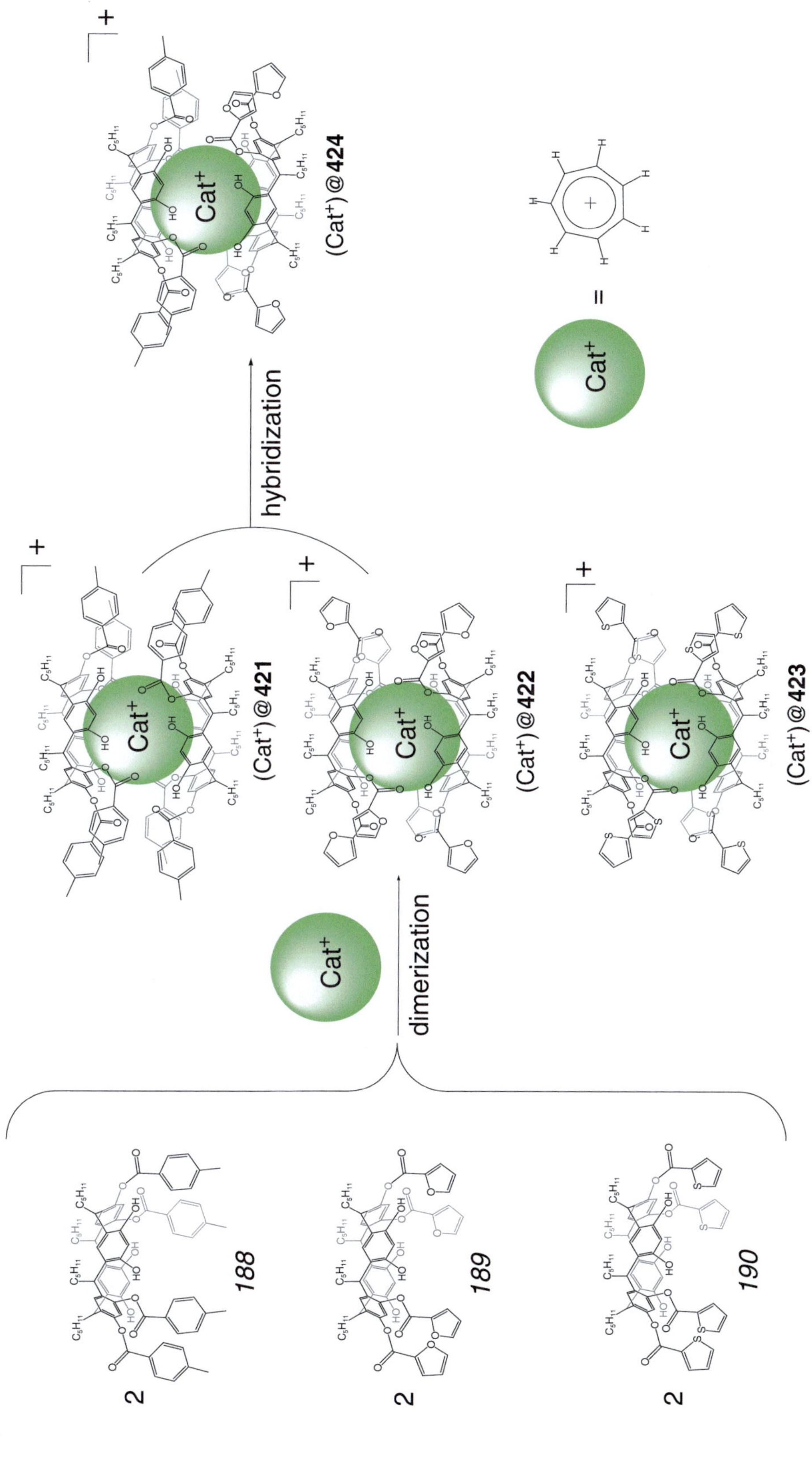

2
188
2
189
2
190
dimerization
Cat+
(Cat+)@421
(Cat+)@422
(Cat+)@423
hybridization
(Cat+)@424
Cat+ =

Scheme 3.82

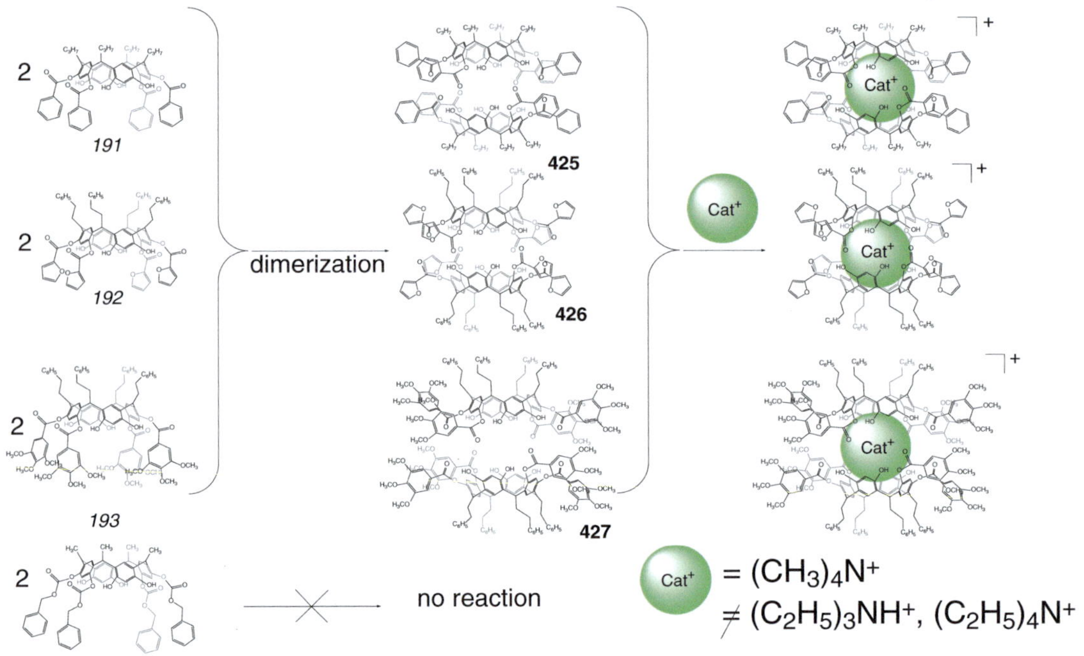

Scheme 3.83

The hexameric resorcinarene caging ligand **401** with a cavity volume of approximately 1400 A^3 encapsulates calix[4]arene or thiacalix[4]arene molecules in the presence of tetramethylammonium or trimethylsulfonium cations by Scheme 3.85, giving the highly stable "*Matreshka*" heteroguest 1:1:1 assemblies [81]. This synergetic co-encapsulation is size selective: the cavity of this supramolecular capsule can adopt thiacalix[4]arene molecule but not *tert*-butylcalix[4]arene or calix[4]arene tetramethyl ethers. Similarly, cage complexes of this type are not formed in the presence of the more bulky tetraethylammonium cation as potent guest [81].

The hexameric arene capsules **397** and **401** (Scheme 3.86) have been studied in [82] as potent caging ligands for the encapsulation of cobaltocenium cation. According to NMR data, these ligands form 1:1 cage complexes that result in substantial changes in the CVs: the reversible wave characteristic of the free cobaltocenium cation disappears after caging. Tetraalkylammonium halides do not affect this encapsulation process. In the case of hexafluorophosphate, tetrafluoroborate, and perchlorate salts of these cations, no encapsulation of this organometallic guest has been observed in [82]. The stability of the cage complex of **401** with the encapsulated cobaltocenium cation is substantially lower than that of **397** and increases in the presence of tetraalkylammonium cations in a row $(C_4H_9)_4N^+ < (C_6H_{13})_4N^+ < (C_{12}H_{25})_4N^+$ [82].

3.2 Cavitand-Based Caging Ligands

3.2.1 Free Cages and Encapsulation of Neutral Molecules

For the first time, the synthesis and characterization of a nanosized bis-cavitand hydrogen-bonded capsule **429**, which is large enough to encapsulate two different guest molecules by Scheme 3.87, are reported by J. Rebek, Jr. and coworkers in [83]; its size- and shape-selective binding has been used for exclusive encapsulation of the heteroguest pair benzene–*para*-xylene.

Formation of the same social isomers by α-, β-, and γ-picolines and toluene as co-guests by

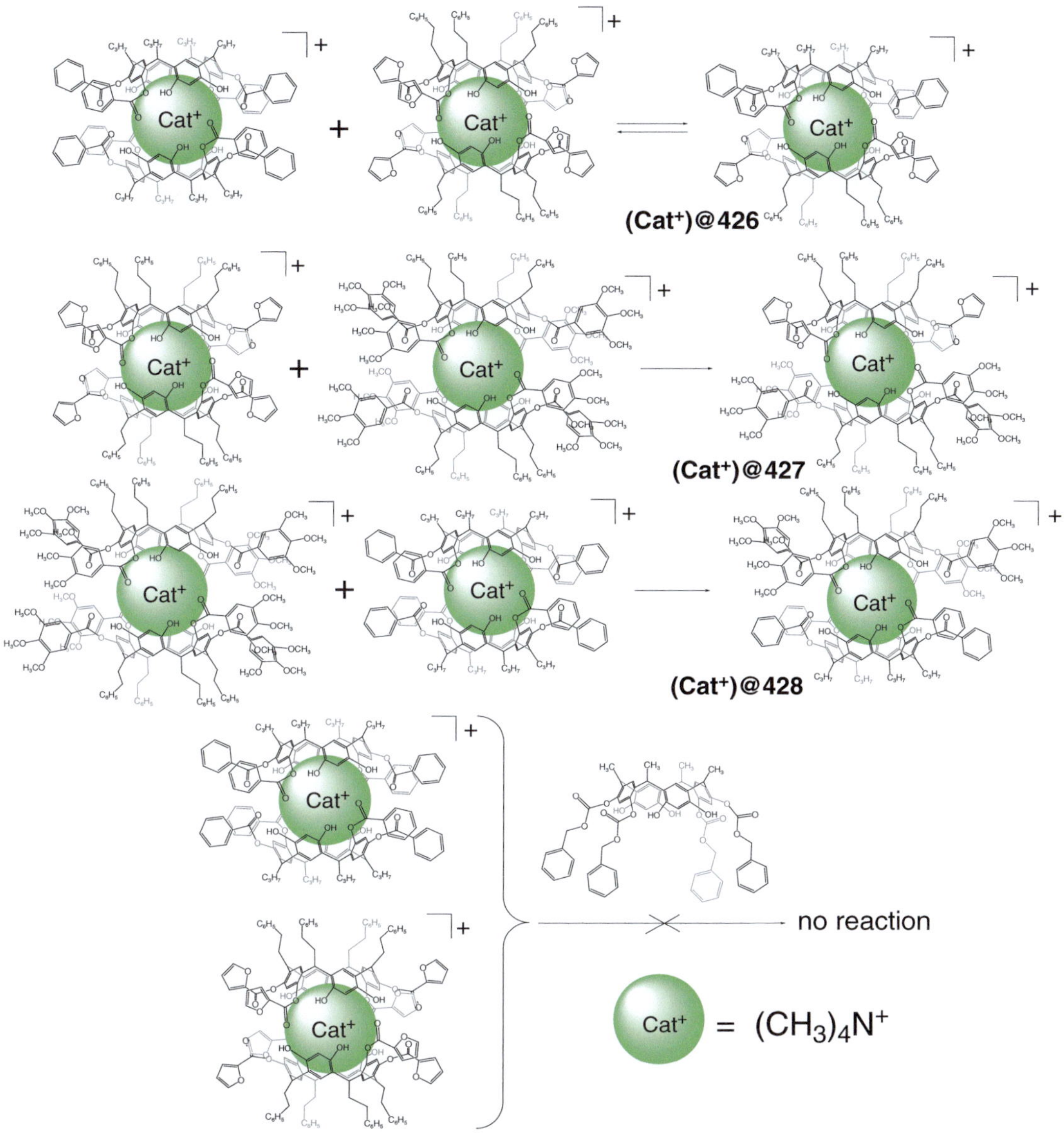

Scheme 3.84

Scheme 3.88 within the cavity of a bis-tetraimide cylindrical hydrogen-bonded capsule **429** has been observed in [84] to proceed reversibly on time scales that range from milliseconds to hours. These guests occupy various microenvironments along its cylinder's axis: the central part of **429** formed by eight imide fragments attracts their more polar parts, while the more hydrophobic sites were found in [84] to be near the capsule's ends.

Molecular recognition of various rigid, flexible, and complex guests (Scheme 3.89) within the cavity of the self-assembled dimeric cylindrical host **429** is reported in [85] to show a range of rates of their uptake, release, and exchange. Small solvent guests such as benzene, toluene, and *para*-xylene that give 1:2 cage complexes rapidly undergo these transformations, while large molecules (in particular, anilide and stilbene) also replace poorly accommodated

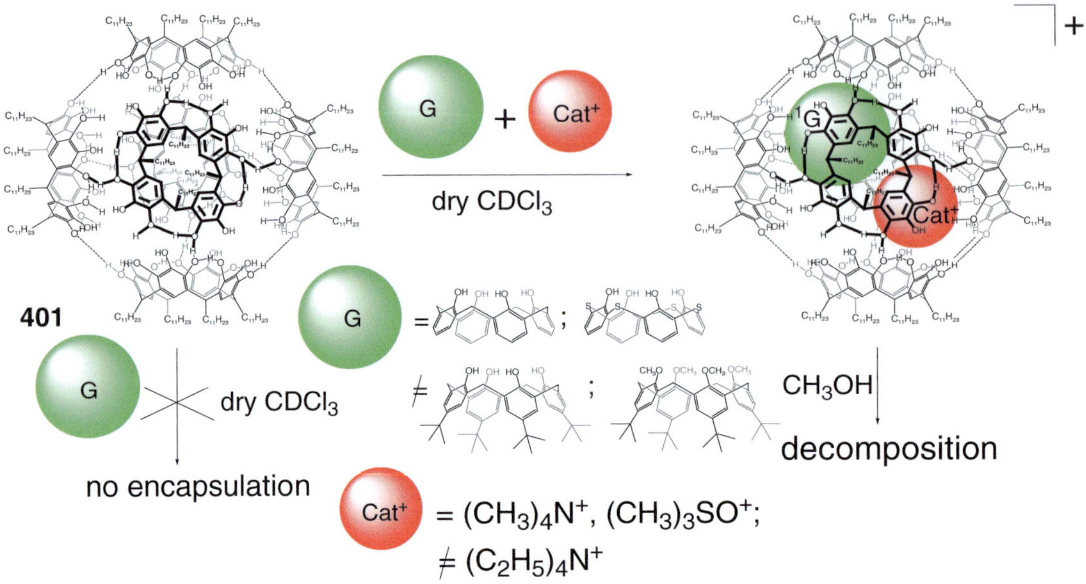

Scheme 3.85

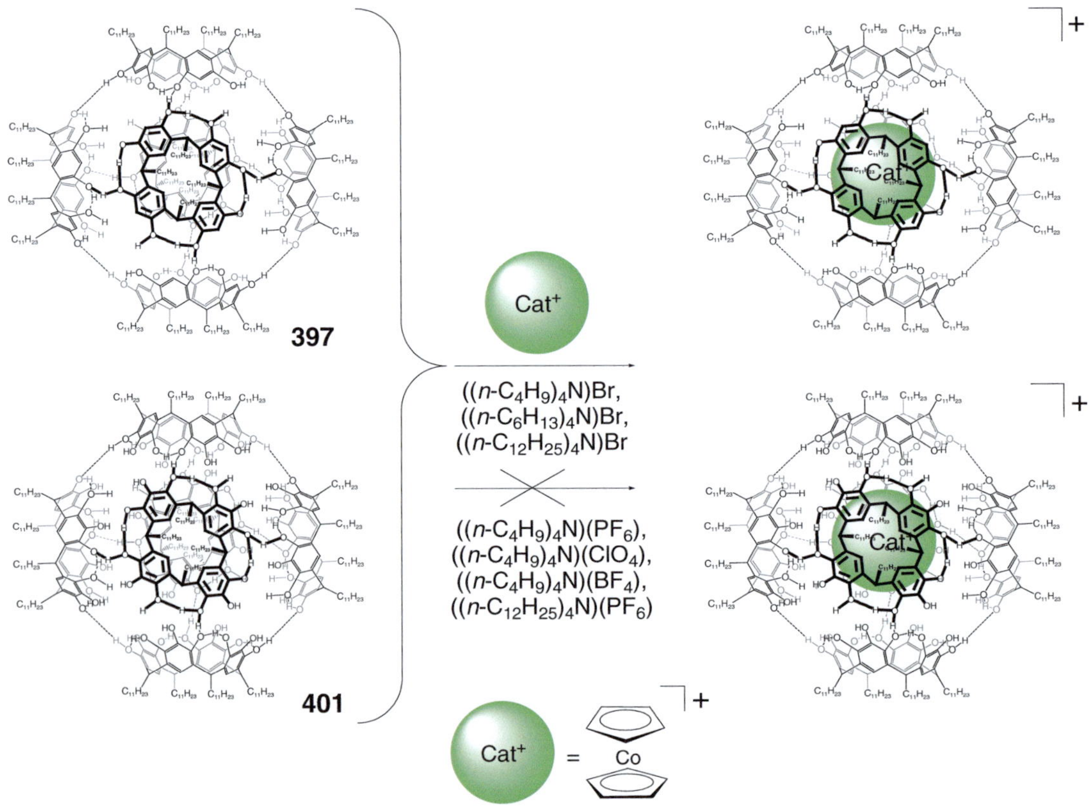

Scheme 3.86

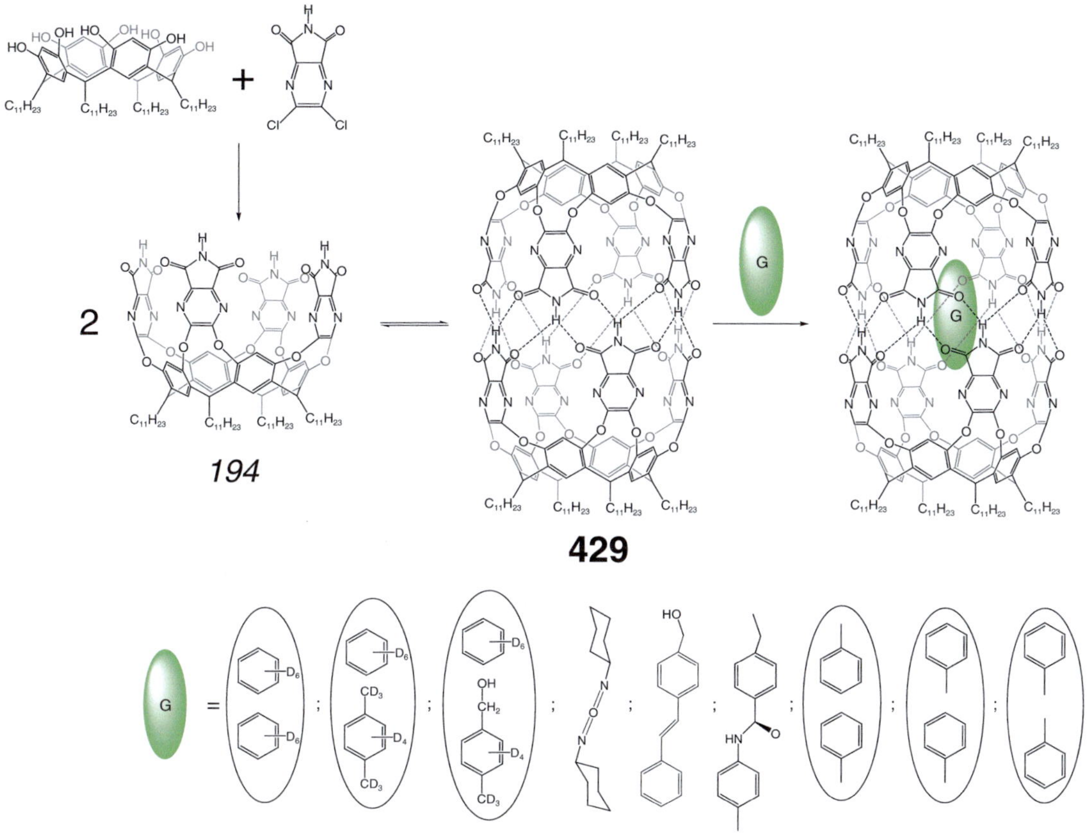

Scheme 3.87

Scheme 3.88

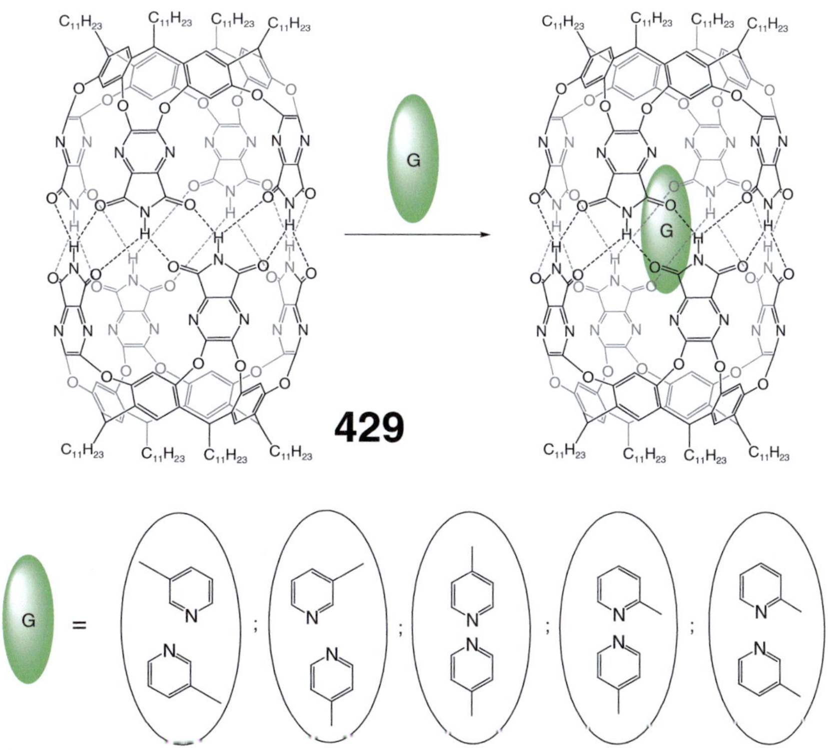

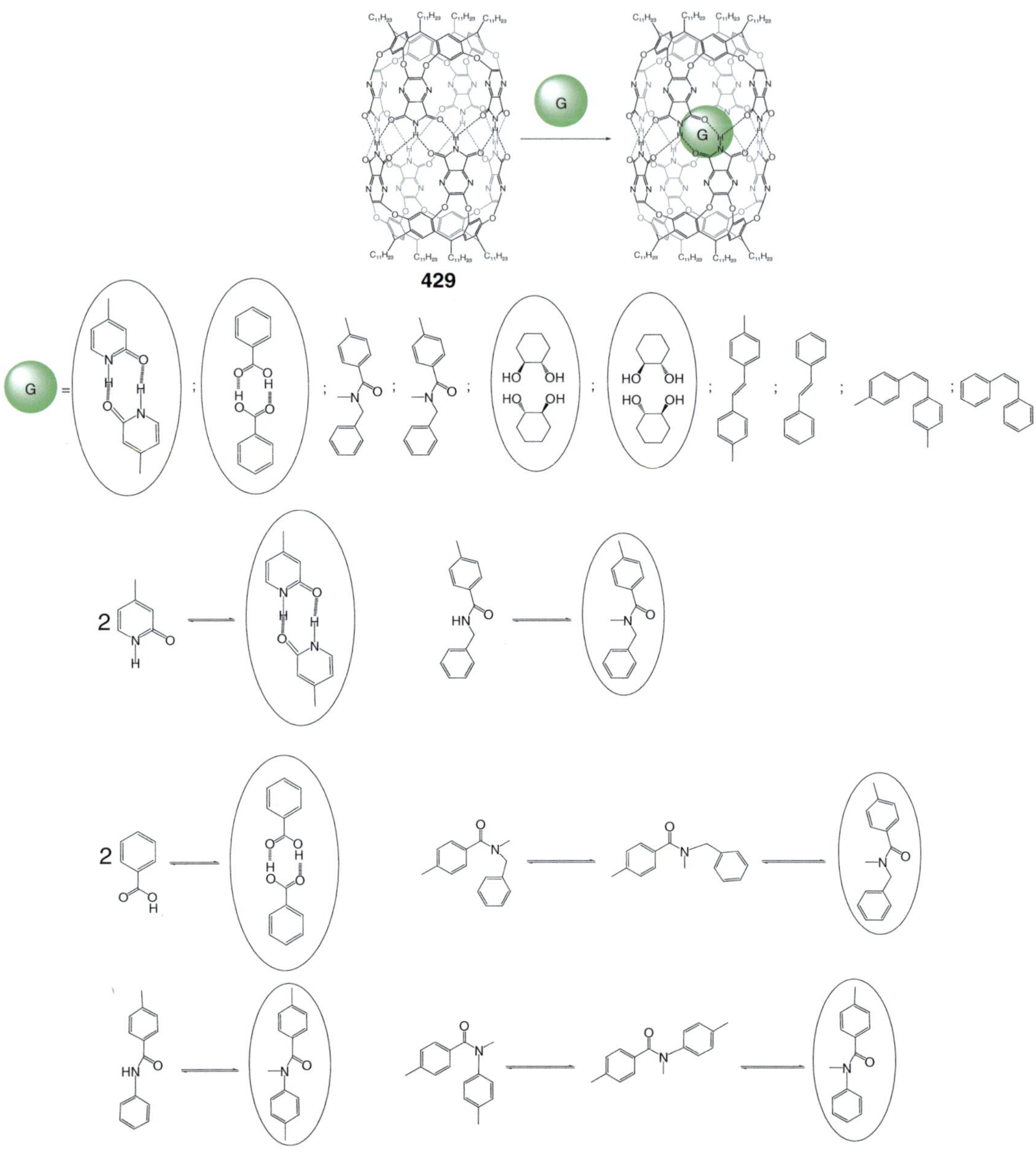

Scheme 3.89

mesitylene on the same time scale. The exchange rates between these large caged molecules are, however, slow (up to days). The sequence of the encapsulation behavior of this hydrogen-bonded caging ligand **429** arises from the considerable size of its cavity and elongated shape thus determining its selectivity for congruent guests [85]. In particular, this capsule is described in [86] to encapsulate large organic guests shown in Scheme 3.90, allowing for ^{1}H NMR spectroscopical estimation of the internal cavity dimensions with the use of a series of homologous molecular "rulers" (e.g., the corresponding aromatic amides). The available space within **429** is estimated to be 5.7 × 14.7 Å, and it readily encapsulates dibenzoyl peroxide in [D_{12}] mesitylene solution, which does not decompose for a long time at 70° C. Moreover, the cavity of **429**

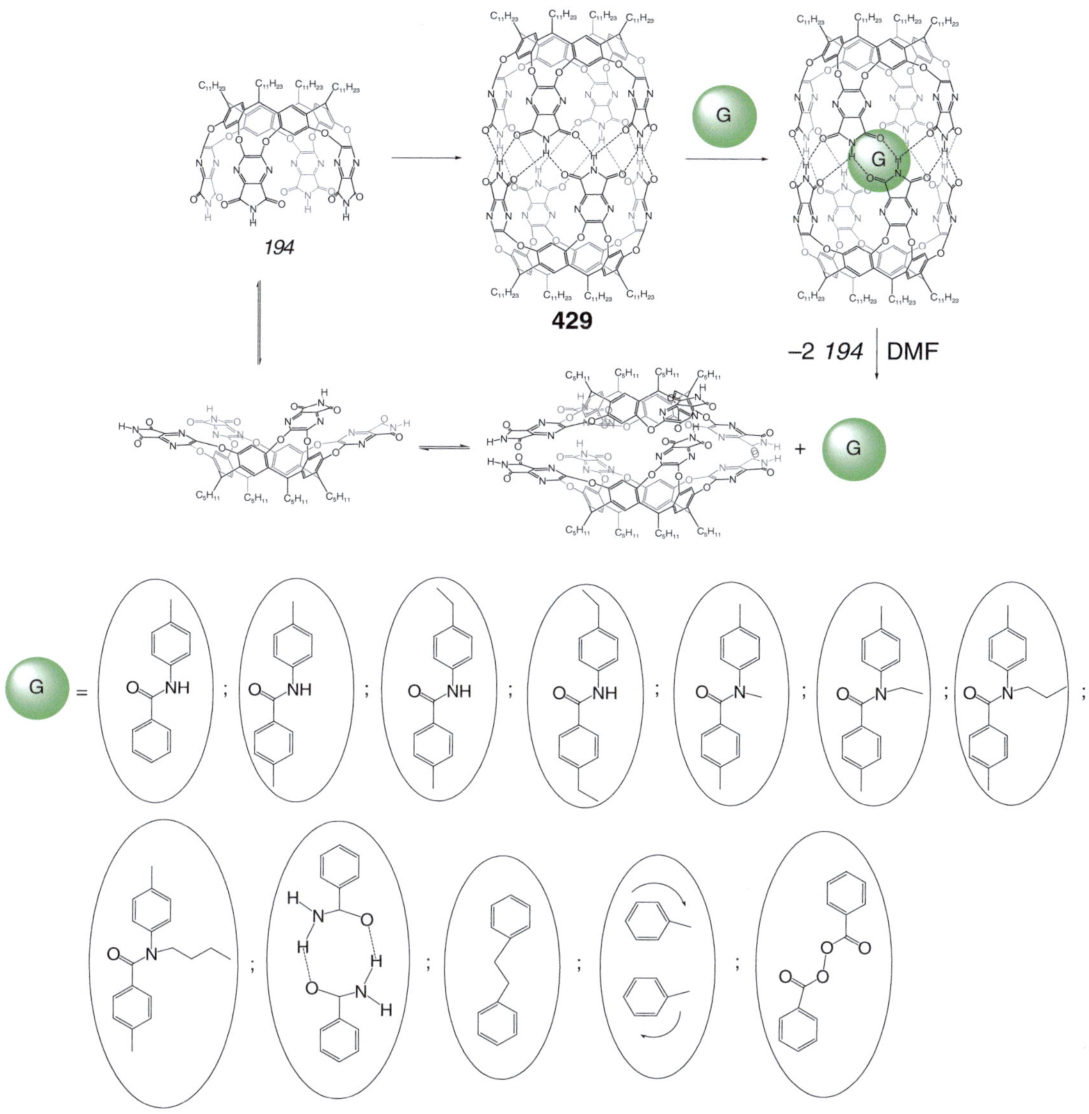

Scheme 3.90

prevents this encapsulated peroxide guest from oxidizing triphenylphosphine and diphenylcarbazide as reductants. The chemical reactivity of this reactive molecule can be restored by its release from the capsule in the presence of DMF, which competes for hydrogen bonds with such caging ligand [86].

The encapsulated solvent molecules shown in Scheme 3.91 are described in [87] to exchange their positions slowly on the NMR time scale within the cavity of the hydrogen-bonded capsule **429**. Translational motion of gas molecules within this cavity and co-encapsulation of these guests by Scheme 3.91 – 1308 have been studied in [88] using various NMR techniques. The caged cyclopropane and butane exchange their locations slowly on the NMR time scale. When those were co-encapsulated with long flexible solvent molecules such as *n*-heptane, their slower motion has been detected.

The hydrogen-bonded caging ligand **429** also encapsulates a series of *N*-protected amino acid

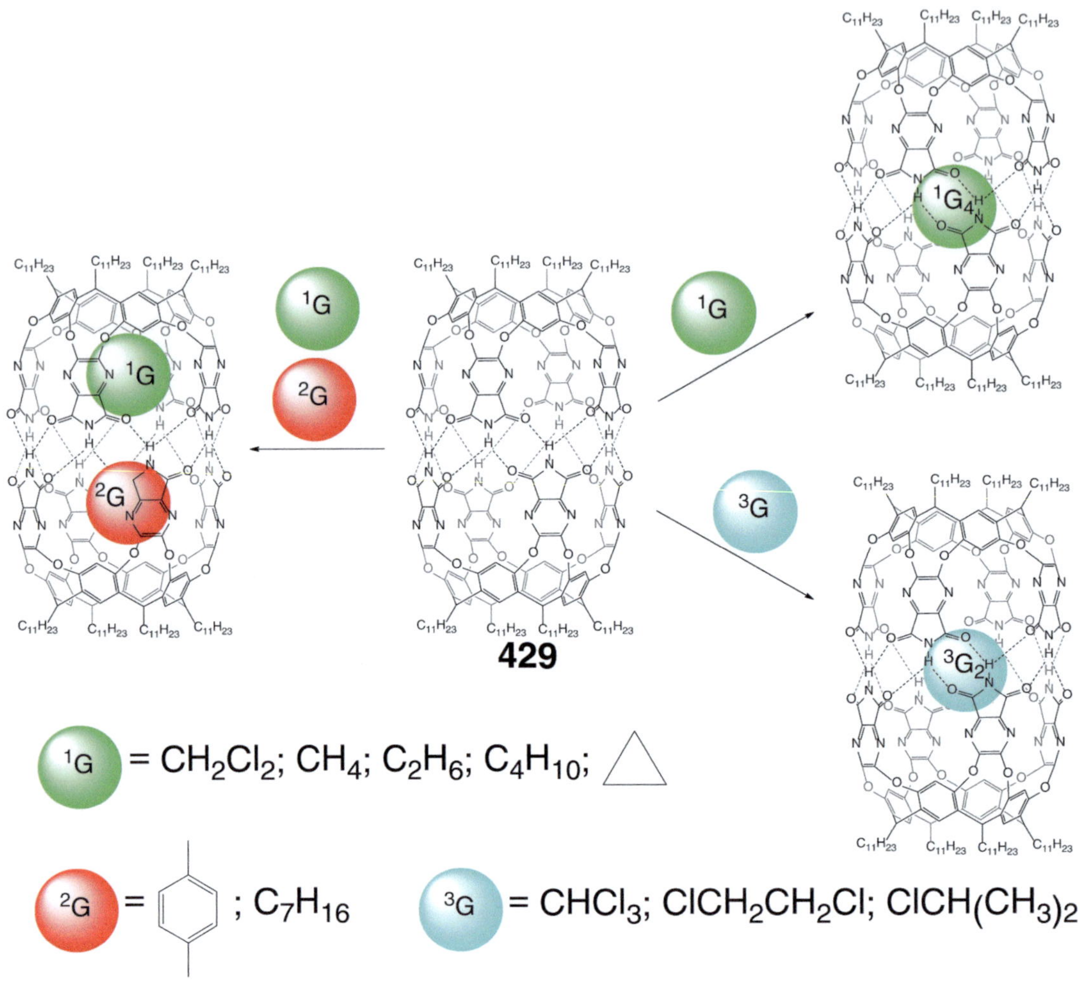

Scheme 3.91

esters and alkanes shown in Scheme 3.92 [89]. As follows from ^{1}H NMR data, these guests that are slightly longer than the inner dimensions of the cavity of **429** have been accommodated by this capsule through adopting of their compacted conformations (in particular, helical folding [90]). Encapsulation of *N*-nitrosoamides also shown in this Scheme by **429** is reported [91] to prevent the rearrangement reactions of these caged guests.

Reinhoudt's social isomerism (see Sect. 3.1.1) is described in [92] to appear in the cage complexes of **429** in the case of co-encapsulation by Scheme 3.93 of the solvent guests with *para*-substituted toluene derivatives. In the case of chloroform and *para*-ethyltoluene, two isomeric capsules with these co-guests have been detected by ^{1}H NMR method. The orientational preference of the guest molecule of one sort depends on the nature of its co-guest; in particular, the presence of a large co-guest gave the isomer with the methyl substituent near ends of the cavity of **429** owing to better fit of this group into that space or more attractive interactions of the ethyl substituent with the co-guest [92].

So-called "isomeric constellations" (arrangements of several molecules) within this cylindrical host that encapsulates two different solvent guests (chloroform and isopropyl chloride) have been studied in [93], and almost all their

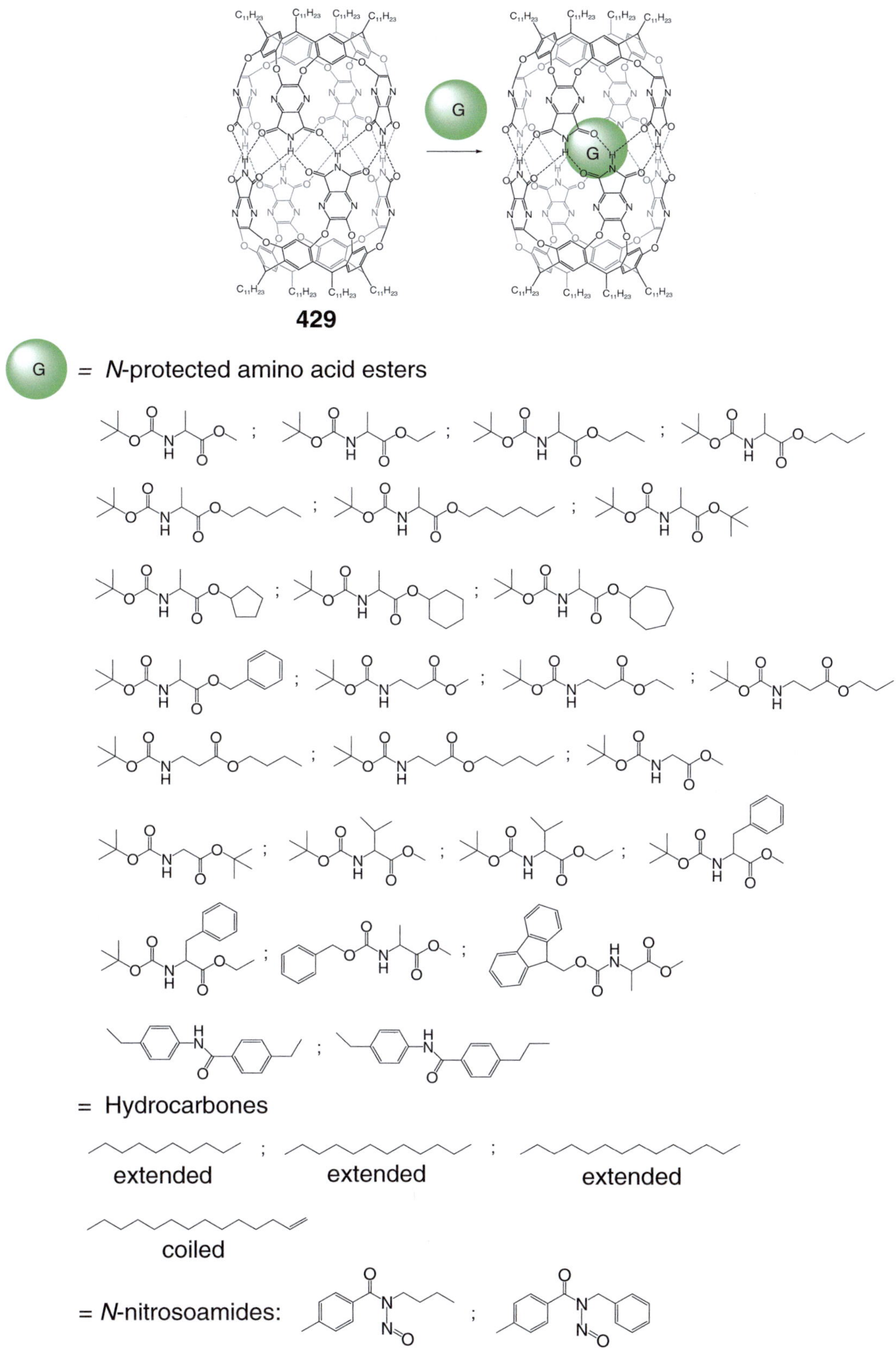
429
G
= N-protected amino acid esters
= Hydrocarbones
extended
extended
extended
coiled
= N-nitrosoamides:

Scheme 3.92

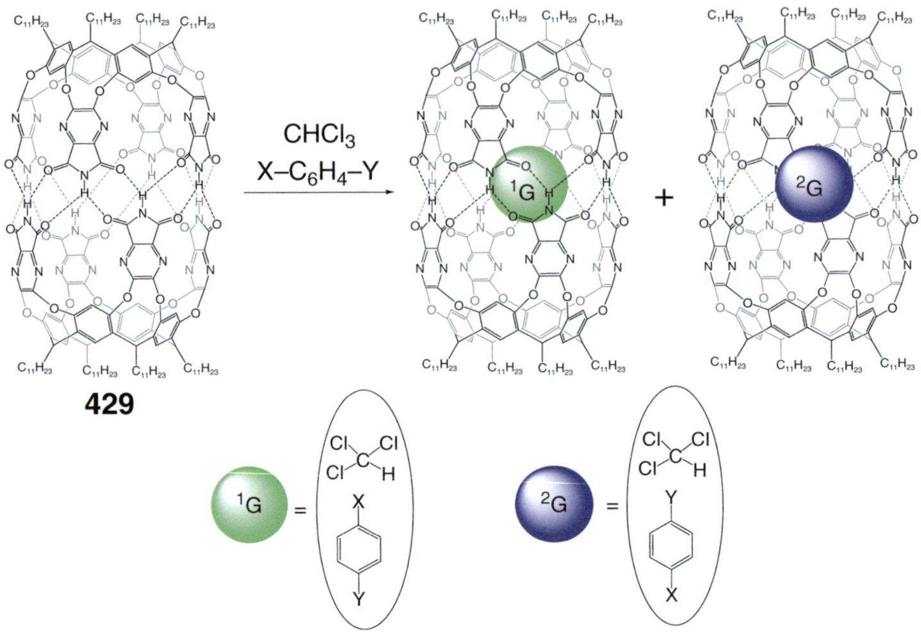

Scheme 3.93

combinations (i.e., five of the possible six) were detected within its cavity by NMR method. Other solvent pairs showed fewer isomeric constellations. This type of isomerism is characteristic of guest assemblies shown in Scheme 3.94 rather than of their individual molecules owing to nearly identical shapes, dimensions, and volumes of these two solvent molecules.

Co-encapsulation of small gaseous hydrocarbons with sizable aromatic molecules by this hydrogen-bonded capsule is described in [94]. The presence of the large aromatic co-guest slows *in–out* exchange of these gases with the rates becoming detectable by NMR method: although no evidence of methane co-encapsulation with *para*-xylene as a co-guest was obtained, this gaseous guest has been detected within the cavity in the presence of a larger anthracene co-guest. With appropriately substituted large co-guests, these co-encapsulated gases showed social isomerism. In particular, *para*-ethyltoluene and cyclopropane gave two different heteroguest cage complexes of **429**. This isomerism [94] results from interactions between the guests and their limited motion within the cavity of the caging ligand. The *para*-ethyltoluene guest is too long to tumble within this capsule, and either its methyl group or ethyl substituent is co-encapsulated with cyclopropane co-guest. Moreover, naphthalene and azulene do not undergo encapsulation with **429**, but their co-encapsulation takes place with ethane or propene as co-guests. For propene–azulene guest system, only the single social isomer shown in Scheme 3.95 has been detected by ^{1}H NMR method. The space not occupied by a larger molecule is appropriate for a smaller one [94].

The social isomerism of three solutes and various solvents also shown in Scheme 3.95 within the cavity of **429** and individual solvent–solute interactions between these guests are described in [95]. The size and shape of the caging ligand limits the mobility of the encapsulated molecules within its cavity. They are too large to slip past each other or to exchange places, and the encapsulated toluene derivative is too long to tumble within the capsule to give a social isomer. This suggests that intermolecular interactions between the caged guests are constrained to two contact areas and are attenuated by the inner surface (i.e., the lining and shape of this hydrogen-bonded

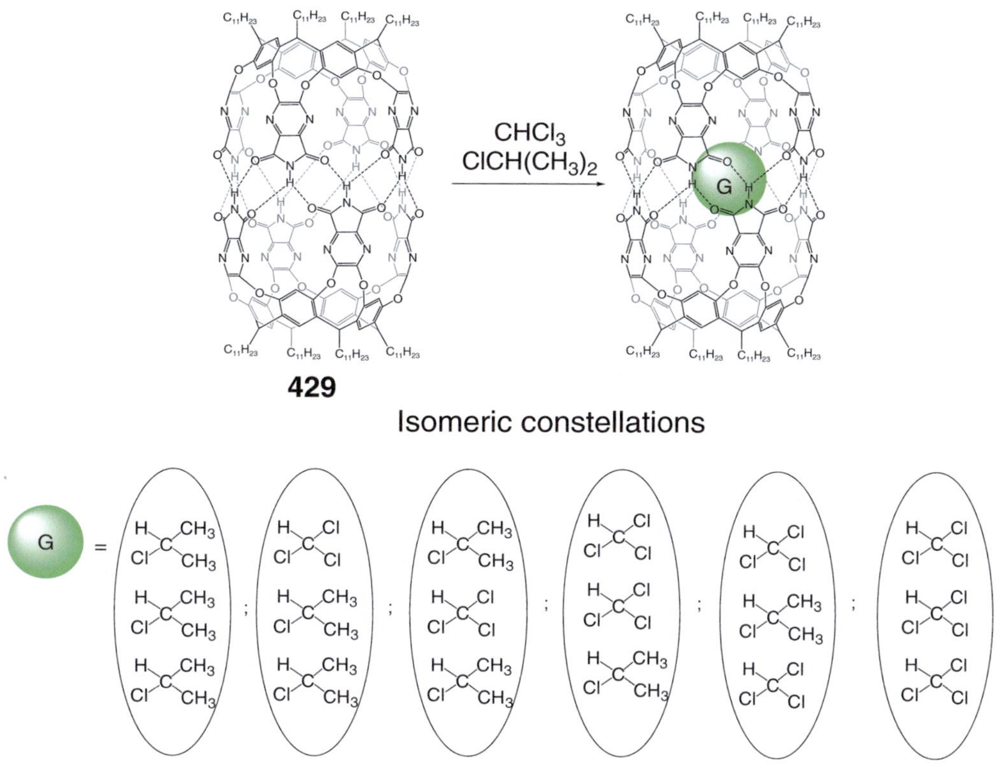

Scheme 3.94

caging ligand): the tapered ends of its cavity are able to accommodate smaller functional groups, while the seam of hydrogen bonds in the center favors the polar fragments of the guests [95].

Cylindrical hydrogen-bonded capsules such as **429** are proposed in [96] for encapsulation of two different guest molecules, thus offering a mode for enantioselection via co-encapsulation: if one of these guests is chiral, their inner spaces also become chiral. The dimensions of these spaces allow selecting appropriately sized combinations of co-encapsulated guests, while their shape prevents tumbling of the rigid caged molecules by orienting their polar groups within the cavity. Chiral mandelic and *R*-*Br*-butyric acids are reported to be promising guests for diastereoselection, which is, however, relatively small (<25 %). Encapsulation of a chiral guest (in particular, *R*-mandelic acid) by a cylindrical caging ligand of appropriate size allows for only two possibilities (**A** and **B**, Scheme 3.96), and the asymmetric element near the middle of its capsule and closer to the other guest (type **A**) is described in [96] to be more effective in distinguishing enantiomers than one near the end of the caging ligand **429** (type **B**). Therefore, the asymmetric recognition is determined by (*i*) multiguest encapsulation, (*ii*) preferential caging of two different molecules as shown in Scheme 3.96, and (*iii*) localization (orientation) of these species within the cavity of a caging ligand that allows the asymmetric elements to be near one another. In the case of *R*-styrene oxide as guest and the cylindrical capsule **429**, one caged molecule fills too little space within the cavity of this host, while two molecules of this type fill too much. Addition of isopropyl chloride as a co-guest resulted in quantitative formation of the corresponding heteroguest 1:1:1 cage complex [96].

Chiral guests within these achiral capsules create their chiral inner spaces. To determine the ability of the chiral element outside the capsule to affect the environment of its inner cavity, the authors of [97] synthesized a series of hydrogen-bonded capsules

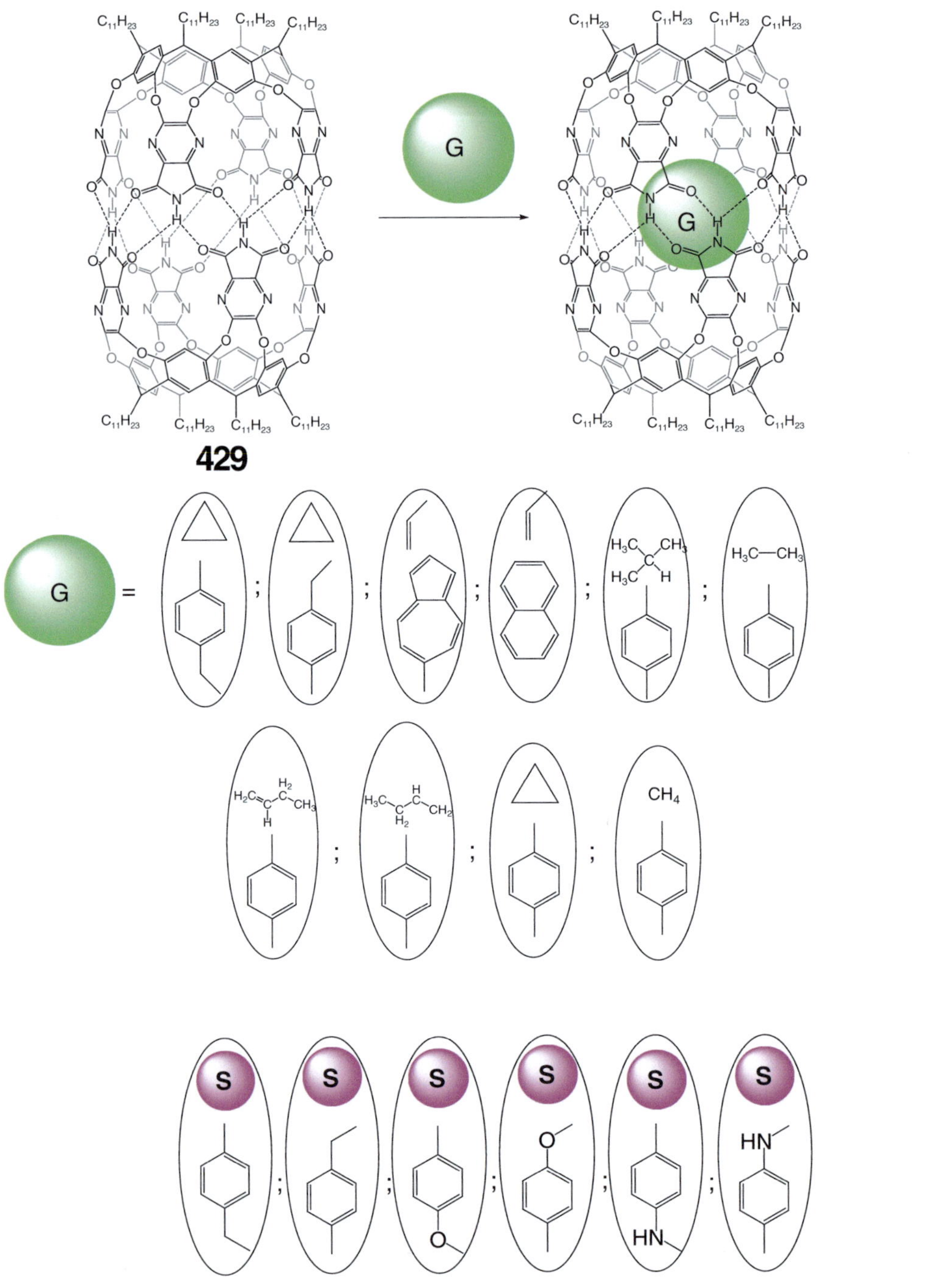

solvent

S = CH_2Cl_2; $CHCl_3$; $(CH_3)_2CHCl$; $(CH_3)_2CCl_2$; CCl_4
$(CH_3)_2CHOH$; $(CH_3)_2CO$; C_6H_6; C_6H_5N; $C_2H_5(CH_3)CHOH$
$C_2H_5CH_2OH$; $C_3H_7CH_2OH$; $C_2H_5CH(CH_3)_2$; n-C_5H_{12}
c-C_6H_{12}; $CH_3CH{=}CH_2$; CH_3CH_3

Scheme 3.95

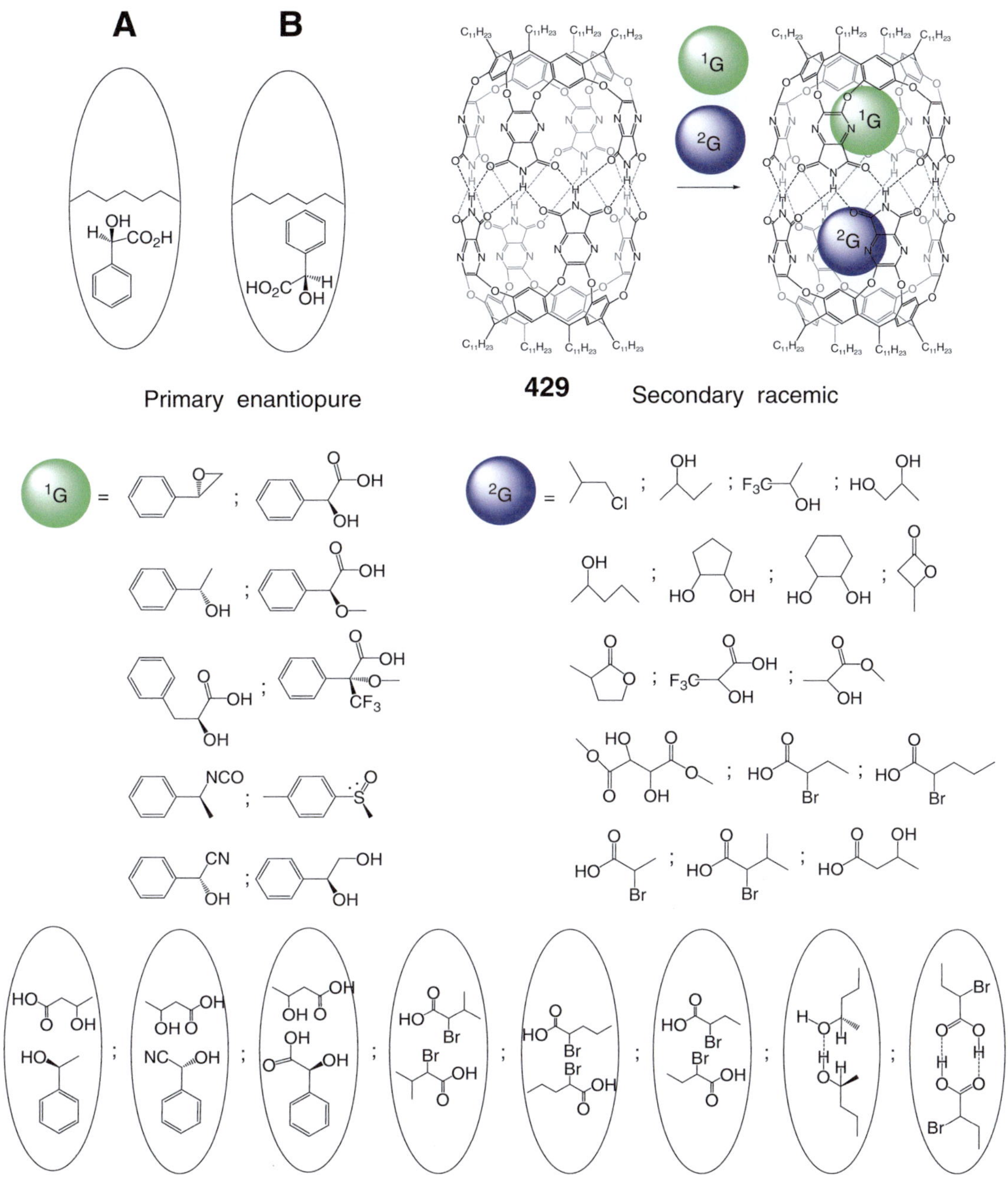

Scheme 3.96

429–434 (Schemes 3.97) from the corresponding ligand syntones by varying the number of asymmetric centers and the distance from the cage framework. The shorter syntone of **430** had the array of asymmetric centers with one methylene group closer to the end of the capsule than its homolog in the cage of **431**, while the longer syntone of **432** had the array with one methylene chain being further away. The influence of the chiral centers outside these hydrogen-bonded capsules on guests encapsulated within their cavities has been clearly shown. The distance between the external asymmetric center and the

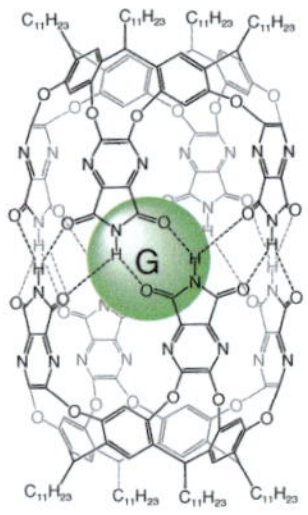

429

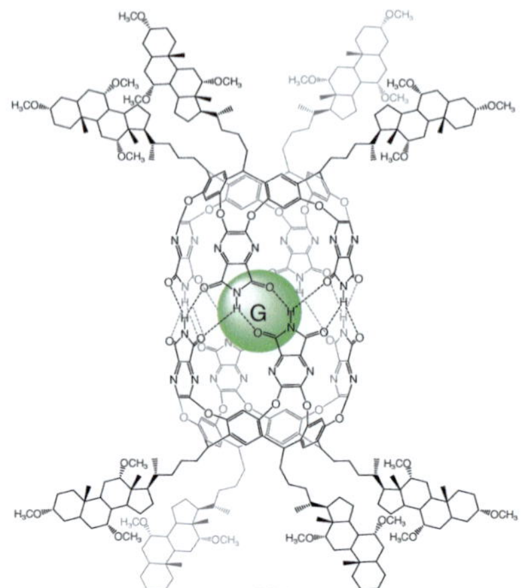

432

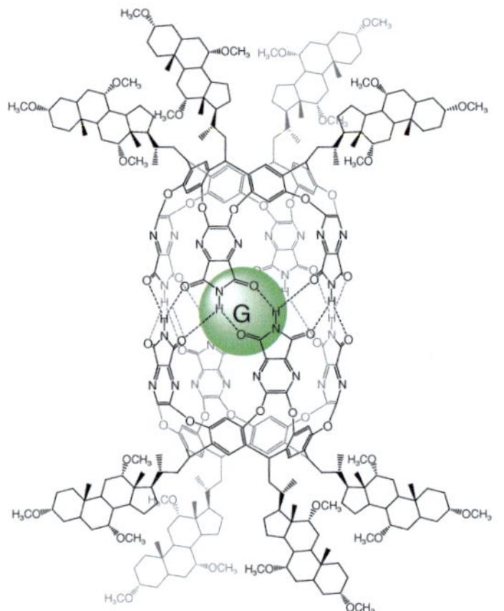

430

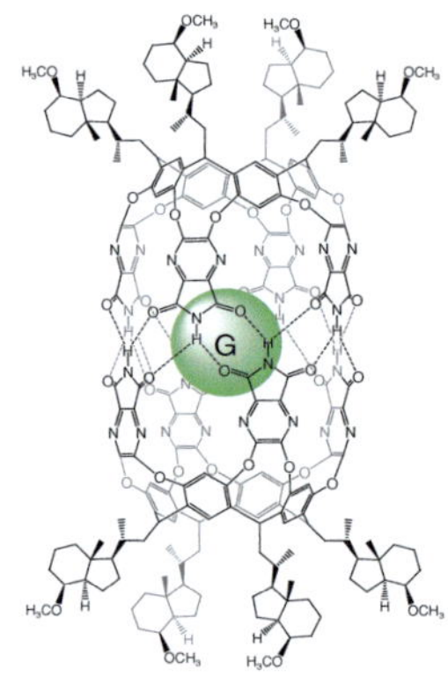

433

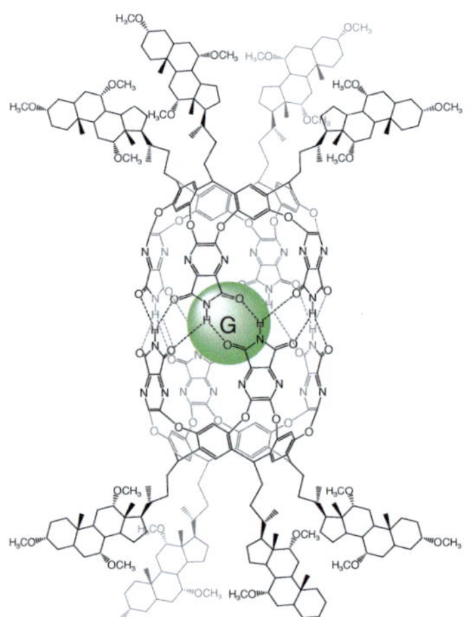

431

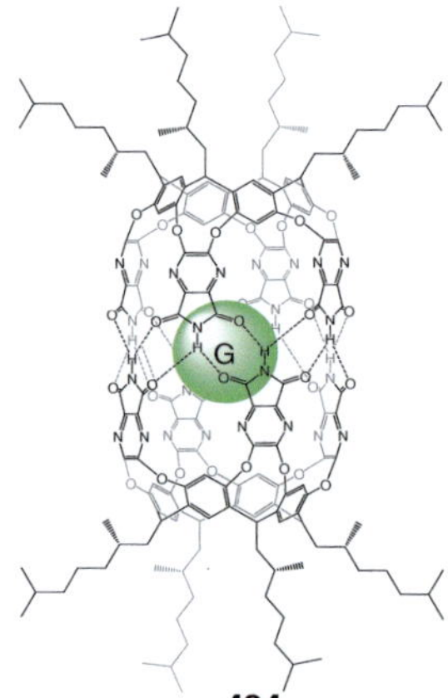

434

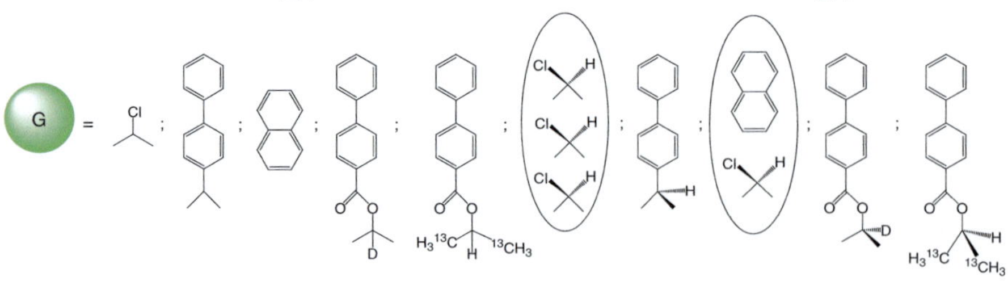

Scheme 3.97

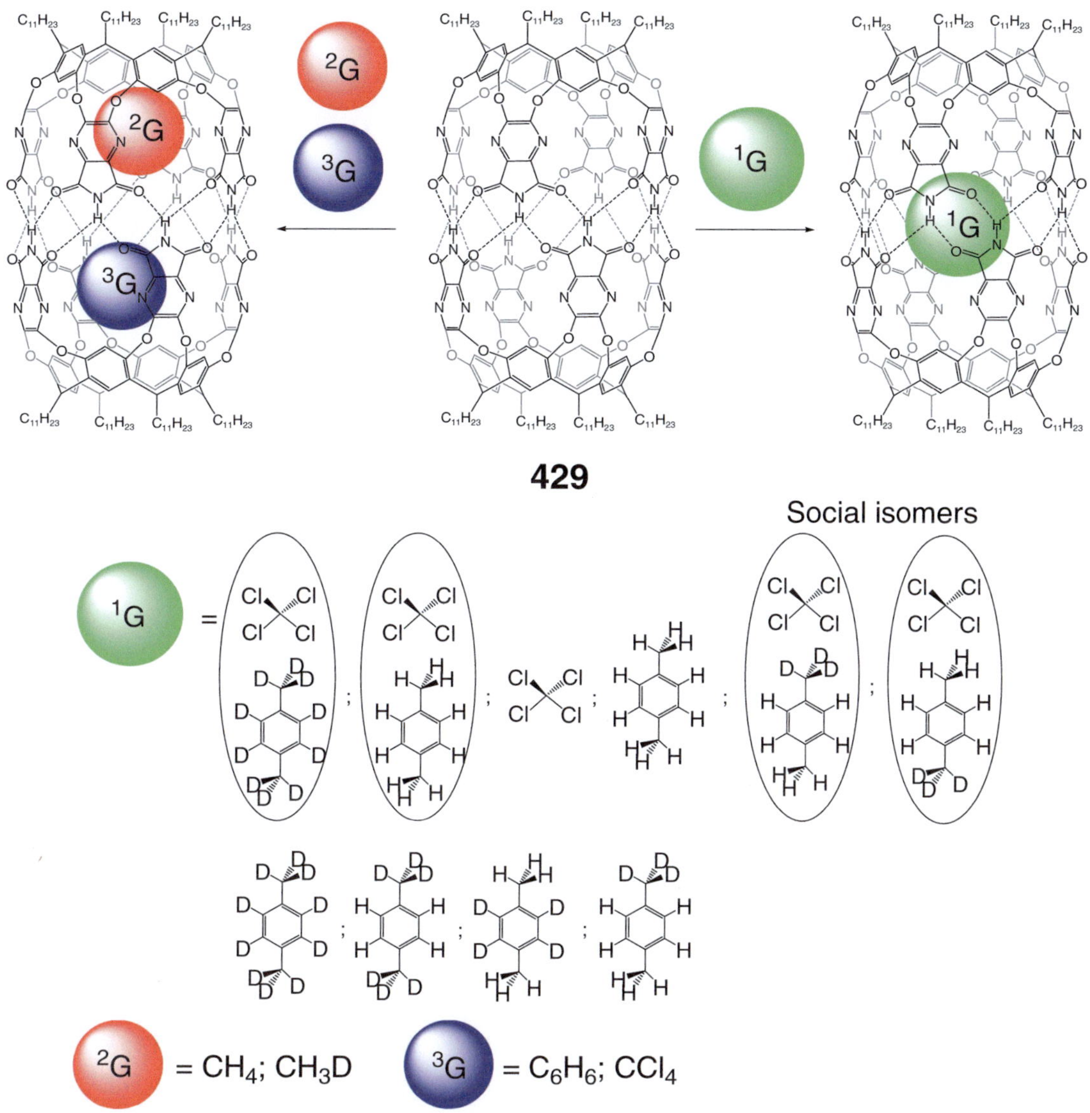

Scheme 3.98

cavity of the caging ligand and the mobility of caged guests within it are reported [97] to be the factors of this influence.

Isotope effects on guest encapsulation by the capsule **429** have been studied in [98] by competition NMR experiments for its cage complexes shown in Scheme 3.98. A preference for encapsulation of CD_3 rather than CH_3 near the resorcinarene fragment of this caging ligand has been observed owing to larger weight of deuterium causing a smaller vibration amplitude and frequency for C–D relative to C–H bonds. Interactions of the methyl hydrogen atoms and the π-systems of the hydrogen-bonded capsule increase the force constant for stretching of these bonds in its proximity [98].

Two model methane—benzene and methane—tetrachloromethane systems have been tested in [99] to determine the effect of isotopic substitution on the stabilities of the corresponding weakly bonded social isomers of these co-guests (Scheme 3.98) within the cavity of **429**. Attraction between the methyl group and benzene is reported to be relatively large due to strong dispersion

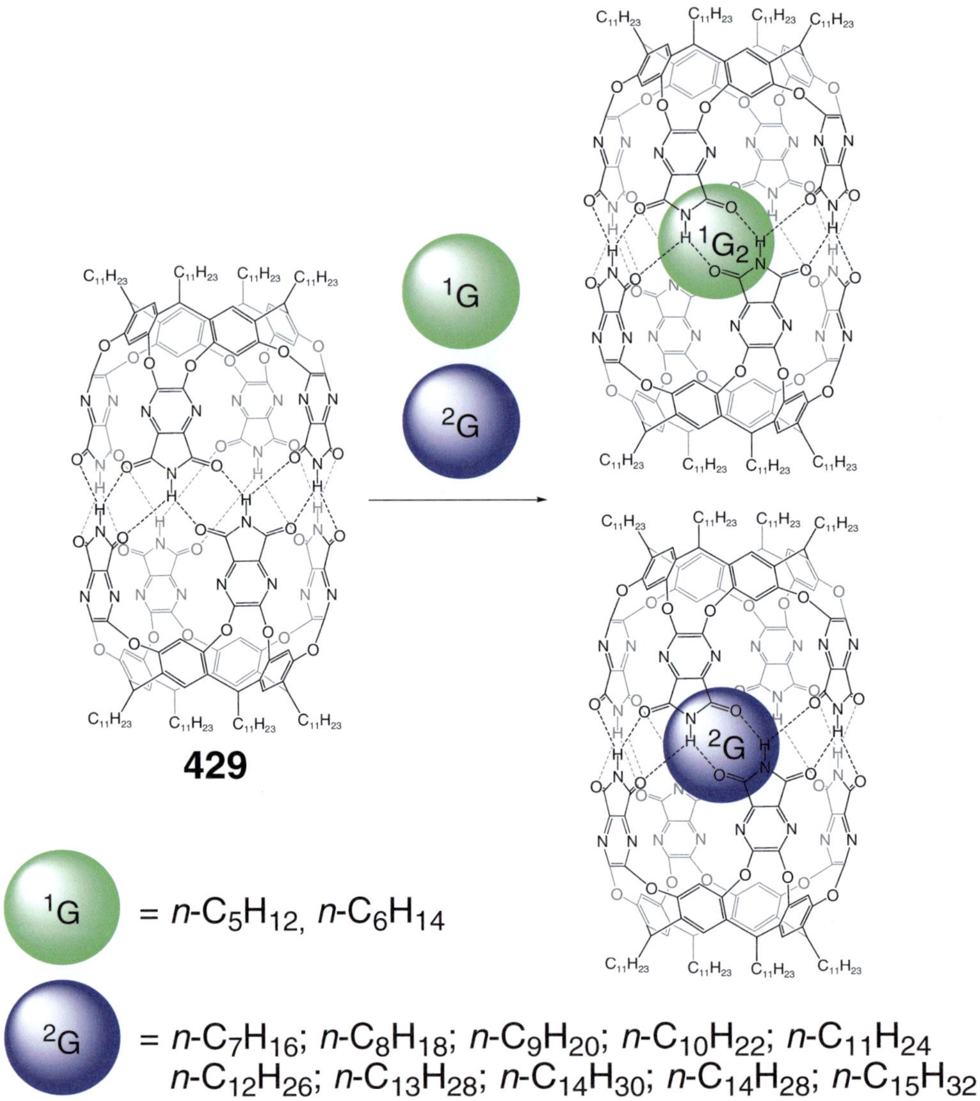

Scheme 3.99

forces between methane and benzene, while the methane–tetrachloromethane interaction is negligible because of low polarizability of the later co-guest. Thus, the cavity of the capsule **429** provides suitable environment to probe these differential nonbonding interactions [98].

Reversible encapsulation of a series of normal alkanes shown in Scheme 3.99 by this cylindrical host has been studied in [100] by NMR methods. For small hydrocarbons, such as *n*-pentane and *n*-hexane, 1:2 cage complexes with two encapsulated alkane molecules have been detected. These guests moved freely within the cavity of **429**, whereas this ligand does not encapsulate *n*-heptane. Longer alkanes, such as *n*-decane, form 1:1 cage complexes with **429**, and its aromatic walls twist to avoid empty spaces and to increase favorable interactions with the encapsulated hydrocarbon molecule. As a best guest, *n*-undecane adopts the conformation with a minimum of *gauche*-interactions; the longest alkane (*n*-tetradecane) adopts helical conformation to fit the cavity of **429**, as this shape maximizes supramolecular host–guest interactions. Therefore, the

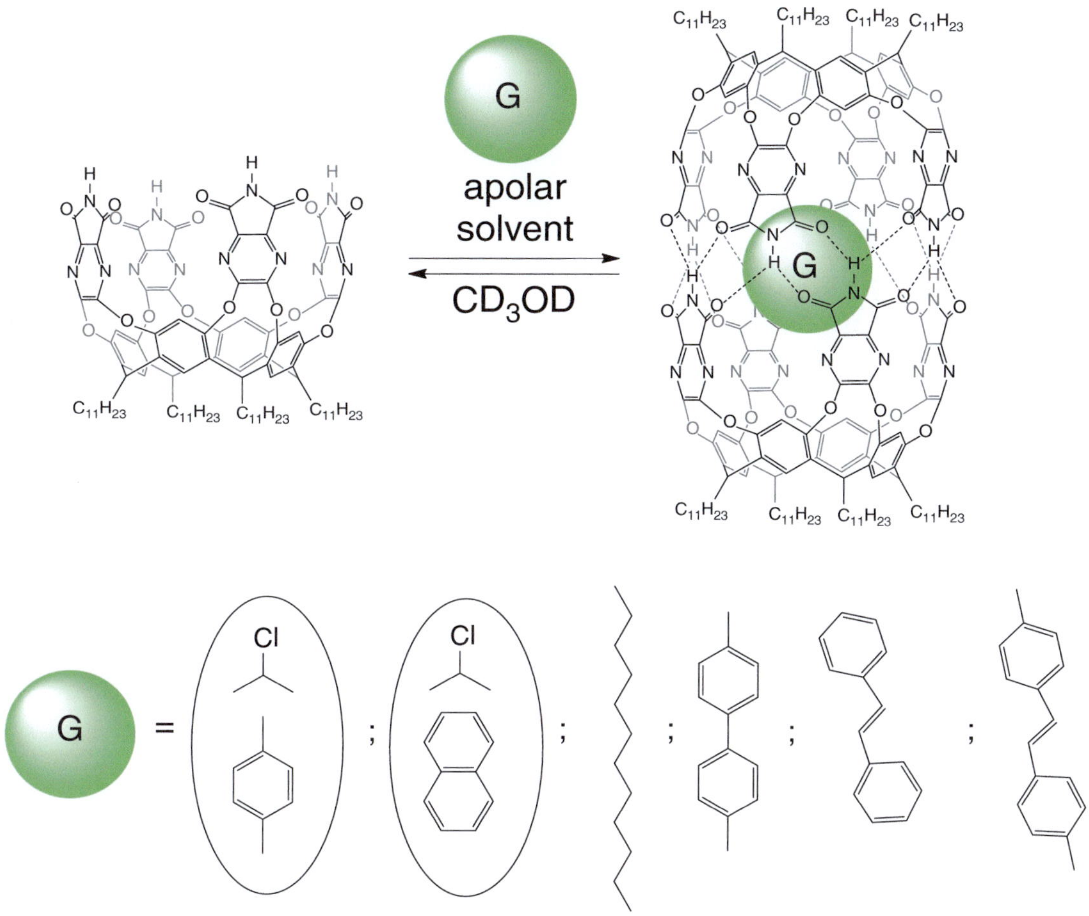

Scheme 3.100

encapsulation of alkanes by the hydrogen-bonded capsule **429** resulted in conformational changes of both the guest and the host, which are driven by the combination of size and shape complementarity with supramolecular interactions between them [100].

Kinetics of encapsulation and thermodynamic stability of hydrogen-bonded homo- and heteroguest cage complexes of **429** shown in Scheme 3.100 in protic media have been studied in [101] using conventional ^{1}H NMR methods. Positive enthalpies and entropies of the encapsulation are indicative of liberation of caged solvent molecules. The rates of dissociation–association of these cage complexes were found to be comparable to those for *in–out* exchange of large guests, suggesting that the guest exchange occurs in protic solvents by complete dissociation of the hydrogen-bonded capsule **429**. Its stability in such media depends strongly on the nature of the encapsulated guests, and the best of them is reported in [101] to be dimethylstilbene.

Encapsulation of oligoethylene glycols and perfluoro-*n*-alkanes (Scheme 3.101) by a caging ligand **429** is described in [102]. The following driving forces for this encapsulation have been noted: (*i*) attractive interactions between this capsule with a cavity volume of approximately 425 Å^3 and the caged guest molecule that includes CH…π interactions and van der Waals contacts, and (*ii*) entropical effects of the formation of hydrogen-bonded seam of the host (its self-assembly does not occur in the absence of suitable guests that can form sufficiently favorable

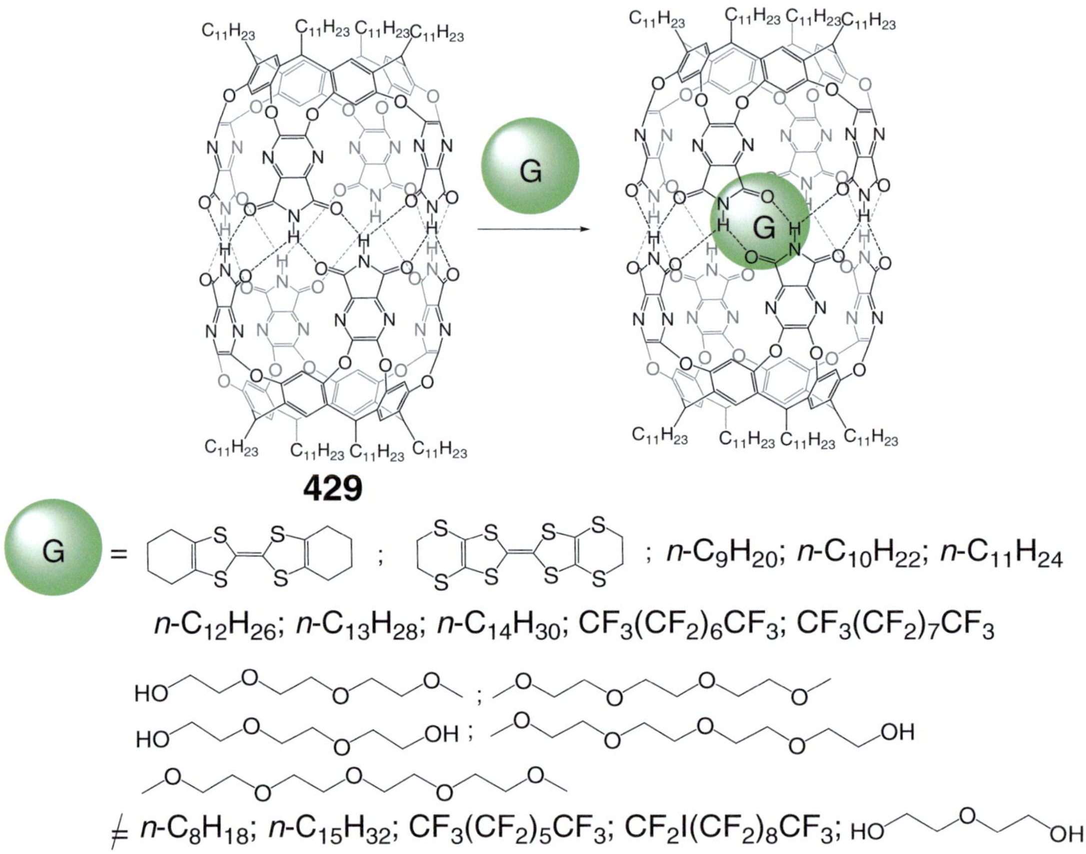

Scheme 3.101

contacts with the interior of **429** to allow them to fill its cavity) [102]. This hydrogen-bonded capsule also effectively binds tetrathiafulvalene derivatives (Scheme 3.101), showing the CT within its cavity [103]. Co-encapsulation of three suitable guest species (including the examples of a highly selective binding) by the caging host **429** giving the anionic heteroguest 1:1:1:1 cage complexes by Scheme 3.102 is reported in [104].

Co-encapsulation of [2,2]-paracyclophane with other guests shown in Scheme 3.103 and its mechanically regulated rotation within the cavity of the hydrogen-bonded capsule **429** has been studied in [105]; as follows from ^{1}H NMR data, these co-guests are able to affect rotational freedom of the caged paracyclophane.

When three chiral guests (*R*- and/or *S*-propylene sulfide) have been co-encapsulated by this cylindrical capsule, the formation of six diastereomeric cage complexes by Scheme 3.104 has been observed in [106]; four of them are reported to demonstrate a diastereomeric constellation. The hydrogen-bonded capsule **429** also selectively bind chiral guests shown in Scheme 3.105 to give the corresponding 1:2 cage complexes [107].

Encapsulation of tertiary amides (Scheme 3.106) by **429** is described in [108] to affect their rotational barriers: for amides *195–200*, the rotation rates decrease or increase within this capsule by up to an order of magnitude as compared with those of free amides in solution depending on their chemical constitution. Such acceleration (deceleration) of rotation is governed by selective destabilization (stabilization) of the ground or transition states of these 1:1 cage complexes. In the case of tertiary amide *201*, this rotation generates and discriminates its two

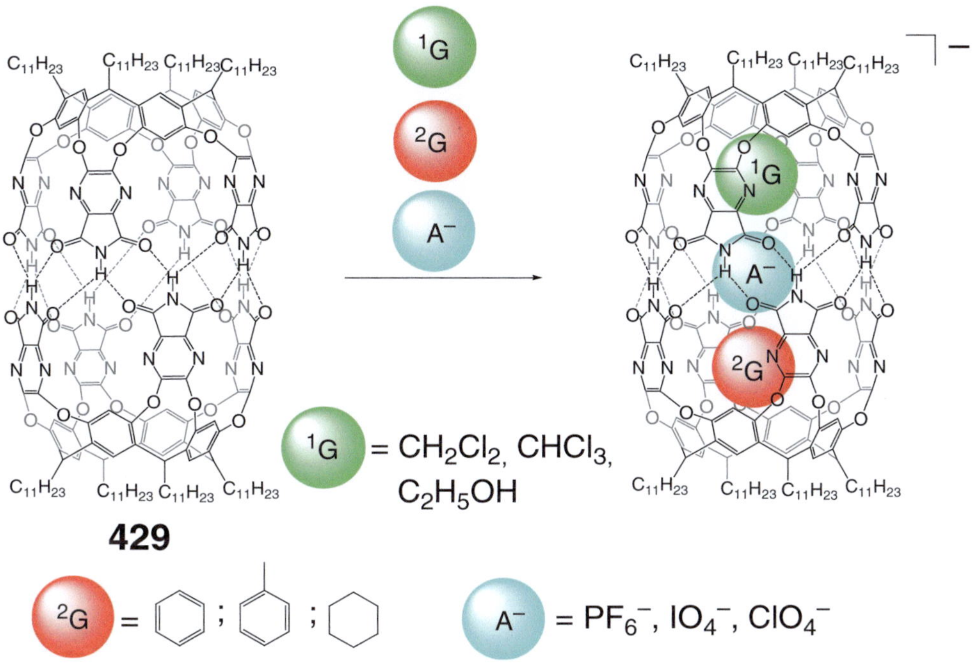

Scheme 3.102

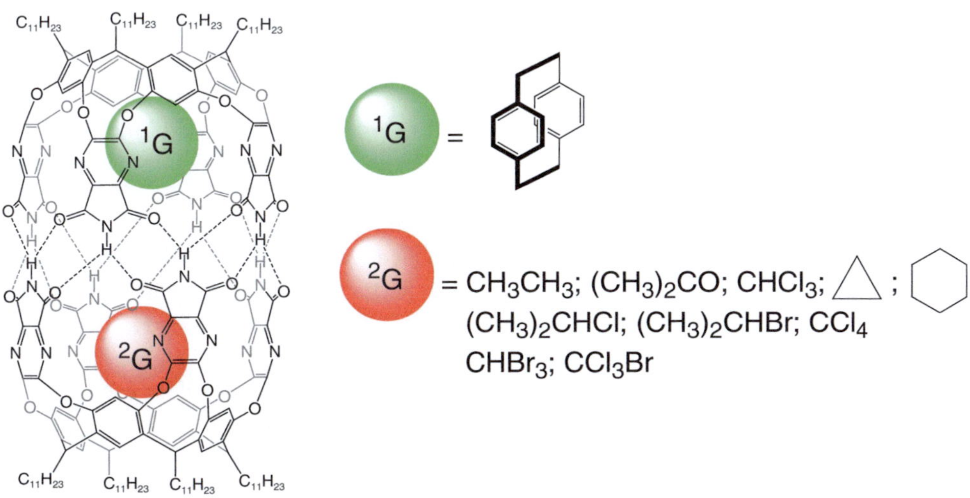

Scheme 3.103

isomers: those are equimolar in bulk solution, but only one of them has been detected within the cavity, and its rotation rate was slowed by several orders of magnitude as a result of such encapsulation. A competition experiment performed in [59] for amide *200* showed the acceleration of its rotation resulted from the destabilization of the ground state of the guest. The caging ligand **429** also encapsulates a series of solid adamantane derivatives shown in this Scheme thus forming their 1:1 cage complexes [109].

Co-encapsulation of suitable guests by the hydrogen-bonded capsule **429** (Scheme 3.107) has been used [110] for direct observation of

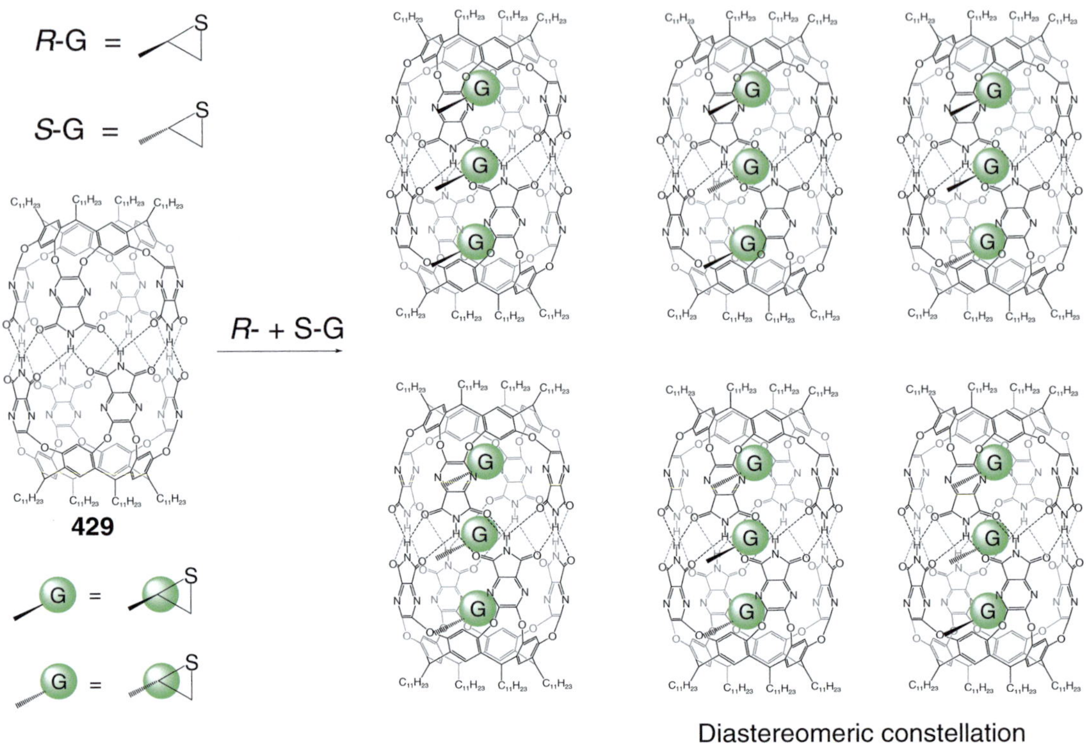

Scheme 3.104

halogen bonding by various NMR methods. The co-caging amplifies the concentrations of both the donor and acceptor of such bonds, while the shape of **429** permits the proper alignment and the extended lifetime of its heteroguest cage complexes [110].

The capsule **429** has been used in [111] to alter the photoluminescence of guest benzyl derivatives shown in Scheme 3.108 by tuning the conformational space available for the aromatic guest within its cavity. This has been used for elucidation of internal dynamics of these cage complexes by luminescence methods.

The use of **429** and its extended analog **435** (Scheme 3.109) allowed the authors of [112] to elucidate coiling (uncoiling) cycles of tetradecane guest. The latter is encapsulated as a helical coil within the cavity of **429**, but upon addition of spacer glycoluril fragments to the solution longer self-assembled capsule **435** appears with the guest relaxes to its extended conformation. Protonation of the aniline sites of these spacers caused this system to revert to coiled tetradecane molecule within the parent capsule **429**, while further addition of triethyl amine to the reaction mixture regenerates the longer capsule **435** with extended guest [112]. NMR experiments have been performed in [113] for the capsule **435** and a series of normal alkanes $C_{16}H_{24}$ to $C_{19}H_{40}$. They showed the coiled forms for the caged $C_{17}H_{36}$ – $C_{19}H_{40}$ molecules. This coiling exerts pressure on the cage framework of **435**: its hydrogen-bonded seams were loosened, and rotation of the capsule's ligand syntones occurred on the NMR time scale, resulting in racemization of the cage complexes. The racemization rates increase with the length of the alkane because longer guests exert more pressure. At the same time, hexadecane guest has been encapsulated in its fully extended conformation exerting no pressure within the cavity of **435** [112].

Encapsulation by a regular hydrogen-bonded capsule **429** of *trans*-stilbene caused its fluorescence quenching due to the distortion of the

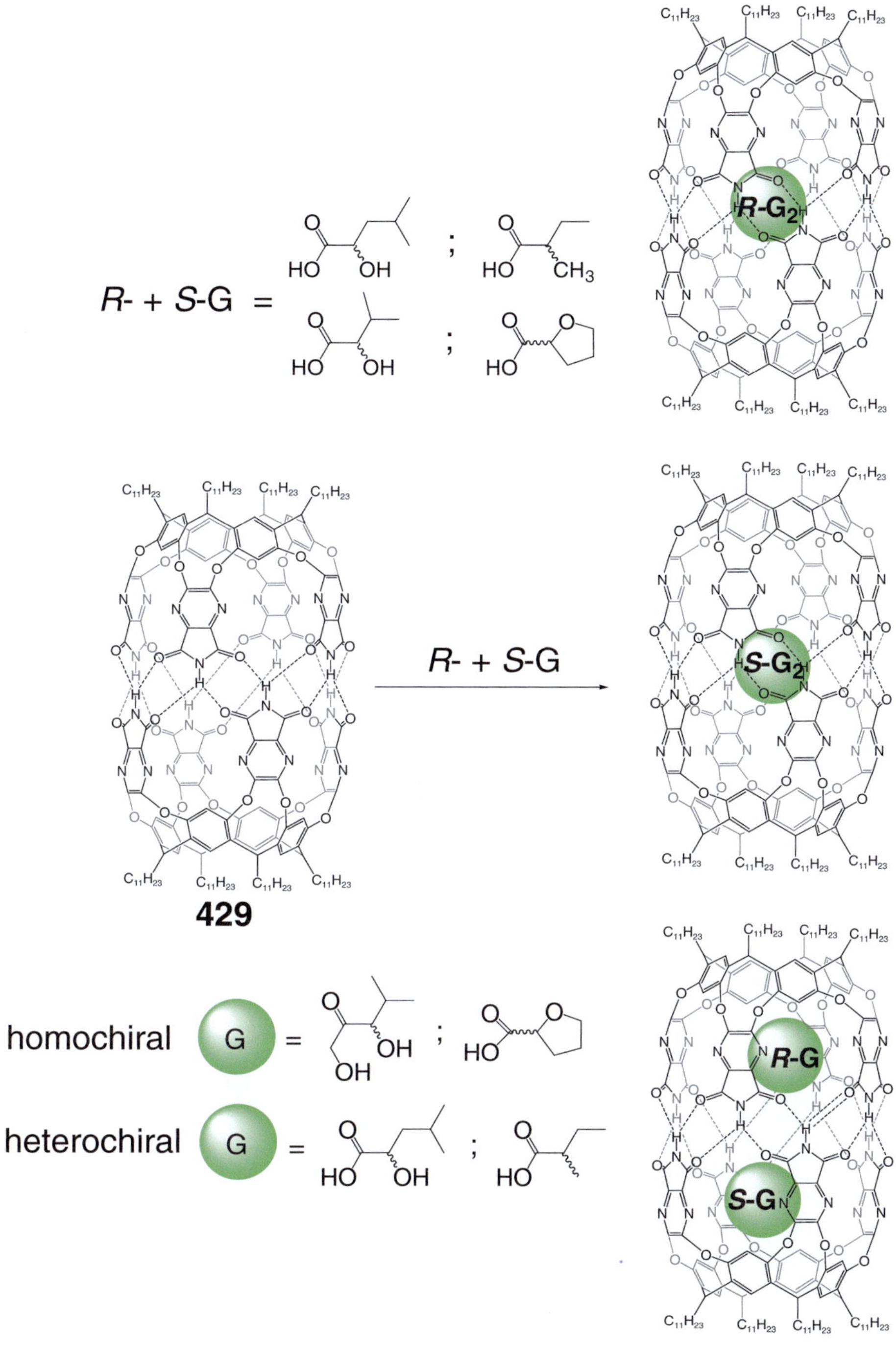

Scheme 3.105

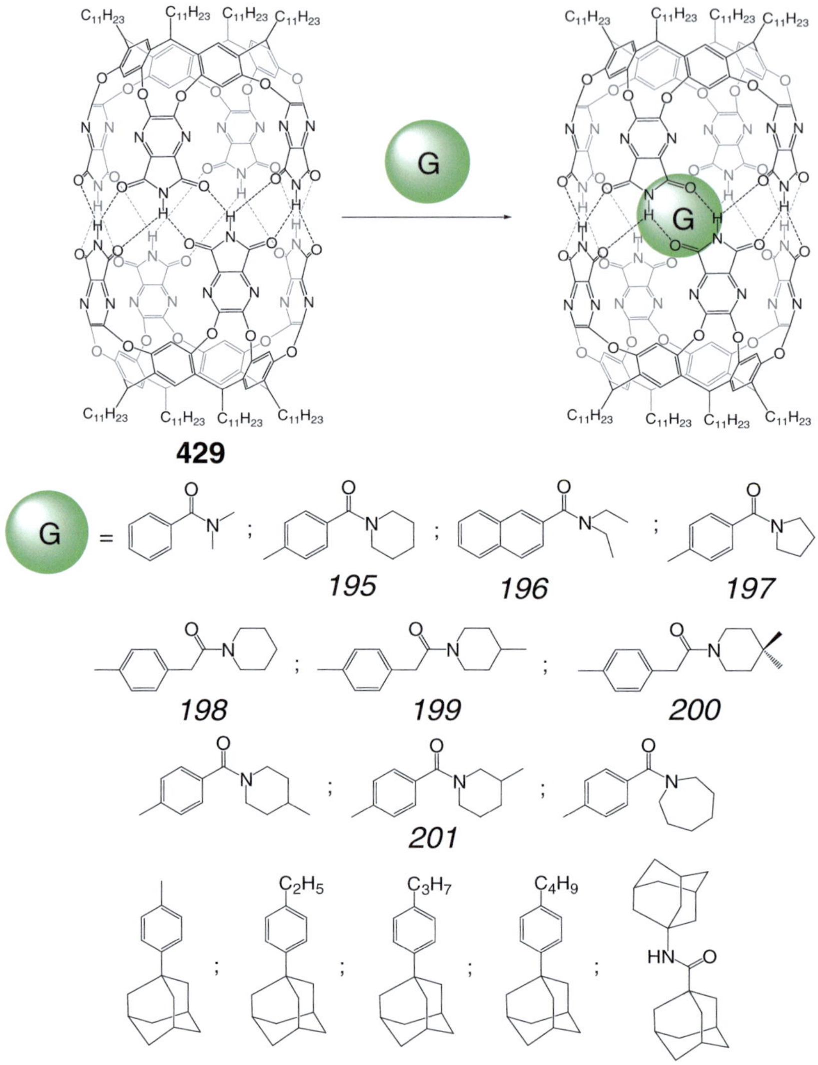

Scheme 3.106

ground-state geometry [114]. When this cage framework has been elongated by incorporation of four glycoluril spacer fragments to give an extended capsule **435**, the caged *trans*-stilbene molecule adapts fully coplanar arrangement that displays fluorescence.

Regular and extended hydrogen-bonded capsules **429** and **435** are also able to encapsulate gas molecules shown in Scheme 3.110, thus forming multiguest 1:n and 1:m cage complexes, respectively [115]. These capsules also bind a series of alkenes and alkynes shown in Scheme 3.111 in contorted conformations to reduce their lengths in order to fit the cavities of these encapsulating ligands [116].

The encapsulated hydrogen-bonded dimers of carboxyl benzoic acids (Scheme 3.112) are reported in [117] to undergo compression within the extended capsule **435**. Caged hydrogen-bonded hetero- and homodimers of benzoic acid and benzamide guests shown in Scheme 3.113 within the cavities of regular and extended hydrogen-bonded capsules **429** and **435** have been investigated in [118] using DFT approach.

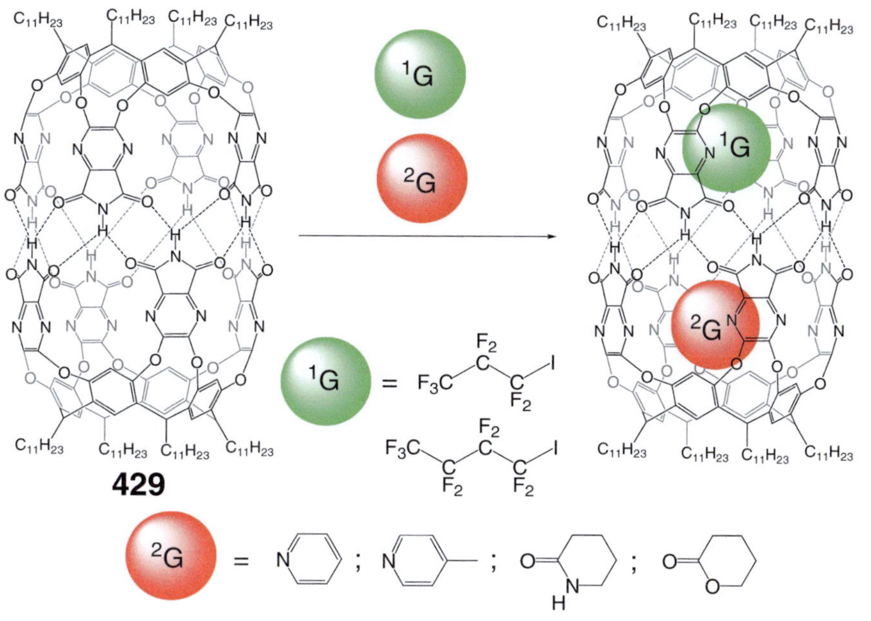

Scheme 3.107

Scheme 3.108

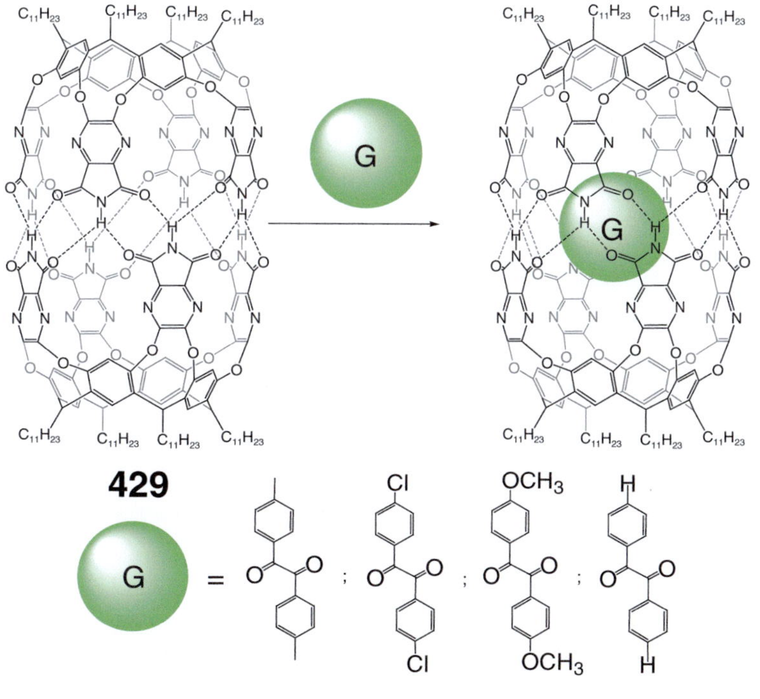

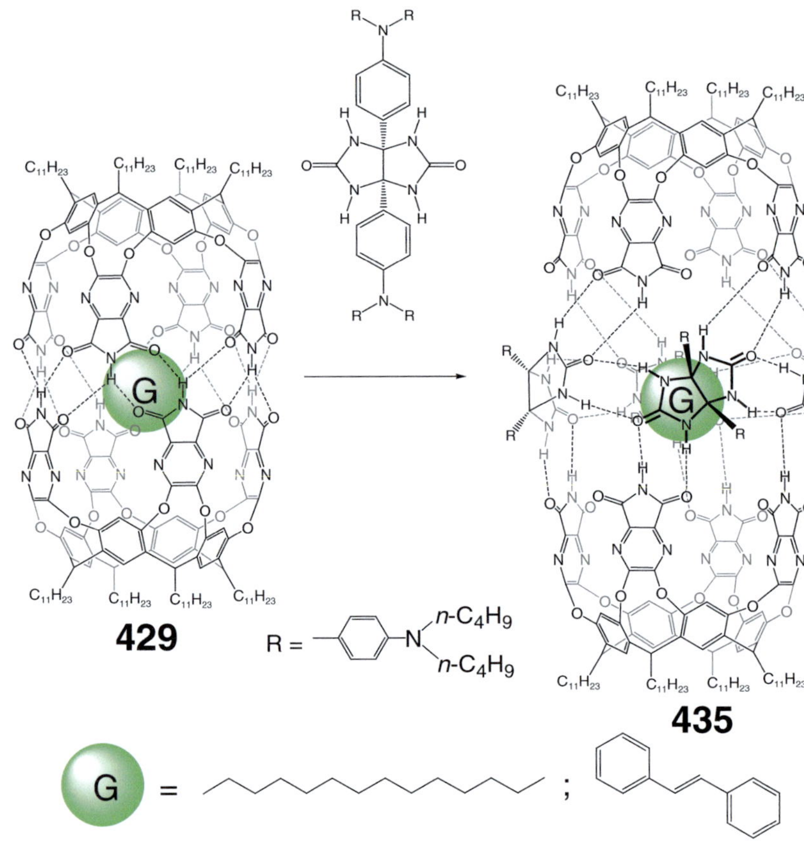

Scheme 3.109

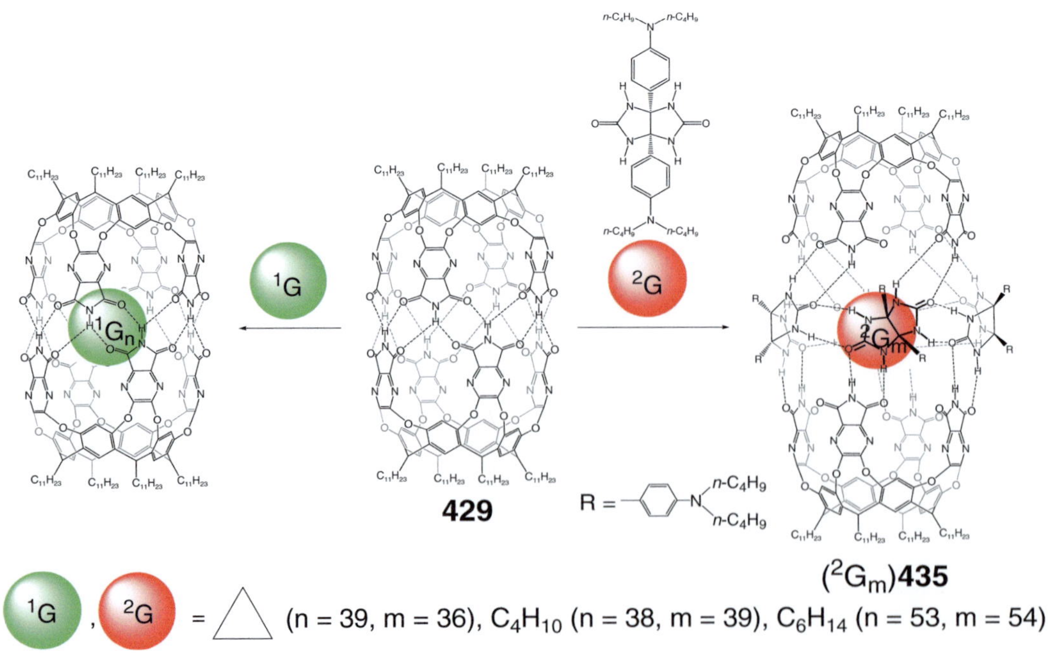

Scheme 3.110

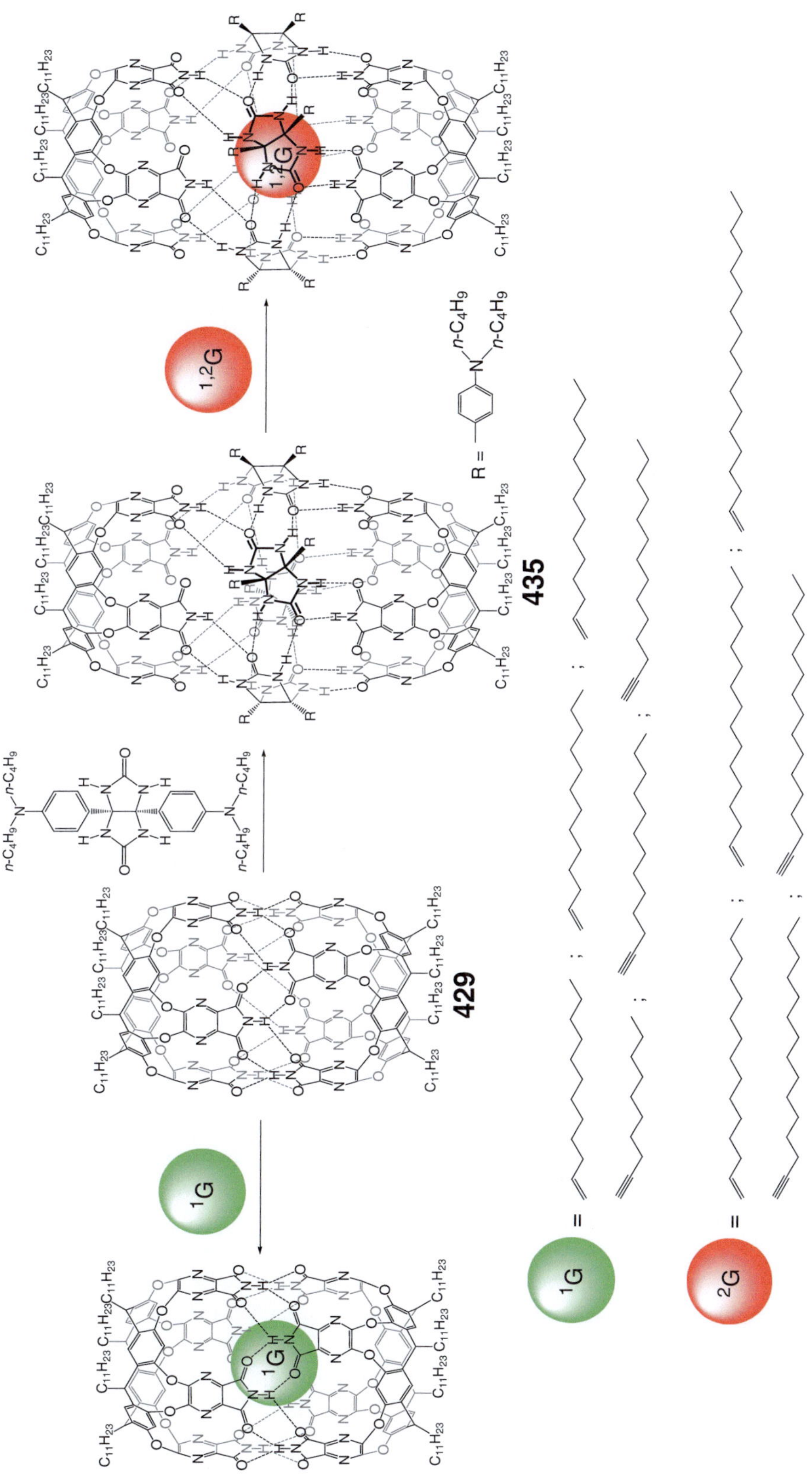

Scheme 3.111

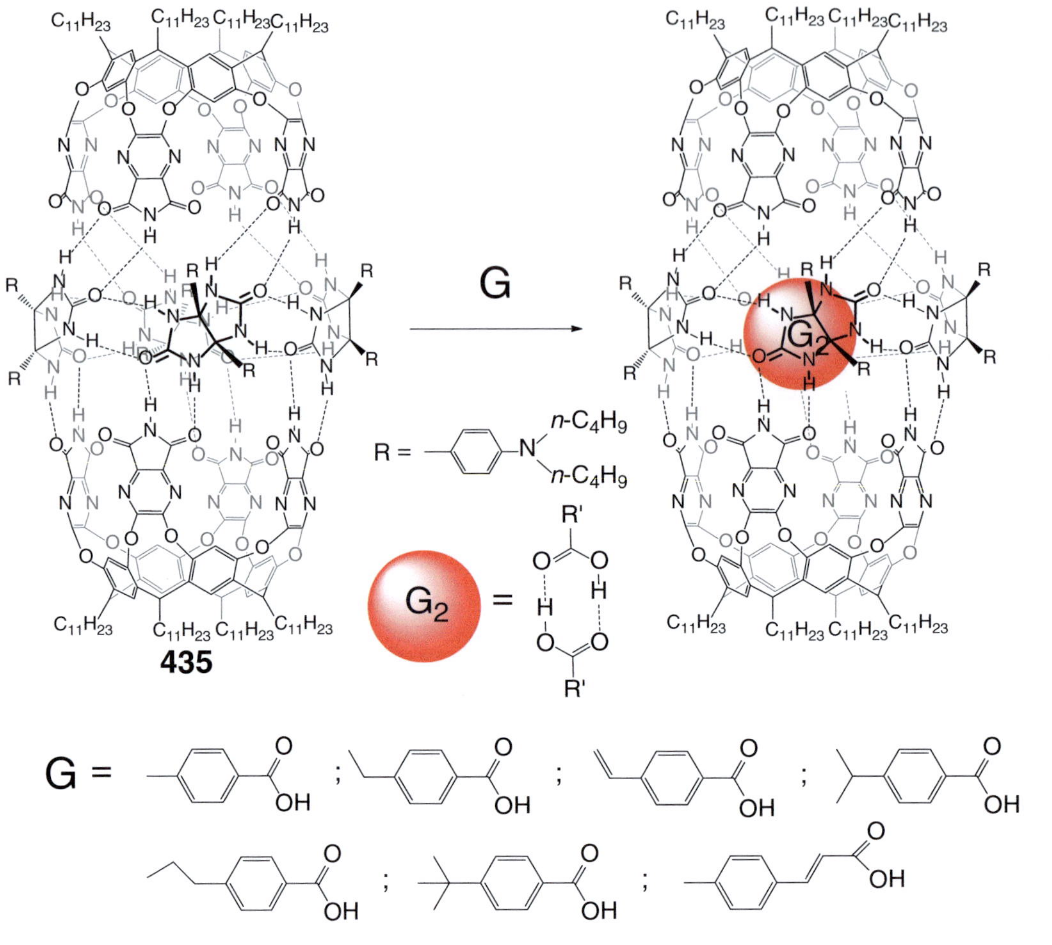

Scheme 3.112

Assembly of two cavitand ligand syntones of different nature shown in Scheme 3.114, each of which is able to form homohexameric- and homodimeric capsules **397** and **429** and to encapsulate solvent chloroform molecules, has been studied in [119]. The hybrid cylindrical framework of **436** is stabilized by eight hydrogen bonds and encapsulates suitable guests that properly fill its inner cavity. Screening of the guests performed in [119] showed that this hybrid capsule can efficiently bind aromatic and metallocene guests as well as the derivatives of cyclohexane. This efficient encapsulation with a proper space filling is the main driving force for the formation of the corresponding hydrogen-bonded capsules that prevent the self-sorting of their cavitand syntones [119].

A deep-cavitand caging ligand **437** with cavity dimensions of 23 × 10 Å and a cavity volume of approximately 800 Å^3, formed by two resorcinarene-based ligand syntones with extended aromatic spacer between them, has been self-assembled in [120] by Scheme 3.115. As follows from ^{1}H NMR data, this ligand encapsulates rigid and long (*ca.* 18 Å) guest molecules *202* and *203* forming kinetically stable 1:1 cage complexes. In the case of more flexible guest *204* and/or shorter (<14 Å) adamantane derivatives *205* and *206*, slow exchange between free and caged guest species has been observed at low temperatures. The small guest *206* is reported in [120] to form a host–guest 1:2 cage complex.

The self-assembled in [121] dimeric water-soluble capsule **438** with ionic pendant substituents

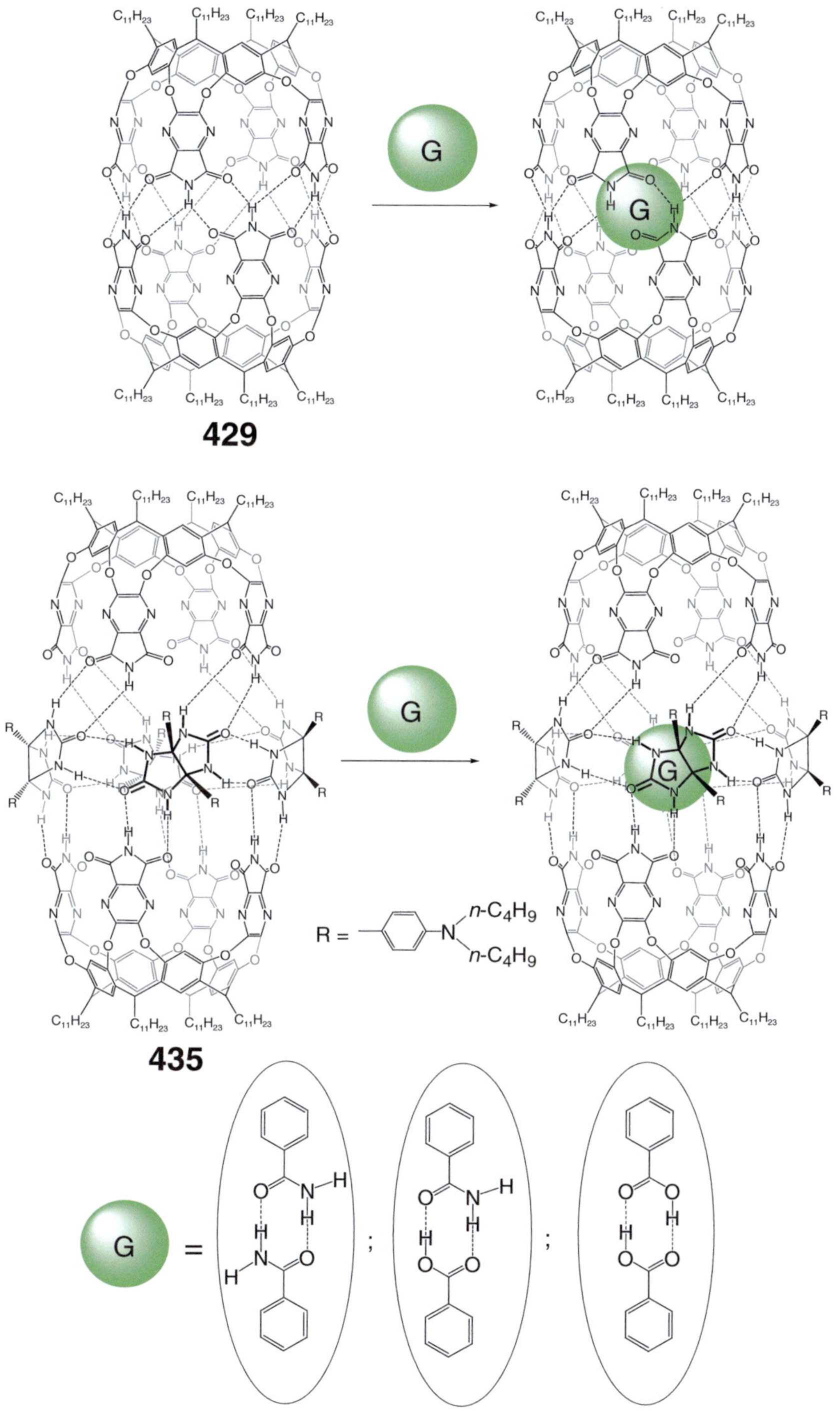

Scheme 3.113

encapsulates long-chain hydrocarbons and other lipophilic guests shown in Scheme 3.116.

Alkane guest-driven reversible assembly of extended and hyperextended cylindrical supramolecular capsules **435** and **439–446** by Scheme 3.117 has been studied in [122] and [123]. These capsules contain from 4 and 8 to 12 glycoluril spacer fragments and have the cavity

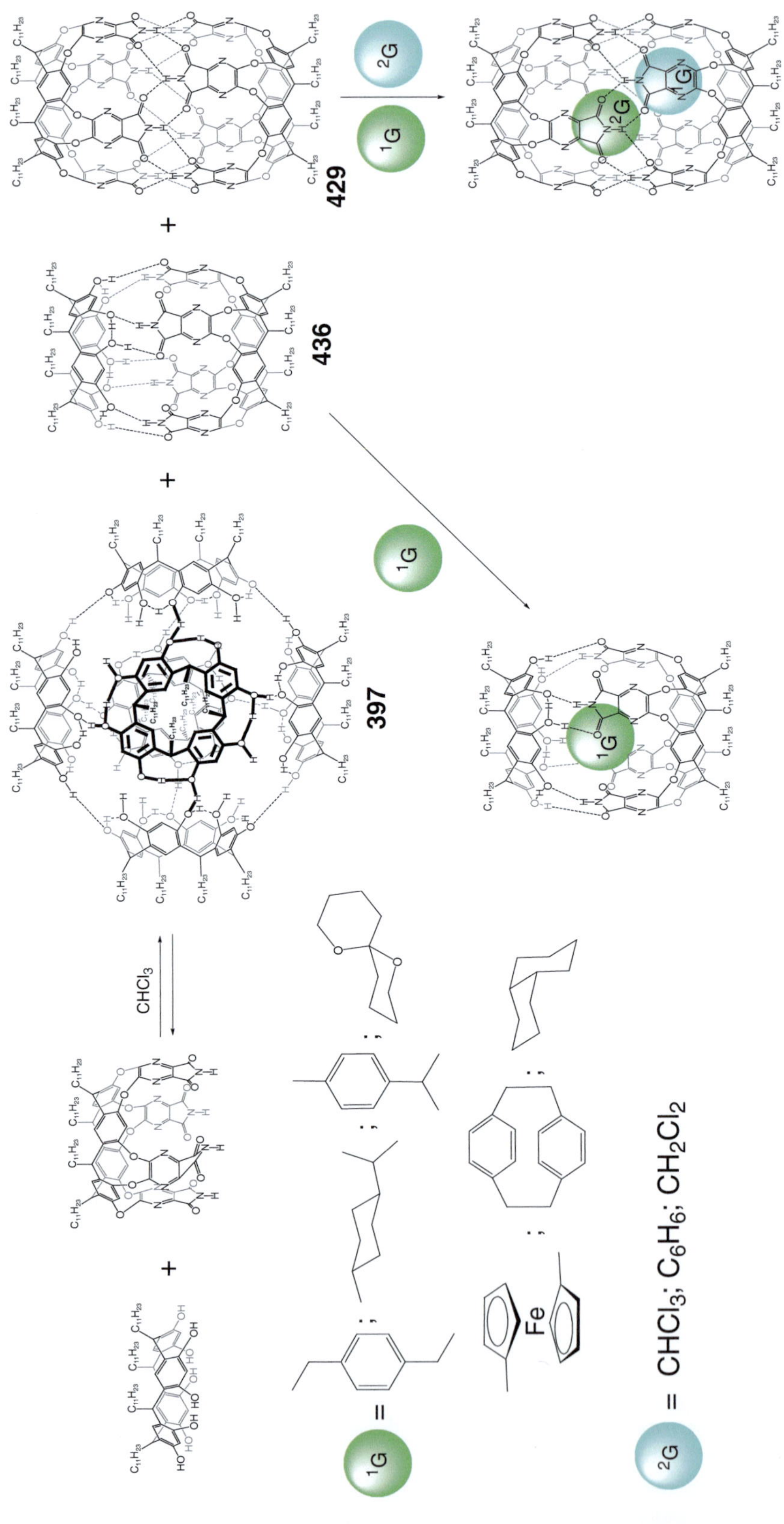

Scheme 3.114

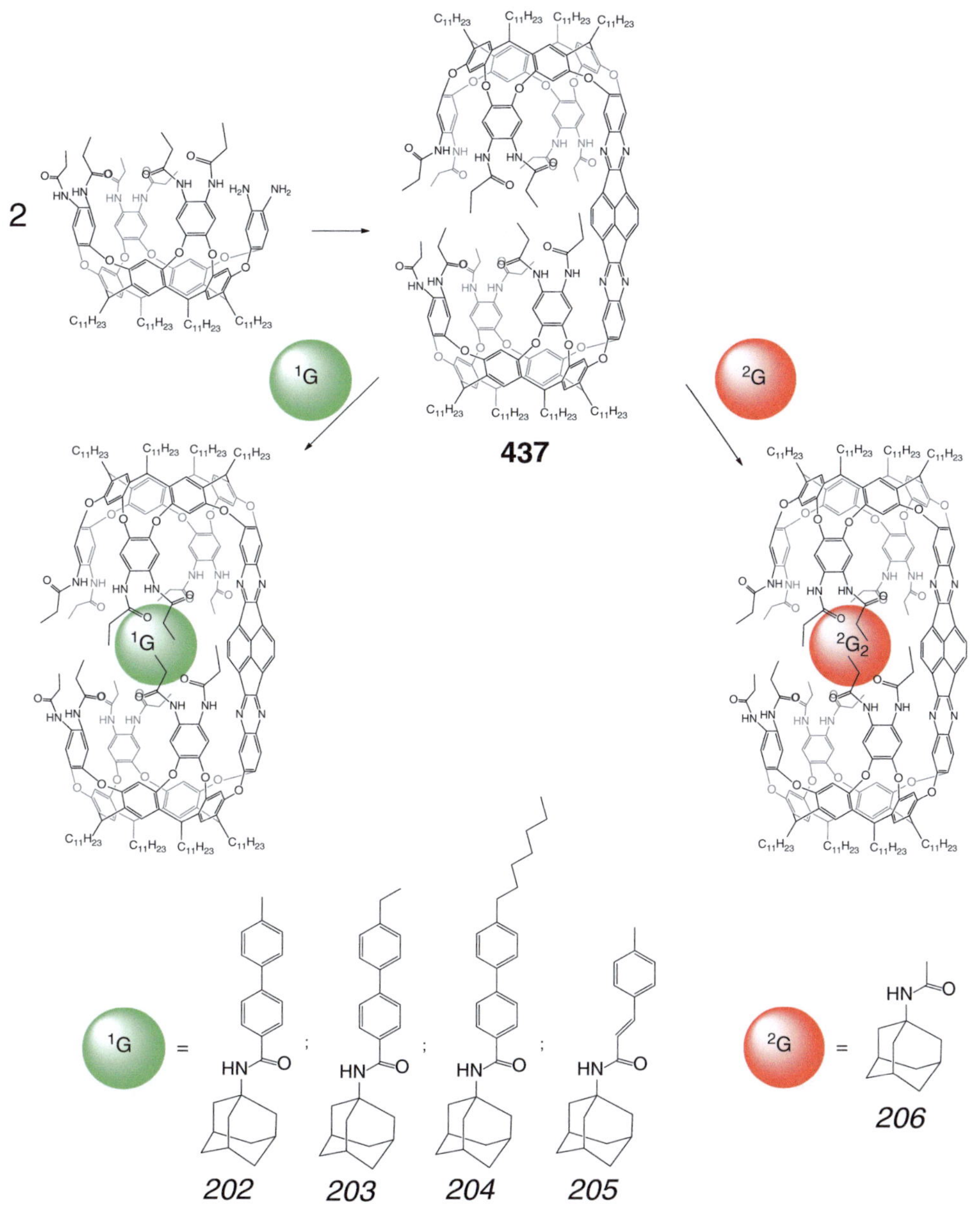

Scheme 3.115

volumes increased up to 530 Å and the lengths up to approximately 21 Å [123].

Self-assembly dynamics of the hybrid bis-cavitand capsule **447** and its cage complexes with different guests shown in Scheme 3.118, the ligand syntones of which were labeled with donor and acceptor fluorophores, has been studied in [124] using FRET. For these guests, a remarkable range of exchange rates (more than four orders of magnitude) has been observed. The same approach has been used in [125] to monitor the self-assembly and the guest exchange for the arene capsules (see Sect. 3.1.1).

The influence of remote asymmetric centers in one of the two co-encapsulated molecules shown in Scheme 3.119 on its partner guest within the cavity of hydrogen-bonded capsule **429** has been studied in [126] by ^{1}H NMR method. Due to the

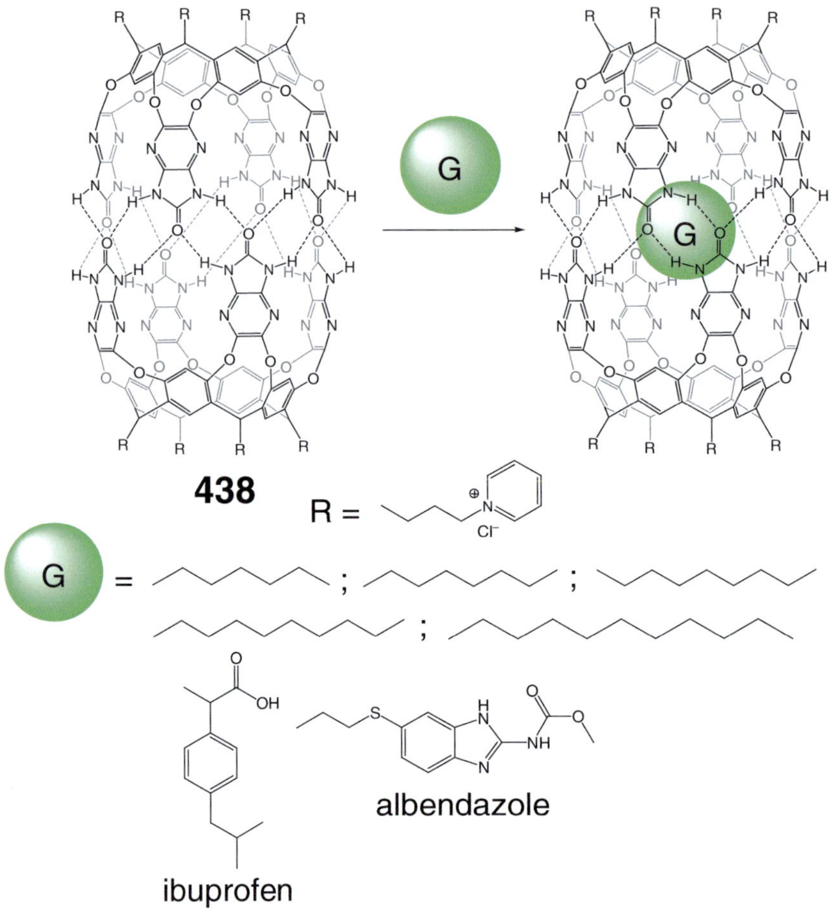

Scheme 3.116

remoteness of asymmetric functions of the chiral guest on its co-encapsulated partner, this result cannot be explained by steric repulsions: their centers are separated by the benzene ring and the proximal asymmetric function as well. So, such remote effects were assigned in [126] to the strictly magnetic effects.

Effects of remote chiral centers of the resorcinarene and cavitand ligand syntones on encapsulated guests have been observed in [127] for their hybrid supramolecular capsules **448–454** giving 1:1 cage complexes with encapsulated neutral guest *202* by Scheme 3.120. These distal effects are additive or subtractive, depending on the relative configurations of the chiral elements.

Assembly of phthalimide deep-cavitand precursor with complimentary resorcinarene syntones by Scheme 3.121 has been studied in [128]. Ethane can be easily accommodated by the cavity of the caging ligand **455** and co-encapsulated by Scheme 3.121 with [2,2]-paracyclophane. Mixing of these cage complexes gave only a mixture of enantiomers of the hybrid 1:1 cage complex with the encapsulated [2,2]-paracyclophane. This bulky guest can be also accommodated by hybrid capsules **456** and **457**. According to NMR data, it undergoes fast rotation within the cavity of these ligands. Fast exchange reaction between their enantiomers is also detected by NMR spectroscopy even at low temperatures; however, this method could not be used to investigate the dynamics of the hybrid capsules that is too fast. The latter has been thus studied [128] by FRET using the perylene-labeled resorcinarene

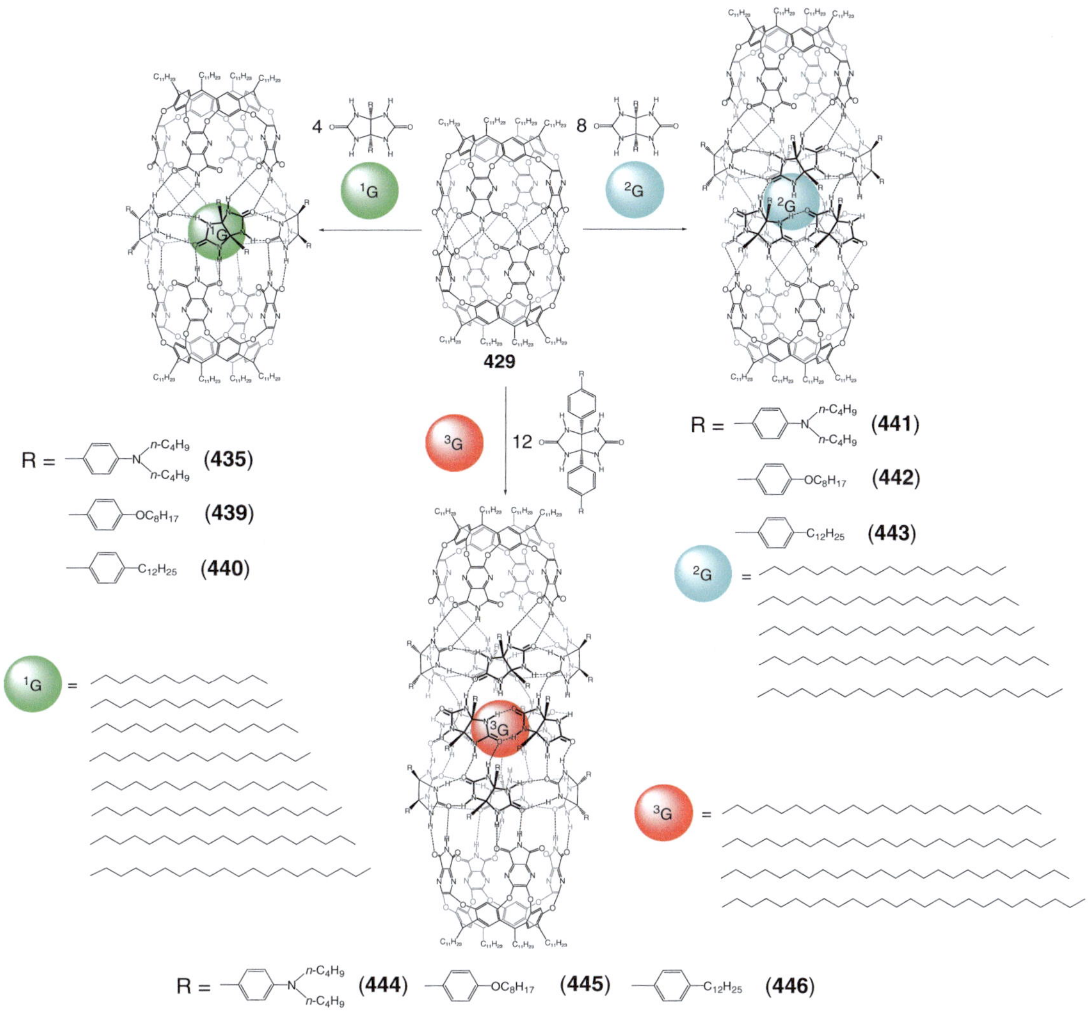

Scheme 3.117

syntone as an acceptor fluorophore and the pyrene-labeled deep-cavitand *194* as a donor fluorophore. No FRET signal has been detected in the absence of guests, whereas in the case of [2,2]-paracyclophane, an intensive signal appears after few days, which indicates the formation of the hybrid cage complex of **456** with the encapsulated [2,2]-paracyclophane [128]. Among the guests that cannot be encapsulated by the homodimers of these ligand syntones but can be accommodated by their hybrid **457**, the nonsymmetrical guests (in particular, *para*-ethyltoluene) form two carceroisomers. In the case of 1,7-dioxaspiro[5.5]undecane and menthol as chiral guests, diastereomeric cage complexes have not been detected due to their free rotation within the cavity, while 2-methylheptane and 2,2-dimethylhexane can be encapsulated by **457**. The openings and seams of hydrogen bonds at the "upper" part of this hybrid capsule are described in [128] to favor bulky and polar guest fragments.

Dimerization of heteroditopic bis-cavitand ligand syntone (Scheme 3.122), containing both self-folding and dimer-forming fragments, has been used in [129] to obtain a polytopic hydrogen-bonded capsule **458**. This host encapsulates guests at the distal binding sites without their direct interaction, and the guests can be released

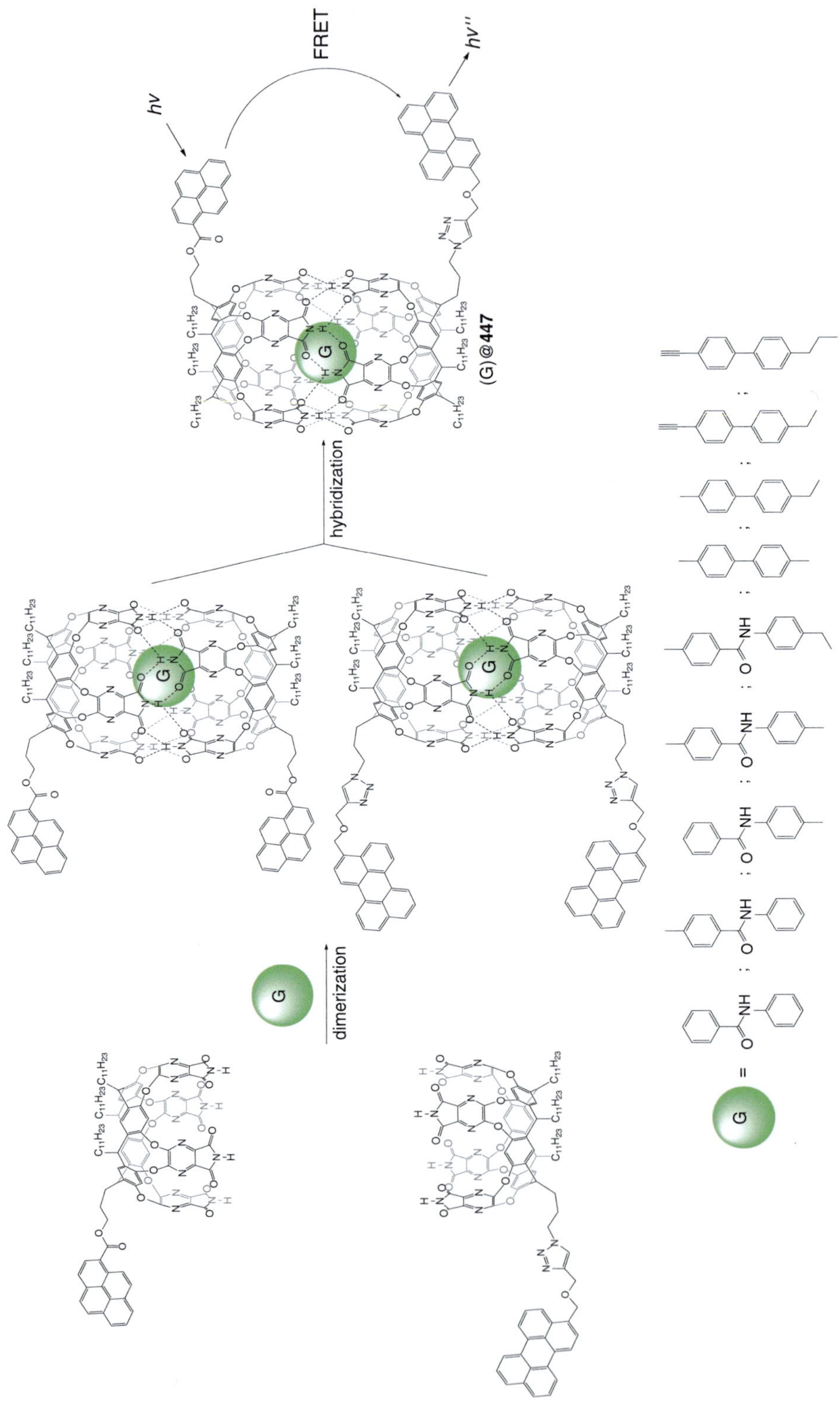

Scheme 3.118

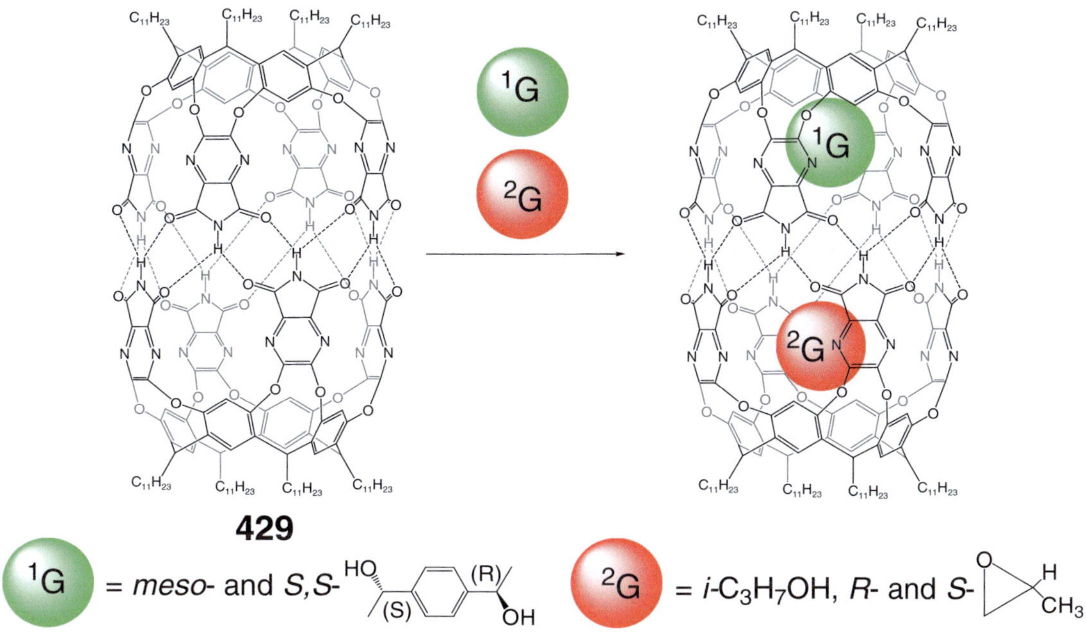

Scheme 3.119

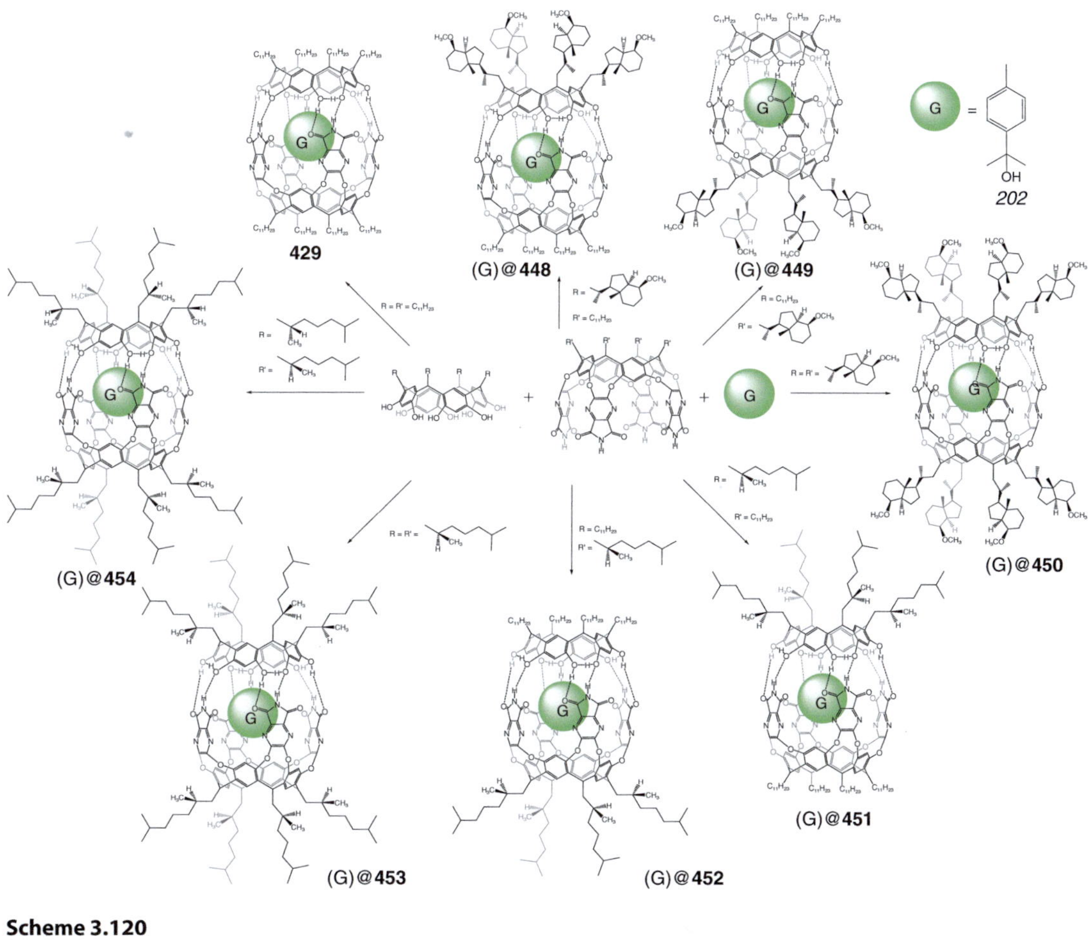

Scheme 3.120

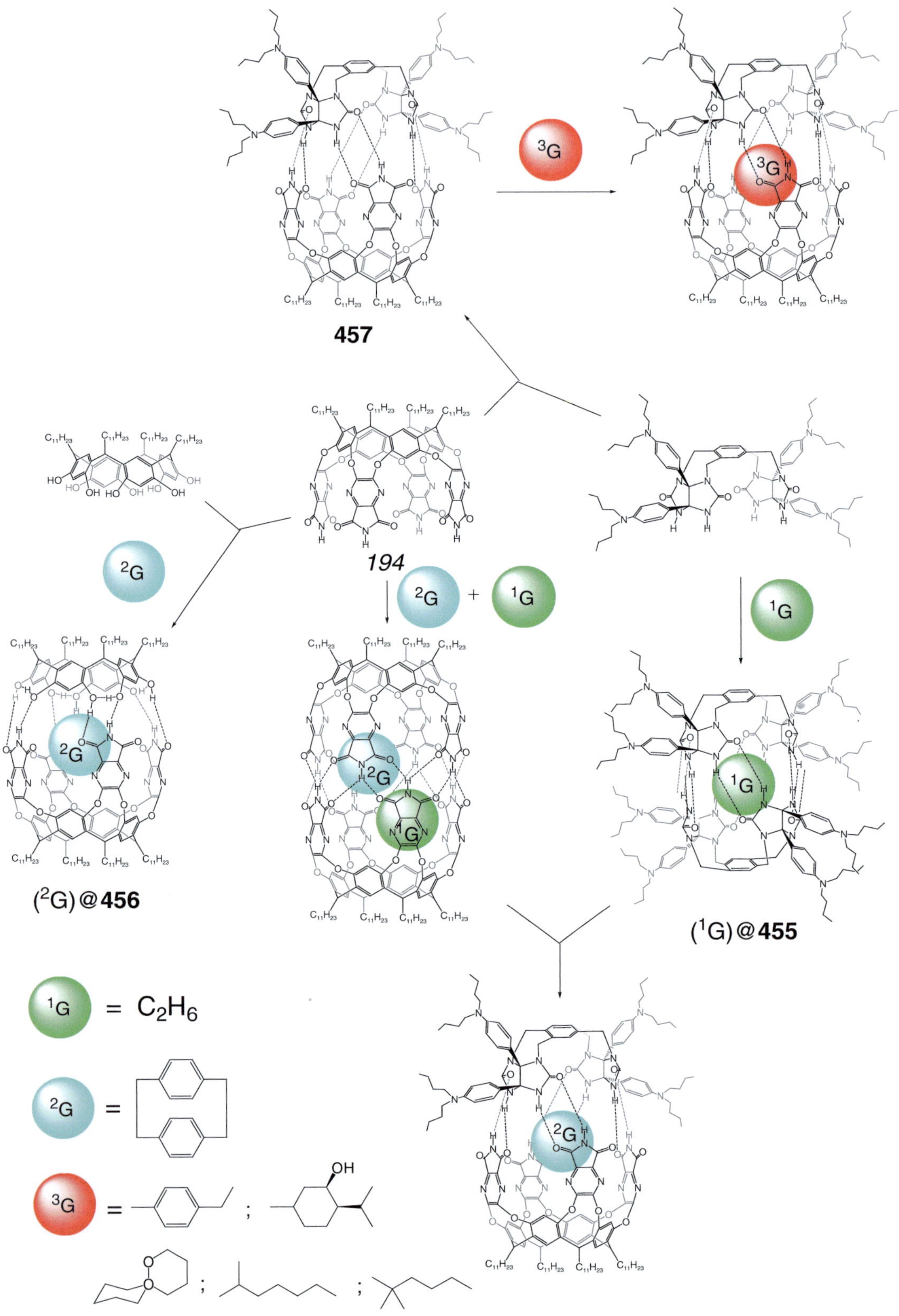

Scheme 3.121

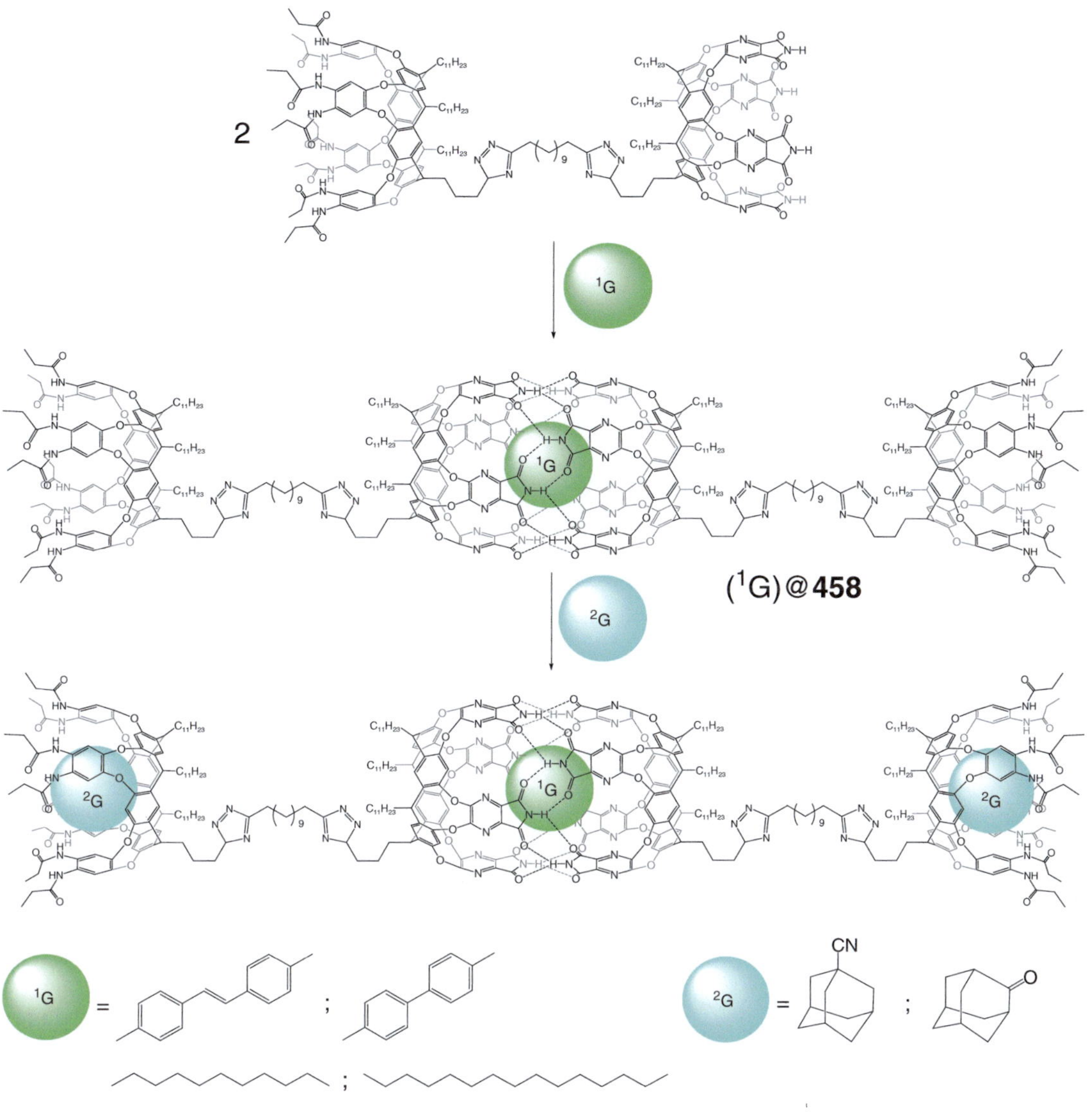

Scheme 3.122

independently from such sites with appropriate chemical reagents. Their orthogonality is reported in [129] to extend the guest exchange dynamics occurring in this case in well-separated time frames.

Hybrid and extended chiral capsules **459** and **460** have been prepared in [130] by Scheme 3.123 using the corresponding methylated glycoluril syntone and parent capsule **429**. The interconverting hydrogen-bonded caging ligands allow tuning the dimensions and conformations of normal alkanes within their cavities.

Hydrogen bonding-directed pairwise co-encapsulation of *para*-ethylphenylboronic acid and its amide- and carboxyl-containing analogs as partners by Scheme 3.124 leading to 1:2 homo- and 1:1:1 heteroguest cage complexes of **435** has been detected in [131] by ^{1}H NMR spectroscopy.

The use of thiourea-containing capsule **461** as a host allowed encapsulation of inconveniently shaped long and bent molecules such as normal $C_{10} - C_{15}$ alkanes by Scheme 3.125 [132].

Extended capsules **462** and **463** with a cavity volume increased by 200 Å and the lengths of

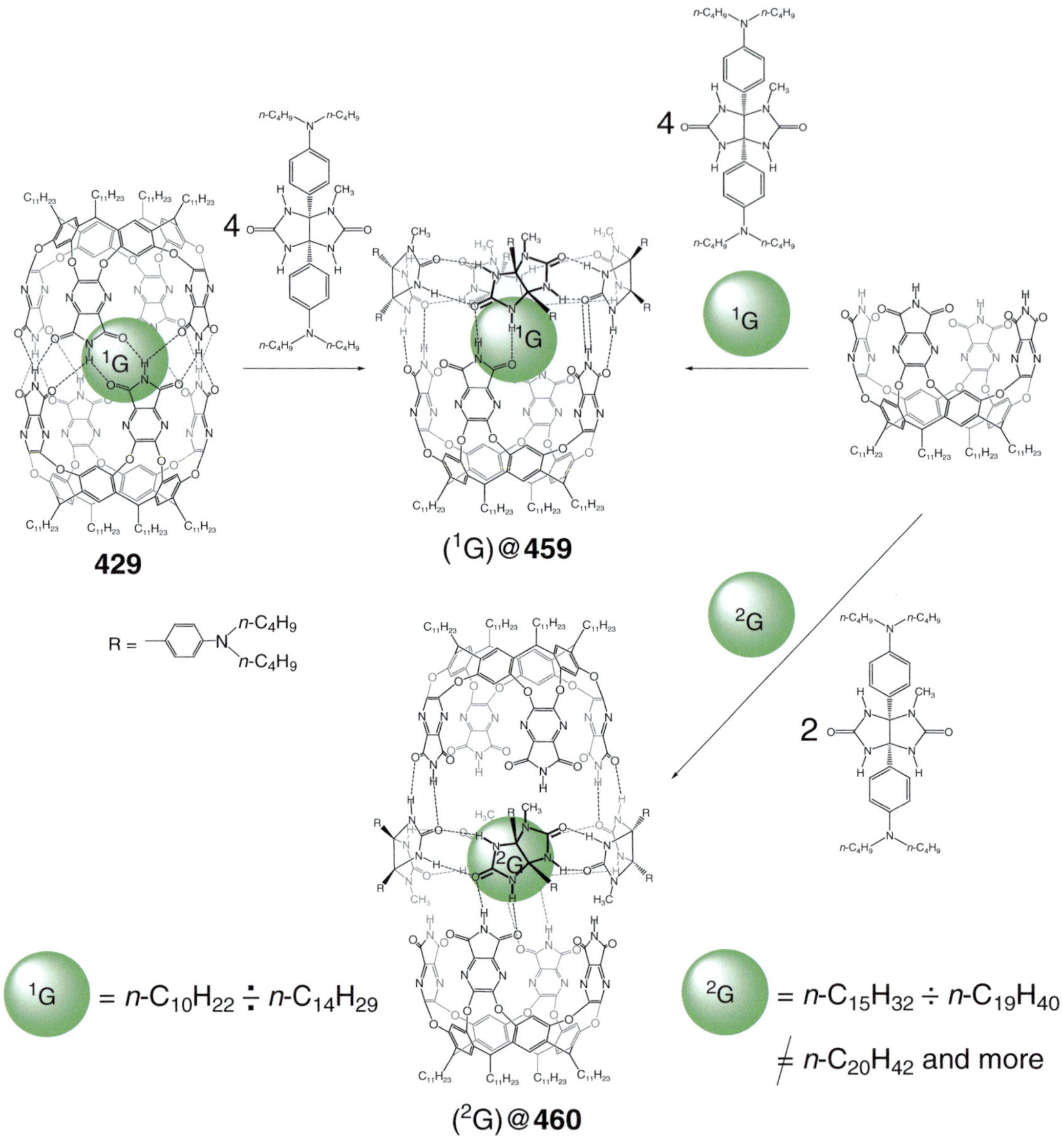

Scheme 3.123

approximately 7 Å, are reported in [133] to encapsulate long-chain alkanes in the their *gauche*-conformation by Scheme 3.126.

Binding properties of the extended hydrogen-bonded capsules **435**, **462**, and **463** with the cavity volumes of approximately 620 Å have been studied in [134]. These caging ligands can encapsulate capsaicin and two *para*-methylstyrene molecules as guests; slow tumbling of caged *para*-disubstituted benzene molecule on the NMR time scale within their cavities leads to their social isomerism depictured in Scheme 3.127. Gases such as propane, butane, isobutene, propylene, 2-methylpropene, 1,3-butadiene, and xenon are reported in [134] to be co-encapsulated with other guests shown in this Scheme.

The use of propane–diurea ligand syntones allowed [135] obtaining of *S*- and banana-shape capsules **464**–**467** by Scheme 3.128. The banana-shaped capsule is reported to selectively

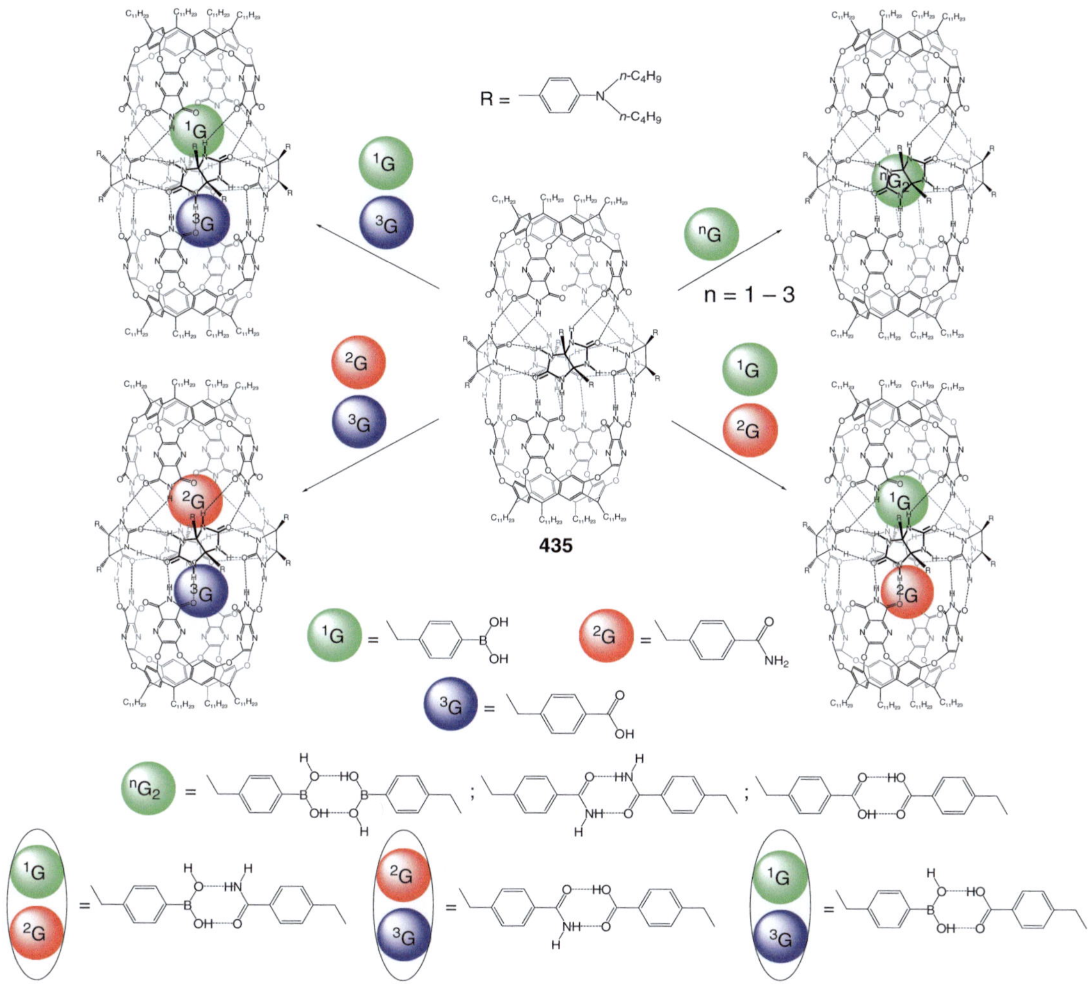

Scheme 3.124

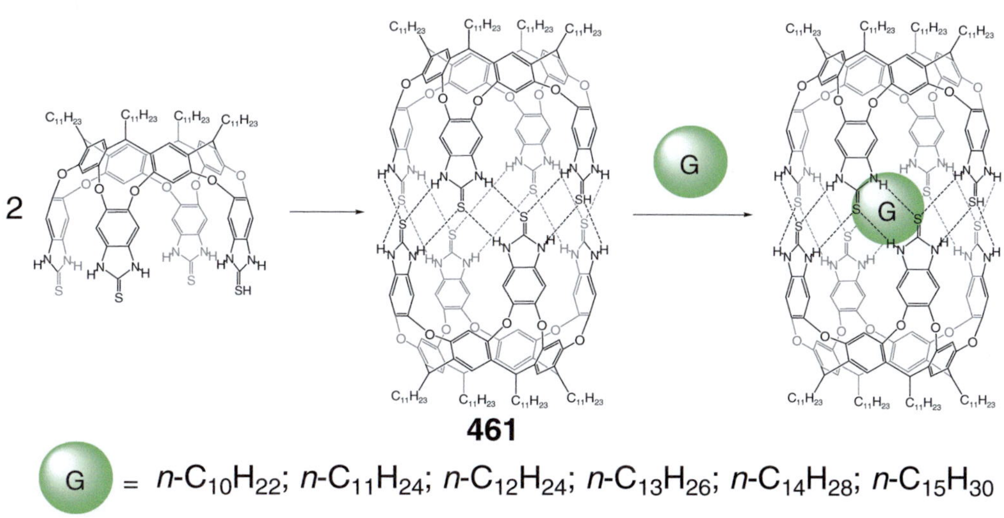

Scheme 3.125

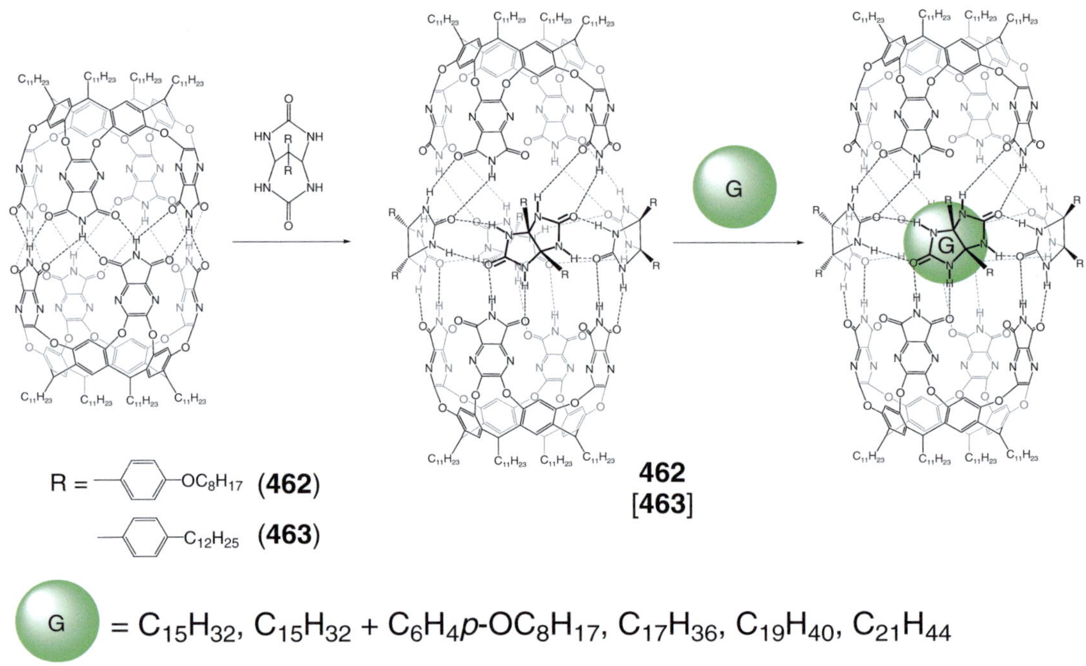

Scheme 3.126

encapsulate complementary shaped alkane guests. π–π interactions on the outside of these capsules have been used in [136] to selectively trigger changes in their cage frameworks.

The redox-active ferrocene and stilbene molecules are described in [137] to form bis-cavitand 1:1 cage complexes of supramolecular capsule **468** by Scheme 3.129 in the presence of *tert*-amine-containing ferrocenyl cation but not with the anionic ferrocenecarboxylate. Such octaacid caging ligand also encapsulates tetrathiafulvalene to give the corresponding 1:1 cage complex but discriminates its more bulky derivatives (as follows from ^{1}H NMR spectra). The caging of the tetrathiafulvalene guest causes dramatic changes in its redox properties; the hindered voltammetric response of these species is explained in [138] mainly by kinetic factors (first of all, by very low ET from an electrode surface to the caged redox-active guest). Relatively free rotation of this guest within the capsule **468** causes a decrease in the stability of the cage complex leading to broadening of its proton signals in ^{1}H NMR spectrum by the exchange of the trapped guest among its several possible locations [138].

Hyperextended hydrogen-bonded capsule **469** is described in [139] to encapsulate two γ-aromatic cations and two carboxylate anions as ion-pairing guests within its cavity; the anionic species are located in the tapered ends of its cage framework, while two cations are near the glycoluril spacers in the capsule's center [139]. In the case of a heteroguest 1:1:1 cage complex of the parent capsule **429** with encapsulated picoline molecules, the strong social isomerism has been observed. A detailed study of the interactions of α-, β-, and γ-picoline guests within the cavity of the cylindrical capsule **469** leading to this type of isomerism has been performed in [140]. In the case of α- and γ-picolines, about 95 % of their cage complexes correspond to the isomer with the methyl group of the α-component inwardly directed and that of the γ-component outwardly directed, respectively. The β-picoline gave predominantly (by approximately 70 %) the 1:2 host–guest complexes (Scheme 3.130) with an outward orientation of its methyl groups [140].

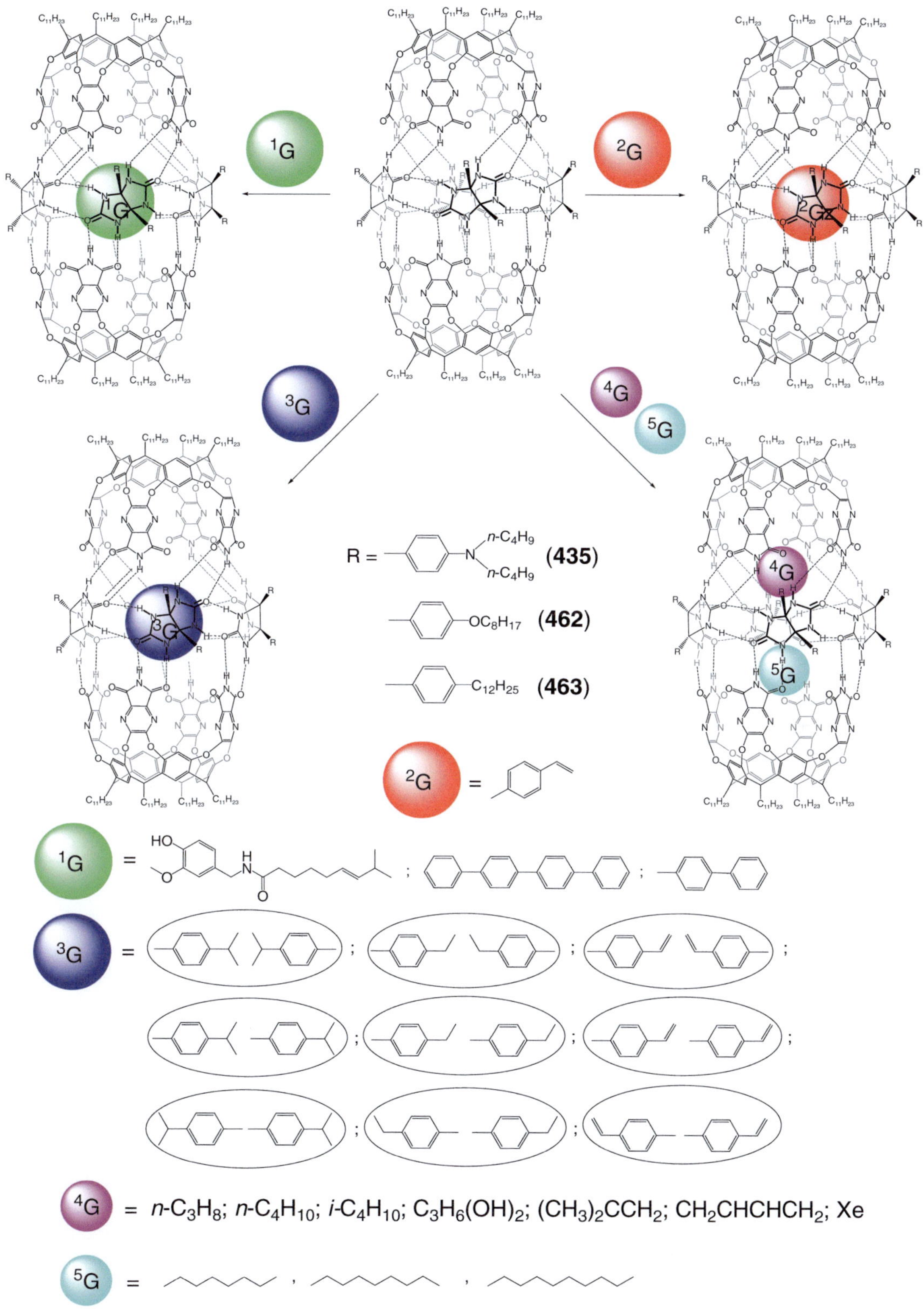

1G
2G
3G
4G
5G
R = n-C4H9 N n-C4H9 (435)
OC8H17 (462)
C12H25 (463)
2G =
1G =
3G =
4G = n-C3H8; n-C4H10; i-C4H10; C3H6(OH)2; (CH3)2CCH2; CH2CHCHCH2; Xe
5G =

Scheme 3.127

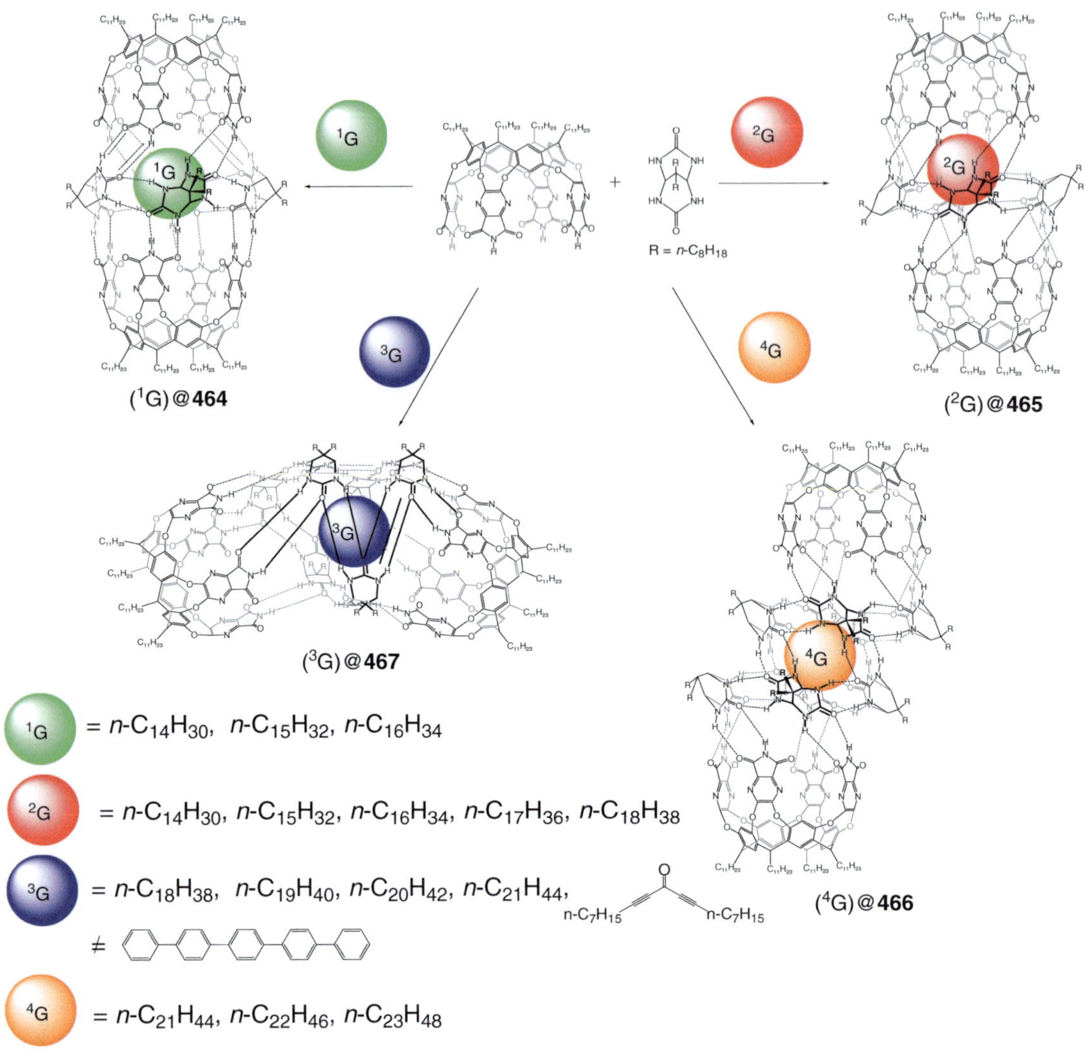

Scheme 3.128

Dimerization of the nonsymmetric cavitand ligand syntone with one short and three long walls in the presence of an appropriate guest molecule by Scheme 3.131 allowed the authors of [141] to obtain two constitutional isomers **470** and **471** of the corresponding hydrogen-bonded capsule. The use of such guests induces predominant formation of only one of these isomers due to the different hydrogen-bonding patterns of the capsules **470** and **471** that affect size, shape, and dynamics of their cavities.

Guest-induced reversible assembly of the water-soluble tetracationic cavitand syntone *207* led to robust 1:1 cage complexes of supramolecular capsule **472** with encapsulated aromatic dications and neutral molecules by Scheme 3.132 [142].

A deep-cavity cavitand octaacid ligand syntone *208* is described by B. C. Gibb and coworker [143] to form a dimeric capsule **473** that is stabilized by hydrophobic effects. As follows from ^{1}H NMR data, this ligand with hydrophobic concave surface and hydrophilic convex surface selectively encapsulates several steroidal guests by Scheme 3.133. The corresponding binding constants decrease in a row (+)-dehydroisoandrosterone *210* > progesterone *213* > estradiol *209* > 17α-ethynylestradiol *212* > estriol *211* > cortisone *214*, cholesterol *216*, spironolactone *215*; so

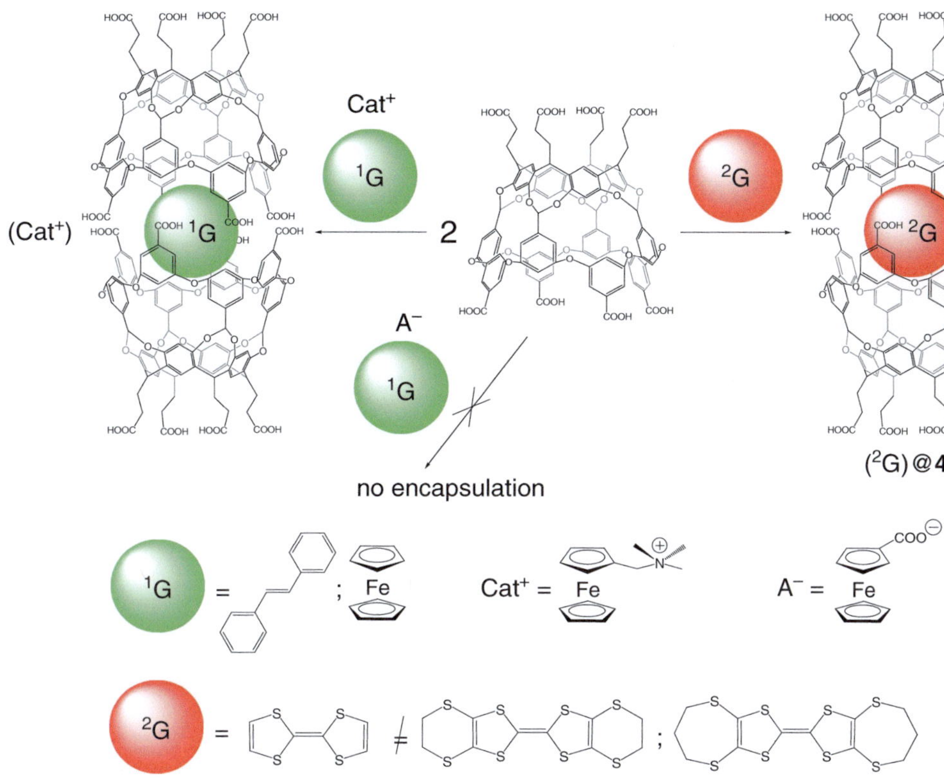

Scheme 3.129

the caging ligand discriminates these steroids by their size and polarity [143]. Encapsulation of guest molecules with a covalently attached ^{15}N-labeled nitroxide substituent (Scheme 3.133) by this bis-cavitand self-complimentary capsule has been performed in [144, 145] for further NMR, EPR, and ESP studies (see Sect. 6.3).

Capsule **473** binds neutral aromatic molecules in different coordination modes to give 1:1 or 1:2 cage complexes depending on the size of these guests [146]. As follows from NMR and luminescence spectra, this caging ligand can adopt two molecules of naphthalene, and both the excimer and monomer emissions have been observed in [146]. Therefore, these caged guests are not constrained within its cavity. At the same time, two encapsulated anthracene molecules form a sandwich excimer via stacking interactions. In the case of more bulky tetracene, the ligand **473** encapsulates only one guest molecule giving a 1:1 cage complex and only monomeric emission has been detected [146]. According to NMR, cyclic, and square-wave voltammetry data, the capsule **473** also efficiently binds the hydrophobic ferrocene guest to give a 1:1 cage complex by Scheme 3.134 but does not encapsulate doubly positively charged aromatic dications [147]. This caging ligand has been also used in [148] as a molecular flask for photochemical reactions (see Sect. 5.2) of encapsulated α-alkyl dibenzyl ketones (Scheme 3.134). According to ^{1}H NMR data, these ketones form 1:1 cage complexes that can be divided into three main groups depending on the orientation of the guest within the cavity of such dimeric encapsulating host. In the case of the caged ketones *217–229*, their two aromatic fragments occupy both the semisphere entities of **473**, while the propyl group of the guest *219* competes with the neighboring phenyl substituent; the alkyl groups and distal phenyl substituents of the encapsulated ketones *220* and *221* occupy the above semisphere fragments. The

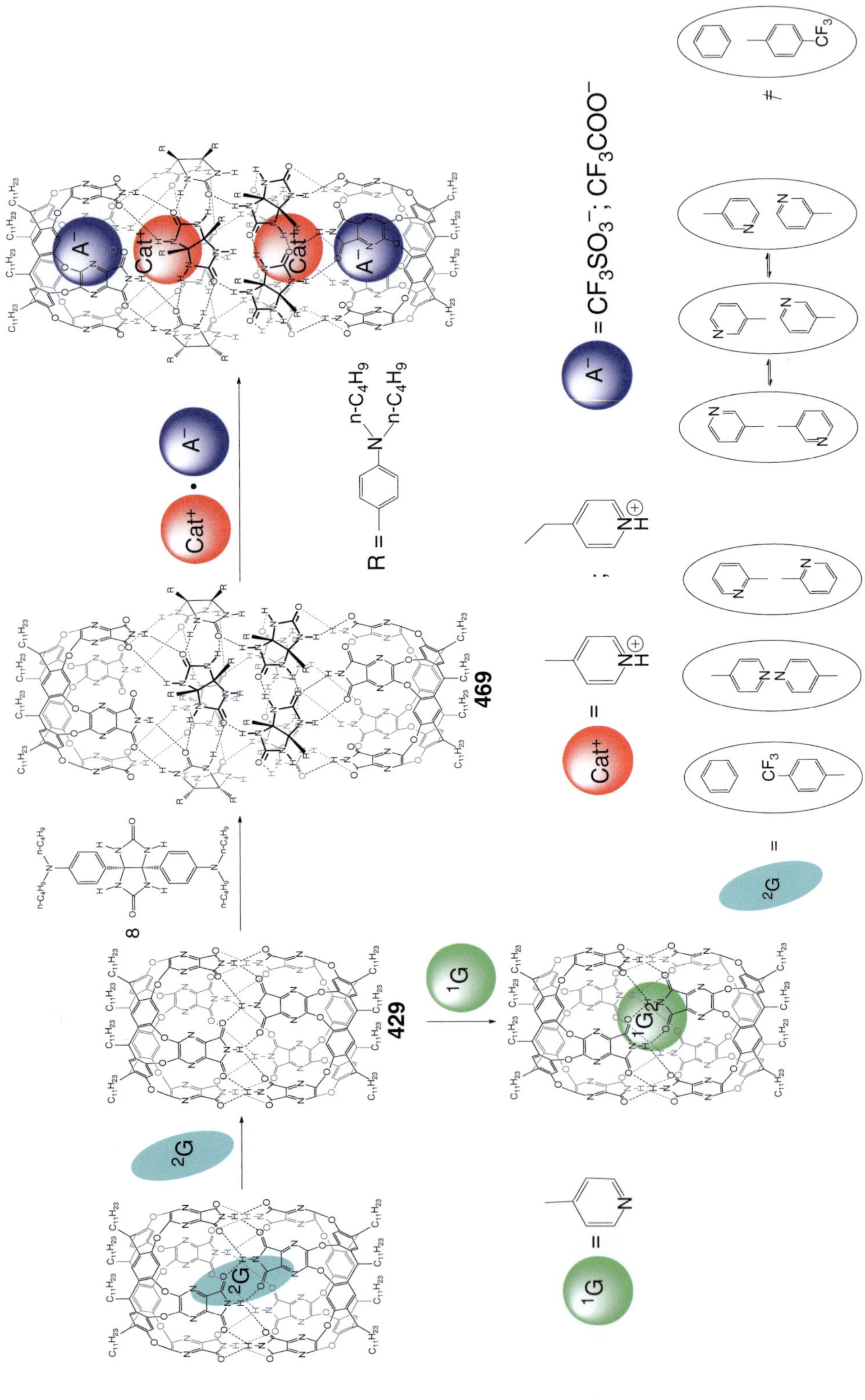

Scheme 3.130

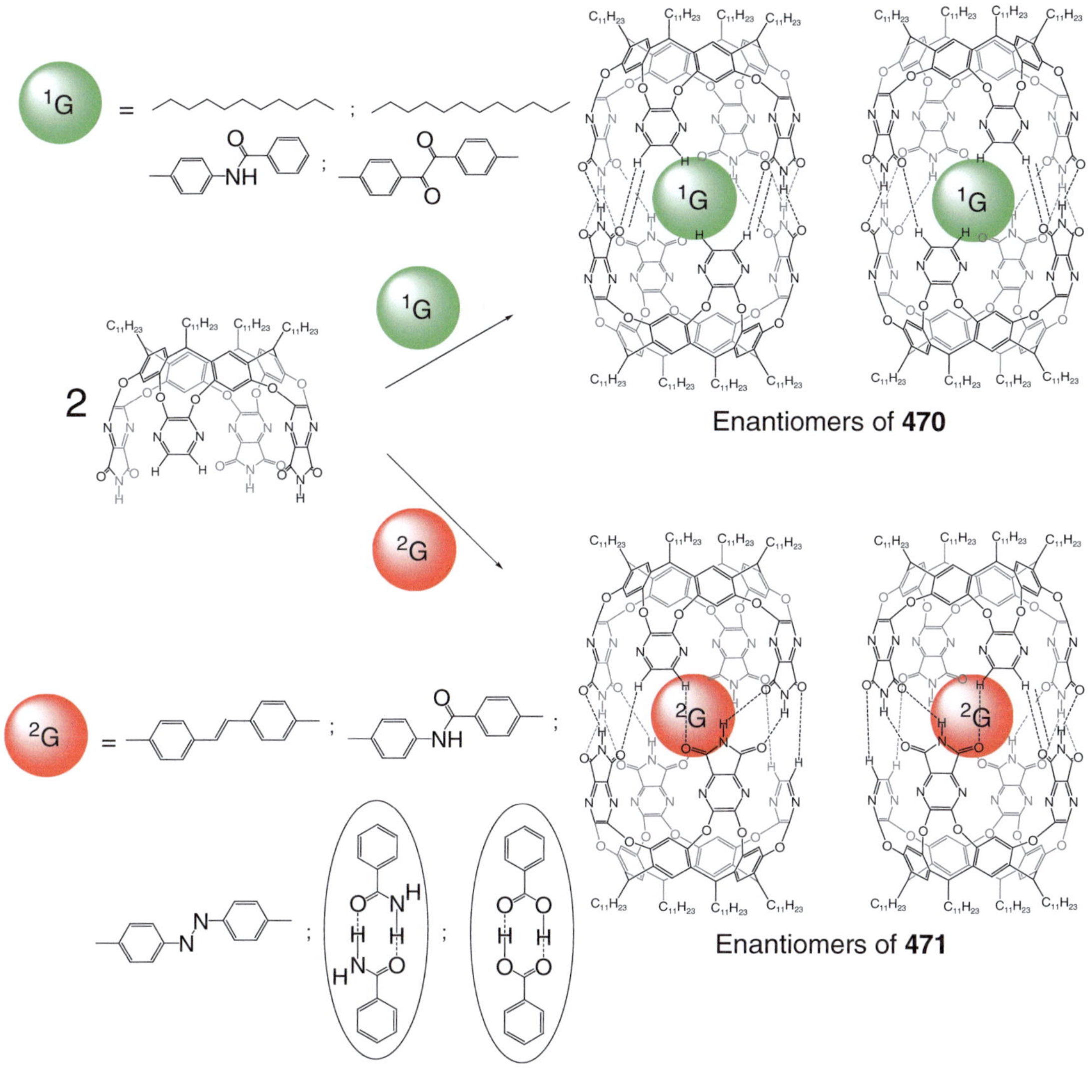

Scheme 3.131

proximal phenyl substituent of the guest *221* occupies one of the two probable isoenergetic positions. In the case of the bulky ketones *222–224*, the corresponding alkyl group and the phenyl substituent occupy semisphere entities of the caging ligand, while the distal substituent of this type is in an equatorial position [148].

The bis-cavitand ligand **473** also encapsulates long-chain esters *225–234* forming 1:1 cage complexes by Scheme 3.134 [149]. The guests with 13 non-hydrogen atoms have a volume of approximately 228 $Å^3$ and are the best for filling of its cavity, while the esters with 16 non-hydrogen atoms have been anticipated in [149] to nearly fill the inner space of **473**. These guests are slightly too long to adopt a fully extended conformation, but the authors of [149] observed no evidence from NOESY NMR that any of them adopted well-defined helical structures allowing for the efficient packing in the ligand cavity [149]. In the first series of these esters, their binding constants increase in a row *227*<*226*<*228*<*229*<*225* according to two main principles. Those are (*i*) a shift of the ester group from the center of a guest molecule to its terminus leading to an increase in steric hindrances between the encapsulated

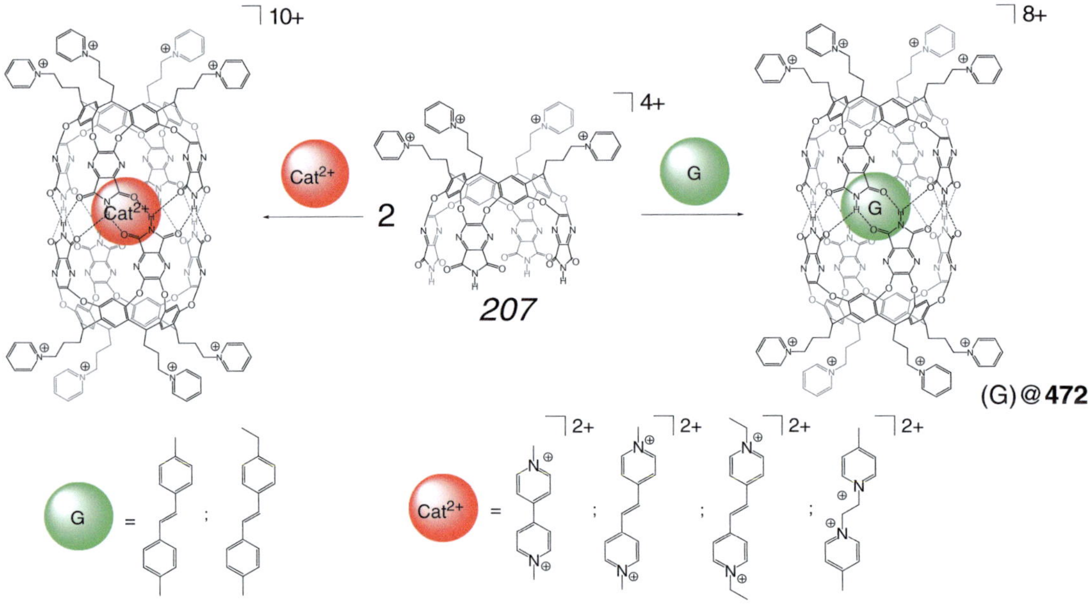

Scheme 3.132

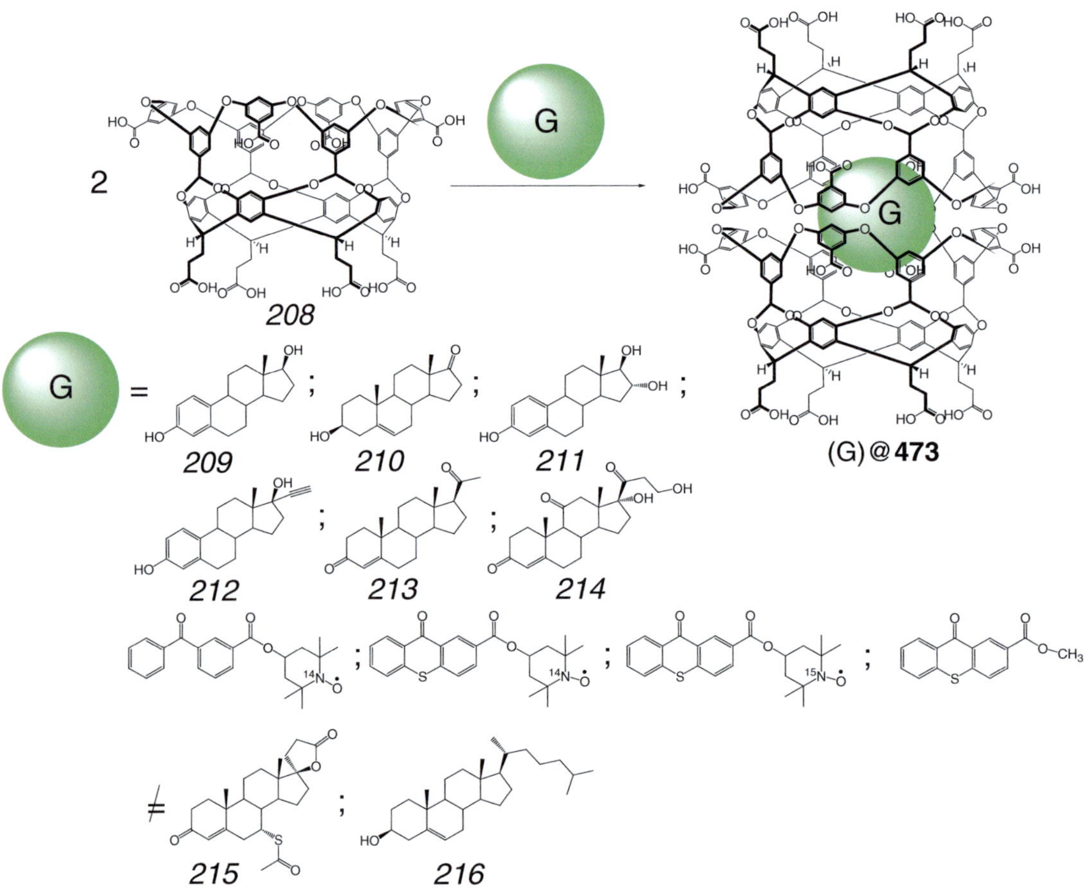

Scheme 3.133

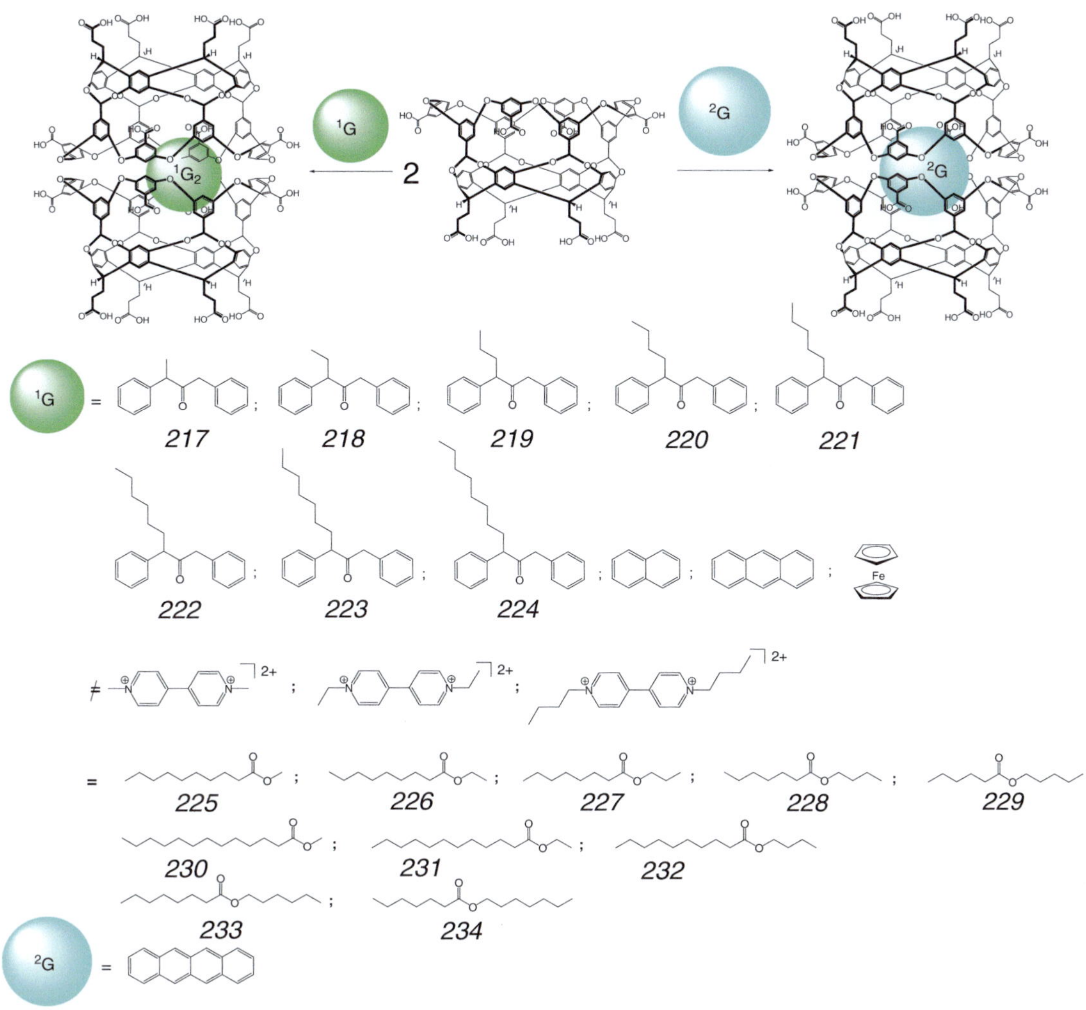

Scheme 3.134

ester and the caging host ligand, and (*ii*) these hindrances being compensated by C–H…π interactions with the energy increasing for more terminal alkoxy group and causing the polarization of the C–H(Me) bonds [149]. If these esters have the similar binding constants, the kinetic resolution does not take place; if they are different in an order or more, it did (in particular, for the esters *225* and *227*). The esters *230–234* containing 16 non-hydrogen atoms have a volume of approximately 283 Å^3 and have been anticipated in [149] to nearly fill the inner space of **473**. They also form 1:1 cage complexes, the stability of which increases in a row *232*<*231*<*234*<*233*<*230* and, in general, is substantially lower than those for the esters of the first series. The preferential binding of the methyl ester *230* is explained in [149] by the formation of strong C–H(Me)…π interactions; the kinetic resolution in this series of esters is very bad and does not exceed 19 %.

The Gibb's octaacid deep-cavity cavitand syntone *208* has been used in [150] as a precursor for the synthesis of its amine- and ammonium-functionalized derivatives. According to ^{1}H NMR data, the latter syntones form caging ligands **474** and **475** in aqueous solutions by Scheme 3.135; those encapsulate neutral organic guests to give 1:1 and 1:2 cage complexes. These complexes were not found in chloroform and methanol media; therefore, the hydrophobic effect is very

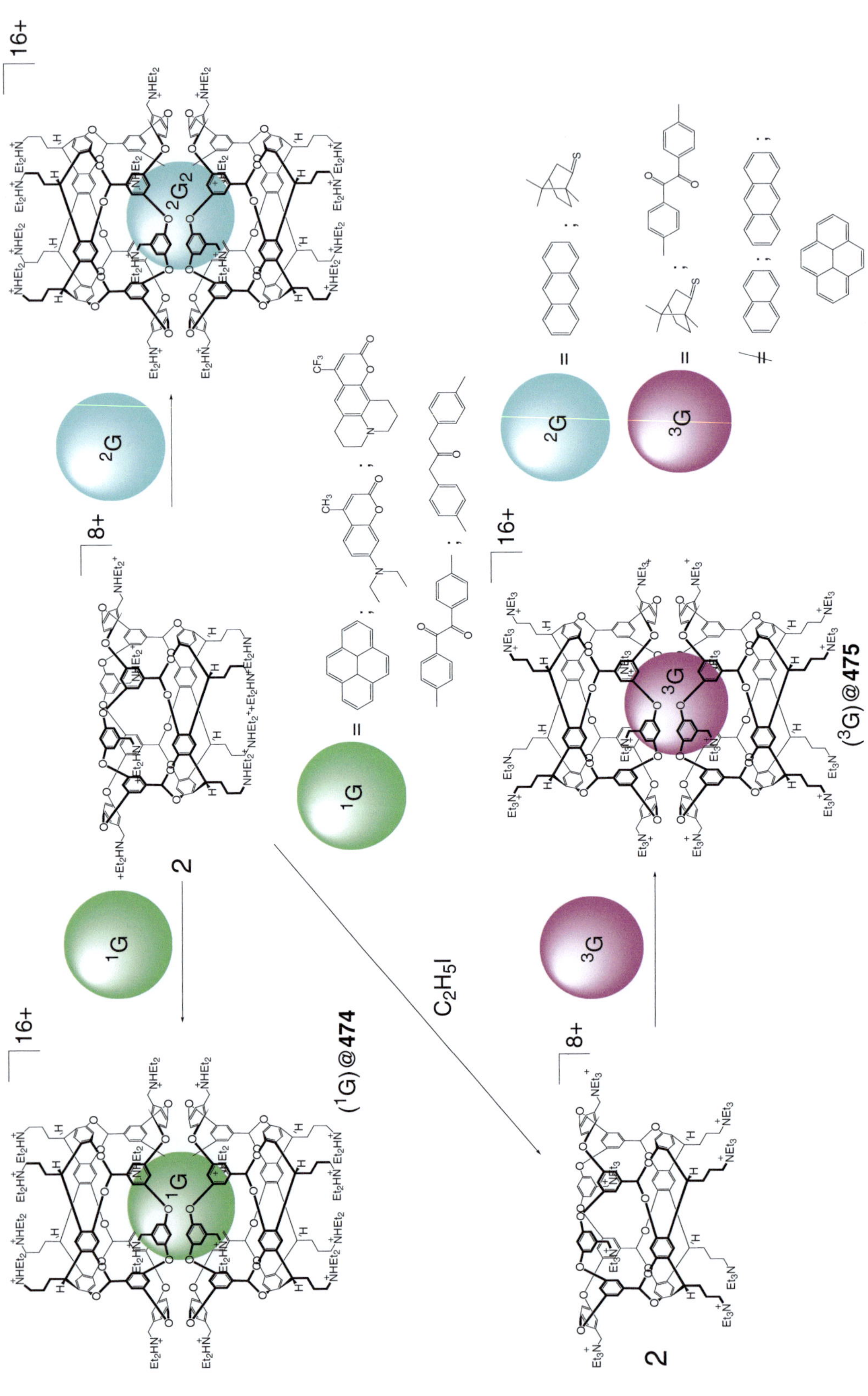

Scheme 3.135

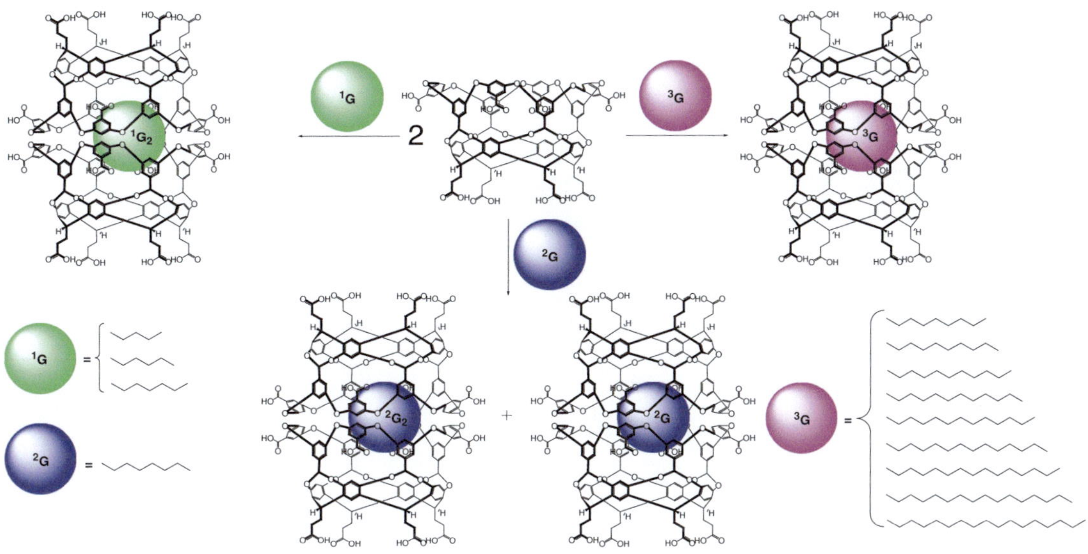

Scheme 3.136

important for their formation. The caging process in the case of the 1:1 complex of **474** with pyrene is confirmed by observation of only the monomer emission in its fluorescence spectrum. The spectra for such cage complexes of **474** suggest that no water molecules are encapsulated within its cavity along with organic guests. The encapsulated anthracene dimer also showed strong excimer emission that was absent in its aqueous solution [150]. Encapsulation also allows controlling the phosphorescent properties of organic guests. In particular, the caged camphorthione and 4,4′-dimethylbenzyl molecules showed intensive phosphorescence after their exciting at room temperature, whereas both self- and oxygen-quenching of phosphorescence of their free molecules in solution was observed at room temperature. The caging ligand **474** can isolate these encapsulated neutral molecules from ambient water or from atmospheric oxygen. Alkylation of this capsule with ethyl iodide gave its analog **475** with eight triethylammonium groups. It encapsulates camphorthione and 4,4′-dimethylbenzyl in aqueous solutions but do not form the cage complexes with aromatic guests [150].

Normal C_5–C_{18} alkanes are described in [151] to form cage complexes of different stoichiometry depending on the size of the guest by dimerization of the deep-cavity cavitand syntone *208* by Scheme 3.136. The C_5–C_7 alkanes gave host–guest 1:2 cage complexes of **473**, and *n*-octane forms both 1:2 and 2:2 complex species. In the case of its more bulky homologs, only 1:1 cage complexes have been detected in [151] by ^{1}H NMR spectroscopy. The terminal methyl groups of these guests are deeply anchored in each of the two cavitand semispheres, and the guests from *n*-decane to *n*-hexadecane adopt their folded conformations to fit into the cavity of an encapsulating ligand **473** [151].

The bis-cavitand capsule **473** is also able to encapsulate trioxolane guests *235–238* in aqueous solutions thus forming 1:1 cage complexes by Scheme 3.137 [152].

A bis-cavitand analog of **473** with methyl substituents, which are partially included in the ligand cavity as well as into its rim, has been synthesized in [153] by Scheme 3.138. The presence of these substituents decreases the ability of the deep-cavitand ligand syntone to form a dimer giving the encapsulating ligand **476**. According to ^{1}H NMR and PGSE diffusion experiments, it forms a host–guest 1:2 cage complex with *n*-pentane and those of host–guest stoichiometry 1:1 with normal C_9–C_4 alkanes by Scheme 3.138. As follows from NMR data [153], the guest in the

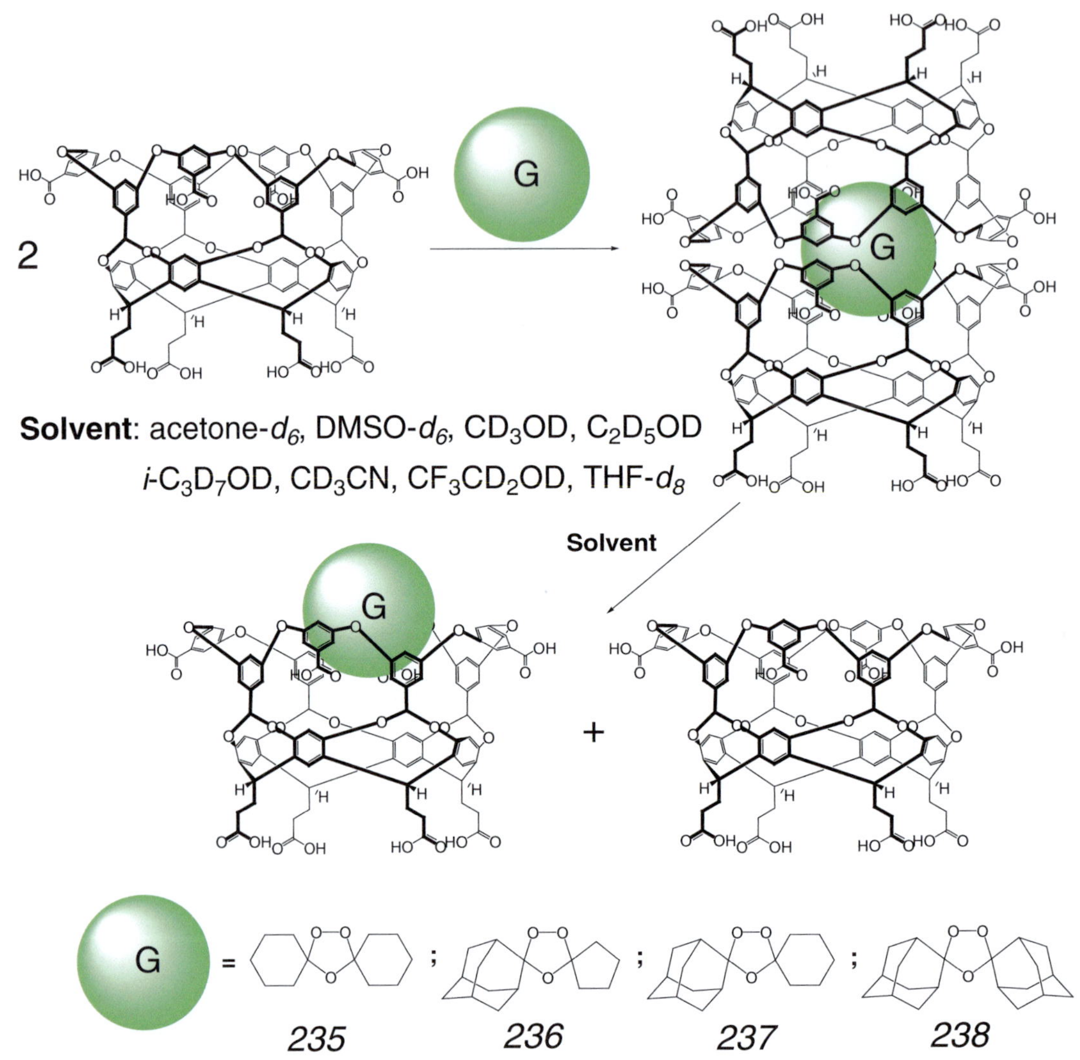

Scheme 3.137

cage complex (n-$C_{14}H_{30}$)@**476** adopts a pole-to-pole conformation with terminal methyl groups located close to the apical fragments of the caging ligand **476**.

In the case of n-hexadecane as a guest, formation of its 1:2 cage complex with tetrameric D_{2d}/*pseudo*-T_d caging ligand **477** having a cavity volume of approximately 1500 Å^3 by Scheme 3.139 has been detected [153] using NMR diffusion experiment; the use of tetra-(n-hexyl ester) pentaerythritol as a guest gave a 1:1 cage complex [154]. The T_d-symmetric 1:2 cage complexes were detected for n-octadecane, n-nonadecane, and n-eicosane, while n-tetracosane formed a hexacavitand 1:3 cage complex of supramolecular capsule **478** having an O_h symmetry. The same octahedral cage complexes were also detected [154] for n-pentacosane and n-hexacosane.

Encapsulation of these long-chain alkanes by the bis-cavitand caging ligand **473** and its diamine-linked hexalene analog **479** has been studied in details [155]. All long-chain guests (Scheme 3.140) starting from n-undecane cannot adopt their fully intended conformation within the cavity of **473** and should be in compressed

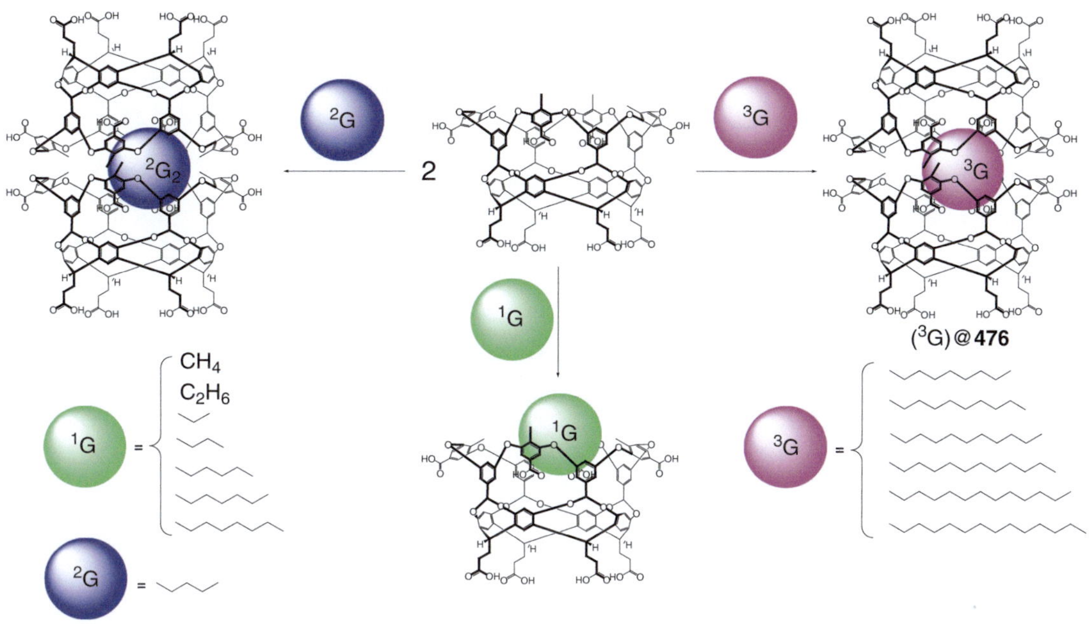

Scheme 3.138

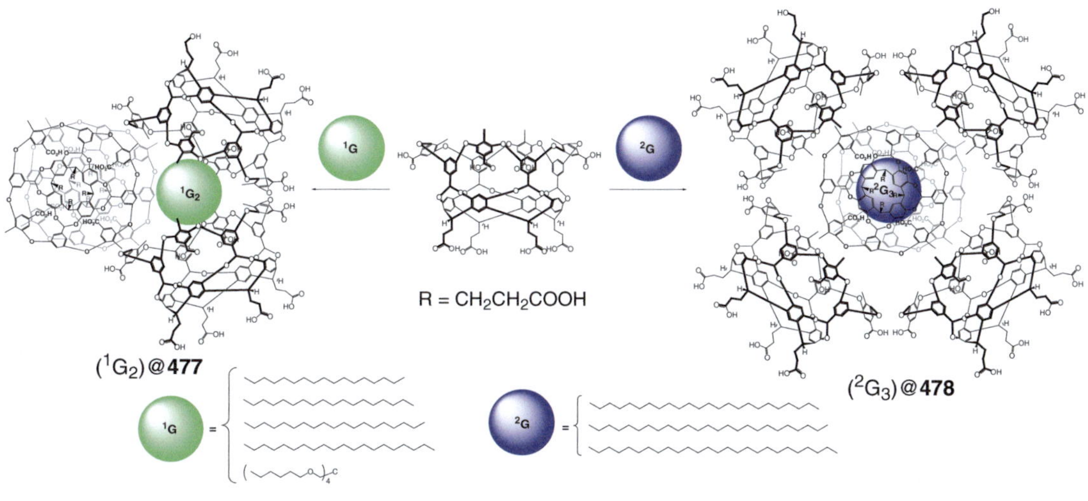

Scheme 3.139

or helical conformations. The guests from *n*-octadecane to tricosane adopt U-shape conformation, while normal C_{24}–C_{26} alkanes exhibit a "spinning-top" motif. A supramolecular capsule **473** is described in [155] to form both the host–guest 1:1 and 1:2 cage complexes with molecular D_{2d} and D_{2h} symmetry, respectively. The *n*-hexane and *n*-octane as guests form 1:2 complexes with this caging ligand (or their mixture with the corresponding 1:1 complexes), while normal C_9–C_{17} alkanes gave 1:1 stoichiometry only. Encapsulation of normal C_{18}–C_{23} alkanes by capsule **479** shown in Scheme 3.141 afforded the cage complexes of two types. According to ^1H NMR data, the major products are 1:1 complexes with an encapsulated guest molecule in a

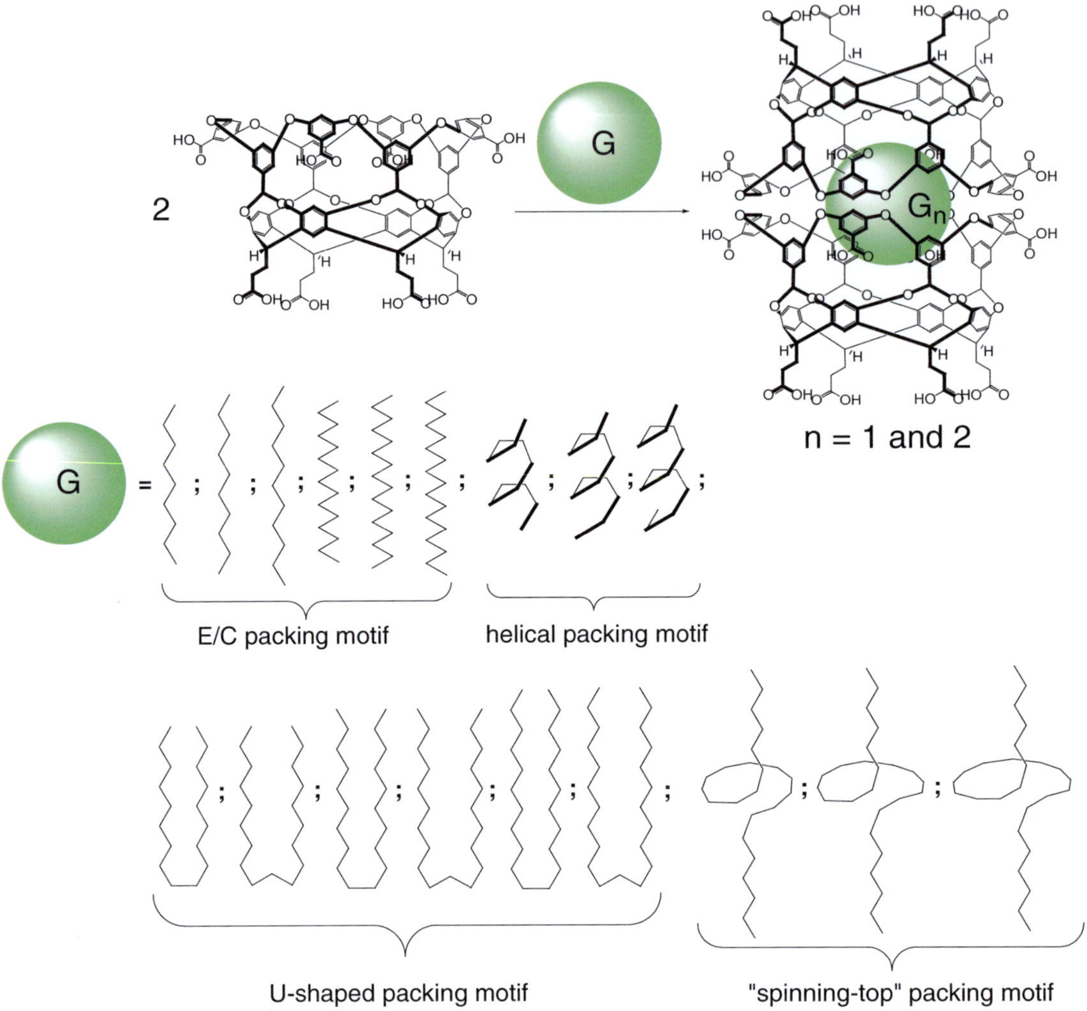

Scheme 3.140

U-shape conformation, while the minor products with a relatively large D_{2d}-symmetric guest molecule are host–guest 2:1 cage complexes [155].

^{1}H NMR titration studies performed in [156] suggest the formation of a cage complex of the hydrogen-bonded capsule **473** with two encapsulated polyaromatic guests shown in Scheme 3.142; the photodimerization of such encapsulated acenaphthylene molecules is described in Sect. 5.2.

Conformational preference of phenyl-substituted alkanes, alkenes, and alkynes guests and their piperidine analogs shown in Scheme 3.142 within the cavity of octaacid bis-cavitand caging ligand **473** has been studied in [157] using ^{1}H NMR and MD methods. Small hydrocarbons adopt a linear conformation, while longer alkanes prefer a folded conformation. The extent of folding and the location of the end groups (methyl and phenyl) are affected by the nature of the group ($H_2C–CH_2$, HC = CH, or C≡C) adjacent to the phenyl substituent.

As follows from ^{1}H NMR and fluorescence data, bis-cavitand caging ligands **480** and **481**, which are formed by dimerization of the corresponding 6-*O*-modified *β*-cyclodextrin ligand syntones *239* and *240*, encapsulate one pyrene molecule in deuterobenzene and deuterocyclohexane solutions by Scheme 3.143 thus giving 1:1

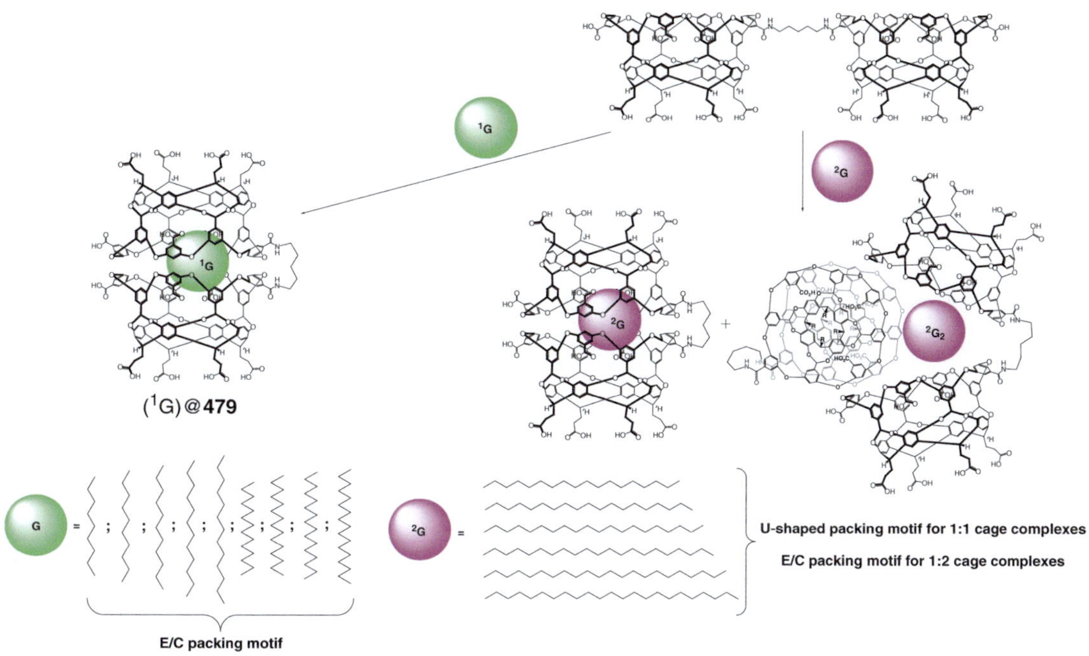

Scheme 3.141

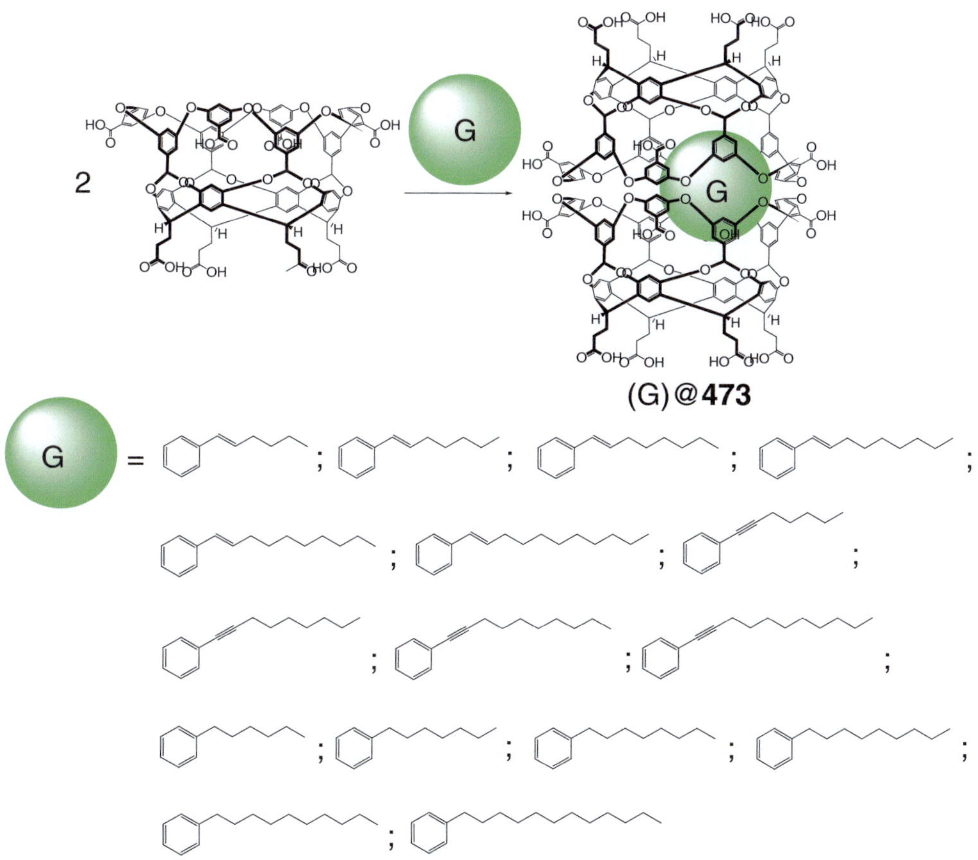

Scheme 3.142

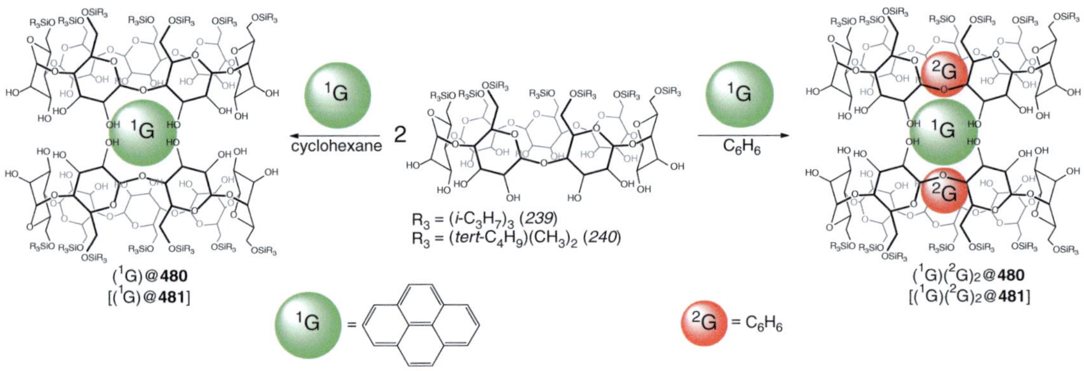

Scheme 3.143

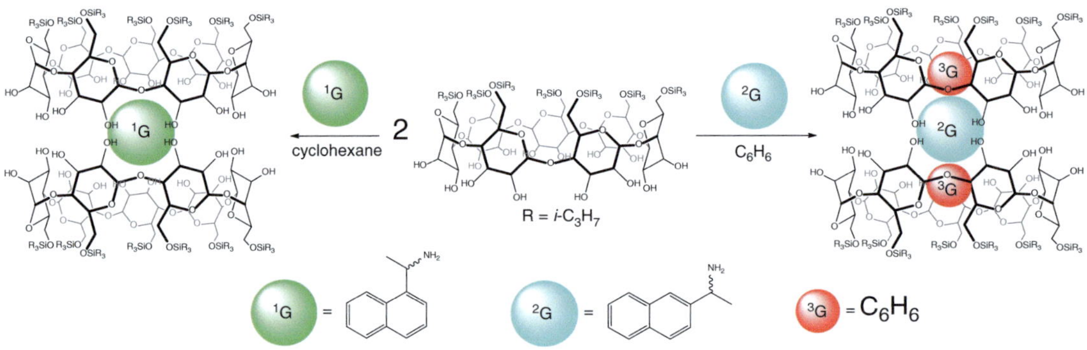

Scheme 3.144

cage complexes. The orientation of the guest molecule within their cavities in solution has been deduced in [158] from two-dimensional NMR spectra. According to X-ray diffraction data, the bis-cavitand caging ligand **480** formed through head-to-head hydrogen-bonding interactions between terminal hydroxyl groups of its two β-cyclodextrin entities encapsulates one pyrene guest molecule. The latter is located in the center of the cavity and forms a sandwich-like supramolecular associate with two benzene molecules through stacking interactions [158].

Chiral recognition of aromatic amines in nonpolar solvents by the bis-cavitand caging ligand **480** has been studied in [159] using various NMR techniques. Both the *R*- and *S*-isomers of 1-(1-naphthyl)ethylamine and 1-(2-naphthyl) ethylamine form 1:1 cage complexes with this ligand by Scheme 3.144 (in deuterocyclohexane for both and in deuterobenzene for *R*- and *S*-isomers of 1-(1-naphthyl)ethylamine). Thus, the position of the 1-aminoethyl substituent in the guest aromatic molecule plays an important role in the encapsulation. According to 2D NMR data, the long axes of the naphthalene guests are perpendicular to that of the cavity of the ligand that passes through the centers of its cyclodextrin entities in their deuterobenzene solutions; those are tilted toward this axis in deuterocyclohexane solutions [159]. As in the case of the cage complex (pyrene)@**480** (see above), the encapsulated guest in the molecule of (*S*-1-(1-naphthyl)ethylamine)@**480** is located in the center of the cavity of the capsule **480** and forms stacking interactions with two solvate benzene molecules [159].

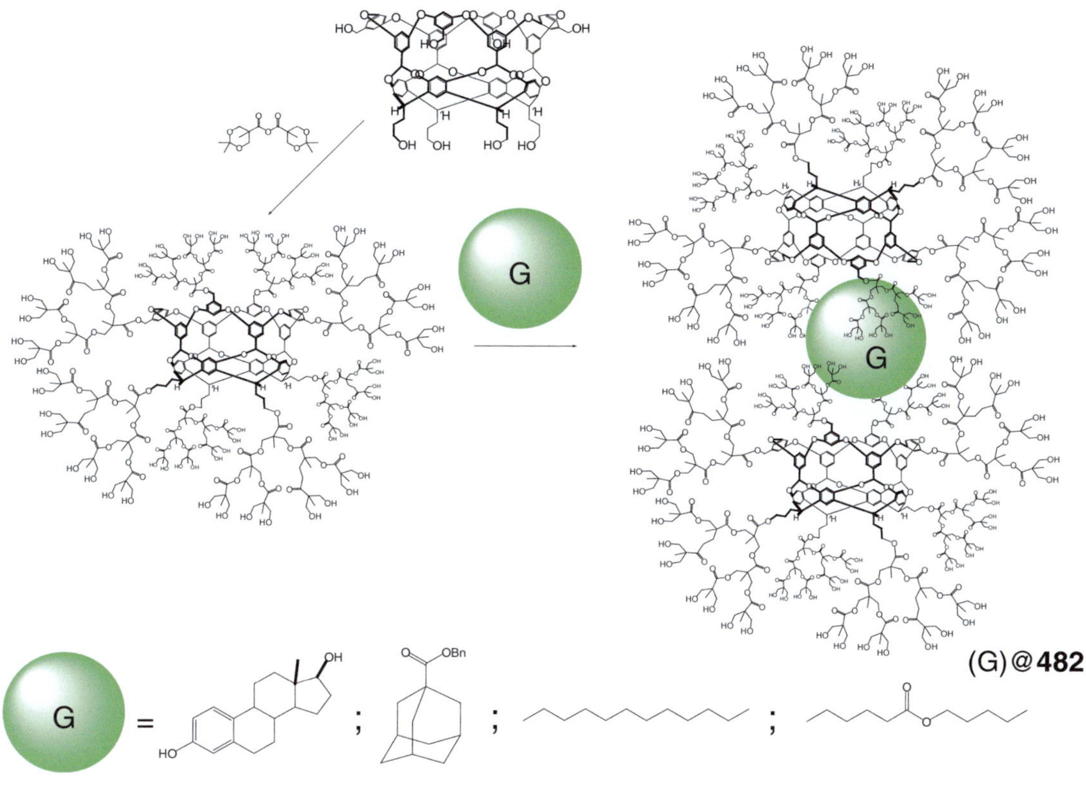

Scheme 3.145

A dendrimeric deep-cavity cavitand capsule **482** has been prepared in [160] by dendrimerization of the corresponding hexahydroxyl-terminated ligand syntone. At neutral pH, this caging ligand easily forms 1:1 cage complexes with various neutral guests shown in Scheme 3.145 [160].

References

1. Timmerman P, Verboom W, Van Veggel FCJM, Van Duynhoven JPM, Reinhoudt DN (1994) A novel type of stereoisomerism in calix[4]arene-based carceplexes. Angew Chem Int Ed 33:2345–2348
2. Koh K, Araki K, Shinkai S (1994) Self-assembled molecular capsule based on the hydrogen-bonding interaction between two different calix[4]arenes. Tetrahedron Lett 35:8255–8258
3. Shimizu KD, Rebek J (1995) Synthesis and assembly of self-complementary calix[4]arenes. Proc Natl Acad Sci U S A 92:12403–12407
4. Mogck O, Böhmer V, Vogt W (1996) Hydrogen bonded homo- and heterodimers of tetra urea derivatives of calix[4]arenes. Tetrahedron 52:8489–8496
5. Mogck O, Paulus EF, Bohmer V, Thondorf I, Vogt W (1996) Hydrogen-bonded dimers of tetraurea calix[4]arenes: unambiguous proof by single crystal X-ray analysis. Chem Commun 2533–2534
6. Mogck O, Pons M, Bolhmer V, Vogt W (1997) NMR studies of the reversible dimerization and guest exchange processes of tetra urea calix[4]arenes using a derivative with lower symmetry. J Am Chem Soc 119:5706–5712
7. Scheerder J, Vreekamp RH, Engbersen JFJ, Verboom W, Van Duynhoven JPM, Reinhoudt DN (1996) The pinched cone conformation of calix[4] arenes: noncovalent rigidification of the calix[4] arene skeleton. J Org Chem 61:3476–3481
8. Vreekamp RH, Verboom W, Reinhoudt DN (1996) Lower rim-upper rim hydrogen-bonded adducts of calix[4]arenes. J Org Chem 61:4282–4288
9. Corbellini F, Van Leeuwen FWB, Beijleveld H, Kooijman H, Spek AL, Verboom W, Crego-Calama M, Reinhoudt DN (2005) Multiple ionic interactions for noncovalent synthesis of molecular capsules in polar solvents. New J Chem 29:243–248
10. Brody MS, Schalley CA, Rudkevich DM, Rebek J Jr (1999) Synthesis and characterization of a unimolecular capsule. Angew Chem Int Ed 38(11): 1640–1644

11. Wyler R, de Mendoza J, Rebek J Jr (1993) A synthetic cavity assembles through self-complementary hydrogen bonds. Angew Chem Int Ed 32(12): 1699–1701
12. Branda N, Grotzfeld RM, Valdes C, Rebek J Jr (1995) Control of self-assembly and reversible encapsulation of xenon in a self-assembling dimer by acid-base chemistry. J Am Chem Soc 117(1): 85–88
13. Meissner RS, Rebek J Jr, de Mendoza J (1995) Autoencapsulation through intermolecular forces: a synthetic self-assembling spherical complex. Science 270(5241):1485–1488
14. Kang J, Rebek J Jr (1996) Entropically driven binding in a self-assembling molecular capsule. Nature 382(6588):239–241
15. Valdes C, Spitz UP, Toledo LM, Kubik SW, Rebek J Jr (1995) Synthesis and self-assembly of pseudo-spherical homo- and heterodimeric capsules. J Am Chem Soc 117(51):12733–12745
16. Valdes C, Spitz UP, Kubik SW, Rebek J Jr (1995) Pseudo-spherical host molecules: synthesis, dimerization, and nucleation effects. Angew Chem Int Ed 34(17):1885–1887
17. Valdes C, Toledo LM, Spitz U, Rebek J Jr (1996) Structure and selectivity of a small dimeric encapsulating assembly. Chem Eur J 2(8):989–991
18. Garcias X, Rebek J Jr (1996) Synthesis and encapsulation behavior of new redox-active dimeric assemblies. Angew Chem Int Ed 35(11):1225–1228
19. Grotzfeld RM, Branda N, Rebek J Jr (1996) Reversible encapsulation of disk-shaped guests by a synthetic, self-assembled host. Science 271(5248): 487–489
20. O'Leary BM, Grotzfeld RM, Rebek J Jr (1997) Ring inversion dynamics of encapsulated cyclohexane. J Am Chem Soc 119(48):11701–11702
21. O'Leary BM, Szabo T, Svenstrup N, Schalley CA, Luetzen A, Schaefer M, Rebek J Jr (2001) "Flexiball" toolkit: a modular approach to self-assembling capsules. J Am Chem Soc 123(47):11519–11533
22. Hamann BC, Shimizu KD, Rebek J Jr (1996) Reversible encapsulation of guest molecules in a calixarene dimmer. Angew Chem Int Ed 35(12): 1326–1329
23. Castellano RK, Kim BH, Rebek J Jr (1997) Chiral capsules: asymmetric binding in calixarene-based dimers. J Am Chem Soc 119(51):12671–12672
24. Thondorf I, Rudzevich Y, Rudzevich V, Böhmer V (2007) Reasons for the exclusive formation of heterodimeric capsules between tetra-tolyl and tetra-tosylurea calix[4]arenes. Org Biomol Chem 5: 2775–2782
25. Castellano RK, Rebek J Jr (1998) Formation of discrete, functional assemblies and informational polymers through the hydrogen-bonding preferences of calixarene aryl and sulfonyl tetraureas. J Am Chem Soc 120(15):3657–3663
26. Ma S, Rudkevich DM, Rebek J Jr (1998) "Deep-Cavity" resorcinarenes dimerize through hydrogen bonding and self-inclusion. J Am Chem Soc 120(20):4977–4981
27. Cho YL, Rudkevich DM, Rebek J Jr (2000) Expanded calix[4]arene tetraurea capsules. J Am Chem Soc 122(40):9868–9869
28. Meissner R, Garcias X, Mecozzi S, Rebek J Jr (1997) Synthesis and assembly of new molecular hosts: solvation and the energetics of encapsulation. J Am Chem Soc 119(1):77–85
29. Kang J, Rebek J Jr (1997) Acceleration of a Diels-Alder reaction by a self-assembled molecular capsule. Nature 385(6611):50–52
30. Kang J, Santamaria J, Hilmersson G, Rebek J Jr (1998) Self-assembled molecular capsule catalyzes a Diels-Alder reaction. J Am Chem Soc 120(29): 7389–7390
31. Kang J, Hilmersson G, Santamaria J, Rebek J Jr (1998) Diels-Alder reactions through reversible encapsulation. J Am Chem Soc 120(15):3650–3656
32. Szabo T, Hilmersson G, Rebek J Jr (1998) Dynamics of assembly and guest exchange in the tennis ball. J Am Chem Soc 120(24):6193–6194
33. Tokunaga Y, Rudkevich DM, Rebek J Jr (1997) Solvation and the synthesis of self-assembled capsules. Angew Chem Int Ed 36(23):2656–2659
34. Rivera JM, Martin T, Rebek J Jr (1998) Chiral spaces: dissymmetric capsules through self-assembly. Science 279(5353):1021–1023
35. Rivera JM, Martin T, Rebek J Jr (1998) Structural rules governing self-assembly emerge from new molecular capsules. J Am Chem Soc 120(4): 819–820
36. Tokunaga Y, Rudkevich DM, Santamaria J, Hilmersson G, Rebek J Jr (1998) Solvent controls synthesis and properties of supramolecular structures. Chem Eur J 4(8):1449–1457
37. Mecozzi S, Rebek J Jr (1998) The 55% solution: a formula for molecular recognition in the liquid state. Chem Eur J 4(6):1016–1022
38. Kang J, Meissner RS, Wyler R, De Mendoza J, Rebek J Jr (2000) Development of synthetic self-assembling molecular capsule: from flexible spacer to rigid spacer. Bull Kor Chem Soc 21(2):221–227
39. Tokunaga Y, Rebek J Jr (1998) Chiral capsules. 1. Softballs with asymmetric surfaces bind camphor derivatives. J Am Chem Soc 120(1):66–69
40. Rivera JM, Craig SI, Martin T, Rebek J Jr (2000) Chiral guests and their ghosts in reversibly assembled hosts. Angew Chem Int Ed 39(12):2130–2132
41. Szabo T, O'Leary BM, Rebek J Jr (1999) Self-assembling sieves. Angew Chem Int Ed 37(24):3410–3413
42. Cho YL, Rudkevich DM, Shivanyuk A, Rissanen K, Rebek J Jr (2000) Hydrogen-bonding effects in calix[4]arene capsules. Chem Eur J 6(20):3788–3796
43. Rivera JM, Martin T, Rebek J Jr (2001) Chiral softballs: synthesis and molecular recognition properties. J Am Chem Soc 123(22):5213–5220
44. Shivanyuk A, Friese JC, Doering S, Rebek J Jr (2003) Solvent-stabilized molecular capsules. J Org Chem 68(17):6489–6496

45. Corbellini F, Di Costanzo L, Crego-Calama M, Geremia S, Reinhoudt DN (2003) Guest encapsulation in a water-soluble molecular capsule based on ionic interactions. J Am Chem Soc 125(33): 9946–9947
46. Letzel MC, Decker B, Rozhenko AB, Schoeller WW, Mattay J (2004) Encapsulated guest molecules in the dimer of octahydroxypyridine[4]arene. J Am Chem Soc 126(31):9669–9674
47. Corbellini F, Knegtel RMA, Grootenhuis PDJ, Crego-Calama M, Reinhoudt DN (2005) Water-soluble molecular capsules: self-assembly and binding properties. Chem Eur J 11:298–307
48. Kuberski B, Szumna A (2009) A self-assembled chiral capsule with polar interior. Chem Commun 1959–1961
49. Szumna A (2009) Water co-encapsulation in an inverted molecular capsule. Chem Commun 4191–4193
50. Martin AD, Boulos RA, Hubble LJ, Hartlieb KJ, Raston CL (2011) Multifunctional water-soluble molecular capsules based on *p*-phosphonic acid calix[5]arene. Chem Commun 47:7353–7355
51. Martin T, Obst U, Rebek J Jr (1998) Molecular assembly and encapsulation directed by hydrogen-bonding preferences and the filling of space. Science 281(5384):1842–1845
52. Johnson DW, Hof F, Iovine PM, Nuckolls C, Rebek J Jr (2002) Solid-state and solution studies of a tetrameric capsule and its guests. Angew Chem Int Ed 41(20):3793–3796
53. Nuckolls C, Hof F, Martin T, Rebek J Jr (1999) Chiral microenvironments in self-assembled capsules. J Am Chem Soc 121(44):10281–10285
54. Hof F, Nuckolls C, Craig SL, Martin T, Rebek J Jr (2000) Emergent conformational preferences of a self-assembling small molecule: structure and dynamics in a tetrameric capsule. J Am Chem Soc 122(44):10991–10996
55. Hiraoka S, Harano K, Shiro M, Shionoya M (2008) A self-assembled organic capsule formed from the union of six hexagram-shaped amphiphile molecules. J Am Chem Soc 130(44):14368–14369
56. Hiraoka S, Harano K, Nakamura T, Shiro M, Shionoya M (2009) Induced-fit formation of a tetrameric organic capsule consisting of hexagram-shaped amphiphile molecules. Angew Chem Int Ed 48(38):7006–7009
57. Hiraoka S, Nakamura T, Shiro M, Shionoya M (2010) In-water truly monodisperse aggregation of gear-shaped amphiphiles based on hydrophobic surface engineering. J Am Chem Soc 132(38):13223–13225
58. MacGillivray LR, Atwood JL (1997) Achiral sphericalmolecular assembly held together by 60 hydrogen bonds. Nature 389:469–472
59. Shivanyuk A, Rebek J Jr (2003) Assembly of resorcinarene capsules in wet solvents. J Am Chem Soc 125(12):3432–3433
60. Yamanaka M, Shivanyuk A, Rebek J Jr (2004) Kinetics and thermodynamics of hexameric capsule formation. J Am Chem Soc 126(9):2939–2943
61. Avram L, Cohen Y (2004) Self-recognition, structure, stability, and guest affinity of pyrogallol[4]arene and resorcin[4]arene capsules in solution. J Am Chem Soc 125:11556–11562
62. Dalgarno SJ, Bassil DB, Tucker SA, Atwood JL (2006) Cocrystallization and encapsulation of a fluorophore with hexameric pyrogallol[4]arene nanocapsules: structural and fluorescence studies. Angew Chem Int Ed 45:7019–7022
63. Antesberger J, Cave GWV, Ferrarelli MC, Heaven MW, Raston CL, Atwood JL (2005) Solvent-free, direct synthesis of supramolecular nano-capsules. Chem Commun 892–894
64. Palmer LC, Rebek J Jr (2005) Hydrocarbon binding inside a hexameric pyrogallol[4]arene capsule. Org Lett 7(5):787–789
65. Palmer LC, Shivanyuk A, Yamanaka M, Rebek J Jr (2005) Resorcinarene assemblies as synthetic receptors. Chem Commun 7:857–858
66. Barrett ES, Dale TJ, Rebek J Jr (2007) Assembly and exchange of resorcinarene capsules monitored by fluorescence resonance energy transfer. J Am Chem Soc 129:3818–3819
67. Barrett ES, Dale TJ, Rebek J Jr (2008) Stability, dynamics, and selectivity in the assembly of hydrogen-bonded hexameric capsules. J Am Chem Soc 130:2344–2350
68. Bassil DB, Dalgarno SJ, Cave GWV, Atwood JL, Tucker SA (2007) Spectroscopic investigations of ADMA encapsulated in pyrogallol[4]arene nanocapsules. J Phys Chem B 111:9088–9092
69. Iyer KS, Norret M, Dalgarno SJ, Atwood JL, Raston CL (2008) Loading molecular hydrogen cargo within viruslike nanocontainers. Angew Chem Int Ed 47:6362–6366
70. Mileo E, Yi S, Bhattacharya P, Kaifer AE (2009) Probing the inner space of resorcinarene molecular capsules with nitroxide guests. Angew Chem Int Ed 48:5337–5340
71. Kvasnica M, Chapin JC, Purse BW (2011) Efficient loading and kinetic trapping of hexameric pyrogallolarene capsules in solution. Angew Chem Int Ed 50:2244–2248
72. Slovak S, Cohen Y (2012) The effect of alcohol structures on the interaction mode with the hexameric capsule of resorcin[4]arene. Chem Eur J 18:8515–8520
73. Zhang Q, Tiefenbacher K (2013) Hexameric resorcinarene capsule is a brønsted acid: investigation and application to synthesis and catalysis. J Am Chem Soc 135:16213–16219
74. Bonaccorso C, Sgarlata C, Grasso G, Zito V, Sciotto D, Arena G (2011) A gemini guest triggers the self-assembly of a calixarene capsule in water at neutral pH. Chem Commun 47:6117–6119
75. Schalley CA, Rivera JM, Martin T, Santamaria J, Siuzdak G, Rebek J Jr (1999) Structural examination of supramolecular architectures by electrospray ionization mass spectrometry. Eur J Org Chem 6: 1325–1331

76. Schalley CA, Castellano RK, Brody MS, Rudkevich DM, Siuzdak G, Rebek J Jr (1999) Investigating molecular recognition by mass spectrometry: characterization of calixarene-based self-assembling capsule hosts with charged guests. J Am Chem Soc 121(19):4568–4579
77. Luetzen A, Renslo AR, Schalley CA, O'Leary BM, Rebek J Jr (1999) Encapsulation of ion-molecule complexes: second-sphere supramolecular chemistry. J Am Chem Soc 121(32):7455–7456
78. Schalley CA, Martin T, Obst U, Rebek J Jr (1999) Characterization of self-assembling encapsulation complexes in the gas phase and solution. J Am Chem Soc 121(10):2133–2138
79. Shivanyuk A, Paulus EF, Böhmer V (1999) Guest-controlled formation of a hydrogen-bonded molecular capsule. Angew Chem Int Ed 38:2906–2909
80. Shivanyuk A (2005) Selective formation of heterodimeric resorcinarene capsules. Tetrahedron 61: 349–352
81. Shivanyuk A (2007) Nanoencapsulation of calix[4]arene inclusion complexes. J Am Chem Soc 129:14196–14199
82. Philip I, Kaifer AE (2005) Noncovalent encapsulation of cobaltocenium inside resorcinarene molecular capsules. J Org Chem 70(5):1558–1564
83. Heinz T, Rudkevich DM, Rebek J Jr (1998) Pairwise selection of guests in a cylindrical molecular capsule of nanometer dimensions. Nature 394(6695):764–766
84. Tucci FC, Rudkevich DM, Rebek J Jr (1999) Stereochemical relationships between encapsulated molecules. J Am Chem Soc 121(20):4928–4929
85. Heinz T, Rudkevich DM, Rebek J Jr (1999) Molecular recognition within a self-assembled cylindrical host. Angew Chem Int Ed 38(8): 1136–1139
86. Korner SK, Tucci FC, Rudkevich DM, Heinz T, Rebek J Jr (2000) A self-assembled cylindrical capsule: new supramolecular phenomena through encapsulation. Chem Eur J 6(1):187–195
87. Shivanyuk A, Rebek J (2002) The inner solvation of a cylindrical capsule. Chem Commun 20: 2326–2327
88. Ajami D, Schramm MP, Rebek J (2009) Translational motion inside self-assembled encapsulation complexes. Tetrahedron 65(35):7208–7212
89. Hayashida O, Sebo L, Rebek J Jr (2002) Molecular discrimination of N-protected amino acid esters by a self-assembled cylindrical capsule: spectroscopic and computational studies. J Org Chem 67(24): 8291–8298
90. Scarso A, Trembleau L, Rebek J Jr (2003) Encapsulation induces helical folding of alkanes. Angew Chem Int Ed 42(44):5499–5502
91. Sather AC, Berryman OB, Ajami D, Rebek J Jr (2011) Reactivity of N-nitrosoamides in confined spaces. Tetrahedron Lett 52(17):2100–2103
92. Shivanyuk A, Rebek J Jr (2002) Social isomers in encapsulation complexes. J Am Chem Soc 124(41): 12074–12075
93. Shivanyuk A, Rebek J Jr (2003) Isomeric constellations of encapsulation complexes store information on the nanometer scale. Angew Chem Int Ed 42(6): 684–686
94. Shivanyuk A, Scarso A, Rebek J Jr (2003) Coencapsulation of large and small hydrocarbons. Chem Commun 11:1230–1231
95. Scarso A, Shivanyuk A, Rebek J Jr (2003) Individual solvent/solute interactions through social isomerism. J Am Chem Soc 125(46):13981–13983
96. Scarso A, Shivanyuk A, Hayashida O, Rebek J Jr (2003) Asymmetric environments in encapsulation complexes. J Am Chem Soc 125(20):6239–6243
97. Amaya T, Rebek J Jr (2004) Steric and magnetic asymmetry distinguished by encapsulation. J Am Chem Soc 126(20):6216–6217
98. Rechavi D, Scarso A, Rebek J Jr (2004) Isotopomer encapsulation in a cylindrical molecular capsule: a probe for understanding noncovalent isotope effects on a molecular level. J Am Chem Soc 126(25): 7738–7739
99. Zhao Y-L, Houk KN, Rechavi D, Scarso A, Rebek J Jr (2004) Equilibrium isotope effects as a probe of nonbonding attractions. J Am Chem Soc 126(37): 11428–11429
100. Scarso A, Trembleau L, Rebek J Jr (2004) Helical folding of alkanes in a self-assembled, cylindrical capsule. J Am Chem Soc 126(41):13512–13518
101. Amaya T, Rebek J Jr (2004) Hydrogen-bonded encapsulation complexes in protic solvents. J Am Chem Soc 126(43):14149–14156
102. Purse BW, Rebek J Jr (2005) Encapsulation of oligoethylene glycols and perfluoro-n-alkanes in a cylindrical host molecule. Chem Commun 6: 722–724
103. Van Anda H, Myles AJ, Rebek J Jr (2007) Charge transfer and encapsulation in a synthetic, self-assembled receptor. New J Chem 31(5):631–633
104. Amaya T, Rebek J Jr (2004) Coencapsulation of three different guests in a cylindrical host. Chem Commun 16:1802–1803
105. Scarso A, Onagi H, Rebek J Jr (2004) Mechanically regulated rotation of a guest in a nanoscale host. J Am Chem Soc 126(40):12728–12729
106. Yamanaka M, Rebek J Jr (2004) Constellational diastereomers in encapsulation complexes. Chem Commun 15:1690–1691
107. Palmer LC, Zhao Y-L, Houk KN, Rebek J Jr (2005) Diastereoselection of chiral acids in a cylindrical capsule. Chem Commun 29:3667–3669
108. Salvio R, Moisan L, Ajami D, Rebek J Jr (2007) Tertiary amide rotation in a nanoscale host. Eur J Org Chem 16:2722–2728
109. Ajami D, Rebek J Jr (2010) Solid guests in reversible encapsulation hosts. Heterocycles 80(1): 109–113
110. Sarwar MG, Ajami D, Theodorakopoulos G, Petsalakis ID, Rebek J Jr (2013) Amplified halogen bonding in a small space. J Am Chem Soc 135(37):13672–13675

111. Ams MR, Ajami D, Craig SL, Yang J-S, Rebek J Jr (2009) "Too small, too big, and just right" – optical sensing of molecular conformations in self-assembled capsules. J Am Chem Soc 131(37):13190–13191
112. Ajami D, Rebek J Jr (2006) Coiled molecules in spring loaded devices. J Am Chem Soc 128(47): 15038–15039
113. Ajami D, Rebek J Jr (2009) Compressed alkanes in reversible encapsulation complexes. Nat Chem 1(1):87–90
114. Ams MR, Ajami D, Craig SL, Yang J-S, Rebek J Jr (2009) Control of stilbene conformation and fluorescence in self-assembled capsules. Beilstein J Org Chem 5:79. doi:10.3762/bjoc.5.79
115. Ajami D, Rebek J Jr (2008) Gas behavior in self-assembled capsules. Angew Chem Int Ed 47(32):6059–6061
116. Ajami D, Rebek J Jr (2008) Reversible encapsulation of terminal alkenes and alkynes. Heterocycles 76(1):169–176
117. Ajami D, Tolstoy PM, Dube H, Odermatt S, Koeppe B, Guo J, Limbach H-H, Rebek J Jr (2011) Encapsulated carboxylic acid dimers with compressed hydrogen bonds. Angew Chem Int Ed 50(2):528–531
118. Tzeli D, Petsalakis ID, Theodorakopoulos G, Ajami D, Jiang W, Rebek J Jr (2012) Encapsulated hydrogen-bonded dimers of amide and carboxylic acid. Chem Phys Lett 548:55–59
119. Ajami D, Schramm MP, Volonterio A, Rebek J Jr (2007) Assembly of hybrid synthetic capsules. Angew Chem Int Ed 46:242–244
120. Tucci FC, Renslo AR, Rudkevich DM, Rebek J Jr (2000) Nanoscale container structures and their host-guest properties. Angew Chem Int Ed 39(6):1076–1079
121. Zhang K-D, Ajami D, Gavette JV, Rebek J (2014) Complexation of alkyl groups and ghrelin in a deep, water-soluble cavitand. Chem Commun 50(38): 4895–4897
122. Ajami D, Rebek J Jr (2007) Longer guests drive the reversible assembly of hyperextended capsules. Angew Chem Int Ed 46(48):9283–9286
123. Ajami D, Rebek J Jr (2009) Expanding capsules. Supramol Chem 21(1–2):103–106
124. Barrett ES, Dale TJ, Jr. Rebek J (2007) Self-assembly dynamics of a cylindrical capsule monitored by fluorescence resonance energy transfer. J Am Chem Soc 129(28):8818–8824
125. Barrett ES, Dale TJ, Rebek J (2007) Synthesis and assembly of monofunctionalized pyrogallolarene capsules monitored by fluorescence resonance energy transfer. Chem Commun 41:4224–4226
126. Schramm MP, Restorp P, Zelder F, Rebek J Jr (2008) Influence of remote asymmetric centers in reversible encapsulation complexes. J Am Chem Soc 130(8): 2450–2451
127. Schramm MP, Rebek J Jr (2008) Effects of remote chiral centers on encapsulated molecules. New J Chem 32(5):794–796
128. Ajami D, Hou J-L, Dale TJ, Barrett E, Rebek J Jr (2009) Disproportionation and self-sorting in molecular encapsulation. Proc Natl Acad Sci U S A 106(26):10430–10434
129. Lledo A, Kamioka S, Sather AC, Rebek J Jr (2011) Supramolecular architecture with a cavitand-capsule chimera. Angew Chem Int Ed 50(6):1299–1301
130. Yamauchi Y, Ajami D, Lee J-Y, Rebek J (2011) Deconstruction of capsules using chiral spacers. Angew Chem Int Ed 50(39):9150–9153
131. Ajami D, Dube H, Jr. Rebek J (2011) Boronic acid hydrogen bonding in encapsulation complexes. J Am Chem Soc 133(25):9689–9691
132. Asadi A, Ajami D, Rebek J (2011) Bent alkanes in a new thiourea-containing capsule. J Am Chem Soc 133(28):10682–10684
133. Ajami D, Rebek J Jr (2006) Expanded capsules with reversibly added spacers. J Am Chem Soc 128(16):5314–5315
134. Ajami D, Rebek J (2009) Multicomponent, hydrogen-bonded cylindrical capsules. J Org Chem 74(17):6584–6591
135. Tiefenbacher K, Ajami D, Rebek J (2011) Self-assembled capsules of unprecedented shapes. Angew Chem Int Ed 50:12003–12007
136. Tiefenbacher K, Rebek J (2012) Selective stabilization of self-assembled hydrogen-bonded molecular capsules through – interactions. J Am Chem Soc 134(6):2914–2917
137. Podkoscielny D, Gadde S, Kaifer AE (2009) Mediated electrochemical oxidation of a fully encapsulated redox active center. J Am Chem Soc 131:12876–12877
138. Qiu Y, Yi S, Kaifer AE (2011) Encapsulation of tetrathiafulvalene inside a dimeric molecular capsule. Org Lett 13:1770–1773
139. Taira T, Ajami D, Rebek J (2012) Encapsulation of ion pairs in extended, self-assembled structures. J Am Chem Soc 134(29):11971–11973
140. Ajami D, Theodorakopoulos G, Petsalakis ID, Rebek J Jr (2013) Social isomers of picolines in a small space. Chem Eur J 19(50):17092–17096
141. Jiang W, Rebek J (2012) Guest-induced, selective formation of isomeric capsules with imperfect walls. J Am Chem Soc 134(42):17498–17501
142. Zhang K-D, Ajami D, Rebek J (2013) Hydrogen-bonded capsules in water. J Am Chem Soc 135(48):18064–18066
143. Gibb CLD, Gibb BC (2004) Well-defined, organic nanoenvironments in water: the hydrophobic effect drives a capsular assembly. J Am Chem Soc 126:11408–11409
144. Chen JY-C, Jayaraj N, Jockusch S, Ottaviani MF, Ramamurthy V, Turro NJ (2008) An EPR and NMR study of supramolecular effects on paramagnetic interaction between a nitroxide incarcerated within a nanocapsule with a nitroxide in bulk aqueous media. J Am Chem Soc 130:7206–7207
145. Jockusch S, Zeika O, Jayaraj N, Ramamurthy V, Turro NJ (2010) Electron spin polarization transfer

from a nitroxide incarcerated within a nanocapsule to a nitroxide in the bulk aqueous solution. J Phys Chem Lett 1:2628–2632
146. Kaanumalle LS, Gibb CLD, Gibb BC, Ramamurthy V (2005) A hydrophobic nanocapsule controls the photophysics of aromatic molecules by suppressing their favored solution pathways. J Am Chem Soc 127:3674–3675
147. Podkoscielny D, Philip I, Gibb CLD, Gibb BC, Kaifer AE (2008) Encapsulation of ferrocene and peripheral electrostatic attachment of viologens to dimeric molecular capsules formed by an octaacid, deep-cavity cavitand. Chem Eur J 14:4704–4710
148. Gibb CLD, Sundaresan AK, Ramamurthy V, Gibb BC (2008) Templation of the excited-state chemistry of α-(n-alkyl) dibenzyl ketones: how guest packing within a nanoscale supramolecular capsule influences photochemistry. J Am Chem Soc 130: 4069–4080
149. Liu S, Gan H, Hermann AT, Rick SW, Gibb BC (2010) Kinetic resolution of constitutional isomers controlled by selective protection inside a supramolecular nanocapsule. Nat Chem 2:847–852
150. Kulasekharan R, Ramamurthy V (2011) New water-soluble organic capsules are effective in controlling excited-state processes of guest molecules. Org Lett 13:5092–5095
151. Gibb CLD, Gibb BC (2007) Straight-chain alkanes template the assembly of water-soluble nanocapsules. Chem Commun 1635–1637
152. Liu S, Gibb BC (2011) Solvent denaturation of supramolecular capsules assembled via the hydrophobic effect. Chem Commun 47:3574–3576
153. Gan H, Benjamin CJ, Gibb BC (2011) Nonmonotonic assembly of a deep-cavity cavitand. J Am Chem Soc 133:4770–4773
154. Gan H, Gibb BC (2013) Guest-mediated switching of the assembly state of a water-soluble deep-cavity cavitand. Chem Commun 49:1395–1397
155. Liu S, Russell DH, Zinnel NF, Gibb BC (2013) Guest packing motifs within a supramolecular nanocapsule and a covalent analogue. J Am Chem Soc 135:4314–4324
156. Kaanumalle LS, Ramamurthy V (2007) Photodimerization of acenaphthylene within a nanocapsule: excited state lifetime dependent dimer selectivity. Chem Commun 1062–1064
157. Choudhury R, Barman A, Prabhakar R, Ramamurthy V (2013) Hydrocarbons depending on the chain length and head group adopt different conformations within a water-soluble nanocapsule: ^{1}H NMR and molecular dynamics studies. J Phys Chem B 117(1):398–407
158. Kida T, Iwamoto T, Fujino Y, Tohnai N, Miyata M, Akashi M (2011) Strong guest binding by cyclodextrin hosts in competing nonpolar solvents and the unique crystalline structure. Org Lett 13(17):4570–4573
159. Kida T, Iwamoto T, Asahara H, Hinoue T, Akashi M (2013) Chiral recognition and kinetic resolution of aromatic amines via supramolecular chiral nanocapsules in nonpolar solvents. J Am Chem Soc 135:3371–3374
160. Giles MD, Liu S, Emanuel RL, Gibb BC, Grayson SM (2008) Dendronized supramolecular nanocapsules: pH independent, water-soluble, deep-cavity cavitands assemble via the hydrophobic effect. J Am Chem Soc 130:14430–14431

4 Encapsulation by Coordination Capsules

Coordination capsules are the species that typically self-assembled from ligand syntones with appropriate terminal donor groups and the coordinate transition and non-transition metal ions or their labile and coordination-unsaturated complexes. The resulting cage frameworks can be divided into three main structural types: (*i*) tetrahedral capsules and their analogs, such as cuboid frameworks, containing 4n end-capping metal ion, (*ii*) octahedral and TP capsules and their analogs with 6n end-capping metal ions, and (*iii*) coordination capsules with bridging (cross-linking) metal ions. Most of these caging ligands are positively or negatively charged (cationic and anionic coordination capsules, respectively); templation of their formation by anionic or cationic guests of appropriate size and shape has been observed in many cases.

4.1 Caging Ligands with 6n End-Capping Metal Ions

4.1.1 Free Cages and Encapsulation of Neutral Molecules

For the first time, the coordination-driven self-assembly of a rigid tris-pyridyl tridentate ligand syntone *241* with ethylenediamine palladium(II) nitrate by Scheme 4.1 has been used by M. Fujita and coworkers in [1] to obtain a nanosized Pd_6L_4 coordination capsule **483**. Its expanded analogs **484** and **485** have been synthesized in [1] by Scheme 4.2 using mono- and diphenylene-containing homologs of *241*.

The caging ligand **483** encapsulates hydrophobic organic molecules by Scheme 4.3 forming host–guest 1:1 and 1:4 cage complexes depending on the size of such guest molecules, but not with 1,4-dibromo-, 1,3,5-tribromo-, and perfluorobenzenes. Its hydrophobic and electrophilic cavity thus discriminates these electron-deficient organic molecules [2]. This coordination capsule is also able to encapsulate polyaromatic guests shown in Scheme 4.3 giving 1:2 cage complexes (as follows from ^{1}H NMR data); the photochemical reactions of these caged species performed in [3] are described in Sect. 5.2.

The coordination capsule **483** has been used in [4] for AND/OR bimolecular recognition of neutral organic guests (Scheme 4.4). In the case of AND, this ligand encapsulates the corresponding guests and their mixtures giving the homoguest and heteroguest cage complexes, respectively. In particular, it binds both *cis*-decalin and perylene, thus forming a host–heteroguest 1:1:1 cage complex with a visible CT between its caging ligand and perylene guest (according to ^{1}H NMR and UV-vis data), but does not give the cage complexes with their individual molecules. This type of bimolecular recognition has been observed in [4] for large aromatic guests and suitable aliphatic co-guests. In the case of OR bimolecular recognition, the caging ligand can adopt either these individual components or their mixture;

Y. Voloshin et al., *The Encapsulation Phenomenon: Synthesis, Reactivity and Applications of Caged Ions and Molecules*, DOI 10.1007/978-3-319-27738-7_4

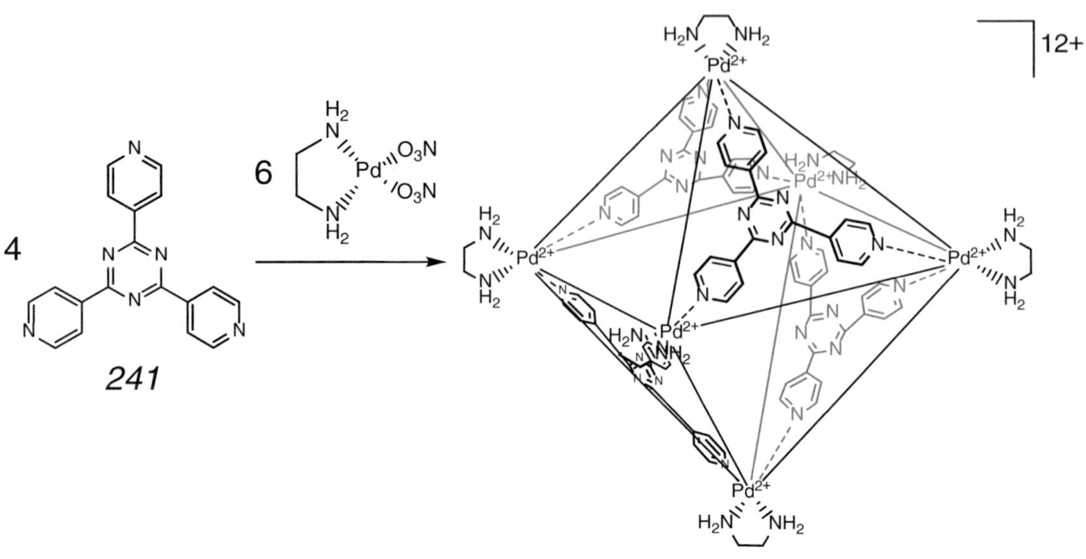

Scheme 4.1

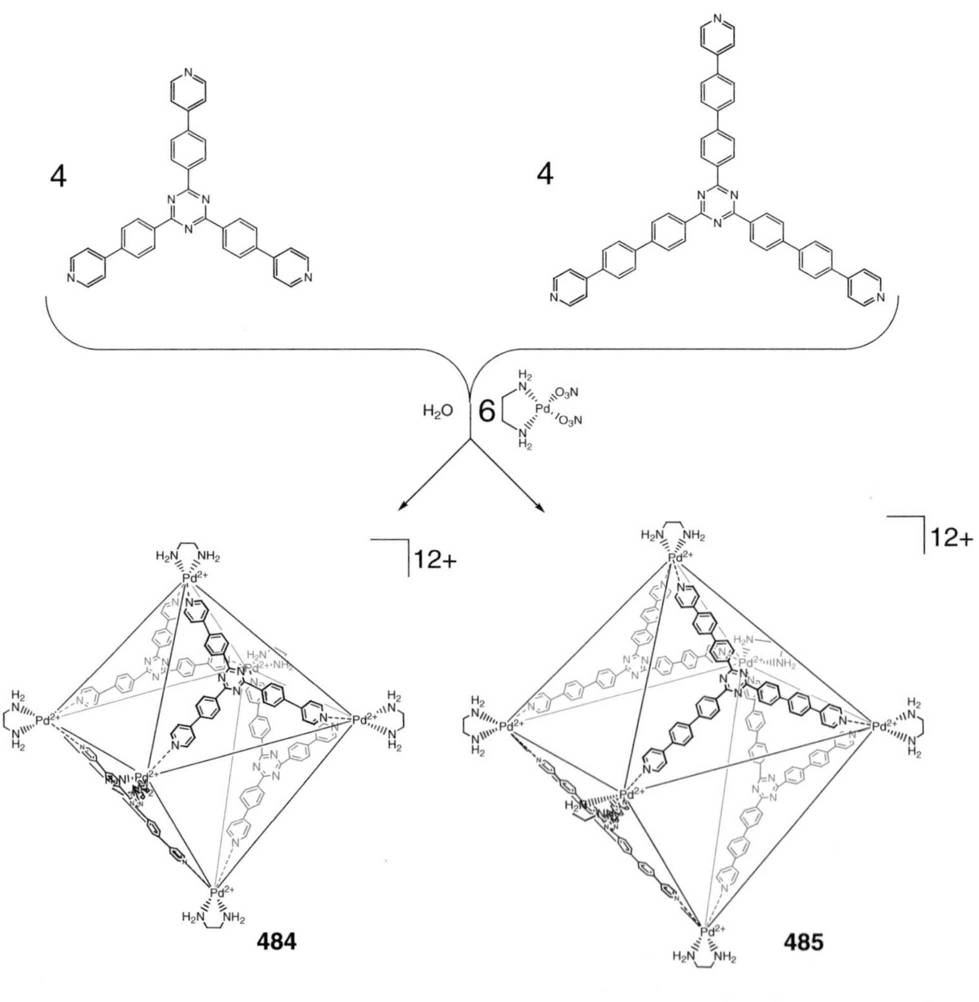

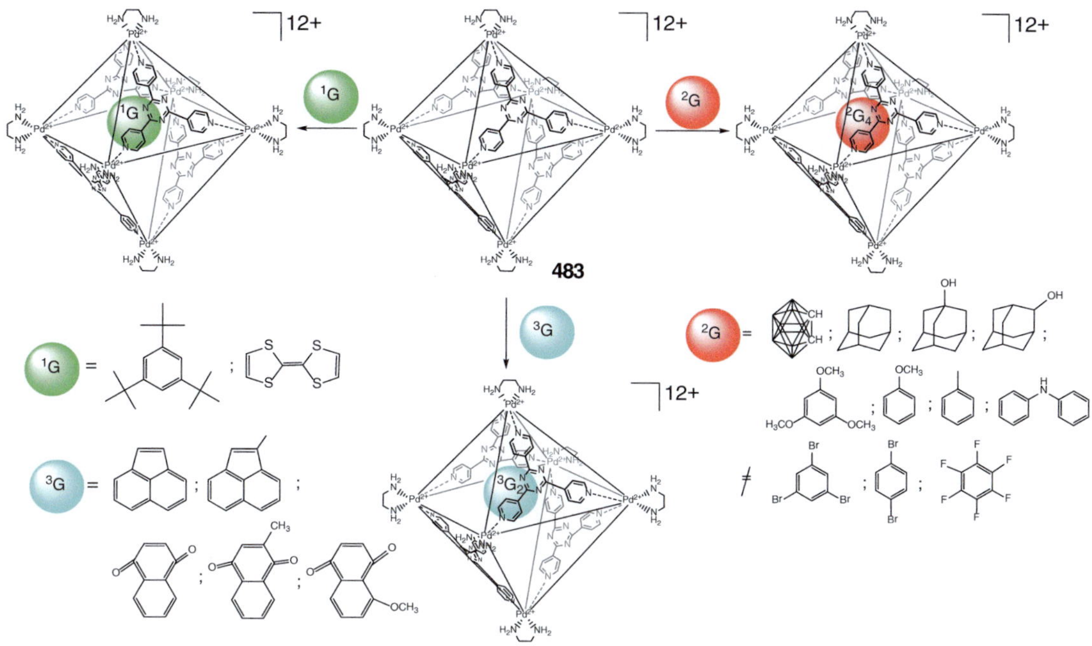

Scheme 4.3

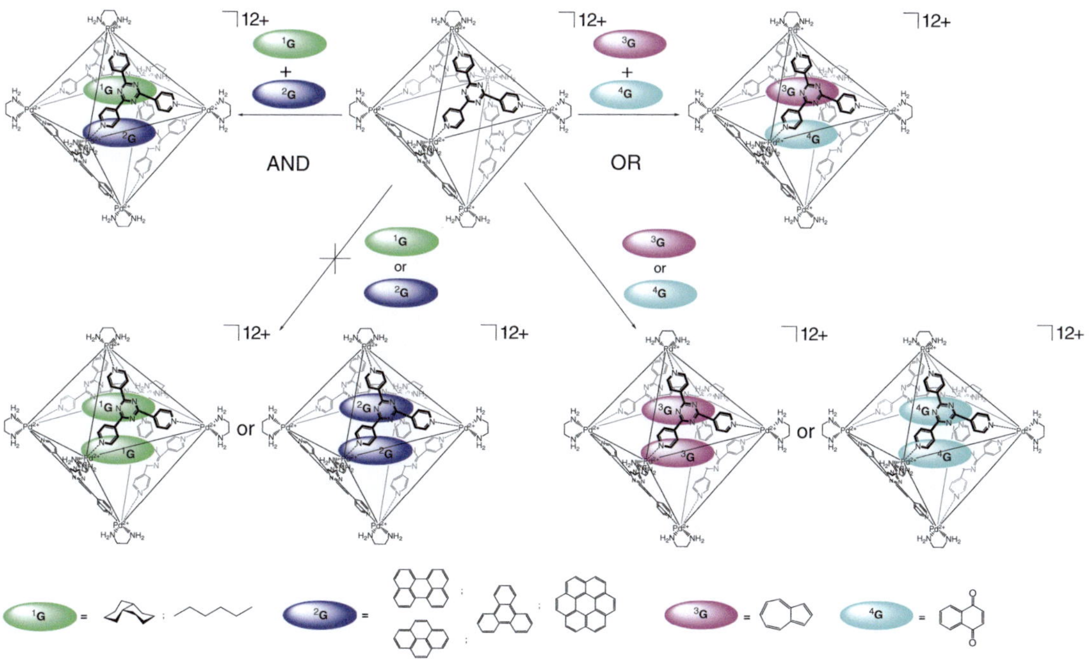

Scheme 4.4

Scheme 4.2

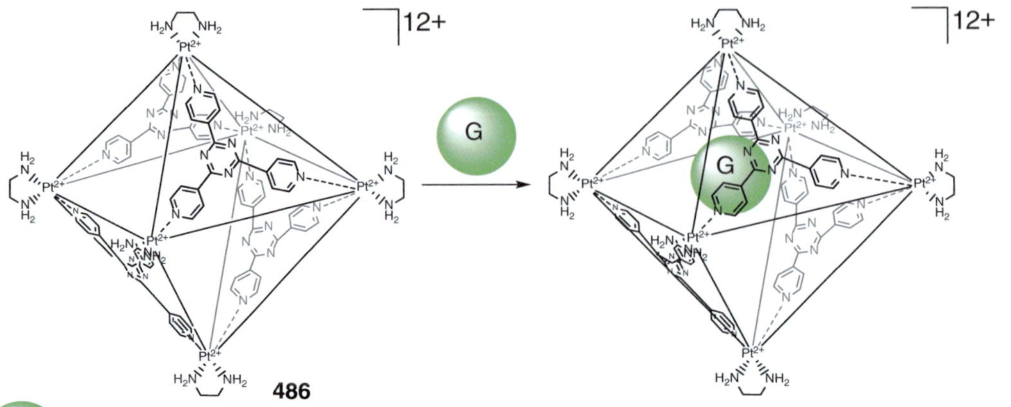

G = Ac-Trp-Trp-Ala-NH_2 (*242*), Ac-Trp-Ala-Trp-NH_2 (*243*) >> Ac-Ala-Trp-Trp-NH_2 (*244*), Ac-Trp-Trp-Gly-NH_2 (*245*), Ac-Ser-Gly-Ala-Trp-Trp-Ala-NH_2 (*247*), Ac-Ala-Trp-Trp-Ala-Gly-Ser-NH_2 (*248*), H-Trp-Trp-Ala-OH (*249*), Ac-Ala-Trp-Trp-Ala-Gly-Lys-NH_2 (*250*), Ac-Ala-Trp-Trp-Ala-Gly-Glu-NH_2 (*251*), Ac-Ala-Tyr-Tyr-NH_2 (*252*)

≠ Ac-Gly-Gly-Ala-NH_2 (*246*)

Scheme 4.5

this type of bimolecular recognition has been observed for relatively small aromatic guests. In particular, both azulene and 1,4-naphthoquinone form homoguest 1:2 cage complexes with **483**, while the use of their mixture resulted in the formation of a corresponding heteroguest 1:1:1 assembly that has been detected in [4] by ^1H NMR method.

The following advanced properties of Pd_6L_4 caging ligand **483** for molecular recognition of different guests are summarized in [5]: (*i*) its inner cavity, formed by four aromatic ligand syntones, can effectively bind aromatic molecules through hydrophobic van der Waals interactions; (*ii*) electron-rich guest molecules form CT interactions with triazine coordination capsule **483**, which is electron deficient due to coordination to platinum(II) ions; (*iii*) its relatively rigid cage framework discriminates guests by their size and shape; (*iiii*) it has more affinity to anionic guests than to neutral molecules or to cationic species, and (*iiiii*) its platinum-containing analog **486** has substantially higher chemical stability in both the acidic and basic media than the palladium(II)-capped cage framework **483** [5].

The capsule **486** selectively recognizes aromatic oligopeptides *242–245* (Scheme 4.5) but discriminates their aliphatic analog *246*. In the case of longer guests *247* and *248*, steric effects of this ligand cause preferential binding of hexapeptide *247* with an additional peptide chain at its *N*-terminus over its analog *248* having this sequence at the *C*-terminus. The latter cage complex is less stable due to steric "mismatches" between the cavity of the host and the guest molecule. According to the X-ray diffraction data for a 1:1 cage complex of capsule **486** with *242*, the guest *C*-terminus is more deeply buried in its cavity than *N*-terminus. Due to the cationic character of this caging ligand, it binds anionic (such as *251*) and neutral peptides much more (up to two orders) efficiently than their cationic analogs (such as *250*). In the case of charged peptides, this binding is affected by pH. Encapsulation of aromatic peptides becomes more efficient in the basic conditions and causes an intensive coloration of a solution as a result of strong CT interactions (such as in the cage complexes of the guests *244* and *252*) [5].

The Pd_6L_4 coordination capsules **483** and **487** are reported [6] to form heteroguest 1:1:1 cage

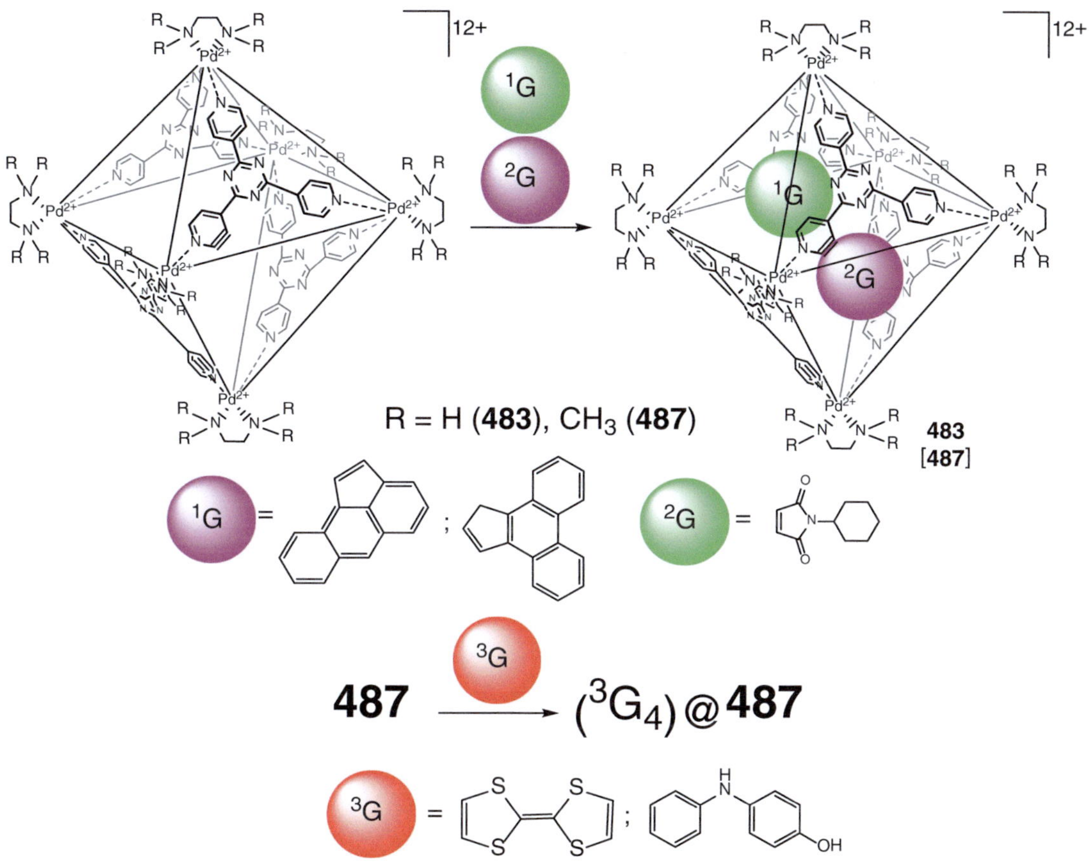

Scheme 4.6

complexes with the encapsulated polyaromatic molecules shown in Scheme 4.6 and the derivative of maleimide; the chemical reactivity of these caged guests is described in Sect. 5.1.

The coordination capsule **487** is reported in [7] to form a host–guest 1:4 cage complex with four encapsulated TTF molecules, which has been characterized both in solution and in solid state by ^{1}H NMR and single-crystal X-ray diffraction data, respectively. The crystals of **487** extract 4-hydroxydiphenylamine from its toluene solution, thus forming 1:4 host–guest complex [8]. These guest molecules within the cavity of **487** lie along its S_4-symmetry axis in such a manner that their secondary amino groups are completely isolated from external factors, whereas hydroxyl substituents are directed to interstitial pores of the crystals [8].

A caging ligand **488** has been self-assembled in [9] by coordination of tris-pyridyl ligand syntone *241* to palladium(II) bipyridinate; this ligand encapsulates neutral organic guests shown in Scheme 4.7. In the case of 4,4′-dimethoxydibenzoyl, the formation of host–guest 1:2 cage complex caused desymmetrization of its cage framework from T_d-symmetric to S_4-symmetric due to a specific orientation of two guest molecules. According to NMR and X-ray diffraction data, they aggregate in an orthogonal fashion and adopt a chiral twisted conformation with opposite screw axes. In the case of asymmetric analogs of this guest, an equimolar mixture of two diastereomeric forms of the corresponding cage complexes has been detected in [9] using single-crystal X-ray diffraction experiment. This coordination capsule also encapsulates ten solvent water molecules by

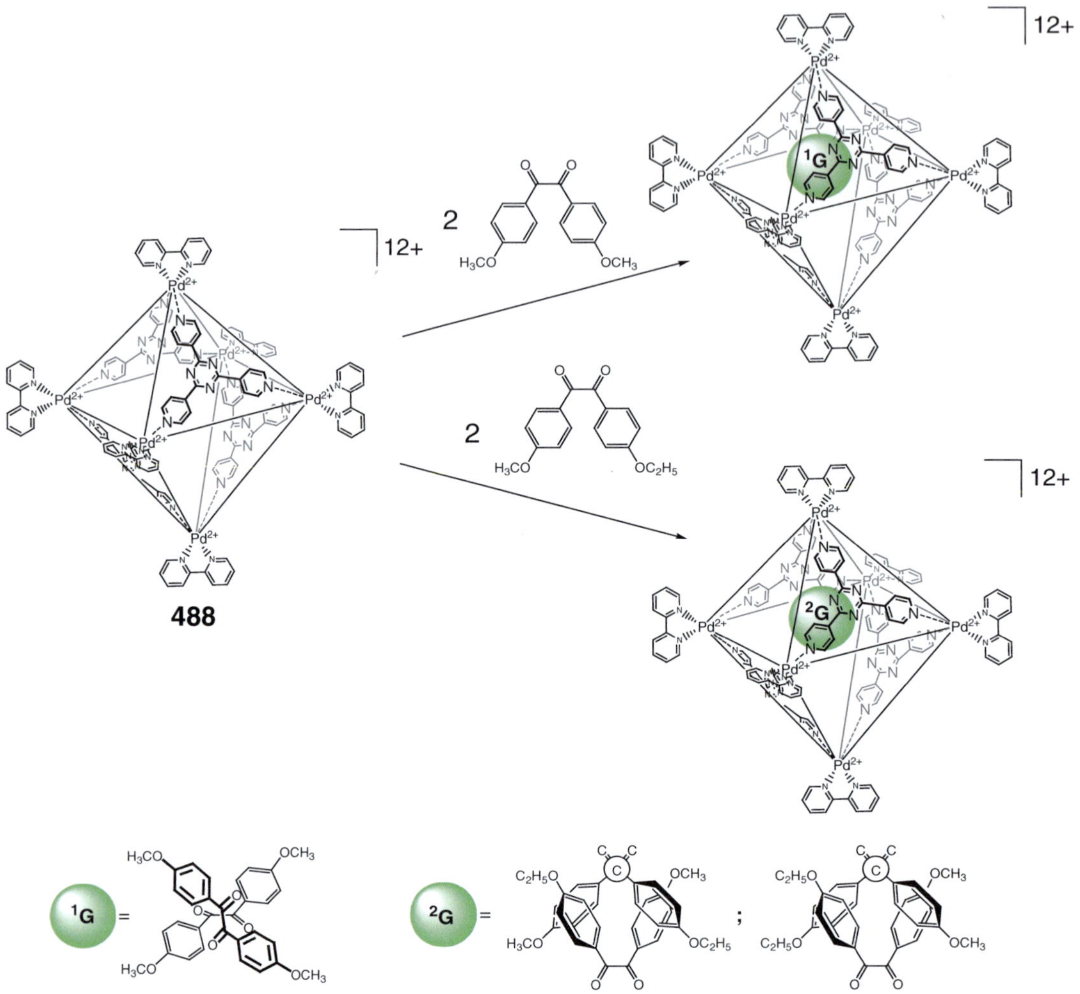

Scheme 4.7

Scheme 4.8 – 649, thus forming an adamantanoid cluster $(H_2O)_{10}$ (so-called "molecular ice") within its cavity. Such "molecular ice" does not melt even at room temperature and is not sustained by any metal cation or anion coordination [10]; it fills the void of the cavity of **488** forming OH…π and H_2O…π interactions. A single-crystal neutron diffraction study of [10] showed that the cluster $(H_2O)_{10}$ donated the electrons of a lone pair of the bridgehead water molecule.

The capsules **483** and **488** are reported in [11] to simultaneously encapsulate *ortho*-quinones *253* and *254* and an adamantane derivative *255*, thus giving heteroguest 1:1:1 cage complexes by Schemes 4.9 and 4.10. The encapsulated aromatic guest forms stacking interactions with facial azine fragments of **488**, while that with bulky adamantyl substituent occupies the remaining free space within its cavity; the photomediated 1,4-radical reaction of the caged molecules is described in Sect. 5.2. The coordination capsule **483** also gave the corresponding homoguest 1:1 and 1:2 cage complexes shown in Scheme 4.10 [11].

Cavity-induced spin–spin interactions between organic radicals encapsulated within the Pd_6L_4 coordination capsule **488** have been studied in [12] using EPR method. The organic radical *256* forms a stable 1:2 cage complex with this host by

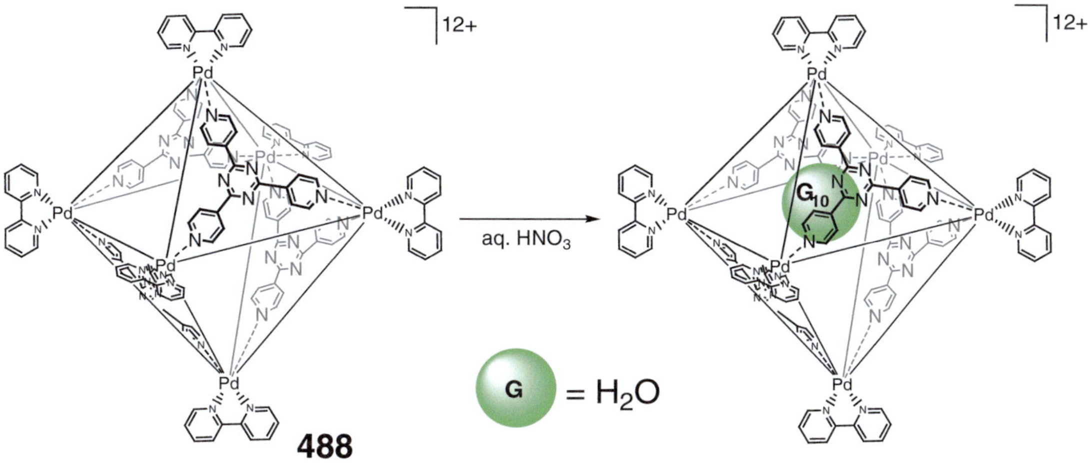

488

Scheme 4.8

Scheme 4.11 and two radical centers of the caged guest molecules are enforced to be close to each other within its cavity. As a result, the through-space interaction between them has been detected both in solution and in solid state [12].

Guanidinium nitrate, DMF, DMA, and DMSO have been used in [13] as efficient ionization-promoting reagents for CSI-MS study of encapsulation abilities of its platinum-based analog **489** in aqueous solutions. This allowed the authors to characterize caged guests shown in Scheme 4.12 (including the direct observation of their molecular behavior) and to elucidate the structures of various labile cage complexes.

The Pd_6L_4 coordination capsule **487** forms host–guest 1:1 and 1:2 cage complexes with encapsulated acenaphtylene molecule(s) by Scheme 4.13, which were characterized in [14] by single-crystal X-ray diffraction data; the olefin photodimerization reactions of these caged guests are described in Sect. 5.2. A sequence-selective recognition of peptides also shown in this Scheme by **487** has been performed in [15]. The ligand **487** selectively binds a peptide Ac–Trp–Trp–Ala–NH_2, giving a stable 1:1 cage complex, while the tripeptides with other sequences as well as those with one changed amino acid residue form the corresponding cage complexes with substantially lower stability constants. In the former cage complex, the encapsulated tripeptide guest having two electron-rich indole fragments forms strong stacking interactions with the electron-deficient triazine fragment of **487** (i.e., the ET *indole* → *triazine*) [15]. At the same time, this ligand is described to discriminate aromatic-free and cationic peptides; in the latter case, this feature is explained by electronic repulsions between the positively charged cage framework and the cationic guest. The same sequence-specific recognition has been also observed in [15] for larger oligopeptides.

The coordination capsule **487** has been used in [16] to stabilize reactive transition metal complexes such as manganese cyclopentadienyl $MeCpMn(CO)_3$, giving a host–guest 1:4 cage complex with this coordinately unsaturated guest shown in Scheme 4.13.

Self-assembly of a triazine ligand syntone *241* with bulky capping ruthenium(II) thiomacrocyclic precursor by Scheme 4.14 is reported in [17]. According to the single-crystal X-ray diffraction data, the donor pyridyl groups of the formed capsule **490** are twisted due to the steric restriction at its vertex metallocenters. This ligand encapsulates four guest molecules of adamantanol isomers, thus forming host–guest 1:4 cage complexes. Such binding causes coplanarity of the triazine facial fragments of **490** and a substantial longwave shift of its metal-to-ligand CT bands [17].

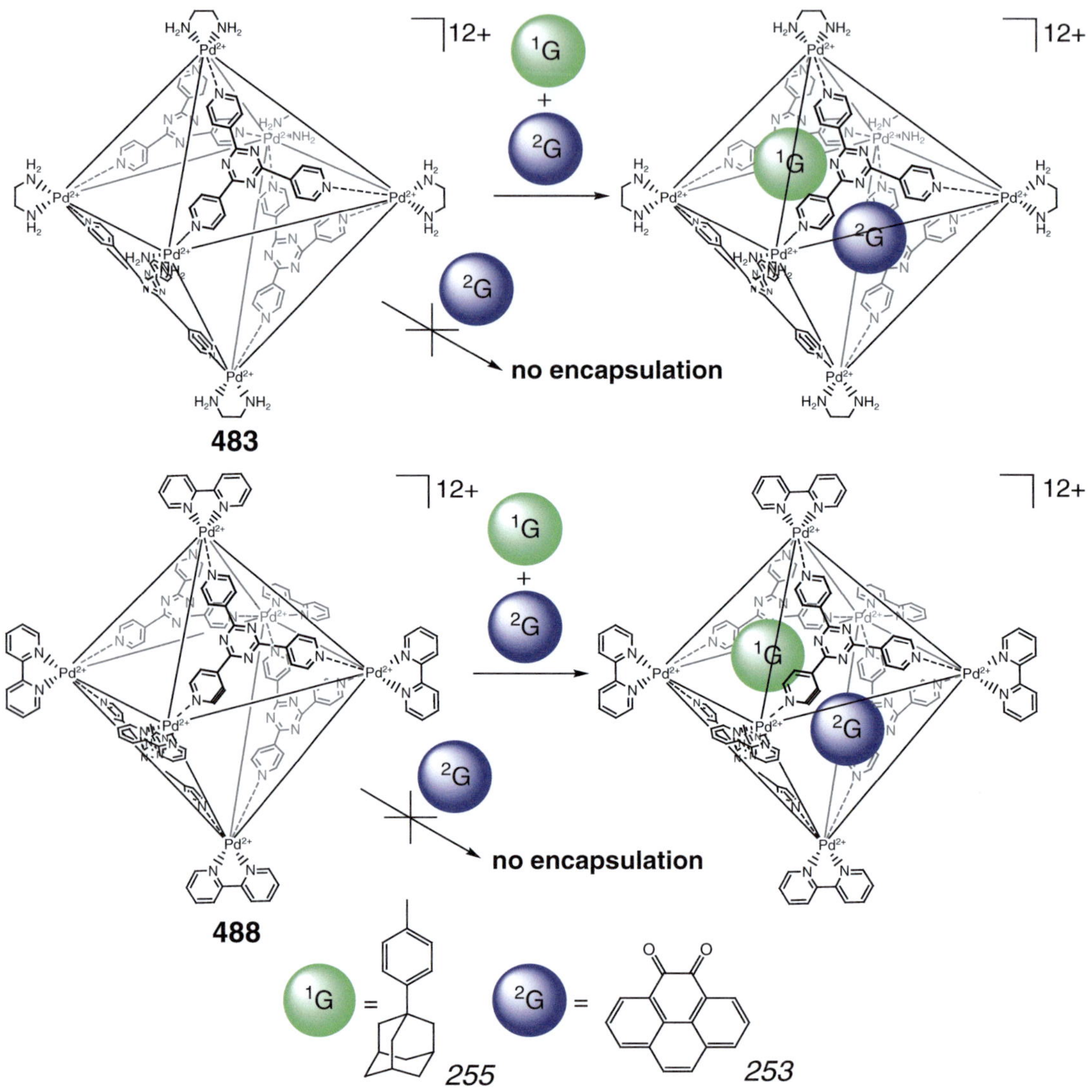

Scheme 4.9

Self-assembly of a verdazyl radical tris-pyridyl ligand syntone *257* with palladium(II) 2,2′-bipyridinate by Scheme 4.15 afforded a M_6L_4 capsule **491** with four spin centers, which showed intramolecular antiferromagnetic interactions [18]. The obtained paramagnetic caging ligand encapsulates nitroxyl radical *258*, giving a 1:1 cage complex with more pronounced interactions of this type. Therefore, the spin state of **491** can be controlled by encapsulation of appropriate guests; in particular, the radical guest *259* forms a host–guest 1:2 cage complex with multiple *H–G–H–G* spin–spin interactions between its six radical centers [18].

The photoinduced self-assembly of platinum(II)-capped coordination capsule **492** has been performed in [19] by Scheme 4.16.

A robust, porous crystalline network **493** (Scheme 4.17) that arises from an infinite array of the Co_6L_4 cage entities occupying approximately

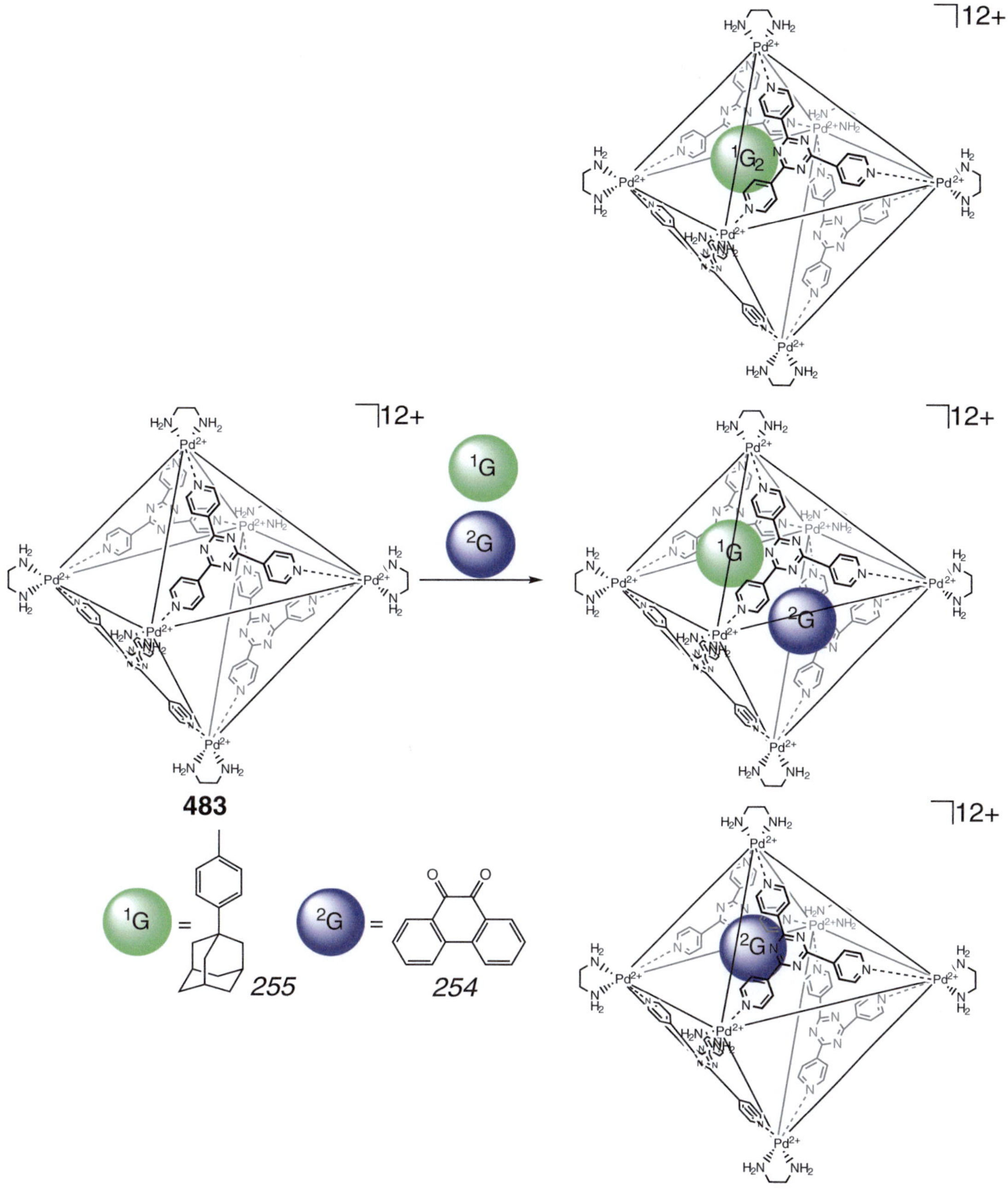

Scheme 4.10

20 % of its crystal volume has been prepared in [7] by self-assembly of cobalt(II) thiocyanate with the ligand syntone *241*. In the crystal, the octahedral coordination capsules are formed by six cross-linking cobalt(II) ions and four paneling syntones, and each vertex metal ion is shared by two adjacent cage entities. Its coordination polyhedron is formed by four pyridyl donor groups of the four syntones *241* and two apical thiocyanate anions. The adjacent and opposite Co...Co

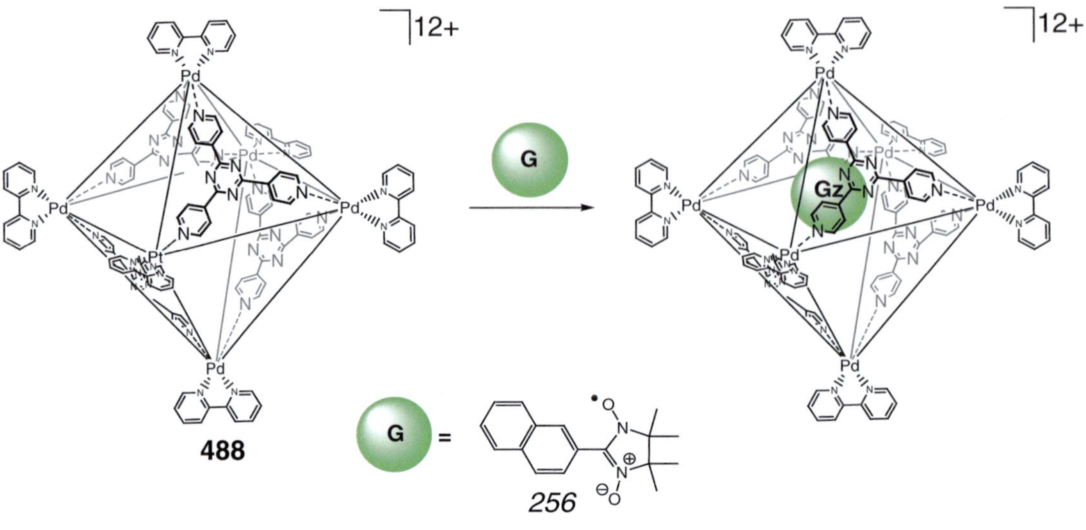

Scheme 4.11

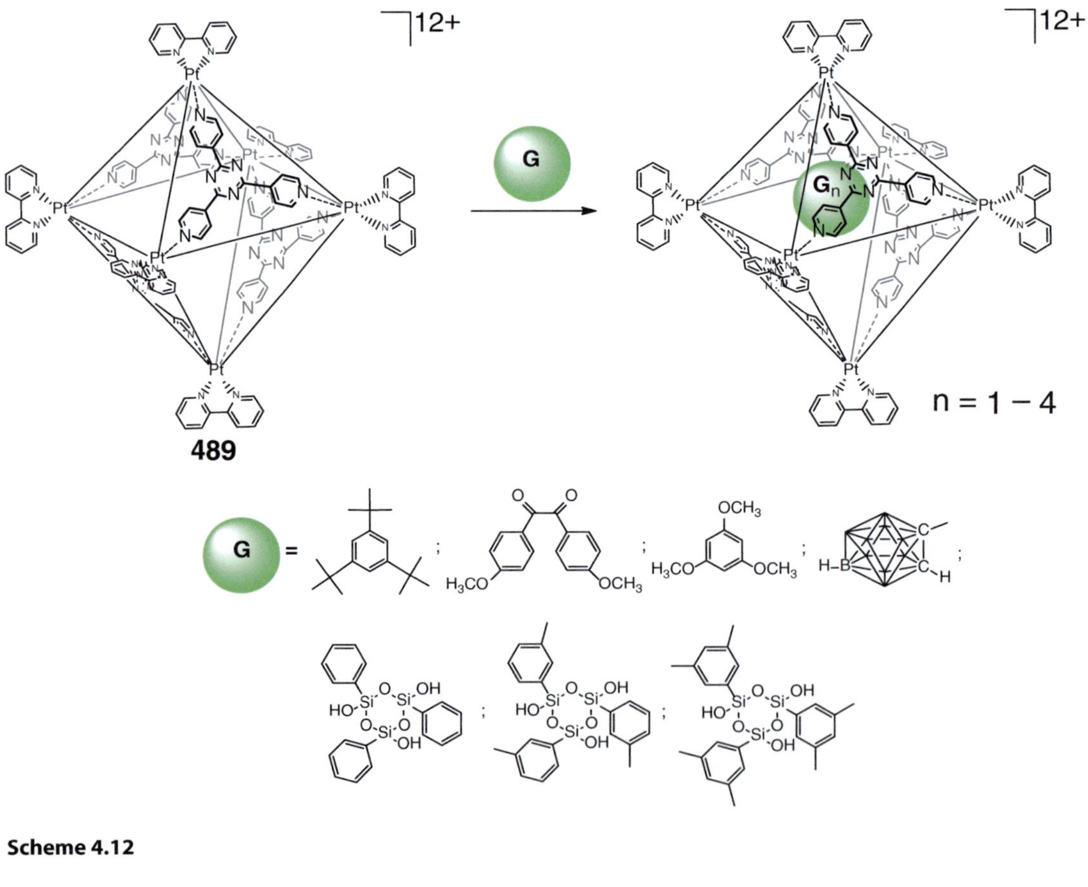

Scheme 4.12

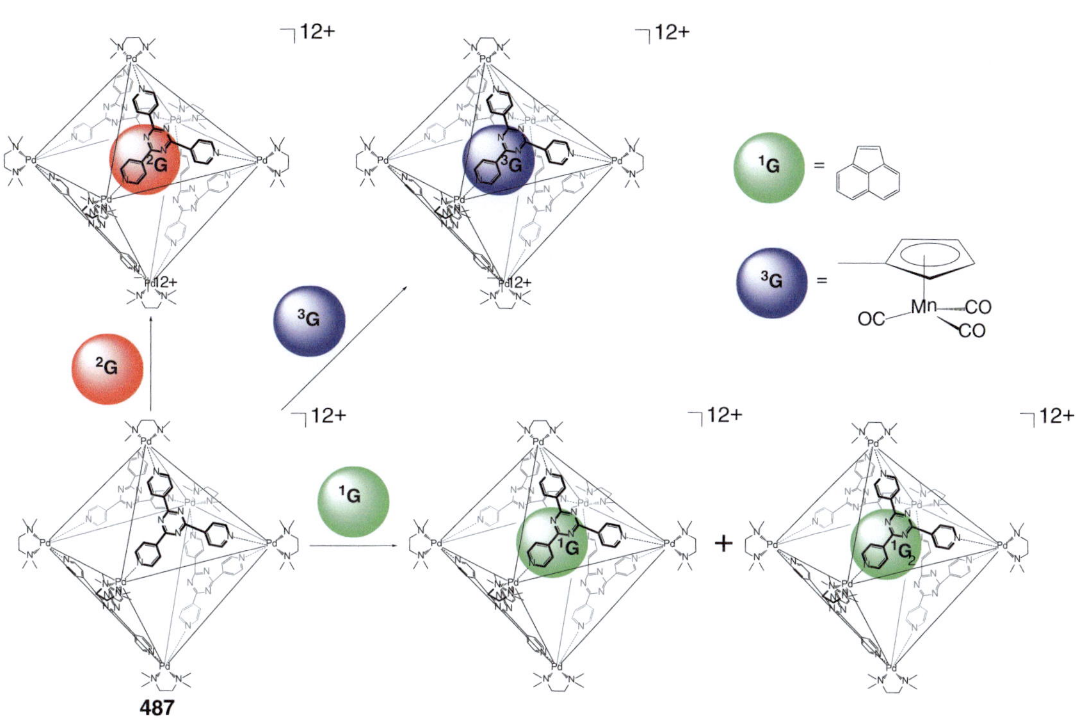

= Ac-Trp-Trp-Ala-NH_2, Ac-Trp-Ala-Trp-NH_2, Ac-Ala-Trp-Trp-NH_2, Ac-Trp-Trp-Gly-NH_2
Ac-Trp-Tyr-Ala-NH_2, Ac-Tyr-Tyr-Ala-NH_2, Ac-Ala-Trp-Trp-Ala-NH_2
Ac-Ser-Gly-Ala-Trp-Trp-Ala-NH_2
≠ Ac-Gly-Gly-Ala-NH_2, Ac-Tyr-His-Ala-NH_2

Scheme 4.13

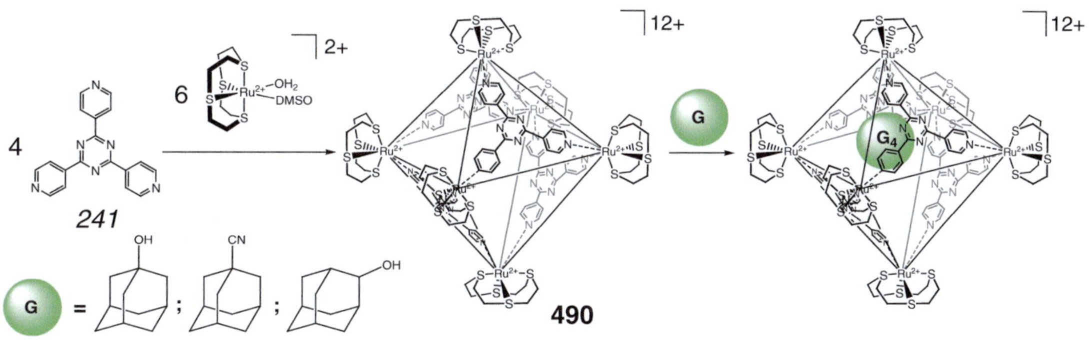

Scheme 4.14

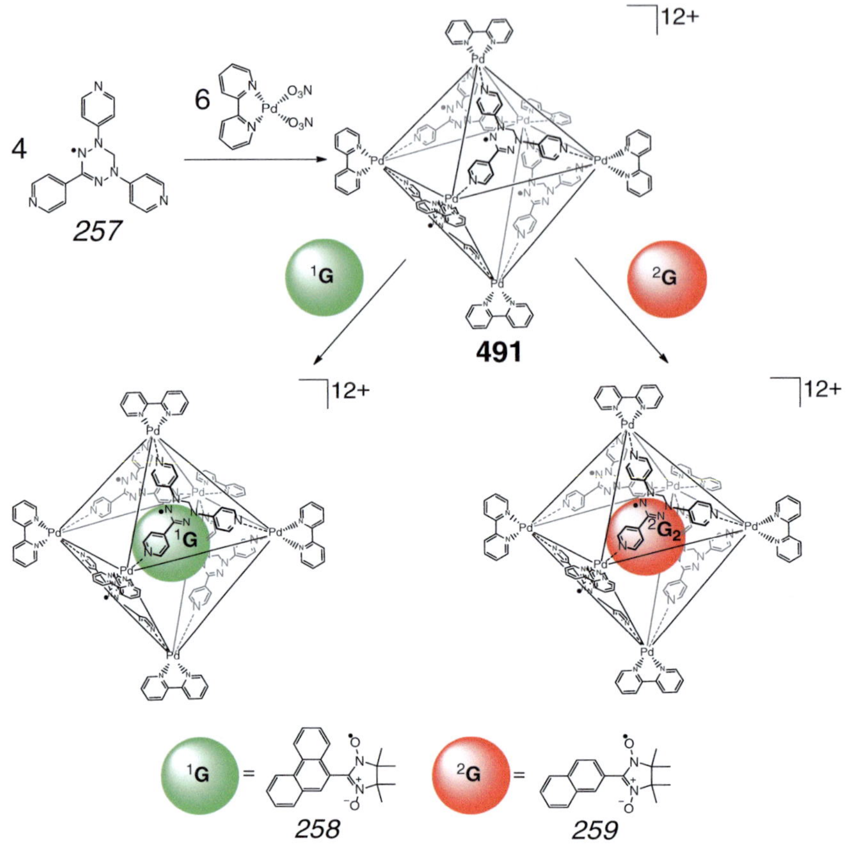

Scheme 4.15

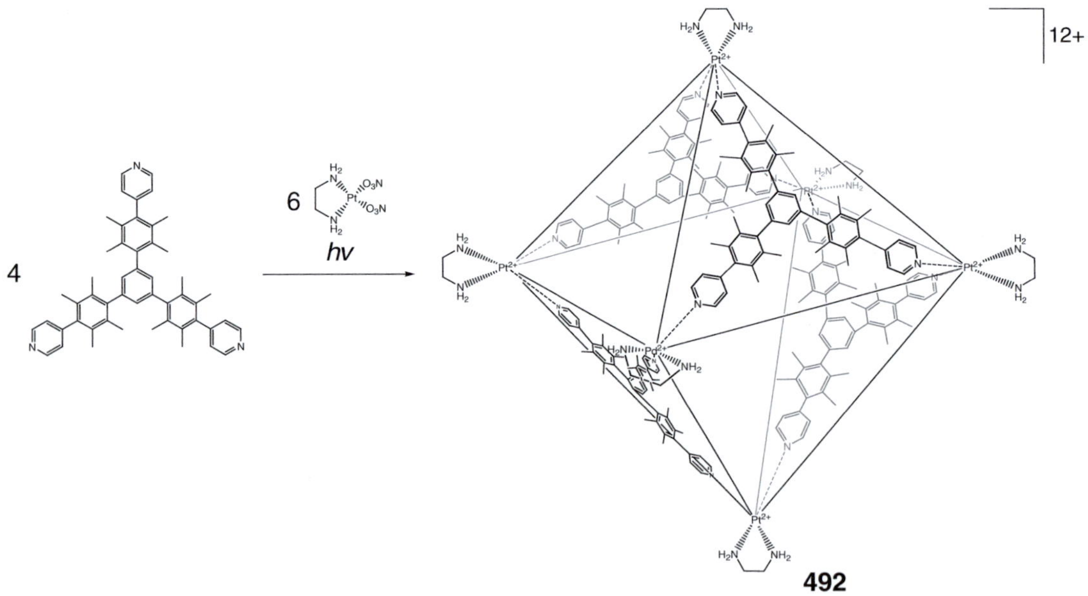

Scheme 4.16

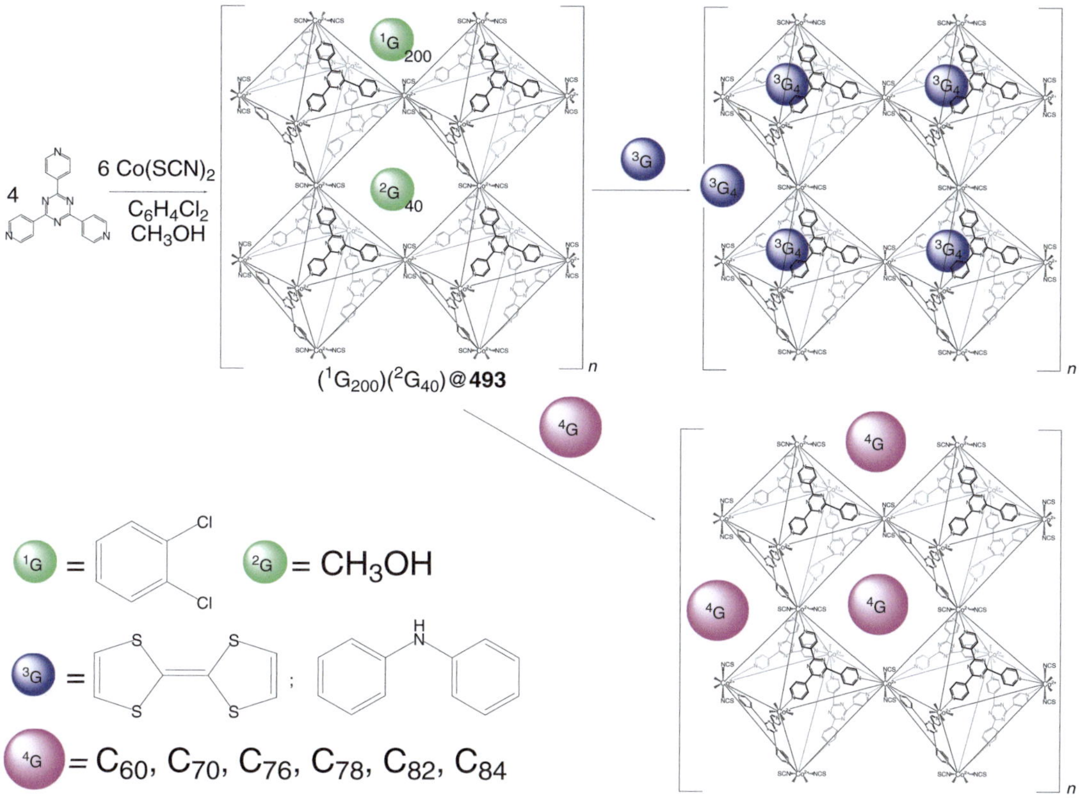

Scheme 4.17

distances in the capsule are equal to 13.24 and 18.73 Å, respectively. This caging host is also reported [7] to bind organic guests, such as TTF and diphenylamine. In the crystal of a host–guest 1:4 cage complex of **493** with TTF, four tightly packed guest molecules in each M_6L_4 entity are stacked with its paneling syntones at the distance of approximately 3.7 Å. Large interstitial voids in the crystal of **493** occupy approximately 60 % of its total lattice volume and contain two types of cuboctahedral $M_{12}L_8$ and $M_{12}L_{24}$ cage frameworks. The first one is surrounded by and shares the paneling ligand syntones and the cobalt(II) vertices with eight octahedral M_6L_4 capsules, and the second one is formed by the edges of 24 bridging syntones *241*, each of which joins two metallocenters. Thus, the porous crystals of **493** contain a small octahedral cage framework, having the narrow windows with a van der Waals diameter of 4.8 Å and two big cuboctahedral frameworks with substantially wider windows (11.5 Å); those allow encapsulating bulky organic guests. In particular, the nanosized cuboctahedral capsules bind fullerene guest molecules and display their molecular recognition by preferentially extracting the fullerene C_{70} from its 1:1 mixture with C_{60}.

Coordination of a tris-pyridyl ligand syntone *260* to palladium(II) ions by Scheme 4.18 has been used in [20] for the synthesis of the first spherical M_6L_8 caging ligand **494**. All eight bridging fragments of this ligand that encapsulate DMSO molecules are equivalent, and C_4-symmetry axis passes through its apical metallocenters. These six centers with PdN_4 square-planar coordination polyhedra form a regular octahedron.

The Pd_6L_{12} cubic coordination capsule **495** has been formed in a quantitative yield by coordination-driven self-assembly of a bis-pyridyl ligand syntone *261* having a bend angle

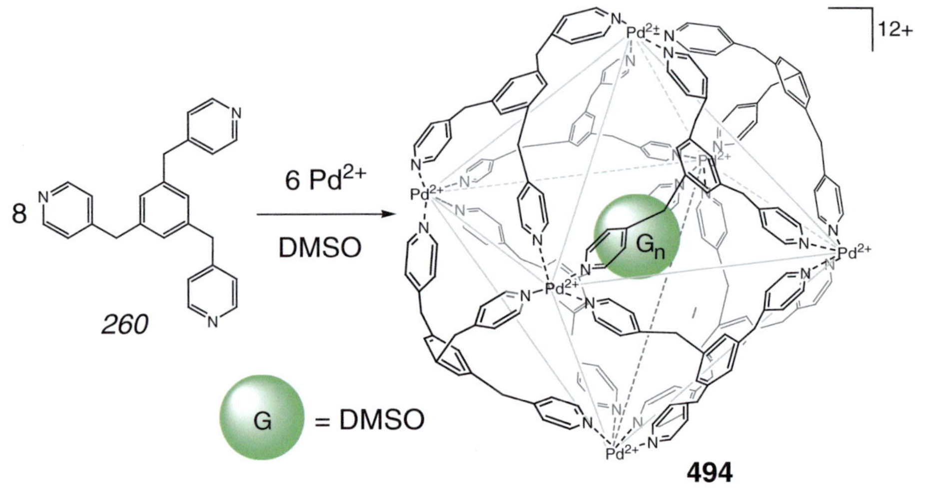

Scheme 4.18

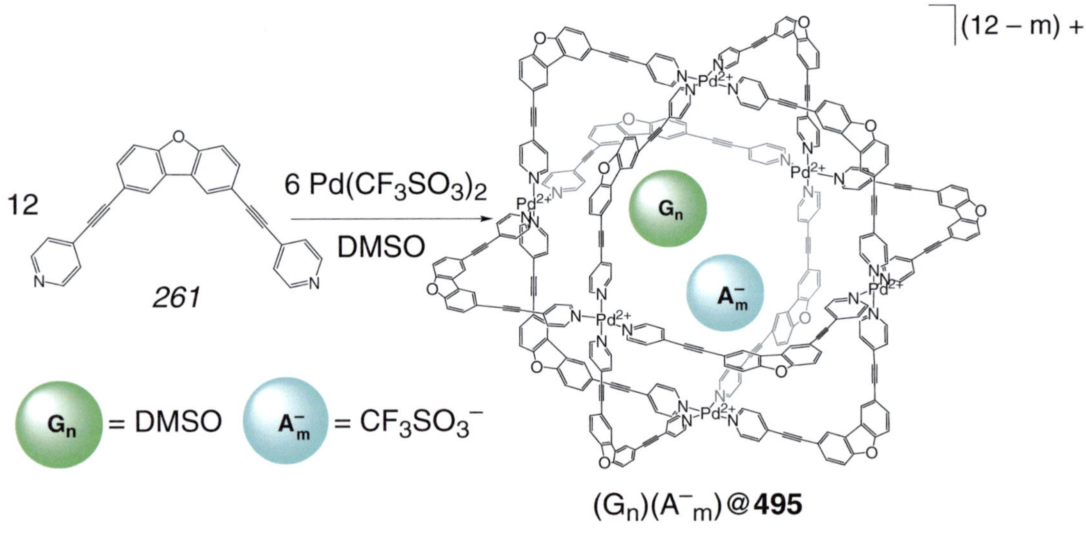

Scheme 4.19

of 90° with palladium(II) ions by Scheme 4.19 [21]. This capsule has highly symmetric 30×30×30 Å cage framework with a cavity volume of approximately 27,000 Å^3 and a Pd…Pd distance between the opposite PdN_4 square-planar metallocenters of approximately 26 Å; it encapsulates solvent DMSO molecules and triflate anions [21].

Self-assembly of a C_3-symmetric tris-*meta*-pyridyl ligand syntone *262* and palladium(II) ions has been used in [22] for the synthesis of a Pd_6L_8 coordination capsule **496** as the only product of the reaction by Scheme 4.20. The ^1H NMR and ESI-MS data suggest an equivalence of all the C_3-symmetric facial entities in its cage framework with a diameter of approximately 24 Å and a truncated octahedral geometry. The combination of a donor nitrogen atom in a *meta*-position to the carboxamidopyridinyl fragment and the tilt of a pyridyl ring versus the facial mean plane of this syntone provided the needed curvature for the formation of such octahedral coordination

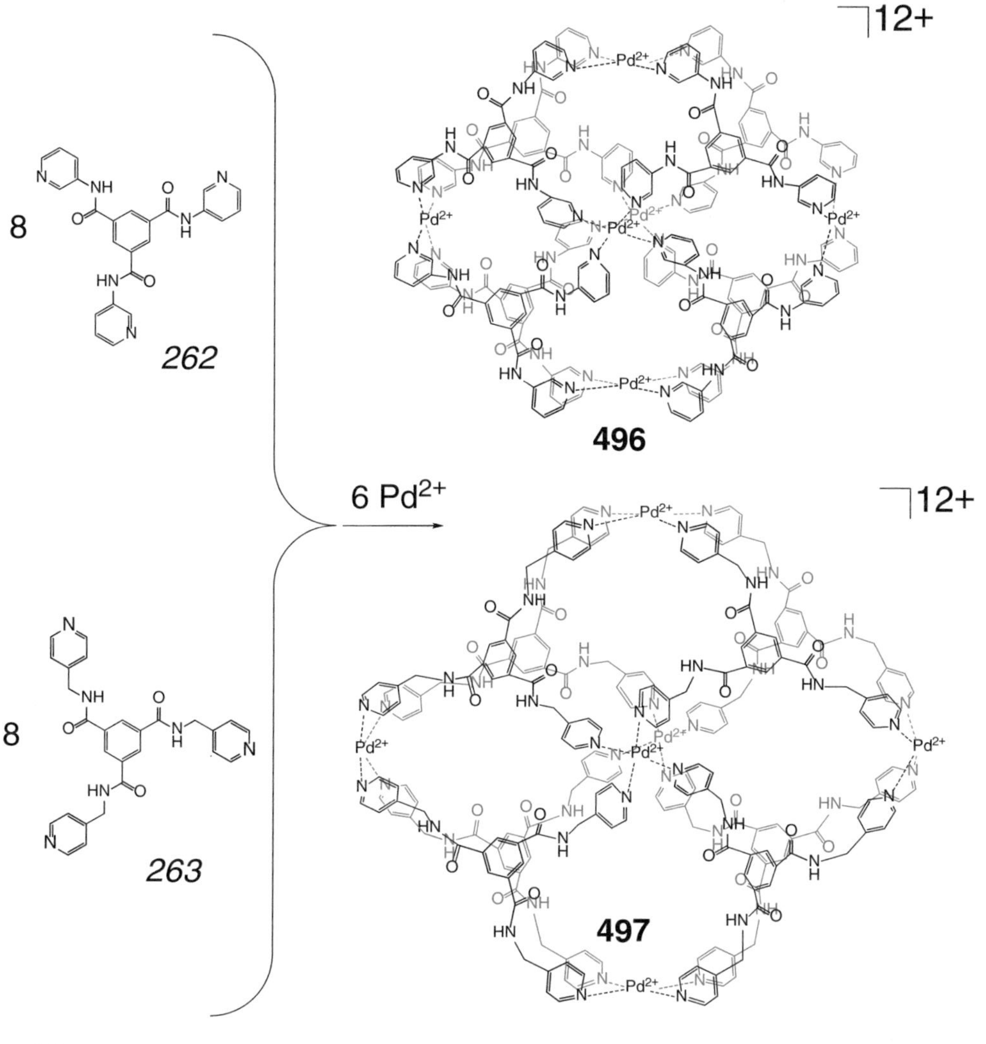

Scheme 4.20

capsule; the cross-linking *PdN₄* fragments have a square-planar geometry. Two statistically disordered conformational isomers occurring in the ratio of 3:2 have been observed in [22]; the major *syn*-isomer has a cavity volume of approximately 1600 Å^3, and that of its *trans*-analog is above 1900 Å^3. At the same time, only the *syn*-conformer of **496** has been detected in solution [22]. The analogous synthetic approach allowed obtaining O_h-symmetric truncated octahedral coordination capsule **497** starting from tris-*para*-pyridyl ligand syntone *263*. Its truncated octahedral cage framework has a diameter of approximately 26 Å and is slightly larger than that for **496**; a cavity volume of **497** (approximately 2200 Å^3) is also larger than those for the above *syn*- and *anti*-conformers of **496** [22].

A similar truncated octahedral Pd_6L_8 coordination capsule **498** with pendant monodentate quinoxalyl donor groups has been synthesized in [23] by Scheme 4.21. As follows from ESI-MS data, this ligand encapsulates four solvent DMSO-d_6 molecules, thus giving a host–guest 1:4 cage complex.

Coordination-driven self-assembly of a rigid angular bis-pyridyl ligand syntone *264* with palladium(II) ions by Scheme 4.22 afforded a cuboctahedral $M_{12}L_{24}$ capsule **499** in quantitative

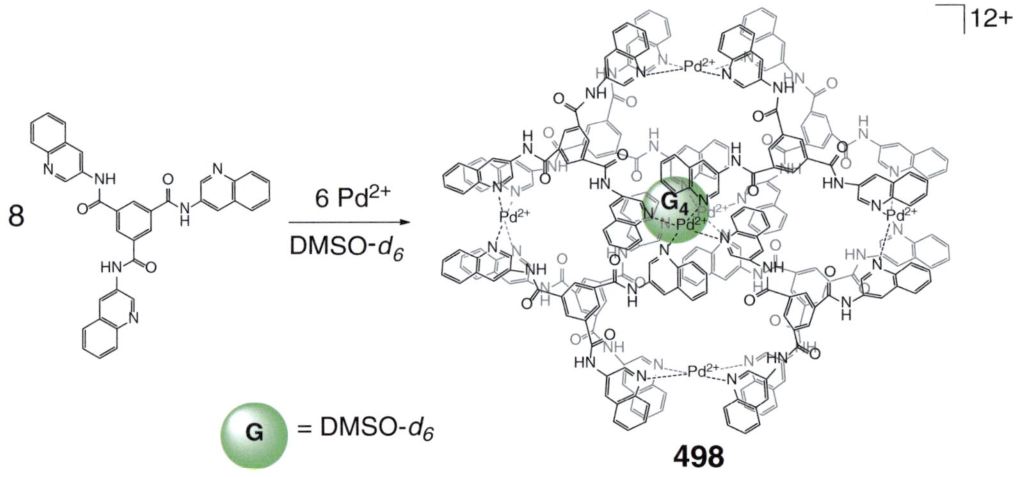

Scheme 4.21

yield [24]. This spherical capsule has a uniformed dimension of 35 Å, while its analog **500**, the derivative of 2,5-(bis-pyridyl)furane ligand syntone *265*, has a diameter of approximately 34 Å; the latter encapsulates nitrate anions and solvent DMSO molecules. An extended angular bis-pyridyl syntone *266* with phenylene spacer fragments forms highly symmetric $M_{12}L_{24}$ coordination capsule **501** as the only product of its coordination-driven self-assembly with Pd^{2+} ions. The use of monofunctionalized syntones *267* and *268* with porphyrin and fullerene pendants allowed obtaining spherical capsules **502** and **503** (Scheme 4.23) with 24 peripheral functionalizing groups [24].

A bowl-shaped Pd_6L_3 precursor *269* undergoes dimerization in the presence of a nine-residue peptide in $NaNO_3$ aqueous solution by Scheme 4.24 to give a 1:1 cage complex of the $Pd_{12}L_6$ coordination capsule **504** with large hydrophobic cavity; this capsule is in equilibrium with its cavitand analog [25]. According to the NMR data, the guest peptide molecule adopts α-helical conformation within the hydrophobic cavity of **504** [25].

Functionalized angular bis-pyridyl diacetylene syntones *265–268* have been used in [26] for the synthesis of spherical $M_{12}L_{24}$ coordination capsules **505–508** (Scheme 4.25) decorated with 24 endohedral substituents. The rigid acetylene spacers in this case not only increase the size of their cage frameworks but also prevent their collapse. The bis-pyridyl ligand's fragments have a planar conformation, and each of four pyridyl donor groups is perpendicular to *a* mean plane of the square-planar PdN_4 coordination polyhedron of the corresponding metallocenter [26].

Self-assembly of 24 analogous perfluoroalkyl-functionalized bis-pyridyl ligand syntones *269–272* with 12 palladium(II) ions by Scheme 4.26 afforded in quantitative yields the fluorophilic caging ligands **509–512** with an inner superhydrophobic interior able to encapsulate perfluoroalkanes [27]. As follows from ^{19}F NMR data, the coordination capsule **509** can accommodate up to six of these molecules to give mostly a host–guest 1:6 complex. The complex with encapsulated disordered perfluorooctane molecules has a rigid oval $Pd_{12}L_{14}$ cage framework of approximate size of 49×42 Å with disordered perfluoroalkyl chains in its inner cavity. The coordination capsules **510** and **511** with more bulky inner perfluoroalkyl substituents can accommodate up to 2.5 guest molecules per one cage framework. Their analog **512** does not form such a cage complex with encapsulated perfluorooctane molecule(s) [27].

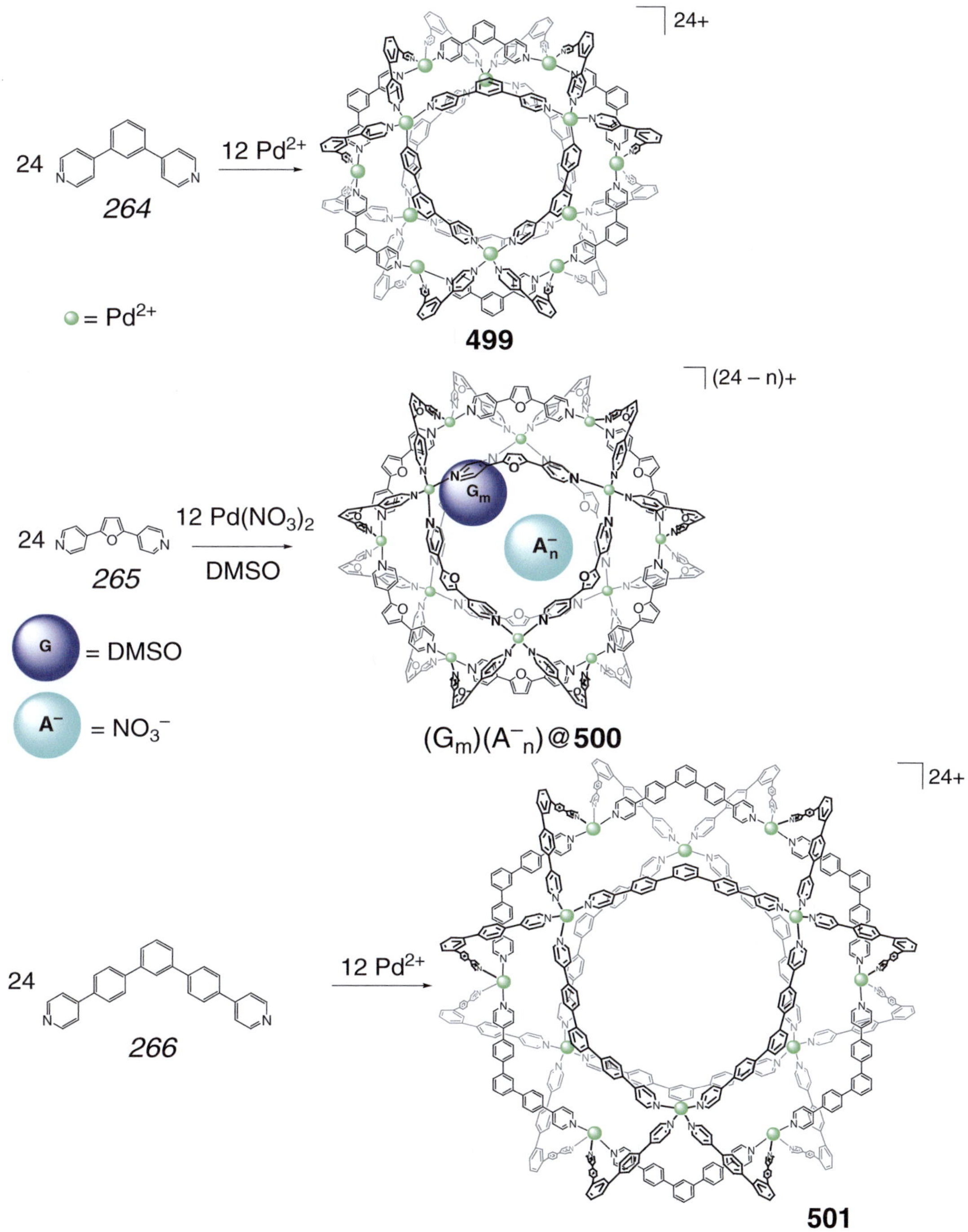
24
264
12 Pd2+
= Pd2+
499
24+
24
265
12 Pd(NO3)2
DMSO
G = DMSO
A− = NO3−
(Gm)(A−n)@500
(24 – n)+
24
266
12 Pd2+
501
24+

Scheme 4.22

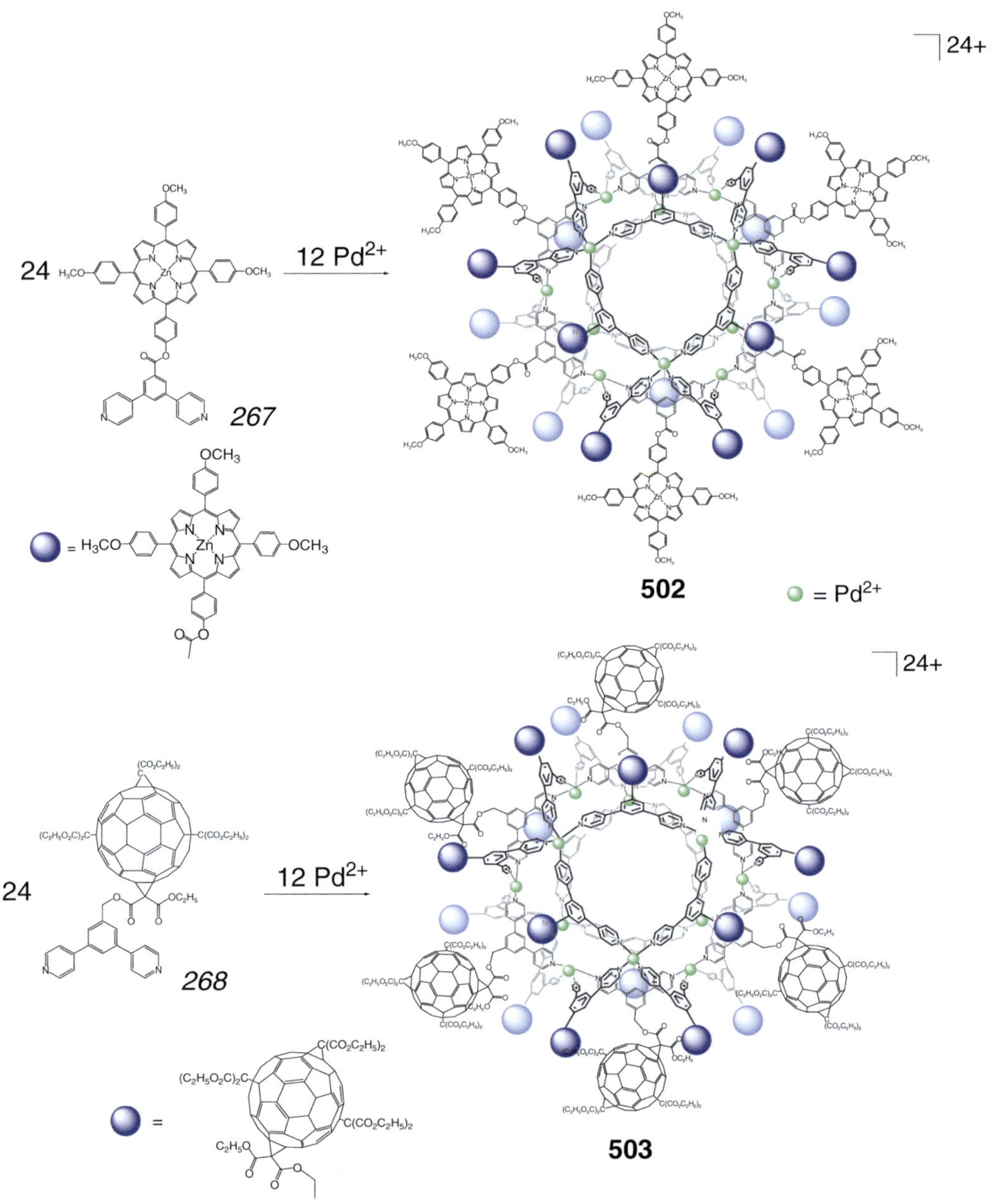

Scheme 4.23

Similar $Pd_{12}L_{24}$ long-chain alkyl-functionalized caging ligands **513–515** with a diameter of approximately 50 Å have been synthesized in [28] by Scheme 4.27. These spherical coordination capsules have hydrophobic inner cavities that are formed by alkyl chains and are able to encapsulate hydrophobic guests (such as dye Nile red) by Scheme 4.27 giving host–guest 1:2, 1:10, and 1:12 cage complexes, respectively [28].

The use of angular amino acid- and peptide-functionalized bis-pyridinyl ligand syntones allowed preparing nanosized spherical $Pd_{12}L_{24}$ coordination capsules **516–524** with peptide-lined

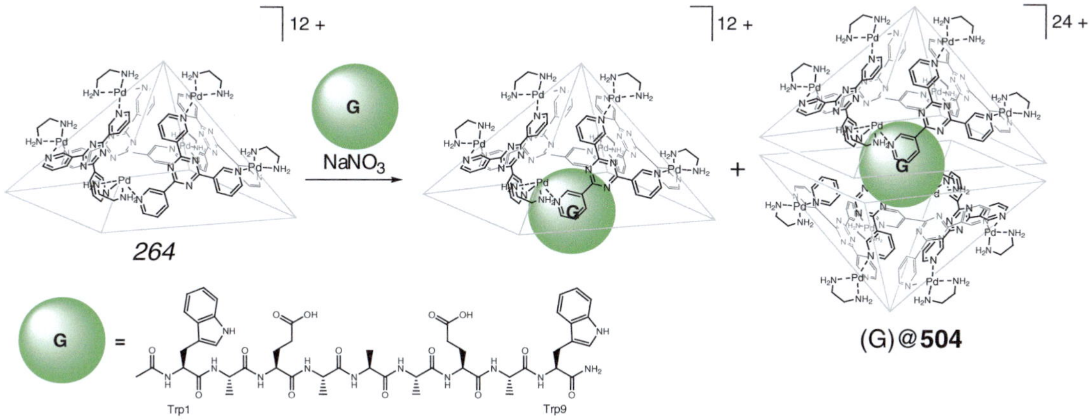

Scheme 4.24

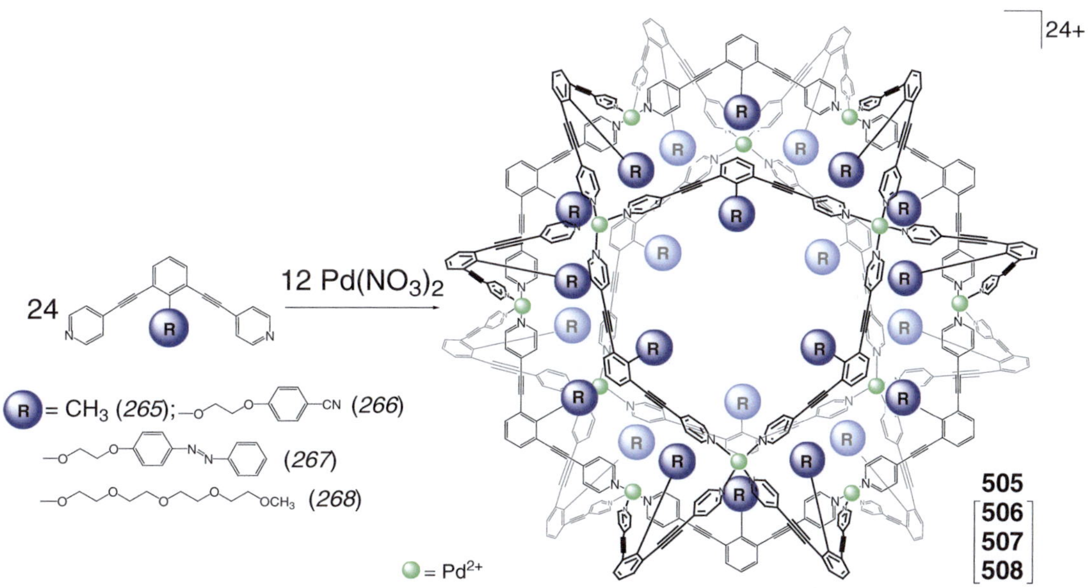

Scheme 4.25

chiral cavities by Scheme 4.28 [29]. While the yield of the corresponding caging ligand with one-residue peptide substituents is quantitative, in the case of five-residual functionalizing groups, it is substantially lower due to steric restrictions within the inner cavity of the capsule. The ligand **516** has a spherical cage framework with a diameter of approximately 46 Å and 24 inner asymmetrical peptide substituents that form chiral environment. As follows from CD data, the chirality of these pendants is conducted through this framework. The use of a combinatorial library of such bipyridyl syntones afforded a mixture of $Pd_{12}L^1{}_nL^2{}_{24-n}$ coordination capsules with different inner peptide substituents [29].

A caging ligand **525** with a polycationic quaternary ammonium interior and its neutral analogs **526** and **527** have been synthesized in [30] by self-assembly of angular aromatic syntones *273–275* with palladium(II) ions, giving by Scheme 4.29 the corresponding $M_{12}L_{24}$ coordination capsules in quantitative yields.

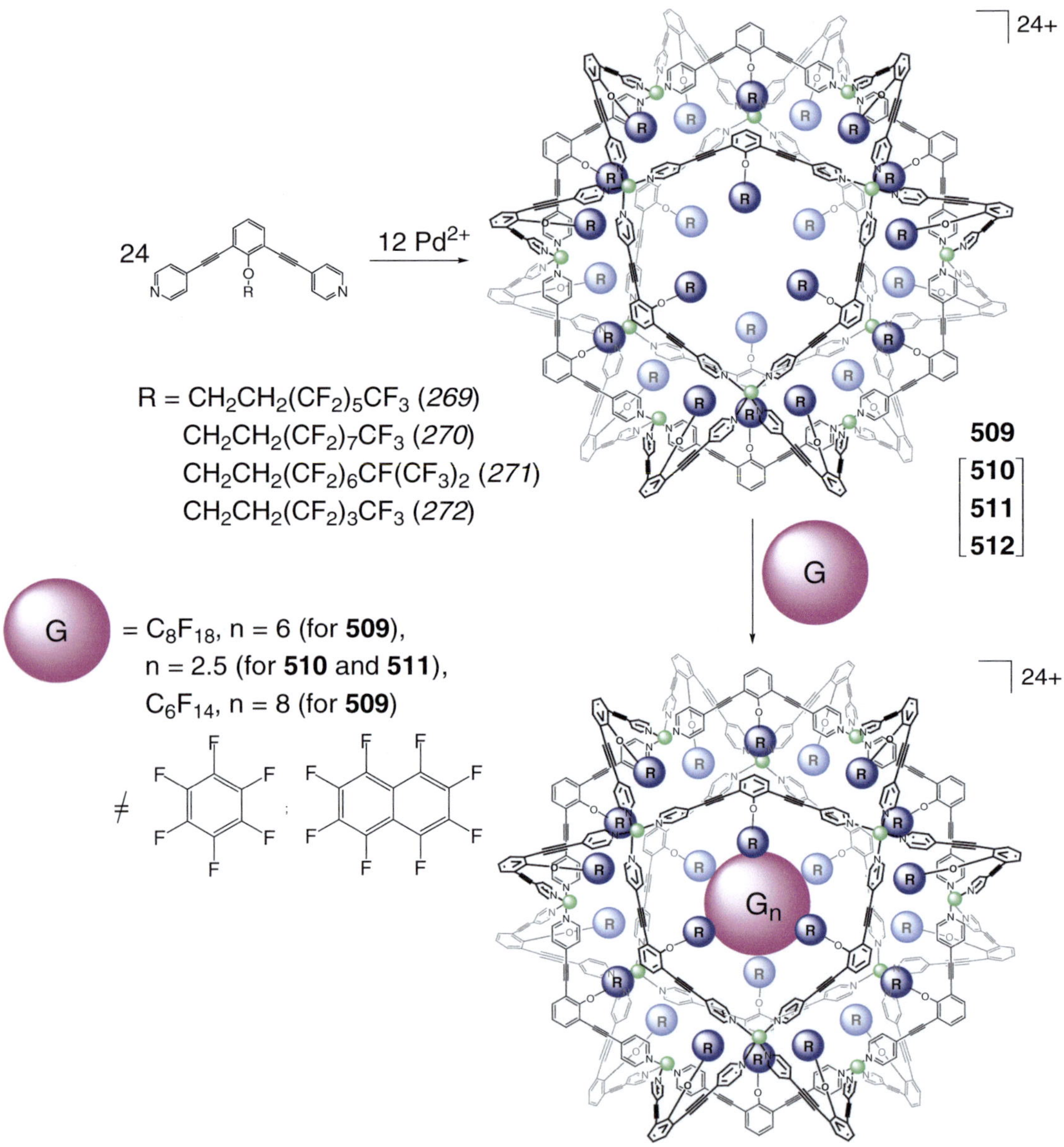

Scheme 4.26

According to MM data, the cage framework of **525** with a diameter of approximately 46 Å has cuboctahedral symmetry, and its inner cavity with a diameter of approximately 23 Å is decorated by 24 cationic residues allowing for multicenter interactions within the cationic microenvironment. Indeed, encapsulation of anionic guests shown in Scheme 4.30 by this spherical caging ligand has been performed. While the carboxylate anions caused complete destruction of its cage framework due to their coordination to palladium(II) ions, SO_3^--containing anions (especially the corresponding polyanionic guests) form 1:1 cage complexes. These caged anions interact with multicationic interior in a multipoint fashion, and such electrostatic interactions are more pronounced in the case of multivalent anions.

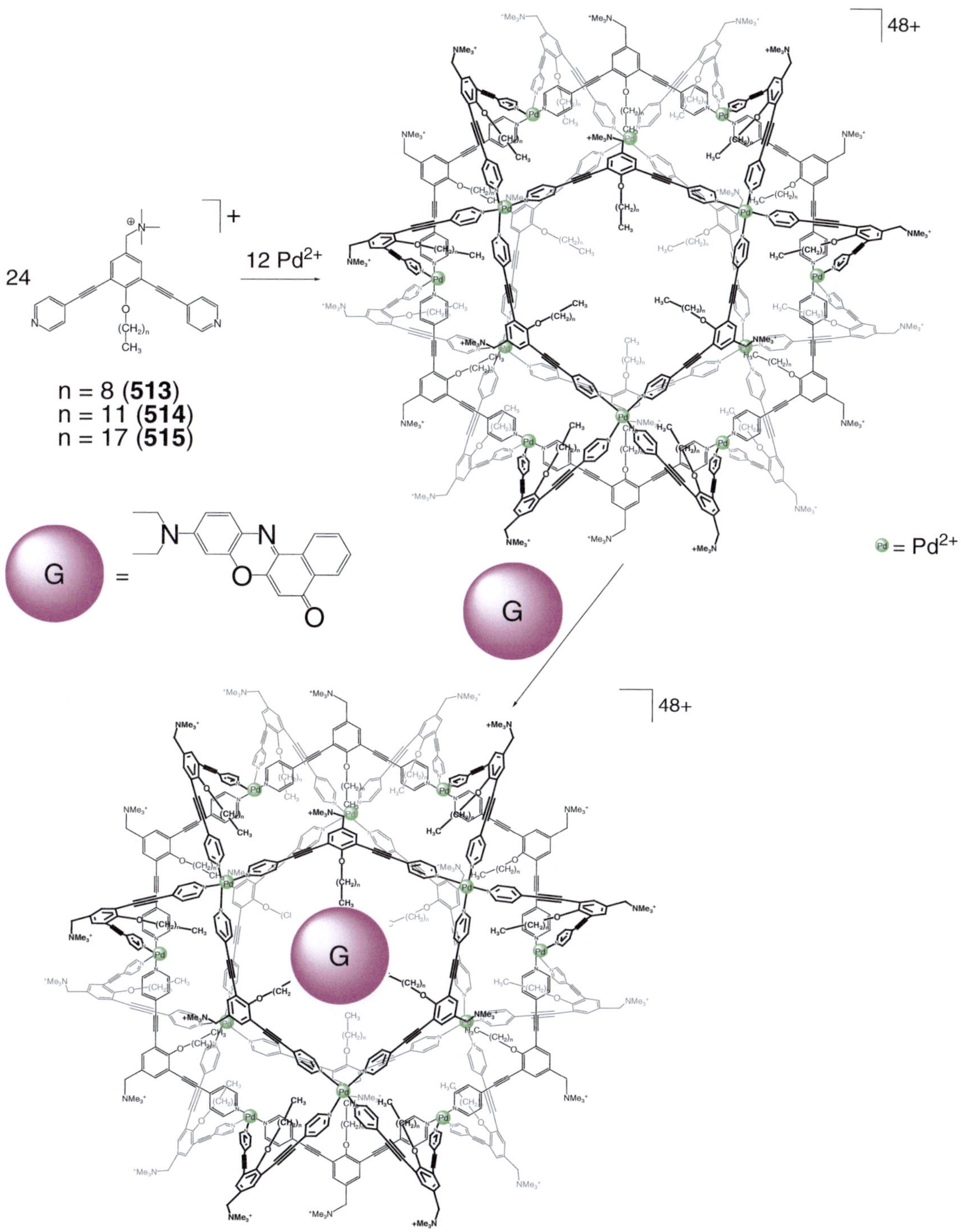

Scheme 4.27

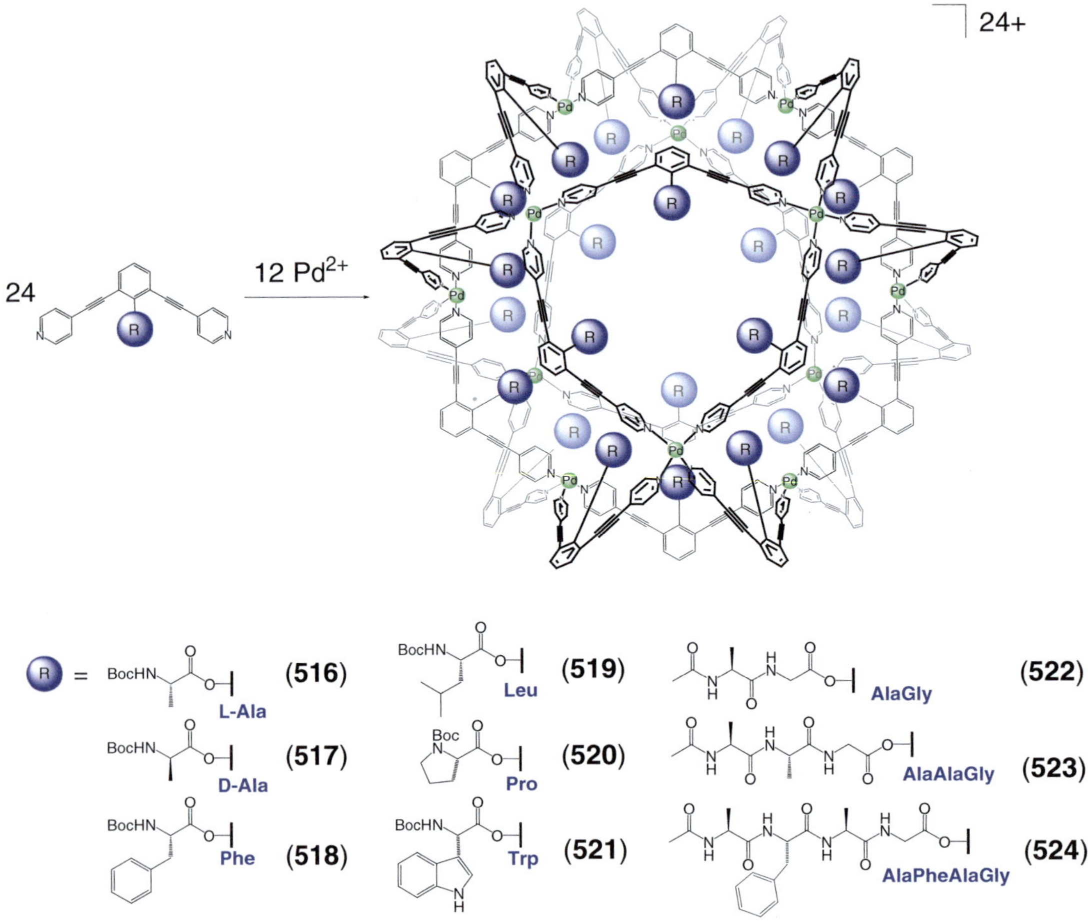

Scheme 4.28

As follows from DOSY NMR experiments, supramolecular interactions of these anionic guests within the cavities of capsules **526** and **527** with neutral pendant substituents are substantially weaker [30].

Nanosized $Pd_{12}L_{24}$ coordination capsules **528**–**530** have been prepared in [31] by coordination-driven self-assembly of rigid angular bis-bipyridyl ligand syntones *276–278* by Schemes 4.30. The caging ligands **528** and **530** encapsulate tetramethoxysilane molecules; their further reactivity within the cavities of these coordination capsules is described in Sect. 5.1.

Coordination-driven self-assembly of 18 palladium(II) ions and six triangular 3,5-pyrimidyl ligand syntones *279* by Scheme 4.31 has been used in [32] to obtain the first hexahedral $M_{18}L_6$ coordination capsule **531** with a cavity volume of approximately 900 Å^3.

Distorted TP – TAP $M_6L^1_2L^2_3$ capsules **532**–**534** have been prepared in [33] by Scheme 4.32 using *one-pot* self-assembly of aromatic tris-carboxylate ligand syntones *280–282* with hexaazamacrocycle *283* and copper(II) ions. Their cage frameworks are formed by three copper(II) azamacrocyclic fragments linked by two η^1-*O*-bridging tris-carboxylate donor fragments, and each of them is coordinated to three *CuN_3O*-metallocenters having a rigid square-planar geometry. TP – TAP framework of **532**, stabilized by stacking interactions, is distorted around its C_3-symmetry axis with the distortion angle φ of approximately 58° [33].

Dimerization of a tris-ethynyl-containing organocobalt precursor *284* in the presence of

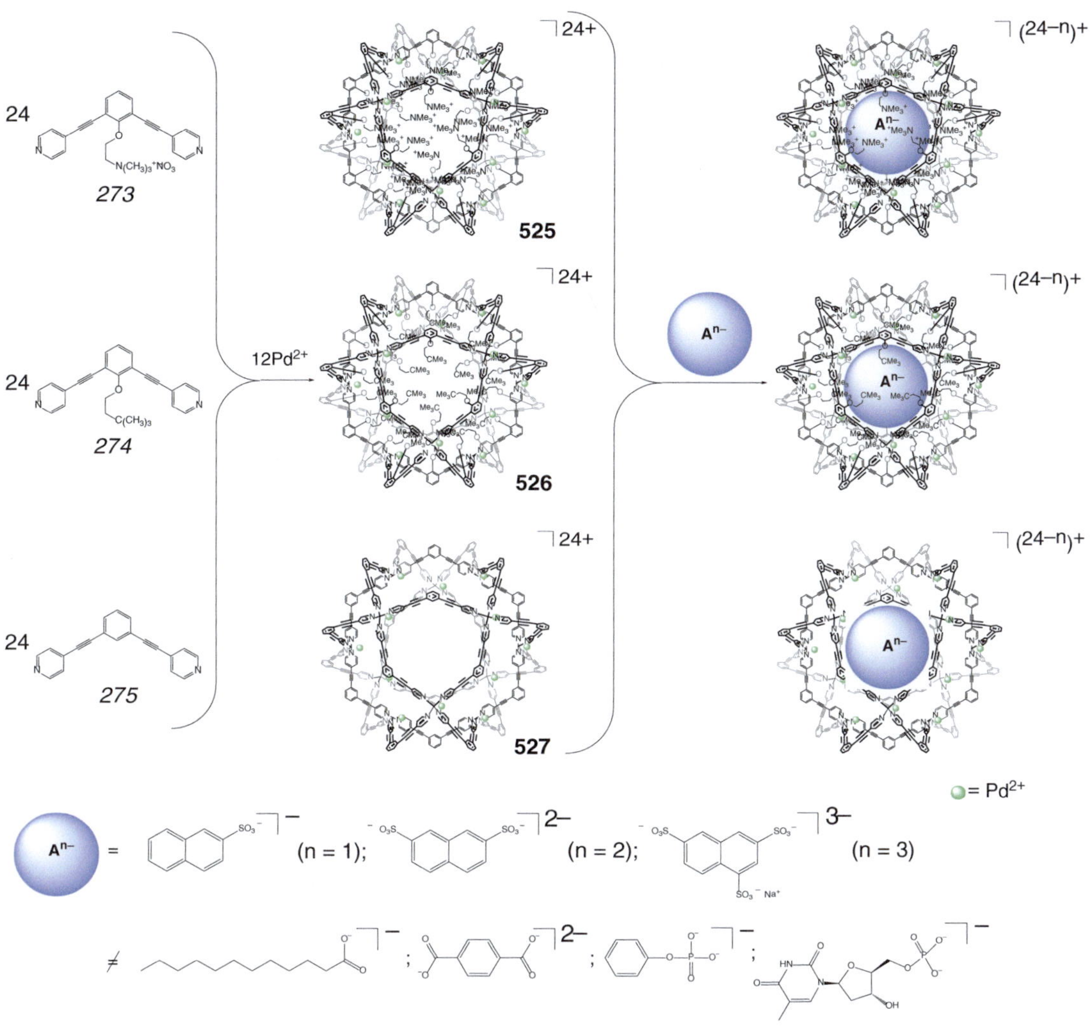

Scheme 4.29

copper(II) acetate and copper(I) chloride by Scheme 4.33 gave the hexanuclear TP capsule **535** [34].

Coordination-driven self-assembly of the ligand syntone *241* with organoruthenium precursors by Scheme 4.34 has been used in [35] for the synthesis of ruthenium-based TP hexacationic capsules **536–539**. Their assembly in the presence of large aromatic molecules, which is also shown in this Scheme, gave the corresponding 1:1 cage complexes. As follows from single-crystal X-ray diffraction data, the capsule **536** encapsulates two pyrene or benzo[e]pyrene guests that form stacking interactions with the aromatic rings of its caging ligand at the distance of 3.42 Å. The average Ru… Ru distance is 7.93 Å for the pillar fragments and becomes 13.2 Å for the trigonal bases of this TP framework. The cavity volumes of the hexanuclear coordination capsules **536–539** are approximately 700 Å^3 [35].

Self-assembly of a diruthenium precursor *285* as a bimetallic connector with tris-pyridyl ligand syntone *241* in the presence of silver(I) ions by Scheme 4.35 afforded a cationic TP (arene)ruthenium $Ru_6{}^1L_3{}^2L_2$ coordination capsule **540** [36]. This capsule encapsulates palladium and platinum(II) acetylacetonates to give the corresponding 1:1 cage complexes [36]. In the platinum-

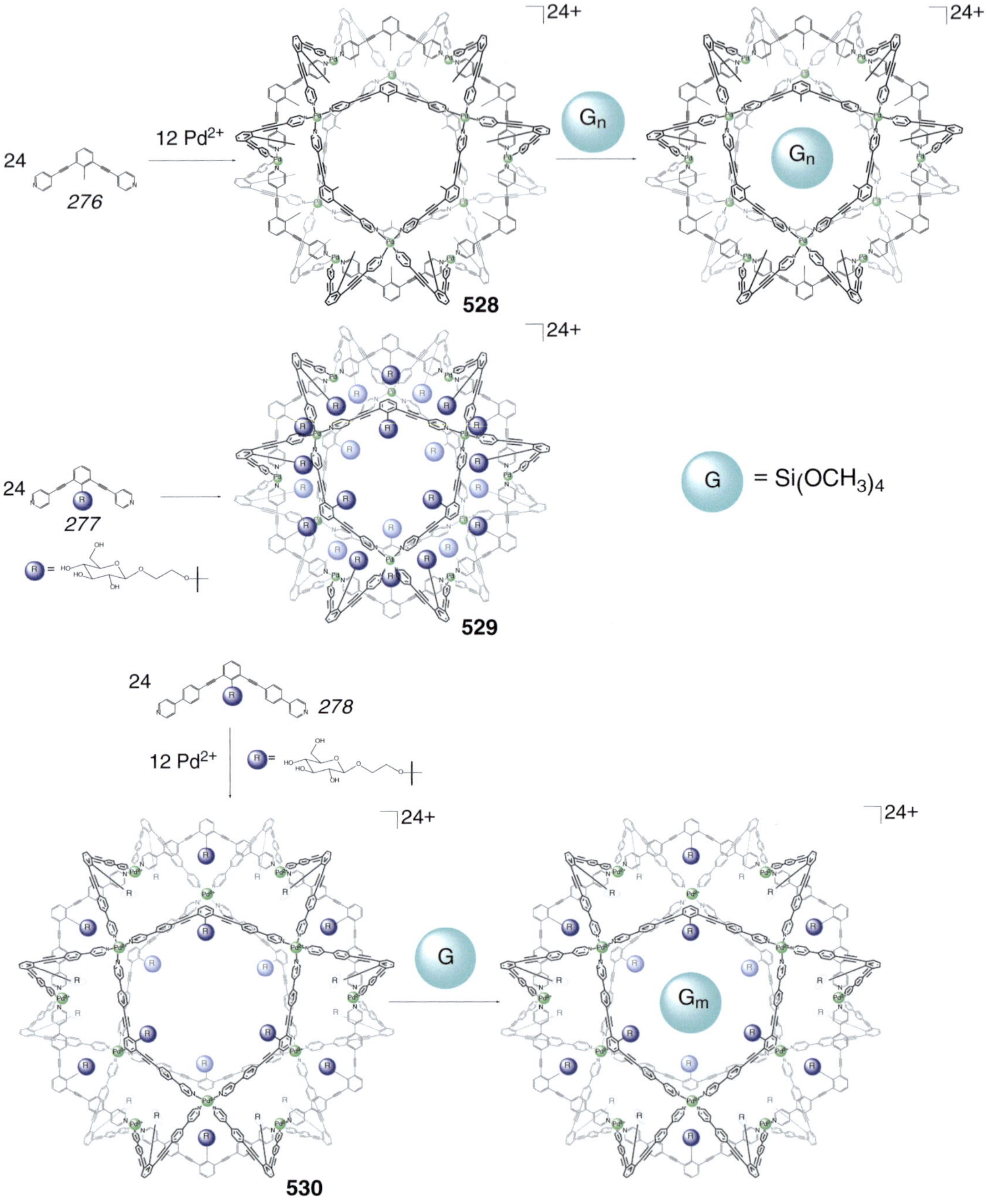

Scheme 4.30

containing complex of this type, the guest molecule is located between the planar base triazine fragments of **540**, thus forming a "*Matreshka*" complex-in-complex cage assembly [36]. The ligand **540** has been used in [37] for encapsulation of a triazine-containing pyrene derivative *286* as a fluorescent aromatic guest by Scheme 4.35; this caging process is reported in [37] to quench its luminescence. The coordination capsule **540** also efficiently binds other planar aromatic guests

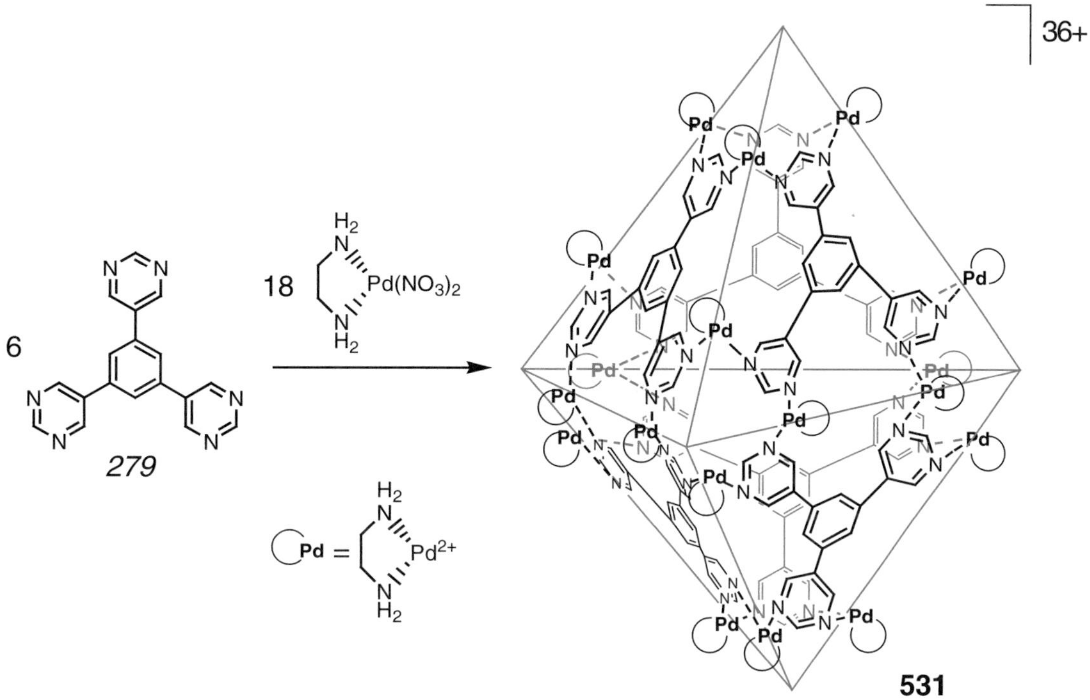

Scheme 4.31

shown in Scheme 4.35 [38]. Encapsulation of monosubstituted pyrenyl-containing guest proceeds in two steps: (*i*) the reaction of an (arene) ruthenium complex syntone with silver(I) ions to give a reactive intermediate, and (*ii*) reaction of the formed intermediate with a tris-pyridyltriazine ligand syntone and a functionalized pyrene derivative to give a cage complex (as follows from the solution ^{1}H NMR spectra). The observed upfield shift of the signals of protons of pyrene guests suggests that these encapsulated molecules are sandwiched between two tris-pyridyl panels of **540** through their stacking interactions [38].

Similar dimetallic (cyclopendadienyl)iridium linkers (connectors) have been used in [39] to obtain the flexible coordination capsules **541** and **542**. Self-assembly of six organometallic capping fragments, two tridentate paneling tris-pyridyl ligand syntones and three *Z-like* pillaring syntones by Scheme 4.36 afforded a mixture of a symmetric caging ligand **541** with the same orientation of methyl substituents and its non-symmetric analog **542**. In the molecule of **541**, its parallel aromatic panels are situated very close each to other on the distance of approximately 3.3 Å due to strong stacking interactions between them, and this causes a substantial (by 33°) deviation of the iridium-based metallocenters from their mean planes. As follows from ^{1}H NMR data, these capsules encapsulate aromatic guest molecules by Scheme 4.36 with a "closing-to-opening" conversion of the corresponding cage frameworks, allowing to accommodate such guests as a result of the face-to-face and edge-to-edge interactions with an interior of their cavities [39]. In particular, in the molecule of 1:1 cage complex of **541** with an encapsulated coronene, this guest forms face-to-face stacking interactions with 1,3,5-triazine bases at the distance of 3.3 Å. As a result, the distance between them increases up to 6.8 Å, while the deviation of iridium metallocenters goes down to 13°. At the same time, the caging ligands **541** and **542** are described in [39] do not encapsulate more bulky aromatic molecules such as hexamethoxytriphenylene.

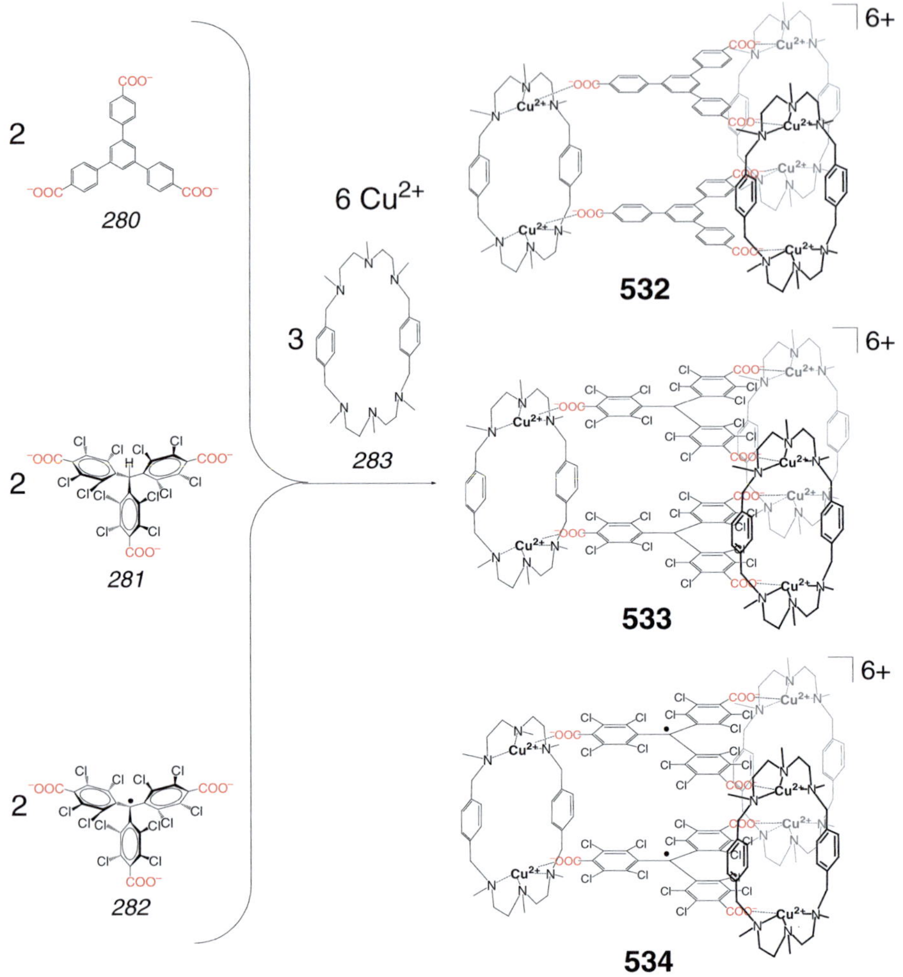

Scheme 4.32

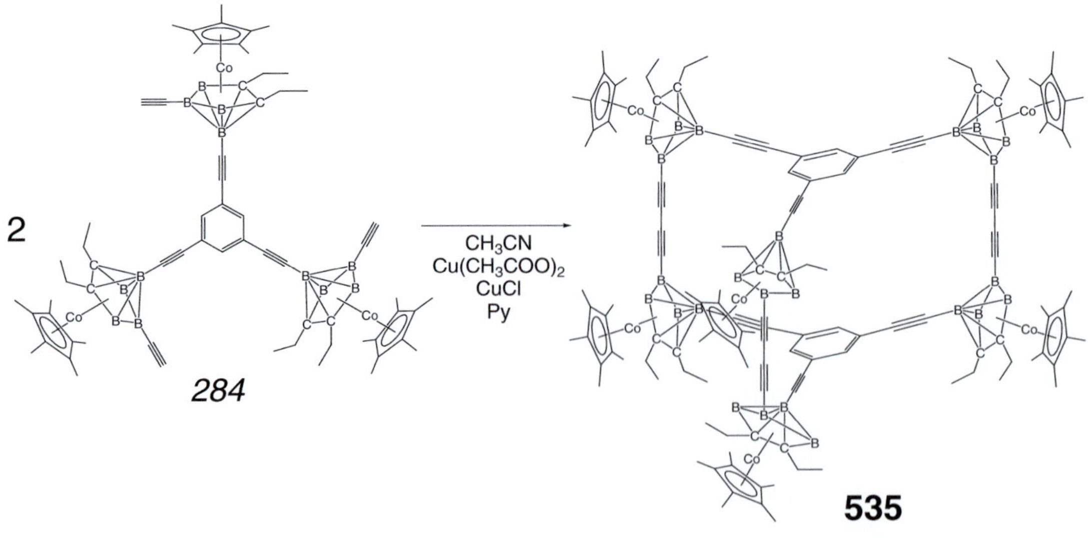

Scheme 4.33

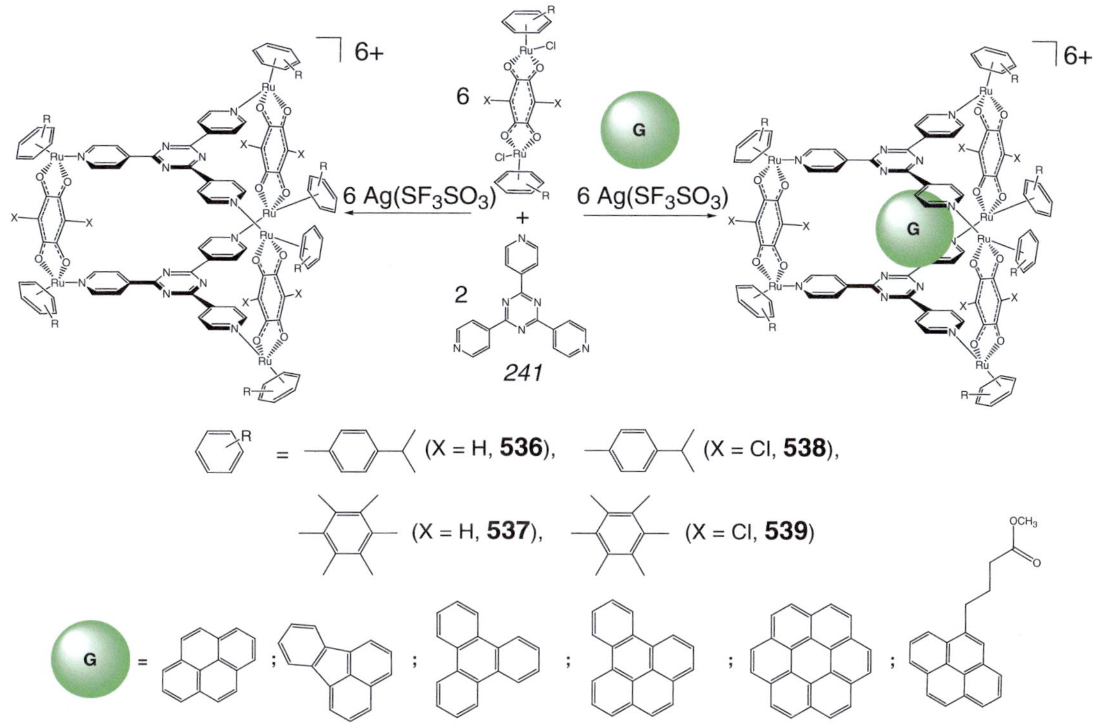

Scheme 4.34

Guest-template effect of hexamethoxytriphenylene molecules in the formation of a pyrazine-pillared TP capsules **543** and **544** has been observed in [40]. According to the ^{1}H NMR data, these ligands forms the corresponding 1:1 cage complexes by Scheme 4.37. In their D_{3h}-symmetric molecules, the encapsulated aromatic guest is oriented within the ligand's cavity to maximize its stacking interactions with the aromatic fragments of the corresponding caging ligand. Extraction of hexamethoxytriphenylene from such cage complexes gave the free coordination capsules **543** and **544**, which efficiently encapsulate other aromatic guests shown in Scheme 4.37. The ligand **544** has been also used to control an equilibrium between planar and non-planar tautomers of these aromatic guests. In particular, only the enol form of the corresponding *β*-diketone *287* is found in [40] to be encapsulated by this caging ligand.

Encapsulation of minimal hydrogen-bonded nucleotide pairs (duplexes) by the ligand **544** in aqueous solution has been studied in [41] using ^{1}H NMR spectroscopy. In the heteroguest 1:1:1 cage complex shown in Scheme 4.38, nucleobase fragments of the encapsulated nucleotides are located within the cavity of the coordination capsule being encaged by its aromatic ribbed panels. The ribose moieties of these co-guests remain outside the cage framework of **544**. The ^{1}H NMR signals of its protons are substantially broadened as compared with those of the corresponding homoguest cage complexes also shown in this Scheme. Thus, the duplex guest is described in [41] to give a more stable coordination capsule with this ligand than its homodimeric analogs. All these nucleotide anionic species form hydrophobic, stacking, and electrostatic interactions with the cationic aromatic host, but only in the case of their heterodimeric pair formation of the hydrogen-bonded duplex caused the additional stabilization of the corresponding cage complex. The X-ray diffraction data of [41] for the cage complex of its *S*,*S*-1,2-diaminocyclohexane analog **545** (Scheme 4.39) confirmed the *anti*-Hoogsteen hydrogen bonding between the encapsulated

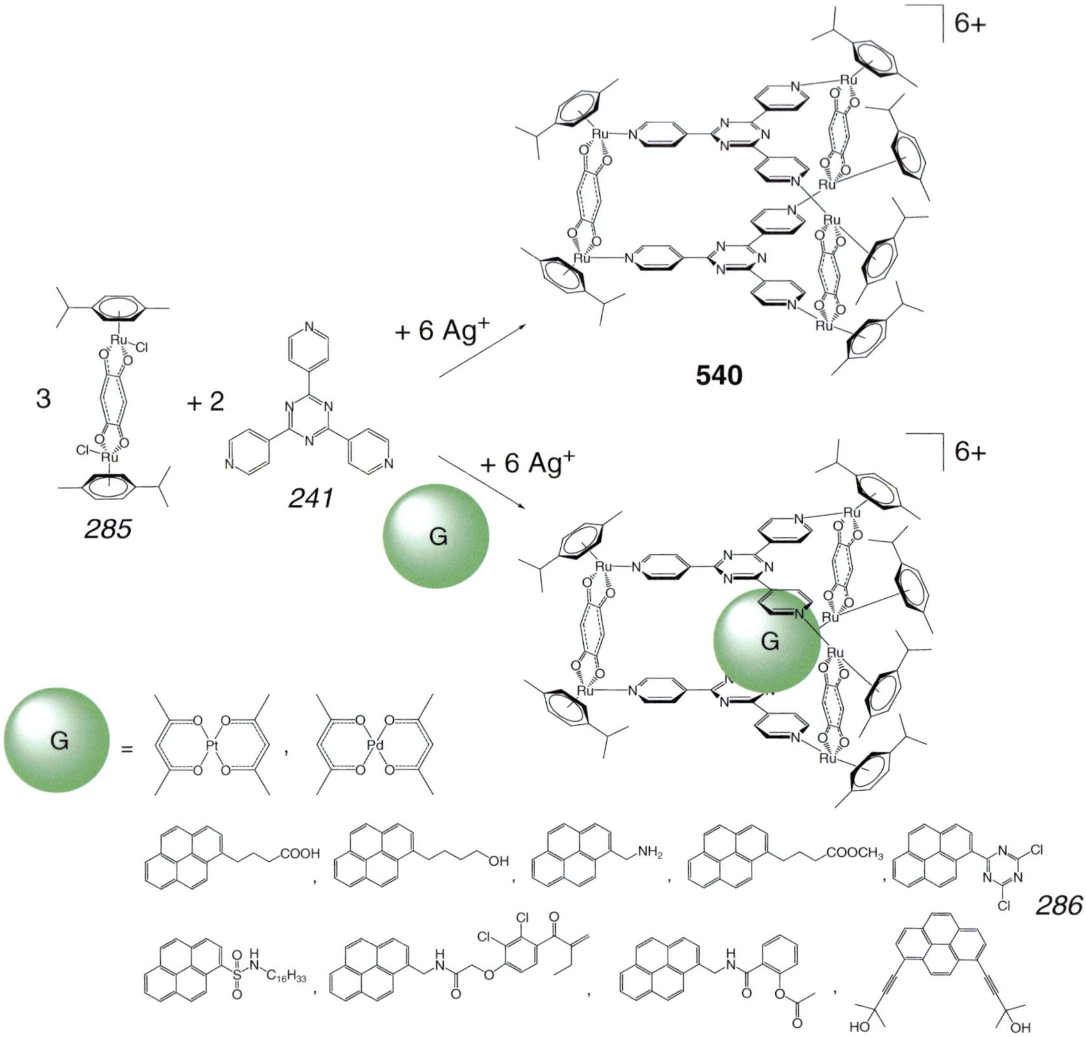

Scheme 4.35

nucleotides. Stacking interactions of a planar *A–U* base-pair guest with triazine fragments of this caging ligand stabilize their heteroguest 1:1:1 cage complex; the formation of an *anti*-Hoogsteen-type nucleobase pair is also confirmed by ^{15}N NMR experiments for the ^{15}N-labeled nucleotides. Encapsulation of dinucleoside monophosphate duplexes in their aqueous solutions by the caging ligand **546** with an extended cavity (Scheme 4.40) has been also studied in [41]. In this case, thymidylyl-(3′-5′)-2′-deoxyadenosine forms the corresponding host–guest 1:2 cage complex. As follows from ^{1}H NMR data, two caged dinucleotides adopt U conformations within its cavity, thus giving two *A–T* hydrogen-bonded base pairs. Formation of these pairs has been confirmed in [41] by X-ray diffraction data for the cage complex of the

Scheme 4.37

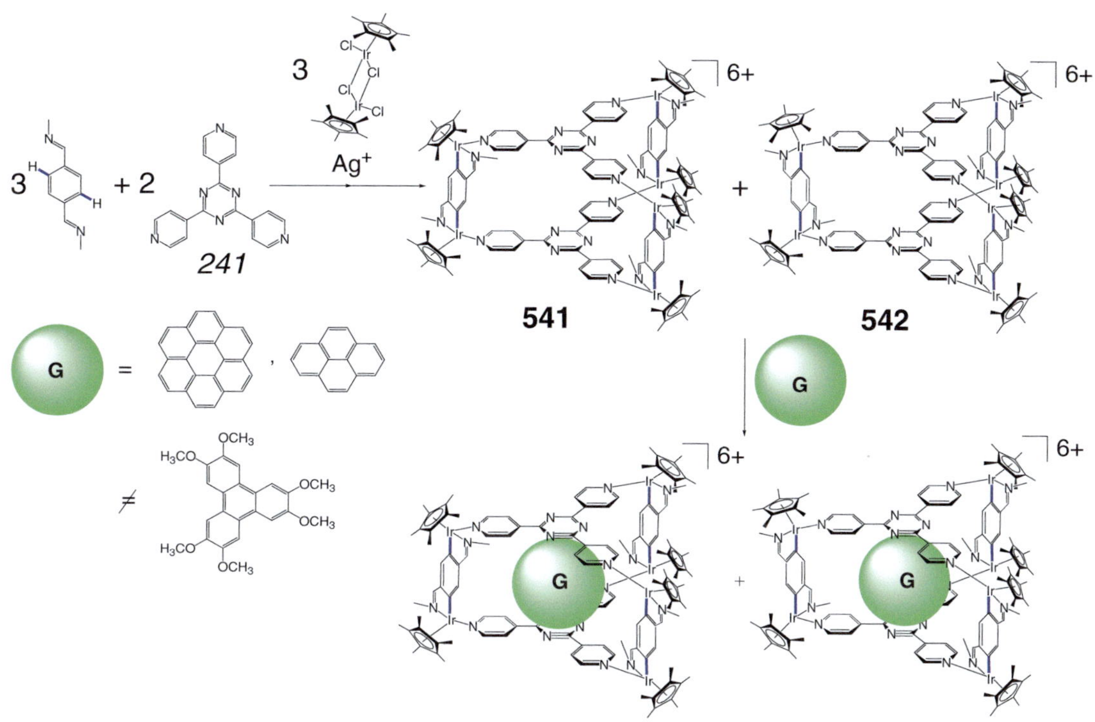
3
3
+ 2
241
Ag+
6+
541
542
+
G
=
≠
G

Scheme 4.36

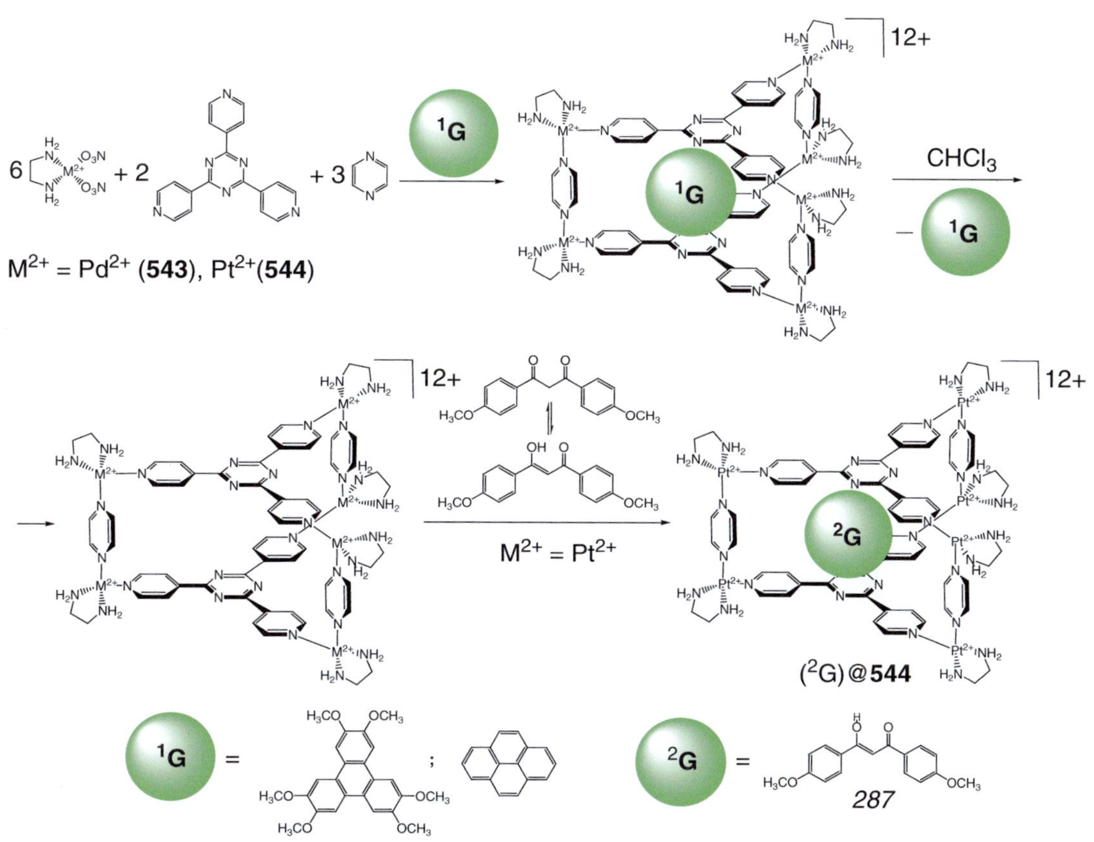
6
+ 2
+ 3
M^{2+} = Pd^{2+} (**543**), Pt^{2+}(**544**)
^{1}G
12+
CHCl3
− ^{1}G
M^{2+} = Pt^{2+}
(^{2}G)@**544**
^{1}G =
^{2}G =
287

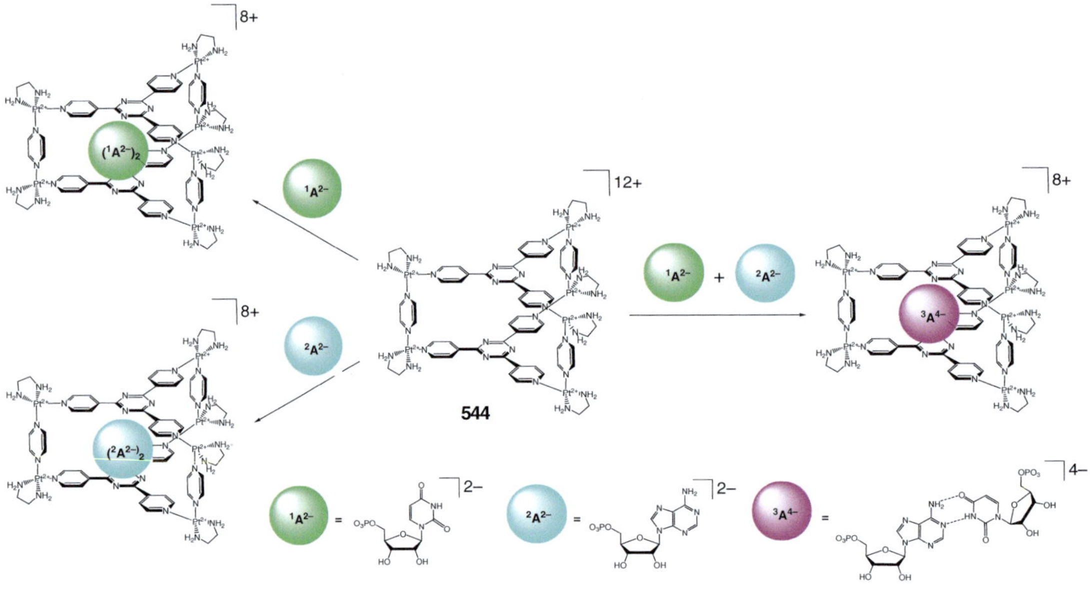

Scheme 4.38

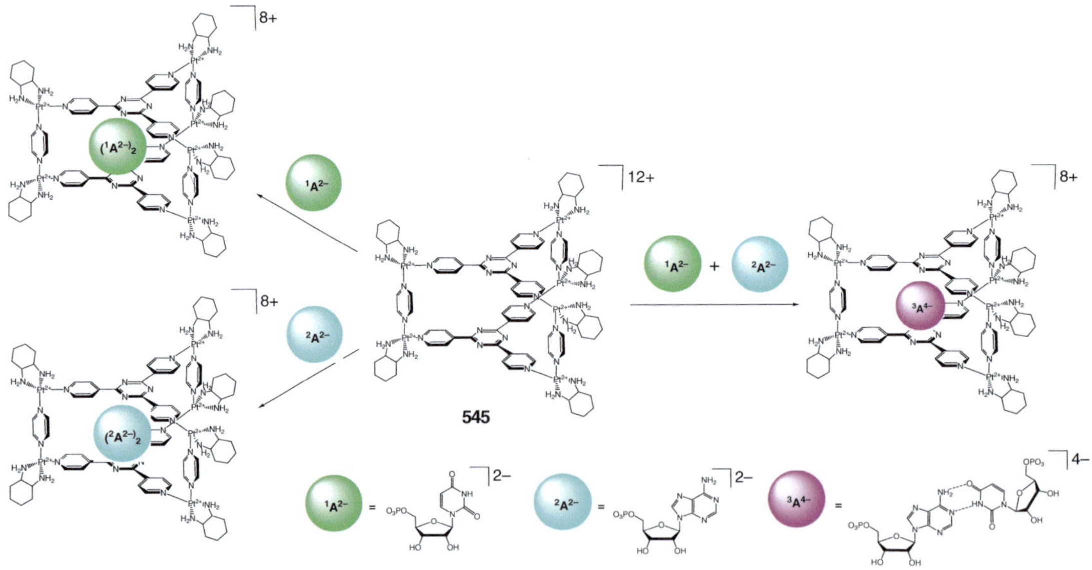

Scheme 4.39

S,*S*-1,2-diaminocyclohexane coordination capsule **547** (Scheme 4.41), which showed the *anti*-Hoogsteen-type pairing between its encapsulated *A–T* species that also form stacking interactions with the triazine bases of this caging ligand.

Duplex binding of the complementary *G–C* nucleotides by the TP coordination capsule **544** (Scheme 4.42) has been detected in [42] using ^{1}H NMR method in aqueous solutions of disodium 5′-guanosine monophosphate and disodium 5′-cytidine monophosphate. In their heteroguest 1:1:1 cage complex, the encapsulated nucleobase fragments are shielded by the aromatic panels of the caging host. Its *S*,*S*-1,2-diaminocyclohexane-

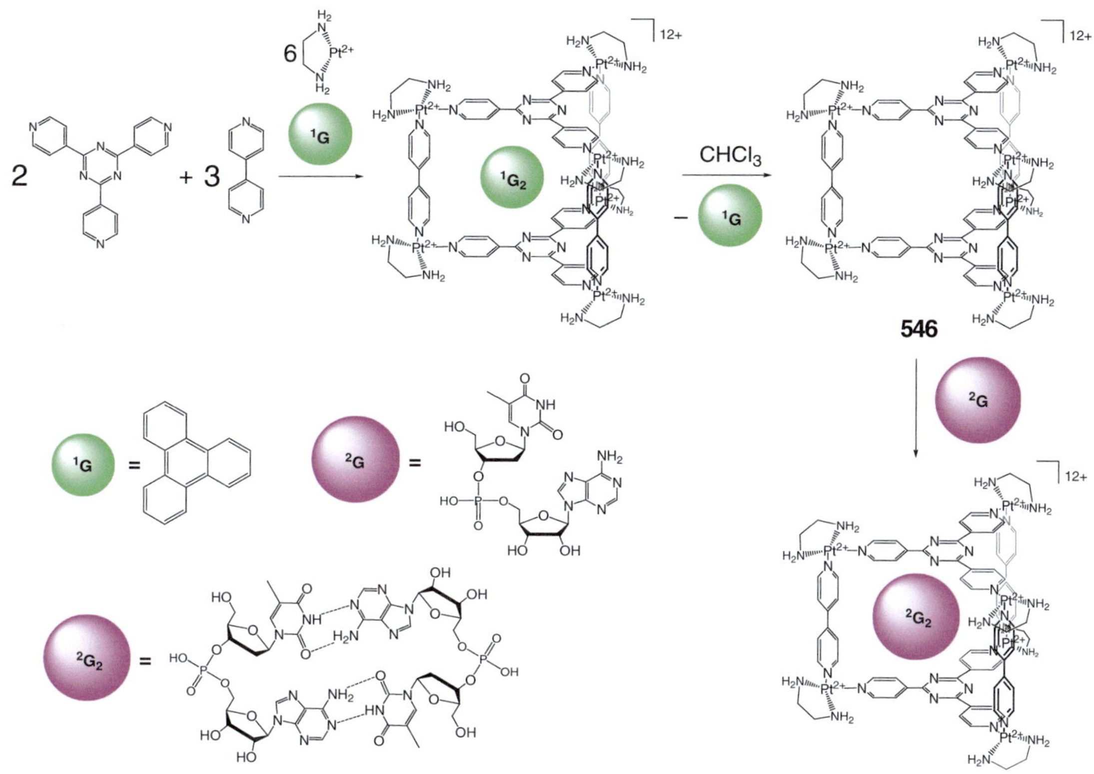

Scheme 4.40

containing analog **545** (Scheme 4.43) encapsulates these complementary guests with the Watson–Crick hydrogen bonds between them; the *G–C* base pair stacks with the electron-deficient triazine bases of the coordination capsule **545** stabilizing their cage complex. The hydrophobic cavity of **544** also contains one nitrate anion and one water molecule. As a result, the caged nucleobase pair is hydrated as in a DNA structure and retains its selectivity and stability in aqueous solutions even without templating DNA helix or additional stabilizing interactions [42]. In the absence of the complementary nucleotide, each of these nucleotides gives the corresponding homoguest 1:2 cage complexes. The first one is more stable owing to its extended aromatic fragment forming strong hydrophobic and aromatic–aromatic interactions between their electron-rich guanine moieties and the electron-deficient triazine bases of the caging ligand. This caused host–guest CT bands to appear in the UV-vis spectrum of an adenosine-containing complex, which are absent in the spectrum of its cytidine-containing analog; this suggestion has been confirmed in [42] by the competition ^{1}H NMR experiments.

Tetraazaporphine *288* is reported in [43] to form a 1:1 cage complex with a TP coordination capsule **544** as host (Scheme 4.44). This caged guest, in contrast to the free molecule, exhibits strong fluorescence within the highly cationic caging ligand; the encapsulation also enhanced the acidity of its interior protons, and addition of triethylamine quenched its emission by deprotonation. Unlike the aromatic hydrocarbons, an electron-deficient macrocyclic tetraazaporphine guest does not form a CT complex with **544** [43]. The reactivity of analogous palladium(II) ethylenediamine-capped capsules **548–550** also shown in this Scheme is described in Chap. 5.

The verdazyl tris-pyridyl radical syntone *257* has been used in [44] for the guest-templated synthesis of a host–guest 1:1 complex of TP

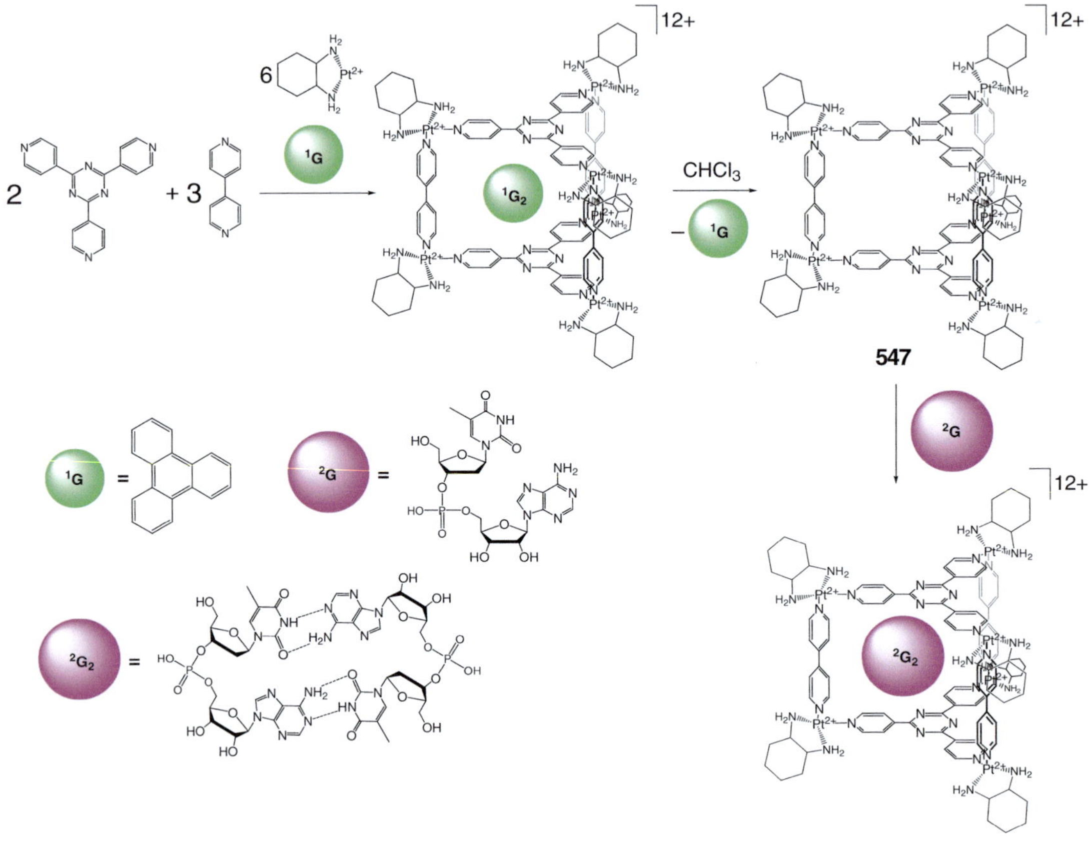

Scheme 4.41

caging ligand **551** by Scheme 4.45 on the triphenylene guest molecule as a matrix. In this coordination capsule, the parallel verdazyl-radical bases showing an EPR-detected magnetic interaction at the distance of 6.8 Å are tilt by 120°. The caging ligand **551** also encapsulates planar copper(II) β-diketonates *289* and *290* (Scheme 4.45) to give the corresponding 1:1 cage complexes with through-space spin–spin interactions [44].

Guest-template self-assembly of tris- and bis-pyridyl ligand syntones *241* and *291* with palladium(II) ions by Scheme 4.46 on the corresponding aromatic molecules as a matrix afforded 1:1 cage complexes of the heterotopic $M_6{}^1L_2{}^2L_3$ ligand **552** with encapsulated coronene and pyrene molecules in quantitative yield [45]. These D_{3h}-symmetric coordination capsules have been characterized by NMR and CSI-MS methods; the same coordination-driven but template-free reaction gave the mixture of **552**, a homotopic $M_6{}^1L_4$ capsule **483** and some oligomeric by-products. 1:2 cage complex of **552** with pyrene has a quadruple-stacking structure with the efficiency of stacking interactions governed by a substantial (approximately 36°) distortion of its cage framework. Use of an extended bis-pyridyl ligand syntone *292* allowed obtaining the coordination capsule **553** with an expanded cavity by Scheme 4.47 [45]. This ligand accommodates two molecules of coronene and one molecule of tris-pyridyl syntone *241*, thus forming a heteroguest 1:2:1 cage complex. Efficient *A–D–A–D–A* stacking within its cavity is explained by the donor–acceptor CT interactions between the electron-deficient *241* and two electron-rich molecules of coronene. The even-odd-number effect of this stacking appears in the UV-vis spectra: the

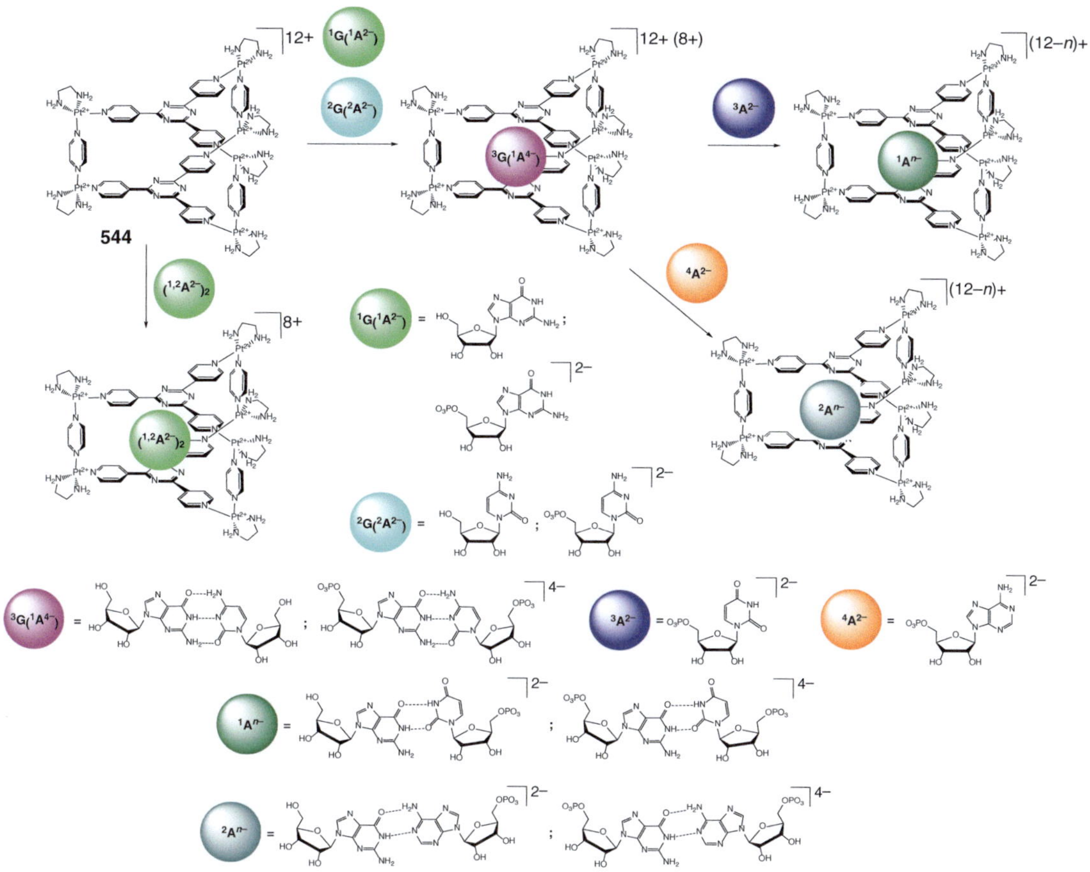

Scheme 4.42

corresponding CT bands undergo a longwave shift going from the *A–D–D–A* cage complexes to their *A–D–A*-containing analogs [45]. The caged pyrene molecule also forms strong stacking interactions with a TP ligand **552**, and the methoxylated pyrene derivatives have been used in [46] as appropriate guests to obtain host–guest 1:2 cage complexes by Scheme 4.46; their chemical reactivity within the cavity of **552** is described in Sect. 5.1.

A $Pd_6{}^1L_2{}^2L_3$ caging ligand **552** encapsulates planar platinum, palladium, and copper(II) acetylacetonates by Scheme 4.48 to give host–guest 1:2 cage complexes in quantitative yields [47]. The metal–metal interactions between their efficiently stacked guest molecules have been studied in solution using UV-vis and ^{195}Pt NMR spectroscopies and in solid state by X-ray diffraction. In the platinum-containing cage compound, two platinum(II) ions are located on the C_2-axis of the stacked complex dimer; the Pt…Pt distance of approximately 3.32 Å is characteristic of the platinum(II)…platinum(II) *d–d* interactions (*ca.* < 3.5 Å). Its encapsulated complex molecules are tilt by 24.8° to decrease steric hindrances between their methyl substituents. Square-planar guest molecules of $Cu(acac)_2$ in its cage complex with **552** showed *d–d* interactions between the copper metallocenters with $s = 1$ within the cavity of this coordination capsule; the average Cu…Cu distance of 3.6 Å is substantially longer than that in its platinum(II)-containing analog. This result has been explained in [47] by different types of their *d–d* interactions: in the case of platinum(II) acetylacetonate, those are caused by an overlap of the d_{z2} and p_z orbitals, while the Cu^{2+}…Cu^{2+}

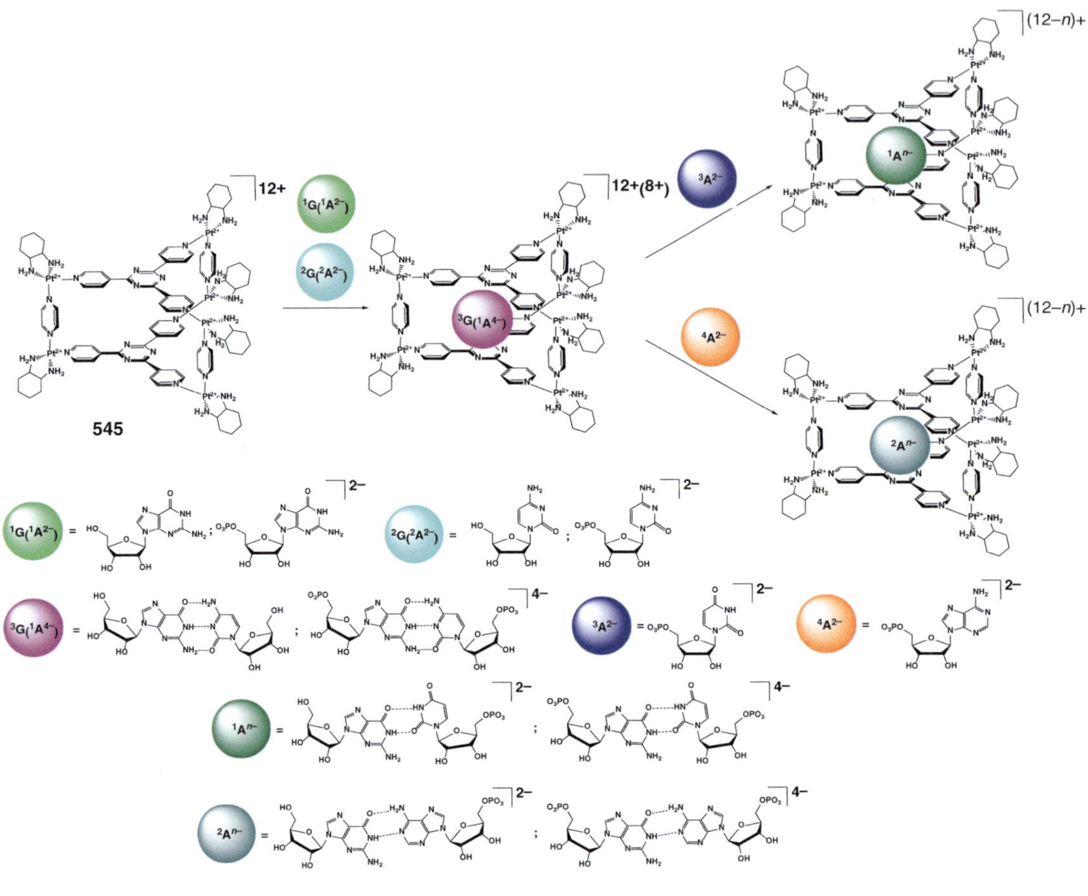

Scheme 4.43

interactions involve non-bonding stacking of two d_{x2-y2} orbitals.

The coordination capsule **552** is reported in [48] to encapsulate also two tetrathiafulvalene molecules, thus giving a host–guest 1:2 cage complex. Its UV-vis spectrum contains an intensive band assigned to the CT from electron-rich caged TTF molecules to the electron-deficient triazine bases of **552**, thus suggesting the cation-radical nature of its encapsulated guests that causes substantial broadening of their ^{1}H NMR signals [48].

The same capsule **552** has been also prepared [49] by guest-templated self-assembly on pyrene as a matrix. The methyl substituents in its pillaring ligand syntone are reported to be crucial for the self-assembly of this 1:2 cage complex with two encapsulated stacked pyrene molecules by Scheme 4.49. Use of its homolog without such substituents afforded the mixture of homotopic M_4L_4 square-planar metallomacrocycle and octahedral $M_6L'_4$ cage complex. The caging ligand **552** is also able to encapsulate a cofacial stacked porphine dimer forming a 1:2 cage complex by Scheme 4.50; its extended analog **553** is described in [50] to give a heteroguest 1:2:1 compound with triple-decker encapsulated species containing two cofacial porphine molecules that sandwich one tris-pyridyl aromatic syntone *241* (Scheme 4.50).

The host–guest and guest–guest stacking interactions have been used in [51] for the synthesis of an aromatic heteroguest 1:3:2, 1:4:2, and 1:5:2 cage complexes of double-cage coordination capsules **554**$_2$–**556**$_2$, respectively, by Schemes 4.51 and 4.52; the formed encapsulated aromatic towers provide vertical π-conjugations.

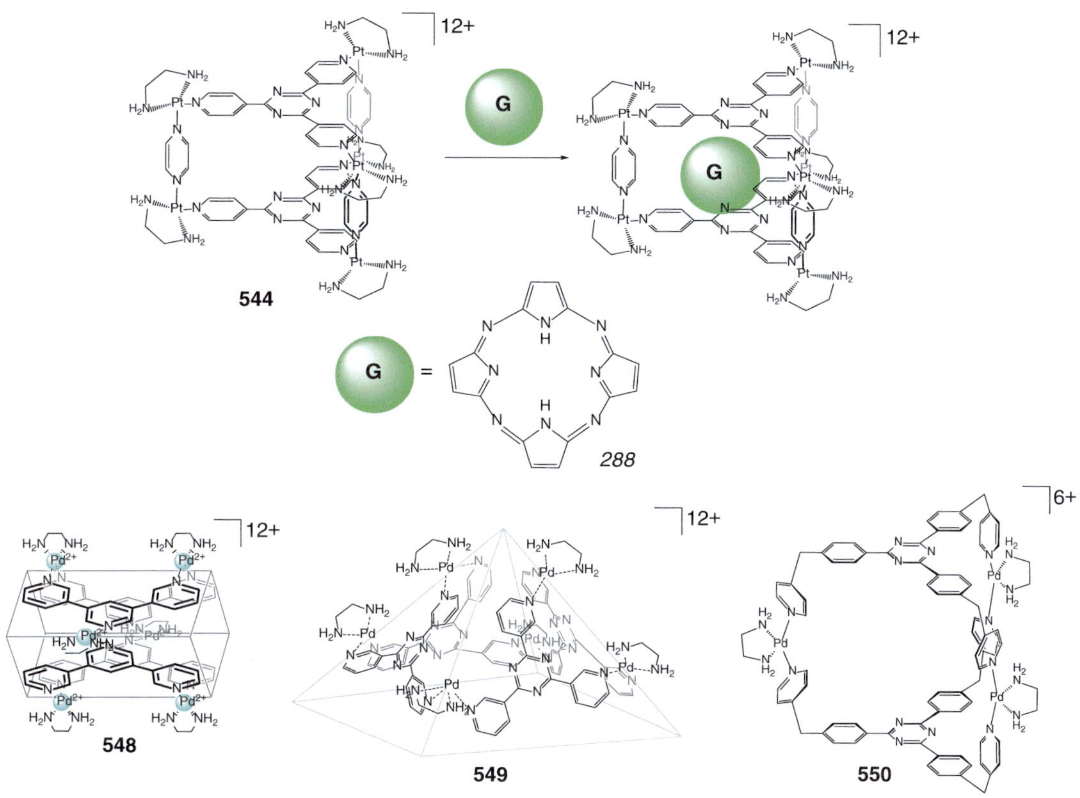

Scheme 4.44

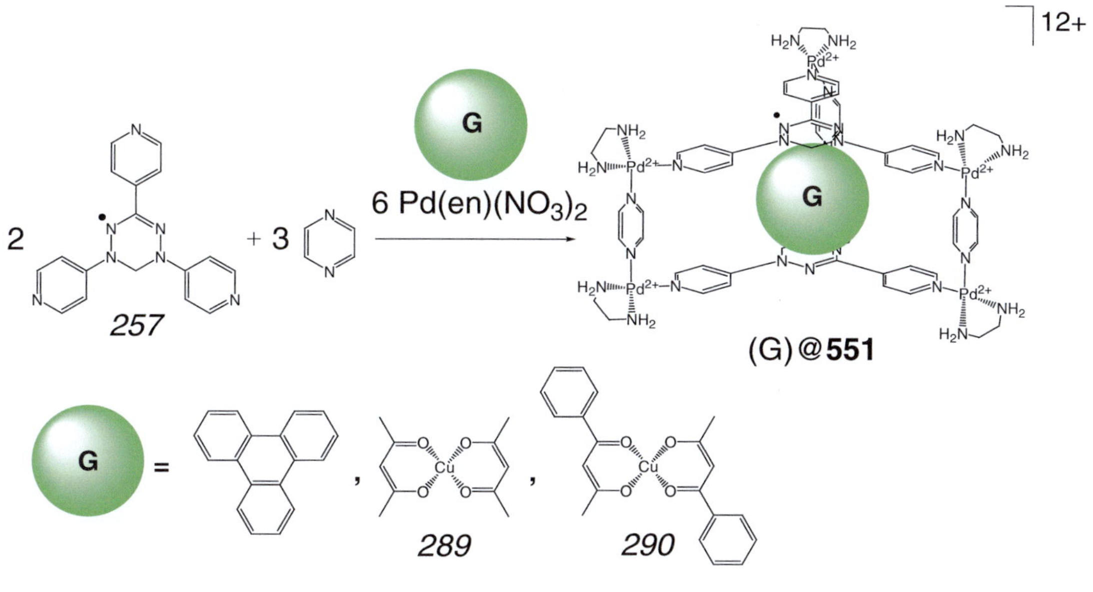

Scheme 4.45

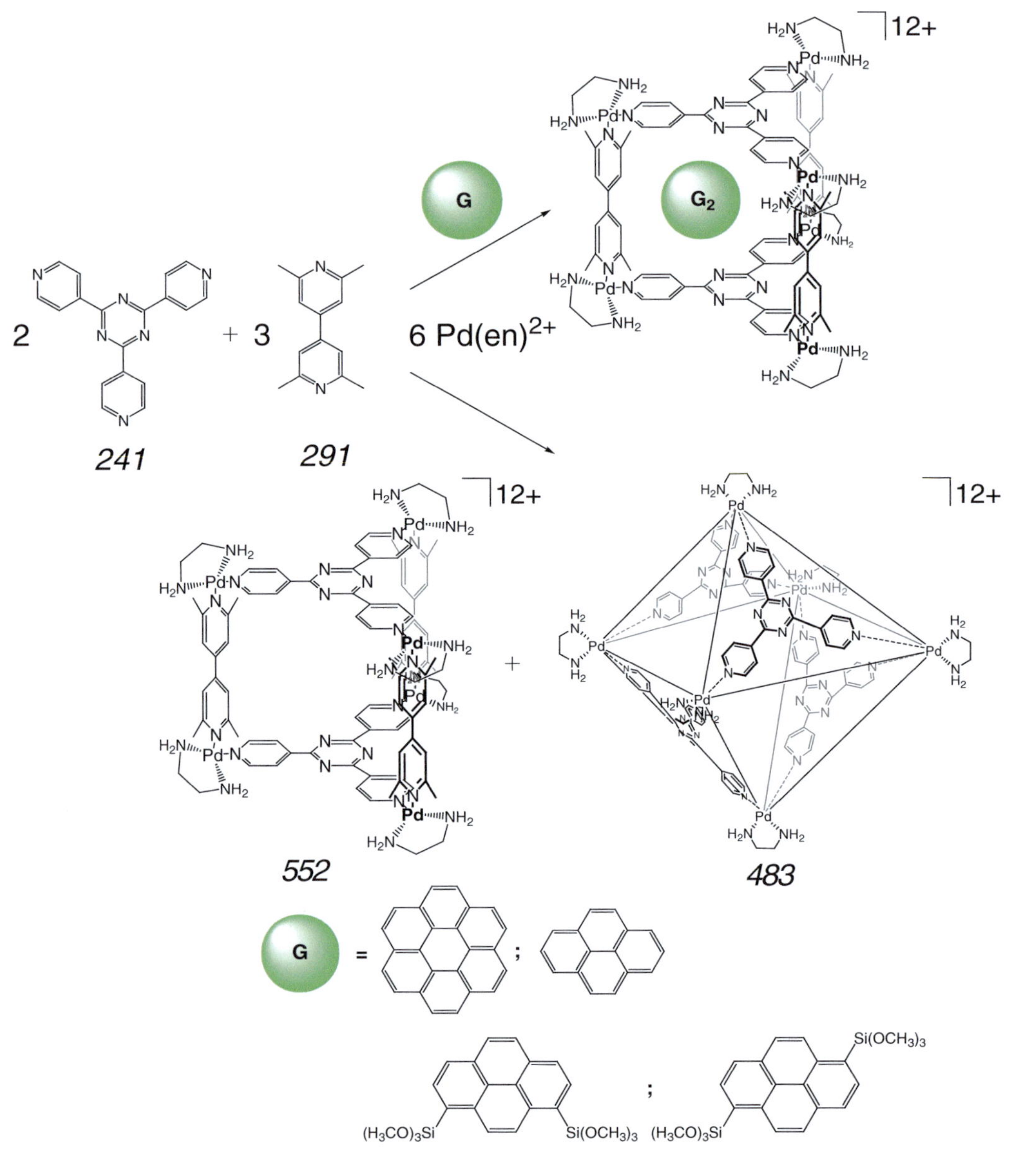

Scheme 4.46

Discrete linear Cu…Cu…Cu, Cu…Pd…Cu, and Cu…Co…Cu metal arrays have been obtained in [52] by Scheme 4.53 within the coordination capsule **553** encapsulating metal(II) azaporphine(porphine) guests (where M^{2+} are Cu^{2+}, Pd^{2+}, or Co^{2+} cations); such metal arrays showed unique magnetic interactions in the EPR spectra.

Encapsulation of square-planar *3d*-metal complexes (in particular, nickel(II) acetylacetoneiminate and cobalt(II) tetraazaporphine) by TP caging ligands **544**, **552**, and **553** (Scheme 4.54) having aromatic bases is reported in [53] to induce an SCO in these guests. This effect was explained by the confined cavity of these coordination capsules inhibiting the changes in the coordination polyhedra of *3d*-metal ions and promoting their configurational changes through metal d_{z^2}-orbital and ligand π electronic interactions.

A coordination-driven guest-templated self-assembly of rigid triazine and bis-pyridyl ligand

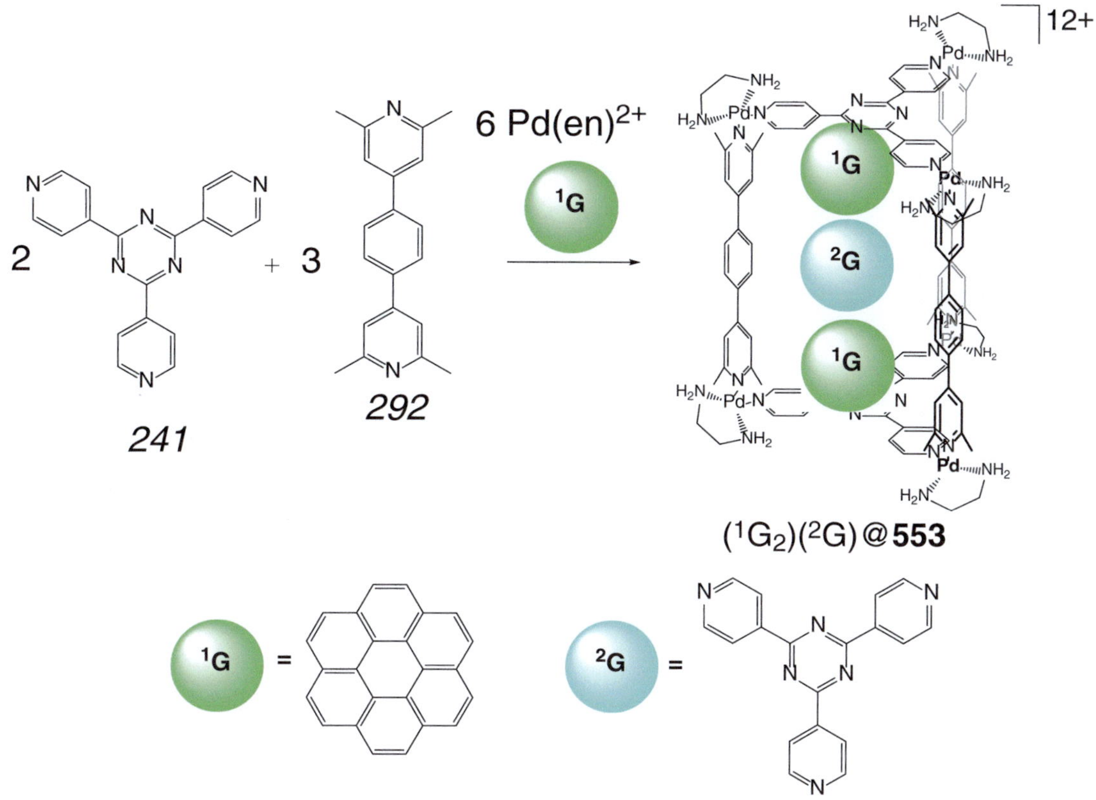

Scheme 4.47

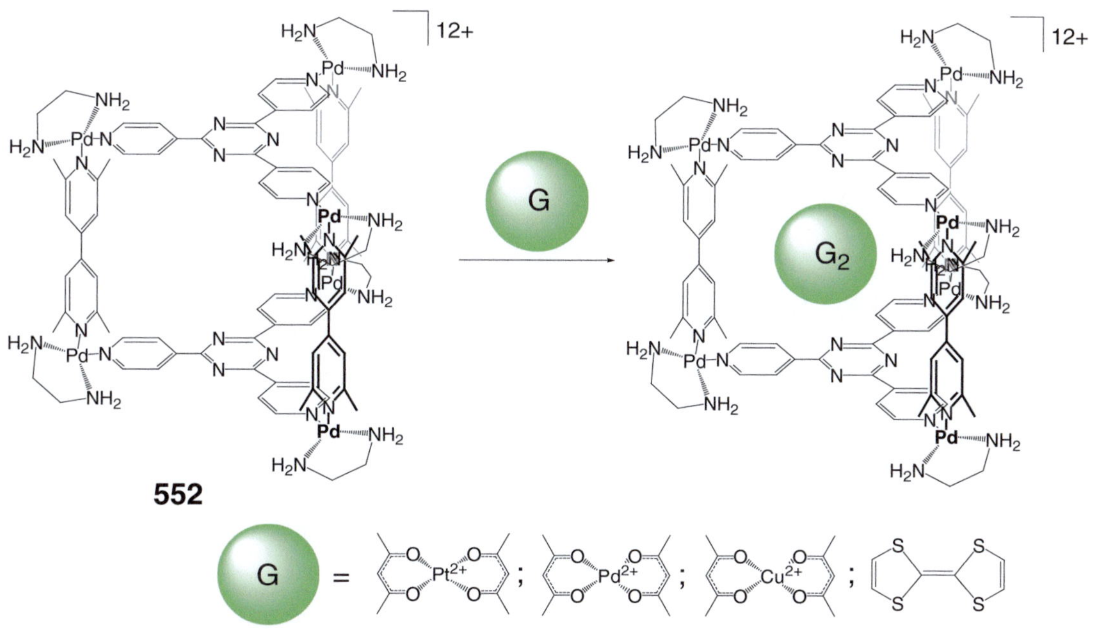

Scheme 4.48

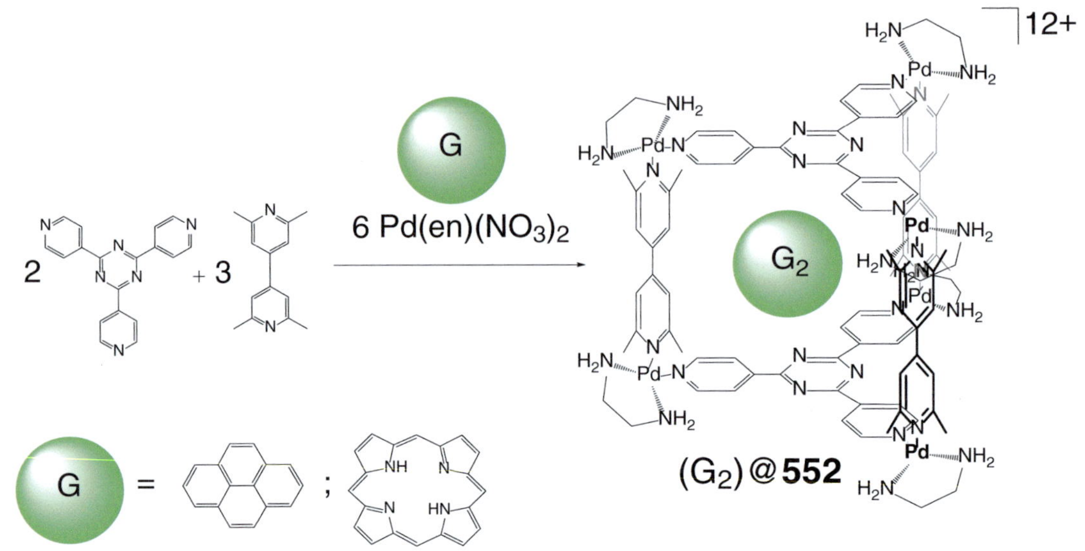

Scheme 4.49

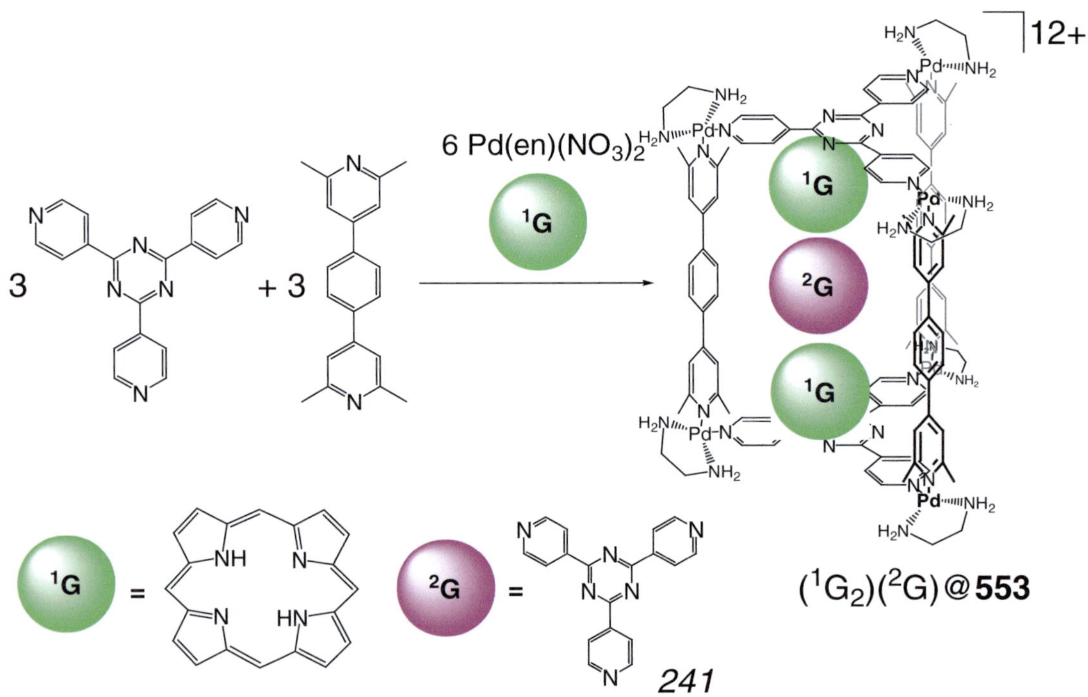

Scheme 4.50

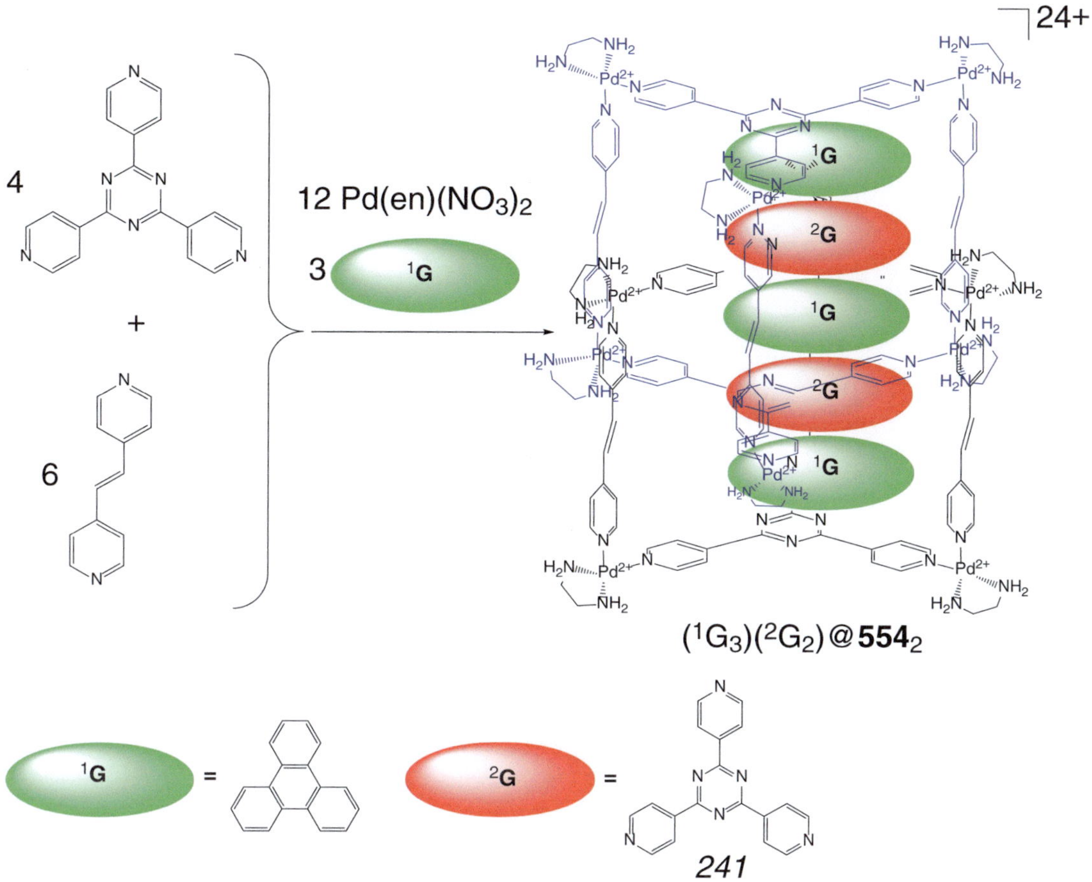

Scheme 4.51

syntones *291*, *293*, and *294* with palladium(II) ions by Scheme 4.55 in the presence of pyrene-4,5-dione as a guest with high dipole moment afforded the corresponding host–guest 1:2, 1:3, and 1:4 cage complexes of coordination capsules **552**, **554**, and **557** [54, 55]. These complexes were used to study discrete stacks of the polarized guests and their preferred conformations. The above caged molecules within these capsules are rotated by approximately 120° with respect to each other, thus canceling the net dipole moment rather than the local dipole moment [54].

Guest-induced transformation of a TP Ru_6L_6 coordination capsule **558** to its pseudoicosahedral dodecanuclear derivative **559** is reported in [56]. The coordination capsule **558** has been self-assembled from the ligand syntone *295* and the organometallic precursor *296* by Scheme 4.56. Its six ruthenium vertices are connected by six 3,5-pyridinedicarboxylate ligands, and each of them is coordinated to three different metal ions through two carboxylate oxygen atoms (the average Ru–O distance is equal to 2.10 Å) and the pyridine nitrogen atom (the Ru–N distance is approximately 2.12 Å). The vertex ruthenium metallocenters form a TAP polyhedron with the lowest Ru…Ru distances of 8.00 and 8.61 Å. This capsule, having an inversion center, encapsulates a disordered methanol molecule. Half of the carbonyl oxygen atoms of **558** are positioned in close proximity to each other at the average distance of approximately 4.25 Å, thus giving O_3-binding sites for metal cations (see insert in this Scheme). A K^+-induced rearrangement of this hexanuclear capsule gave the $Pd_{12}L_{12}$ cage framework **559** with the coordination modes similar to

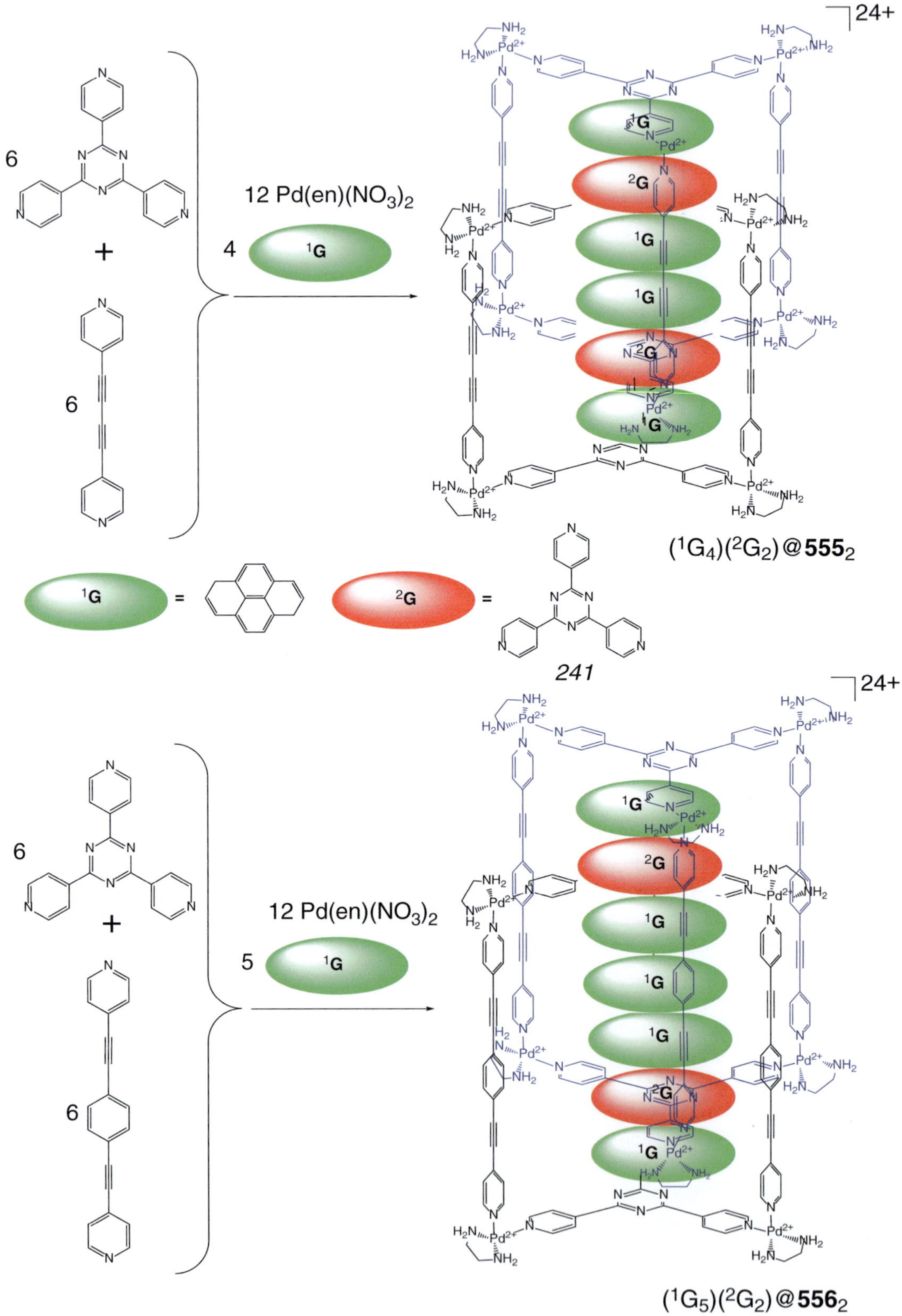

6
+
6
12 Pd(en)(NO3)2
4 1G
24+
(1G4)(2G2)@5552
1G =
2G =
241
6
+
6
12 Pd(en)(NO3)2
5 1G
24+
(1G5)(2G2)@5562

Scheme 4.52

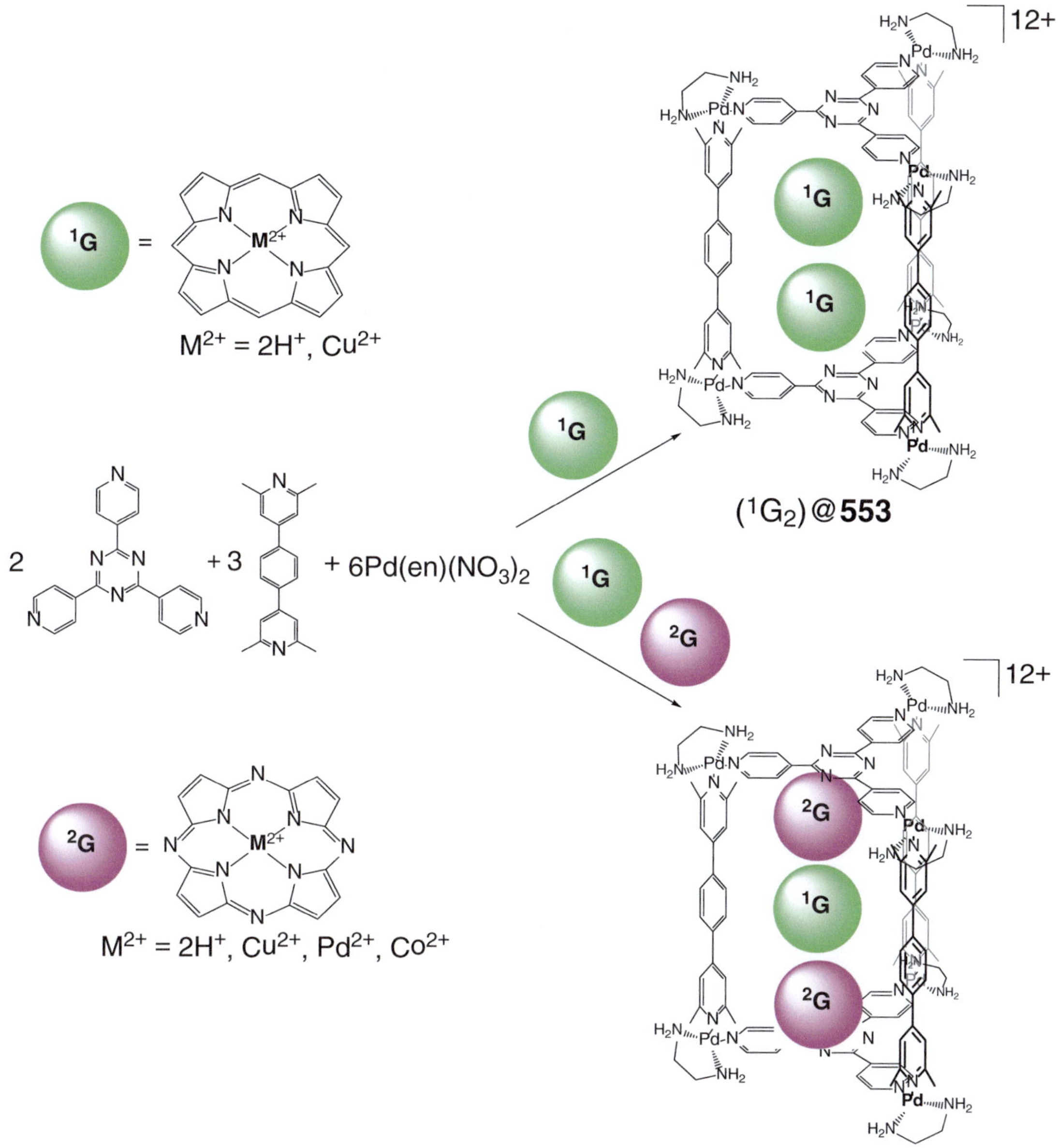

Scheme 4.53

those for **558** and also contains eight potassium(I) cations. The vertex ruthenium metallocenters of **559** form an icosahedral polyhedron with the lowest Ru…Ru distances from 8.2 to 8.7 Å. The Ru…Ru distances for symmetry-related ruthenium centers are approximately 16 Å, and those for the corresponding nitrogen atoms are equal to 13.8 Å. A cavity volume of **559**, having the outside diameter of 23.8 Å, is approximately 1100 Å^3, and this coordination capsule encapsulates 18 disordered water molecules. Twelve faces of its icosahedral cage framework are occupied by bridging 3,5-pyridinecarboxylate ligand syntones, while the remaining eight faces, having a small opening, are surrounded by three carbonyl groups. These groups form a metal binding site (like those in the case of **558**; see above) occupied by potassium(I) ions. Such a rearrangement of the

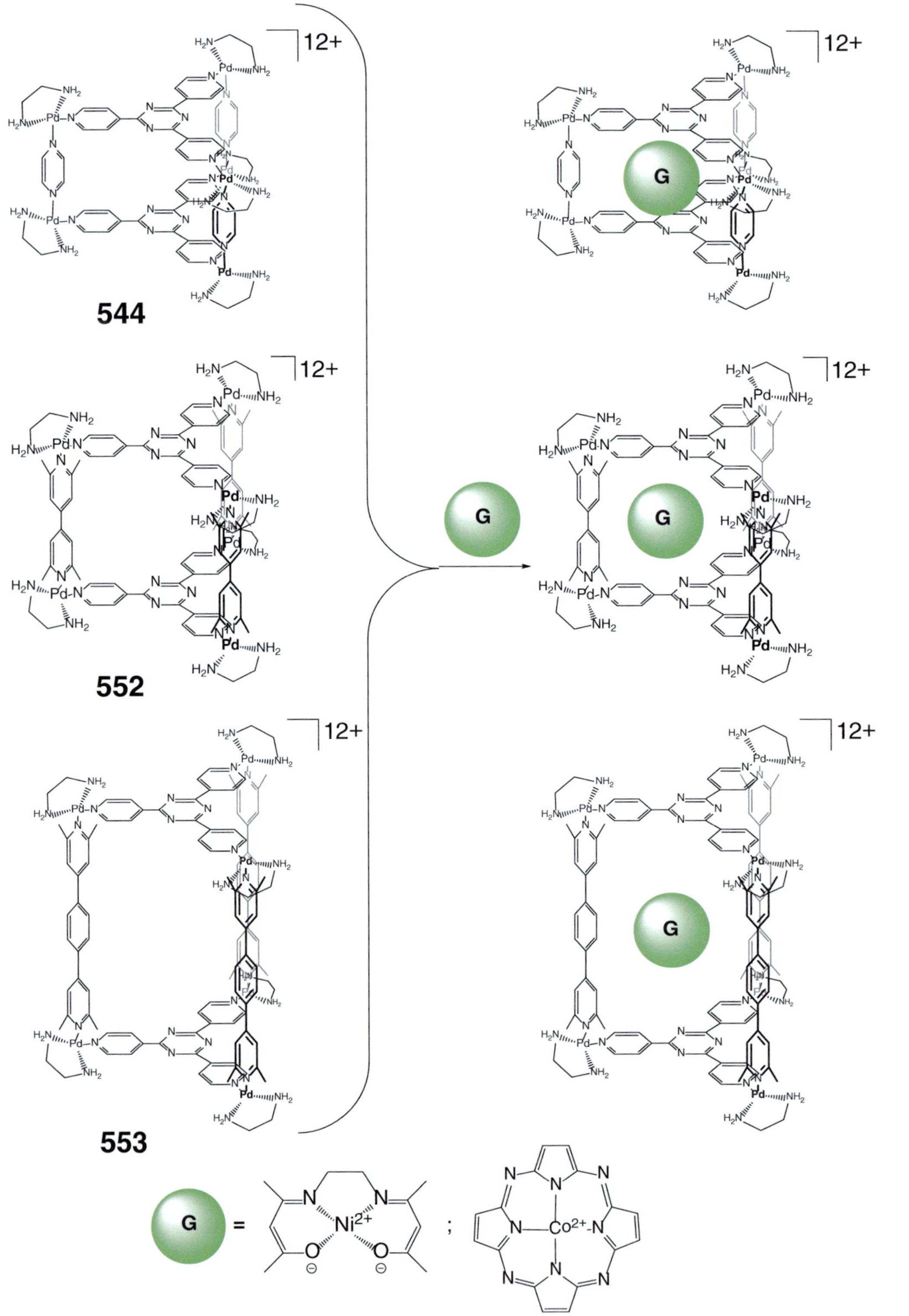

Scheme 4.54

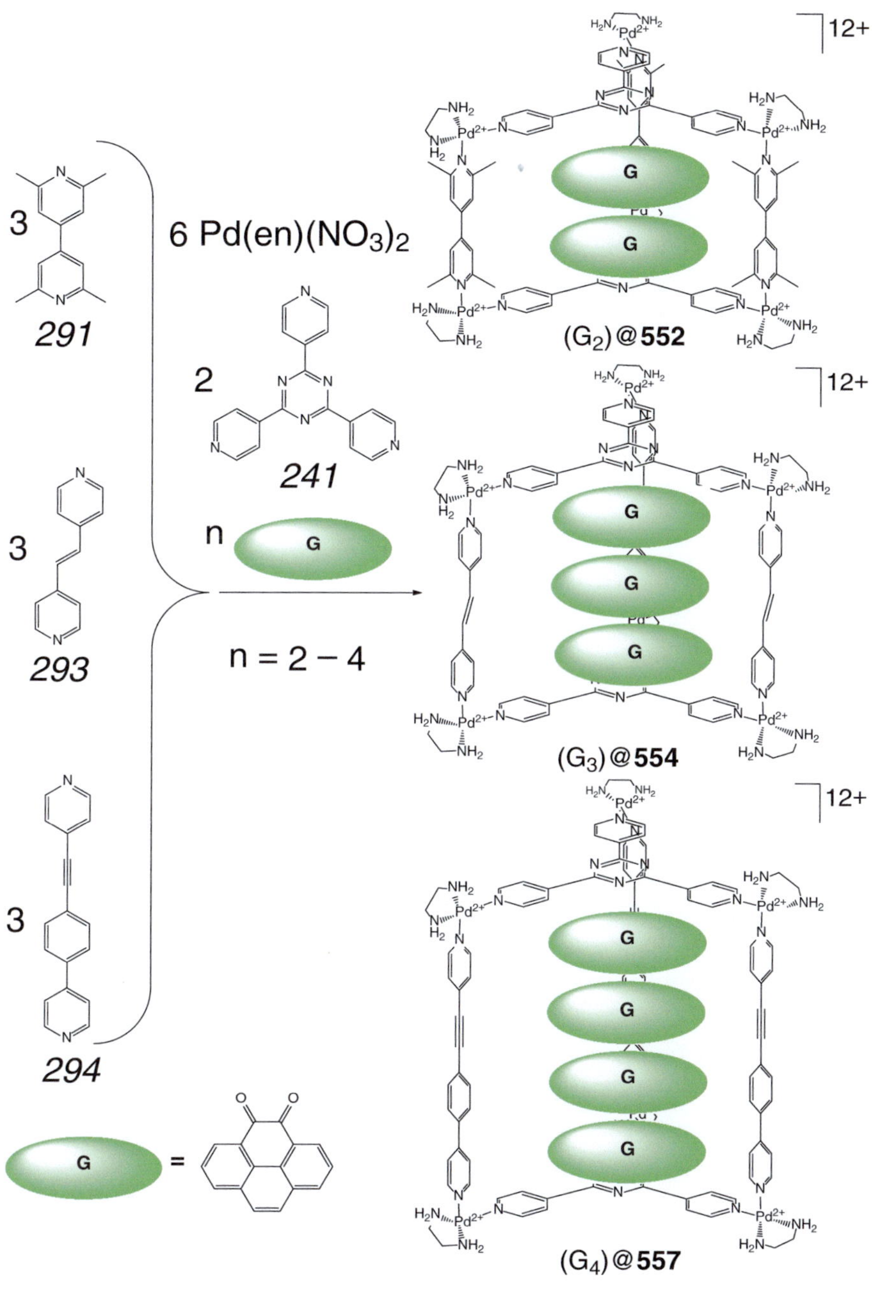

Scheme 4.55

hexanuclear capsule **558** into a dodecanuclear framework **559** is reported in [56] to be an entropically disfavored process, which is compensated by the formation of more potassium(I)–oxygen interactions in the case of **559** and by more favorable geometry of its binding sites for the coordination to alkali metal ions as compared with **558**: the O…O distances in this Ru_6L_6 caging ligand are approximately 4.25 Å, thus being rather long for K^+ and Cs^+ ions, whereas those in the dodecanuclear capsule **559** are approximately 3.8 Å [56].

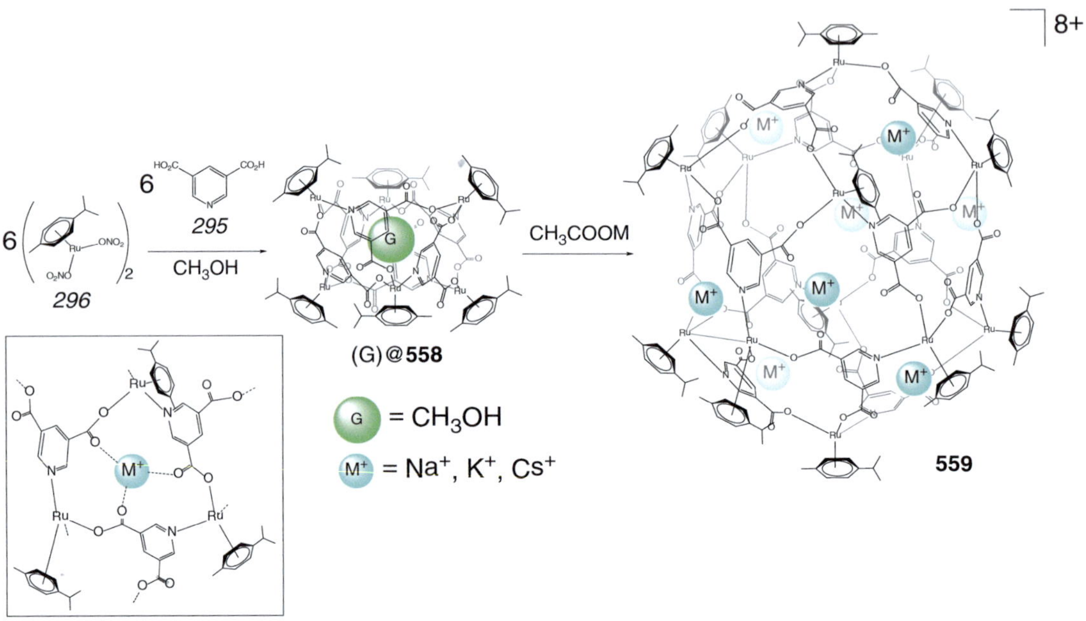

Scheme 4.56

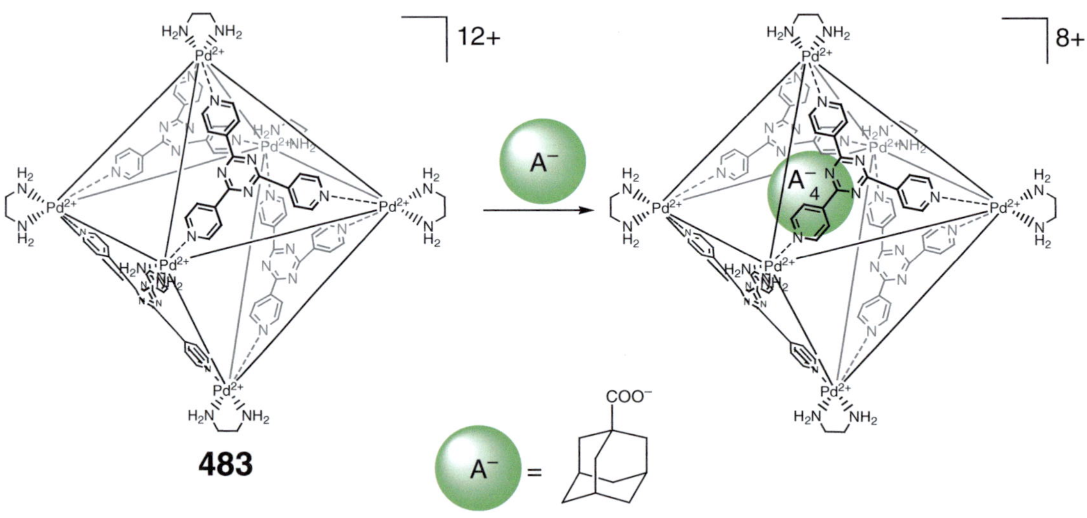

Scheme 4.57

4.1.2 Encapsulation of Anions

A 12-cationic caging ligand **483** (see Sect. 4.1.1) is described in [1] to encapsulate adamantylcarboxylate anions to give a host–guest 1:4 cage complex by Scheme 4.57. In this complex, hydrophobic carbocyclic fragments of the caged anionic species are located within the cavity of **483**, and their hydrophilic carboxyl groups are directed to the palladium-containing vertices, thus stabilizing this octahedral coordination capsule [1].

An *exo*-hexadentate ligand syntone *279* has been used in [32] for the synthesis of a trigonal bipyramidal $Pd_{18}L_6$ caging ligand **531** by Scheme 4.58. In its host–guest 1:5 complex with

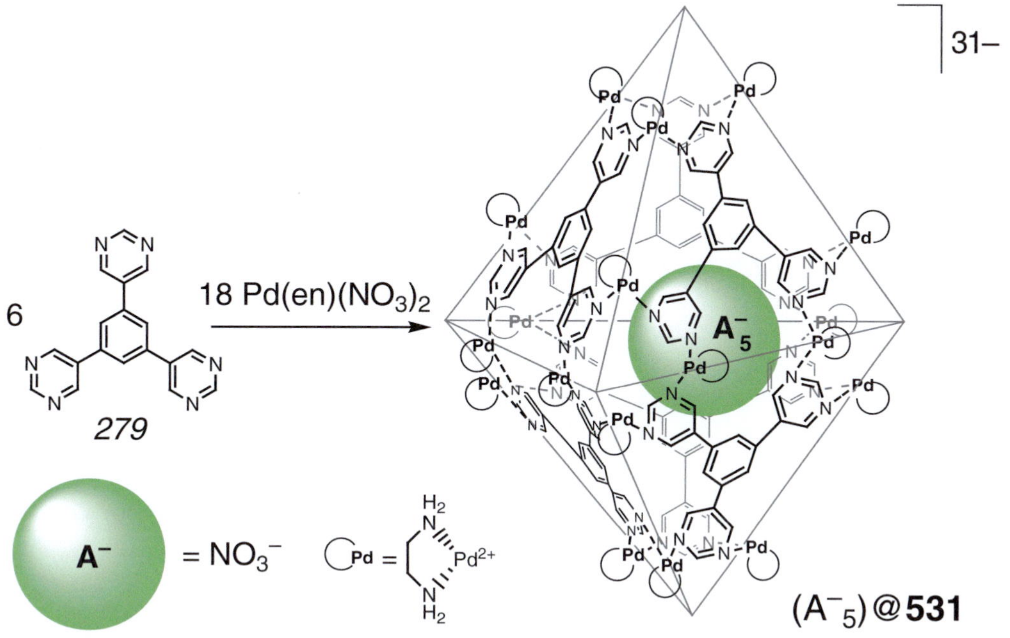

Scheme 4.58

encapsulated nitrate ions, each equatorial corner of this hexahedral coordination capsule with a cavity volume of approximately 900 Å^3 is formed by four triangular PdN_2 fragments, giving a cyclic moiety with a small "pinhole" of approximate size 2×2 Å [32].

4.1.3 Encapsulation of Cations

An octahedral coordination capsule **560** with triangular faces covered by threefold chelating synthones has been prepared in [57]. *One-pot* coordination-driven self-assembly of tris(2-hydroxybenzylidene)triaminoguanidinium chloride with palladium(II) chloride and sodium 5,5-diethylbarbiturate by Scheme 4.59 in the presence of triethylamine and tetraethylammonium chloride gave the crystals of a cage complex of **560**. This *T*-symmetric chiral complex contains four encapsulated sodium cations and caged solvent water molecules as well [57].

Use of an analogous bromine-containing ligand syntone *296* is described in [58] to prevent the formation of the same octahedral coordination capsule due to steric effects of its bulky bromine substituents. *One-pot* condensation of this syntone with the same co-reagents by Scheme 4.60 in the same reaction conditions gave a trigonal bipyramidal host–heteroguest 1:4:1 cage complex of the capsule **561**. Its chiral framework contains six tridentate syntones in the trigonal faces; each of them is coordinated to three palladium(II) ions. This dodecaanionic capsule with a cavity volume of approximately 1600 Å^3 encapsulates four triethylammonium and one tetraethylammonium cations and solvent water molecules. The self-assembly by Scheme 4.60 also gave a tetrahedral coordination pseudocage **562** with one open bottom face. Its closed faces are formed by three palladium(II) ions and one tridentate bromoguanidinium ligand syntone. Four of the 11 palladium(II) metallocenters coordinate the monoprotonated barbiturate anion, and these four anions are hydrogen-bonded with each other. The resulting chiral pseudocaging ligand with a cavity volume of approximately 800 Å^3 encapsulates three tetraethylammonium cations [58].

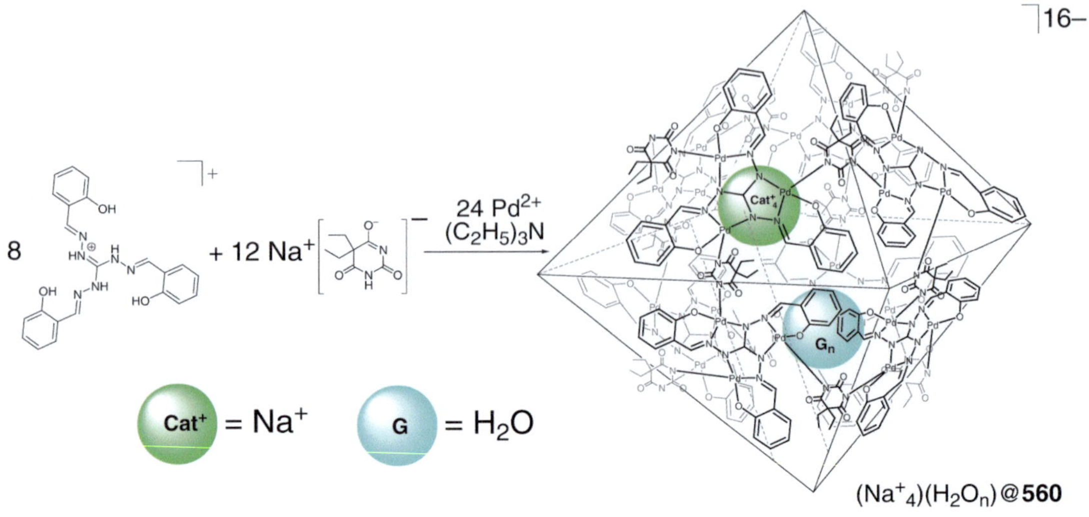

Scheme 4.59

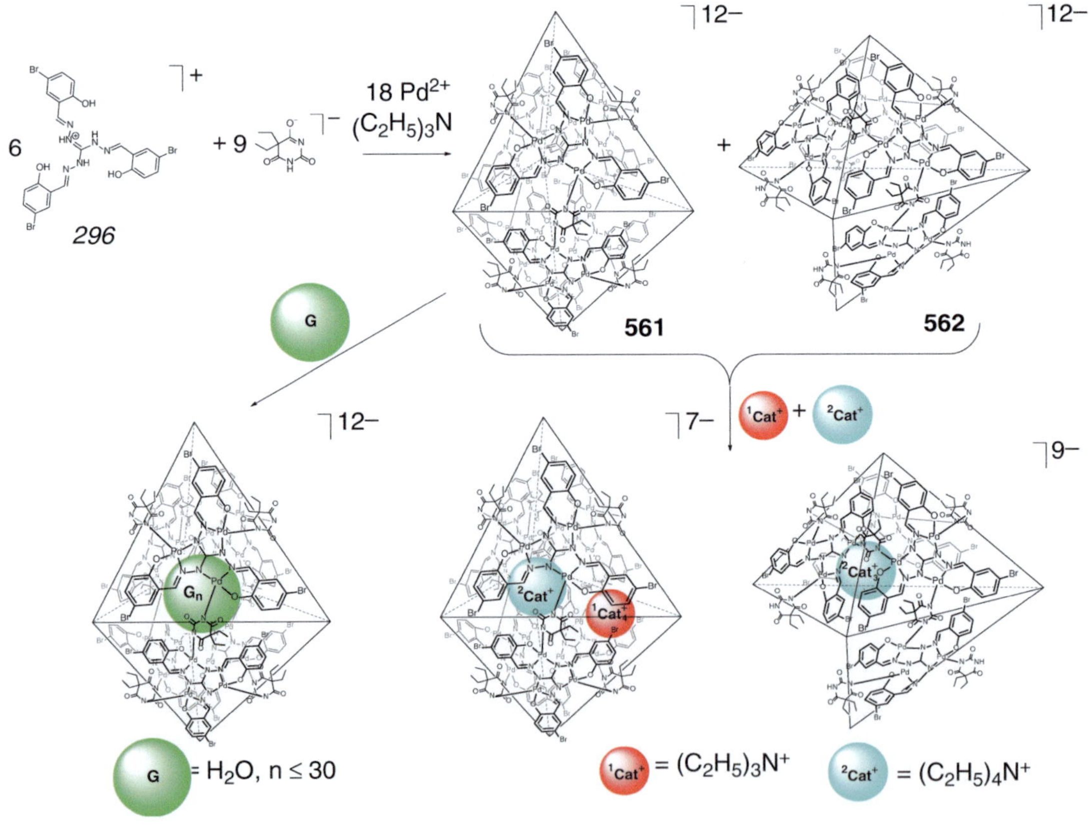

Scheme 4.60

4.2 Caging Ligands with 4n End-Capped Metal Ions

4.2.1 Free Cages and Encapsulation of Neutral Molecules

The first tetrahedral Mg_4L_6 coordination capsule **563** has been synthesized by R. Saalfrank [59] using *one-pot* coordination-driven self-assembly of malonic ester, oxalyl chloride, and methylmagnesium iodide (Scheme 4.61); the corresponding tetradentate ligand syntone could not be isolated, as it readily undergoes acid cleavage. The formation of analogous "adamantoid" compounds **564–567** with vertex manganese, zinc, cobalt, and nickel(II) ions is reported in [60]. Although the Co_4L_6 tetraanionic capsule **565** has an exact C_2 symmetry in crystal, it can be described as having nearly T symmetry with three C_2 and four C_3 axes, and its tetrahedral cage framework is therefore chiral. All four metallocenters (i.e., Mg^{2+}, Mn^{2+}, Co^{2+}, Ni^{2+}, and Zn^{2+} cations) in these chiral racemic T-symmetric tetraanions have identical $\Delta\Delta\Delta\Delta$-*fac* or $\Lambda\Lambda\Lambda\Lambda$-*fac* coordination polyhedra [61]. The use of iron(III) ions for cross-linking of the same ligand syntones allowed obtaining the first neutral "adamantanoid" coordination capsule using similar *one-pot* self-assembly procedures (Scheme 4.62). The neutral $Fe^{III}{}_4L_6$ capsule **568** has an almost ideal tetrahedral geometry with iron(III) ions in all of its vertices. Six edges of this tetrahedral cage framework are each formed by the corresponding bis-bidentate doubly negatively charged bridging ligand syntone and six oxygen atoms surrounding each of the four cross-linking iron(III) ions in

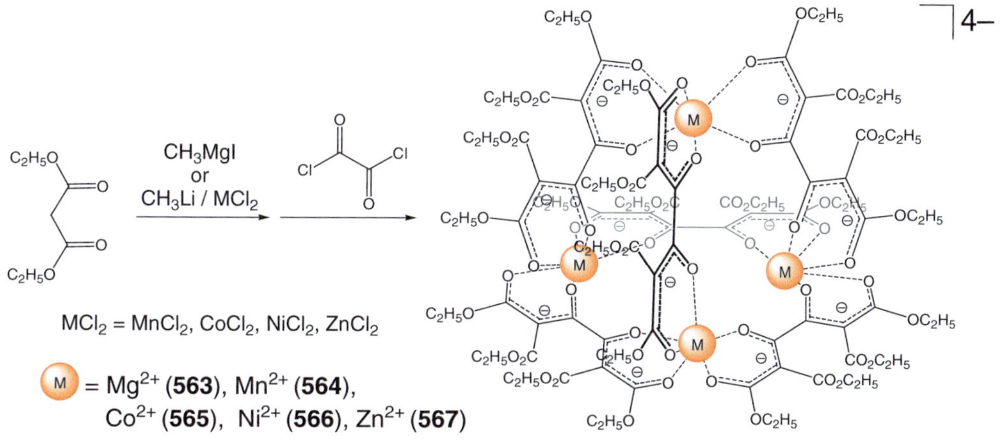

Scheme 4.61

Scheme 4.62

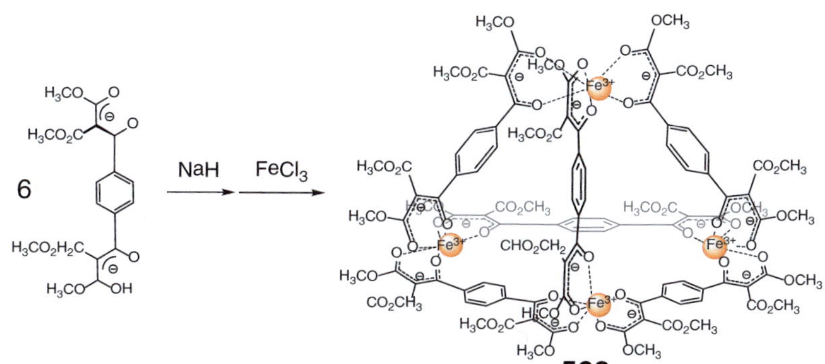

568

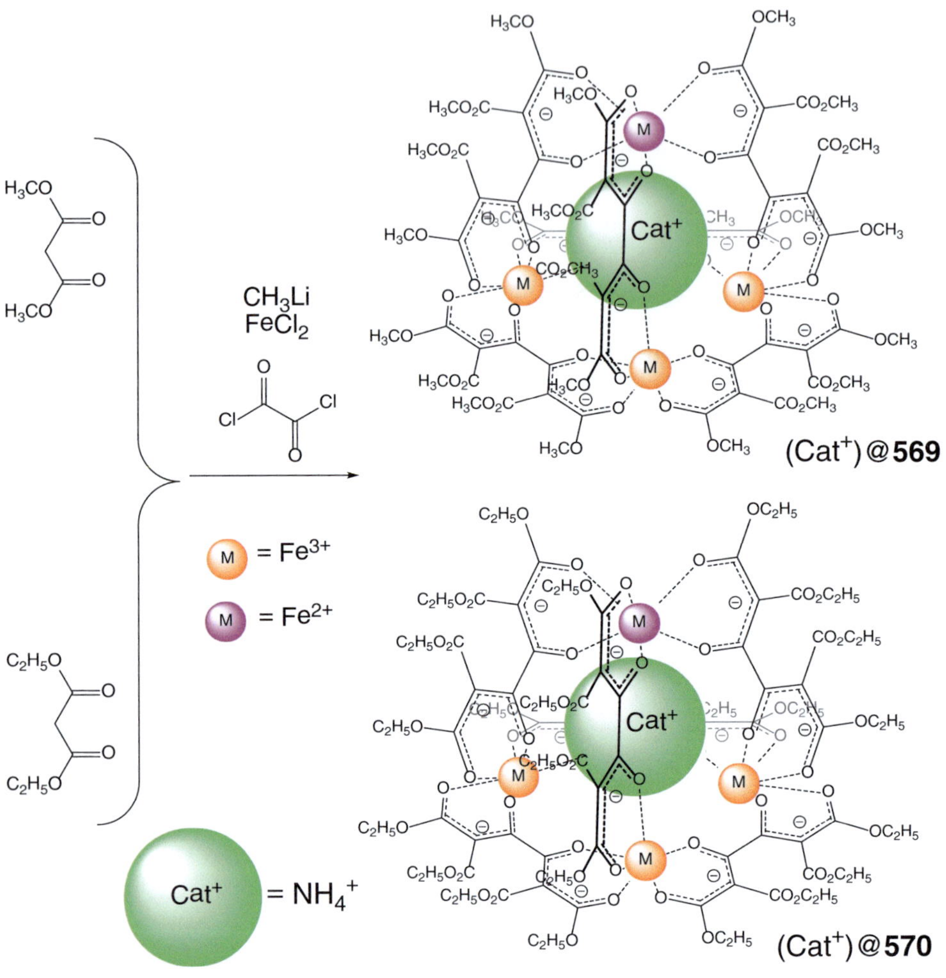

Scheme 4.63

an octahedral arrangement [61]. Two enantiomeric pairs (*Δ*-, *Λ*-*fac*, and *Δ*-, *Λ*-*mer*) are possible, in general, for octahedral tris-bidentate complexes. The Fe_4L_6 cage framework **568** is described in [61] to have an exact S_4 symmetry in crystal and is thus achiral (*meso*-form); its ligand syntones around the four iron(III) metallocenters are arranged facially, and two of the four iron centers have the same (*Δ*,*Δ*)-/(*Λ*, *Λ*)-*fac* configurations.

The mixed-valence $Fe_3^{III}Fe^{II}L_6$ 1:1 cage complexes with an encapsulated ammonium cation have been isolated in [62] when iron(II) chloride was used as a source of cross-linking metal ions (Scheme 4.63). The encapsulated ammonium cation serves as a template for their coordination-driven self-assembly. This caged cationic guest also compensates the negative charge of the encapsulating ligands, thus allowing to isolate the tetranuclear mixed-valence intracomplexes. The coordination capsules **569** and **570** were characterized using ^{57}Fe Mössbauer, EPR and CV methods, and single-crystal X-ray diffraction data as well. Their cage frameworks have a geometry of regular tetrahedron with four vertex metallocenters formed by one iron(II) and three iron(III) ions. Six edges of these coordination capsules are each formed by the corresponding bis-bidentate, doubly negatively charged bridging ligand syntone, and each of the four iron ions

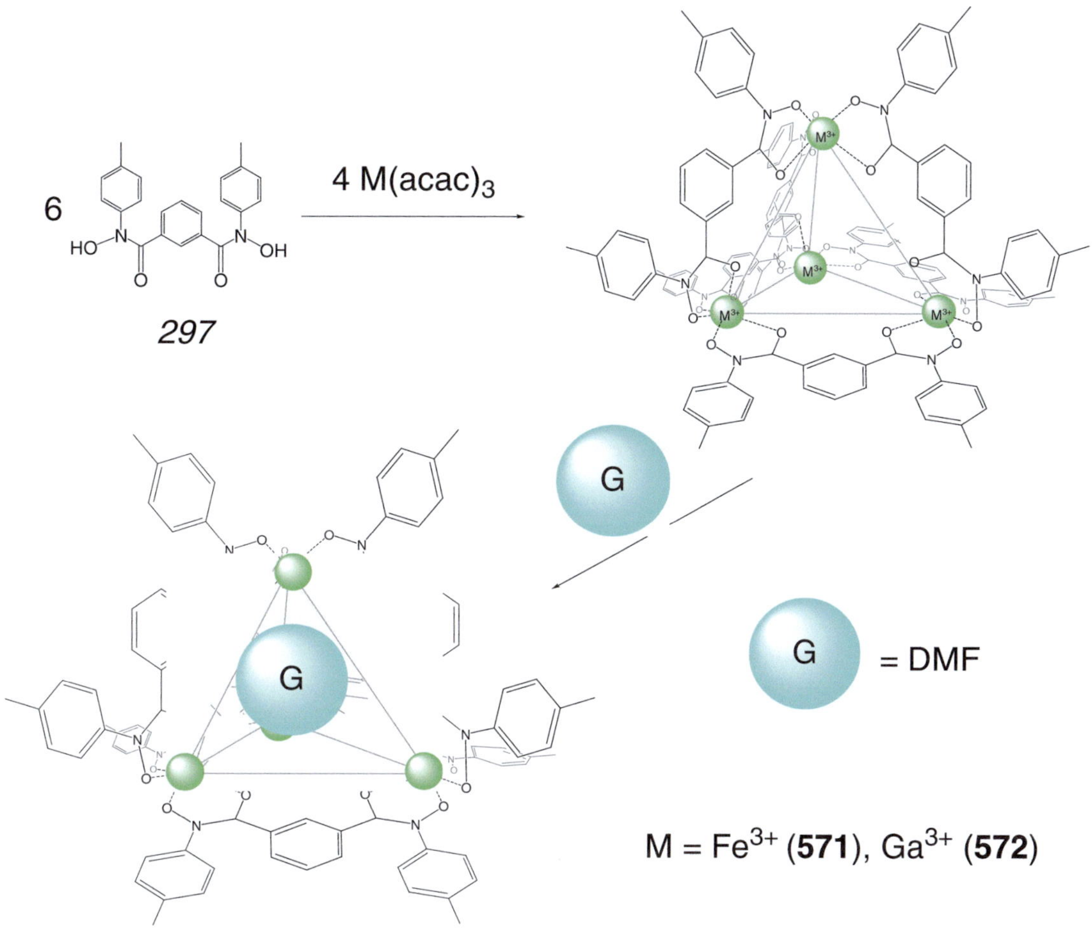

Scheme 4.64

coordinates six oxygen atoms forming the octahedral $Fe^{II}O_6$ or $Fe^{III}O_6$ coordination polyhedra. All four cross-linking metallocenters in these chiral racemic coordination capsules **569** and **570** have identical coordination environments, so they are present as either *ΔΔΔΔ-fac-* or as *ΛΛΛΛ-fac* stereoisomers. In an idealized representation, those are regarded [62] as having an approximate *T* symmetry. Note that their phenylene-containing analog **568** has an exact S_4 symmetry (see above) and thus is achiral (*meso*-form).

For the first time, tetrahedral M_4L_6 coordination capsules **571** and **572** have been synthesized by K. Raymond and coworkers in [63] by coordination-driven self-assembly of Ga^{3+} and Fe^{3+} cations with tetradentate hydroxamate ligand syntone *297* by Scheme 4.64. As follows from single-crystal X-ray diffraction data for **572**, this caging ligand encapsulates four solvent DMF molecules within its hydrophobic cavity [63]. Dynamic isomerization of its cage framework both in aqueous solution and in solid state has been studied in [64] using single-crystal X-ray diffraction and variable-temperature ^{1}H NMR methods, respectively. In this framework, each vertex metallocenter is chiral (*Δ* or *Λ*), resulting in the caging ligands with *T* (*ΔΔΔΔ* or *ΛΛΛΛ*), C_3 (*ΔΔΔΛ* or *ΛΛΛΔ*), or S_4 (*ΔΔΛΛ*) symmetry. The rigid ligand syntone *297* is bimodal (accommodate either mixed or homochiral metal centers at either end), but locks in the chirality of the cage framework is formed. All three isomers are seen in solution, and their interconversion, although still on the

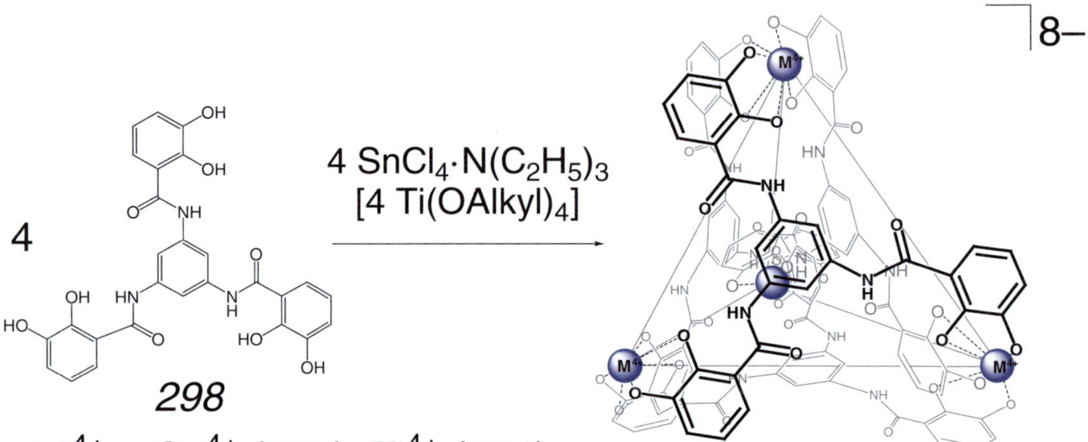

$M^{4+} = Sn^{4+}$ (**573**), Ti^{4+} (**574**)

Scheme 4.65

NMR time scale, is significantly slower than isomerization of similar unimolecular hydroxamate complexes. The distribution of the *T*, C_3, and S_4 isomers in aqueous solution is 4, 58, and 38 %, respectively. In the crystal of complex (DMF_4)@**572** with the symmetry-related four guest solvent molecules, the Ga…Ga distances between the metallocenters with the same and the opposite chiralities are 9.0 and 8.8 Å, respectively. Two different ligand conformations are observed, with one bridging homochiral metal center and with mixed chiral centers; their nearly equal stability explains the mix (*T*, C_3, or S_4) of cluster isomers. This ligand couples the metal vertices in the cluster to significantly increase the transition-state free energy for Δ–Λ interconversion but does not couple the chirality for the Δ or Λ ground state. The rigid ligand is bimodal in the four metal cluster: it can accommodate equally either Δ or Λ chirality at one end when the chirality at the other end is fixed. Quantitative NMR studies have shown that all of the possible isomers are present in solution with nearly statistical distribution. Various stereoisomers form and interconvert with the identical activation barrier for the interconversion. The cage framework, however, inhibits a dissociative or trigonal twist path required for the $\Delta \leftrightarrow \Lambda$ interconversion [64].

The octaanionic tetrahedral M_4L_4 capsules **573** and **574** have been obtained in [65] by coordination-driven self-assembly of the hexadentate tris-catecholate ligand syntone *298* with oxophilic Sn^{4+} and Ti^{4+} cations (Scheme 4.65). In **574**, all vertex titanium(IV) ions have the same chirality (all Δ or Λ). No evidence of encapsulation of any guest by its small cavity was found in this work, whereas its analogs were recognized as efficient cation receptors (see Sect. 4.2.3).

A dodecaanionic tetrahedral Ga_4L_6 capsule **575** (see Sect. 4.2.3) is described in [66] to form 1:1 cage complexes with normal C_5–C_9 alkanes, C_5–C_{10} cycloalkanes, and related hydrocarbons by Scheme 4.66; this encapsulation is explained by hydrophobic effect and C–H…π interactions between the caging ligand **575** and a caged alkane. The ligand, however, has a limited cavity size and cannot encapsulate *n*-decane due to steric reasons. Relative binding affinities of **575** to C_5–C_8 normal and cycloalkanes have been studied in [66] by ^{1}H NMR method. These affinities increase from C_5 to C_8 hydrocarbons and from normal alkanes to their alicyclic analogs in accordance with the corresponding hydrophobic effects as the entropy-driven force of this supramolecular binding [66].

Encapsulation by this coordination capsule **575** with hydrophobic interior has been also used

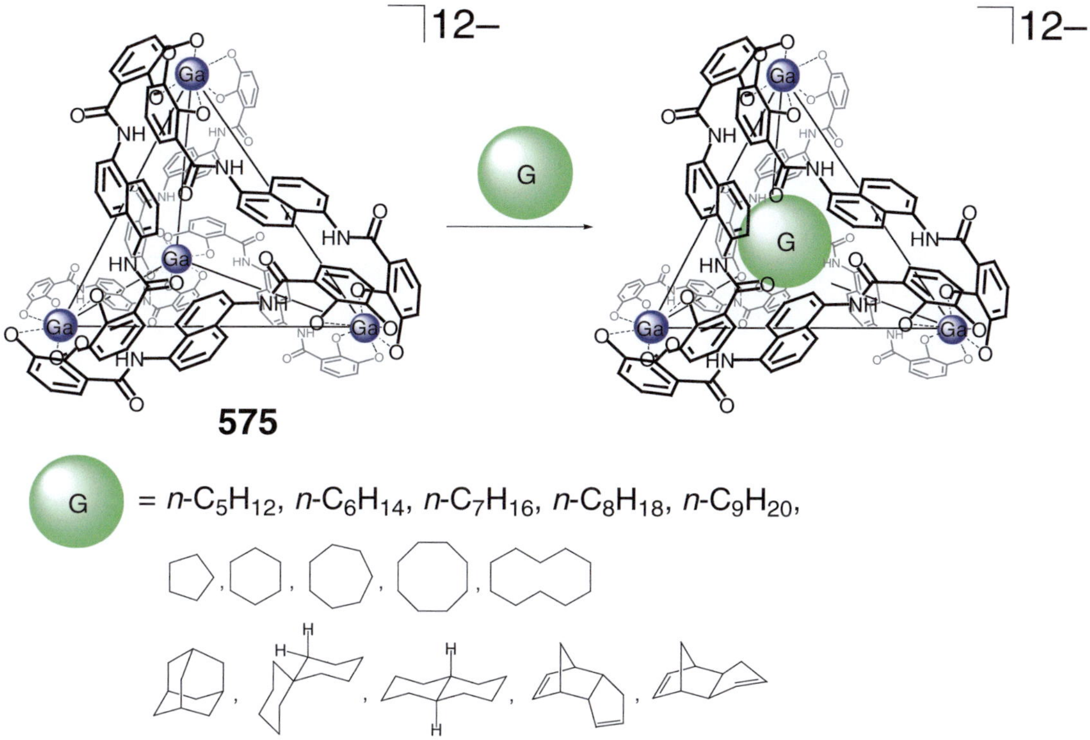

Scheme 4.66

for simultaneous binding of hydrophobic aromatic guests and for resolution of chiral natural products shown in Scheme 4.67 in aqueous solutions [67]. Small aromatic guests underwent encapsulation, and simultaneous caging of various multiple species has been observed in many cases. This host is reported in [67] to be able to recognize different substitutional isomers of disubstituted benzene guests: *ortho*-substituted guest forms a host–guest 1:2 compound, whereas *meta*- or *para*-isomers gave 1:1 cage complexes. Upon encapsulation of chiral molecules of natural products by this racemic capsule, diastereomeric host–guest complexes have been formed with diastereoselectivities of up to 78:22 [67].

Selective guest-induced formation of a tetrahedral capsule **576** has been accomplished in [68] by Scheme 4.68 with the use of stabilizing effect of encapsulated guest molecules such as CBr_4, giving a host–guest 1:1 cage complex. The coordination-driven self-assembly of triangular ligand syntones is reported to be controlled by sterically demanding substituent(s) in appropriate positions.

Dynamic coordination-driven self-assembly of bis-3-pyridyl ligand syntones *299* and *300* with palladium(II) ions is reported in [69] to give M_3L_6 metallomacrocycles and M_4L_8 tetrahedral coordination capsules **577** and **578** (Schemes 4.69 and 4.70); interconversion between such metallomacrocycles and capsules has been also observed in this work. These processes are affected by the nature of a solvent and counterion as well as by concentrations of the reaction components. In aqueous solution, the M_2L_2 trench-like complex *301* (Scheme 4.69) has been formed. In DMSO, a mixture of this complex and a M_3L_3 metallomacrocycle *302* were in equilibrium; at higher concentration of these reagents, the M_4L_4 metallomacrocycle *303* has been mostly observed. The tetrahedral M_4L_8 coordination capsule **577** has been prepared in quantitative yield

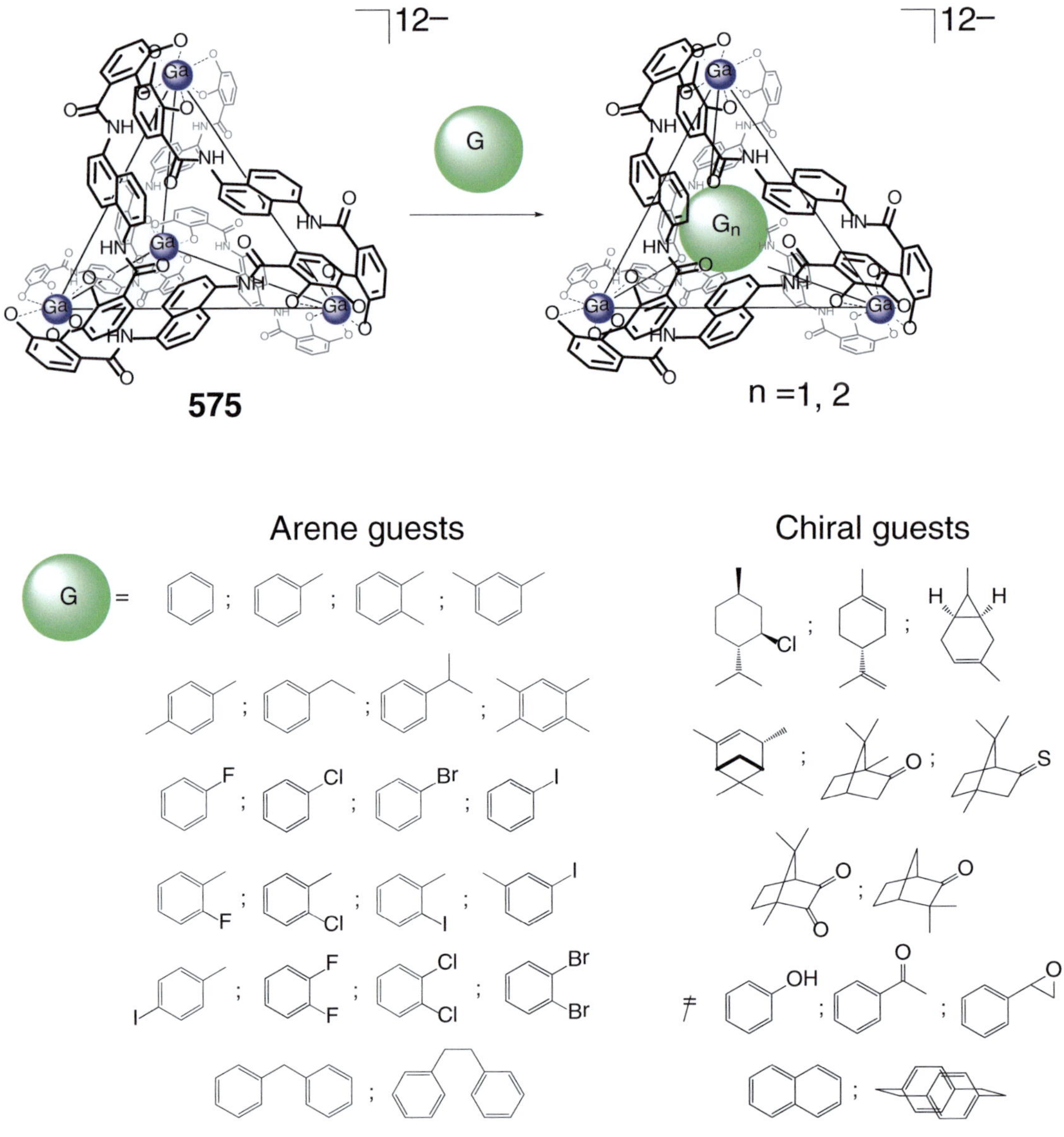

Scheme 4.67

by self-assembly of *299* with $Pd(NO_3)_2$, whereas that with $Pd(CF_3SO_3)_2$ gave a corresponding M_3L_6 double-walled triangular complex *304*. The anion is thus an important template in the formation of these coordination assemblies. Self-assembly of the extended ligand syntone *300* with $Pd(NO_3)_2$ by Scheme 4.70 gave mostly a tetrahedral Pd_4L_8 coordination capsule **586**, along with a corresponding double-walled triangular complex *305*. Reaction of *300* with $Pd(CF_3SO_3)_2$ afforded the latter complex with a negligible amount of the coordination capsule (such products are in dynamic equilibrium). The Pd_4L_8 coordination capsule **578** has been also prepared in [69] using palladium(II) *para*-tosylate as an initial Pd^{2+} salt.

A tris-pyridyl ligand syntone *306* has been used in [70] for the synthesis of Ag_4L_4 and Ag_6L_4 coordination capsule **579** and pseudocapsule **580** by Scheme 4.71. As follows from ^1H NMR titration

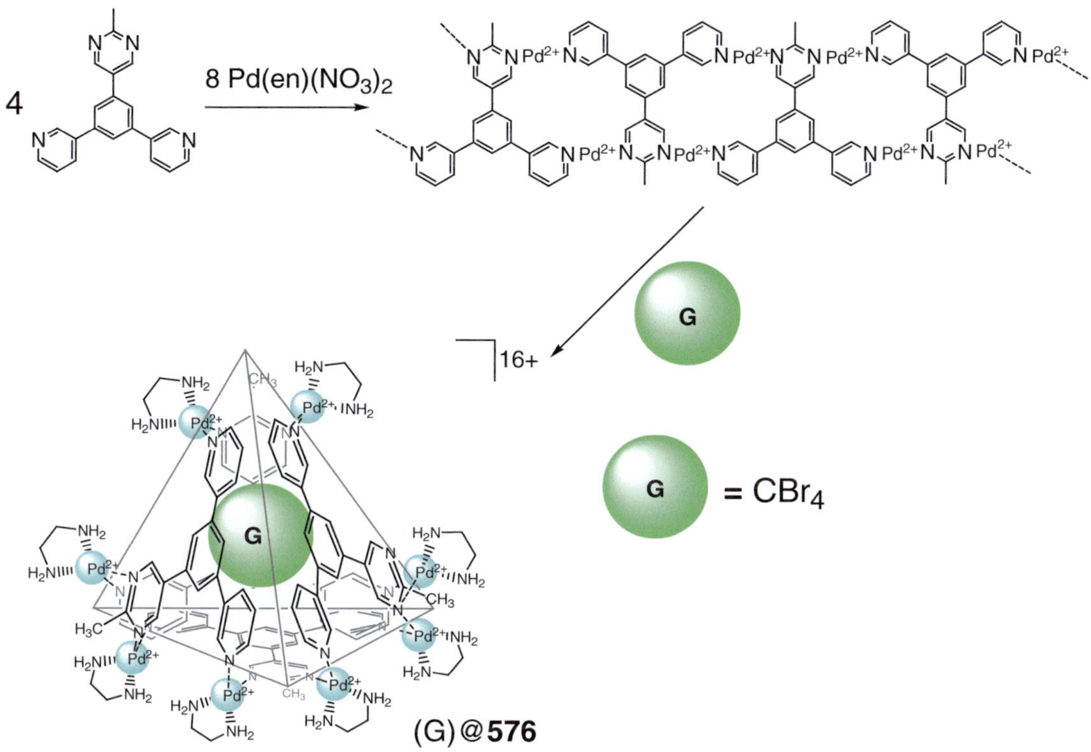

Scheme 4.68

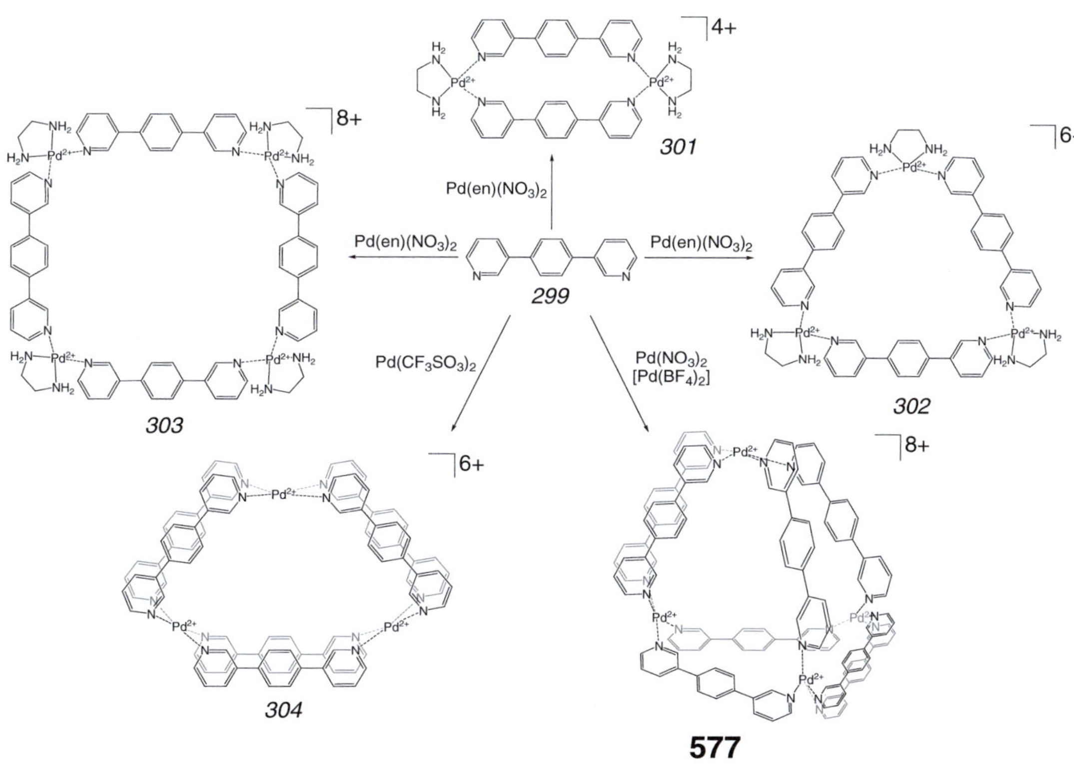

Scheme 4.69

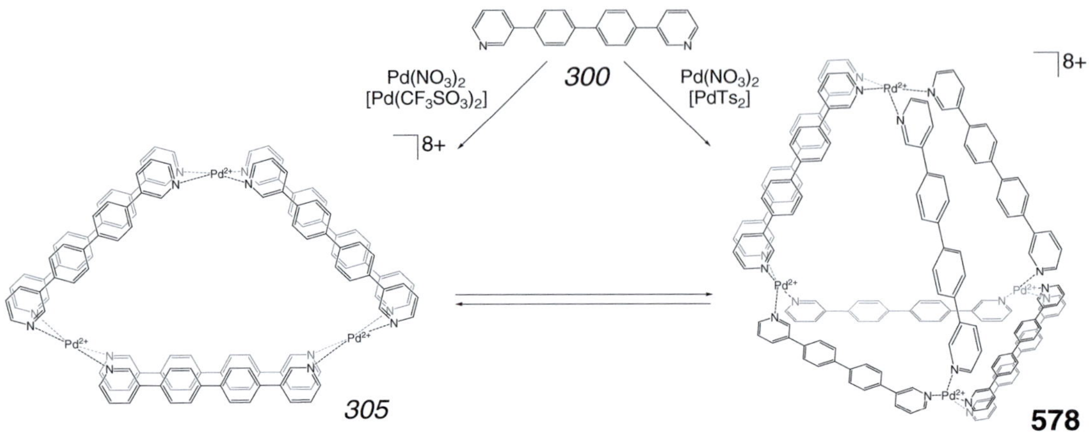

Scheme 4.70

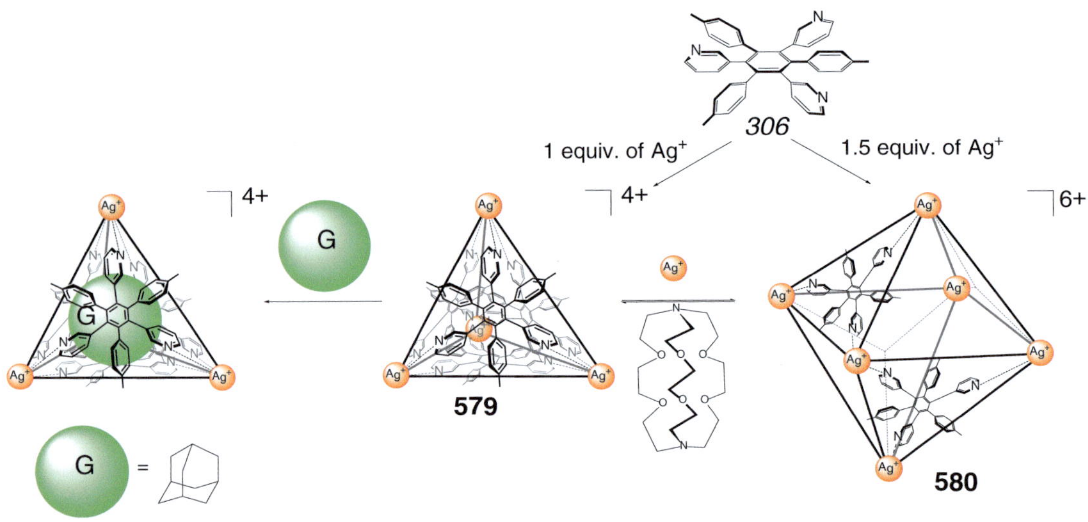

Scheme 4.71

experiments, self-assembly of these compounds depends on the molar ratio of components (i.e., Ag^+ ions and the above syntone): they are quantitatively formed at the component's ratios 1:1 and 3:2, respectively. These thermodynamically stable compounds easily undergo interconversion with a change in this molar ratio [70]. As follows from ^{1}H NMR data, **579** has a trigonal pyramidal cage framework with four tetracoordinate vertex Ag^+ ions; each of them coordinates three pyridyl groups of different tris-pyridyl syntones. This Ag_4L_4 caging ligand efficiently encapsulates adamantane by Scheme 4.71, giving a 1:1 cage complex. The Ag_6L_4 pseudocapsule **580** has a distorted octahedral (tetragonal bipyramidal) geometry in which doubly coordinate Ag^+ cations occupy vertex positions, and half of its faces are formed by these syntones with others being opened [70].

A water-soluble tetrahedral Fe_4L_6 coordination capsule **581** has been isolated in [71] by Scheme 4.72 from a dynamic library of linear diamine and aromatic aldehyde ligand syntones and iron(II) sulfate in the presence of tetramethylammonium hydroxide as organic base. The caging ligand **581** with an approximate cavity

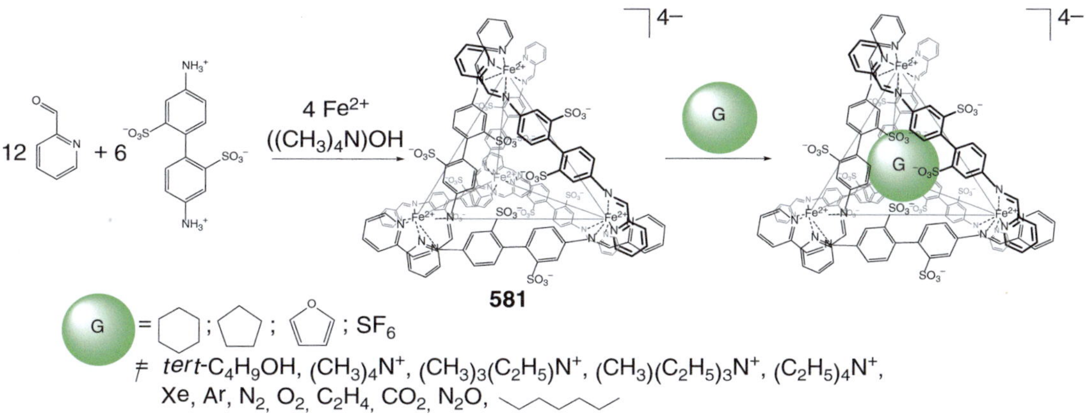

Scheme 4.72

volume of 141 Å[3] encapsulates cyclohexane or cyclopentane as guests to give 1:1 cage complexes with occupations of 61 and 55 %, respectively; those are close to the optimal Rebek's value of 55 % (see Chap. 3). At the same time, this tetraanionic caging ligand does not encapsulate *n*-butanol and alkylammonium cations. In the latter case, this fact has been explained in [71] by screening of the potent cationic guest from externally directed anionic sulfonate groups, along with the presence of four positively charged vertex iron(II) ions within the ligand's cavity. The caging ligand **581** is also reported in [72] to selectively encapsulate sulfur hexafluoride to give a 1:1 cage complex by Scheme 4.72. In the latter complex, this guest is located in the geometrical center of the cavity of **581** and forms van der Waals interactions with its hydrophobic interior. Such encapsulating ligand, however, does not form analogous cage complexes with other gas molecules shown in Scheme 4.72 [71].

At the same time, the coordination capsule **581** efficiently binds air-sensitive white phosphorus by Scheme 4.73, thus giving an air-stable 1:1 cage complex with an encapsulated P_4 guest molecule. This complex has been characterized in [73] using ^{1}H and ^{31}P NMR spectra and by X-ray diffraction data. The P_4 molecule within the cavity of a coordination capsule **581** is stabilized by van der Waals interactions with the hydrophobic phenylene ribbed fragments of its cage framework and is air stable for a long time, even in aqueous solutions. This result is explained by a possible oxygen-containing intermediate that is large enough to be accommodated by the cavity of **581**. This explanation has been confirmed by irreactivity of this cage complex against single-oxygen atom transfer reagents (e.g., nitrous oxide or pyridine N-oxide) [73]. According to ^{1}H NMR data of [74], a tetraanionic capsule **581** also encapsulates furan, preventing its Diels–Alder reaction with maleimide; the exchange reactions and reactivity of this 1:1 cage complex are described in Sect. 5.4. A detailed kinetic and thermodynamic study of encapsulation of a wide range of various neutral guest molecules shown in Scheme 4.74 by **581** has been performed in [75].

To explain the stereochemical coupling (i.e., the influence of chirality of an iron(II)-based vertex on the chirality of neighboring apical moieties), the authors of [76] have used tetramethyl-substituted and isomeric dimethyl-containing aromatic diamines *307–310*. These C_2-symmetric diamines form diastereomeric Fe_4L_6 capsules **582–585** by Scheme 4.75. In the case of 2,2″-dimethyl-substituted diaminoterphenyl, this reaction gave a mixture of the T-, S_4-, and C_3-diastereomers of **583** (67, 10, and 23 %, respectively). The X-rayed crystal of **583** contains in its asymmetric unit two homochiral molecules with $\Lambda\Lambda\Lambda\Lambda$ or $\Delta\Delta\Delta\Delta$ configurations at their metallocenters. As follows from ^{1}H NMR data, the acetonitrile solution of these crystals contains only T-diastereomer of the coordination

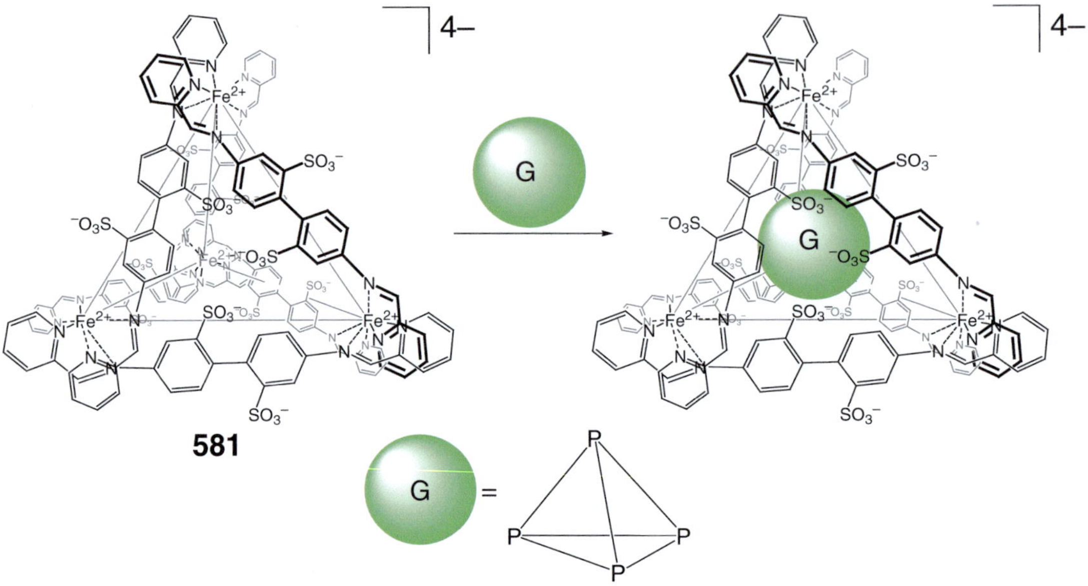

Scheme 4.73

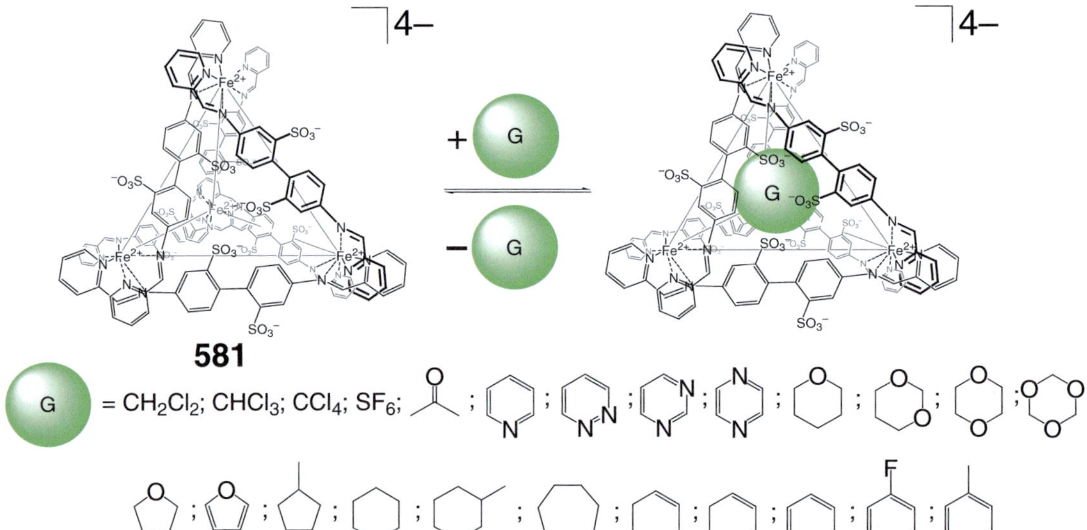

Scheme 4.74

capsule, and the reversible mixture of its diastereomers appears in 5 days [76]. The use of tetramethyl-substituted aromatic diamine as a ligand syntone in the coordination-driven imine condensation gave only S_4- and C_3-diastereomers of the caging ligand **584** (88 and 12 %, respectively), whereas the dissolution of its single crystals, suitable for X-ray diffraction experiment, gave an S_4-diastereomer only just after this procedure. The coordination capsule **585** exists in solution as a mixture of T-, S_4-, and C_3-diastereomers (44, 35, and 21 %, respectively). Thus, (*i*) the parent aromatic diamine gave the mixture of diastereomers in nearly equal amounts, (*ii*) the presence of two methyl substituents in terminal aromatic groups causes the T-diastereomer

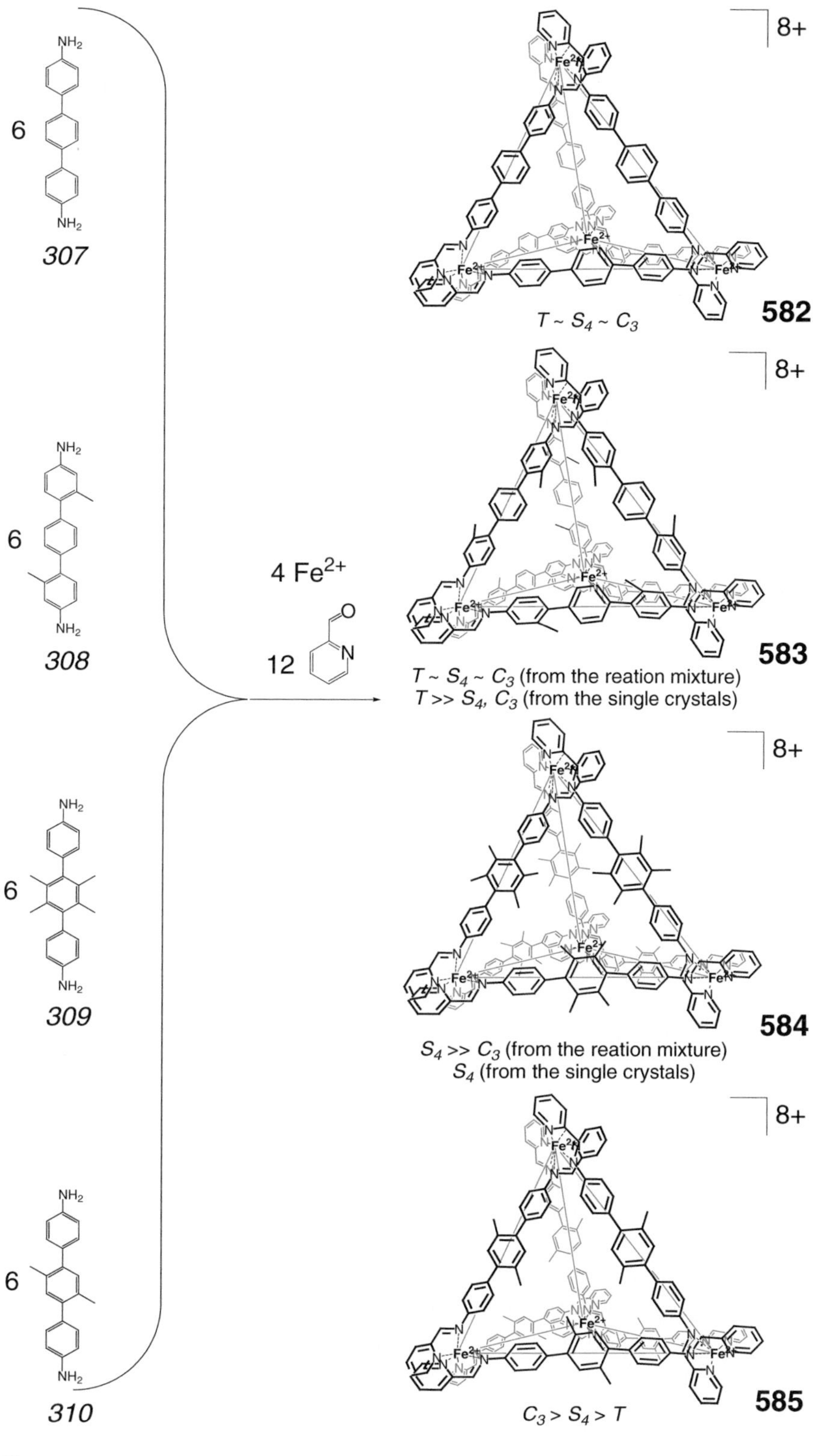

6
307
6
308
6
309
6
310
4 Fe^{2+}
12
8+
582
$T \sim S_4 \sim C_3$
8+
583
$T \sim S_4 \sim C_3$ (from the reation mixture)
$T >> S_4, C_3$ (from the single crystals)
8+
584
$S_4 >> C_3$ (from the reation mixture)
S_4 (from the single crystals)
8+
585
$C_3 > S_4 > T$

Scheme 4.75

to dominate, and (*iii*) the use of the tetrasubstituted diamine ligand syntone generates predominantly the S_4-diastereomer, whereas that with two methyl substituents in the central benzene fragment gave mainly the C_3-diastereomer of the corresponding Fe_4L_6 coordination capsule [76].

Cooperative stereochemical communication between iron(II) metallocenters up to 20 Å apart has been studied in [77] using Fe_4L_6 coordination capsules of various sizes. The large, optically active Fe_4L_6 caging ligands **586–595** have been prepared by Scheme 4.76 using self-assembly of linear 5,5′-bis(2-formylpyridinate) non-, mono-, di-, and tris-*para*-phenylene-containing ligand syntones *311–315*, chiral amines, and iron(II) ions. When the coordination capsule has been prepared from the syntone bridged by one *para*-xylene spacer and a bulky chiral amine, both a homochiral Fe_2L_3 helicate and an Fe_4L_6 cage framework coexist in solution. In the case of a less bulky chiral amine, only the Fe_4L_6 capsule has been formed [77]. The larger bis- and tris-*para*-phenylene capsules showed a long-range stereochemical coupling between their metallocenters: stereochemical information from the less bulky chiral amine is transmitted to each vertex iron(II) ion inefficiently, although this transfer is amplified by cooperative stereochemical communication between these four metallocenters. This communication is reported in [77] to be mediated by the syntone's geometries and rigidity, as opposed to gearing effects between xylene methyl groups, and takes place even in the larger coordination capsules with the shortest Fe…Fe distances more than 20 Å (independent of the ligand's axial configuration). Therefore, the axial chirality of these ligand syntones does not affect the chiral information transfer between the vertex metallocenters. The transmission of stereochemical information within their cage frameworks does not depend the length of their edge fragments in helical conformations and bridging vertex iron(II) stereocenters. The long-range cooperative stereochemical communication originates from three-dimensional geometric effects within the tetrahedral M_4L_6 framework as a whole. The energetic preference of such coordination capsules for adopting approximate *T* symmetry is due to the distortion of their cage frameworks that incorporate metallocenters of opposite handedness, the latter being relevant to the tetrahedral polyhedra of various sizes [77].

Interconverting tetrahedral Fe_4L_6 and pentagonal prismatic $Fe_{10}L_{15}$ coordination capsules **596** and **597** have been self-assembled in [78] by Scheme 4.77. The tetrahedral capsule **596** with the average Fe…Fe distance of 12.8 Å has one bis-bidentate bridging syntone along each of its six edges; all the vertex metallocenters have *fac*-coordination polyhedra. The cage framework of **597** has a geometry of a twisted pentagonal prism and is formed by two parallel staggered Fe_5L_5 pentagonal metallomacrocycles, connected to each other at the vertex iron(II) ions with a *mer*-arrangement by five ligand syntones. The Fe…Fe distances are from 12.23 to 12.40 Å in the Fe_5L_5 cycles and from 12.14 to 12.20 Å between the pentagonal bases [78].

A long diaminoterphenylene ligand syntone *316* bearing chiral glyceryl groups has been used in [79] for the enantioselective formation of a large water-soluble Fe_4L_6 coordination capsule **598** (Scheme 4.78) that encapsulates a wide range of guests (e.g., chiral natural products). The suitable guests for this caging host include (*i*) large hydrophobic molecules that bind weakly (those that can fit within its cavity demonstrate a slow exchange on the NMR time scale), (*ii*) medium-sized hydrophobic guests that bind strongly (those appear to be suitably sized for the cavity of *ΔΔΔΔ*-**598**, forming host–guest 1:1 cage complexes, and they also show a slow exchange on the NMR time scale), and (*iii*) smaller molecules such as cyclopentane, dichlorvos (famous acetylcholine esterase inhibitor insecticide), and benzene for which the fast exchange on this time scale has been observed in [79].

A face-capped tetrahedral coordination capsules **599** and **600**, the derivatives of the C_3-symmetric trialdehyde ligand syntone *317*, formed in situ by metal-templated imine condensation of tris(formylpyridyl) benzene and the corresponding chiral amine, have been diastereoselectively

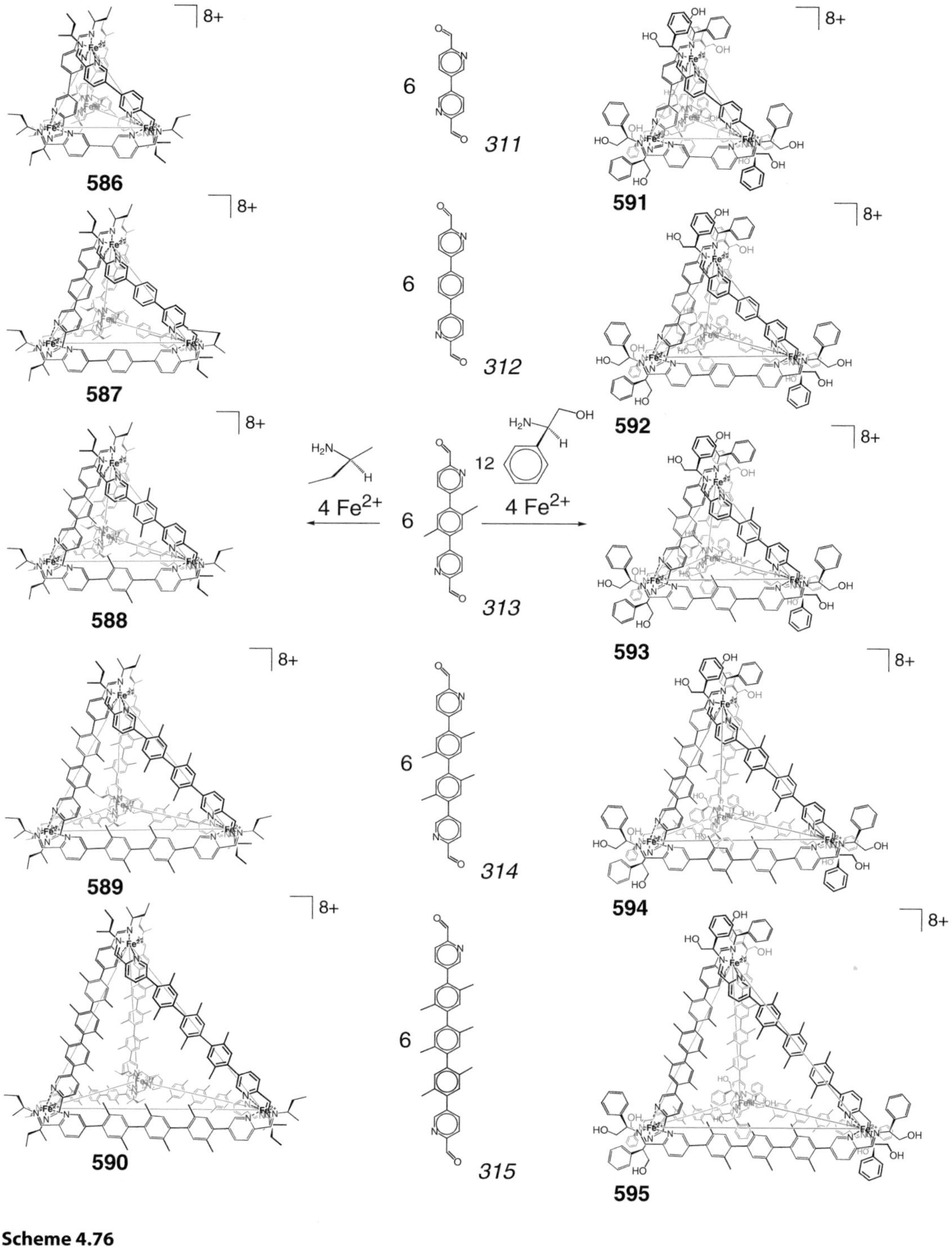
8+
586
587
588
589
590
6
311
312
313
314
315
H_2N
4 Fe^{2+}
12
H_2N
OH
H
591
592
593
594
595

Scheme 4.76

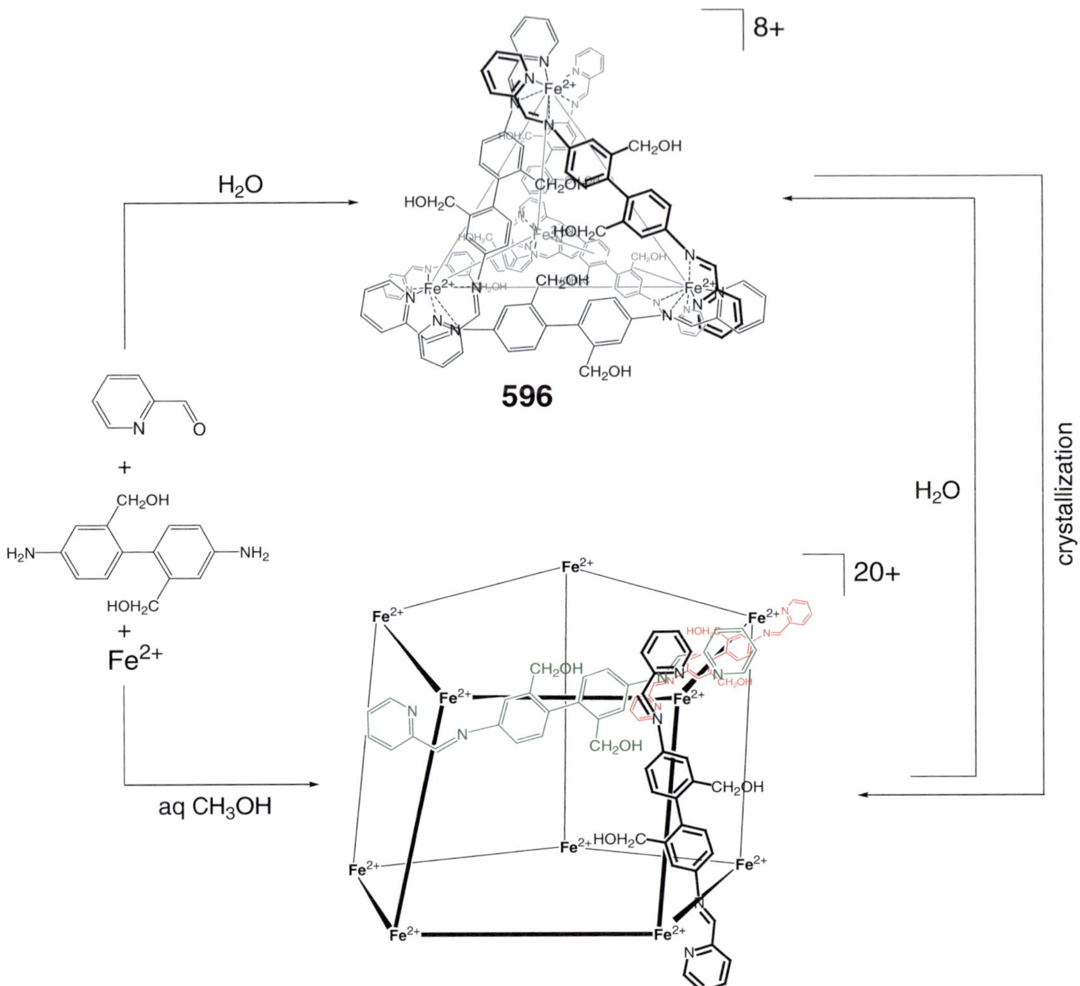

Scheme 4.77

self-assembled in [80] by Scheme 4.79. Strong cooperative stereochemical coupling between the vertex iron(II) stereocenters allows retaining the configuration upon replacement of its chiral amine precursors to their achiral analogs (Pathway II). In contrast, the direct self-assembly of the same aldehyde, achiral amines, and iron(II) ions (Pathway I) afforded a mixture of $\Delta\Delta\Delta\Delta$- and $\Lambda\Lambda\Lambda\Lambda$-isomers of these coordination capsules. Strong communication of handedness between the vertex metallocenters effectively mediated by the helical sense of the C_3-symmetric ligand syntones, robustness of the face-capped cage frameworks, and conduction of stereochemical information through chiral amine syntones are described in [80] to be the key features that enable its stereochemical memory: it prevents racemization of these capsules, thus allowing long-lasting storage of chiral information.

Face-capped tetrahedral Fe_4L_4 coordination capsules **601** and **602** have been prepared in [81] using self-assembly of iron(II) ions, 2-formyl-*N*-methylimidazole, and the corresponding triamine ligand syntone by Scheme 4.80. The caging ligand **601** has a larger cavity than its analog **602** and is able to encapsulate guests, whereas the latter is not. Both these Fe_4L_4 capsules are reported to undergo temperature-induced SCO in solution and in solid state; the binding of either carbon tetrabromide or carbon disulfide as guests substantially affects this behavior.

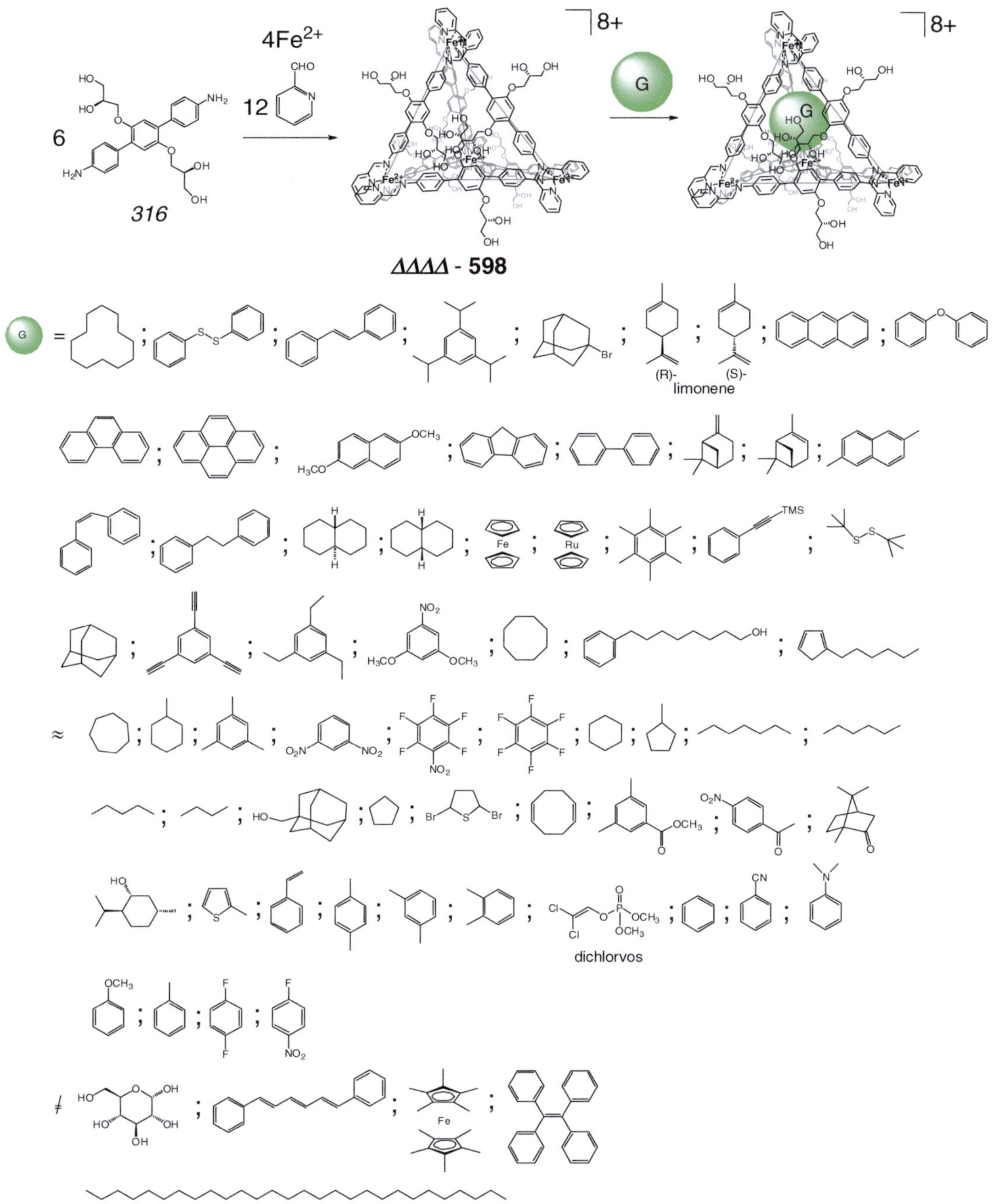

Scheme 4.78

Cobalt- and nickel-containing analogs **603** and **604** of the Fe_4L_6 coordination capsule **581** have been prepared in [82] by Scheme 4.81. In the case of cobalt(II) ion, the self-assembly gave a mixture of the capsule **603** and its 1:1 cage complex with an encapsulated 2-formylpyridine guest. This difference results from the larger cavity of the Co_4L_6 framework that allows it to encapsulate more bulky guests [82]. All these iron-, nickel-, and cobalt(II)-capped capsules have crystallographic and approximate non-crystallographic *T* symmetry; their solvent-filled

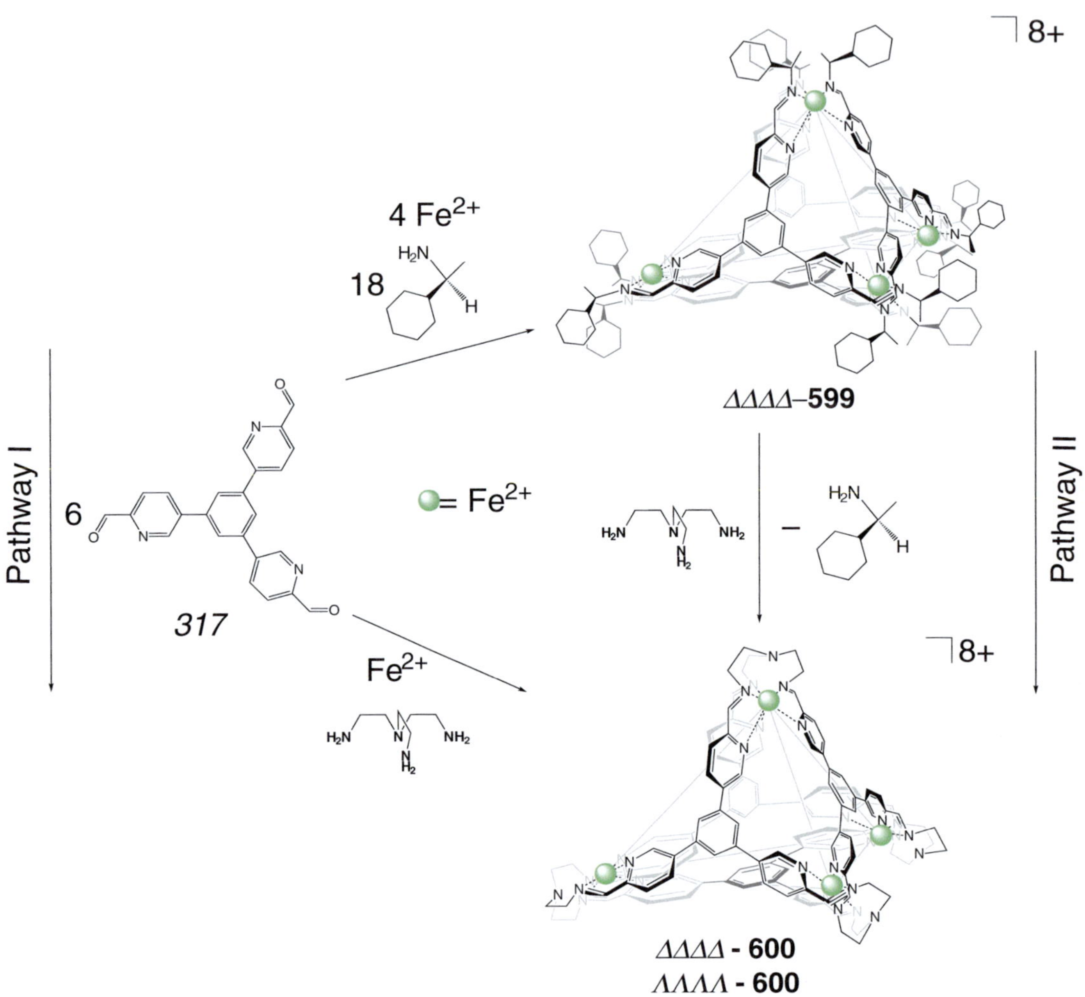

Pathway I : racemic mixture of *ΔΔΔΔ* - **600** and *ΛΛΛΛ* - **600**
Pathway II : *ΔΔΔΔ* - **600**

Scheme 4.79

cavities have the volumes of approximately 118, 140, and 149–153 Å^3, respectively. The caging ligand **603** encapsulates a wide range of neutral organic guests shown in Scheme 4.81–1089a but discriminates bulky molecules of cyclodecane, *n*-heptane, *n*-octane, and naphthalene. In 1:1 cage complexes of **581** and **604** with di-, tri-, and tetrachloromethanes and tetrabromomethane, the cavity volumes are substantially (by approximately 20 Å^3) larger than those of the initial capsules. This results [82] from the large size of the encapsulated guests giving rise to a distortion of cage frameworks, as they occupy from 60 to approximately 100 % of those cavities, respectively.

Self-assembly of a diformyl ligand syntone *318*, tripodal amine *trpn*, and cadmium (II) ions by Scheme 4.82 has been used in [83] for the synthesis of a tetrahedral Cd_4L_6 caging ligand **605**. It has four tris(3-aminopropyl) amine fragments in its vertex metallocenters and six ligand syntones that form its edge fragments. Each cross-linking cadmium(II) ion coordinates seven nitrogen donor atoms and adopt an approximately face-capped octahedral coordination geometry. This coordination capsule has a cavity volume of approximately 200 Å^3, the latter being larger than that of its Fe_4L_6 analog by 60 Å^3 [83].

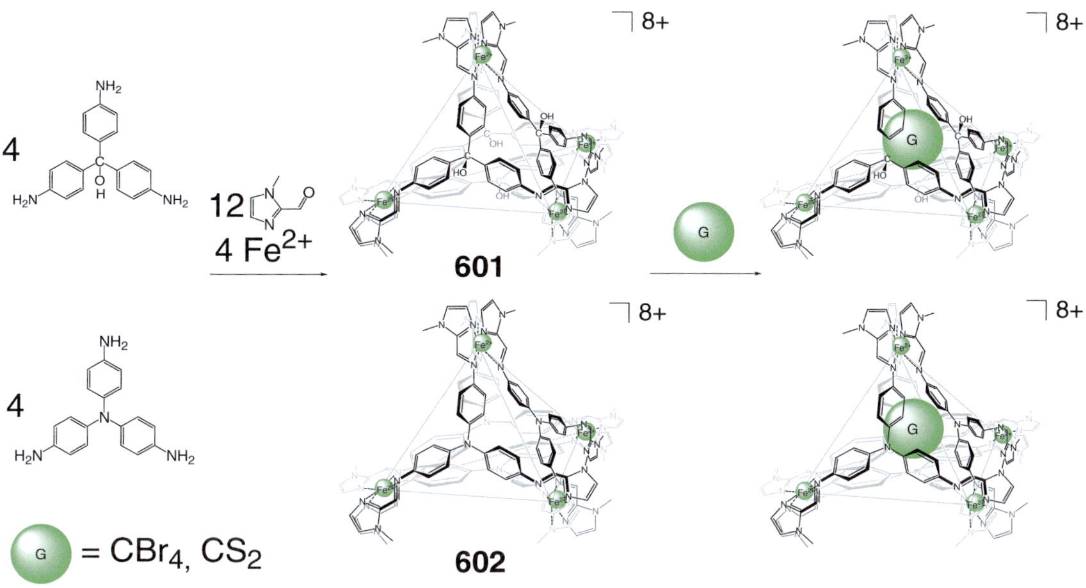

Scheme 4.80

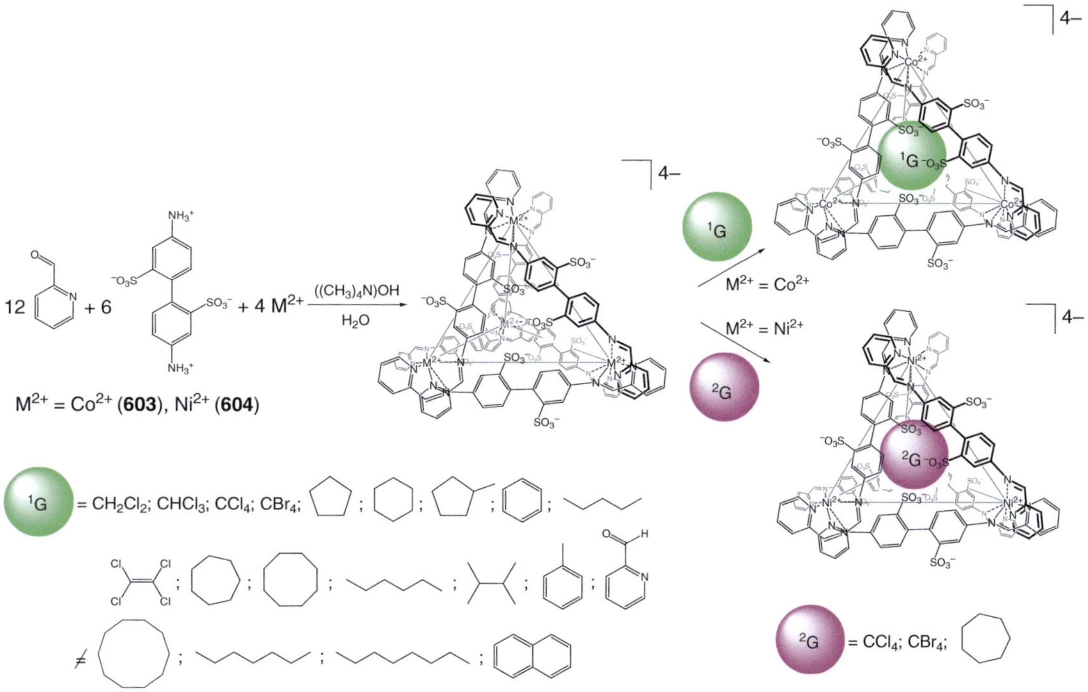

Scheme 4.81

A heterometallic $Fe_8Pt_6L_{24}$ cubic coordination capsule **606** has been prepared in [84] using two synthetic approaches shown in Scheme 4.83, one including *one-pot* self-assembly of four different components (of 62 building blocks in total) and the other being a more efficient two-step procedure that involves isolation of a square-planar platinum-based intermediate *319*.

Self-assembly of 12 3,3′-diformyl-4,4′-bipyridine ligand syntones, 24 primary aromatic

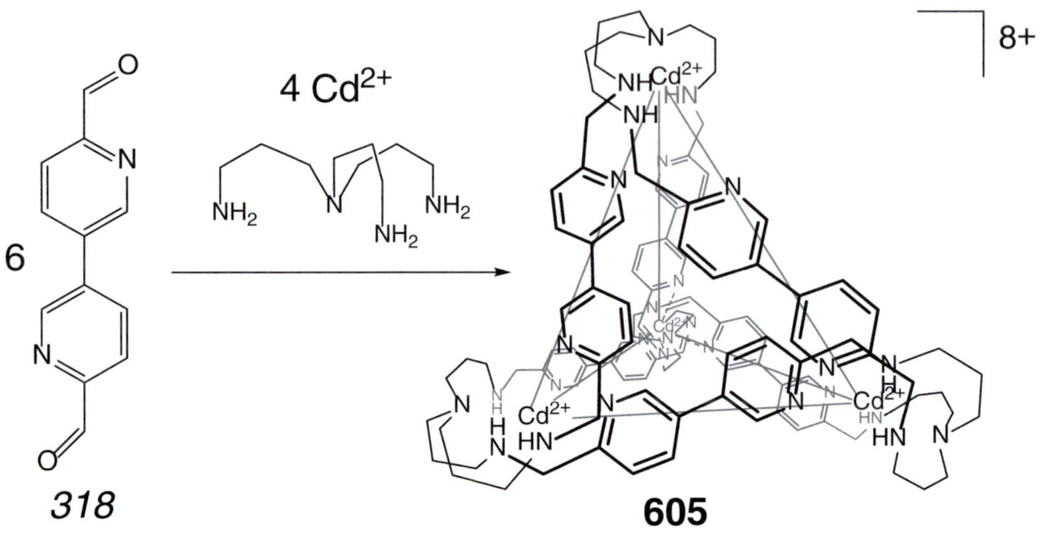

Scheme 4.82

Scheme 4.83

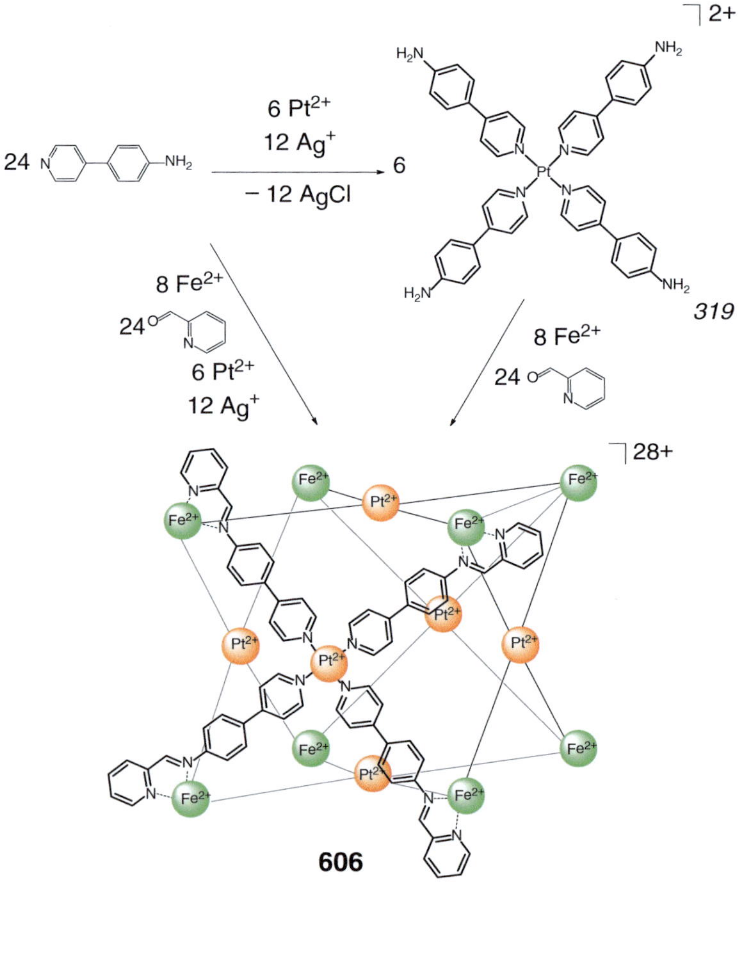

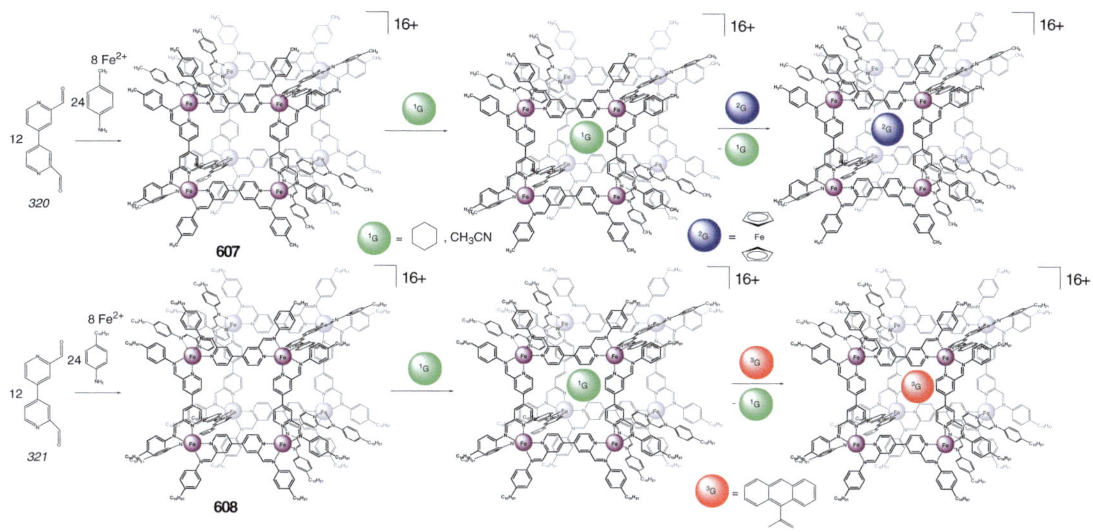

Scheme 4.84

amines *320* or *321*, and 8 iron(II) ions by Scheme 4.84 has been used in [85] to obtain Fe_8L_{12} cubic coordination capsules **607** and **608**. In the cage framework of **607**, each of its metallocenters of Λ-stereochemistry is adjacent to three Δ-centers and *vice versa*, thus giving the coordination capsule with an almost *T* symmetry. The average Fe…Fe distance along the edges is approximately 11 Å, and a cavity volume of this caging ligand (approximately 1000 $Å^3$) is reported in [85] to be enough to encapsulate various guests also shown in Scheme 4.84.

A rigid ligand syntone obtained by imine condensation of *para*-toluidine and 6,6′-diformyl-3,3′-bipyridine has been used in [86] for the study of synergistic and competing effects of metal ions and anion templation on the formation of coordination capsules shown in Schemes 4.85. Coordination-driven self-assembly of this syntone in the presence of iron, cobalt, nickel, and zinc(II) cations and triflimide, tetrafluoroborate, triflate, perchlorate, and nitrate anions gave six main types of products: discrete tetragonal *T*-symmetric M_4L_6, cuboid S_4-symmetric M_8L_{12} and pentagonal prismatic $M_{10}L_{15}$ coordination capsules, metallopolymers, or their dynamic combinatorial library. The tetrahedral iron-based caging ligand **609** (Scheme 4.85) encapsulates BF_4^-, OTf^-, or PF_6^- monoanions within its inner cavity but not the large NTf_2^- ion. Therefore, an anion is not a necessary template for this coordination-driven self-assembly. The 1:1 cage complexes of this Fe_4L_6 capsule with encapsulated tetrafluoroborate and perchlorate anions undergo the structural rearrangements upon heating into their pentagonal prismatic $Fe_{10}L_{15}$ analog **610**. This D_5-symmetric barrel-like ligand with a distorted pentagonal prismatic polyhedron is formed by two Fe_5L_5 circular helicates with iron(II) ions having *mer*-coordination environment. These helicates are bridged by five axial ligand syntones, thus saturating the coordination polyhedra of these ten vertex metallocenters. Such cage framework is stabilized by strong stacking interactions between its electron-rich toluidine and electron-poor pyridine rings [86]. More wide range of the reaction products shown in Scheme 4.86 in the case of the high-spin cobalt(II) cation as compared with those for the low-spin iron(II) ion are explained in [86] by the increased Shannon radius of Co^{2+} ion. Nickel(II) cation in octahedral coordination environment, having a smaller Shannon radius than the Co^{2+} ion, also forms the various coordination capsules, depending on the nature of the anionic component (Schemes 4.86 and 4.87). The S_4-symmetric complex of **615** has a distorted cuboid Ni_8L_{12} cage framework, and each of its two binding pockets

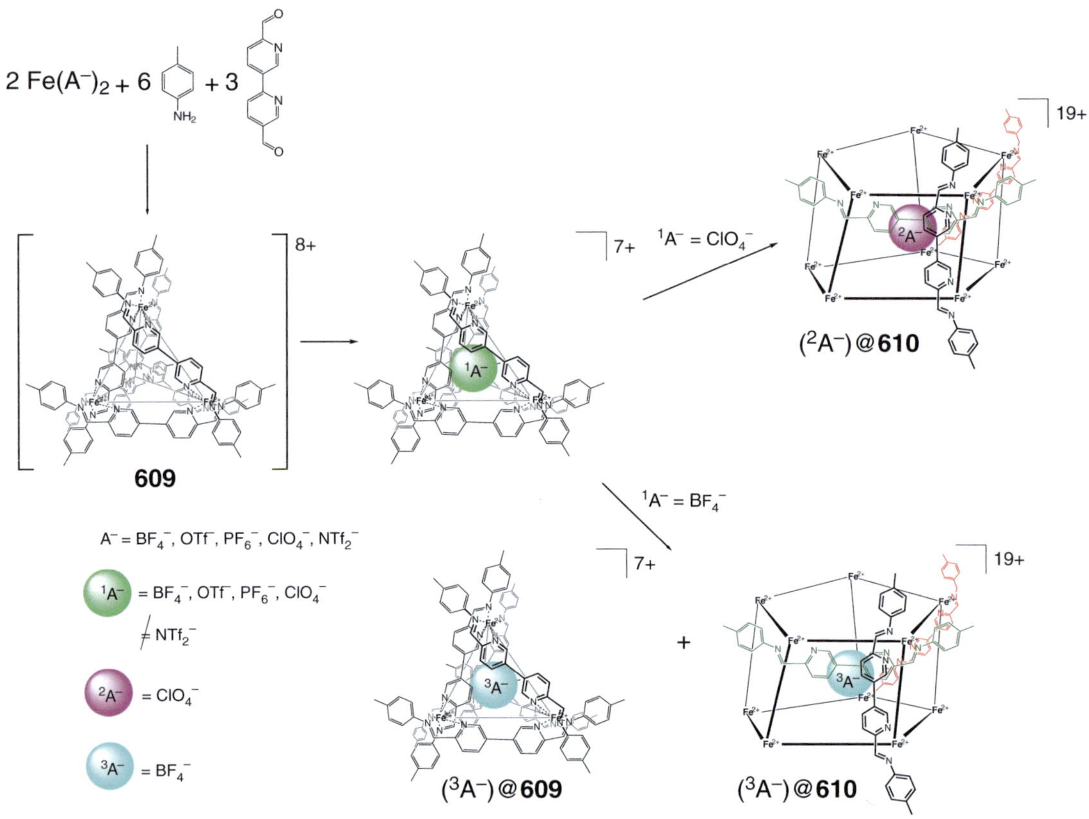

Scheme 4.85

with the cavity volumes of approximately 155 Å^3 encapsulates two nitrate anions. The tetrahedral Ni_4L_6 coordination capsule **616** and its cobalt-based analog **611** encapsulate acetonitrile solvate molecule; the Ni…Ni distances in the capsule **616** (approximately 9.7 Å) is intermediate between those for its iron(II)- and cobalt(II)-based analogs **609** and **611** (9.5 and 9.8 Å, respectively).

Self-assembly of the same ligand syntone with zinc(II) cation, which has a larger Shannon radius than the low-spin iron(II) ion, caused the formation of a wide range of the products shown in Scheme 4.88. Such reactions in the presence of appropriate templating anions gave $Zn_{10}L_{15}$ and Zn_8L_{12} coordination capsules **617** and **618**, but not the tetrahedral Zn_4L_6 cage framework. This result is explained in [86] by fast ligand exchange kinetics in the case of zinc(II) complexes and thus by rapid equilibration of the reaction mixtures. Anionic templation has been confirmed by the formation of the dynamic library of the reaction products in the case of zinc(II) triflimide.

The following main conclusions were made by the authors of [86]: (*i*) larger metal ions enhance structural diversity of the coordination capsules that they form with the same ligand syntone, (*ii*) subtle differences in the size of the templating anions substantially affect the structure of these capsules, and (*iii*) both *fac*-and *mer*-coordination environments of their capping metallocenters must be considered.

A C_4-symmetric tetrakis-bidentate ligand syntone *322*, a derivative of the molybdenum(II) “paddle wheel” complex, has been self-assembled in [87] with iron(II) cations by Scheme 4.89 to give a polycationic Fe_8Mo_{12} cubic capsule **619**. Its coordinatively unsaturated molybdenum centers are located on each face of the cage framework and are able to coordinate donor species. Eight C_3-symmetric tris-pyridylimine iron(II) vertex metallocenters have *fac*-coordination polyhedra with the same Δ or Λ stereochemistry; both enantiomers are presented in its X-rayed crystal. The faces of this capsule, encapsulating three disordered acetonitrile

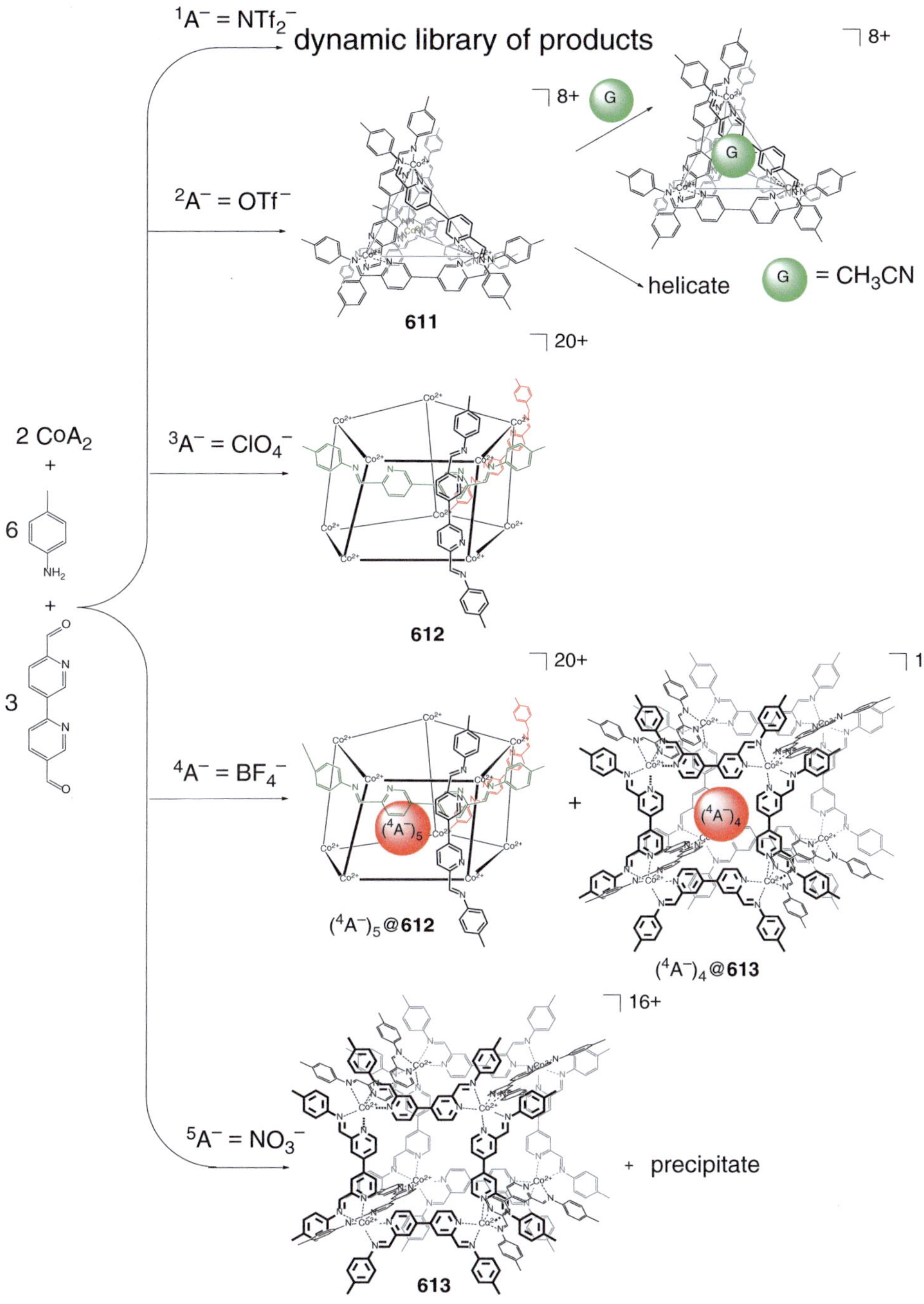

Scheme 4.86

molecules, with a cavity volume of 560 Å^3 are formed by six dimolybdenum(II) "paddle wheels". The average distance between the interior molybdenum centers of these faces is equal to 11.3 Å, while the average Fe…Fe distance diagonally across them is approximately 18.7 Å [87]. This highly positively charged caging ligand encapsulates halogenide anions and neutral guests to give 1:1 cage complexes. It can also form heteroguest 1:1:1 host–guest compounds by Scheme 4.90; if ammonium is encapsulated by **619**, the affinity of this capsule to iodide ion [87] increases eightfold.

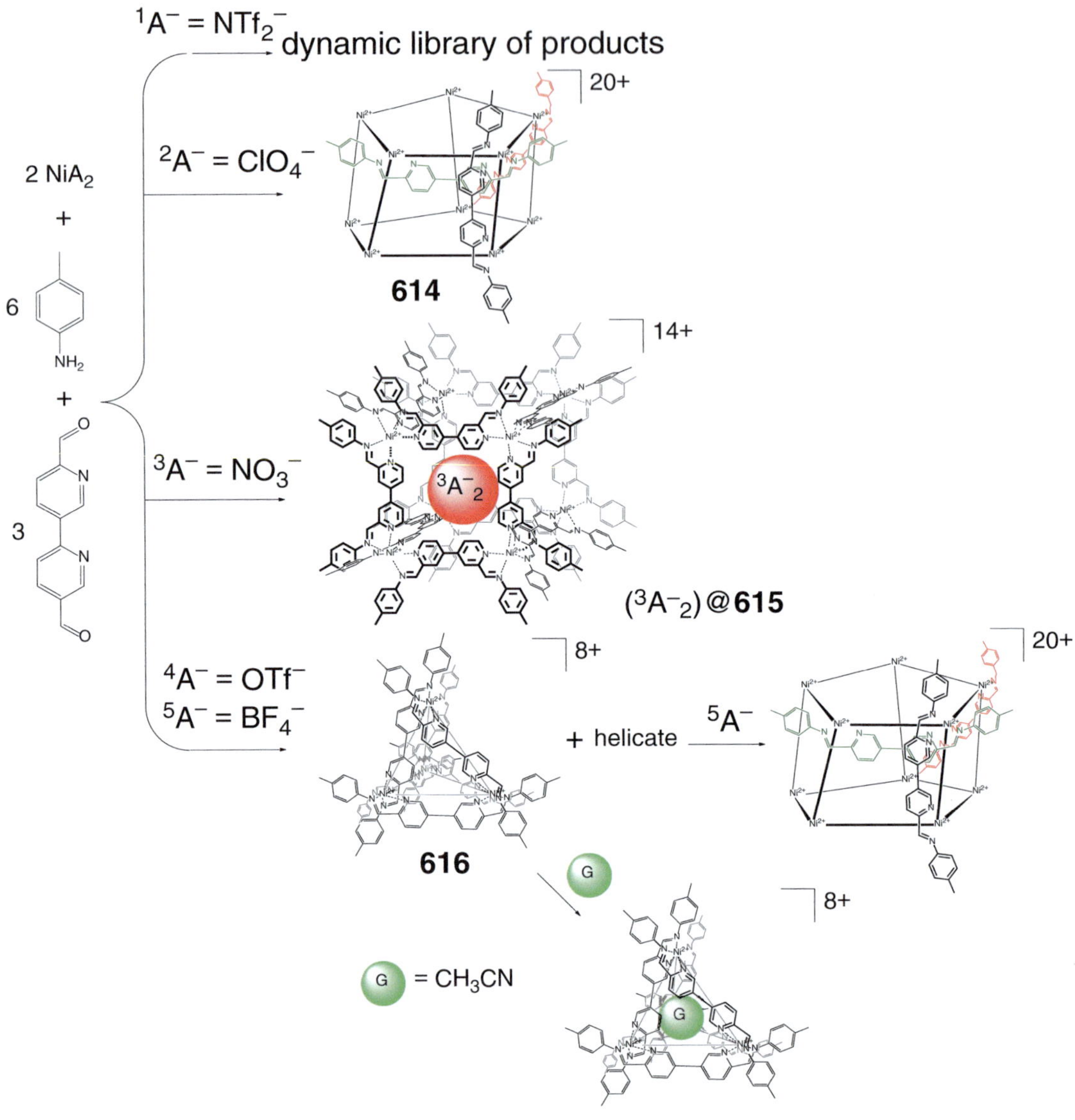

Scheme 4.87

A more weakly binding trimethylamine guest results in a further increase in this affinity, while more strongly binding guest CH_3PO shows the smallest increase in the iodide binding strength. In contrast, the presence of an excess of trifluoroacetate anion competitively inhibits the encapsulation of iodide anion. Therefore, the lining of the hosting cavity of **619** with well-defined binding sites allowed co-guest species to directly influence each other's binding affinities through their mutual steric and electronic interactions [87].

Binding properties of the Fe_8Mo_{12} caging ligand **619** have been tuned in [88] by allosteric binding to its molybdenum sites shown in Scheme 4.91: the externally binding donors (such as tricyclohexylphosphine, tri-*n*-octylphosphine, or tetraphenylborate anion) are described to allosterically reduce its encapsulating ability. In particular, trialkylphosphines that are coordinated to the exterior of the coordination capsule **619** at its molybdenum sites inhibit the encapsulation of neutral and anionic guests; the binding of anionic species has been allosterically inhibited by tetraphenylborate anion due to electrostatic repulsions. As the two allosteric active sites of its cage framework act independently of each other, they

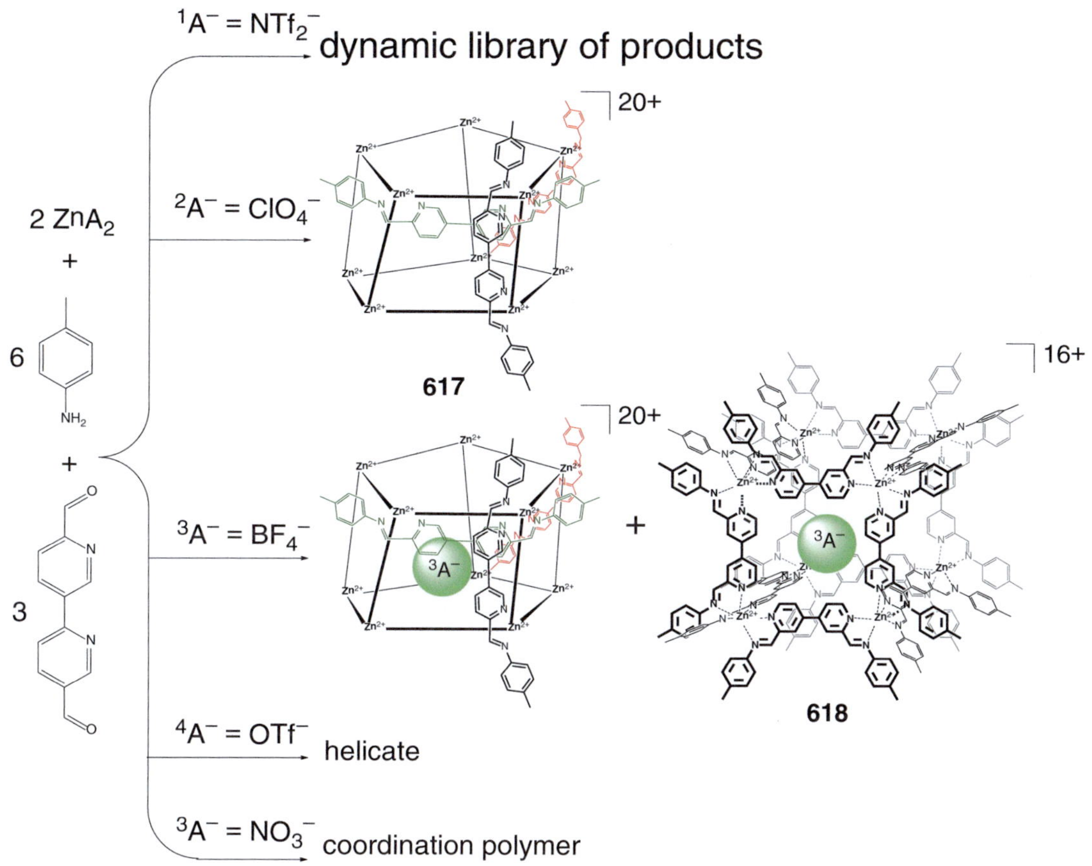

Scheme 4.88

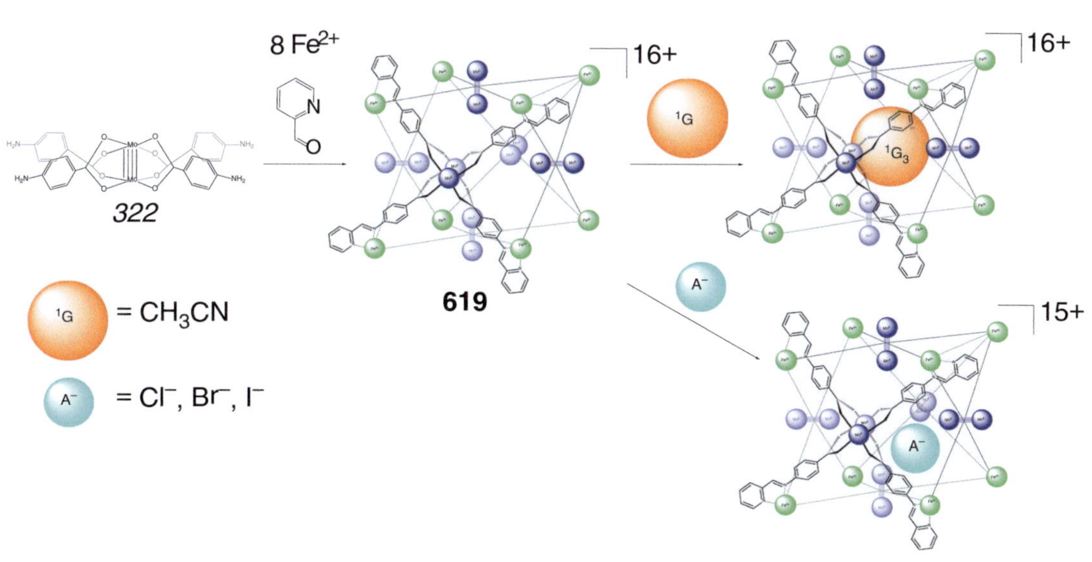

Scheme 4.89

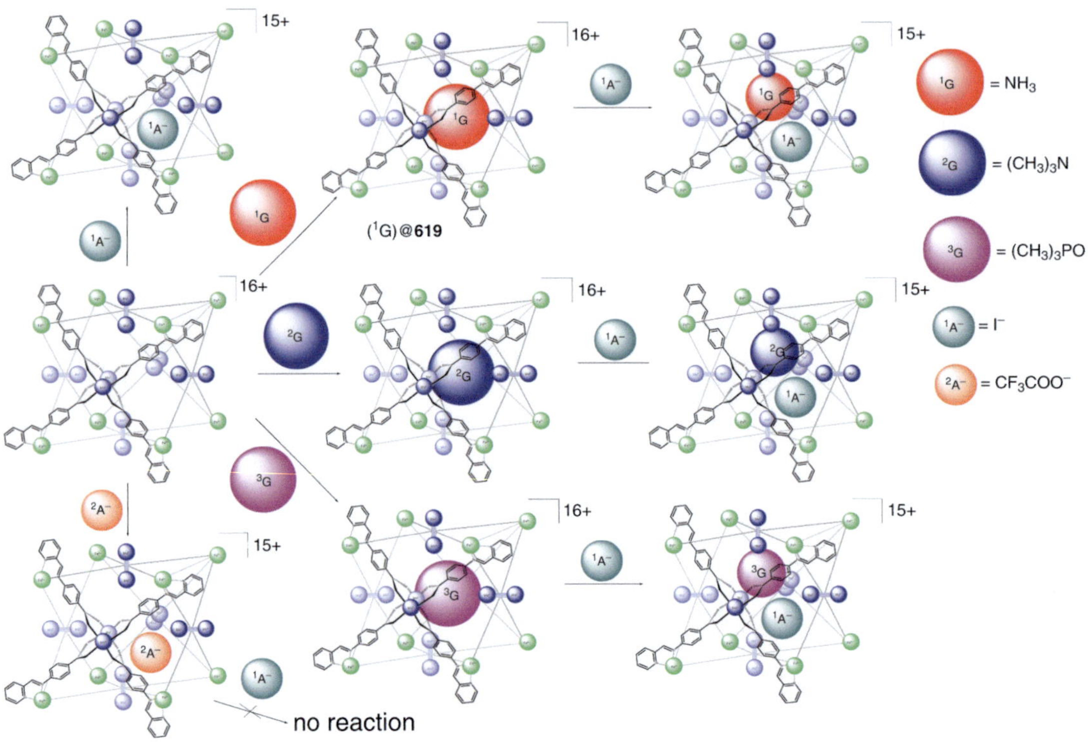

Scheme 4.90

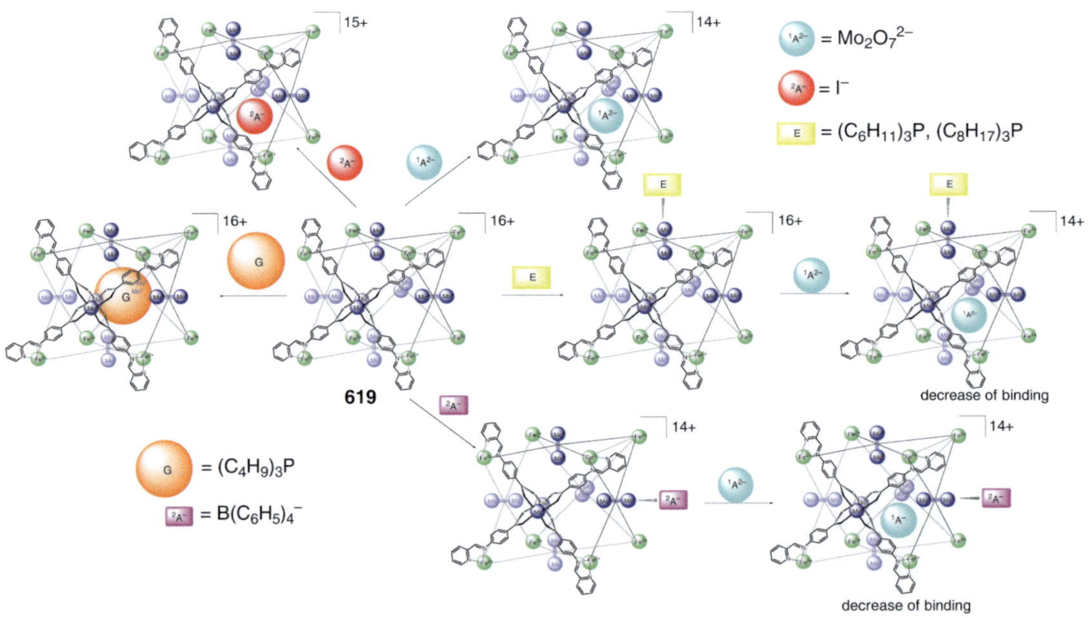

Scheme 4.91

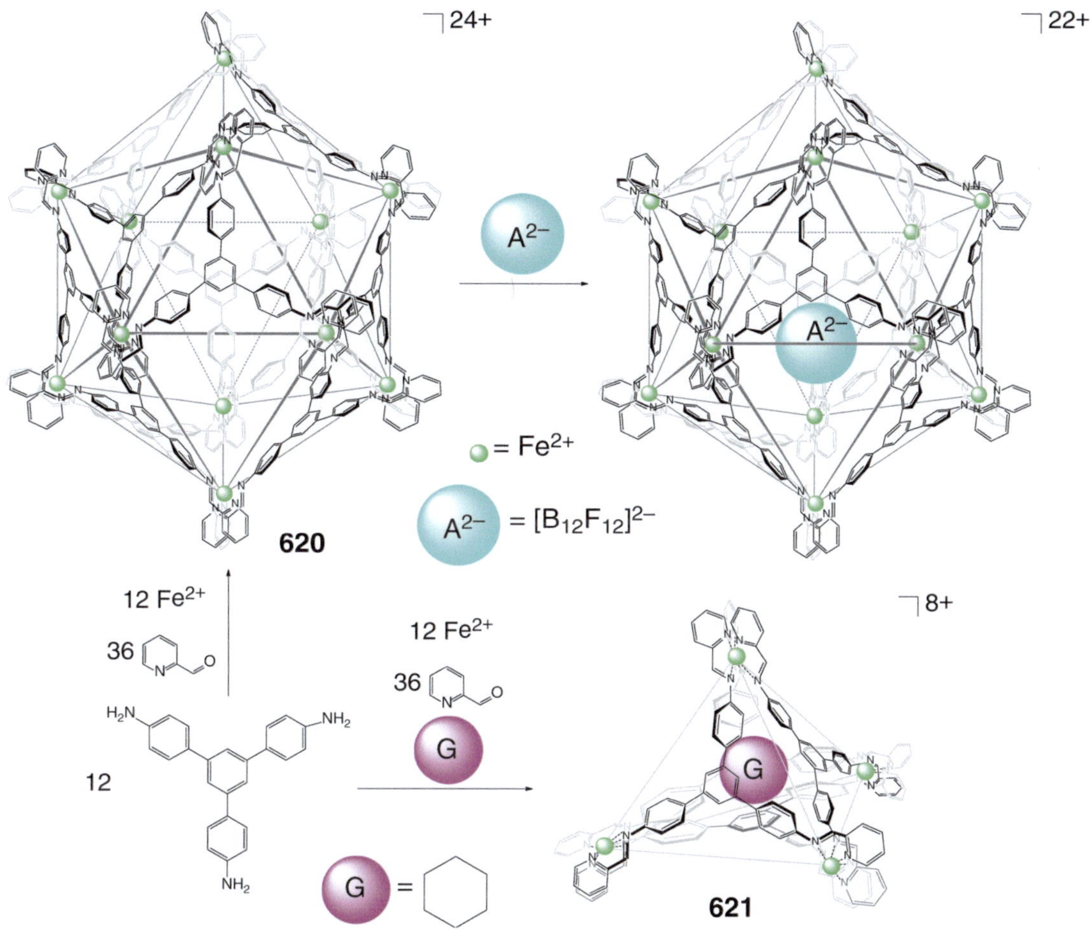

Scheme 4.92

provide parallel modes of modulating the binding of $Mo_2O_7^{2-}$ or iodide anionic guests [88].

An $Fe_{12}L_{12}$ coordination capsule **620** with an icosahedral cage framework has been self-assembled in [89] from appropriate ligand syntones and iron(II) ions by Scheme 4.92. This capsule is preferentially formed from iron(II) triflate in a 1:1 methanol–acetonitrile mixture at 343 K, while an Fe_4L_4 tetrahedral capsule **621** has been isolated from iron(II) triflimide in acetonitrile at 323 K in the presence of cyclohexane as a template. The longest Fe…Fe distances between the antipodal vertices of **620** are approximately 27 Å and its overall diameter is approximately 36 Å; the average Fe…Fe distance along its edges is 14.3 Å. The caging ligand **620** encapsulates dodecafluoro-*closo*-dodecaborate dianion to give a 1:1 cage complex; this binding occurs through electrostatic attraction and van der Waals supramolecular interactions [89].

A tetraaminoporphyrin *323* and its metallocomplexes have been successfully used in [90] for the design and preparation of Schiff-base M_8L_6 cubic coordination capsules **622–624** by Scheme 4.93. In this *one-pot* coordination-driven imine condensation, 2-formylpyridine and iron(II) ions play the roles of an active aldehyde component and a coordinate metallocenter, respectively. The nickel(II)-based coordination capsule **623** has an O_h symmetry, and each of its six low-spin capping NiN_6 metallocenters has the same (either Λ or Δ) configuration. Therefore, its crystal contains both $\Lambda,\Lambda,\Lambda,\Lambda,\Lambda,\Lambda$ and $\Delta,\Delta,\Delta,\Delta,\Delta,\Delta$ enantiomeric cage frameworks with opposite Ni…Ni distances of approximately 15 Å

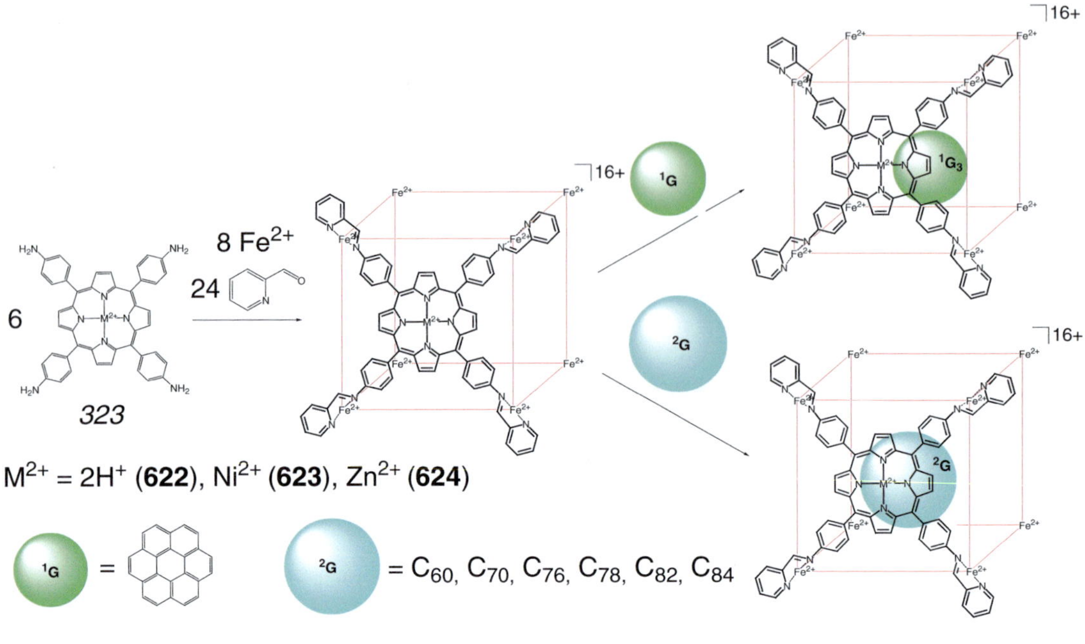

Scheme 4.93

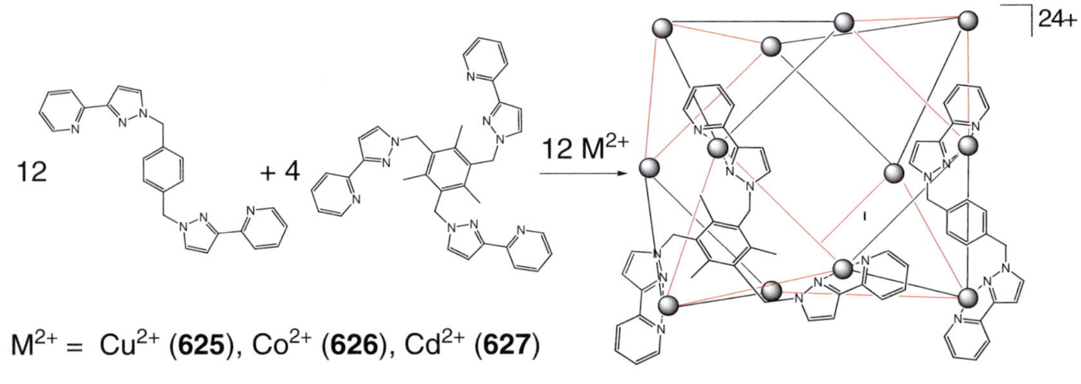

Scheme 4.94

and a cavity volume of 1340 Å^3. These cubic polyaromatic caging ligands are described in [90] to encapsulate planar aromatic molecules of coronene, giving host–guest 1:3 complexes. The ligands **622–624** preferably bind non-spherical (i.e., forming more stacking interactions) C_{70}–C_{84} fullerenes rather than their spherical analog C_{60}; all these bulky guests form 1:1 cage complexes with such hosts [90].

Dodecanuclear cuboctahedral $M_{12}(\mu_3\text{-}L)_4(\mu_2\text{-}L')_{12}$ coordination capsules **625–627** (Scheme 4.94) have been self-assembled in [91] from the corresponding bidentate edge-bridging and triface-capping ligand syntones and characterized using ^{1}H and ^{113}Cd NMR and ESI-MS data. The dodecacadmium cage framework **625** with fourfold D_2 symmetry is reported in this work to remain intact in solution.

A bis-bidentate pyrazolyl-pyridine ligand syntone *324* with a luminescent naphthalene spacer has been used in [92] to design polyhedral compounds bearing photoactive peripheral pendants with high-energy excited states. Its coordination-driven self-assembly with $Ni(BF_4)_2$ in their molar ratio 3:2 by (Scheme 4.95) gave a Ni_8L_{12} coordination capsule **628** with an unusual octanuclear "cuneane"-like framework, which was characterized by single-crystal X-ray diffraction experiment.

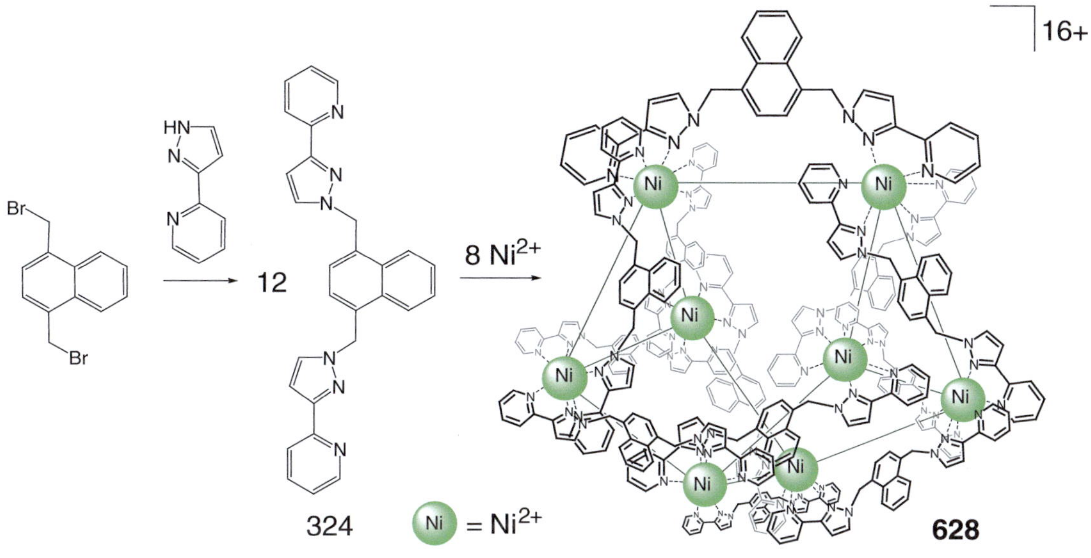

Scheme 4.95

This C_{2v}-symmetric framework is a topological isomer of the cube with a capping hexacoordinate nickel(II) ion at each of its eight vertices that are bridged by bis-bidentate ligand syntones forming each of 12 edges of this cage framework; the Ni… Ni distances along these edges vary in a wide range from 9.77 to 11.21 Å. The Ni_8L_{12} cage framework of **628** is stabilized by strong stacking interactions between the electron-rich naphthalene linkers and the bridging pyrazolyl-pyridine fragments, which are electron deficient due to their coordination to the nickel(II) ions. The authors of [92] noted that among 257 possible eight-vertex polyhedra, only in cubic and cuneane frameworks the capping fragments are three-connected.

Dynamic equilibrium between the bis-ruthenium metallomacrocycle *325*, ligand syntone *241*, and a Ru_6L_6 coordination capsule **629** (Scheme 4.96) is reported in [93] to be shifted in the direction of the corresponding host–guest 1:2 cage complex by addition of coronene as an appropriate guest. In this coordination capsule, the triazine ligand syntones are stacked on top of each other, and two coronene molecules are sandwiched between them.

Self-assembly of a rigid ligand syntone *241* and its spacer-containing analog *326* with palladium and platinum(II) ions by Scheme 4.97 afforded $M_6{}^1L_2{}^2L_2$ catenanes **630** and **631**. These coordination capsules contain two interlocked cage frameworks with an efficient quadruple stacking interaction between their aromatic fragments [94].

4.2.2 Encapsulation of Anions

A tetrahedral Co_4L_6 caging ligand **632** (Scheme 4.98) has been prepared by M.D. Ward and coworkers [95] by coordination-driven self-assembly of a bis-bidentate *ortho*-phenylene pyrazolyl-pyridine ligand syntone *327* with cobalt(II) ions. Each of four cross-linking metallocenters of **632** with pseudooctahedral coordination polyhedron coordinates three bidentate donor fragments of three bis-bidentate syntones and those form a tetrahedral cage framework with six ribbed bridging fragments *327*; the latter is stabilized by intramolecular stacking interactions of their aromatic groups. This positively charged ligand encapsulates tetrafluoroborate anion to give a 1:1 cage complex. Each fluorine atom of the guest is directed to the center of a Co_3 face of this tetragonal ligand **632**; the same is also observed in solution. The efficient encapsulation of BF_4^- anion by it is explained [95] by a good fit of its size to that of the cavity of **632** and by their electrostatic (Coulombic) interactions.

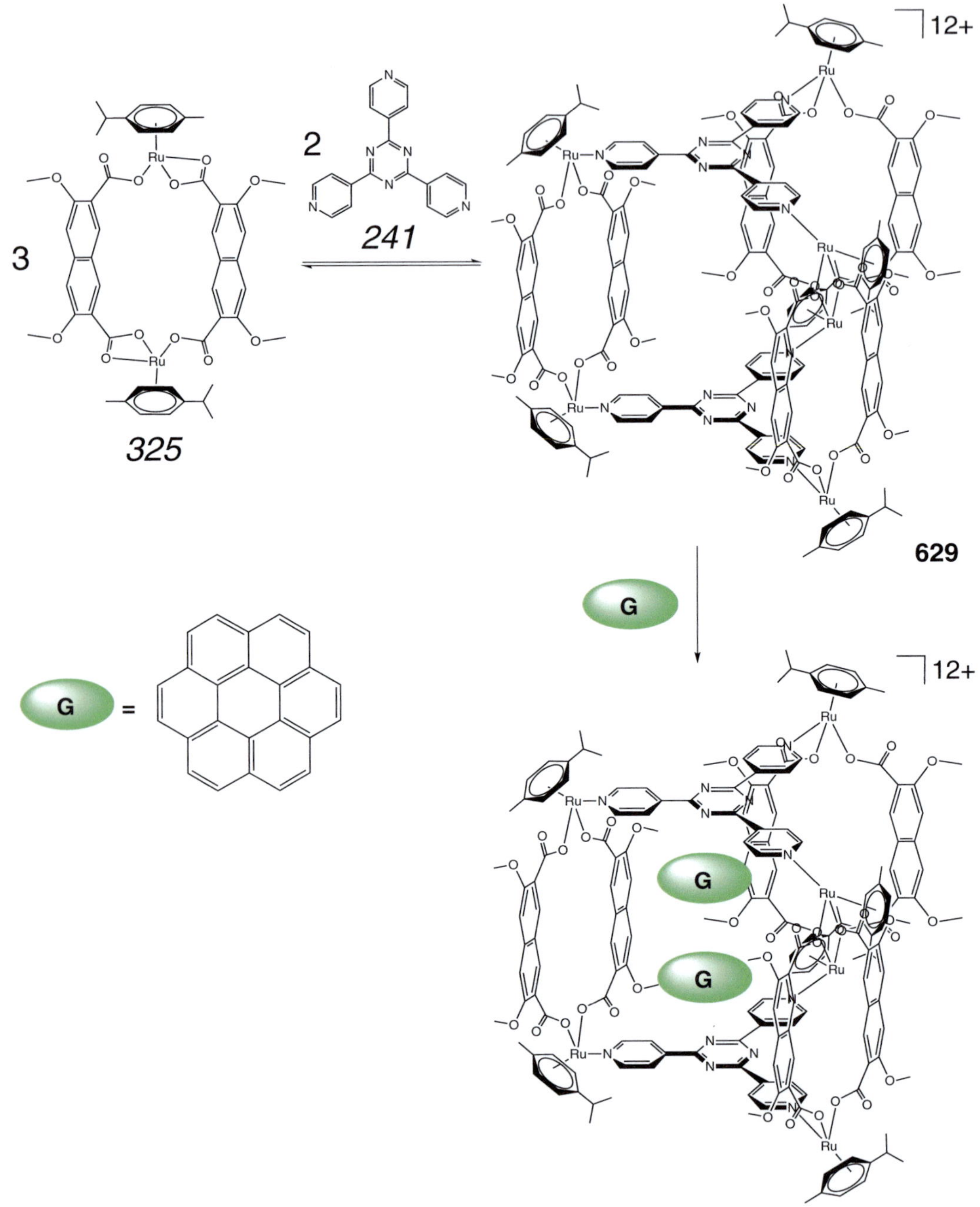

Scheme 4.96

The use of an analogous chiral bis-bidentate syntone *328* allowed diastereoselectively obtain M_4L_6 pseudotetrahedral 1:1 cage complexes of coordination capsules **633** and **634** (Scheme 4.99) with vertex cobalt and zinc(II) cations, respectively, and encapsulated tetrafluoroborate anion [96]. Their only diastereomers showing more than 30-fold increase in the optical rotation per mol have a *T* symmetry. The conformational changes at this diastereoselective formation of a cage complex are

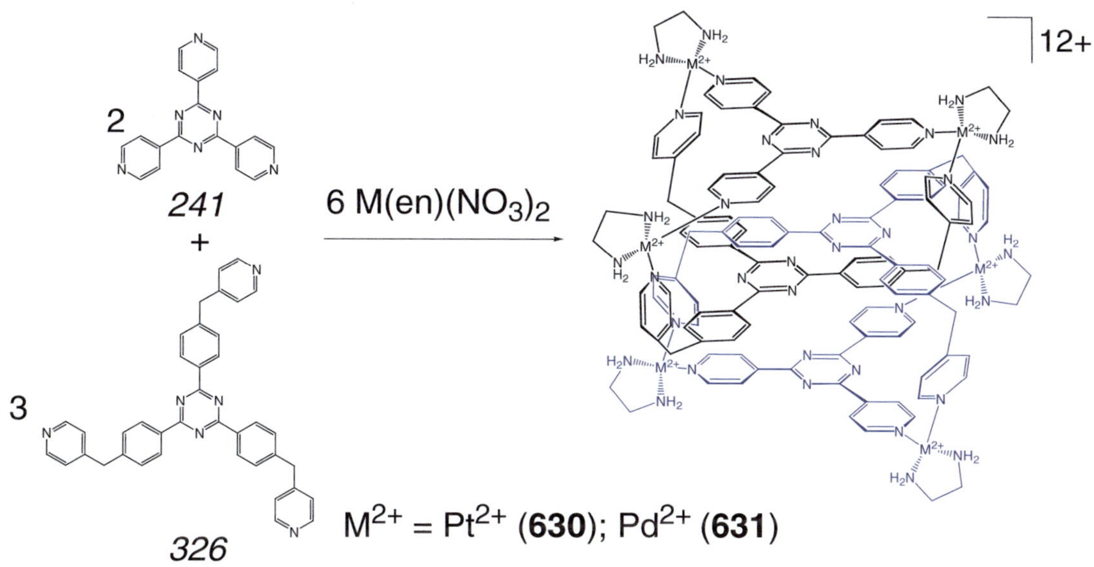

Scheme 4.97

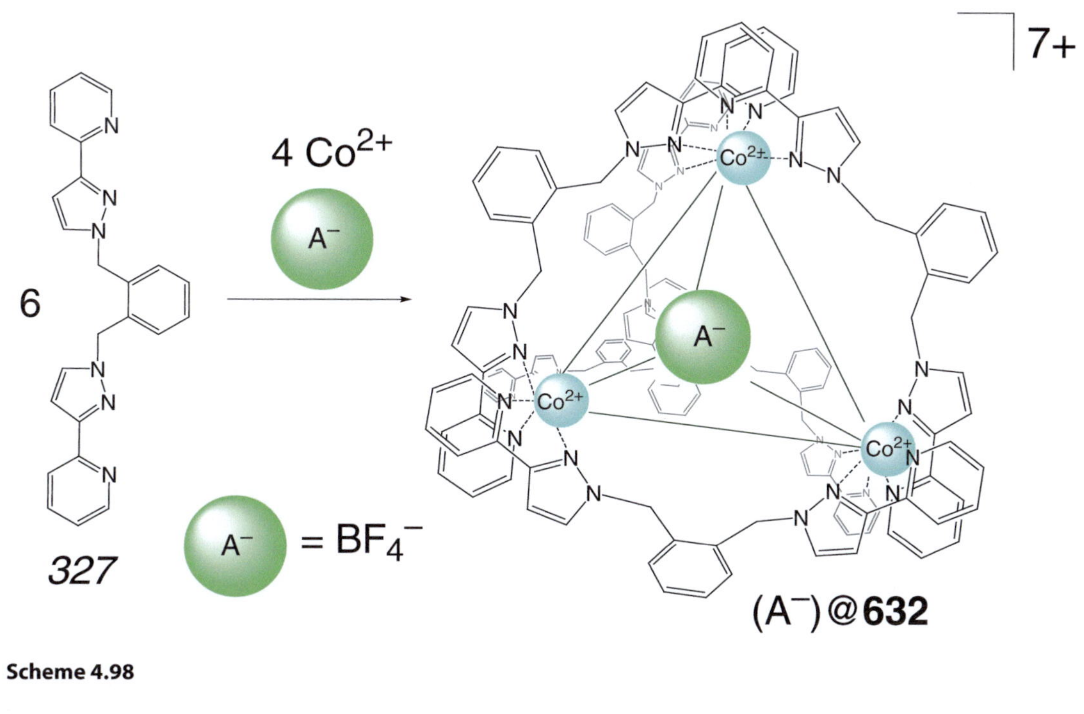

Scheme 4.98

reported to cause an increase in the optical rotation of each chiral bis-bidentate ligand syntone *328* by a factor of 5 upon its coordination to a cross-linking metal ion [96].

Anion-template effect of BF_4^- ion on self-assembly of a *T*-symmetric Co_4L_6 *ortho*-naphthalene capsule **635**, the derivative of ligand syntone *329* (Scheme 4.100), has been confirmed in [97] by ^1H NMR studies; it has been also observed for ClO_4^-anion. In the Co_4L_6 cage complex with encapsulated BF_4^- anion, its vertex cobalt ions have *fac*-pseudooctahedral

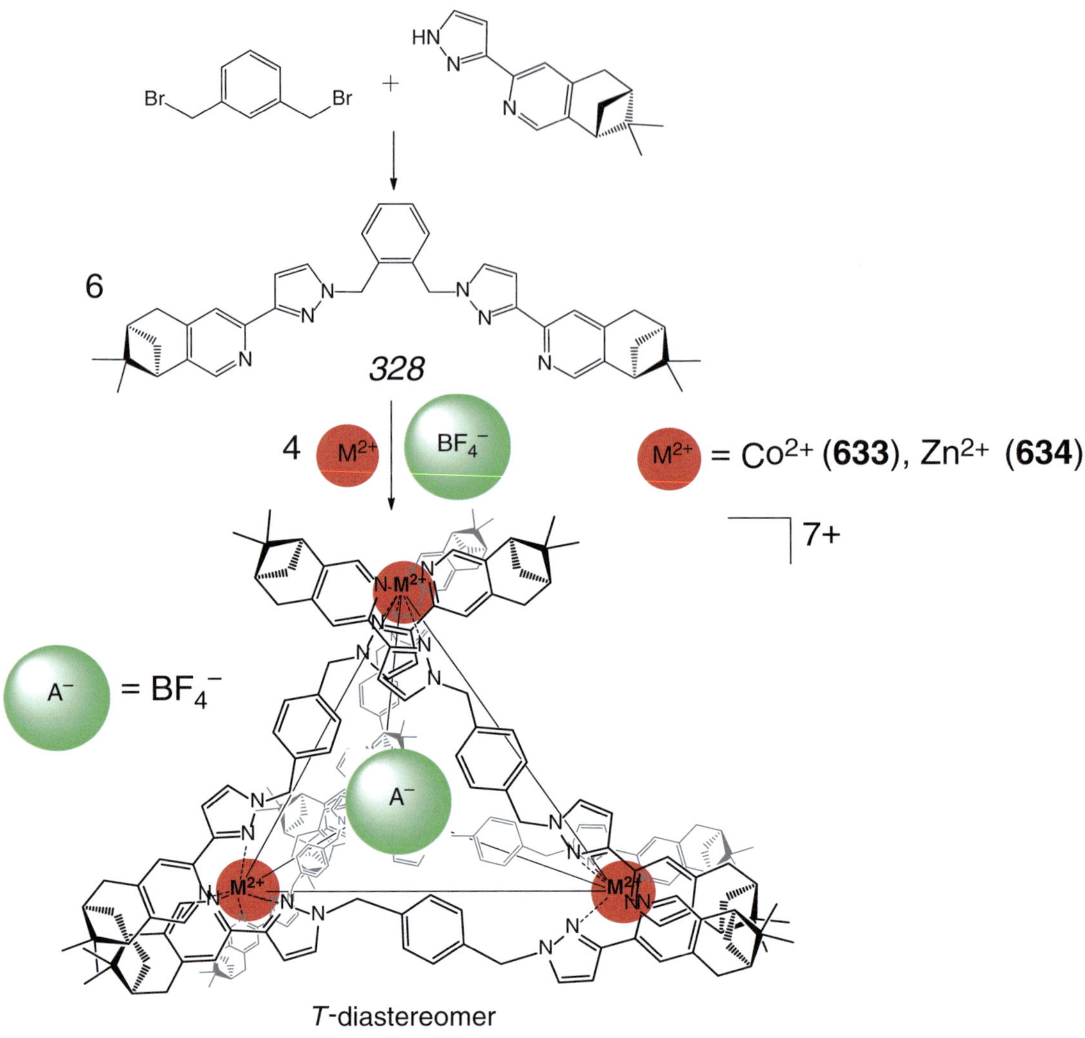

Scheme 4.99

tris-chelate coordination polyhedra, and four fluorine atoms of the encapsulated tetrahedral guest are directed to the centers of four face fragments of this coordination capsule [97].

The same *ortho*-naphthalene pyrazolyl-pyridine ligand syntone *329* has been self-assembled in [98] with zinc(II) ions by Scheme 4.101 to give a similar tetrahedral 1:1 cage complex with encapsulated BF_4^- ion, forming C–H…F hydrogen bonds with methylene fragments of the coordination capsule **636**. Such complex also remains in solution [98].

Self-assembly of a more extended 2,6-pyridine pyrazolyl-pyridine ligand syntone *330* with zinc and cobalt(II) ions by Scheme 4.102 has been performed in [99]. The obtained M_8L_{12} caging ligands **637** and **638** encapsulate BF_4^- and ClO_4^- anions to give the corresponding 1:1 cage complexes. In the Zn_8L_{12} capsule **637** with encapsulated ClO_4^- anion, its S_6-symmetric cubic framework is formed by eight vertex zinc(II) ions with the Zn…Zn separations of 9.75–10.15 Å, linked by bridging ligand syntone fragments. Two of these eight metallocenters have *fac*-tris-chelate coordination polyhedra, while other six Zn^{2+} ions has *mer*-N_6-polyhedra. A non-crystallographic C_3-symmetry axis of this cage framework passes through the opposite zinc(II)

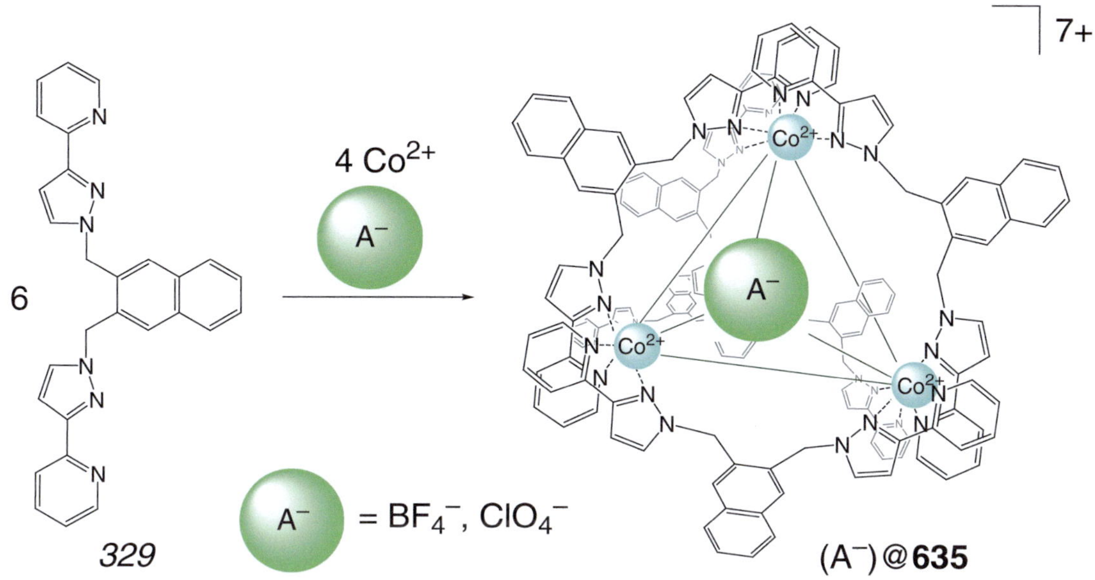

Scheme 4.100

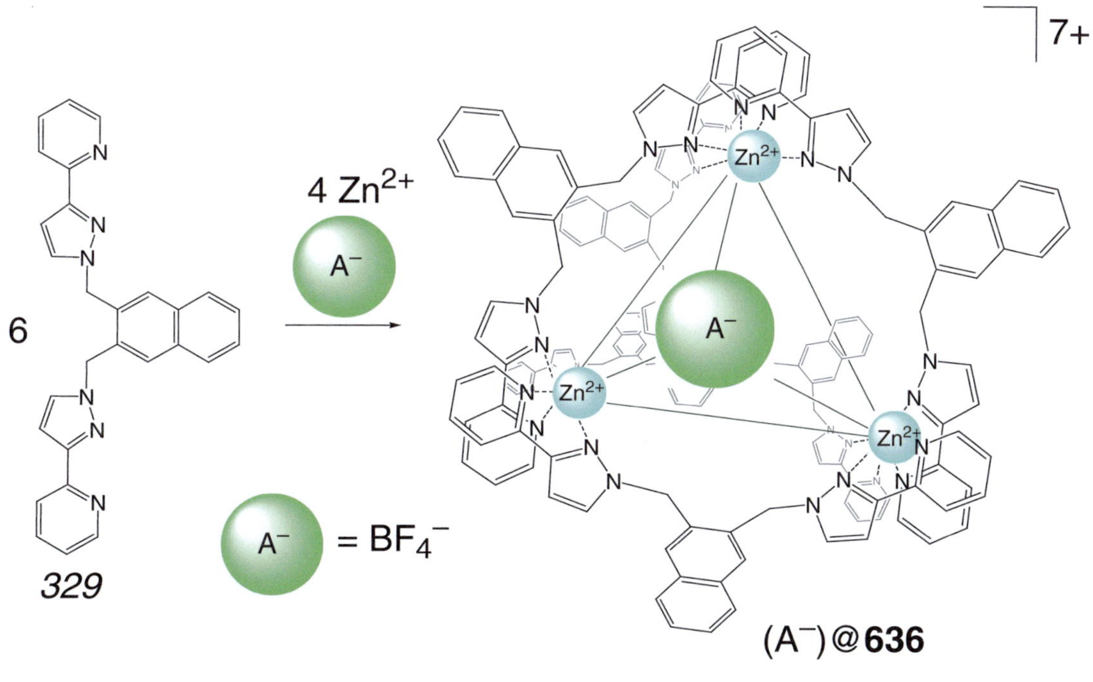

Scheme 4.101

ions with *fac*-coordination polyhedra; the encapsulated perchlorate anion is located in its cavity center [99].

The analogous *meta*-phenylene bis-bidentate ligand syntone *331* also forms the cubic host–guest 1:2 cage complexes **639** and **640** with zinc(II) ions by Scheme 4.103 on tetrafluoroborate anion as a matrix [100]. Their Zn_8L_{12} cage frameworks with two symmetry-related encapsulated anionic guests have almost cubic geometry, and S_6-symmetry axis

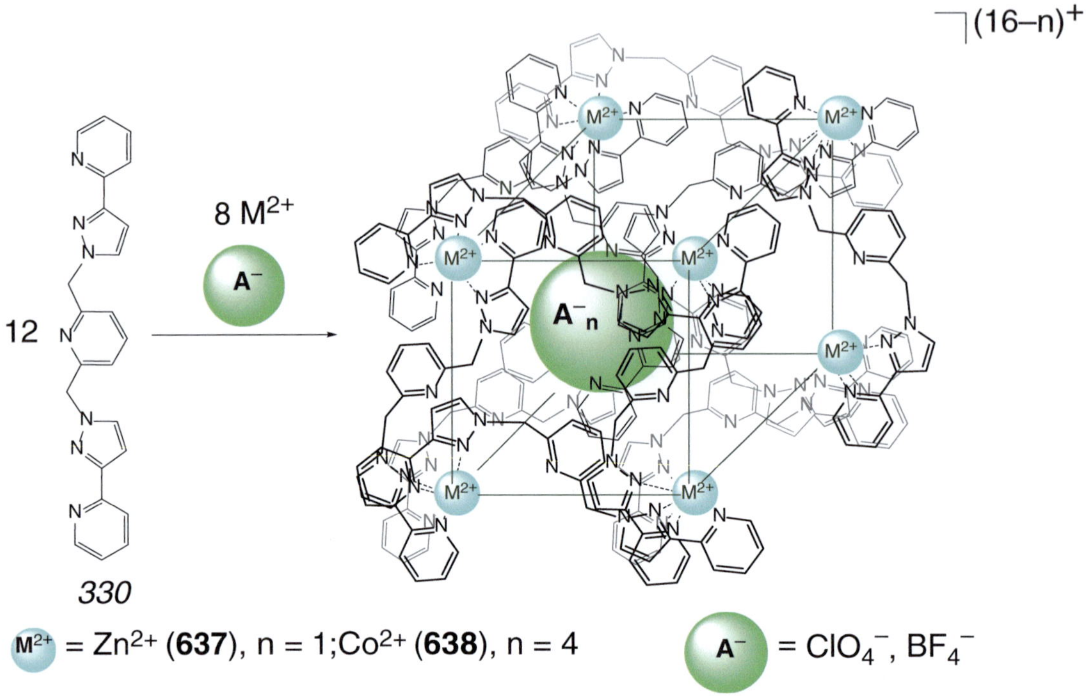

Scheme 4.102

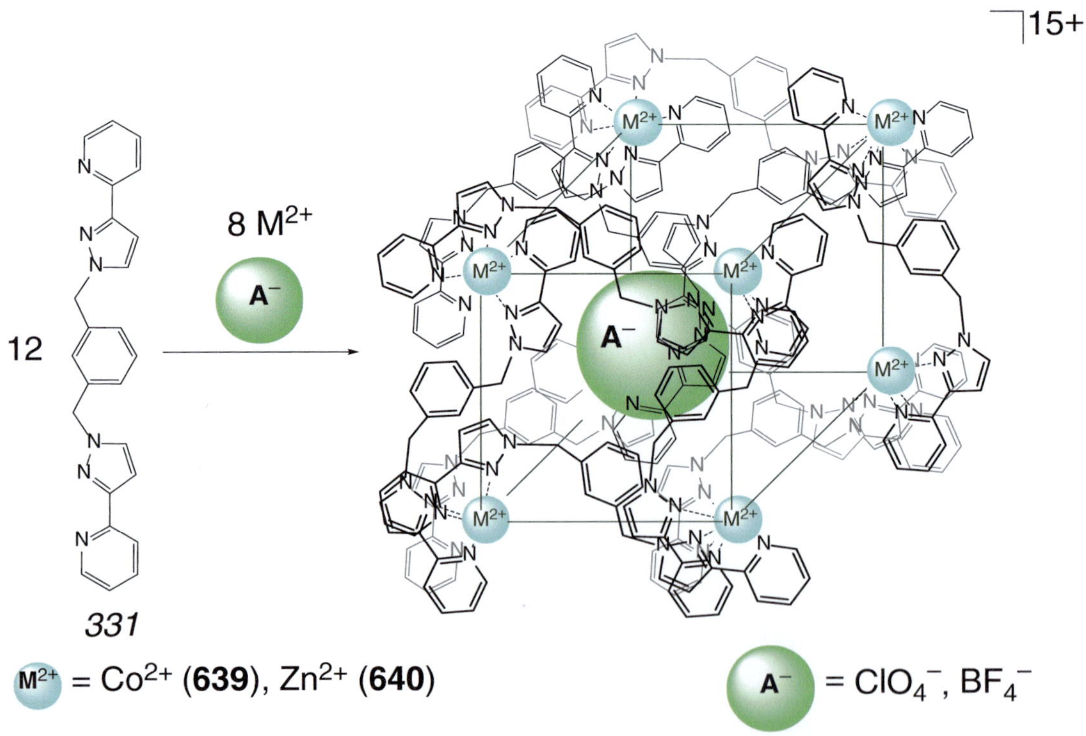

Scheme 4.103

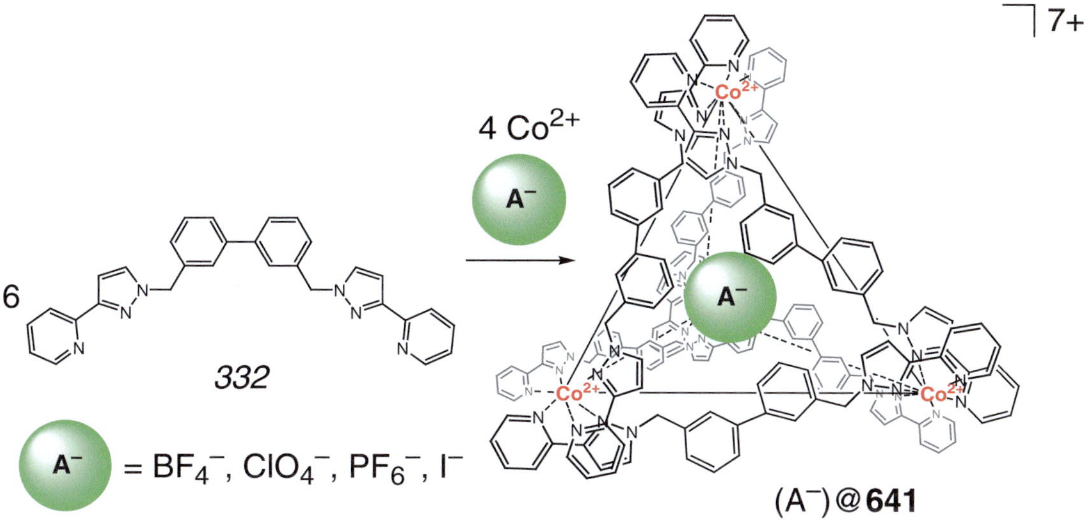

Scheme 4.104

of these frameworks with strong stacking interactions between their aromatic fragments passes through their long diagonal [100].

The detailed X-ray diffraction study of the above M_8L_{12} capsules has been performed in [101]. In *meta*-phenylene Co_8L_{12} cage complex of **638**, eight vertex metal ions form an approximate cubic framework with 11 bridging bis-bidentate ribbed fragments between the hexacoordinate metallocenters. The *meta*-phenylene spacer of each of these linkers is sandwiched between the coordinated pyrazolyl-pyridine donor fragments of two other linkers of this type. Stacking interactions between the electron-deficient and the electron-rich aromatic entities at the distance of approximately 3.5 Å additionally stabilize this coordination capsule that encapsulates four BF_4^- ions. Going to the pyridine-containing cage complexes of **639** and **640** causes a decrease in the number of encapsulated tetrafluoroborate anions: only one BF_4^- ion is caged in their cavities [101].

Anion-templated self-assembly in a system of cobalt(II) ion–aromatic bis-bidentate ligand syntone *332* by Scheme 4.104 in the presence of BF_4^-, ClO_4^-, PF_6^-, and I^- monoanions gave paramagnetic Co_4L_6 pseudotetrahedral 1:1 cage complexes of coordination capsule **641** with an encapsulated monoanion [102]. In this capsule, one of its four vertex cobalt(II) ions has a *fac*-tris-pyrazolyl-pyridine coordination polyhedron with a molecular C_3-pseudoaxis passing through its cross-linking fragment and the caged anion; other three metallocenters have *mer*-tris-chelate polyhedra. The encapsulated $BF_4^-$and PF_6^- ions are shifted in the direction of one of these vertex cobalt(II) metallocenters. Fluorine atoms of large PF_6^- anion are directed to the hydrogen atoms of the encapsulating ligand **641**, thus forming weak C–H…F hydrogen bonds; the average Co…Co distance in this case is similar to those for tetrafluoroborate- and perchlorate-encapsulating complexes (11.92, 11.80 и 11.88 Å, respectively). The same cage structures are also found in solution [102]. Template effect of a caged anion in the case of this cobalt-based encapsulating ligand is less probable than for their zinc-based analogs: the faces of the Co_4L_6 framework contain holes that are large enough for the anions to exchange (as follows from variable-temperature ^{19}F NMR experiments of [102] for BF_4^- and PF_6^- ions).

Coordination-driven self-assembly of a *para*-phenylene bis-bidentate pyrazolyl-pyridine syntone *333* with labile hexacoordinate metal ions such as Zn^{2+} and Cd^{2+} by Scheme 4.105 afforded $M_{16}L_{24}$ coordination capsules **642** and **643**, respectively, having tetracapped truncated tetrahedral cage frameworks with the cavity volumes

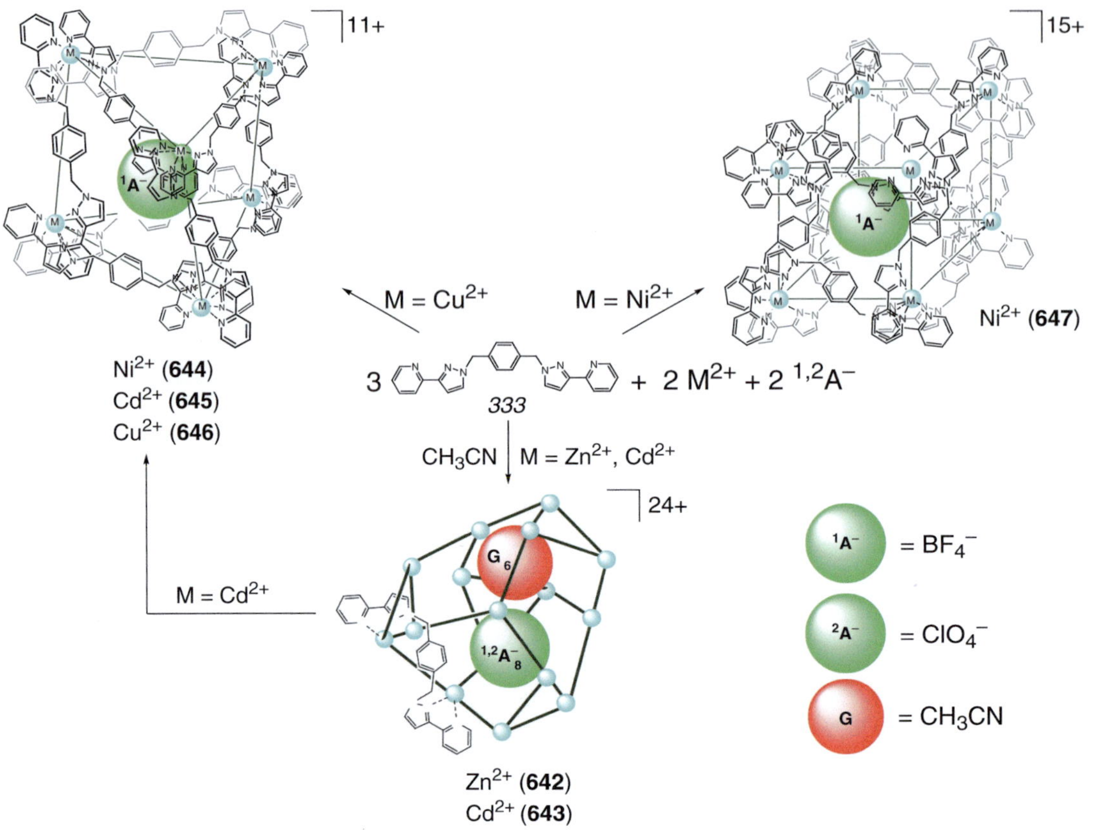

Scheme 4.105

of approximately 700 Å^3 [103, 104]. The cadmium(II)-based ligand **643** encapsulates eight perchlorate anions and six solvate acetonitrile molecules. In this cage complex, each apical metallocenter has a truncated tetrahedral CdN_6-coordination polyhedron formed by three bridging bis-bidentate fragments of its framework. Electron-rich *para*-phenylene spacers as donors (D) interact with coordinated to a metal ion, electron-deficient pyrazolyl-pyridine acceptors (A) forming 12 five-component *A–D–A–D–A* stacks. These interactions stabilize the cage framework with an effective diameter of its cavity of 11 Å and a cavity volume of approximately 700 Å^3; this large cavity allows encapsulating at least eight perchlorate anions and six solvent acetonitrile molecules [104]. While the analogous zinc(II)-based cage complex of **642** has the same crystal and molecular structure, in the case of copper(II) ion a hexanuclear 1:1 cage complex **646** with encapsulated BF_4^- ion has been isolated in [104]. As follows from X-ray diffraction data, the guest anion is included in C–H…F hydrogen bonds. Self-assembly of the same ligand syntone and $Ni(BF_4)_2$ in the molecular ratio 3:2 gave an octanuclear 1:1 cage complex of **647** with eight cross-linking metallocenters forming a distorted cubic polyhedron (Scheme 4.105) that contains encapsulated BF_4^- anion. This capsule has S_6 symmetry, and extensive interligand aromatic stacking with electron-rich phenyl ring sandwiched between a pair of coordinated electron-deficient pyrazolyl-pyridine unit gives a total of eight five-layer alternating *A–D–A–D–A* stacks, each one associated with a face of the cubic cage framework. As follows from ^1H NMR and ESI-MS data, the $Cd_{16}L_{24}$ capsule **643** is only a kinetic intermediate product of a self-assembly reaction that easily converts to the thermodynamically more stable

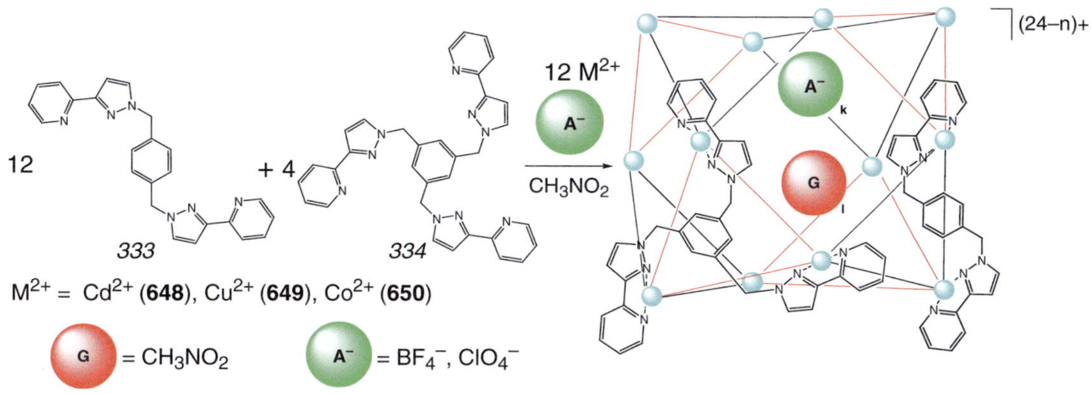

Scheme 4.106

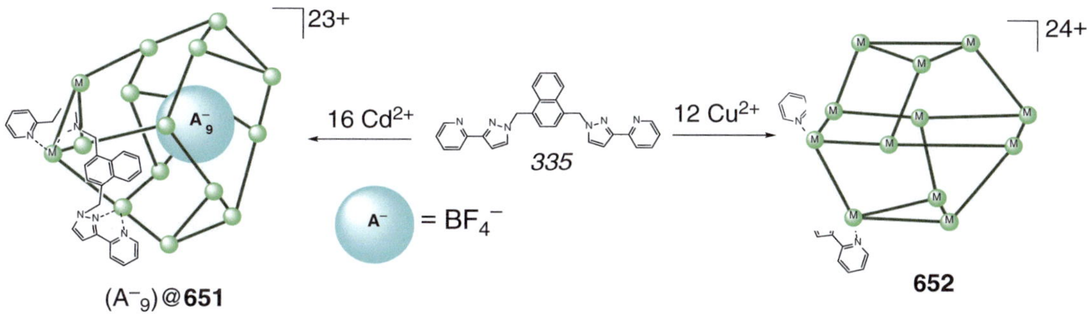

Scheme 4.107

hexanuclear cage framework **645**. In the case of nickel(II) ions, both the hexanuclear and octanuclear capsules **644** and **647** have been detected in [104].

Coordination-driven self-assembly of a mixture of an edge-bridging bis-bidentate ligand syntone *333* and a face-capping tris-bidentate syntone *334* with cadmium(II) ions allowed obtaining by Scheme 4.106 heteroligand $Cd_{12}L_{12}L'_4$ coordination capsules **648** with an unusual cuboctahedral cage framework formed by eight trigonal and six square faces [103]. The formation of the corresponding homoligand capsules has been also observed in this work. The copper(II)-based caging ligand **649** with a cavity volume of approximately 450 Å^3 encapsulates disordered tetrafluoroborate anions and solvent nitromethane molecules [103].

Naphthalene-containing analogs of **642** and **643** have been prepared in [105] from the corresponding rigid bis-pyrazolyl-pyridine ligand syntone *335*. Its self-assembly with cadmium(II) ions by Scheme 4.107 gave a tetracapped truncated tetrahedral $Cd_{16}L_{24}$ coordination capsule **651** with Cd…Cd distances of 9.17–10.46 Å. Each of 24 ribbed fragments of its *T*-symmetric framework with a cavity volume of 1300 Å^3 is formed by these rigid syntones, which are cross-linked by 16 chiral hexacoordinate metallocenters. The caging ligand **651** with stacking intramolecular interactions between its electron-rich naphthalene groups and electron-deficient pyrazolyl-pyridine fragments encapsulates nine tetrafluoroborate anions to give a host–guest 1:9 cage complex. As follows from ^{1}H and ^{113}Cd NMR data, this complex is stable even in its diluted solution for a long time due to the above extensive stacking interactions. Its cage framework contains 12 *mer*- and 4 *fac*-tris-bidentate cadmium(II)-based metallocenters, as confirmed [105] by X-ray diffraction study. A

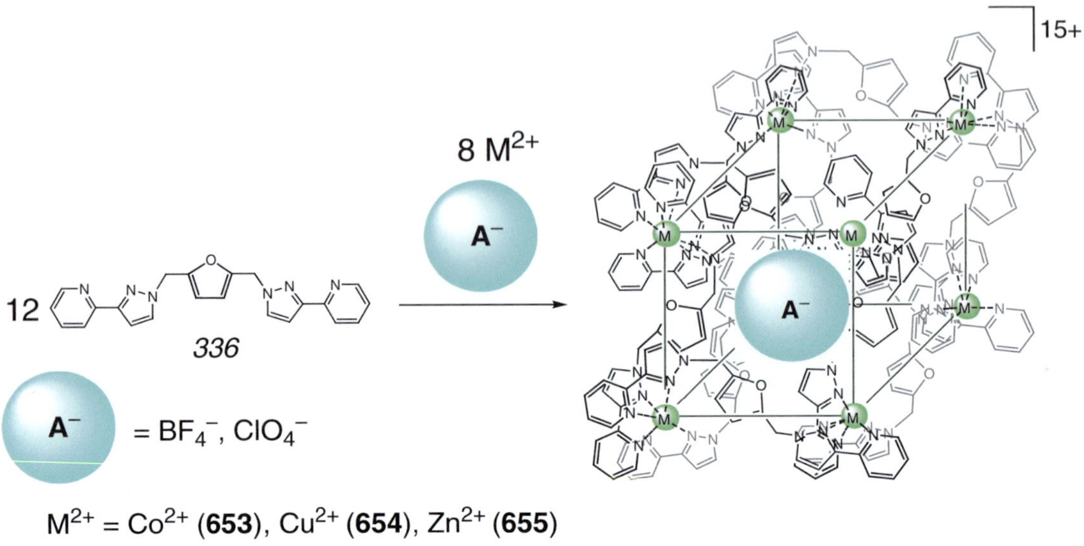

Scheme 4.108

fluorescence solution spectrum of the above complex in the visible range also contains a broad low-energy band characteristic of an excited state of the naphthalene fragments stabilized by their interactions with neighboring electron-deficient groups in a stacked ligand array [105]. Self-assembly of the ligand syntone *335* with copper(II) ions proceeds much less efficiently: the authors of [105] failed to obtain an analogous $Cu_{16}L_{24}$ coordination capsule and its cage complexes by Scheme 4.107 with a reasonable yield; only crystals of the $Cu_{12}L_{15}$ framework **652** with an uncomplete and irregular polyhedron have been isolated in a low yield and then characterized by X-ray diffraction data [105].

The pyrazolyl-pyridine ligand syntone *336* is reported in [106] to undergo self-assembly by Scheme 4.108 with *3* metal (II) ions that prefer octahedral coordination to give D_4-symmetric M_8L_{12} pseudocubic capsules **653–655**; their two cyclic tetranuclear helicate M_4L_4 units are connected by four additional pillar ligand syntones. As follows from X-ray diffraction data of [106], these 16-cationic coordination capsules encapsulate tetrahedral BF_4^- and ClO_4^- anions, thus forming 1:1 cage complexes.

Self-assembly of a 1,8-naphthalene pyrazolyl-pyridine syntone *337* with copper(II) ions by Scheme 4.109 gave 1:4 cage complex of a dodecanuclear $Cu_{12}L_{18}$ ligand **656** [107]. This *T*-symmetric coordination capsule with a non-crystallographic C_3 symmetry has a truncated tetrahedral framework with the vertex hexacoordinate Jahn–Teller distorted copper(II) ions. The naphthalene spacers and pyrazolyl-pyridine fragments form strong stacking interactions on a periphery of this cage complex with four encapsulated perchlorate ions forming weak O…H–C hydrogen bonds with the interior of the inner cavity of **656**. All tris-chelate metallocenters of this framework have the same configurations that results in its chirality. Cobalt and cadmium(II) ions are described to give the same structurally labile $M_{12}L_{18}$ capsules **657** and **658**. In contrast, self-assembly of zinc(II) ions with an optically active pyrazolyl-pyridine ligand syntone *338* afforded a tetrahedral Zn_4L_6 cage framework **659** by Scheme 4.110. This framework, which encapsulates disordered BF_4^- anion, contains four vertex zinc(II) ions and ribbed bridging bis-bidentate fragments. It is the single enantiomer in which the chirality of its pinene fragments controls the chirality of coordination capsule **659** [107].

A 1:1 cage complex of the tetrahedral Hg_4L_6 coordination capsule **660** with encapsulated perchlorate anion has been prepared in [108] by

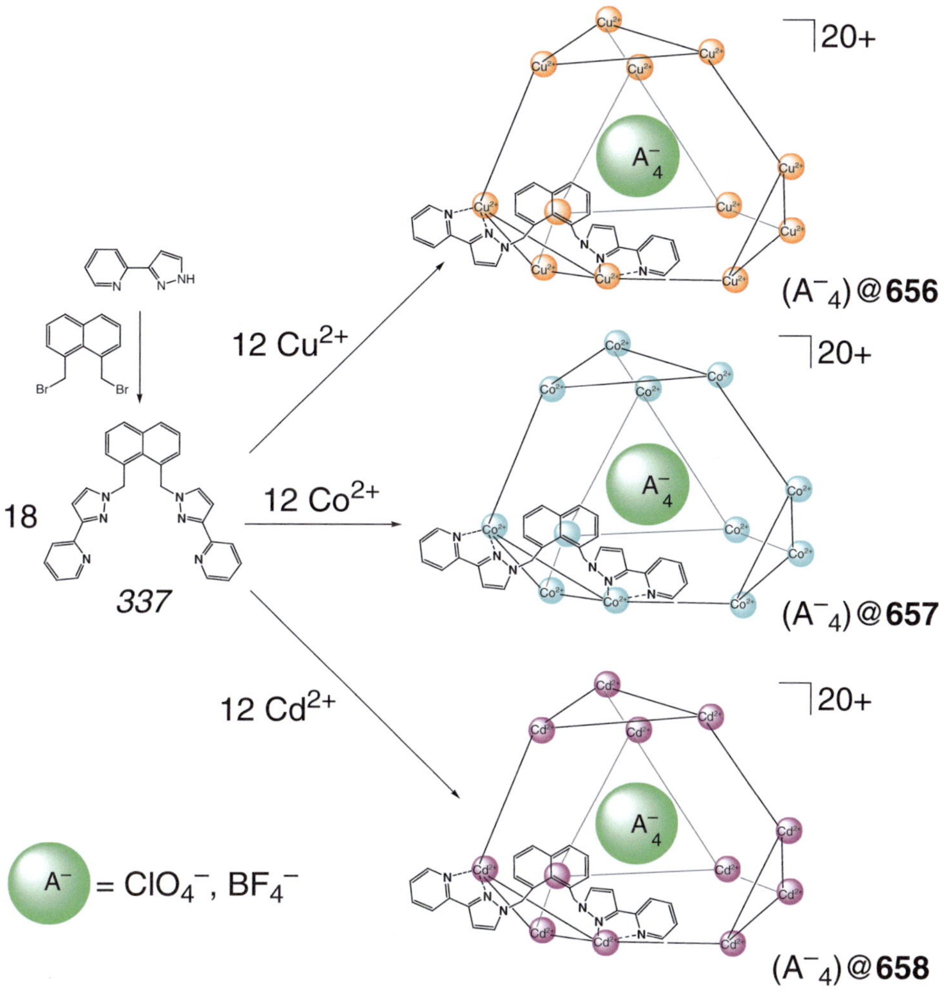

Scheme 4.109

Scheme 4.111. It has a C_3-symmetric cage framework with Hg…Hg separations along the edges varying in a narrow range of 11.64–12.23 Å. Two types of metal coordination geometry are present in this framework: one of Hg^{2+} ions has a *fac*-tris-chelate geometry, when the three others have a *mer*-geometry. All metal centers have the same optical configuration (all Δ or all Λ) with both enantiomers present in the unit cell. Intramolecular stacking interactions are reported in [108] to be an important feature of **660**.

Extensive stacking interactions between the adjacent ligand syntones resulted in red-shifted "excimer-like" luminescence in the $Zn_4L'_6$ and $Cu_{12}L''_{18}$ cage complexes of naphthalene-containing coordination capsules **636** and **656** (Scheme 4.112), allowed the monitoring of their formation [109].

A 1,8-naphthalene-containing Cd_4L_6 coordination capsule **661** having a tetrahedral cage framework with an average Cd…Cd distance of 9.28 Å is reported in [110] to encapsulate BF_4^- anion. The use of zinc(II) hexafluorosilicate in the self-assembly gave a 1:1 cage complex of its zinc-based analog **636** with encapsulated SiF_6^{2-} dianion. The average Zn…Zn distance in its tetrahedral framework (approximately 9.73 Å) is very close to that in the analogous complex with caged BF_4^- anion (approximately 9.72 Å) [107].

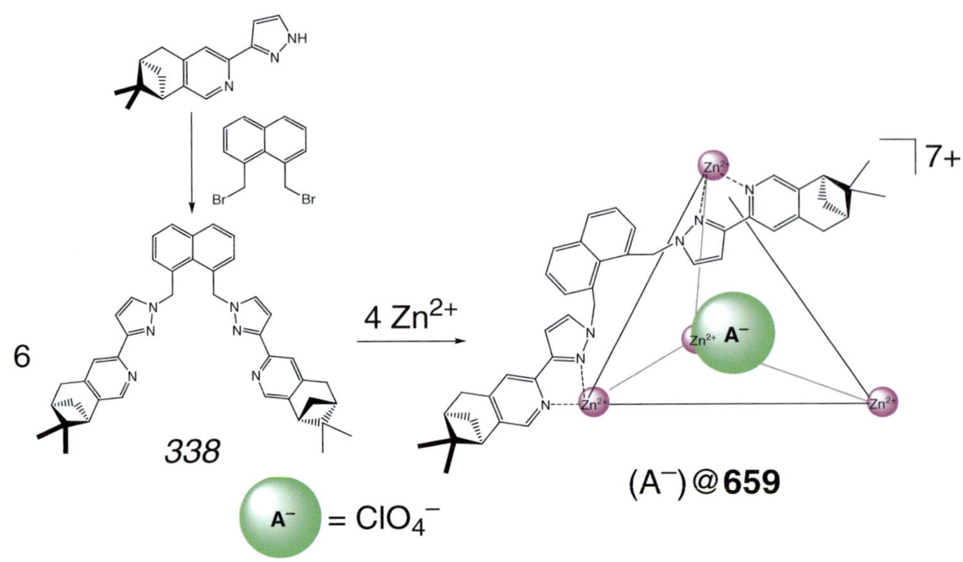

Scheme 4.110

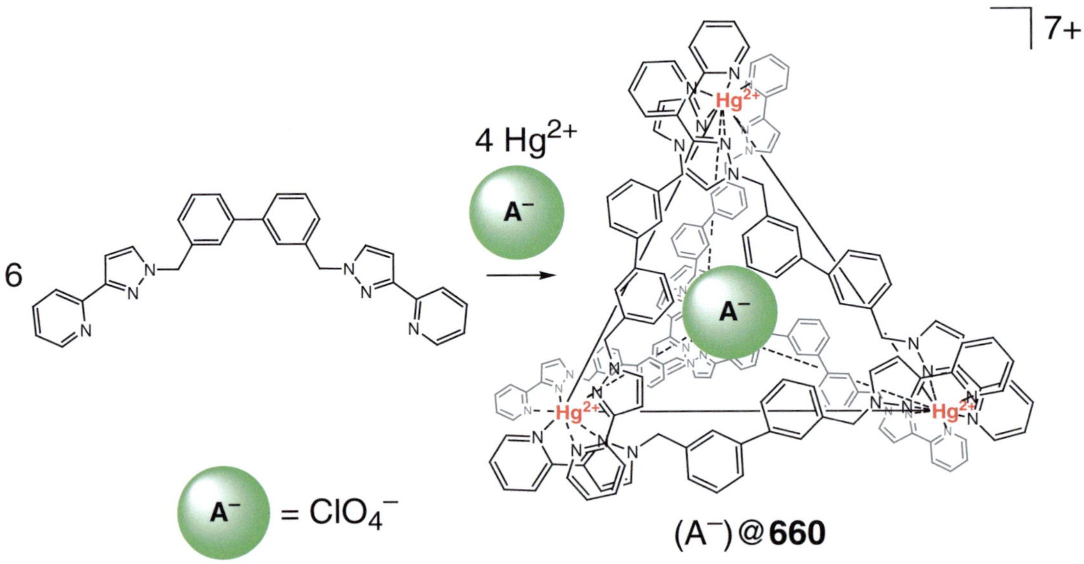

Scheme 4.111

A similar ligand syntone *339* with a 1,10-anthracene spacer fragment has been assembled with zinc and cobalt(II) ions by Scheme 4.113 to give M_4L_6 capsules **662** and **663** [110]. The cage complex of **662** with encapsulated tetrafluoroborate anion has a non-crystallographic *T* symmetry caused by the same chirality of its vertex tris-chelate metallocenters. The anthracene spacers are sandwiched between the coordinated pyrazolyl-pyridine fragments of two neighboring ribbed bridging fragments; the Zn…Zn distances along them are approximately 9.7 Å. Its cage framework retains a twofold symmetry in solution. The isostructural cobalt(II)-based capsule

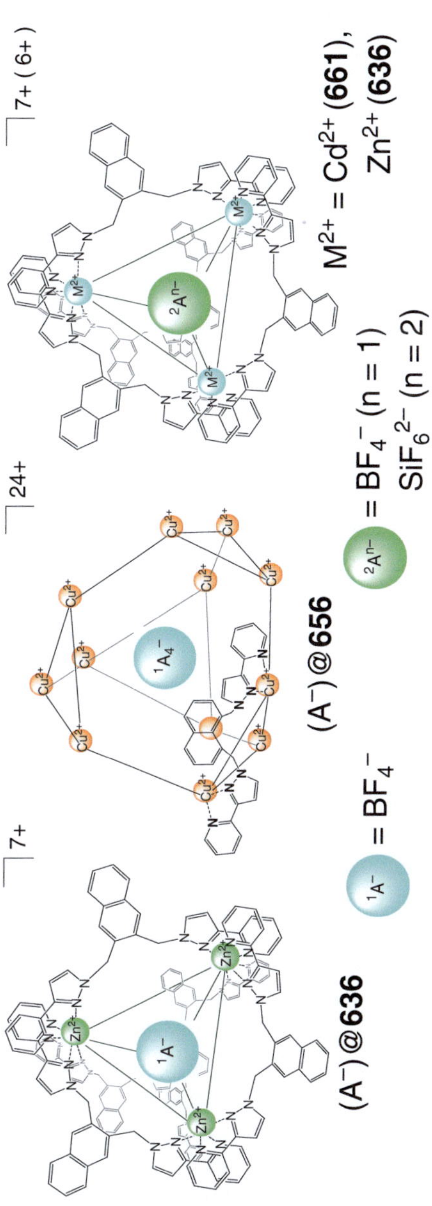

Scheme 4.112

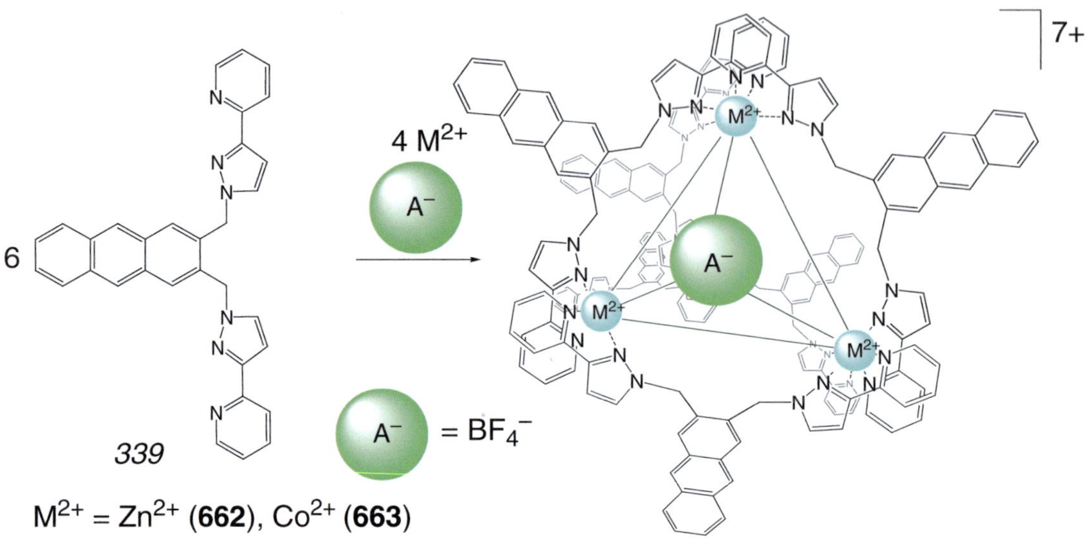

Scheme 4.113

663 has been prepared in [110] using a solvothermal method.

Fe_4L_4 face-capped tetrahedral coordination capsules **664–667** have been synthesized in [111] using imine condensation (including a guest-templated procedure) of the corresponding C_3-symmetric triamines *340–343*, 2-formylpyridine, and iron(II) ions by Scheme 4.114; three of these amine syntones also form the corresponding Fe_2L_3 helicates.

An eight-cationic Fe_4L_6 coordination capsule **668** has been self-assembled in [112] by Scheme 4.115 from 4,40-diaminobiphenyl, 2-formylpyridine, and iron(II) ions. Its cage framework exists in solution as a mixture of interconverting *T*-, C_3-, and S_4-diastereomers, and encapsulation of guest anions resulted in a redistribution between them [112].

A tripodal ligand syntone *344*, obtained by condensation of pyridyl-containing donor components with an appropriate rigid triptycene precursor, is reported in [113] to undergo coordination-driven self-assembly with europium(III) ions by Scheme 4.116, giving a tetrahedral Eu_4L_4 1:1 cage complex with encapsulated perchlorate anion. An average Eu…Eu distance in coordination capsule **669** is approximately 12.0 Å; therefore, its cavity is large enough to encapsulate also tetrafluoroborate and triflate anions [113].

Rigid 3-pyridyl-terminated iron(II) clathrochelates *345* and *346* have been used in [114] as ditopic *N*-donor angular ligand syntones for the synthesis of large heterometallic Pd_6L_{12} coordination capsules **670–673** by their coordination-driven self-assembly with palladium(II) ions by Scheme 4.117. The cavities of their octahedral cage frameworks **670**, **672**, and **673** are almost completely occupied by guest solvent molecules and NO_3^- or BF_4^- anions. These anions are located in interior binding pockets close to the vertex palladium(II) ions. A cavity volume of these frameworks is approximately 5200 Å^3, but much of it is occupied by the ribbed substituents of their macrobicyclic ligand syntones. As a result, these coordination capsules have the void cavity volumes of approximately 1400/1800, 1300, and 1100 Å^3, respectively. The conformation of their clathrochelate syntones in crystal affects the internal space within these dodecacationic coordination capsules having central cavity with a hydrophobic internal environment formed by methyl or cyclohexane substituents at the cage framework. The coordination capsule **670** gave 1:1 cage complex with encapsulated lipophilic tetraphenylborate anion by Scheme 4.119; this encapsulation has been confirmed in [114] by single-crystal synchrotron X-ray diffraction data.

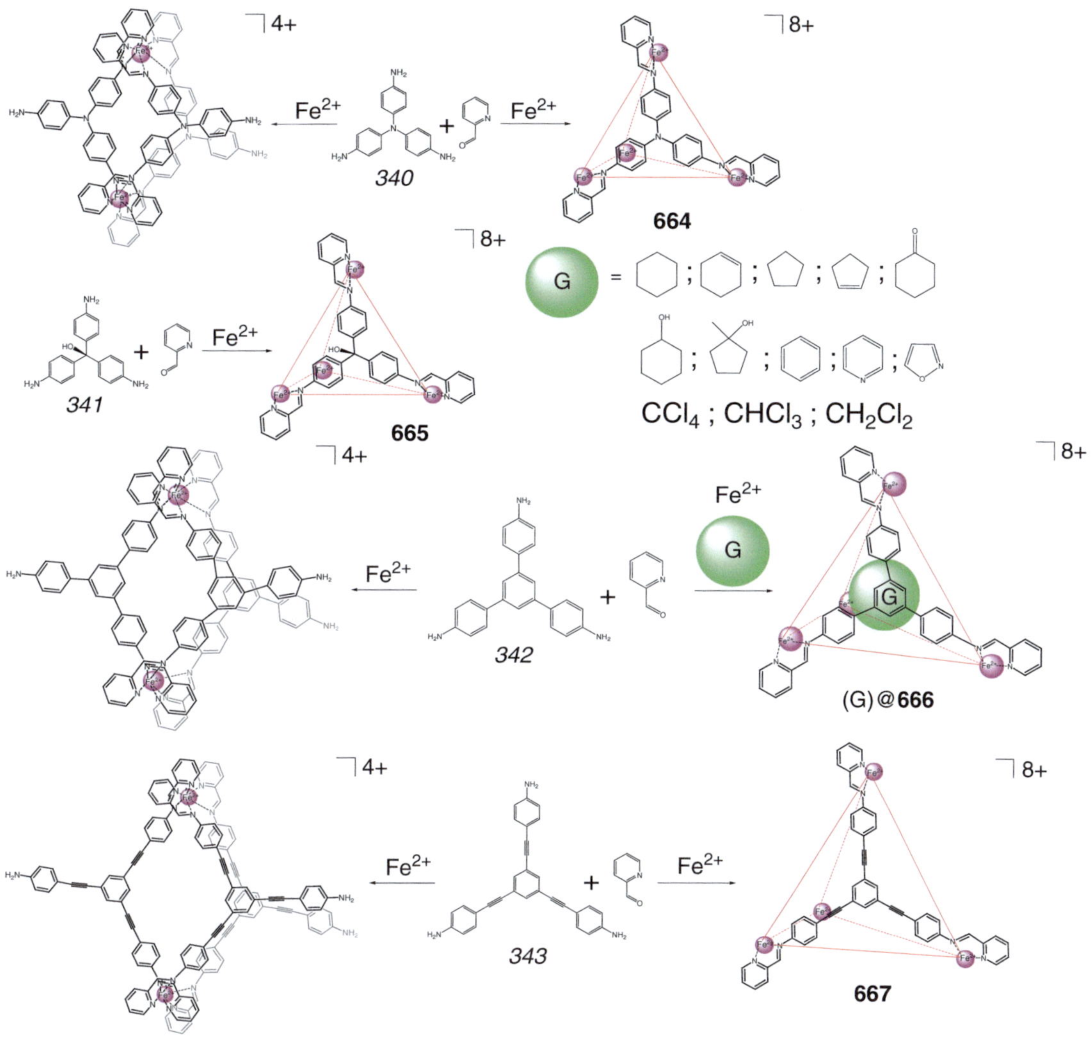

Scheme 4.114

4.2.3 Encapsulation of Cations

Encapsulation-driven interconvertion of anthracene-containing M_2L_3 triple helicates into the corresponding M_4L_6 cage complexes by Scheme 4.118 in the presence of tetramethylammonium cation as a guest has been performed by K. Raymond and coworkers in [115]. Self-assembly of the ligand syntone *347* with $TiO(acac)_2$ and $Ga(acac)_3$ gave these helicates in the absence of this cation and the M_4L_6 coordination capsules **674** and **675** in its presence. These tris-helicates have been converted into their tetrahedral cage derivatives by addition of tetramethylammonium chloride. A highly symmetric Ti_4L_6 1:1 cage complex with encapsulated $(CH_3)_4N^+$ cation is reported to be a racemic mixture of homochiral tetrahedral frameworks ($\Delta\Delta\Delta\Delta$ or $\Lambda\Lambda\Lambda\Lambda$ at its metal vertices) with an average Ti…Ti distance of 16.1 Å.

Five bis-catecholate ligand syntones (Scheme 4.119) have been used [116] to evaluate combinatorial libraries of their coordination self-assemblies with metal ions as hosts for various guests. As follows from ESI-MS and NMR data, the dynamic combinatory library with many assemblies simultaneously being present in a solution is made from labile ligand syntones and

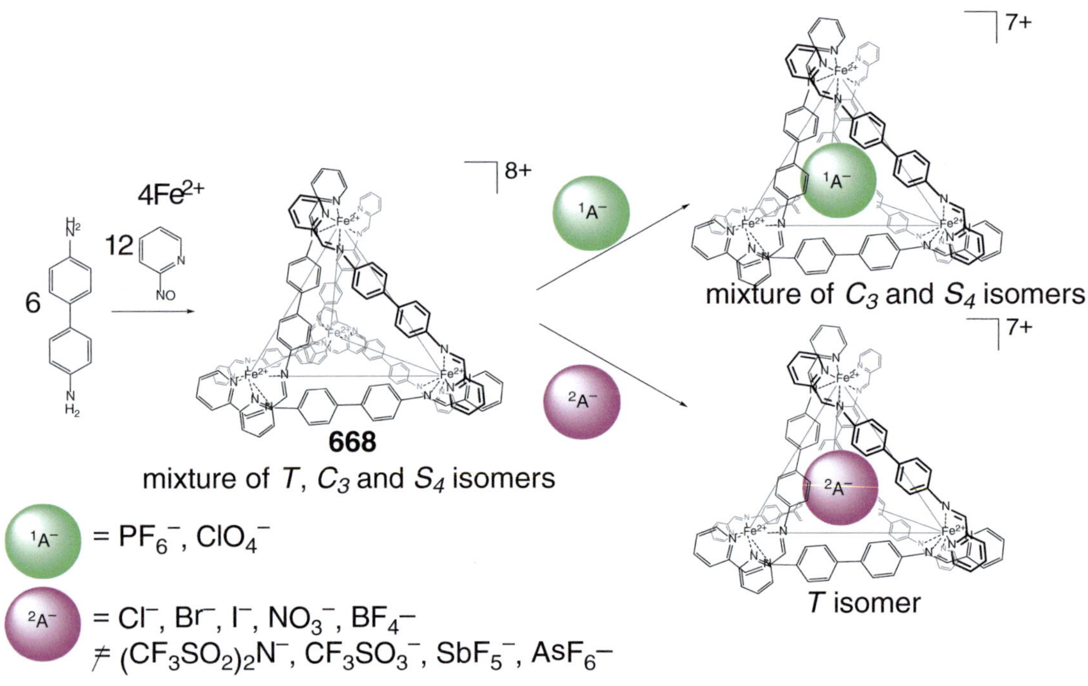

Scheme 4.115

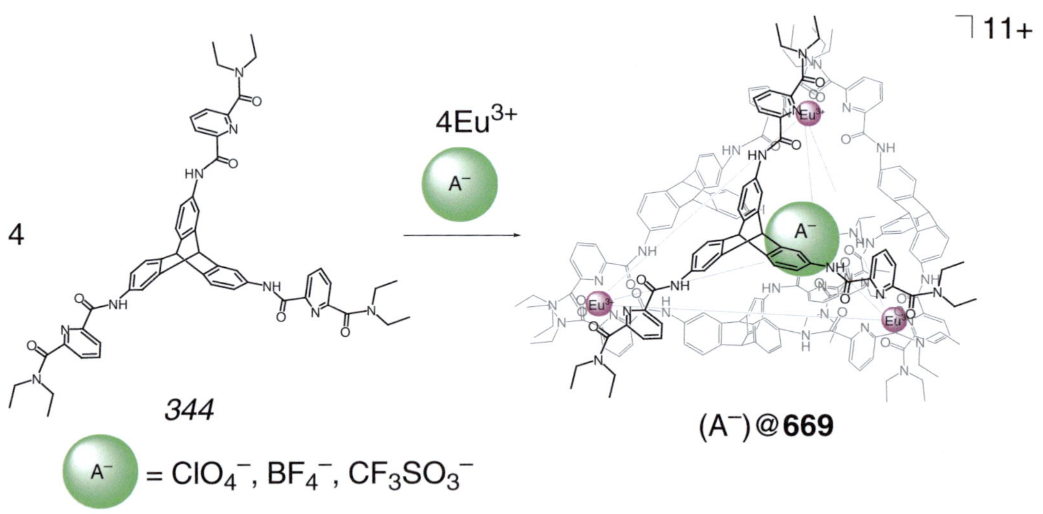

Scheme 4.116

coordinate metal ions forming kinetically inert coordination capsules.

To avoid the formation of M_2L_3 complexes as to favor the tetrahedral M_4L_6 capsule, K. Raymond and coworkers used [117] a rigid ligand syntone *348* with a naphthalene spacer. The latter gave a 1:1 cage complex of capsule **575** (see Sect. 4.2.1) with encapsulated tetrabutylammonium cation by Scheme 4.120; this complex was characterized by ^1H NMR and X-ray diffraction data. The caging ligand **575** is also able to encapsulate various organic cations, such as $(CH_3)_2(C_3H_7)_2N^+$, $(C_3H_7)_4N^+$, and

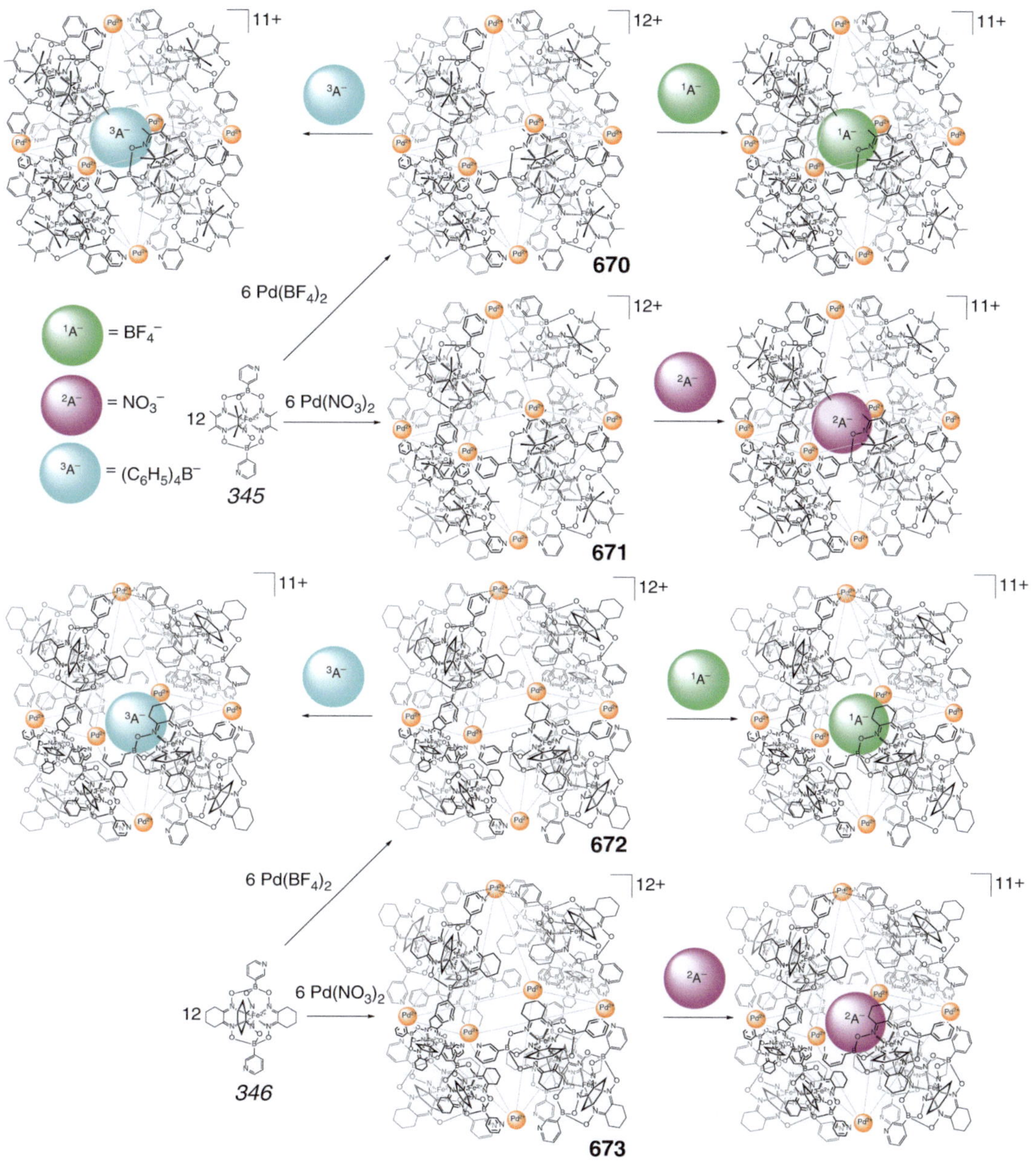

Scheme 4.117

N,N,N′,N′-tetramethyl-1,3-propanediammonium, in their aqueous solutions [118].

Complete resolution of isomers of a chiral anionic capsule **577** has been performed in [119] with the chiral *N*-methylnicotinium cation as a guest. The latter is an ideal probe for determining the diastereomeric excess of the resolved capsule, as it readily undergoes further exchange reaction with tetraethylammonium cation with complete retention of chirality (such enantiopurity retains in alkaline aqueous solution for a long time). It gives the enantiopure *ΔΔΔΔ* and *ΛΛΛΛ* isomers of the corresponding 1:1 cage complexes (Scheme 4.121) the absolute configuration and enantiopurity of which was determined in [119] from their CD spectra; the absolute configuration

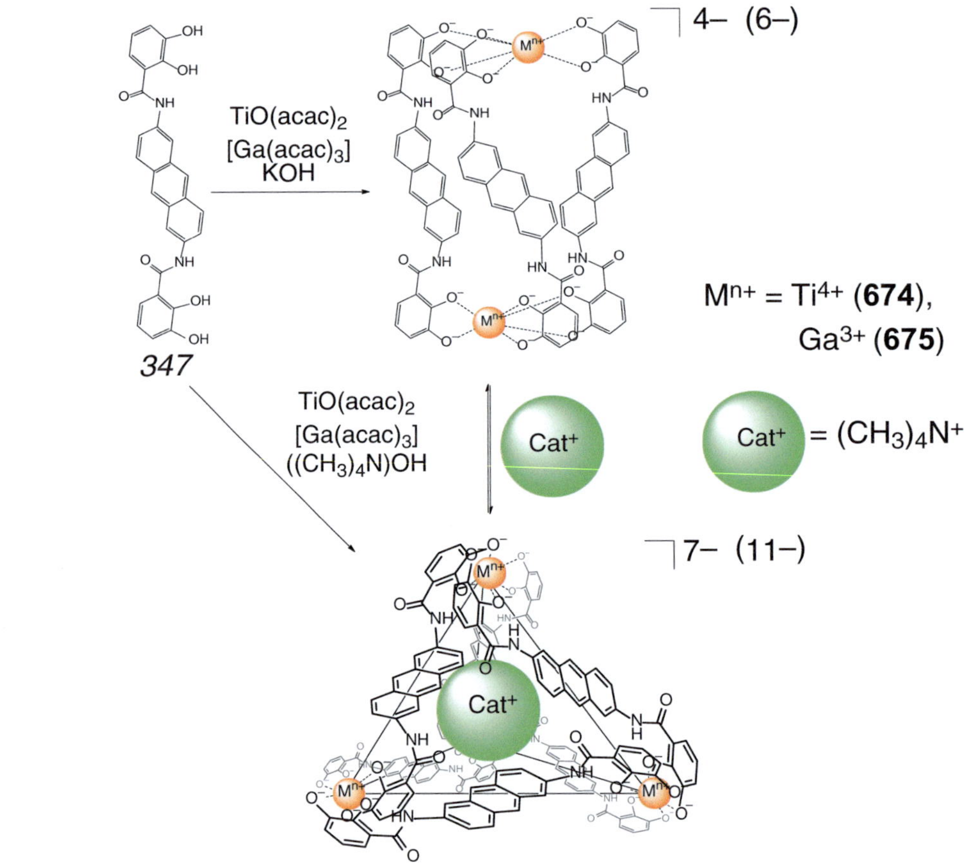

Scheme 4.118

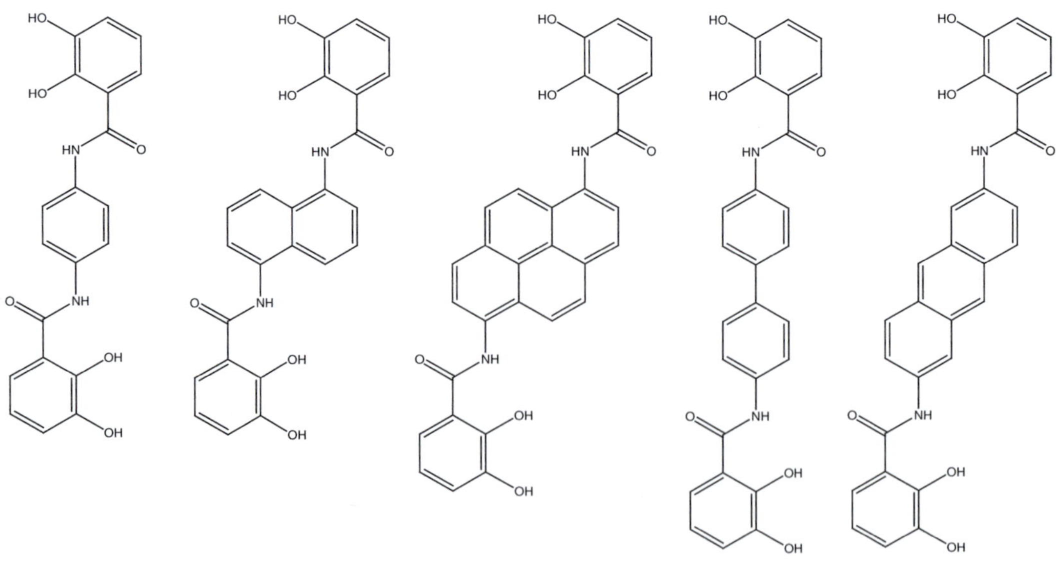

Scheme 4.119

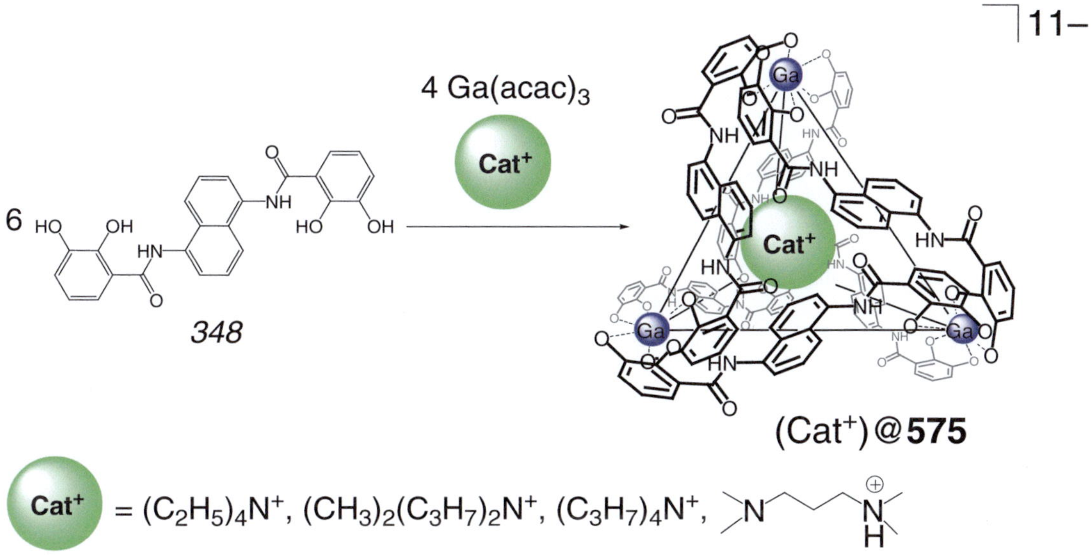

Scheme 4.120

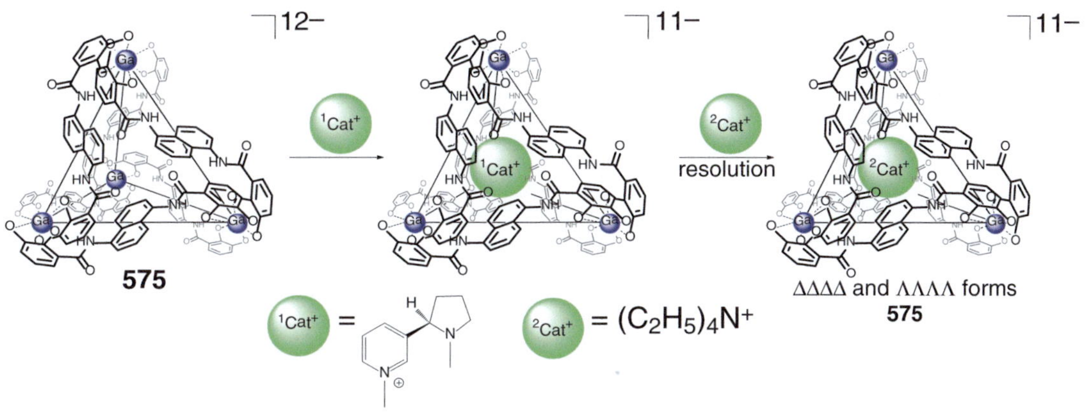

Scheme 4.121

of its *ΔΔΔΔ* isomer was also confirmed by X-ray diffraction data. The vertex gallium(III) ions have a distorted octahedral O_6-coordination polyhedra with average Ga–O distances equal to 1.966 Å with the distortion angle of 40.2° and Ga…Ga distances of approximately 12.66 Å [119].

This caging ligand is also reported in [120] to be able to encapsulate not only alkali metal cations but also their complexes with crown ethers, thus forming the corresponding 1:1 "*Matreshka*" complex-in-complex assemblies by Scheme 4.122. As follows from the multinuclear NMR data, these coordination capsules are in a dynamic equilibrium in their aqueous solutions [120].

A similar Fe_4L_4 capsule **676** has been prepared in [121] by a *one-pot* synthetic procedure (Scheme 4.123) from iron(II) chloride and the corresponding ligand syntone. The use of bis-*para*-tolyl malonate instead of bis-*tert*-butyl syntone under the same reaction conditions resulted in the formation of a Fe_6L_6 cage framework of **677**. As follows from their ^{57}Fe Mössbauer spectra, both these coordination capsules contain only the vertex iron(III) ions [121].

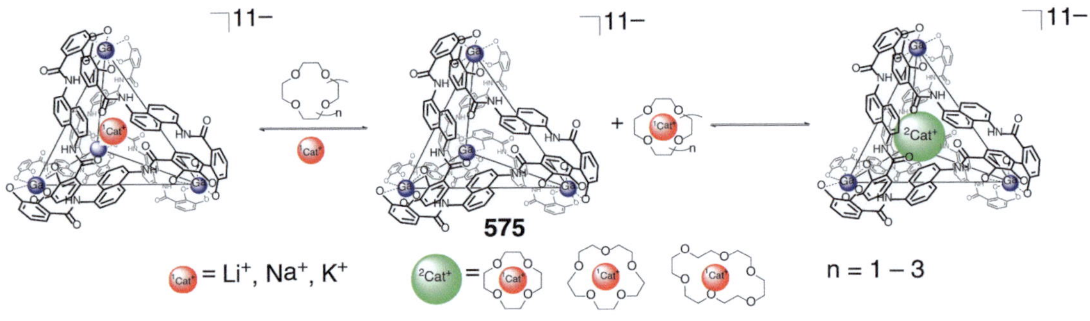

Scheme 4.122

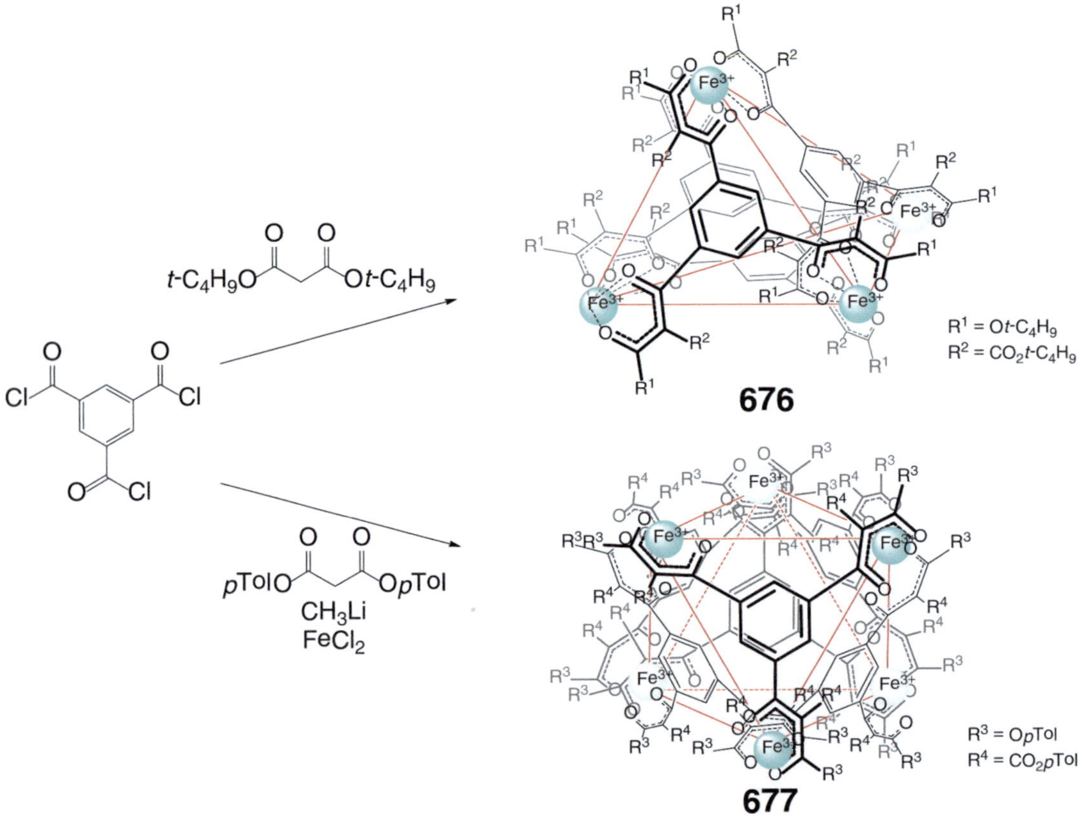

Scheme 4.123

Bis- and tris-catecholate syntones *298*, *348*, and *349* (Schemes 4.124, 4.125, and 4.126) have been used [122] for the synthesis of M_4L_6 and M_4L_4 capsules, respectively, by their coordination to oxophilic and strongly Lewis-basic gallium(III), tin, and titanium(IV) ions; these polyanionic caging ligands are reported to be prospective cation receptors. Self-assembly of gallium(III) acetylacetonate with a tris-catecholate ligand syntone *349* in the presence of triethylamine by Scheme 4.124 gave a highly symmetric Ga_4L_4 capsule **678** as the only product of this reaction. In the case of tin(IV) tetrachloride, it afforded a Sn_4L_4 cage framework **573** with

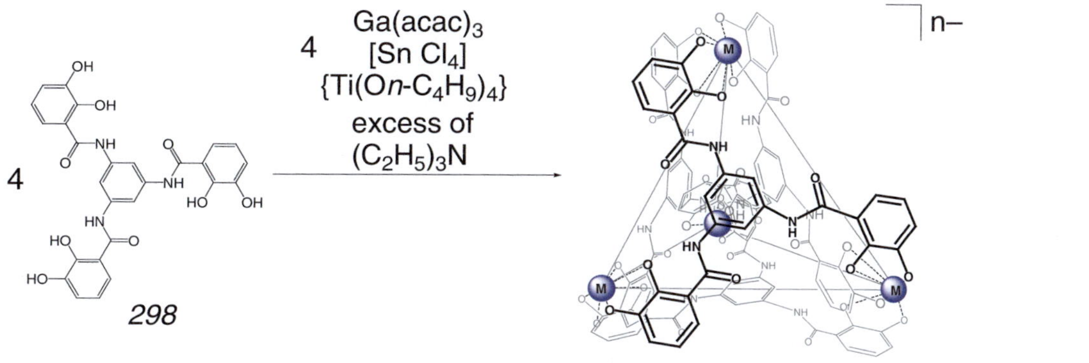

= Ga (**678**) [n = 12], Sn (**573**) [n = 8], Ti (**574**) [n = 8]

Scheme 4.124

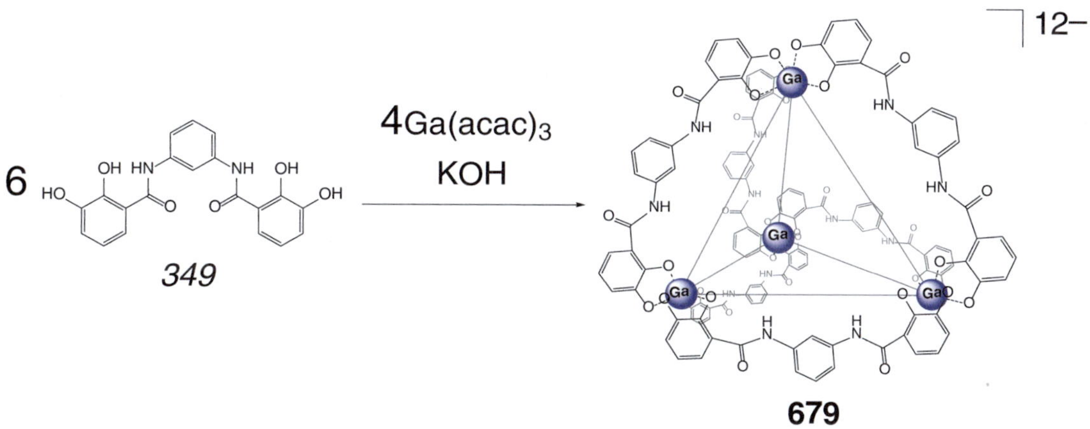

Scheme 4.125

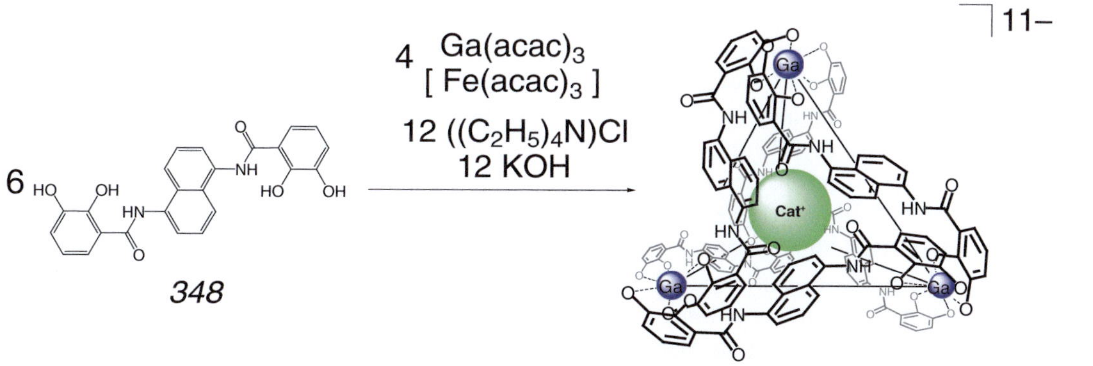

= Ga^{3+} (**575**), Fe^{3+} (**680**) Cat+ = $(C_2H_5)_4N^+$

Scheme 4.126

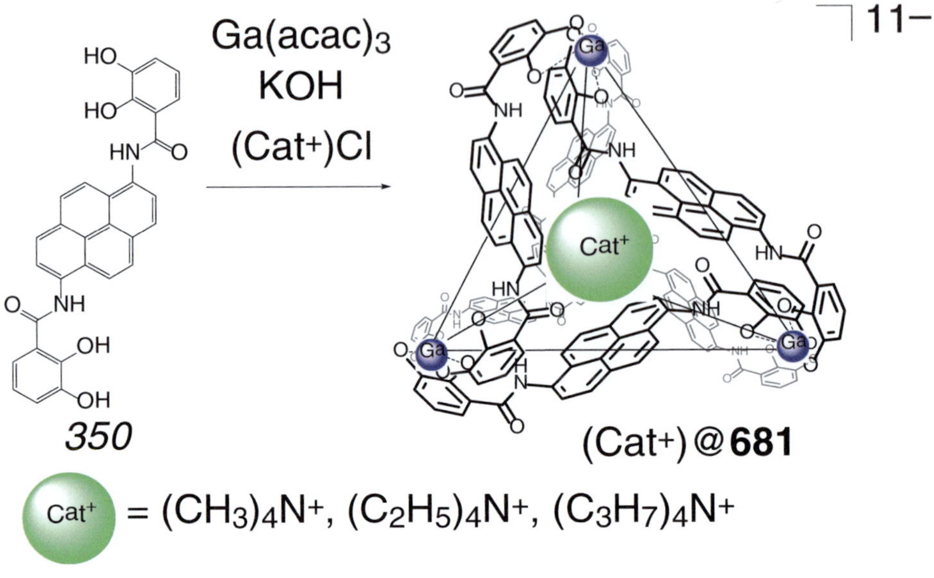

Scheme 4.127

C_3-symmetry axes that pass through its facial tris-catecholate fragments. The slightly modified self-assembly with titanium(IV) tetrabutylate gave a Ti_4L_4 coordination capsule **574**. Its crystal contains a racemic mixture of homoconfigurational tetrahedra (all Δ or all Λ at the metallocenters within a given *T*-symmetric capsule), while four C_3-symmetric tris-catecholate fragments form the faces of this tetrahedral cage framework and are cross-linked by four vertex titanium(IV) ions. Self-assembly of the bis-catecholate ligand syntones *350* and *348* with gallium(III), tin, and titanium(IV) cations by Schemes 4.125 and 4.126 afforded M_4L_6 coordination capsules **575**, **679**, and **680**. As follows from their ^{1}H NMR spectra, the *T*-symmetric ligands **575** and **680** in the course of this reaction encapsulate tetraethylammonium cation, giving the corresponding 1:1 cage complexes. The authors of [122] noted that the caged cation began to experience the chiral environment of their *T*-symmetric inner cavities. These coordination capsules thus seem to be potent chiral molecular flasks (see Sect. 6.3).

The cation-induced synthesis of tetrahedral cage complexes by Scheme 4.127 is reported in [123]; the formation of such derivatives of the rigid pyrenocatecholate ligand syntone *350* is induced by the interactions between an appropriate cationic guest species and the anionic caging host **681**. This coordination-driven self-assembly is described in [123] to be size selective: small cationic species do not form such cage complexes, whereas the tetraalkylammonium cations $(CH_3)_4N^+$, $(C_2H_4)_4N^+$, and $(C_3H_7)_4N^+$ do. The binding constant for $(C_2H_5)_4N^+$ ion is higher by approximately 300 times than that for tetrapropylammonium cation. In the host – guest 1:1 complex $((C_2H_5)_4N^+)$@**681**, all the vertex metallocenters of its cage framework have the same (Δ or Λ) configurations; thus, it exists as a racemic mixture of the homoconfigurational ligand frameworks with the tetraethylammonium cation encapsulated within their cavities. The space-filling model of the resulting solid-state structure showed this cationic guest to be completely isolated from external factors by the extended π-system of the pyrene rings [123] and the host–guest interactions play a key role in the formation of coordination capsule **681** [123].

The example of reactive guest stabilization within the cavity of a caging ligand **575** is described in [124]. Cationic species $[(CH_3)_2C(OH)P(C_2H_5)_3]^+$ easily decomposes in aqueous solutions, but its encapsulation

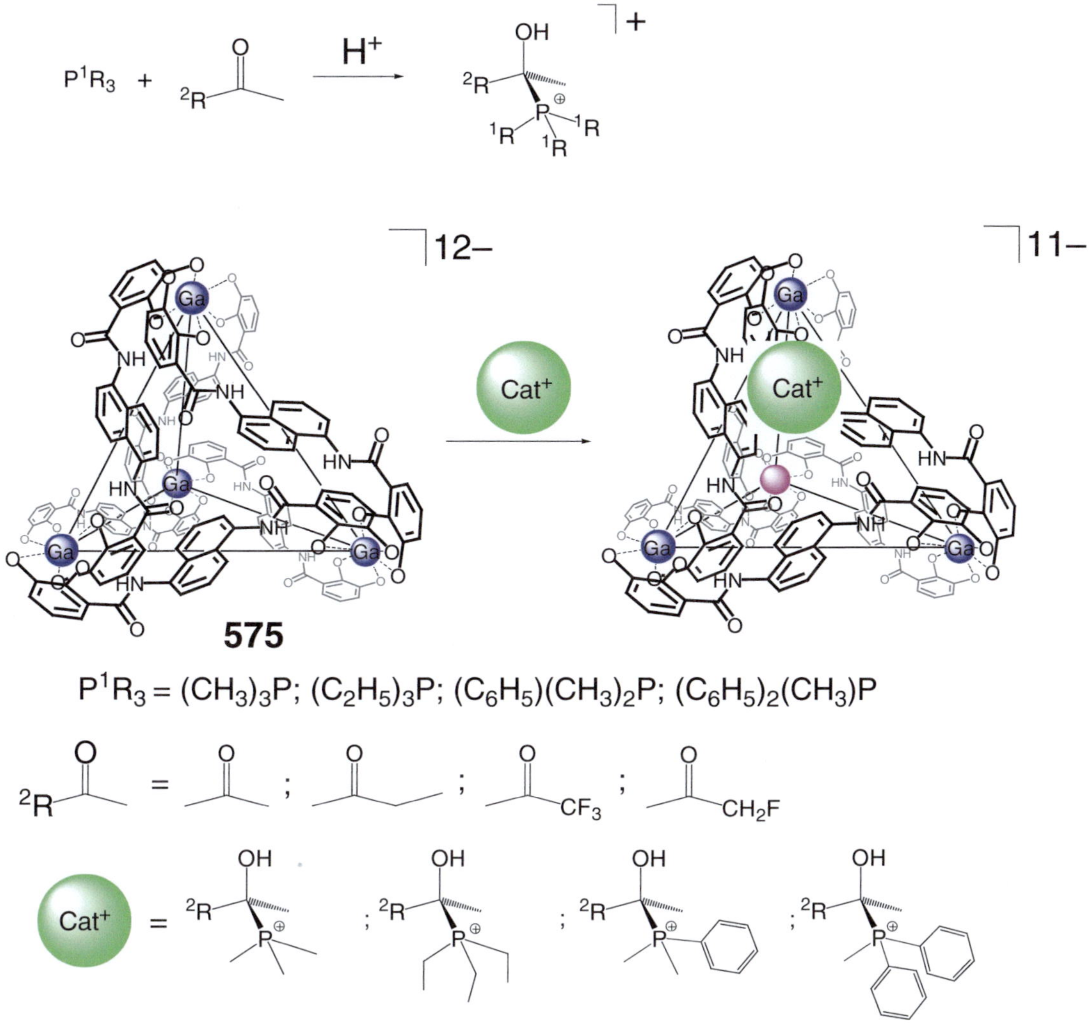

Scheme 4.128

by this caging ligand results in the stabilization of a phosphine–acetone adduct within the ligand's cavity. The resulting cage complex with the encapsulated adduct is reported in [124] to be stable in aqueous and wet methanolic solutions. A detailed study of stereoselective encapsulation of phosphonium guests has been performed in [125] for a wide range of the phosphines {$(CH_3)_3P$, $(C_2H_5)_3P$, $C_6H_5(CH_3)_2P$, and $(C_6H_5)_2CH_3P$)} and ketones (acetone, methyl ethyl ketone, 1,1,1-trifluoroacetone, and fluoroacetone) shown in Scheme 4.128. Stabilization of reactive phosphonium–ketone adducts of a general formula $[R^1CH_3C(OH)PR_3]^+$ using their encapsulation by a Ga_4L_6 tetrahedral capsule **575** is reported in [125]; although these cations decompose in aqueous solution, in its hydrophobic cavity they have substantially longer lifetimes, in some cases, up to weeks. By varying the phosphines and ketones that form the adducts and the pD of the solutions, it was shown that the size and shape of the guest cations and acidity (basicity) of the aqueous solution play an important role in the stability of these host–guest complexes. Encapsulation of chiral guests by the chiral capsule **575** results in the formation of diastereomers, which were characterized in [125] using ^1H, ^{19}F, and ^{31}P NMR spectra. Although

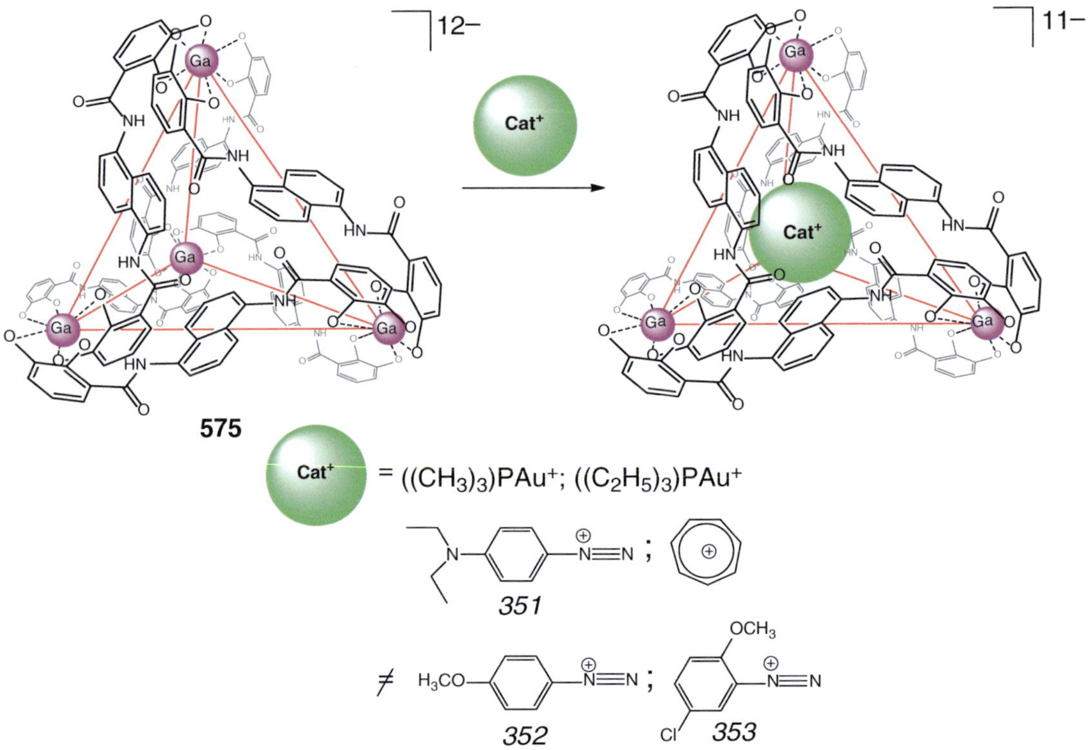

Scheme 4.129

this capsule is formed from non-chiral ligand syntones, it has *ΔΔΔΔ* or *ΛΛΛΛ* chirality around its vertex metallocenters. Due to the chirality of **575**, diastereomeric selectivity is observed upon initial guest encapsulation (typical diastereomeric excesses in this case are 30–50 %). The initial diastereomeric selectivity decreases over time to reach an equilibrium but does not become 1:1, indicating that both the kinetic and thermodynamic processes promote this selective encapsulation. The stability of the caged phosphonium–ketone adducts increases with decreasing pH, and fit of the guest molecule into the cavity of **575** is an important consideration for these cage complexes; dramatic increases in stability of caged reactive species are observed upon encapsulation by this capsule [125].

The ligand **575** is reported in [126] to stabilize reactive aromatic diazonium (*351*) and tropylium cations, forming 1:1 cage complexes by Scheme 4.129. In contrast, its methoxyl-containing analogs *352* and *353* do not form such complexes (as follows from ^{1}H NMR data). This result is explained in [126] by the difference in the hydrophobicity and in the size of the guest molecules; encapsulation prevents these reactive aromatic cations from solvolysis or this process is slowed by their caging [126]. This ligand also encapsulates dialkylphosphinegold(I) cations to give cage complexes of the same stoichiometry by Scheme 4.129 [127].

Two tris-catecholate ligand syntones *354* and *355* (Scheme 4.130) have been tested in silico as potent tris-bidentate chelators for the synthesis of very large M_4L_4 capsules **682**–**685** [128]. In the Ga_4L_4 cage framework of **682**, the distance Ga… Ga is approximately 19 Å, while the estimated volume of its cavity is approximately 450 Å^3. In contrast, the bis-phenylene spacers in *355* increase its lability, causing a distortion from a planar geometry that is necessary for the design of M_4L_4 capsules; according to the MM data, this ligand cannot form such capsules. Indeed, self-assembly of the designed M_4L_4 derivatives **682**–

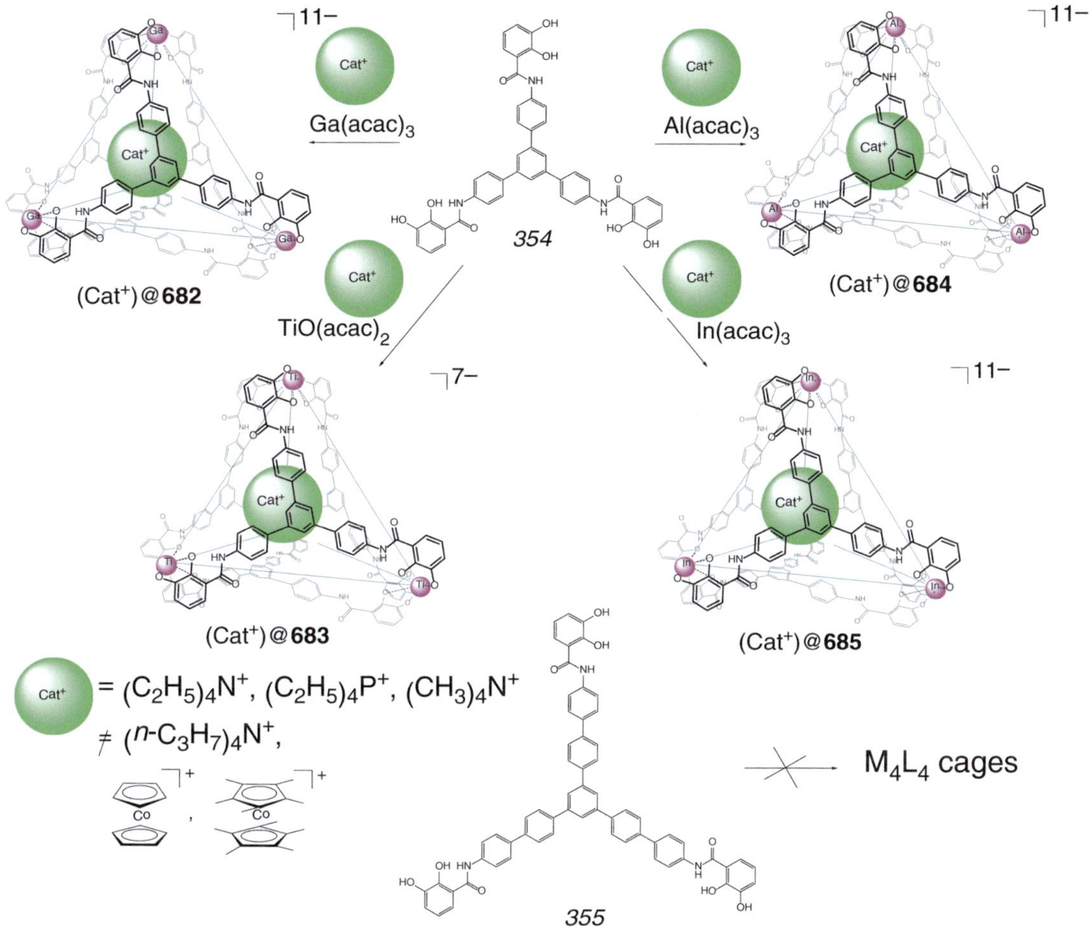

Scheme 4.130

685 of the tris-catecholate ligand syntone *354* has been performed in [128] by Scheme 4.130. According to the NMR data, their highly symmetric tetrahedral cage frameworks have four C_3-symmetry axes. The anionic Ga_4L_4 and Ti_4L_4 ligands **682** and **683** encapsulate organic monocations of suitable size and shape but do not form the cage complexes with anions, neutral molecules, or polycations. As follows from the NMR data, these capsules bind tetraethylammonium and tetraethylphosphonium cations with approximate volumes of 360 and 390 Å^3, respectively, giving 1:1 cage complexes. In contrast, tetrapropylammonium and tetramethylammonium cations with an approximate volume of 770 and 150 Å^3, respectively, do not form cage complexes with these caging ligands. Besides, such tetrahedral hosts are selective for the guest's shape: they also do not encapsulate cylindrical molecules of bis(cyclopentadienyl)cobaltocenium and bis(decamethyl cyclopentadienyl)cobaltocenium ions with the approximate volumes of 190 and 450 Å^3, respectively [128].

Titanium- and germanium(IV)-containing analogs of the tetrahedral Fe_4L_4 and Ga_4L_4 capsules **575** and **680** have been prepared in [129] under more rigorous reaction conditions (at higher temperature and with more prolonged reaction time) than used for the gallium(III) complex due to the reduced lability of their titanium and germanium(IV) catecholate precursors. In contrast to the metal(III) cations, the formation of such metal(IV)-based coordination capsules demands suitable guest templates; the

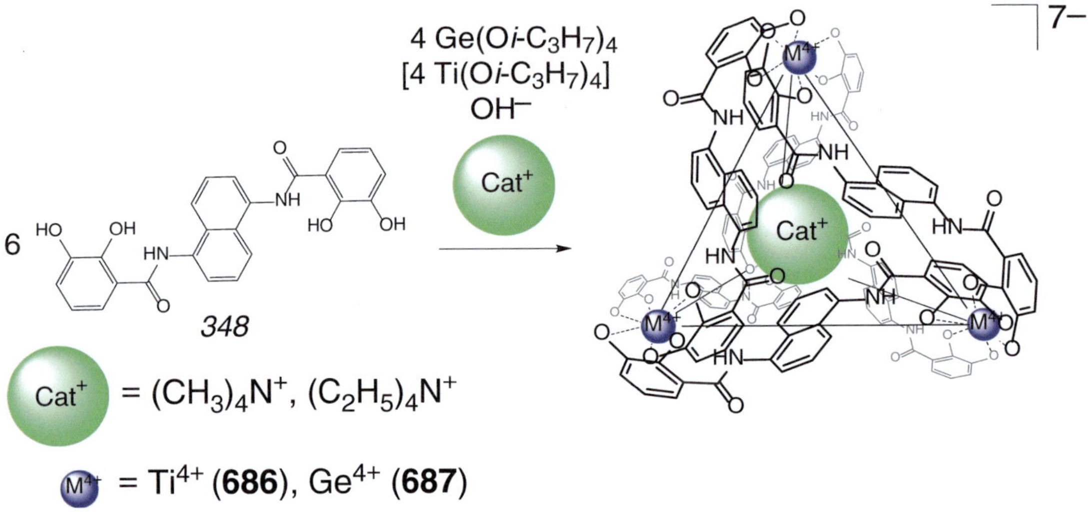

Scheme 4.131

guest-templated self-assembly by Scheme 4.131 gave 1:1 cage complexes of coordination capsules **686** and **687** with encapsulated tetramethyl- and tetraethylammonium cations [129].

The Ga_4L_4 caging ligand **575** is reported in [130] to interact with an organometallic compound [CpRuCl(cod)], giving two diastereomeric forms of 1:1 cage complex with an encapsulated cationic species $[CpRu(\textit{cis}\text{-}1,3,7\text{-octatriene})]^+$ by Scheme 4.132. In contrast to the free form of this cation that easily and irreversibly hydrolyses, the encapsulated one is stable in aqueous solutions for a long time. This organometallic precursor $[CpRu(\textit{cis}\text{-}1,3,7\text{-octatriene})]^+$ also reacts with carbon monoxide to give an organometallic cation $[CpRu(cod)(CO)]^+$ that undergoes encapsulation in its aqueous solution by **575**, thus giving 1:1 cage complex. The same coordination capsule has been prepared by a reaction of the cage complex of a parent organometallic guest cation with carbon monoxide in aqueous solution; the Cp*-containing analog of this guest gave a stable 1:1 cage complex with the tetrahedral coordination capsule **575** [130]. This ligand is reported in [131] also to form 1:1 host–guest complexes with very reactive iridium metallocomplex cations shown in Scheme 4.133.

The catecholate Ti_4L_4 coordination capsule **574** and the cage complexes of naphthalene-containing Ga_4L_4 ligand **575** with encapsulated $(CH_3)_4N^+$, $(C_2H_5)_4N$, $(C_3H_7)_4N^+$, and Cp^*Co^+ cations have been thoroughly characterized in [132] and [133] by ESI-ITMS and ESI-MS, respectively.

Resolution of racemic mixtures of the tetrahedral Ga_4L_4, Fe_4L_4, and Al_4L_4 capsules **575**, **680**, and **688** onto their homochiral *ΛΛΛΛ* and *ΔΔΔΔ* forms has been performed by Scheme 4.134 in [134] with chiral *S*-nicotinium cation as a guest. This resolution can also be accomplished using the ionic associates of this cation with the corresponding polyanionic 1:1 cage complexes with encapsulated $(C_2H_5)_4N^+$ cation blocking the cavity interior. This suggests that the external binding sites of the cage frameworks are responsible for the difference in their enantiomer interactions with the above chiral cationic guest [134].

A 1:1 cage complexes of **575** with various encapsulated reactive organometallic rhodium cations have been prepared in [135] by Scheme 4.135. To understand the factors driving the molecular recognition process by the tetrahedral M_4L_6 caging ligands, the authors of [136] determined standard thermodynamic parameters for the encapsulation of a series of organometallic cations also shown in this Scheme. Such caging process for these guests is reported in [135] to have enthalpy–entropy compensation effects.

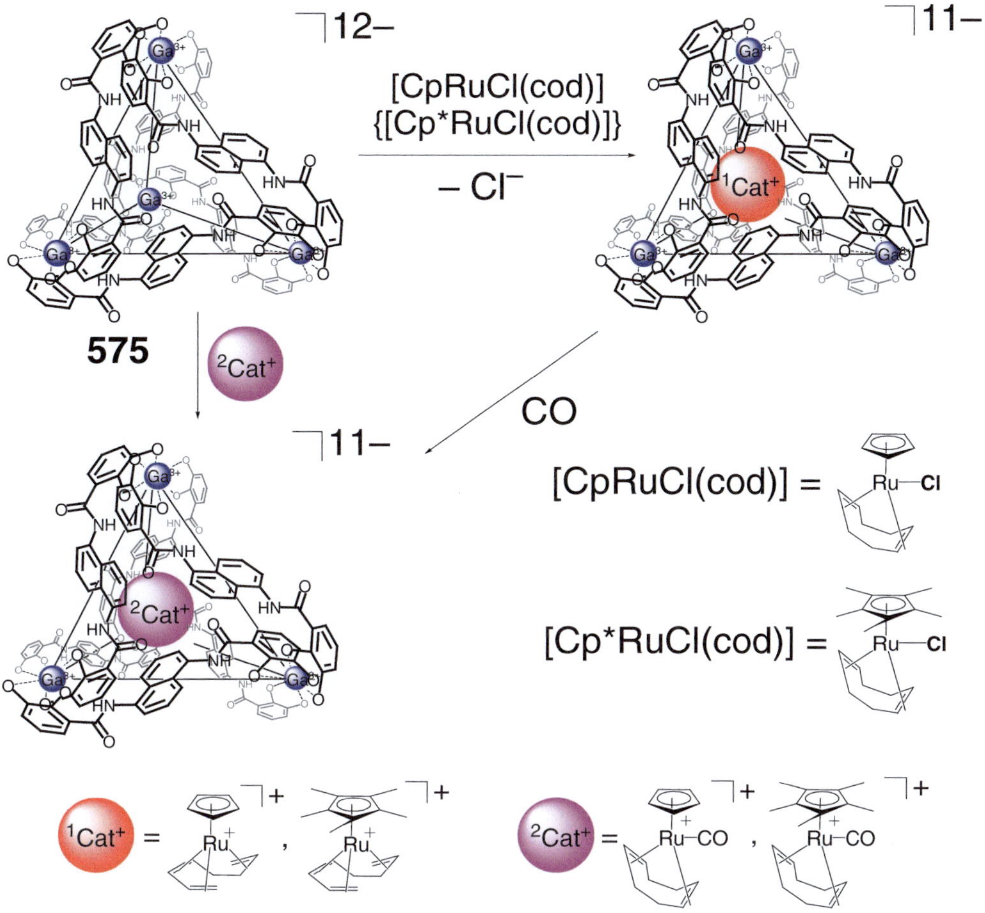

Scheme 4.132

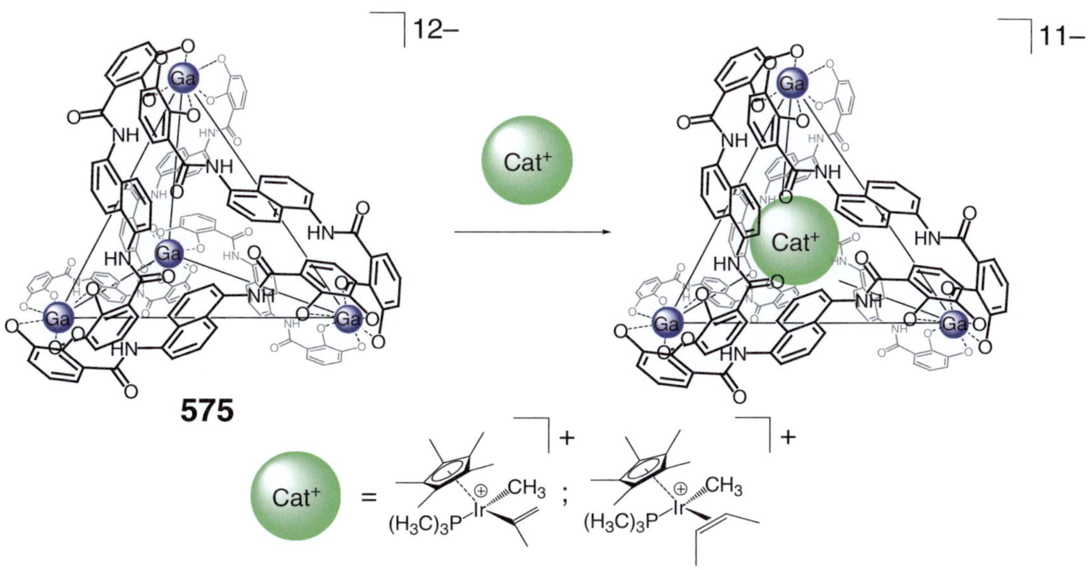

Scheme 4.133

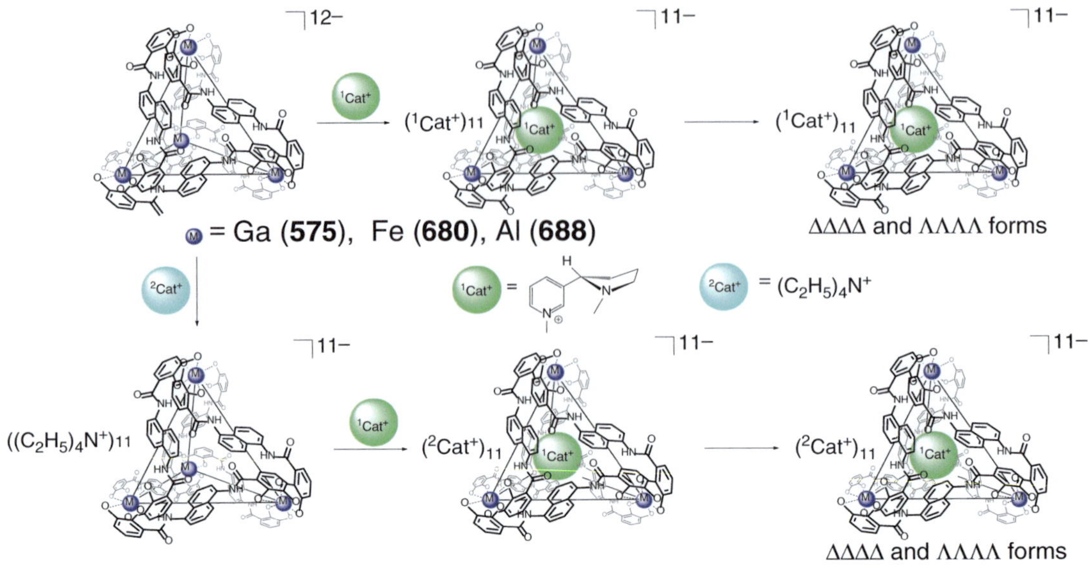

Scheme 4.134

The caging ligand **575** enantioselectively binds organoruthenium cations shown in Scheme 4.137 to give 1:1 cage complexes. As follows from ^{1}H NMR data, the encapsulation of these chiral cations proceed with diastereoselectivities up to 70 %, and the resolved assembly is exploited for a dynamic resolution of one of the organometallic species [137]. Cage complexes with encapsulated cationic ruthenium sandwich guests have been characterized in [138] using ESI-MS and NMR methods. The reactivity of these caged organometallic cations is altered by their encapsulation, and the binding of the achiral guest [CpRu(*para*-cymene)]$^+$ within the cavity of such chiral ligand **575** caused its enantiotopic methyl protons to become diastereotopic [138].

Cage complexes of **575** with encapsulated ruthenocene guest cations, beared with long-chain alkyl spacer substituents, are reported in [139] to undergo fluxional structural changes by Scheme 4.137. These changes are caused by the dynamic second-order Jahn–Teller effect: at low temperatures, this system does not have enough thermal energy to populate its appropriate excited states sufficiently to overcome the activation energy, and the alkyl chain of the guest is extruded through a single face of a caging ligand **575**, but at higher temperatures, its greater statistical population of excited electronic states permits the distortion of the transition state toward the *T*-symmetric intermediate. This system can easily overcome a smaller activation barrier to return to one of the C_3 ground states. At ambient temperatures, this distortion is very fast, allowing the system to interconvert rapidly between the four degenerate C_3 states through the *T* intermediate one, leading to the time-averaged *T* symmetry on the NMR time scale [139] .

The ligand **575** has been also used in [140] for stabilization of iminium ions formed in situ by imine condensation of the corresponding secondary amines and ketones in aqueous solution. This binding (Scheme 4.138) is size selective: encapsulation of too small or too large iminium guests is inefficient. Thus, caging by **575** allowed generating iminium ions in basic aqueous solutions and stabilizing these species at room temperature [140].

Encapsulation of protonated amine and phosphine guests (Scheme 4.139) by coordination capsule **575** is described in [141] to make them more basic; the process leads to a dramatical shift in the effective basicity of their caged protonated forms. For encapsulated protonated

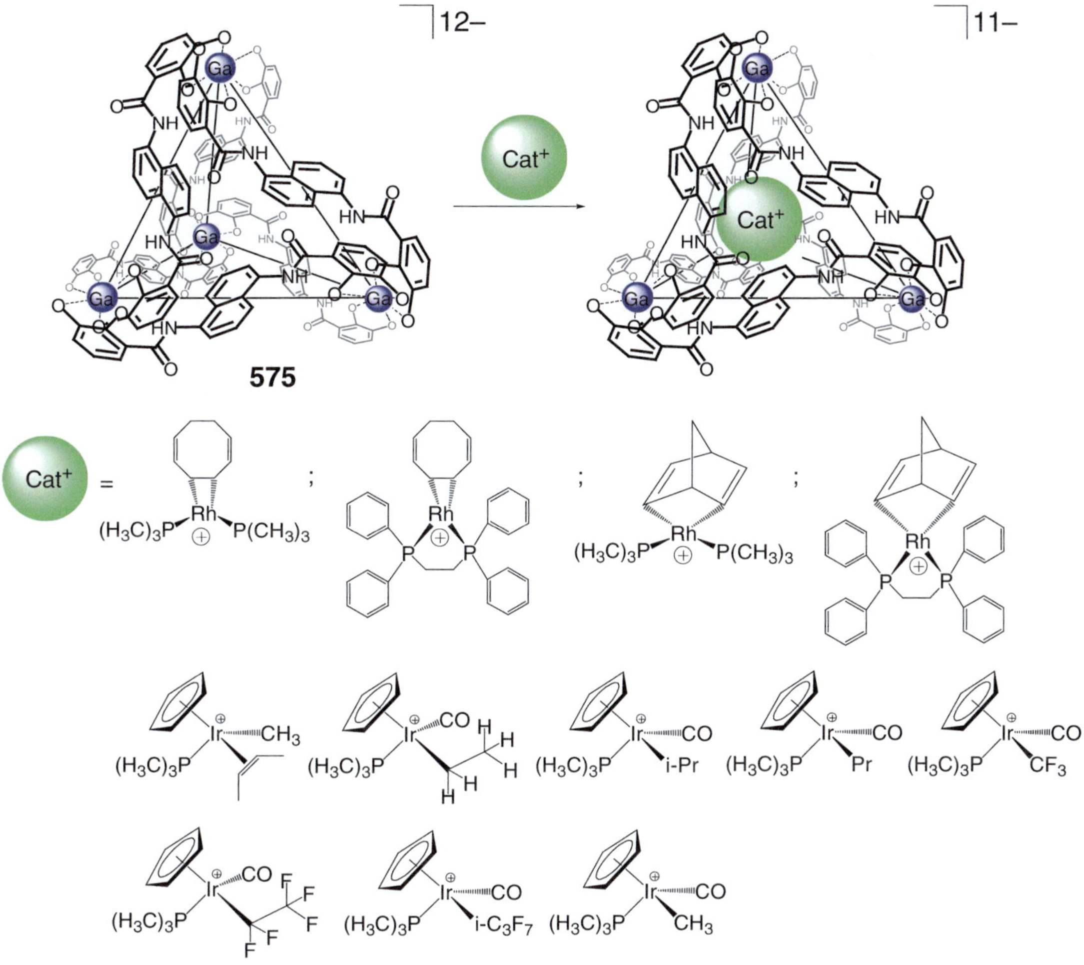

Scheme 4.135

amine and phosphine guests, the size selectivity is consistent with a constrained binding environment of **575**. Protonation of the encapsulated guests has been confirmed by ^{31}P NMR and MS data and a pH dependence of the encapsulation. The binding by this caging ligand dramatically alter the effective basicities of caged amines by over 4 pKa units for a wide variety of such guests; those are exchanging freely to the exterior of the capsule **575**, eliminating the possibility of encapsulation acting as a kinetic trap [141].

1:1 cage complexes of coordination capsule **689** with an extended cavity have been synthesized in [142] by Scheme 4.140 using an appropriate 1,1′-binaphtyl ligand syntone *357*. Such cavity allowed efficient encapsulation of relatively large tetraalkylammonium and tetraaryl(alkyl)ammonium cations shown in this Scheme.

A coordination-driven self-assembly of tris-pyrocatecholate ligand syntones *358* and *359* with titanoyl bis(acetylacetone) by Schemes 4.141 and 4.142 afforded truncated tetrahedral M_4L_4 capsules **690** and **691** with more "close" and more "open" cage frameworks, respectively [143]. In the host–guest 1:4 cage complex of **690** with an encapsulated cationic guest $[K(DMF)_3]^+$, the distance Ti…Ti between its vertex metallocenters (approximately 17 Å) is substantially shorter than that in the case of its analog **691** formed by the more extended tris-pyrocatecholate

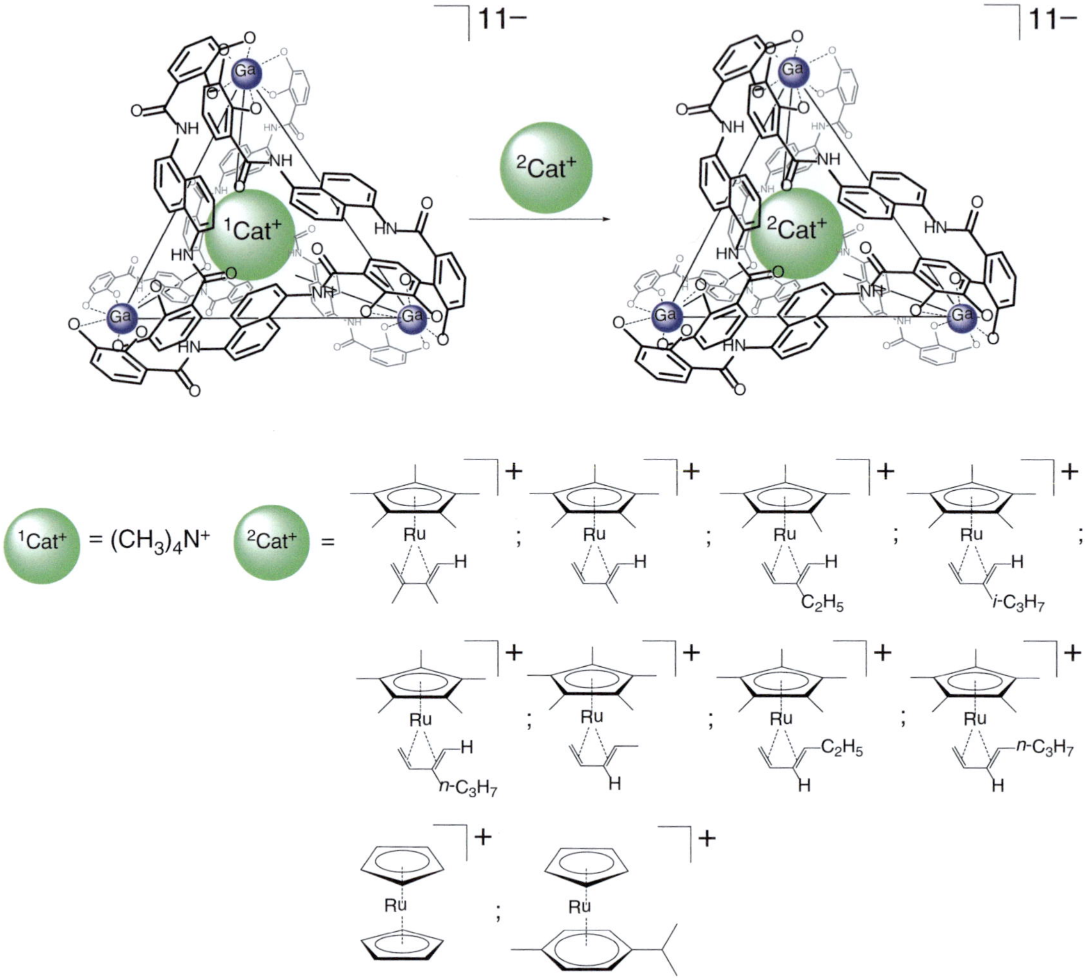

Scheme 4.136

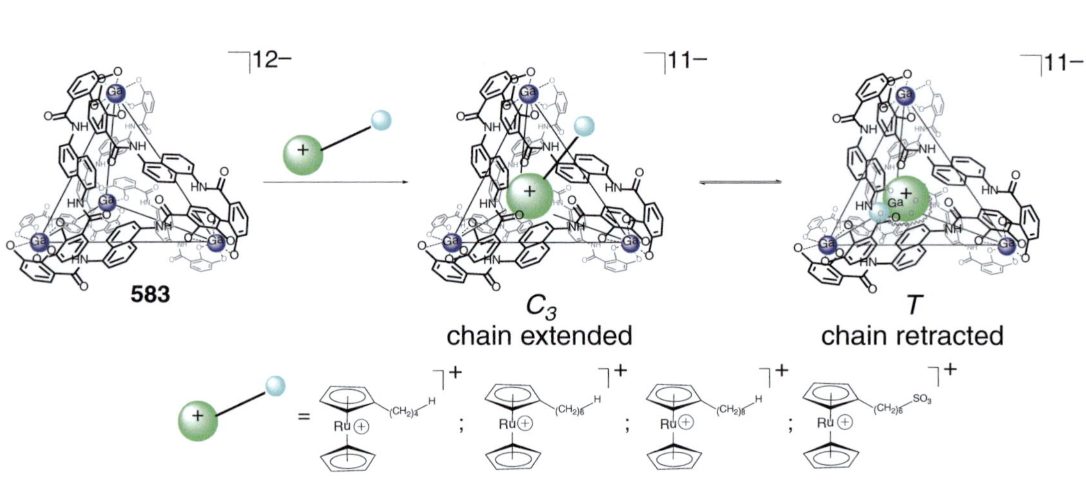

Scheme 4.137

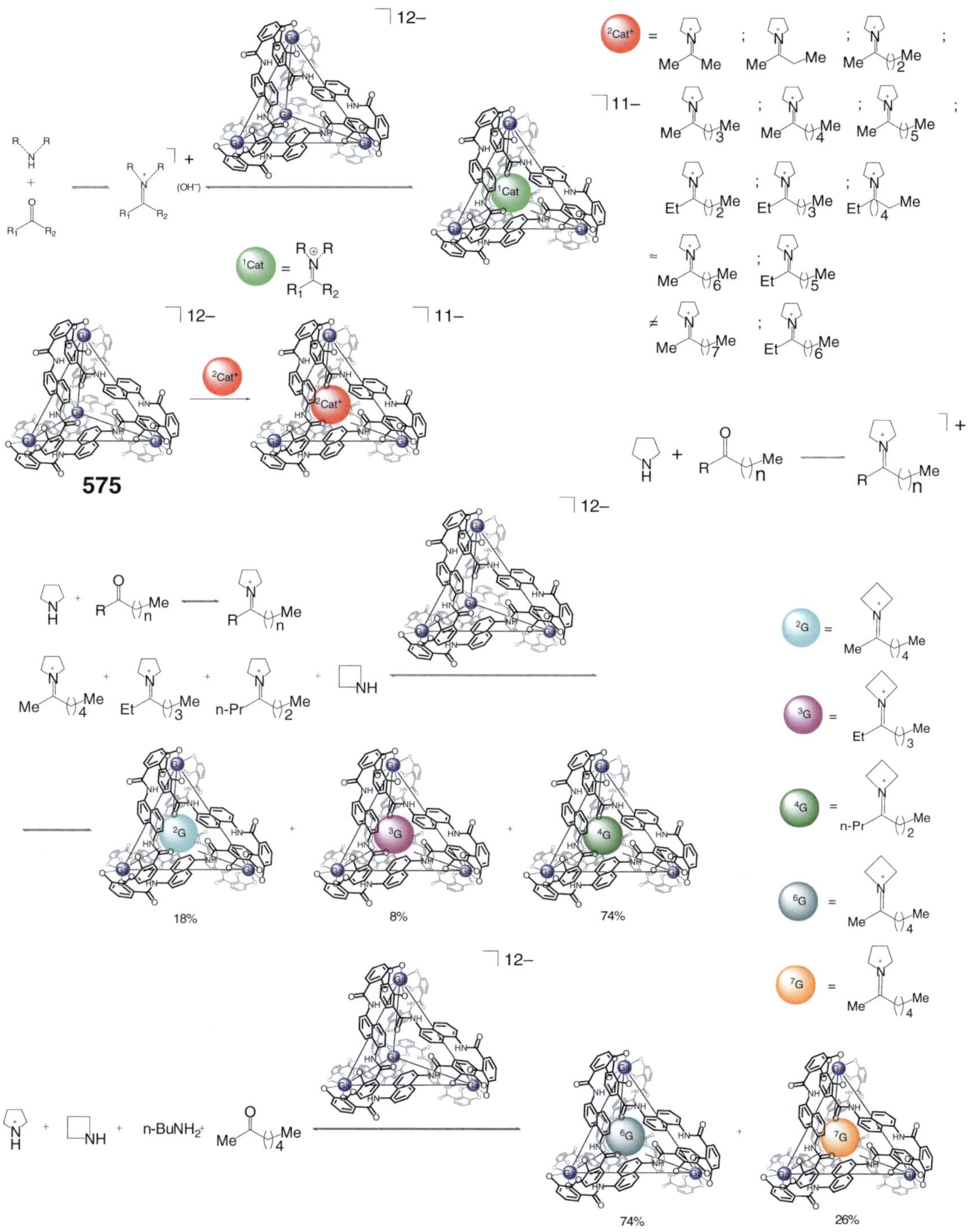

Scheme 4.138

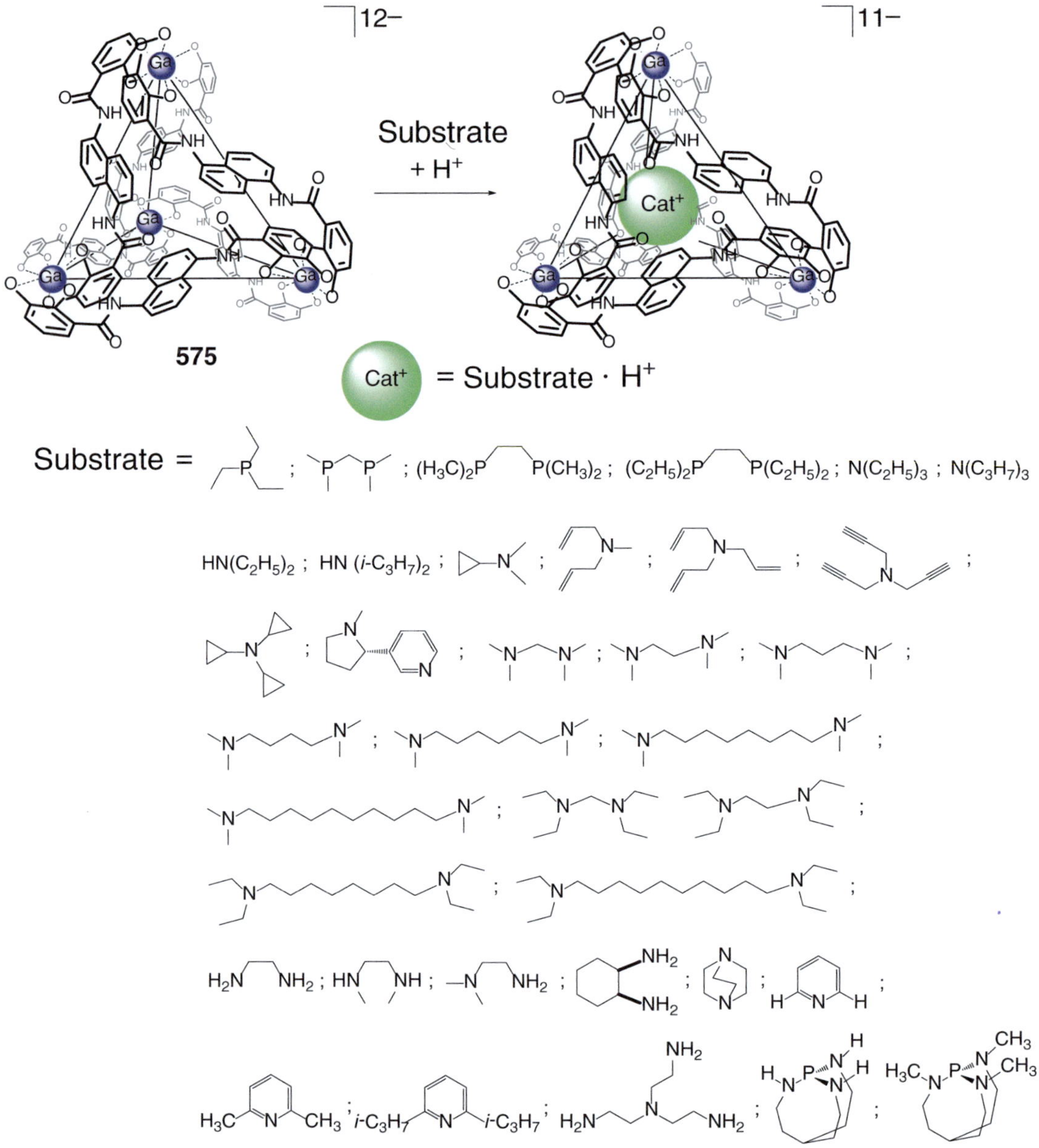

Scheme 4.139

syntone *359* (approximately 23.5 Å). Therefore, the latter capsule can easier encapsulate and release the guest molecules [143].

Detailed analysis of the cavity shapes and volumes and the effect of charge on the cage framework properties for a series of M_4L_6 coordination capsules **575**, **680**, and **686** with encapsulated ammonium and organometallic cations shown in Scheme 4.143 has been performed in [144]. Single-crystal X-ray diffraction data showed that intermolecular π…π, cation…π, and CH…π interactions depend on the nature of the cationic guests. The cation…π interactions are formed near the electron-rich catechol fragments of the caging ligand, while the π…π interactions are localized near its naphthalene rings. Changing of

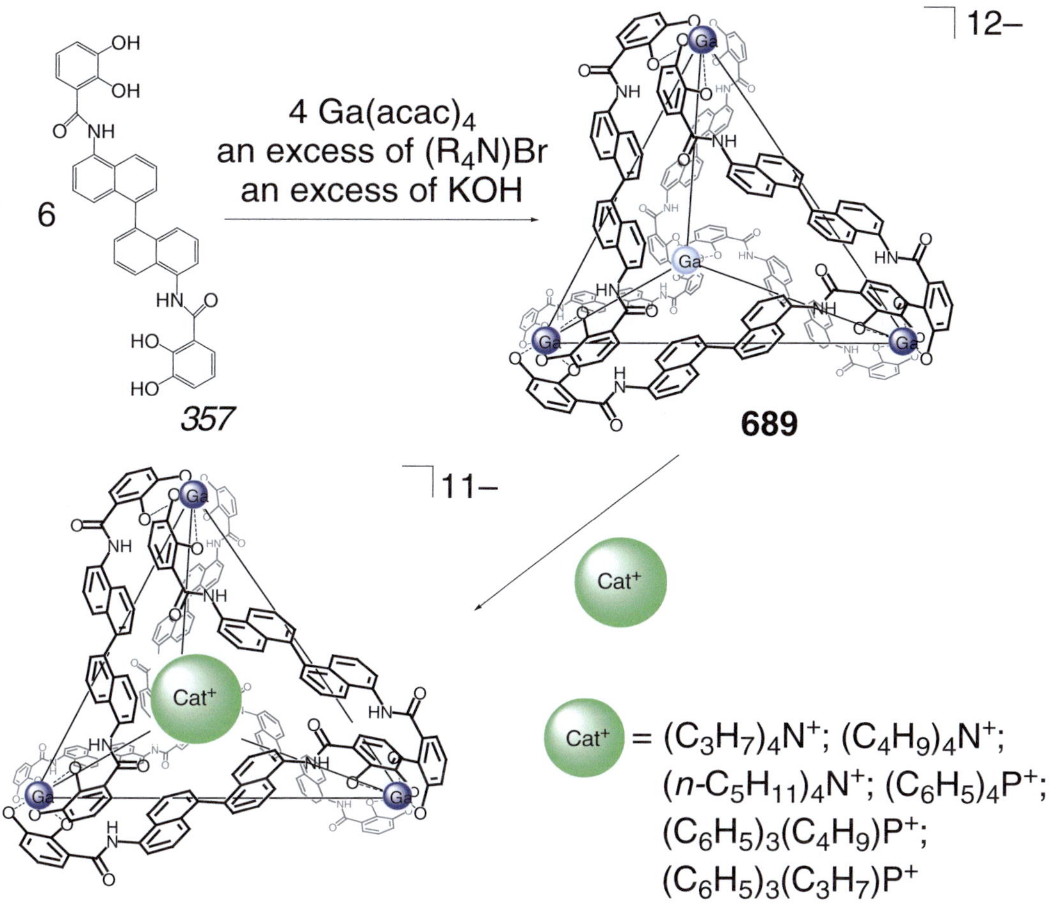

Scheme 4.140

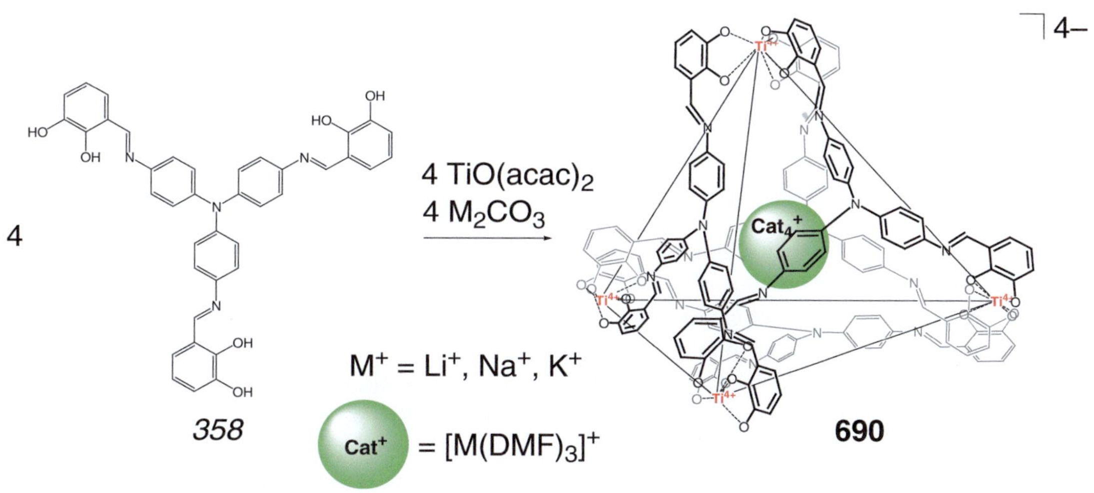

Scheme 4.141

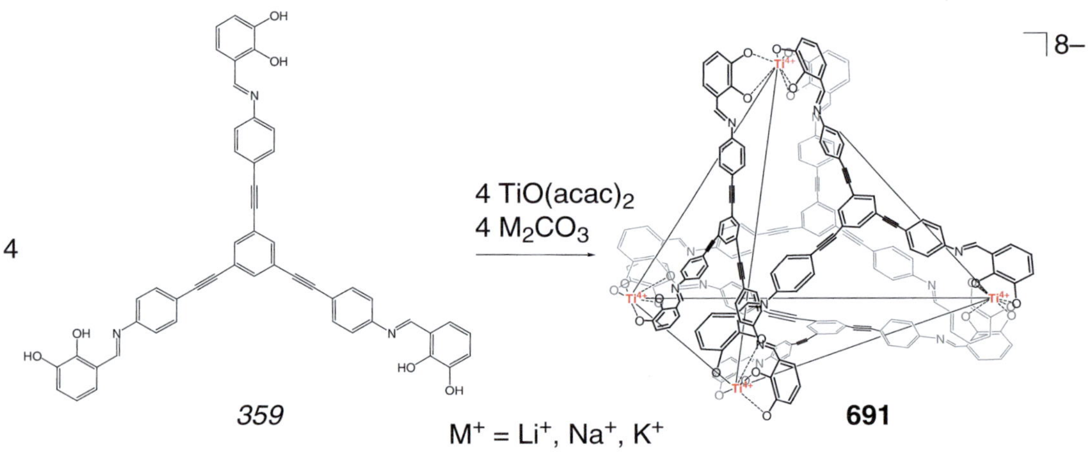

Scheme 4.142

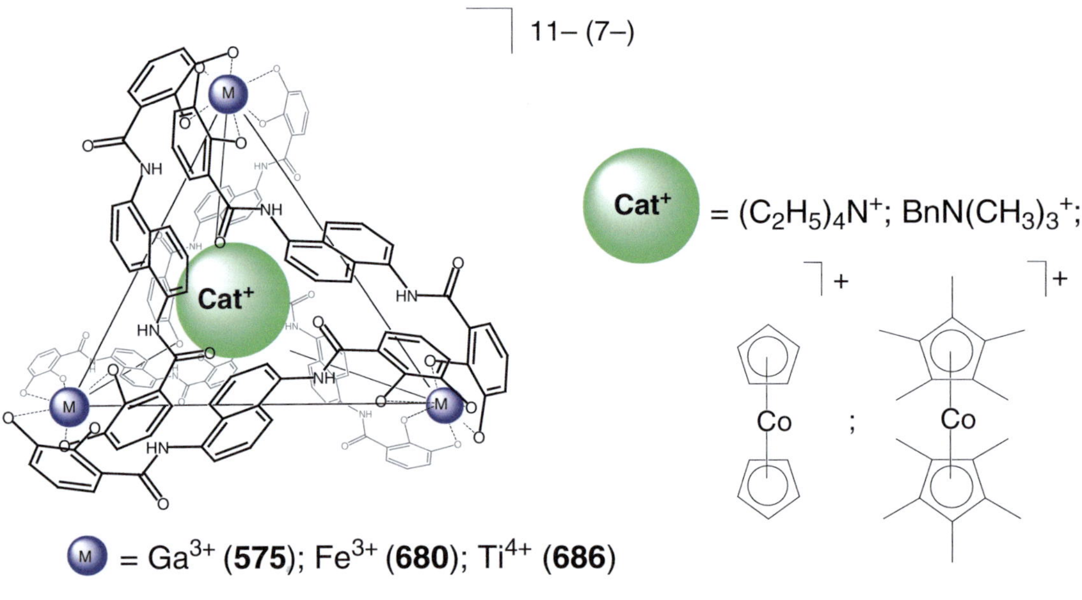

Scheme 4.143

the vertex metal ions and their charges does not affect the geometry and shape of their cage framework, whereas the interior cavities undergo distortion to accommodate a wide range of guest molecules; the volume of their cavities falls from 253 to 434 Å^3 depending on the nature of the caged guest cation [144].

^{1}H NMR chemical shifts for the encapsulated tetraalkylphosphonium, tetraalkylammonium, and trimethylarylphosphonium cations and neutral methane molecule within the cavity of a coordination capsule **575** have been calculated in [145] using DFT approach, allowing accurately probing "a caging host-caged guest" geometry and mapping a magnetic environment of the interior of its cage framework, which discriminates between the different host–guest geometries. The unexpected downfield NMR chemical shifts experimentally detected in [145] has been thus explained for these caged species. Substantial differences between the internal and external binding of mono- and tetraalkylammonium guest cations by this Ga_4L_6 caging

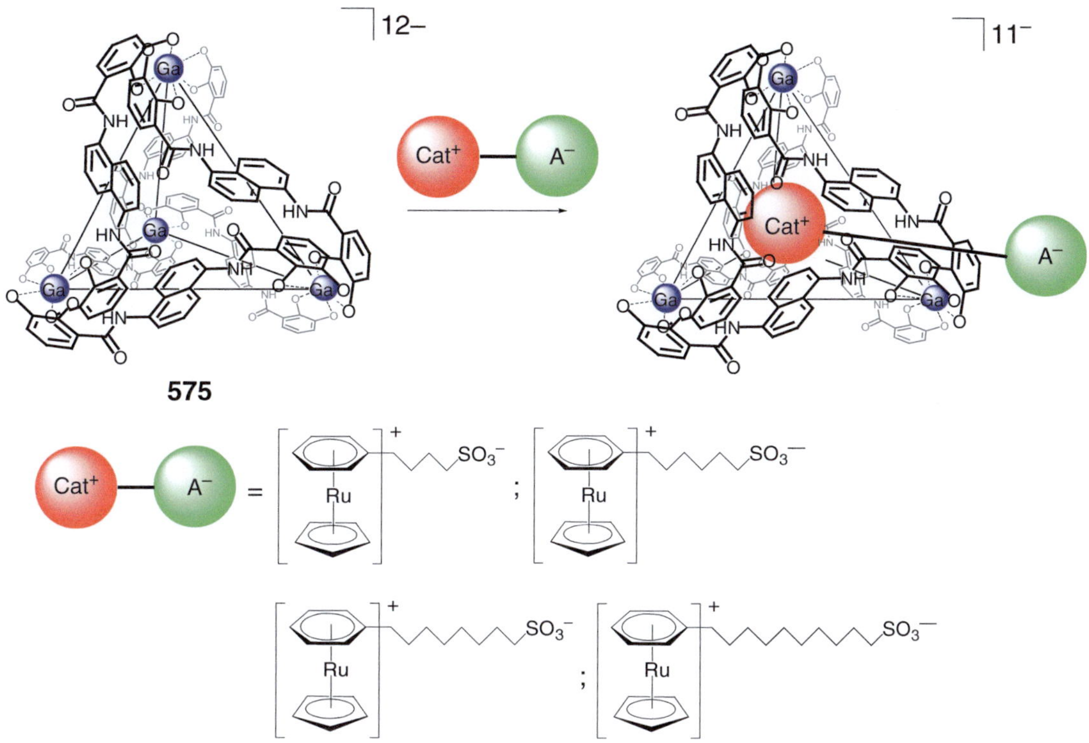

Scheme 4.144

ligand have been detected in [146] using NMR, UV-vis, and ITC methods. External binding due to attractive interactions between the guests and the exterior surface of this dodecaanionic coordination capsule is enthalpy driven, while the encapsulation, which results in desolvation and release of solvent molecules from its cavity, is entropy driven. Tetraethylammonium cation binds more strongly to both the interior and exterior of the coordination capsule than its methyl-containing homologs. These different internal and external host–guest interactions of **575** are explained in [146] by high negative charge and hydrophilic outer surface of this coordination capsule in contrast to its hydrophobic inner cavity.

This coordination capsule **575** is reported in [147] to only partially encapsulate ruthenocenyl-based zwitterionic guests by Scheme 4.144. Whereas their cationic "heads" are situated within its cavity, the spacer to the anionic terminal group passes through one of the small openings at the centers of the triangular faces of its dodecaanionic cage framework.

4.3 Coordination Capsules with Bridging (Cross-Linking) Metal Ions

Edge-directed coordination-driven self-assembly of 50 predesigned precursors (30 diplatinum linear complexes *360* and *361* and 20 tris-pyridyl ligand syntones *362* with directing angle of approximately 108°) by Scheme 4.145 afforded dodecahedral coordination capsules **692** and **693**. As follows from TEM and ^{1}H, ^{31}P, and PGSE NMR data, these highly symmetric cage frameworks have the diameters along their C_3-symmetry axes of 52 and 75 Å, respectively. The self-assembly includes the formation of 60 new coordination bonds from 50 ligand syntones; their rigidity and fitting in size are described in [148] to play a key role.

A similar approach has been used in [149] for the synthesis of a cuboctahedral $Pt_{24}{}^{1}L_{12}{}^{2}L_{8}$ coordination capsules **694** and **695**. It allows controlling the positioning of their either ferrocene or crown ether functionalizing pendant substituents.

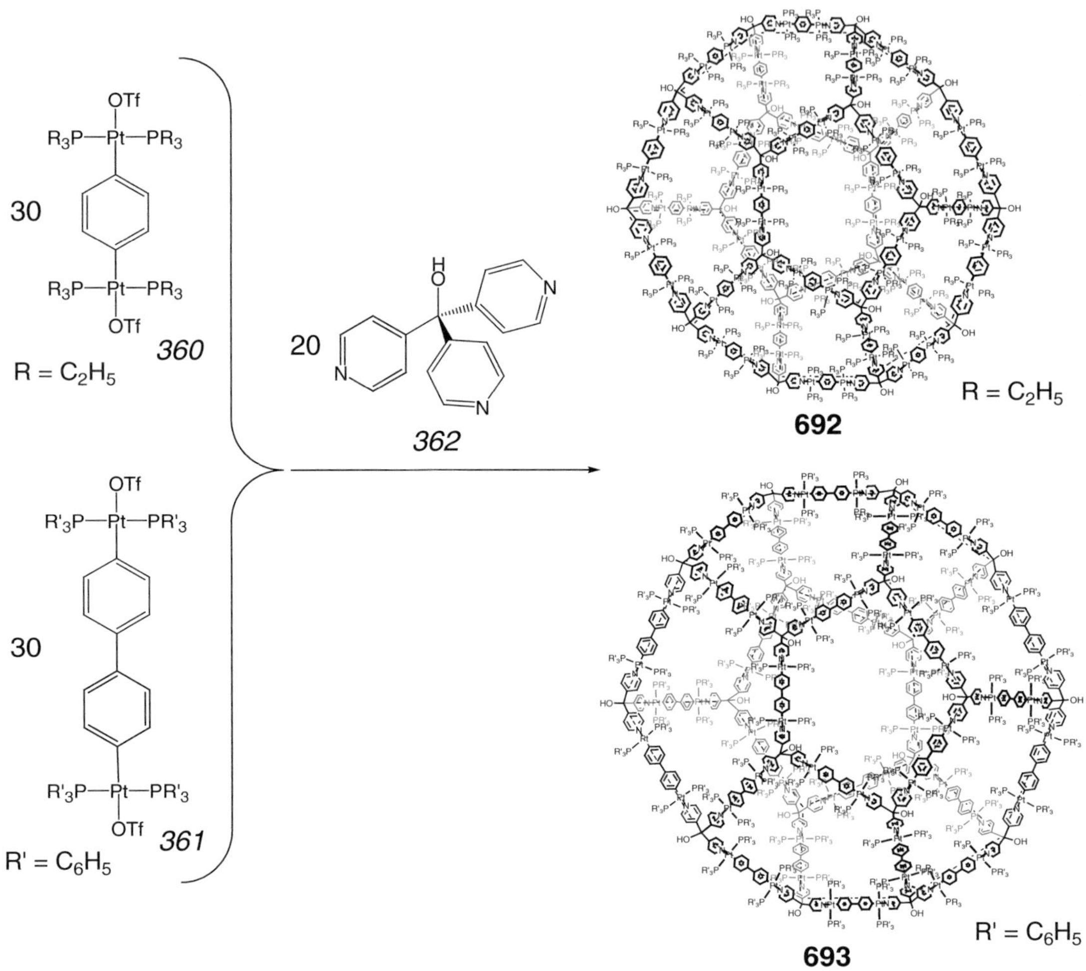

Scheme 4.145

These coordination-driven reactions by Scheme 4.146 include a self-assembly of the diplatinum complexes *362* or *363* as an acceptor with a directing angle of 120° and the tris-pyridyl ligand syntone *364* as a donor. The multinuclear NMR data suggest the formation of only these capsules with highly symmetric cuboctahedral rigid frameworks and the external diameters of approximately 67 and 84 Å; those of their inner cavities are approximately 50 Å. According to the CV data, **694** and **695** undergo redox processes that are reversible on the CV time scale; the authors of [149] report a fast ET between the pendant redox-active ferrocenyl substituents in **694**.

The S_3-symmetric Ag_2L_3 and S_4-symmetric Cu_2L_4 cage complexes with encapsulated monoanions have been self-assembled in [150] by Scheme 4.147. In the coordination capsule **696**, each silver(I) ion triangularly coordinates three imine nitrogen atoms of three C_3-symmetric ligand syntones adopting a *cis*-conformation. Their two benzimidazole substituents are perpendicular to the arene ribbed fragment of the caging ligand, thus giving the trigonal prismatic coordination capsule **696** with one encapsulated triflate anion. Its C_3-symmetry axis passes through the two capping silver(I) ions, and the capsule has a S_3 symmetry. The encapsulated triflate ion interacts with two capping silver(I) ions; the distance O–Ag and Ag…Ag are 2.543 and 7.5 Å, respectively. In the coordination capsule **697**, the four ligand syntones adopt a *cis*-conformation and

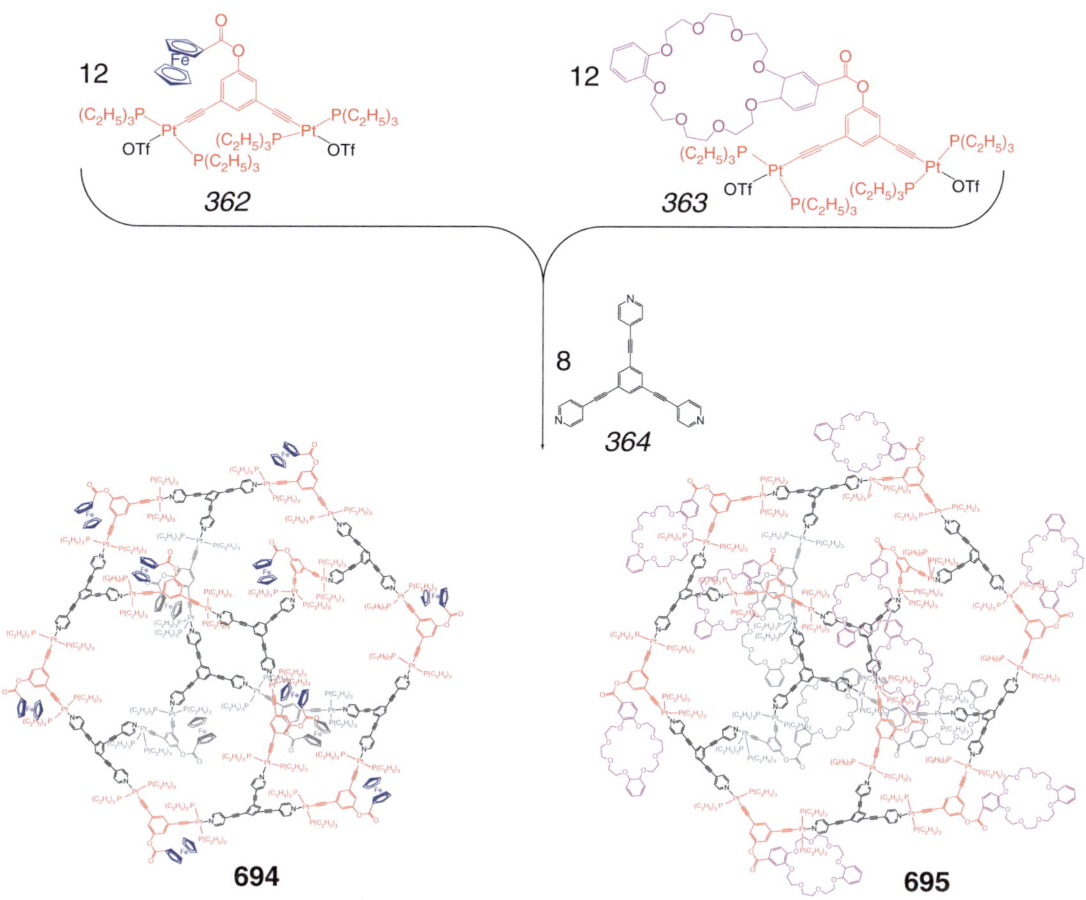

Scheme 4.146

are cross-linked by two copper(II) cations; each of these ions coordinates nitrogen donor atoms of the benzimidazole groups from four different ligand syntones, forming a square-planar CuN_4 coordination polyhedron. The aromatic bases of **697** are reported in [150] to "turn up," thus allowing an encapsulation of tetrahedral perchlorate anion. The C_4-symmetry axis of **697** passes through two cross-linking copper(II) ions, and it thus has a S_4 symmetry. The Cu…Cu distance of approximately 7.4 Å and the distance of 10.3 Å between the two parallel arene ribbed fragments of the cage framework are reported in [150] to define the void cavity volume of this C_4-symmetric coordination capsule.

Coordination-driven self-assembly of the arene-based ligand syntone having three benzimidazolyl pendant substituents with an excess $AgBF_4$ and CuI by Scheme 4.148 gave M_3L_2 cage complexes **698** and **699**, respectively. In the coordination capsule **698**, three doubly coordinate silver(I) ions almost linearly bind two face-to-face positioned ligand syntones, giving the Ag_3L_2 framework. The six benzimidazolyl groups are nearly perpendicular to its arene bases, thus forming a TP cage framework with the height *h* of approximately 11 Å and the distance between its edges of approximately 6 Å. This caging ligand encapsulates one tetrafluoroborate anion, one acetonitrile, and one water molecules, thus giving a heteroguest 1:1:1:1 cage complex. A similar Cu_3L_2 cage complex **699** has been prepared by reaction of the same ligand syntone with CuI. The obtained TP coordination capsule framework with slightly bent N–Cu–N fragments encapsulates CuI_3^{2-} dianion. The template effect

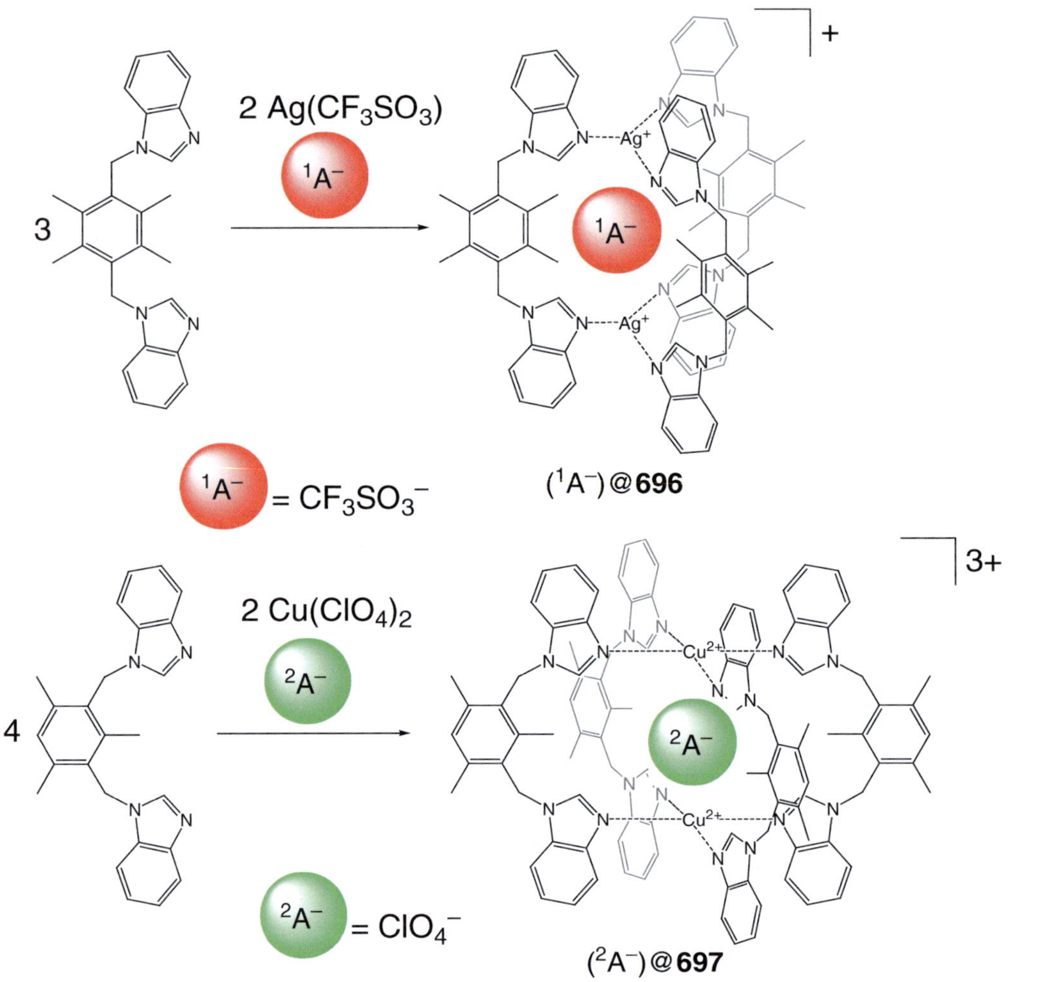

Scheme 4.147

of these anionic guests is reported [151] not to be specific because they can vary from the tetrahedral F_4^- monoanion to the planar CuI_3^{2-} dianion.

A tris-imidazolyl ligand syntone *364* has been used in [152] to obtain by Scheme 4.149 host–guest 1:1 cage complexes of a TP coordination capsule **700** with encapsulated tetrahedral and octahedral monoanions.

Tetragonal prismatic M_2L_4 (**701** and **702**) and TP M_2L_3 and M_3L_2 (**703**–**705**) coordination capsules with bridging and capping silver(I) and nickel(II) ions, respectively, have been self-assembled in [153] by Scheme 4.150 from the corresponding angular ditopic ligand syntones *365–367*.

Aliphatic and aromatic multidentate *N*-donor bridging ligand syntones and silver(I) acetylide $Ag_2(C{\equiv}C)_2$ have been used in [154] to obtain different types of monomeric and polymeric coordination capsules **706**–**710** with encapsulated acetylide dianions. In the case of the rigid aromatic pyrazine linker, the self-assembly by Scheme 4.151 gave a cage-based polymer **706** with pyrazine-linked silver columns, which was characterized by X-ray diffraction data. Its monomeric unit contains a double-cage framework, and each of the distorted TP and TAP entities encapsulates one acetylide dianion. The use of more Lewis-acidic and bulky aliphatic 1,8-diazabicyclooctane allowed obtaining a monomeric capsule **707**. Its crystal contains Ag_{14} double-cage frameworks with a C_2 pseudoaxis that passes through their shared edges; each of its

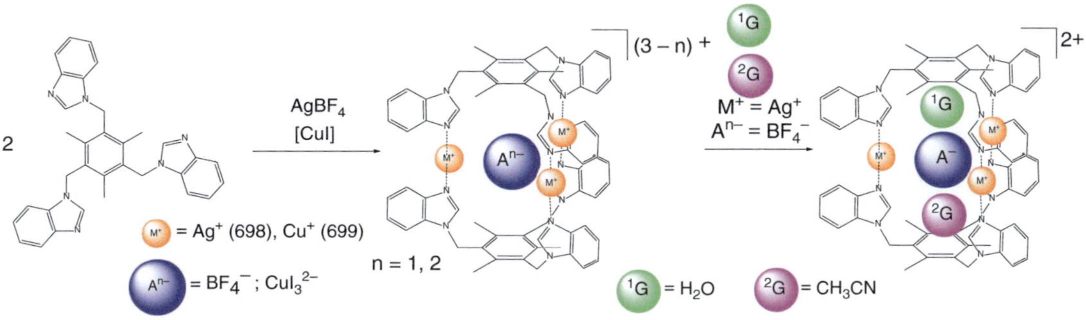

Scheme 4.148

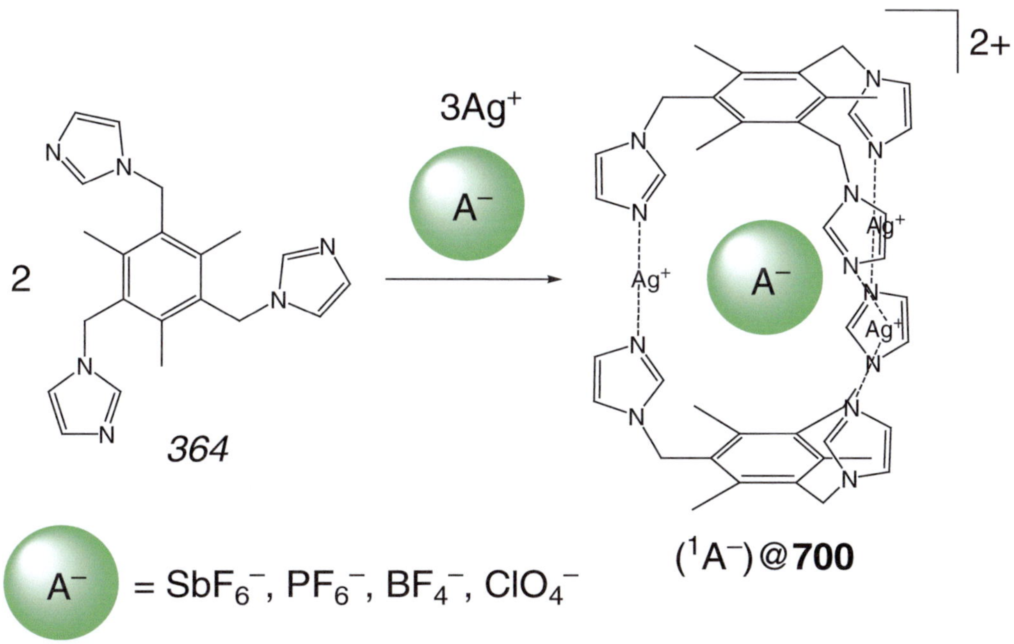

Scheme 4.149

distorted triangulated dodecahedral entities also contains one caged acetylide dianion. The more labile bis-pyridyl linker *368* forms two types of coordination capsules: the major product **708** contains the monoprotonated linkers, while in the minor product **709** this bis-pyridyl fragment is a neutral connector between its two cage fragments. The capsule **708** has a branched-tree architecture based on the distorted square-antiprismatic cage frameworks sharing their edges. These frameworks are bonded through additional Ag…Ag interactions, giving chain-like silver columns. The monoprotonated bis-pyridyl linker has a *gauche*-conformation and forms hydrogen bonds with a carboxylate anion of a neighboring coordination chain that gives an undulated layer in this crystal. In contrast, the bidentate linker in the crystal of **709** has both *gauche-* and *anti*-conformations. Its building blocks, the decanuclear coordination capsules

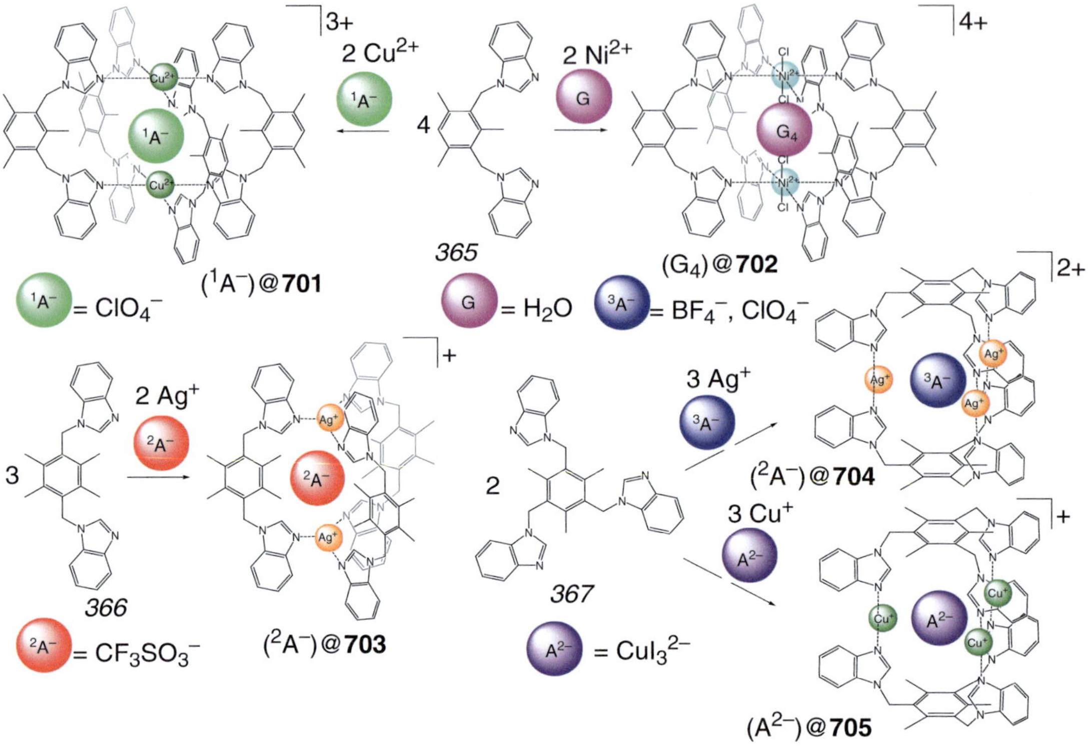

Scheme 4.150

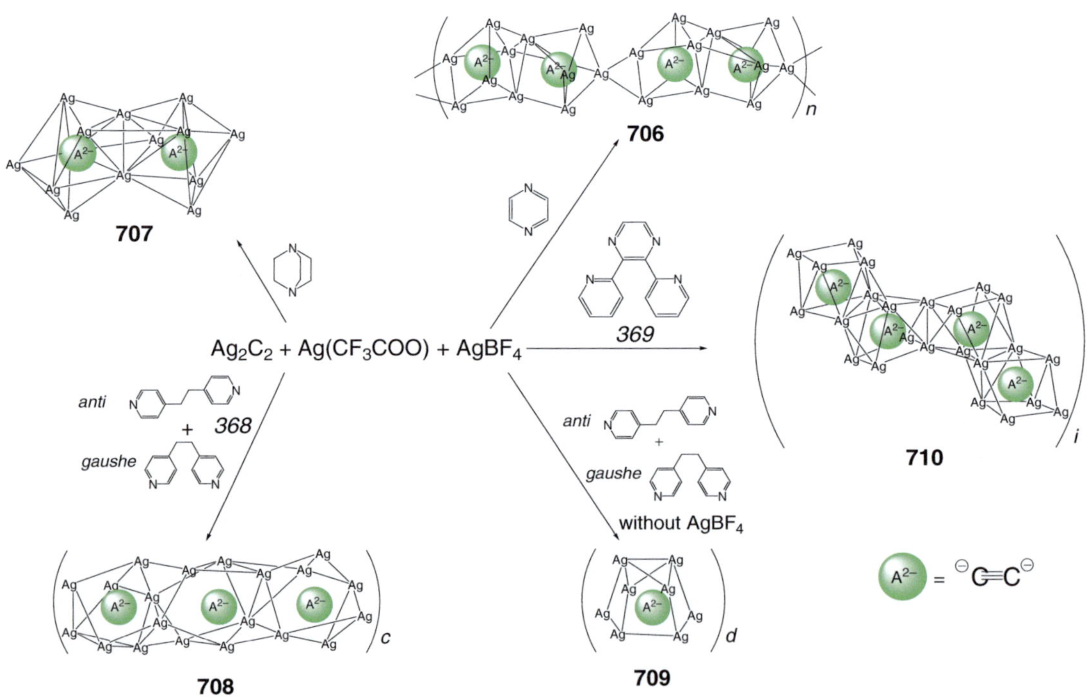

Scheme 4.151

$[(Ag_2C_2)(AgCF_3CO_2)_8]$, are linked by *anti*-conformers in the chains, forming a 2D network through its *gauche*-conformers. The same 2D network has been observed by the authors of [154] in the crystal of capsule **710** that is formed by an angular aromatic bridging ligand syntone *364*. In this case, the tetracage building block framework is formed by two types of the Ag_7 monocages. The C≡C bond length for the encapsulated acetylide dianions falls in the range 1.16–1.23 Å in all these coordination capsules; i.e., the triple character of this bond persists within their cavities. The Ag…Ag distances in these cage frameworks (2.821–3.368 Å, within twice the van der Waals radius of Ag^+) suggest the presence of argentophilic interactions [154].

A hydrothermal synthesis of the analogous coordination capsules **711–715**, the derivatives of *para*- and *meta*-cyanopyridines and the corresponding pyridinecarboxamides as the products of their hydrolysis, has been performed in [155]. Coordination-driven self-assembly of silver(I) salts by Scheme 4.152 with *meta*- and *para*-cyanopyridines at different temperatures afforded mono- and tetracage complexes **711** and **712** at 108 °C and their tetracage analogs **713** and **714** at 60 °C; *meta*-cyanopyridine also forms a coordination capsule **715** at 85 °C. The branched-tree motif of the capsule **711** is formed by square-antiprismatic Ag_8 polyhedral subunits encapsulating an acetylide anion. These cages form a columnar structure throughout the trifluoroacetate anion, which is stabilized by coordinated *meta*-pyridinecarboxamide ligand syntones. Hydrogen bonding between their amino groups and solvent water molecules as H-donors and oxygen atoms of trifluoroacetate anions as H-acceptors gave a 3D columnar network. In the crystal of **712**, the distorted triangulated-dodecahedral polyhedral entities that are conjugated by two Ag–Ag bonds form *zigzag* composite chains. These chains are linked by *para*-pyridylcarboxamide ligand syntones into a 2D network. 3D network in the crystal of **713** is formed by bis-capped TP Ag_8 cage frameworks, which are linked by *meta*-pyridyl-

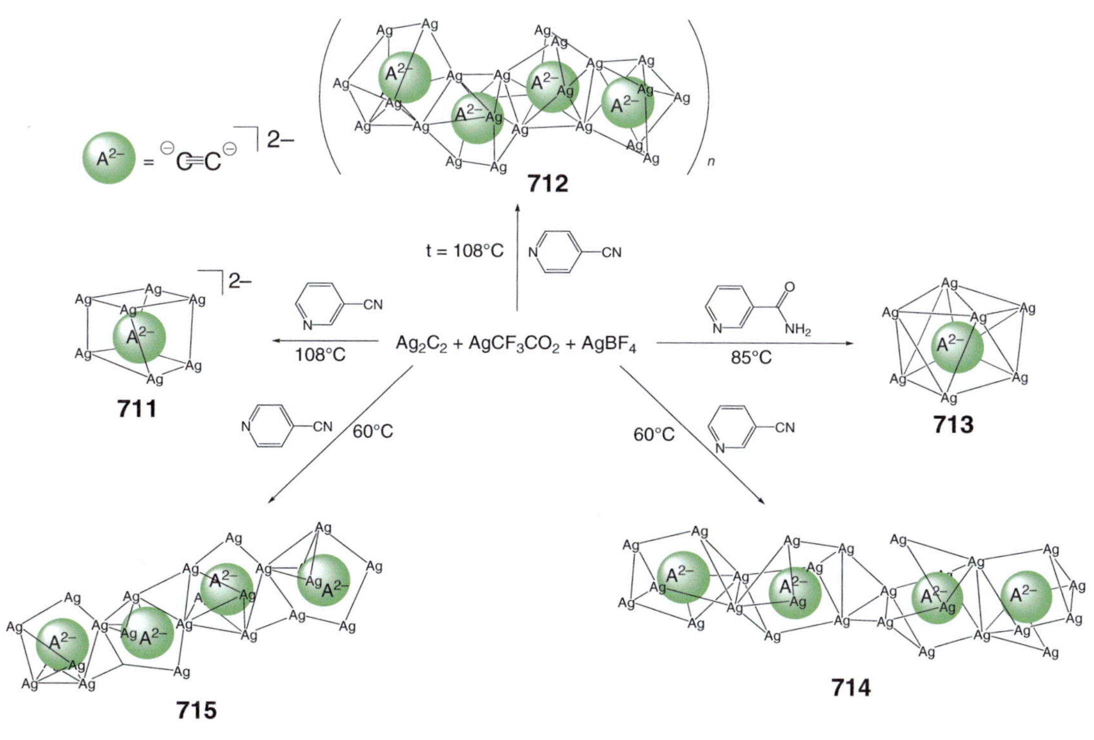

Scheme 4.152

carboxamide ligand syntones. The coordination capsules **714** and **715** also have the same 3D architectures [155].

In the case of tripodal tris-aminopyrrolyl ligand syntones, metal dithiocarbamate moieties can play a role of linkers in the ribbed fragments of the corresponding coordination capsules. *One-pot* reactions of these capping syntones with CS_2 and metal(II) halides by Scheme 4.153 in high dilution conditions gave nickel-, zinc-, and copper(II)-based M_3L_2 capsules **716–719**. UV-vis study performed in [156] for the capsule **717** showed that this copper(II)-based caging ligand efficiently binds only benzoate anion and discriminates chloride and dihydrophosphate ions, while its copper(III)-containing oxidized derivative **720** preferably encapsulates chloride ion. Therefore, the encapsulation stabilizes such neutral copper(II, III)-containing capsule (Cl^-)@**720** [156].

An 11-nuclear 21-vertex alcohol capsule **721**, formed by 11 sodium cations and 10 *tert*-butanolate anions, has been prepared in [157] by Scheme 4.154. This capsule contains one encapsulated hydroxide anion and retains its cage framework in solution [157].

The M_6L_6 coordination capsules **722** and **723** have been self-assembled in [158] by Scheme 4.155 from copper and silver(I) ions and an alkyne tetrapyridyl ligand syntone *371*. Each of their metallocenters coordinates four nitrogen atoms of three tetrapyridyl fragments. These caging ligands encapsulate BF_4^- and SbF_6^- monoanions both in solution and in solid state but discriminate more bulky $[Co(C_2H_{11}B_9)_2]^-$ anion [158]; the preparation of their analog **724** (Scheme 4.156) is described in Chap. 5.

The *meta-* and *para*-tris-pyridyl cyclotriveratrylene ligand syntones *372* and *373* have been used in [159] for the synthesis of dimeric M_2L_2 and tetrahedral prismatic M_4L_4 coordination capsules **725** and **726** by Scheme 4.157. The caging ligand **725** encapsulates solvent acetonitrile molecules, giving a host–guest 1:2 cage complex. In this C_2-symmetric capsule, the caged molecules are coordinated to cross-linking silver(I) metallocenters. The tetranuclear Ag_4L_4 capsule **726** also encapsulates acetonitrile molecules to give a 1:5 cage complex. As follows from X-ray diffraction data of [160], capping (cross-linking) Ag+ cations in its vertices form a tetrahedral Ag_4L_4 framework with Ag…Ag distances of 8.591 and 9.034 Å. Each of the ligand syntones is coordinated to three silver cations forming a face of the coordination capsule **726**. Its 2-quinolyl-containing analog **727** has been prepared in [161] by Scheme 4.158. Cross-linking silver(I) cations in its vertices have a linear geometry with Ag–N distances of 2.21 and 2.17 Å and N–Ag–N angle of 176.0°. These four crystallographically equivalent Ag^+ cations are arranged in a distorted tetrahedron manner with Ag…Ag distances of 9.31 and 10.34 Å. While in the above tetrahedral cage framework **726** the capping silver(I) ions form the vertices of the tetrahedron and each ligand syntone is coordinated to three metallocenters forming a face of this tetrahedral capsule, each ligand syntone in the cage framework of **727** is coordinated to two Ag^+ cations through only two of its three quinolyl fragments of *374* [161]. Detailed synthetic procedure, NMR and X-ray diffraction characterization of this coordination capsule and its dimeric analog **725** (Scheme 4.159) are reported in [162]. In the coordination capsule **725**, two ligand syntones *372* are arranged in a face-to-face manner by two tetrahedral bridging silver(I) ions; each Ag^+ cation coordinates two pyridyl groups from one ligand syntone and one pyridyl nitrogen donor atom from the other. The encapsulated acetonitrile molecule completes its coordination polyhedron.

A similar ligand syntone *375* has been self-assembled in [163] with silver(I) cations to give a Ag_3L_3 coordination capsule **728** by Scheme 4.160. It has a trigonal bipyramidal polyhedron with two ligand syntones as an enantiomeric pair; this capsule encapsulates solvent acetonitrile molecules [163].

A starburst prismatic Pd_6L_8 coordination capsule **729** with a stella octangular geometry has been prepared in [164] by Scheme 4.161. In it, the Pd…Pd distances along the edges of its octahedral cage framework range from 16.58 to 16.65 Å, and the longest one between the vertices

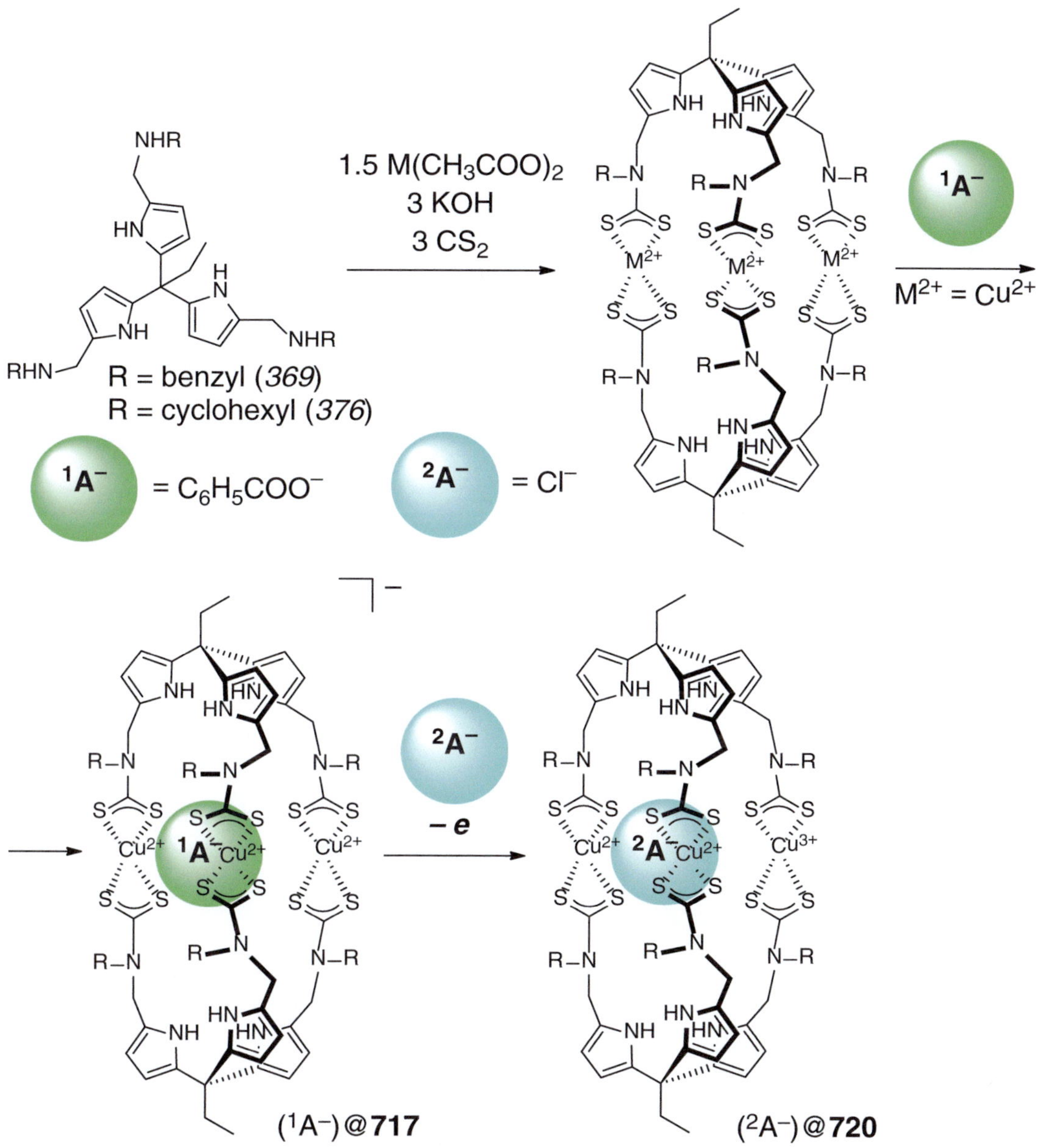

R = benzyl (*369*), M = Ni^{2+} (**716**) R = benzyl (*369*), M = $Zn(pyr)^{2+}$ (**718**)
R = benzyl (*369*), M = Cu^{2+} (**717**) R =cyclohexyl (*376*), M = Ni^{2+} (**719**)

Scheme 4.153

is approximately 23.54 Å. The ligand syntones *376* occupy eight faces of **729**, all the cavities of which face the center of this coordination capsule. Each 12-cationic Pd_6L_8 framework is self-assembled with eight ligand syntones of the same chirality; so this 3.1 nm-sized capsule **729** is chiral but exists as a racemate with both the enantiomers present in its X-rayed crystal [164].

Analogous [2]-catenane M_6L_4 and monocage M_3L_2 coordination capsules **730–732** with bridging silver(I) and copper(II) ions (Schemes 4.162

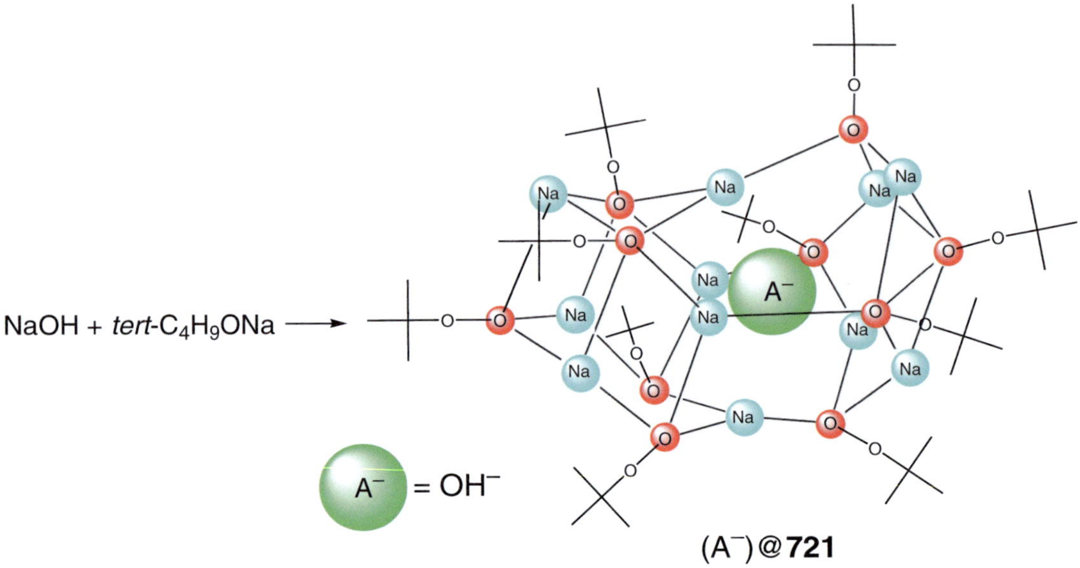

Scheme 4.154

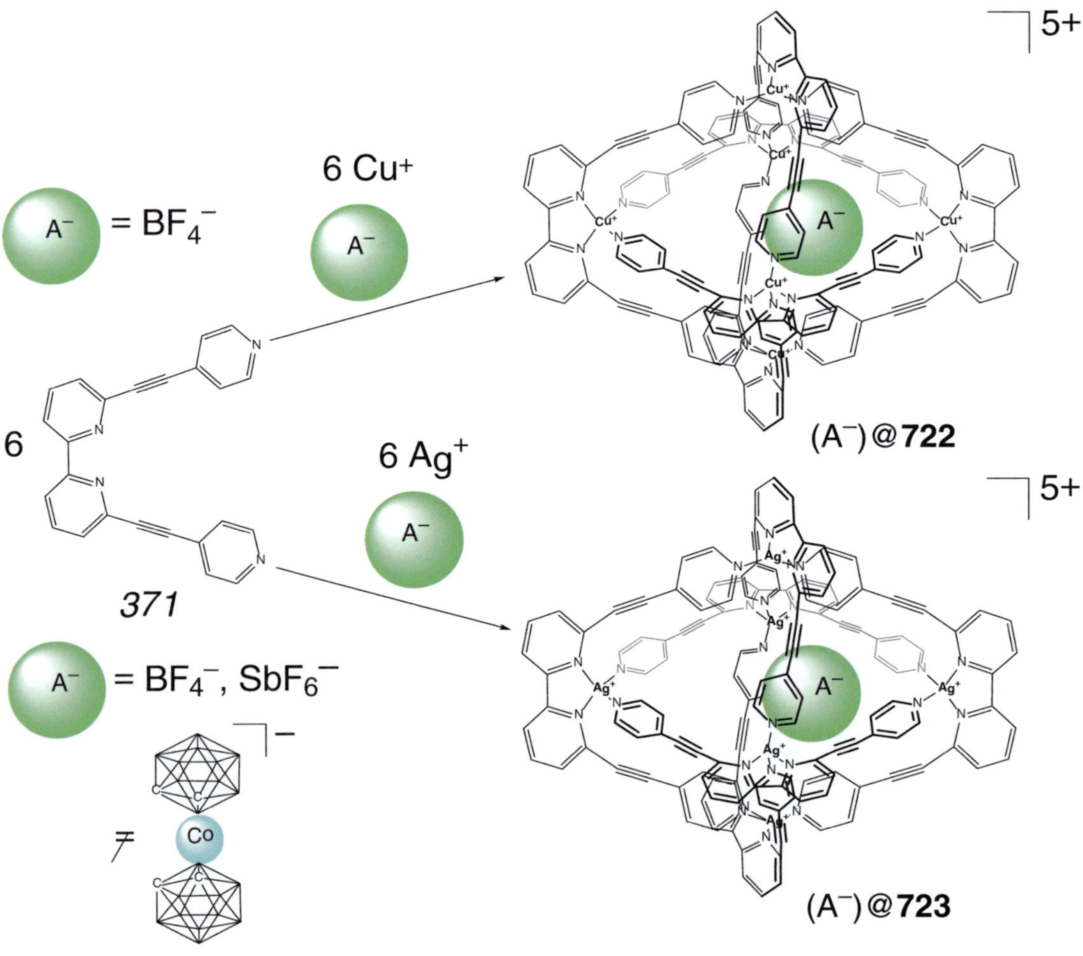

Scheme 4.155

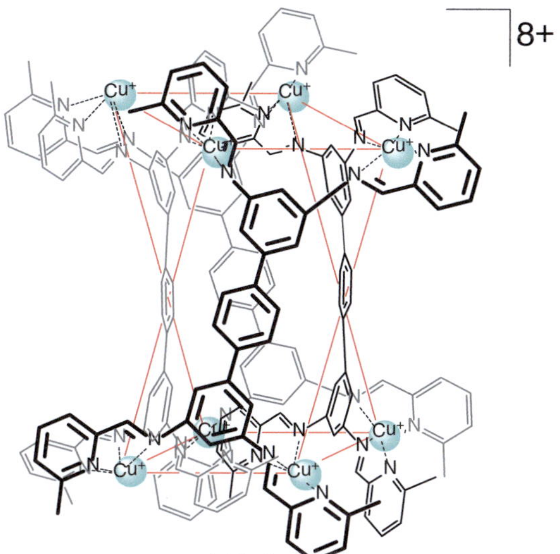

Scheme 4.156

and 4.163) have been isolated and structurally characterized in [165].

Self-assembly of four polyaromatic bis-pyridyl ligand syntone *377* with two palladium(II) ions by Scheme 4.164 gave a C_4-symmetric macrotricyclic coordination capsule **733** encapsulating one nitrate anion to give a 1:1 cage complex [166].

Anion-templated synthesis of a 1:1 cage complex of a rigid Co_2L_4 coordination capsule **734** with two tripodal cross-linking cobalt(II) ions and four ribbed bridging bis-benzimidazolyl ligand syntones *378* by Scheme 4.165 has been performed in [167]. In it, these coordinatively unsaturated cross-linking cations form the donor–acceptor bonds Co–F (r_i=2.045 Å) with encapsulated BF_4^- anion, and this anion does not undergo complete exchange reaction with PF_6^- ion but gives only a mixture of the corresponding cage complexes. The host–guest 1:1 compound with encapsulated hexafluorophosphate anion has been prepared in [168] in a straightforward manner by self-assembly of a solvato-complex $[Co(CH_3CN)_4](PF_6)_2$ with the same bis-benzimidazolyl syntone *378*. Its cobalt(II) metallocenters with a solvent-containing square-pyramidal polyhedra also coordinate caged PF_6^- anion, thus forming the donor–acceptor Co–F bonds [168].

A tetranuclear octaanionic iron(II)-bridged bis-resorcene[4]arene capsule **735** has been prepared in [169] by Scheme 4.166 using coordination-driven self-assembly of iron(II) chloride with the iminodiacetic acid-functionalized ligand syntone *379*. The tetranuclear cage framework of **735** is formed by four iron(II) cations linking together two hemispherical syntones *379* by coordination of their two iminodiacetate moieties in an *N,N-cis-fac*-manner. Coordination polyhedra of these metallocenters that form a belt around the outside of the cavity of **735** have a TP geometry. The interior of this capsule is surrounded by phenyl, ether, and hydrocarbon groups, and it encapsulates six water molecules, giving a distorted octahedron with the distances O…O of approximately 3 Å through the hydrogen-bonded network. The caging ligand **735** is reported in [169] to be suitable in size (11.0×9.71 Å) for encapsulation of small organic molecules, such as benzene, THF, and hexane. As follows from solution NMR data [169], this capsule binds guests shown in Scheme 4.166; such caging of bromobenzene in solid state is confirmed by X-ray diffraction study of the corresponding host–guest 1:1 complex that contains one disordered guest molecule.

Analogous tetranuclear cobalt(II)-based coordination capsules **736** and **737** have been self-assembled in [170] by Scheme 4.167. Bridging cobalt ions in **736** with one principle C_2 and two perpendicular Co…Co C_2 axes have an alternating $\Delta\Lambda\Delta\Lambda$ arrangement. The cage framework of its 1:1 cage complex with one encapsulated disordered ethylbenzene molecule has the same structure and consists of two hemispherical resorcinarene syntones *380* bridged by four cobalt(II) ions. These cations are positioned at the edge of its resorcinarene rims and are close to the equator of this D_{2d}-symmetric capsule. Two iminodiacetate moieties are coordinated in an *N,N-cis-fac*-manner to each such metallocenter having a distorted TP coordination polyhedron, thus giving an octaanionic capsule. The dimensions of its cavity are 10.9×9.7 Å, and it encapsulates a variety of organic guests shown in Scheme 4.167 [170].

Coordination-driven self-assembly of six pyrogallol[4]arene ligand syntones *382* with 12

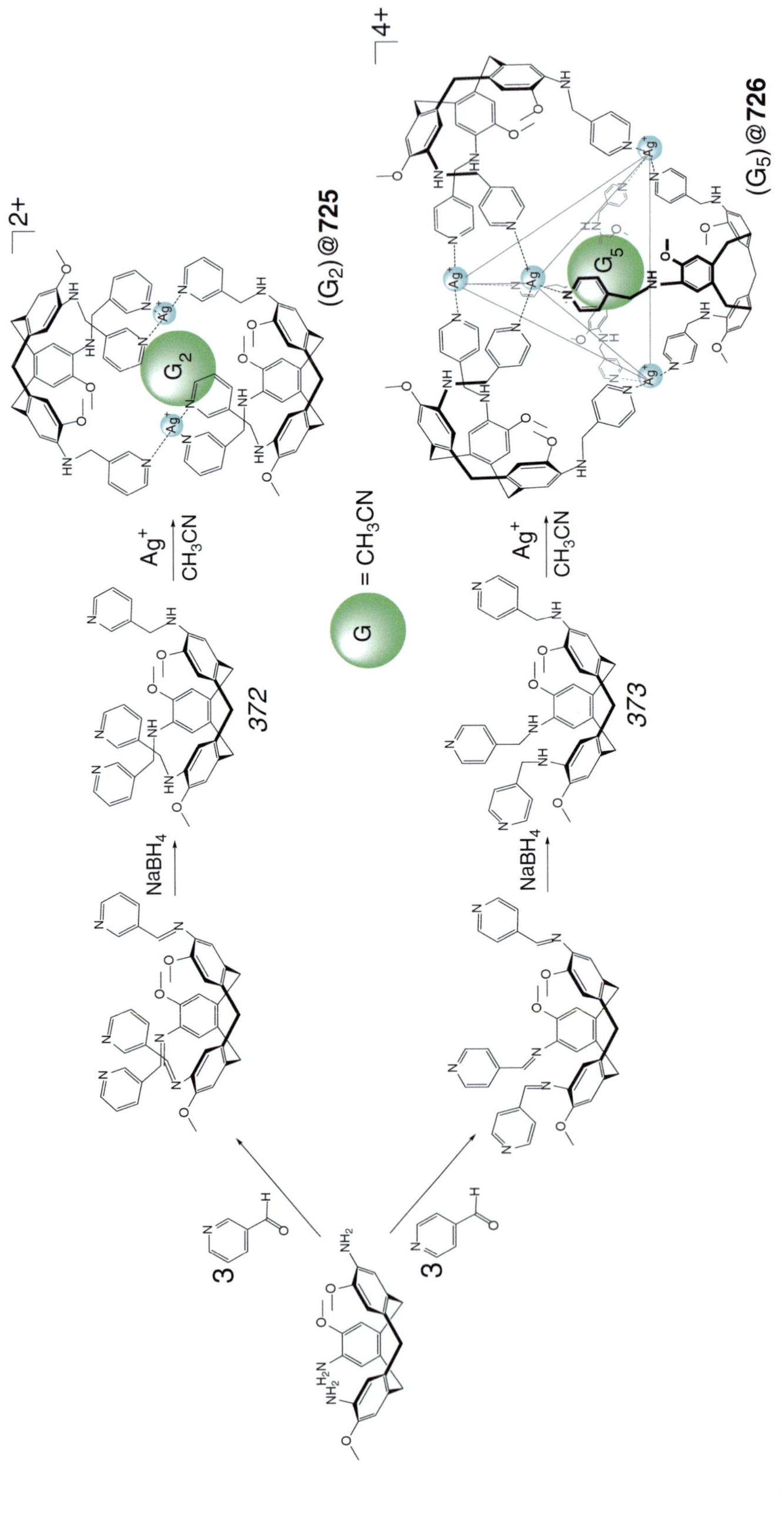
3
3
NaBH4
NaBH4
372
373
Ag+
CH3CN
Ag+
CH3CN
G = CH3CN
(G2)@725
(G5)@726
2+
4+

Scheme 4.157

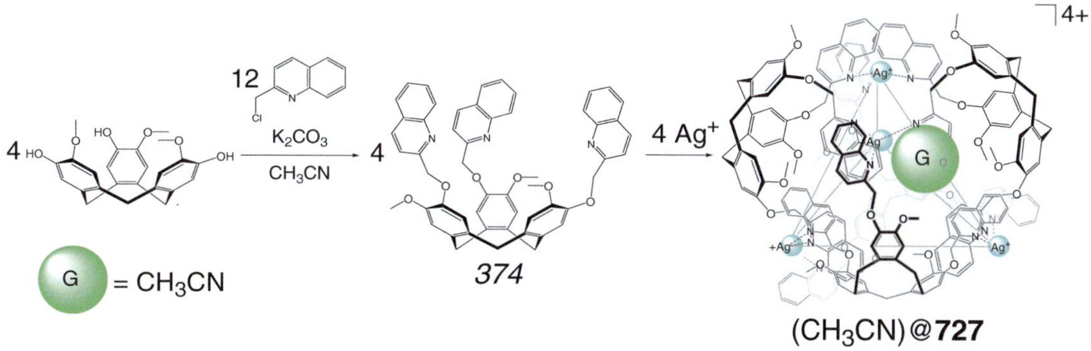

Scheme 4.158

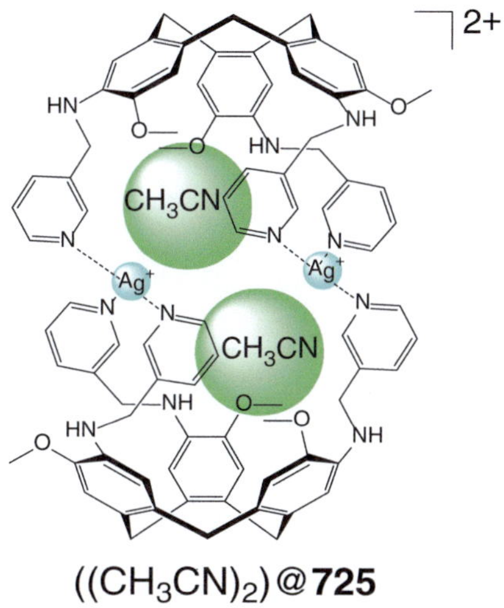

Scheme 4.159

gallium(III) cations by Scheme 4.168 gave a large coordination capsule **738** with a "rugby ball" shape and a cavity volume of approximately 1150 Å^3. Its cage framework is formed by four cross-linking Ga_2O_3 arrays, 16 intramolecular OH…O$^-$ hydrogen bonds, and four intramolecular OH…OH hydrogen bonds. This ligand encapsulates 12 solvent water molecules coordinated to its metallocenters, six uncoordinated water and eight acetone molecules. Due to spatial limitations, encapsulated water molecules are reported in [171] to form two pseudolinear hydrogen-bonded $(H_2O)_5$ chains.

Encapsulation of large aromatic guests (Scheme 4.169) by a D_4-symmetric bis-cavitand Ag_4L_2 coordination capsule **739** has been studied in [172]. This capsule provides a highly rigid cavity in which guest molecules with the length of approximately 14 Å can be selectively accommodated, thus making it possible to distinguish slight structural differences in flexible alkyl-chain molecules and rigid aromatic guests. Detailed thermodynamic studies revealed that not only the CH…π interactions between the terminal methyl groups of these guests and the aromatic cavity walls play a key role in the encapsulation but also does the desolvation of the cavity. It selects hydrogen-bonded heterodimers of a mixture of two or three carboxylic acids. This chiral capsule **739** is also reported in [172] to diastereoselectively encapsulate chiral guests.

The Mn_{26} coordination capsules **740** and **741** with the encapsulated complex cation obtained by Scheme 4.170 are reported in [173, 174] to show single-molecule magnet behavior. The caged guest of **740** [173] is formed by 16 Mn^{III} ions, 12 methoxide, and 16 oxide bridges and is surrounded by the strands of 12 manganese(II) ions and ten linking ligand syntones *383*. The adamantanoid cage framework of **740** is formed by four Mn^{II} ions found at the threefold axes that are bridged by *383*–Mn^{III}–*383* chains [173]. In its analog **741** [174], the analogous adamantanoid Mn^{II}_4-capped coordination capsule also encapsulates the same manganese oxide guest.

"*Matreshka*" complex-in-complex assemblies of the metallacrown caging ligand **742** have been prepared [175, 176] by Scheme 4.171. In the terbium(III) complex of **742** [175], an eight-

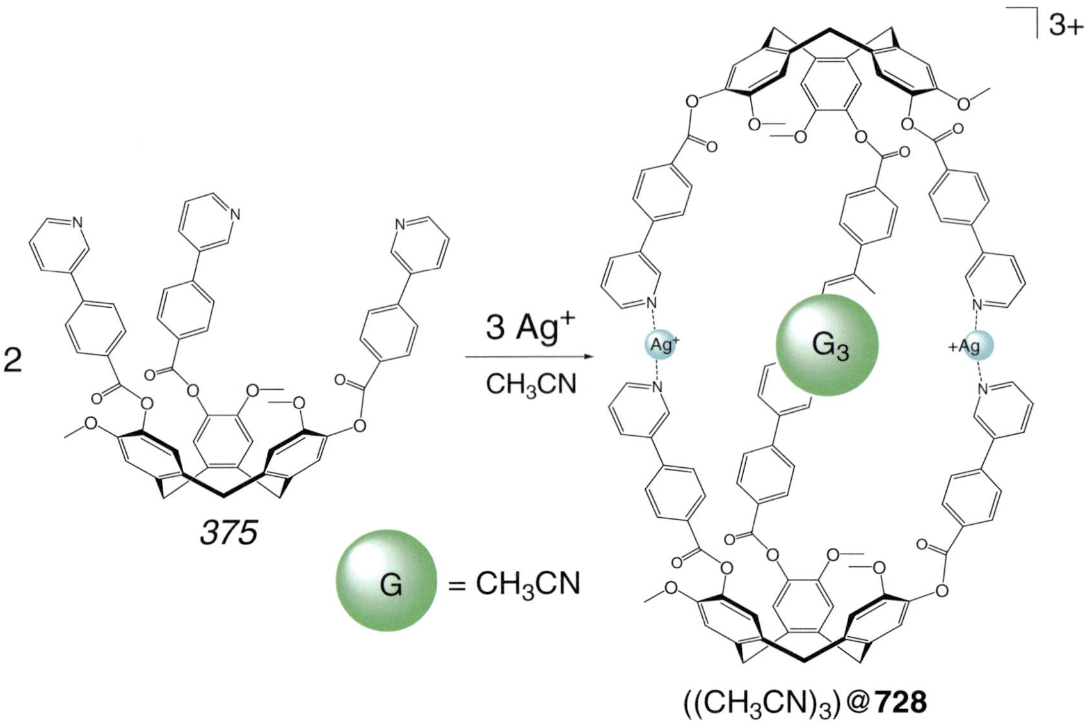

Scheme 4.160

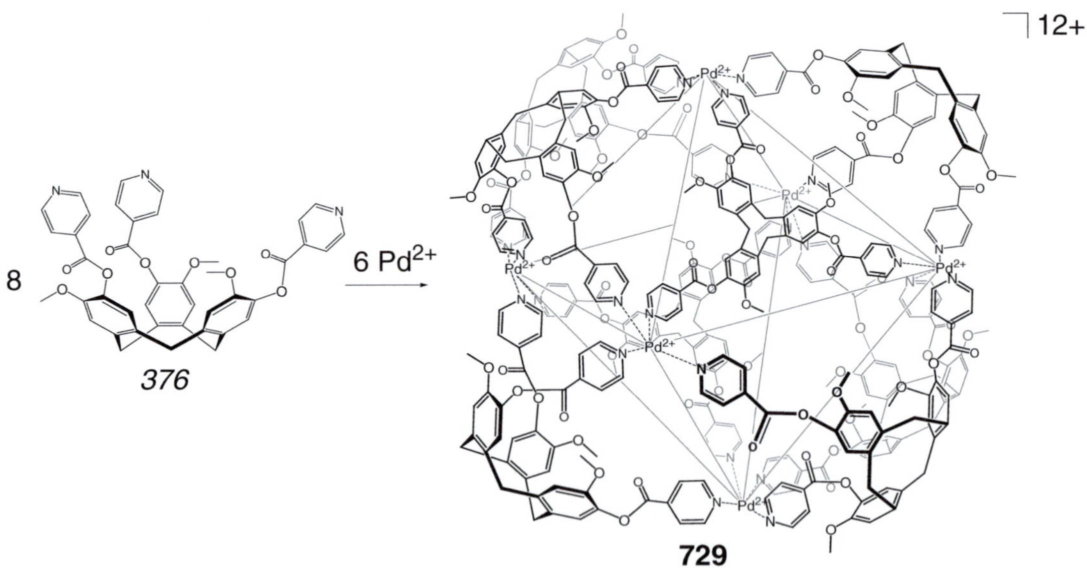

Scheme 4.161

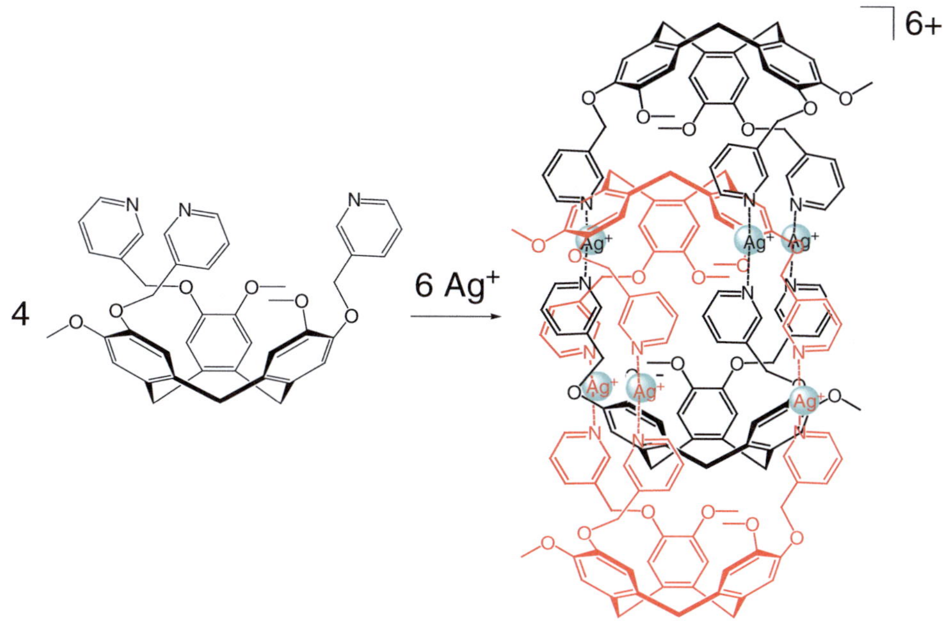

730

Scheme 4.162

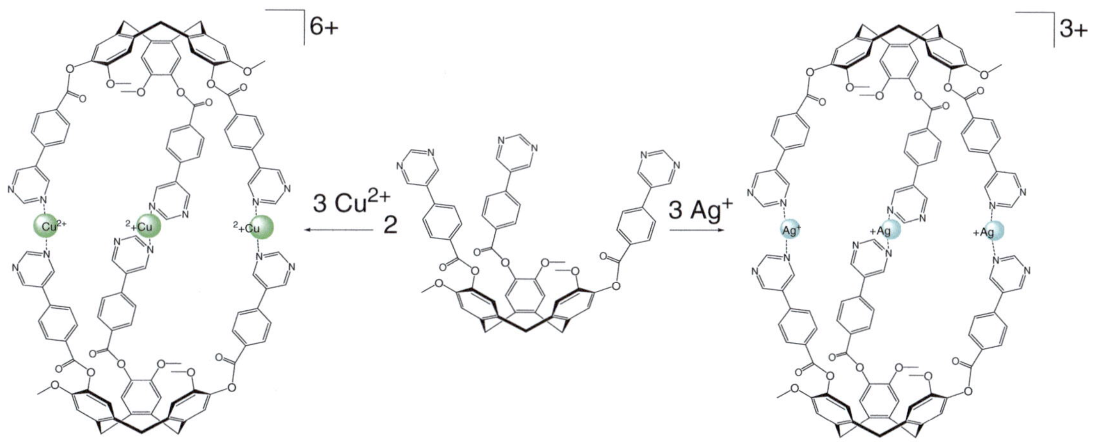

732 **731**

Scheme 4.163

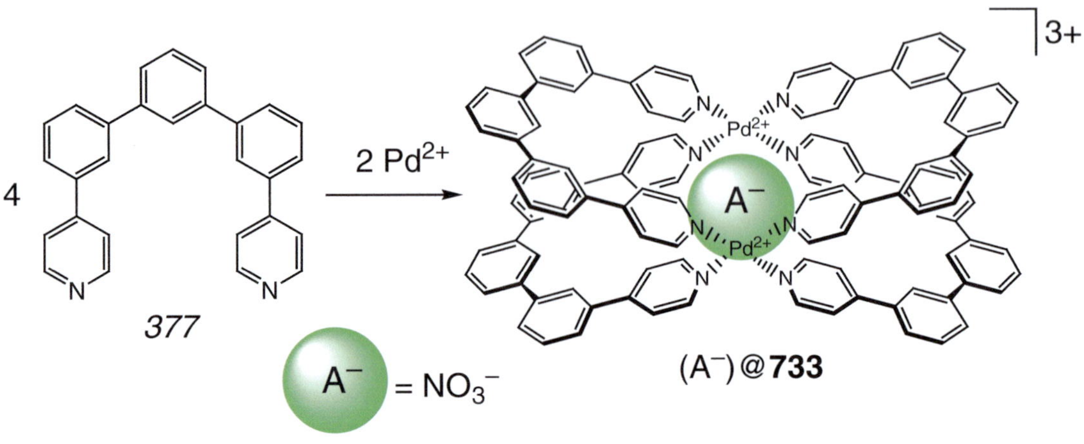

Scheme 4.164

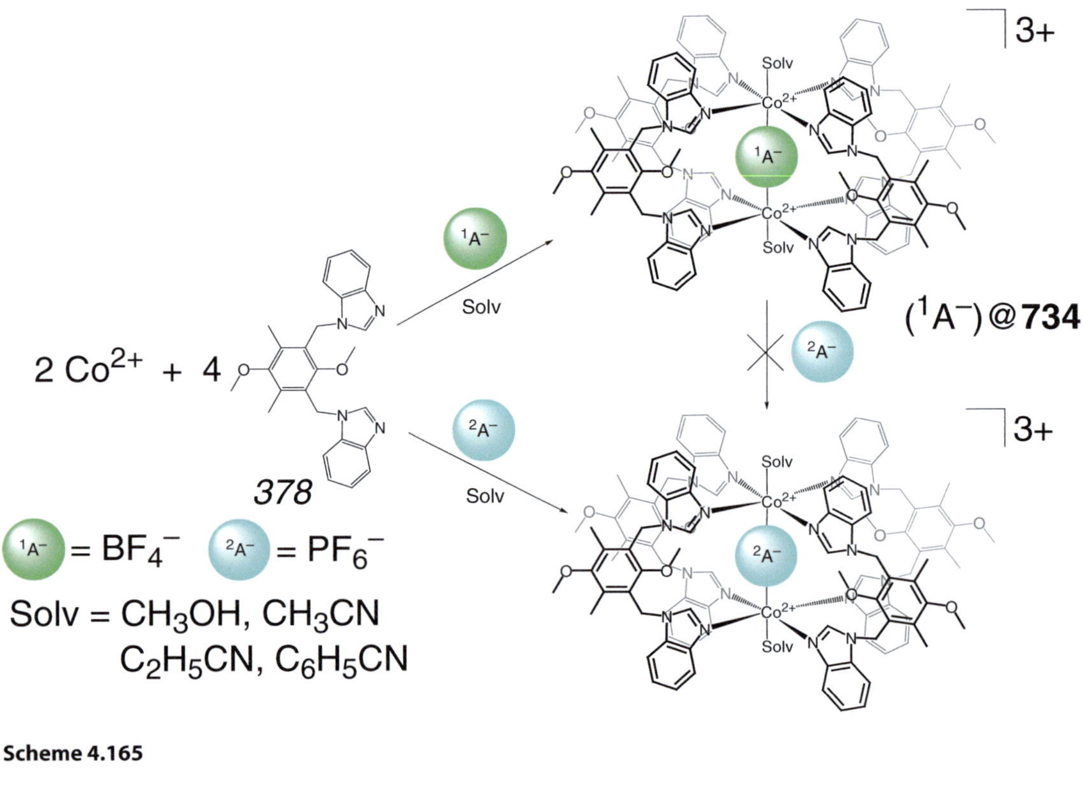

Scheme 4.165

coordinate central metal ion is sandwiched between the two concave metallacrown syntones. This sandwich complex is encapsulated within the cavity of the $Zn_{24}L_{24}$ cage framework of **742** having an overall S_8 symmetry. Its metallamacrocyclic fragments are formed by octacoordinate zinc(II) ions that possess alternating Λ and Δ absolute stereochemical configurations (see insert in Scheme 4.171). In its dysprosium(III)-containing analog [176] with the crystallographic fourfold axis that passes through the Dy^{3+} ions, the cationic coordination capsule also has an overall S_8 symmetry and consists of three layers: a large $[24\text{-}MC_{Zn(II),quinHA}\text{-}8](py)_8$ ring is sand-

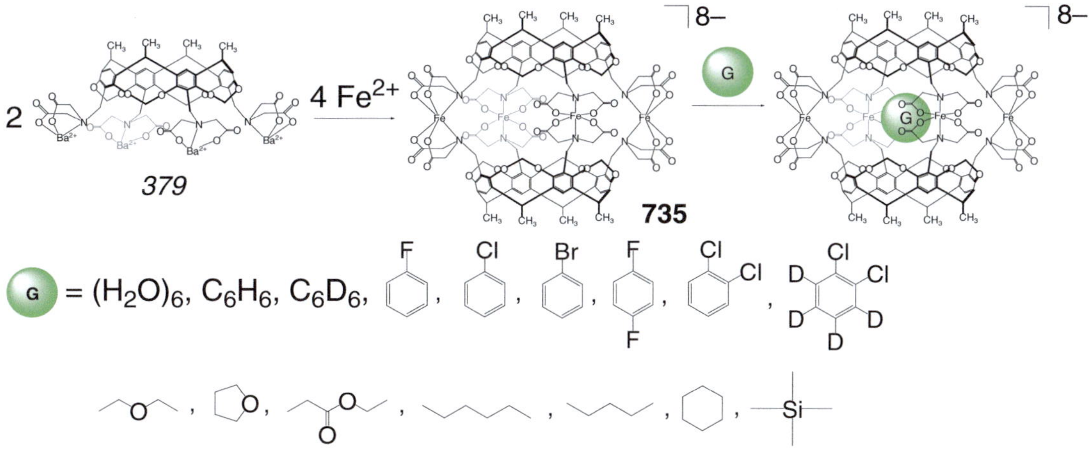

Scheme 4.166

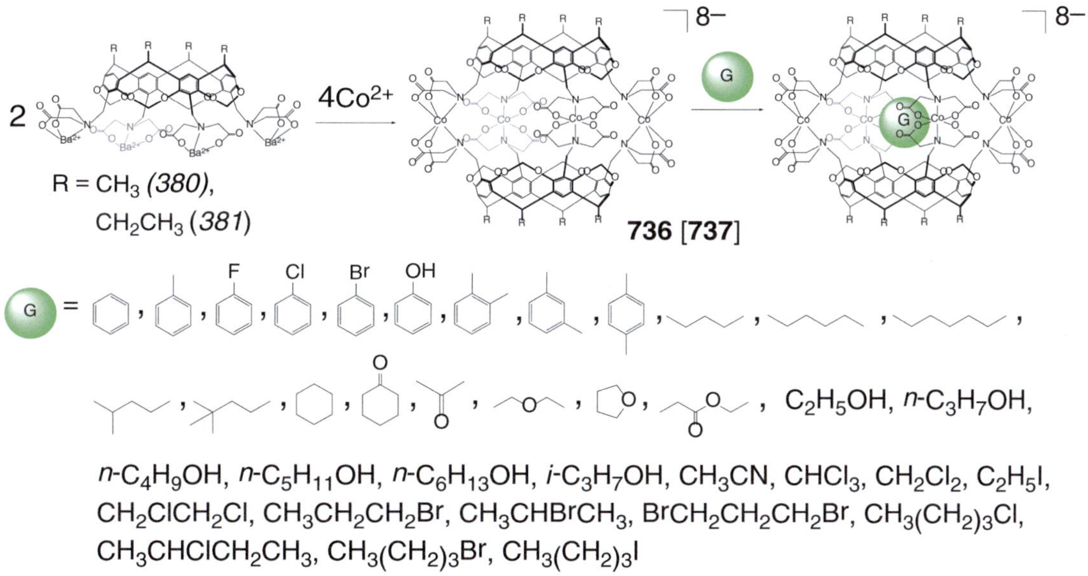

Scheme 4.167

wiched between two smaller concave [12-$MC_{Zn(II),quinHA}$-4] rings. Unusual photophysical properties of these highly emitting near-infrared lanthanide metallacrown cage complexes with excitation shifted toward lower energy are explained [176] by their encapsulated sandwich metallacrown structure, which allows one to put lanthanide ions at a predetermined and relatively shielded position. As a result, the closest C–H group in these capsules is located at a fixed and constant distance of approximately 7 Å from the central lanthanide metal ion [176].

Hexanuclear $[Cu_6(\mu_3\text{-}O)_2L^1{}_6L^2{}_3]^-$ coordination capsules **743–746** have been prepared in [177] by two synthetic approaches shown in Scheme 4.172: one of them is based on direct self-assembly of copper(II) ions with the corresponding pyrazolate ligand syntones and bis(triphenylphosphine)iminium chloride in the presence of NaOH and the second approach uses

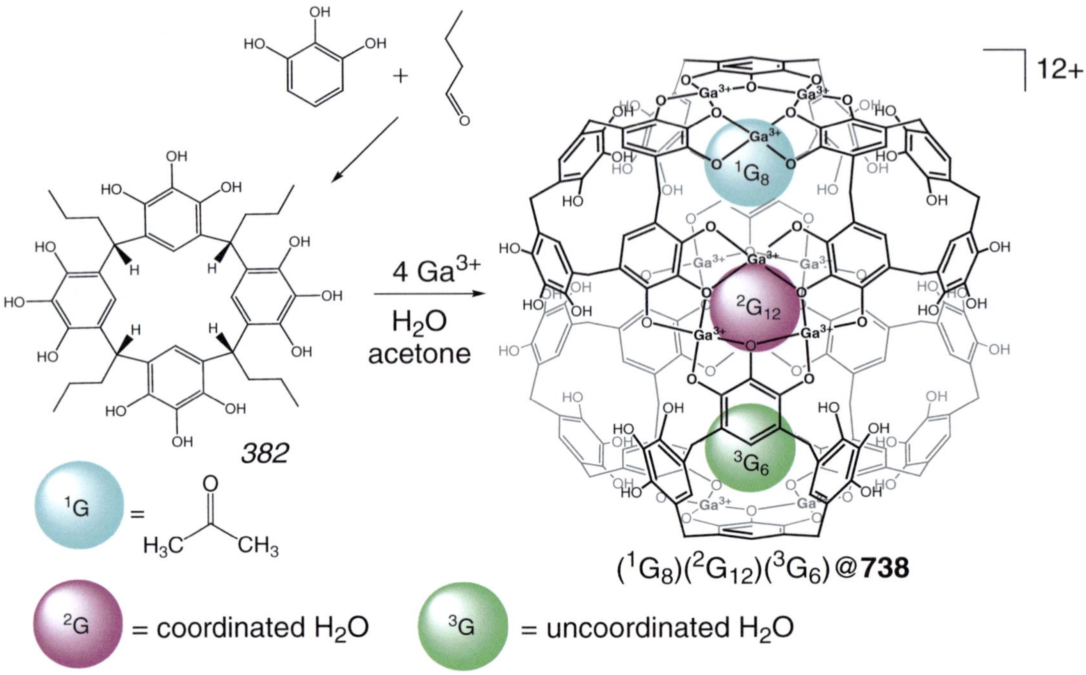

Scheme 4.168

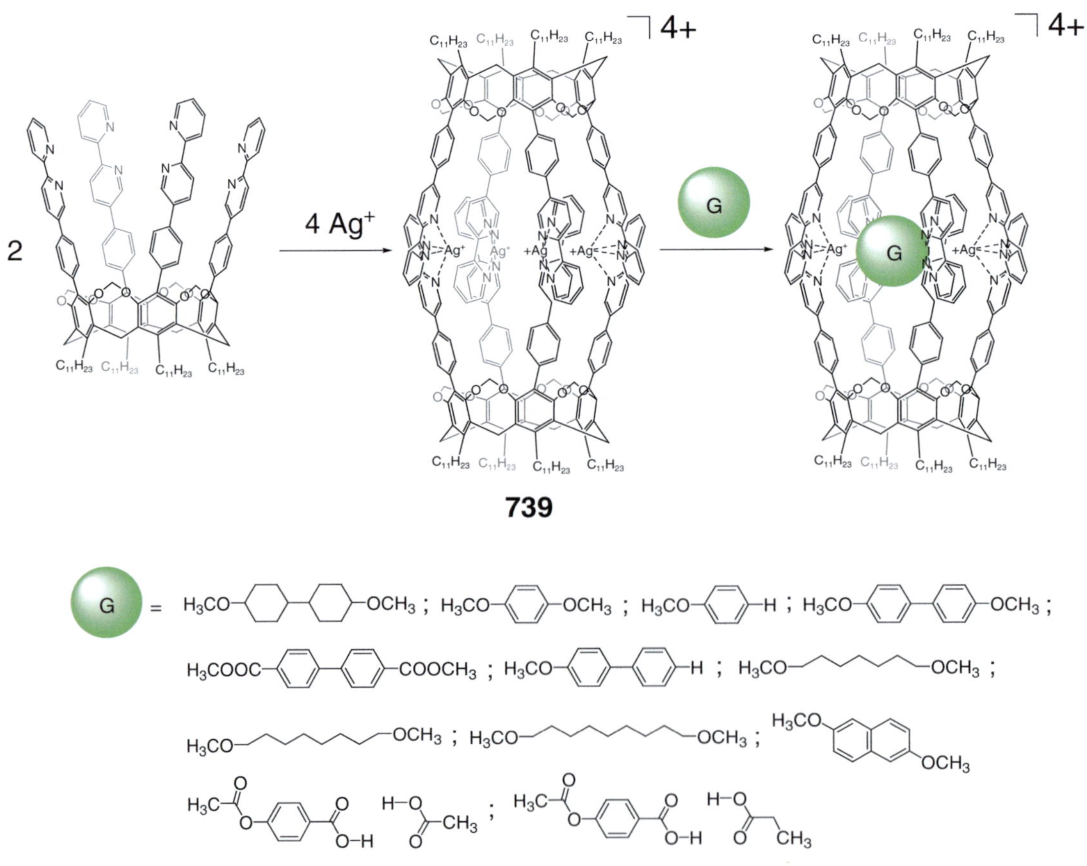

Scheme 4.169

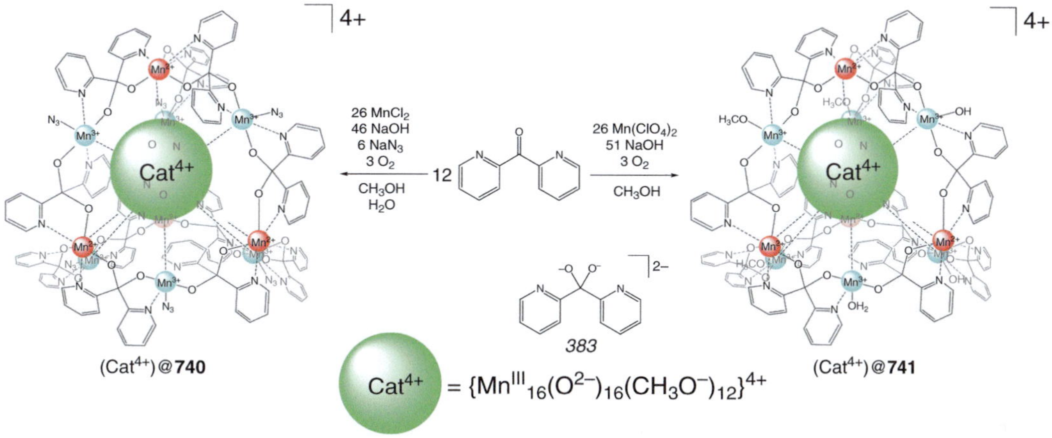

Scheme 4.170

condensation of initially prepared trinuclear Cu_3- and Ag_3-metallomacrocycles as complex syntones. As follows from the CV data, their two redox-active $Cu_3(\mu_3\text{-}O)$ cross-linking metallocenters demonstrate strong electronic interactions through rigid conjugated cage frameworks. No template effect of anions on the formation of these coordination capsules has been observed in [177], but they are able to encapsulate them as guests. In the framework of **743**, six cross-linking copper(II) ions form a TP $Cu_6L'_2L''_3$ cage framework that is contracted along its C_3-symmetry pseudoaxis.

Self-assembly of an excess of copper(II) ions with pyrogallol[4]arene ligand syntone *384* by Scheme 4.173 gave $M_{24}L_6$ coordination capsule **747**, which encapsulate water (approximately 35) and methanol (approximately 20) solvate molecules. The same reaction of a propyl-containing pyrogallol ligand syntone with gallium(II) ions afforded a Ga_{16}-containing pseudocapsule **748**, which forms a heterometallic $Ga_{16}Cu_8$ coordination capsule **749** [178] by Scheme 4.174 with copper(II) ions. In the case of a mixture of the pyrogallol[4]arene C_6- and C_{11}-containing ligand syntones, non-discriminate nature of their self-assembly with copper(II) ions to give the corresponding heteroarene coordination capsules with each possible permutation has been observed in [178].

Coordination of a deprotonated tripodal tris-*β*-diketone ligand syntones *385* to copper(II) ions by Scheme 4.175 has been used in [179] for the synthesis of a neutral Cu_3L_2 coordination capsule **750**.

A heterometallic $^1M_6{}^2M_8L_8$ capsule **751** (Scheme 4.176) has been prepared in [180] by a stepwise coordination-driven synthetic procedure based on the "bottom-to-up" principle. Its tris-pyrazolyl-pyridine ligand syntone *386* with two types of nonequivalent donor centers underwent two-step transformation. First one gave a copper(I) hydroborato-tris-pyrazolate (scorpionate) complex precursor *387* as a building block with terminal pyridyl donor groups for their further coordination to a metal ion preferring an octahedral coordination (in particular, iron(II) ion). Therefore, the self-assembly of this complex precursor with Fe^{2+} and NCS^- ions gave a nanoball capsule **751**, the distorted rombododecahedral framework of which with the diameter of approximately 30 Å is formed by *Fe_6*-octagedral and *Cu_8*-cubic polyhedra; this coordination capsule encapsulates disordered acetonitrile solvent molecules. The use of 0 ↔ 5/2 SCO iron(II) ions for cross-linking of the scorpionate precursors *387* is explained in [180] by their well-known potential magnetic switching ability. Indeed, the SQUID magnetometry data suggests this SCO for approximately 50 % of total iron(II) ions per cage framework containing two types of coordination polyhedra of these cross-linking ions $Fe^{2+}{}_6Cu^+{}_8L_8$ {$[Fe^{II}(NCS)_2(py)_4]$ (Fe–A, 56 %) and $[Fe^{II}(NCS)(CH_3CN)(py)_4]$ (Fe–B, 44 %}. As follows from ^{57}Fe Mössbauer spectra, two crystallographically nonequivalent Fe–A sites with a total occupancy of 57 % are in a high-spin

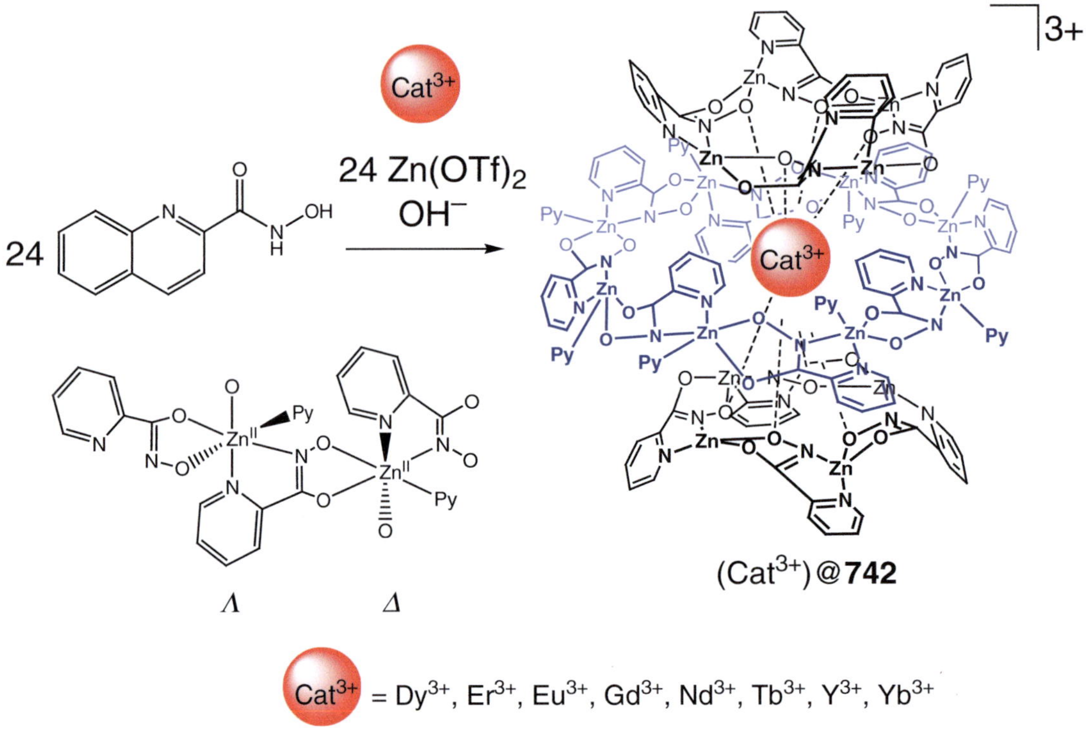

Scheme 4.171

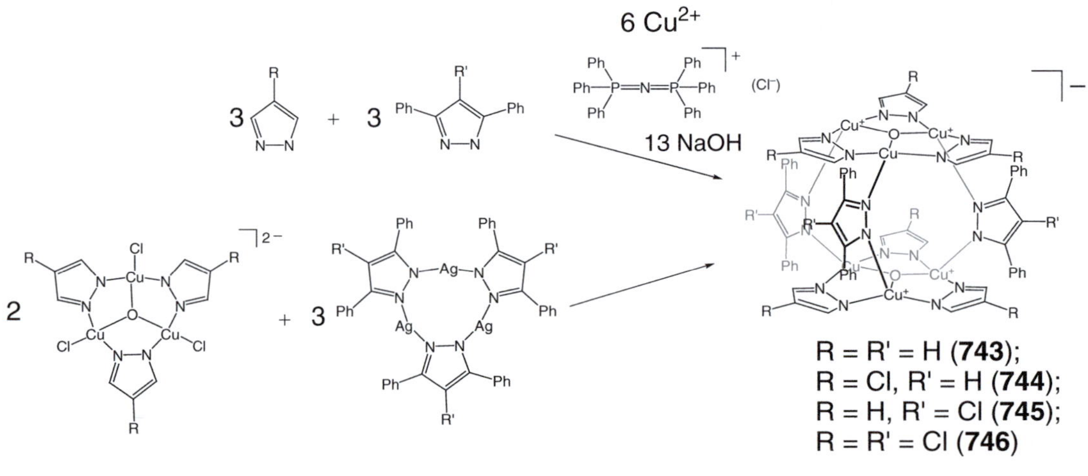

Scheme 4.172

state at 4.2 K. Therefore, the Fe–B sites undergo a spin transition because their acetonitrile ligands have a higher ligand field strength than thiocyanate anion [180]. The capsule **751** also undergoes a reverse light-induced SCO by the LIESST effect, and this allowed it to be a thermal- and light-operated molecular switch (see Sect. 6.5). The above SCO behavior [180] is affected by encapsulation of solvate molecules, which undergo temperature desorption giving the cage framework with a vacant cavity and only high-spin metallocenters. This coordination capsule does not undergo SCO but is able to

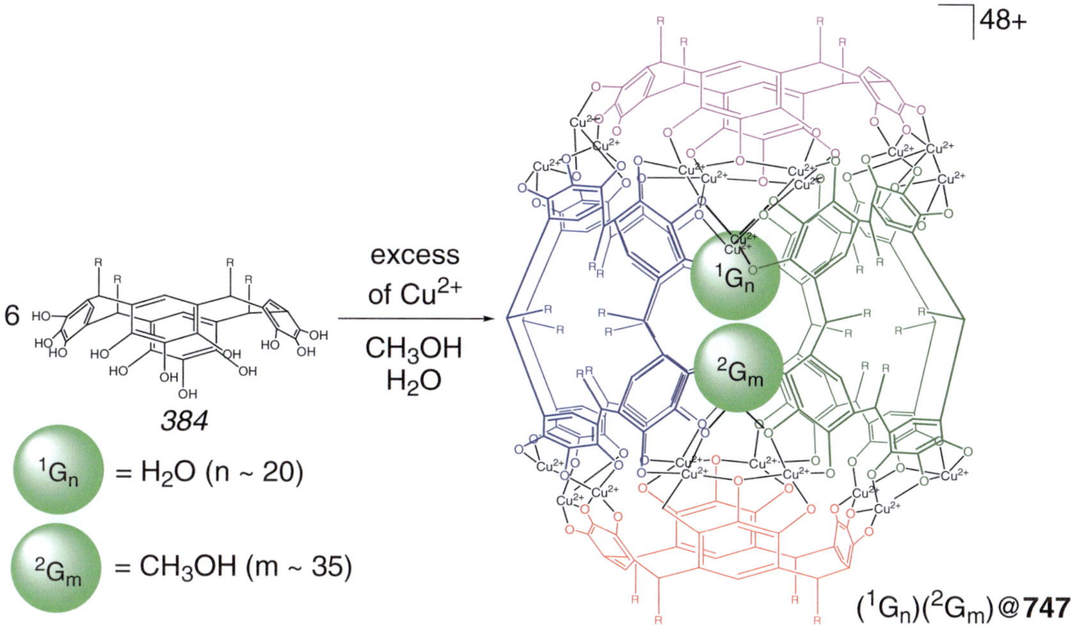

R = C_2H_5, *n*-C_3H_7, *n*-C_4H_9, *n*-C_5H_{11}, *n*-C_6H_{13}, *n*-C_7H_{15}, *n*-C_8H_{17}, *n*-C_9H_{19}, *n*-$C_{10}H_{21}$, *n*-$C_{11}H_{23}$, *n*-$C_{12}H_{25}$, *n*-$C_{13}H_{27}$, *iso*-C_4H_9

Scheme 4.173

encapsulate acetonitrile molecules after soaking in this solvent. The resulting cage complex with the encapsulated solvent molecule that undergoes thermal SCO. Thus, the capsule **761** demonstrates the guest-dependent SCO properties [180].

Analogous heterometallic $M_6^{2+}Cu^+_8L_8$ coordination capsules **752–757** (where M^{2+} are Cu^{2+}, Mn^{2+}, Zn^{2+}, Cd^{2+}, and Fe^{2+} ions) with a diameter of approximately 28 Å have been prepared in [181] by Scheme 4.177 using a generated in situ copper(I) scorpionate *387* as a complex precursor. In the copper(II)-containing capsule **752**, each of its tetracoordinate copper(I) ions is located in a pocket of tripodal scorpionate fragment and forms three donor–acceptor Cu–N bonds and one Cu–N bond a with solvate acetonitrile molecule. These complex fragments, solvate acetonitrile molecules, and counterions form a pseudooctahedral (tetragonal bipyramidal) coordination polyhedra of the cross-linking M^{2+} ions; approximately half of its copper(II) metallocenters coordinate nitrogen atoms of four pyridyl donor groups, perchlorate anion, and encapsulated solvate acetonitrile molecule. Six caged acetonitrile molecules are arranged within its cavity, forming a hydrogen-bonded polyhedral motif of the same geometry through short C–H…O contacts [181].

Two Cu^IFe^{II}-heterometallic capsules **758** and **759** with encapsulated acetonitrile molecules, the derivatives of selenocyanate and cyanoborohydride anions, have been synthesized in [182] by Scheme 4.178. As follows from the magnetometry data, those and its analog **751** undergo thermal and light-induced SCO between their high- and low-spin states with $T_{1/2}$ of 173, 162, and 124 K, respectively; no hysteresis or small cooperativity between the bridging iron(II) cations have been observed in all these coordination capsules. The inrising transition temperatures in a row **751** < **758** < **759** [182] govern their diverse photomagnetic behavior: they exhibited a LIESST effect at different degrees and at different temperatures as well.

Tris-*para*-pyridyl ligand syntone *373* (see above) underwent coordination-driven self-assembly with

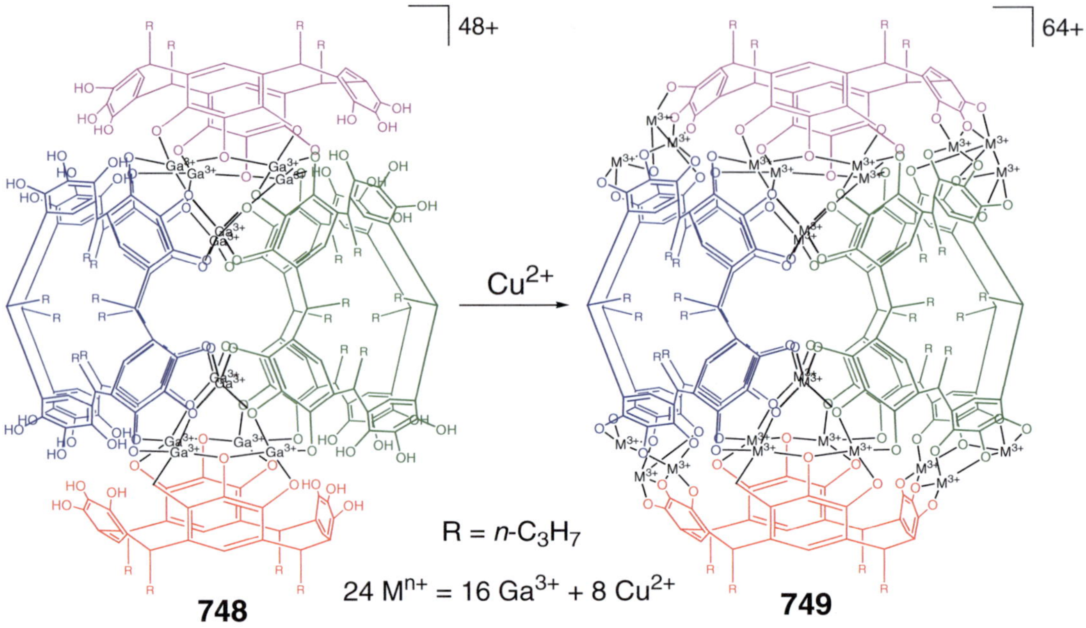

Scheme 4.174

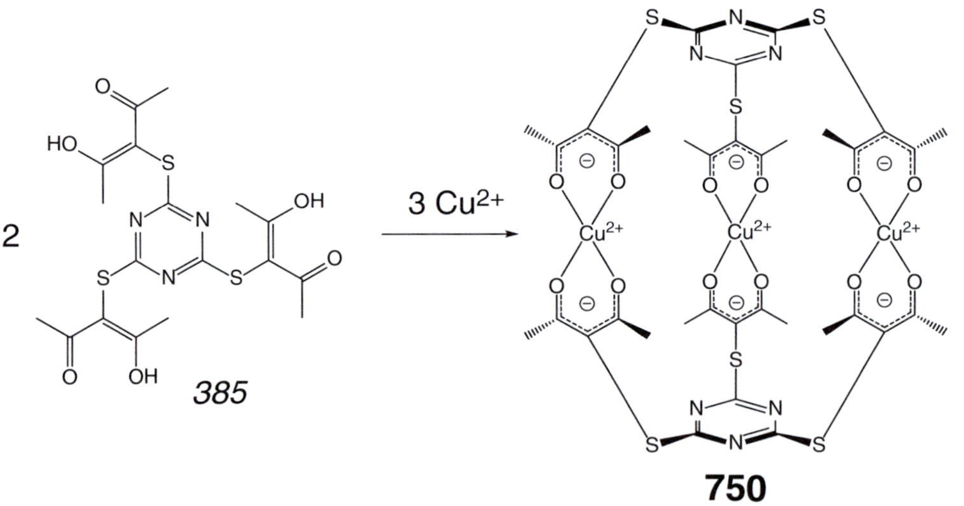

Scheme 4.175

cadmium(II) acetate by Scheme 4.179, giving the M_6L_2 cage framework **780** [183].

A hexadigallium-containing $(M_2)_6L_6$ coordination capsule **761** with six tricarboxylate fragments has been self-assembled in [184] by Scheme 4.180. Analogous diindium and dialuminum complexes have not been isolated due to side reactions of their metal–metal bonds with this protonogenic ligand syntone. The framework of **761** is substantially flattened but is able to encapsulate solvate THF molecule that does not undergo exchange reactions. The resulting 1:1 cage complex also retains in solution [184].

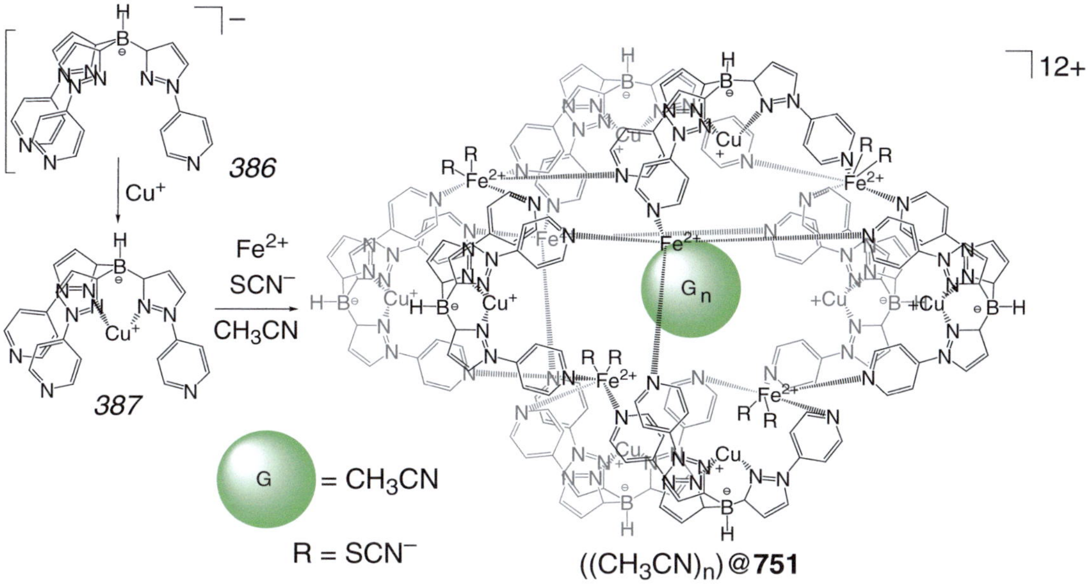

Scheme 4.176

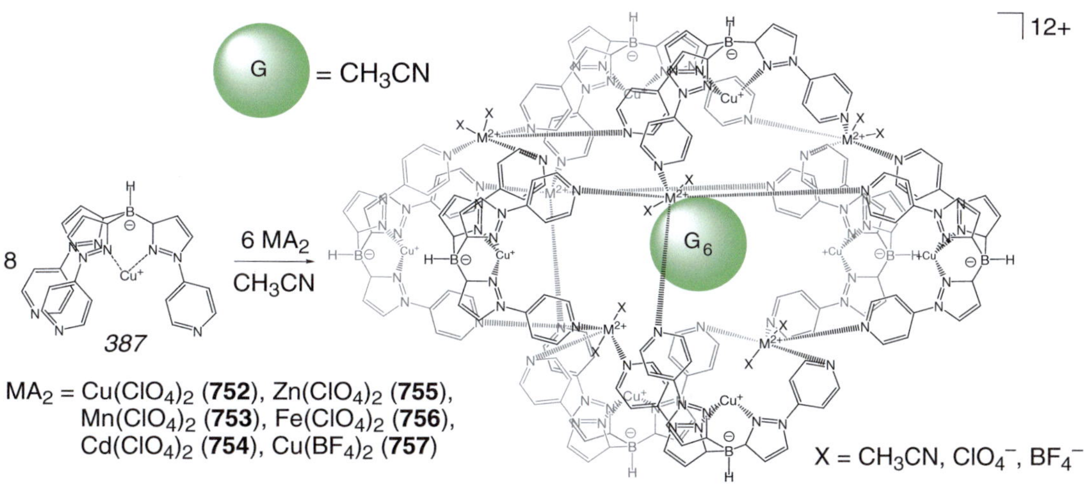

Scheme 4.177

Anionic tetrahedral M_6L_4 capsules **762–764** have been prepared in [185] by self-assembly of a cyclotricatechylene ligand syntone *388* with vanadyl(I), manganese, and cobalt(II) cations by Scheme 4.181. A 1:2 cage complex of the vanadyl-based capsule **762** has a tetrahedral framework formed by three-connecting cyclotricatechylene fragments cross-linked by two-connecting vanadyl cations. Six metallocenters of **762** form a tetragonal bipyramidal polyhedron that is elongated along its C_2-symmetry axis passing through its two vanadium-containing vertices (with four encapsulated disordered 1,4-dioxane molecules); the manganese-based capsule **763** has the similar structure [185]. An attempt to obtain such anionic Mn_6L_4 cage complexes with encapsulated Rb^+ and Cs^+ ions unexpectedly allowed isolating a highly symmetric 1:1 cage complex with one encapsulated $[Na(H_2O)_6]^+$

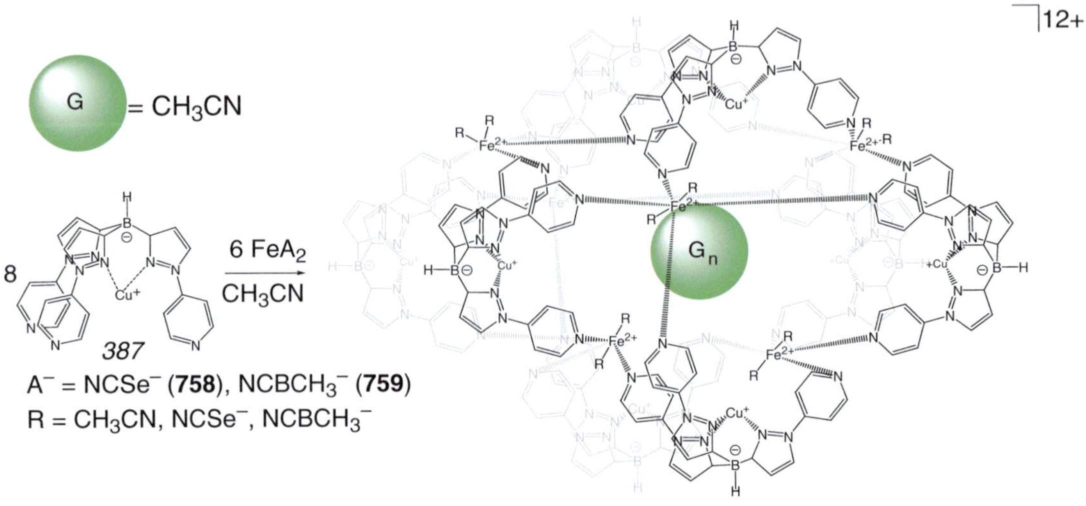

Scheme 4.178

cation and four outside Cs^+ cations; the same crystal and molecular structure is observed for its cobalt(II)-based analog **764** [185].

Four such cyclotricatechylene ligand syntones *388* and six copper(II) ions have been self-assembled in [186] by Scheme 4.182 to give a tetrahedral Cu_6L_4 coordination capsule **765**. Its highly symmetric cage framework with four C_3-symmetry axes and six crystallographically equivalent metallocenters forming an octahedral polyhedron encapsulates sodium cations, iodide anion, and solvate water molecules. The cross-linking copper(II) ions have a square-planar CuO_4-coordination polyhedra formed by two chelate catecholate donor fragments. Each of these capsules is associated "edge by edge" with other four cage entities by both hydrogen bonding and electrostatic (Coulombic) interactions at the distance of approximately 16 Å. Such supramolecular binding gave in the crystal a diamond-like network with unusually large adamantanoid chambers having the volumes of approximate 2900 $Å^3$ [186].

A Pd_2L_3 coordination capsule **766** has been prepared in [187] in a quantitative yield by self-assembly of palladium(II) ions with a banana-shaped ligand syntone *389* by Scheme 4.183. This rigid capsule is formed by two cross-linking palladium(II) ions with square-planar coordination polyhedra at the distance of 17 Å and four ribbed banana-shaped ligand fragments. As follows from ^1H DOSY NMR data, the cage framework of **766** has a radius of approximately 11 Å in solution, in a good agreement with X-ray diffraction data showing its spherical shape and an average diameter of 20 Å in crystal [187]. The caging ligand **766** encapsulates guest molecules such as 1,1′-ferrocenyl-bis-sulfonate dianion through its four large portals. The electrochemically generated reduced form of this caged redox-active guest is stabilized due to electron-withdrawing effect of two positively charged palladium-containing cross-linking fragments [187].

Coulombic interactions between the tetracationic Pd_2L_3 and Pt_2L_3 caging ligands **766** and **767** as hosts and bis-sulfonate dianionic guests of a suitable size are reported by G.H. Clever and M. Shionoya [188] to be a driving force for their encapsulation, leading to the corresponding host–guest 1:1 cage complexes and pH-switchable pseudorotaxanes by Scheme 4.184.

A discrete stretch of stacked dianionic and dicationic platinum complexes has been generated in [189] through their encapsulation by the above coordination capsule **767**: two $PtPy_4$ mean planes of its cage framework are parallel to each other and are approximately 17 Å apart. The concave shape of this capsule creates a spherical cavity co-encapsulating an alternately stacked dianionic trimeric guest $\{[PtX_4][PtY_4][PtX_4]\}^{2-}$

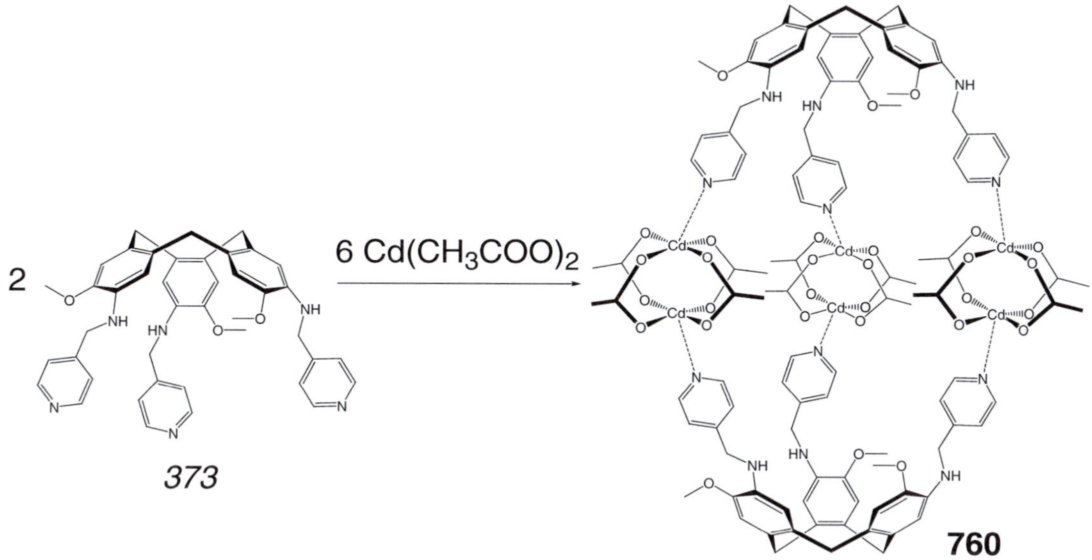

Scheme 4.179

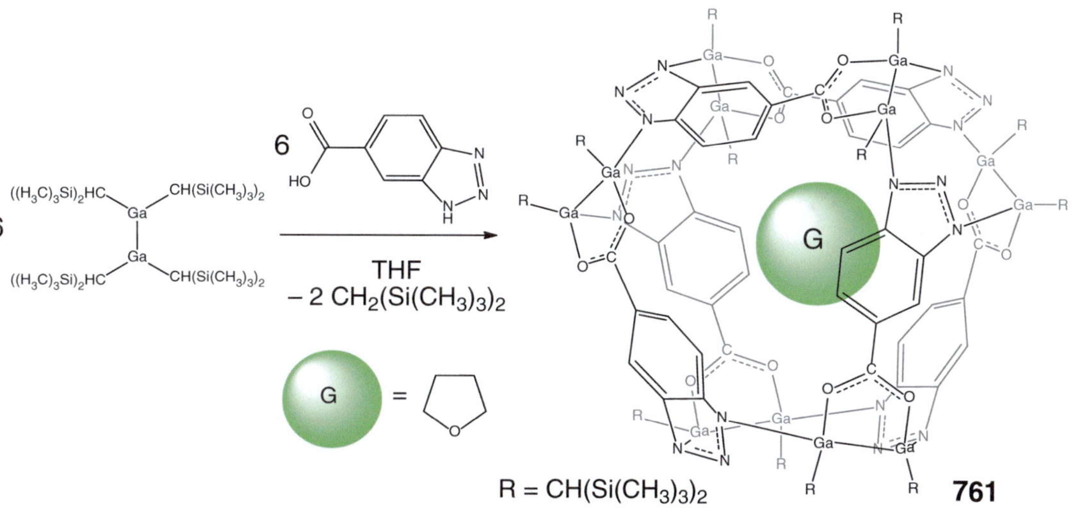

Scheme 4.180

between the doubly positively charged planes of **767**, thus forming a heteroguest 1:2:1 cage complex with an approximate C_4 symmetry by Scheme 4.185. All four ligand syntones of its framework are twisted in a helical manner around the Pt…Pt axis, and two enantiomers of this chiral capsule in solid state have the same amounts [189]. At the same time, the caging ligand **766** is reported in [190] to quantitatively encapsulate a hexamolybdate $[Mo_6O_{19}]^{2-}$ dianion in solution, forming a 1:1 cage complex by Scheme 4.186. This complex undergoes interconversion during its recrystallization into the corresponding hydrogen-bonded cyclic complex *390*.

Self-assembly of angular dibenzosuberone ligand syntones *390*, *391*, and *392* with palladium(II) ions by Scheme 4.187 under mild reaction conditions afforded M_2L_4 coordination capsules **768** and **769** characterized in [191] by NMR and ESI-MS data. Their double-cage

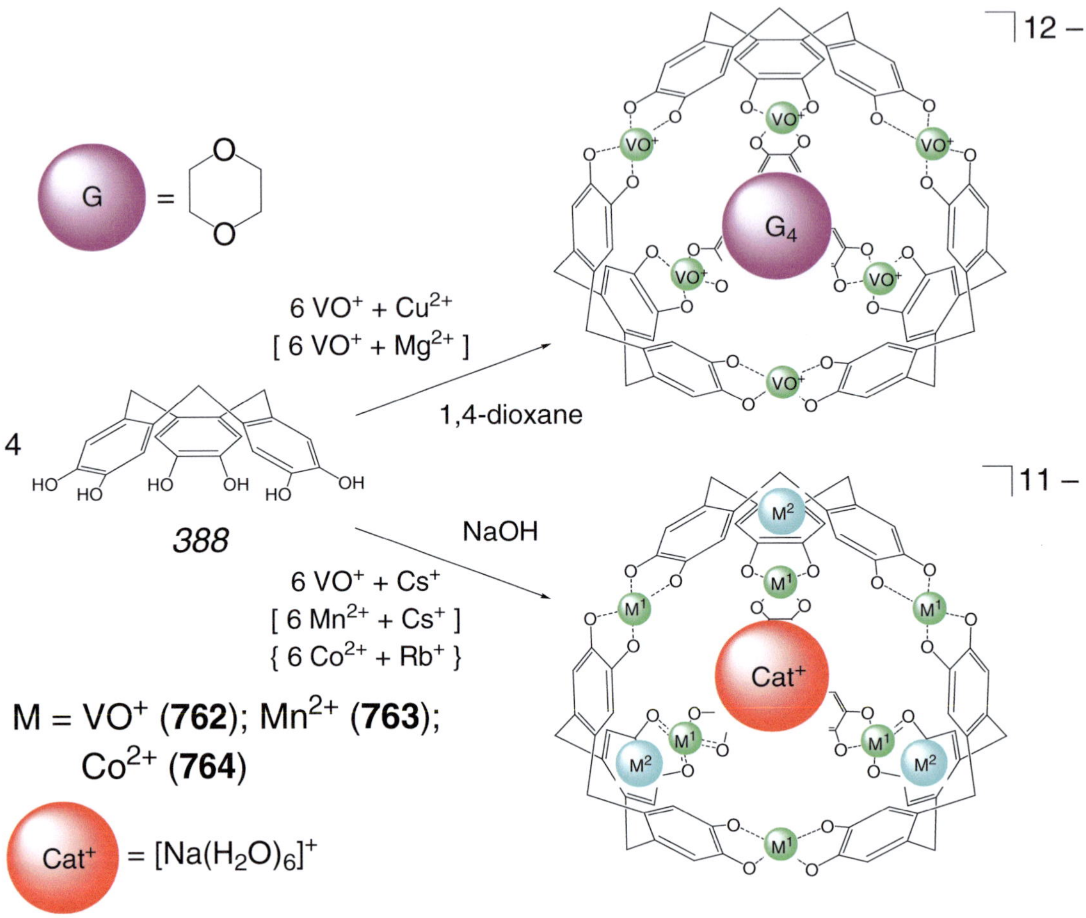

Scheme 4.181

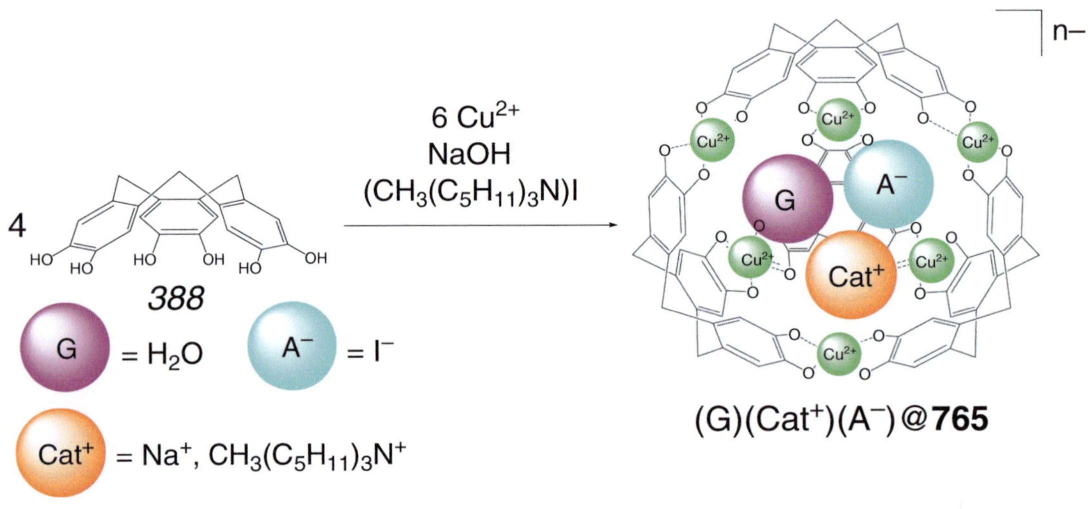

Scheme 4.182

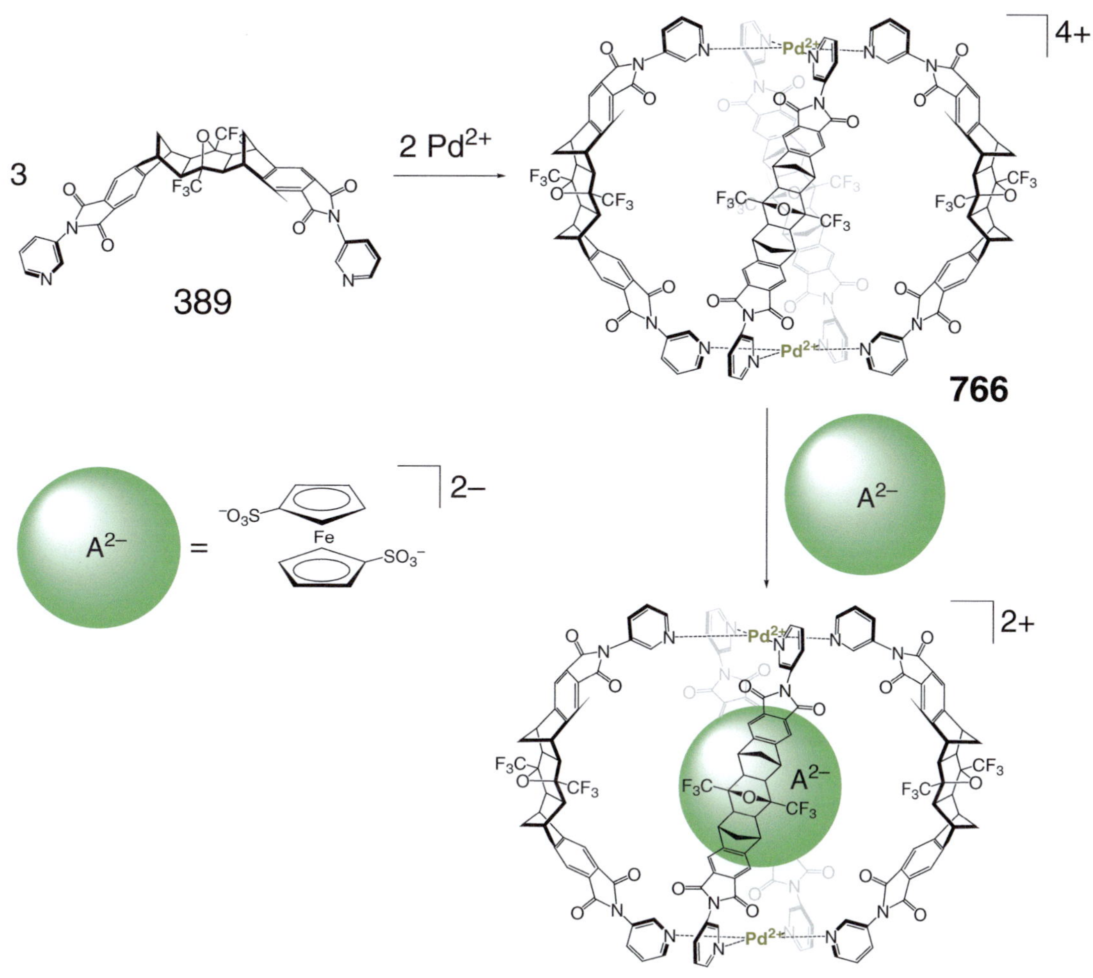

Scheme 4.183

analog **770** with an extended framework has three vacant cavities, and it forms a highly symmetrical host–guest 1:3 cage complex with one tetrafluoroborate anion tightly encapsulated in its inner cavity and two anionic guests caged in outer pockets. This capsule has been prepared [191] in quantitative yield after a prolonged heating of the same reaction mixture. In the double-cage framework **770** with Pd…Pd distances of 24.436, 8.093, and 8.251 Å between terminal, inner, and inner–terminal metallocenters, respectively, four cross-linking palladium(II) ions occupy its fourfold symmetry axis.

Coordination-driven self-assembly of homologous bis-phenothiazine sulfide, sulfoxide, and sulfone ligand syntones *393–395* by Scheme 4.188 has been used in [192] for the synthesis of Pd_4L_8 eight-phenothiazine double-cage capsules **771–773** with three encapsulated tetrafluoroborate anions. The sulfide capsule **771** undergoes eightfold mono- and dioxygenation in solution and in crystal, giving its oxidized forms **772** and **773**, respectively. The relative affinity of these capsules and their dibenzosuberon-based analog **770** toward Cl^- and Br^-anions has been studied in [193]. Structural changes between the double cages **771–773**, which are a consequence of the oxidation state of the ligand's sulfur atoms, are reported affect the size of their cavities. The binding affinity of these encapsulating hosts toward halide anions decreases with an increase in the cavity size. The dibenzosuberone-based

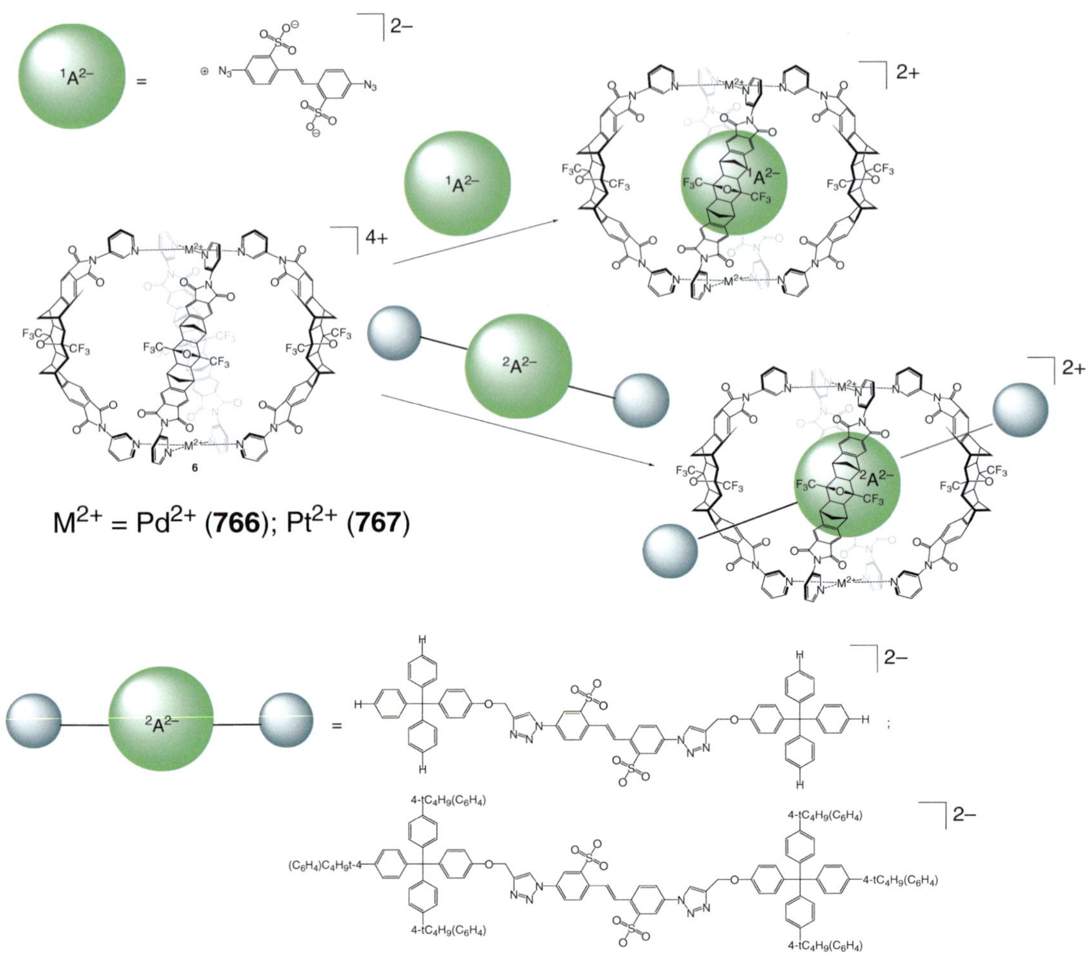

Scheme 4.184

capsule **770** is described in [193] to bind Cl^- anion substantially stronger than Br^- ion.

Single-crystal X-ray diffraction study of 1:3 double-cage complex of **773** with three encapsulated tetrafluoroborate anions (Scheme 4.188) and systematic structural comparison of all the X-ray structurally characterized Pd_4L_8 frameworks of this type with that of their monocage analog **774** (Scheme 4.189) have been performed in [194]; general scheme for the formation of these palladium(II)-based coordination capsules by various angular ligand syntones is also reported. A *push*-and-*pull* model for the allosteric anion binding in such double-cage coordination capsules with three encapsulated monoanions has been evaluated in [195].

Quantum chemical calculations and a simple *push*-and-*pull* model have been used in [195] to explain the allosteric effect in anion binding by the capsule **770**. For this purpose, the double-cage framework has been divided into three subsystems as shown in Scheme 4.187, and the DFT-calculated geometrical and thermodynamic parameters are reported in [195] to be in an excellent agreement with the experimental data.

Use of structural differences between the two isomeric ligand syntones *396* and *397* allowed obtaining [196] both the non-intertwined Pd_2L_4 coordination capsule **775** and its intertwined Pd_3L_6 analog **776**, respectively (Scheme 4.190). The double trefoil knot-mimicking capsule **776**

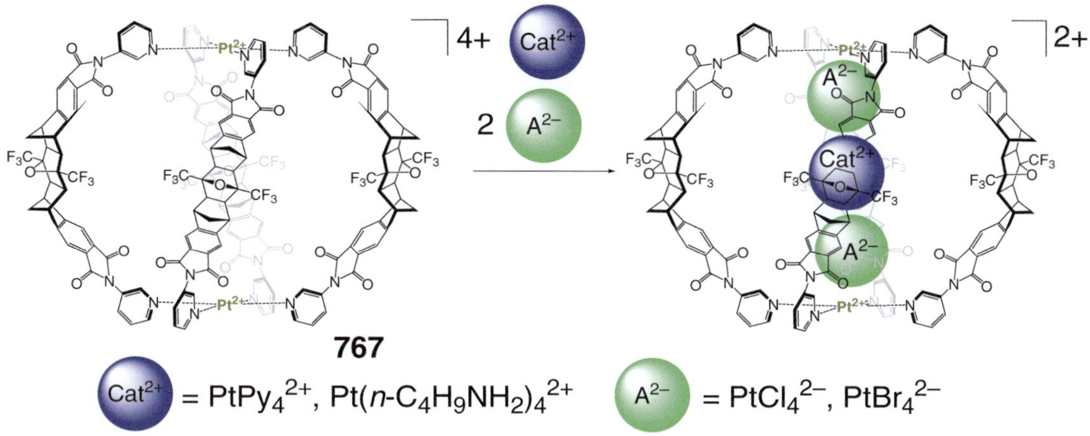

Scheme 4.185

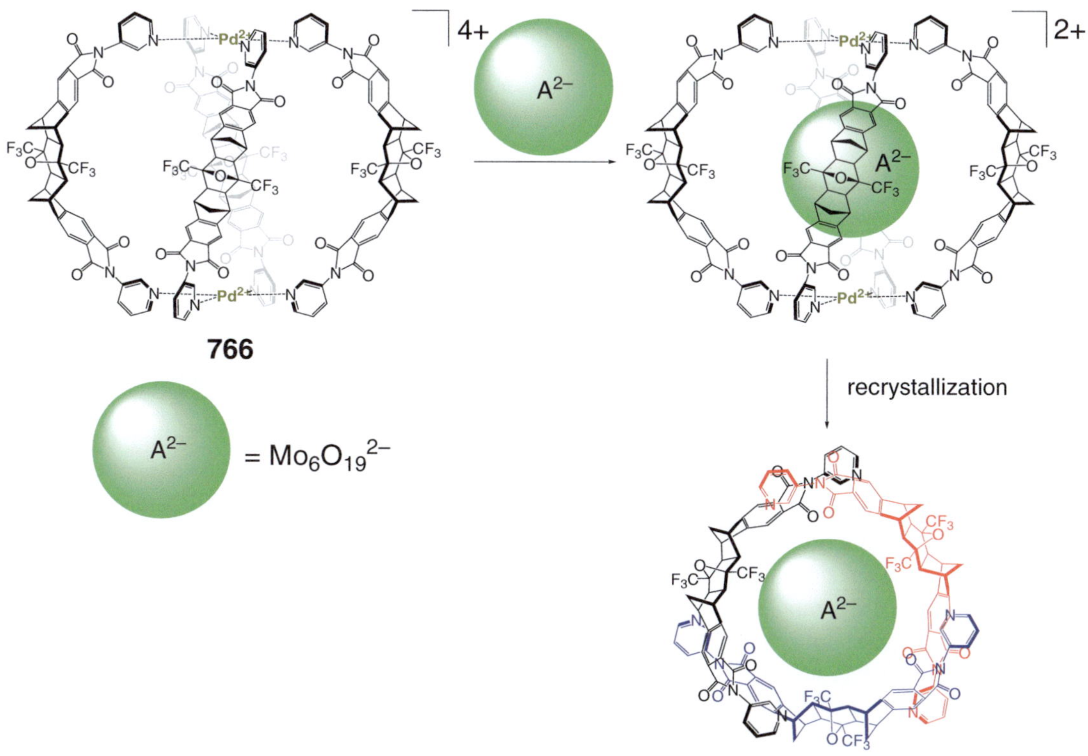

Scheme 4.186

gave 1:1 cage complexes with an encapsulated chiral camphorsulfonate anion [196].

Both the yellowish flexible opened and intensively blue-colored forms of the angular acetylene-bis-pyridinate syntones *398* and *399* have been used in [197] for the synthesis of Pd_2L_4 coordination capsules **777** and **778** by Scheme 4.191. These ligands as well as their angular syntones can be reversibly interconverted without photodegradation using UV and white irradiation. As follows from ^{1}H and DOSY NMR data, labile molecules of **777** are smaller than

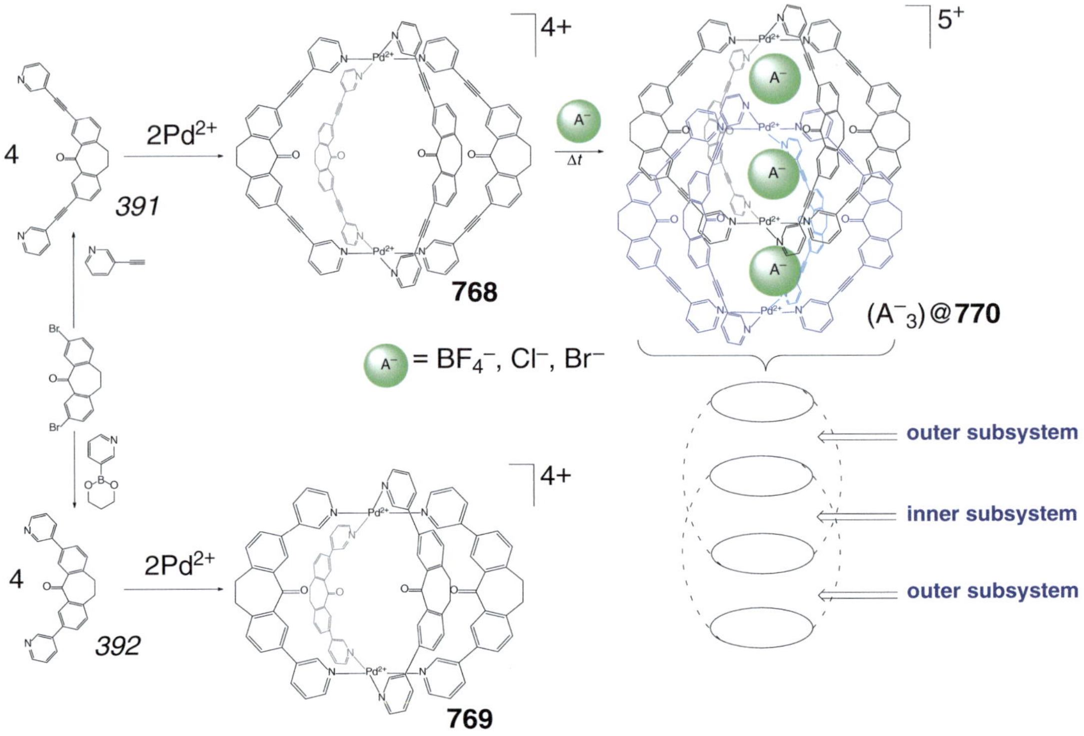

Scheme 4.187

those of its rigid analog **778**. These caging ligands encapsulate a dodecafluorododecacarborate dianion giving host–guest 1:1 cage complexes as a result of entropy-driven and endothermic processes: the effect of solvent release from the interior of these capsules governs the driving force of such encapsulation [197]. Their binding constants are, however, substantially (by approximately 500 times) different. This result has been explained in [197] by lability of **777** that allowed closely surrounding this anionic guest in the fashion of an induced fit, whereas its rigid analog **778** cannot adopt this guest in the same manner.

The coordination capsules **766** and its polyethylene glycol-containing analog **779** are reported in [198] to encapsulate most of the tested dianionic guests shown in Scheme 4.192, thus giving 1:1 cage complexes (as follows from the NMR and ESI-MS data). Only oversized *para*-substituted aromatic dianion *400* does not form the corresponding cage complex, and its outside binding mode has been detected. In the case of the capsule **766**, the composition of its cage complexes in solid state and in solution can be controlled by the order in which their guest molecules are added. Addition of one equivalent of the dye Japan Red 1 *401* resulted in the formation of red-colored 1:1 dicationic cage species by Scheme 4.193, giving the red precipitate with an aromatic dianion *400* [198]. In contrast, addition of this dianion to a solution of the capsule **766** gave the corresponding colorless precipitate with two outside-coordinated dianions.

Use of an uncoordinating solvent allowed obtaining 1:1 cage complex of Co_2L_4 coordination capsule **780** (Scheme 4.194) with one encapsulated and two outside-coordinated tetrafluoroborate anions [199]; its caged BF_4^- anion forms a donor–acceptor Co–F bond of 2.31 Å, and the Co…Co distance is approximately 6.9 Å. The use of other cobalt(II) salts (in particular, its nitrate or chloride) gave either a 1D coordination polymer or a metallomacrocyclic complex, respectively; no template effect of these anions in

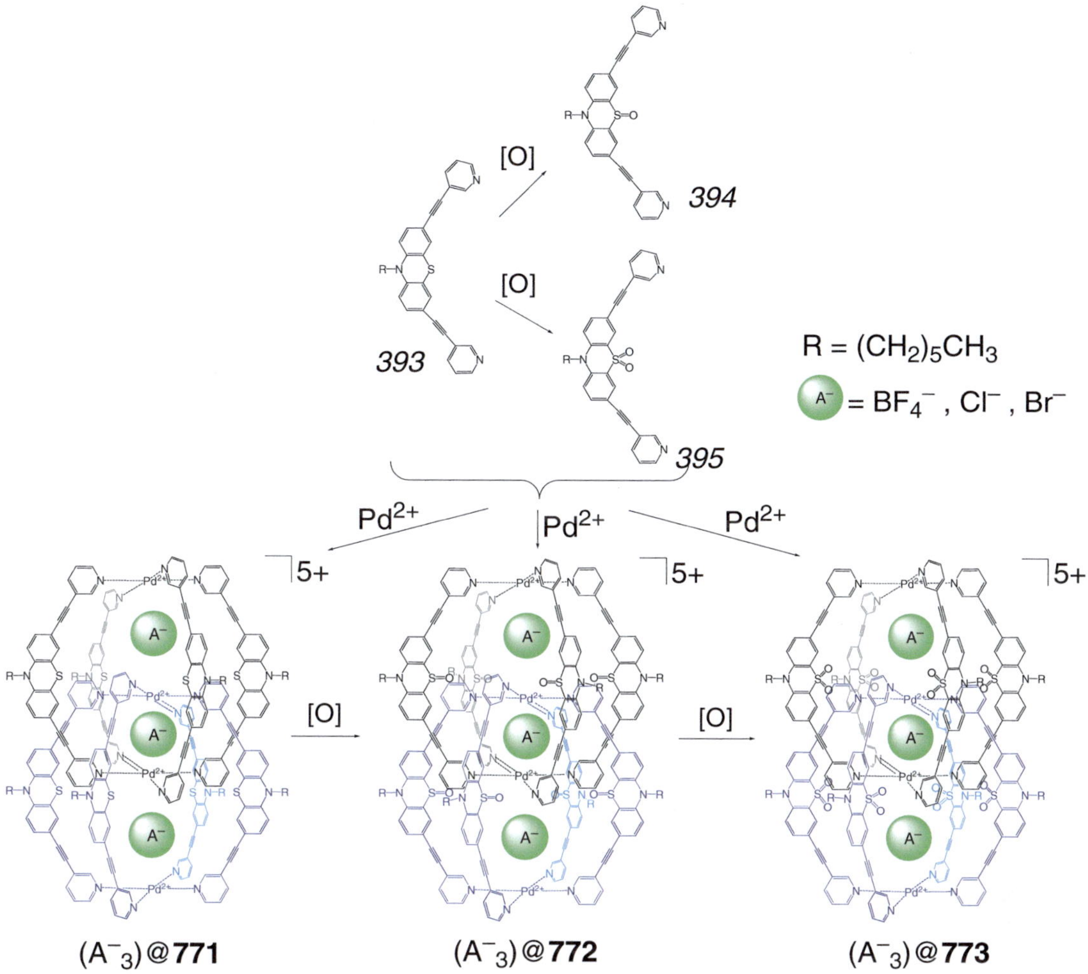

Scheme 4.188

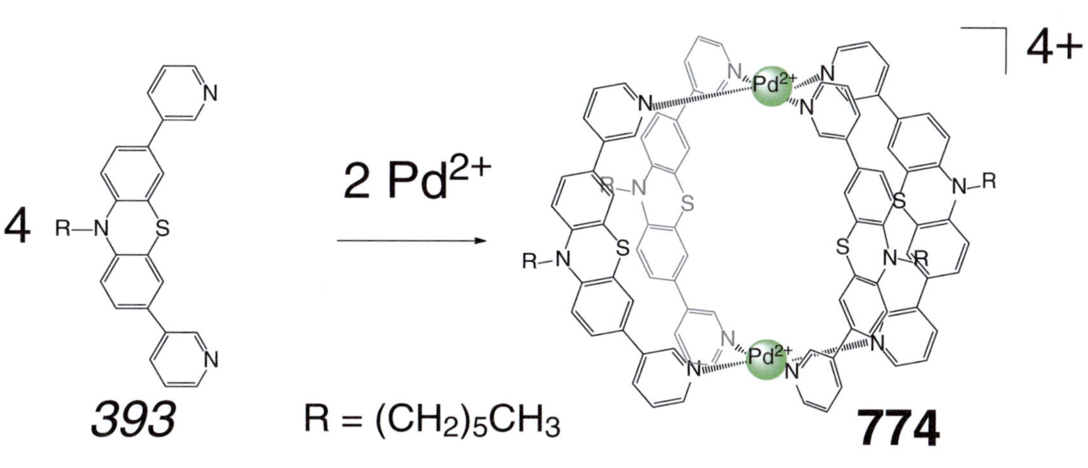

Scheme 4.189

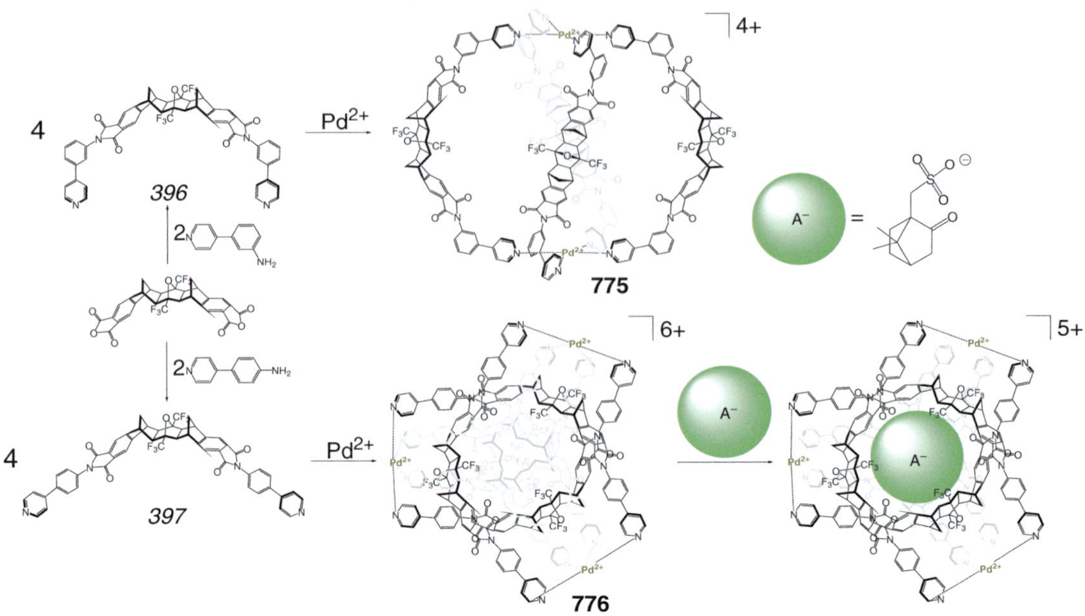

Scheme 4.190

this coordination-driven self-assembly has been observed in [199].

Self-assembly of a tris-pyridyl ligand syntone *402* with palladium(II) ions in the molar ratio 2:1 by Scheme 4.195 gave a cage complex of macrotricyclic Pd_2L_4 caging ligand **781** with encapsulated disordered solvent methanol and water molecules as the only product of this reaction [200]. In its coordination-saturated quadruply stranded cage framework; each of the cross-linking palladium(II) ions coordinates four pyridyl donor groups of four tridentate ligand syntones forming a square-planar PdN_4-coordination polyhedron. The capsule **781** encapsulates two cisplatin molecules to give a host–guest 1:1 cage complex. As follows from the X-ray diffraction data, the caged guests form NH…N and CH…Cl hydrogen bonds with the interior of **781**, and the platinum(II) ions are aligned, suggesting a Pt…Pt interaction between them at the distance of approximately 3.32 Å [200].

A M_2L_4 coordination capsule **782** with an aromatic shell has been prepared in [201] and [202] by Scheme 4.196. Its D_4-symmetric cage framework is formed by four bis-anthracene ligand syntones *403* having a twisted conformation; they are cross-linked by two palladium(II) ions possessing a square-planar geometry of their coordination polyhedra. The cavity of **782** with a diameter of approximately 10 Å and a cavity volume of 580 Å^3 is defined by the Pd…Pd distance of 13.6 Å and by the diagonal distance between the opposite *meta*-phenylene hydrogen atoms of 11.2 Å. The solvent guests (seven methanol and one water molecules) and nitrate anion are caged within this cavity. In a 1:2 cage complex with two encapsulated 1-methylpyrene molecules, their stacked dimer with an interplane distance of 3.5 Å interacts with the capsule **782** through complementary CH (guest)…π (anthracene host) and CH (pyridine host)…π (guest) interactions [201]. Spherical (such as paracyclophane, adamantane derivatives, and fullerene C_{60}), planar (such as pyrenes, phenanthrene, and triphenylene), and bowl-shaped molecules (such as corannulene) can also be encapsulated within the large spherical cavity of **782**, giving host–guest 1:1 and 1:2 cage complexes by Scheme 4.196 [202]. The caged planar guests adopt a stacked-dimer structure, while the bowl-shaped guests form a con-

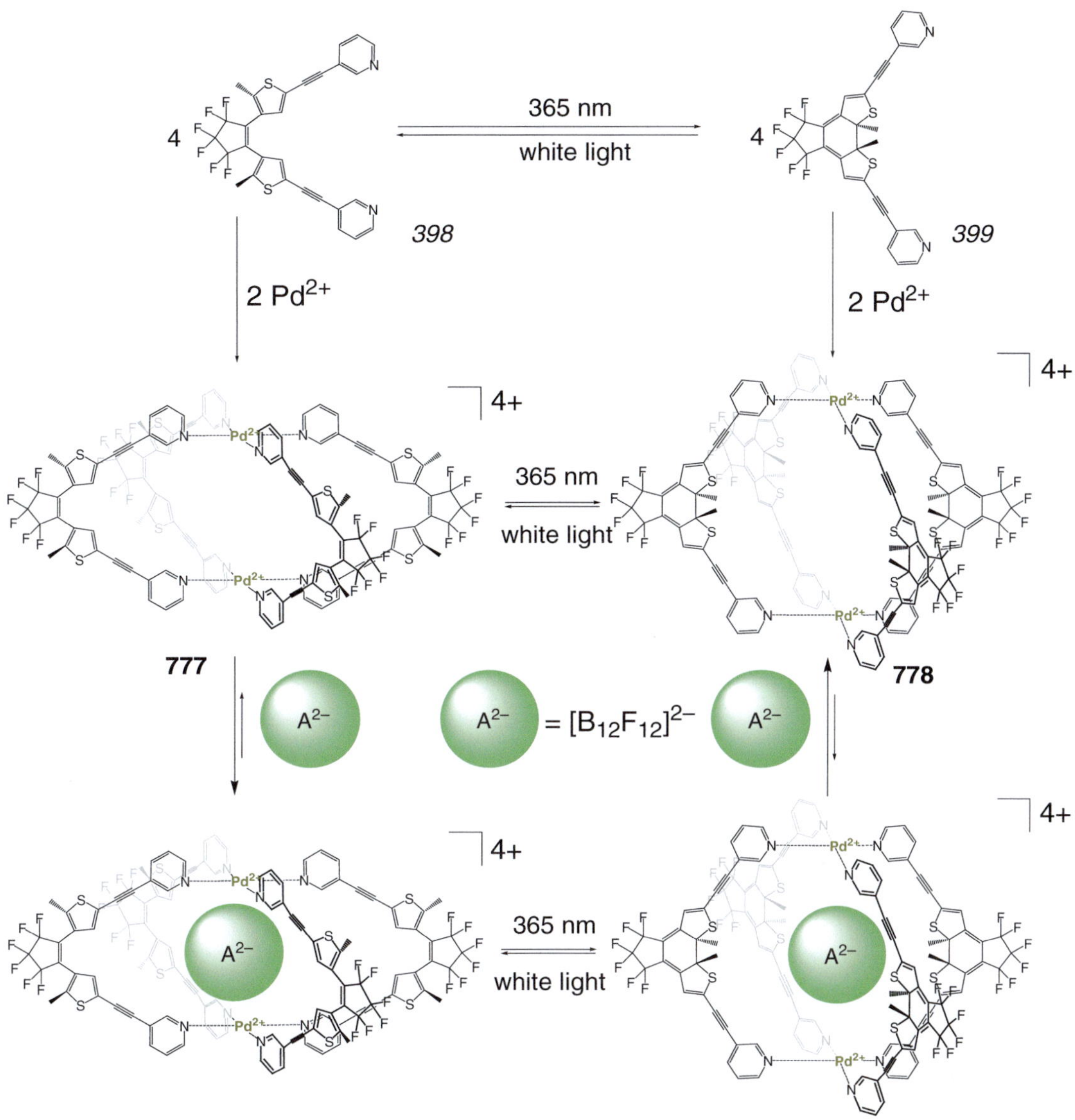

Scheme 4.191

cave-to-concave capsular geometry. Competitive binding experiments performed in [201] for a set of planar guests showed a decrease in the binding affinity in a row *pyrenes* ≈ *phenanthrene* > *triphenylene*. The coordination capsule **782** also selectively forms a heteroguest 1:1:1 cage complex shown in Scheme 4.196 with encapsulated triphenylene and corannulene as co-guests [202].

Anthracene-based M_2L_4 coordination capsules **783–785** have been self-assembled in [203] by Scheme 4.197 from the corresponding metal(II) ions ($M^{2+} = Zn^{2+}$, Ni^{2+}, and Pd^{2+}) and bent bidentate ligand syntone *404* with anthracene fluorophoric fragments. The distorted spherical D_4-symmetric cage framework of the nickel(II)-based capsule **784** is formed by four syntones and two cross-linking nickel(II) ions with N_6-coordination polyhedra possessing octahedral geometries: four pyridyl groups of these ligand syntones are coordinated in the square-planar positions, and the apical positions

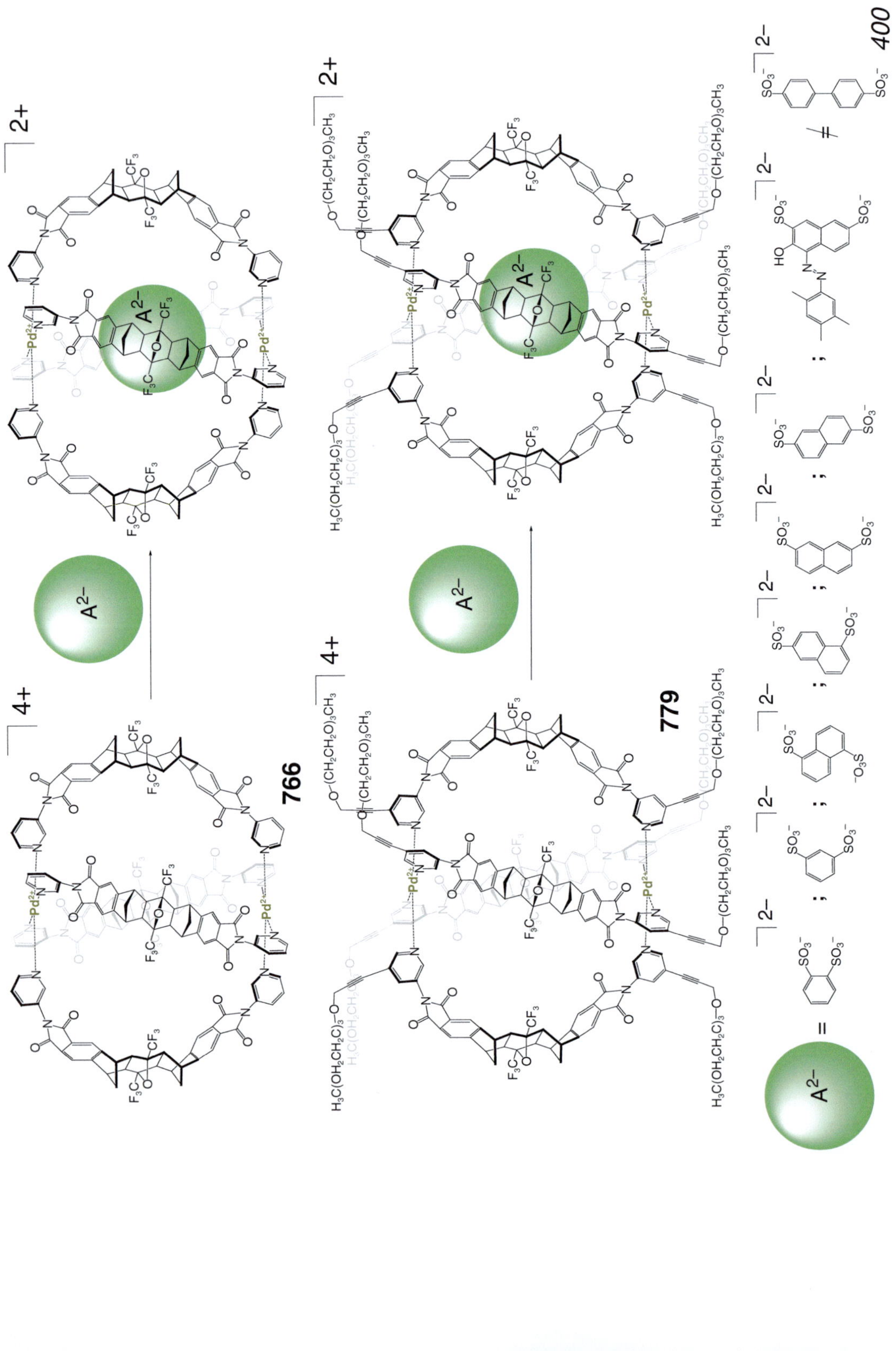

Scheme 4.192

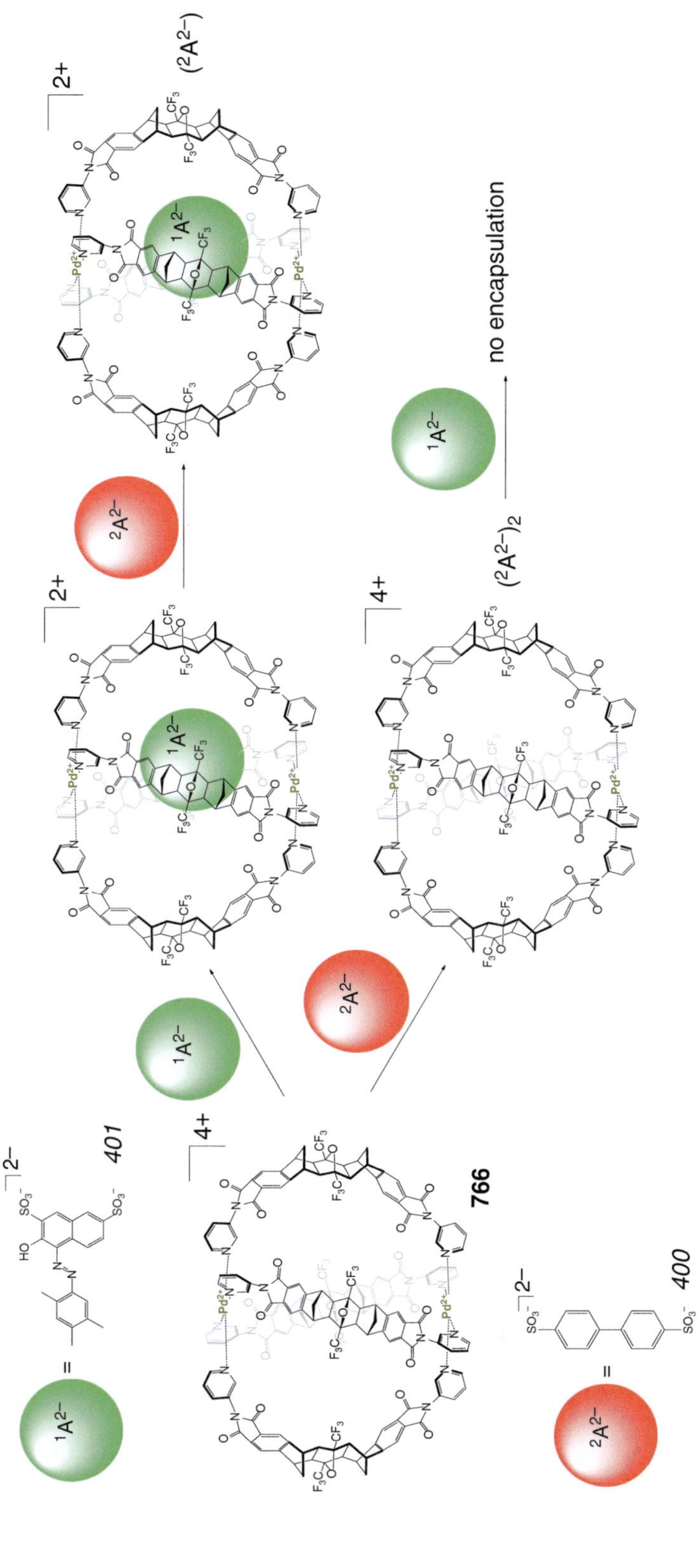

Scheme 4.193

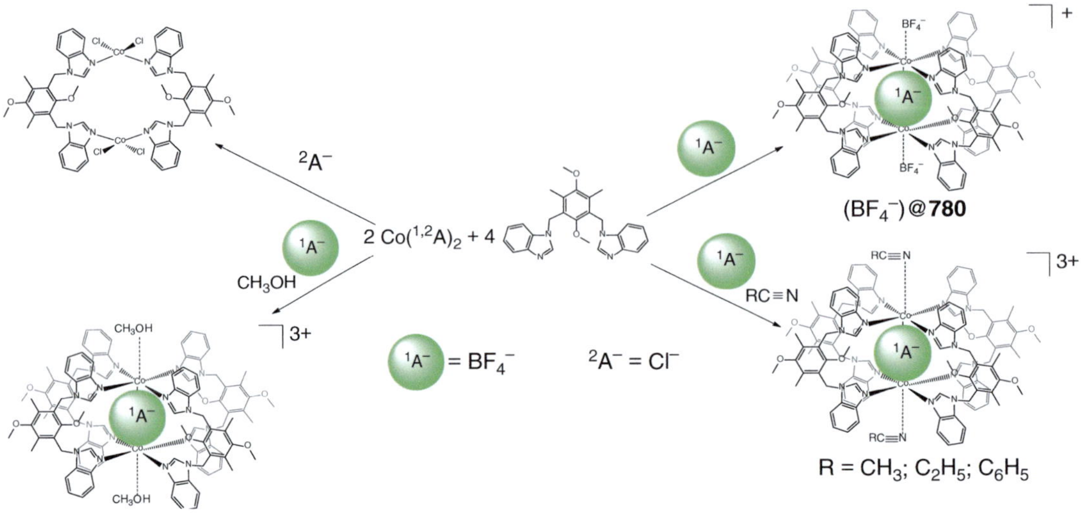

Scheme 4.194

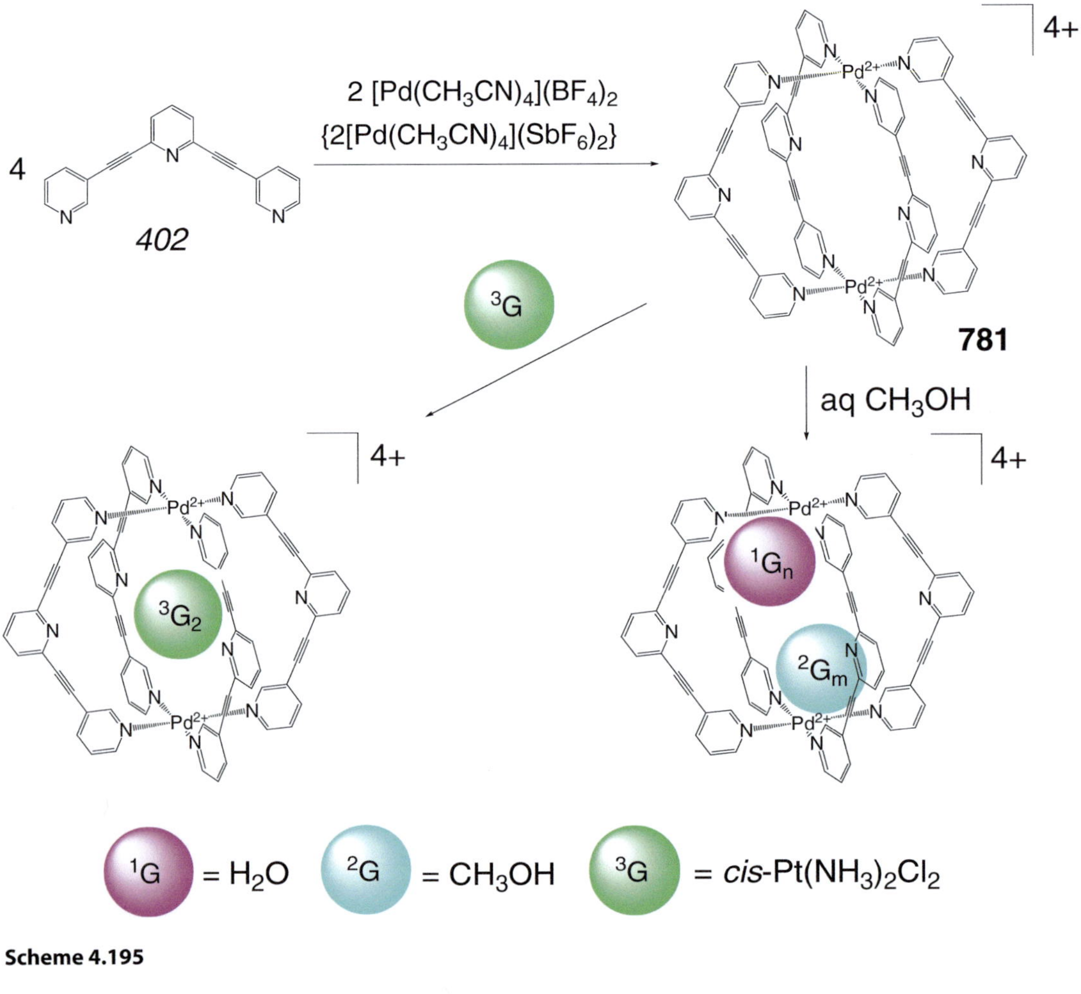

Scheme 4.195

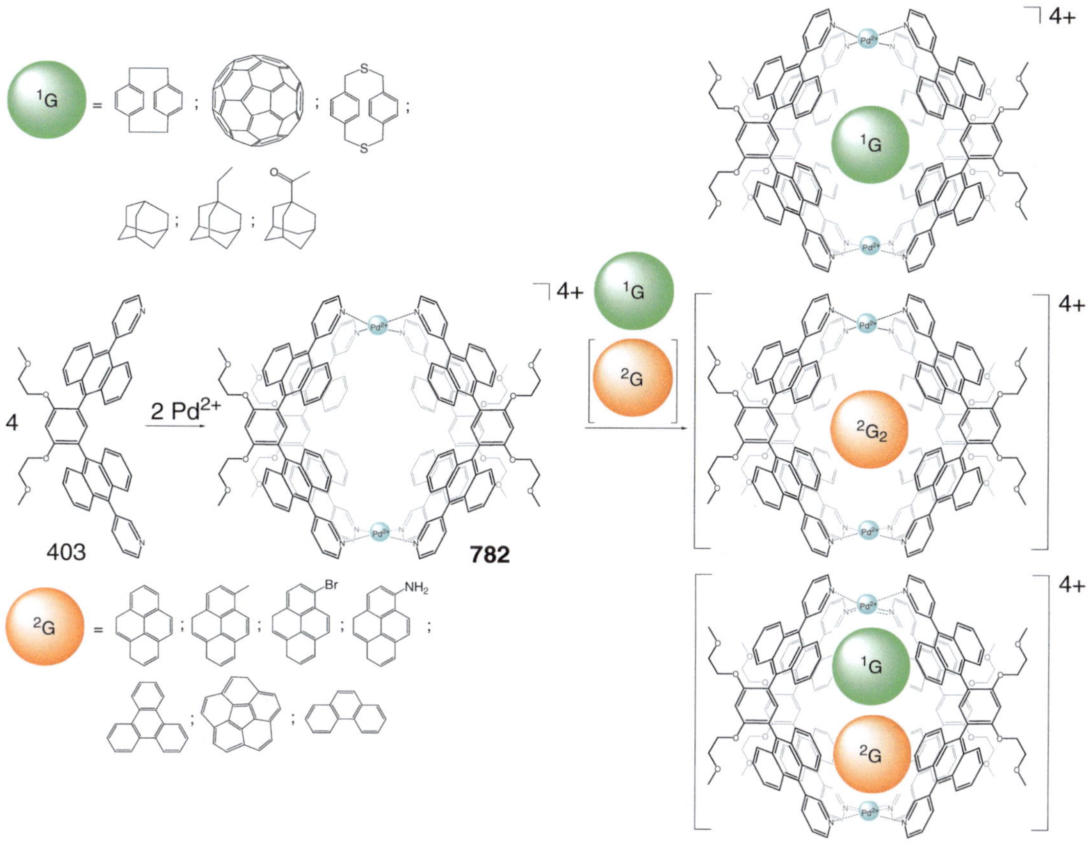

Scheme 4.196

are occupied by water and/or acetonitrile molecules. The zinc- and nickel(II)-capped capsules **783** and **784** are reported in [203] to be fluorescent emissive with high to moderate quantum yields. The detailed X-ray diffraction, NMR, ESI-TOF, and fluorescent studies of these coordination capsules and their platinum-, cobalt-, manganese-, and copper(II)-capped analogs **786–789** have been performed in [204]. The fluorescence properties of these isostructural coordination capsules depend on the nature of cross-linking metal ions. The zinc-capped capsule **783** emits strong blue fluorescence with a high quantum yield; its nickel- and manganese-based analogs **784** and **788** are weakly emissive, and palladium-, platinum-, and cobalt(II)-based capsules **785–787** are completely non-emissive. The copper(II)-capped coordination capsule **789** is reported in [204] to exhibit solvatochromism and a solvent-dependent emission behavior.

The disk-shaped tridentate ligand syntone *405* is reported in [205] to form structurally equivalent dodecacationic M_6L_8 coordination capsules **790–799** with a series of d^5–d^{10} metal dications by Scheme 4.198. In the mercury-based coordination capsule **799**, eight ligand syntones are arranged in an octahedral cage framework with six cross-linking mercury(II) ions in its vertices. Each of these ions coordinates four pyridyl nitrogen donor atoms of four ligand syntones and two oxygen atoms of axial triflate anions from inside and outside of this capsule, giving a N_4O_2-coordination polyhedron. The distances between two neighboring mercury(II) ions and two diagonal ions of this type are 17.9 Å and 25.4 Å. The cavity volume of this Hg_6L_8 coordination capsule

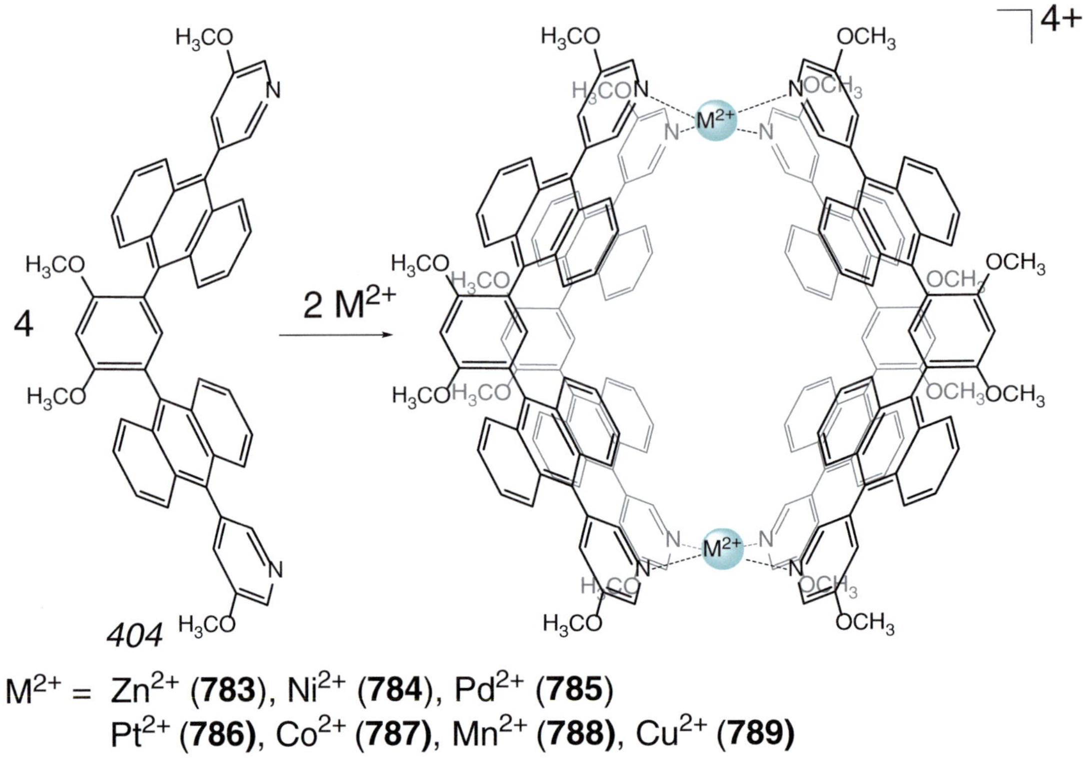

Scheme 4.197

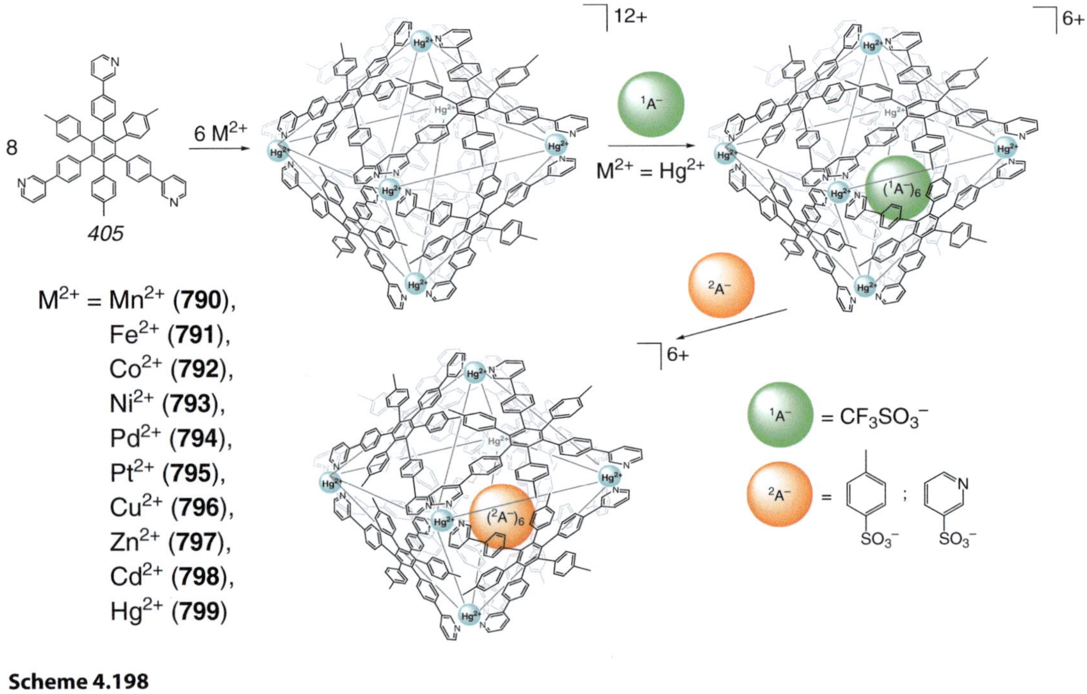

Scheme 4.198

has been estimated in [205] as equal to approximately 2700 Å^3, which allows encapsulating six triflate anions to give a hexacationic 1:6 host–guest cage complex. These encapsulated guests have been site-selectively replaced in [205] by *para*-tolyl- and 3-pyridinesulfonate anions. The authors of [206] have designed a rod-shaped bis-sulfonate dianion *406* (Scheme 4.199) as a ditopic dianionic guest. In this dianion, the distance between the two negatively charged oxygen atoms is approximately 20 Å, and it is suitable for bridging two apical cross-linking mercury(II) ions of the coordination capsule **799**.

A 4 nm-sized covalent analog **801** of the metal(II)-based coordination capsules **790–799** has been obtained in [207] by Scheme 4.200, starting from its Pd_6L_8 cage precursor **800** by a three-step synthetic procedure that included (*i*) olefin metathesis reaction between nearest-neighbor olefin groups in this precursor, (*ii*) removal of the cross-linking palladium(II) ions, and (*iii*) reduction of internal olefin groups. The 24 pyridyl substituents of the covalent capsule **801** underwent methylation giving its 5 nm-sized derivative **802**.

Quantitative self-assembly of a Hg_6L_4 coordination capsule **803** and its reversible interconversion with a Hg_6L_8 cage framework **799** by Scheme 4.201 in response to Hg^{2+}/-ligand syntone ratio are reported in [208]. This structural rearrangement is described as elimination of four ligand syntones from the closely packed octahedral capsule accompanied by changes in the coordination polyhedra of mercury(II) ions from N_6-octahedral to N_2-linear geometry. These dynamic structural changes caused the fluores-

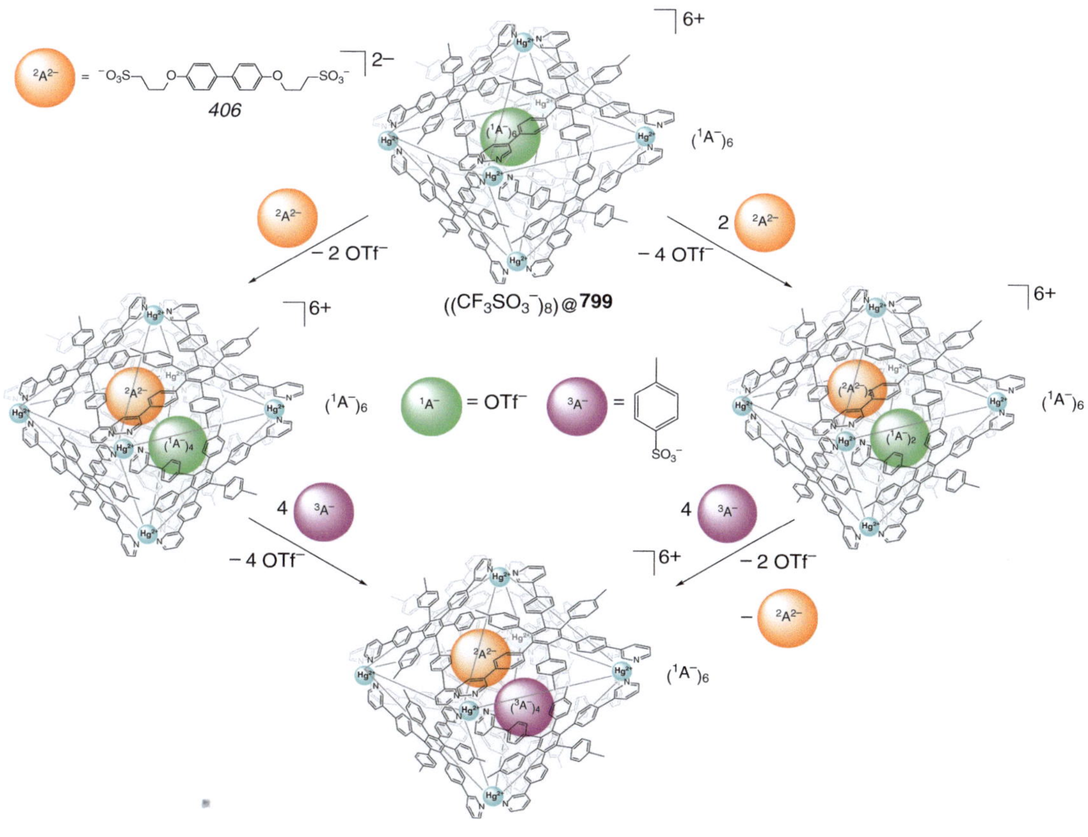

Scheme 4.199

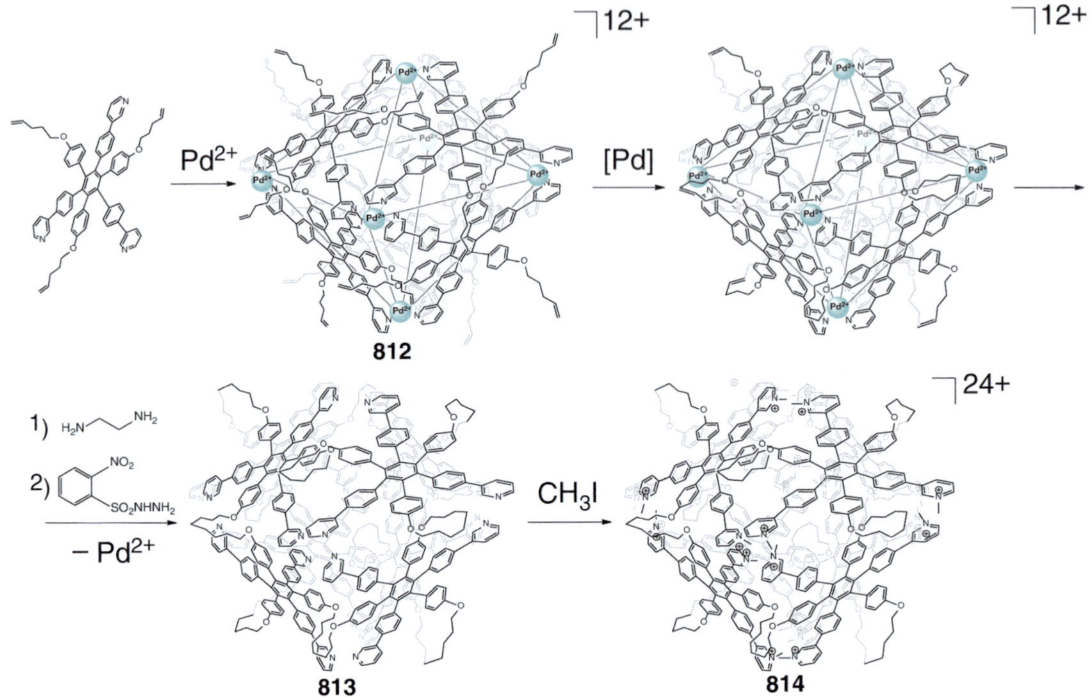

Scheme 4.200

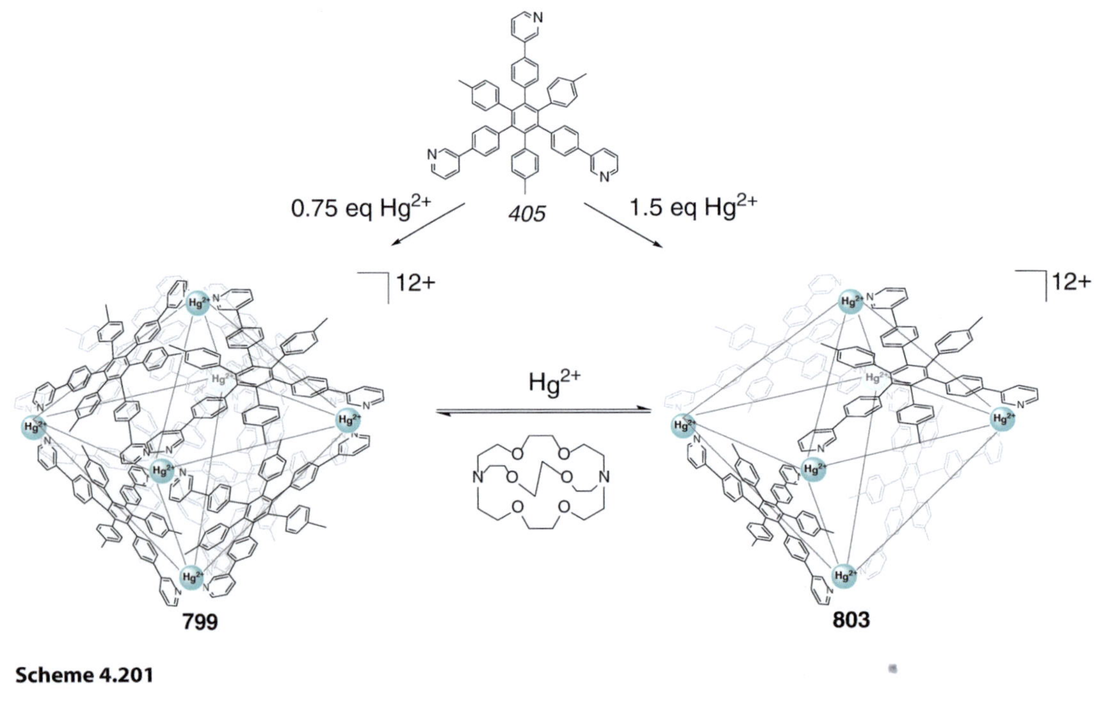

Scheme 4.201

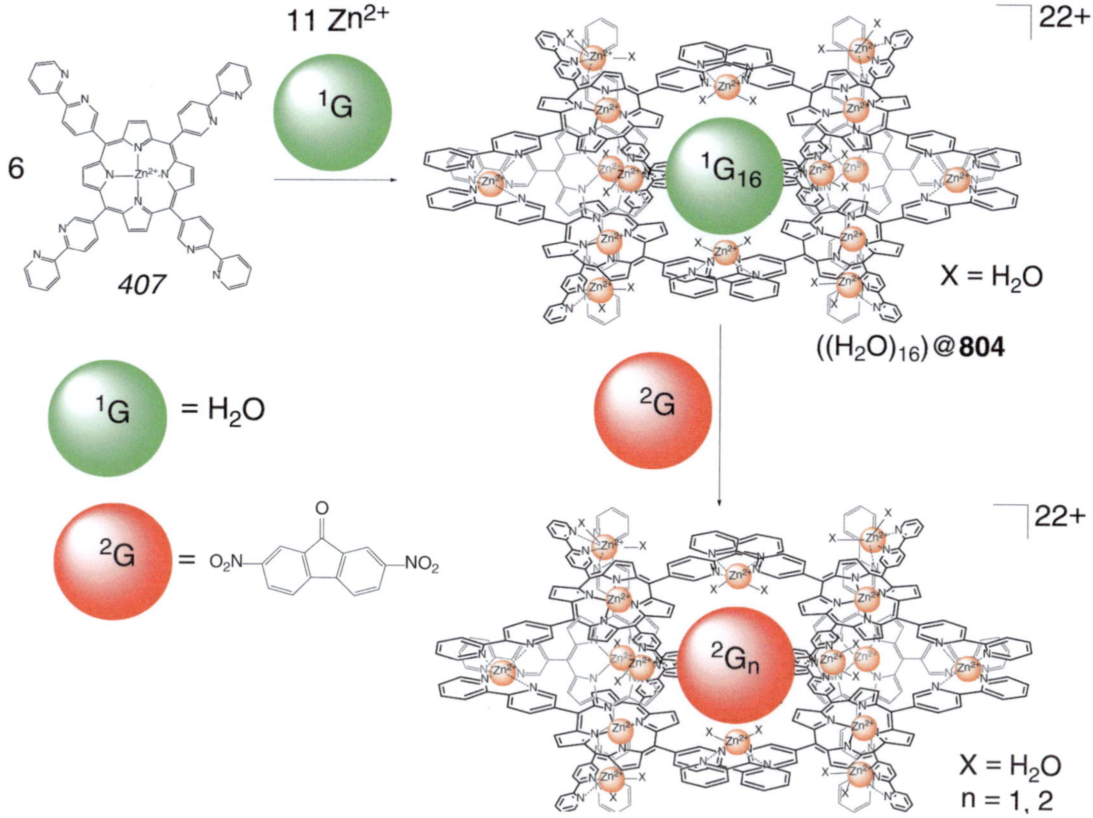

Scheme 4.202

cence switching between the fluorescent Hg_6L_8 capsule **799** and the non-fluorescent Hg_6L_4 cage framework **803** [208].

Coordination-driven self-assembly of a multiporphyrin $Zn_{11}L_6$ coordination capsule **804** with different types of cross-linking zinc(II) ions has been performed in [209] by Scheme 4.202. Its cage framework possess a triangular bipyramidal-shaped cavity of an approximate size of 9×18 Å with a cavity volume of 730 $Å^3$; this cavity is surrounded by six metalloporphyrin rings *407*. The caging ligand **804** has three windows with a diameter of approximately 4 Å in the middle of its framework, which allow it to encapsulate up to two symmetric π-acceptor guests (Scheme 4.202), thus giving 1:1 and 1:2 host–guest cage complexes [209].

An analogous tetraporphyrin 1:1 cage complex of coordination capsule **805** has been prepared in [210] by guest-templated self-assembly (Scheme 4.203) of labile tetrakis(bipyridyl)porphyrin ligand syntones *407*, zinc(II) ions, and guest C_{60}. In this coordination capsule, four such zinc(II) porphyrin syntones are wrapped around the fullerene guest, resulting in a barrel-like shape with a S_4 pseudosymmetry of the encapsulating ligand **805**. Its metallocomplex fragment adopt a C_1 symmetry, so this cage complex is achiral [210].

The hybrid diphenylphosphine-functionalized catecholate ligand syntone has been used in [211] and [212] for preparation of $M_2M_3'L_6$ coordination capsules **806** and **807** with the cross-linking titanium and tin(IV) ions by

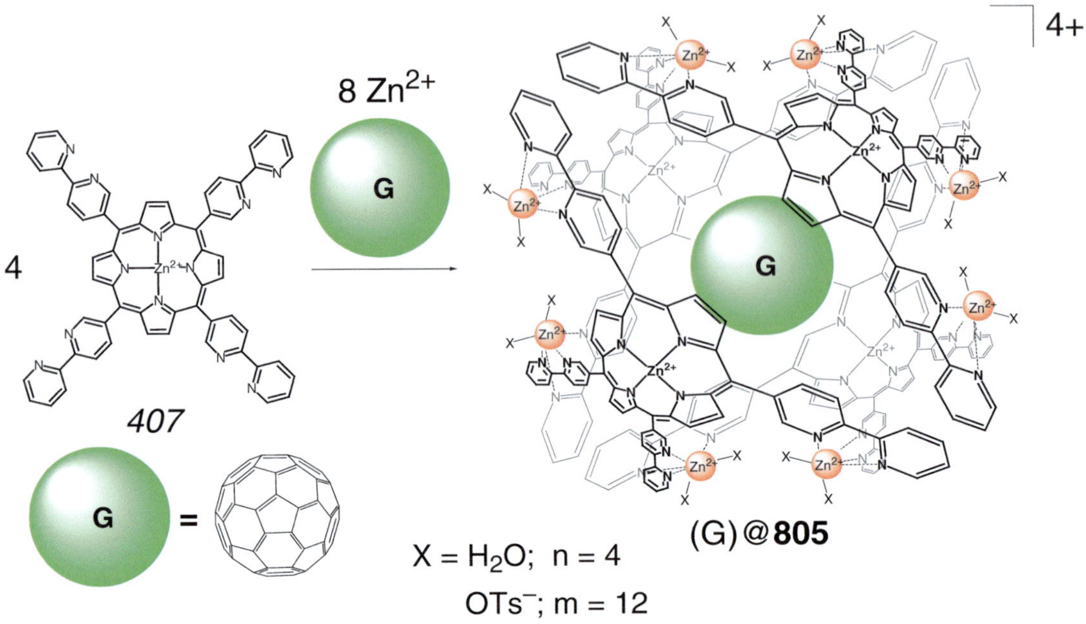

Scheme 4.203

Scheme 4.204. As follows from X-ray diffraction data [211] for their cesium salts, these cage frameworks have a crystallographically imposed C_{3h} symmetry and are achiral with one cross-linking titanium or tin ion having Δ and the other having Λ configuration; the intramolecular Ti… Ti distance is equal to 6.76 Å. The three bridging palladium(II) ions of this framework have distorted square-planar coordination polyhedra formed by *trans*-coordinated phosphine moieties; three cesium cations are encapsulated in its clefts. The Sn…Sn distance is approximately 6.88 Å being 0.12 Å larger than that for the titanium-based coordination capsule. This higher separation is explained [211] by larger ionic radius of tin(IV) ion compared to that of titanium(IV) (r_i=0.830 and 0.745 Å, respectively). This allowed observing Cs…Br interactions that occurred within the clefts of the tin-based caging ligand, which are absent in the case of its titanium-containing analog. The exclusive formation of these coordination capsules is reported in [212] to be sensitive to the nature of their counterions: those are formed in high yields with alkali metal cations such as K^+, Rb^+, and Cs^+, while $(C_2H_5)_3NH^+$ and $(CH_3)_4N^+$ ions gave none of the corresponding cage complexes. The synthetic procedure by Scheme 4.204 with DABCO or NH_4OH as bases afforded 1:3 cage complexes of the corresponding cations; these guests are encapsulated into the deep clefts of the coordination capsules. As follows from X-ray diffraction data [212] for the DABCO·H^+-containing complex of **807**, its cage framework also has a C_{3h} symmetry with the twist angle around its tin center of 41.9°; an average Sn–O distance is approximately 2.05 Å. The significantly larger Sn…Sn distance (7.47 Å) compared to one observed in the $Sn_2Pd_3L_6$ cage framework of **807** (6.88 Å) is explained in [212] by necessity to accommodate three bulky organic DABCO·H^+ cations as guests encapsulated within its expanded ligand's clefts.

The titanium(IV) tris-catecholate *408* has been used in [213] as a complex syntone for the synthesis of a heterometallic $M_2^1M_3^2L_6$ cage framework **808** by its reaction with palladium(II) ions (Scheme 4.205). The coordination capsule **808** has been also obtained by reversible conversion from the metallamacrocyclic precursor *409*

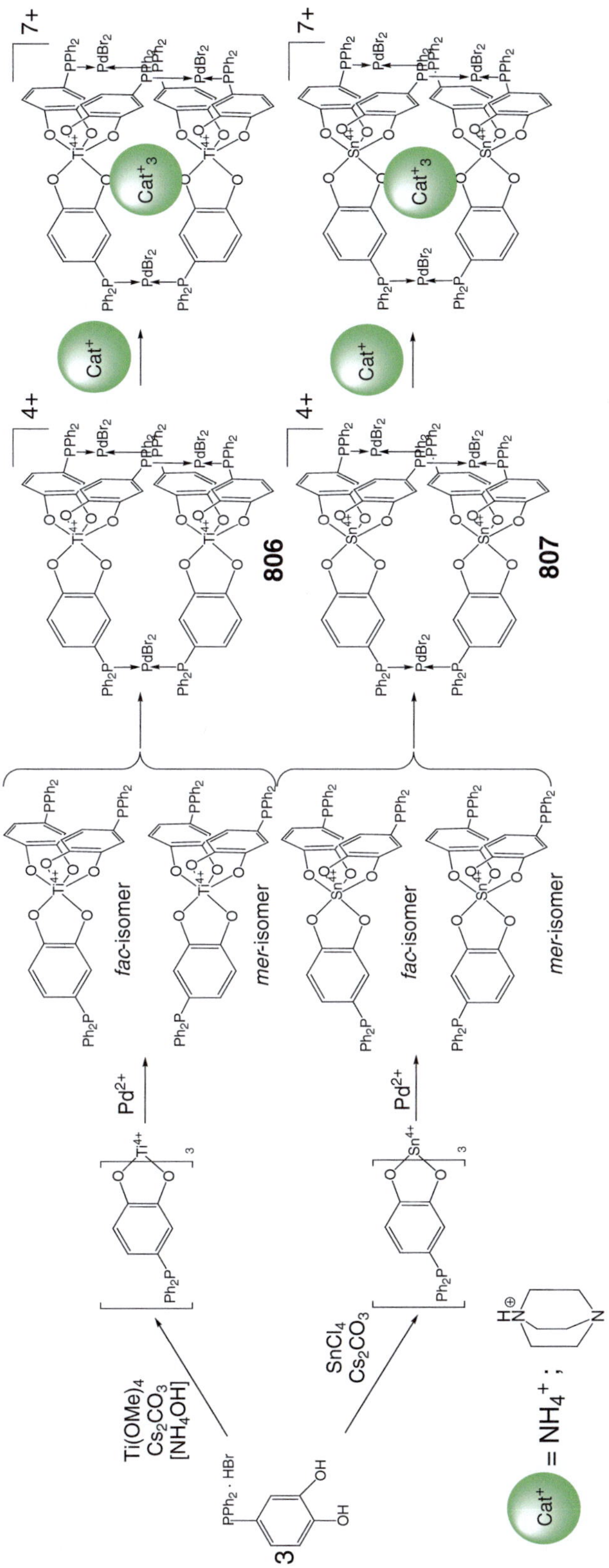
Ti(OMe)4
Cs2CO3
[NH4OH]
SnCl4
Cs2CO3
Pd2+
fac-isomer
mer-isomer
806
807
4+
7+
Cat+
Cat+3
Cat+ = NH4+ ;

Scheme 4.204

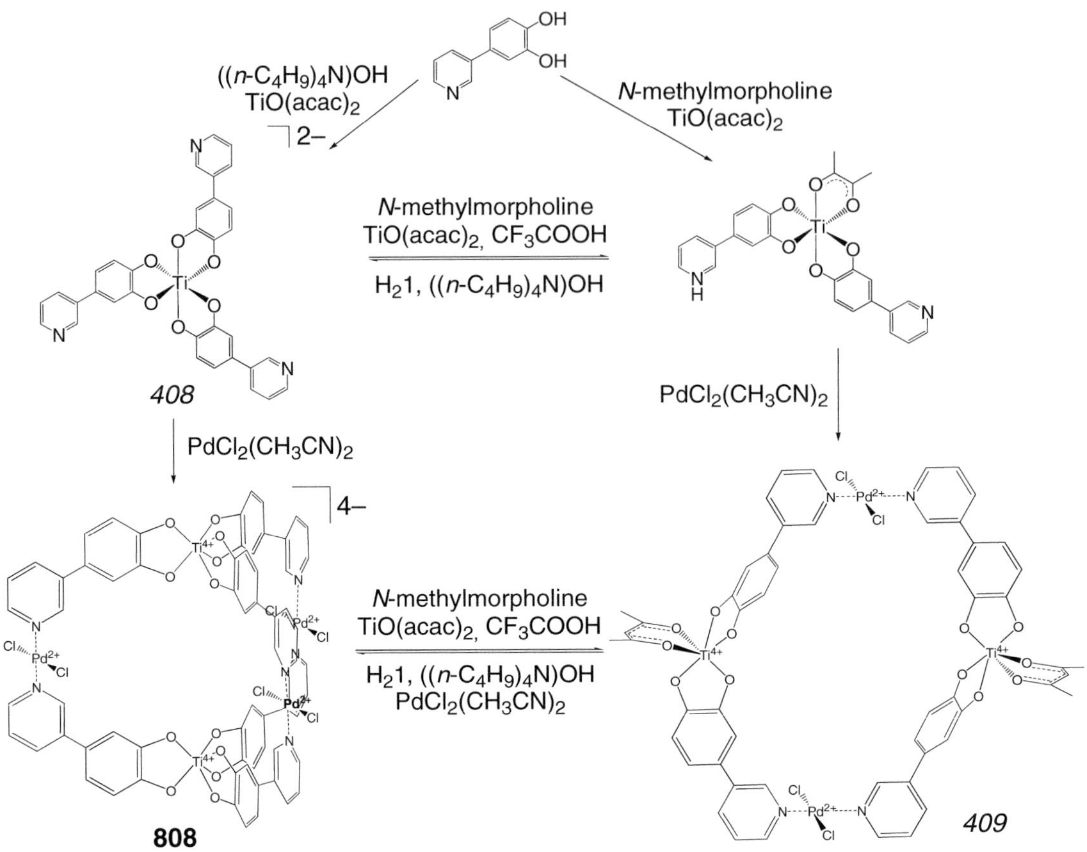

Scheme 4.205

by changing the acidity–basicity of the reaction mixture and the ratio of components. This dynamic molecular system is therefore based on interconvertion between palladium(II)–titanium(IV) heteronuclear metallomacrocyclic and cage complexes, which is triggered by Ti^{4+}-centered coordination changes between homoligand titanium(IV) tris-catecholate and heteroligand catecholato-acetylacetonate species [213].

Dithiocarbamate building blocks are reported in [214] to form coordination capsules **809–820** (Scheme 4.206) with appropriate octahedral transition metal cations, such as cobalt and iron(III) ions, by *one-pot* reaction of the corresponding secondary amine, carbon disulfide, organic (triethylamine), or inorganic (KOH) base and the corresponding metal(III) halide. As follows from ESI-MS, CV, and square-wave voltammetry data, cobalt(III)-containing polyether caging ligands encapsulate alkali metal cations, and this process is regulated by the fitting of these ions to the cavity size, while in the case of their iron(III)-containing analogs, the encapsulation of these cations is insignificant. The oxidized forms of the amide capsules **815–820** efficiently encapsulate $H_2PO_4^-$ anion as a donor of hydrogen bonds, giving 1:1 cage complexes. The nickel(IV)-based capsule **821** of this type has been prepared by an alternative pathway using chemical oxidation of its square-planar nickel(II) bis-dithiocarbamate precursor *410* (Scheme 4.207). According to the CV data, this oxidized capsule **821** is able to encapsulate chloride and nitrate anions [214].

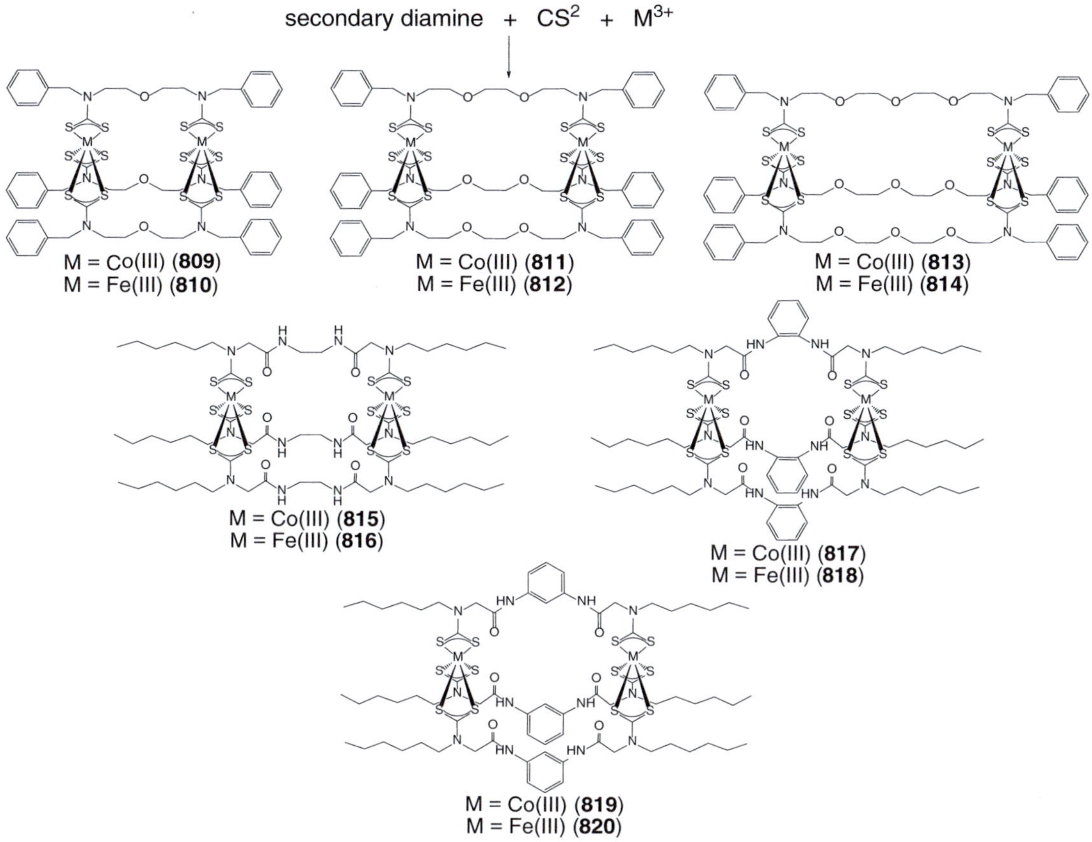

Scheme 4.206

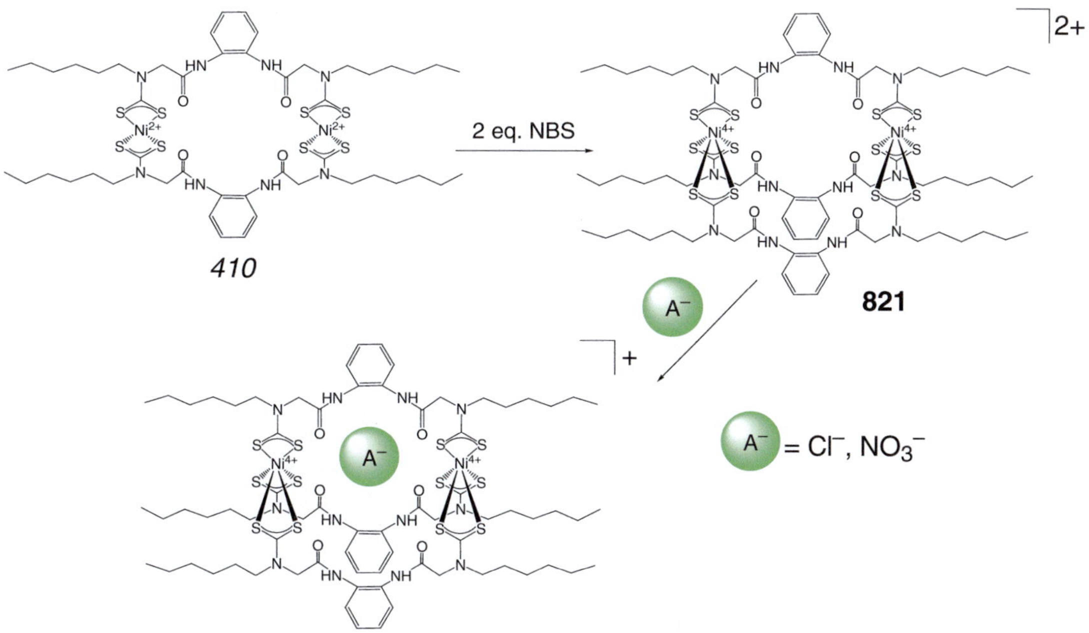

Scheme 4.207

References

1. Fujita M, Oguro D, Miyazawa M, Oka H, Yamaguchi K, Ogura K (1995) Self-assembly of ten molecules into nanometre-sized organic host frameworks. Nature 378:469–471
2. Kusukawa T, Fujita M (1998) Encapsulation of large, neutral molecules in a self-assembled nanocage incorporating six palladium(II) ions. Angew Chem Int Ed 37:3142–3144
3. Yoshizawa M, Takeyama Y, Kusukawa T, Fujita M (2002) Cavity-directed, highly stereoselective [2+2] photodimerization of olefins within self-assembled coordination cages. Angew Chem Int Ed 41: 1347–1349
4. Yoshizawa M, Tamura M, Fujita M (2004) AND/OR bimolecular recognition. J Am Chem Soc 126: 6846–6847
5. Tashiro S, Fujita M (2006) Selective recognition of Trp- and Tyr-rich oligopeptides by self-assembled coordination hosts. Bull Chem Soc Jpn 79:833–837
6. Horiuchi S, Nishioka Y, Murase T, Fujita M (2010) Both [2+2] and [2+4] additions of inert aromatics via identical ternary host-guest complexes. Chem Commun 46:3460–3462
7. Inokuma Y, Arai T, Fujita M (2010) Networked molecular cages as crystalline sponges for fullerenes and other guests. Nat Chem 2:780–783
8. Inokuma Y, Kojima N, Arai T, Fujita M (2011) Bimolecular reaction via the successive introduction of two substrates into the crystals of networked molecular cages. J Am Chem Soc 133:19691–19693
9. Kusukawa T, Yoshizawa M, Fujita M (2001) Probing guest geometry and dynamics through host–guest interactions. Angew Chem Int Ed 40:1879–1884
10. Yoshizawa M, Kusukawa T, Kawano M, Ohhara T, Tanaka I, Kurihara K, Niimura N, Fujita M (2005) Endohedral clusterization of ten water molecules into a "Molecular Ice" within the hydrophobic pocket of a self-assembled cage. J Am Chem Soc 127:2798–2799
11. Yamaguchi T, Fujita M (2008) Highly selective photomediated 1,4-radical addition to o-quinones controlled by a self-assembled cage. Angew Chem Int Ed 47:2067–2069
12. Nakabayashi K, Kawano M, Yoshizawa M, Ohkoshi S, Fujita M (2004) Cavity-induced spin-spin interaction between organic radicals within a self-assembled coordination cage. J Am Chem Soc 126(51): 16694–16695
13. Sakamoto S, Yoshizawa M, Kusukawa T, Fujita M, Yamaguchi K (2001) Characterization of encapsulating supramolecules by using CSI-MS with ionization-promoting reagents. Org Lett 3(11):1601–1604
14. Takaoka K, Kawano M, Ozeki T, Fujita M (2006) Crystallographic observation of an olefin photodimerization reaction that takes place via thermal molecular tumbling within a self-assembled host. Chem Commun 1625–1627
15. Tashiro S, Tominaga M, Kawano M, Therrien B, Ozeki T, Fujita M (2005) Sequence-selective recognition of peptides within the single binding pocket of a self-assembled coordination cage. J Am Chem Soc 127:4546–4547
16. Kawano M, Kobayashi Y, Ozeki T, Fujita M (2006) Direct crystallographic observation of a coordinatively unsaturated transition-metal complex in situ generated within a self-assembled cage. J Am Chem Soc 128:6558–6559
17. Yamashita K, Kawano M, Fujita M (2007) Ru(II)-cornered coordination cage that senses guest inclusion by color change. Chem Commun 40: 4102–4103
18. Nakabayshi K, Ozaki Y, Kawano M, Fujita M (2008) A self-assembled spin cage. Angew Chem Int Ed 47:2046–2048
19. Yamashita K, Sato K, Kawano M, Fujita M (2009) Photo-induced self-assembly of Pt(II)-linked rings and cages via the photolabilization of a Pt(II)–py bond. New J Chem 33:264–270
20. Chand DK, Biradha K, Fujita M, Sakamoto S, Yamaguchi K (2002) A molecular sphere of octahedral symmetry. Chem Commun 2486–2487
21. Suzuki K, Tominaga M, Kawano M, Fujita M (2009) Self-assembly of an M_6L_{12} coordination cube. Chem Commun 1638–1640
22. Moon D, Kang S, Park J, Lee K, John RP, Won H, Seong GH, Kim YS, Kim GH, Rhee H, Lah MS (2006) Face-driven corner-linked octahedral nanocages: M6L8 cages formed by C3-symmetric triangular facial ligands linked via C4-symmetric square tetratopic PdII ions at truncated octahedron corners. J Am Chem Soc 128:3530–3531
23. Liu Y, Wu X, He C, Zhang R, Duan C (2008) A truncated octahedral nanocage for fluorescent detection of nucleoside. Dalton Trans 5866–5868
24. Tominaga M, Suzuki K, Kawano M, Kusukawa T, Ozeki T, Sakamoto S, Yamaguchi K, Fujita M (2004) Finite, spherical coordination networks that self-organize from 36 small components. Angew Chem Int Ed 43:5621–5625
25. Tashiro S, Tominaga M, Yamaguchi Y, Kato K, Fujita M (2006) Peptide recognition: encapsulation and α-helical folding of a nine-residue peptide within a hydrophobic dimeric capsule of a bowl-shaped host. Chem Eur J 12:3211–3217
26. Tominaga M, Suzuki K, Murase T, Fujita M (2005) 24-Fold endohedral functionalization of a self-assembled $M_{12}L_{24}$ coordination nanoball. J Am Chem Soc 127:11950–11951
27. Sato S, Iida J, Suzuki K, Kawano M, Ozeki T, Fujita M (2006) Fluorous nanodroplets structurally confined in an organopalladium sphere. Science 313: 1273–1276
28. Suzuki K, Iida J, Sato S, Kawano M, Fujita M (2008) Discrete and well-defined hydrophobic phases confined in self-assembled spherical complexes. Angew Chem Int Ed 47:5780–5782

29. Suzuki K, Kawano M, Sato S, Fujita M (2007) Endohedral peptide lining of a self-assembled molecular sphere to generate chirality-confined hollows. J Am Chem Soc 129:10652–10653
30. Kikuchi T, Murase T, Sato S, Fujita M (2008) Polymerisation of an anionic monomer in a self-assembled $M_{12}L_{24}$ coordination sphere with cationic interior. Supramol Chem 20:81–94
31. Suzuki K, Sato S, Fujita M (2010) Template synthesis of precisely monodisperse silica nanoparticles within self-assembled organometallic spheres. Nat Chem 2:25–29
32. Takeda N, Umemoto K, Yamaguchi K, Fujita M (1999) A nanometre-sized hexahedral coordination capsule assembled from 24 components. Nature 398: 794–796
33. Company A, Roques N, Guell M, Mugnaini V, Gomez L, Imaz I, Datcu A, Sola M, Luis JM, Veciana J, Ribas X, Costas M (2008) Nanosized trigonal prismatic and antiprismatic CuII coordination cages based on tricarboxylate linkers. Dalton Trans 1679–1682
34. Yao H, Grimes RN (2003) Small cobaltacarborane clusters in synthesis. Peralkylation, perhalogenation, and macrocycle construction. J Org Chem 680(1–2): 51–60
35. Mattsson J, Govindaswamy P, Furrer J, Sei Y, Yamaguchi K, Süss-Fink G, Therrien B (2008) Encapsulation of aromatic molecules in hexanuclear arene ruthenium cages: a strategy to build up organometallic carceplex prisms with a dangling arm standing out. Organometallics 27:4346–4356
36. Therrien B, Süss-Fink G, Govindaswamy P, Renfrew AK, Dyson PJ (2008) The "complex-in-a-complex" cations $[(acac)_2M{\subset}Ru_6(p\text{-}iPrC_6H_4Me)_6(tpt)_2(dhbq)_3]^{6+}$: a trojan horse for cancer cells. Angew Chem Int Ed 47:3773–3776
37. Zava O, Mattsson J, Therrien B, Dyson PJ (2010) Evidence for drug release from a metalla-cage delivery vector following cellular internalisation. Chem Eur J 16:1428–1431
38. Mattsson J, Zava O, Renfrew AK, Sei Y, Yamaguchi K, Dyson PJ, Therrien B (2010) Drug delivery of lipophilic pyrenyl derivatives by encapsulation in a water soluble metalla-cage. Dalton Trans 39: 8248–8255
39. Han Y-F, Jin G-X (2011) Flexible organometallic cages: efficient formation by C–H activation-directed muticomponent assembly, isomerization, and host–guest properties. Chem Asian J 6: 1348–1352
40. Kumazawa M, Biradha K, Kusukawa T, Okano T, Fujita M (2003) Multicomponent assembly of a pyrazine-pillared coordination cage that selectively binds planar guests by intercalation. Angew Chem Int Ed 42:3909–3913
41. Sawada T, Yoshizawa M, Sato S, Fujita M (2009) Minimal nucleotide duplex formation in water through enclathration in self-assembled hosts. Nat Chem 1:53–56
42. Sawada T, Fujita M (2010) A single watson–crick G·C base pair in water: aqueous hydrogen bonds in hydrophobic cavities. J Am Chem Soc 132:7194–7201
43. Ono K, Klosterman JK, Yoshizawa M, Sekiguchi K, Tahara T, Fujita M (2009) ON/OFF red emission from azaporphine in a coordination cage in water. J Am Chem Soc 131(35):12526–12527
44. Ozaki Y, Kawano M, Fujita M (2009) Engineering noncovalent spin-spin interactions in an organic-pillared spin cage. Chem Commun 4245–4247
45. Yoshizawa M, Nakagawa J, Kumazawa K, Nagao M, Kawano M, Ozeki T, Fujita M (2005) Discrete stacking of large aromatic molecules within organic-pillared coordination cages. Angew Chem Int Ed 44:1810–1813
46. Yamauchi Y, Fujita M (2010) Self-assembled cage as an endo-template for cyclophane synthesis. Chem Commun 46:5897–5899
47. Yoshizawa M, Ono K, Kumazawa K, Kato T, Fujita M (2005) Metal–metal d–d interaction through the discrete stacking of mononuclear M(II) complexes (M = Pt, Pd, and Cu) within an organic-pillared coordination cage. J Am Chem Soc 127:10800–10801
48. Yoshizawa M, Kumazawa K, Fujita M (2005) Discrete stacking of large aromatic molecules within organic-pillared coordination cages. J Am Chem Soc 127:13456–13457
49. Yoshizawa M, Nagao M, Kumazawa K, Fujita M (2005) Side chain-directed complementary cis-coordination of two pyridines on Pd(II): selective multicomponent assembly of square-, rectangular-, and trigonal prism-shaped molecules. J Org Chem 690(23):5383–5388
50. Ono K, Yoshizawa M, Kato T, Watanabe K, Fujita M (2007) Porphine dimeric assemblies in organic-pillared coordination cages. Angew Chem Int Ed 46(11):1803–1806
51. Yamauchi Y, Yoshizawa M, Fujita M (2008) Engineering stacks of aromatic rings by the interpenetration of self-assembled coordination cages. J Am Chem Soc 130(18):5832–5833
52. Ono K, Yoshizawa M, Kato T, Fujita M (2008) Three-metal-center spin interactions through the intercalation of metal azaporphines and porphines into an organic pillared coordination box. Chem Commun 20:2328–2330
53. Ono K, Yoshizawa M, Akita M, Kato T, Tsunobuchi Y, Ohkoshi S, Fujita M (2009) Spin crossover by encapsulation. J Am Chem Soc 131(8): 2782–2783
54. Yamauchia Y, Yoshizawa M, Akita M, Fujita M (2009) Discrete stack of an odd number of polarized aromatic compounds revealing the importance of net vs. local dipoles. Proc Natl Acad Sci U S A 106(26):10435–10437
55. Kiguchi M, Takahashi T, Takahashi Y, Yamauchi Y, Murase T, Fujita M, Tada T, Watanabe S (2011) Electron transport through single molecules comprising aromatic stacks enclosed in self-assembled cages. Angew Chem Int Ed 50:5708–5711

56. Brasey T, Scopelliti R, Severin K (2006) Guest-induced formation of an icosahedral coordination cage. Chem Commun 3308–3310
57. Müller IM, Spillmann S, Franck H, Pietschnig R (2004) Rational design of the first closed coordination capsule with octahedral outer shape. Chem Eur J 10:2207–2213
58. Müller IM, Möller D (2005) Rational design of a coordination cage with a trigonal-bipyramidal shape constructed from 33 building units. Angew Chem Int Ed 44:2969–2973
59. Saalfrank RW, Hörner B, Stalke D, Salbeck J (1993) The first neutral adamantanoid iron(III)-chelate complex: spontaneous formation, structure, and electrochemistry. Angew Chem Int Ed 32:1179–1182
60. Saalfrank RW, Stark A, Bremer M, Hummel H-U (1990) Formation of tetranuclear chelate(4-) ions of divalent metals (Mn, Co, Ni) with idealized T symmetry by spontaneous self-assembly. Angew Chem Int Ed 29:311–314
61. Saalfrank RW, Stark A, Peters K, Von Schnering HG (1988) The first "adamantoid" alkaline earth metal chelate complex: synthesis, structure, and reactivity. Angew Chem Int Ed 27:851–853
62. Saalfrank RW, Burak R, Breit A, Stalke D, Herbst-Irmer R, Daub J, Porsch M, Bill E, Müther M, Trautwein AX (1994) Mixed-valence, tetranuclear iron chelate complexes as endoreceptors: charge compensation through inclusion of cations. Angew Chem Int Ed 33:1621–1623
63. Beissel T, Powers RE, Raymond KN (1996) Symmetry-based metal complex cluster formation. Angew Chem Int Ed 35(10):1084–1086
64. Beissel T, Powers RE, Parac TN, Raymond KN (1999) Dynamic isomerization of a supramolecular tetrahedral M_4L_6 cluster. J Am Chem Soc 121(17): 4200–4206
65. Bruckner C, Powers RE, Raymond KN (1998) Symmetry-driven rational design of a tetrahedral supramolecular Ti_4L_4 cluster. Angew Chem Int Ed 37(13/14):1837–1839
66. Biros SM, Bergman RG, Raymond KN (2007) The hydrophobic effect drives the recognition of hydrocarbons by an anionic metal–ligand cluster. J Am Chem Soc 129:12094–12095
67. Hastings CJ, Pluth MD, Biros SM, Bergman RG, Raymond KN (2008) Simultaneously bound guests and chiral recognition: a chiral self-assembled supramolecular host encapsulates hydrophobic guests. Tetrahedron 64(36):8362–8367
68. Yoshizawa M, Nagao M, Umemoto K, Biradha K, Fujita M, Sakamoto S, Yamaguchi K (2003) Side chain-directed assembly of triangular molecular panels into a tetrahedron vs. open cone. Chem Commun 15:1808–1809
69. Chand DK, Biradha K, Kawano M, Sakamoto S, Yamaguchi K, Fujita M (2006) Dynamic self-assembly of an M_3L_6 molecular triangle and an M_4L_8 ,etrahedron from naked Pd^{II} ions and bis(3-pyridyl)-substituted arenes. Chem Asian J 1:82–90
70. Hiraoka S, Harano K, Shiro M, Shionoya M (2005) Quantitative dynamic interconversion between agI-mediated capsule and cage complexes accompanying guest encapsulation/release. Angew Chem Int Ed 44:2727–2731
71. Mal P, Schultz D, Beyeh K, Rissanen K, Nitschke JR (2008) An unlockable–relockable iron cage by subcomponent self-assembly. Angew Chem Int Ed 47:8297–8301
72. Riddell IA, Smulders MMJ, Clegg JK, Nitschke JR (2011) Encapsulation, storage and controlled release of sulfur hexafluoride from a metal-organic capsule. Chem Commun 47:457–459
73. Mal P, Breiner B, Rissanen K, Nitschke JR (2009) White phosphorus is air-stable within a self-assembled tetrahedral capsule. Science 324:1697–1699
74. Smulders MMJ, Nitschke JR (2012) Supramolecular control over Diels-Alder reactivity by encapsulation and competitive displacement. Chem Sci 3:785–788
75. Smulders MMJ, Zarra S, Nitschke JR (2013) Quantitative understanding of guest binding enables the design of complex host-guest behaviour. J Am Chem Soc 135(18):7039–7046
76. Meng W, Clegg JK, Thoburn JD, Nitschke JR (2011) Controlling the transmission of stereochemical information through space in terphenyl-edged Fe_4L_6 cages. J Am Chem Soc 133:13652–13660
77. Ousaka N, Grunder S, Castilla AM, Whalley AC, Stoddart JF, Nitschke JR (2012) Efficient long-range stereochemical communication and cooperative effects in self-assembled Fe_4L_6 cages. J Am Chem Soc 134(37):15528–15537
78. Zarra S, Clegg JK, Nitschke JR (2013) Selective assembly and disassembly of a water-soluble $Fe_{10}L_{15}$ prism. Angew Chem Int Ed 52(18):4837–4840
79. Bolliger JL, Belenguer AM, Nitschke JR (2013) Enantiopure water-soluble $[Fe_4L_6]$ cages: host-guest chemistry and catalytic activity. Angew Chem Int Ed 52(31):7958–7962
80. Castilla AM, Ousaka N, Bilbeisi RA, Valeri E, Ronson TK, Nitschke JR (2013) High-fidelity stereochemical memory in a $FeII_4L_4$ tetrahedral capsule. J Am Chem Soc 135(47):17999–18006
81. Bilbeisi RA, Zarra S, Feltham HLC, Jameson GNL, Clegg JK, Brooker S, Nitschke JR (2013) Guest binding subtly influences spin crossover (SCO) in an $FeII_4L_4$ capsule. Chem Eur J 19(25):8058–8062
82. Ronson TK, Giri C, Beyeh NK, Topić F, Holstein JJ, Rissanen K, Nitschke JR (2013) Size-selective encapsulation of hydrophobic guests by self-assembled M_4L_6 cobalt and nickel cages. Chem Eur J 19:3374–3382
83. Meng W, Ronson TK, Clegg JK, Nitschke JR (2013) Transformations within a network of cadmium architectures. Angew Chem Int Ed 52(3):1017–1021
84. Smulders MMJ, Jimenez A, Nitschke JR (2012) Integrative self-sorting synthesis of a $Fe_8Pt_6L_{24}$ cubic cage. Angew Chem Int Ed 51(27):6681–6685
85. Browne C, Brenet S, Clegg JK, Nitschke JR (2013) Solvent-dependent host-guest chemistry of an Fe_8L_{12} cubic capsule. Angew Chem Int Ed 52(7):1944–1948

86. Riddell IA, Hristova YR, Clegg JK, Wood CS, Breiner B, Nitschke JR (2013) Five discrete multinuclear metal-organic assemblies from one ligand: deciphering the effects of different templates. J Am Chem Soc 135(7):2723–2733
87. Ramsay WJ, Ronson TK, Clegg JK, Nitschke JR (2013) Bidirectional regulation of halide binding in a heterometallic supramolecular cube. Angew Chem Int Ed 52(50):13439–13443
88. Ramsay WJ, Nitschke JR (2014) Two distinct allosteric active sites regulate guest binding within a $Fe_8Mo_{12}^{16+}$ cubic receptor. J Am Chem Soc 136(19): 7038–7043
89. Bilbeisi RA, Ronson TK, Nitschke JR (2013) A self-assembled $[Fe^{II}_{12}L_{12}]$ capsule with an icosahedral framework. Angew Chem Int Ed 52(34):9027–9030
90. Meng W, Breiner B, Rissanen K, Thoburn JD, Clegg JK, Nitschke JR (2011) A self-assembled M_8L_6 cubic cage that selectively encapsulates large aromatic guests. Angew Chem Int Ed 50:3479–3483
91. Al-Rasbi NK, Tidmarsh IS, Argent SP, Adams H, Harding LP, Ward MD (2008) Mixed-ligand molecular paneling: dodecanuclear cuboctahedral coordination cages based on a combination of edge-bridging and face-capping ligands. J Am Chem Soc 130:11641–11649
92. Stephenson A, Ward MD (2011) An octanuclear coordination cage with a 'cuneane' core-a topological isomer of a cubic cage. Dalton Trans 40: 7824–7826
93. Mirtschin S, Slabon-Turski A, Scopelliti R, Velders AH, Severin K (2010) A coordination cage with an adaptable cavity size. J Am Chem Soc 132:14004–14005
94. Fujita M, Fujita N, Ogura K, Yamaguchi K (1999) Spontaneous assembly of ten components into two interlocked, identical coordination cages. Nature 400:52–55
95. Fleming JS, Mann KLV, Carraz C-A, Psillakis E, Jeffery JC, McCleverty JA, Ward MD (1998) Anion-templated assembly of a supramolecular cage complex. Angew Chem Int Ed 37:1279–1281
96. Argent SP, Riis-Johannessen T, Jeffery JC, Harding LP, Ward MD (2005) Diastereoselective formation and optical activity of an M_4L_6 cage complex. Chem Commun 4647–4649
97. Paul RL, Bell ZR, Jeffery JC, McCleverty JA, Ward MD (2002) Anion-templated self-assembly of tetrahedral cage complexes of cobalt(II) with bridging ligands containing two bidentate pyrazolyl-pyridine binding sites. Proc Natl Acad Sci U S A 99:4883–4888
98. Paul RL, Bell ZR, Jeffery JC, Harding LP, McCleverty JA, Ward MD (2003) Complexes of a bis-bidentate ligand with d^{10} ions: a mononuclear complex with Ag(I), and a tetrahedral cage complex with Zn(II) which encapsulates a fluoroborate anion. Polyhedron 22:781–787
99. Bell ZR, Harding LP, Ward MD (2003) Self-assembly of a molecular M_8L_{12} cube having S_6 symmetry. Chem Commun 2432–2433
100. Argent SP, Adams H, Harding LP, Ward MD (2006) A closed molecular cube and an open book: two different products from assembly of the same metal salt and bridging ligand. Dalton Trans 542–544
101. Najar AM, Adams H, Tidmarsh IS, Ward MD (2009) Cubes, squares, and books: a simple transition metal/bridging ligand combination can lead to a surprising range of structural types with the same metal/ligand proportions. Inorg Chem 48:11871–11881
102. Paul RL, Argent SP, Jeffery JC, Harding LP, Lynam JM, Ward MD (2004) Structures and anion-binding properties of M_4L_6 tetrahedral cage complexes with large central cavities. Dalton Trans 3453–3458
103. Argent SP, Adams H, Riis-Johannessen T, Jeffery JC, Harding LP, Ward MD (2005) High-nuclearity homoleptic and heteroleptic coordination cages based on tetra-capped truncated tetrahedral and cuboctahedral metal frameworks. J Am Chem Soc 128:72–73
104. Stephenson A, Argent SP, Riis-Johannessen T, Tidmarsh IS, Ward MD (2010) Structures and dynamic behavior of large polyhedral coordination cages: an unusual cage-to-cage interconversion. J Am Chem Soc 133:858–870
105. Stephenson A, Sykes D, Ward MD (2013) Cu_{12} and Cd_{16} coordination cages and their Cu_3 and Cd_3 subcomponents, and the role of inter-ligand π-stacking in stabilising cage complexes. Dalton Trans 42:6756–6767
106. Stephenson A, Ward MD (2011) Molecular squares, cubes and chains from self-assembly of bis-bidentate bridging ligands with transition metal dications. Dalton Trans 40:10360–10369
107. Argent SP, Adams H, Riis-Johannessen T, Jeffery JC, Harding LP, Mamula O, Ward MD (2006) Coordination chemistry of tetradentate N-donor ligands containing two pyrazolyl–pyridine units separated by a 1,8-naphthyl spacer: dodecanuclear and tetranuclear coordination cages and cyclic helicates. Inorg Chem 45:3905–3919
108. Argent SP, Adams H, Riis-Johannessen T, Jeffery JC, Harding LP, Clegg W, Harrington RW, Ward MD (2006) Complexes of Ag(I), Hg(I) and Hg(II) with multidentate pyrazolyl-pyridine ligands: from mononuclear complexes to coordination polymers via helicates, a mesocate, a cage and a catenate. Dalton Trans 4996–5013
109. Al-Rasbi NK, Sabatini C, Barigelletti F, Ward MD (2006) Red-shifted luminescence from naphthalene-containing ligands due to π-stacking in self-assembled coordination cages. Dalton Trans 4769–4772
110. Tidmarsh IS, Tailor BF, Hardie MJ, Russo L, Clegg W, Ward MD (2009) Further investigations into tetrahedral M_4L_6 cage complexes containing guest anions: new structures and NMR spectroscopic studies. New J Chem 33:366–375
111. Bilbeisi RA, Clegg JK, Elgrishi N, de Hatten X, Devillard M, Breiner B, Mal P, Nitschke JR (2011) Subcomponent self-assembly and guest-binding properties of face-capped $Fe_4L_4^{8+}$ capsules. J Am Chem Soc 134:5110–5119
112. Clegg JK, Cremers J, Hogben AJ, Breiner B, Smulders MMJ, Thoburn JD, Nitschke JR (2013) A stimuli responsive system of self-assembled anion-binding $Fe_4L_6^{8+}$ cages. Chem Sci 4(1):68–76

113. Hamacek J, Poggiali D, Zebret S, El Aroussi B, Schneider MW, Mastalerz M (2012) Building large supramolecular nanocapsules with europium cations. Chem Commun 48(9):1281–1283
114. Wise MD, Holstein JJ, Pattison P, Besnard C, Solari E, Scopelliti R, Bricogne G, Severin K (2015) Large, heterometallic coordination cages based on ditopic metallo-ligands with 3-pyridyl donor groups. Chem Sci 6:1004–1010
115. Scherer M, Caulder DL, Johnson DW, Raymond KN (1999) Coordination number incommensurate cluster formation. Part 11. Triple helicate-tetrahedral cluster interconversion controlled by host-guest interactions. Angew Chem Int Ed 38(11):1588–1592
116. Ziegler M, Miranda JJ, Andersen UN, Johnson DW, Leary JA, Raymond KN (2001) Combinatorial libraries of metal-ligand assemblies with an encapsulated guest molecule. Angew Chem Int Ed 40(4):733–736
117. Caulder DL, Powers RE, Parac TN, Raymond KN (1998) The self-assembly of a predesigned tetrahedral M_4L_6 supramolecular cluster. Angew Chem Int Ed 37(13/14):1840–1843
118. Parac TN, Caulder DL, Raymond KN (1998) Selective encapsulation of aqueous cationic guests into a supramolecular tetrahedral $[M_4L_6]^{12-}$ anionic host. J Am Chem Soc 120(31):8003–8004
119. Terpin AJ, Ziegler M, Johnson DW, Raymond KN (2001) Resolution and kinetic stability of a chiral supramolecular assembly made of labile components. Angew Chem Int Ed 40(1):157–160
120. Parac TN, Scherer M, Raymond KN (2000) Host within a host: encapsulation of alkali ion- crown ether complexes into a $[Ga_4L_6]^{12-}$ supramolecular cluster. Angew Chem Int Ed 39(7):1239–1242
121. Saalfrank RW, Glaser H, Demleitner B, Hampel F, Chowdhry MM, Schunemann V, Trautwein AX, Vaughan GBM, Yeh R, Davis AV, Raymond KN (2002) Self-assembly of tetrahedral and trigonal antiprismatic clusters $[Fe_4(L_4)_4]$ and $[Fe_6(L_5)_6]$ on the basis of trigonal tris-bidentate chelators. Chem Eur J 8(2):493–497
122. Caulder DL, Brückner C, Powers RE, König S, Parac TN, Leary JA, Raymond KN (2001) Design, formation and properties of tetrahedral M_4L_4 and M_4L_6 supramolecular clusters. J Am Chem Soc 123:8923–8938
123. Johnson DW, Raymond KN (2001) The self-assembly of a $[Ga_4L_6]^{12-}$ tetrahedral cluster thermodynamically driven by host-guest interactions. Inorg Chem 40(20):5157–5161
124. Ziegler M, Brumaghim JL, Raymond KN (2000) Stabilization of a reactive cationic species by supramolecular encapsulation. Angew Chem Int Ed 39(22):4119–4121
125. Brumaghim JL, Michels M, Raymond KN (2004) Hydrophobic chemistry in aqueous solution: stabilization and stereoselective encapsulation of phosphonium guests in a supramolecular host. Eur J Org Chem 22:4552–4559
126. Brumaghim JL, Michels M, Pagliero D, Raymond KN (2004) Encapsulation and stabilization of reactive aromatic diazonium ions and the tropylium ion within a supramolecular host. Eur J Org Chem 24:5115–5118
127. Wang ZJ, Brown CJ, Bergman RG, Raymond KN, Toste FD (2011) Hydroalkoxylation catalyzed by a gold(I) complex encapsulated in a supramolecular host. J Am Chem Soc 133:7358–7360
128. Yeh RM, Xu J, Seeber G, Raymond KN (2005) Large M4L4 (M = Al(III), Ga(III), In(III), Ti(IV)) tetrahedral coordination cages: an extension of symmetry-based design. Inorg Chem 44:6228–6239
129. Davis AV, Raymond KN (2005) The big squeeze: guest exchange in an M4L6 supramolecular host. J Am Chem Soc 127:7912–7919
130. Fiedler D, Bergman RG, Raymond KN (2006) Stabilization of reactive organometallic intermediates inside a self-assembled nanoscale host. Angew Chem Int Ed 45:745–748
131. Leung DH, Bergman RG, Raymond KN (2006) Scope and mechanism of the C-H bond activation reactivity within a supramolecular host by an iridium guest: a stepwise ion pair guest dissociation mechanism. J Am Chem Soc 128(30):9781–9797
132. Konig S, Bruckner C, Raymond KN, Leary JA (1998) Electrospray ionization ion trap mass spectrometry of a tetrahedral supramolecular Ti_4L_4 cluster. J Am Soc Mass Spectrom 9(10):1099–1103
133. Andersen UN, Seeber G, Fiedler D, Raymond KN, Lin D, Harris D (2006) Characterization of self-assembled supramolecular $[Ga_4L_6]$ host-guest complexes by electrospray ionization mass spectrometry. J Am Soc Mass Spectrom 17(3):292–296
134. Davis AV, Fiedler D, Ziegler M, Terpin A, Raymond KN (2007) Resolution of chiral, tetrahedral M_4L_6 metal-ligand hosts. J Am Chem Soc 129(49): 15354–15363
135. Leung DH, Bergman RG, Raymond KN (2007) Highly selective supramolecular catalyzed allylic alcohol isomerization. J Am Chem Soc 129(10): 2746–2747
136. Leung DH, Bergman RG, Raymond KN (2008) Enthalpy-entropy compensation reveals solvent reorganization as a driving force for supramolecular encapsulation in water. J Am Chem Soc 130(9):2798–2805
137. Fiedler D, Leung DH, Bergman RG, Raymond KN (2004) Enantioselective guest binding and dynamic resolution of cationic ruthenium complexes by a chiral metal- ligand assembly. J Am Chem Soc 126(12):3674–3675
138. Fiedler D, Pagliero D, Brumaghim JL, Bergman RG, Raymond KN (2004) Encapsulation of cationic ruthenium complexes into a chiral self-assembled cage. Inorg Chem 43(3):846–848
139. Tiedemann BEF, Raymond KN (2007) Second-order Jahn-Teller effect in a host-guest complex. Angew Chem Int Ed 46(26):4976–4978
140. Dong VM, Fiedler D, Carl B, Bergman RG, Raymond KN (2006) Molecular recognition and

stabilization of iminium ions in water. J Am Chem Soc 128(45):14464–1446
141. Pluth MD, Bergman RG, Raymond KN (2007) Making amines strong bases: thermodynamic stabilization of protonated guests in a highly-charged supramolecular host. J Am Chem Soc 129(37): 11459–11467
142. Biros SM, Yeh RM, Raymond KN (2008) Design and formation of a large tetrahedral cluster using 1,1′-binaphthyl ligands. Angew Chem Int Ed 47(32): 6062–6064
143. Albrecht M, Janser I, Burk S, Weis P (2006) Self-assembly and host-guest chemistry of big metallosupramolecular M_4L_4 tetrahedra. Dalton Trans 2875–2880
144. Pluth MD, Johnson DW, Szigethy G, Davis AV, Teat SJ, Oliver AG, Bergman RG, Raymond KN (2009) Structural consequences of anionic host-cationic guest interactions in a supramolecular assembly. Inorg Chem 48(1):111–120
145. Mugridge JS, Bergman RG, Raymond KN (2011) ^{1}H NMR chemical shift calculations as a probe of supramolecular host-guest geometry. J Am Chem Soc 133(29):11205–11212
146. Sgarlata C, Mugridge JS, Pluth MD, Tiedemann BEF, Zito V, Arena G, Raymond KN (2010) External and internal guest binding of a highly charged supramolecular host in water: deconvoluting the very different thermodynamics. J Am Chem Soc 132:1005–1009
147. Tiedemann BEF, Raymond KN (2005) Dangling arms: a tetrahedral supramolecular host with partially encapsulated guest. Angew Chem Int Ed 45:83–86
148. Olenyuk B, Levin MD, Whiteford JA, Shield JE, Stang PJ (1999) Self-assembly of nanoscopic dodecahedra from 50 predesigned components. J Am Chem Soc 121:10434–10435
149. Ghosh K, Hu J, White HS, Stang PJ (2009) Construction of multifunctional cuboctahedra via coordination-driven self-assembly. J Am Chem Soc 131:6695–6697
150. Su C-Y, Cai Y-P, Chen C-L, Zhang H-X, Kang B-S (2001) Coordination-directed assembly of trigonal and tetragonal molecular boxes encapsulating anionic guests. J Chem Soc Dalton Trans 359–361
151. Su C-Y, Cai Y-P, Chen C-L, Lissner F, Kang B-S, Kaim W (2002) Self-assembly of trigonal-prismatic metallocages encapsulating BF_4^- or CuI_3^{2-} as anionic guests: structures and mechanism of formation. Angew Chem Int Ed 41:3371–3375
152. Liu H-K, Huang X, Li T, Wang X, Sun W-Y, Kang B-S (2008) Discrete and infinite ^{1}D, ^{2}D/^{3}D cage frameworks with inclusion of anionic species and anion-exchange reactions of Ag_3L_2 type receptor with tetrahedral and octahedral anions. Dalton Trans 3178–3188
153. Su C-Y, Cai Y-P, Chen C-L, Smith MD, Kaim W, zur Loye H-C (2003) Ligand-directed molecular architectures: self-assembly of two-dimensional rectangular metallacycles and three-dimensional trigonal or tetragonal prisms. J Am Chem Soc 125: 8595–8613
154. Wang Q-M, Mak TCW (2003) Assembly of discrete, one-, two-, and three-dimensional silver(I) supramolecular complexes containing encapsulated acetylide dianion with nitrogen-donor spacers. Inorg Chem 42:1637–1643
155. ZhaoX-L, Mak TCW (2004) Silver cages with encapsulated acetylenediide as building blocks for hydrothermal synthesis of supramolecular complexes with n-cyanopyridine and pyridine-n-carboxamide (n = 3, 4). Dalton Trans 3212–3217
156. Beer PD, Cheetham AG, Drew MGB, Fox OD, Hayes EJ, Rolls TD (2003) Pyrrole-based metallomacrocycles and cryptands. Dalton Trans 603–611
157. Geier J, Grutzmacher H (2003) Synthesis and structure of $[Na_{11}(OtBu)_{10}(OH)]$: ^{1}H NMR shift of a hydroxide ion encapsulated in a 21-vertex alcoholate cage. Chem Commun 2942–2943
158. Dolomanov OV, Blake AJ, Champness NR, Schroder M, Wilson C (2003) A novel synthetic strategy for hexanuclear supramolecular architectures. Chem Commun 682–683
159. Sumby CJ, Hardie MJ (2005) Capsules and starburst polyhedra: an $[Ag_2L_2]$ capsule and a tetrahedral $[Ag_4L_4]$ metallosupramolecular prism with cyclotriveratrylene-type ligands. Angew Chem Int Ed 44:6395–6399
160. Sumby CJ, Carr MJ, Franken A, Kennedy JD, Kilner CA, Hardie MJ (2006) Crystal-packing motifs of $[Ag_4L_4]^{4+}$ star-burst tetrahedra. New J Chem 30: 1390–1396
161. Carruthers C, Ronson TK, Sumby CJ, Westcott A, Harding LP, Prior TJ, Rizkallah P, Hardie MJ (2008) The dimeric "hand-shake" motif in complexes and metallo–supramolecular assemblies of cyclotriveratrylene-basedligands.ChemEurJ14:10286–10296
162. Sumby CJ, Fisher J, Prior TJ, Hardie MJ (2006) Tris(pyridylmethylamino)cyclotriguaiacylene cavitands: an investigation of the solution and solid-state behaviour of metallo-supramolecular cages and cavitand-based coordination polymers. Chem Eur J 12:2945–2959
163. Ronson TK, Hardie MJ (2008) Extended 3^6 and 6^3 arrays of capsule motifs using ligand tris{4-(3-pyridyl)phenylester}cyclotriguaiacylene. Cryst Eng Commun 10:1731–1734
164. Ronson TK, Fisher J, Harding LP, Hardie MJ (2007) Star-burst prisms with cyclotriveratrylene-type ligands: a $[Pd_6L_8]^{12+}$ stella octangular structure. Angew Chem Int Ed 46:9086–9088
165. Henkelis JJ, Ronson TK, Harding LP, Hardie MJ (2011) M_3L_2 metallo-cryptophanes: [2]catenane and simple cages. Chem Commun 47:6560–6562
166. Chand DK, Biradha K, Fujita M (2001) Self-assembly of a novel macrotricyclic Pd(II) metallocage encapsulating a nitrate ion. Chem Commun 1652–1653

167. Amouri H, Mimassi L, Rager MN, Mann BE, Guyard-Duhayon C, Raehm L (2005) Host–guest interactions: design strategy and structure of an unusual cobalt cage that encapsulates a tetrafluoroborate anion. Angew Chem Int Ed 44:4543–4546
168. Desmarets C, Poli F, Le Goff XF, Muller K, Amouri H (2009) A unique type of a dicobalt cage templated by a weakly coordinated hexafluorophosphate anion: design, structure and solid-state NMR investigations. Dalton Trans 10429–10432
169. Fox OD, Dalley NK, Harrison RG (1999) Structure and small molecule binding of a tetranuclear iron(II) resorc[4]arene-based cage complex. Inorg Chem 38(25):5860–5863
170. Fox OD, Leung JFY, Hunter JM, Dalley NK, Harrison RG (2000) Metal-assembled cobalt(II) resorc[4]arene-based cage molecules that reversibly capture organic molecules from water and act as nmr shift reagents. Inorg Chem 39(4):783–790
171. McKinlay RM, Thallapally PK, Cave GWV, Atwood JL (2005) Hydrogen-bonded supramolecular assemblies as robust templates in the synthesis of large metal-coordinated capsules. Angew Chem Int Ed 44:5733–5736
172. Haino T, Kobayashi M, Fukazawa Y (2006) Guest encapsulation and self-assembly of a cavitand-based coordination capsule. Chem Eur J 12:3310–3319
173. Dendrinou-Samara C, Alexiou M, Zaleski CM, Kampf JW, Kirk ML, Kessissoglou DP, Pecoraro VL (2003) Synthesis and magnetic properties of a metallacryptate that behaves as a single-molecule magnet. Angew Chem Int Ed 42(32):3763–3766
174. Zaleski CM, Depperman EC, Dendrinou-Samara C, Alexiou M, Kampf JW, Kessissoglou DP, Kirk ML, Pecoraro VL (2005) Metallacryptate single-molecule magnets: effect of lower molecular symmetry on blocking temperature. J Am Chem Soc 127(37):12862–12872
175. Jankolovits J, Andolina CM, Kampf JW, Raymond KN, Pecoraro VL (2011) Assembly of near-infrared luminescent lanthanide host(host-guest) complexes with a metallacrown sandwich motif. Angew Chem Int Ed 50(41):9660–9664
176. Trivedi ER, Eliseeva SV, Jankolovits J, Olmstead MM, Petoud S, Pecoraro VL (2014) Highly emitting near-infrared lanthanide "encapsulated sandwich" metallacrown complexes with excitation shifted toward lower energy. J Am Chem Soc 136(4):1526–1534
177. Mezei G, Rivera-Carrillo M, Raptis RG (2007) Trigonal-prismatic Cu^{II}_6-pyrazolato cages: structural and electrochemical study, evidence of charge delocalization. Dalton Trans 37–40
178. Dalgarno SJ, Power NP, Warren JE, Atwood JL (2008) Rapid formation of metal-organic nanocapsules gives new insight into the self-assembly process. Chem Commun 1539–1544
179. Bray DJ, Antonioli B, Clegg JK, Gloe K, Gloe K, Jolliffe KA, Lindoy LF, Wei G, Wenzel M (2008) Assembly of a trinuclear metallo-capsule from a tripodal tris(β-diketone) derivative and copper(II). Dalton Trans 1683–1685
180. Duriska MB, Neville SM, Moubaraki B, Cashion JD, Halder GJ, Chapman KW, Balde C, Létard J-F, Murray KS, Kepert CJ, Batten SR (2009) A nanoscale molecular switch triggered by thermal, light, and guest perturbation. Angew Chem Int Ed 48:2549–2552
181. Duriska MB, Neville SM, Li J, Iremonger SS, Boas JF, Kepert CJ, Batten SR (2009) Systematic metal variation and solvent and hydrogen-gas storage in supramolecular nanoballs. Angew Chem Int Ed 48:8919–8922
182. Duriska MB, Neville SM, Moubaraki B, Murray KS, Balde C, Létard J-F, Kepert CJ, Batten SR (2012) A family of discrete magnetically switchable nanoballs. Chem Plus Chem 77:616–623
183. Carruthers C, Fisher J, Harding LP, Hardie MJ (2010) Host–guest influence on metallo-supramolecular assemblies with acyclotriveratrylene-type ligand. Dalton Trans 39:355–357
184. Uhl W, Vob M, Layh M, Rogel F (2010) Gallium-gallium bonds as efficient templates for the generation of a large cage containing twelve gallium atoms. Dalton Trans 39:3160–3162
185. Abrahams BF, FitzGerald NJ, Robson R (2010) Cages with tetrahedron-like topology formed from the combination of cyclotricatechylene ligands with metal cations. Angew Chem Int Ed 49:2896–2899
186. Abrahams BF, Boughton BA, FitzGerald NJ, Holmes JL, Robson R (2011) A highly symmetric diamond-like assembly of cyclotricatechylene-based tetrahedral cages. Chem Commun 47:7404–7406
187. Clever GH, Tashiro S, Shionoya M (2009) inclusion of anionic guests inside a molecular cage with palladium(II) centers as electrostatic anchors. Angew Chem Int Ed 48:7010–7012
188. Clever GH, Shionoya M (2010) A pH switchable pseudorotaxane based on a metal cage and a bis-anionic thread. Chem Eur J 16(39):11792–11796
189. Clever GH, Kawamura W, Tashiro S, Shiro M, Shionoya M (2012) Stacked platinum complexes of the magnus' salt type inside a coordination cage. Angew Chem Int Ed 51(11):2606–2609
190. Han M, Hey J, Kawamura W, Stalke D, Shionoya M, Clever GH (2012) An inclusion complex of hexamolybdate inside a supramolecular cage and its structural conversion. Inorg Chem 51(18):9574–9576
191. Freye S, Hey J, Torras-Galán A, Stalke D, Herbst-Irmer R, John M, Clever GH (2012) Allosteric binding of halide anions by a new dimeric interpenetrated coordination cage. Angew Chem Int Ed 51: 2191–2194
192. Frank M, Hey J, Balcioglu I, Chen Y-S, Stalke D, Suenobu T, Fukuzumi S, Frauendorf H, Clever GH (2013) Assembly and stepwise oxidation of interpenetrated coordination cages based on phenothiazine. Angew Chem Int Ed 52(38):10102–10106
193. Frank M, Dieterich JM, Freye S, Mata RA, Clever GH (2013) Relative anion binding affinity in a series of interpenetrated coordination cages. Dalton Trans 42(45):15906–15910

194. Frank M, Krause L, Herbst-Irmer R, Stalke D, Clever GH (2014) Narcissistic self-sorting vs. statistic ligand shuffling within a series of phenothiazine-based coordination cages. Dalton Trans 43(11): 4587–4592
195. Dieterich JM, Clever GH, Mata RA (2012) A push-and-pull model for allosteric anion binding in cage complexes. Phys Chem Chem Phys 14(37): 12746–12749
196. Engelhard DM, Freye S, Grohe K, John M, Clever GH (2012) NMR-Based structure determination of an intertwined coordination cage resembling a double trefoil knot. Angew Chem Int Ed 51(19):4747–4750
197. Han M, Michel R, He B, Chen Y-S, Stalke D, John M, Clever GH (2013) Light-triggered guest uptake and release by a photochromic coordination cage. Angew Chem Int Ed 52:1319–1323
198. Clever GH, Kawamura W, Shionoya M (2011) Encapsulation versus aggregation of metal–organic cages controlled by guest size variation. Inorg Chem 50:4689–4691
199. Amouri H, Desmarets C, Bettoschi A, Rager MN, Boubekeur K, Rabu P, Drillon M (2007) Supramolecular cobalt cages and coordination polymers templated by anion guests: self-assembly, structures, and magnetic properties. Chem Eur J 13:5401–5407
200. Lewis JEM, Gavey EL, Cameron SA, Crowley JD (2012) Stimuli-responsive Pd_2L_4 metallosupramolecular cages: towards targeted cisplatin drug delivery. Chem Sci 3:778–784
201. Kishi N, Li Z, Yoza K, Akita M, Yoshizawa M (2011) An M_2L_4 molecular capsule with an anthracene shell: encapsulation of large guests up to 1 nm. J Am Chem Soc 133(30):11438–11441
202. Kishi N, Li Z, Sei Y, Akita M, Yoza K, Siegel JS, Yoshizawa M (2013) Wide-ranging host capability of a PdII-linked M_2L_4 molecular capsule with an anthracene shell. Chem Eur J 19(20):6313–6320
203. Li Z, Kishi N, Hasegawa K, Akita M, Yoshizawa M (2011) Highly fluorescent M_2L_4 molecular capsules with anthracene shells. Chem Commun 47(30):8605–8607
204. Li Z, Kishi N, Yoza K, Akita M, Yoshizawa M (2012) Isostructural M2L4 molecular capsules with anthracene shells: synthesis, crystal structures, and fluorescent properties. Chem Eur J 18(27):8358–8365
205. Hiraoka S, Harano K, Shiro M, Ozawa Y, Yasuda N, Toriumi K, Shionoya M (2006) Isostructural coordination capsules for a series of 10 different d^5–d^{10} transition-metal ions. Angew Chem Int Ed 45: 6488–6491
206. Hiraoka S, Kiyokawa M, Hashida S, Shionoya M (2010) Site-selective internal cross-linking between mercury(II)-centered vertices of an octahedral mercury(II) capsule by a rod-shaped ditopic ligand. Angew Chem Int Ed 49(1):138–143
207. Hiraoka S, Yamauchi Y, Arakane R, Shionoya M (2009) Template-directed synthesis of a covalent organic capsule based on a 3 nm-sized metallocapsule. J Am Chem Soc 131(33):11646–11647
208. Harano K, Hiraoka S, Shionoya M (2007) 3 nm-scale molecular switching between fluorescent coordination capsule and nonfluorescent cage. J Am Chem Soc 129(17):5300–5301
209. Nakamura T, Ube H, Shiro M, Shionoya M (2013) A self-assembled multiporphyrin cage complex through three different zinc(II) center formation under well-balanced aqueous conditions. Angew Chem Int Ed 52(2):720–723
210. Nakamura T, Ube H, Miyake R, Shionoya M (2013) A C60-templated tetrameric porphyrin barrel complex via zinc-mediated self-assembly utilizing labile capping ligands. J Am Chem Soc 135(50): 18790–18793
211. Sun X, Johnson DW, Caulder DL, Powers RE, Raymond KN, Wong EH (1999) Exploiting incommensurate symmetry numbers: rational design and assembly of $M_2M_3L_6$ supramol. clusters with a C_{3h} symmetry. Angew Chem Int Ed 38(9):1303–1307
212. Sun X, Johnson DW, Caulder DL, Raymond KN, Wong EH (2001) Rational design and assembly of $M_2M'_3L_6$ supramolecular clusters with C3h symmetry by exploiting incommensurate symmetry numbers. J Am Chem Soc 123(12):2752–2763
213. Hiraoka S, Sakata Y, Shionoya M (2008) Ti(IV)-centered dynamic interconversion between Pd(II), Ti(IV)-containing ring and cage molecules. J Am Chem Soc 130(31):10058–10059
214. Beer PD, Berry NG, Cowley AR, Hayes EJ, Oates EC, Wong WWH (2003) Metal-directed self-assembly of bimetallic dithiocarbamate transition metal cryptands and their binding capabilities. Chem Commun 2408–2409

Reactivity of Encapsulated Species 5

Unusual chemical and photochemical reactivity (or, in contrast, incredible stability) of encapsulated species and the unexpected products of such reactions within the confined cavities of molecular capsules have been observed in many cases. This can be explained by steric restrictions due to relative rigidity of covalent and coordination capsules, by isolation of these guests from external factors such as solvent effects, and by relatively strong supramolecular host–guest interactions.

5.1 Chemical Reactions

A cyclic trimer *409* (Scheme 5.1) has been obtained in [1] by condensation of trialkoxysilane guest molecules encapsulated within the nanosized cavity of a self-assembled coordination capsule **488**. If the free trimer *409* is a kinetic and short-lived product rapidly converting to the corresponding thermodynamically favored cyclic tetramer and other condensed products, its caged molecule is remarkably stable and remains intact over a long time at room temperature in both neutral aqueous solutions and acidic media (pH <1) [1].

Selective oligomerization of alkoxysilanes as an example of cavity-directed synthesis (i.e., a reaction that is controlled by encapsulation of reactive species within the caging ligands so that their optimal guests are formed as products) has been performed in [2] using a series of the coordination capsules **483**, **548**, and **549** with the cage frameworks of the same M_6L_4 stoichiometry but with different size and shape of their cavities. A TP caging ligand **548** has the smallest cavity and can accommodate only one rodlike guest molecule; its cavity promotes only hydrolysis of trialkoxysilanes into silanetriol products without their subsequent polycondensation (an example for 2-naphthyltrimethoxysilane giving *410* is shown in Scheme 5.2). In contrast, selective preparation of a silanol dimer *411* from the condensation reaction of 2-naphthyltrimethoxysilane or *meta*-biphenyltrimethoxysilane has been performed within the tetrahedral capsule **549**. The latter has a cavity large enough to accommodate two molecules of the corresponding silane precursor or one molecule of its dimer. Within the spherical cavity of the octahedral coordination capsule **483**, the same guest, 2-naphtyltrimetoxysilane, underwent trimerization into a caged cyclic trimer *412* with all-*cis*-conformation. Thus, the formation of monomeric, dimeric, and trimeric products strongly depends on the size and shape of the cavities of these caging coordination capsules [2].

Encapsulated molecules of trimethoxy(*meta*-tolyl)silane are reported in [3] to easily hydrolyze within the cavity of coordination capsule **488**, and resulting trihydroxyl-containing guests undergo trimerization leading to a 1:1 host–guest complex of this ligand with caged trioxane trimer *413* (Scheme 5.3). C_{3v} symmetry of this cage complex in solution and in solid state has been confirmed in [3] using ^{1}H NMR and X-ray

Y. Voloshin et al., *The Encapsulation Phenomenon: Synthesis, Reactivity and Applications of Caged Ions and Molecules*, DOI 10.1007/978-3-319-27738-7_5

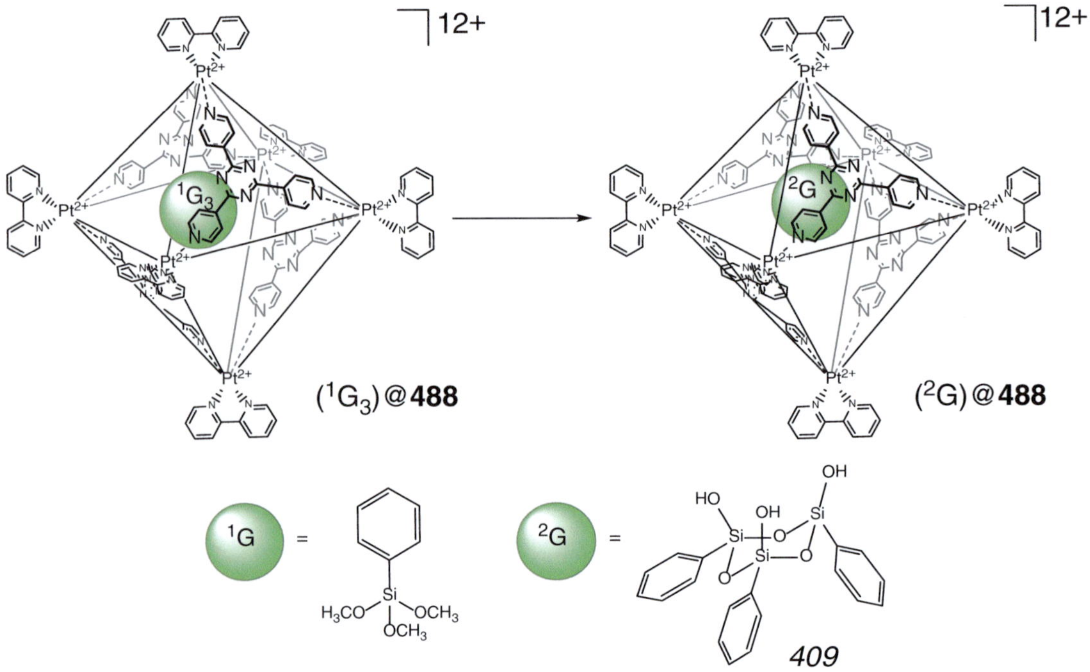

Scheme 5.1

diffraction methods, respectively. The encapsulated molecule of *413* has all-*cis*-configuration of its substituents relative to the almost planar six-membered siloxane ring.

Sol-gel condensation of encapsulated tetramethoxysilane guest molecules within the spherical cavities of coordination capsules **529** and **530** has been used in [4] for preparation of monodisperse silica nanoparticles by Scheme 5.4.

The similar spherical caging ligand **525** with cationic interior has been used in [5] for controlled polymerization of anionic monomers within its cavity. Free radical polymerization of sodium *para*-styrenesulfonate (Scheme 5.5), initiated by the redox system $(NH_4)_2S_2O_8-NaHSO_3$, gave an insoluble cage complex with encapsulated polymeric species. The cationic capsule **525** accelerates the initial stage of the polymerization reaction; however, the final degree of conversion in this case (30 %) is substantially lower than that in the absence of this caging ligand (45 %). Increase in the concentration of the initiator caused an increase in the degree of conversion up to 62 % in the absence of **525**, whereas it did not affect the polymerization process in the presence of such molecular flask. This result is explained in [5] by tight incorporation of a growing polymeric chain within the ligand's cavity that resulted in fast polymerization of only monomeric entities trapped in (around) its polycationic interior through electrostatic (Coulombic) interactions. As follows from GFC data, the number average molecular weight (M_n) and polydispersity (M_w/M_n; where M_w is a weight average molecular weight) in the presence of **525** are equal to 2860 and 1.61; in its absence they are 5800 and 3.53, respectively. Therefore, this capsule allows tuning the number of monomeric species of sodium *para*-styrenesulfonate that undergo polymerization and controlling the molecular weight of the polymeric product [5]. Similar results are reported for the caging ligand **526**: the corresponding M_n and M_w/M_n values are equal to 2100 and 1.41, respectively. There is a clear correlation between the molecular weight of the polymeric product and the number of interior cationic residues: cationic centers govern the number of anionic monomeric species trapped in (around) this capsule. A plausible scheme of the polymerization process includes the electrostatic

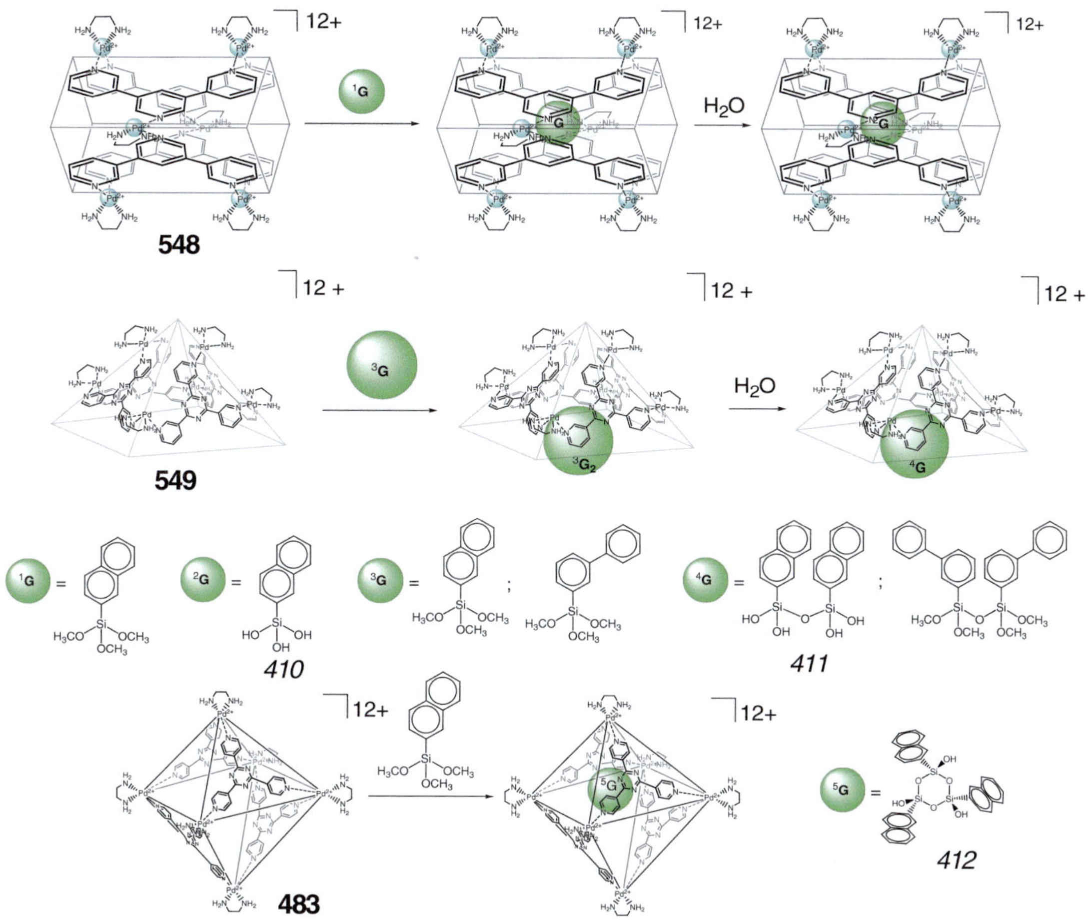

Scheme 5.2

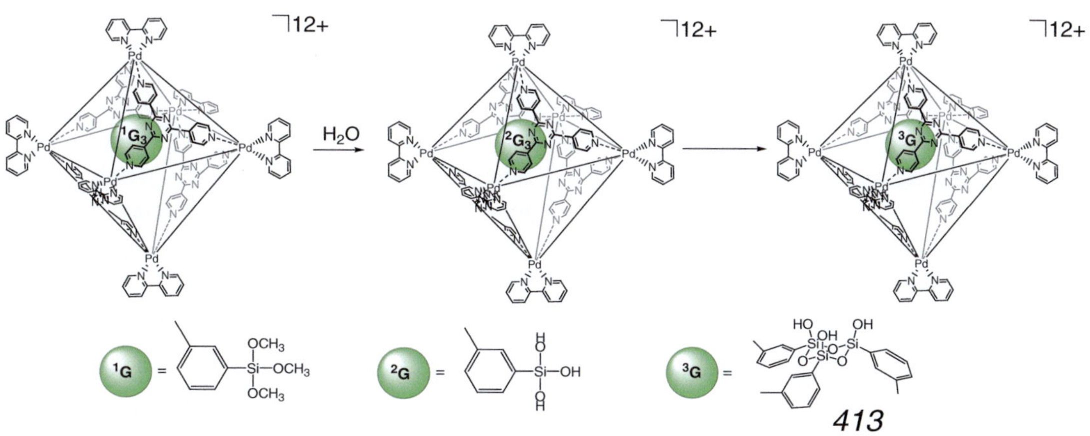

Scheme 5.3

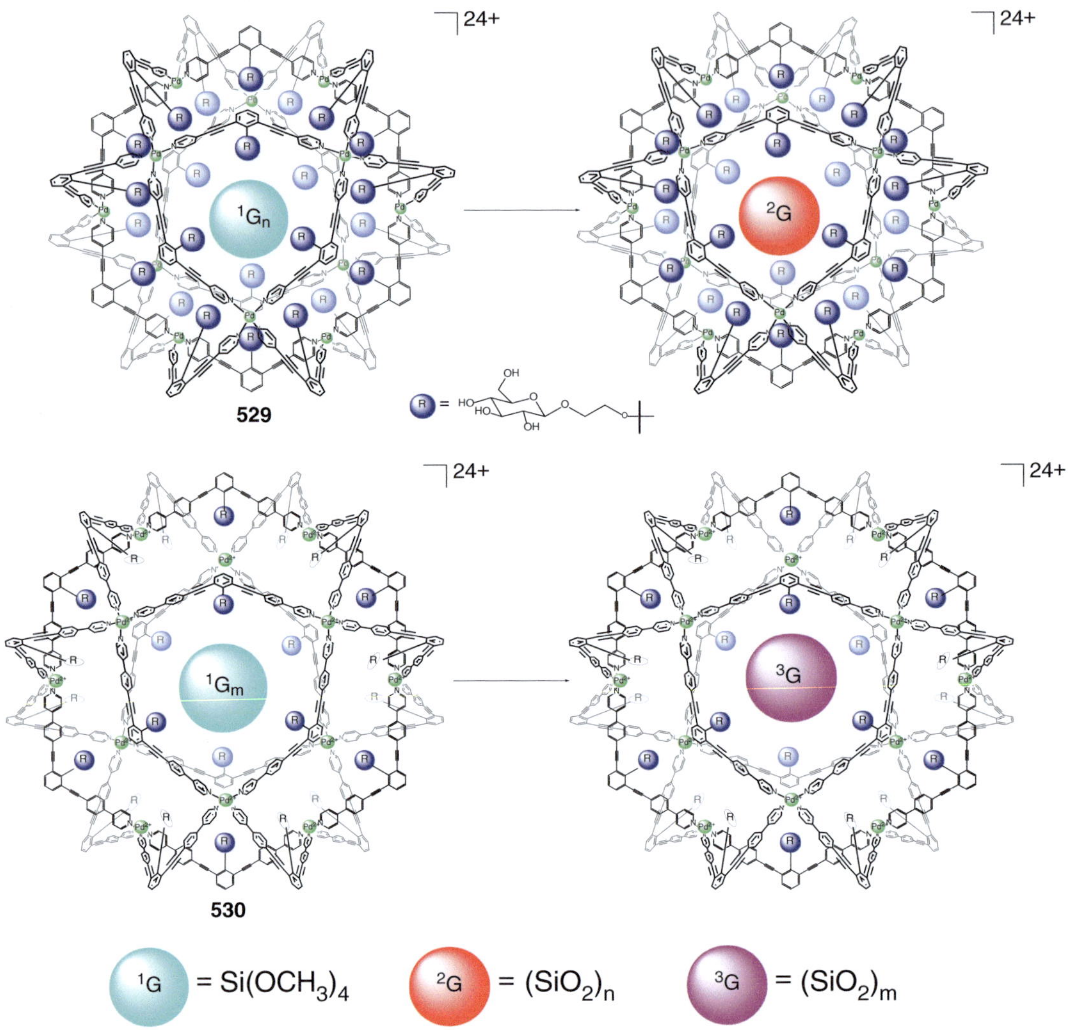

Scheme 5.4

accumulation of anionic monomers within the cavity of a coordination capsule followed by their polymerization. The growing polymer chain having multivalent negative charges is tightly incorporated in its polycationic interior. This resulted both in the increase in the polymerization rate and in the formation of the insoluble counterionic cage complex that prevents further polymerization and limits the degree of conversion [5].

Suitable aromatic guests within the cavities of coordination capsules **483** and **487** undergo both [2+2] and [2+4] pericyclic additions by Scheme 5.6 depending on the reaction conditions [6]. A thermal initiation of these Diels–Alder reactions gave *endo*-[2+4] adducts with *syn*-stereochemistry at the projecting terminal benzene fragment of acenaphthylene, whereas irradiation of the same solution afforded [2+2] cross-adducts. A reactivity of these guest molecules is reported in [6] to be very sensitive to the nature of a caging ligand.

Two encapsulated molecules of the constitutional isomers *414* and *415* of bis(trimethoxysilyl) pyrene undergo dimerization within the cavity of coordination capsule **552** by Scheme 5.7 giving 1:1 cage complex of only one conformational isomer of the resulting dimer *416* having the parallel *syn*-conformation and that with

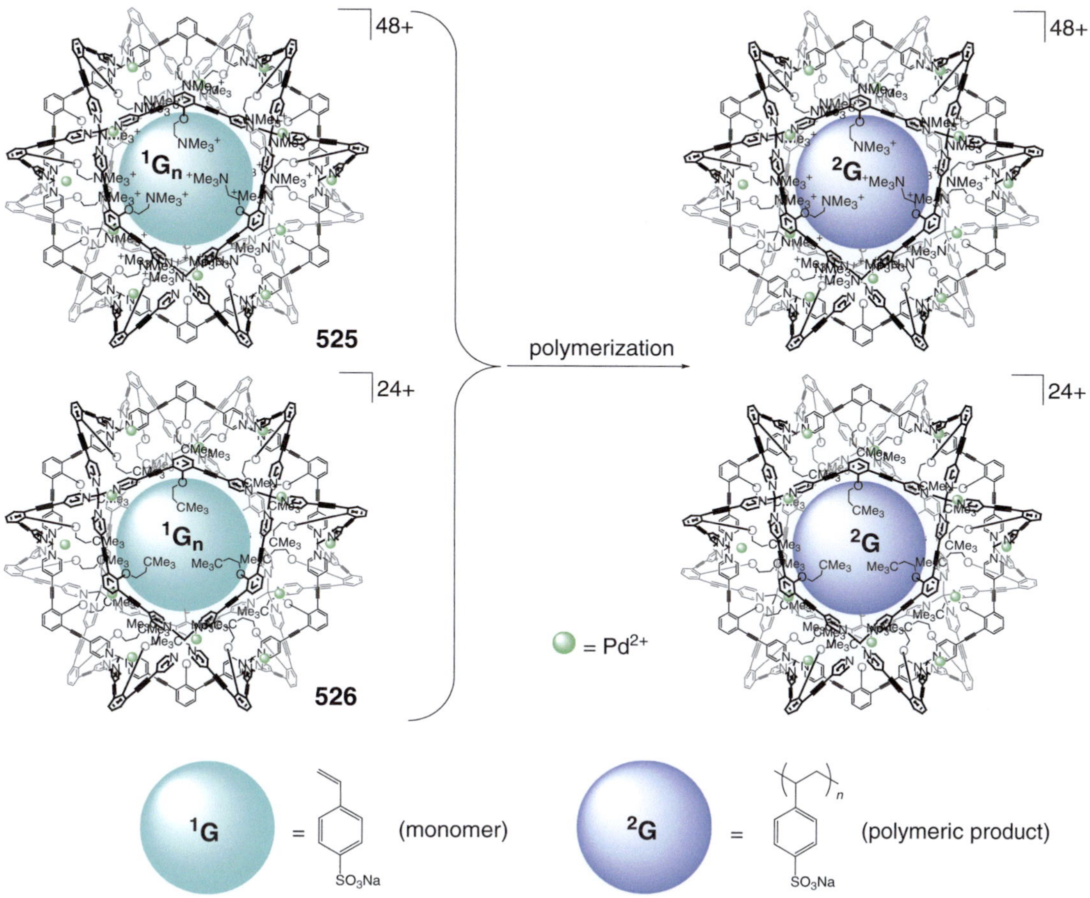

Scheme 5.5

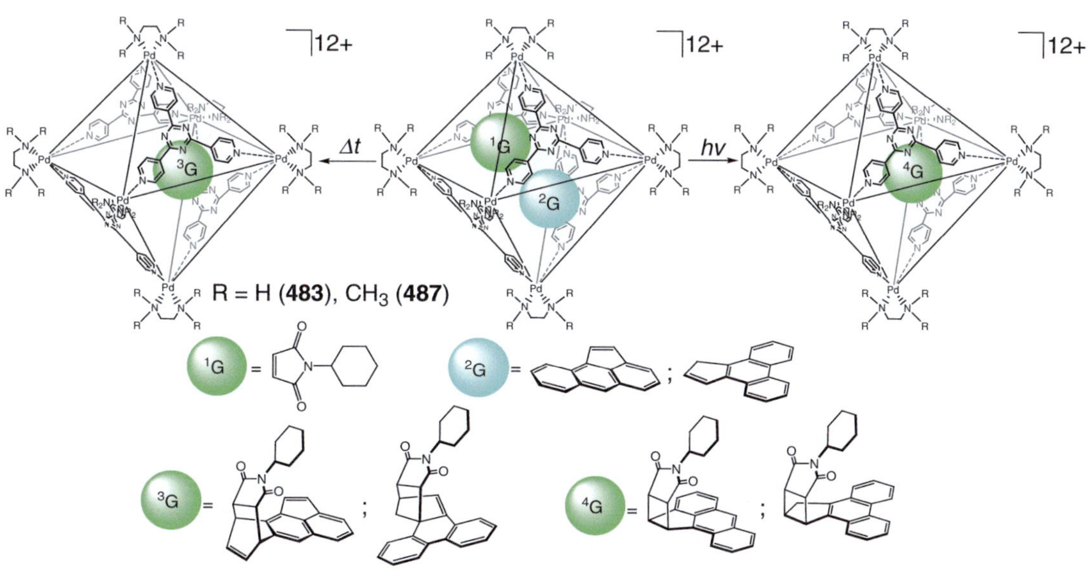

Scheme 5.6

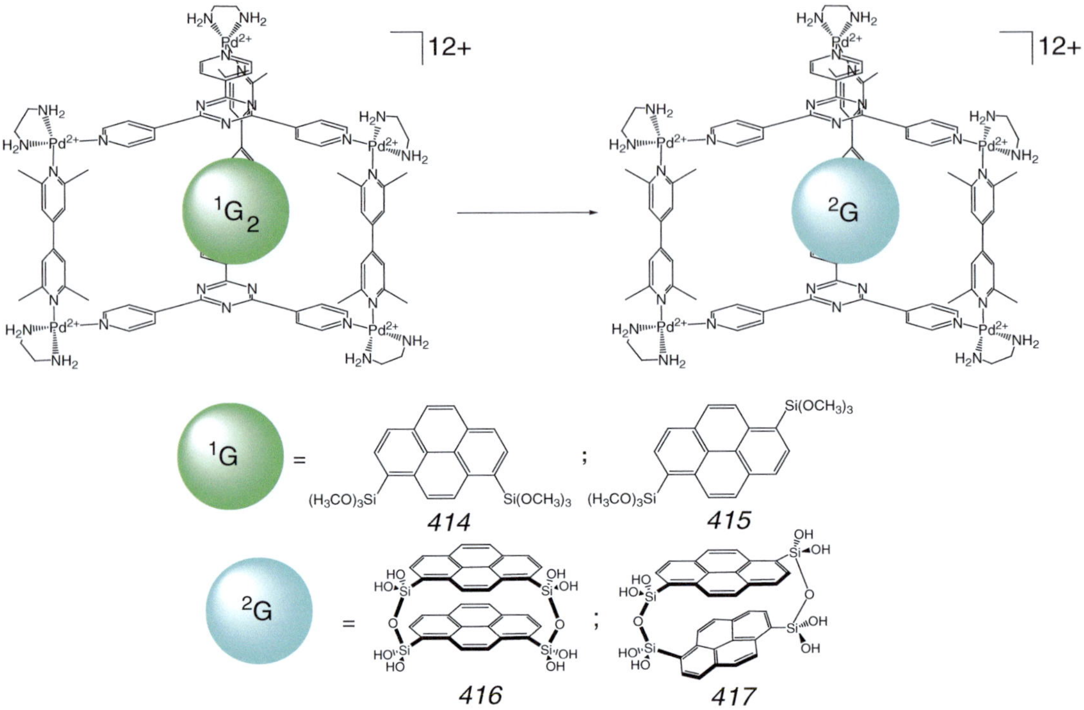

Scheme 5.7

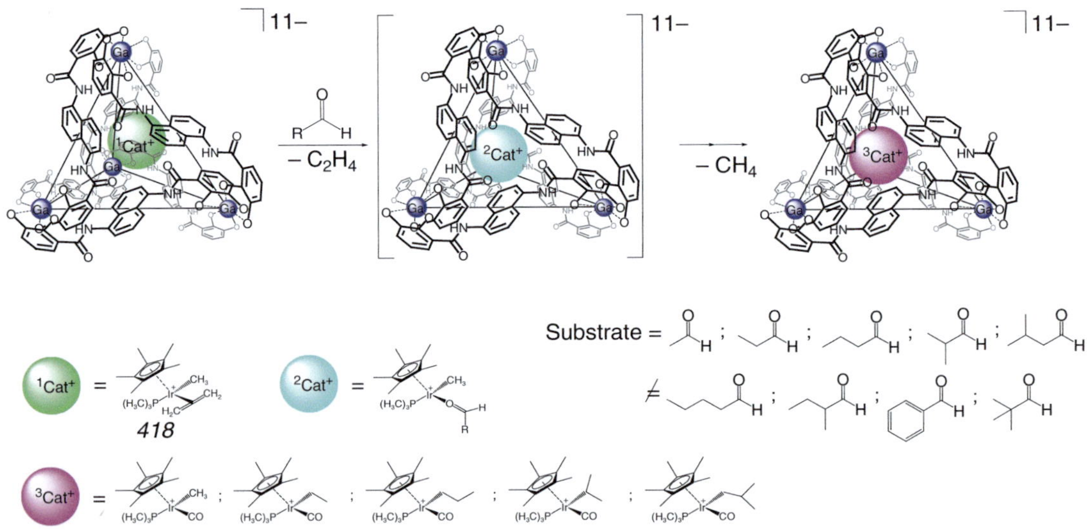

Scheme 5.8

syn-[3, 3] cyclophane molecule *417* as a guest, respectively [7].

Caged organoiridium cation *418* is reported [8] to selectively activate C–H bond in several aldehydes thus giving new encapsulated organometallic products in moderate yields by Scheme 5.8 and discriminating other aldehyde substrates. These C–H bond activation reactions within a Ga_4L_6 capsule **575** depend strongly on the size and shape of the guest during the diastereoselective product formation; the selectivity of activation of aldehyde molecules is explained [8] by occurrence of these reactions within its cavity rather than by extrusion of a metallocomplex guest from the coordination capsule, the further outside reaction with aldehyde and re-encapsulation by **575**. As a result, the C–H bond activation of aldehydes proceeds with highly specific size and shape discrimination and modest diastereoselectivity [8].

Scope and mechanism of this C–H bond activation reaction by an encapsulated iridium guest metallocomplex and its more reactive propene- and *cis*-butane-containing analogs *419* and *420* (Scheme 5.9) have been thoroughly studied in [9] using thermodynamic, mechanistic, and kinetic approaches as well as by ^{1}H NMR techniques. The use of *cis*-butene-containing guest not only allowed providing more efficient activation of aldehyde substrates but also extending it over other organic and inorganic compounds (Scheme 5.9), such as ethers, strained alkanes (in particular, cyclopropane), and molecular hydrogen. All these reactions are reported in [9] to exhibit significant kinetic diastereoselectivity within the capsule; they are largely entropy driven, and the caged iridium complex leaves its cavity by a stepwise mechanism including the formation of a strongly bound ion-pair intermediate. Thus, the result of [9] demonstrated the ability of caging hosts to act as nanoscale molecular flasks for the functional encapsulation of reactive organometallic guests and their further reactions within the ligand's cavity.

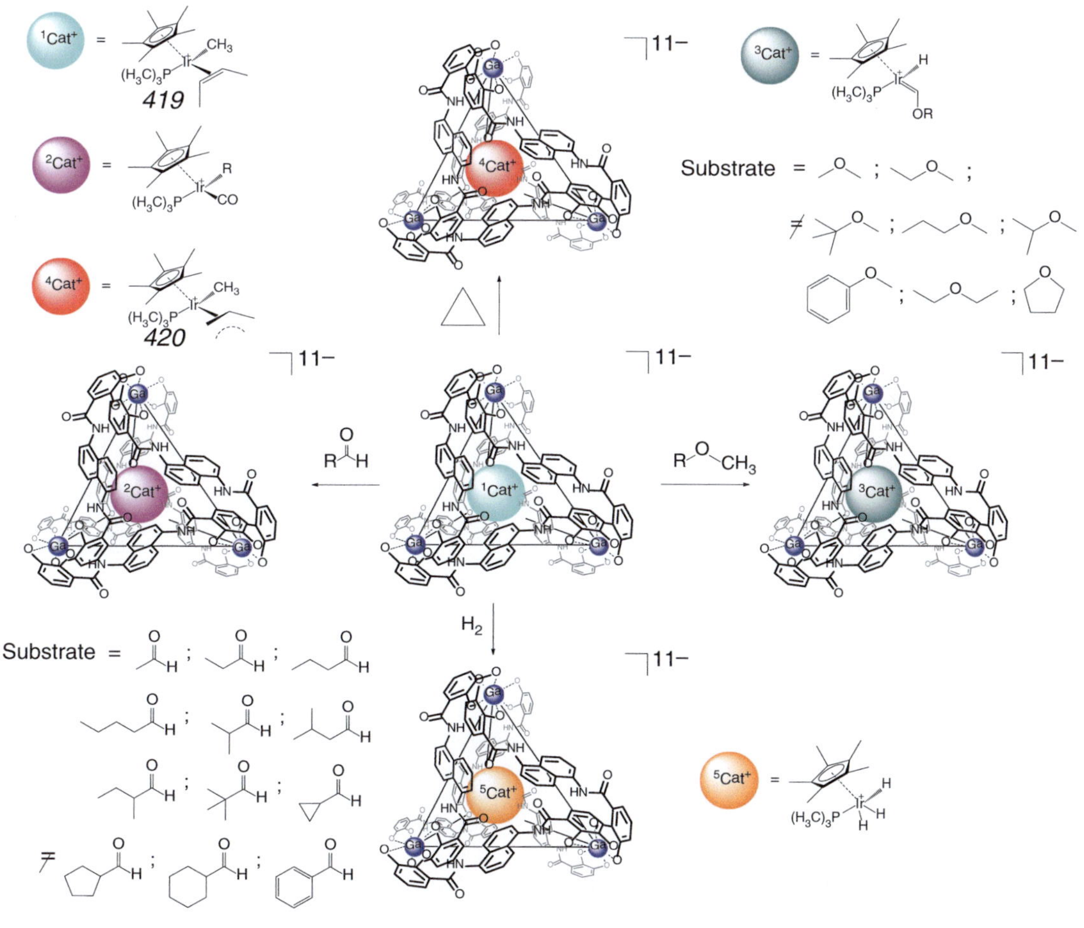

Scheme 5.9

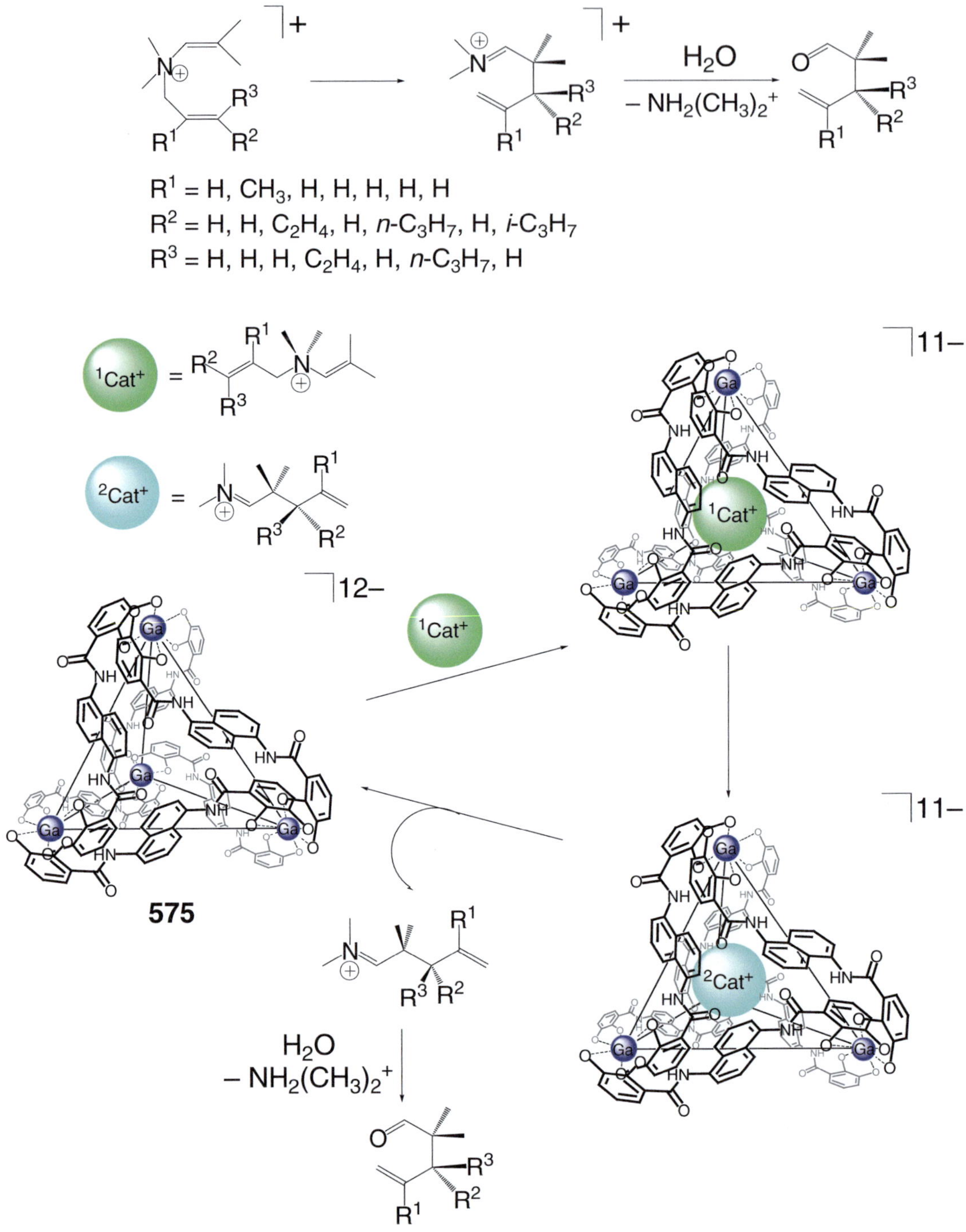

Scheme 5.10

The same caging ligand **575** has been used in [10] for unimolecular transformation (i.e., 3-aza-Cope rearrangement) of cationic guest species within its cavity by Scheme 5.10. In this transformation, the initial enammonium cation undergoes [3, 3]-sigmatropic rearrangement leading to iminium ion that then hydrolyzes to the corresponding aldehyde [10]. Encapsulation of

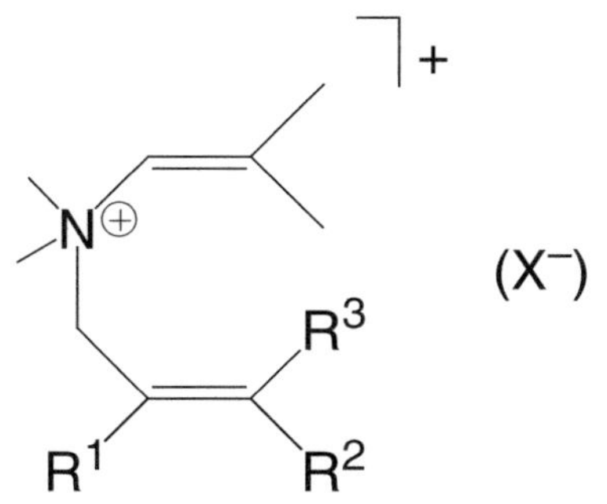

R^1 = H, CH_3, H, H, H, H, H, H, H, H
R^2 = H, H, C_2H_4, H, n-C_3H_7, H, i-C_3H_7, n-C_4H_9, TMA, CH_3
R^3 = H, H, H, C_2H_4, H, n-C_3H_7, H, H, H, CH_3
X^- = Br^-, Br^-, Br^-, OTs^-, OTs^-, OTs^-, OTs^-, OTs^-, OTs^-, Br^-

Scheme 5.11

substrates in their reactive conformations afforded acceleration in the rates of this rearrangement by up to three orders of magnitude thus suggesting the ability of tetrahedral M_4L_6 capsules to provide their size- and shape-defined cavities for the catalysis of different types of unimolecular organic reactions [10] as so-called "nanozymes" [11]. More detailed kinetic and quantitative NMR studies of this reaction with various substrates (Scheme 5.11) have been performed in [12]. Their results suggest that this cavity-catalyzed rearrangement is followed by the release of the iminium product from the cavity to the exterior ion-pairing binding site, and then an ionic associate undergoes hydrolysis [12].

The same rearrangement has been performed in [11] for various propargyl enammonium cations shown in Scheme 5.12. In this case, the initial propargyl derivative undergoes [3, 3]-sigmatropic rearrangement giving an allenyl iminium cation; its hydrolysis results in an allenyl aldehyde. The rates of this reaction for the encapsulated propargylamine substrates increase by a factor of 200 due to a decrease in the entropy of its activation as compared with the background (i.e., encapsulating catalyst-free) reaction, typically requiring a Lewis or Brønsted acid; the plausible catalytic cycle of this rearrangement of propargyl derivatives is also shown in this Scheme. The catalytic reaction within **575** as a molecular flask is enhanced due to entropy factors and obeys the Michaelis–Menten enzymatic kinetics [11]. This tetrahedral coordination capsule, acting as enzyme mimics, has been also used [13] as a chiral caging catalyst to catalyze 3-aza-Cope rearrangement of allyl enammonium cations by Scheme 5.13 through their preorganization into reactive conformations within its cavity.

Moreover, encapsulation by the dodecaanionic caging ligand **575** with a hydrophobic interior cavity, which thermodynamically stabilizes the protonated forms of guest molecules, is reported in [14] to catalyze the acidic hydrolysis of various orthoformates by Scheme 5.14 in basic solution at pH = 11 with rate acceleration of up to three orders of magnitude. The size-selective catalytic reaction obeys the Michaelis–Menten kinetics characteristic of enzymatic reactions and exhibits competitive inhibition [14]. The enzyme-like mechanism of this orthoformate hydrolysis reaction within the cavity of the caging ligand **575** has been studied in [15] using kinetic, thermodynamic, and mechanistic approaches (including ^{13}C-labeling experiments) and solvent isotope effects as well. Their results confirmed an

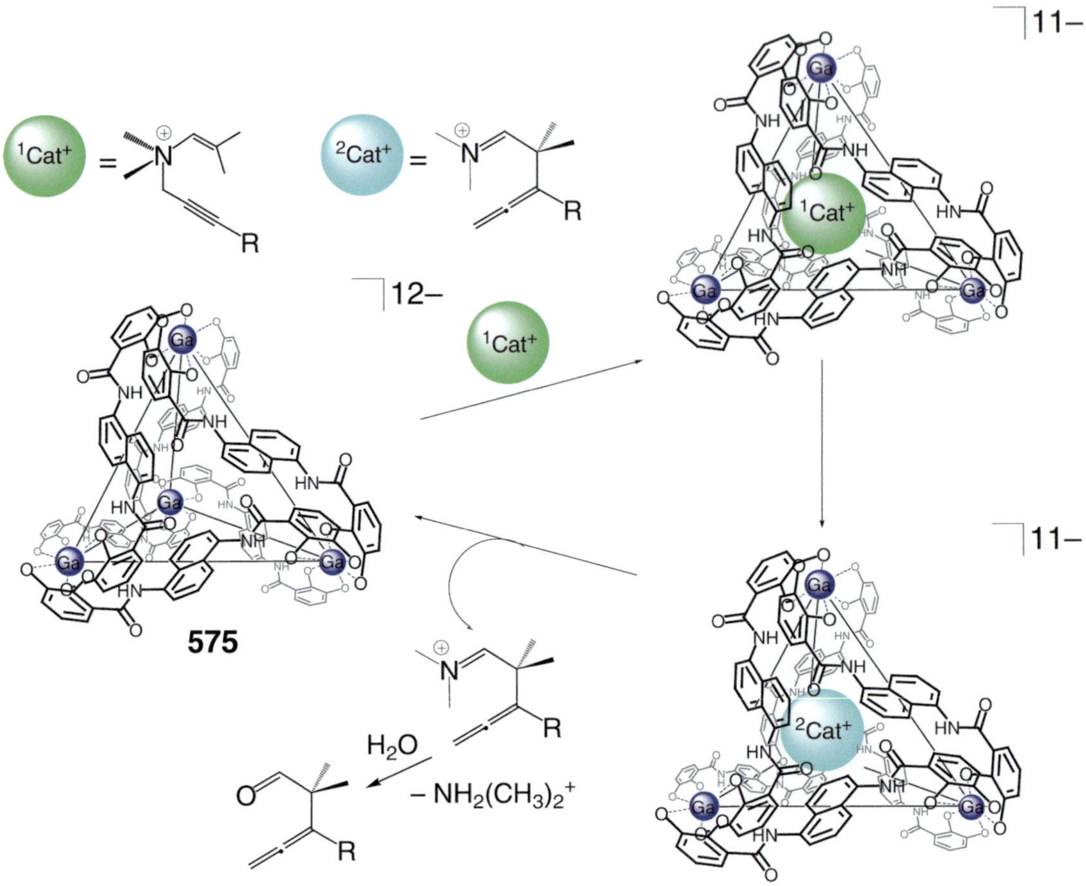

Scheme 5.12

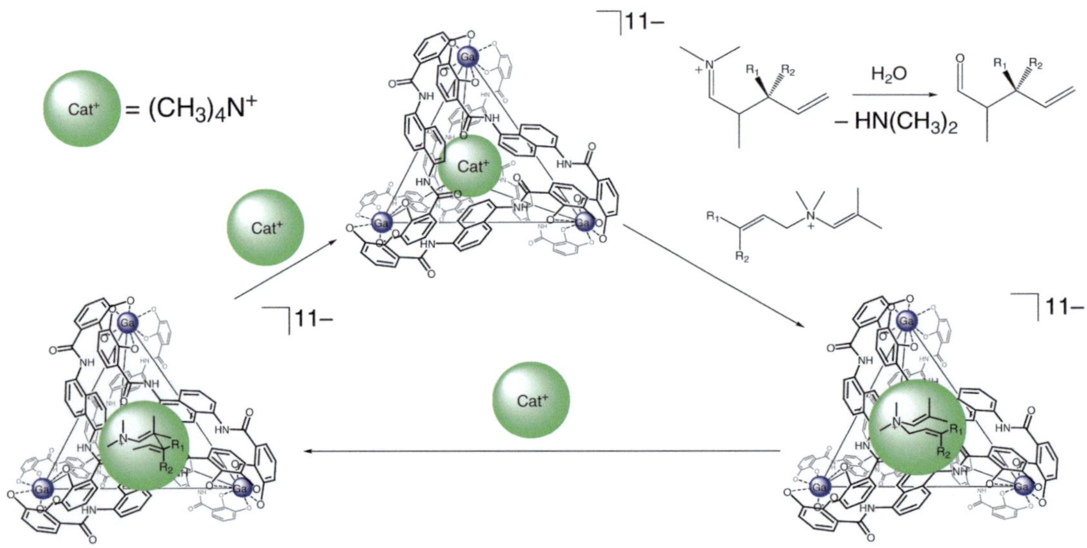

Scheme 5.13

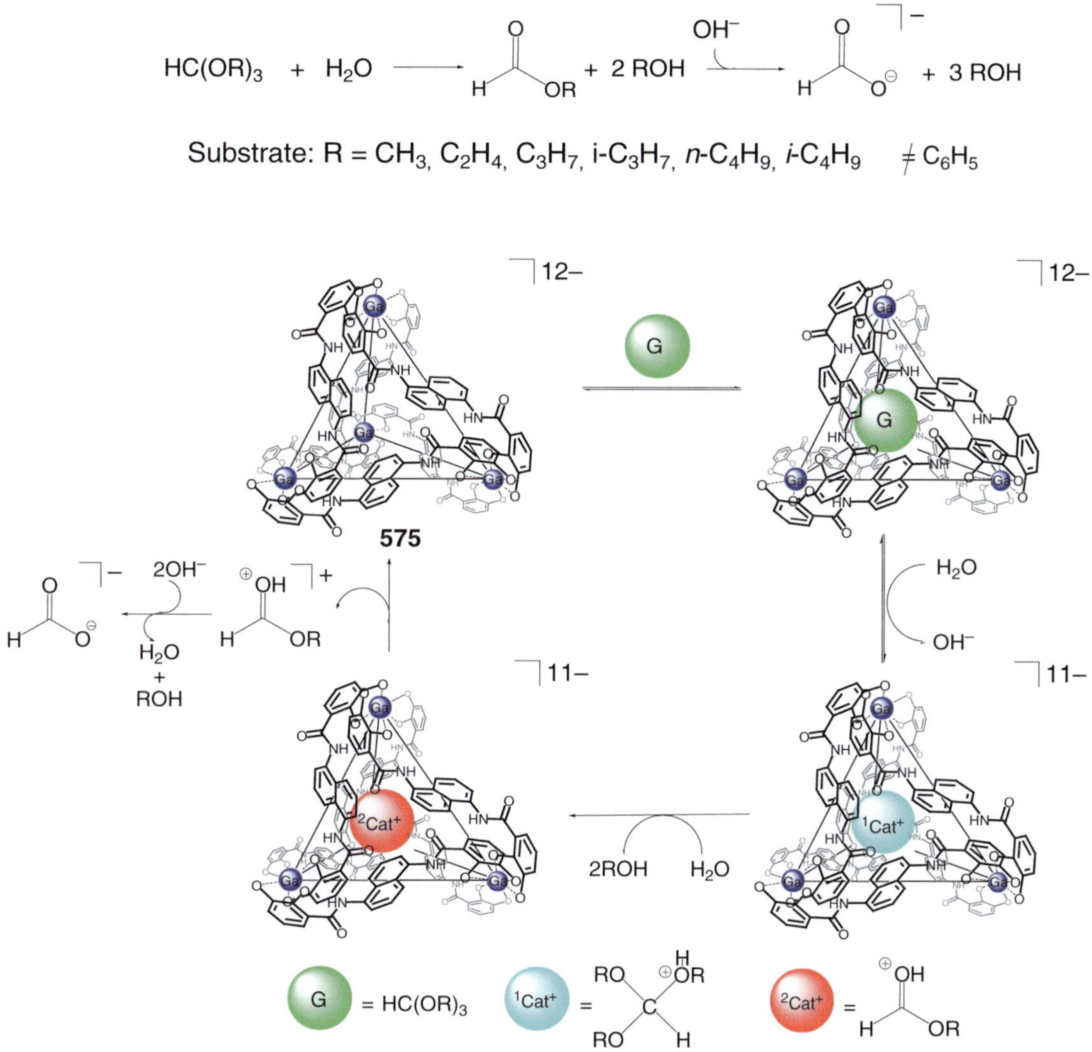

Scheme 5.14

$A-S_{E^2}$ mechanism (Scheme 5.15) within the capsule with a rate-limiting stage being the proton transfer in contrast to the *A–1* mechanism of the uncatalyzed reaction [15]. In the latter case, the decomposition of the protonated substrate is the rate-limiting stage.

This “nanozyme” capsule **575** in basic solution at pH 10 also catalyzes deprotection of various acetals and ketals shown in Scheme 5.16 [16].

The caged organorhodium cation *421*, generated in situ by Scheme 5.17, has been used [17] for catalyzing allylic alcohol isomerization of various allylic substrates. As follows from ^1H NMR data, the encapsulation of this organometallic catalyst by the capsule **575** allows control of the catalytic isomerization of allylic alcohols to give highly specific size and shape substrate selectivity; crotyl alcohols are nonreactive under the same reaction conditions.

Encaged enzyme-like catalysis of the Nazarov cyclization (i.e., the acid-catalyzed reaction of 1,4-dien-3-ol giving cyclopentadiene) by Scheme 5.18 within the tetrahedral coordination capsule **575** as a molecular flask is reported in [18]; Cp*H product is a competitive guest in this

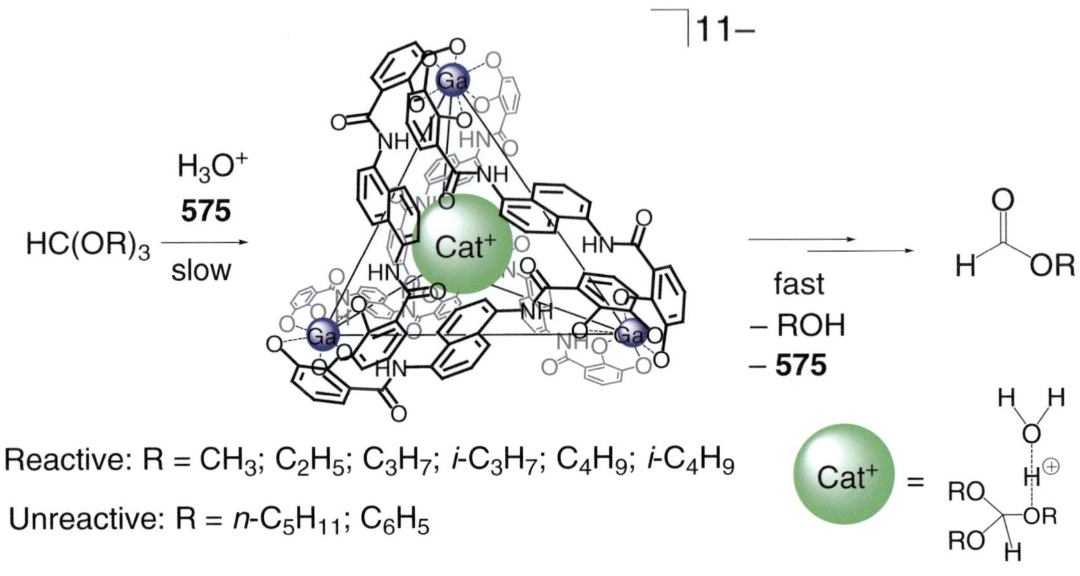

Scheme 5.15

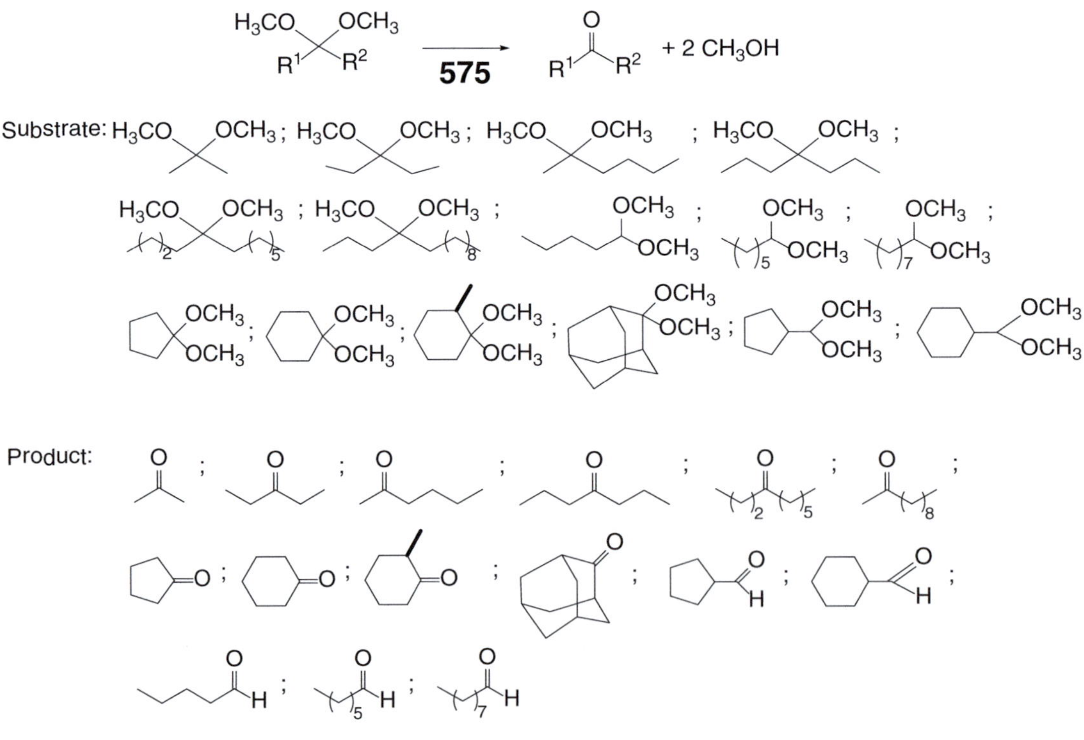

Scheme 5.16

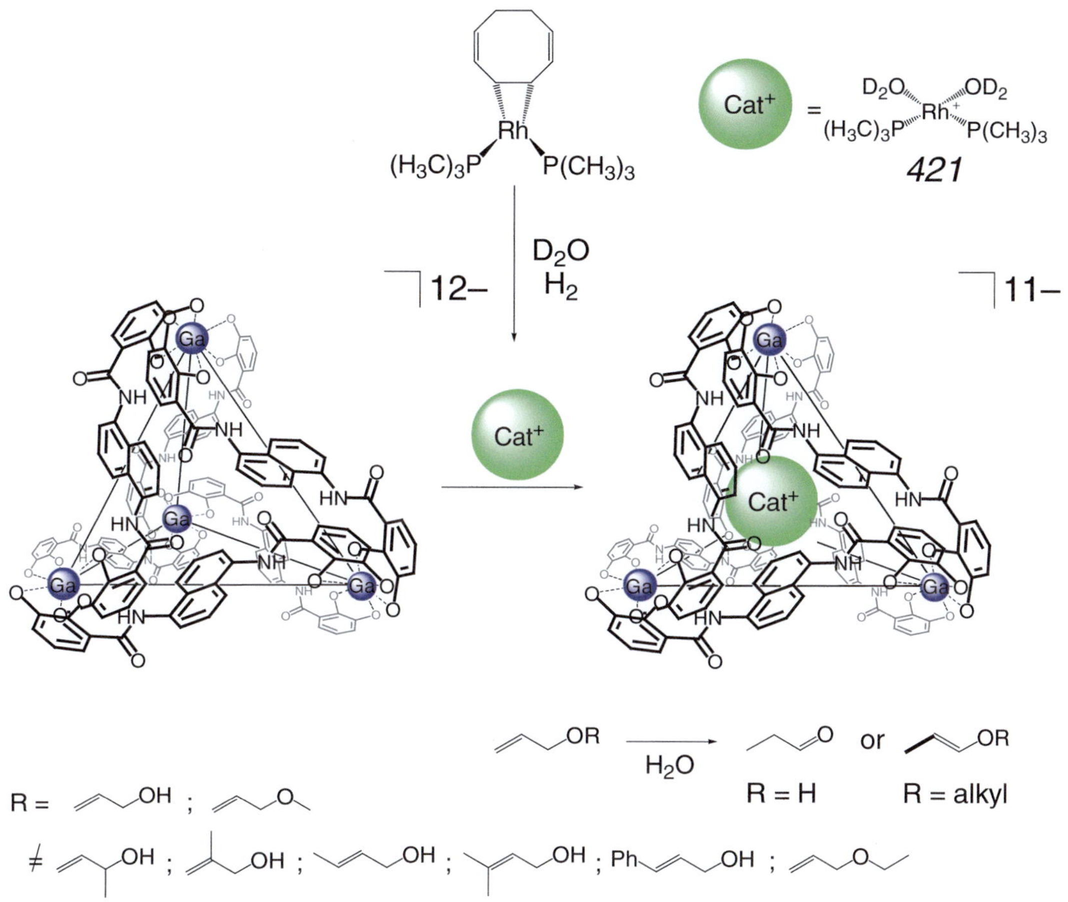

Scheme 5.17

case, and addition of maleimide to the reaction mixture has been used to convert it into the uncompeting Diels–Alder adduct. When this reaction proceeds in DMSO–water mixture at 45 °C in the presence of maleonitrile, the concentration-vs-time plots for cyclization reaction showed no deviation from pseudo-first-order kinetics. This allows observing the efficient catalysis of Nazarov cyclization of three possible stereoisomers of olefin 1,4-pentadien-3-ol. The rate of the catalytic reaction is two million times higher than that of the uncatalyzed cyclization, which is explained in [18] by preorganization of the guest substrate and stabilization of the transition state within the cavity of **575** as well as by an increase in basicity of the alcohol group of the caged molecule. An unexpected transformation has been observed in [19] when performing the capsule-catalyzed Nazarov cyclization by Scheme 5.19 in unbuffered D_2O at room temperature: while in the case of the unsymmetric isomer *424* the catalyzed condensation gave Cp*H as the expected product, in the case of symmetric substrates *422* and *423* under the same reaction conditions the process yielded its isomer – dihydrofulvene *432*. This result is explained in [19] by the stereochemistry of the corresponding encapsulated cyclopentenyl cationic intermediates *426* and *428*: as 4π electrocyclization of pentadienyl cations occurs in a conrotatory fashion, the alkene stereochemistry of the pentadienyl cations governs the stereochemistry of the resulting cyclopentenyl cation. Therefore, the electrocyclization of the pentadienyl cations *430* and *425* yields the cation *431* with methyl groups in a *trans*-conformation, while the *E,Z*-pentadienyl

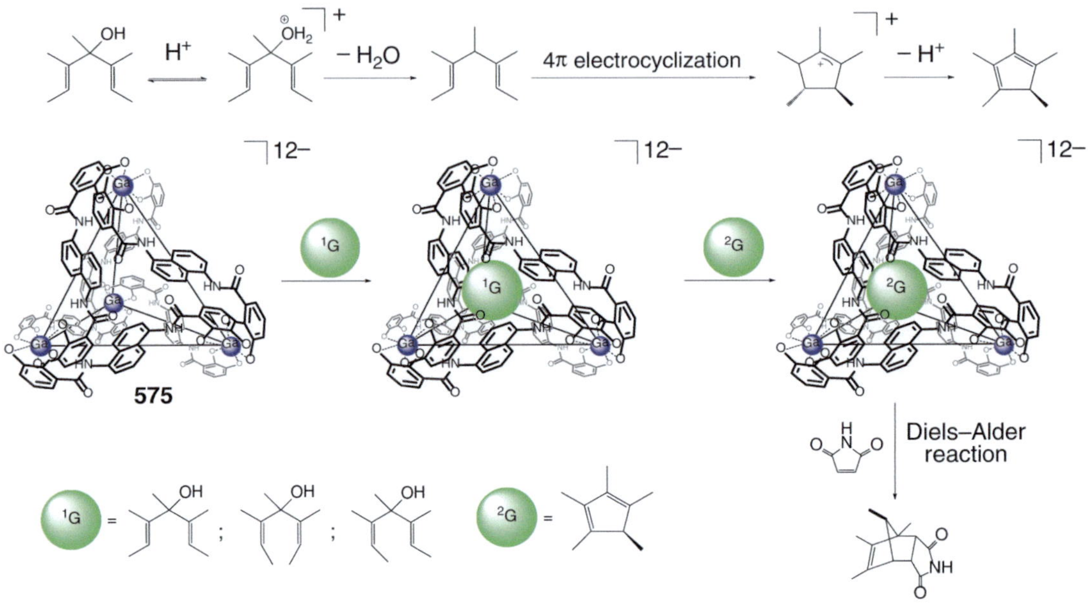

Scheme 5.18

cation *427* undergoes a transformation by Scheme 5.19 leading to the product *428* with methyl groups in a *cis*-orientation. This suggestion has been supported in [19] by the observation of *432* as the only product of the capsule-catalyzed dehydration of the alcohol *424* by Scheme 5.19, which proceeds through an intermediate *trans*-cyclopentenyl cation *431*. Thus, the regiochemistry of deprotonation in this encaged catalytic reaction is determined by the stereochemistry of the intermediate cyclopentenyl cation, the structure of which is governed by the alkene stereochemistry of the guest substrate. Changing the relative stereochemistry of two methyl groups in the caged carbocationic intermediate from *trans* to *cis* has been used in [19] to completely switch the regioselectivity of deprotonation.

Detailed kinetic analysis and ^{18}O-exchange experiments performed in [20] allowed evaluating the mechanism of the encaged Nazarov cyclization shown in Scheme 5.20. It includes reversible encapsulation, protonation, and water loss from substrate, followed by irreversible 4π electrocyclization. Although the electrocyclization is a rate-determining stage in the uncatalyzed reaction, the water loss and the electrocyclization processes have similar reaction rates within the capsule **575**. Analysis of the energetics of the catalyzed and uncatalyzed reactions performed in [19] showed that transition state stabilization through its encapsulation contributed significantly to the higher rate of the capsule-catalyzed cyclization.

The catalytic activity of an organogold(I) cationic guests *434* and *435* (Scheme 5.21), caged by the Ga_4L_6 coordination capsule **575**, in hydroalkylation reaction of allenes has been studied in [21]. These encapsulated cations in its cavity leave enough vacant space for coordination of an organic substrate. In particular, an allenyl alcohol *436* undergoes 50 % catalytic conversion in the target cyclic product with relatively high turnover number and substantially (eight times) higher rate than in the absence of the caged catalyst. In the case of a more bulky allenyl substrate *437*, the observed acceleration of the catalytic cyclization reaction is lower than that for its less bulky analog *436* [21].

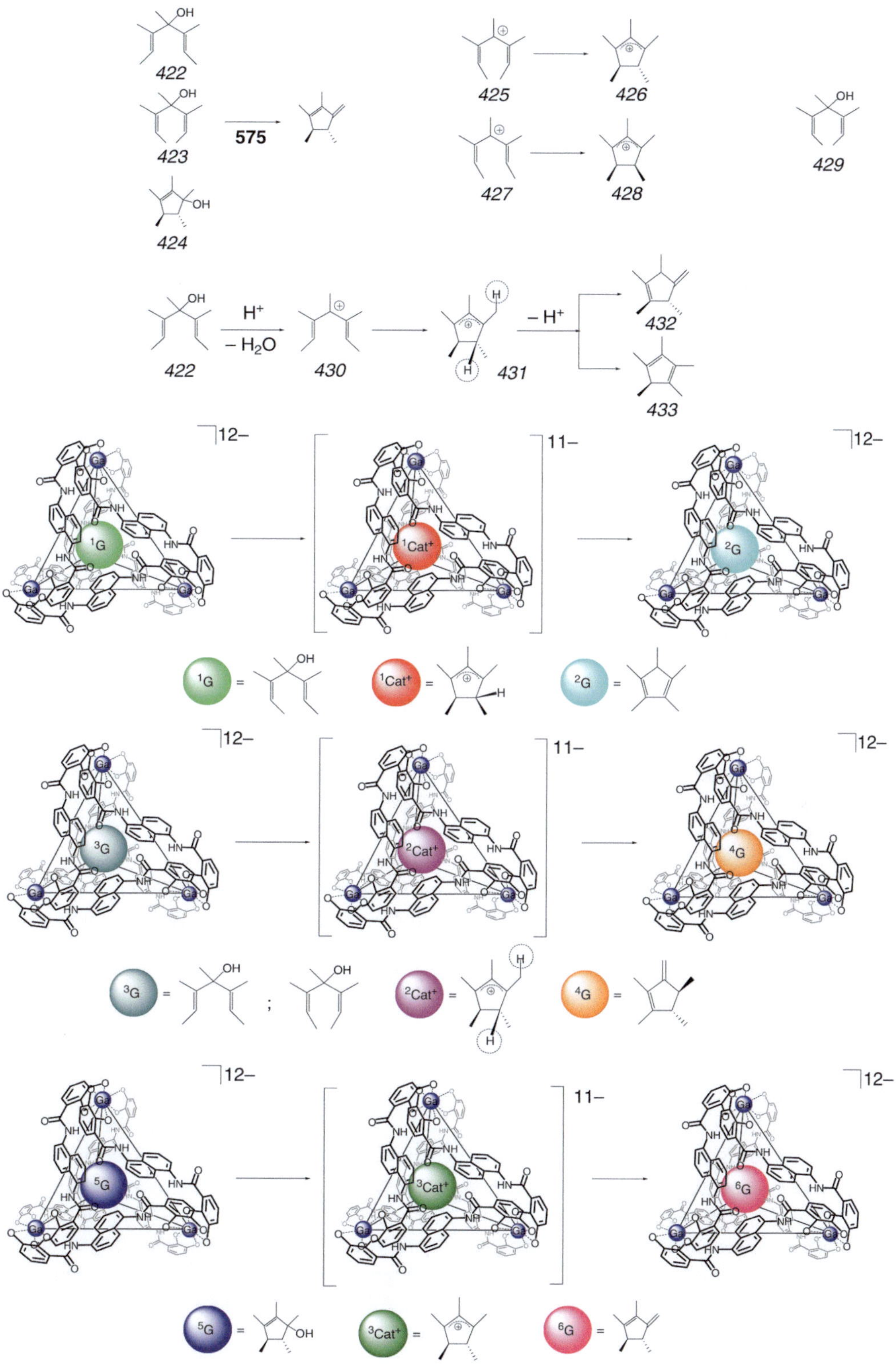

422
423
424
575
425
426
427
428
429
H+
– H2O
430
431
– H+
432
433
12–
11–
1G
1Cat+
2G
3G
2Cat+
4G
5G
3Cat+
6G

Scheme 5.19

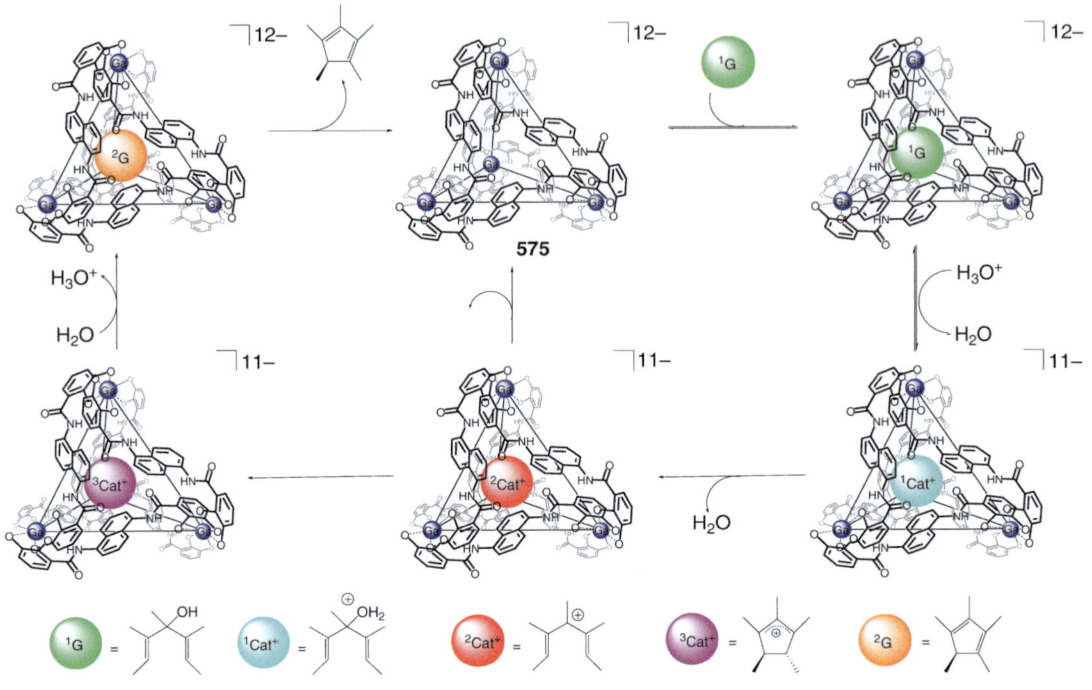

Scheme 5.20

An encapsulated monoterpene citronellal *438* and its two *gem*-dimethyl-containing and -lacking analogs *439* and *440* are described in [22] to undergo selective cyclization by Scheme 5.22 within the cavity of a caging ligand **575**. Such encapsulation prevents nucleophilic capture of reactive intermediates by water, directing the cyclization of monoterpene guest toward deprotonation.

The same effect has been observed in the case of an encapsulated organogold catalyst of cycloisomerization of enyne *441* by Scheme 5.23, and it has been attributed [22] to the hydrophobic environment within the coordination capsule **575**, preventing carbenium-ionic intermediates of the catalytic process from their side reactions with water. The same caging catalyst has been used in [23] to perform the combined enzymatic and transition metal catalysis of tandem reactions by Scheme 5.24. The authors of this work also developed a tandem olefin isomerization–reduction reaction by Scheme 5.25 with the encapsulated organoruthenium cation as a catalyst on its first step and ADH as a reductant on the second step.

The caged $(CH_3)_3PAu^+$ cation also catalyzes with high turnovers of the allyl alcohol isomerization in aqueous solutions by Scheme 5.26, performing more than 1000 catalytic cycles [24].

Oxygenation of furan into fumaraldehydic acid with an 80 % yield by Scheme 5.27 (Pathway I) in the presence of a catalytic amount of the coordination capsule **581** is reported in [25]. In the absence of this caging ligand, hydroxybutenolide has been formed in a 35 % yield by Pathway II along with other oxidation products by Pathway III. Upon addition of this catalyst to the reaction mixture, the target product, fumaraldehydic acid, has been generated. Therefore, functioning of this *one-pot* producing system is based on the catalytic activity of such robust capsule giving the target product in a high yield by Pathway I. This approach is scalable up to 50-fold allowing to avoid intermediate purification procedures and changes in reaction conditions [25].

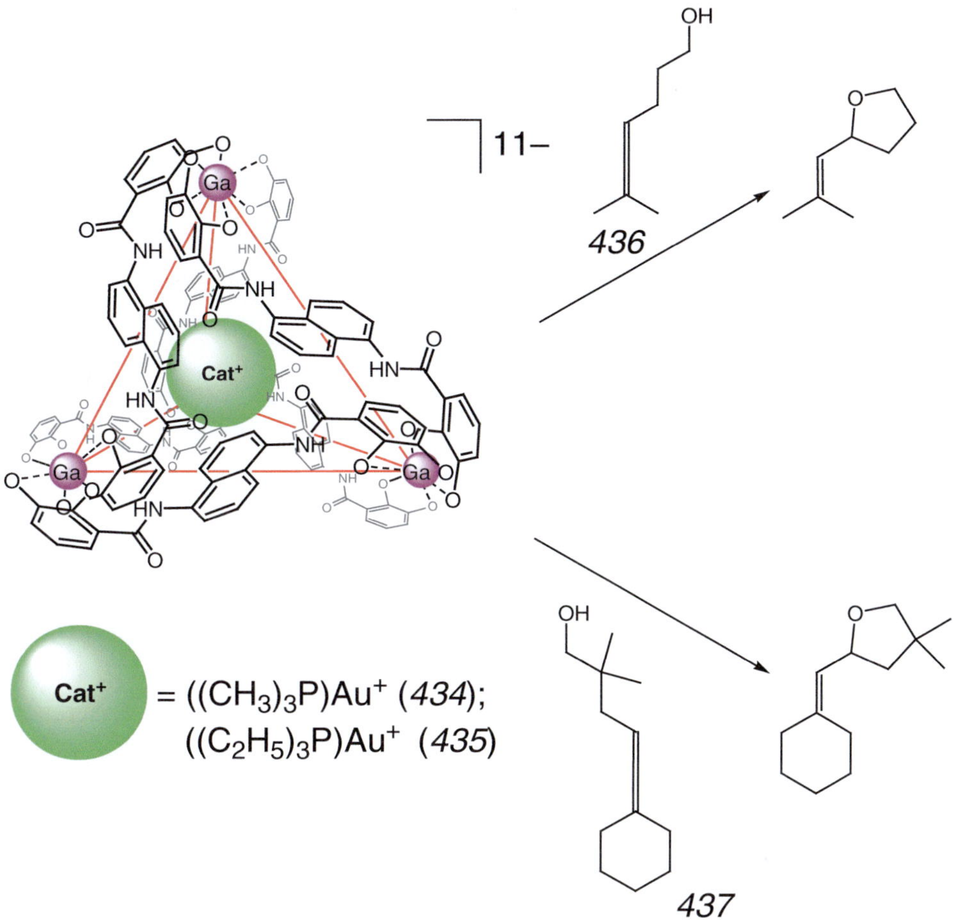

Scheme 5.21

Catalytic activity of oxidovanadium(V)-containing hemicryptophane capsules *P*-(*S*,*S*,*S*)-**207** and *M*-(*S*,*S*,*S*)-**207** and their tripodal analog in a model oxidation of organic sulfides to sulfoxides with cumyl and *tert*-butyl hydroperoxides as oxidants by Scheme 5.28 has been studied in [26]. These caging ligands showed a very high catalytic activity (the yields up to 95 %) together with a high (up to 100 %) sulfoxide–sulfone selectivity and a turnover number up to 180. The different diastereomeric forms of these encapsulating catalysts (i.e., those with Δ and Λ configurations of their vanatrane fragments) have the similar catalytic activity. In the case of their non-caging tripodal analog shown in this Scheme, the yields of the oxidized products are substantially lower, and more time is needed; the corresponding kinetic rate constants are up to six times lower than those for its caging analogs *P*-(*S*,*S*,*S*)-**207** and *M*-(*S*,*S*,*S*)-**207** [26].

Competitive study of the catalytic activity of a 1:1 cage complex of covalent capsule **209** with encapsulated copper(II) ion, its tripodal non-caging analog *441*, and hexahydrated Cu^{2+} ion in oxidation of cyclohexane, cyclooctane, and adamantane with hydrogen peroxide by Scheme 5.29 has been performed in [27]. This cage complex showed high catalytic activity and turnover numbers (from 110 to 130) at various reaction conditions, thus being a substantially more efficient

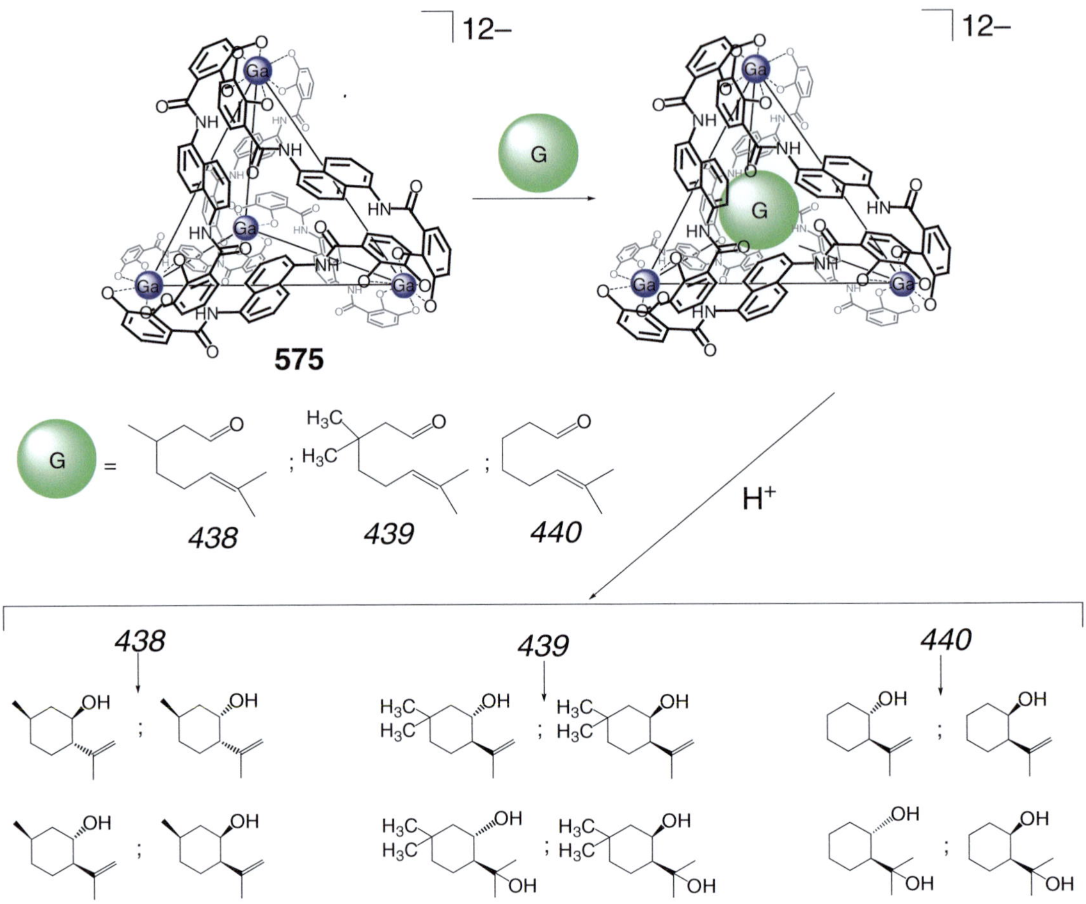

Scheme 5.22

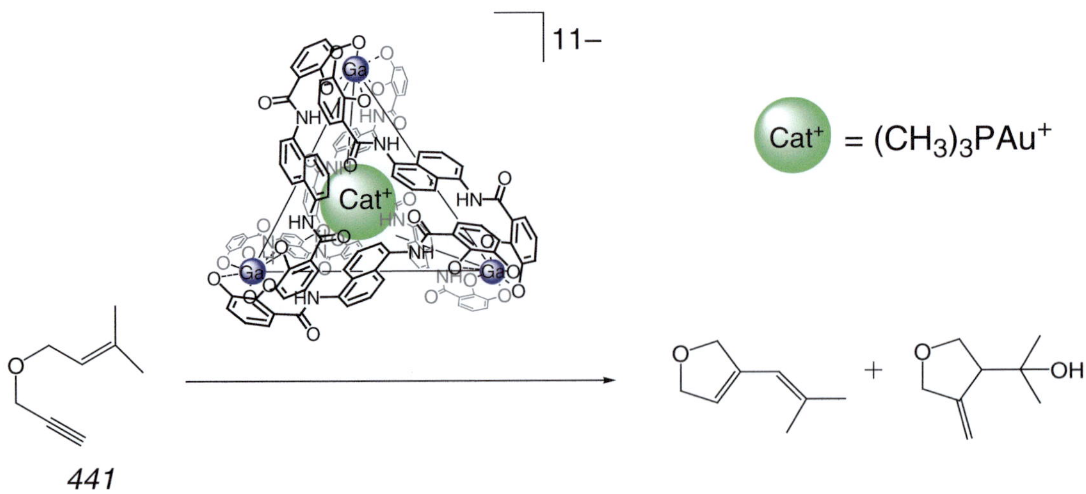

Scheme 5.23

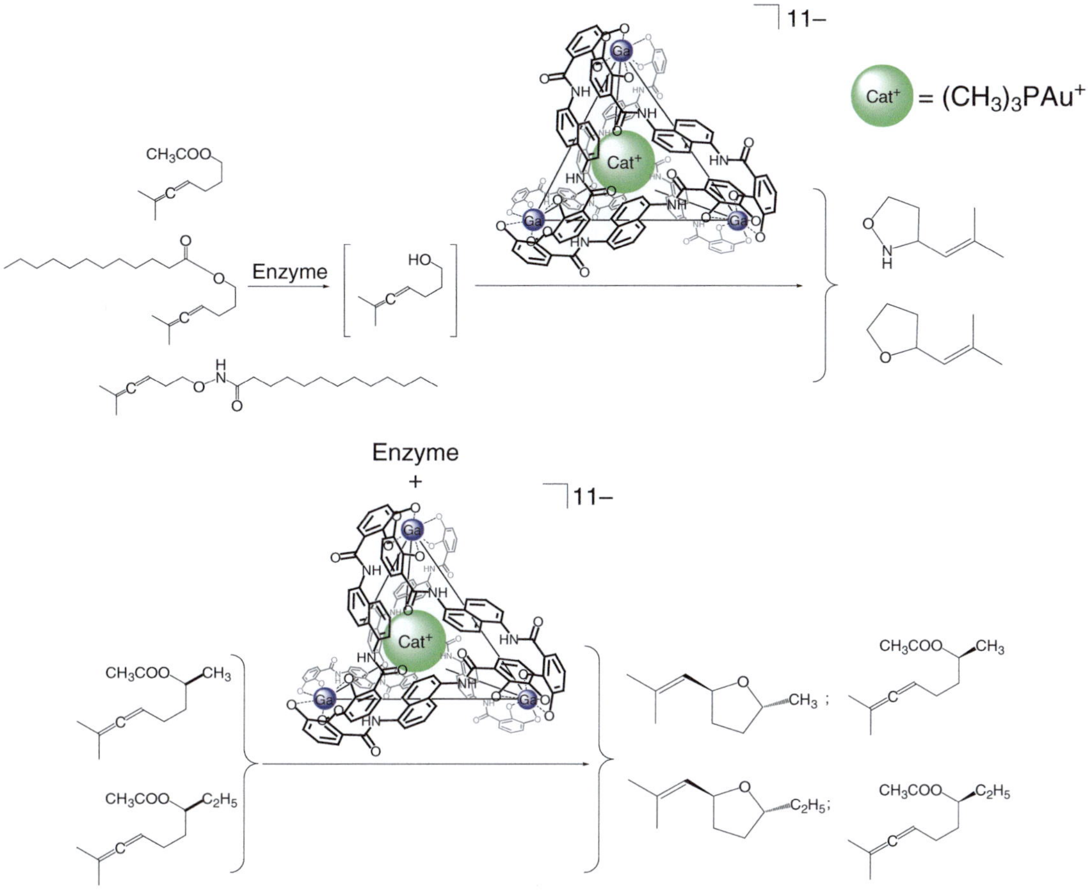

Scheme 5.24

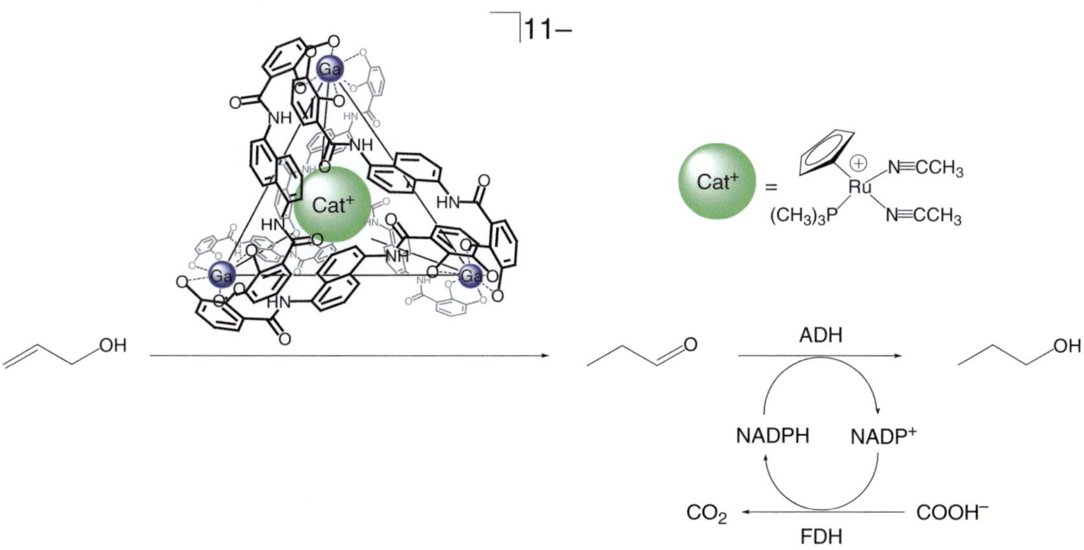

Scheme 5.25

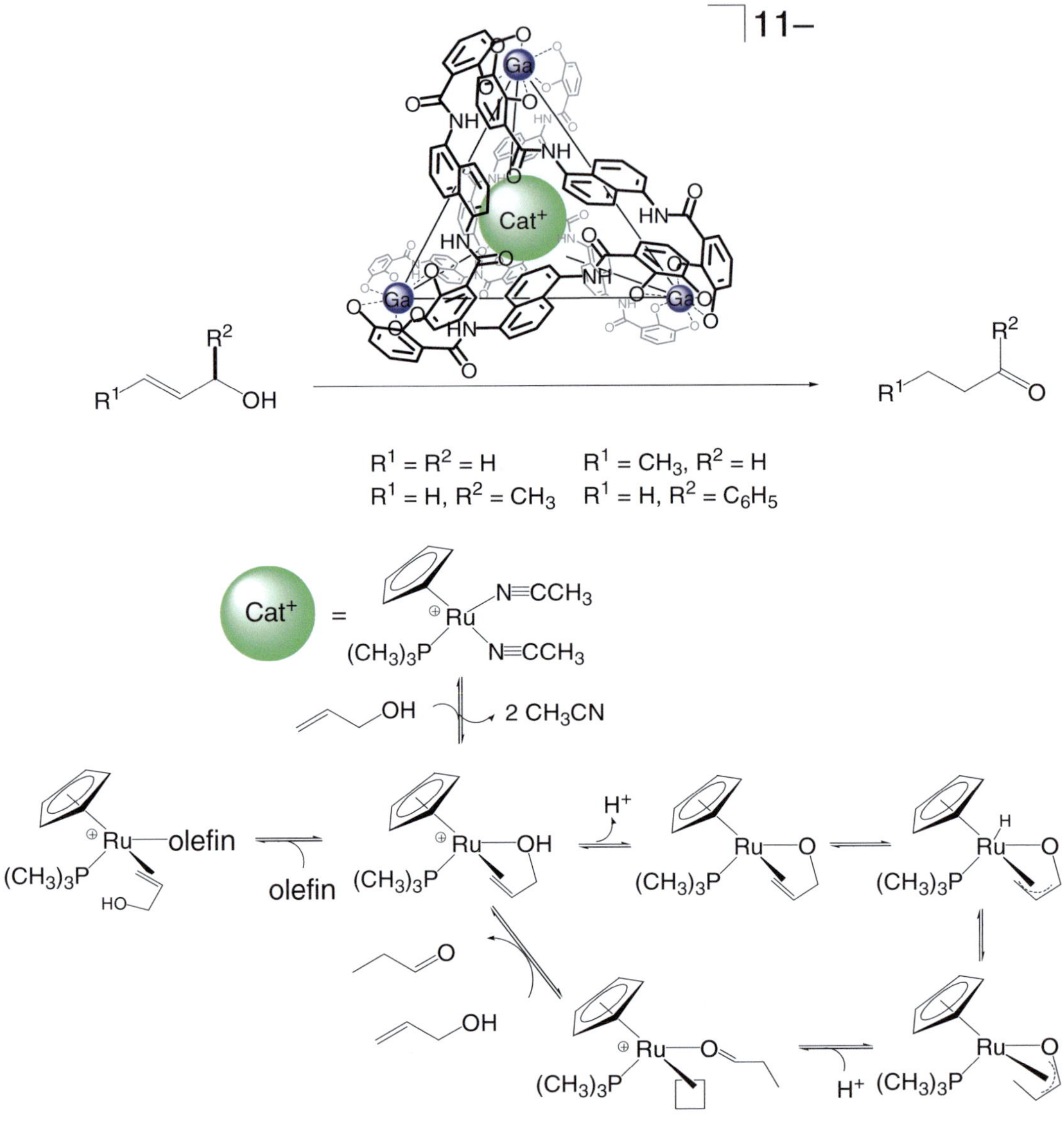

Scheme 5.26

catalyst than the solvato-complex or the non-caging analog. According to concurrent catalytic experimental data, the cage catalyst also discriminates cyclohexane over cyclooctane and adamantane, suggesting that encapsulation of a substrate by covalent capsule **209** plays a key role in this catalytic oxidation reaction [27].

Catalysis of Diels–Alder reactions by a hydrogen-bonded capsule **309** (Schemes 5.30, 5.31, and 5.32) is described in [28–30] and explained by encapsulation of reactants in this "soft" caging ligand. The product inhibition (lack of dissociation) prevents these systems from showing true catalytic behavior, while an increase in the reaction rate of over two orders of magnitude due to the effective enhancement of reagent concentration within the cavity of **309** (Scheme 5.30) has been detected in [28]. Its use as a true catalyst of the specific Diels–Alder reaction of *para*-benzoquinone with the thiophene dioxide derivative by

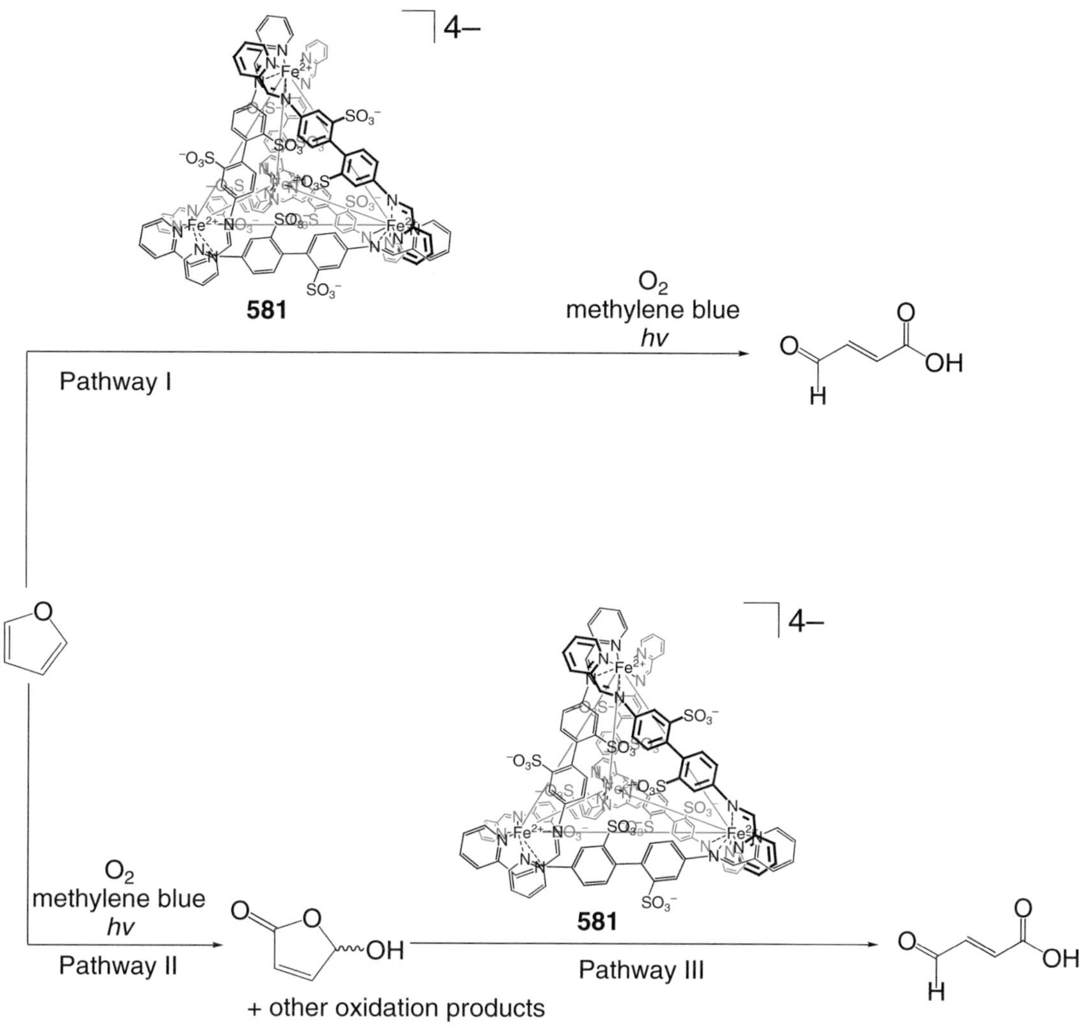

Scheme 5.27

Scheme 5.31 is reported in [29]; the plausible catalytic cycle is also given in this scheme. Bimolecular Diels–Alder reactions between the caged dienes and dienophiles by Scheme 5.32 are reported in [30] to be accelerated by **309**: both the reactants can be encapsulated by this ligand concurrently, and the observed size selectivity, saturation kinetics, and product inhibition studies suggest that these reactions take place within the cavity of its hydrogen-bonded cage framework [30].

A 1,3-dipolar cycloaddition reaction by Scheme 5.33 between the reactants, which are co-encapsulated within the cavity of a hydrogen-bonded capsule **429**, has been performed in [31] to give 1:1 cage complex with an unsymmetrically loaded product molecule.

Reversible cyclization reactions within the cavity of this self-assembled bis-cavitand **429** have been studied in [32] using ^{1}H NMR spectra; amplification and stabilization of otherwise unfavored forms of the caged guest molecules in their ring-to-chain isomerization processes are illustrated by Scheme 5.34.

The reactions of *para*-tolylacetic acid *442* with *n*-butyl isonitrile *443* within the cavity of the hydrogen-bonded capsule **429** (Scheme 5.35)

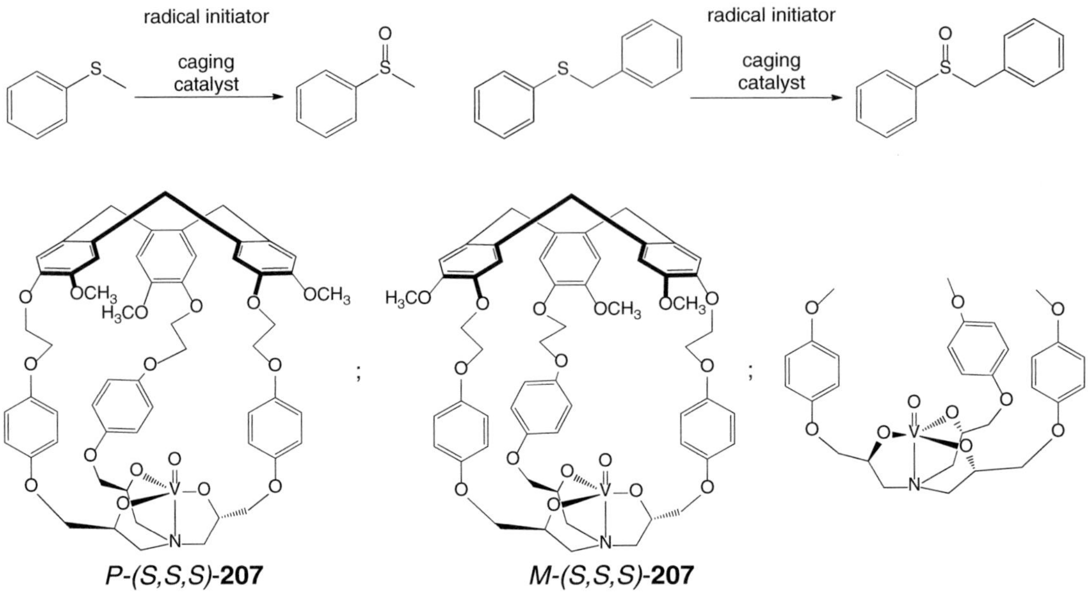

Scheme 5.28

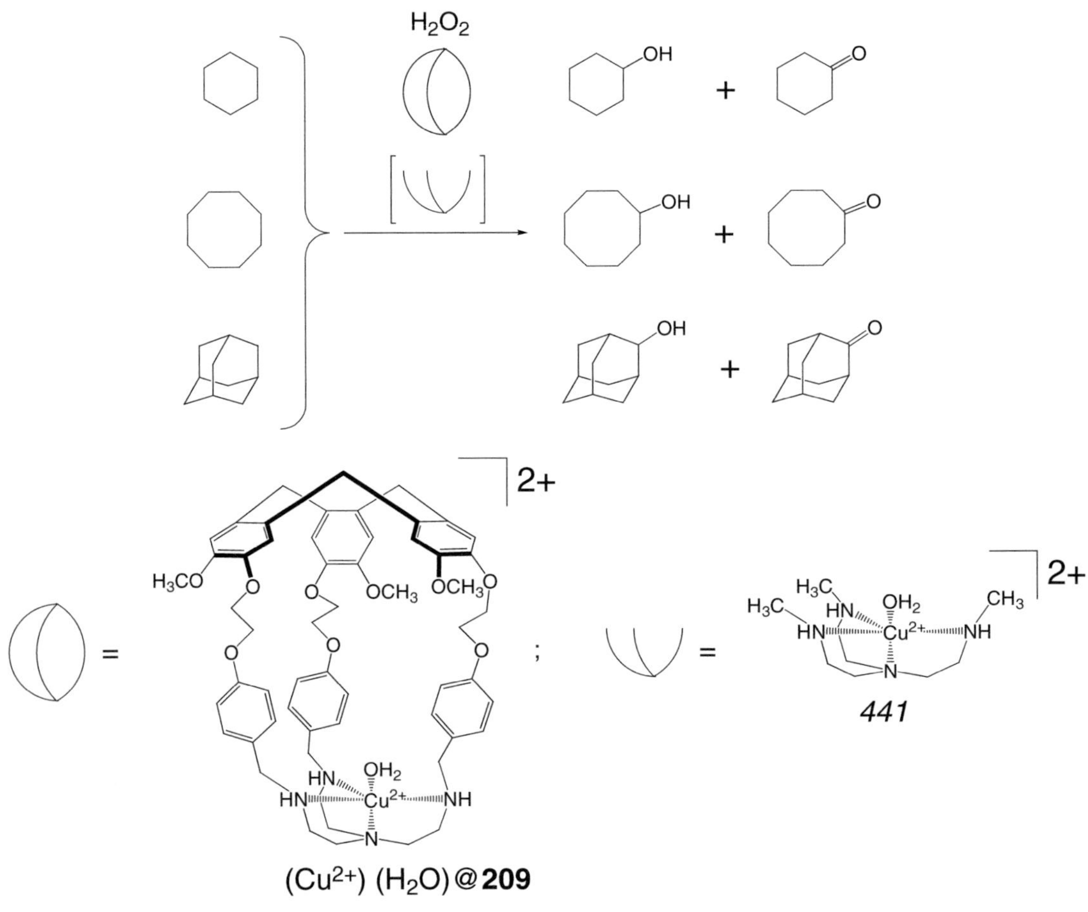

Scheme 5.29

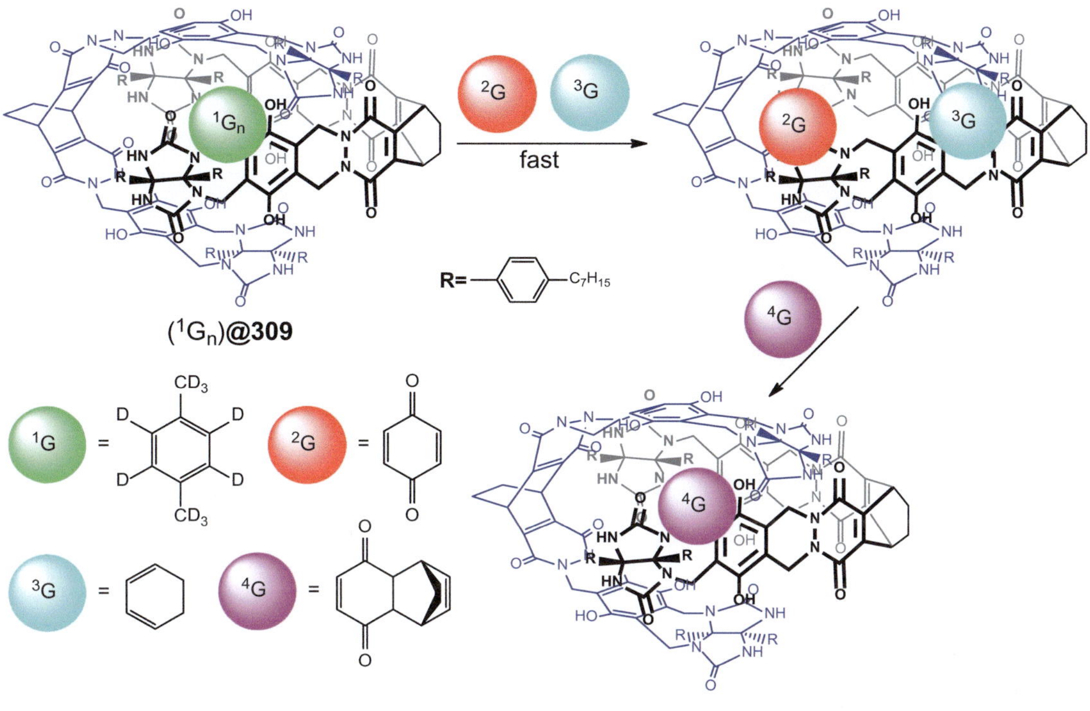

Scheme 5.30

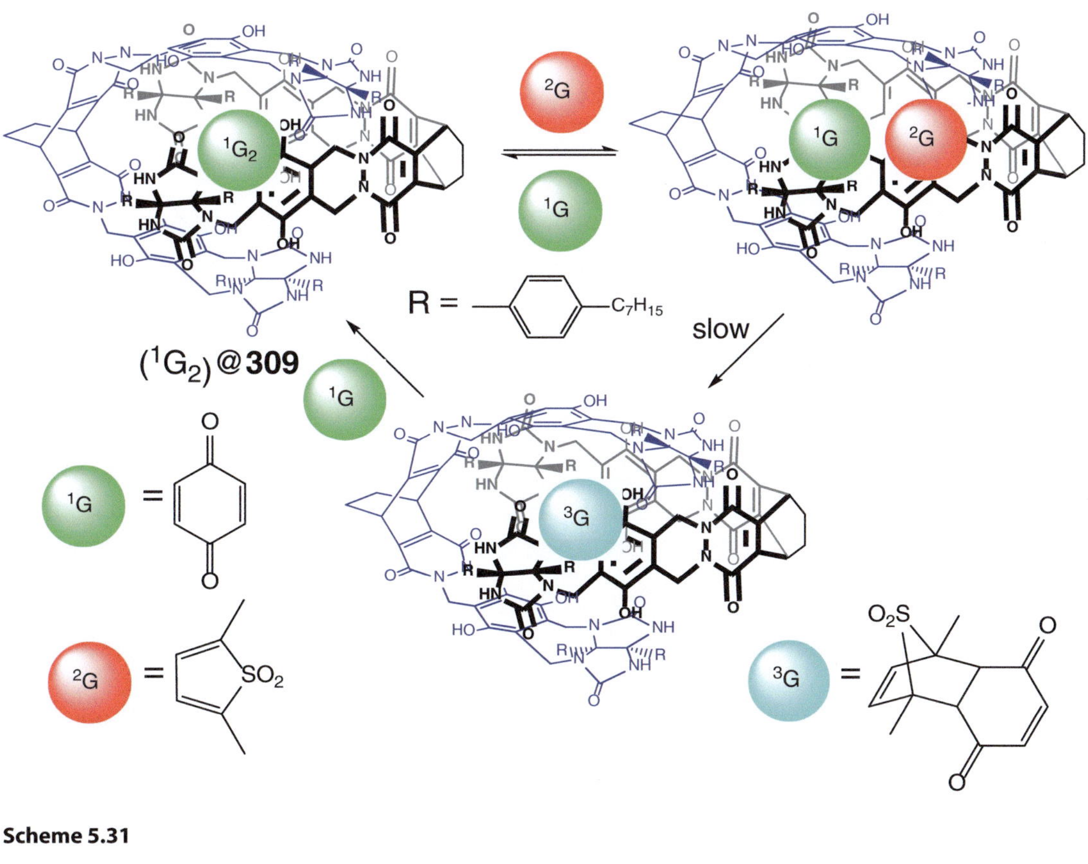

Scheme 5.31

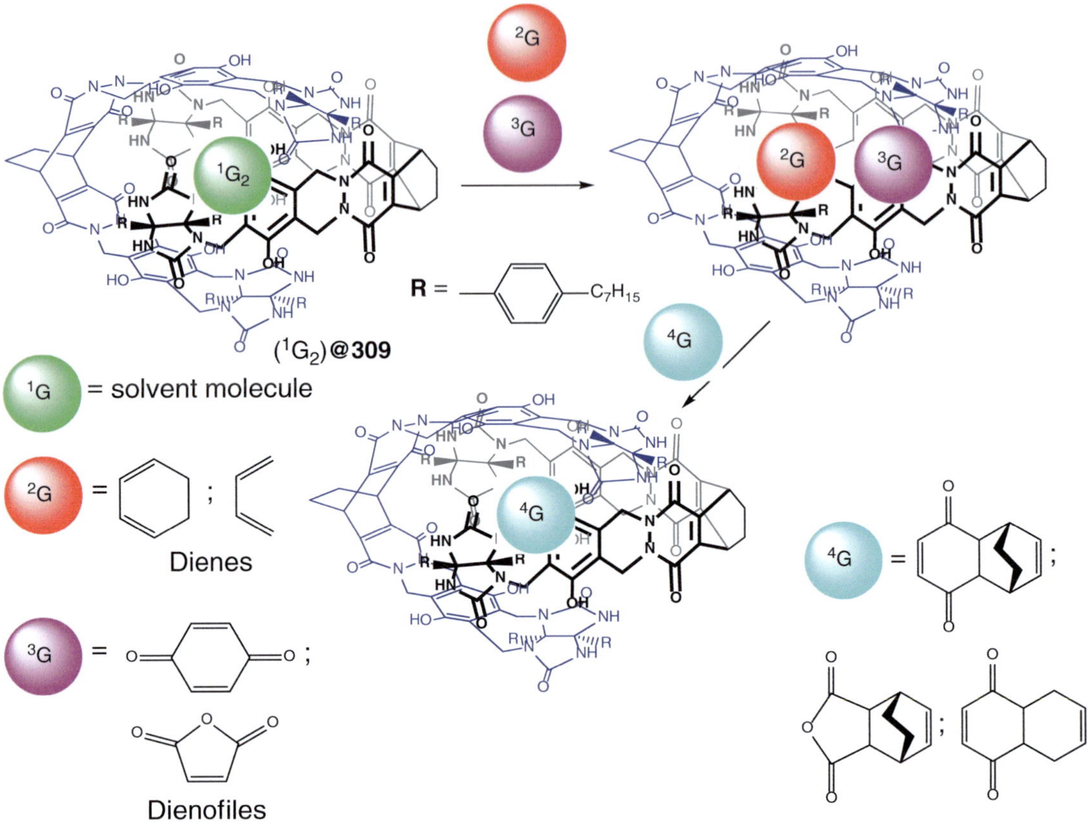

Scheme 5.32

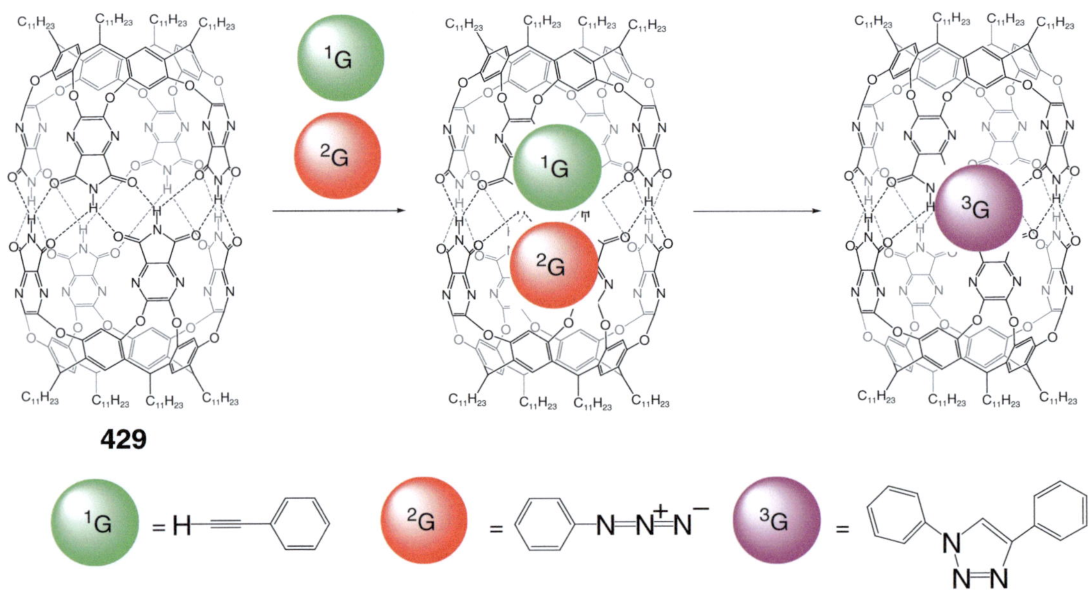

Scheme 5.33

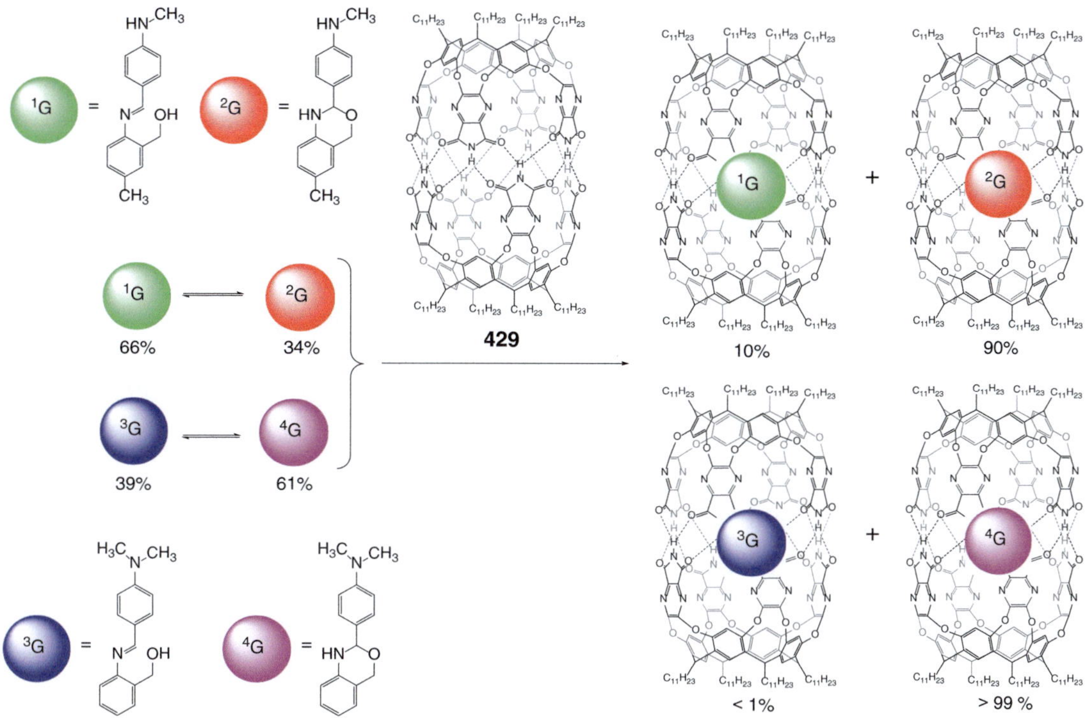

Scheme 5.34

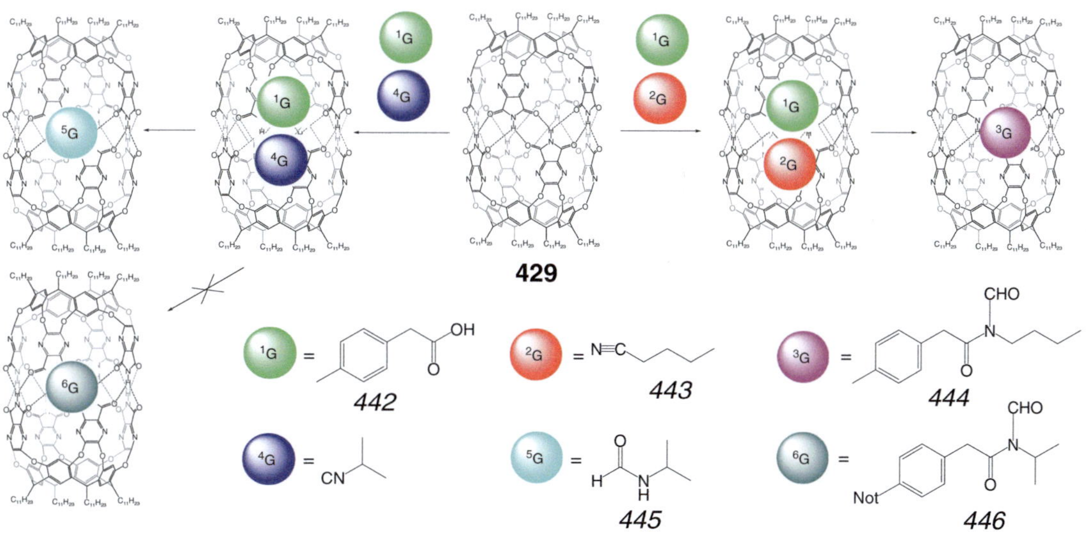

Scheme 5.35

afforded a rearrangement product *444* even at room temperature, while under the same reaction conditions but in the absence of this capsule, the corresponding formamide and anhydride products are formed. Thus, this rearrangement takes a different and accelerated reaction course within **429** [33]. Indeed, the capsule-free reaction of *442* with isopropyl isonitrile gave a formamide in a 40 % yield and only 1 % of a rearrangement product, whereas **429** forms a heteroguest 1:1:1 complex with co-encapsulated reagents by Scheme 5.35, and their isonitrile and acid groups are located near one another in the polar middle of its cage framework. Further transformation of these encapsulated molecules gave a 1:1:1 cage complex with co-encapsulated formamide *445* and acid *442* as co-guests, while the expected rearrangement product *446* has not been detected. The following principles of co-encapsulation of the different guest molecules within the capsule's cavity have been evaluated in [33]: congruence of shapes, compatibility with lengths, conformity with volumes, and complementarity of surfaces.

An extended hydrogen-bonded capsule **435** has been used in [34] to facilitate reactions of encapsulated isocyanates with water to give *N*,*N*-dialkylureas by Scheme 5.36.

A 1:1 cage complex of the ligand **822** (Scheme 5.37) with encapsulated nitrosonium cation has been described in [35] to be a mild nitrosation agent for secondary amides RCONHR': this agent selectively reacts with *N*-methylated amides only due to the steric hindrances between more bulky alkyl substituents of analogous alkylamide

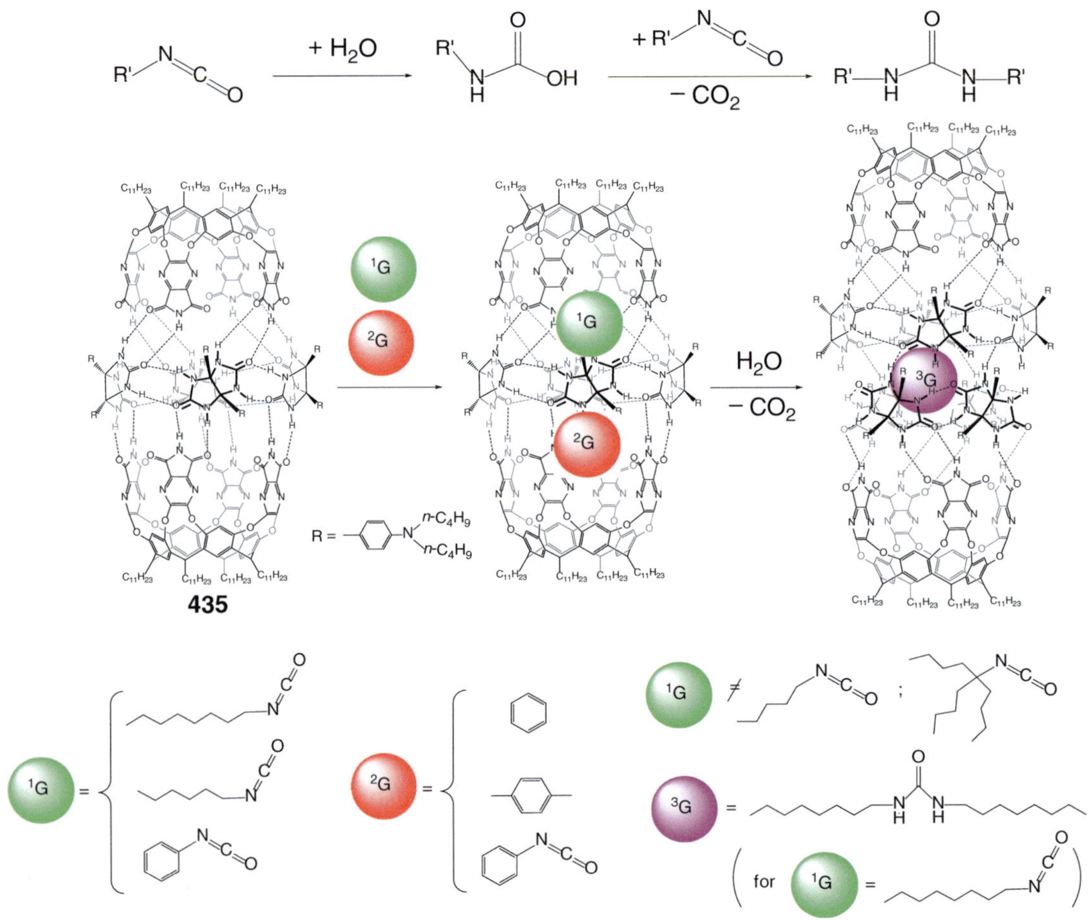

Scheme 5.36

Scheme 5.37

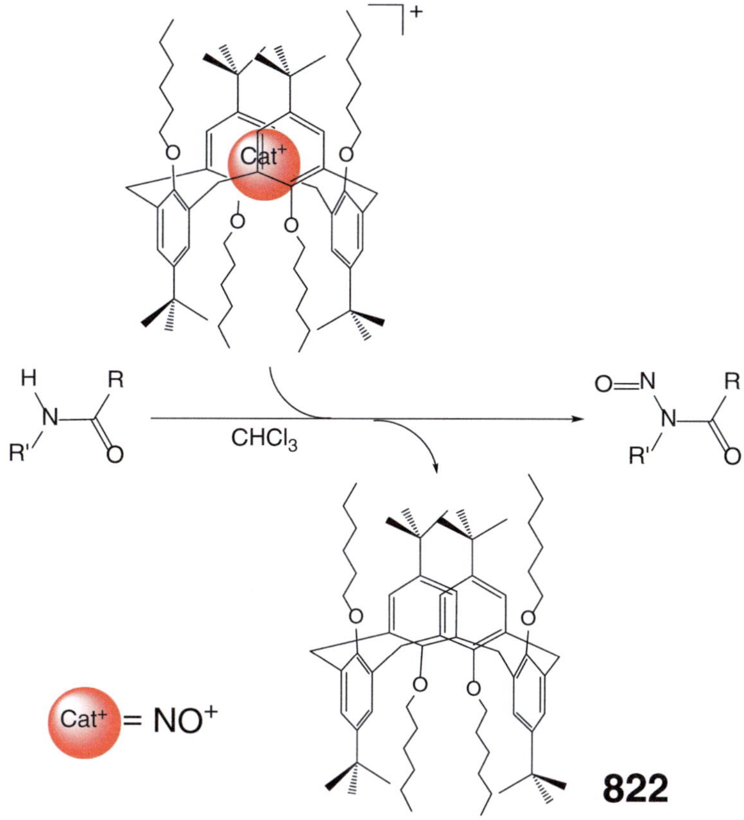

822

substrates. The caged NO^+ cation is described in [36] to be a soft nitrosylating agent for secondary amides also in the case of its encapsulation by bis-calix[4]arene capsule **823**, giving the corresponding *N*-nitroamides and the initial caging ligand by Scheme 5.38.

Kinetics of the hydrolysis of esters *447–451* shown in Scheme 5.39 as guests encapsulated by hydrogen-bonded capsule **473** has been studied in [37] using ^{1}H NMR method. The lowest hydrolysis rates are observed for pentyl- and hexyl-containing guests *450* and *451*, while the highest reaction rate is found to be for their methyl-containing analog *447*, but the observed difference is reported to be not enough for efficient kinetic resolution. The kinetic curves in the absence and in the presence of the encapsulating ligand are substantially different, which results from two plausible mechanisms: either the encapsulated ester molecule releases in to the bulk solution and undergoes the hydrolysis or it is the hydroxide anion entering the cavity of **473** that it does [37].

Encapsulation of organogold catalyst *452* by the hexameric hydrogen-bonded capsule **397** has been used in [38] to control chemo- and regioselectivity of transformations of an alkyne substrate by Scheme 5.40 within the cavity of **397**. The reaction of this substrate in bulk solution quickly leads to an almost exclusive and quantitative formation of the regular product, the corresponding ketone. The same reaction within the cavity of **397** occurs much slowly and gives, in addition to this product, significant amounts of the linear aldehyde (4 %). Moreover, the formation of 1,2-dihydronaphthalene *453* as a result of intramolecular rearrangement has been observed in [38] only in the absence of water: this intramolecular reaction is reported to be favored if taking place within the cavity of **397** due to specific folding of an alkyne substrate.

Enantiomeric *N*-acylation of the racemic guest *454* with acetic or benzoic anhydride in the presence of the cavitand syntone *238* by Scheme 5.41 has been performed in [39]. In all cases, the enantioselectivity of acylation of its

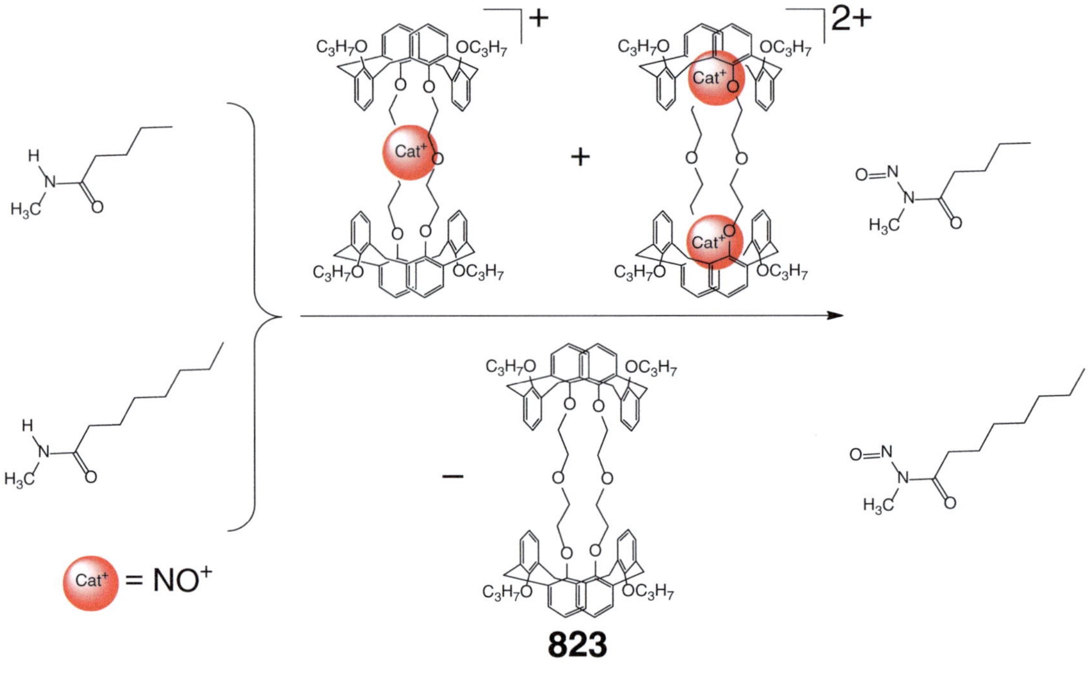

Scheme 5.38

R-isomer exceeds 50 % and increases with the concentration of *238*. This suggests that selective encapsulation of *454* by the corresponding bis-cavitand capsule **480** plays a key role in the process. Moreover, in the case of benzoic anhydride as a substrate with more bulky substituents, its *N*-benzyl alkylation reaction proceeds with higher enantioselectivity [39].

5.2 Photochemical Reactions

Irradiation of a solution of homoguest 1:2 cage complexes of **483** with two encapsulated acenaphthylenes *455* and *456* resulted in regioselective dimerization of these guests by Scheme 5.42 leading to the corresponding 1:1 complexes with caged *syn*-isomeric products only. The regioselectivity in the photodimerization of 2-metylnaphthoquinone within this capsule is reported in [40] to be also very high (96 % of the head-to-tail isomer). These [2+2] addition reactions do not proceed without encapsulation at the same photochemical reaction conditions even at high concentrations of these aromatic monomers, and the coordination capsule **483** acts as a molecular flask to promote this intermolecular photodimerization of large olefins in a highly efficient fashion [40].

Photoreactivity of the encapsulated adamantane molecule within the cavity of coordination capsules **487** has been studied in [41]. Irradiation of the homoguest 1:4 cage complex of **487** afforded a mixture of its heteroguest 1:1:3 derivatives with one caged oxidized 1-adamantylhydroperoxide or 1-adamantanol molecule, whereas three other adamantane guests remained unchanged (Scheme 5.43). Under anaerobic conditions, the photooxidation of this cage complex in solution and in solid state gave blue air-sensitive radical products. Going from the tris-bipyridyltriazine ligand syntone to its 1,3,5-substituted benzene-containing analog prevents this photoreaction; so these radical species are generated by its triazine fragment. The use of ^{18}O labeling on O_2 or H_2O allowed evaluating the photoreaction pathway shown in Scheme 5.43: it includes a photochemical excitation of the triazine fragment and electron transfer from the adamantane guest to the cage framework, giving a 1-adamantyl radical and H^+ ion as well as the

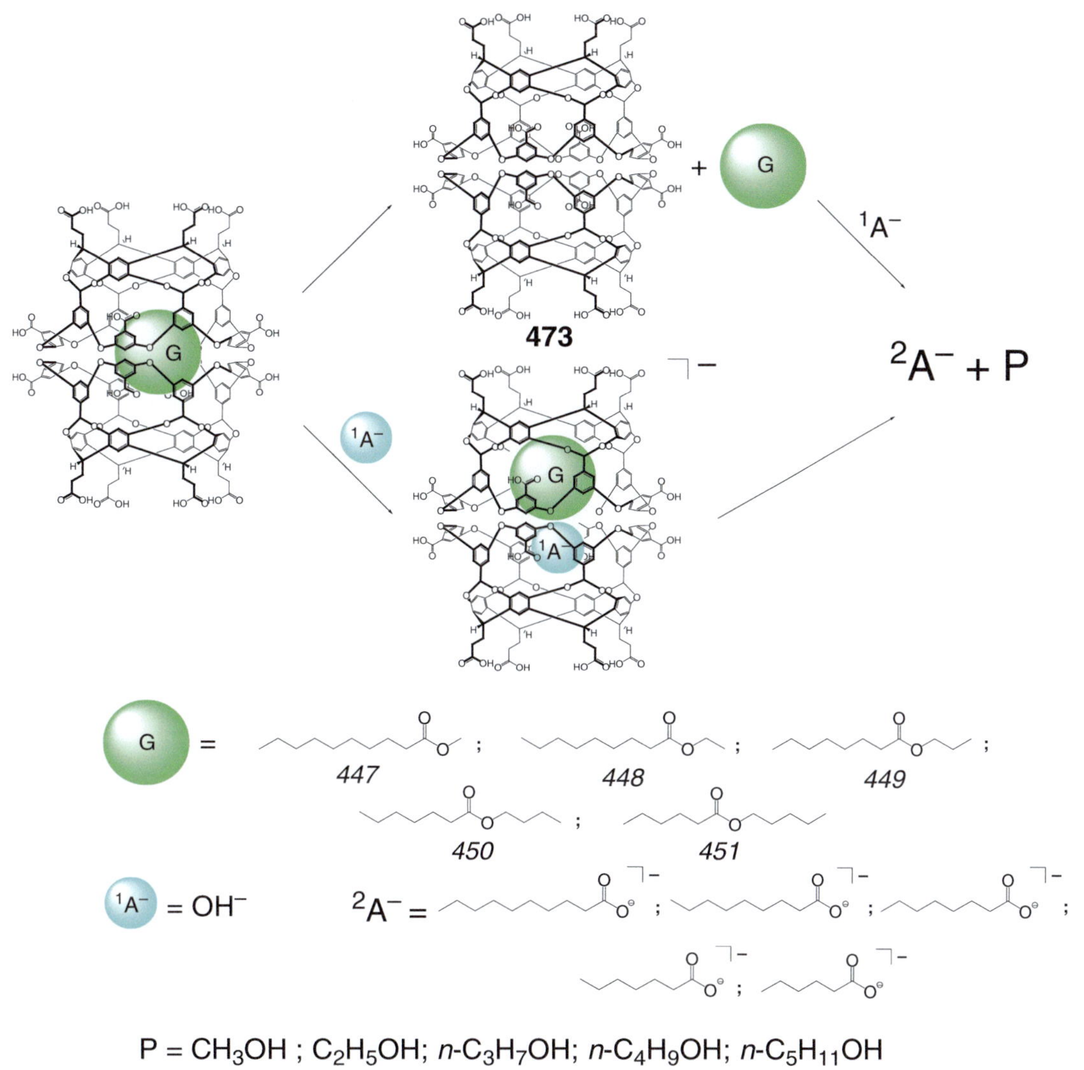

Scheme 5.39

corresponding anion-radical cage species. Oxygen or water molecules under aerobic conditions trap the encapsulated 1-adamantyl radical giving 1-adamantanol or 1-adamantylhydroperoxide. Under anaerobic conditions, these radical species are stable but are rapidly quenched with molecular oxygen under aerobic conditions. The same oxidation reactions, leading to the corresponding hydroperoxides and ketones, have been observed in [41] for encapsulated six-, seven-, and eight-membered alicyclic guest molecules.

Unusual photooxidation of alkanes within the cavity of the coordination capsule **488** has been studied in [42] using in situ IR, UV and electrochemical measurements, and quantum chemical calculations; the X-ray diffraction study of its 1:4 cage complex with encapsulated adamantane molecules [41] revealed the close aggregation of these guests within the cavity of **488**. A plausible reaction mechanism shown in Scheme 5.44 involves generation of host anionic and guest cationic radical species by guest-to-host PET; the former anion radical has been experimentally detected using in situ IR spectra. The electrochemical experiments and theoretical calculations performed in [42] showed strong

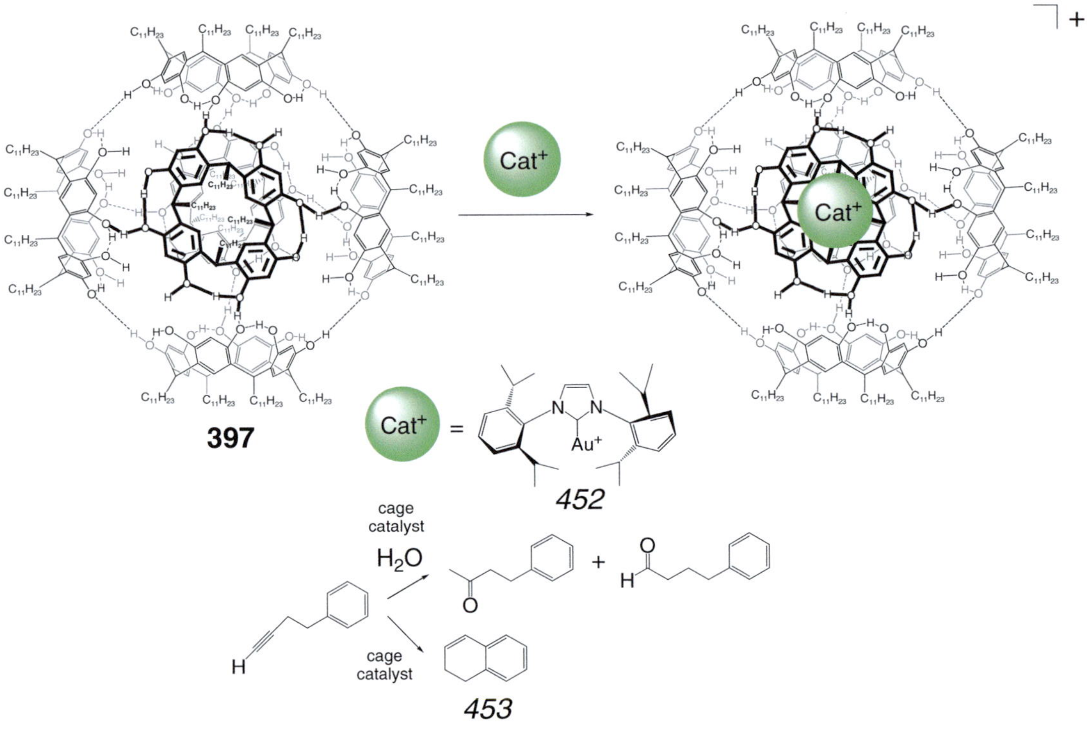

Scheme 5.40

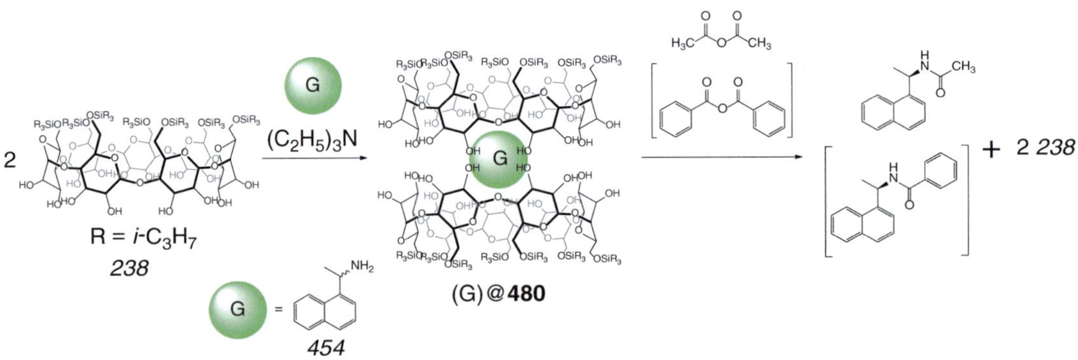

Scheme 5.41

electron-withdrawing effects of the triazine fragments of this capsule. Four guest molecules are tightly bound and packed within its cavity, so that the guest adamantane molecules are located very close to the triazine fragments, thus facilitating this unusual PET [42].

Highly selective [2+2] cross-photodimerization of olefins by Scheme 5.45 within the Pd_6L_6 capsule **483** has been performed in [43]. It encapsulates acenaphthylene and 5-ethoxynaphthoquinone molecules in a pairwise selective fashion, thus giving a heteroguest 1:1:1 cage complex. The encapsulation assists this photoreaction of the caged olefins allowing it to proceed extremely efficiently with high reaction rate, stereoselectivity, and pairwise selectivity. In particular, maleimide derivatives, which themselves are photochemically inactive

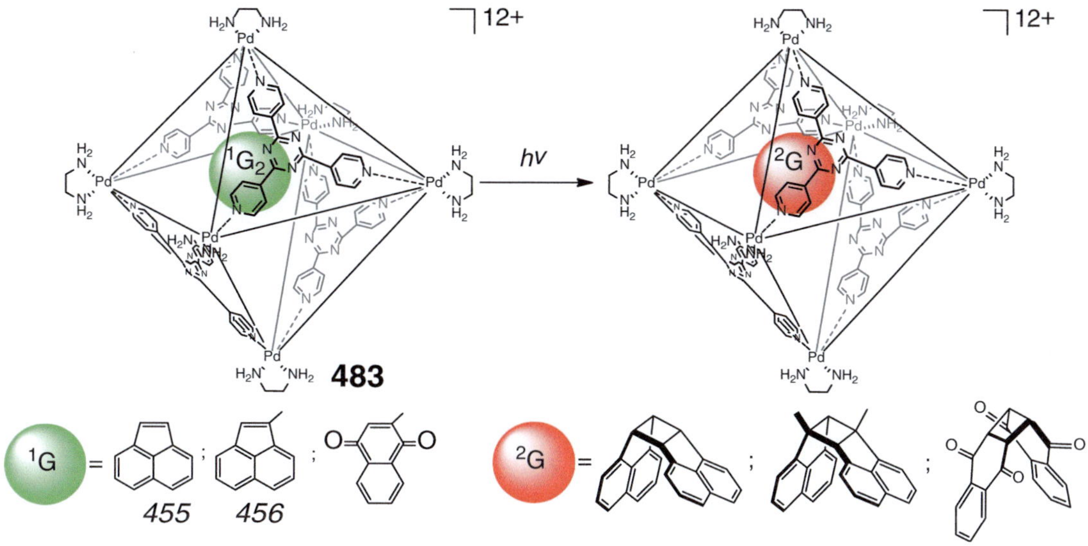

Scheme 5.42

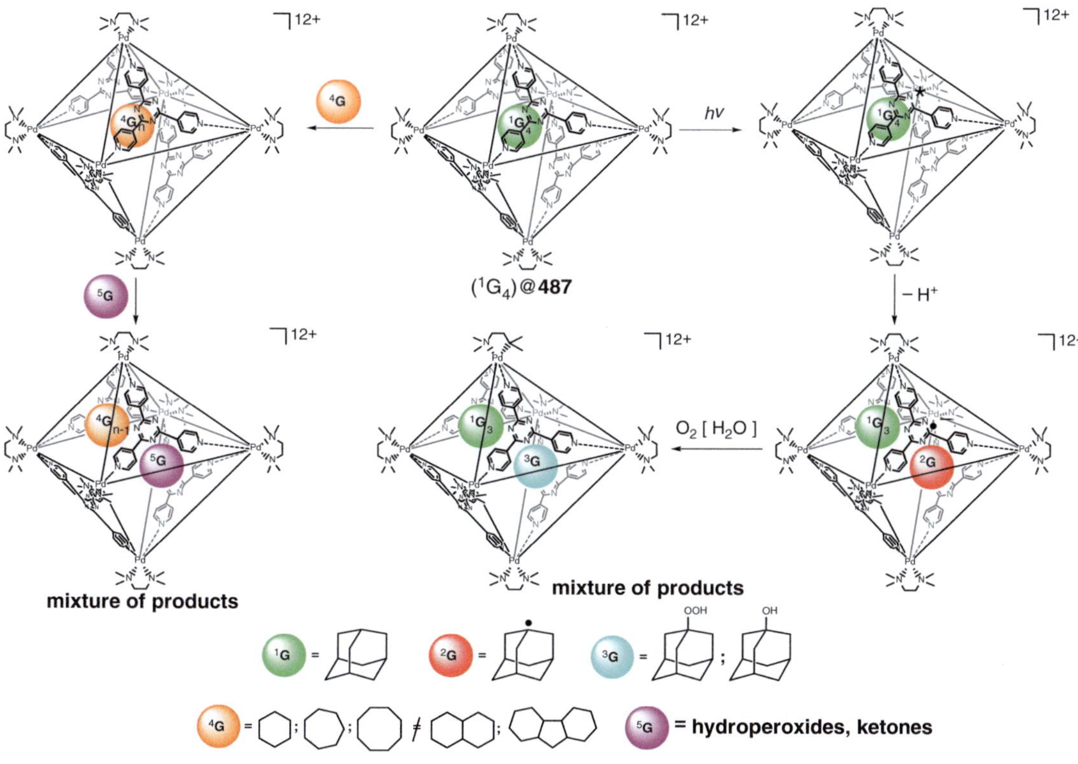

Scheme 5.43

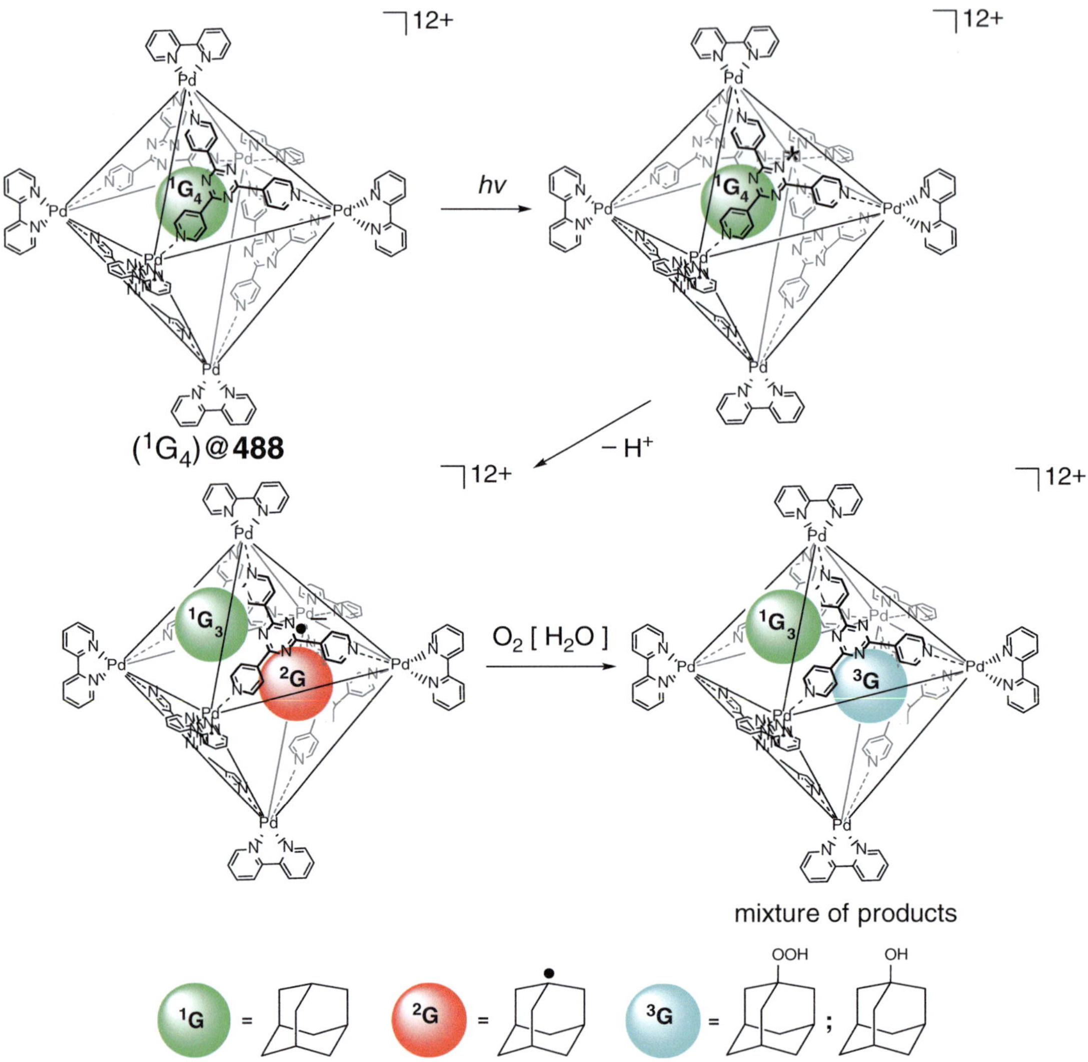

Scheme 5.44

under ordinary reaction conditions, quite efficiently undergo cross-dimerization with acenaphthylene and dibenzosuberenon within the cavity of **483**. It facilitates [2+2] cross-photodimerization of large olefins giving the corresponding *syn*-dimers in high yields. In such a cavity-directed synthesis, the capsule's confined space allows strictly controlling the encapsulation process of the co-guests by steric effects; the subsequent chemical transformation of these co-caged reagents may also be strictly controlled by this coordination capsule playing a role of a molecular flask [43]. The encapsulated arene molecules are reported in [44] to also undergo [2+2] photoaddition and [2+4] thermoadditions by Scheme 5.46 within the cavities of **483** and its Pt_6L_6 and Pd_6L_6' analogs **486** and **487**. Highly efficient, *syn*-selective [2+2] photoadditions of *N*-cyclohexylmaleimide to phenanthrene and of maleimide to fluoranthene have been also performed under similar reaction conditions [44].

Encapsulated olefin guests undergo photodimerization within the cavity of the coordination capsule **487** even in crystals [45]. In particular, two caged acenaphthylene molecules have been thermally tumbled: their [2+2] photoaddition

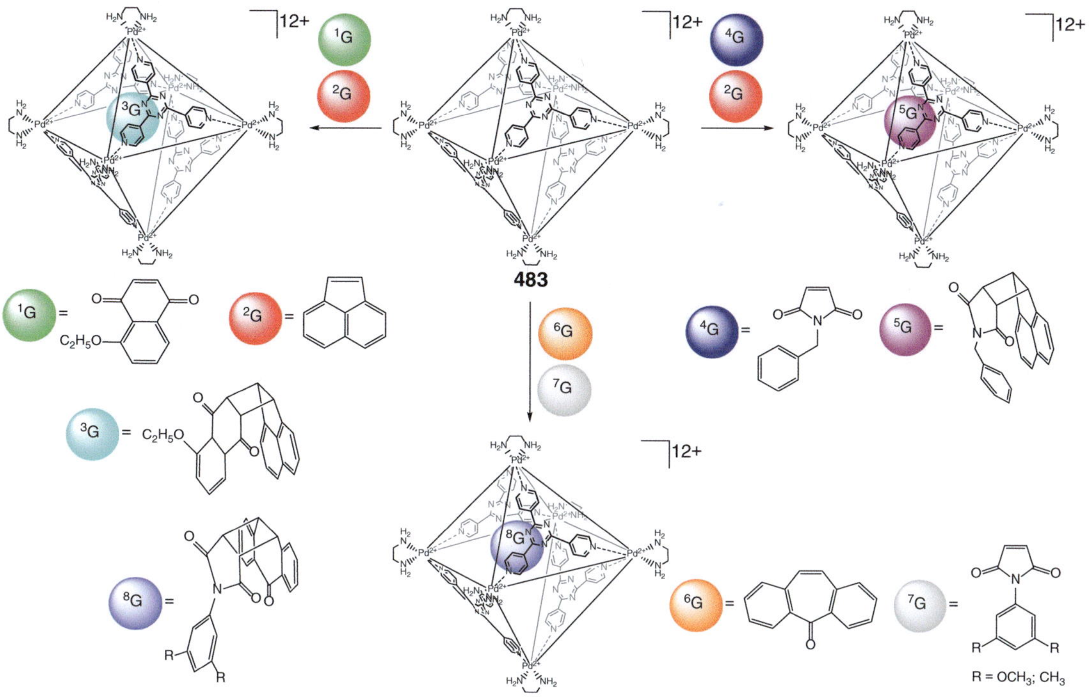

Scheme 5.45

reaction by Scheme 5.47 gave the *syn*-isomer of the corresponding dimeric guest in a quantitative yield. This photoaddition reaction within the cavity of **487** involves the disordered acenaphthylene monomers that are located rather far (up to 9 Å) from each other; such a distance substantially exceeds those for regular organic crystals (approximately 4.2 Å). The rigidity of the coordination capsule prevents destruction of these crystals in the course of this reaction; the high cavity volume of **487** is described in [45] to be a reason for the above non-topochemical photoreactions without a loss of its single crystallinity [45].

One of the four encapsulated coordination-saturated complex guests $(CH_3cp)Mn(CO)_3$ underwent photodissociation by Scheme 5.48 within the cavity of the coordination capsule **487**, giving its coordination-unsaturated analog $(CH_3cp)Mn(CO)_2$. In the formed heteroguest 1:1:3 cage complex, this 16-electron molecule has a pyramidal geometry and undergoes no recombination with CO molecules [46].

Irradiation of an aqueous solution of the heteroguest 1:1:1 cage complex of **483** with encapsulated pyrene-4,5-quinone and *para*-adamantyl toluene molecules highly selectively gave a 1:1 cage complex with an encapsulated 1,4-adduct *457* shown in Scheme 5.49 [47]. The plausible mechanism of this photoaddition reaction proposed in [47] includes radical recombination process: the photoexcited pyrene-4,5-quinone molecule abstracts hydrogen atom from the *para*-adamantyl toluene co-guest giving the corresponding benzyl and semiquinone radicals, the recombination of which resulted only in the *O*-coupled 1,4-adduct *457*. Due to the proximity of the reactive centers of these intermediate radical species, the formation of their 1,2-adduct is not observed [47]. In the case of its analog with an encapsulated 9,10-phenanthrenequinone, the caged guest molecule also selectively undergoes photomediated radical coupling within the cavity of the coordination capsule **483**. On the initial stage of the encapsulation reaction, the formation of the corresponding homoguest cage complexes by Scheme 5.49 together with their heteroguest

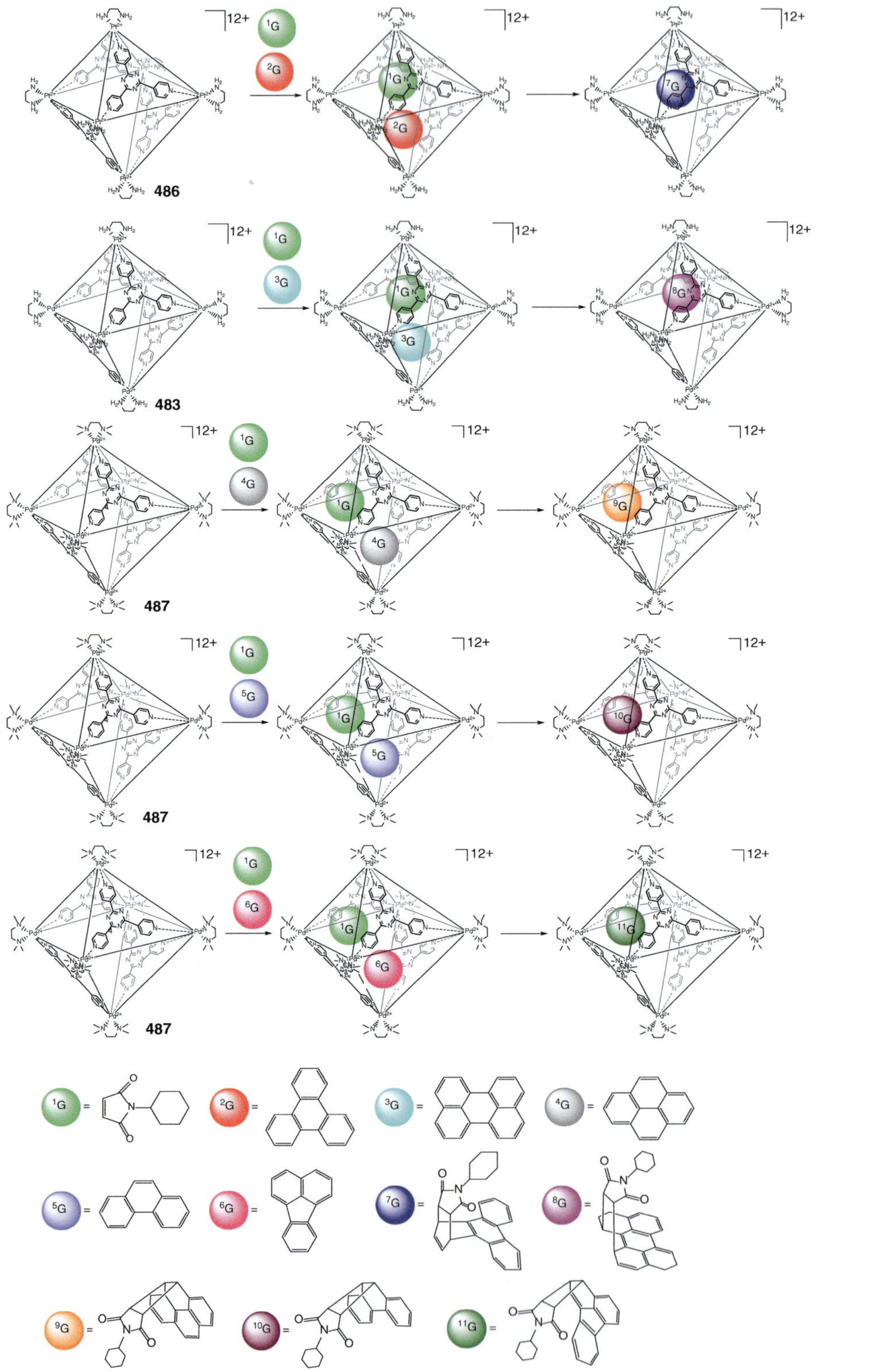

Scheme 5.46

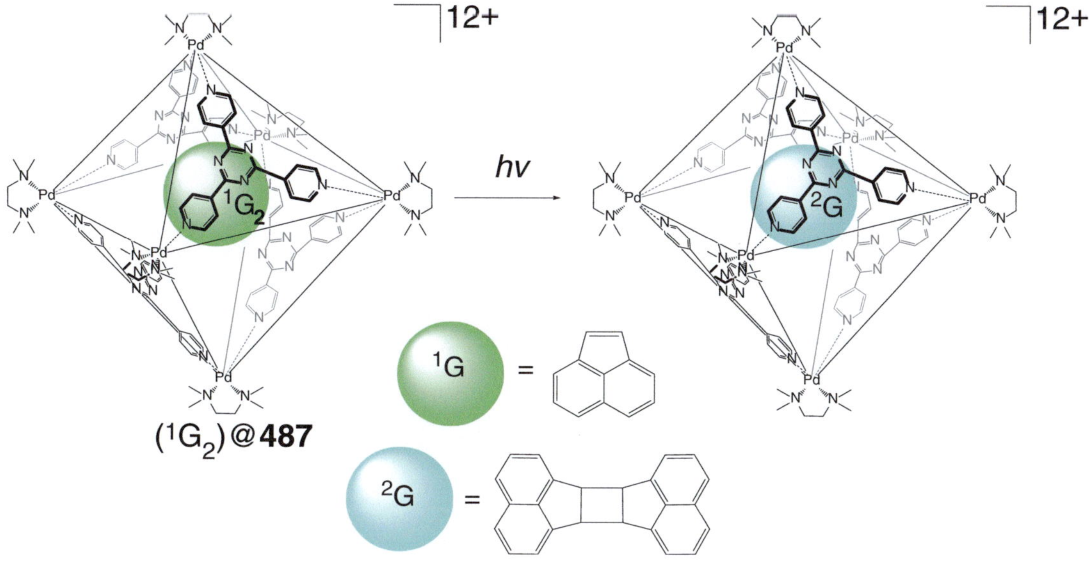

Scheme 5.47

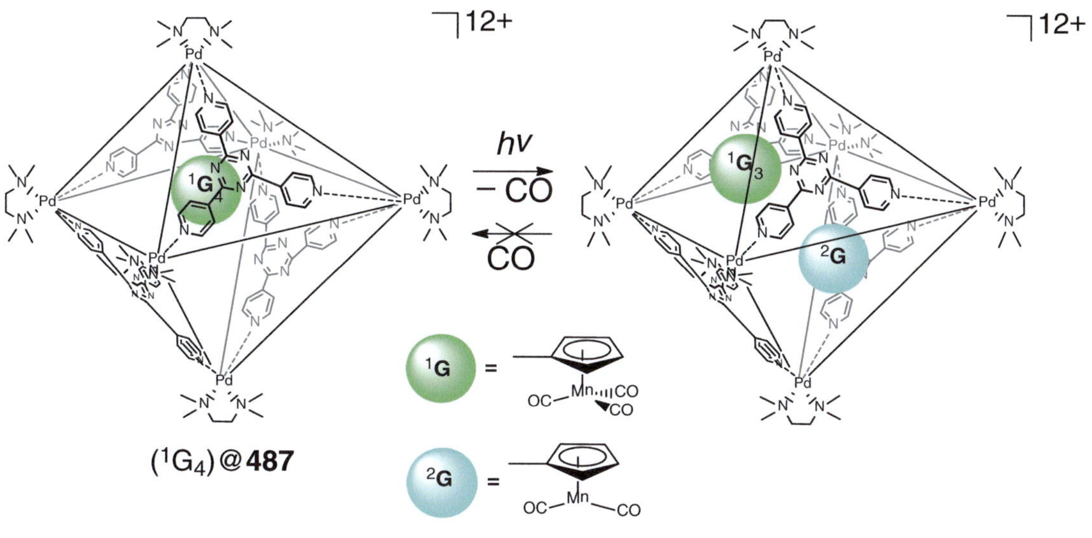

Scheme 5.48

analog has been detected. Irradiation of the equilibrium mixture gave a 1:1 cage complex with an encapsulated 1,4-adduct in a moderate yield; so the equilibrium between the homo- and heteroguest cage complexes of the encapsulating ligand **483** is reached quickly. In contrast, the uncaged substrate molecules do not form any *O*-coupled 1,4-adducts. Therefore, an encapsulation of these reactants by the coordination capsule promotes an *O*-coupling reaction preventing other side processes [47].

Immersing the crystals of the host–guest 1:4 cage complex of polymeric capsule **493** with encapsulated 4-hydroxyphenylamine *458* in *n*-decan solution of ethyl isocyanate has been used in [48] for chemoselective *O*- and *N*,*O*-acylation of these guests by Scheme 5.50. The simple *N*-acylated product has been not detected due to

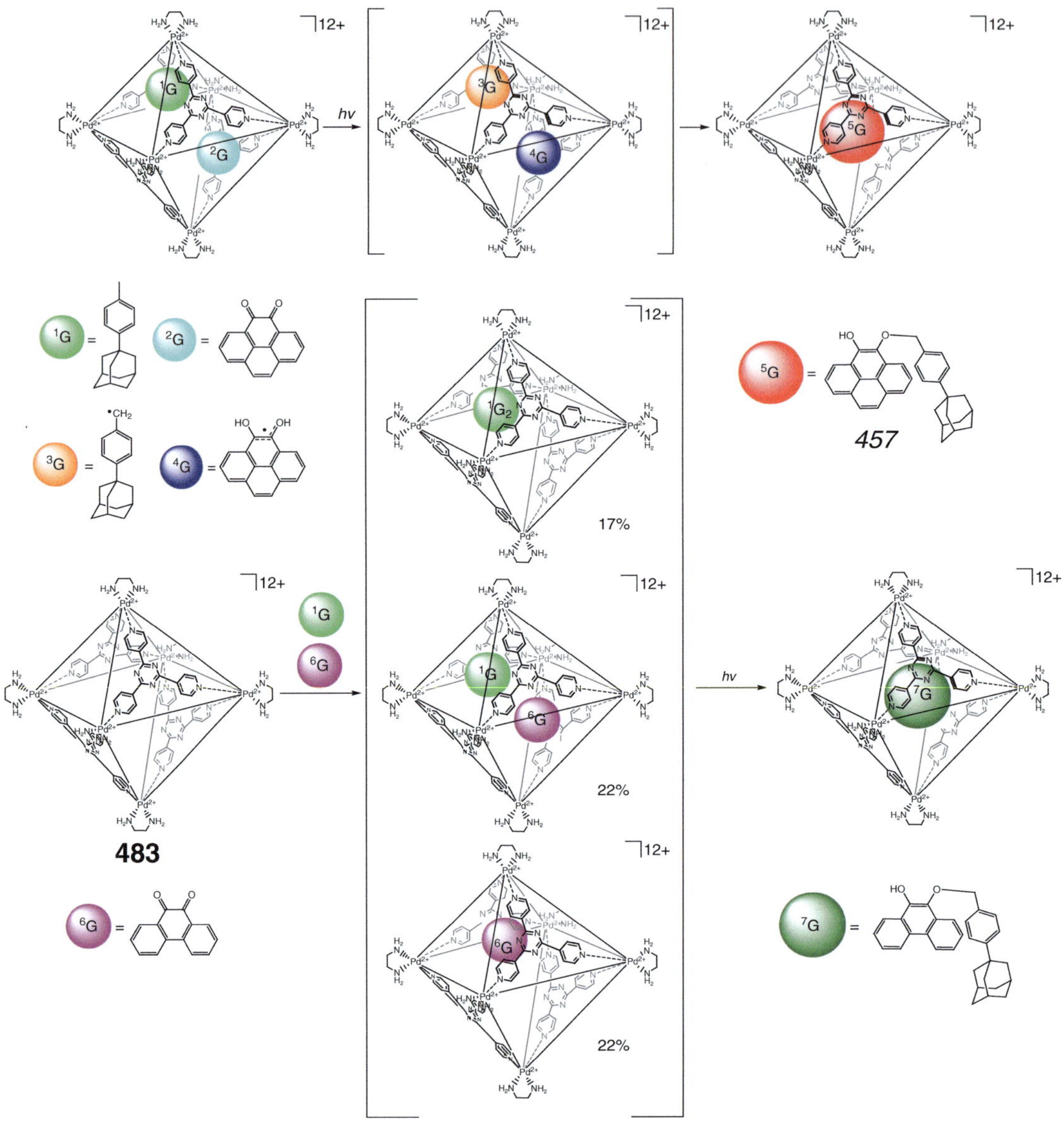

Scheme 5.49

the protection of its secondary amino groups by **493**: this intermediate formed on the early stage of the reaction undergoes further *O*-acylation giving the *N,O*-acylated guest *460*. Thus, two parallel reactions occur in the reaction mixture: (*i*) *N*-acylation of 4-hydroxydiphenylamine followed by *O*-acylation of a product in large interstitial pores, where such substrates are mobile, and (*ii*) *O*-acylation within the cavity of **493** without subsequent chemical reactions of capsule-protected amino groups (the product *459*). This protection has been confirmed in [48] in a study of analogous reactions within the molecular capsule **483**, also encapsulating 4-hydroxydiphenylamine that undergoes *O*-acylation with ethyl isocyanate within its cavity. The *O*-acylated product is not a suitable guest for the capsule **483** releasing such product from its cavity [48].

Selective thiocarbamoylation of aromatic amines by Scheme 5.51 using encapsulated methyl thiocyanate molecules of a host–guest 1:4 cage complex of the metallopolymeric cap-

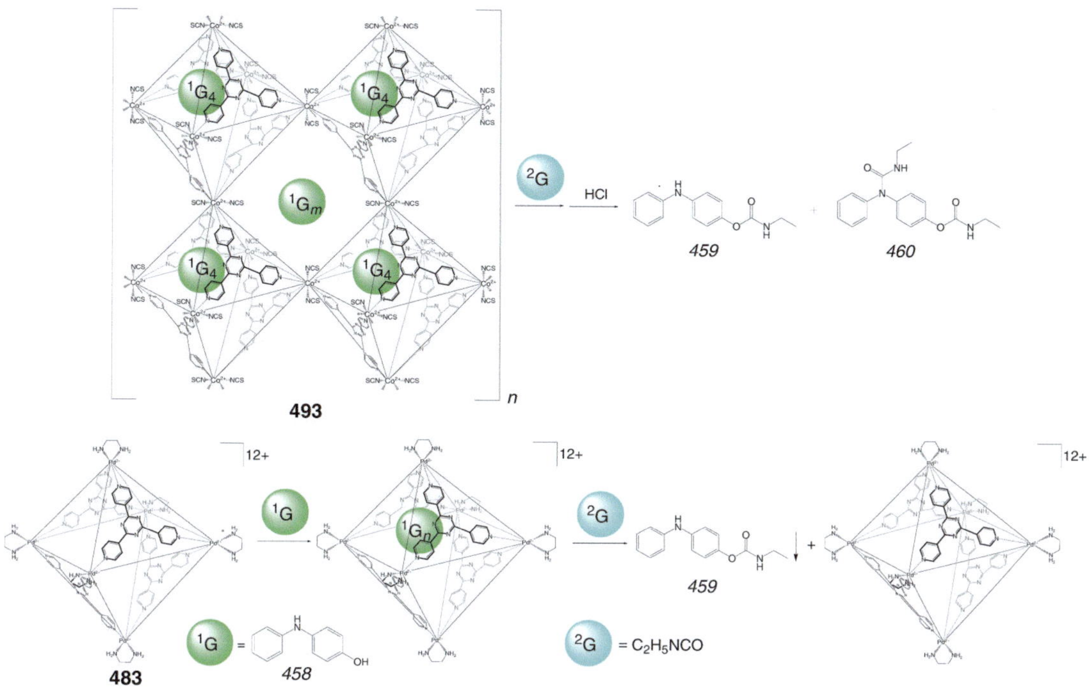

Scheme 5.50

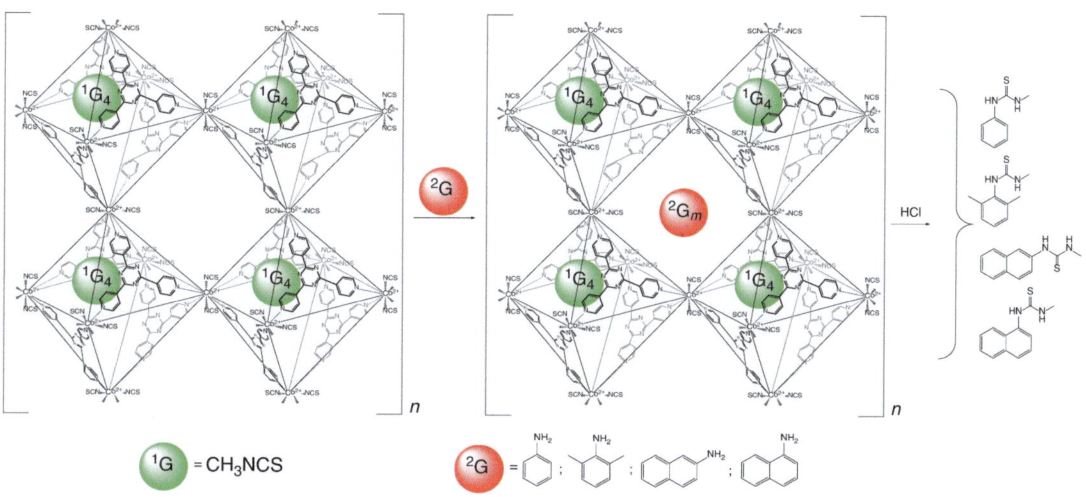

Scheme 5.51

sule **493** has been performed in [49]. The caged reactant discriminates aniline over its more bulky substituted analogs, in particular, 2,6-dimethylaniline: a competitive experiment showed a substantial (by approximately one order) increase in the reaction rate. It also discriminates 2- and 1-naphthylamines in this thiocarbamoylation reaction with a fivefold product selectivity [49].

1:1 cage complexes of Pd_2L_4 coordination capsules **779** and **766** with encapsulated *cis*-4,4′-azobenzene bis-sulfonate dianion are reported in [50] to release the guest after its photoisomerization in a *trans*-isomer by Scheme 5.52–1446. In

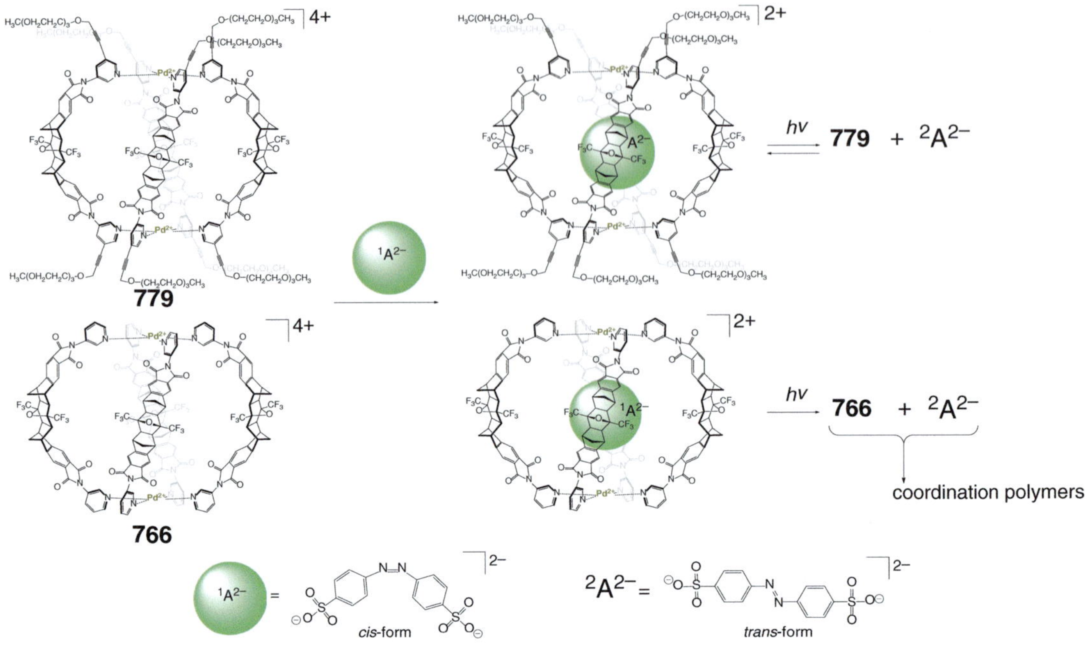

Scheme 5.52

the case of the highly soluble caging ligand **779**, the full reversibility of this light-driven encapsulation/release process has been observed. The analogous cage complex of **766** resulted in immediate crystallization upon photoisomerization reaction with a release of this caged dianion from this capsule, thus giving coordination polymers with its outside binding to cross-linking palladium(II) metallocenters [50].

Analogous coordination capsules **777** and **778** formed by four photoswitchable bis-monodentate pyridyl ligand syntones are reported in [51], allowing the light-triggered guest uptake and release by Scheme 5.53: irradiation of host–guest 1:1 cage complexes with encapsulated spherical fully fluorinate dodeca-*closo*-borate dianion resulted in reversible uptake and release of this guest.

Encapsulated α-pyrone molecule of 1:1 hemicarceplex of the covalent capsule **262** undergoes thermal and photochemical transformations by Scheme 5.54 within its cavity [52]. A key stage of these processes is the formation of the corresponding cage complex with encapsulated cyclo butadiene molecule by decarboxylation of the photoproduct of transformations of this α-pyrone. This observation stimulated further research [53] into the encapsulation of *ortho*-benzyne and its analogs by the caging ligand **264** (Scheme 5.55), which has been selected for the following reasons: MM data suggested that this capsule has the cavity large enough to accommodate benzocyclobutenedione but prevented *ortho*-benzyne from leaving through one of its four equatorial portals at ambient temperature. Sharp decrease of UV absorption of this capsule above 275 nm allowed the photolysis of the guest molecules to occur at larger wave lengths. As follows from ^{1}H and ^{13}C NMR experiments [53] with deuterated and ^{13}C-enriched guests, the parent 1:1 cage complex of **264** with encapsulated benzocyclobutenedione molecule underwent photochemical transformations by Scheme 5.55 giving the target capsule with the caged *ortho*-benzyne molecule.

Two modes of photodimerization of the acenaphthylene guests within the octaacid hydrogen-bonded capsule **473** have been studied in [54]. The photoreaction of the singlet excited state of

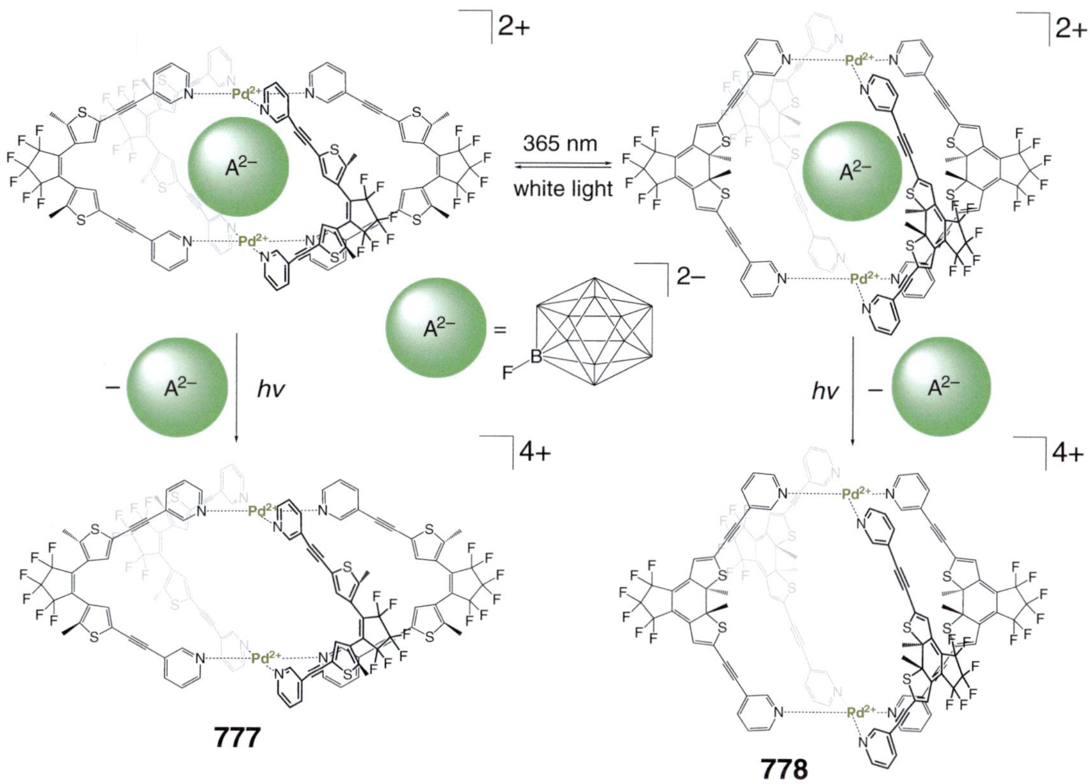

Scheme 5.53

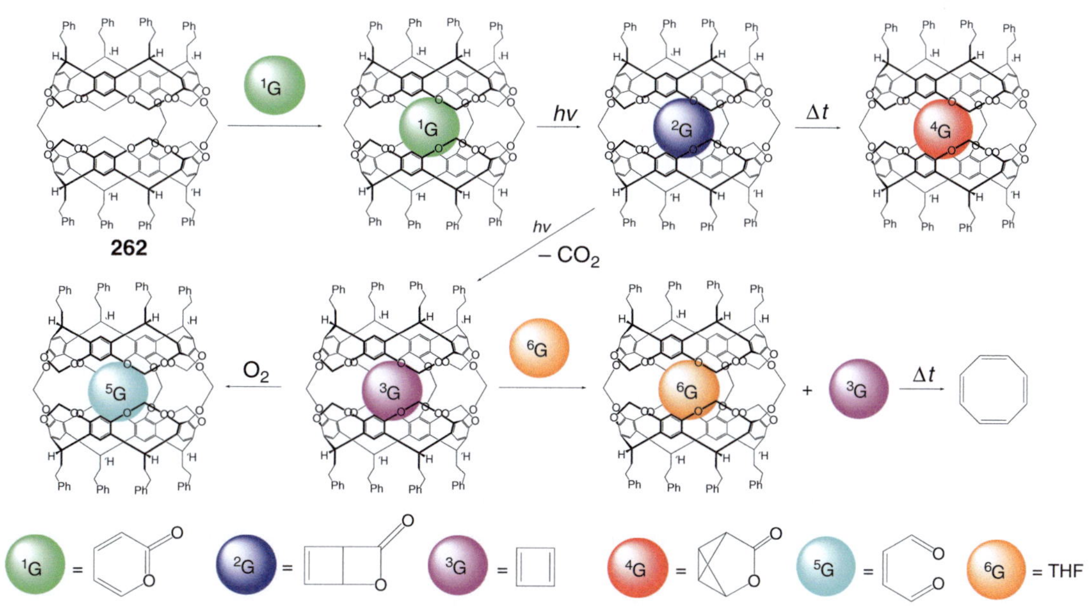

Scheme 5.54

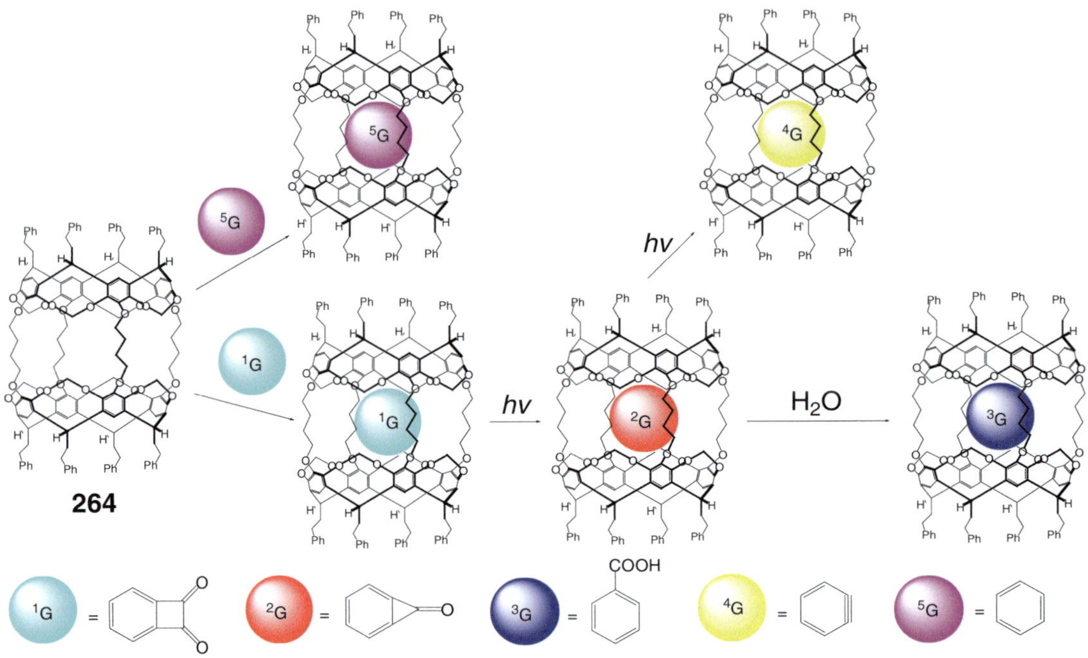

Scheme 5.55

acenaphthylene by Scheme 5.56 gave only *syn*-dimer, but in the case of its triplet state in the presence of eosin Y, it afforded a mixture of *syn*- and *anti*-forms; the former form has been encapsulated by **473**, while the latter has been detected by NMR spectroscopy in a bulk solution. On the time scale of S_1 state (0.35 ns), the guest acenaphthylene molecules are preorganized in a *syn*-form, the photodimerization of which gave only the caged *syn*-dimer. Irradiation of an aqueous solution of this 1:2 cage complex in the visible region ($\lambda > 500$ nm) in the presence of a triplet sensitizer (the lifetime of T_1 state is 6 ms) gave a mixture of the dimeric forms. It seems that the excited acenaphthylene molecules in their triplet state leave the hydrogen-bonded capsule as an excimer that then undergoes rearrangement with the formation of a biradical intermediate giving its *anti*-isomer by Scheme 5.56 [54].

The encapsulated molecules of the alkyl-substituted benzyl ketones are described in [55] to undergo unusual photoreactions within the cavity of a hydrogen-bonded capsule **473**. The products of these reactions suggest that the intermediate radical pairs are also encapsulated by the capsule during their transformation. C_1–C_8 α-alkyl dibenzyl ketones undergo Norrish type II and/or type I photoreactions by Scheme 5.57 in their bulk solutions and within the cavity of **473** [55]. Their cage complexes are kinetically stable and fall into three packing motifs modulated by the length of their α-alkyl chain with the capsule **473** acting as an external template to promote the formation of distinct guest conformers. The major products from all these ketones upon irradiation in bulk solutions resulted from their Norrish type I reaction. Within the cavity of **473**, the photoreactions of their caged molecules depend on the length of the alkyl chain. The α-hexyl-, α-heptyl-, and α-octyl-substituted dibenzyl ketones as guests mainly gave the products of Norrish type II transformations. The caged α-butyl- and α-pentyl-substituted dibenzyl ketones yielded equimolar amounts of their two rearranged isomers, while encapsulated α-methyl-, α-ethyl-, and α-propyl-substituted dibenzyl ketones yielded only one isomer of the initial ketone [55].

Encapsulation of optically pure α-alkyldeoxybenzoins within the cavity of the

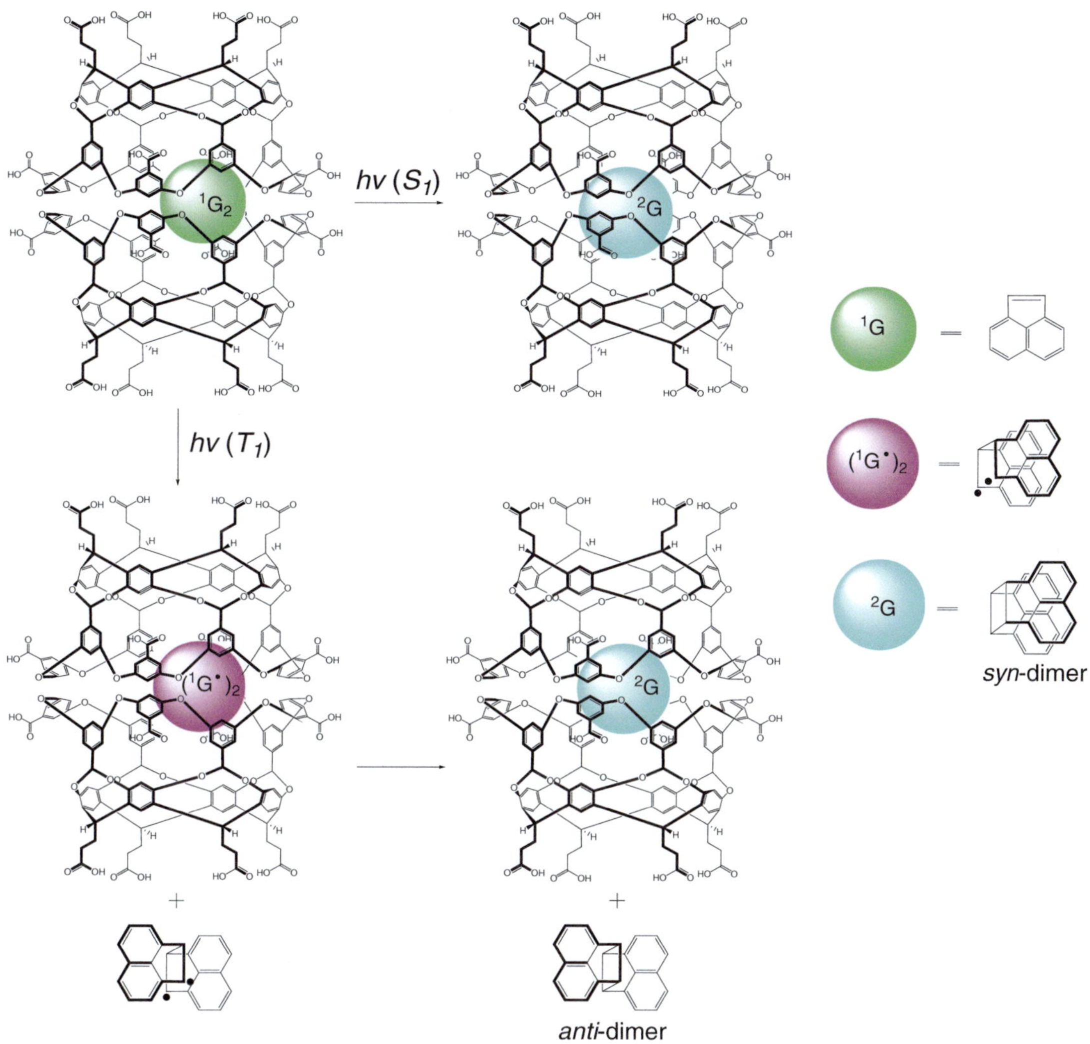

Scheme 5.56

octaacid capsule **473** [56] also affects the products of the Norrish type I and type II photoreactions of the caged molecules shown in Scheme 5.58.

Photodimerization of appropriate reactive guests (Scheme 5.59) giving 1:2 cage complexes with the supramolecular capsule **473** has been performed in [57]. This caging ligand plays the role of a molecular flask that not only solubilizes organic molecules in water and brings two molecules close to each other but also preorganizes them to yield one or two encapsulated dimeric products.

Reversible light-induced and thermoinduced *trans*-to-*cis* isomerization of encapsulated 4,4′-dimethylazobenzene is reported in [58] to affect its binding abilities, allowing one to control co-encapsulation of a co-guest molecule by the hydrogen-bonded capsule **429** (Scheme 5.60). This remote control of the reversible binding is prospective for the design of a supramolecular fluorescence *on–off* switching devices: when *trans*-4-ethyl-4′-methylstilbene has been used as a co-guest, its fluorescence is altered depending on whether it is in bulk solution or encapsulated within a cavity of **429**. As a result, such encapsulation (and, therefore, the guest fluorescence intensity) has been controlled in [58] by light-triggered

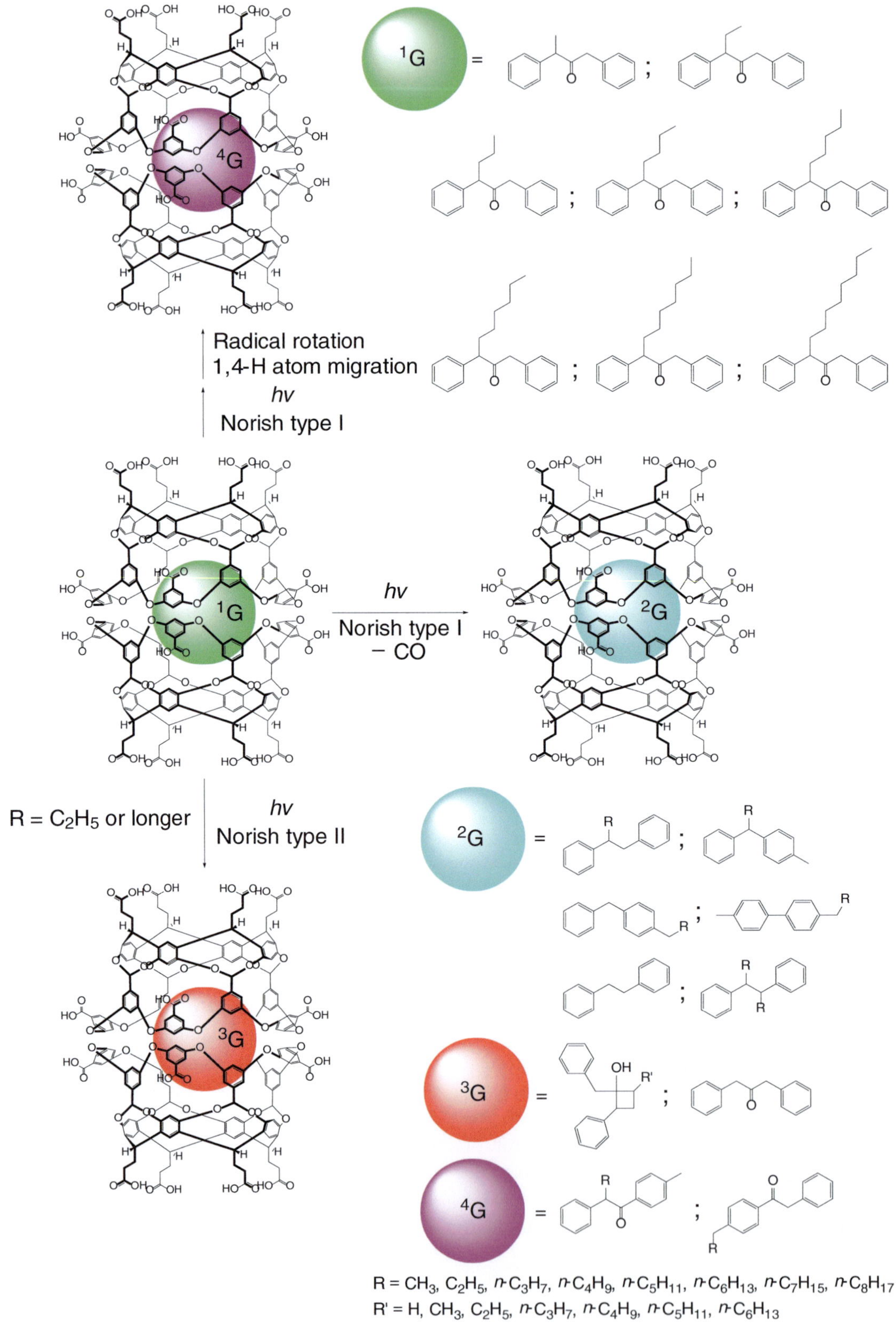

^{1}G
^{4}G
^{2}G
^{3}G
Radical rotation
1,4-H atom migration
hv
Norish type I
hv
Norish type I
− CO
R = C2H5 or longer
hv
Norish type II
R = CH3, C2H5, n-C3H7, n-C4H9, n-C5H11, n-C6H13, n-C7H15, n-C8H17
R' = H, CH3, C2H5, n-C3H7, n-C4H9, n-C5H11, n-C6H13

Scheme 5.57

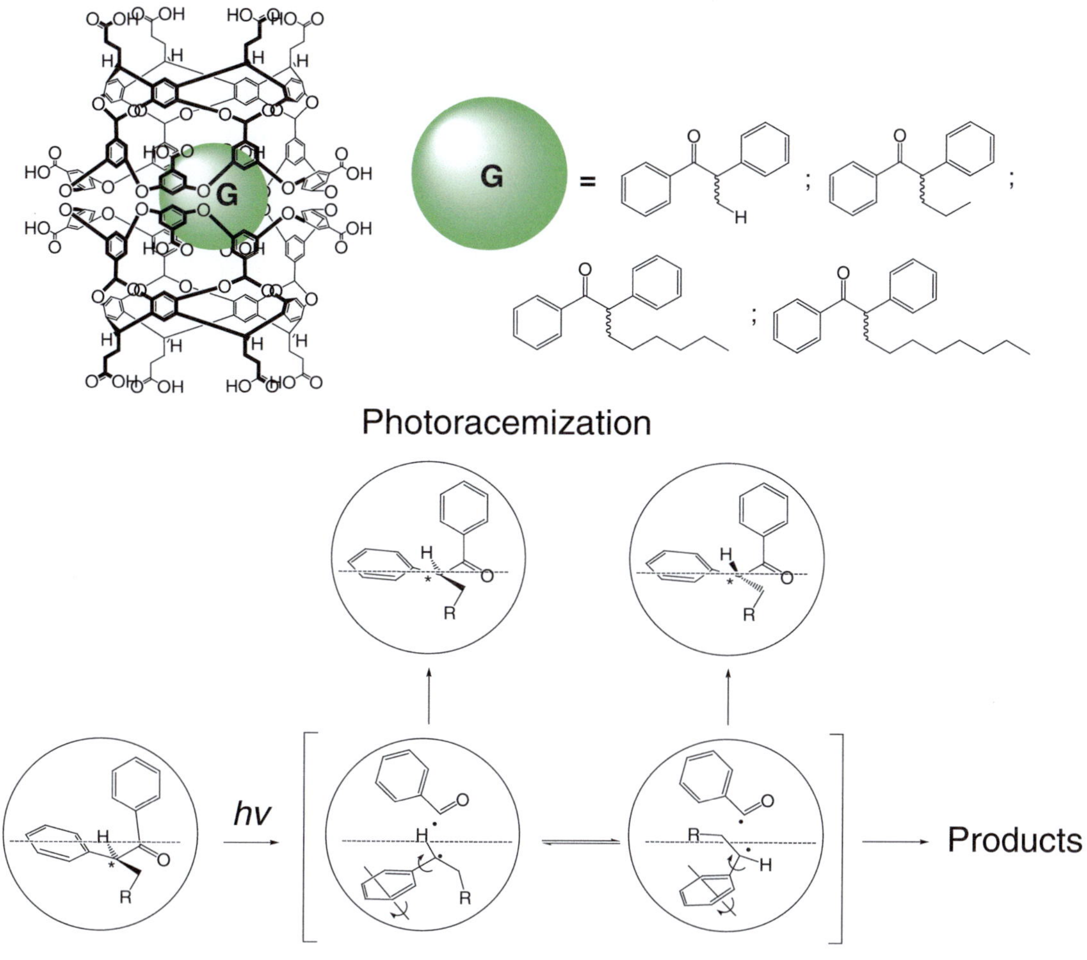

Scheme 5.58

trans-to-*cis* isomerization of 4,4′-dimethylazobenzene as a concurrent guest.

The same approach (i.e., *cis*-to-*trans* photoisomerization of azobenzenes shown in Scheme 5.61) has been used in [59] for photochemical control of reversible encapsulation by hydrogen-bonded capsules **429**, **435**, and **440**. Photoexcitation of the guest *trans*-4,4′-dimethylazobenzene molecule caused the release of its formed *cis*-isomer, which no longer fits within this capsule, thus making it possible to encapsulate other guests. On heating its bulk solution, *cis*-4,4′-dimethylazobenzene undergoes a reverse thermal isomerization to its *trans*-isomer, easily and rapidly releasing caged molecules in to the bulk solution; the release–exchange cycle can be repeated many times [59]. The same wavelength-dependent and sequential light-triggered processes by Scheme 5.62 for the hydrogen-bonded capsules **429** and **440** have been performed in [60]. A photo- and thermochemically operated self-assembled system shown in Scheme 5.63 allowed the authors of [61] to reversibly control the switching process between the homodimeric bis-cavitand capsule **429** and its hybrid analog **436**. Undesired cage complexes have been eliminated from the reaction mixture using appropriate guests with high selectivity and affinity for this encapsulating ligand.

Encapsulation of ruthenium(II) tris-bipyridinate by a hydrogen-bonded hexameric resorcin[4]arene capsule **397** (Scheme 5.64) is described in [62] to turn off photocatalytic aerobic oxidation of an aliphatic sulfide with this photosensitizer.

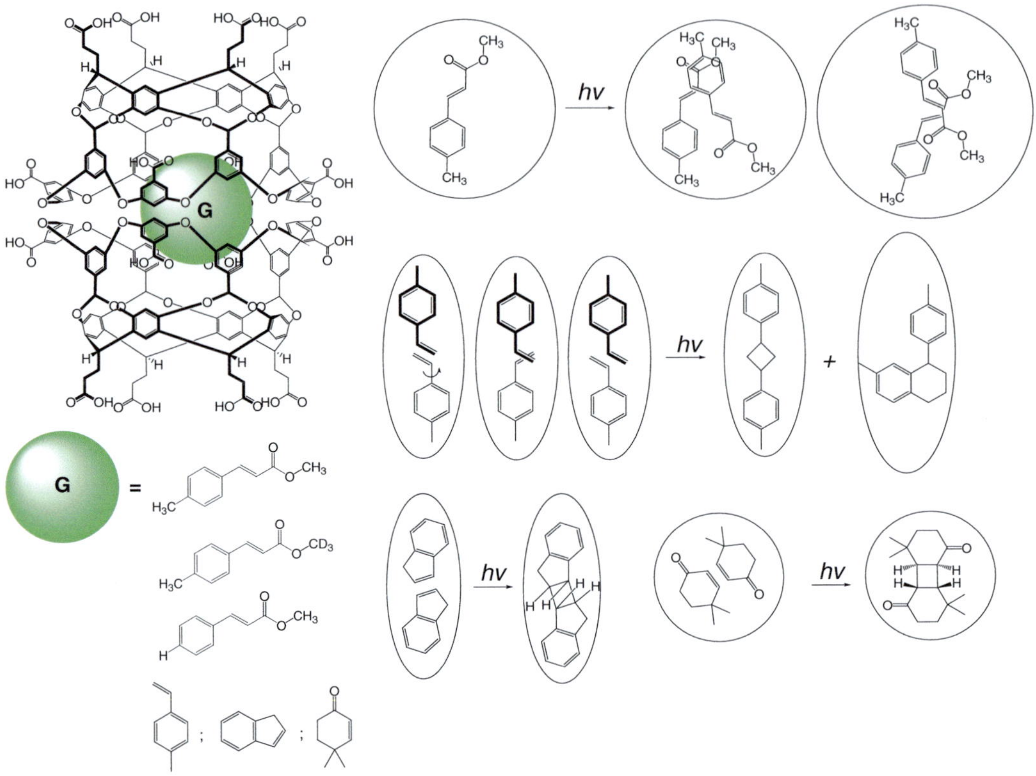

Scheme 5.59

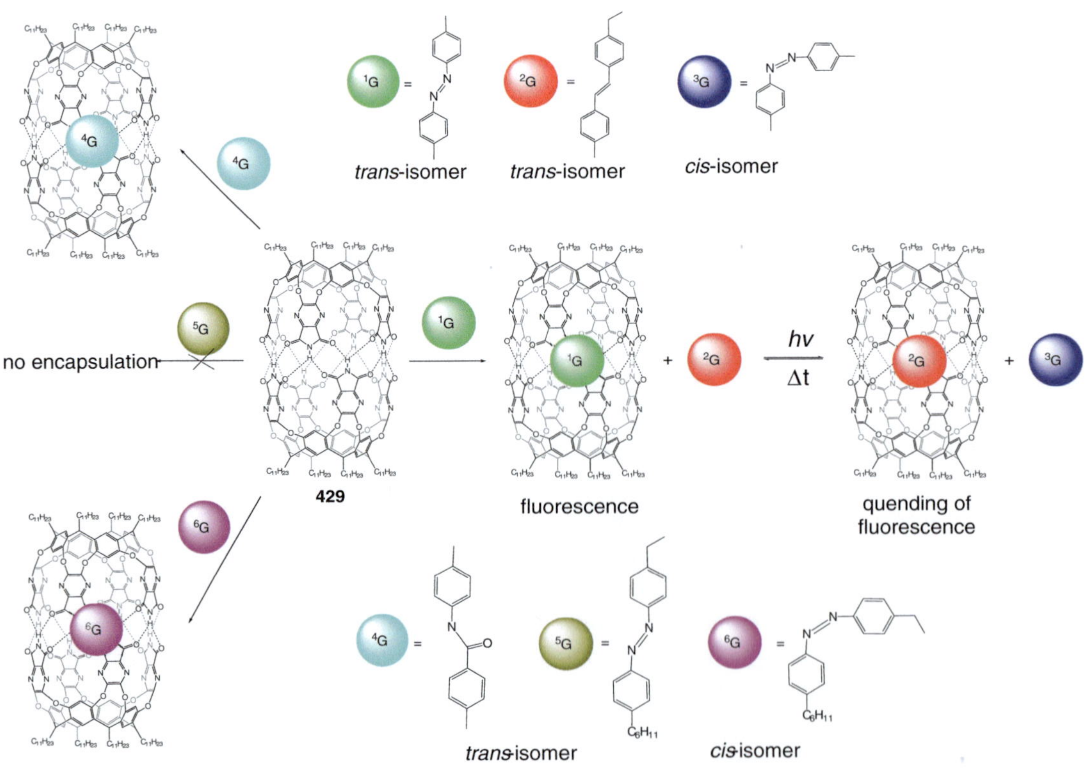

Scheme 5.60

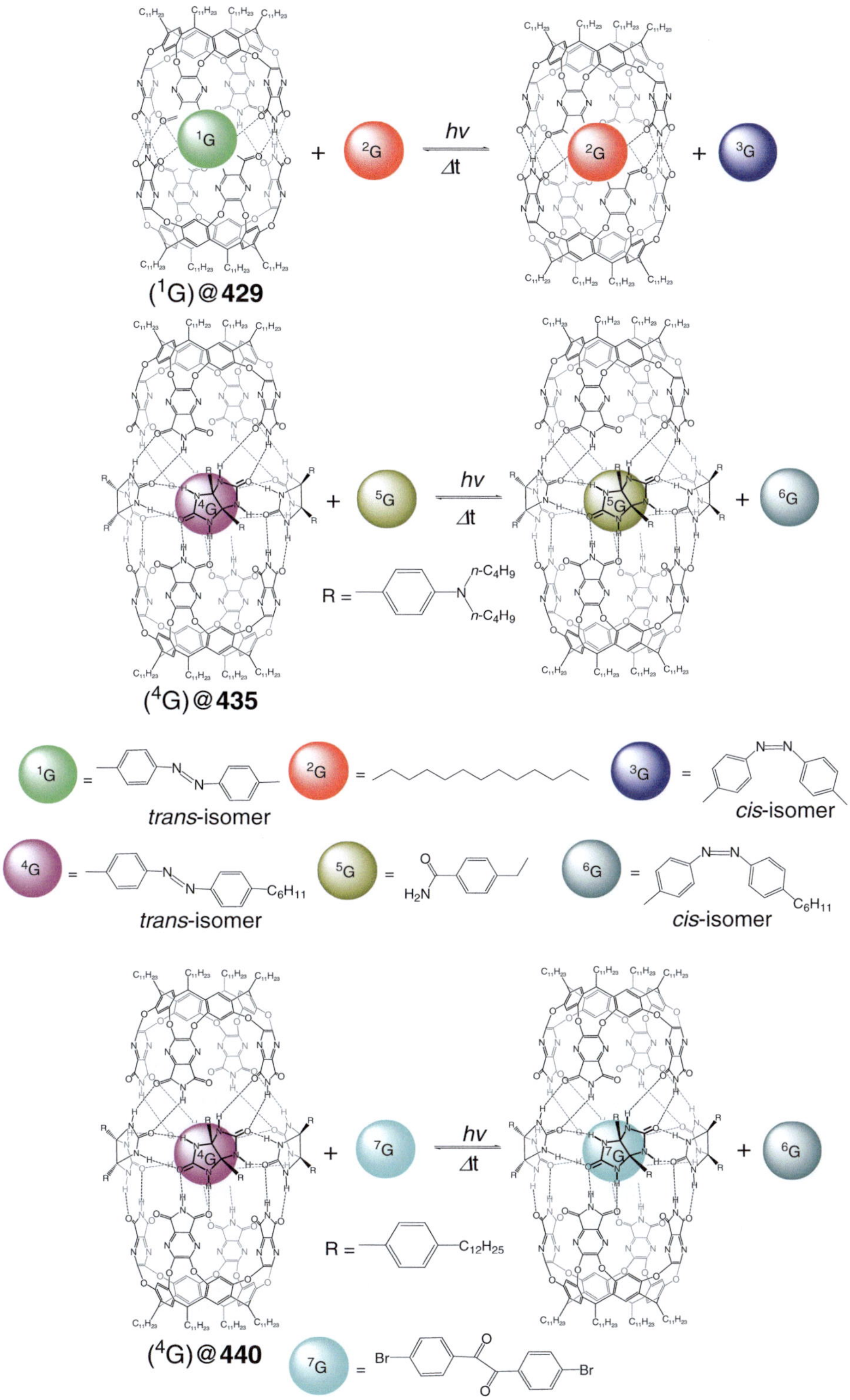

(¹G)@429
hν
Δt
(⁴G)@435
R = n-C4H9 N n-C4H9
trans-isomer
cis-isomer
trans-isomer
cis-isomer
C6H11
H2N
(⁴G)@440
R = C12H25
Br

Scheme 5.61

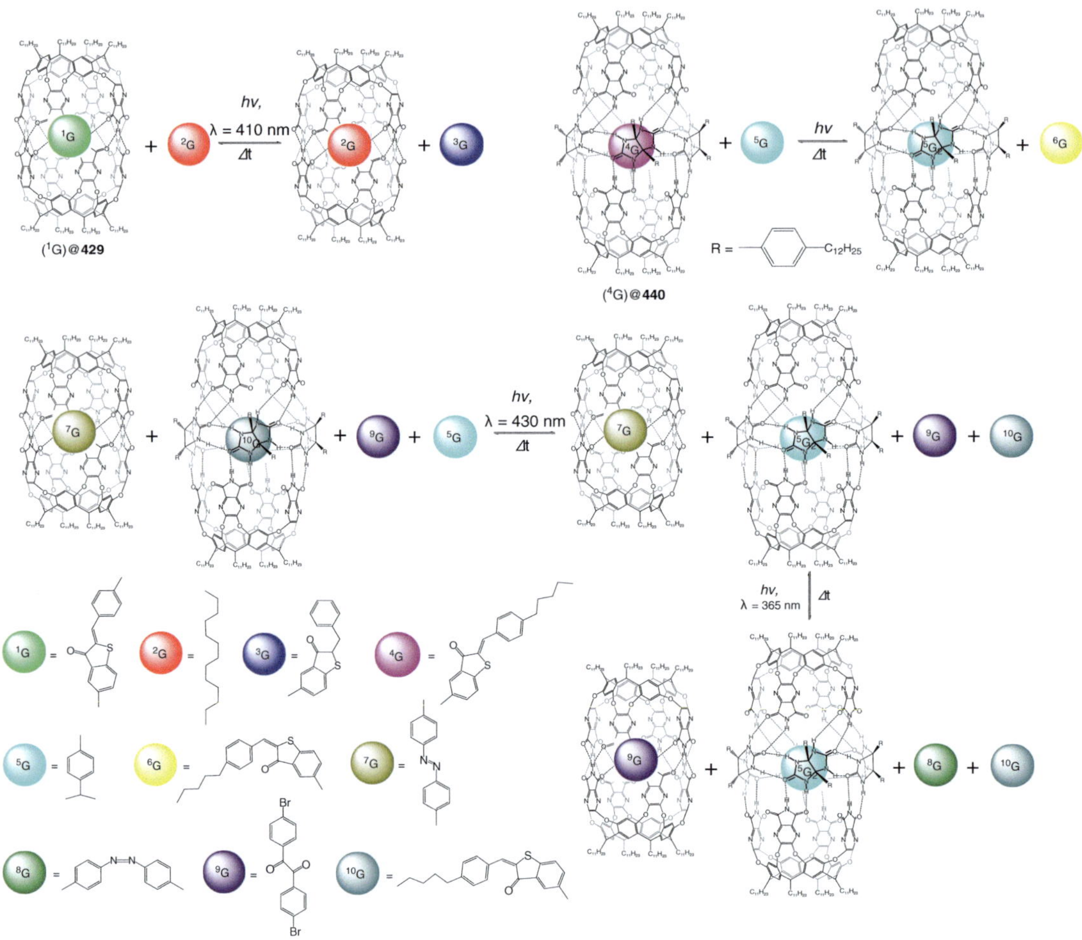

Scheme 5.62

5.3 Redox Reactions

Two encapsulated TTF molecules undergo stepwise oxidation by Scheme 5.65 within the cavity of a coordination capsule **552**, which results in mixed-valence monocation-radical species and bis(cation-radical) form and then a bis-dicationic cage complex [63]. This complex is unstable on the CV time scale, as it releases the guests, whereas the monooxidized caged radical species, which do not form in a bulk solution, are stable for a relatively long time. Their stability is explained [63] by efficient stacking interactions between two caged TTF molecules [63]. Slow electrochemical oxidation of the same guest within the cavity of a bis-cavitand hydrogen-bonded capsule **468** resulted in disassembly of its cage framework and in the release of less hydrophobic cation-radical species (Scheme 5.66) [64].

Redox reactions of the encapsulated ferrocene within the cavity of the octaacid caging ligand **468** have been studied in [65]. The oxidized dicationic ferrocene derivative undergoes intramolecular ET from the guest molecule with its re-reduction to ferrocenyl monocation giving its cage complex. This complex releases the monocation, thus electrochemical reduction of which resulted in the ferrocene re-encapsulation by Scheme 5.67 [65]. When ferrocenylcarboxylate anion has been tested instead of Fc^{+} ion, the ET process did not occur due to the absence of any strong interactions between the ferrocenyl-based electron mediator and the exterior of the molecule (Fc)@**468**. Indeed, the study of this

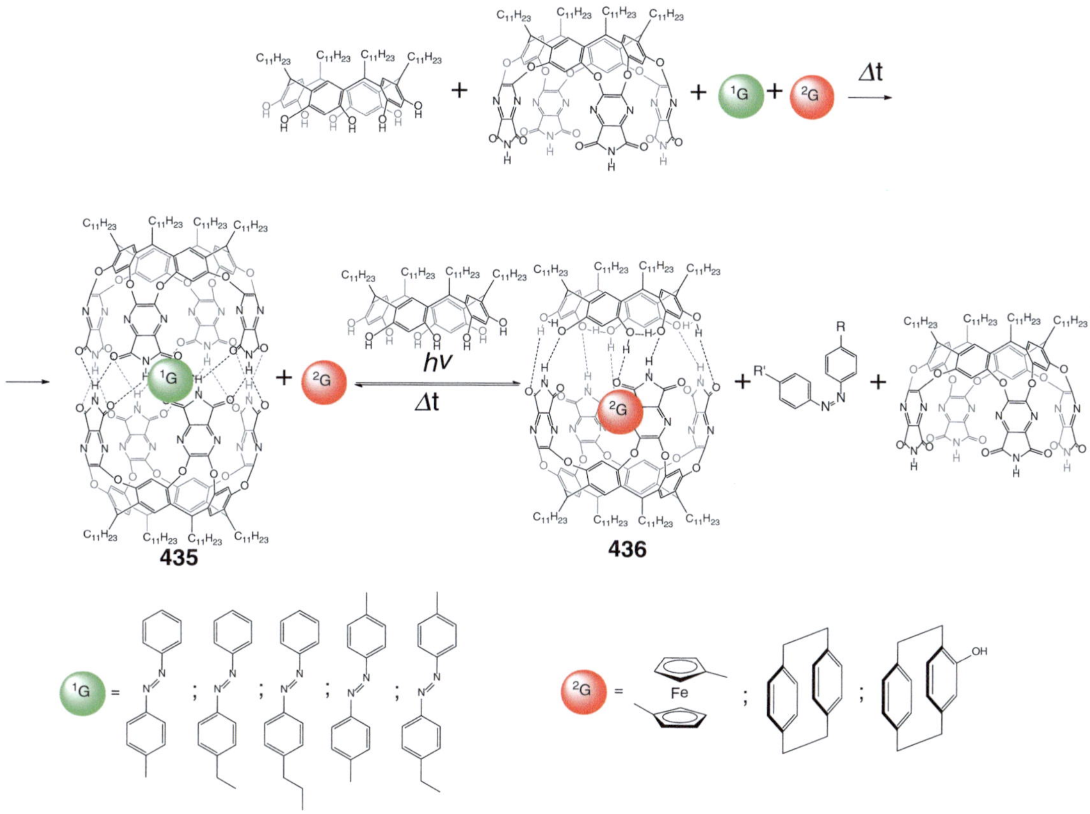

Scheme 5.63

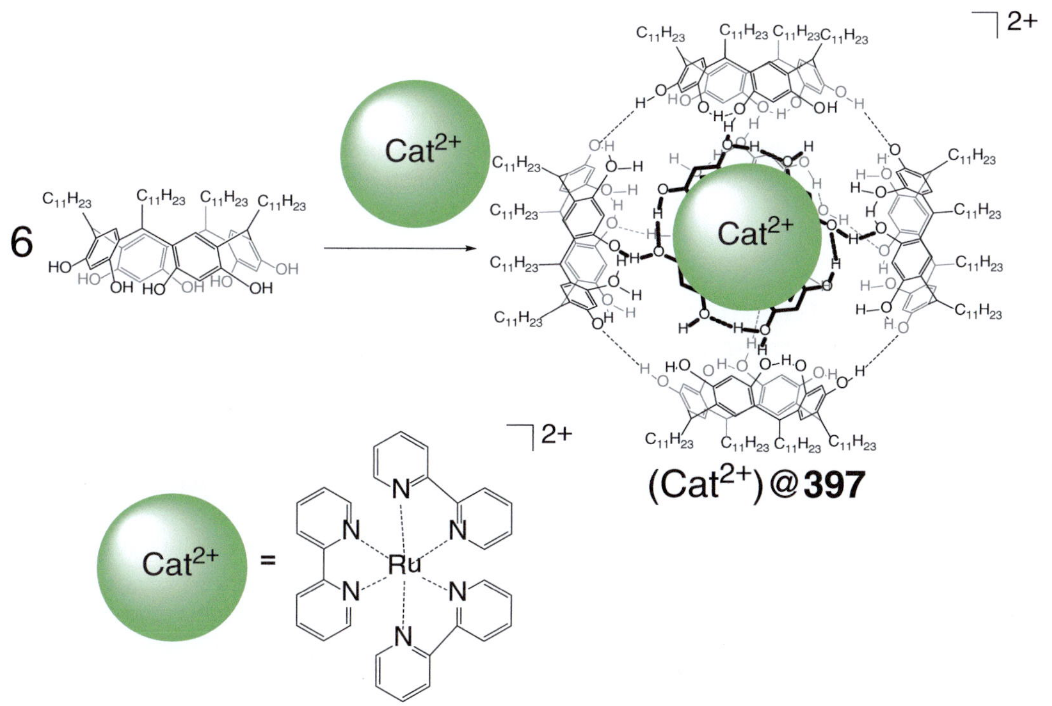

Scheme 5.64

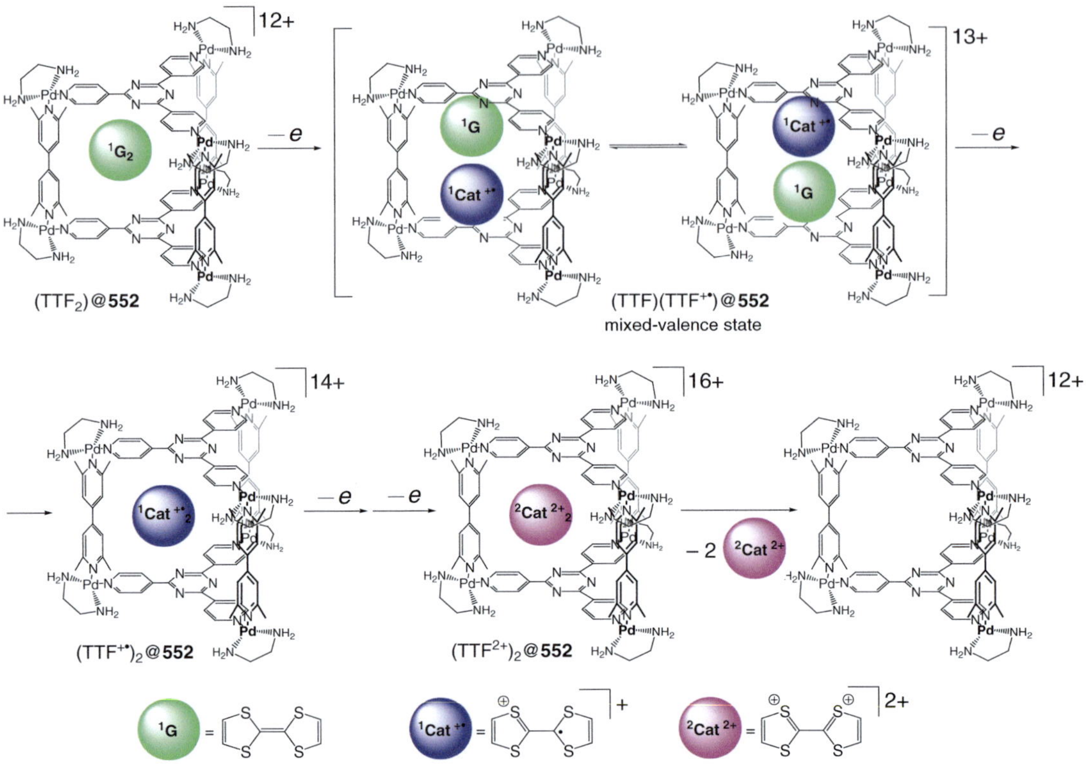

Scheme 5.65

mediated electrochemical oxidation process in the presence of cucurbit[7]uril as a host for the ferrocenyl cation showed decrease in the rate of this process due to the formation of their host–guest complex. Its occurrence at measurable rates thus needs strong outer-sphere binding between the ferrocene-based mediator and the hydrogen-bonded capsule [65].

5.4 Exchange, Extraction (Guest Release), Rearrangement, and Decomposition Reactions

Mixing of tetrahedral $M_6L^1_4$ and TP $M_3L^2_2$ coordination capsules **483** and **550** is reported in [66] to give a hybrid cage framework **631** by Scheme 5.68.

Encapsulation by molecular and polymeric coordination capsules **483** and **493** is reported in [67] to be reversible: organic solvents (such as toluene) extract the guests from the cavities of these capsules, thus giving the empty cage frameworks by Scheme 5.69.

As the Ag_4L_4 coordination capsule **579** is able to encapsulate one adamantane molecule to give a 1:1 host–guest cage complex in contrast to its Ag_6L_4 analog **580**, this system has been used in [68] for reversible encapsulation/release of such guest by addition of two equivalents of $AgPF_6$ that resulted in the formation of **580** (Scheme 5.70). The following addition of two equivalents of the cryptand *461* that traps these Ag^+ ions caused re-encapsulation of the adamantane molecule with the formation of the initial cage complex of **579** [68].

A tetrahedral Ga_4L_6 coordination capsule **575**, a derivative of the labile achiral ligand syntone *348*, is reported in [69] to exhibit structural memory (i.e., retention of chirality of a cage framework). Although its molecular composition has been altered, chirality of the coordination capsule serves as a structural reporter, preserving the

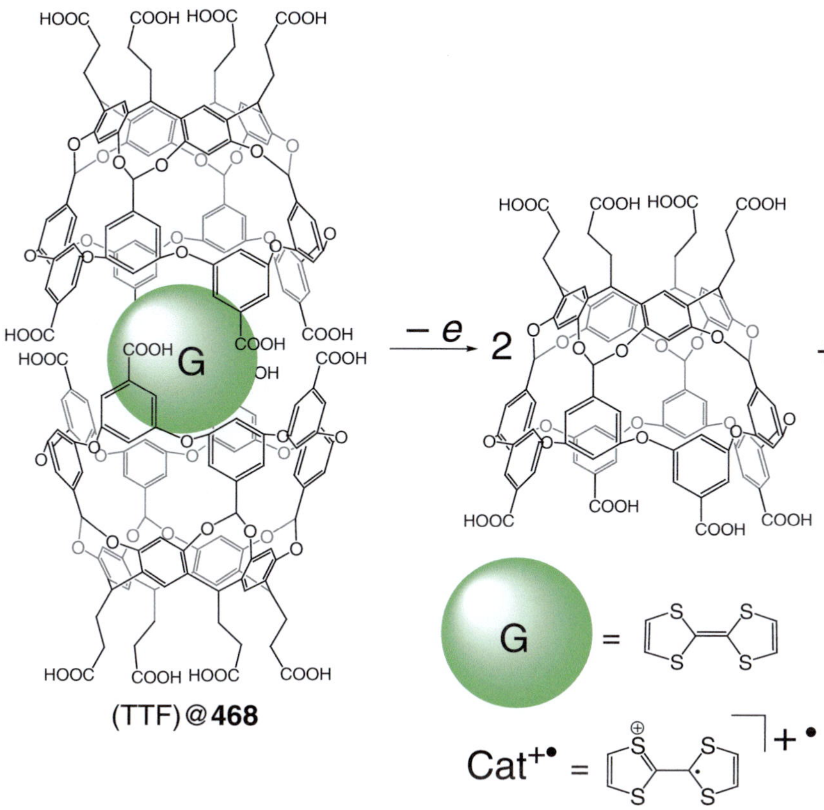

Scheme 5.66

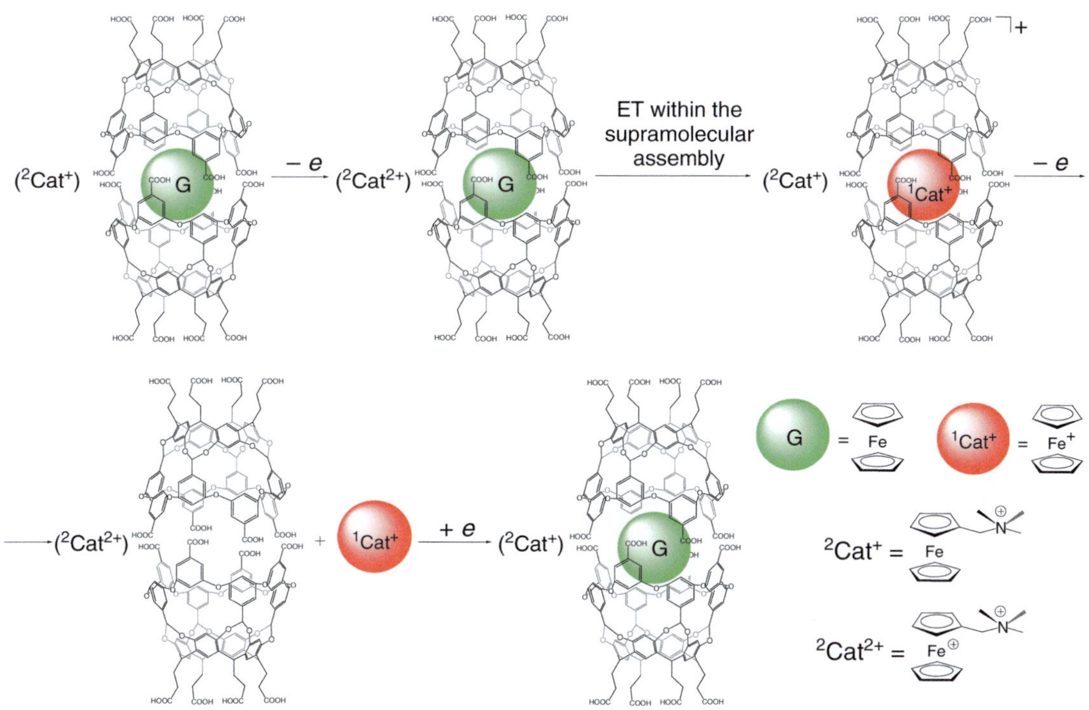

Scheme 5.67

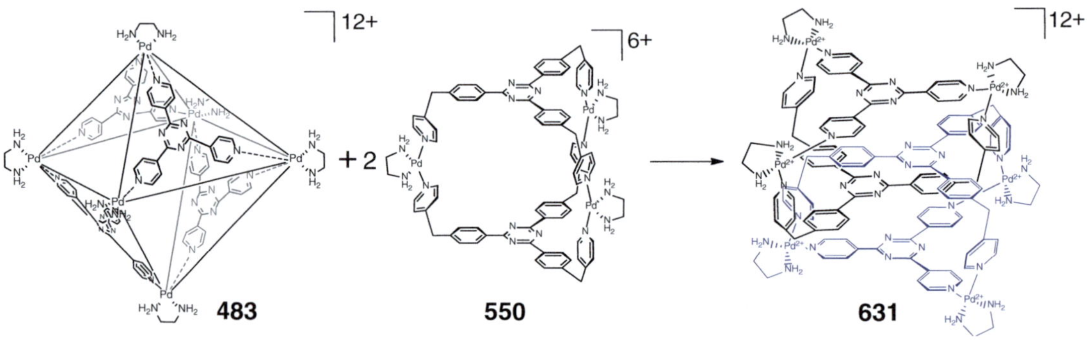

Scheme 5.68

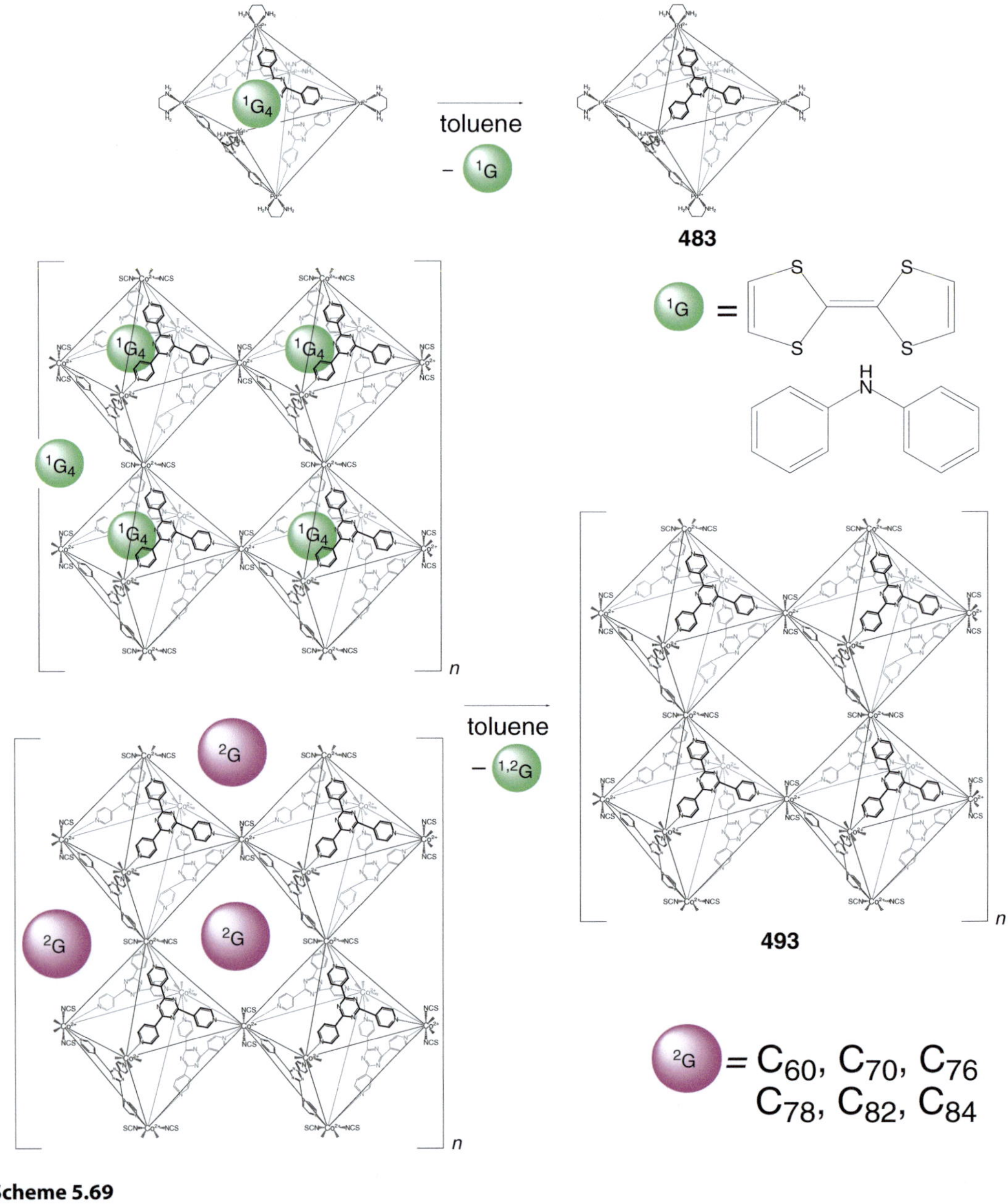

Scheme 5.69

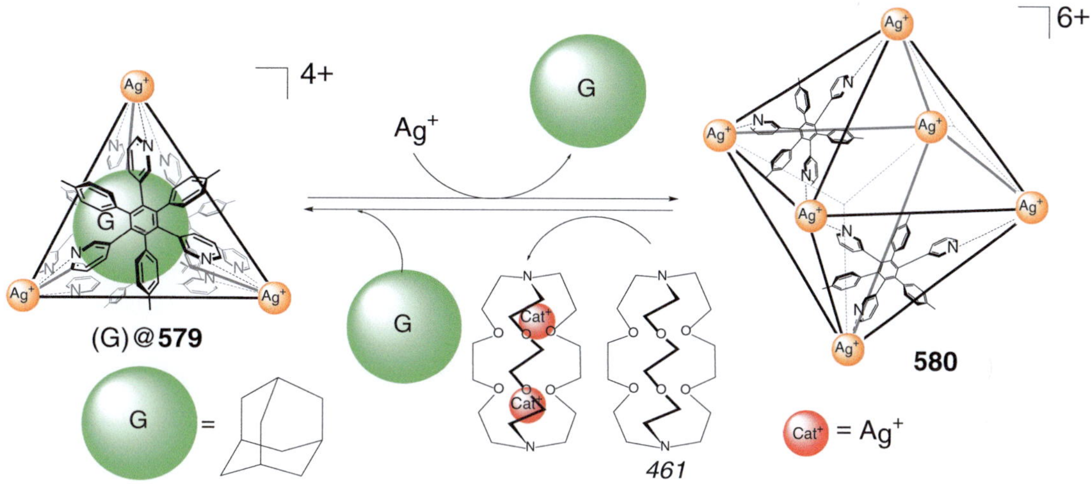

Scheme 5.70

initial cage structure. As follows from ^{1}H NMR and CD data, the interaction by Scheme 5.71 of tetrahedral 1:1 cage complex of the naphthalene *ΔΔΔΔ* ligand syntone *348* with encapsulated $(C_2H_5)_4N^+$ cation and the corresponding M_2L_3 helicate **824** of its *para*-phenylene analog *354*, which does not form a tetrahedral Ga_4L_6 coordination capsule, gave heteroligand $Ga_4L_{6-n}L'_n$ cage complexes of the same chirality [69].

Host–guest exchange kinetics for the Ga_4L_6 coordination capsule **575** has been studied in [70] using various ^{1}H NMR techniques. Cationic guest self-exchange in this host by Scheme 5.72 proceeds through a non-dissociative mechanism in which an aperture in its cage framework enlarges to accommodate guest passage. Note that the guest displacement reaction proceeds through dissociation of an initially caged cation, and then encapsulation of a second positively charged guest takes place. The cations showing binding are reported in [70] to produce higher rates of the guest exchange. As follows from PGSE ^{1}H NMR data [71], the diffusion coefficient of this capsule also depends on the nature of cationic counterions (Scheme 5.73) in solution. Saturation studies revealed a small number of ion-association sites on the exterior of **575** and provided direct observation of ionic association in water. Addition of an excess of alkali metal cations caused displacement of the associated hydrophobic tetraalkylammonium cations, while the comparison between $(C_2H_5)_4N^+$ and $(n\text{-}C_3H_7)_4N^+$ cations as guests showed a preference for ionic association with more hydrophobic guests suggesting a stepwise guest encapsulation process. On the first stage, the freely solvated cationic guest forms an ionic associate(s) with the dodecaanionic host **575**, and this process is followed by its encapsulation [71].

Deuterium substitution in cationic guests, isotopologues of the $[D_n][CpRu(\eta^6\text{-}C_6H_6)]^+$ cation (Scheme 5.73), has been used in [72] to probe the sensitivity limits of the guest exchange process with the highly negatively charged caging host. Kinetics of their exchange in the presence of an excess of $(C_2H_5)_4P^+$ cation as a more strongly binding guest has been monitored by ^{1}H NMR spectroscopy, which showed the deuteration of the guests to result in an inverse kinetic isotope effect (i.e., faster displacement of the deuterated guest from the cavity of **575**). Deuteration at cyclopentadienyl and benzene rings has a measurable impact on the guest exchange kinetics and the calculated isotope effect per deuterium atom values (up to 11 %). This effect has been attributed to increases in the relative spacing of C–H/D vibrational zero-point energy levels of the guest. It arises from steepening of the vibrational potential energy functions at the sterically strained transition state [72].

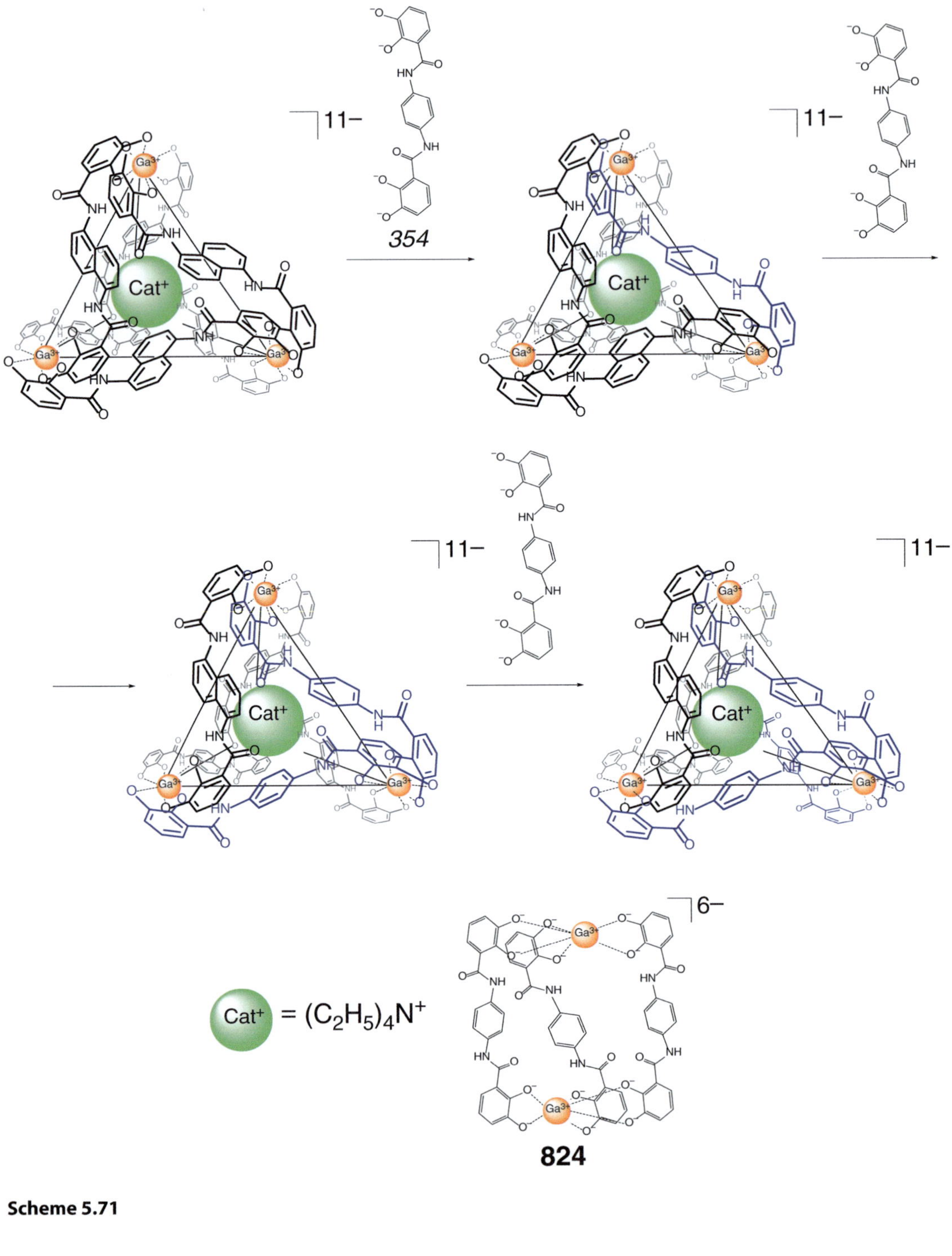

Scheme 5.71

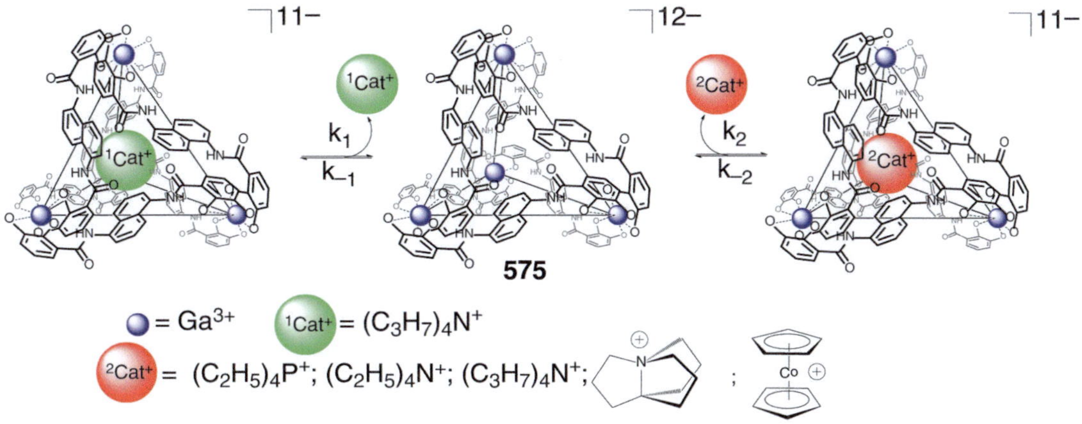

Scheme 5.72

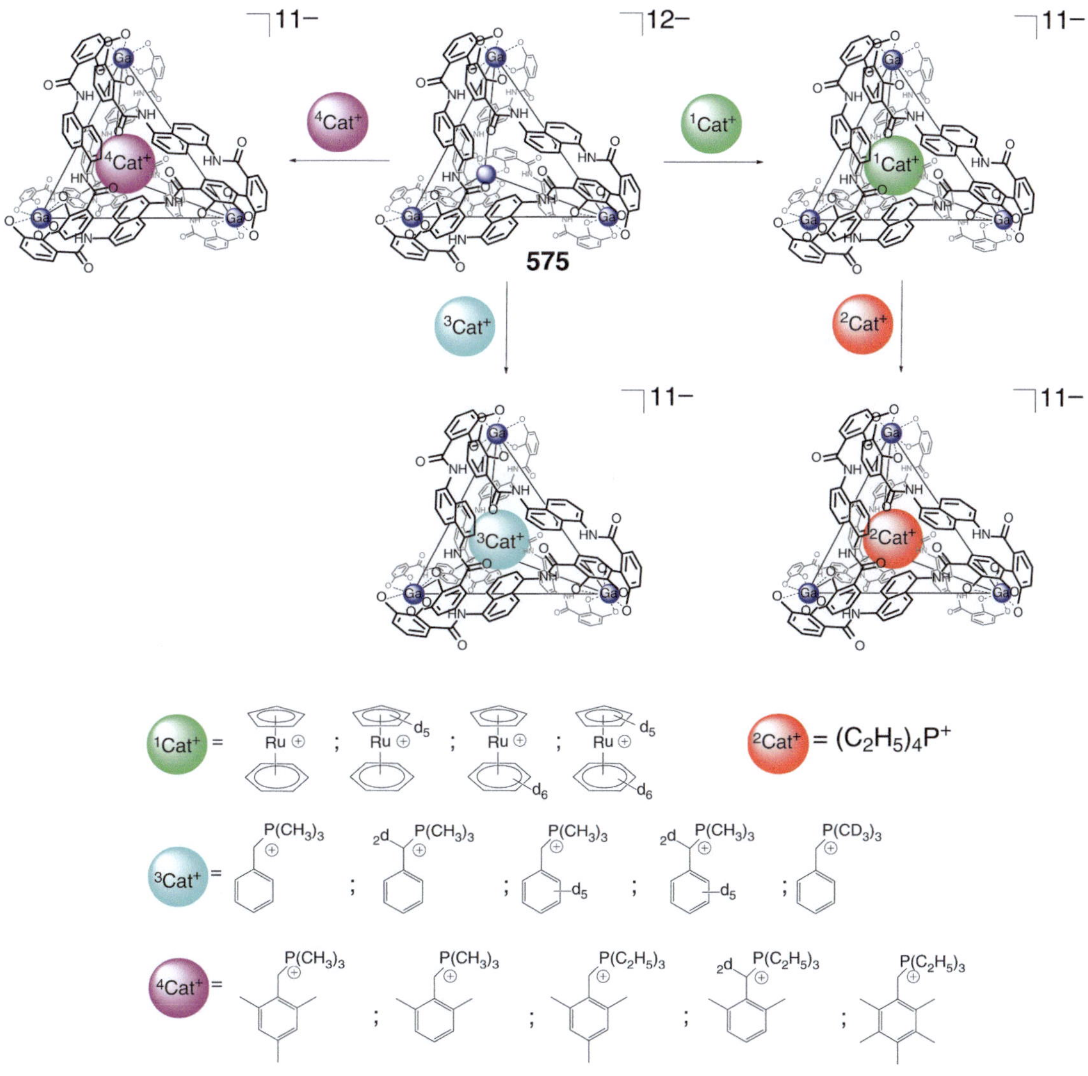

Scheme 5.73

NMR titration method has been used in [73] to measure secondary EIEs on the binding affinity of cationic guests shown in Scheme 5.73, which are isotopologues of benzyltrimethylphosphonium cation, toward the cavity of **575**. Deuteration of the methyl and benzyl C–H groups of the guest had larger equilibrium isotope effects than that of its aromatic C–H groups. This suggests small differences in the interactions of these C–H/D groups with the caging ligand [73]; the details of such high-precision NMR experiments are reported in [74]. The obtained EIE values have been explained by considering changes in vibrational force constants and zero-point energies, while DFT calculations confirmed the origins of these EIEs and suggested that changes in low-frequency C–H/D vibrational motions were mainly responsible for EIEs.

Motion dynamics (i.e., bond rotation and tumbling) of caged *ortho*-substituted benzyltrialkylphosphonium guest cations also shown in Scheme 5.73 has been used in [75] to probe steric consequences of encapsulation within the cavity of **575** using selective inversion recovery NMR method. The encapsulation increases the Ph–CH_2 bond rotational barrier for these caged cations in the range of 3–6 kcal per mol (depending on both their size and shape). The longer prolate-shaped guests tumble more slowly in the host's cavity than larger but more spherical cations reducing the symmetry of this capsule from T to C_1 in solution at low temperatures; such a distortion has been also observed in solid state [75]. The motion of the guest [76] is affected by the nature of a bulk solvent (water, methanol, or DMF) and applied pressure (up to 150 MPa), allowing one to use their rotational dynamics to probe changes in host's cavity size and flexibility. The NMR experiments demonstrate that an increase in solvent internal pressure or in applied external pressure reduces this size or flexibility, restricting the motion of the encapsulated guests and allowing one to tune their physical properties and reactivity within the cavity of this labile coordination capsule **575** [76].

A detailed study of the mechanism of guest exchange reactions for 1:1 cage complexes of the tetrahedral M_4L_6 coordination capsules with encapsulated organic monocations has been performed in [77] using three different theoretical and experimental strategies. First one tested the caging hosts with different lability of their frameworks, while the second theoretical and third experimental approach used MM calculations and the experimental data on the exchange of guest species with different sizes. These exchange reactions can proceed either by the Pathway I (Scheme 5.74) through partial dissociation of one catecholate fragment or by deformation of the cage framework to create a portal for guest passage by the Pathway II. The M_4L_6 capsules have four hollows on their C_3-symmetry axes leading from an inner cavity to the surrounding solution, and the non-dissociative Pathway II includes an increase in size of one of these hollows for the transport of encapsulated guests. To study a relevance of the dissociative Pathway I, the guest exchange experiments for the labile Ga_4L_6 caging ligand **575** and its inert Ti_4L_6 and Ge_4L_6 analogs **686** and **687** (Scheme 5.75) have been performed in [77]. According to ^{1}H NMR data, the reaction rates are nearly equal, with minor differences explained by different charges of their dodeca- and eight-anionic frameworks, respectively. Therefore, the guest exchange rate does not depend on the nature of metal–ligand interactions. The facile guest exchange for all these M_4L_6 coordination capsules suggests a non-dissociative mechanism by the Pathway II: the guest species squeeze into and out of a cage framework through its faces. High lability of these capsules has been confirmed in [77] by theoretical MM calculations. The extension of their C_3-symmetric hollows is 7.3 Å for tetramethylammonium guest ion and becomes 9.8 Å for pentamethylcyclopentadienyl cobalt cation, with twice the calculated activation energy. In the case of the partial dissociation mechanism (Pathway I), the guest exchange is practically independent from the nature of a M_4L_6 caging ligand. In contrast, the rate of the non-dissociative process strongly depends on these factors. As follows from the experimental data for the 1:1 cage complex with an encapsulated $Cp^*{}_2Co^+$ cation, this large cationic guest cannot freely rotate within the ligand's cavity, so the corresponding 1:1 cage

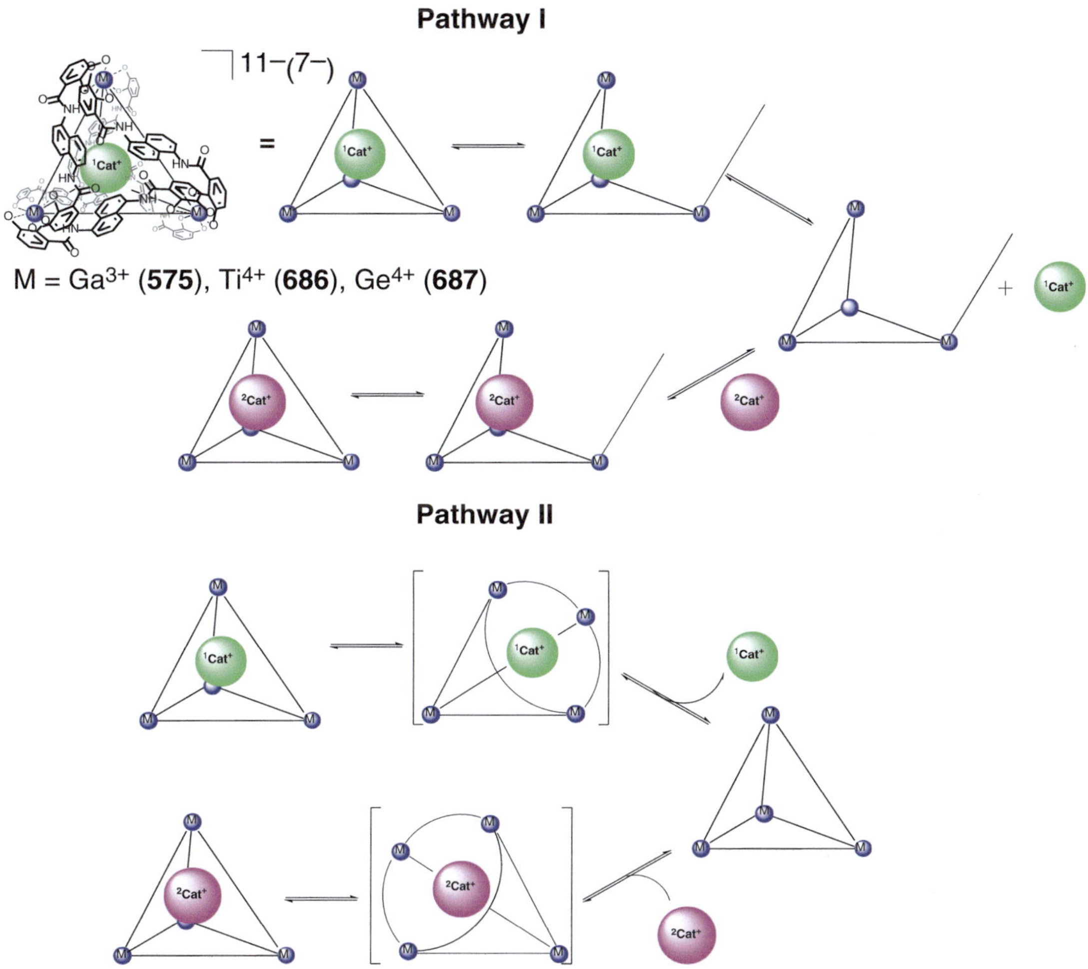

Scheme 5.74

complex loses its C_3-symmetry axis. As a result, it adopts a D_2 symmetry instead of a T symmetry for all small guest cations. This organometallic guest did not undergo an exchange reaction with an excess of tetraethylphosphonium cation even for 21 days at r.t., and it starts at 75 °C only, whereas for other cationic guests, the exchange reaction started in less than 10 min (in particular, half-life for the exchange reaction in the system $((C_2H_5)_4N)^+$–$((C_2H_5)_4P)^+$ is approximately 23 s at r.t.). A dramatic difference in the exchange reaction rates is explained [77] by the non-dissociative mechanism with strong dependence on the size and shape of the encapsulated guests.

Four $[Li(DMF)_3]^+$ cations encapsulated by a Ti_4L_4 coordination capsule **690** are reported in [78] to undergo stepwise substitution by Scheme 5.76 with 4-bromoanilinium and 3,5-dimethylanilinium ions as concurrent cationic guests, which also form host–guest 1:4 cage complexes. In contrast, neutral electron-rich or electron-deficient guests do not substitute the caged $[Li(DMF)_3]^+$ cations under the same reaction conditions [78].

A cage complex of the coordination capsule **581** with cyclohexane decomposes in the presence of amine *tren* through imine exchange by Scheme 5.77 with a following release of the guest [79], and the initial complex does not reform after addition of an acid. In contrast, variation in pH has been used for the reversible release of cyclohexane by its exchange with an

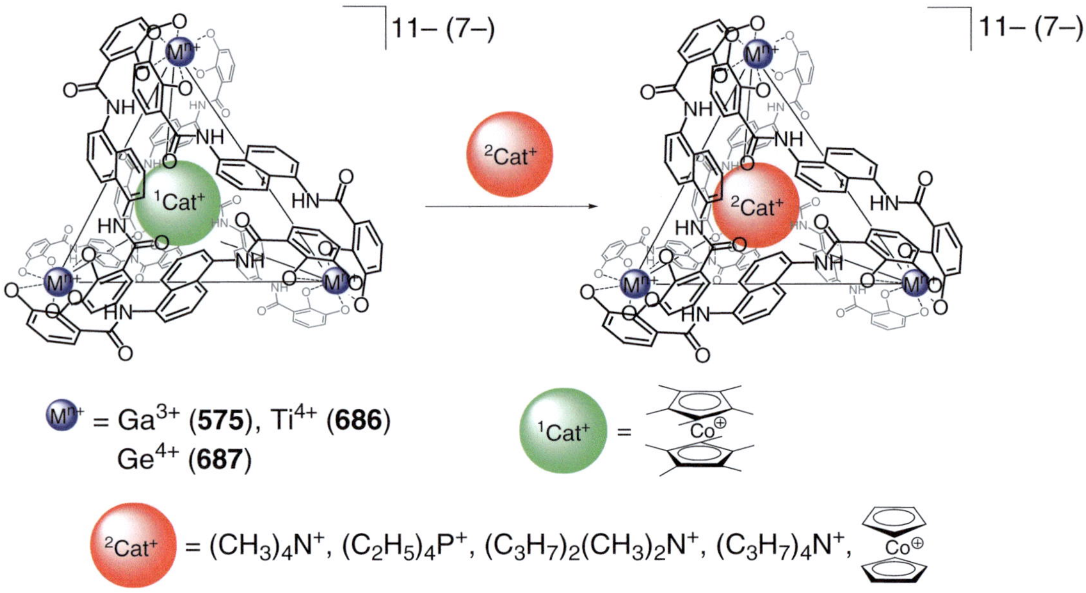

Scheme 5.75

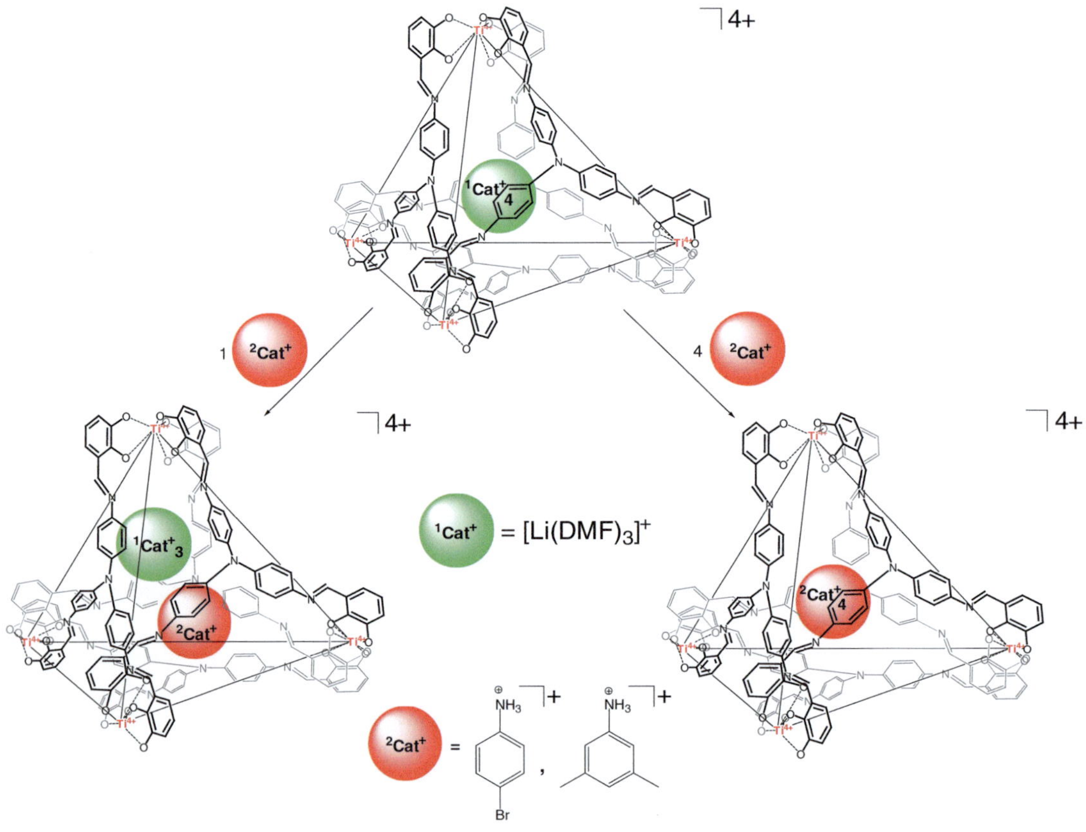

Scheme 5.76

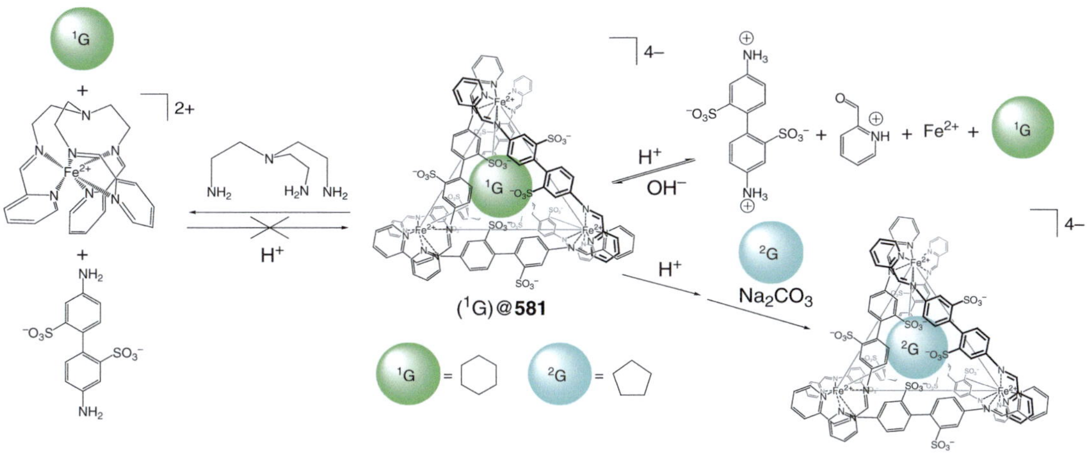

Scheme 5.77

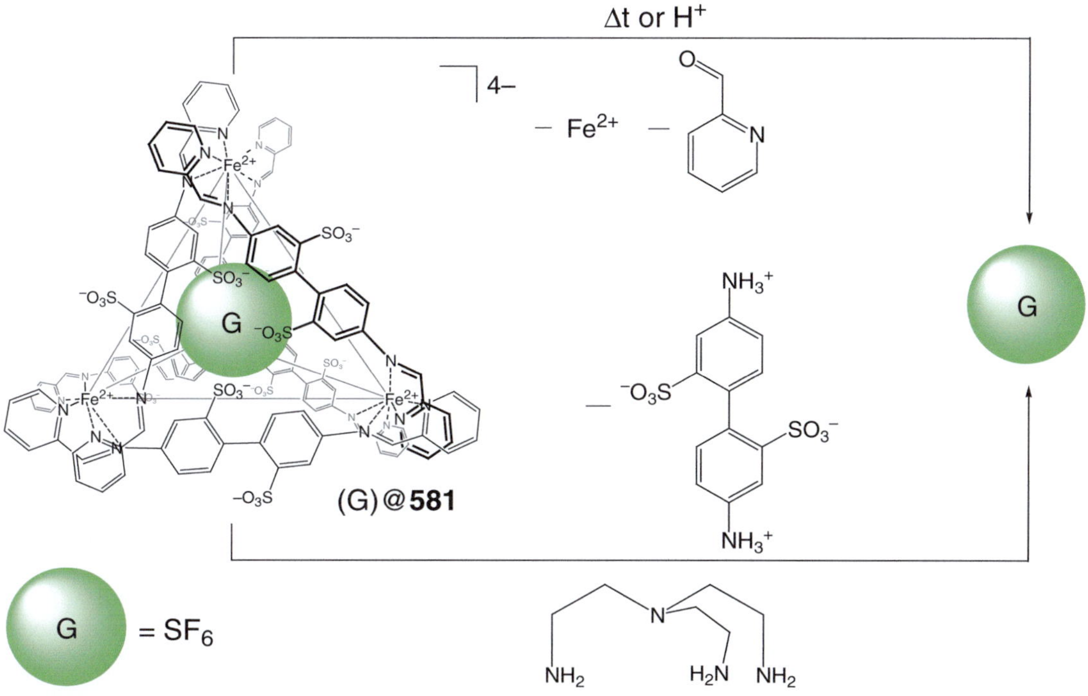

Scheme 5.78

excess of cyclopentane [79]. A 1:1 cage complex of **581** with encapsulated SF_6 molecule [80] releases the guest by Scheme 5.78 via two main pathways: (*i*) while the complex is heated or kept in acidic media or (*ii*) while it is treated with the same tripodal amine to remove iron(II) ions and to form the corresponding Schiff bases with 2-formylpyridine syntones [80].

As follows from the ^{19}F NMR data [81], the tetrahedral coordination capsules **581** and **825** can encapsulate PF_6^- ion that completely substitutes the caged BF_4^- and $CF_3SO_3^-$ species by Scheme 5.79.

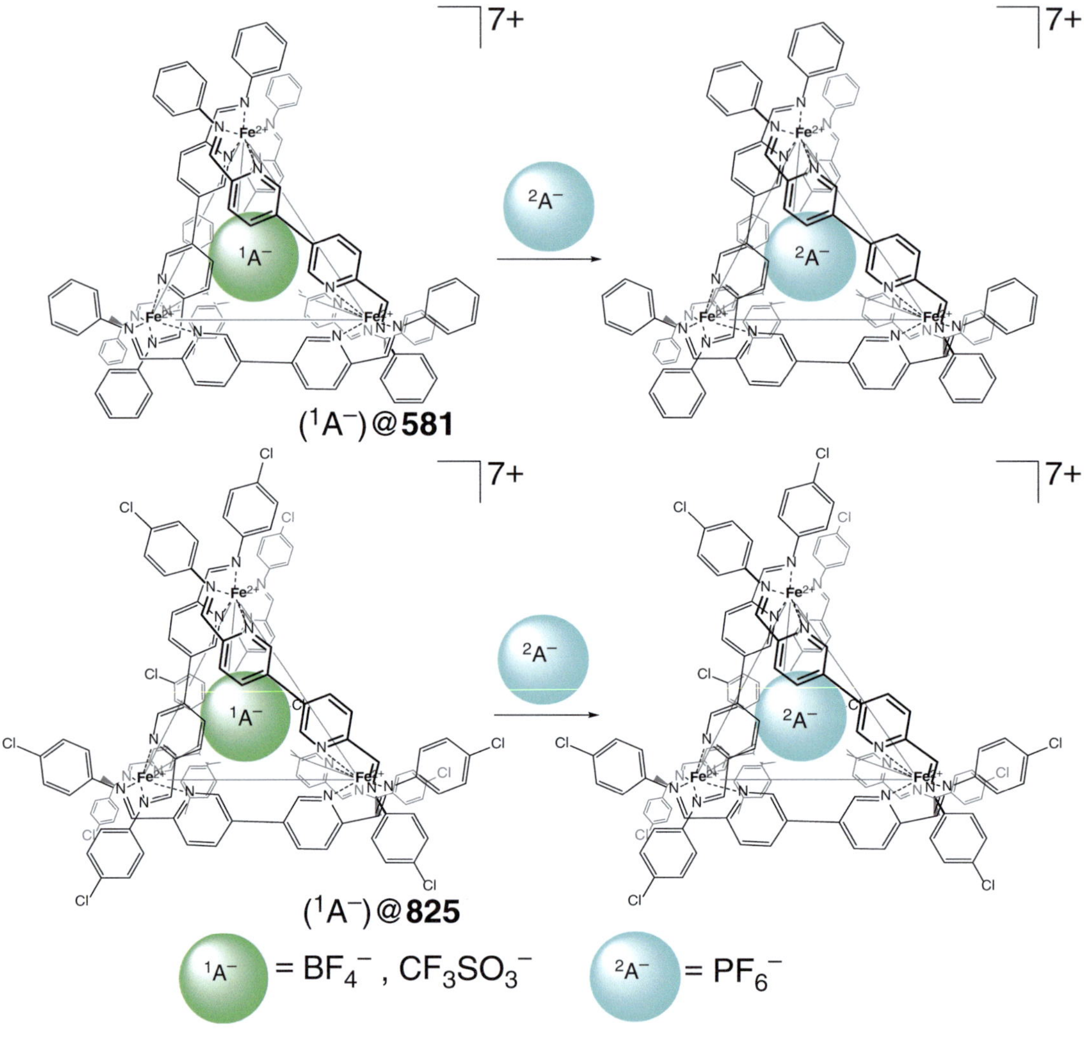

Scheme 5.79

Their *para*-chloroaniline cage complexes undergo exchange reactions with electron-rich anilines by Scheme 5.80 giving the corresponding imine capsules with the same encapsulated anions. As follows from ESI-MS data, the imine condensation of 3,3′-bipyridine-6,6′-dicarboxaldehyde with a mixture of *para*-bromo-, *para*-chloro-, and *para*-iodoanilines by Scheme 5.81 afforded a library of the corresponding imine capsules **826** and **827**. Addition of *para*-methoxyaniline to this mixture gave a homoleptic capsule **827** as the only product of the re-imination reaction [81].

Cubic hexaporphyrin Fe_8L_6 coordination capsules **622–624** are reported in [82] to decompose by Scheme 5.82 in acidic media or in the presence of tripodal amine *tren*. Their 1:1 complexes with an encapsulated fullerene C_{60} molecule easily undergo guest exchange reactions with an excess of coronene and fullerene C_{70}, giving the corresponding 1:1 cage compounds.

A 1:1 cage complex of tetragonal coordination capsule **581** with an encapsulated furane molecule (Scheme 5.83) is described in [83] to slowly dissociate in aqueous solution, whereas in the presence of benzene as concurrent guest, it releases furane into bulk solution during 1 h. Addition of maleimide resulted in its Diels–Alder reaction with furane in this solution [83].

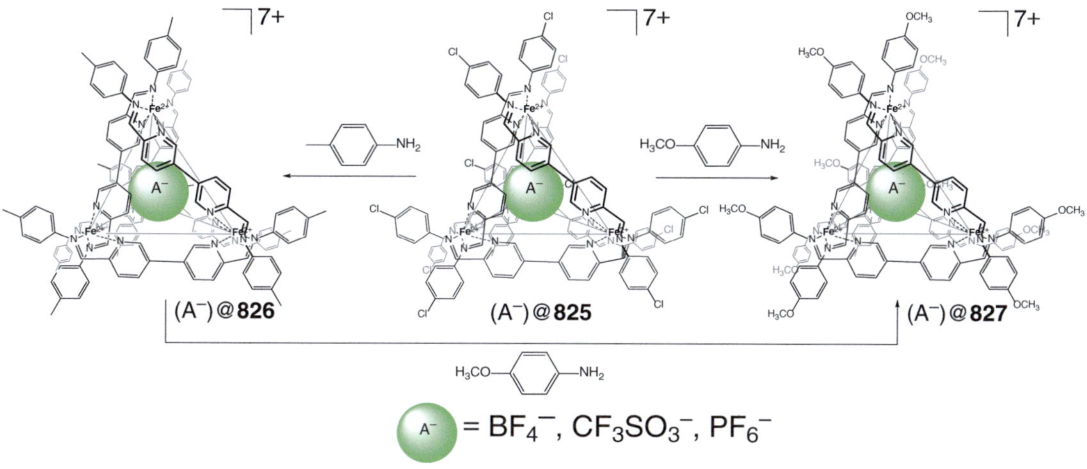

Scheme 5.80

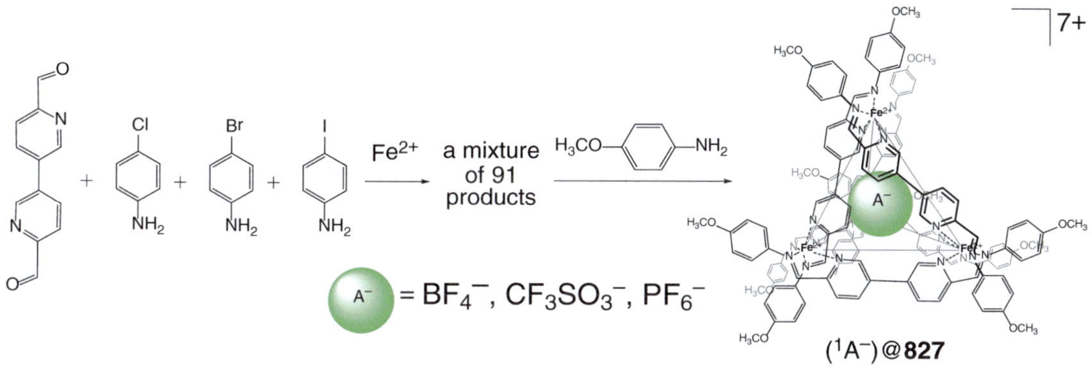

Scheme 5.81

Subcomponent substitution of a racemic mixture of isomers of 1:1 cage complex of **826** with encapsulated triflate anion in the presence of optically active primary amine *462* by Scheme 5.84 allowed to obtain an optically active *ΔΔΔΔ* form of this complex [84]. The use of a racemic mixture of *R*,*S*-isomers of the amine subcomponent afforded *R*,*S*-imine capsules **828** but not its homochiral analogs. Therefore, chiral self-sorting is not observed in this system.

Conversion of a tetrahedral M_4L_6 coordination capsule **829** into a heterometallic cubic cage framework **606** upon addition of a square-planar metallocomplex ligand syntone *319*, followed by the conversion of this cubic capsule into a tetrahedral capsule **668** upon addition of 4,4′-diaminobiphenyl (Scheme 5.85), has been performed in [85].

Guanidinium–sulfonate interactions between the faces of a tetragonal Fe_4L_6 coordination capsule **581** have been used for modulation of its guest exchange kinetics. A correlation between the concentration of guanidinium ions and the rate of the encapsulation process by Scheme 5.86 is described in [86].

Differential binding affinities of the smaller and larger Fe_4L_6 tetrahedral coordination capsules **668** and **830** for a series of guest anions in solution are reported in [87]; the corresponding anion-exchange sequence can be illustrated by Scheme 5.87.

Addition of triethylamine as an allosteric effector to a 1:1 cage complex of the coordination

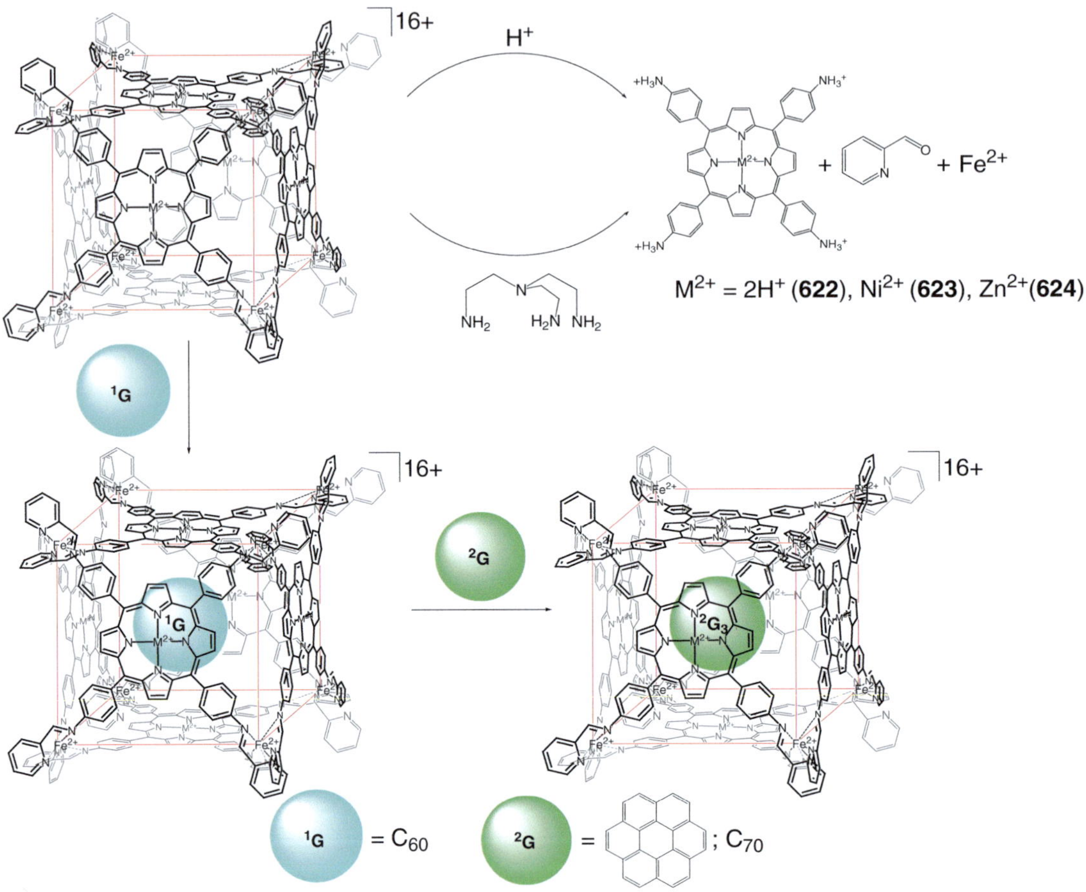

Scheme 5.82

capsule **619** with encapsulated *tert*-butylphosphine molecule has been used in [88] to allosterically release this guest from the cavity of **619** into bulk solution by Scheme 5.88.

An encapsulated anion $Cu(Au(CN)_2)_2^-$ undergoes transmetallation with silver(I) cation by Scheme 5.89 within the cavity of the coordination capsule **724** to give a 1:1 cage complex of monoanionic guest $Ag(Au(CN)_2)_2^-$, which is substantially more stable than its copper-containing analog. This fact is explained in [89] by a larger size of the Ag^+ cation as compared to Cu^{2+}, so it forms strong van der Waals interactions with π-electron-donating cavity of this encapsulating ligand. All the cage complexes of **724** release the encapsulated guest molecules in their DMSO solutions by Scheme 5.90 giving the D_{2d}-symmetric guest-free capsule **724** [89].

Anion-exchange reactions of cage complexes of the Ag_3L_2 ligand **700** with tetrahedral (ClO_4^- and BF_4^-) and octahedral (PF_6^- and SbF_6^-) anions by Scheme 5.91 have been studied in [90]. Anion-exchange experiments showed the anion selectivity to vary in a row $SbF_6^- > PF_6^- > BF_4^-, ClO_4^-$ and this ligand to prefer encapsulating octahedral and tetrahedral anions rather than their linear or triangular analogs.

Cage complexes of a coordination capsule **780** undergo transformation with an excess of tetra-*n*-butylammonium nitrate by Scheme 5.92 resulting in a 1D coordination polymer. This fact has been explained in [91] by the template effect of NO_3^- ion in this metathesis of the Co_2L_4 framework.

Disassembly and reassembly of a 1:1 cage complex of the capsule **781** with an encapsulated cisplatin guest molecule by Scheme 5.93

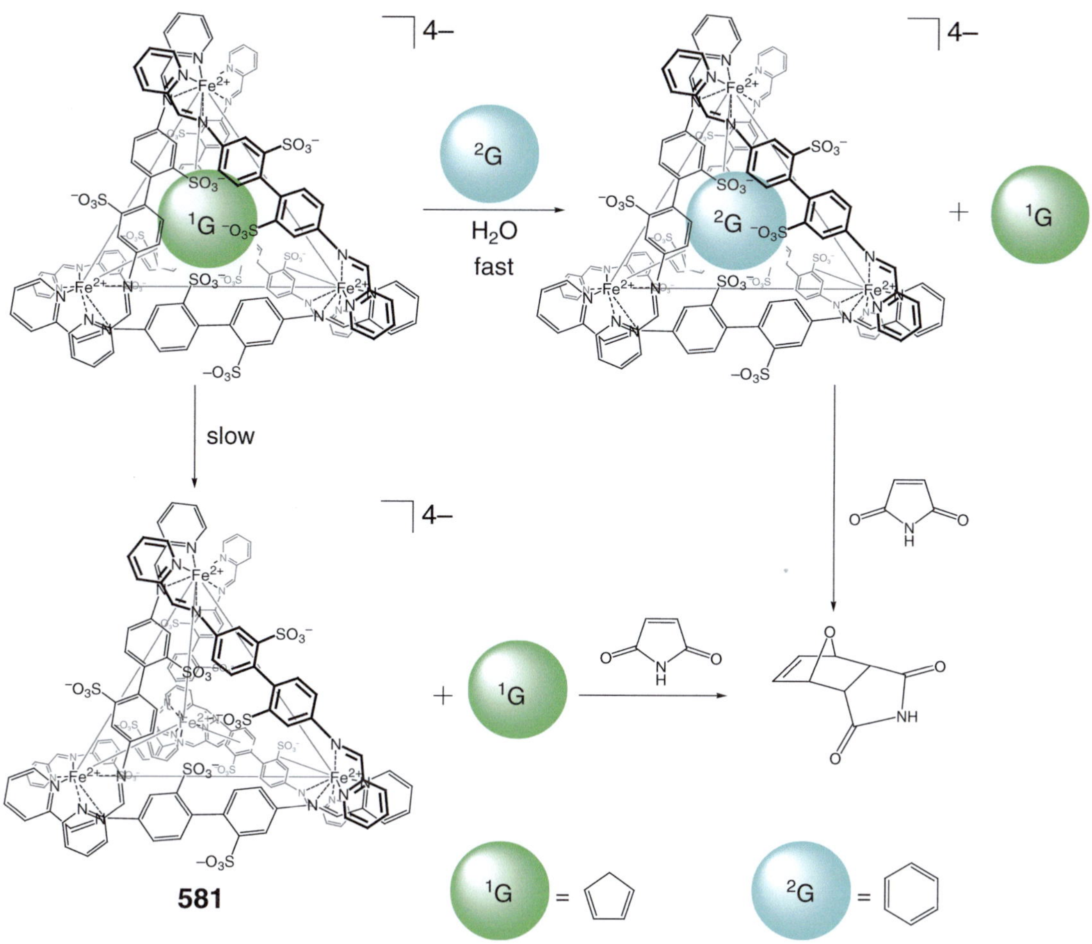

Scheme 5.83

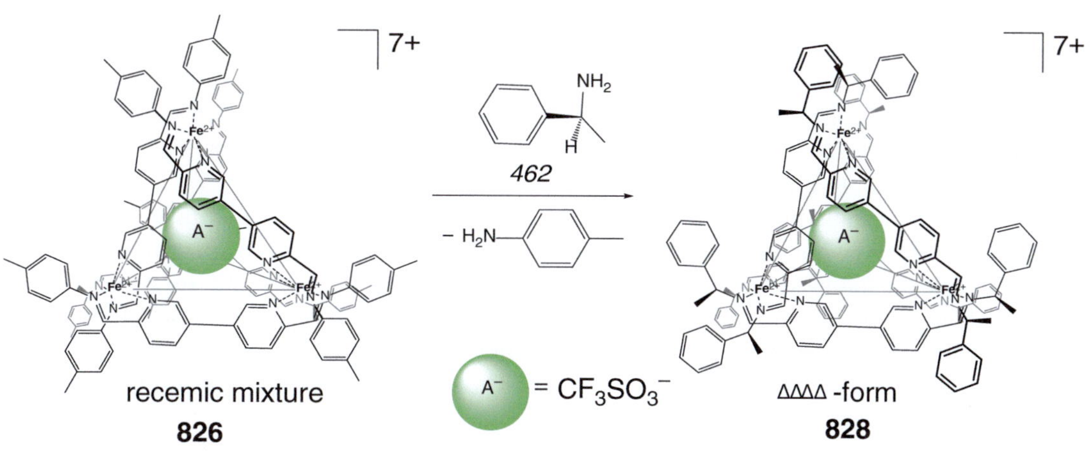

Scheme 5.84

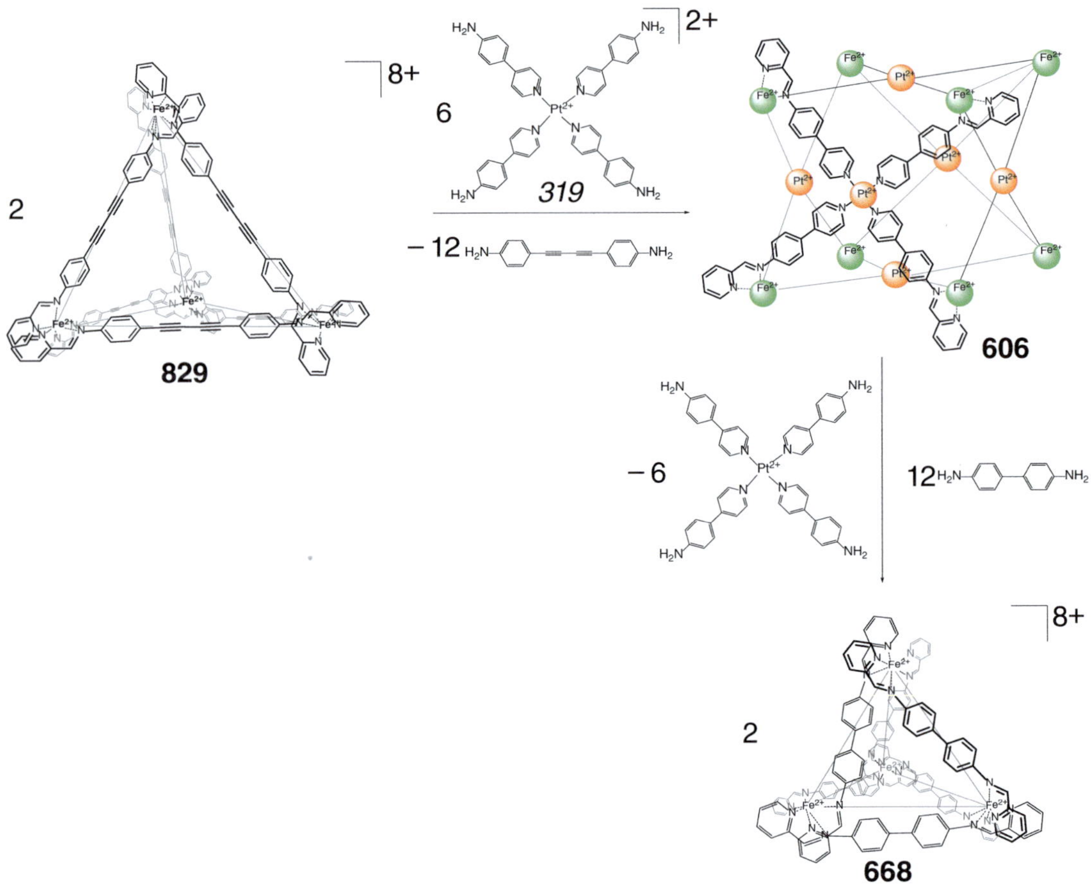

Scheme 5.85

in response to stimuli are reported in [92]. Its interaction with 4-dimethylaminopyridine as a competing ligand caused the complete destruction of this cage framework, but the following addition of *para*-toluenesulfonic acid or (+)-camphor-10-sulfonic acid resulted in a quantitative reassembly of the initial coordination capsule. This caging ligand also undergoes complete destruction with tetrabutylammonium chloride, while the followed addition of $AgSbF_6$ caused its reassembly. The initial cage complex also releases an encapsulated cisplatin molecule with D_2O [92].

Two tetrafluoroborate ions that are encapsulated in the outer cavities of the 1:3 cage complex with caging ligand **770** undergo ion-exchange reactions by Scheme 5.94, not only with the water-soluble tetraalkylammonium chlorides but also with insoluble AgCl giving a heteroguest 1:1:2 double-cage complex $(Cl^-{}_2)(BF_4^-)$@**770**. This process is reversible and gives the initial homoguest capsule with a high excess of BF_4^- anion [93].

An enantiomerically pure carceplex (G)@R_4-**261** (where G is (*S*)-$BrCH_2CH(CH_3)CH_2CH_3$, Scheme 5.95) and its S_4-isomer are reported in [94] to display stereoselectivity in the chiral guest release from their cavities with a factor from 1.5 to 2. The rate constants for release of these guest molecules from the R_4- and S_4-isomers of caging ligand **261** are different by order of magnitude.

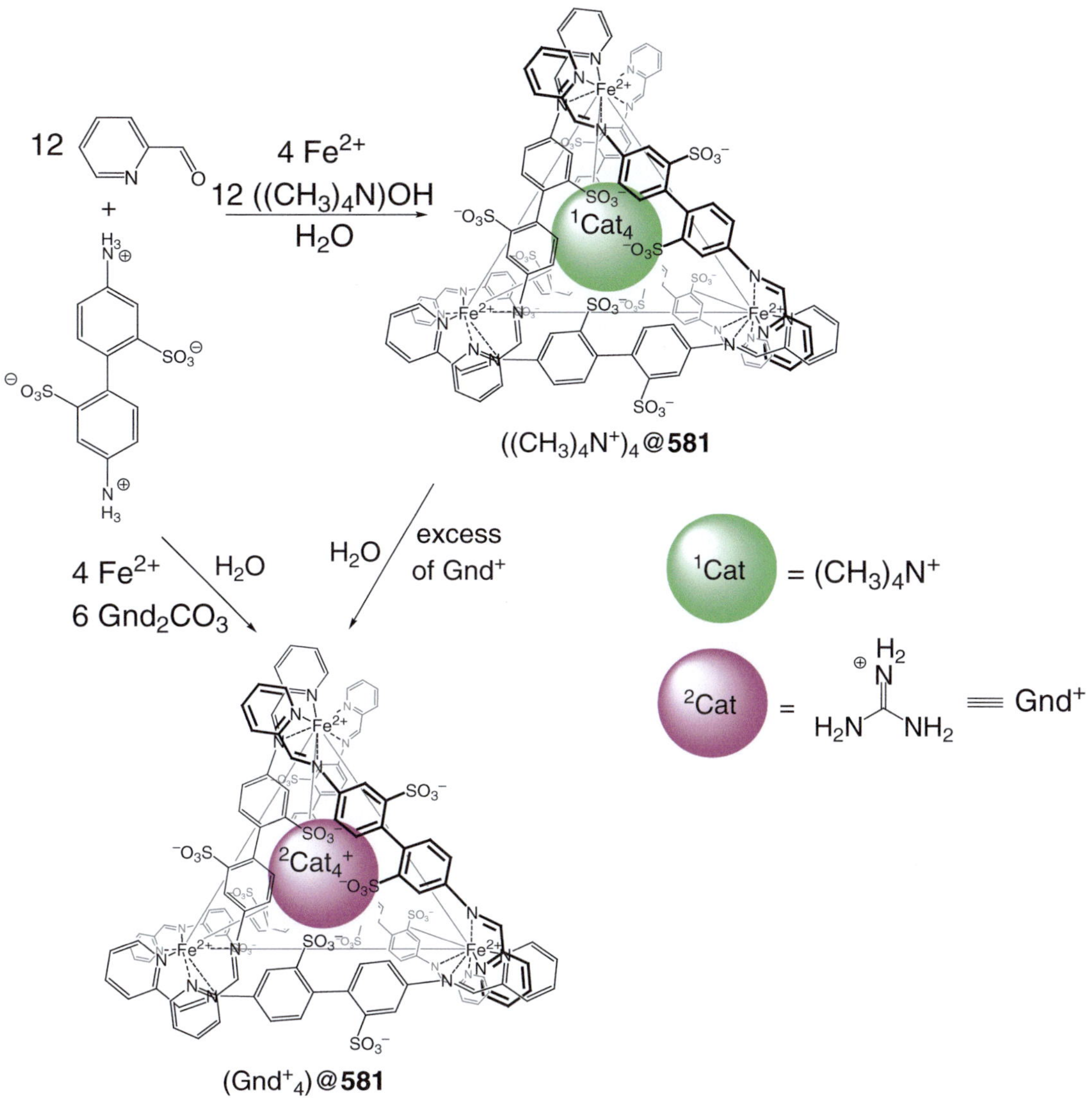

Scheme 5.86

Binding properties of bis-hemicarcerand capsules **268** and **269** toward a wide range of chiral and achiral guests (Scheme 5.96) have been studied in [95] using ^{1}H NMR method. The caged solvent molecules undergo exchange with diastereomeric equilibrated ratios from 2.7:1 for $BrCH_2CH_2C^*HBrCH_3$ as a guest to 1:1 for $ClCH_2C^*HClCH_3$ and $(CH_3)_2CHCH_2CH_2C^*HOHCH_3$ as concurrent guests, entering and departing through the chiral portals of covalent capsule, which are larger than non-chiral portals. Encapsulation of six enantiomeric pairs of guests within the cavity of **268** gave the diastereomeric ratios up to 1.4: the two non-chiral 26-membered ring portals of this caging ligand are described in [95] to be less sterically overloaded than its two chiral 26-membered ring portals. The equilibrium chiral recognition factors for encapsulation of chiral guest molecules by *S*-**269** (Scheme 5.96) are from 1 (for two

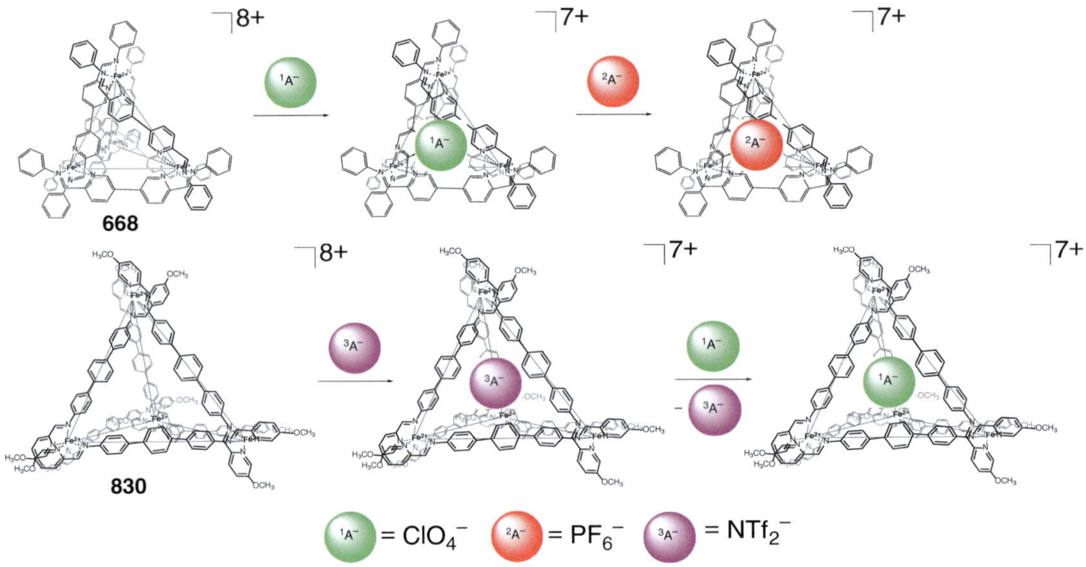

Scheme 5.87

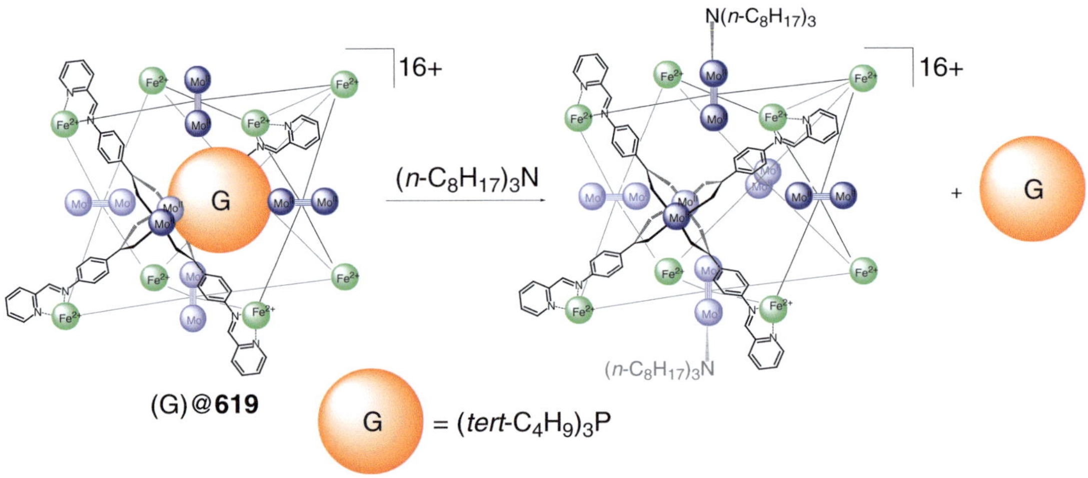

Scheme 5.88

branched alcohols and 1,2-dichloropropane as guests) to more than 20 for *R*-4-$CH_3C_6H_4S(O)CH_3$. Higher chiral recognition has been observed [95] for the covalent capsule *S*-**269** as compared with that for its analog *SS*-**268**.

A bis-hemicarcerand capsule **277** has been reported by D. Rudkevich and coworkers in [96] to be a prospective caging ligand for encapsulation, storage, and release of gases in the solid state. Its cavity volume allows for reversible encapsulation–release of the solvent molecules and gaseous guests such as CO_2, N_2, O_2, Xe, and N_2O in solution (Scheme 5.97). According to NMR data, the molecules of CO_2, N_2O, and N_2 undergo *in–out* exchange processes, which are affected mainly by the distance between the hemispheric entities of a caging ligand **277** rather than the inner size of this covalent capsule. It can also encapsulate two or more small gas molecules such as molecular hydrogen and

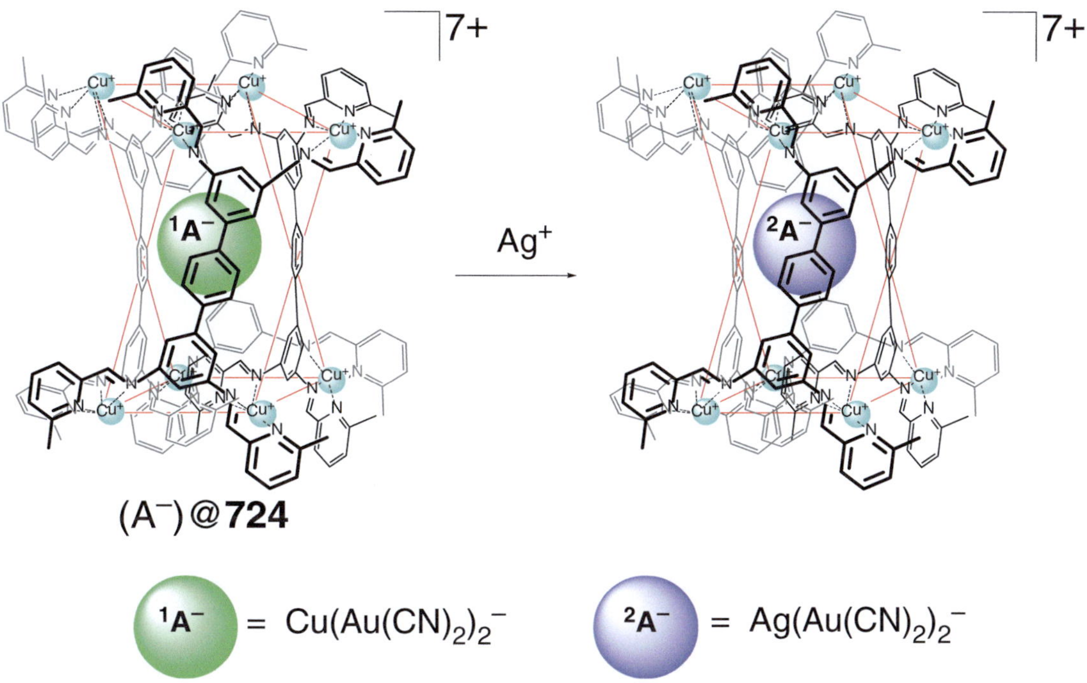

Scheme 5.89

helium; such exchange reactions take place in solution but not in solid state [96].

Encapsulated NO^+ cationic species of their cage complexes with arene capsules **823** and **831–833** are described in [97] to undergo reversible release/re-absorption reactions by Scheme 5.98 in solution with 18-crown-6 and $SnCl_4$, respectively. Addition of water to their solutions caused irreversible release of these guests due to the formation of nitrous acid [97].

D_{3h}-symmetric coordination capsules **743–746** undergo slow self-rearrangement by Scheme 5.99 in their solutions leading to the corresponding polynuclear metallomacrocyclic complexes [98].

Imine bonds in a Schiff-base tetrahedral covalent capsule **82** are reported in [99] to be rather labile in the substitution reaction giving a cylindrical cage framework **87** with an excess of ethylene diamine (Scheme 5.100) by releasing of tripodal amine *tren*.

A caging ligand **248** encapsulates aromatic dications in its two asymmetrical 24-crown-8 cavities, and the caged species undergo reversible exchange reactions with potassium(I) cations by Scheme 5.101 [100].

Rates of dissociation–association reactions for dimeric hydrogen-bonded capsules **350** and **351** and the scheme of guest exchange have been evaluated in [101]. The use of two differently substituted glycoluril functions on the durene spacer of the ligand syntone reduces the symmetry of **350** and **351**, and these self-assembled cage frameworks are pair of enantiomers shown in Scheme 5.102. These enantiomers undergo interconversion (racemization) through dissociation and recombination of their two ligand syntones. The dynamics of both methane and ethane guest exchange for their cage complexes with **350** and **351** have been examined in [101] by NMR experiments. Two plausible schemes of guest exchange in these hydrogen-bonded dimeric capsules involve ring inversion of a seven-membered cyclic fragment connecting the glycoluril to the aromatic spacer. This inversion breaks four hydrogen bonds, and the ring that undergoes inversion may be either proximal or distal to the

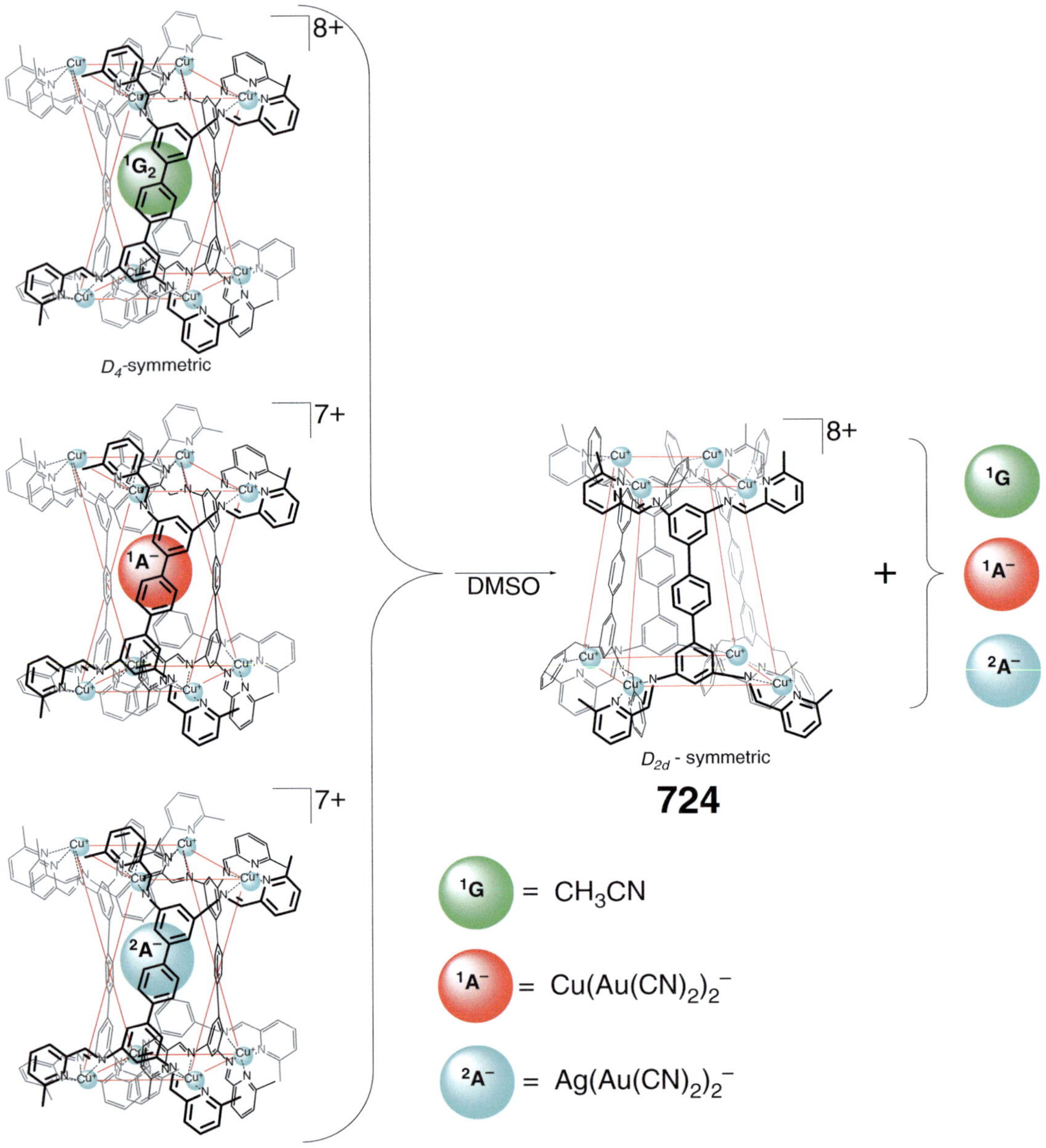
8+
$^1G_2$
D_4-symmetric
7+
$^1A^-$
7+
$^2A^-$
DMSO
8+
D_{2d} - symmetric
724
+
^1G
$^1A^-$
$^2A^-$
^1G = CH_3CN
$^1A^-$ = $Cu(Au(CN)_2)_2^-$
$^2A^-$ = $Ag(Au(CN)_2)_2^-$

Scheme 5.90

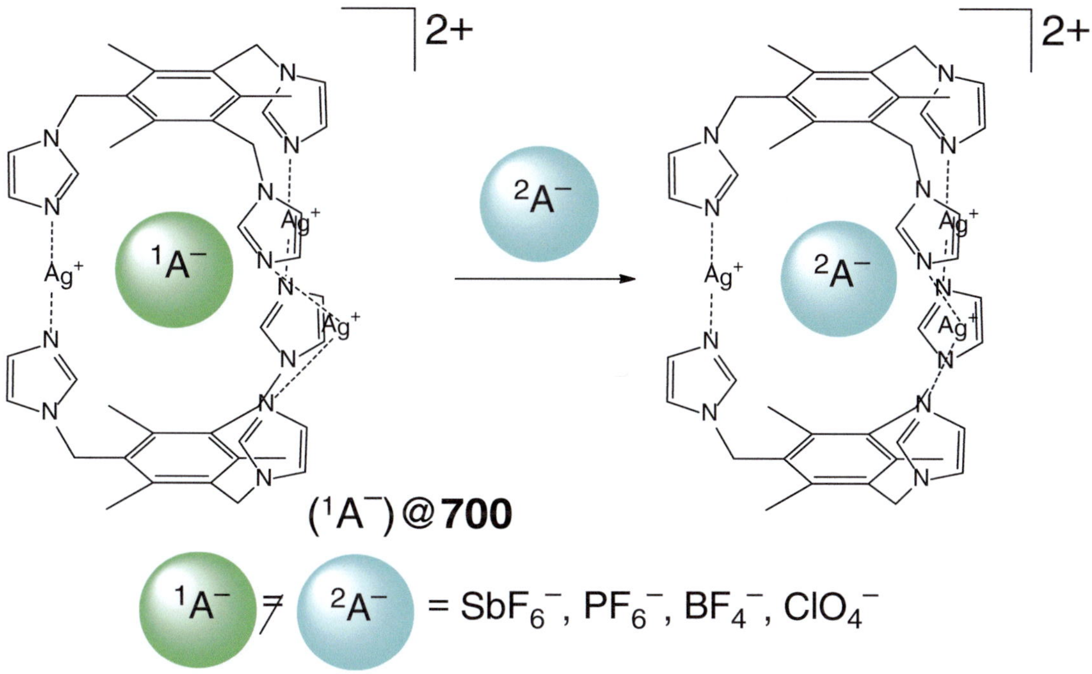

Scheme 5.91

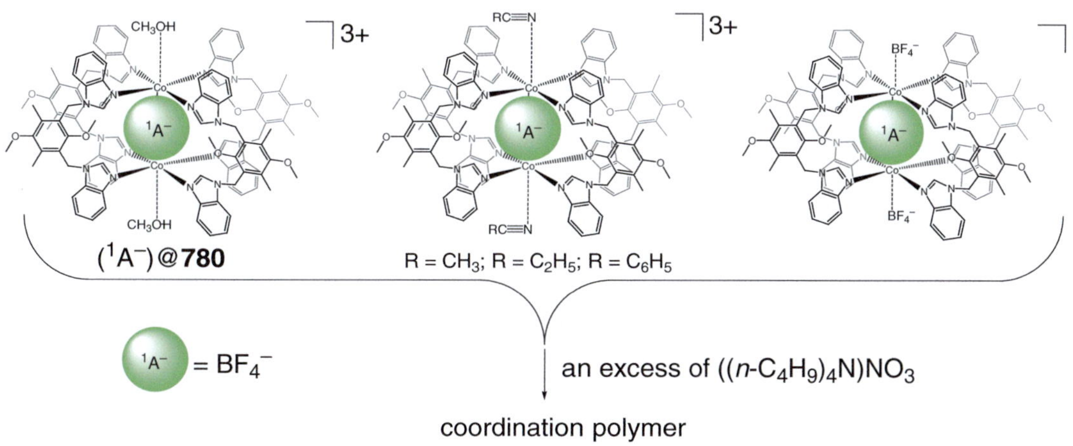

Scheme 5.92

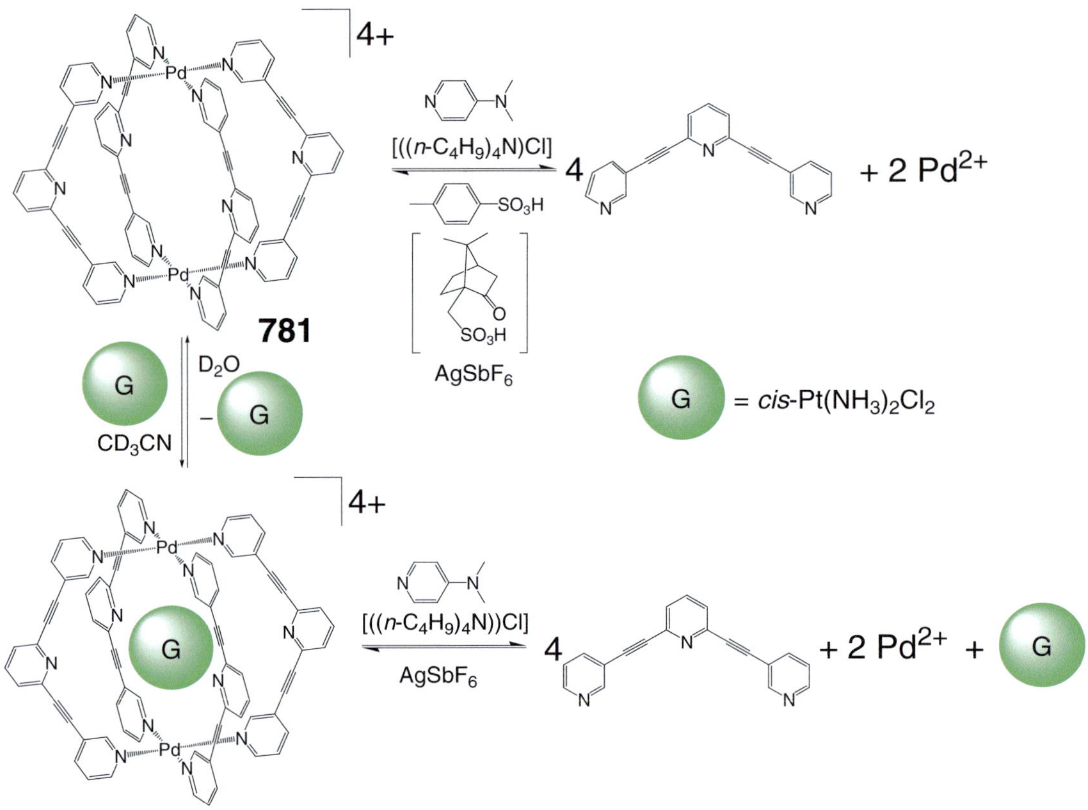

Scheme 5.93

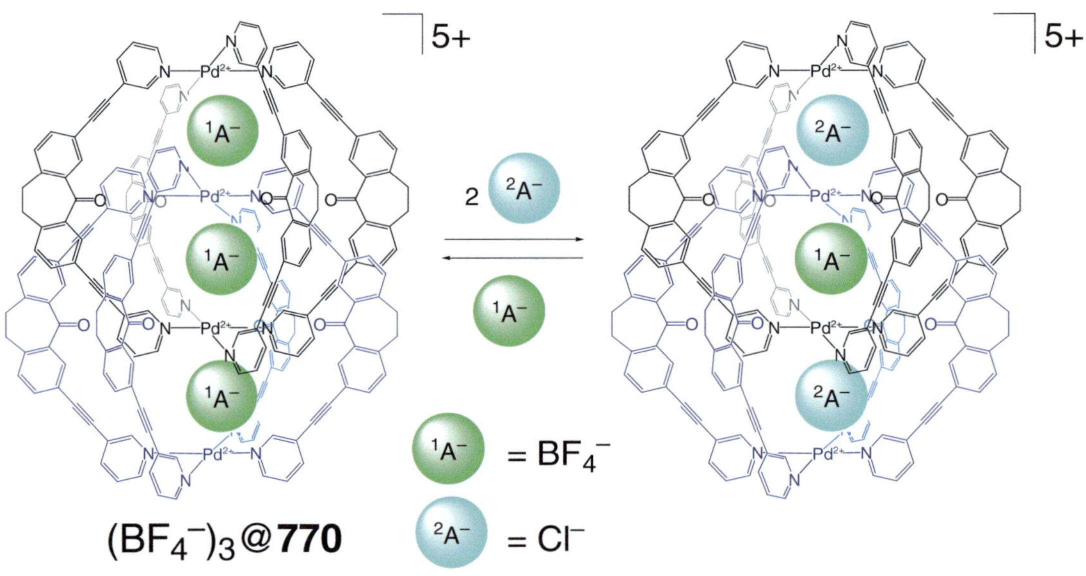

Scheme 5.94

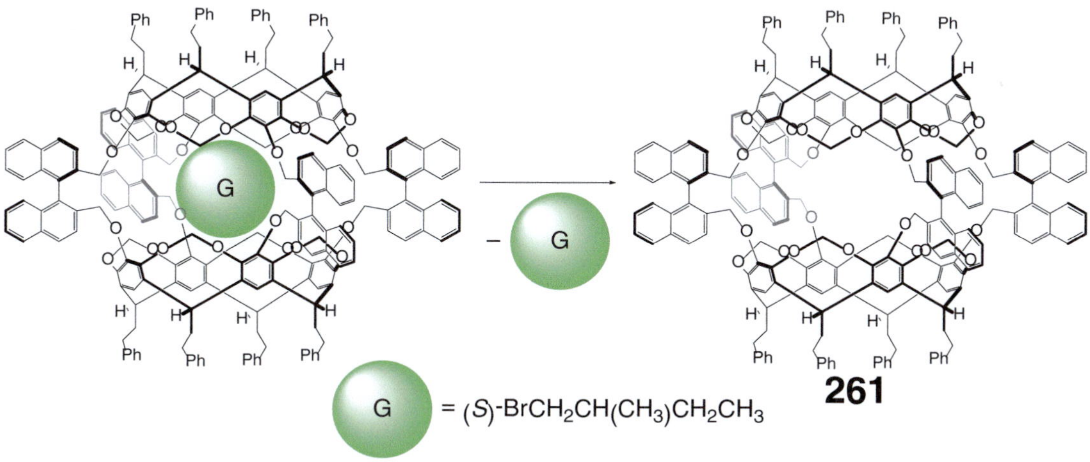

Scheme 5.95

hydrogen bonds being broken, while the remaining four hydrogen bonds hold two ligand syntones of these tennis ball-like capsules together. Such motion provides an opening large enough to allow the guest exchange to occur through an S_N2-like process, the displacement of the outgoing guest by an incoming one [101].

Single-molecule guest exchange from the cavity of a cylindrical hydrogen-bonded capsule **429** (Scheme 5.103) has been studied in [102]; this process depends on the size and shape of the guest molecules but does not involve complete dissociation of the caging ligand or creation of empty cavities. The solvent-bridged intermediates are reported to determine the kinetic parameters of this exchange.

A reaction of a hydrogen-bonded capsule **330** with a urea derivative *463* by Scheme 5.104 is reported in [103] to liberate encapsulated *para*-fluorobenzene guest. The bis-calix[5]arene cage complex of **382** with an encapsulated carboplatin molecule (Scheme 5.105) releases this guest after removing of water giving also the corresponding ligand syntone [104]. Solvent-induced denaturation of the supramolecular bis-cavitand capsule **473** with encapsulate trioxolane guests by Scheme 5.106 has been examined in [105].

Reversible photochemical transformations by Scheme 5.107 between the hemicarcerand **834** and its closed carcerand derivative **835** are reported in [106]. A photoswitchable cycle between the open (hemicarcerand) and closed (carcerand) states of this host has been controlled by using different wavelengths of radiation, and the controlled encapsulation and release of the guest molecules such as 1,4-dimethoxybenzene have been also observed in this work.

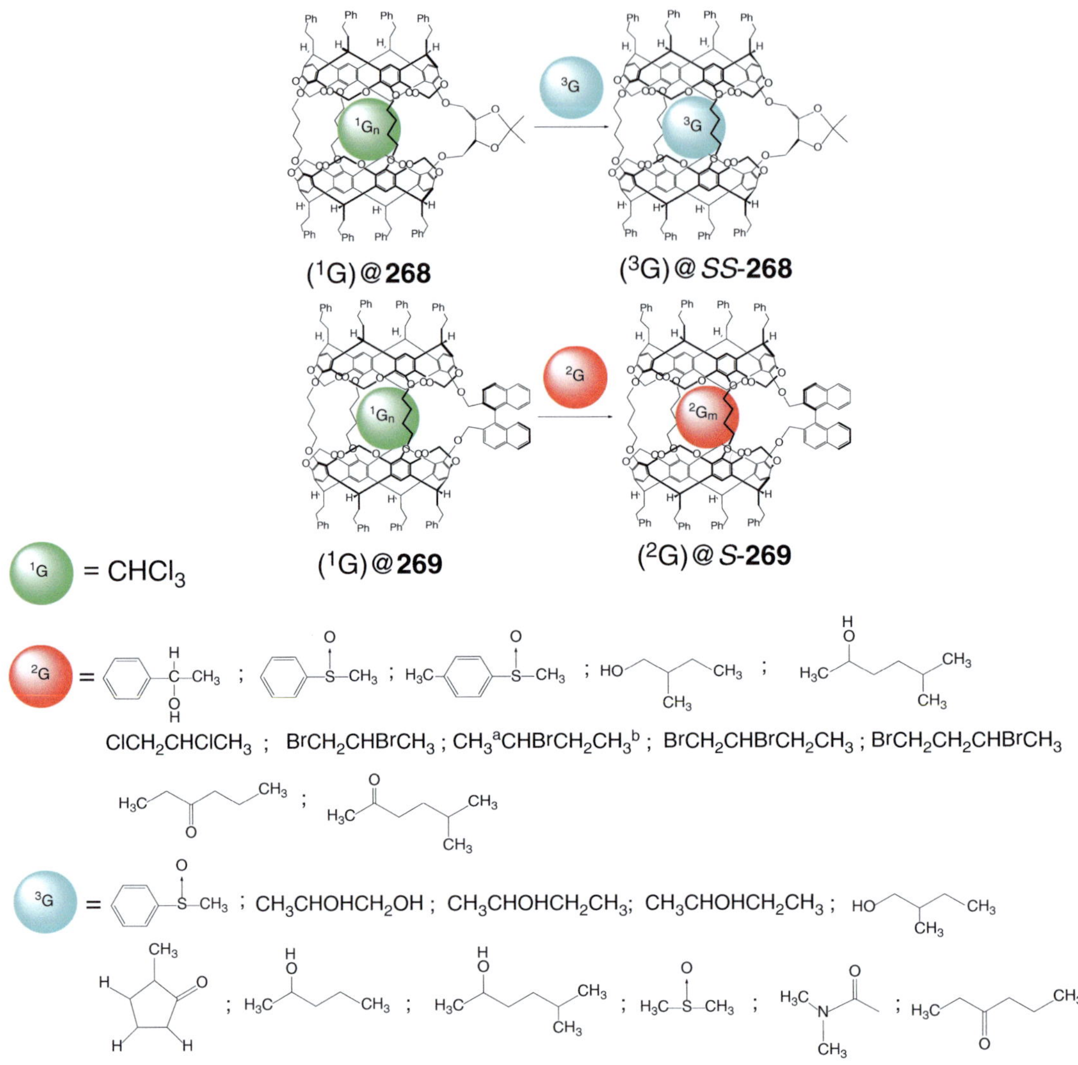
(1G)@268
(3G)@SS-268
(1G)@269
(2G)@S-269
= CHCl3
ClCH2CHClCH3 ; BrCH2CHBrCH3 ; CH3aCHBrCH2CH3b ; BrCH2CHBrCH2CH3 ; BrCH2CH2CHBrCH3
CH3CHOHCH2OH ; CH3CHOHCH2CH3; CH3CHOHCH2CH3 ;

Scheme 5.96

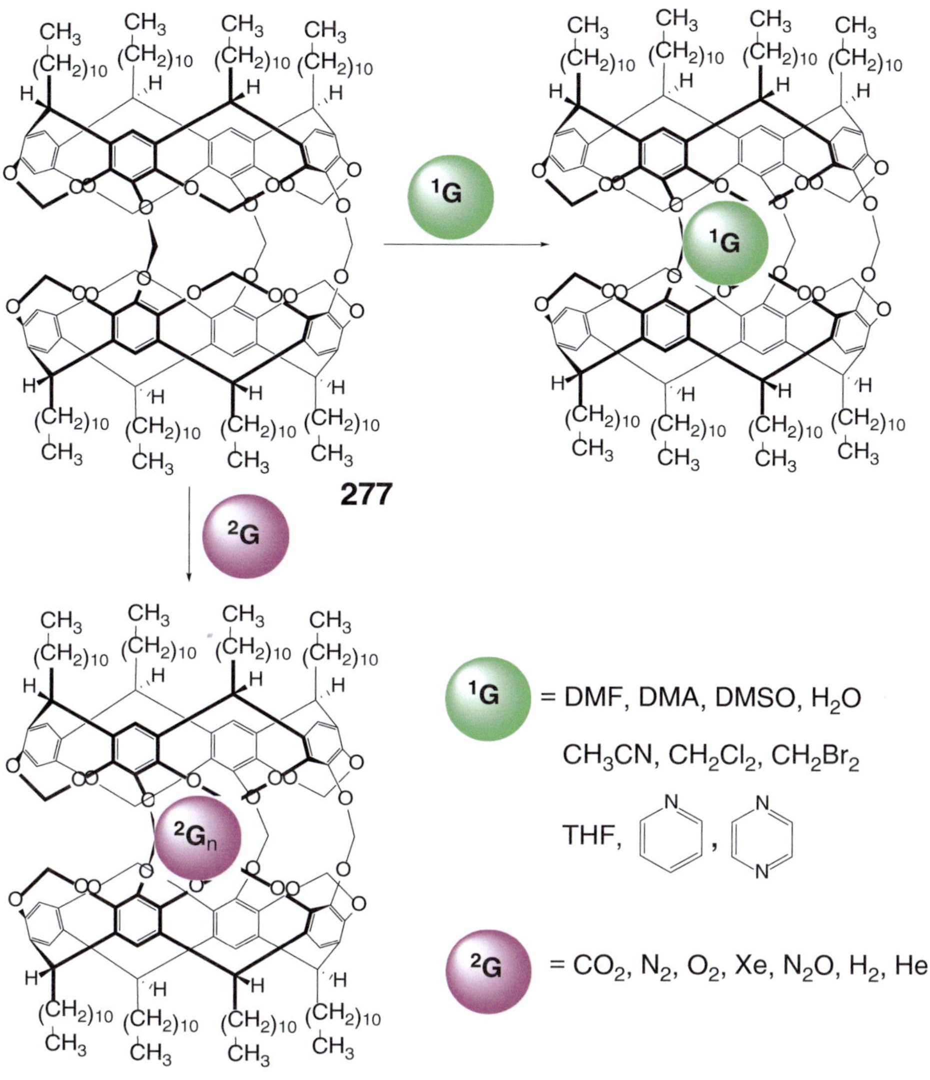
277
^1G
^2G
^2G_n
^1G = DMF, DMA, DMSO, H_2O
CH_3CN, CH_2Cl_2, CH_2Br_2
THF,
^2G = CO_2, N_2, O_2, Xe, N_2O, H_2, He

Scheme 5.97

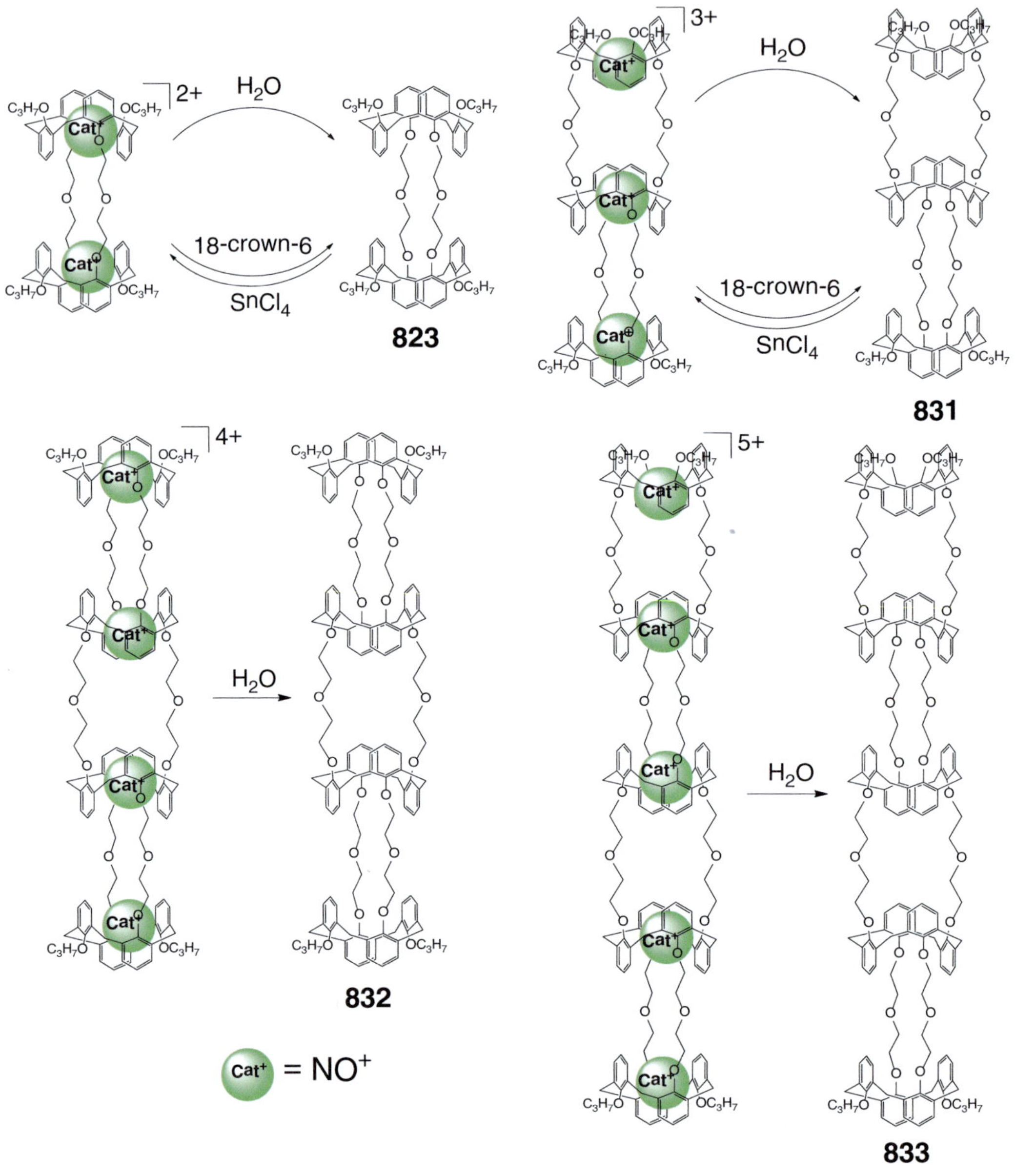
2+
H_2O
18-crown-6
$SnCl_4$
823
3+
H_2O
18-crown-6
$SnCl_4$
831
4+
H_2O
832
5+
H_2O
833
Cat+ = NO+

Scheme 5.98

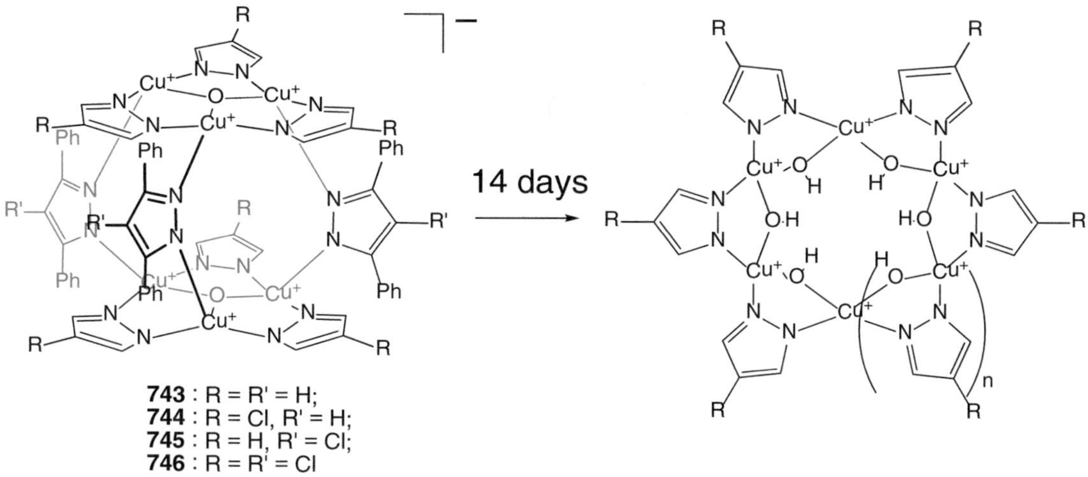

Scheme 5.99

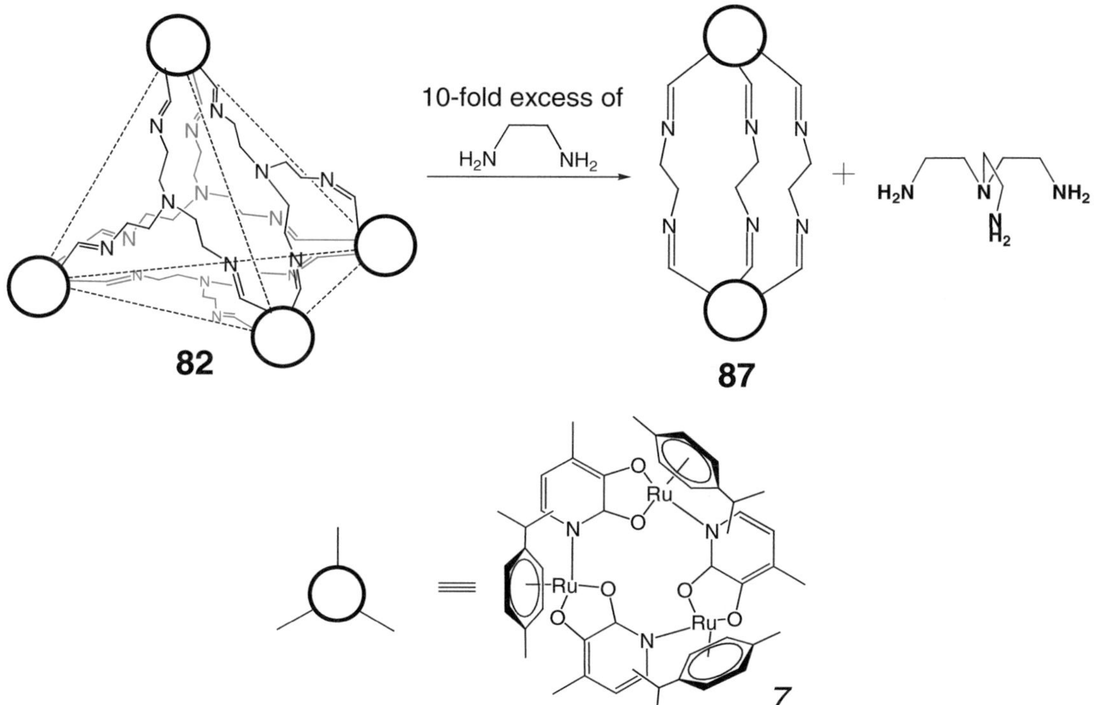

Scheme 5.100

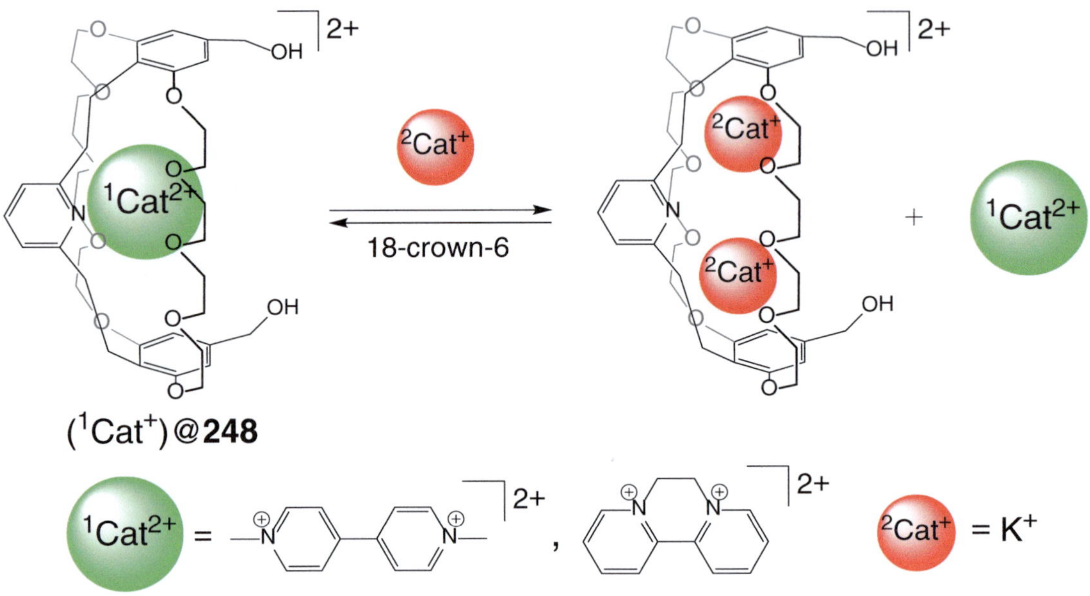

Scheme 5.101

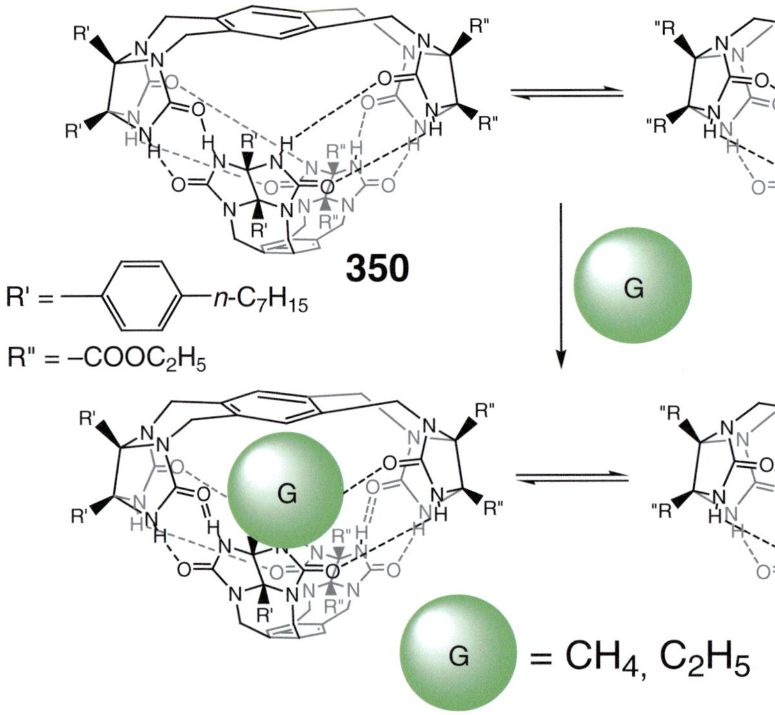

Scheme 5.102

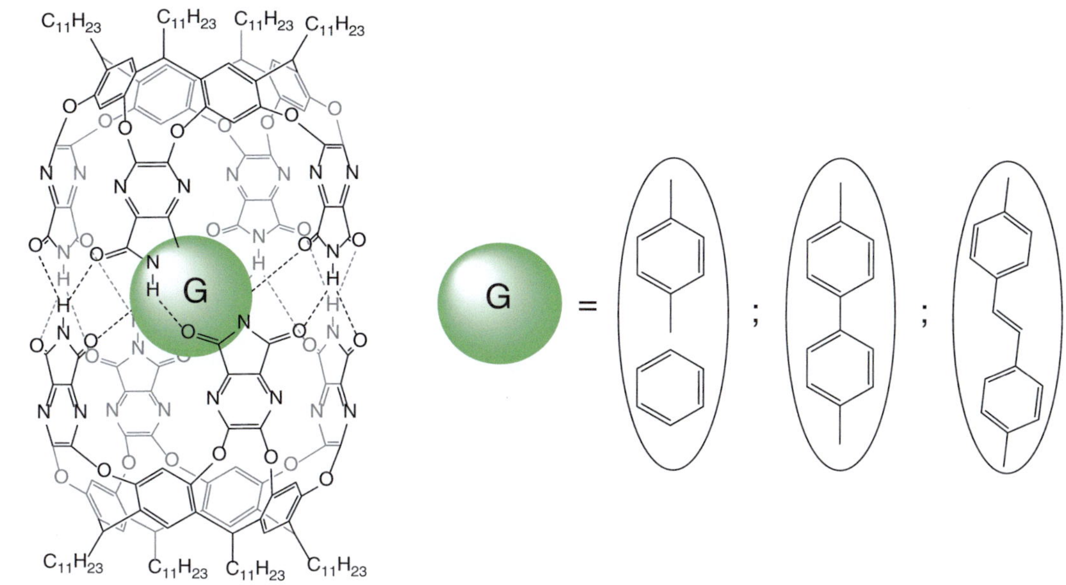

(G)@**429**

Scheme 5.103

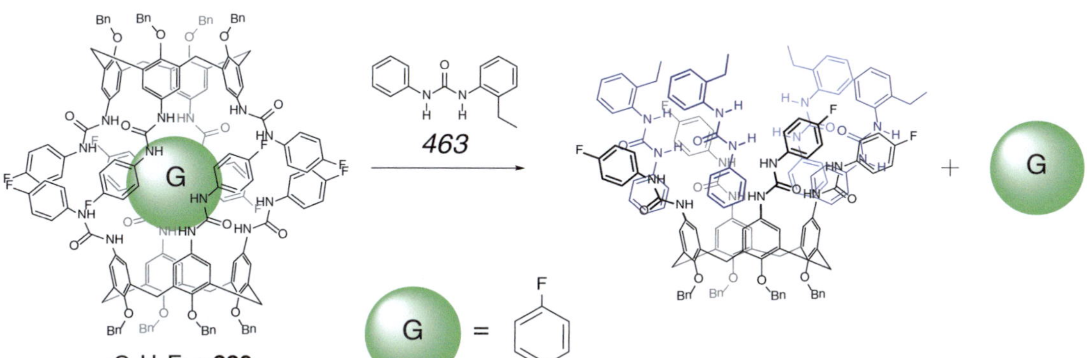

(C_6H_5F)@**330**

Scheme 5.104

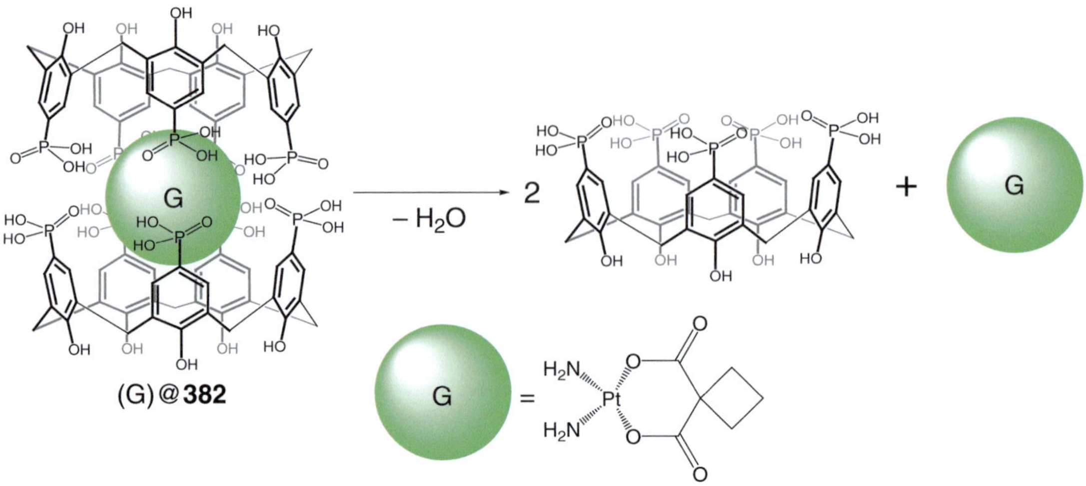

Scheme 5.105

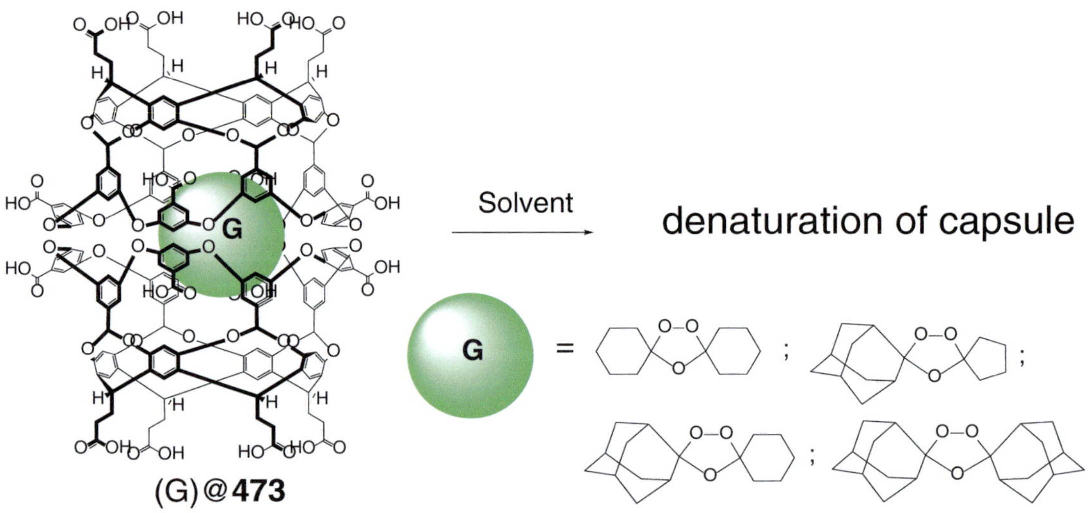

Scheme 5.106

Scheme 5.107

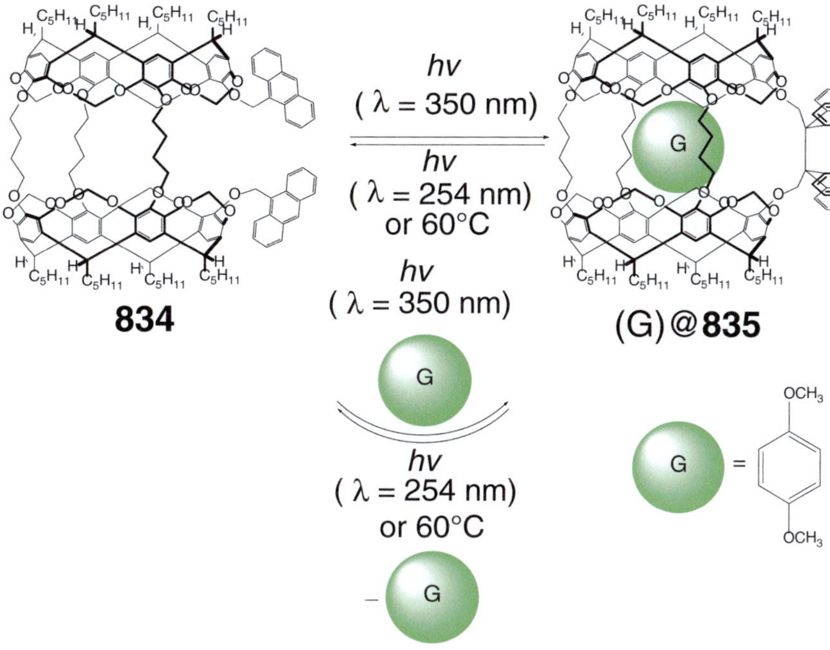

References

1. Yoshizawa M, Kusukawa T, Fujita M, Yamaguchi K (2000) Ship-in-a-bottle synthesis of otherwise labile cyclic trimers of siloxanes in a self-assembled coordination cage. J Am Chem Soc 122(26):6311–6312
2. Yoshizawa M, Kusukawa T, Fujita M, Sakamoto S, Yamaguchi K (2001) Cavity-directed synthesis of labile silanol oligomers within self-assembled coordination cages. J Am Chem Soc 123(43):10454–10459
3. Kusukawa T, Yoshizawa M, Fujita M (2001) Probing guest geometry and dynamics through host–guest. Angew Chem Int Ed 40:1879–1984
4. Suzuki K, Sato S, Fujita M (2010) Template synthesis of precisely monodisperse silica nanoparticles within self-assembled organometallic spheres. Nat Chem 2:25–29
5. Kikuchi T, Murase T, Sato S, Fujita M (2008) Polymerisation of an anionic monomer in a self-assembled $M_{12}L_{24}$ coordination sphere with cationic interior. Supramol Chem 20:81–94
6. Horiuchi S, Nishioka Y, Murase T, Fujita M (2010) Both [2+2] and [2+4] additions of inert aromatics via identical ternary host–guest complexes. Chem Commun 46:3460–3462
7. Yamauchi Y, Fujita M (2010) Self-assembled cage as an endo-template for cyclophane synthesis. Chem Commun 46:5897–5899
8. Leung DH, Fiedler D, Bergman RG, Raymond KN (2004) Selective C-H bond activation by a supramolecular host-guest assembly. Angew Chem Int Ed 43(8):963–966
9. Leung DH, Bergman RG, Raymond KN (2006) Scope and mechanism of the c-h bond activation reactivity within a supramolecular host by an iridium guest: a stepwise ion pair guest dissociation mechanism. J Am Chem Soc 128(30):9781–9797
10. Fiedler D, Bergman RG, Raymond KN (2004) Supramolecular catalysis of a unimolecular transformation: aza-cope rearrangement within a self-assembled host. Angew Chem Int Ed 43(48):6748–6751
11. Hastings CJ, Fiedler D, Bergman RG, Raymond KN (2008) Aza cope rearrangement of propargyl enammonium cations catalyzed by a self-assembled "nanozyme". J Am Chem Soc 130(33):10977–10983
12. Fiedler D, van Halbeek H, Bergman RG, Raymond KN (2006) Supramolecular catalysis of unimolecular rearrangements: substrate scope and mechanistic insights. J Am Chem Soc 128(31):10240–11025
13. Brown CJ, Bergman RG, Raymond KN (2009) Enantioselective catalysis of the aza-cope rearrangement by a chiral supramolecular assembly. J Am Chem Soc 131:17530–17531
14. Pluth MD, Bergman RG, Raymond KN (2007) Acid catalysis in basic solution: a supramolecular host promotes orthoformate hydrolysis. Science 316(5821):85–88
15. Pluth MD, Bergman RG, Raymond KN (2008) Supramolecular catalysis of orthoformate hydrolysis in basic solution: an enzyme-like mechanism. J Am Chem Soc 130(34):11423–11429
16. Pluth MD, Bergman RG, Raymond KN (2007) Catalytic deprotection of acetals in basic solution with a self-assembled supramolecular "nanozyme". Angew Chem Int Ed 46(45):8587–8589

17. Leung DH, Bergman RG, Raymond KN (2007) Highly selective supramolecular catalyzed allylic alcohol isomerization. J Am Chem Soc 129(10):2746–2747
18. Hastings CJ, Pluth MD, Bergman RG, Raymond KN (2010) Enzymelike catalysis of the nazarov cyclization by supramolecular encapsulation. J Am Chem Soc 132(20):6938–6940
19. Hastings CJ, Backlund MP, Bergman RG, Raymond KN (2011) Enzyme-like control of carbocation deprotonation regioselectivity in supramolecular catalysis of the nazarov cyclization. Angew Chem Int Ed 50(45):10570–10573
20. Hastings CJ, Bergman RG, Raymond KN (2014) Origins of large rate enhancements in the nazarov cyclization catalyzed by supramolecular encapsulation. Chem Eur J 20(14):3966–3973
21. Wang ZJ, Brown CJ, Bergman RG, Raymond KN, Toste FD (2011) hydroalkoxylation catalyzed by a gold(I) complex encapsulated in a supramolecular host. J Am Chem Soc 133:7358–7360
22. Hart-Cooper WM, Clary KN, Toste FD, Bergman RG, Raymond KN (2012) Selective monoterpene-like cyclization reactions achieved by water exclusion from reactive intermediates in a supramolecular catalyst. J Am Chem Soc 134(43):17873–17876
23. Wang ZJ, Clary KN, Bergman RG, Raymond KN, Toste FD (2013) A supramolecular approach to combining enzymatic and transition metal catalysis. Nat Chem 5(2):100–103
24. Brown CJ, Miller GM, Johnson MW, Bergman RG, Raymond KN (2011) High-turnover supramolecular catalysis by a protected ruthenium(II) complex in aqueous solution. J Am Chem Soc 133(31):11964–11966
25. Salles AG, Zarra S, Turner RM, Nitschke JR (2013) A self-organizing chemical assembly line. J Am Chem Soc 135(51):19143–19146
26. Martinez A, Dutasta J-P (2009) Hemicryptophane–oxidovanadium(V) complexes: lead of a new class of efficient supramolecular catalysts. J Catal 267:188–192
27. Perraud O, Sorokin AB, Dutasta J-P, Martinez A (2013) Oxidation of cycloalkanes by H_2O_2 using a copper–hemicryptophane complex as a catalyst. Chem Commun 49:1288–1290
28. Kang J, Rebek J Jr (1997) Acceleration of a Diels-Alder reaction by a self-assembled molecular capsule. Nature 385(6611):50–52
29. Kang J, Santamaria J, Hilmersson G, Rebek J Jr (1998) Self-assembled molecular capsule catalyzes a Diels-Alder reaction. J Am Chem Soc 120(29):7389–7390
30. Kang J, Hilmersson G, Santamaria J, Rebek J Jr (1998) Diels-Alder reactions through reversible encapsulation. J Am Chem Soc 120(15):3650–3656
31. Chen J, Rebek J Jr (2002) Selectivity in an encapsulated cycloaddition reaction. Org Lett 4(3):327–329
32. Iwasawa T, Mann E, Rebek J Jr (2006) A reversible reaction inside a self-assembled capsule. J Am Chem Soc 128(29):9308–9309
33. Hou J-L, Ajami D, Rebek J Jr (2008) Reaction of carboxylic acids and isonitriles in small spaces. J Am Chem Soc 130(25):7810–7811
34. Taira T, Ajami D, Rebek J (2012) Hydration of isocyanates in an expandable, self-assembled capsule. Chem Commun 48(68):8508–8510
35. Zyryanov G, Rudkevich DM (2003) Org Lett 5(8):1253–1256
36. Zyryanov GV, Rudkevich DM (2004) Toward synthetic tubes for NO2/N2O4: design, synthesis, and host–guest chemistry. J Am Chem Soc 125:4264–4270
37. Liu S, Gan H, Hermann AT, Rick SW, Gibb BC (2010) Kinetic resolution of constitutional isomers controlled by selective protection inside a supramolecular nanocapsule. Nature Chem 2:847–852
38. Cavarzan A, Scarso A, Sgarbossa P, Strukul G, Reek JNH (2011) Supramolecular control on chemo- and regioselectivity via encapsulation of (NHC)-Au catalyst within a hexameric self-assembled host. J Am Chem Soc 133:2848–2851
39. Kida T, Iwamoto T, Asahara H, Hinoue T, Akashi M (2013) Chiral recognition and kinetic resolution of aromatic amines via supramolecular chiral nanocapsules in nonpolar solvents. J Am Chem Soc 135:3371–3374
40. Yoshizawa M, Takeyama Y, Kusukawa T, Fujita M (2002) Cavity-directed, highly stereoselective [2+2] photodimerization of olefins within self-assembled coordination cages. Angew Chem Int Ed 41:1347–1349
41. Yoshizawa M, Miyagi S, Kawano M, Ishiguro K, Fujita M (2004) Alkane oxidation via photochemical excitation of a self-assembled molecular cage. J Am Chem Soc 126:9172–9173
42. Furutani Y, Kandori H, Kawano M, Nakabayashi K, Yoshizawa M, Fujita M (2009) In situ spectroscopic, electrochemical, and theoretical studies of the photoinduced host-guest electron transfer that precedes unusual host-mediated alkane photooxidation. J Am Chem Soc 131(13):4764–4768
43. Yoshizawa M, Takeyama Y, Okano T, Fujita M (2003) Cavity-directed synthesis within a self-assembled coordination cage: highly selective [2 + 2] cross- photodimerization of olefins. J Am Chem Soc 125(11):3243–3247
44. Nishioka Y, Yamaguchi T, Yoshizawa M, Fujita M (2007) Unusual [2+4] and [2+2] cycloadditions of arenes in the confined cavity of self-assembled cages. J Am Chem Soc 129(22):7000–7001
45. Takaoka K, Kawano M, Ozeki T, Fujita M (2006) Crystallographic observation of an olefin photodimerization reaction that takes place via thermal molecular tumbling within a self-assembled host. Chem Commun 1625–1627
46. Kawano M, Kobayashi Y, Ozeki T, Fujita M (2006) Direct crystallographic observation of a coordinatively unsaturated transition-metal complex in situ generated within a self-assembled cage. J Am Chem Soc 128:6558–6559

47. Yamaguchi T, Fujita M (2008) Highly selective photomediated 1,4-radical addition to o-quinones controlled by a self-assembled cage. Angew Chem Int Ed 47:2067–2069
48. Inokuma Y, Kojima N, Arai T, Fujita M (2011) Bimolecular reaction via the successive introduction of two substrates into the crystals of networked molecular cages. J Am Chem Soc 133:19691–19693
49. Inokuma Y, Ning G-H, Fujita M (2012) Reagent-installed capsule network: selective thiocarbamoylation of aromatic amines in crystals with preinstalled CH_3NCS. Angew Chem Int Ed 51:2379–2381
50. Clever GH, Tashiro S, Shionoya M (2010) Light-triggered crystallization of a molecular host-guest complex. J Am Chem Soc 132(29):9973–9975
51. Han M, Michel R, He B, Chen Y-S, Stalke D, John M, Clever GH (2013) Light-triggered guest uptake and release by a photochromic coordination cage. Angew Chem Int Ed 52(4):1319–1323
52. Cram DJ, Tanner ME, Thomas R (1991) The taming of cyclobutadiene. Angew Chem Int Ed 30:1024–1027
53. Warmuth R (1997) o-Benzyne: strained alkyne or cumulene–NMR characterization in a molecular container. Angew Chem Int Ed 36:1347–1350
54. Kaanumalle LS, Ramamurthy V (2007) Photodimerization of acenaphthylene within a nanocapsule: excited state lifetime dependent dimer selectivity. Chem Commun 1062–1064
55. Gibb CLD, Sundaresan AK, Ramamurthy V, Gibb BC (2008) Templation of the excited-state chemistry of α-(n-alkyl) dibenzyl ketones: how guest packing within a nanoscale supramolecular capsule influences photochemistry. J Am Chem Soc 130:4069–4080
56. Kulasekharan R, Maddipatla MVSN, Parthasarathy A, Ramamurthy V (2013) Role of free space and conformational control on photoproduct selectivity of optically pure α-alkyldeoxybenzoins within a water-soluble organic capsule. J Org Chem 78:942–949
57. Parthasarathy A, Samanta SR, Ramamurthy V (2013) Photodimerization of hydrophobic guests within a water soluble nanocapsule. Res Chem Intermed 39:73–87
58. Dube H, Ams MR, Rebek J Jr (2010) Supramolecular control of fluorescence through reversible encapsulation. J Am Chem Soc 132(29):9984–9985
59. Dube H, Ajami D, Rebek J Jr (2010) Photochemical control of reversible encapsulation. Angew Chem Int Ed 49(18):3192–3195
60. Dube H, Rebek J (2012) Selective guest exchange in encapsulation complexes using light of different wavelengths. Angew Chem Int Ed 51(13):3207–3210
61. Lux J, Rebek J Jr (2013) Reversible switching between self-assembled homomeric and hybrid capsules. Chem Commun 49(21):2127–2129
62. Bianchini G, Scarso A, La Sorella G, Strukul G (2012) Switching the activity of a photoredox catalyst through reversible encapsulation and release. Chem Commun 48:12082–12084
63. Yoshizawa MM, Ono K, Kumazawa K, Kato T, Fujita M (2005) Metal–metal d–d interaction through the discrete stacking of mononuclear M(II) complexes (M = Pt, Pd, and Cu) within an organic-pillared coordination cage. J Am Chem Soc 127:10800–10801
64. Qiu Y, Yi S, Kaifer AE (2011) Encapsulation of tetrathiafulvalene inside a dimeric molecular capsule. Org Lett 13:1770–1773
65. Podkoscielny D, Gadde S, Kaifer AE (2009) Mediated electrochemical oxidation of a fully encapsulated redox active center. J Am Chem Soc 131:12876–12877
66. Fujita M, Fujita N, Ogura K, Yamaguchi K (1999) Spontaneous assembly of ten components into two interlocked, identical coordination cages. Nature 400:52–55
67. Inokuma Y, Arai T, Fujita M (2010) Networked molecular cages as crystalline sponges for fullerenes and other guests. Nat Chem 2:780–783
68. Hiraoka S, Harano K, Shiro M, Shionoya M (2005) Quantitative dynamic interconversion between Ag^{I}-mediated capsule and cage complexes accompanying guest encapsulation/release. Angew Chem Int Ed 44:2727–2731
69. Ziegler M, Davis AV, Johnson DW, Raymond KN (2003) Supramolecular chirality: a reporter of structural memory. Angew Chem Int Ed 42(6):665–668
70. Davis AV, Fiedler D, Seeber G, Zahl A, Van Eldik R, Raymond KN (2006) Guest exchange dynamics in an M_4L_6 tetrahedral host. J Am Chem Soc 128(4):1324–1333
71. Pluth MD, Tiedemann BEF, van Halbeek H, Nunlist R, Raymond KN (2008) Diffusion of a highly charged supramolecular assembly: direct observation of ion association in water. Inorg Chem 47(5):1411–1413
72. Mugridge JS, Bergman RG, Raymond KN (2010) Does size really matter? The steric isotope effect in a supramolecular host-guest exchange reaction. Angew Chem Int Ed 49(21):3635–3637
73. Mugridge JS, Bergman RG, Raymond KN (2010) High-precision measurement of isotope effects on noncovalent host-guest interactions. J Am Chem Soc 132(4):1182–1183
74. Mugridge JS, Bergman RG, Raymond KN (2012) Equilibrium isotope effects on noncovalent interactions in a supramolecular host-guest system. J Am Chem Soc 134(4):2057–2066
75. Mugridge JS, Szigethy G, Bergman RG, Raymond KN (2010) Encapsulated guest-host dynamics: guest rotational barriers and tumbling as a probe of host interior cavity space. J Am Chem Soc 132(45):16256–16264
76. Mugridge JS, Zahl A, van Eldik R, Bergman RG, Raymond KN (2013) Solvent and pressure effects on the motions of encapsulated guests: tuning the flexibility of a supramolecular host. J Am Chem Soc 135(11):4299–4306

77. Davis AV, Raymond KN (2005) The big squeeze: guest exchange in an M4L6 supramolecular host. J Am Chem Soc 127:7912–7919
78. Albrecht M, Janser I, Burk S, Weis P (2006) Self-assembly and host-guest chemistry of big metallosupramolecular M_4L_4 tetrahedra. Dalton Trans 2875–2880
79. Mal P, Schultz D, Beyeh K, Rissanen K, Nitschke JR (2008) An unlockable–relockable iron cage by subcomponent self-assembly. Angew Chem Int Ed 47:8297–8301
80. Riddell IA, Smulders MMJ, Clegg JK, Nitschke JR (2011) Encapsulation, storage and controlled release of sulfur hexafluoride from a metal–organic capsule. Chem Commun 47:457–459
81. Hristova YR, Smulders MMJ, Clegg JK, Breiner B, Nitschke JR (2011) Selective anion binding by a "chameleon" capsule with a dynamically reconfigurable exterior. Chem Sci 2:638–641
82. Meng W, Breiner B, Rissanen K, Thoburn JD, Clegg JK, Nitschke JR (2011) A self-assembled M_8L_6 cubic cage that selectively encapsulates large aromatic guests. Angew Chem Int Ed 50:3479–3483
83. Smulders MMJ, Nitschke JR (2012) Supramolecular control over Diels-Alder reactivity by encapsulation and competitive displacement. Chem Sci 3:785–788
84. Ousaka N, Clegg JK, Nitschke JR (2012) Nonlinear enhancement of chiroptical response through subcomponent substitution in M_4L_6 cages. Angew Chem Int Ed 51:1464–1468
85. Smulders MMJ, Jimenez A, Nitschke JR (2012) Integrative self-sorting synthesis of a $Fe_8Pt_6L_{24}$ cubic cage. Angew Chem Int Ed 51(27):6681–6685
86. Zarra S, Smulders MMJ, Lefebvre Q, Clegg JK, Nitschke JR (2012) Guanidinium binding modulates guest exchange within an [M4L6] capsule. Angew Chem Int Ed 51(28):6882–6885
87. Ma S, Smulders MMJ, Hristova YR, Clegg JK, Ronson TK, Zarra S, Nitschke JR (2013) Chain-reaction anion exchange between metal-organic cages. J Am Chem Soc 135(15):5678–5684
88. Ramsay WJ, Nitschke JR (2014) Two distinct allosteric active sites regulate guest binding within a $Fe_8Mo_{12}^{16+}$ cubic receptor. J Am Chem Soc 136(19):7038–7043
89. Meng W, Clegg JK, Nitschke JR (2012) Transformative binding and release of gold guests from a self-assembled Cu_8L_4 tube. Angew Chem Int Ed 51:1881–1884
90. Liu HK, Huang X, Li T, Wang X, Sun WY, Kang BS (2008) Discrete and infinite 1D, 2D/3D cage frameworks with inclusion of anionic species and anion-exchange reactions of Ag_3L_2 type receptor with tetrahedral and octahedral anions. Dalton Trans 3178–3188
91. Amouri H, Desmarets C, Bettoschi A, Rager MN, Boubekeur K, Rabu P, Drillon M (2007) Supramolecular cobalt cages and coordination polymers templated by anion guests: self-assembly, structures, and magnetic properties. Chem Eur J 13:5401–5407
92. Lewis JEM, Gavey EL, Cameron SA, Crowley JD (2012) Stimuli-responsive Pd_2L_4 metallosupramolecular cages: towards targeted cisplatin drug delivery. Chem Sci 3:778–784
93. Freye S, Engelhard DM, John M, Clever GH (2013) Counterion dynamics in an interpenetrated coordination cage capable of dissolving AgCl. Chem Eur J 19:2114–2121
94. Judice JK, Cram DJ (1991) Stereoselectivity in guest release from constrictive binding in a hemicarceplex. J Am Chem Soc 113:2790
95. Yoon J, Cram DJ (1997) Chiral recognition properties in complexation of two asymmetric hemicarcerands. J Am Chem Soc 119:11796–11806
96. Leontiev AV, Rudkevich DM (2004) Encapsulation of gases in the solid state. Chem Commun 1468–1469
97. Organo VG, Sgarlata V, Firouzbakht F, Rudkevich DM (2007) Long synthetic nanotubes from calix[4]arenes. Chem Eur J 13:4014–4023
98. Mezei G, Rivera-Carrillo M, Raptis RG (2007) Trigonal-prismatic Cu^{II}_6-pyrazolato cages: structural and electrochemical study, evidence of charge delocalization. Dalton Trans 37–40
99. Granzhan A, Schouwey C, Riis-Johannessen T, Scopelliti R, Severin K (2011) Connection of metallamacrocycles via dynamic covalent chemistry: a versatile method for the synthesis of molecular cages. J Am Chem Soc 133:7106–7115
100. Zhang M, Zheng B, Xia B, Zhu K, Wu C, Huang F (2010) Synthesis of a bis(1,2,3-phenylene) cryptand and its dual-response binding to paraquat and diquat. Eur J Org Chem 2010:6804–6809
101. Szabo T, Hilmersson G, Rebek J Jr (1998) Dynamics of assembly and guest exchange in the tennis ball. J Am Chem Soc 120(24):6193–6194
102. Craig SL, Lin S, Chen J, Rebek J Jr (2002) An NMR study of the rates of single-molecule exchange in a cylindrical host capsule. J Am Chem Soc 124(30):8780–8781
103. Hamann BC, Shimizu KD, Rebek J Jr (1996) Reversible encapsulation of guest molecules in a calixarene dimmer. Angew Chem Int Ed 35(12):1326–1329
104. Martin AD, Boulos RA, Hubble LJ, Hartlieb KJ, Raston CL (2011) Multifunctional water-soluble molecular capsules based on p-phosphonic acid calix[5]arene. Chem Commun 47:7353–7355
105. Liu S, Gibb BC (2011) Solvent denaturation of supramolecular capsules assembled via the hydrophobic effect. Chem Commun 47:3574–3576
106. Wang H, Liu F, Helgeson RC, Houk KN (2013) Reversible photochemically gated transformation of a hemicarcerand to a carcerand. Angew Chem Int Ed 52:655–659

6 Practical Applications of Molecular Capsules and Their Cage Complexes

6.1 Chemical Separation, Guest Recognition, Carrying, and Storage

A cascade of dicobalt-containing capsule shown in Scheme 6.1 has been proposed [1] as an oxygen carrier: spectrophotometrical studies of its oxygenation, deoxygenation, and degradation showed to be a suitable encapsulating reagent for separation of O_2 from gaseous mixtures.

Host–guest complexes of a calix[4]arene ligand **822** (Scheme 6.2) with encapsulated multiple nitrosonium guests have been used [2] as precursors of peptide-based NO-releasing pharmaceuticals (including their chiral derivatives) and stable reagents for regioselective activation of peptides through their *N*-nitrosation. Its nanotube calix[4]arene-based analogs **823** and **831–833** are reported [3] to be prospective for reversible storage of NO_x gases, for generating NO gas within the capsules, and for releasing the caged guest. Calix[4]arene caging ligand **277** (Scheme 6.3) and robust polymeric materials based on it have been also used in [4] for gas separation, storage, and sensing as well as for manipulating with and storing of reactive gases.

An octaamide macrotricyclic caging ligand **117** (Scheme 6.4) has been used in [5] as a highly selective synthetic receptor for recognition of disaccharides. A similar macrotricyclic polyamide capsule **118** is reported [6] to be a biomimetic carbohydrate receptor with consistent behavior across a wide range of media. The covalent capsules **119–122** of this type play a role of synthetic lectin for recognition of *β*-glycosyl; the selectivity of this binding is quite low but still strong enough for blood glucose monitoring [7].

A nanoball $M_{12}L_{24}$ coordination capsule **508** (Scheme 6.5) with a poly(ethylene oxide) inner cavity is suitable for reversible storage of metal cations (in particular, of lanthanides), which can be released by addition of highly donor solvents such as DMSO [8].

CT-induced colorization of solutions of an aromatic peptide as a result of their encapsulation by coordination capsule **486** (Scheme 6.6) is reported in [9] to be applicable for the naked-eye recognition of three aromatic residues (Trp, Tyr, and Phe); the selectivity of this binding may be used for site-specific recognition of protein surfaces.

A hydrophobic dimeric bowl-shaped caging ligand **504** (Scheme 6.6) has been used in [10] for encapsulation and α-helical folding of a nine-residue oligopeptide. This ligand stabilizes secondary structure of the guest peptide that mimics biological systems. *β*-strand, *β*-hairpin, and turn structures can be also stabilized within its hydrophobic cavity. Their encapsulation is suitable for studying the pathways of protein folding [10].

Cage complexes with encapsulated cytotoxic guests have been widely used as so-called "Trojan horses" for cancer cells. A comparative study of cytotoxicities of an arene–ruthenium TP coordination capsule **540** (Scheme 6.7) and its 1:1 "Matreshka" complex-in-complex assemblies

Y. Voloshin et al., *The Encapsulation Phenomenon: Synthesis, Reactivity and Applications of Caged Ions and Molecules*, DOI 10.1007/978-3-319-27738-7_6

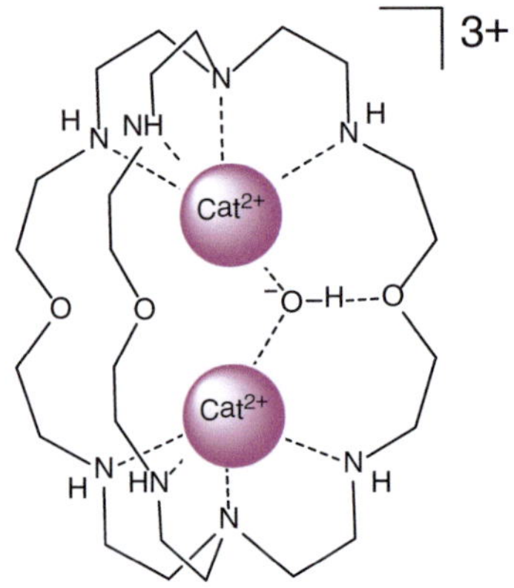

$\{(Co^{2+})_2(OH^-)\}$@**110**

Scheme 6.1

with encapsulated palladium and platinum(II) acetylacetonates against A2780 human ovarian cancer cells has been performed by P.J. Dyson and coworkers in [11]. The cationic capsule **540** plays the role of a carrier that promotes uptake in cancer cell by encapsulating cisplatin, which easily undergoes extrusion from its hydrophobic cavity in aqueous solutions [11]. Palladium and platinum(II) acetylacetonates as more hydrophobic guests form more stable cage complexes, but they themself show no cytotoxic effect against these cancer cells; the encapsulating ligand **540** demonstrates moderate cytotoxicity. Its cage complexes are the most active in this series: 1:1 host–guest compound $(Pt(acac)_2)$@**540** is two

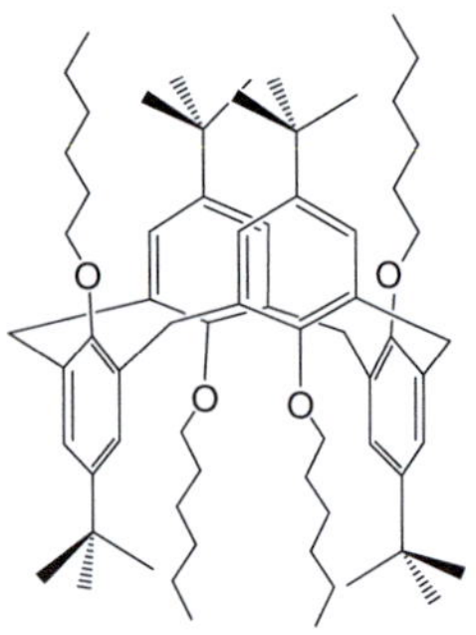

822

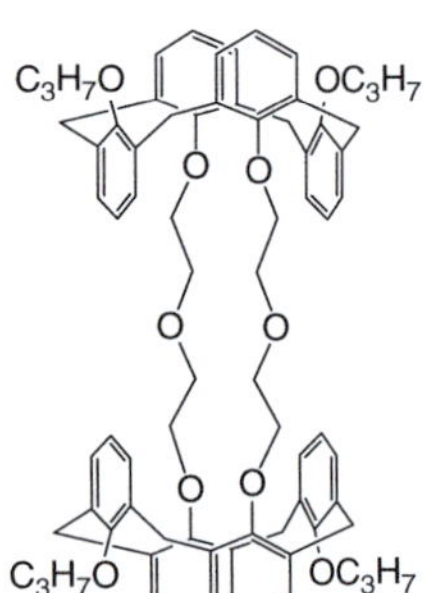

823

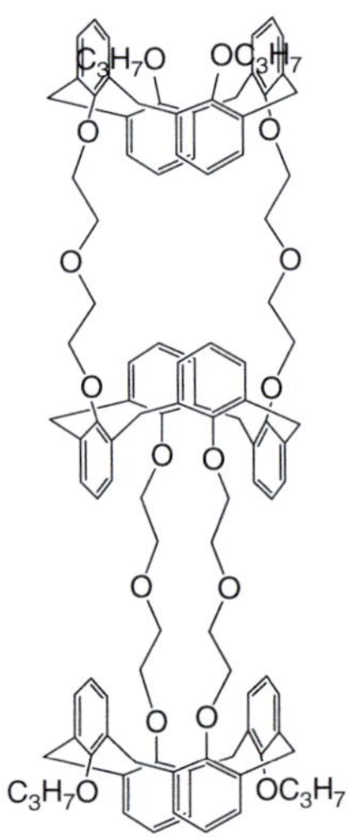

831

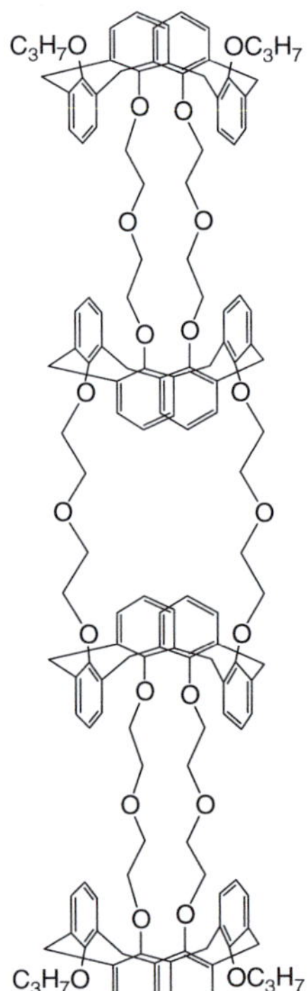

832

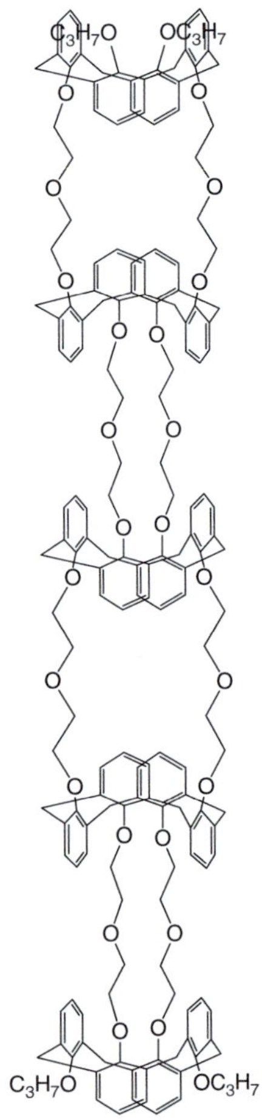

833

Scheme 6.2

times more active than this capsule, while its palladium-containing derivative is ten times more cytotoxic than the platinum-encapsulating complex. This result is explained in [11] by a higher rate of release of caged palladium(II) acetylacetonate from the cavity of **540**.

The similar results have been obtained in [12] for 1:1 cage complex of **540** with encapsulated pyrene-R molecule. In this case, fluorescent flow cytometry suggested more intensive fluorescence of the cells treated with the complex and allowed quantitatively estimating the uptake of the fluorophore, thus providing a direct evidence for the efficient release of this hydrophobic molecule from the cavity of a TP coordination capsule **540** following the uptake into a cell. The use of such a complex with caged fluorescent guest as a cargo allowed increasing the cytotoxicity and cellular uptake and monitoring its accumulation into the cells by microscopy and fluorescent flow cytometry. The entry of coordination capsules of this type into the cell is reported [12] to depend on the assisted diffusion pathway. The existence of a facilitated mode of this process implies certain cell specificity, offering a potential advantage for the use of the cage complexes with encapsulated cytotoxic guests as drug candidates [12]. An in vitro study [13] of the cytotoxicities of water-soluble 1:1 cage complexes of **540** with encapsulated lipophilic pyrenyl guests also showed higher activity of some of them as compared to the initial coordination capsule. For one of these guests, the effect is the most pronounced, and its cage complex is ten times more cytotoxic than **540**. Thus, the nature of the guest pyrenyl derivative strongly affects the cytotoxicity of its cage complex, allowing increasing it by using appropriate functionalizing substituents in the pyrene

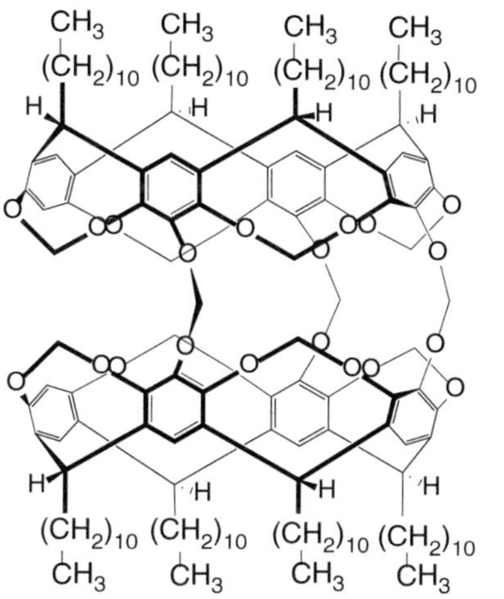

277

Scheme 6.3

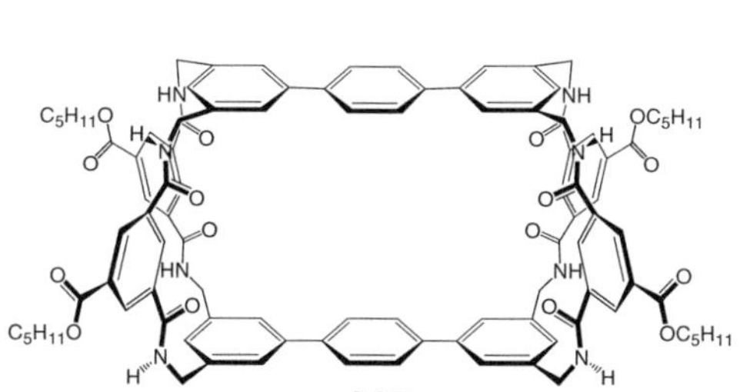

117

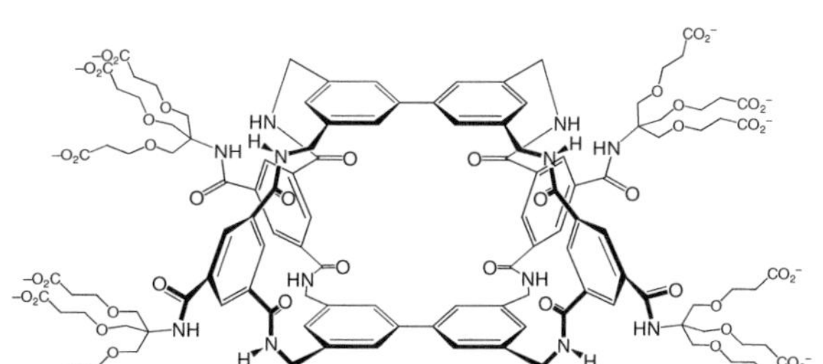

118

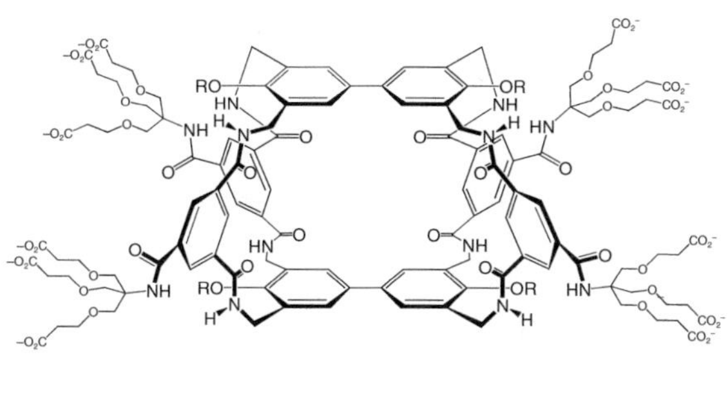

R = OCH_3 (**119**)
OC_2H_5 (**120**)
OC_3H_7 (**121**)
OC_4H_9 (**122**)

Scheme 6.4

Scheme 6.5

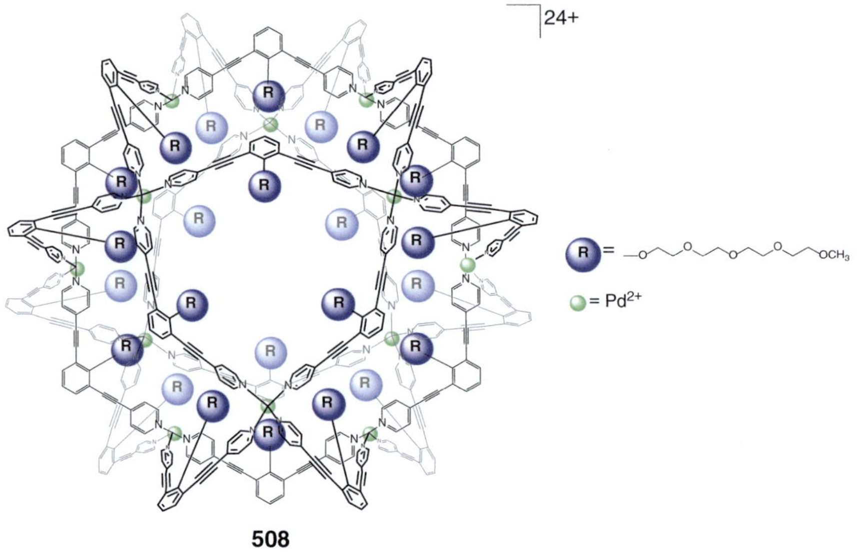

508

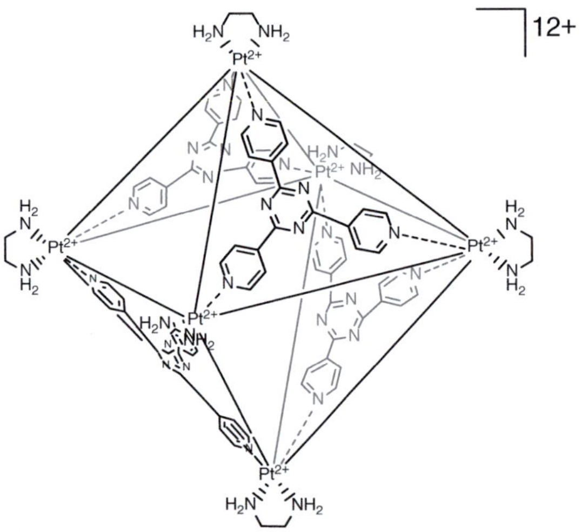

486

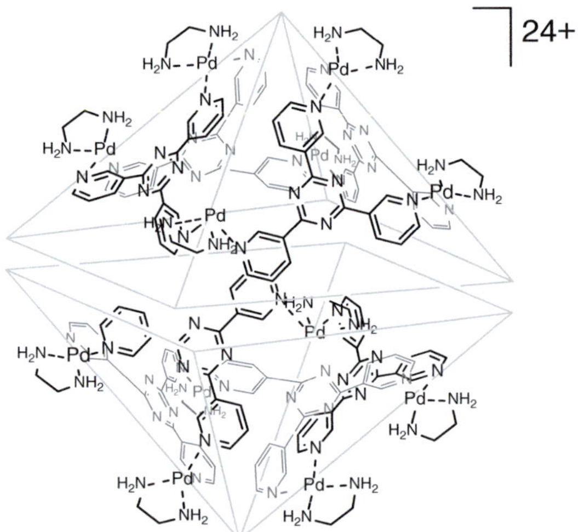

504

Scheme 6.6

guest. In particular, the encapsulation of carbonic anhydrase and glutathione transferase inhibitors for their selective delivery to the tumor environment may be useful to avoid additional side effects of the capsules [13].

A 1:1 complex of *para*-phosphonic acid–beared calix[5]arene-based caging ligand **382** (Scheme 6.8) with an encapsulated carboplatin molecule releases this guest in aqueous solutions, so the ligand can be used for delivery of these drug molecules [14]. A similar approach, developed in [15], used a Pd_2L_4 coordination capsule **781**: its 1:1 cage complexes are reported to have great potential as stimuli-responsive vectors for targeted and controlled drug delivery. Cage complexes of a bis-cavitand capsule **473** (Scheme 6.9) with trioxalane guests are proposed [16] as new class of prospective anti-malarials.

Covalent, supramolecular, and coordination capsules have been also widely used for storage of reactive species. In particular, a paramagnetic caging ligand **551** (Scheme 6.10) is reported in

Scheme 6.7

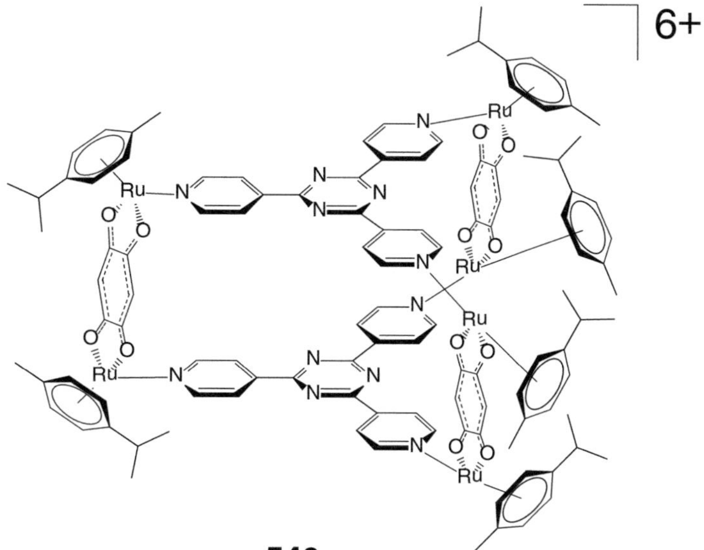

540

Scheme 6.8

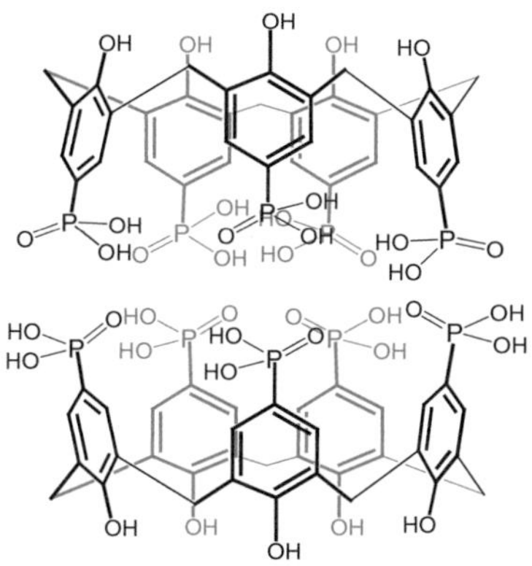

382

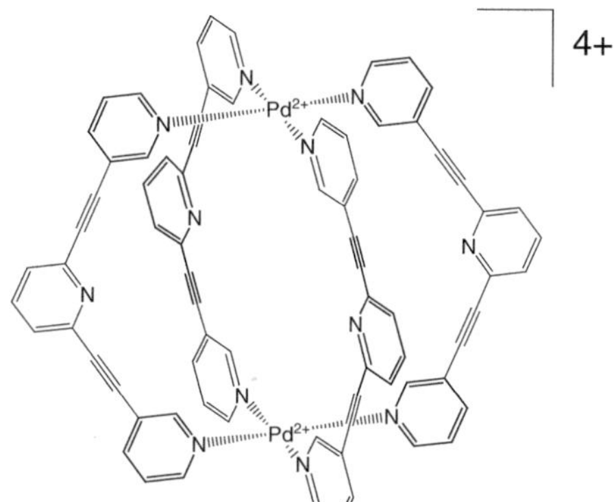

781

[17] to encapsulate radical guests, thus allowing for their further manipulations in aqueous solutions. The coordination capsule **581** is described in [18] to encapsulate the potent greenhouse gas SF_6; under ambient conditions, its caged molecule is significantly compressed and can be readily stored in solution, separated, and recycled. Moreover, the air-sensitive white phosphorus after its encapsulation within the hydrophobic inner cavity of a caging ligand **581** became air stable and water-soluble (see Chaps. 1 and 4). The guest P_4 molecules can be released in a controlled fashion without disrupting the cage frameworks by adding benzene as a competing guest [19].

A tube Cu_8L_4 coordination capsule **724** (Scheme 6.11) has been used in [20] for selective encapsulation and release of dicyanoaurate guest anion; the process is reported to be useful for extraction and refining of gold [20]. A TP caging ligand **552** has been used [21] for stabilization of a mixed-valence state of guest organic dimer $[(TTF)_2]^{+\bullet}$.

Polyamine caging ligands **144** and **836** (Scheme 6.12), their homologs, and solid support-bound derivatives have been patented in [22] as promising chelators for encapsulation of $[^{19}F]$-fluoride anion and as reagents for the preparation of ^{19}F-radiopharmaceuticals suitable for such radiodiagnostics.

The observed template effect of a coordination capsule **14** (Scheme 6.13) on the formation of 2D-layered water clusters is reported in [23] to be very important for correct description of association of water molecules in biological systems [23, 24].

Reactive cationic intermediates of a chemical reaction have been stabilized within the cavity of M_4L_6 coordination capsule **691** (Scheme 6.14), while the final products are released quickly [25]. A boronic ester-based bis-cavitand **278** has been used in [26] as a photosensitized and covalent capsule preventing the photochemical reactions of an encapsulated polyaromatic molecule.

An anthracene-based Pd_2L_4 coordination capsule **837** has been also used in [27] for safe storage of radical initiators shown in Scheme 1480. Their caged molecules have remarkable stability toward light and heat within the well-defined cavity of this polyaromatic caging ligand. The encapsulated and stabilized initiators are suitable for radical polymerization of olefins upon their spontaneous release from the cavity of **837** under reaction conditions (Scheme 6.15) [27].

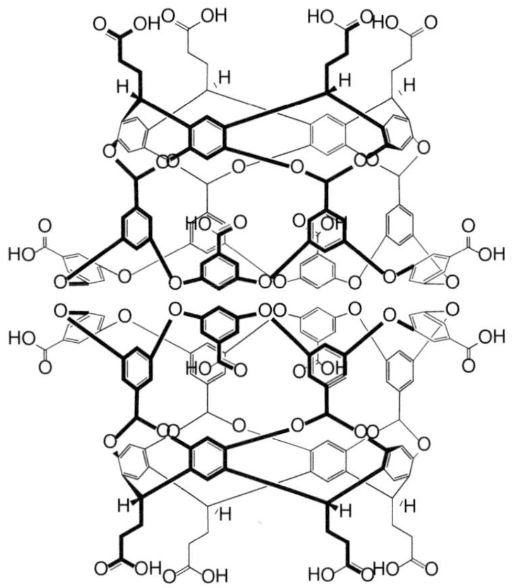

473

Scheme 6.9

6.2 Chiral Separation

Coordination capsules with chiral inner cavities are prospective for chiral recognition and resolution of chiral guests. In particular, diastereoselective formation of tetrahedral M_4L_6 cage complexes of coordination capsules **633** and **634** (Scheme 6.16) from a chiral bridging ligand syntone, which resulted in a substantial chiral amplification caused by adoption by their encapsulating ligands of a suitable conformations upon binding, is reported in [28]. These complexes are of interest not only owing to their chiroptical properties but also to the possible stereoselectivity in the host–guest chemistry associated with their large central cavities. Chiral cationic $M_{16}L_{24}$ caging ligands **642** and

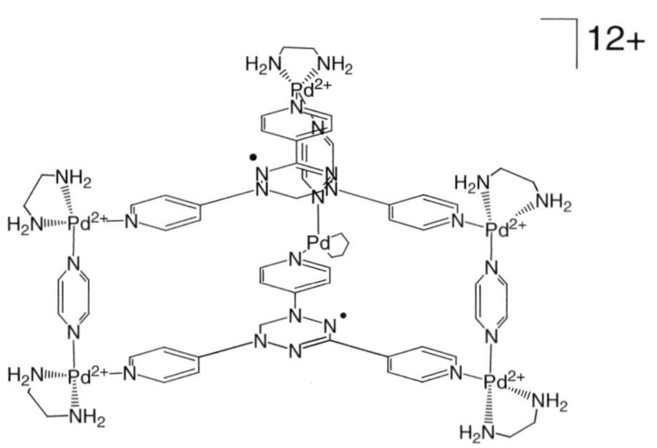

551

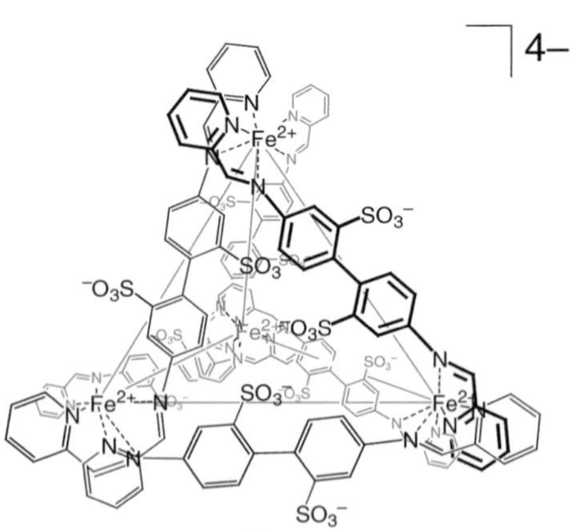

581

Scheme 6.10

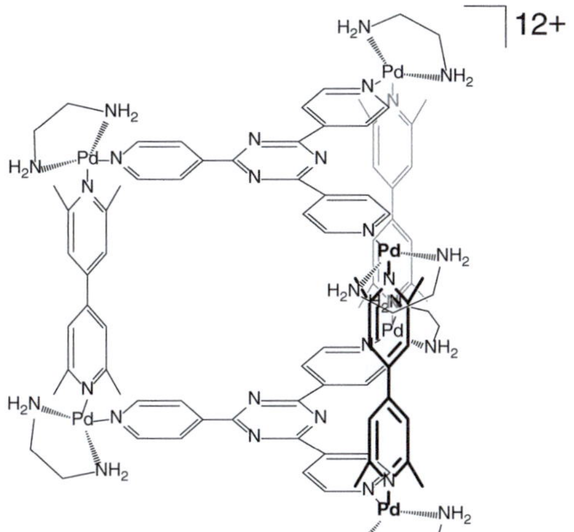

552

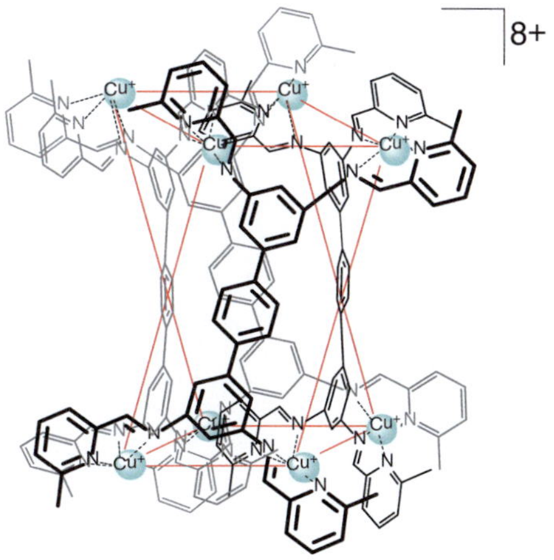

724

Scheme 6.11

Scheme 6.12

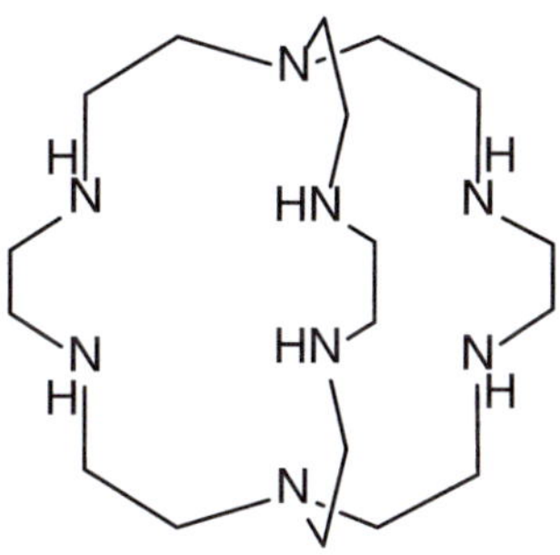

144

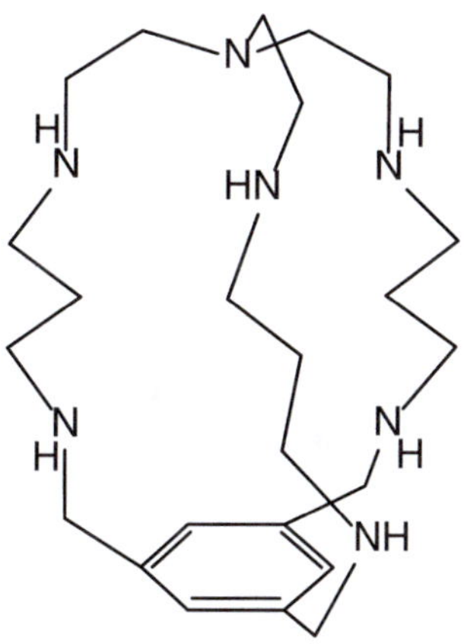

836

Scheme 6.13

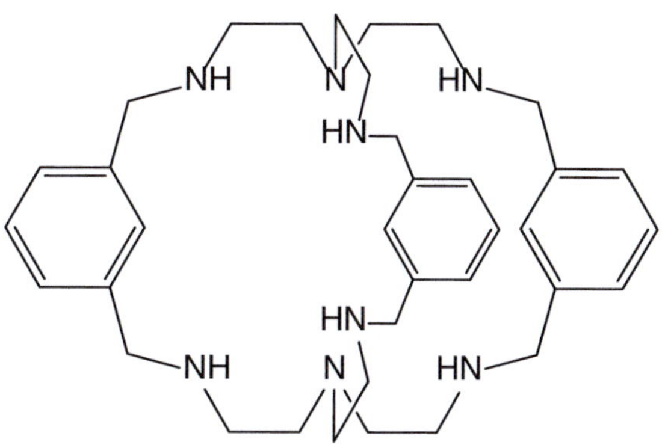

14

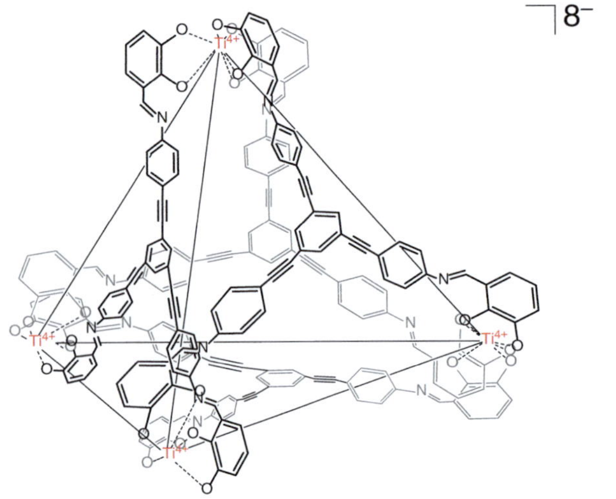

691

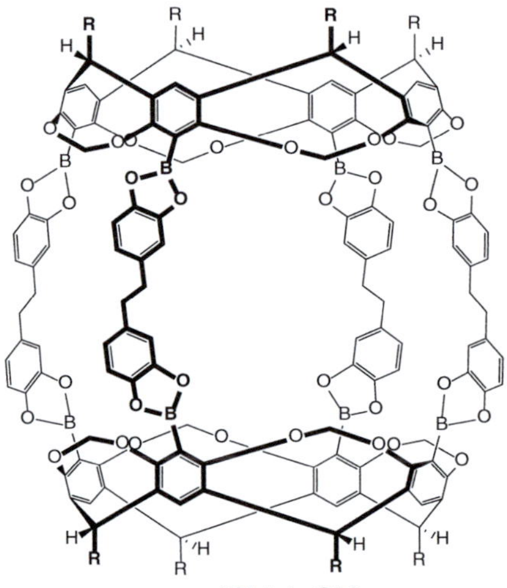

R = $(CH_2)_6CH_3$
278

Scheme 6.14

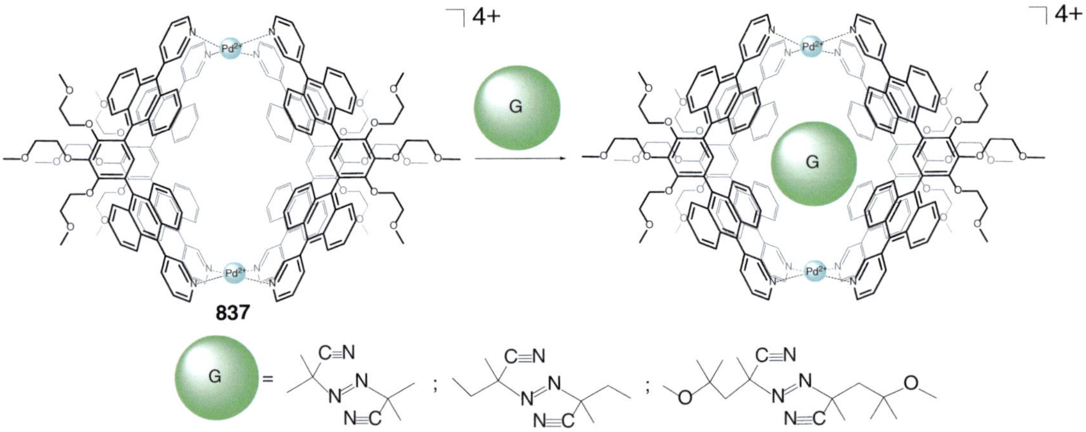

Scheme 6.15

643, $M_{12}L_{12}L'_4$ and coordination capsules **648–650** (Scheme 6.17) with homochiral vertex metallocenters have been proposed in [29] as suitable encapsulating hosts for chiral anionic guests.

Enantiomerically defined tetranuclear coordination capsules **582–585**, **826**, and **828** shown in Scheme 6.18 are described [30, 31] to provide useful chiral spaces for enantioselective catalysis and differential molecular recognition of enantiomers, allowing for the discrimination and transformation of the diastereomeric guests. The observed higher amplification in the case of these tetranuclear capsules than their cage analogs is explained in [31] by the presence of cooperative communication between its vertex metallocenters. This allowed one to elucidate the mechanisms by which stereochemical information has been transmitted between their subunits to provide new methods for the design and synthesis of diastereomerically pure capsules with large inner cavities that allow performing chiral separations and transformations of the caged guests [31].

Scheme 6.16

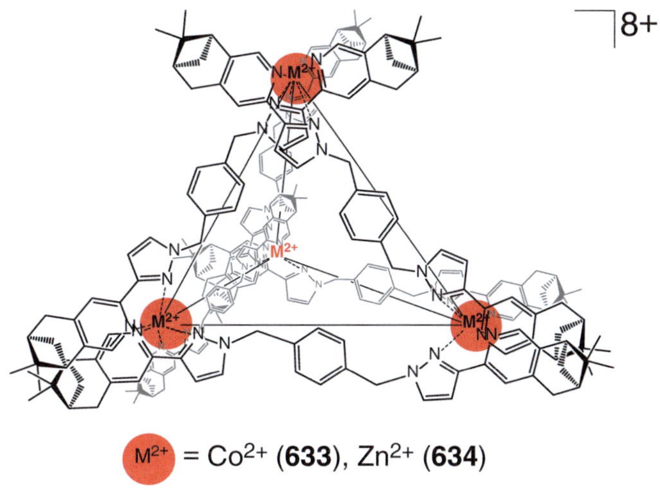

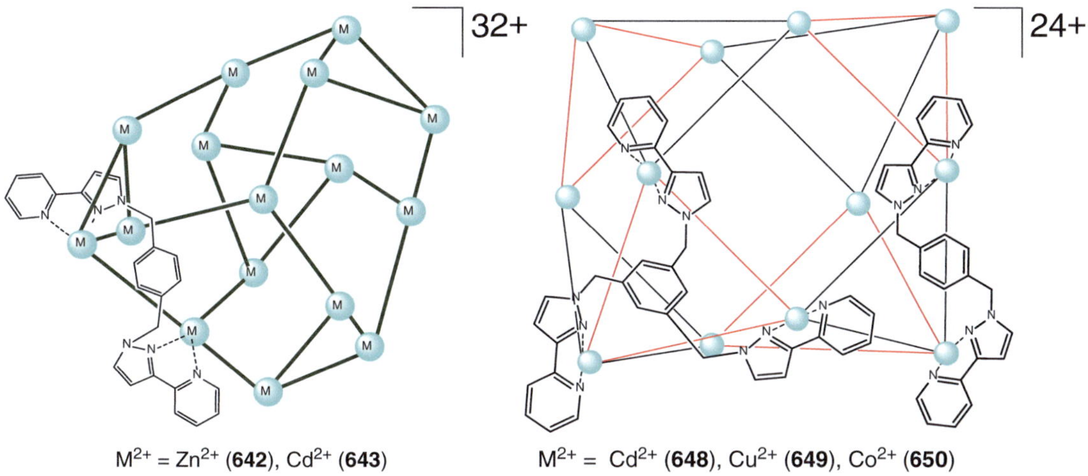

Scheme 6.17

6.3 Chiral and Achiral Caging Catalysts and Cage Flasks

Size-based substrate selectivity of the reaction of C–H-activated substrates with a caged organoiridium complex encapsulated by the Ga_4L_6 dodecaanionic coordination capsule **575** (Scheme 6.19), allowing for a nondissociative *in–out* guest exchange, has been observed in [32].

Two reactive organometallic intermediates of the catalytic dimerization and oligomerization of substituted butadienes have been stabilized in the cavity of a chiral capsule **575**: encapsulation by this size- and shape-complementary host is reported [33] to prevent their decomposition. At the same time, they react stoichiometrically with CO using this coordination capsule as a molecular flask.

Activity of a $(CH_3)_3PAu^+$ cation as a catalyst of intramolecular hydroalkoxylation of allenes became eight times higher upon its encapsulation within the cavity of the Ga_4L_6 coordination capsule **575**. This allowed performing these reactions in water: up to 67 catalytic turnovers by the caged catalyst have been observed in [34].

A peptide-lined coordination capsules **516–524** (Scheme 6.20) is reported [35] to generate chiral vacant cavities that could be used as

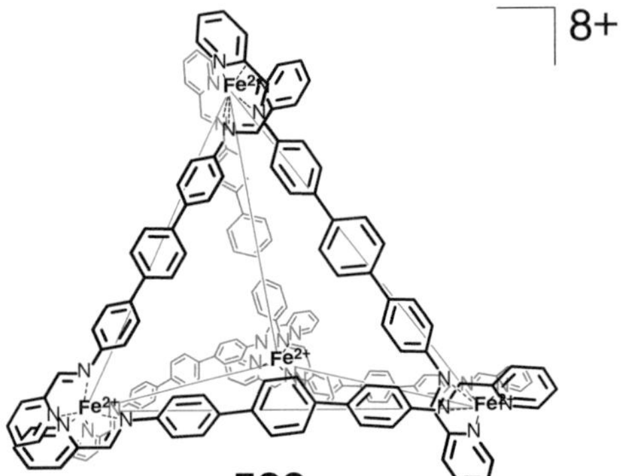

582

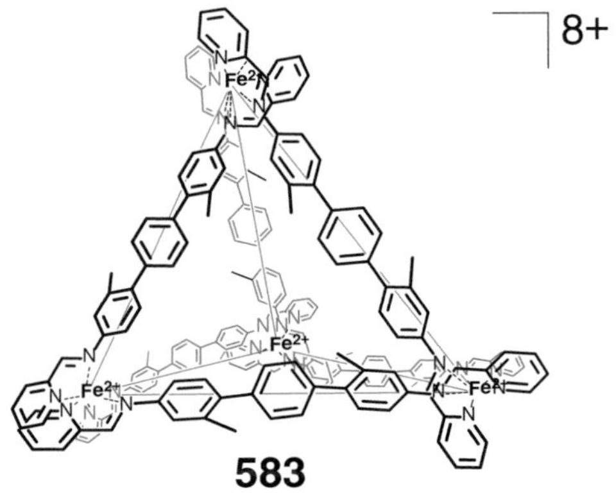

583

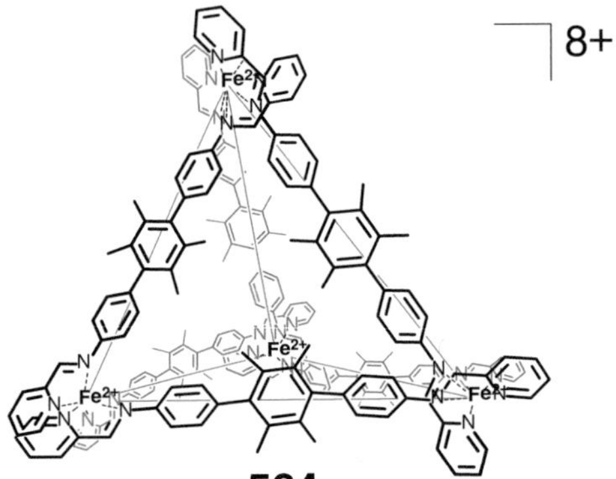

584

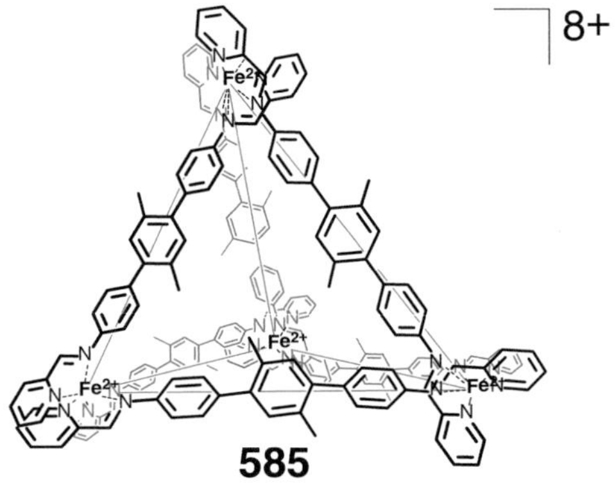

585

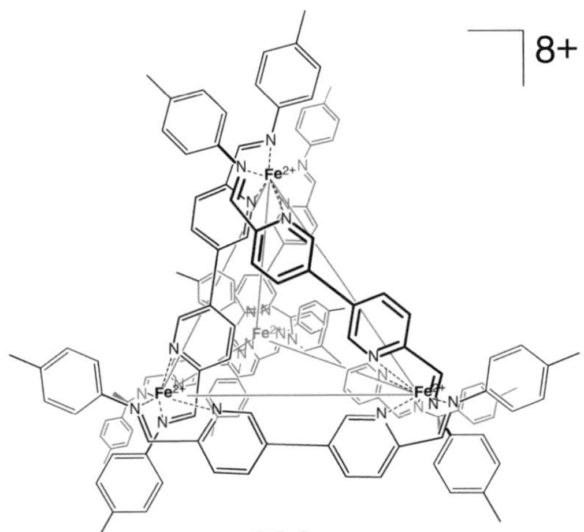

826

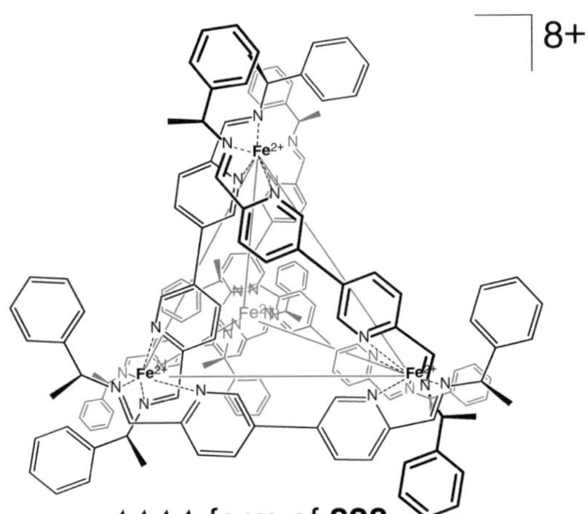

ΔΔΔΔ-form of **828**

Scheme 6.18

Scheme 6.19

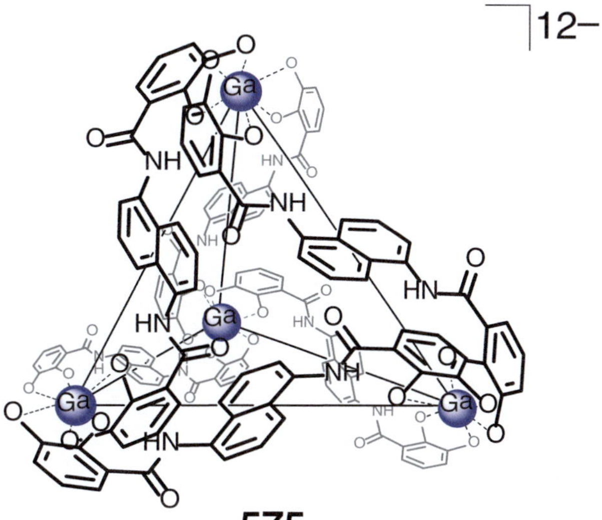

575

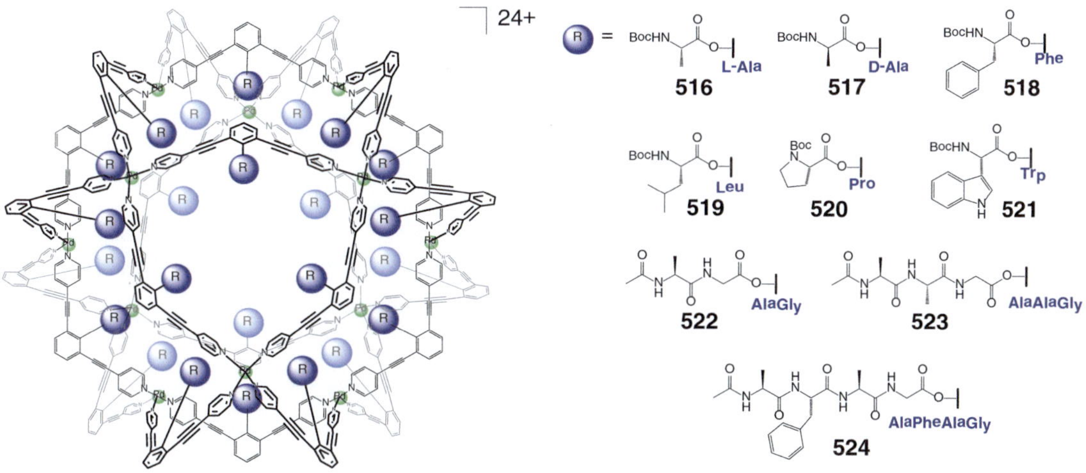

516 517 518 519 520 521 522 523 524

Scheme 6.20

binding pockets for asymmetric molecular recognition and reactions. The ratio and position of amino acid residues within these cavities, playing the role of artificial enzyme pockets, can be controlled by the choice of appropriate ligand syntones and template guests.

A large water-soluble Fe_4L_6 caging ligand **598** (Scheme 6.21) is reported [36] to catalytically accelerate the hydrolysis of acetylcholine esterase inhibitors, such as insecticide dichlorvos (see Sect. 4.1.1).

If free radical polymerization of sodium *para*-styrenesulfonate as an anionic monomer was carried out [37] with a cationic coordination capsules **525** and **526** (Scheme 6.22) as a template, acceleration of the polymerization and control of the molecular weight of the polymeric product, depending on the number of positive charges on the cage template, have been achieved.

A chiral hemicryptophane covalent capsules **205** and **206** and their oxidovanadium-containing derivatives **207** (Scheme 6.23) are proposed [38] as artificial models of enzymes and supramolecular asymmetric catalysts. In particular, the hemicryptophane oxidovanadium(V) caging ligands with the helical chirality (i.e., Δ or Λ con-

Scheme 6.21

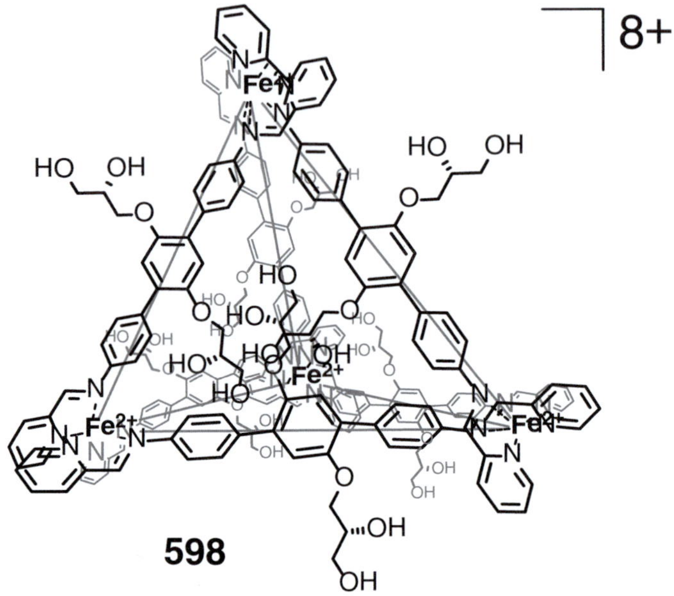

598

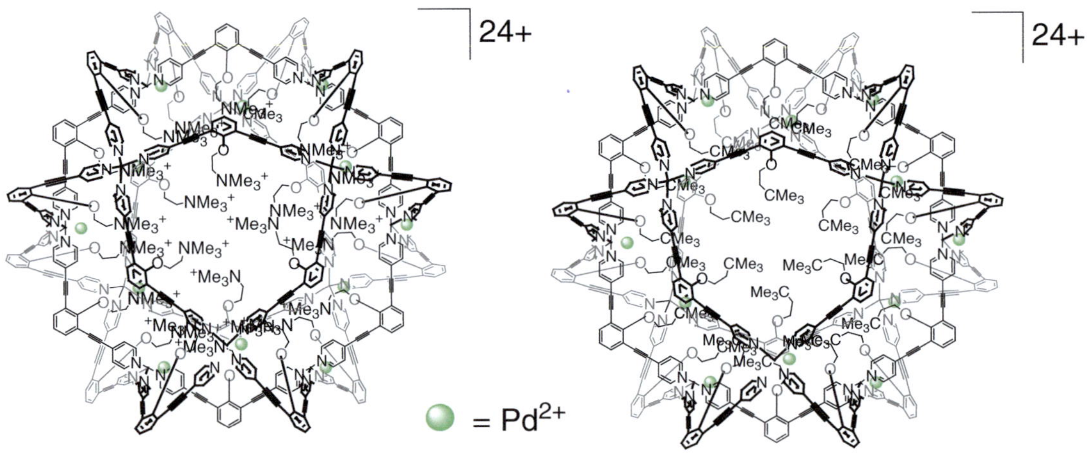

525 **526**

Scheme 6.22

figurations) of the atrane moiety of the capsule allows modeling specific features of vanadate-dependent haloperoxidases, such as the trigonal bipyramidal O_4N–coordination polyhedron of their oxidovanadium(V) metallocenters and the chiral environment of its active center. The chirality is described in [39] to be switched by changing the solvent. Chiral covalent capsules **208** and **209** are also reported [40] to provide a suitable environment for the formation of their cage metal complexes, having a caged catalytically active metallocenter.

6.4 Electrochemical, Paramagnetic, Optical, and X-ray Caging Sensors and Caged Probes

Bimetallic dithiocarbamate coordination capsules **809–820** (Scheme 6.24) with redox-active metal–ligand fragments are proposed in [41] as electrochemical sensors for encapsulated guest molecules.

A radical caging ligand **491** (Scheme 6.25) with four spin centers is reported in [42] to be a

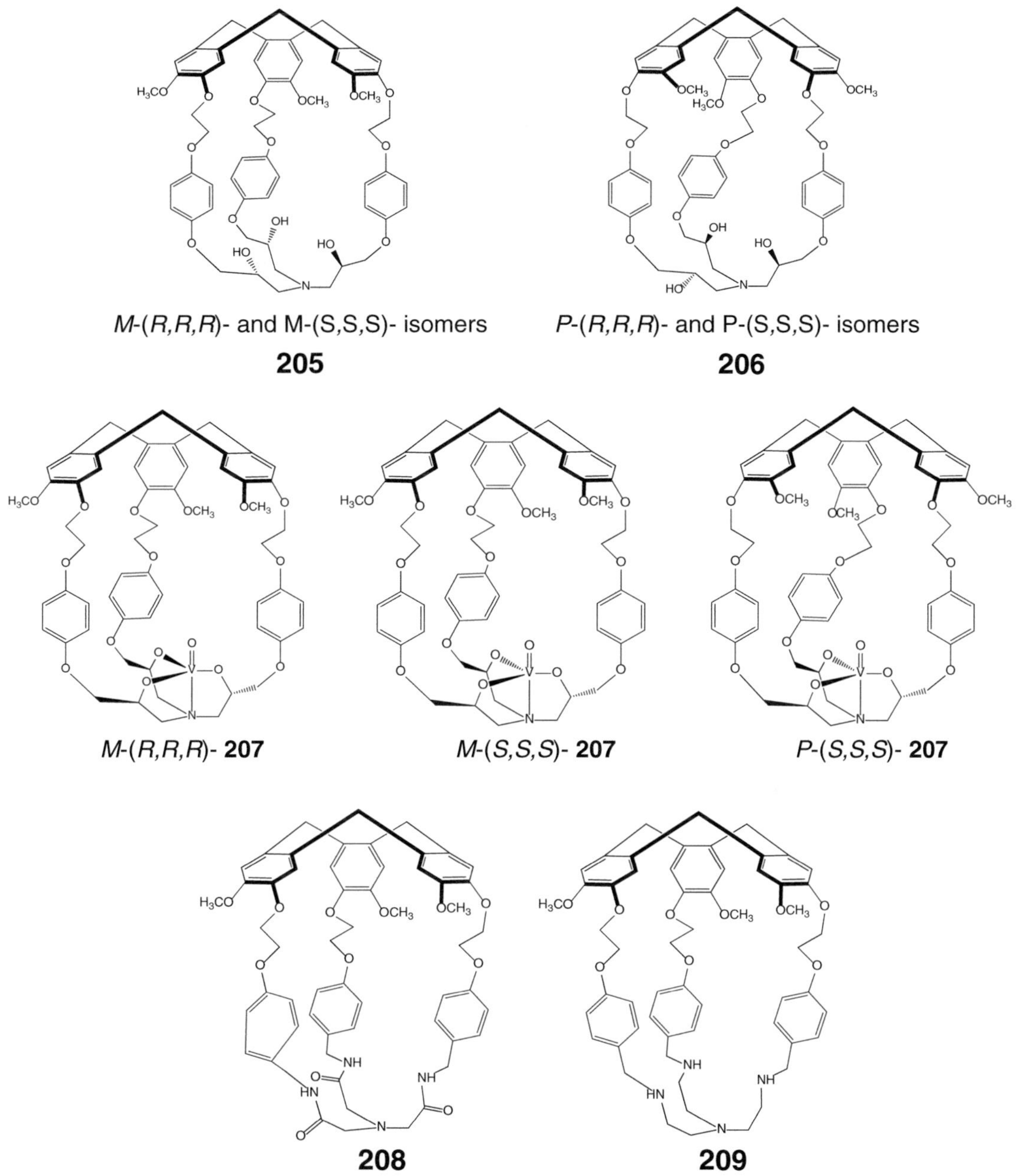

Scheme 6.23

prospective for the design of spin probe reagents and MRI contrast agents.

Simultaneous EPR observation of the electron spin–spin superexchange between the encapsulated paramagnetic nitroxide molecule (Scheme 6.26) within the cavity of a hydrogen-bonded bis-cavitand capsule **473** (Scheme 6.8) and the nitroxide molecule in the external aqueous solution is reported in [43]. The EPR data also provide direct information on the motion and the polarity of the ^{15}N-labeled encapsulated guest and its ^{14}N-labeled free molecule in the bulk solution; communication has been controlled by supramolecular attraction and repulsion between the cage complex of this charged guest and charged nitroxide molecules in the

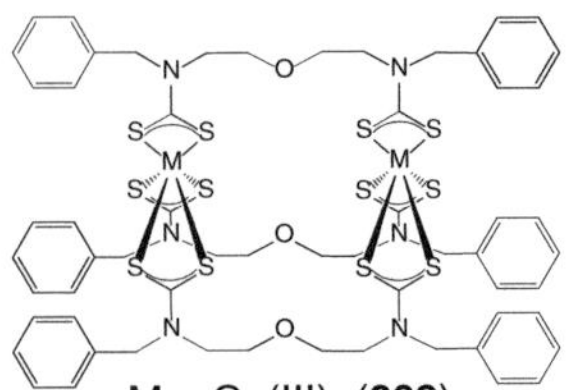

M = Co(III) (**809**)
M = Fe(III) (**810**)

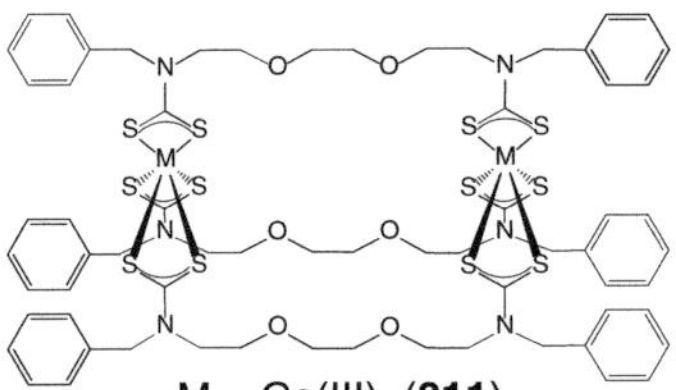

M = Co(III) (**811**)
M = Fe(III) (**812**)

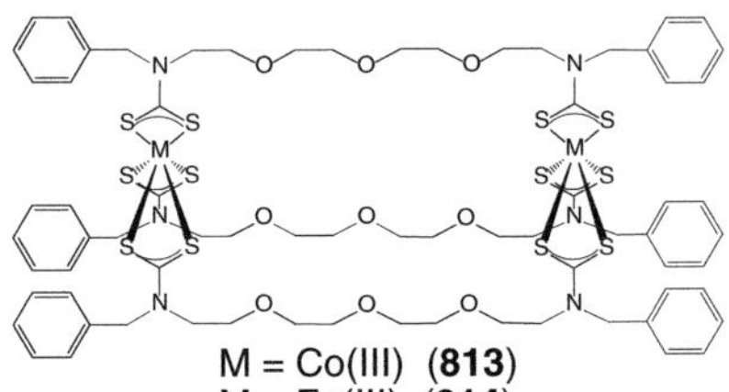

M = Co(III) (**813**)
M = Fe(III) (**814**)

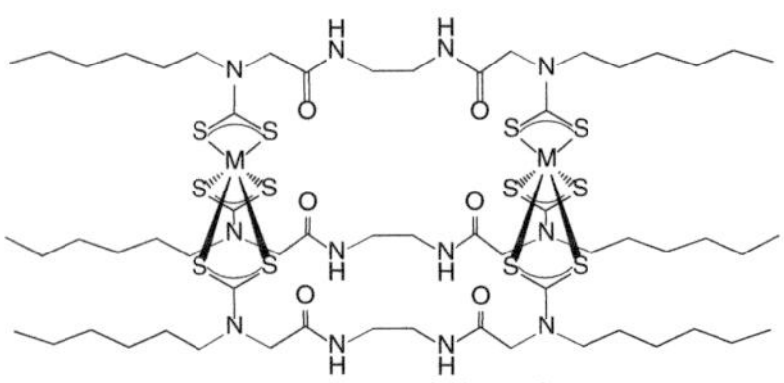

M = Co(III) (**815**)
M = Fe(III) (**816**)

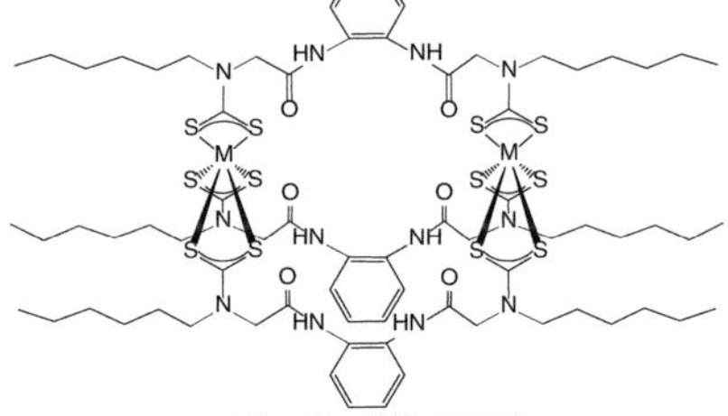

M = Co(III) (**817**)
M = Fe(III) (**818**)

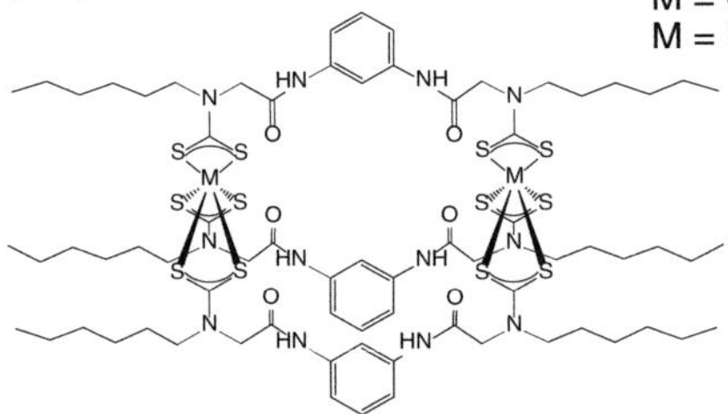

M = Co(III) (**819**)
M = Fe(III) (**820**)

Scheme 6.24

Scheme 6.25

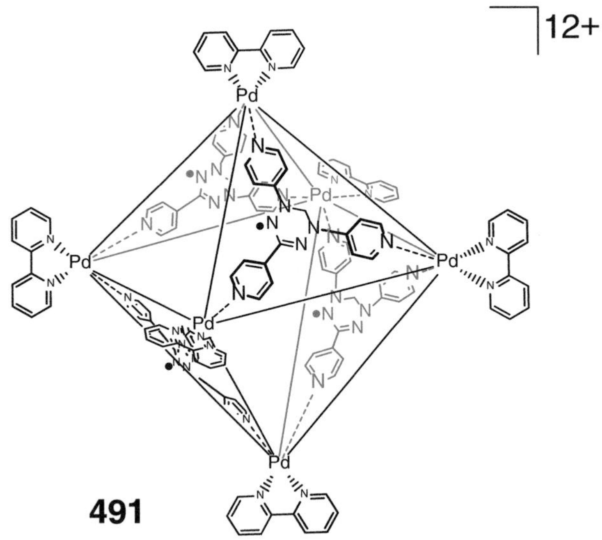

491

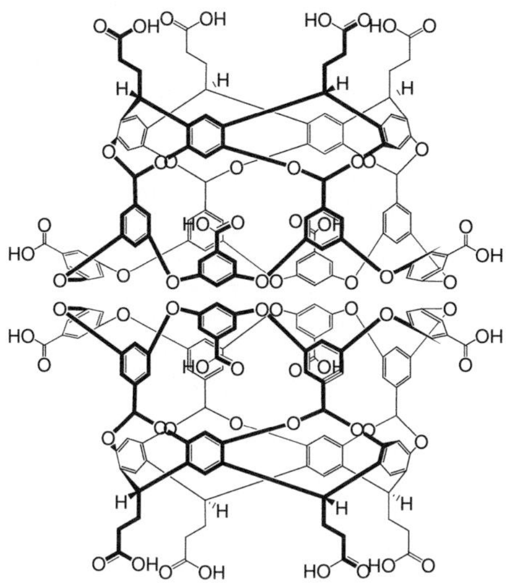

Scheme 6.26

bulk solution. Such encapsulation by **473** has been evidenced [43] through paramagnetic NMR relaxation experiments that involved electron spin–nuclear spin interactions. The following supramolecular and electron spin effects have been observed [43]: (*i*) pronounced spin–spin interactions between two nitroxide species, even with one of them protected by the hydrogen-bonded capsule, (*ii*) they can be increased by supramolecular binding resulting from Coulombic attractions or inhibited by Coulombic repulsions, and paramagnetic nuclear relaxation of an encapsulated guest molecule and internal relaxants can be used to elucidate these supramolecular interactions and structures. Thioxanthone molecule with a covalently attached ^{15}N-labeled nitroxide group encapsulated by the same octaacid capsule is reported in [44] to generate ESP of the nitroxide in the bulk aqueous solution through its cage framework after photoexcitation of the guest.

A 1:1 cage complex of cryptophane covalent capsules **177**, **199**, **203**, and **204** (Scheme 6.27) with encapsulated xenon-129 molecule is reported in [45–47] as ^{129}Xe MRI biosensors owing to the chemical shift sensitivity and large NMR signal of hyperpolarized ^{129}Xe. They are described as excellent candidates for in vivo MRI experiments after chemical transformations, improving their solubility and stability in biological media; the xenon T_1 value of the bound state can be lengthened by deuteration of such capsules. The improvements for sensing applications also include tuning of the *in–out* exchange rate to increase sensitivity and selectivity of xenon encapsulation [46].

A cascade binuclear cage complex of **112** with two encapsulated zinc(II) ions (Scheme 6.28) has been reported [48] to be an efficient fluorescent sensor for anions (first of all, azide ion) in aqueous solutions. An analogous covalent capsule with two encapsulated copper(II) ions, the caging macrobicyclic ligand **10** of which contains functionalizing 2,5-dimethylfuran spacer fragments, has been used in [49] for specific recognition and colorimetric sensing of chloride and bromide anions, forming anion-to-metal CTBs upon their encapsulation.

A hexaresorcinarene caging ligands **402** and **403** (Scheme 6.29) with pendant donor and acceptor fluorophore substituents that show FRET has been used in [50] as a tool to probe the dynamic behavior of hydrogen-bonded hexameric capsules at nanomolar concentrations. The encapsulation of a fluorescent molecule by the supramolecular capsule **561** (Scheme 6.30) is reported [51] to cause FRET that occurs from the caged guest to its labeled caging ligand [51]. A truncate octahedral supramolecular capsule **498** has been used in [52] as an artificial chemosensor for the fluorescence detection of nucleosides. The double hydrogen bonding of its amide group makes this caging ligand an adaptive receptor for selective sensing of uridine over other nucleotides; this amide group works as an efficient communicator converting the recognition information into fluorescent signals.

A covalent capsule **251** (Scheme 6.31) exhibits an ON/OFF caging ability relative to guest 2,7-diazapyrenium derivatives, forming the corresponding cage complexes with only a *cis*-isomer of the ligand. The photocontrolled process

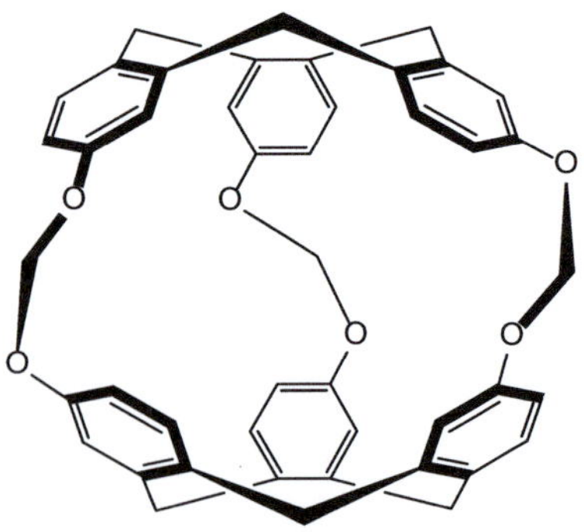

177

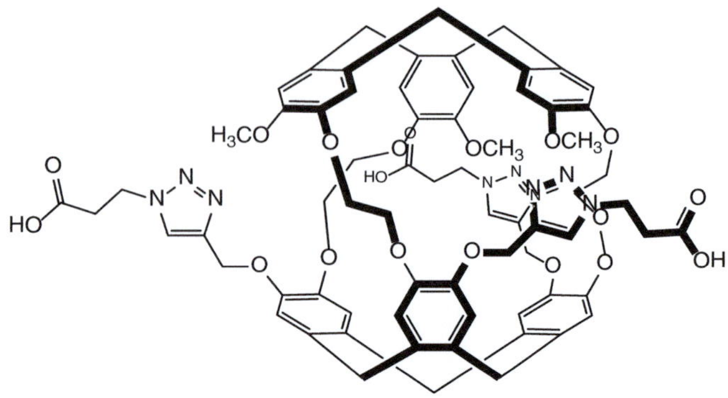

199

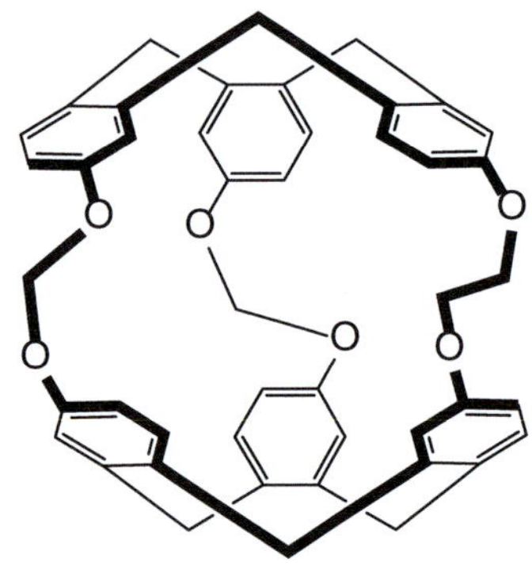

203

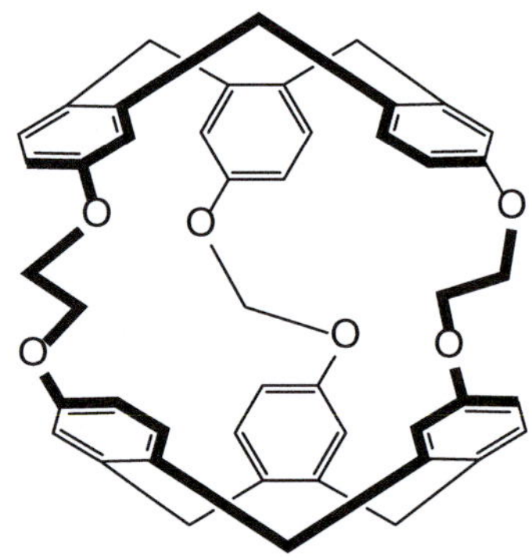

204

Scheme 6.27

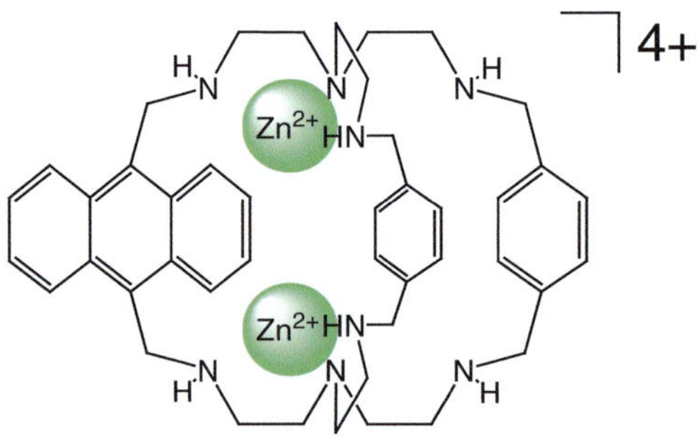

$(Zn^{2+})_2$@**112**

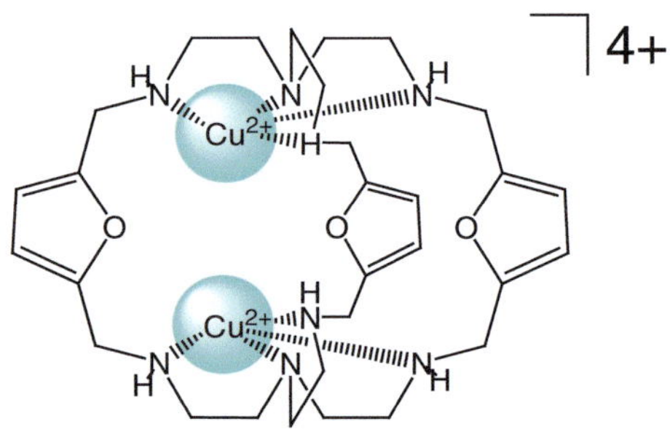

$(Cu^{2+})_2$@**10**

Scheme 6.28

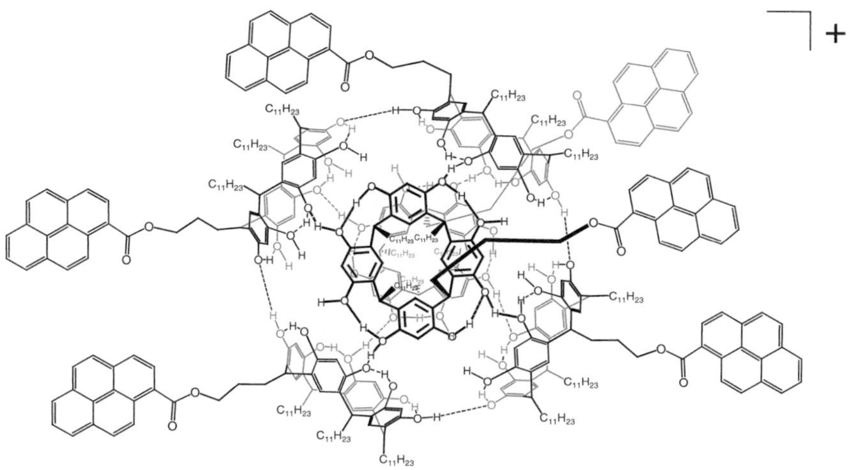

402

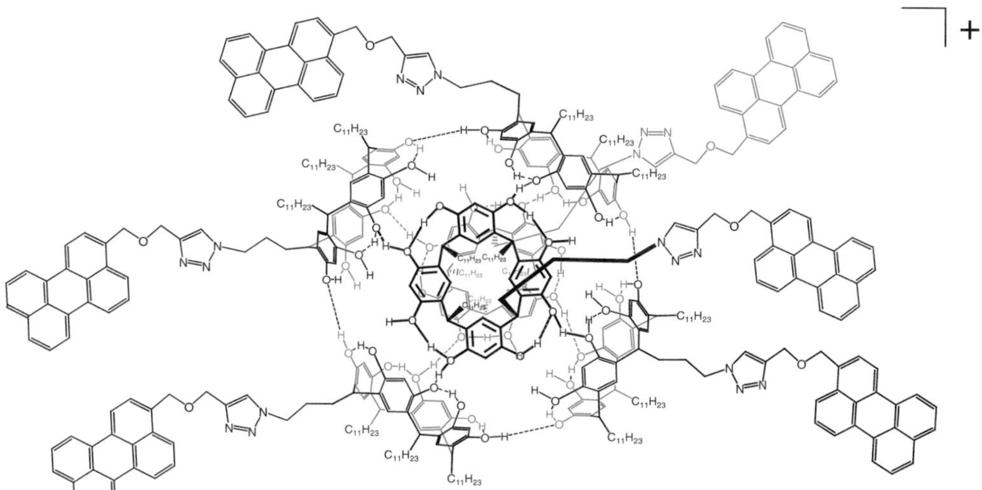

403

Scheme 6.29

is reported [53] to have fluorescence output, allowing for its optical monitoring.

Covalent capsules **50** and **838–841** (Scheme 6.32) are proposed [54] as potent highly affinity organic materials for sensing of γ-butyrolactone: the selectivity of its encapsulation over ethanol and water is determined by the presence of interior functionalizing groups allowing for preferential hydrogen bonding to γ-butyrolactone guest molecule by intramolecular protonation of amino groups of the capsules **838–841**, thus creating a supramolecular system with strong hydrogen bonding of γ-butyrolactone to these ammonium centers. The hydrogen bond donor ability decreases in the range of caging ligands **839–841**: while the proton in **50** is situated between the imine and the phenolate units and is too hindered to form an additional intermolecular hydrogen bond, the capsule **838** forms intramolecular zwitterionic species with an acidic proton that transforms its amine fragment in to a very potent H-donor. The encapsulating ligands **839** and **840** demonstrate the affinities that are intermediate between these two. The covalent capsule **839** is reported [54] as being able to activate the proton on its hydroxyl group, while the

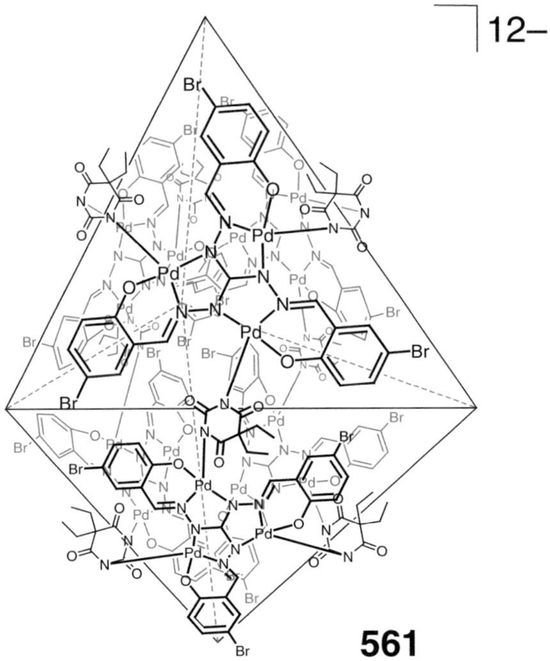

561

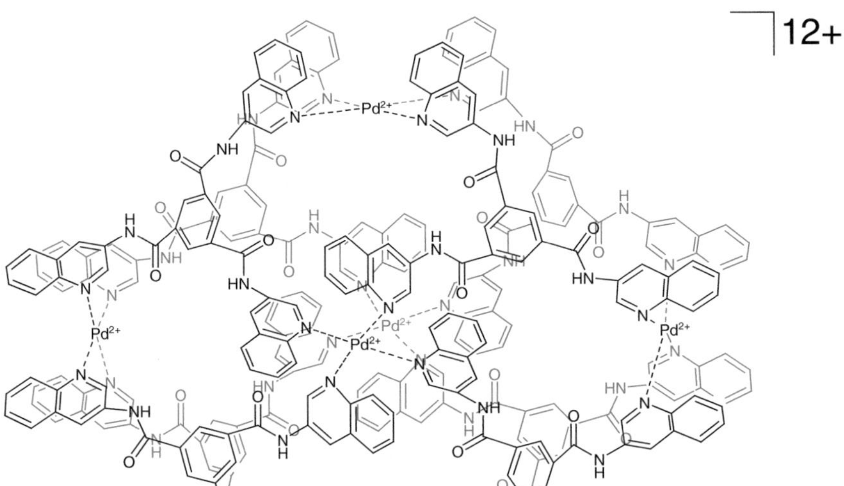

498

Scheme 6.30

cage framework **840** cannot form such supramolecular bonds; so their affinity to γ-butyrolactone as a guest is comparable to that of the caging host **50**. Effects of cavity size and of functioning groups within this cavity allowed designing of triptycene-based caging ligands **50** and **838–841** for the detection of γ-butyrolactone in sensor devices [54].

A method of single-crystal X-ray diffraction not requiring the crystallization of the compound under study has been developed in [55, 56]. The method is based on the use of networked cobalt- and zinc-containing metallopolymers **493** and *464* [57, 58] shown in Scheme 6.33 as so-called "crystalline sponges".

Near-infrared luminescent lanthanide complexes are considered [59] very promising for practical applications, as their optical properties have several complementary advantages over organic fluorophores and semiconductor nanoparticles. The specificity of f-f emission signals and long luminescence lifetimes have made lanthanide ions the key elements for a large number of novel advanced materials for bioanalytical applications and biological imaging, telecommunications, lasers, OLED/LED devices, and energy conversion. The fundamental challenge for lanthanide luminescence is their sensitization through

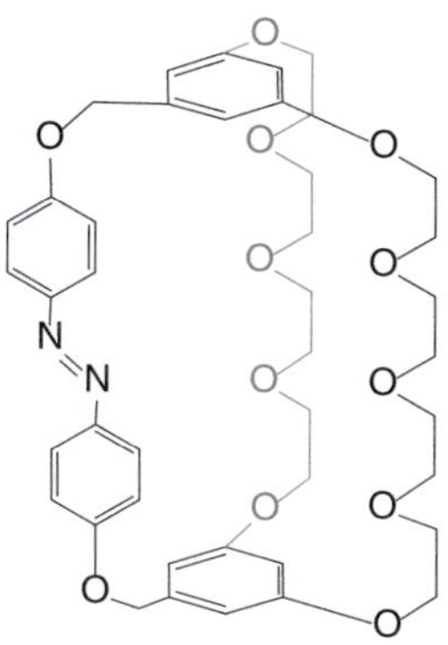

251

Scheme 6.31

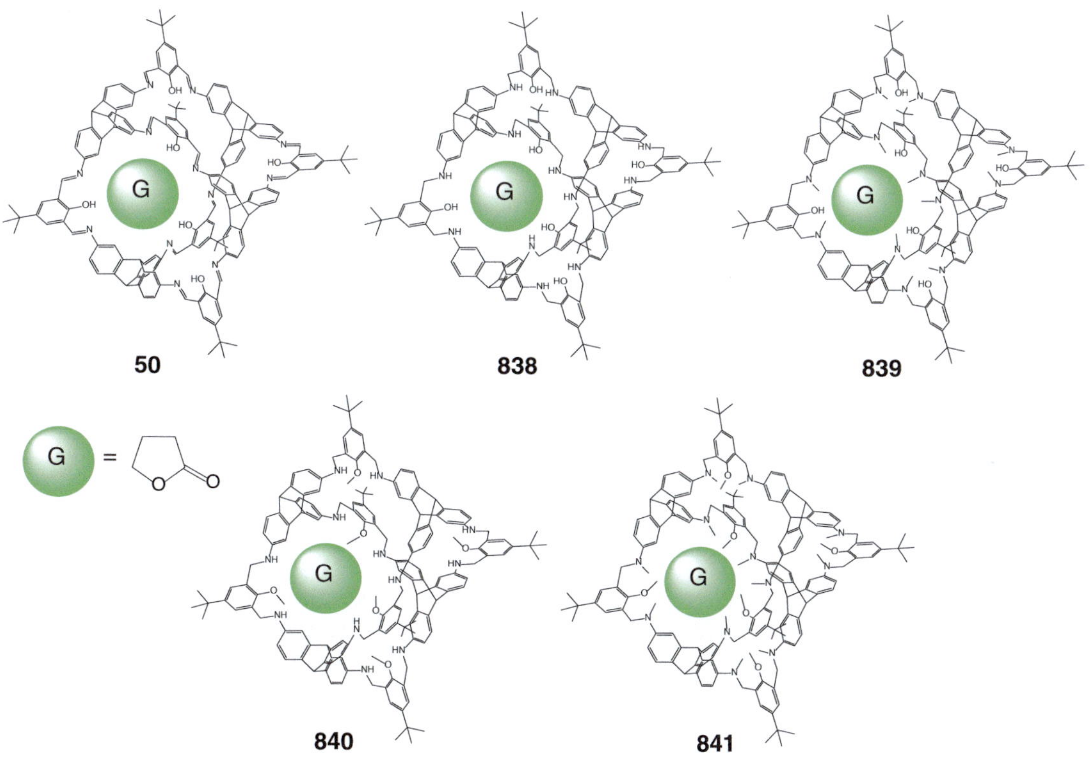

Scheme 6.32

Scheme 6.33

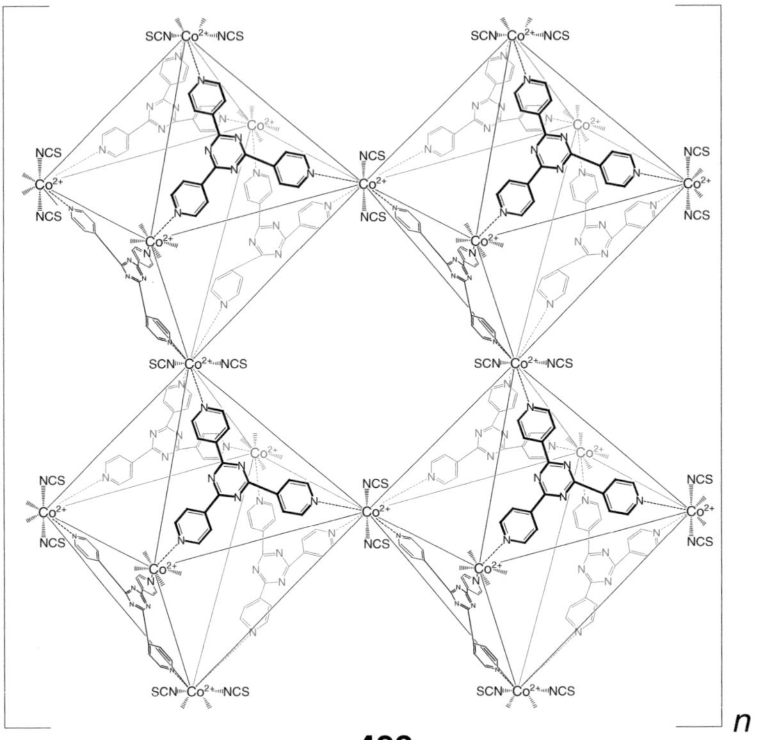

493

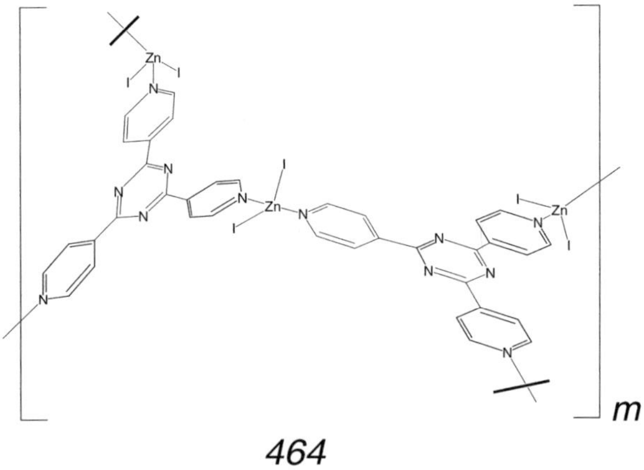

464

suitable chromophores, and the use of metallacrown complexes is proposed [59] as a prospective strategy to arrange several organic sensitizers at a well-controlled distance from the central lanthanide ion. Unusual photochemical properties of highly emitting near-infrared lanthanide-encapsulated sandwich metallacrown capsules (Cat^{3+})@**742** (Scheme 6.34) with excitation shifted toward lower energy, such as long luminescence lifetimes and high quantum yields, allow [59] considering them as novel and promising compounds for a wide variety of applications.

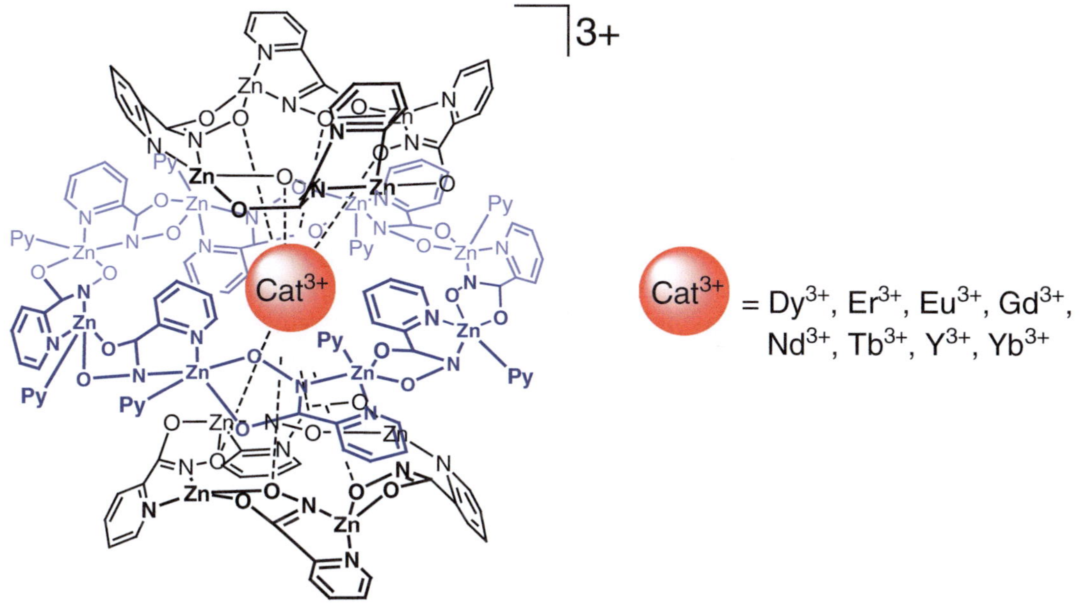

(Cat^{3+})@**742**

Scheme 6.34

6.5 Molecular and Supramolecular Electronic and Mechanic Devices and Building Blocks

AND/OR bimolecular recognition of organic guests by the coordination capsule **483** (Scheme 6.35) providing a prototypical molecular logic gate is described in [60]. The output in this system comes from the intrinsic nature of the input neutral organic molecules and is much less disturbed than in the case of their ion-encapsulating analogs. More importantly, by combining two input molecules with physical or chemical interactions, physical properties of the reaction products can be generated as outputs of their bimolecular recognition [60]. For AND bimolecular recognition, a caging host ligand does not encapsulate guests G_1 and G_2 separately, thus giving no corresponding homoguest cage complexes but only a heteroguest capsule. For OR bimolecular recognition, this caging ligand can adopt two of these guests or each of them separately. In particular, AND bimolecular recognition has been observed in [60] for large aromatic and suitable aliphatic guests, while OR recognition is found in [60] for two smaller aromatic guests.

Nanoball coordination capsules **751–757** (Scheme 6.36) are proposed [61, 62] for the design of a family of metal-varied discrete cage-like materials with a range of supramolecular and material functionalities. In particular, a discrete capsule **751** showed the reversible LIESST between its high- and low-spin states. Its nanosized cage molecules monodispersed on a solid surface in a display device or electronic switch are described in [62] to undergo a solvent-sensitive, physically addressable electronic spin switching. The switching occurs by thermal, light, or solvent perturbation, and its ON and OFF states can be switched by green or red laser irradiation, respectively. Moreover, this coordination capsule undergoes the LIESST: according to SQUID magnetometry data, photoexcitation of

Scheme 6.35

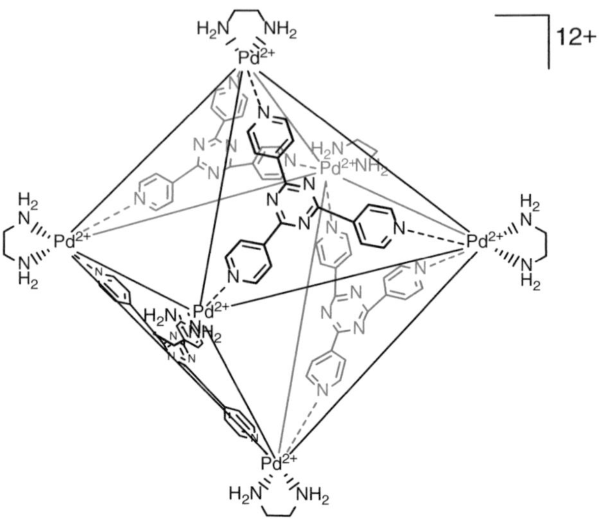

483

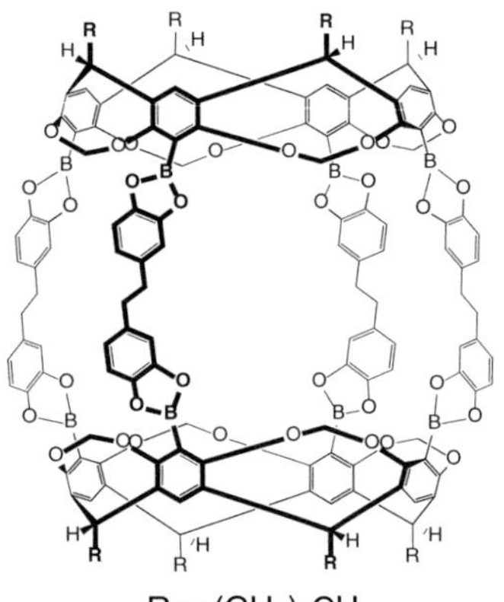

278

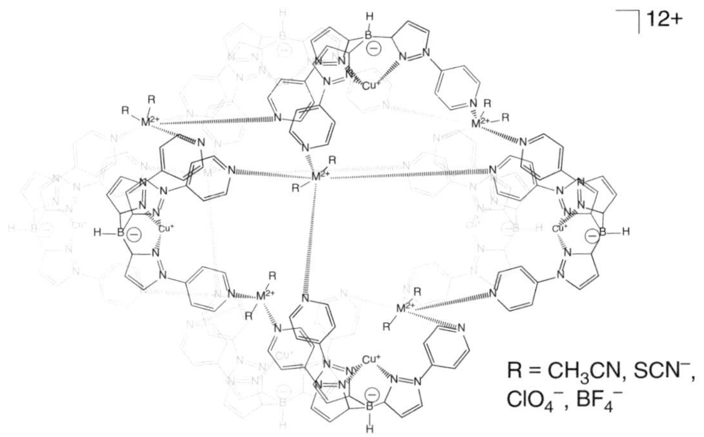

Fe^{2+} / SCN^- (**751**), Cu^{2+} / ClO_4^- (**752**), Zn^{2+} / ClO_4^- (**755**), Mn^{2+} / ClO_4^- (**753**), Fe / ClO_4^- (**756**), Cd^{2+} / ClO_4^- (**754**), Cu^{2+} / BF_4^- (**757**)

Scheme 6.36

low-spin cage species caused their transition to a "metastable" high-spin form. The photoinduced "stored" information can be "erased" by heating of this capsule up to 55 K or by its irradiation with red diode laser with $\lambda = 830$ nm at 10 K by the reverse LIESST effect. This ON/OFF process is reported in [62] to be fully reversible during numerous cyclings.

A chiral caging ligand **278** (Scheme 6.36) is reported [63] to form a supramolecular gyroscope directed toward unidirectional guest rotation. NMR study of rotation of substituted biphenyl guests within the cavity of the capsule showed structure-dependent rotation rate; for one of them, the rotation is very slow on NMR time scale at room temperature [63].

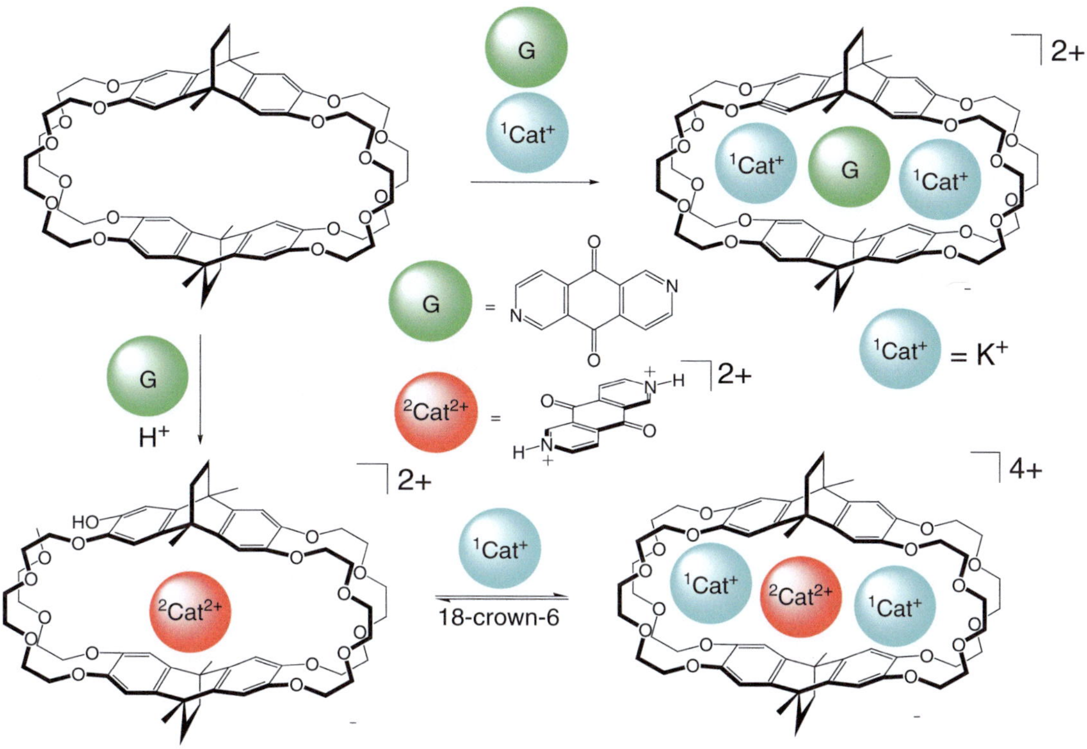

Scheme 6.37

Atrane-based hemicryptophane covalent capsules *M*-(*S,S,S*)-**207** and *P*-(*S,S,S*)-**207** (Scheme 6.23) exhibit reversible change in chirality controlled by the solvent, allowing one to obtain solvent-directed molecular switches and molecular propellers with clockwise and anticlockwise tilting motions [39, 64].

Rotation of an encapsulated aromatic guest within the covalent capsule **240** as a molecular gyroscope has been controlled in [65] by Scheme 6.37 using changes in its binding geometry with respect to the macropolycyclic host upon the addition and extrusion of co-caged potassium(I) cations.

References

1. Motekaitis R, Martell AE (1988) The dioxygen carrier properties of the dicobalt-obistren cryptate in aqueous solution. Chem Commun 1020–1022
2. Zyryanov GV, Rudkevich DM (2003) Encapsulated reagents for nitrosation. Org Lett 5(8):1253–1256
3. Organo VG, Sgarlata V, Firouzbakht F, Rudkevich DM (2007) Long synthetic nanotubes from calix[4]arenes. Chem Eur J 13:4014–4023
4. Leontiev AV, Rudkevich DM (2004) Encapsulation of gases in the solid state. Chem Commun 1468–1469
5. Lecollinet G, Dominey AP, Velasco T, Davis AP (2002) Highly selective disaccharide recognition by a tricyclic octaamide cage. Angew Chem Int Ed 41:4093–4096
6. Klein E, Crump MP, Davis AP (2005) Carbohydrate recognition in water by a tricyclic polyamide receptor. Angew Chem Int Ed 44(2):298–302
7. Barwell NP, Crump MP, Davis AP (2009) A synthetic lectin for β-glucosyl. Angew Chem Int Ed 48:7673–7676
8. Tominaga M, Suzuki K, Murase T, Fujita M (2005) 24-Fold endohedral functionalization of a self-assembled $M_{12}L_{24}$ coordination nanoball. J Am Chem Soc 127:11950–11951
9. Tashiro S, Fujita M (2006) Selective recognition of Trp- and Tyr-rich oligopeptides by self-assembled coordination hosts. Bull Chem Soc Jap 79:833–837
10. Tashiro S, Tominaga M, Yamaguchi Y, Kato K, Fujita M (2006) Peptide recognition: encapsulation and α-helical folding of a nine-residue peptide within a hydrophobic dimeric capsule of a bowl-shaped host. Chem Eur J 12:3211–3217

11. TherrienB,Süss-FinkG,GovindaswamyP,RenfrewAK, Dyson PJ (2008) The "complex-in-a-complex" cations $[(acac)_2M\subset Ru_6(p\text{-}iPrC_6H_4Me)_6(tpt)_2(dhbq)_3]^{6+}$: a trojan horse for cancer cells. Angew Chem Int Ed 47:3773–3776
12. Zava O, Mattsson J, Therrien B, Dyson PJ (2010) Evidence for drug release from a metalla-cage delivery vector following cellular internalization. Chem Eur J 16:1428–1431
13. Mattsson J, Zava O, Renfrew AK, Sei Y, Yamaguchi K, Dyson PJ, Therrien B (2010) Drug delivery of lipophilic pyrenyl derivatives by encapsulation in a water soluble metalla-cage. Dalton Trans 39:8248–8255
14. Martin AD, Boulos RA, Hubble LJ, Hartlieb KJ, Raston CL (2011) Multifunctional water-soluble molecular capsules based on *p*-phosphonic acid calix[5]arene. Chem Commun 47:7353–7355
15. Lewis JEM, Gavey EL, Cameron SA, Crowley JD (2012) Stimuli-responsive Pd_2L_4 metallosupramolecular cages: towards targeted cisplatin drug delivery. Chem Sci 3:778–784
16. Liu S, Gibb BC (2011) Solvent denaturation of supramolecular capsules assembled via the hydrophobic effect. Chem Commun 47:3574–3576
17. Ozaki Y, Kawano M, Fujita M (2009) Engineering noncovalent spin-spin interactions in an organic-pillared spin cage. Chem Commun 4245–4247
18. Riddell IA, Smulders MMJ, Clegg JK, Nitschke JR (2011) Encapsulation, storage and controlled release of sulfur hexafluoride from a metal-organic capsule. Chem Commun 47:457–459
19. Mal P, Breiner B, Rissanen K, Nitschke JR (2009) White phosphorus is air-stable within a self-assembled tetrahedral capsule. Science 324:1697–1699
20. Meng W, Clegg JK, Nitschke JR (2012) Transformative binding and release of gold guests from a self-assembled Cu_8L_4 tube. Angew Chem Int Ed 51:1881–1884
21. Yoshizawa M, Ono K, Kumazawa K, Kato T, Fujita M (2005) Metal–metal d–d interaction through the discrete stacking of mononuclear M(II) complexes (M = Pt, Pd, and Cu) within an organic-pillared coordination cage. J Am Chem Soc 127:10800–10801
22. Jackson A, Bhalla R (2009) Fluoride processing method. US Patent WO2009083530 A2, 9 July 2009
23. Lakshminarayanan PS, Suresh E, Chosh P (2005) Formation of an infinite 2D-layered water of $(H_2O)_{45}$ cluster in a cryptand–water supramolecular complex: a template effect. J Am Chem Soc 127:13132–13133
24. Saeed MA, Wong BM, Fronczek FR, Venkatraman R, Hossain MA (2010) Formation of an amine-water cyclic pentamer: a new type of water cluster in a polyazacryptand. Cryst Growth Design 10:1486–1488
25. Albrecht M, Janser I, Burk S, Weis P (2006) Self-assembly and host–guest chemistry of big metallosupramolecular M_4L_4 tetrahedra. Dalton Trans 2875–2880
26. Nishimura N, Kobayashi K (2010) Self-assembled boronic ester cavitand capsule as a photosensitizer and a guard nanocontainer against photochemical reactions of 2,6-diacetoxyanthracene. J Org Chem 75:6079–6085
27. Yamashina M, Sei Y, Akita M, Yoshizawa M (2014) Safe storage of radical initiators within a polyaromatic nanocapsule. Nat Commun 5:4662
28. Argent SP, Riis-Johannessen T, Jeffery JC, Harding LP, Ward MD (2005) Diastereoselective formation and optical activity of an M_4L_6 cage complex. Chem Commun 4647–4649
29. Argent SP, Adams H, Riis-Johannessen T, Jeffery JC, Harding LP, Ward MD (2005) High-nuclearity homoleptic and heteroleptic coordination cages based on tetra-capped truncated tetrahedral and cuboctahedral metal frameworks. J Am Chem Soc 128:72–73
30. Meng W, Clegg JK, Thoburn JD, Nitschke JR (2011) Controlling the transmission of stereochemical information through space in terphenyl-edged Fe_4L_6 cages. J Am Chem Soc 133:13652–13660
31. Ousaka N, Clegg JK, Nitschke JR (2012) Nonlinear enhancement of chiroptical response through subcomponent substitution in M_4L_6 cages. Angew Chem Int Ed 51:1464–1468
32. Davis AV, Raymond KN (2005) The big squeeze: guest exchange in an M4L6 supramolecular host. J Am Chem Soc 127:7912–7919
33. Fiedler D, Bergman RG, Raymond KN (2006) Stabilization of reactive organometallic intermediates inside a self-assembled nanoscale host. Angew Chem Int Ed 45:745–748
34. Wang ZJ, Brown CJ, Bergman RG, Raymond KN, Toste FD (2011) Hydroalkoxylation catalyzed by a gold(I) complex encapsulated in a supramolecular host. J Am Chem Soc 133:7358–7360
35. Suzuki K, Kawano M, Sato S, Fujita M (2007) Endohedral peptide lining of a self-assembled molecular sphere to generate chirality-confined hollows. J Am Chem Soc 129:10652–10653
36. Bolliger JL, Belenguer AM, Nitschke JR (2013) Enantiopure water-soluble $[Fe_4L_6]$ cages: host-guest chemistry and catalytic activity. Angew Chem Int Ed 52(31):7958–7962
37. Kikuchi T, Murase T, Sato S, Fujita M (2008) Polymerisation of an anionic monomer in a self-assembled $M_{12}L_{24}$ coordination sphere with cationic interior. Supramol Chem 20:81–94
38. Gautier A, Mulatier J-C, Crassous J, Durasta J-P (2005) Chiral trialkanolamine-based hemicryptophanes: synthesis and oxovanadium complex. Org Lett 7:1207–1210
39. Martinez A, Robert V, Gornitzka H, Dutasta J-P (2010) Controlling helical chirality in atrane structures: solvent-dependent chirality sense in hemicryptophane-oxidovanadium(V) complexes. Chem Eur J 16:520–527
40. Dimitrov-Raytchev P, Perraud O, Aronica C, Martinez A, Dutasta J-P (2010) A new class of C_3-symmetrical hemicryptophane hosts: triamide- and tren-hemicryptophanes. J Org Chem 75:2099–2102

41. Beer PD, Berry NG, Cowley AR, Hayes EJ, Oates EC, Wong WWH (2003) Metal-directed self-assembly of bimetallic dithiocarbamate transition metal cryptands and their binding capabilities. Chem Commun 2408–2409
42. Nakabayshi K, Ozaki Y, Kawano M, Fujita M (2008) A self-assembled spin cage. Angew Chem Int Ed 47:2046–2048
43. Chen JY-C, Jayaraj N, Jockusch S, Ottaviani MF, Ramamurthy V, Turro NJ (2008) An EPR and NMR study of supramolecular effects on paramagnetic interaction between a nitroxide incarcerated within a nanocapsule with a nitroxide in bulk aqueous media. J Am Chem Soc 130:7206–7207
44. Jockusch S, Zeika O, Jayaraj N, Ramamurthy V, Turro NJ (2010) Electron spin polarization transfer from a nitroxide incarcerated within a nanocapsule to a nitroxide in the bulk aqueous solution. J Phys Chem Lett 1:2628–2632
45. Hill PA, Wei Q, Eckenhoff RG, Dmochowski IJ (2007) Thermodynamics of xenon binding to cryptophane in water and human plasma. J Am Chem Soc 129:9262–9263
46. Fogarty HA, Berthault P, Brotin T, Huber G, Desvaux H, Dutasta J-P (2007) A cryptophane core optimized for xenon encapsulation. J Am Chem Soc 129:10332–10333
47. Kotera N, Delacour L, Traoré T, Tassali N, Berthault P, Buisson D-A, Dognon J-P, Rousseau B (2011) Design and synthesis of new cryptophanes with intermediate cavity sizes. Org Lett 13:2153–2155
48. Fabbrizzi L, Faravelli I (1998) A fluorescent cage for anion sensing in aqueous solution. Chem Commun 971–972
49. Amendola V, Bergamaschi G, Boiocchi M, Fabbrizzi L, Poggi A, Zema M (2008) Halide ion inclusion into a dicopper(II) bistren cryptate containing 'active' 2,5-dimethylfuran spacers: the origin of the bright yellow colour. Inorg Chim Acta 361:4038–4046
50. Barrett ES, Dale TJ, Rebek J Jr (2007) Assembly and exchange of resorcinarene capsules monitored by fluorescence resonance energy transfer. J Am Chem Soc 129:3818–3819
51. Takeda N, Umemoto K, Yamaguchi K, Fujita M (1999) A nanometre-sized hexahedral coordination capsule assembled from 24 components. Nature 398:794–796
52. Liu Y, Wu X, He C, Zhang R, Duan C (2008) A truncated octahedral nanocage for fluorescent detection of nucleoside. Dalton Trans 5866–5868
53. Liu M, Yan X, Hu M, Chen X, Zhang M, Zheng B, Hu X, Shao S, Huang F (2010) Photoresponsive host–guest systems based on a new azobenzene-containing crytpand. Org Lett 12:2558–2561
54. Brutschy M, Schneider MW, Mastalerz M, Waldvogel SR (2013) Direct gravimetric sensing of GBL by a molecular recognition process in organic cage compounds. Chem Commun 49(75):8398–8400
55. Inokuma Y, Yoshioka S, Ariyoshi J, Arai T, Hitora Y, Takada K, Matsunaga S, Rissanen K, Fujita M (2013) X-ray analysis on the nanogram to microgram scale using porous complexes. Nature 495:461–466
56. Inokuma Y, Yoshioka S, Ariyoshi J, Arai T, Fujita M (2014) Preparation and guest-uptake protocol for a porous complex useful for 'crystal-free' crystallography. Nat Protoc 9:246–252
57. Inokuma Y, Arai T, Fujita M (2010) Networked molecular cages as crystalline sponges for fullerenes and other guests. Nat Chem 2:780–783
58. Biradha K, Fujita M (2002) A springlike 3D-coordination network that shrinks or swells in a crystal-to-crystal manner upon guest removal or readsorption. Angew Chem Int Ed 41:3392–3395
59. Trivedi ER, Eliseeva SV, Jankolovits J, Olmstead MM, Petoud S, Pecoraro VL (2014) Highly emitting near-infrared lanthanide "encapsulated sandwich" metallacrown complexes with excitation shifted toward lower energy. J Am Chem Soc 136(4):1526–1534
60. Yoshizawa M, Tamura M, Fujita M (2004) AND/OR bimolecular recognition. J Am Chem Soc 126:6846–6847
61. Duriska MB, Neville SM, Li J, Iremonger SS, Boas JF, Kepert CJ, Batten SR (2009) Systematic metal variation and solvent and hydrogen-gas storage in supramolecular nanoballs. Angew Chem Int Ed 48:8919–8922
62. Duriska MB, Neville SM, Moubaraki B, Cashion JD, Halder GJ, Chapman KW, Balde C, Létard J-F, Murray KS, Kepert CJ, Batten SR (2009) A nanoscale molecular switch triggered by thermal, light, and guest perturbation. Angew Chem Int Ed 48: 2549–2552
63. Nishimura N, Yoza K, Kobayashi K (2010) Guest-encapsulation properties of a self-assembled capsule by dynamic boronic ester bonds. J Am Chem Soc 132:777–790
64. Martinez A, Guy L, Dutasta J-P (2010) Reversible, solvent-induced chirality switch in atrane structure: control of the unidirectional motion of the molecular propeller. J Am Chem Soc 132:16733–16736
65. Yen M-L, Chen N-C, Lai C-C, Liu Y-H, Peng S-M, Chiu S-H (2011) Controlling the rotation of a complexed guest within a molecular cage. Dalton Trans 40:2163–2166

Appendix

List of capsules

1.

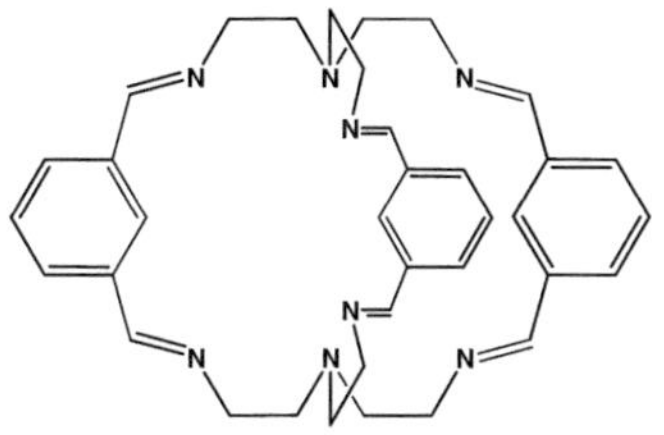

2.

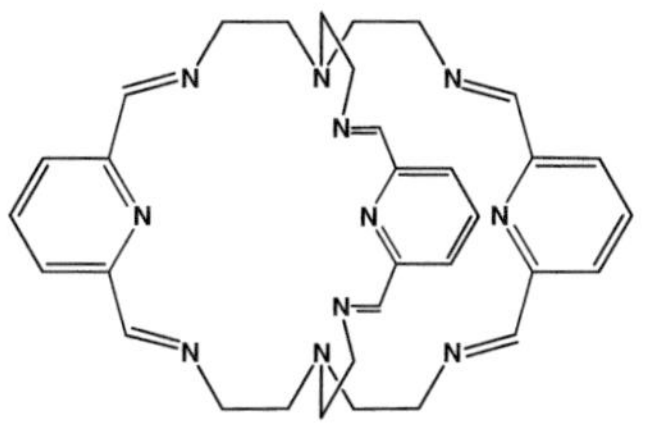

3.

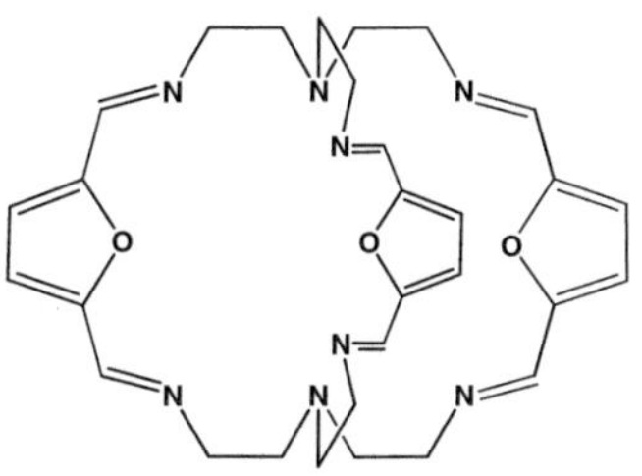

4.

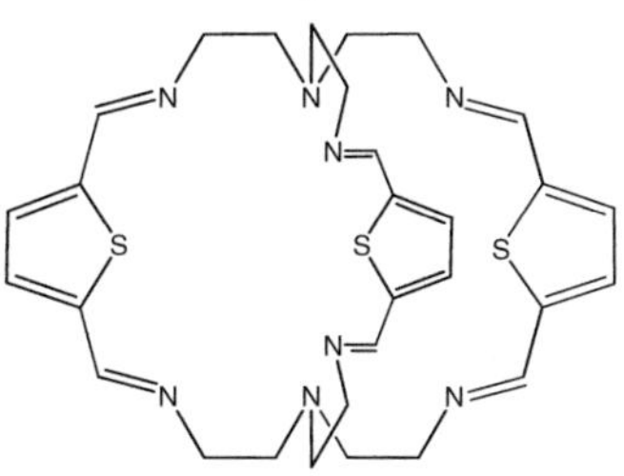

5.

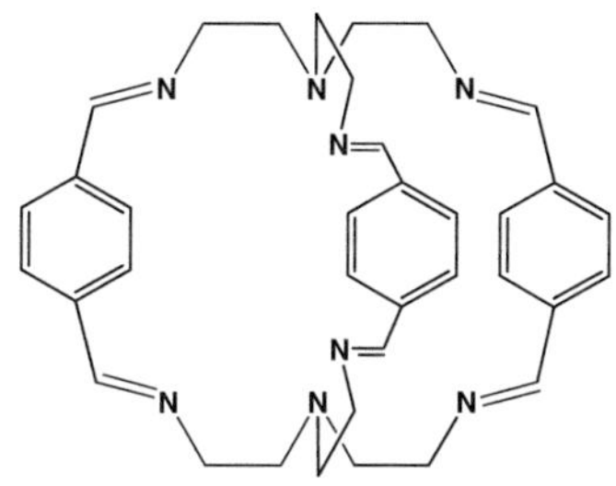

6.

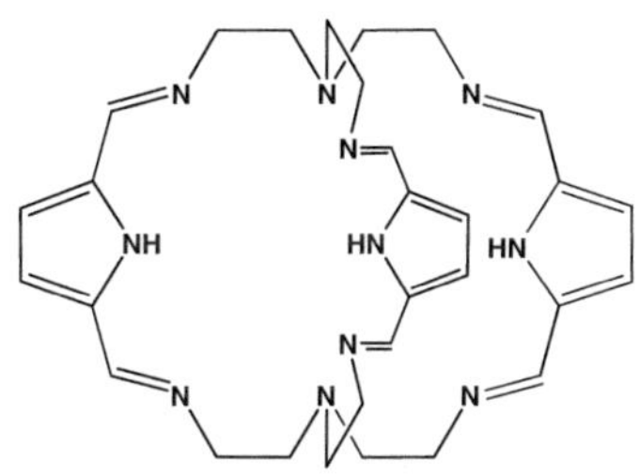

7.

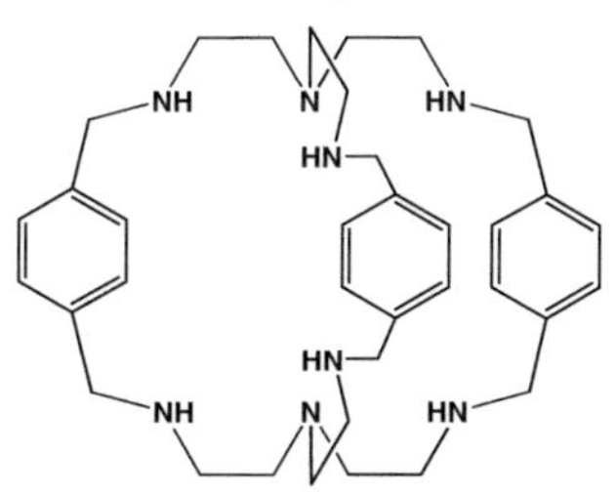

8.

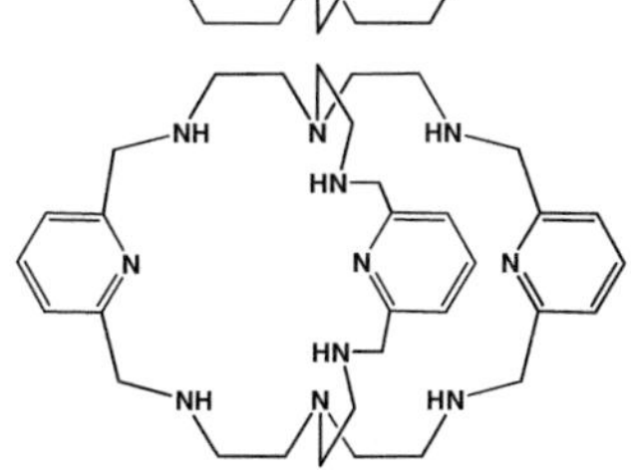

Y. Voloshin et al., *The Encapsulation Phenomenon: Synthesis, Reactivity and Applications of Caged Ions and Molecules*, DOI 10.1007/978-3-319-27738-7

9.

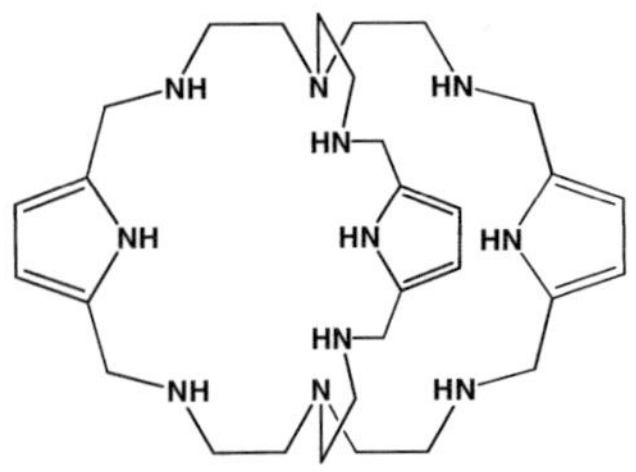

10.

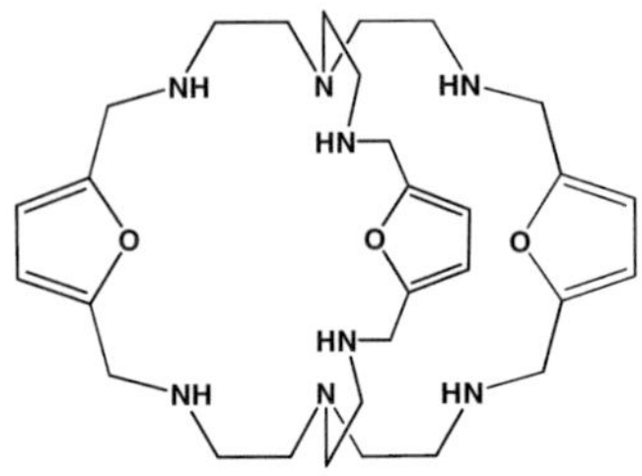

11.

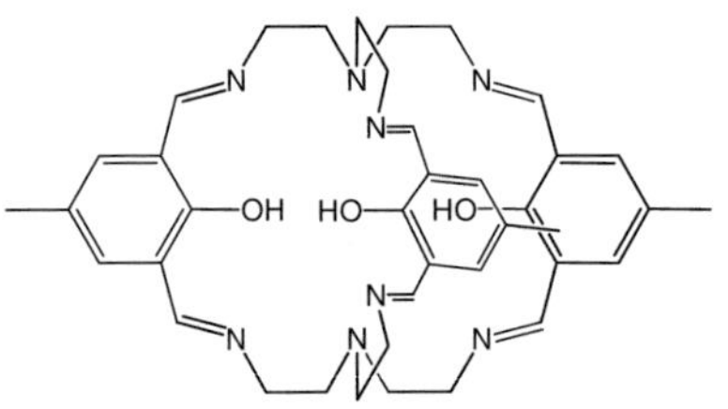

12.

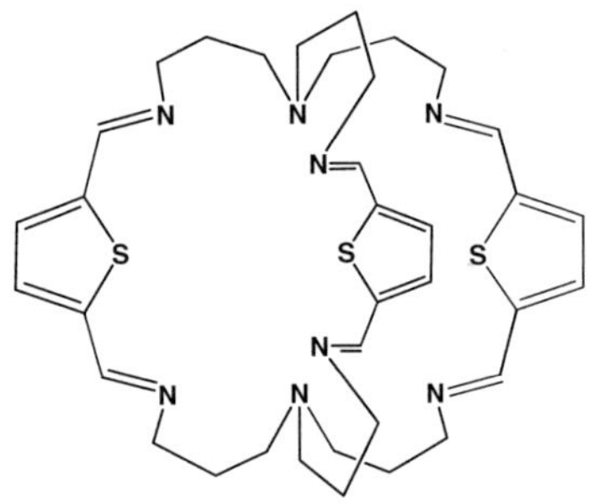

13.

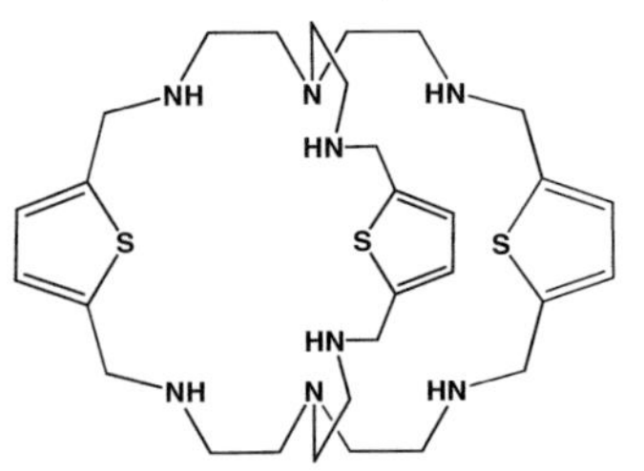

14.

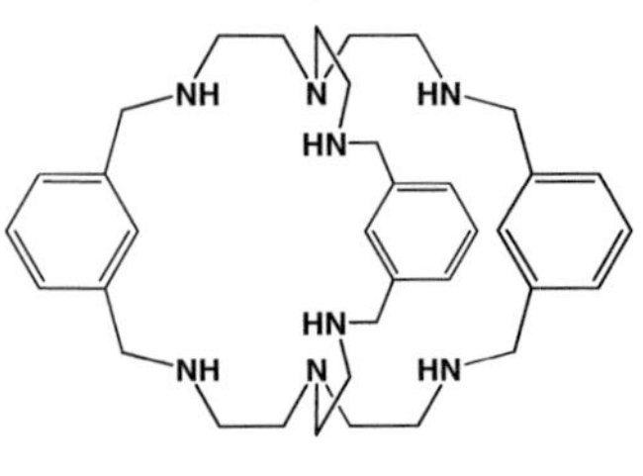

15.

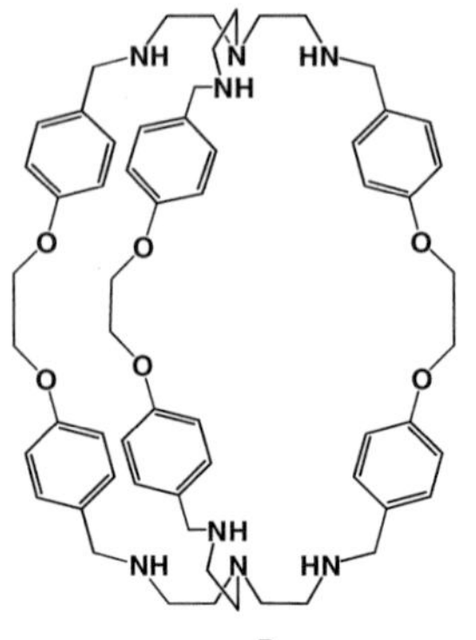

16.

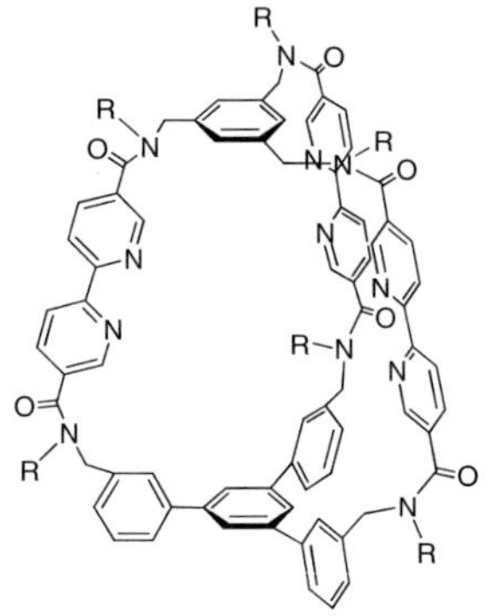

$R = CH_2C_6H_5$

17.

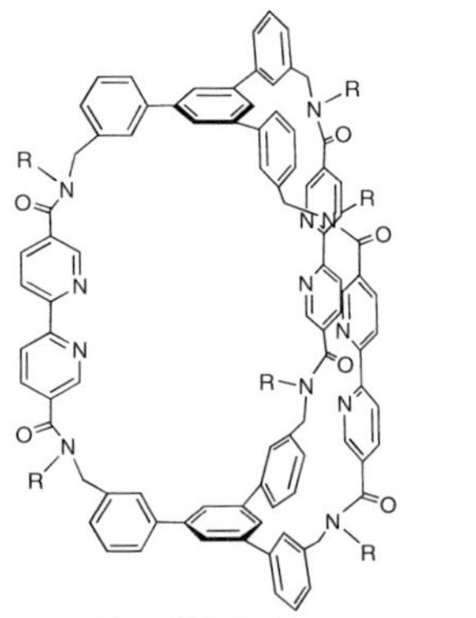

$R = CH_2C_6H_5$

18.

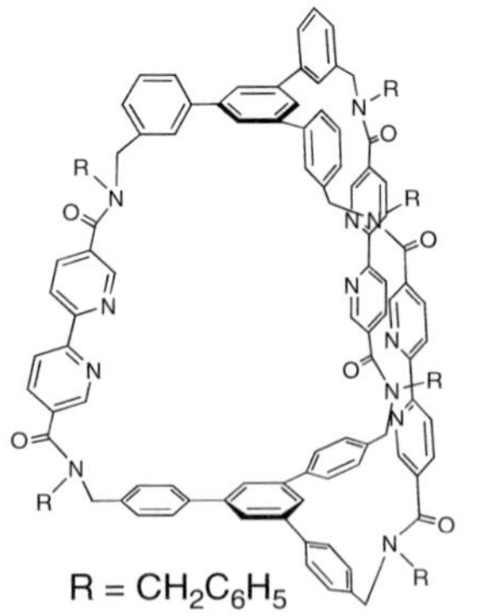

$R = CH_2C_6H_5$

19.

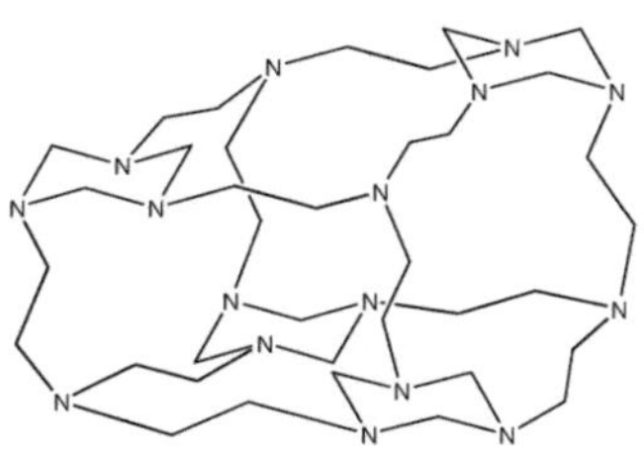

20.

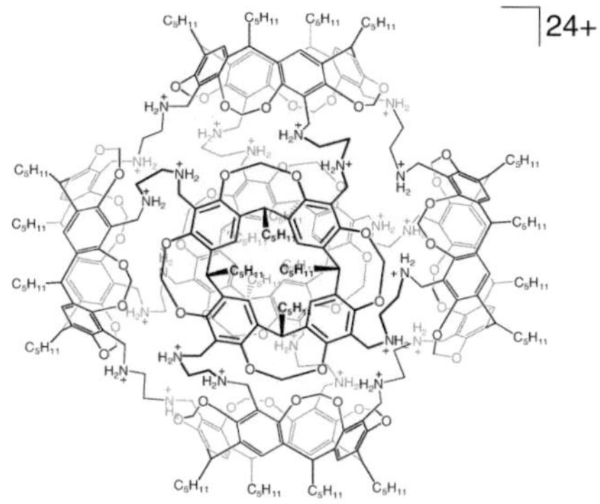

21.

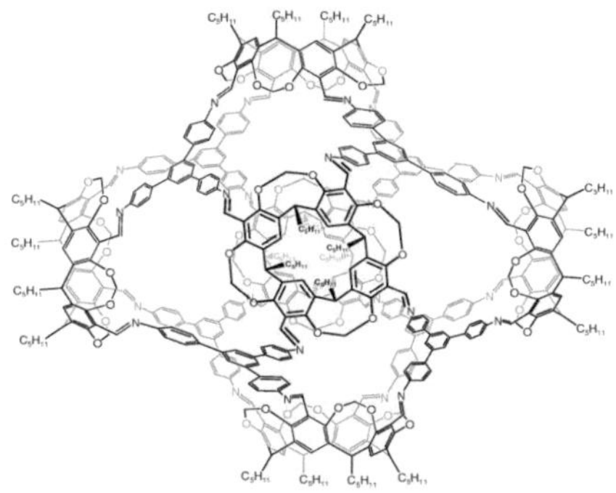

22.

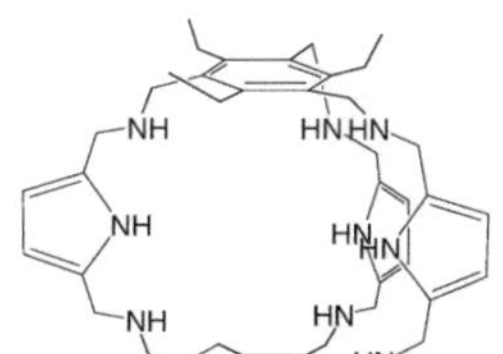

23.

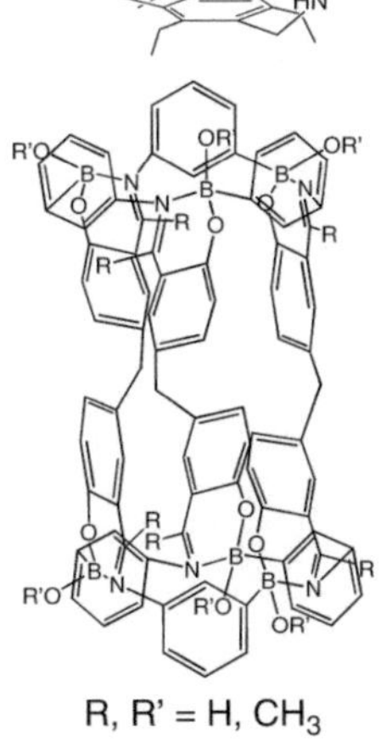

R, R' = H, CH_3

24.

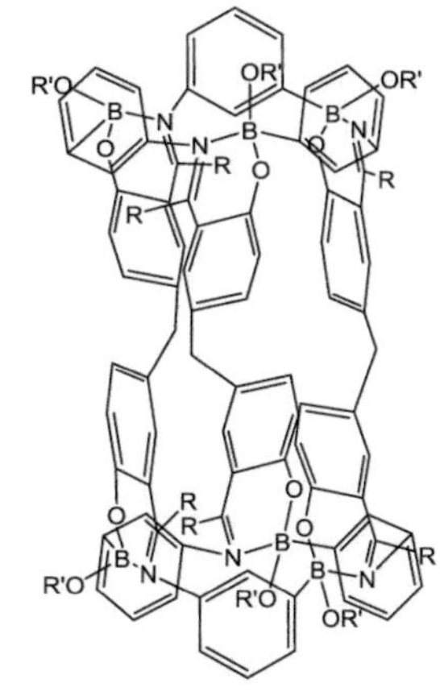

R, R' = H, C_2H_5

25.

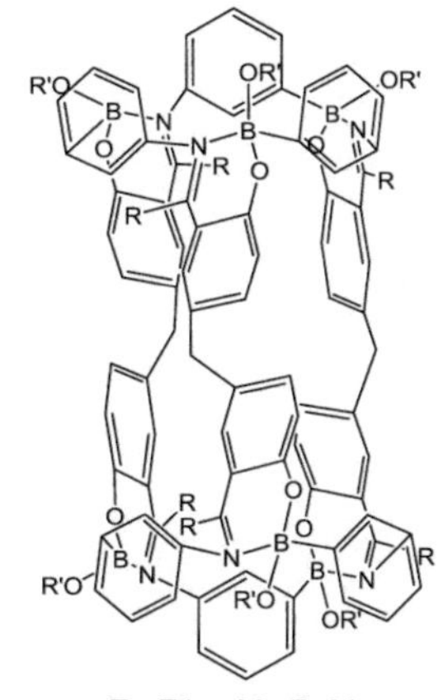

R, R' = H, C_3H_7

26.

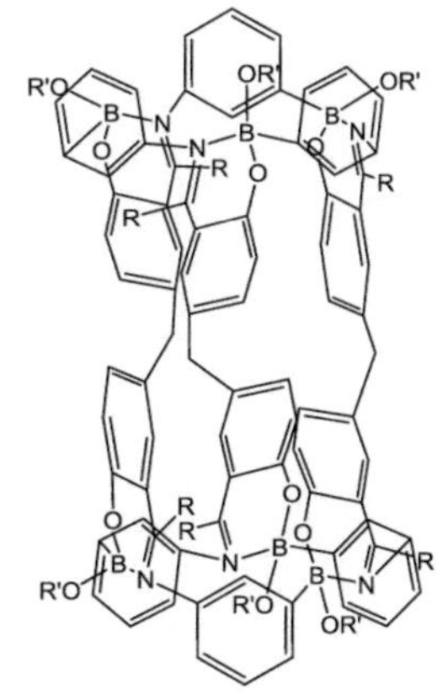

R, R' = CH_3, CH_3

27.

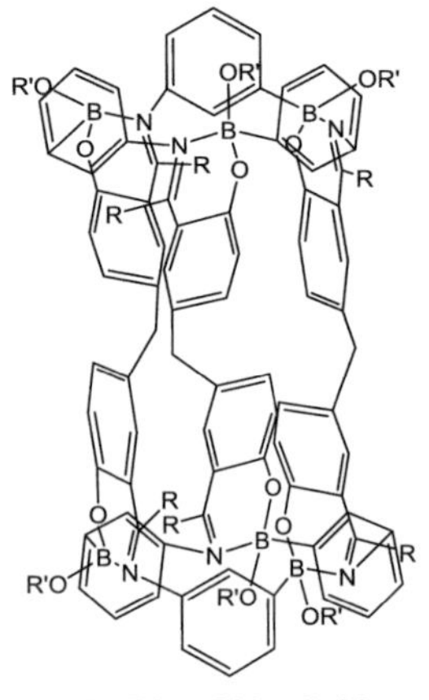

R, R' = CH_3, C_2H_5

28.

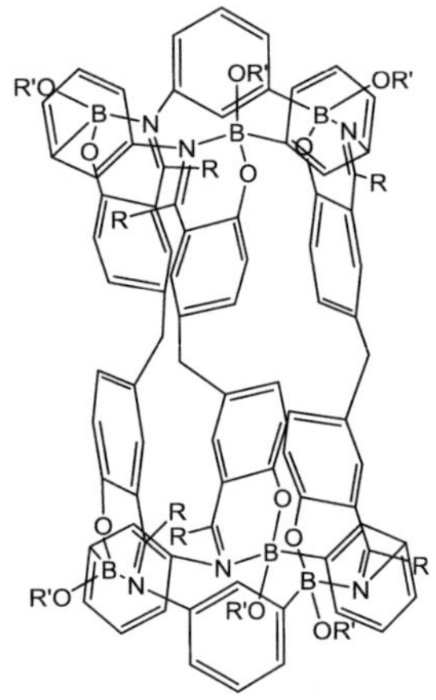

R, R' = CH_3, C_3H_7

29.

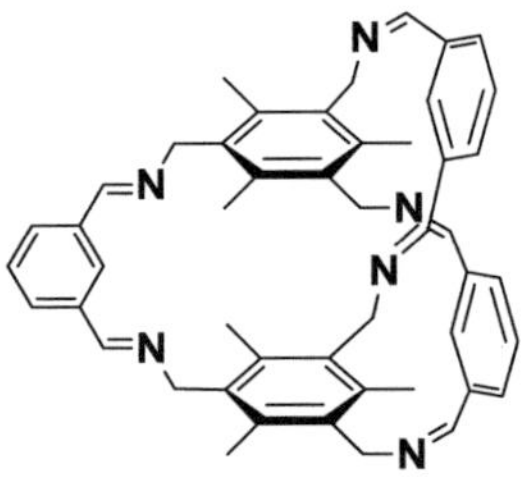

30.

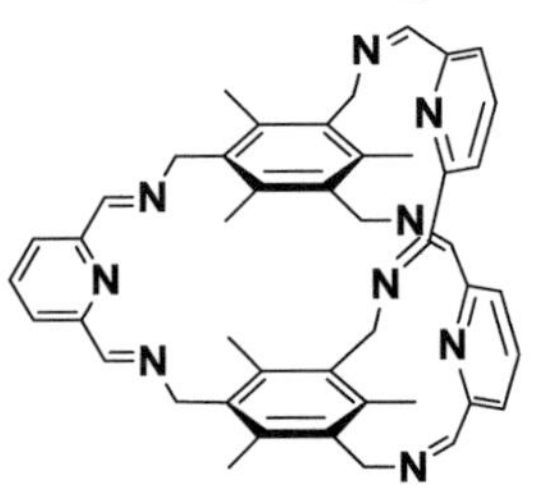

31.

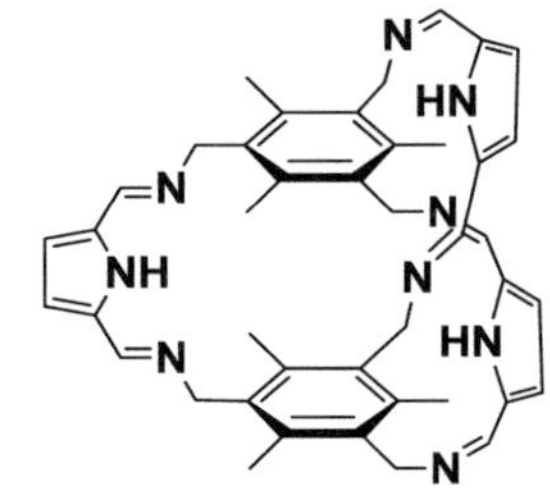

32.

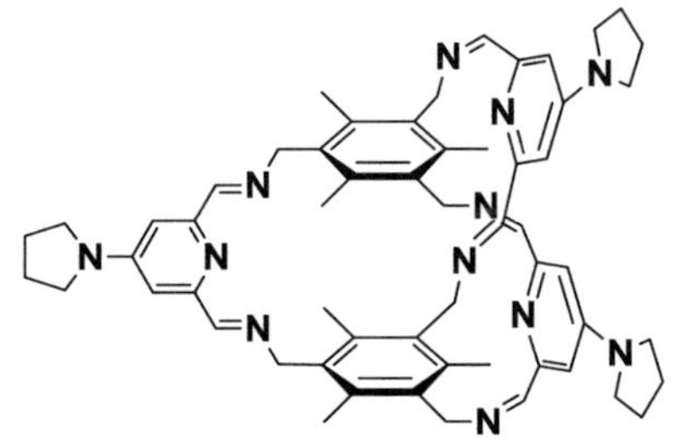

33.

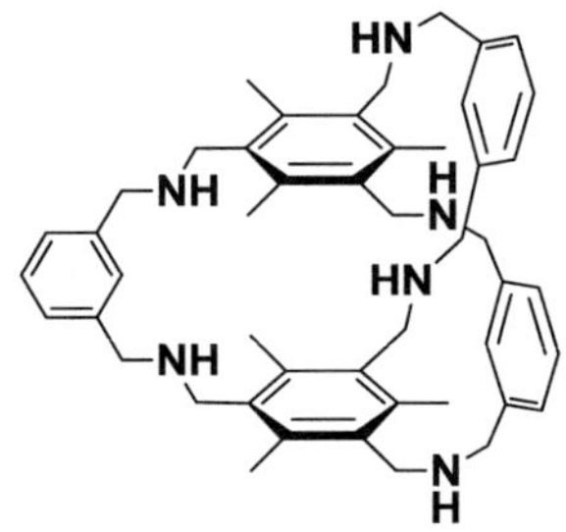

34.

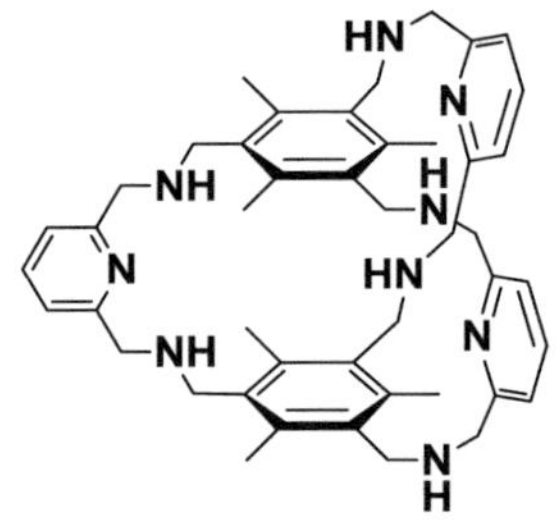

35.

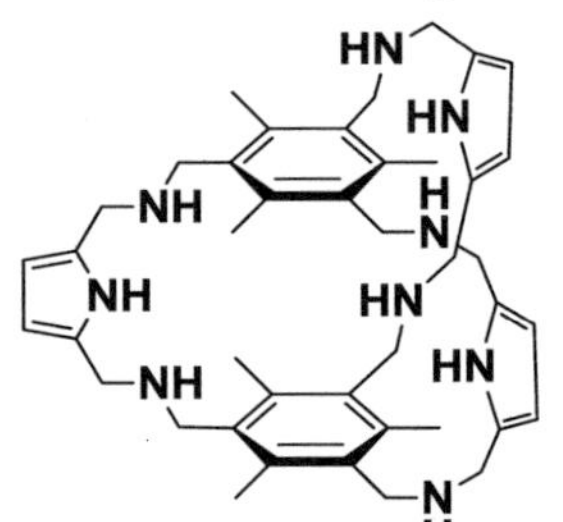

36.

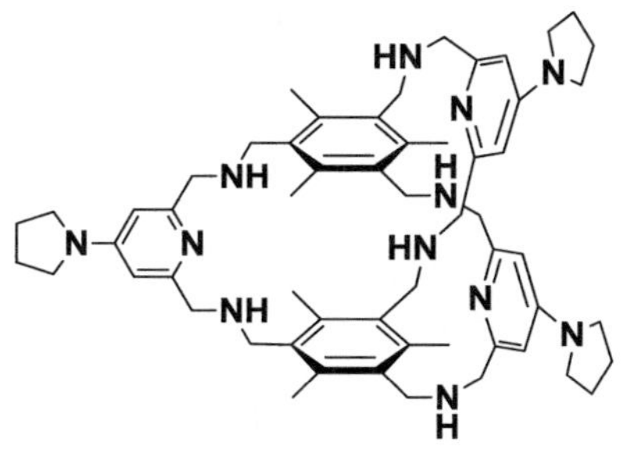

37.

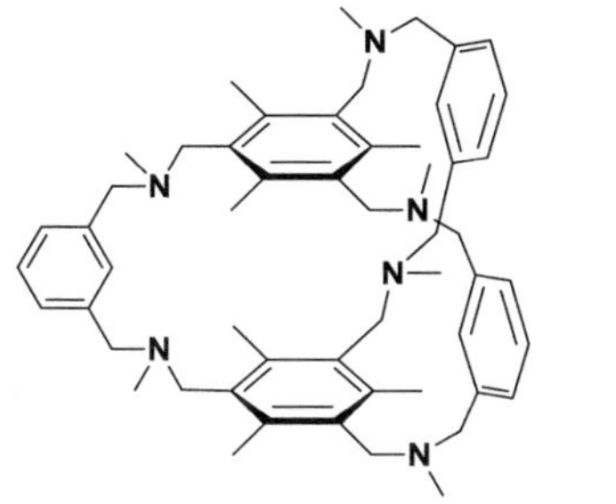

38.

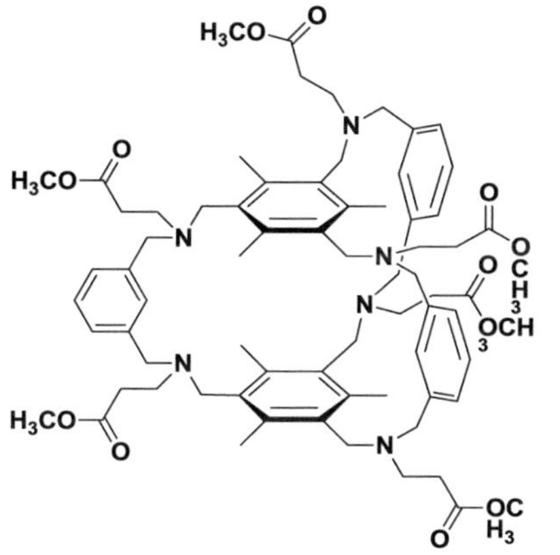

39.

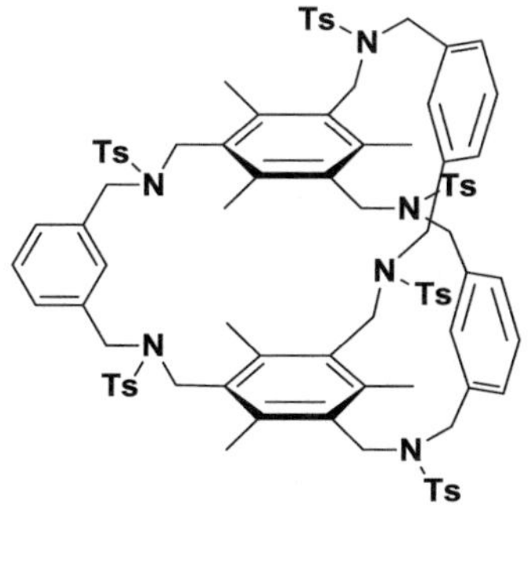

40.

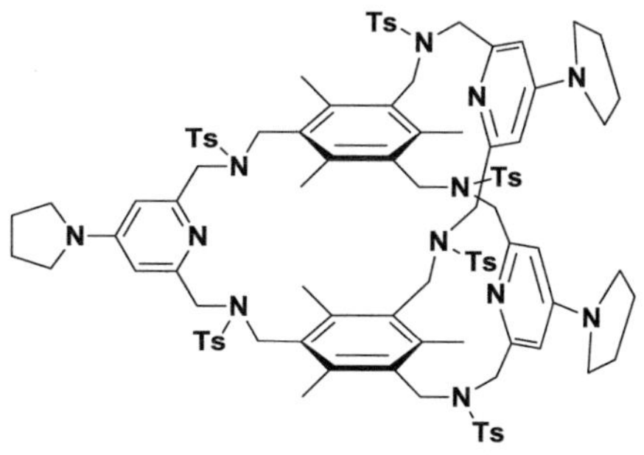

41.

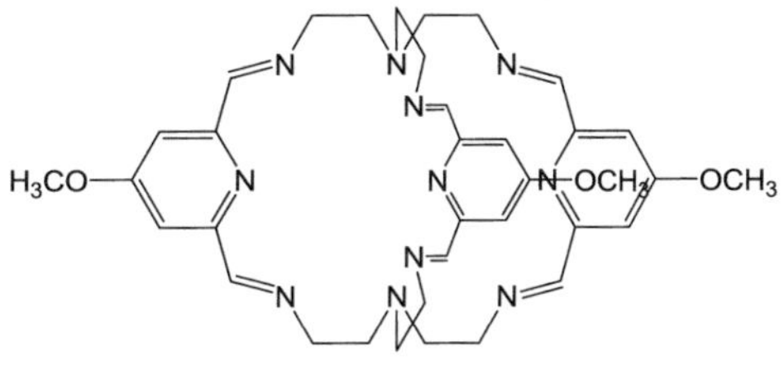

42.

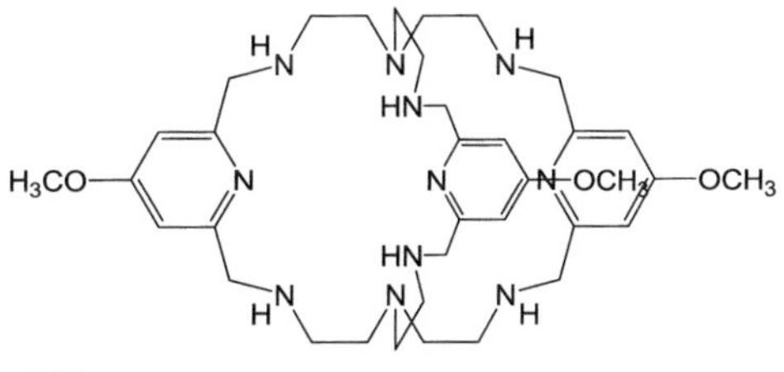

43.

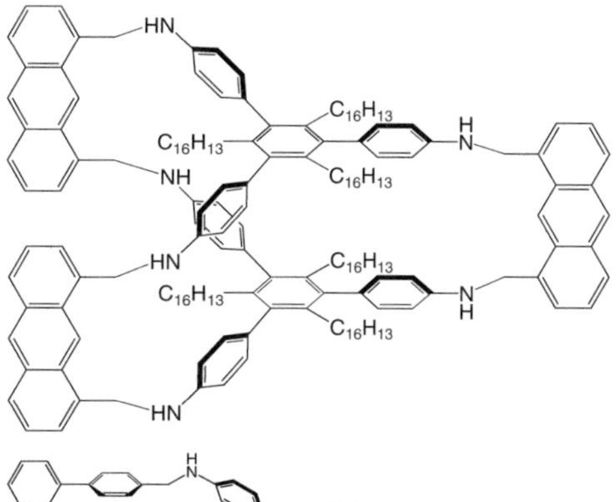

44.

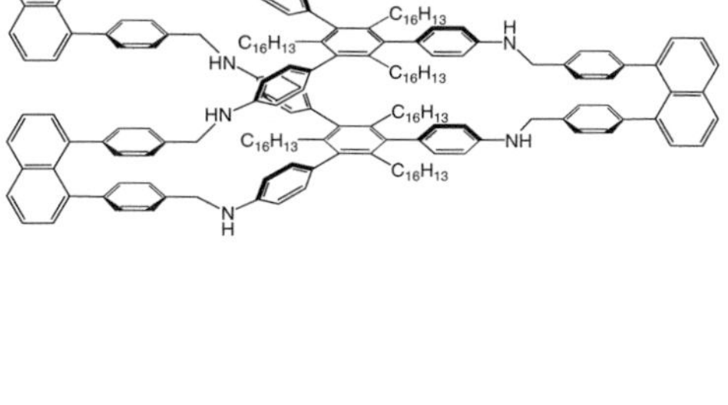

45.

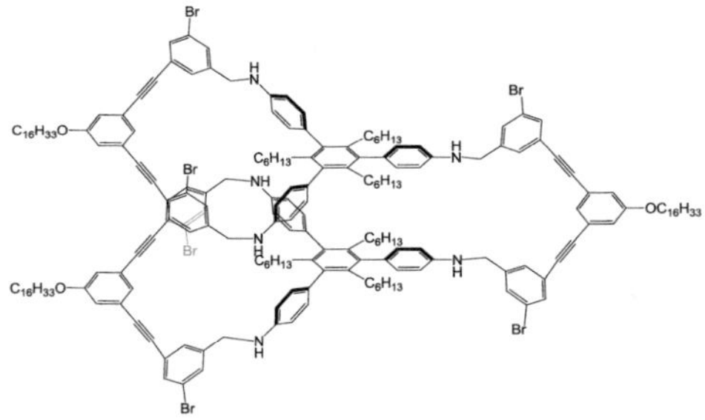

46.

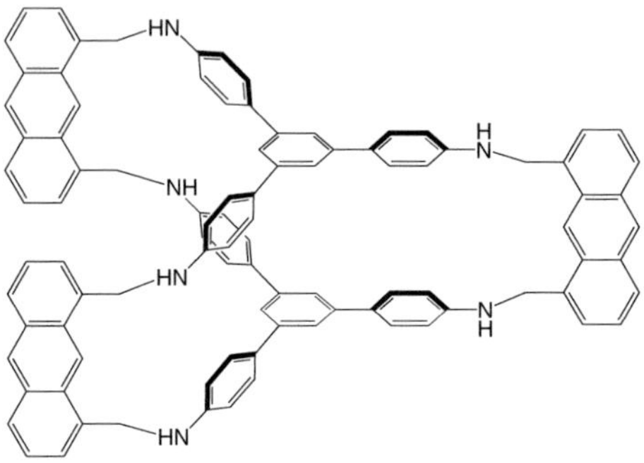

47.

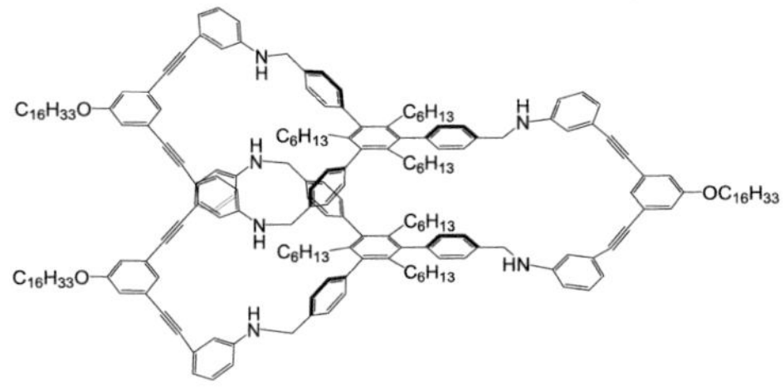

48.

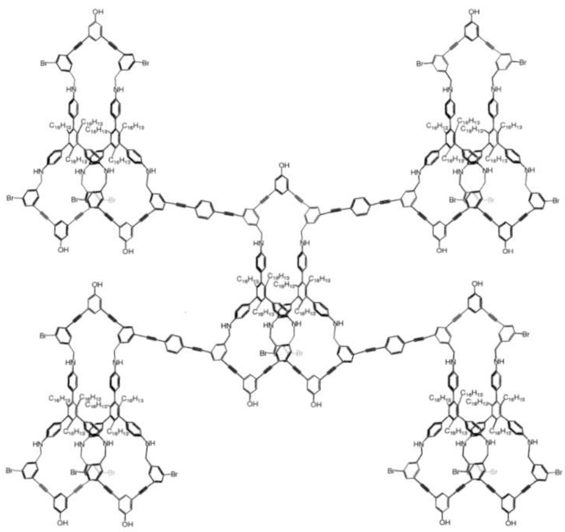

49.

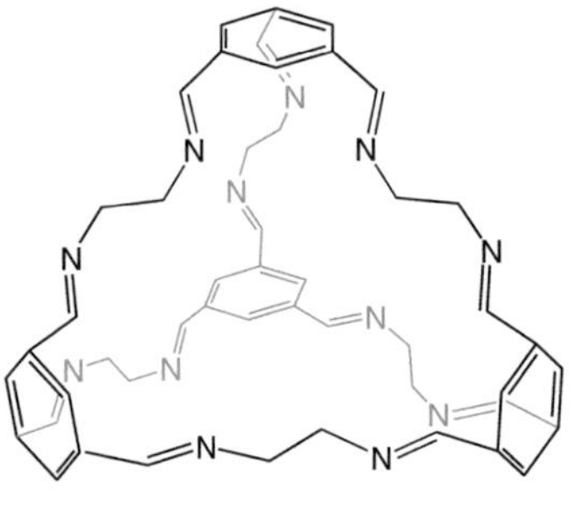

50.

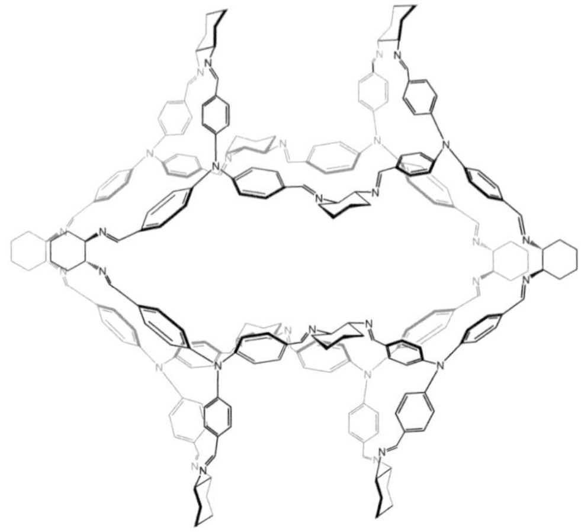

51.

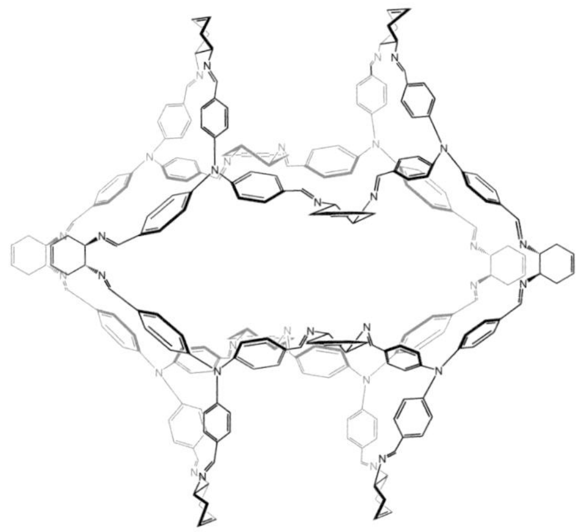

52.

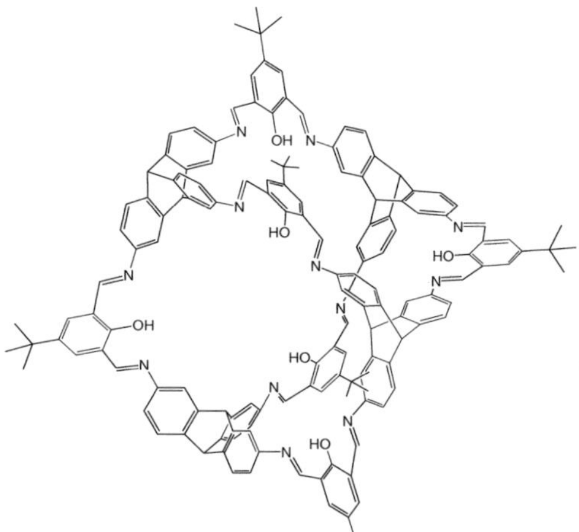

53.

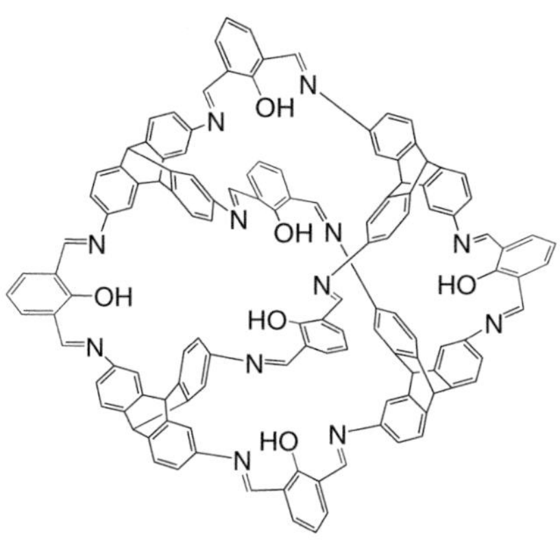

54.

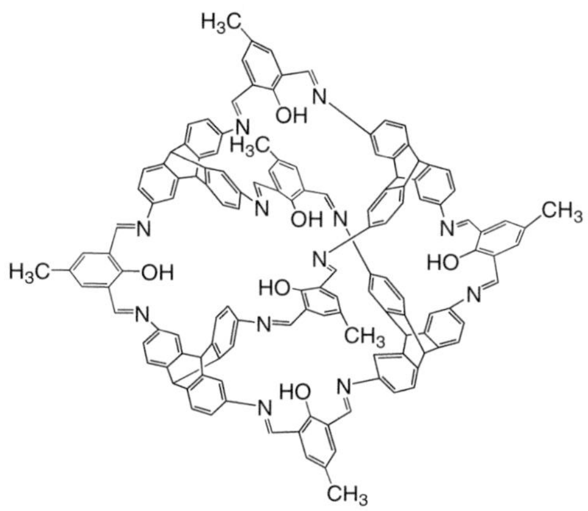

55.

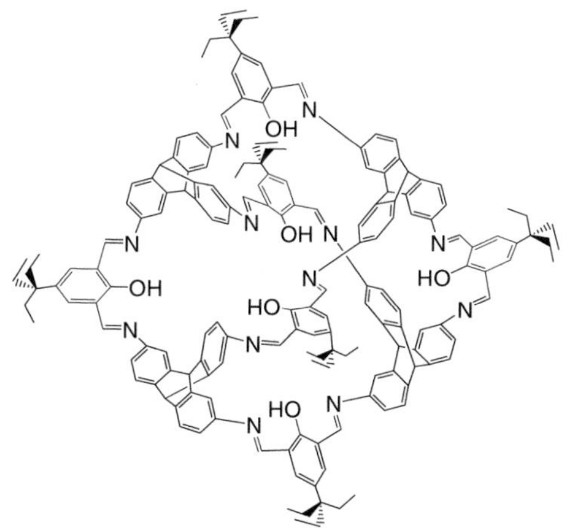

56.

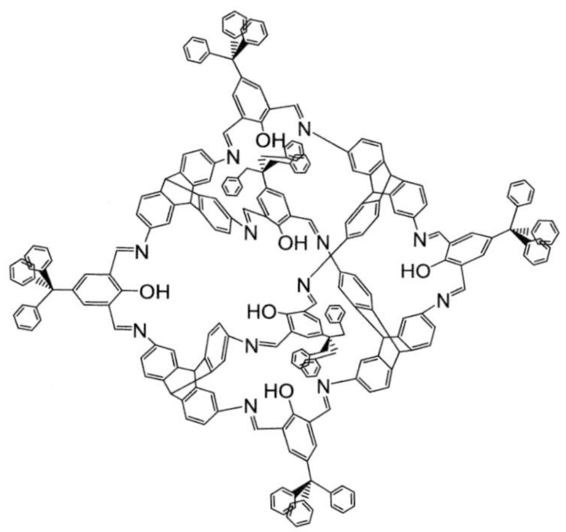

57.

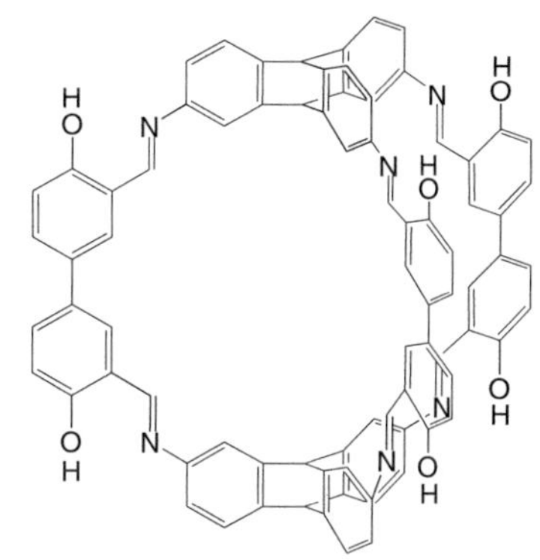

58.

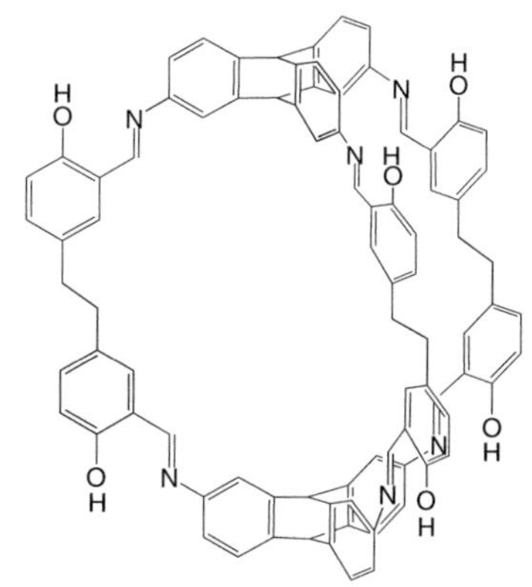

59.

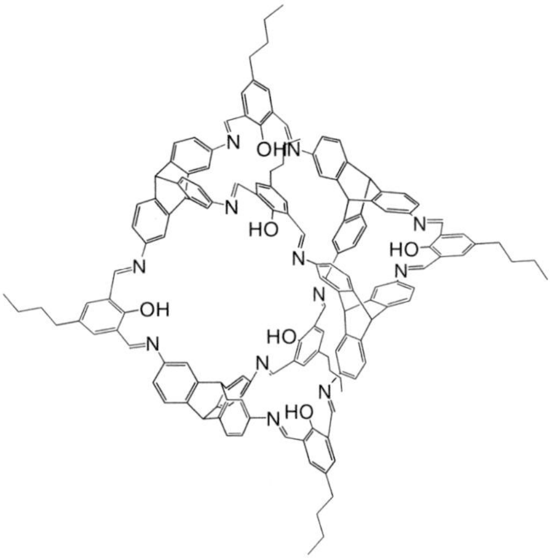

60.

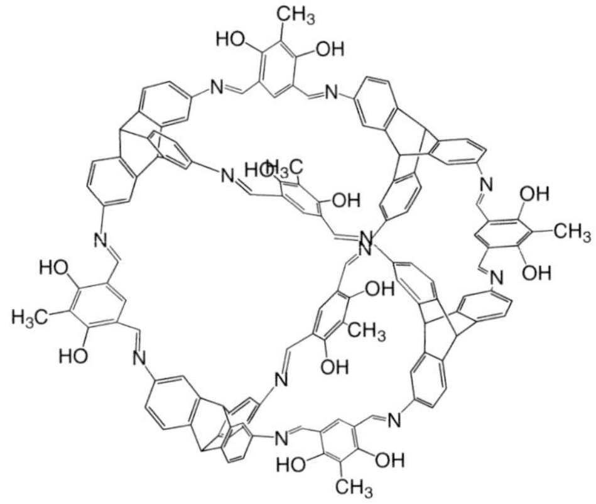

61.

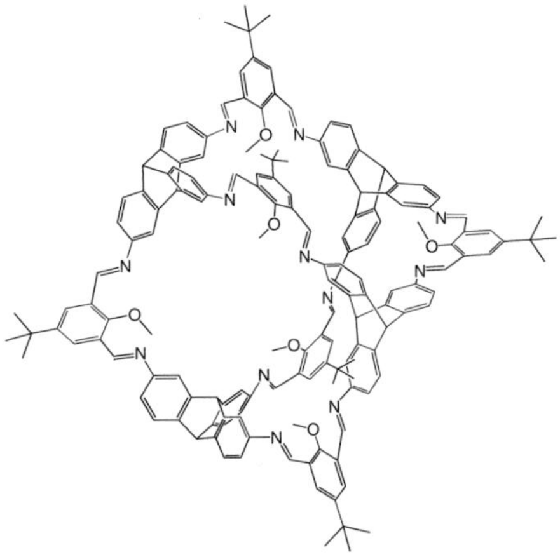

62.

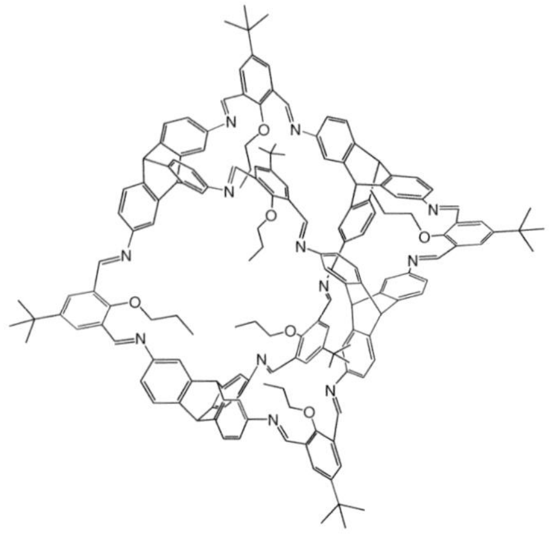

63.

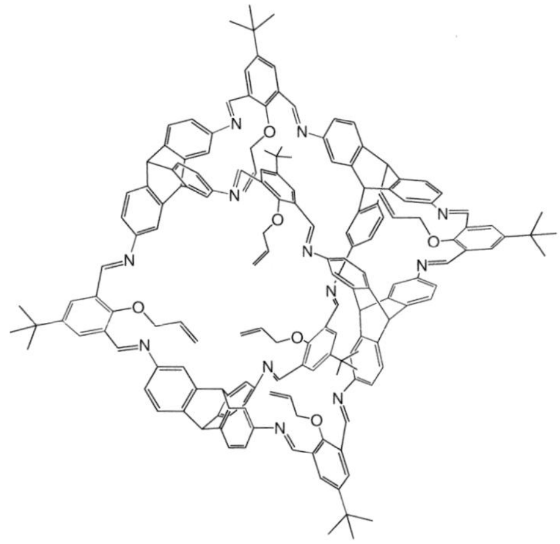

64.

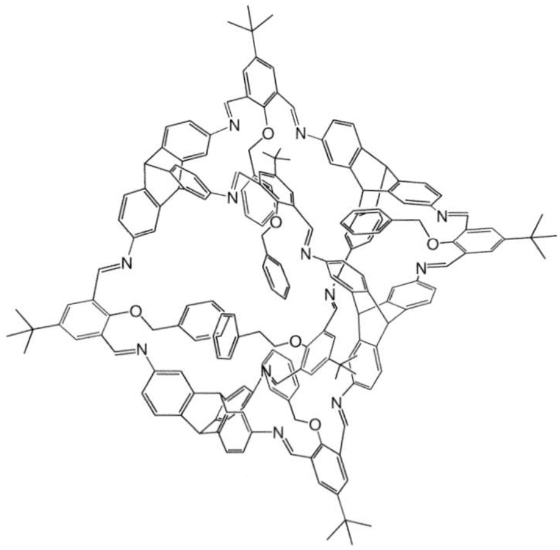

65.

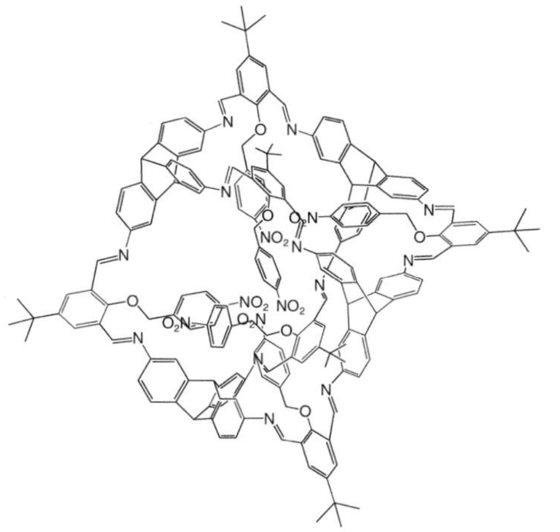

69.

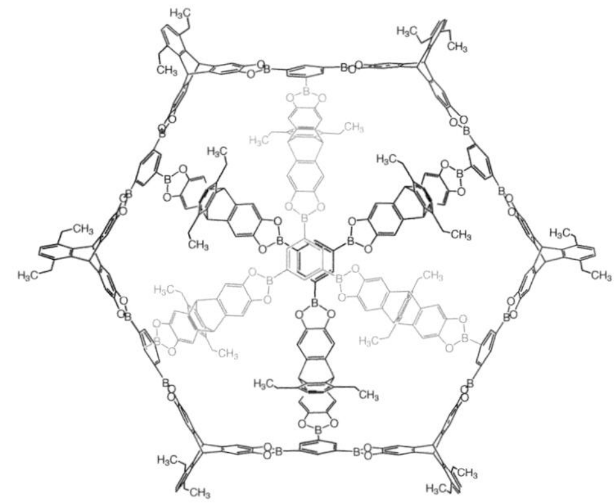

70.

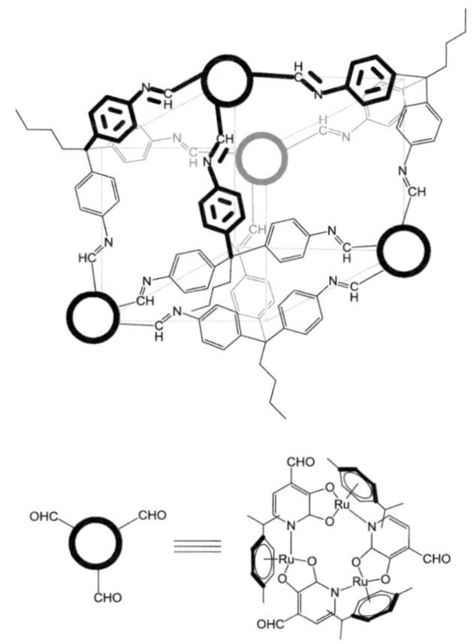

66.

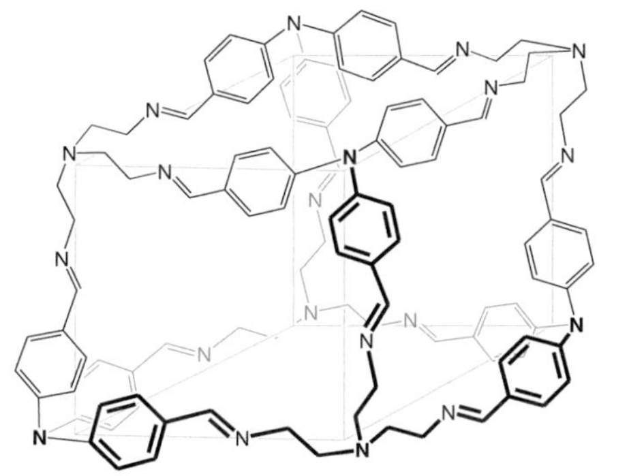

67.

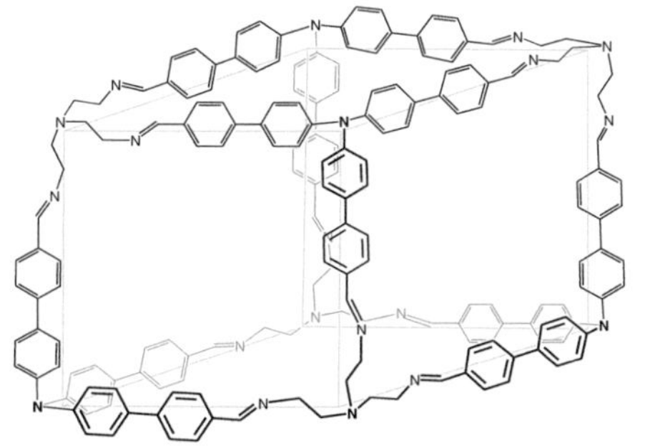

71.

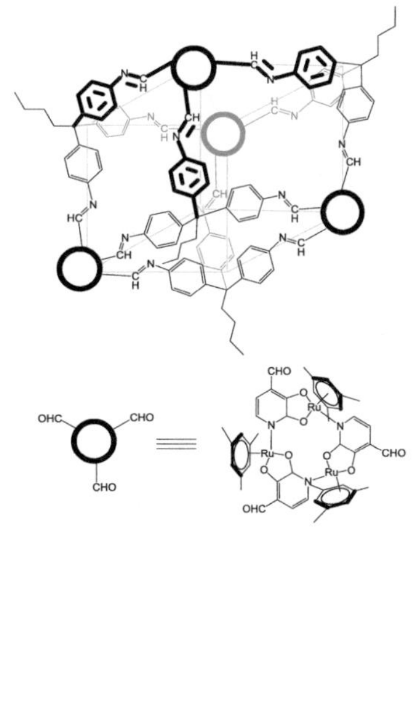

68.

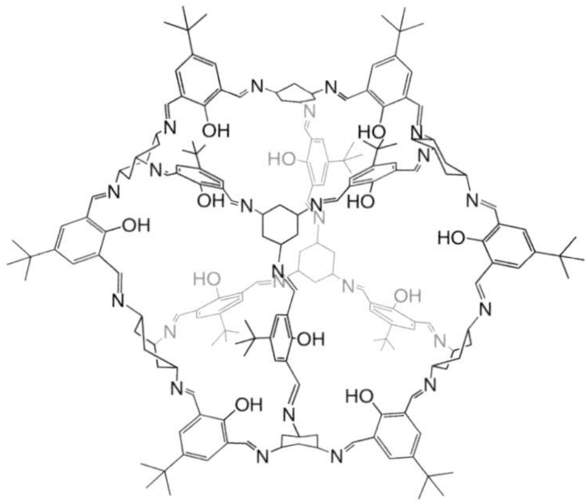

72.

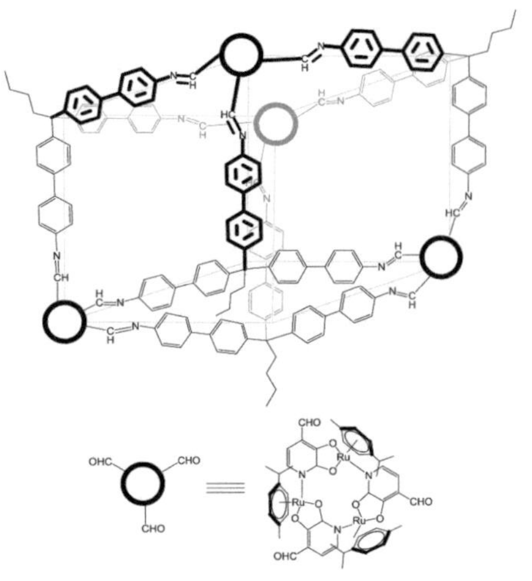

73.

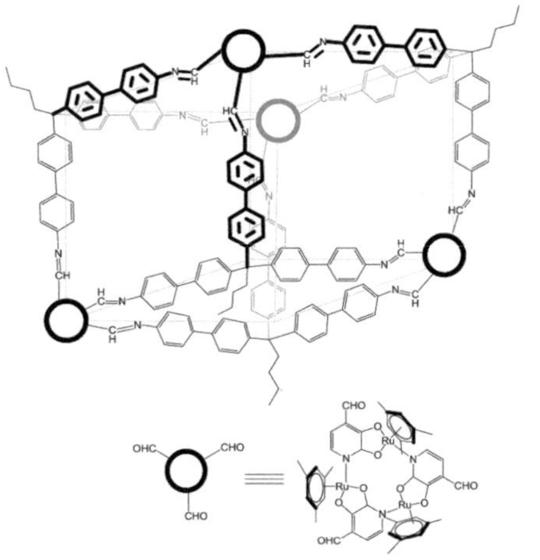

74.

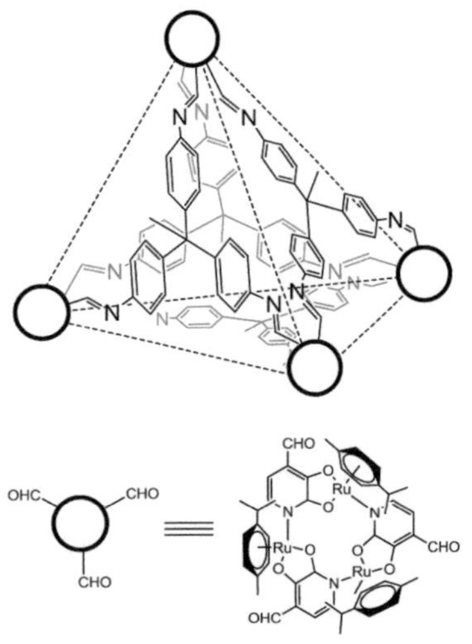

75.

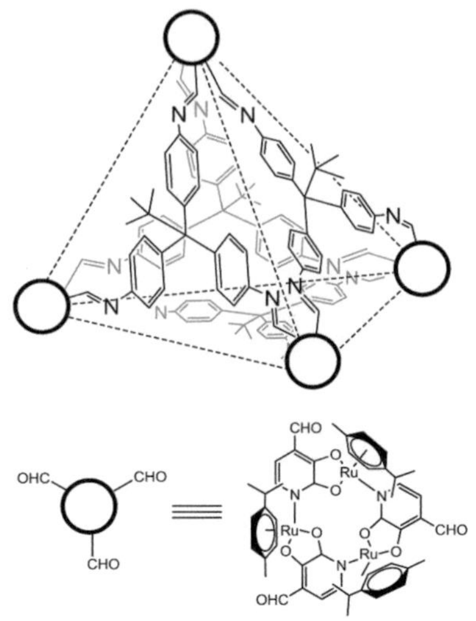

76.

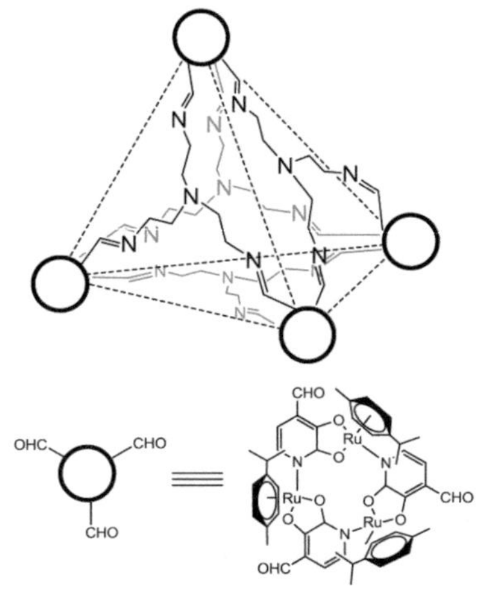

77.

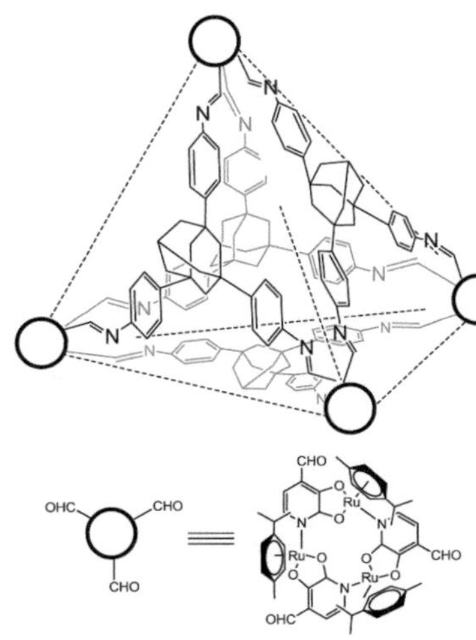

78.

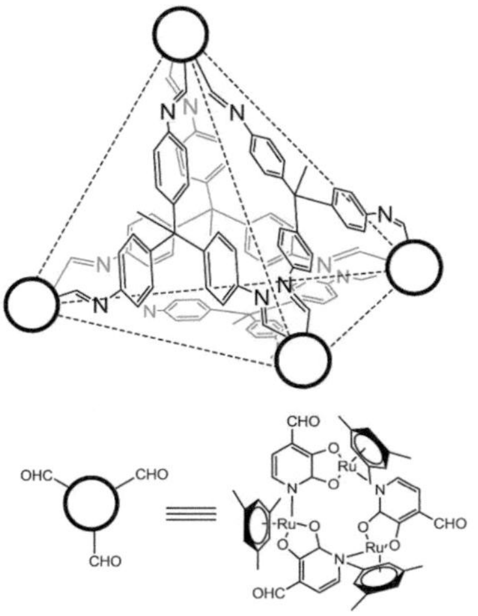

79.

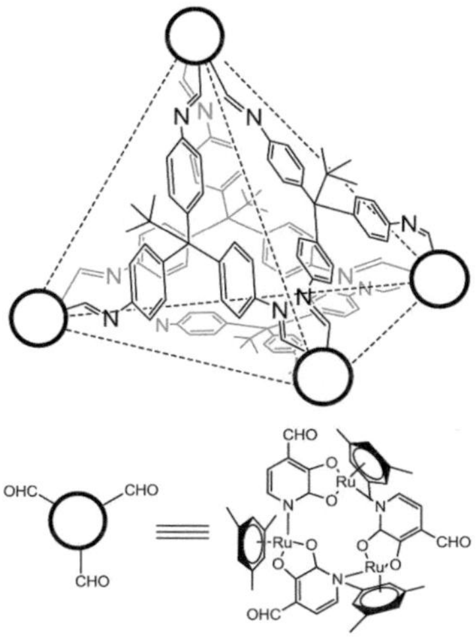

80.

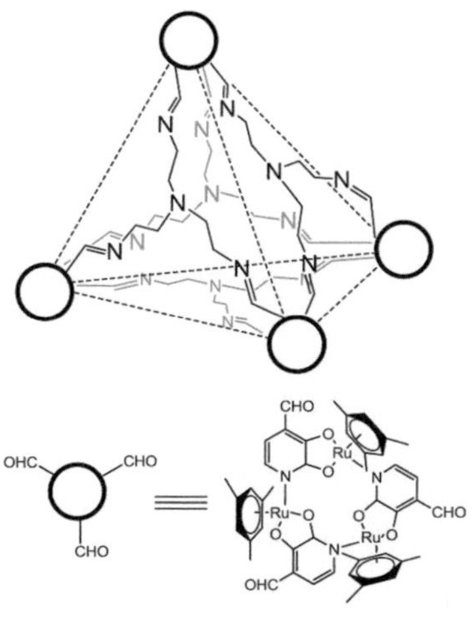

81.

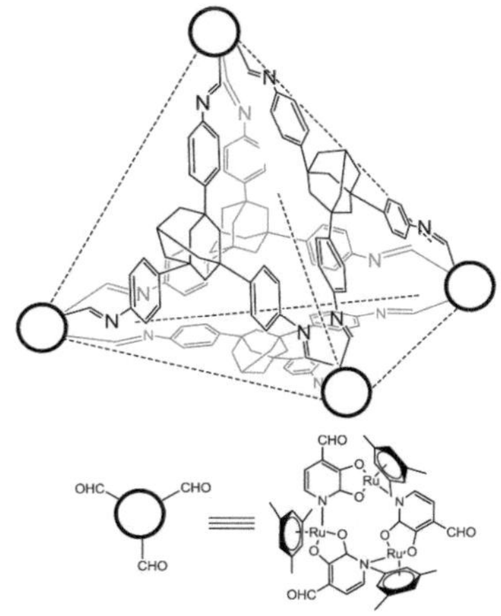

82.

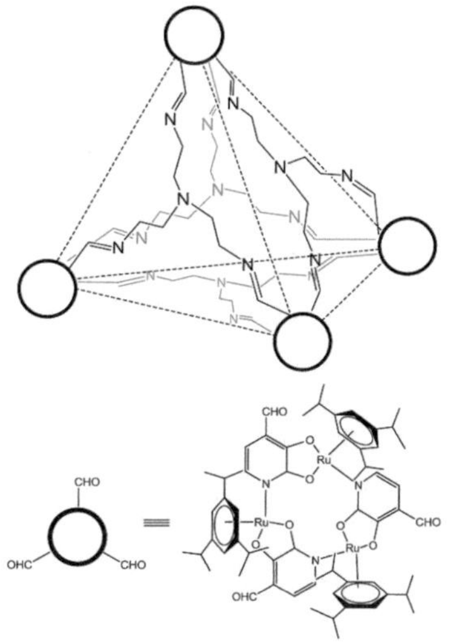

83.

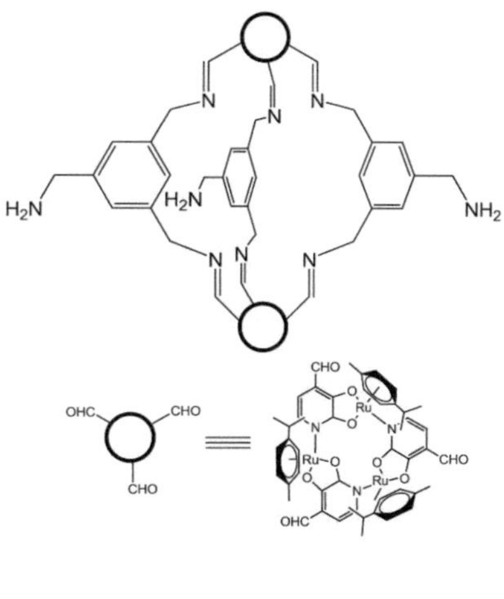

84.

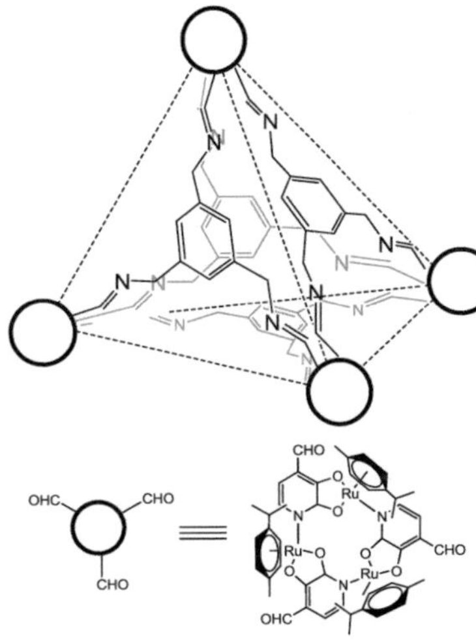

85.

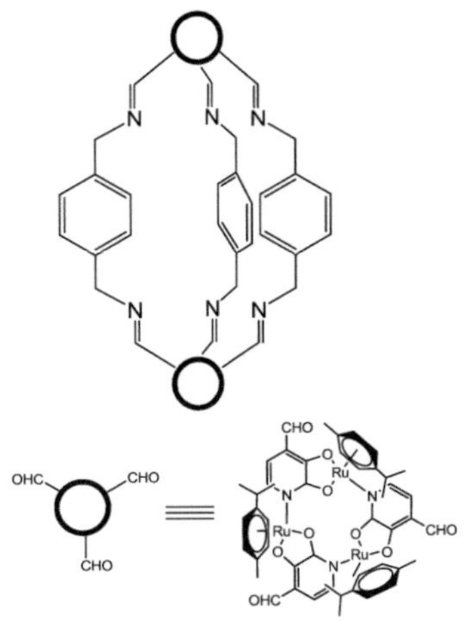

86.

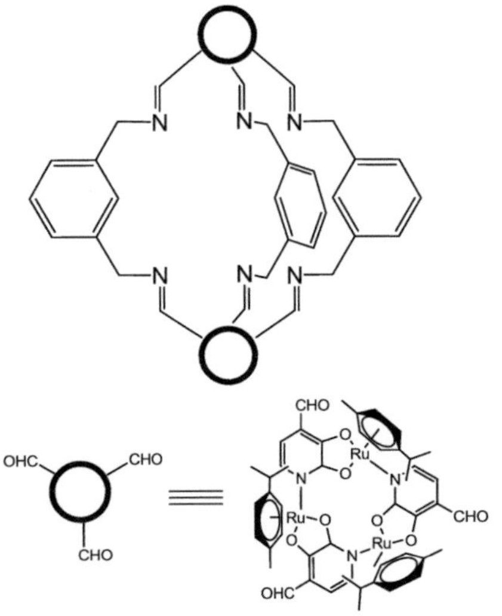

87.

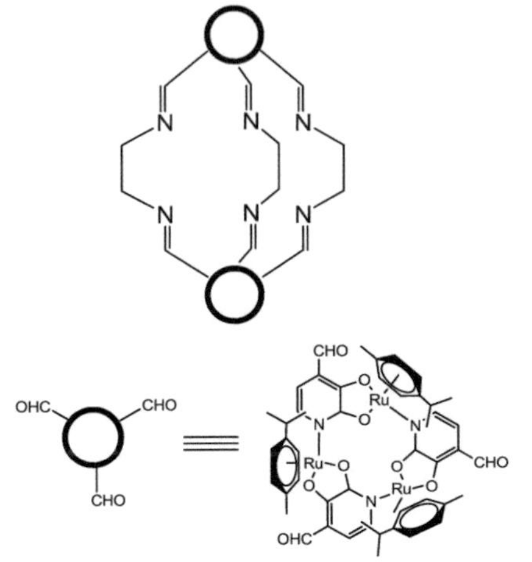

88.

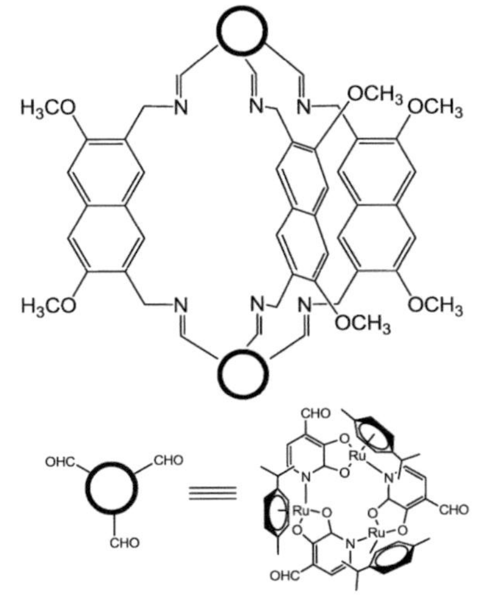

89.

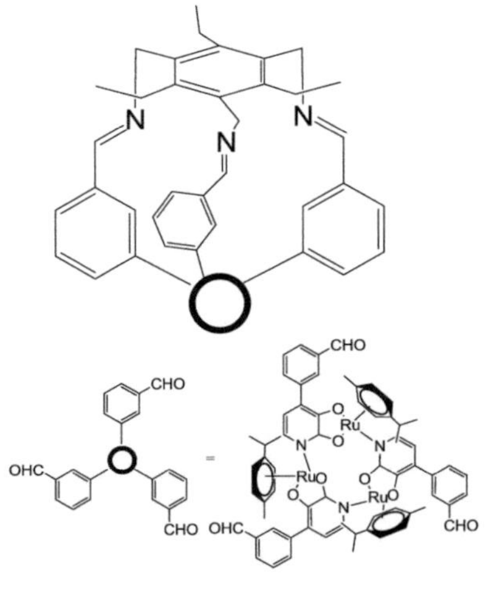

90.

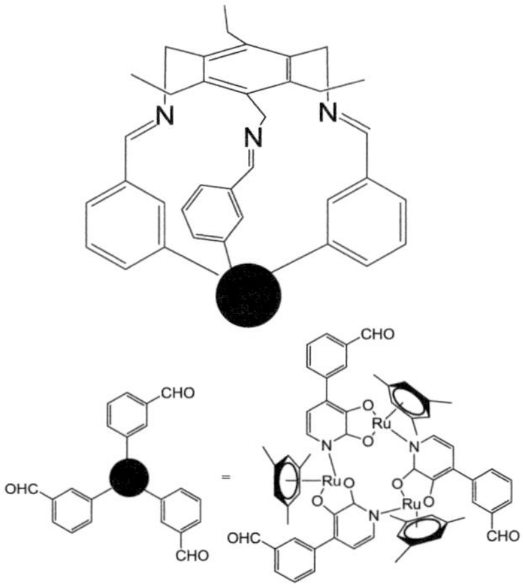

91.

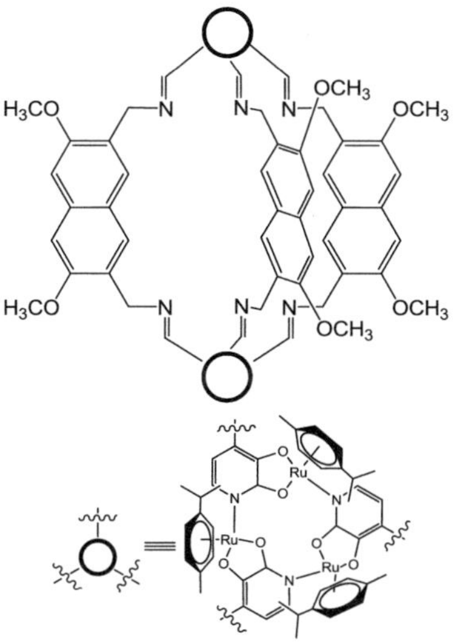

92.

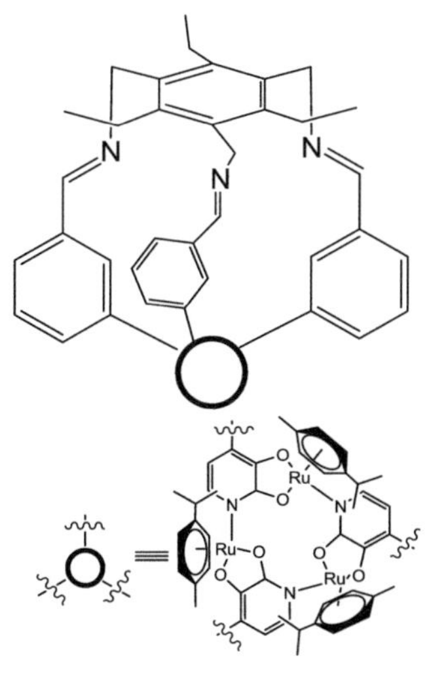

93.

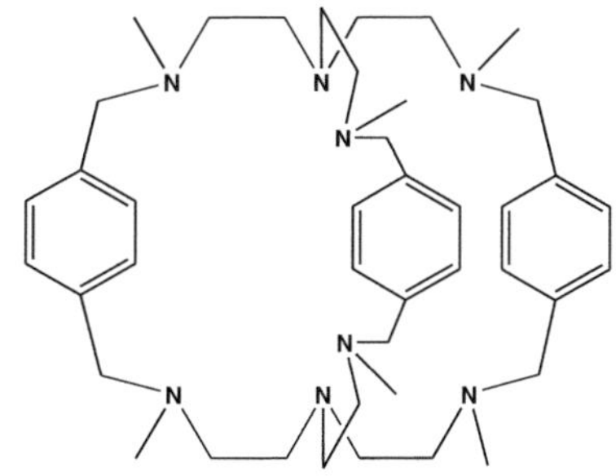

94.

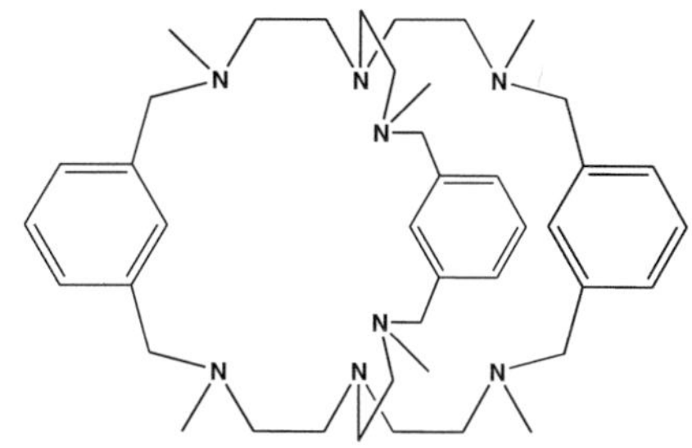

95.

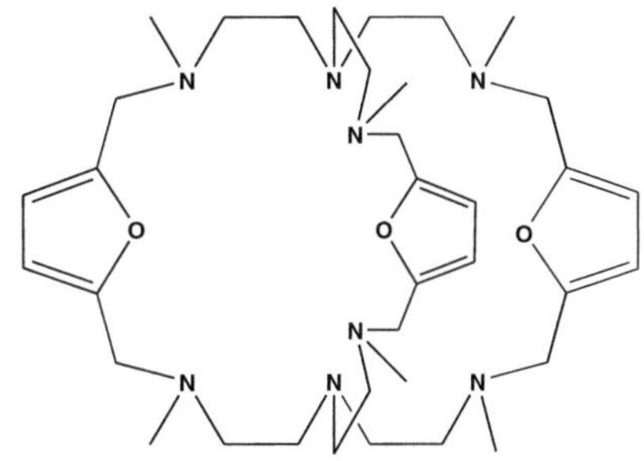

96.

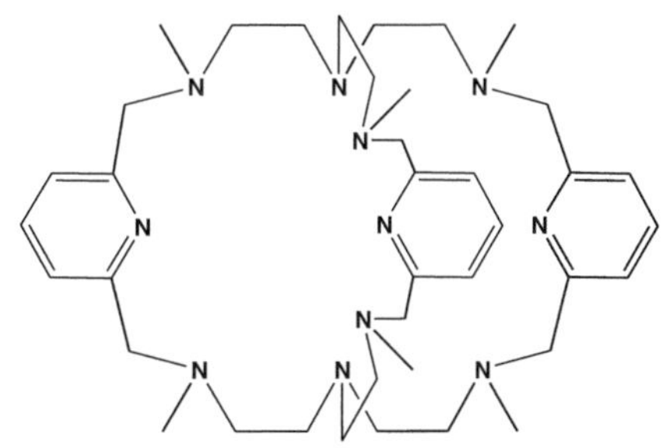

97.

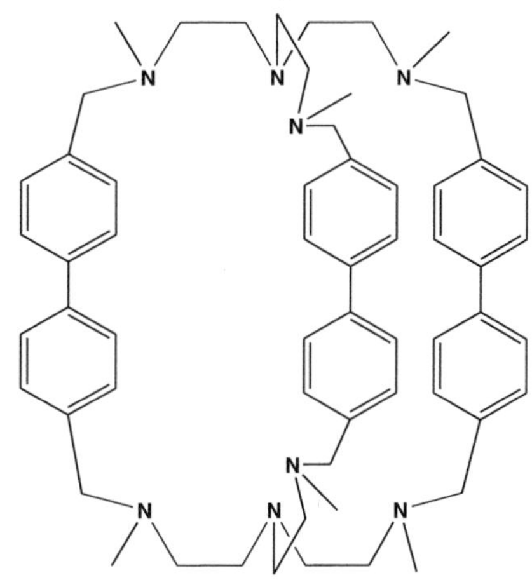

98.

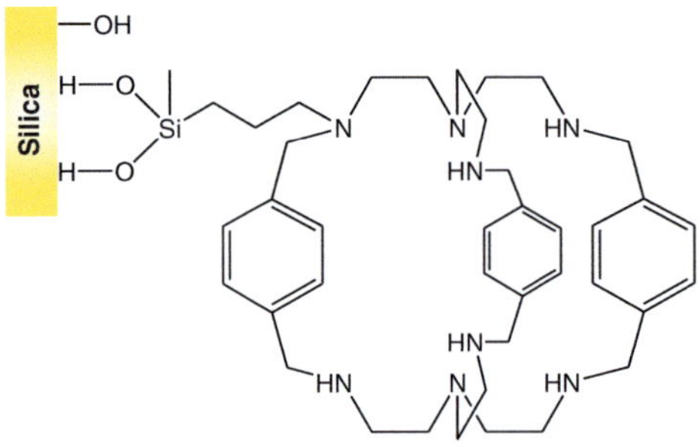

99.

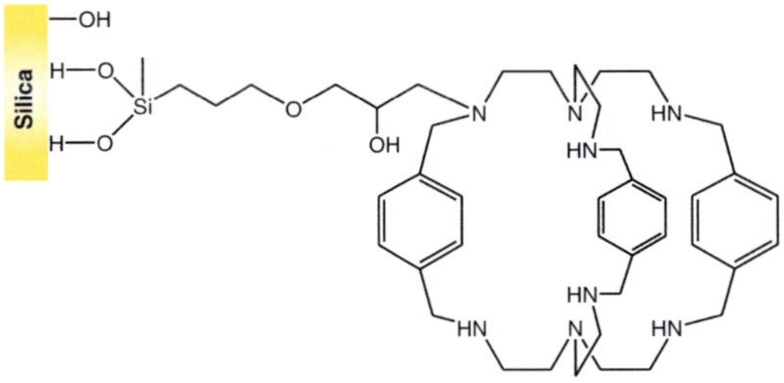

100.

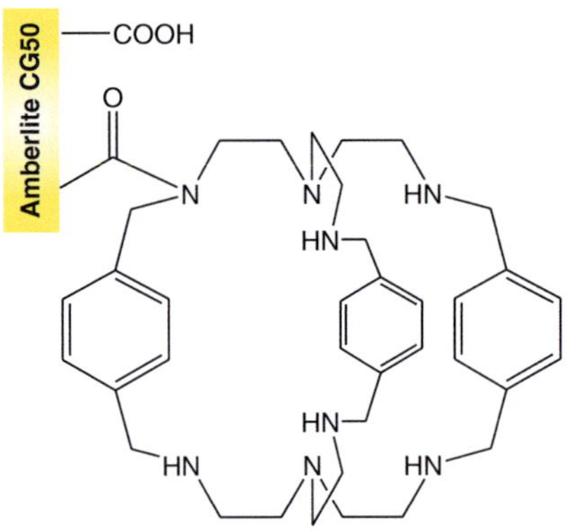

101.

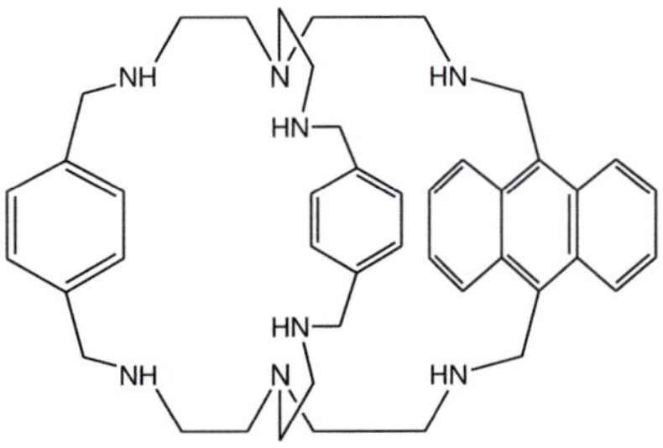

102.

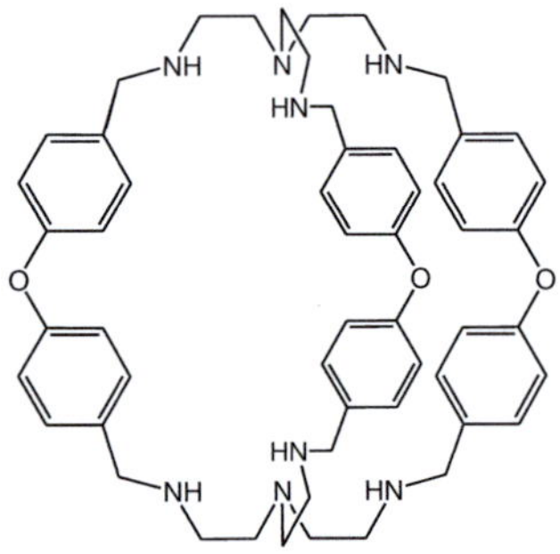

103.

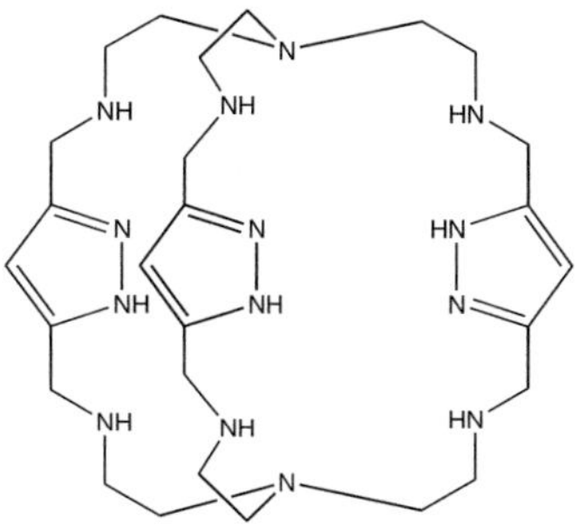

104.

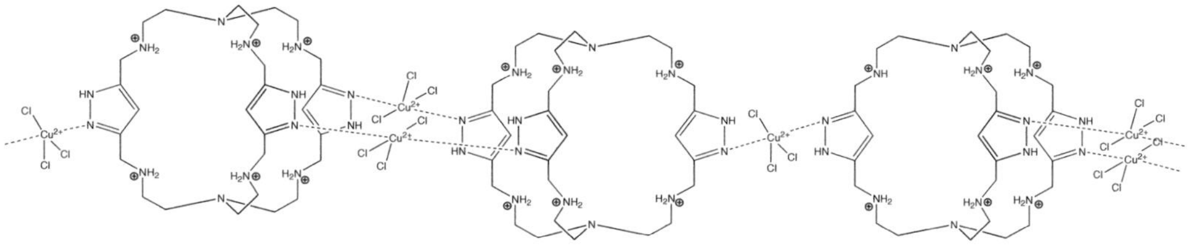

105.

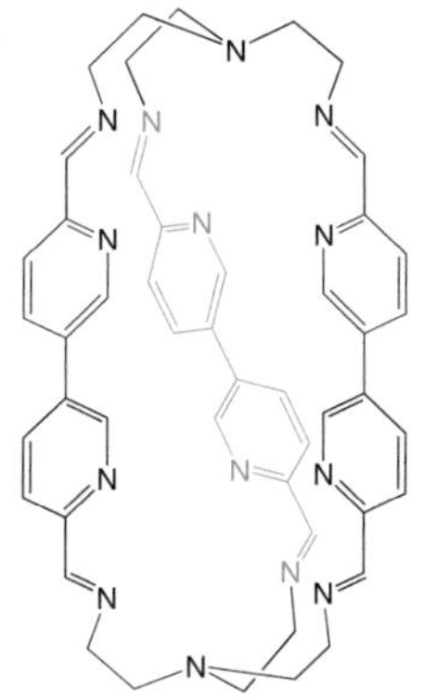

106.

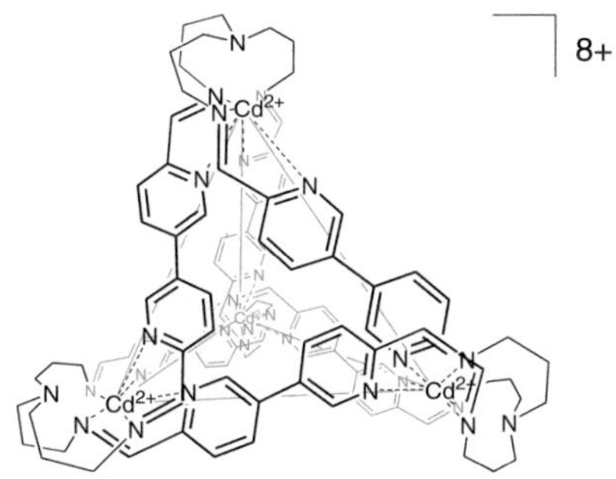

107.

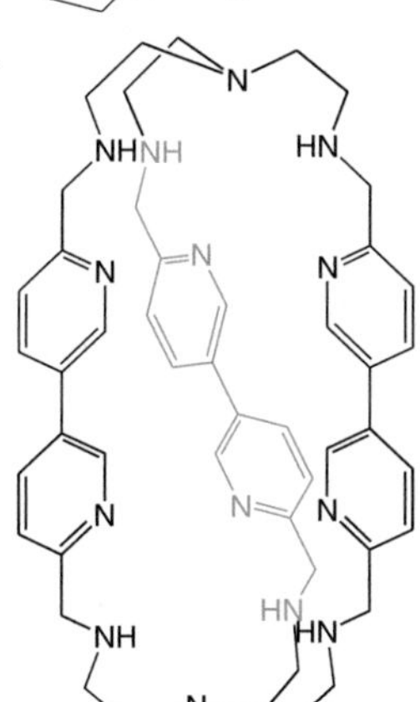

108.

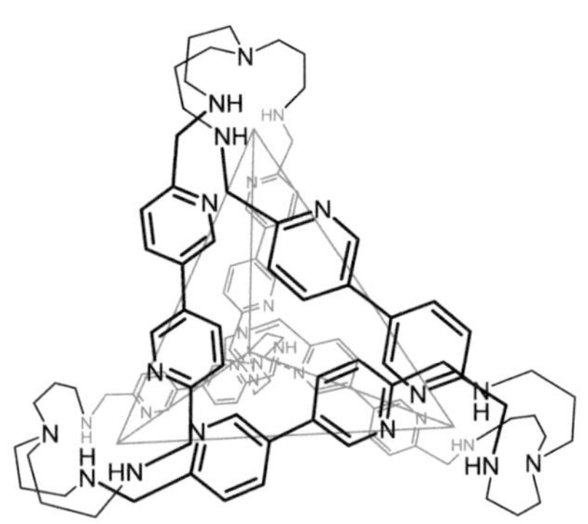

109.

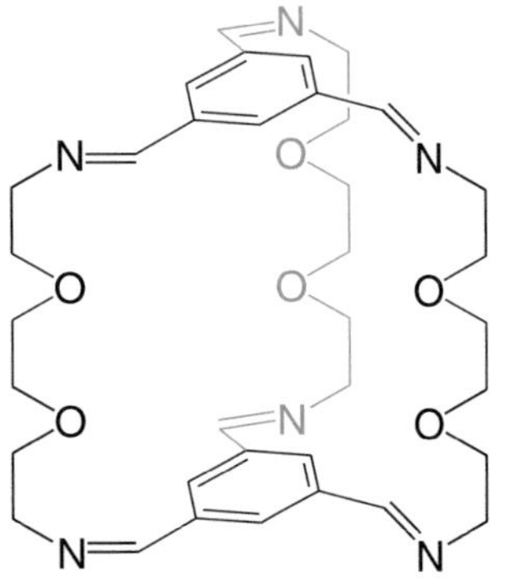

110.

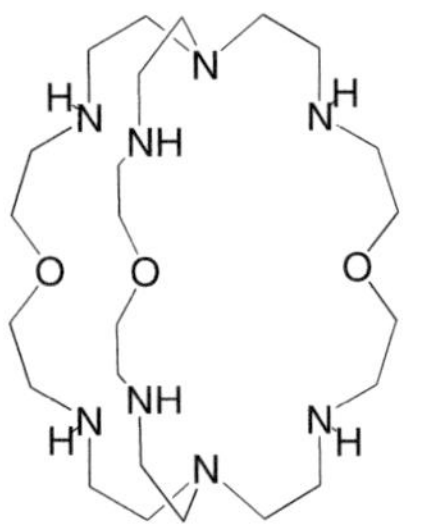

111.

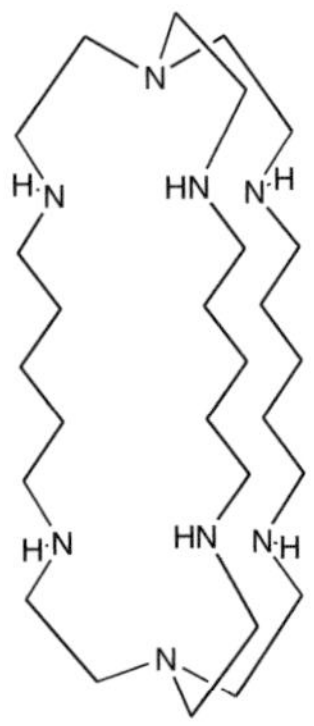

112.

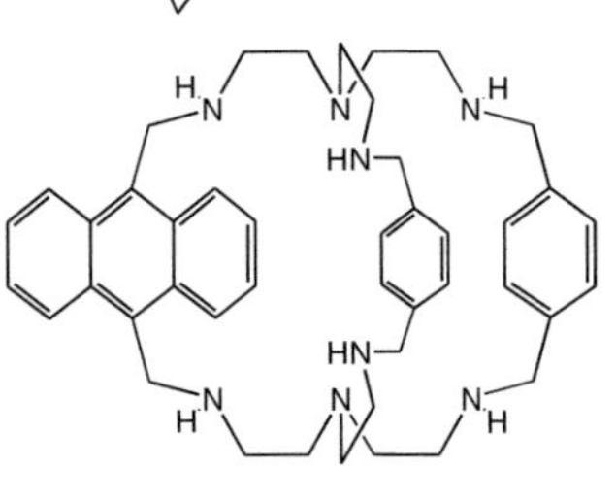

113.

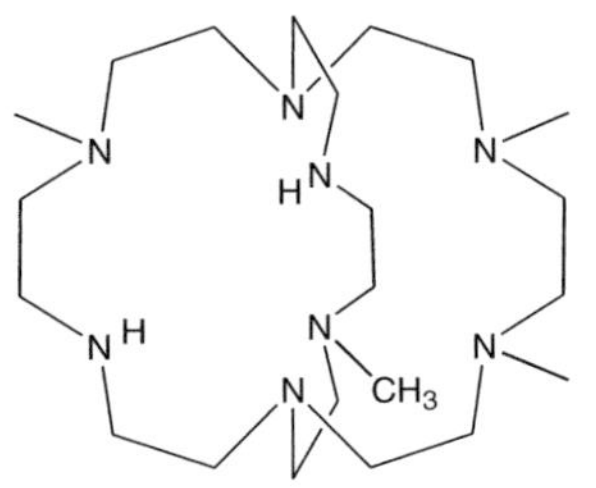

114.

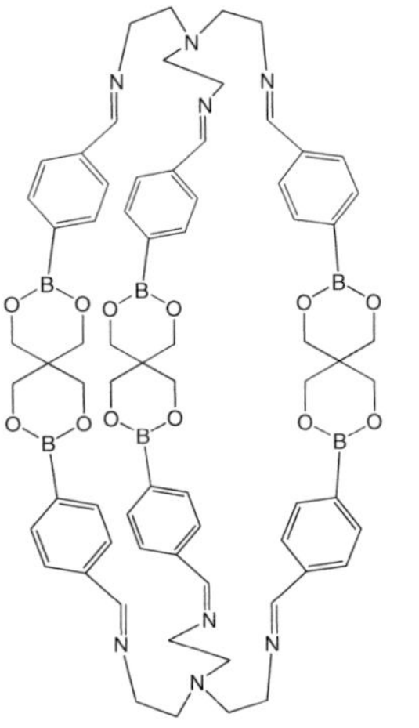

115.

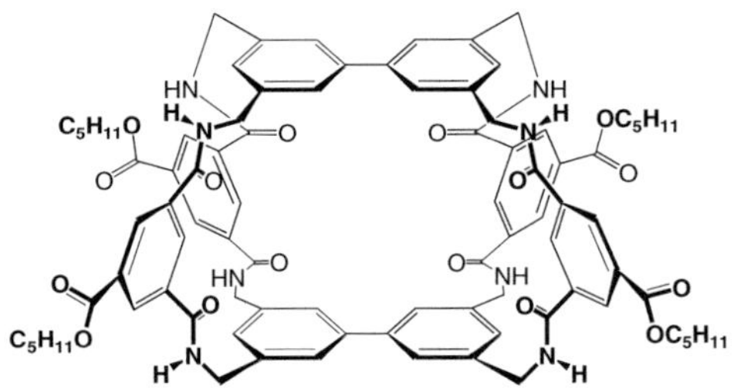

116.

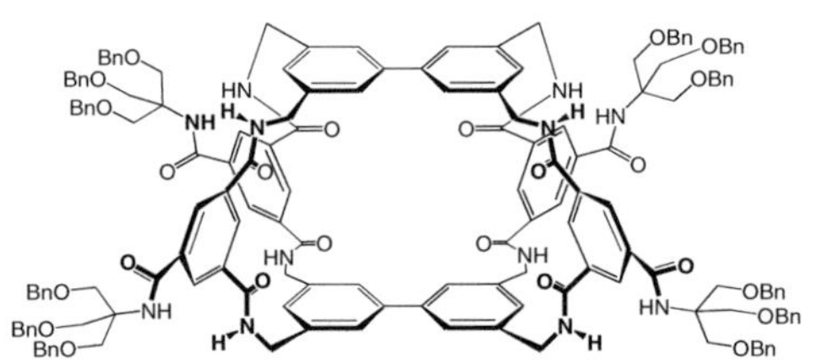

117.

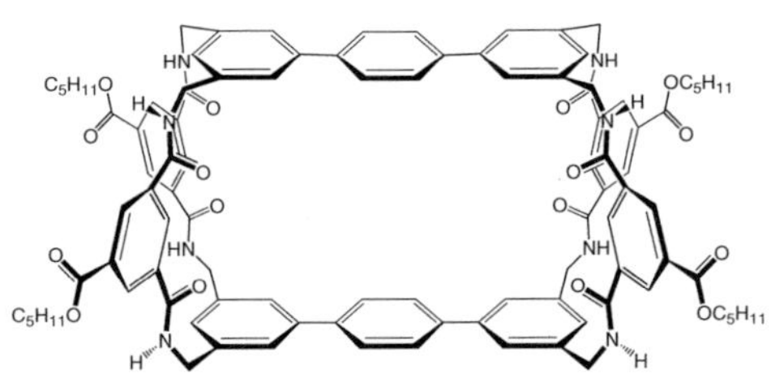

118.

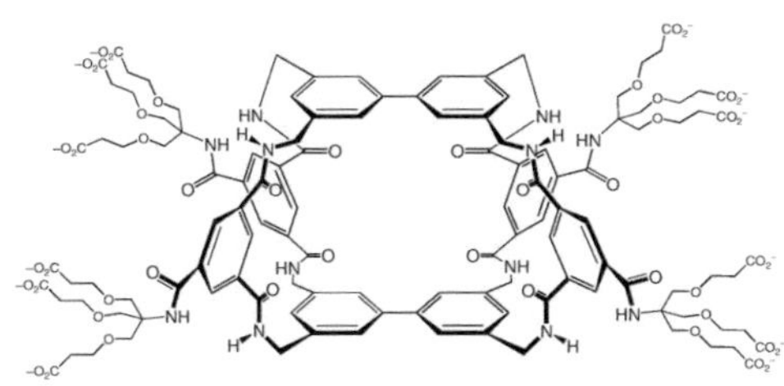

119.

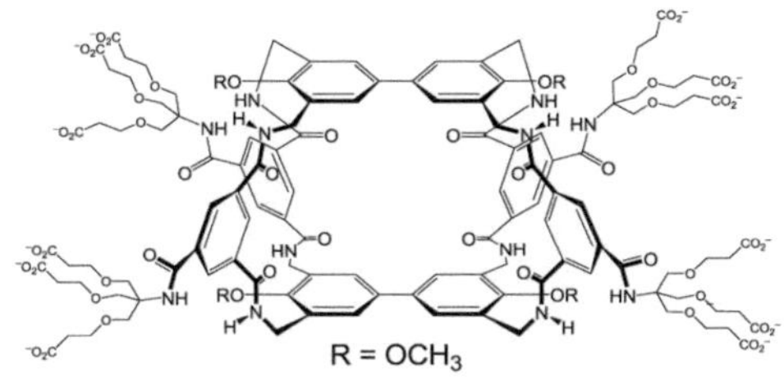

120.

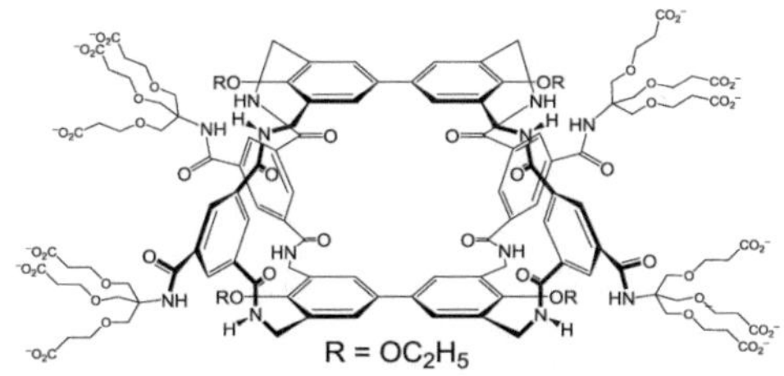

121.

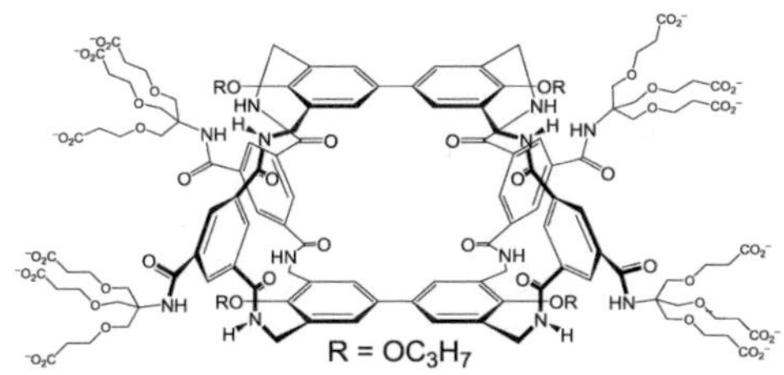

122.

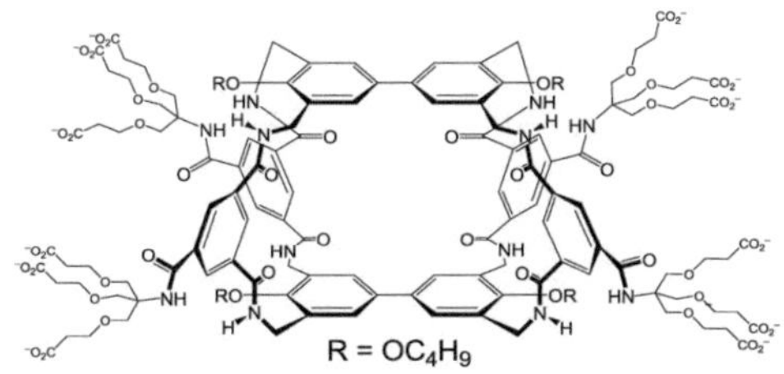

123.

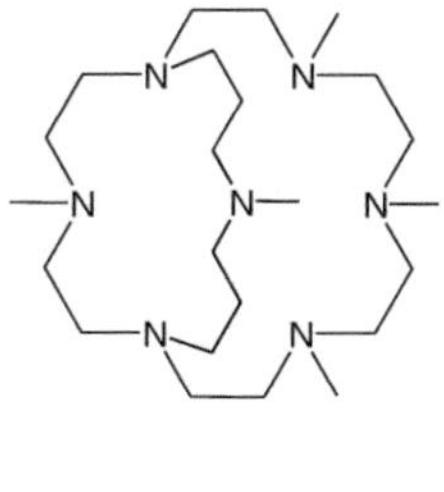

124.

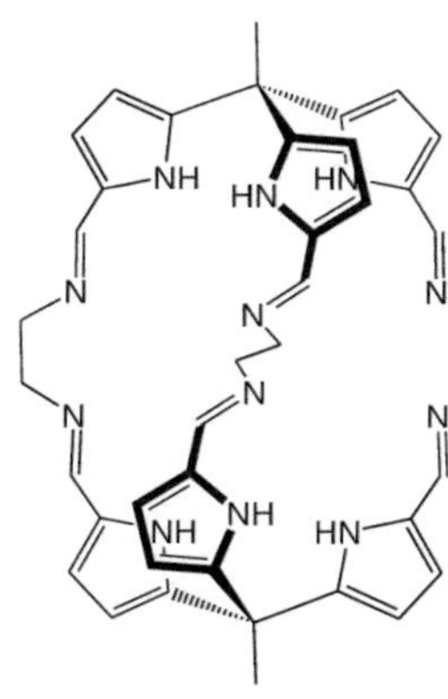

125.

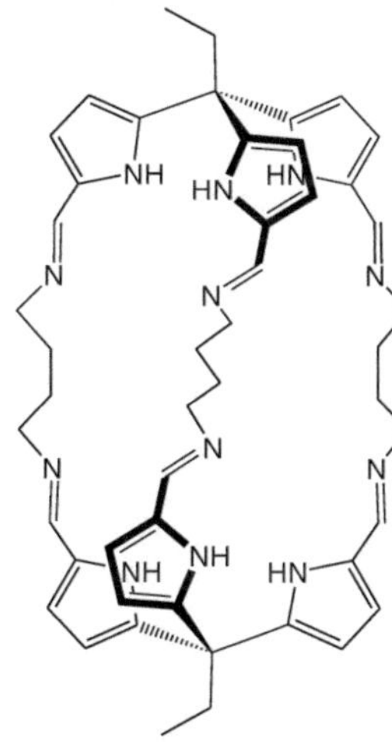

126.

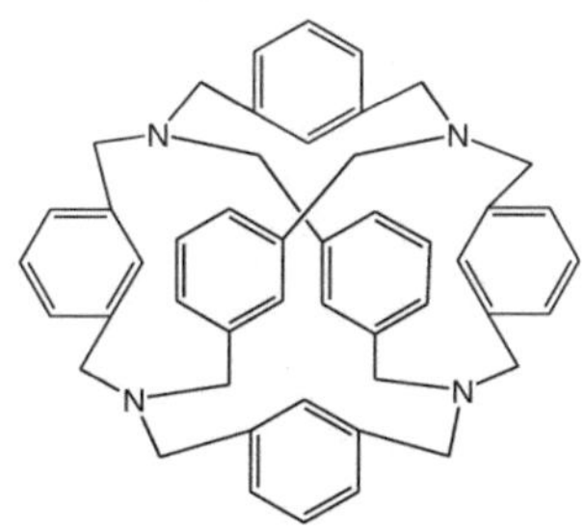

127.

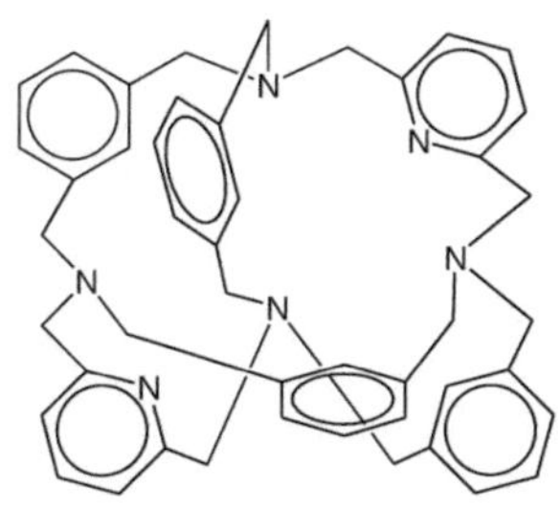

128.

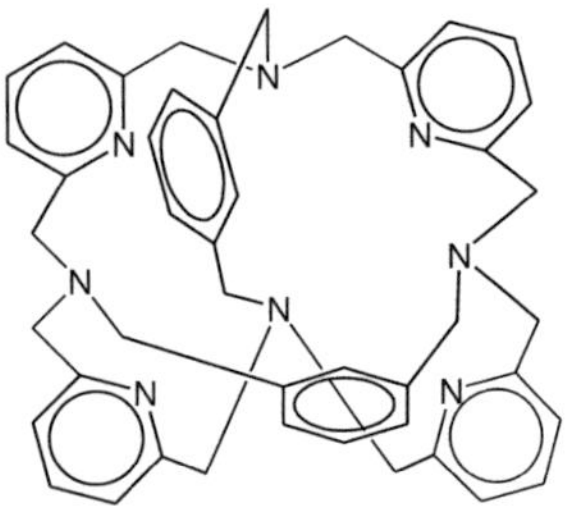

129.

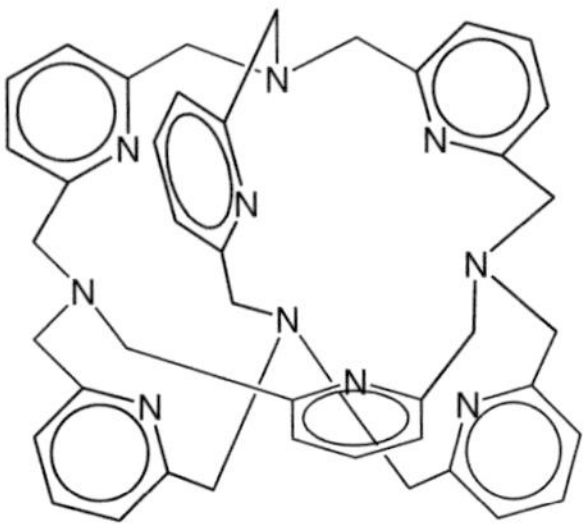

130.

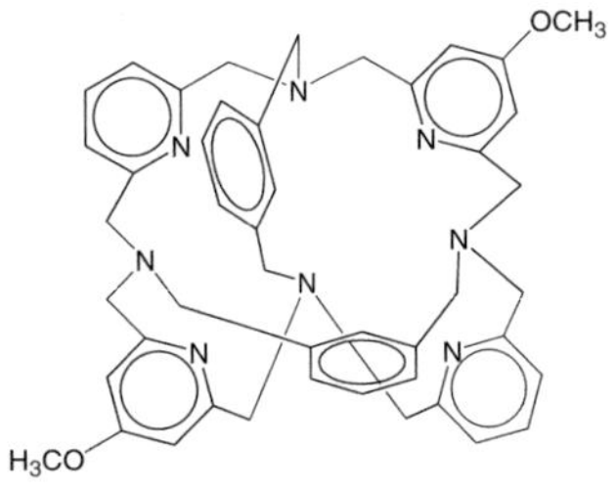

131.

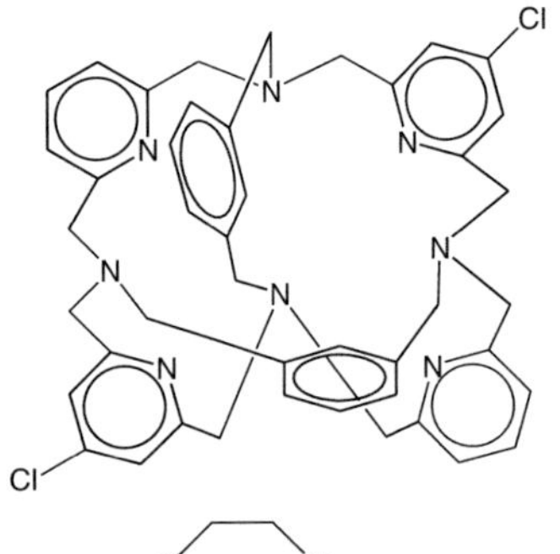

132.

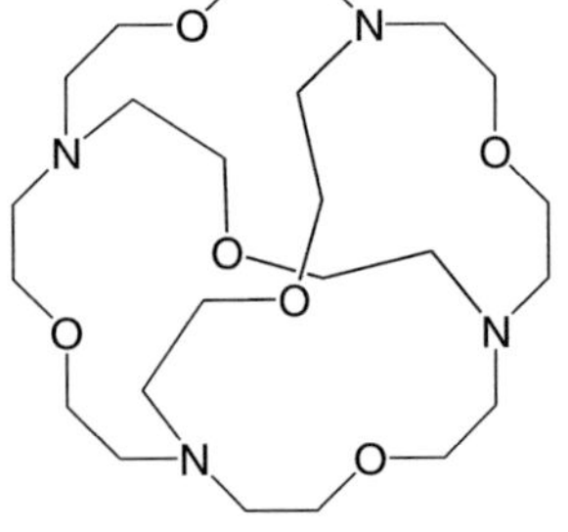

133.

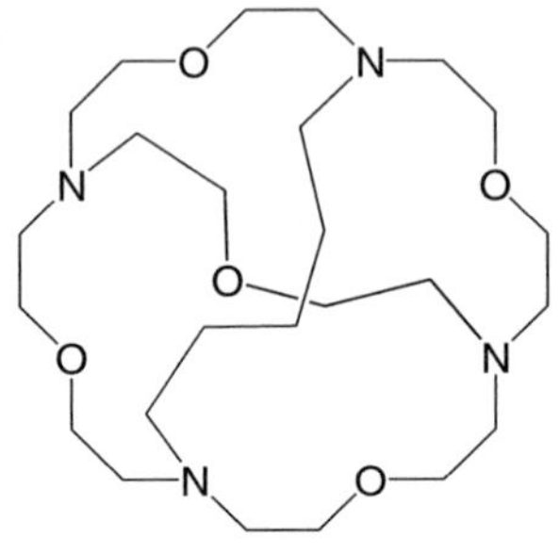

134.

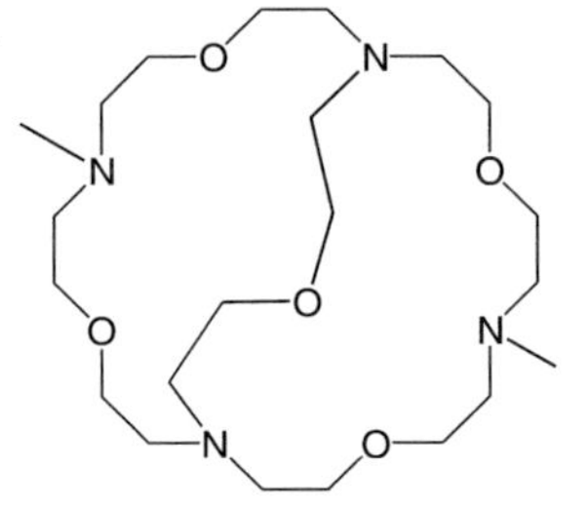

135.

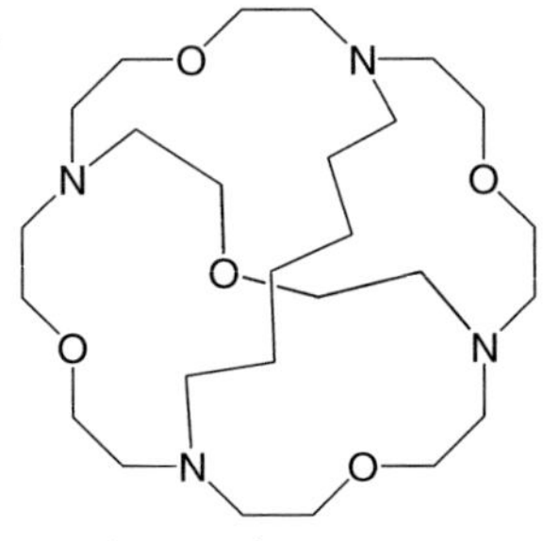

136.

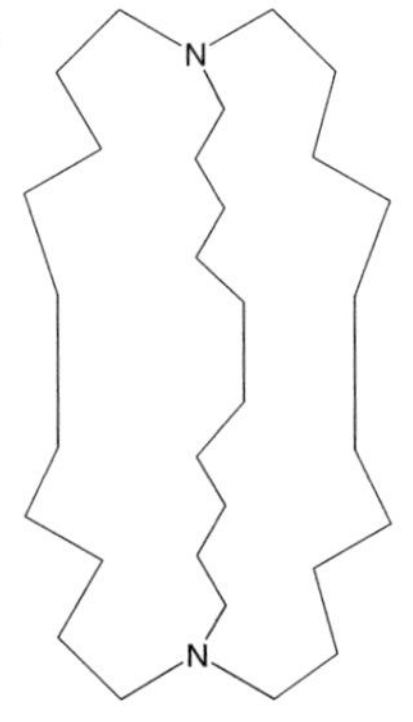

137.

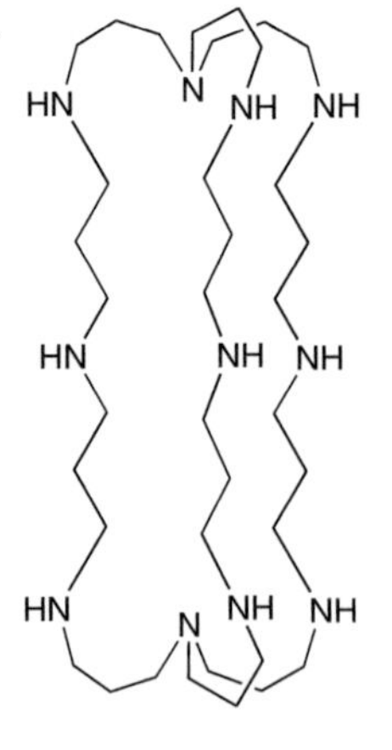

138.

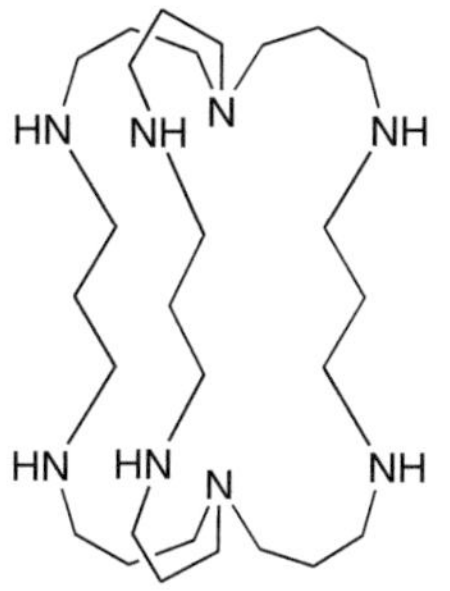

139.

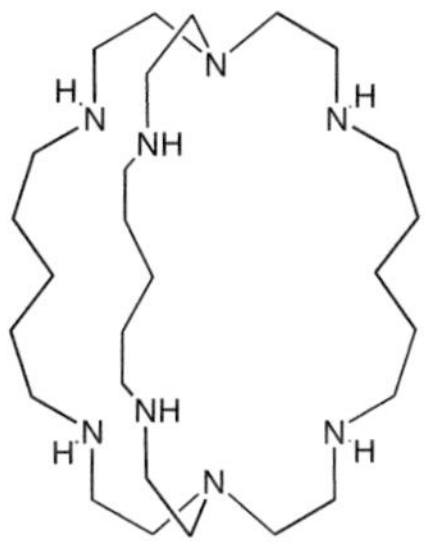

140.

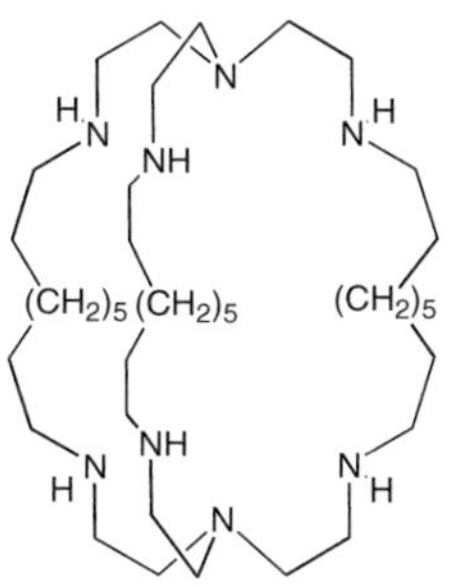

141.

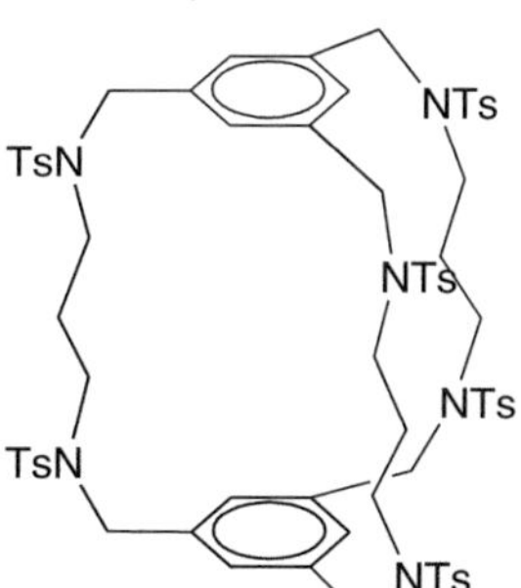

142.

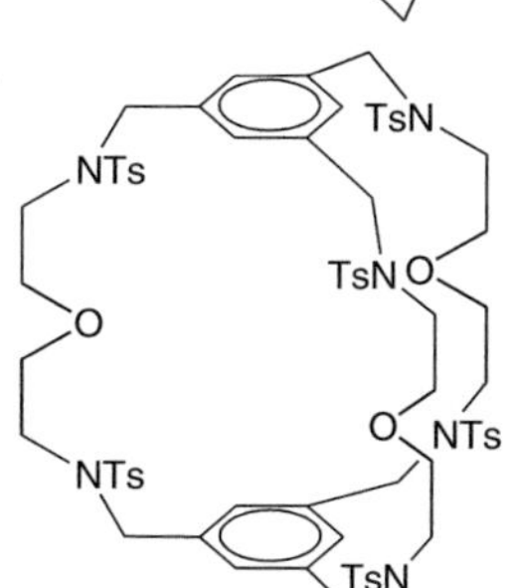

143.

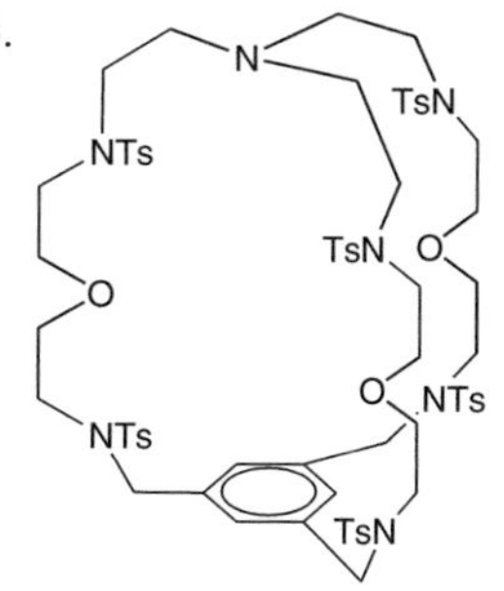

144.

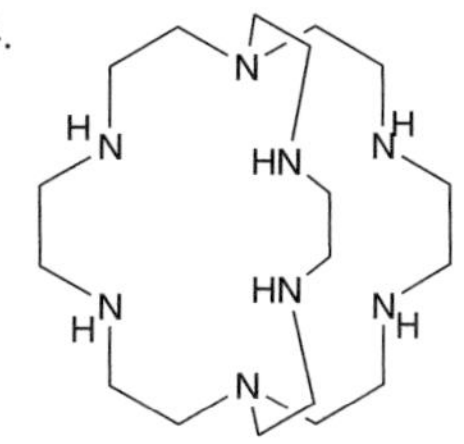

145.

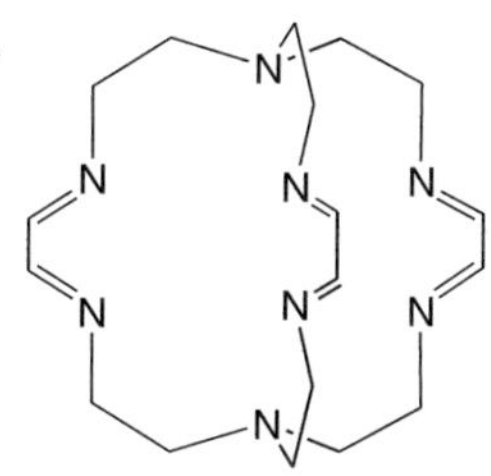

146.

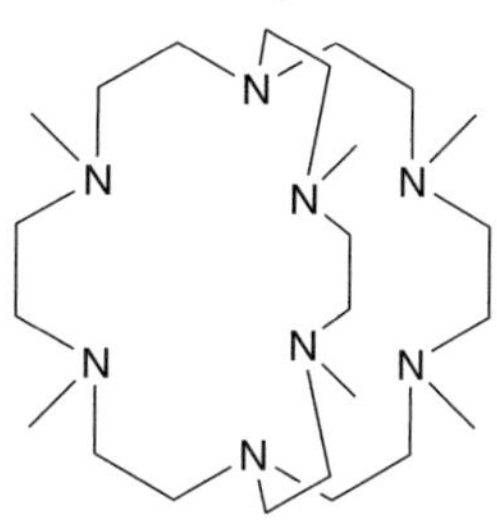

147.

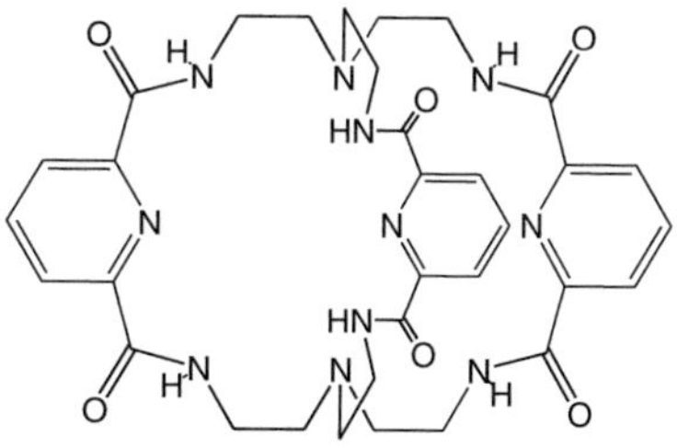

148.

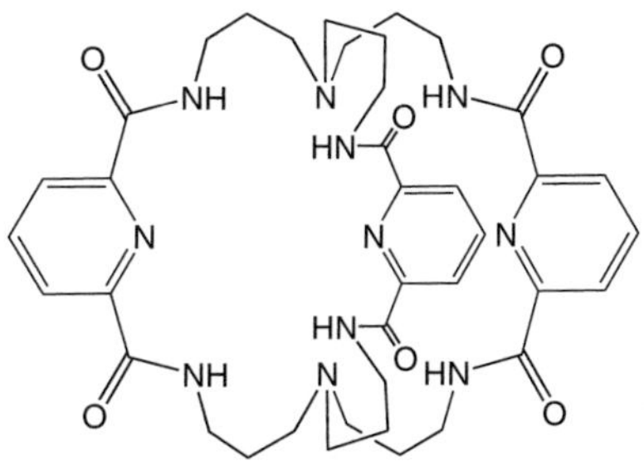

149.
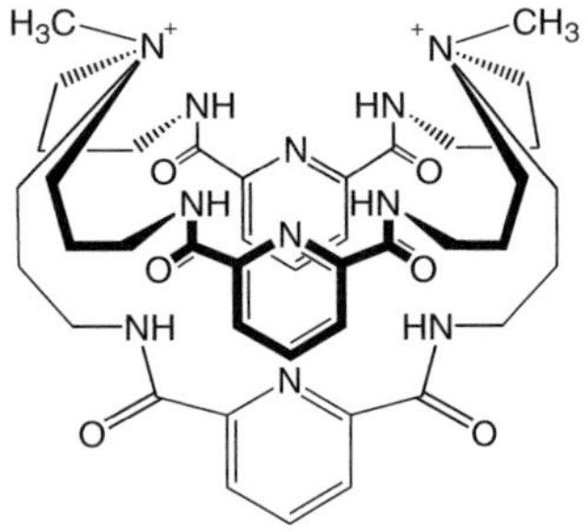

150.
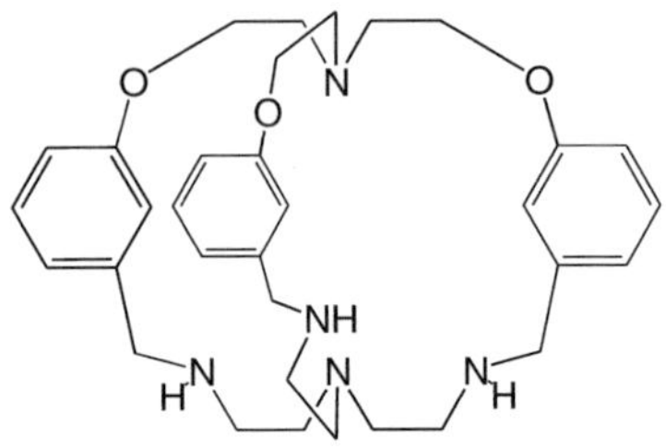

151.
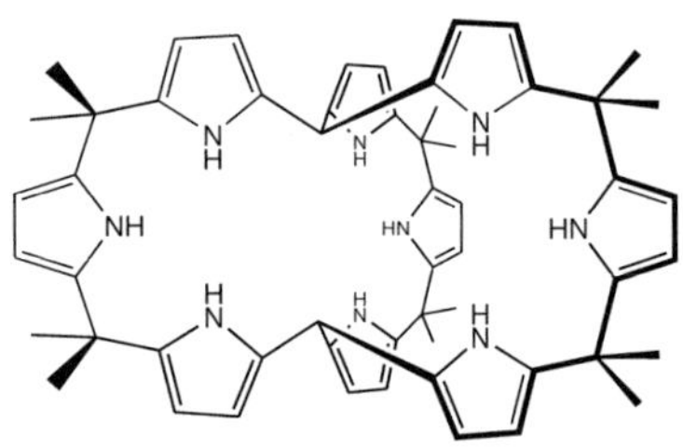

152.
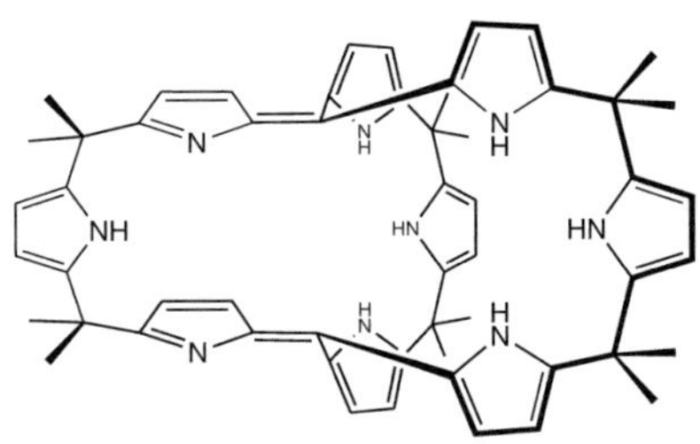

153.
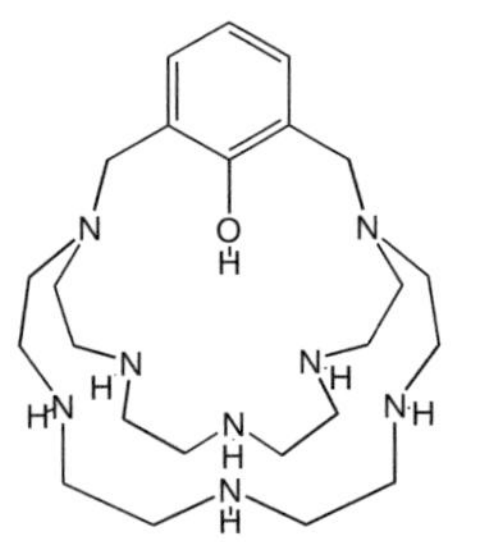

154.
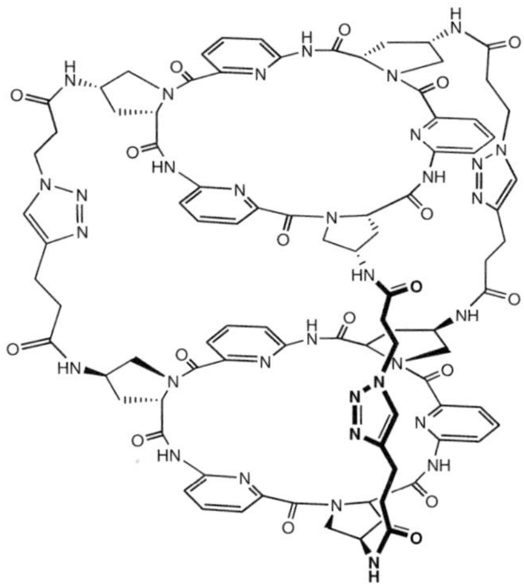

155.
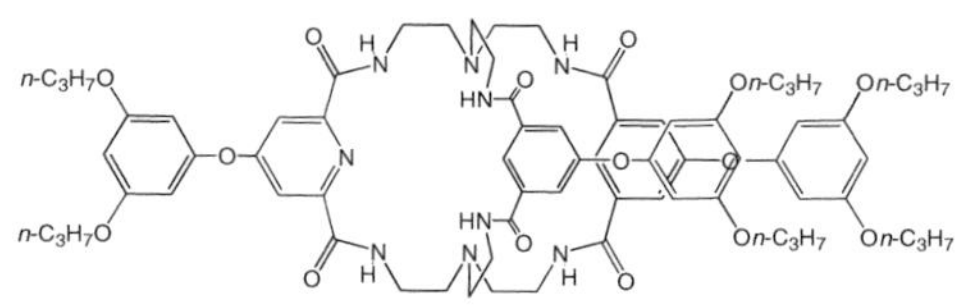

156.
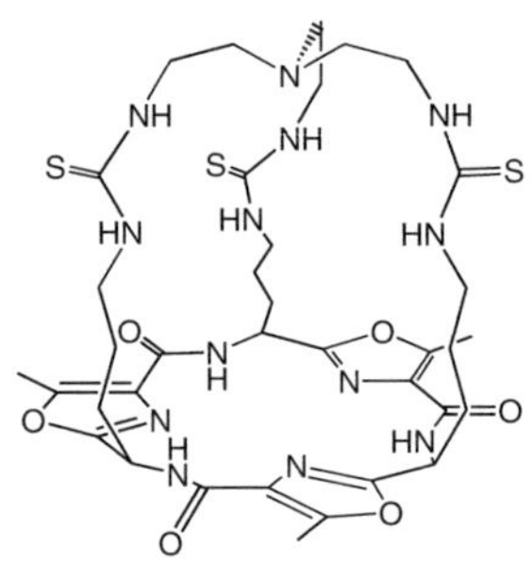

157.
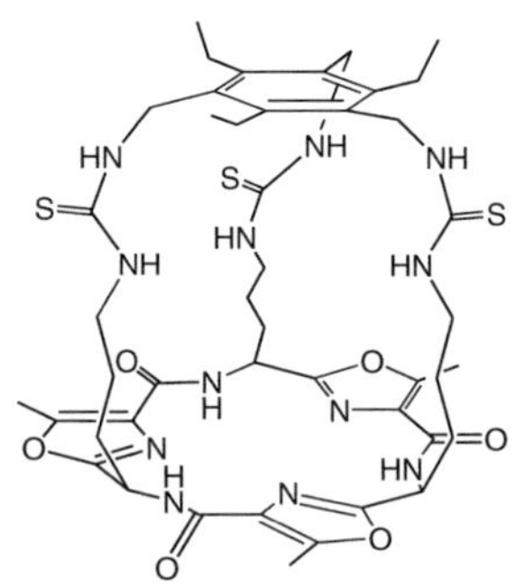

158.
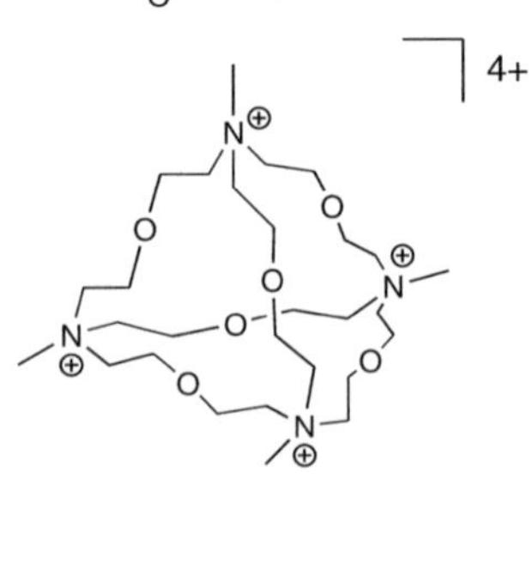

159.

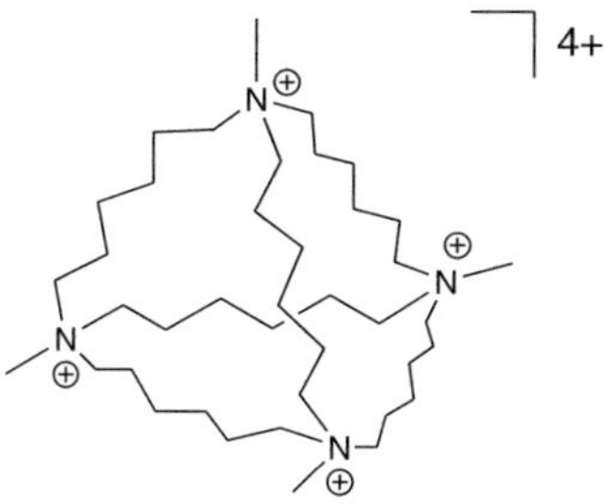

160.

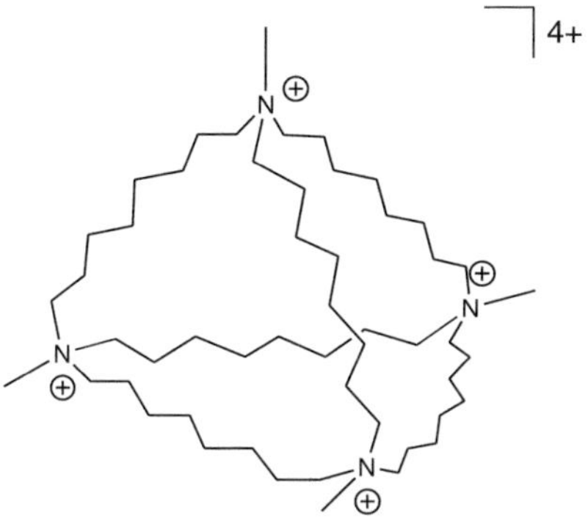

161.

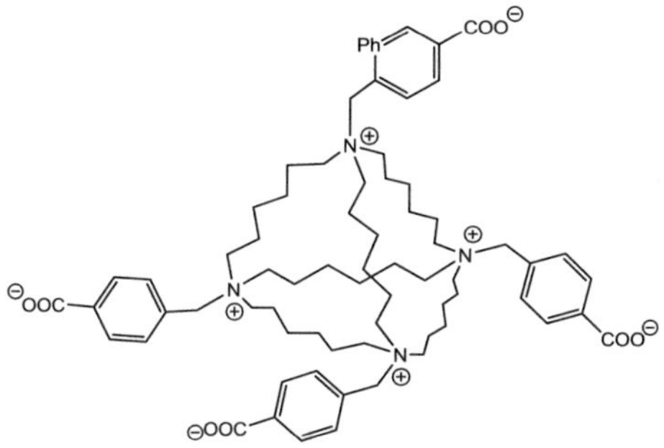

162.

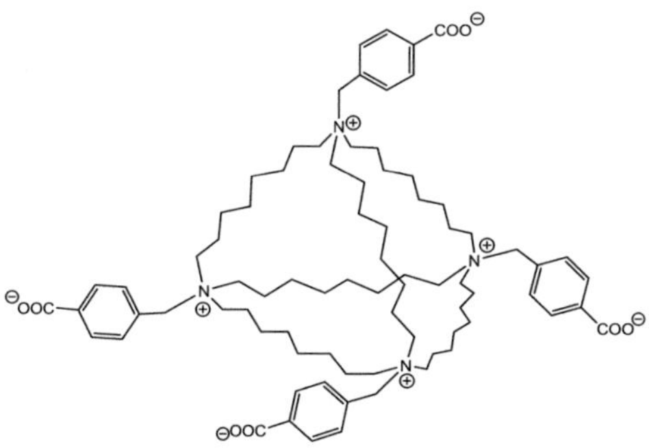

163.

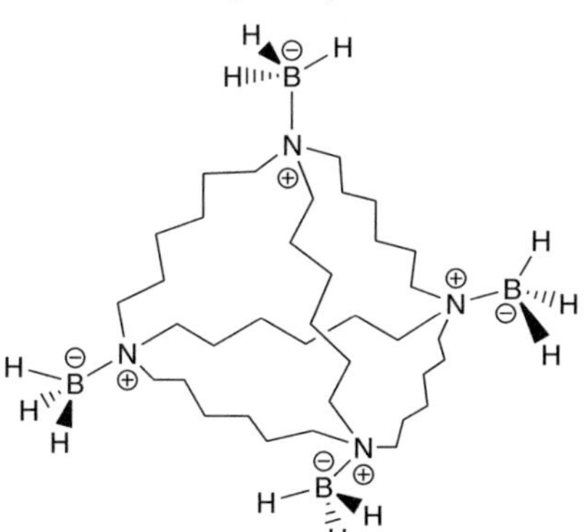

164.

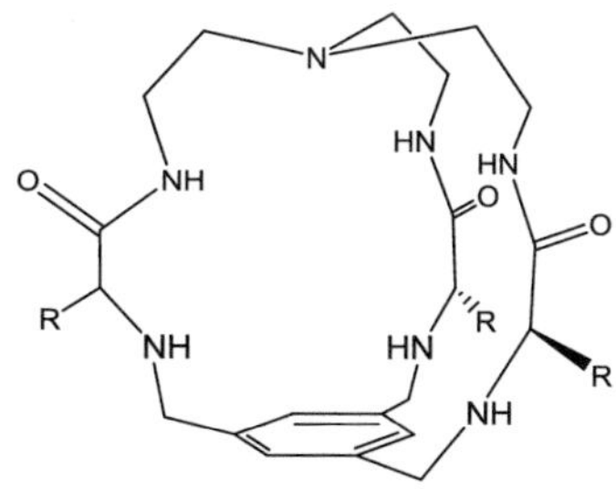

R = $CH_2C_6H_5$, *i*-C_3H_7, *i*-C_4H_9, *sec*-C_4H_9

165.

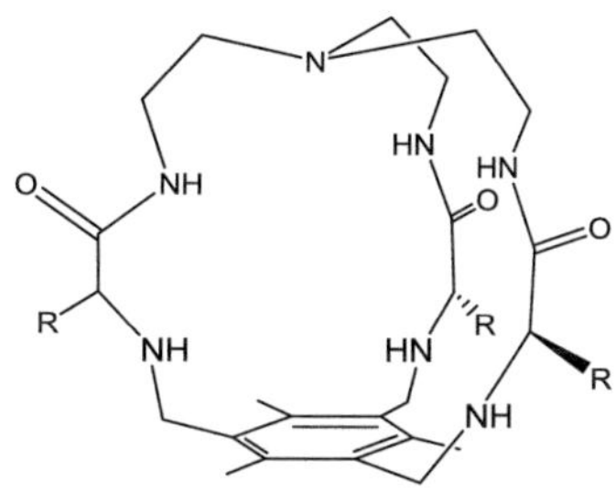

R = $CH_2C_6H_5$, *i*-C_3H_7, *i*-C_4H_9, *sec*-C_4H_9

166.

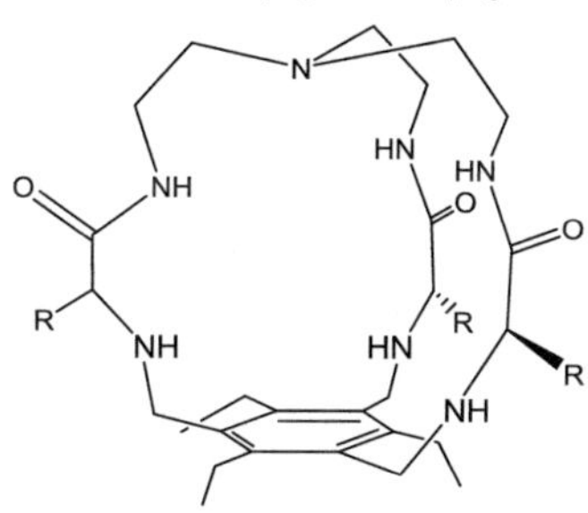

R = $CH_2C_6H_5$, *i*-C_3H_7, *i*-C_4H_9, *sec*-C_4H_9

167.

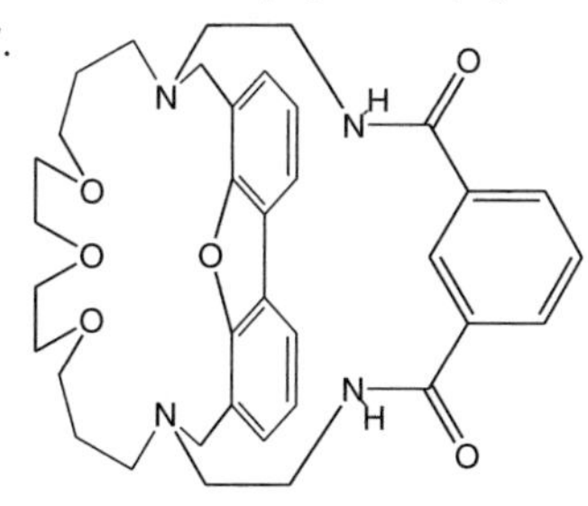

168.

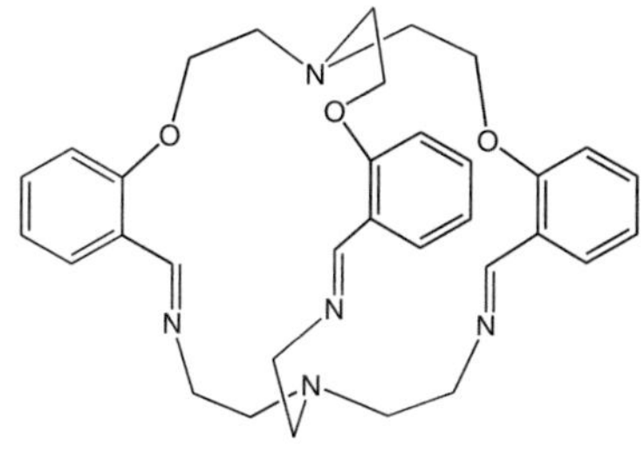

169.

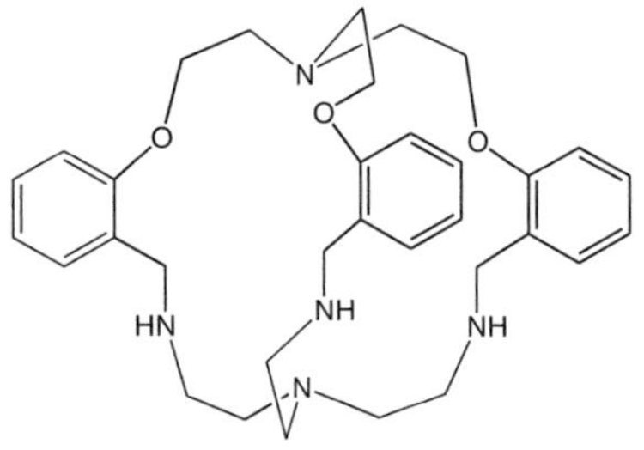

170.

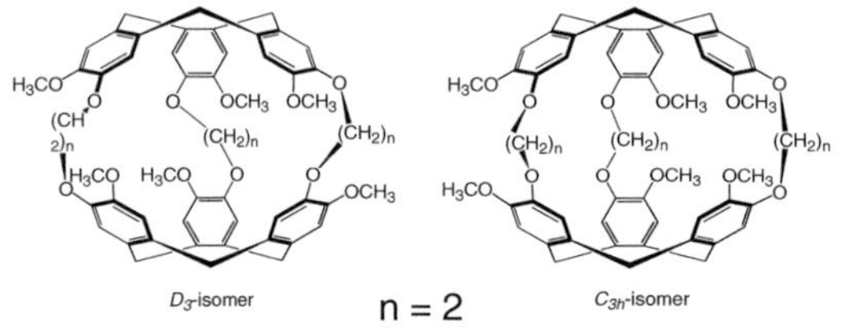

171.

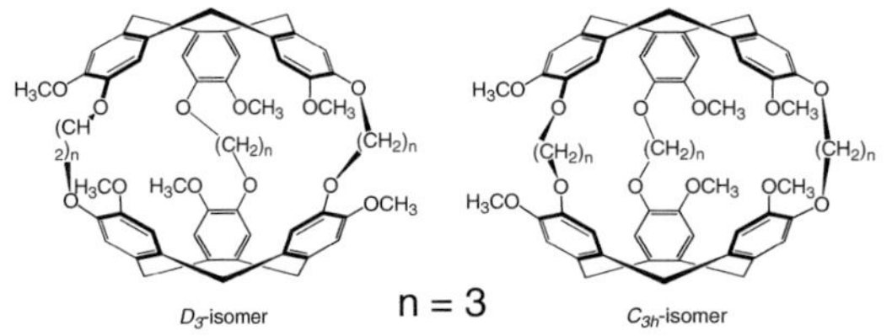

172.

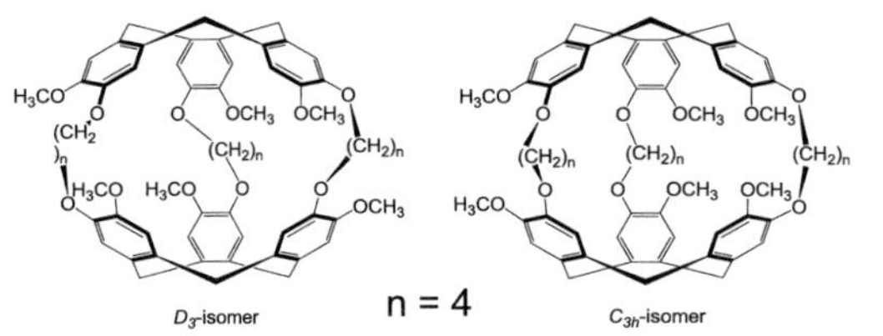

173.

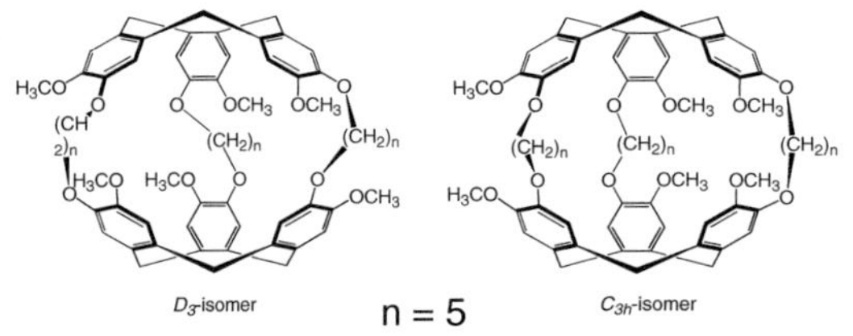

174.

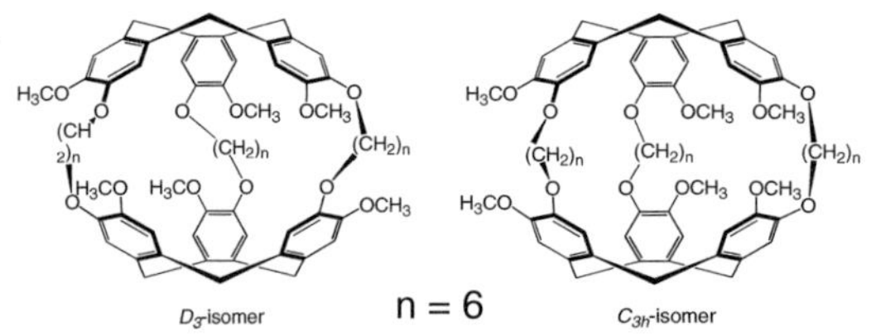

175.

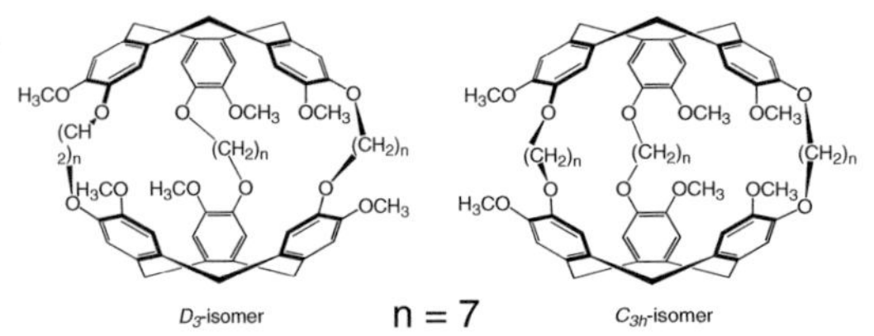

176.

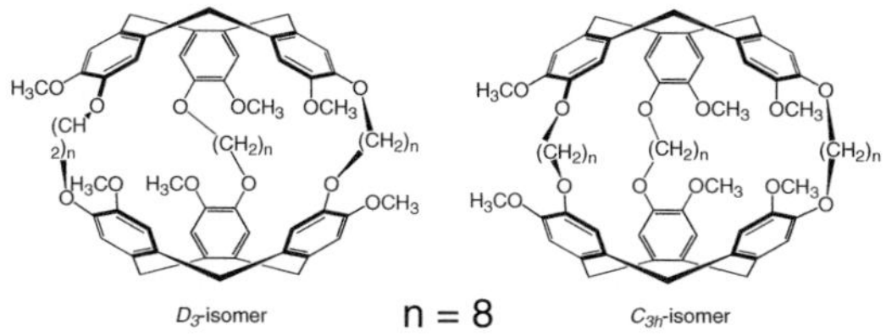

177.

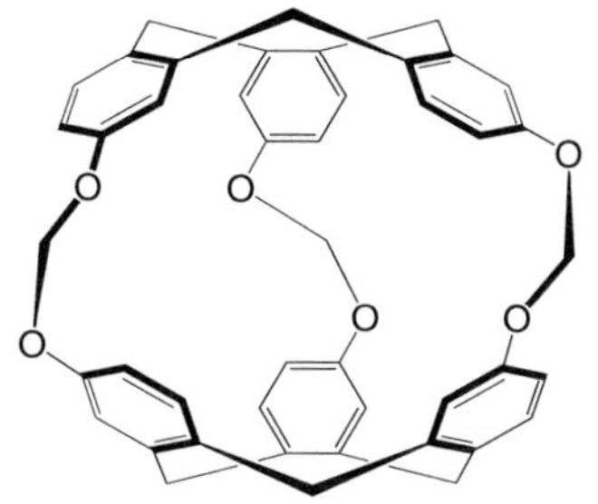

178.

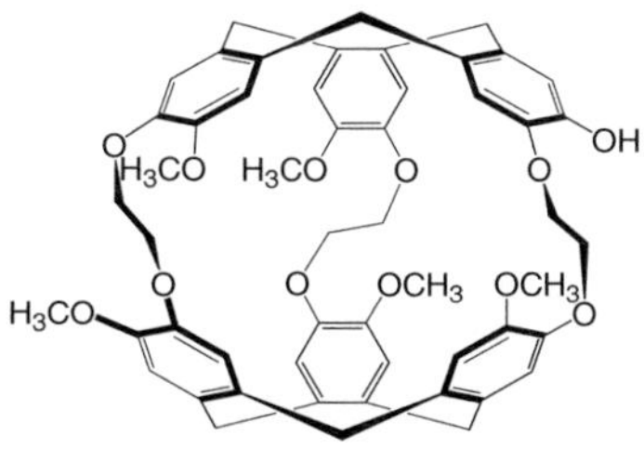

179.

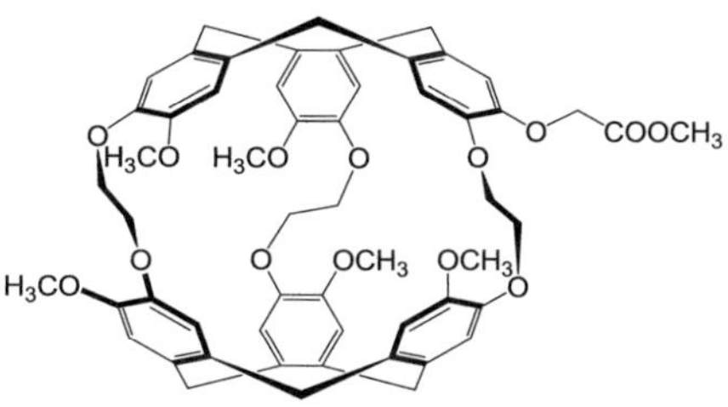

180.

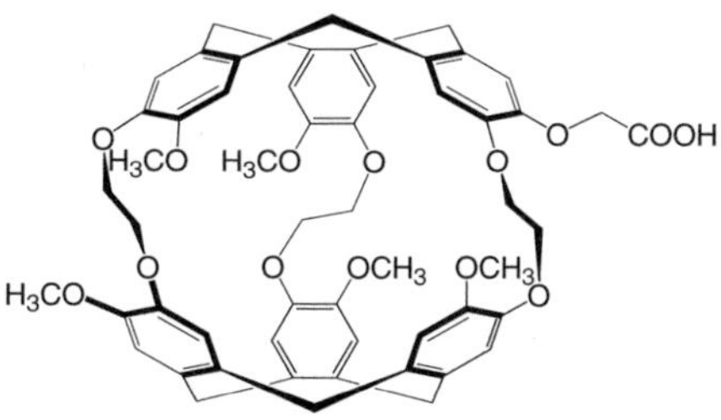

181.

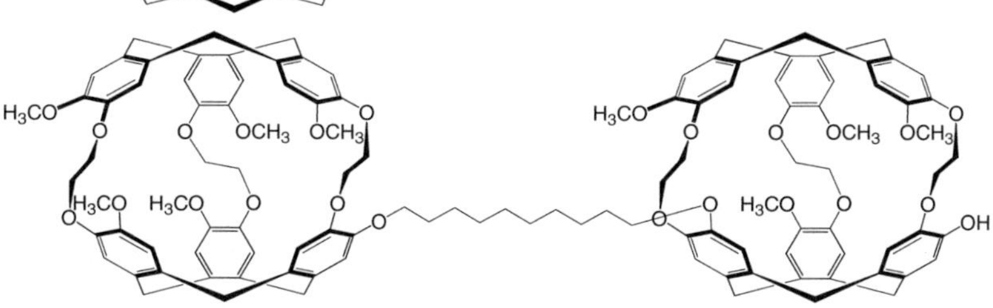

182.

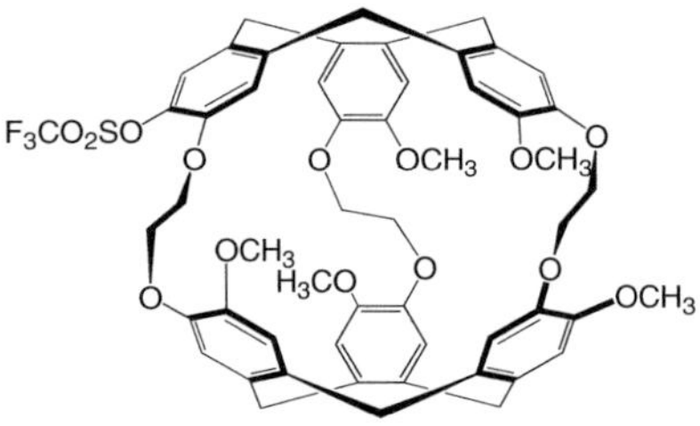

183.

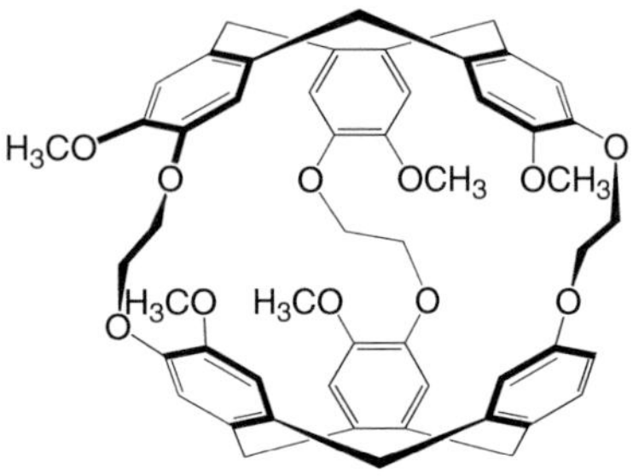

184.

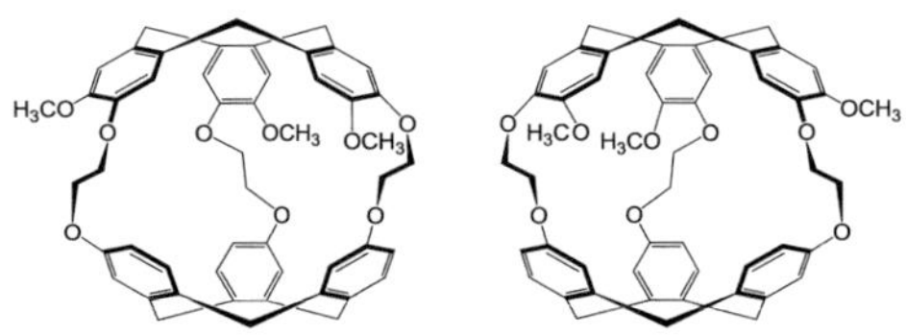

(–)-isomer (+)-isomer

185.

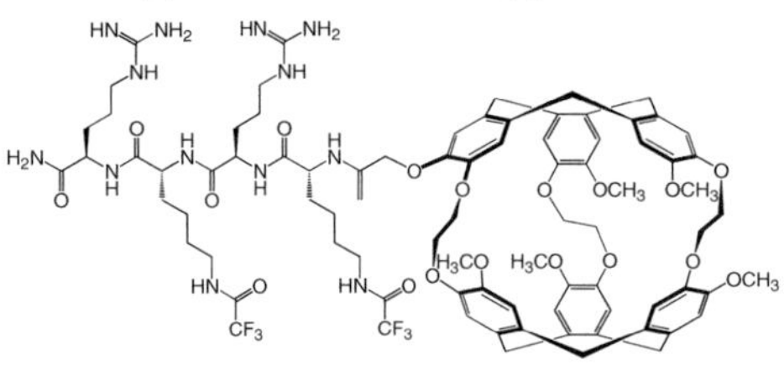

186.

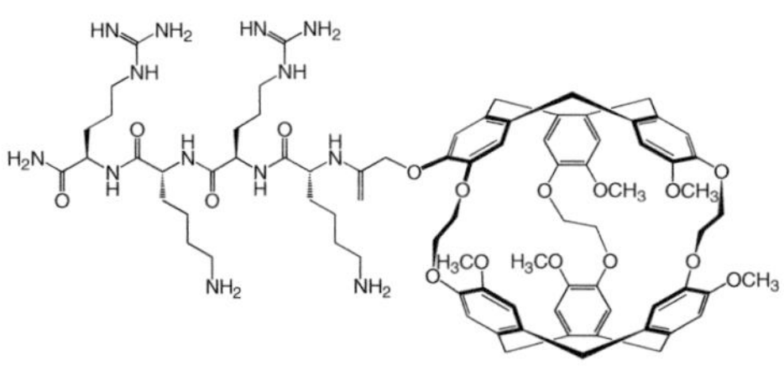

187.

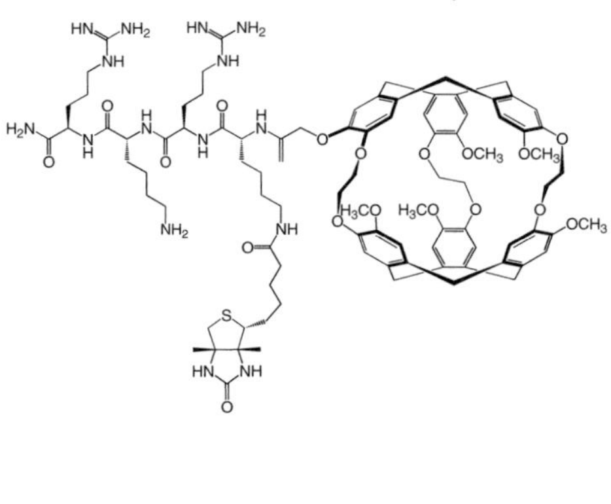

188.

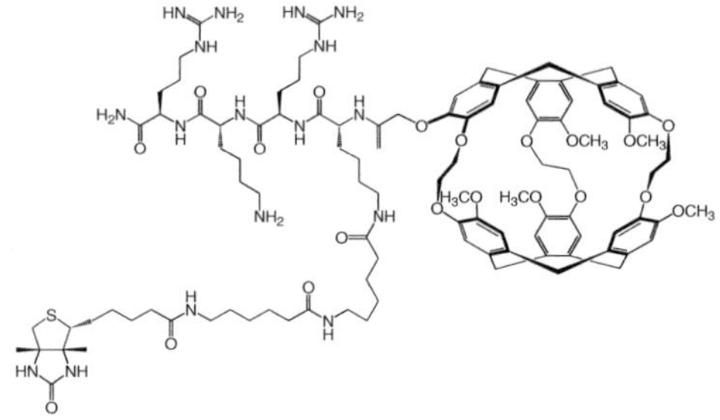

189.

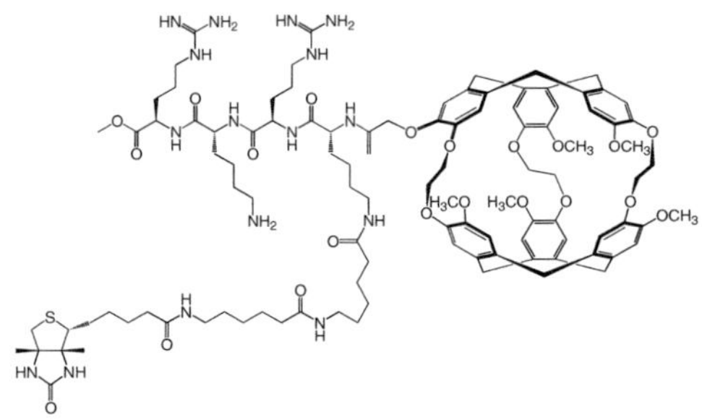

190.

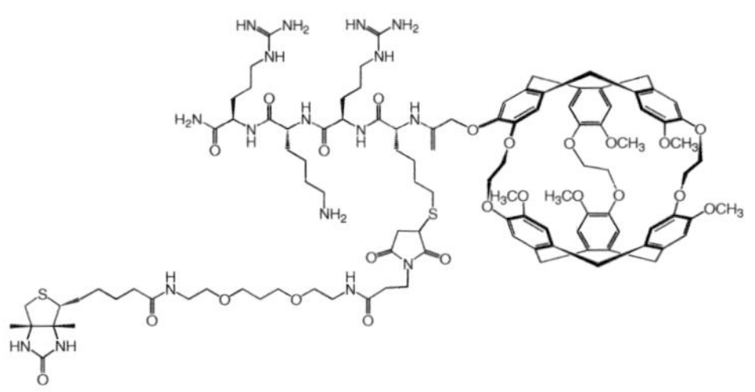

191.

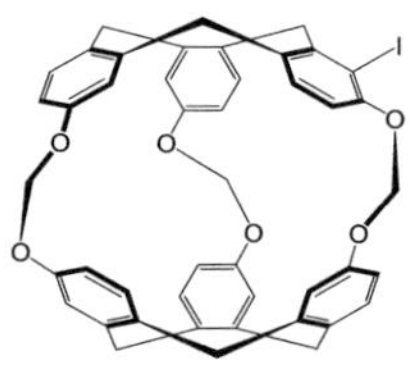

192.

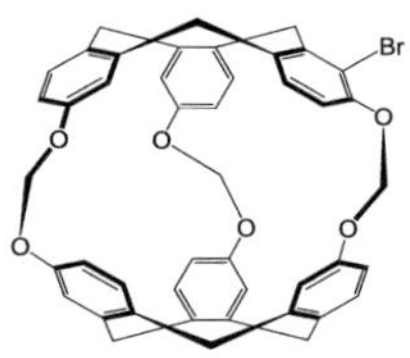

193.

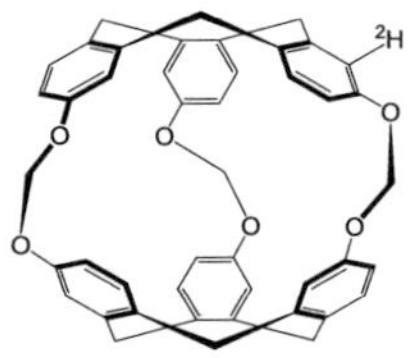

194.

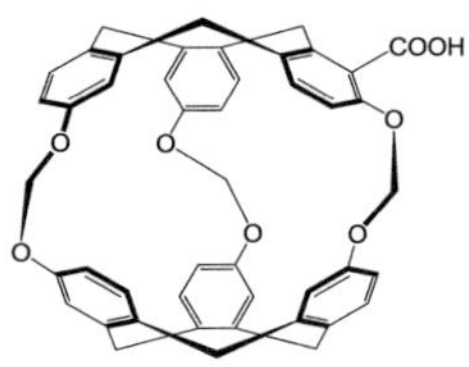

195.

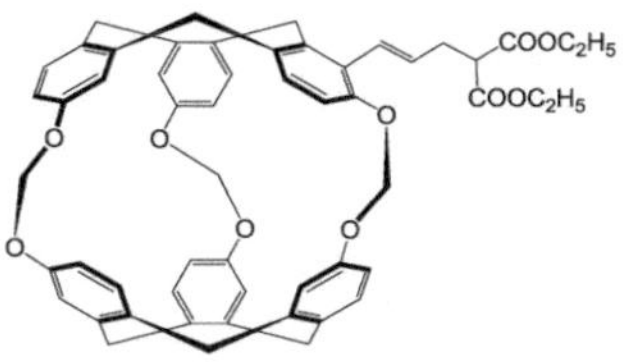

196.

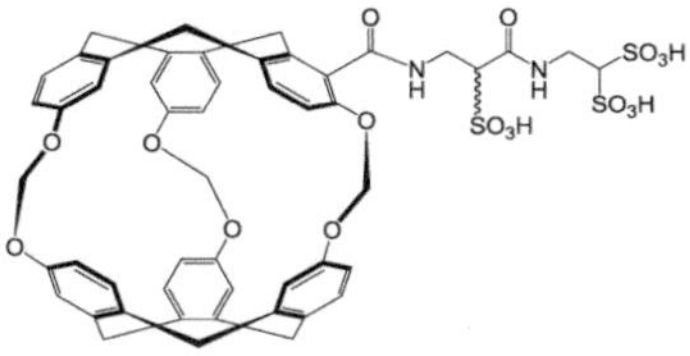

197.

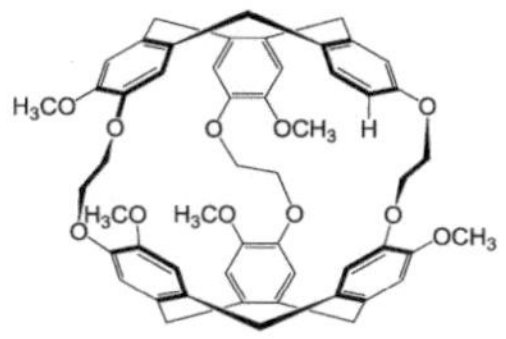

MM–isomer; *PP*–isomer

198.

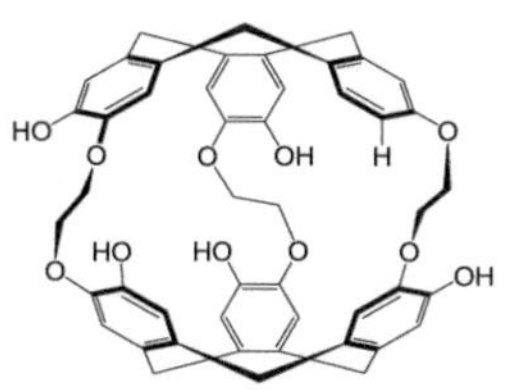

MM–isomer; *PP*–isomer

199.

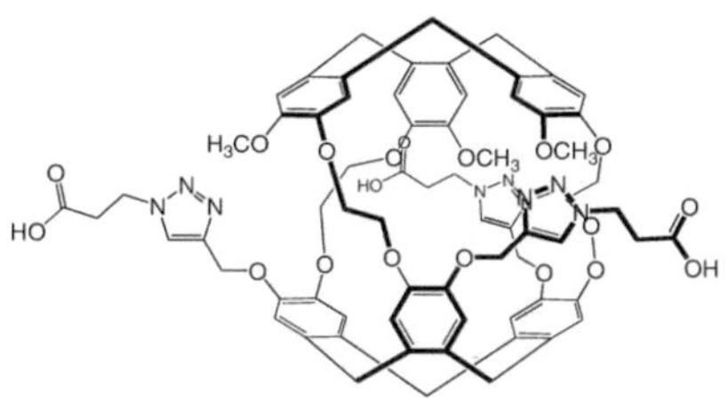

200.

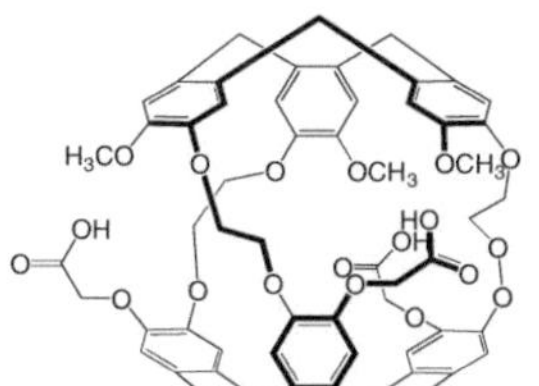

201.

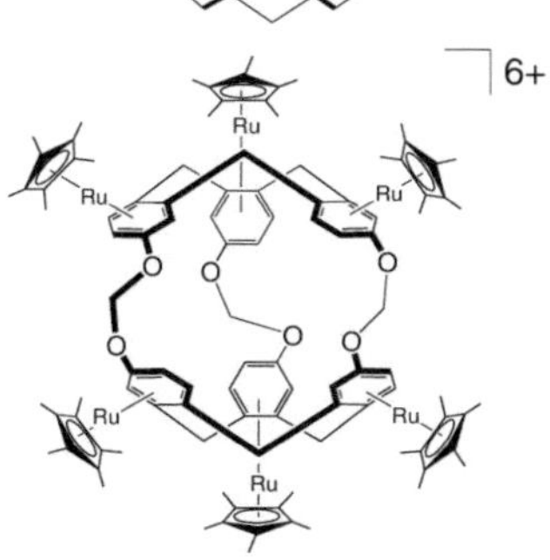

202.

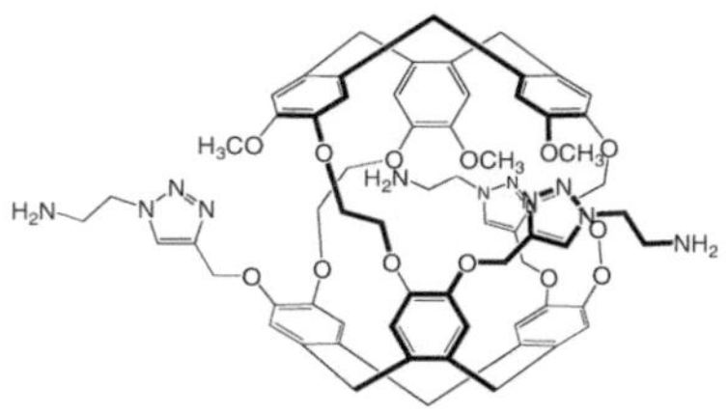

203.

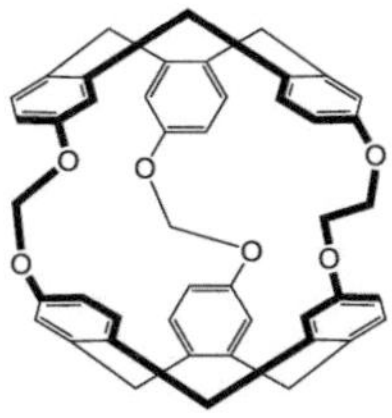

204.

205.

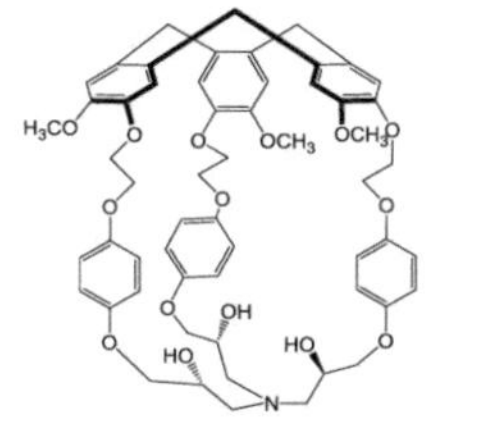

M-(*R,R,R*)- and M-(S,S,S)- isomers

206.

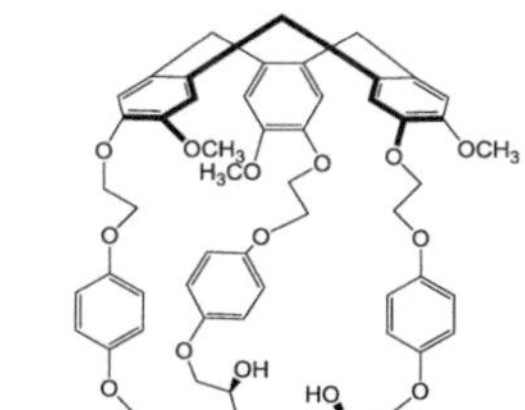

P-(*R,R,R*)- and P-(S,S,S)- isomers

207.

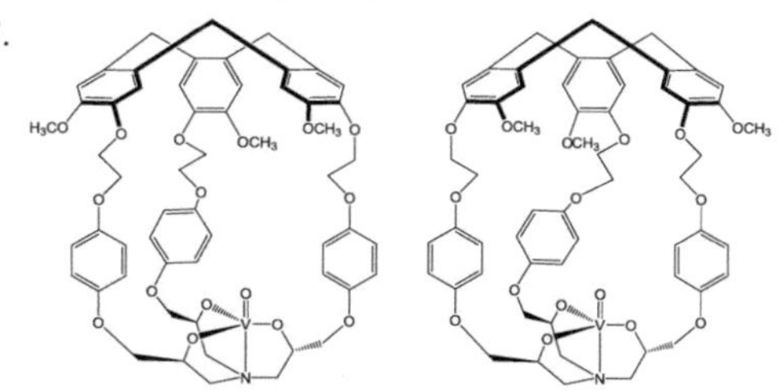

M-(SSS)- and *M*-(*RRR*)-isomers *P*-(SSS)-isomer

208.

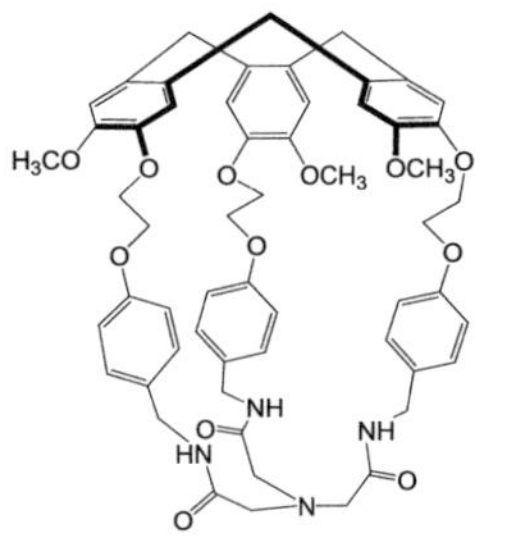

209.

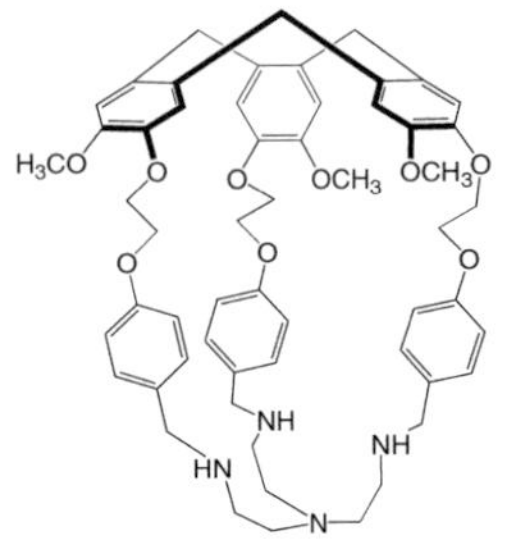

210.

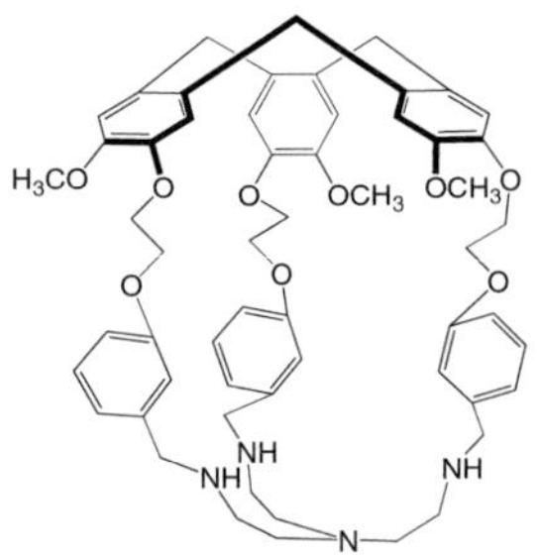

211.

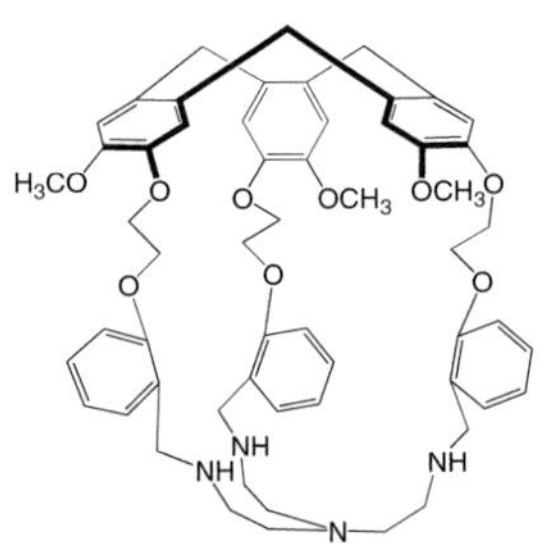

212.

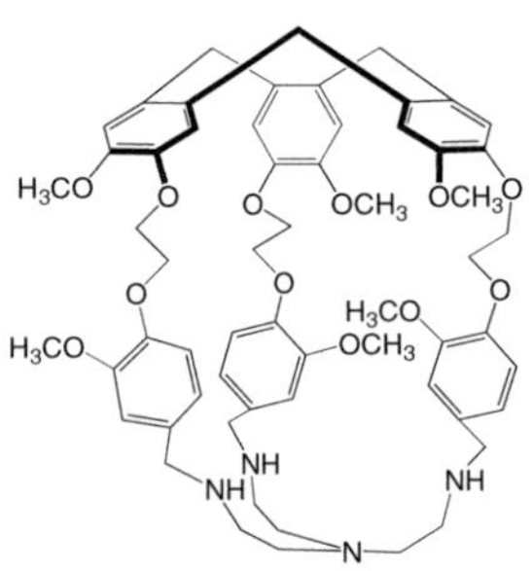

213.

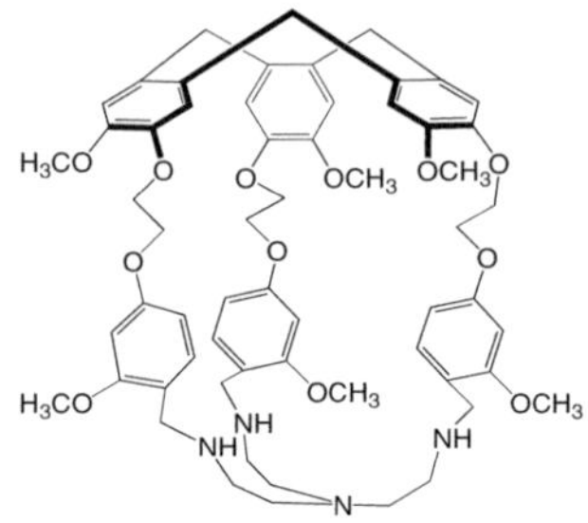

214.

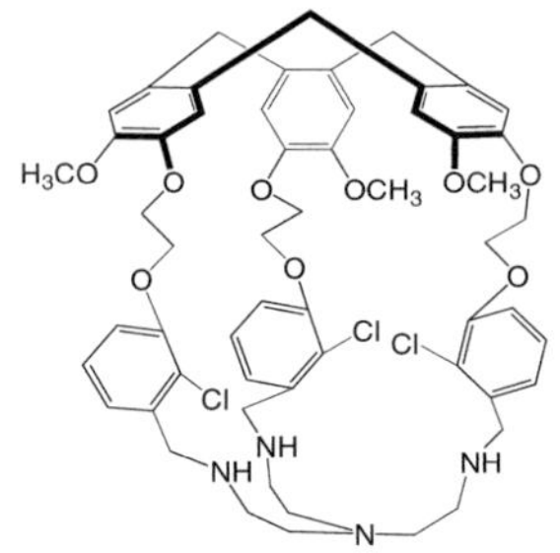

215.

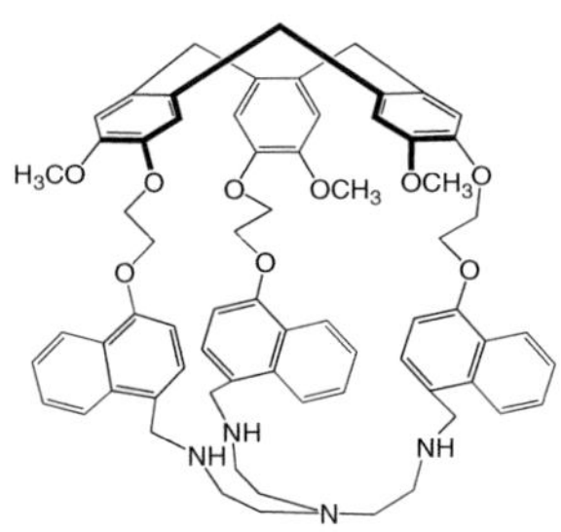

216.

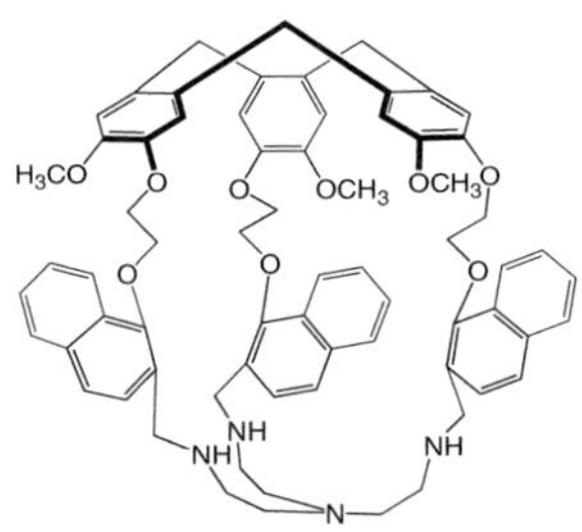

217.

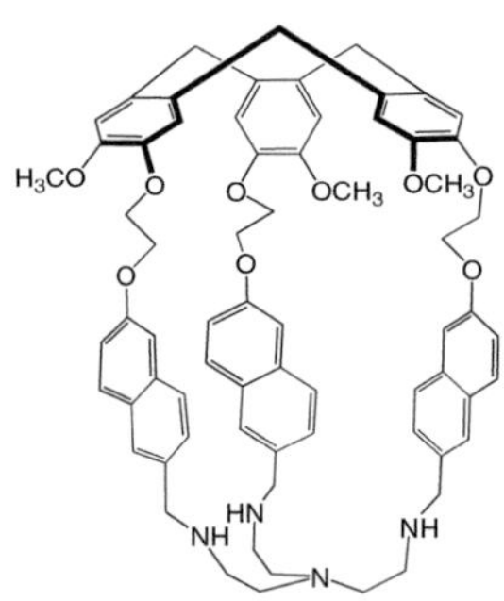

218.

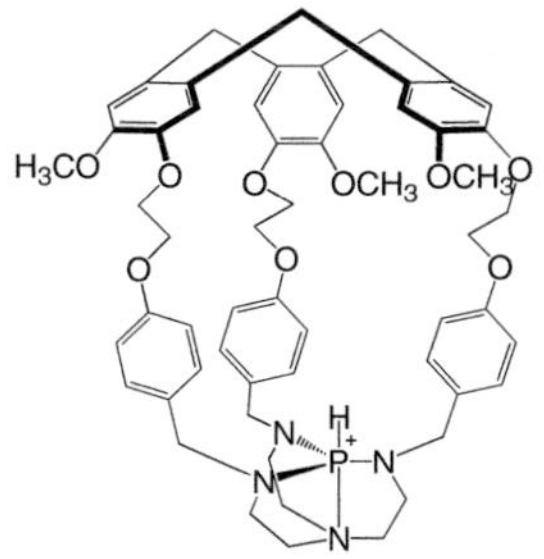

219.

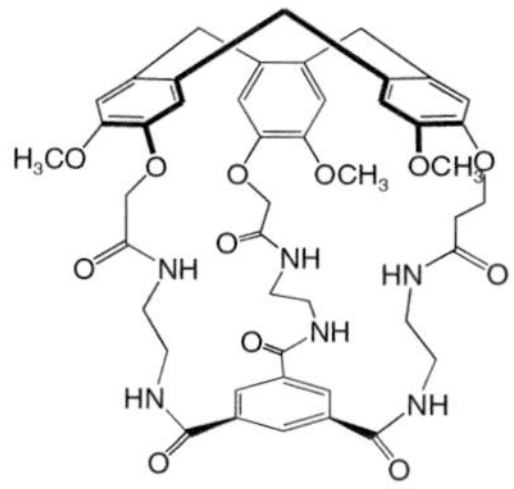

220.

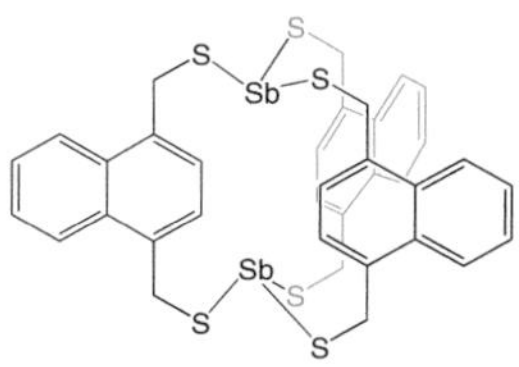

221.

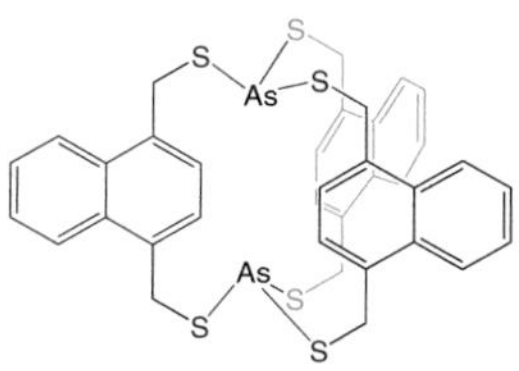

222.

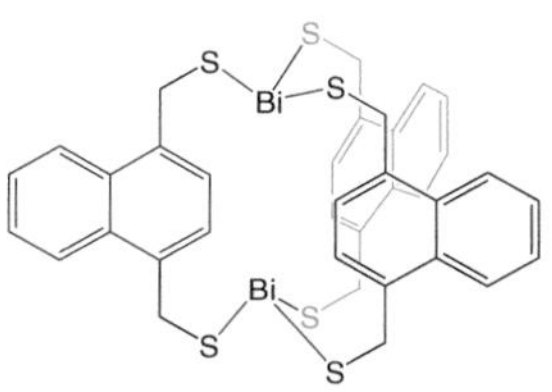

223.

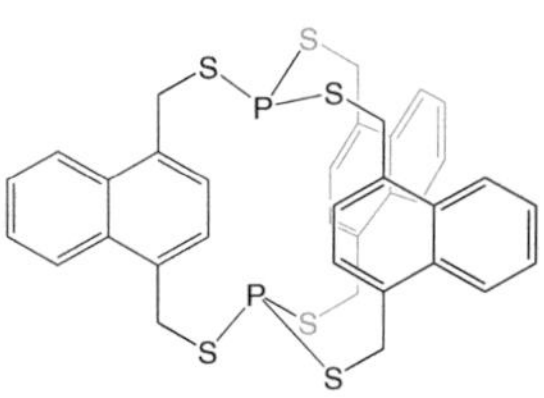

224.

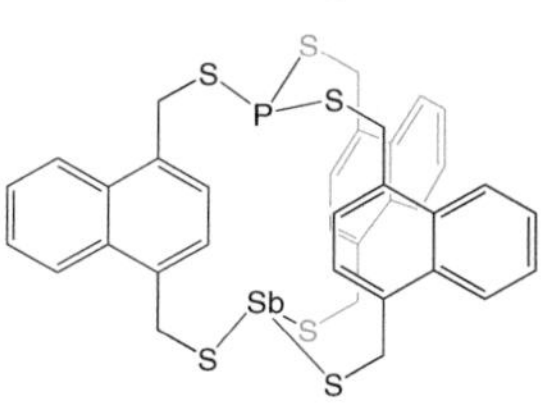

225.

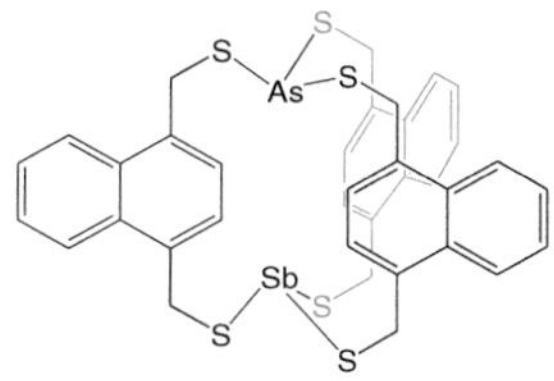

226.

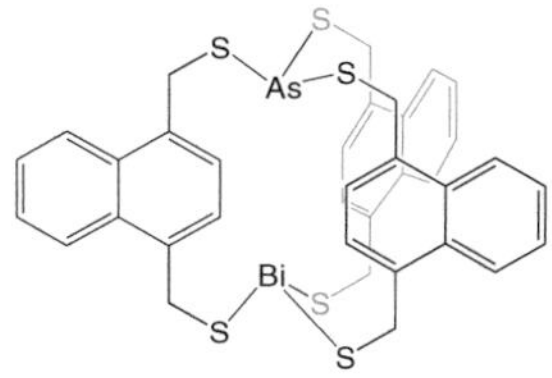

227.

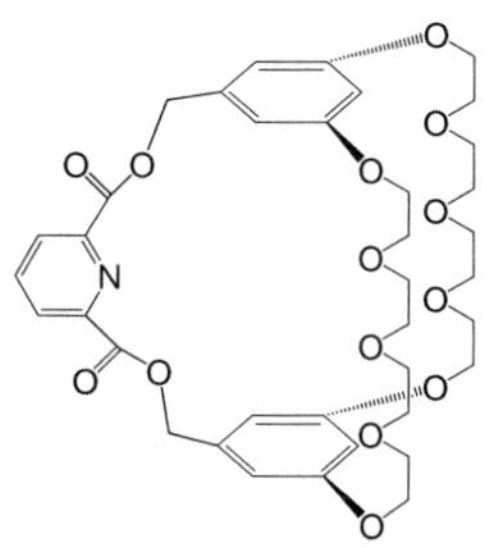

228.

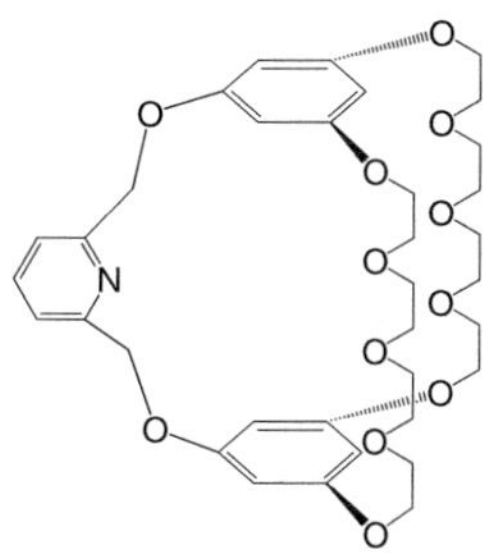

229.

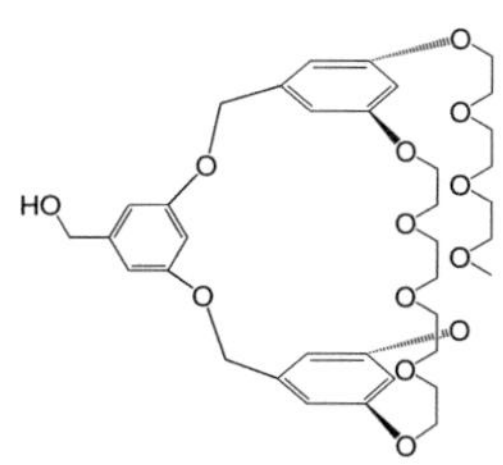

230.

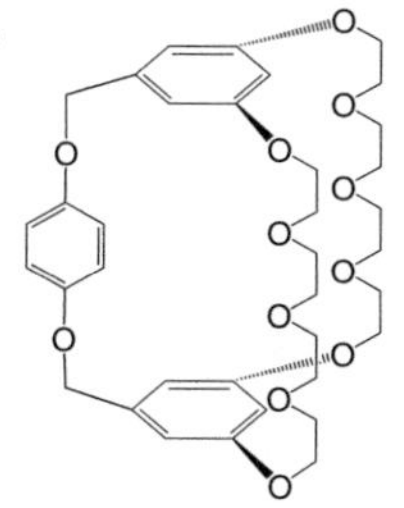

231.

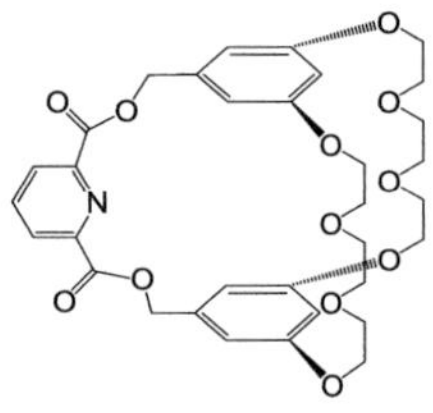

232.

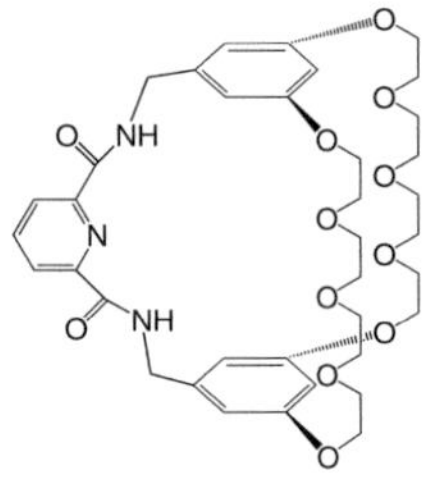

233.

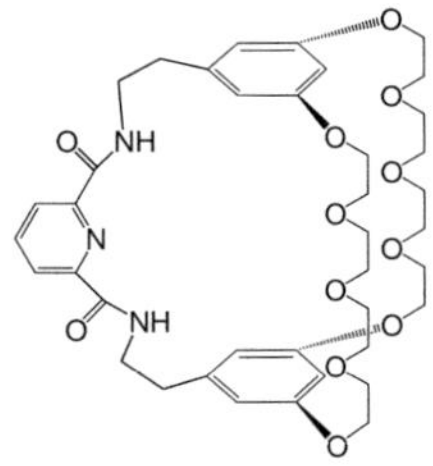

234.

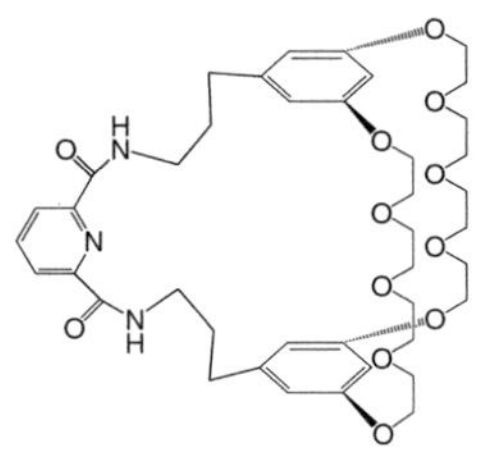

235.

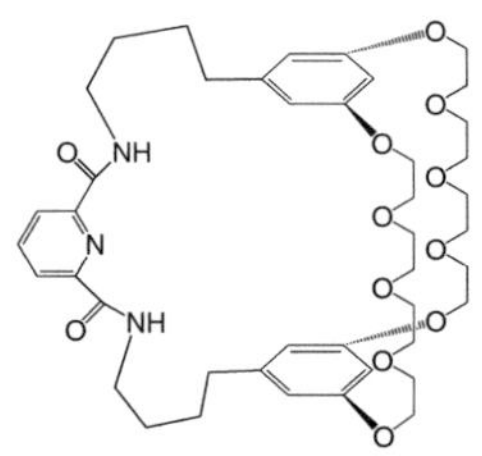

236.

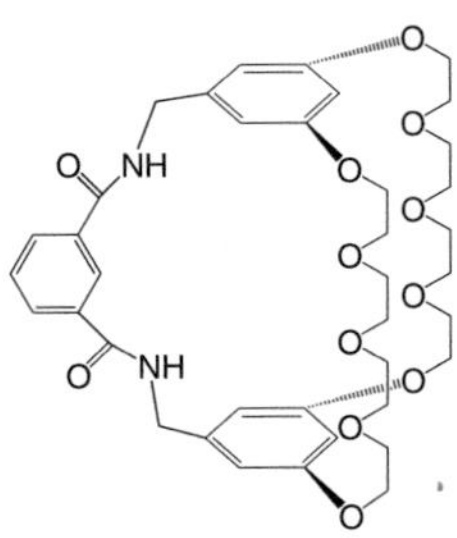

237.

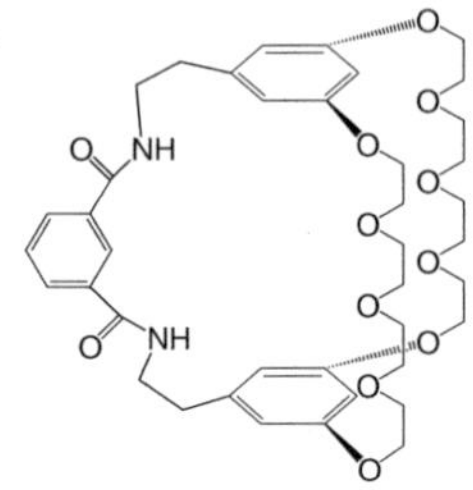

238.

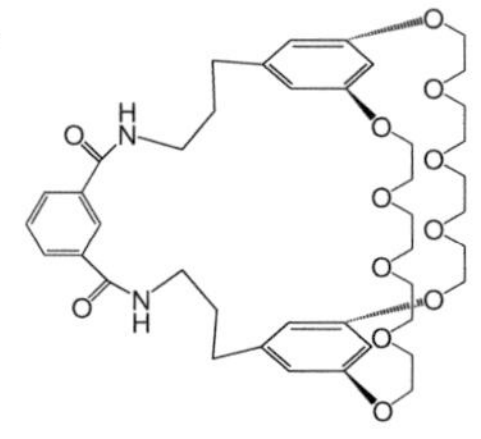

239.

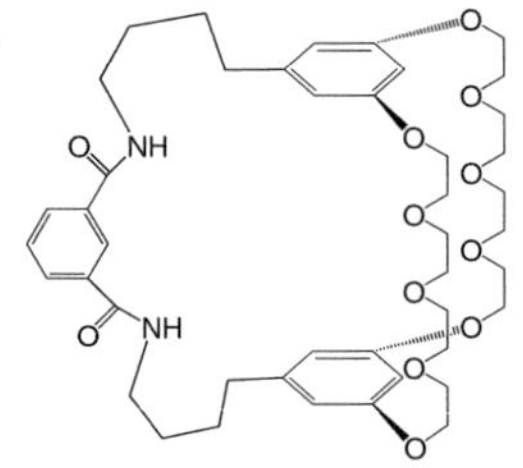

240.

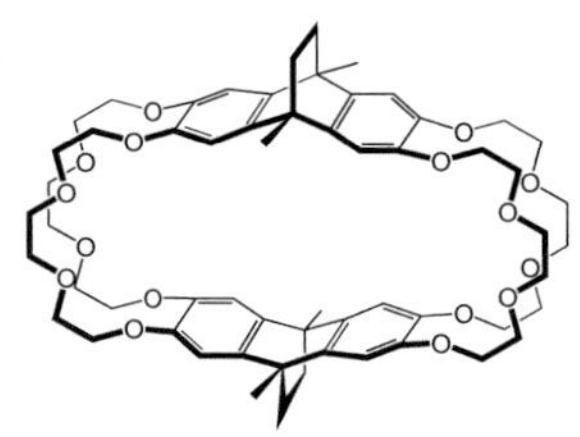

241.

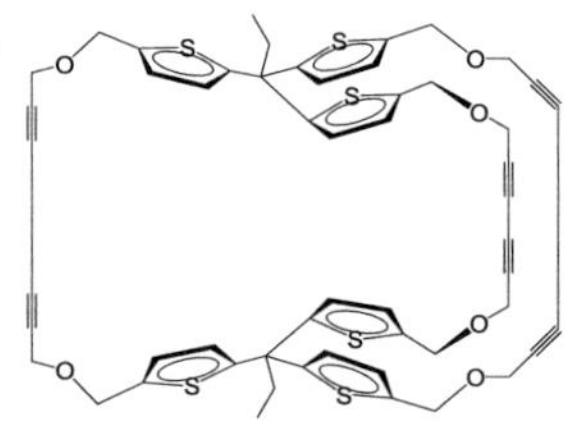

242.

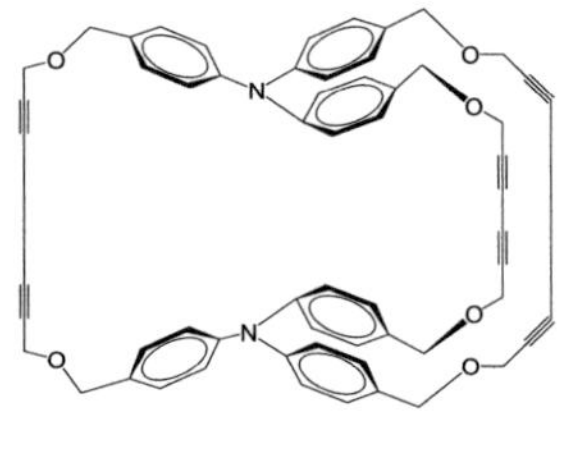

243.

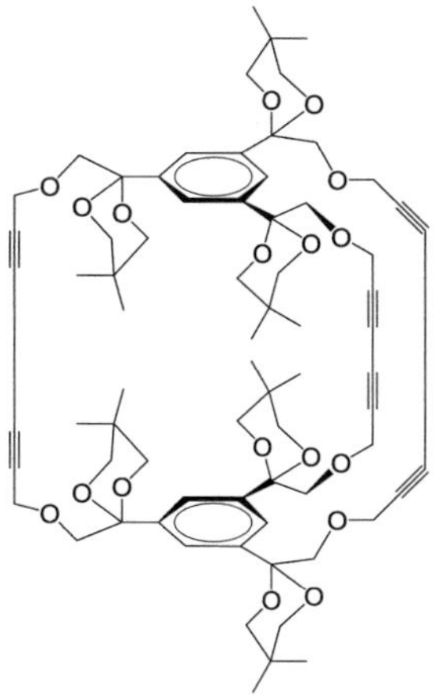

244.

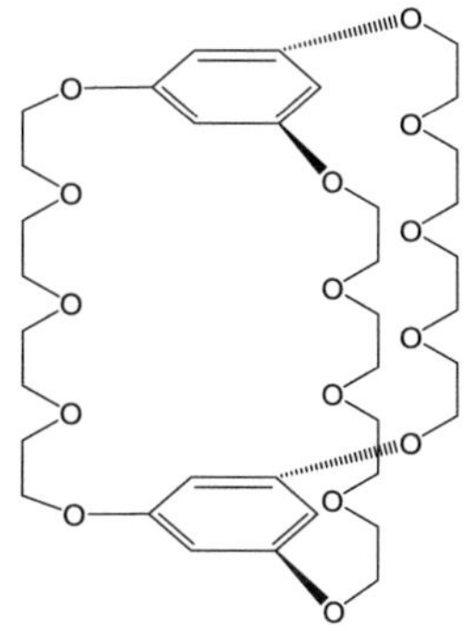

245.

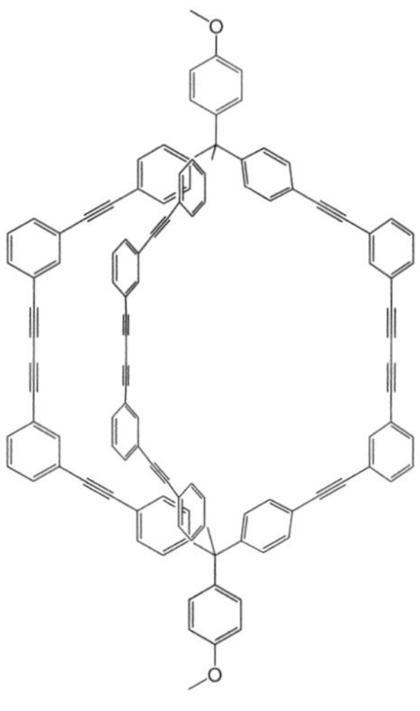

246.

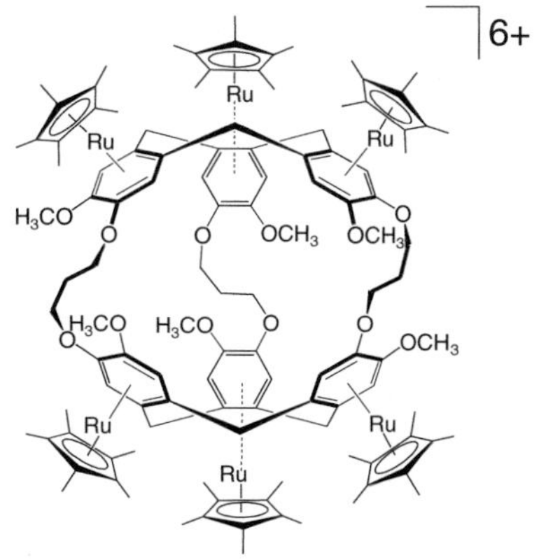

247.

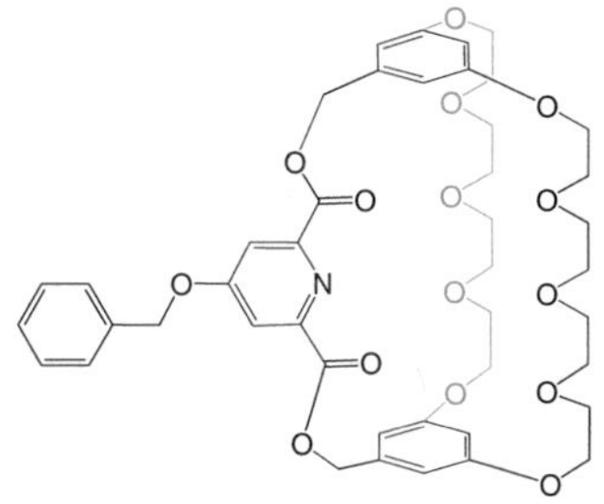

248.

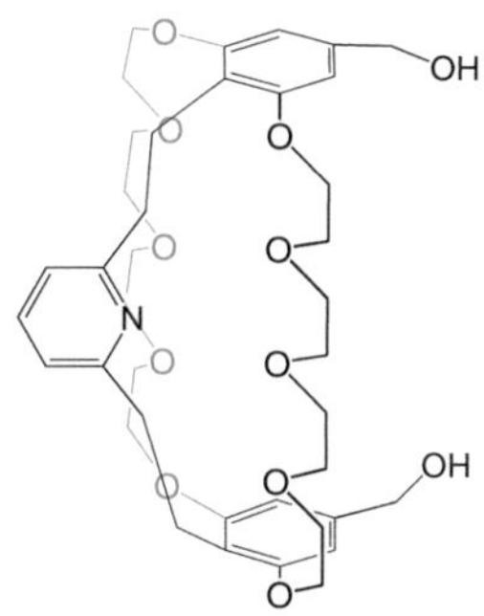

249.

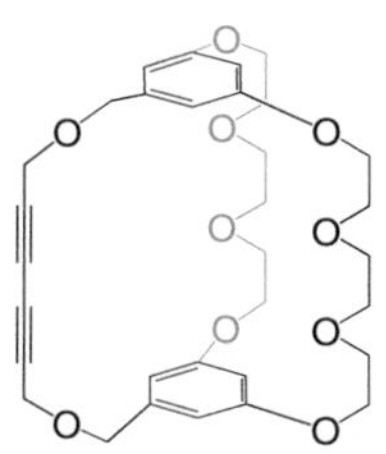

250.

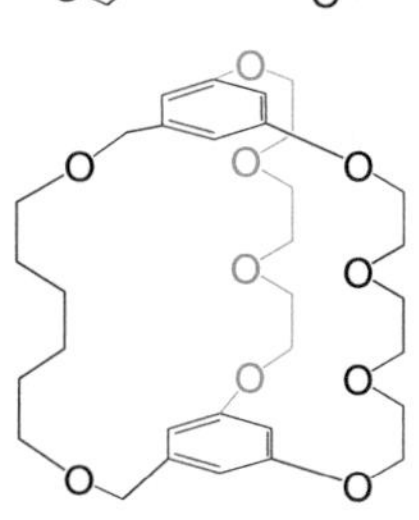

251.

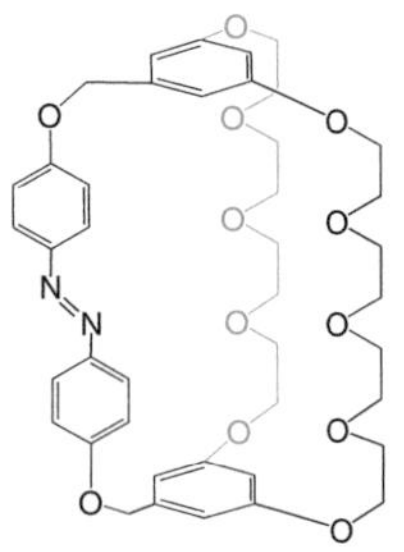

252.

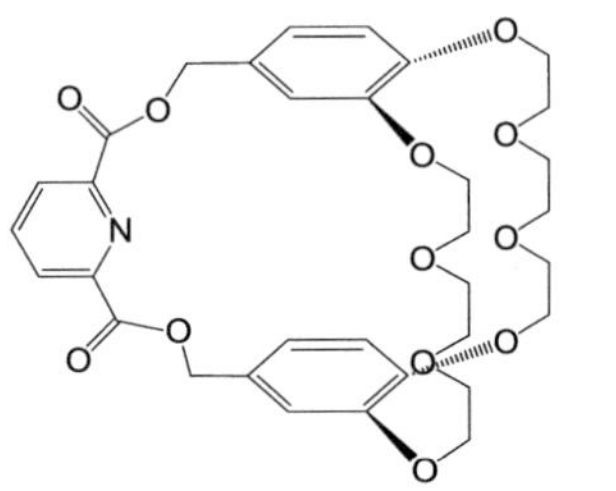

253.

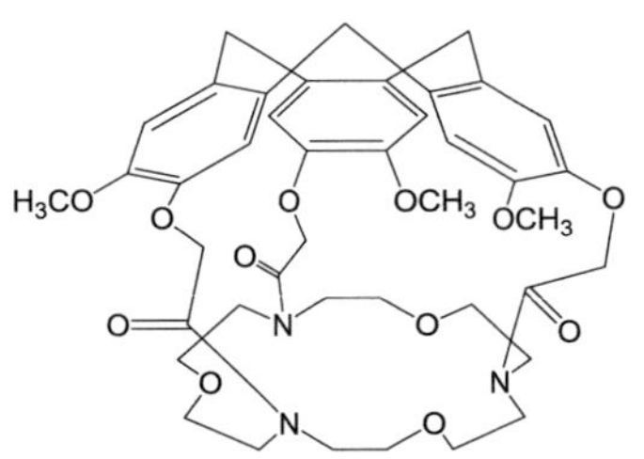

254.

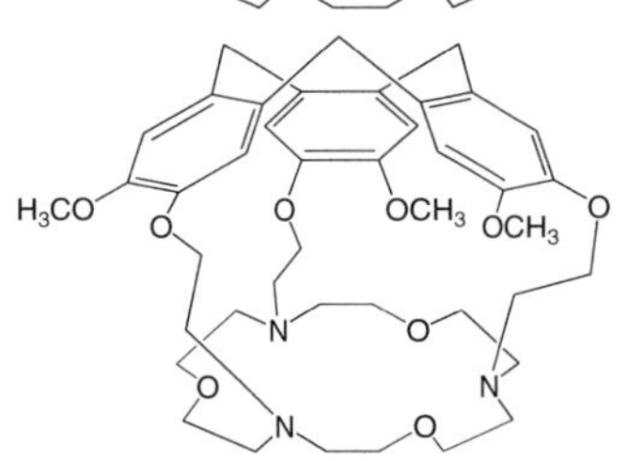

255.

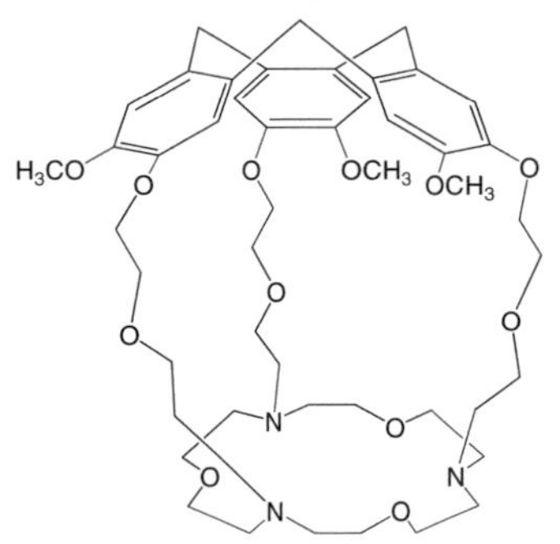

256.

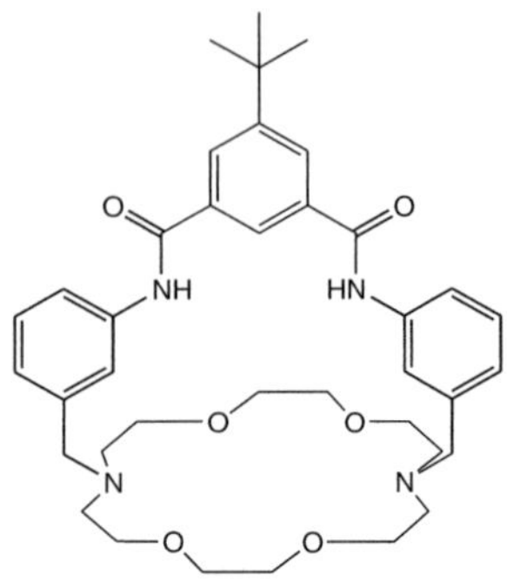

257.

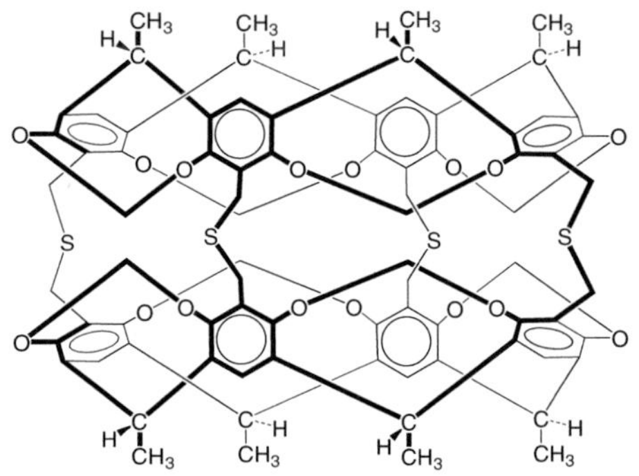

258.

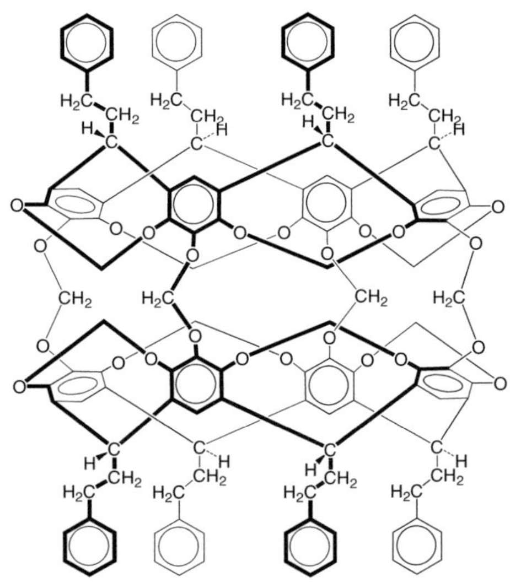

259.

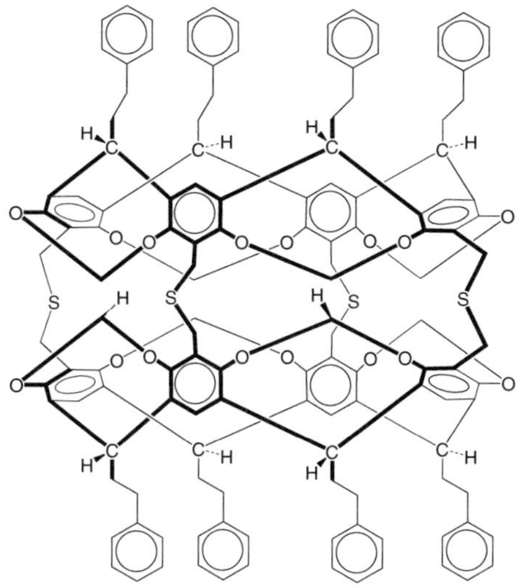

260.

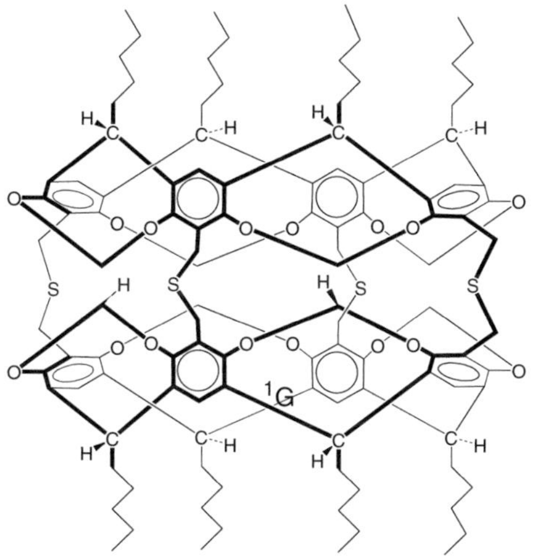

261.

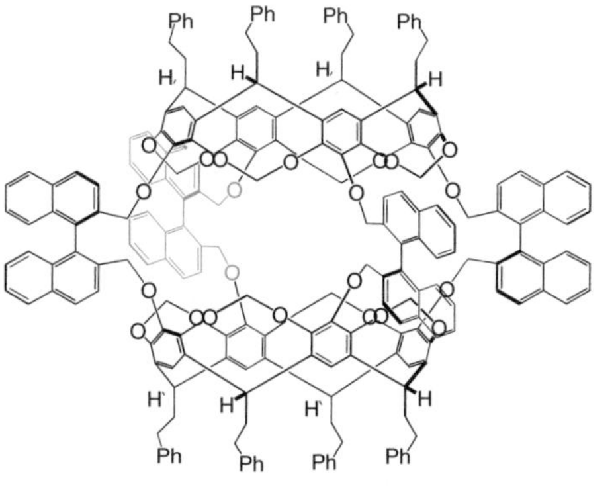

262.

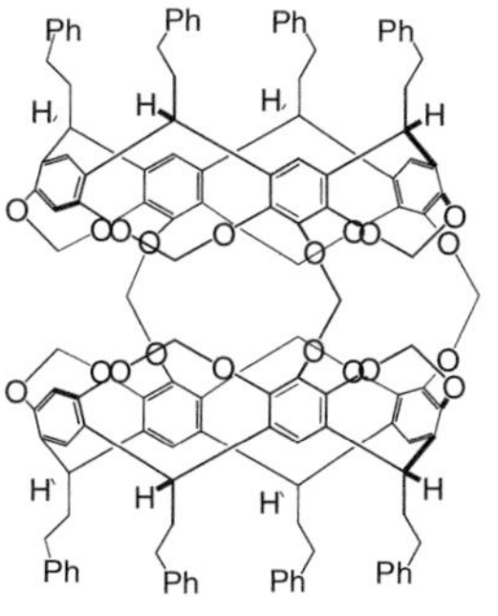

263.

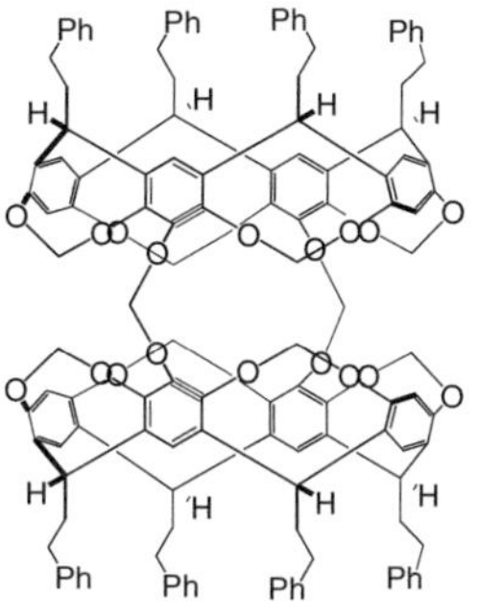

264.

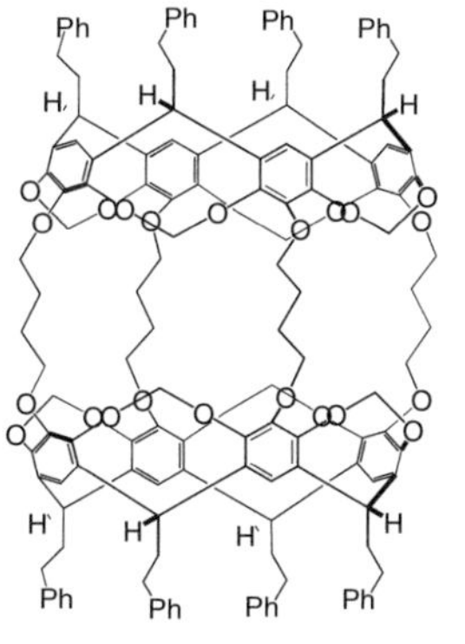

265.

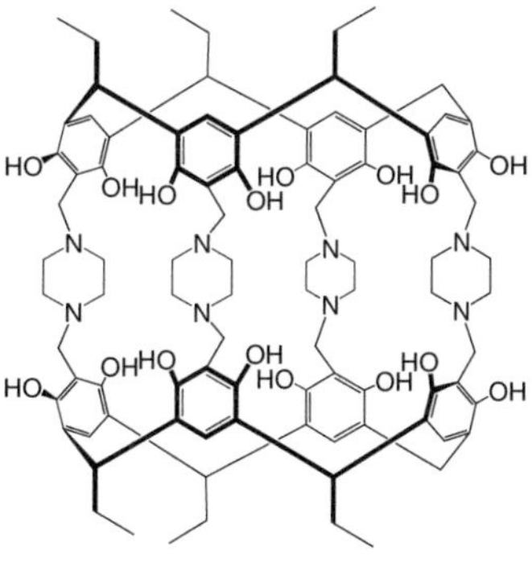

266.

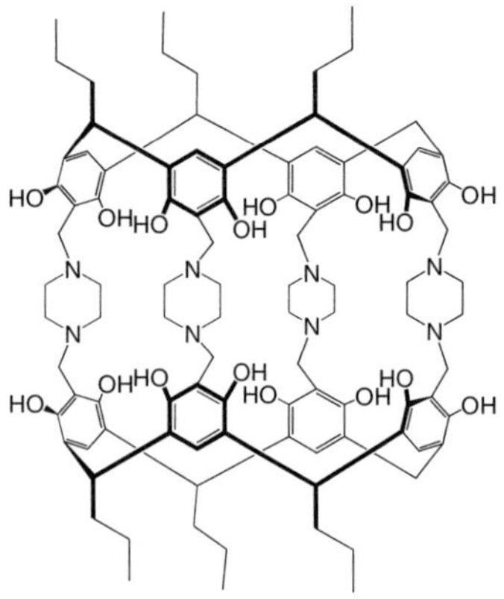

267.

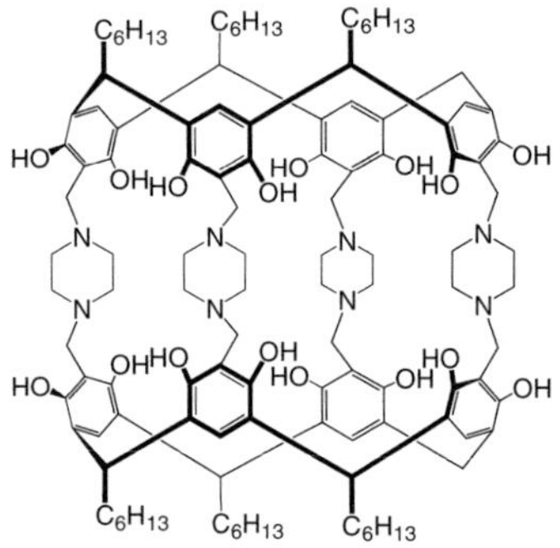

268.

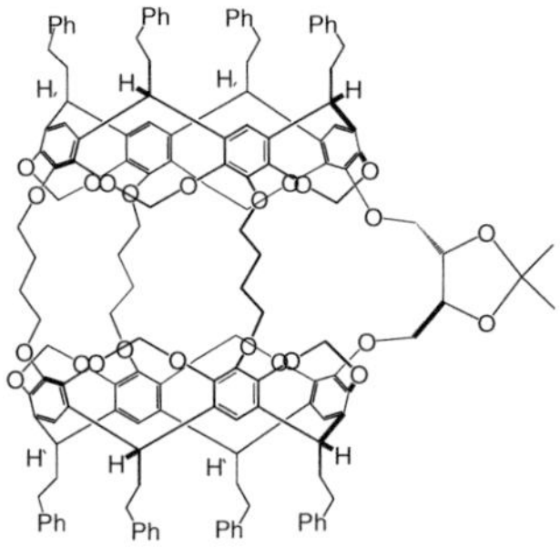

269.

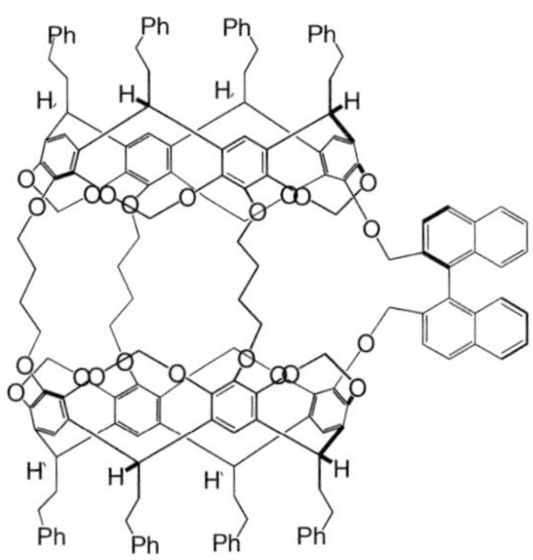

270.

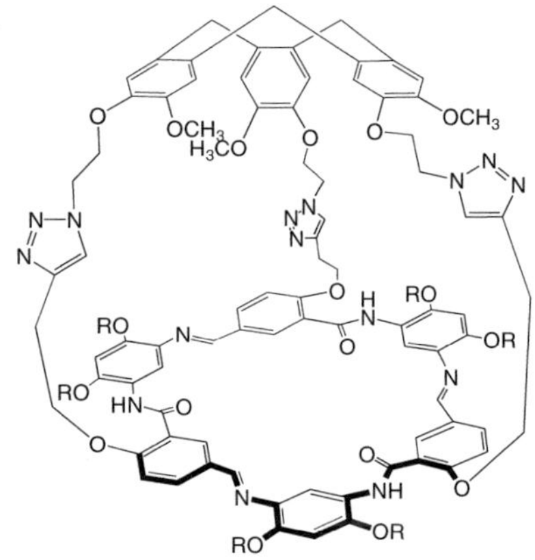

271.

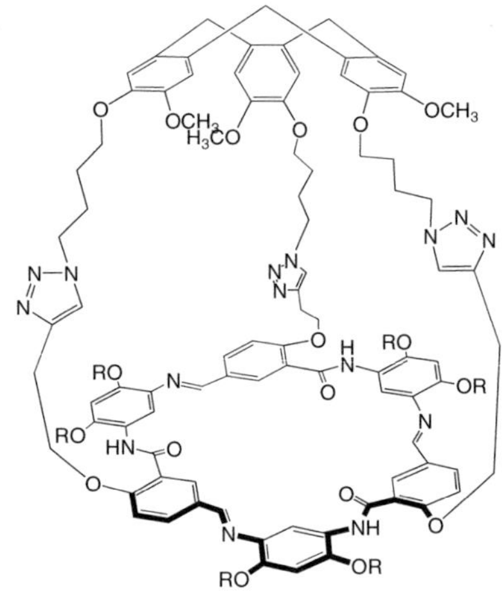

272.

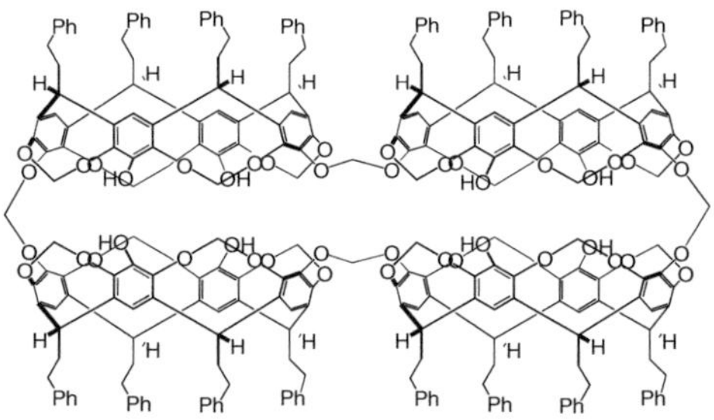

273.

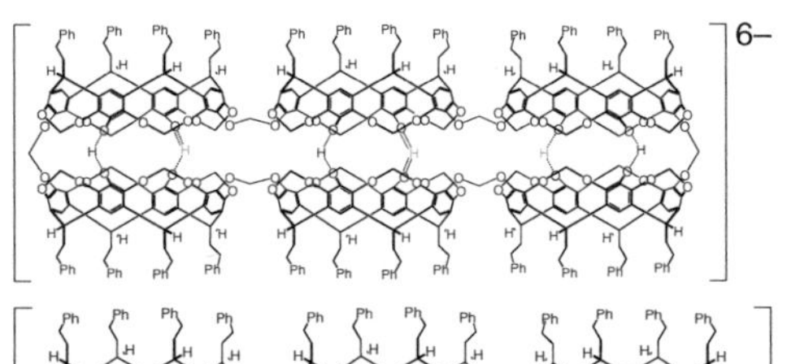

274.

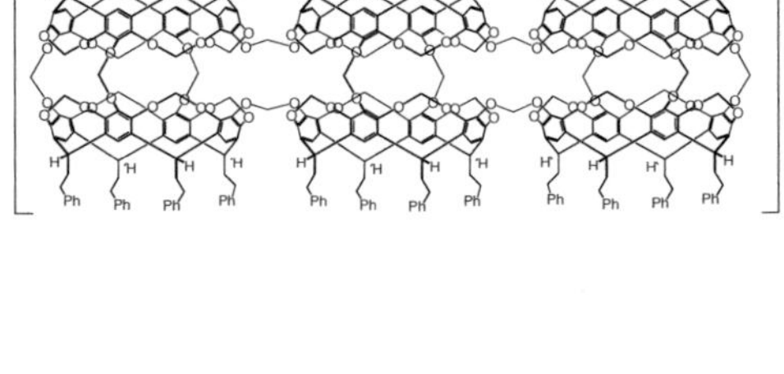

275.

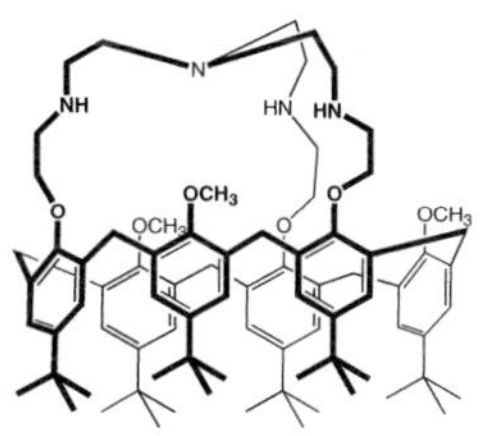

276.

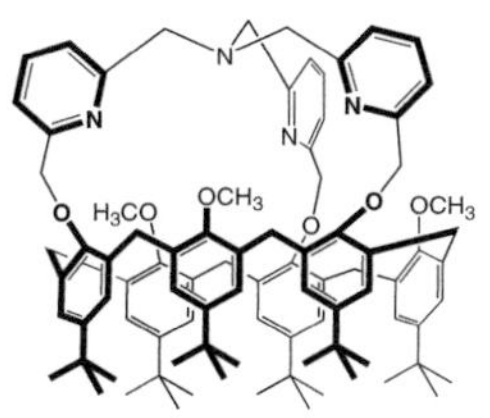

277.

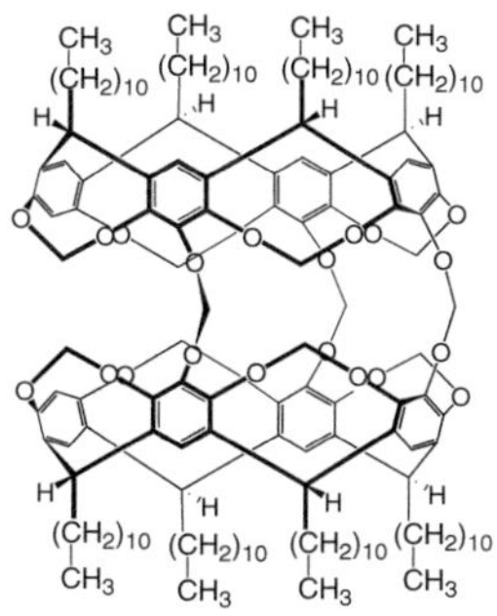

278.

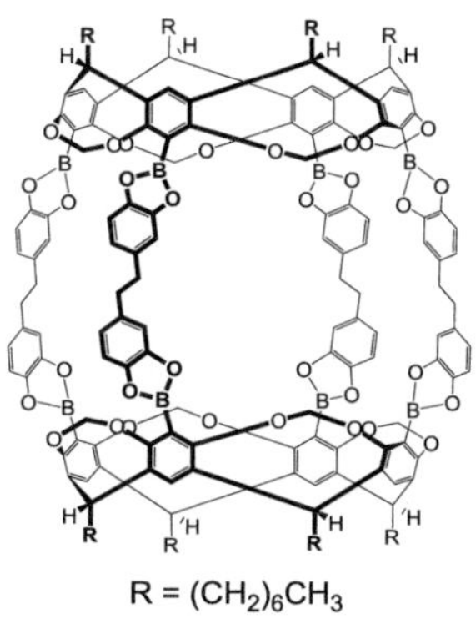

R = $(CH_2)_6CH_3$

279.

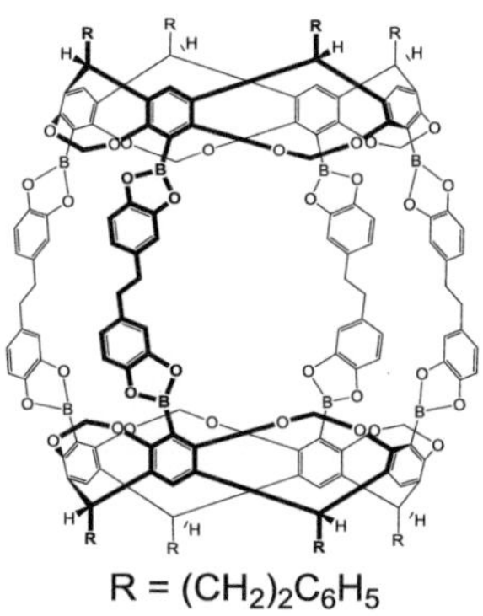

R = $(CH_2)_2C_6H_5$

280.

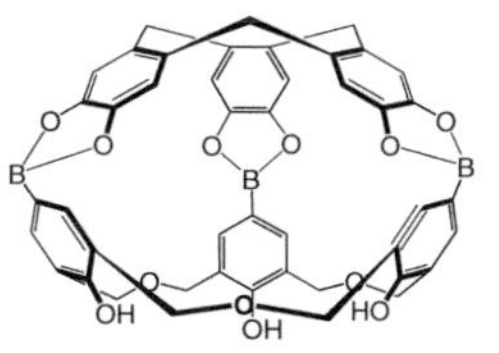

281.

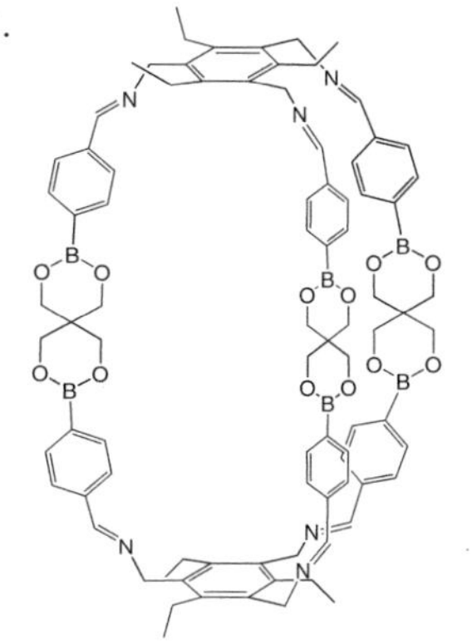

282.

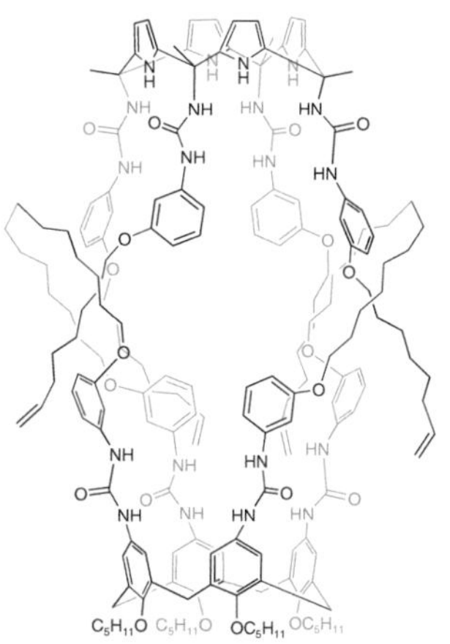

283.

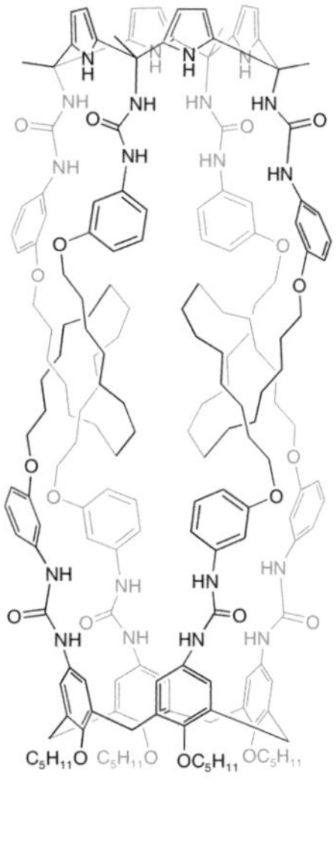

284.

285.

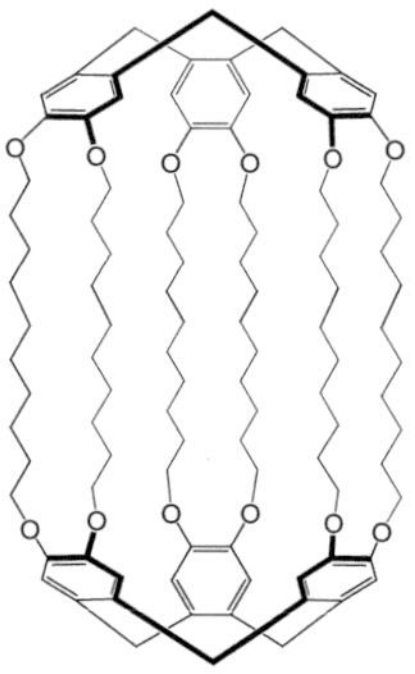

286.

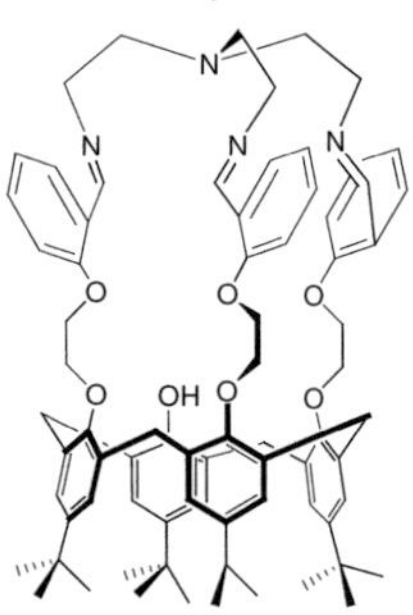

287.

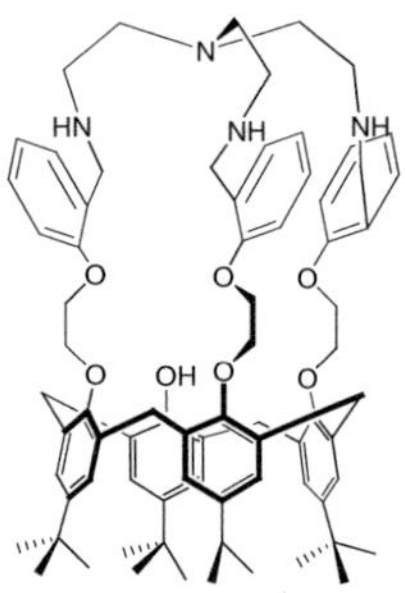

288.

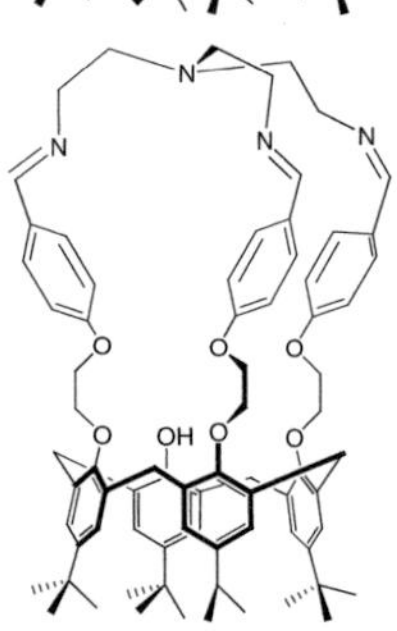

289.

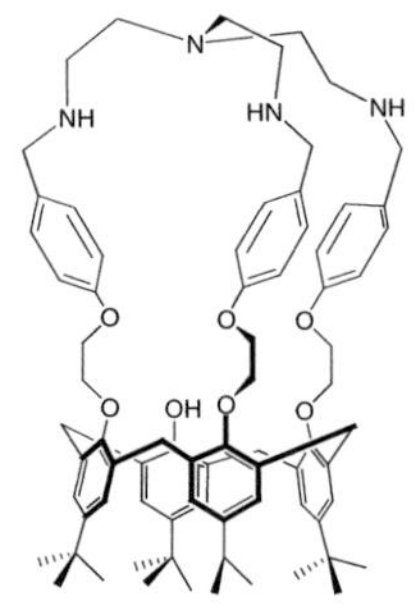

290.

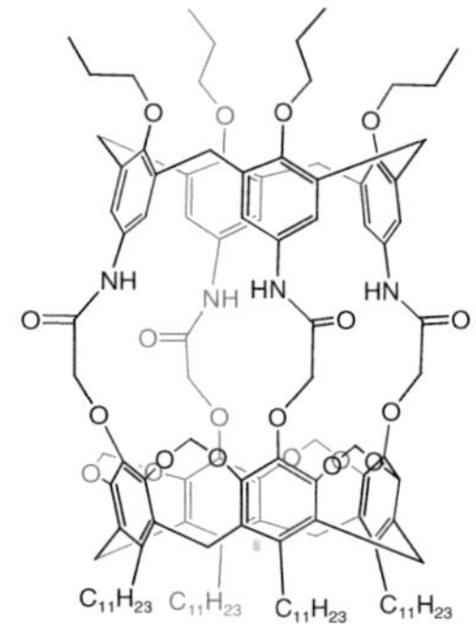

291.

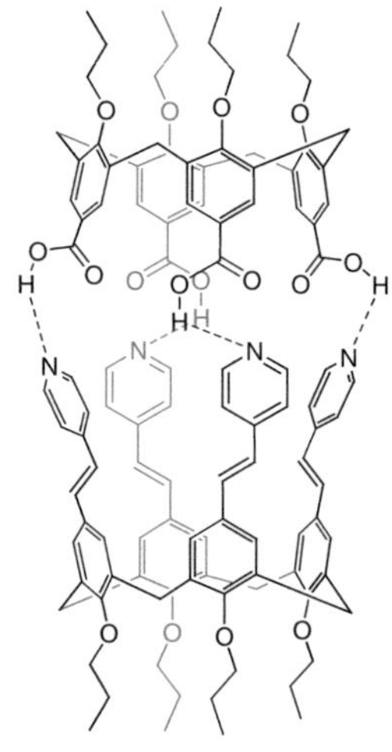

292.

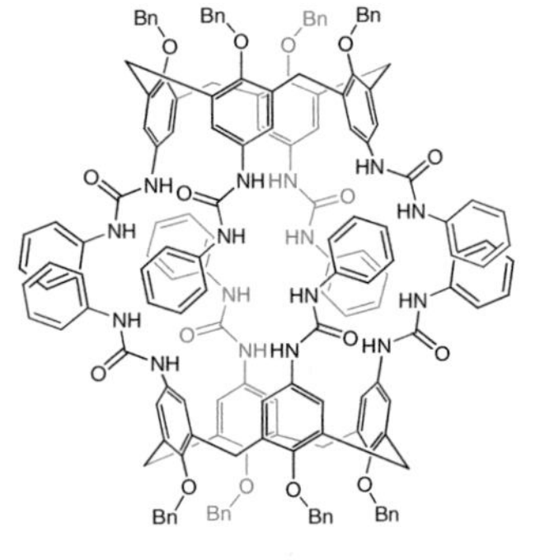

293.

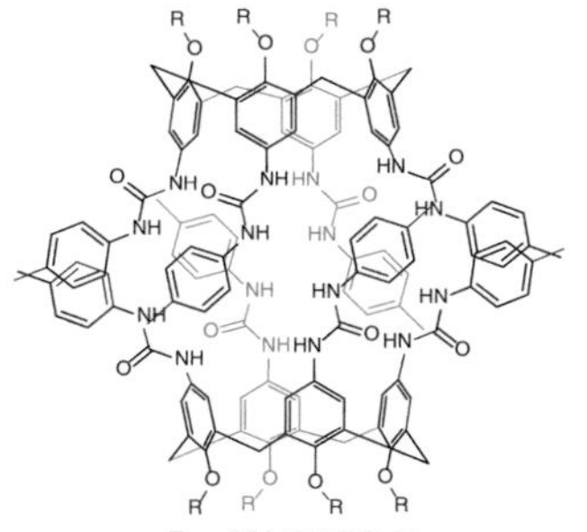

R = $CH_2COOC_2H_5$

294.

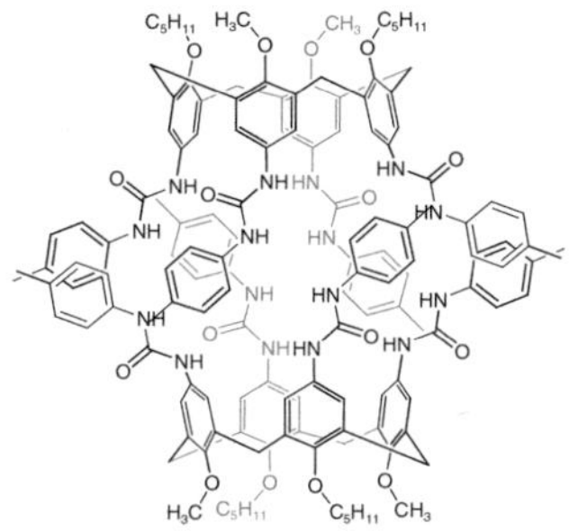

295.

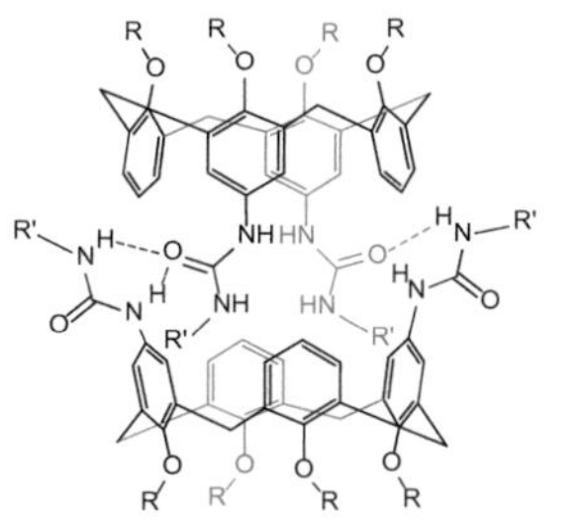

R = $CH_2COOC_2H_5$

R' = 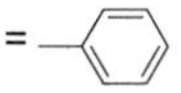

296.

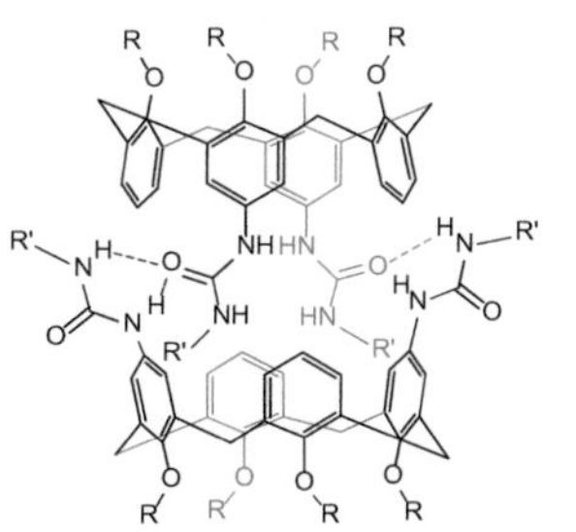

R = $CH_2COOC_2H_5$

R' = *tert*-C_4H_9

297.

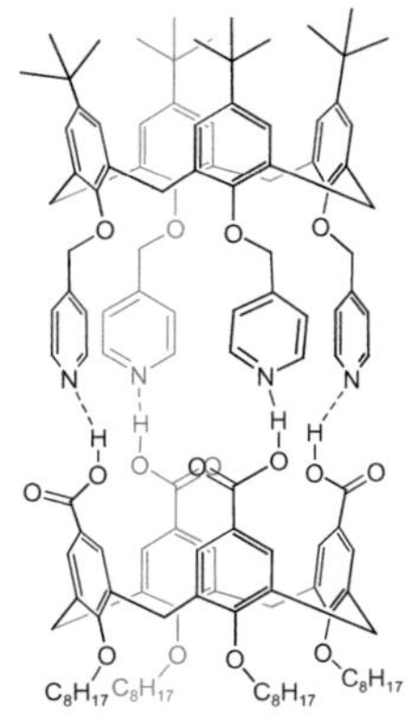

298.

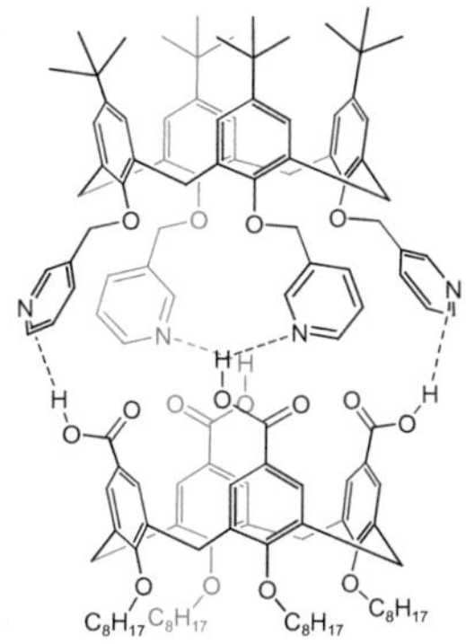

299.

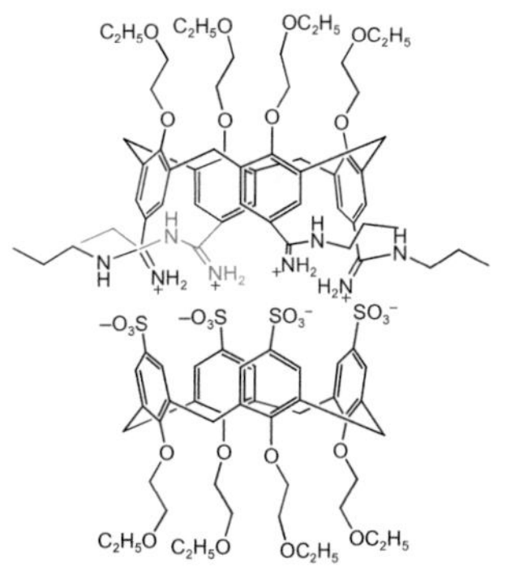

300.

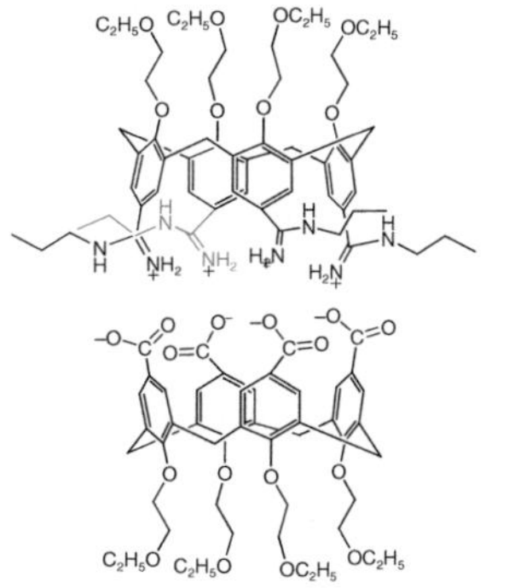

301.

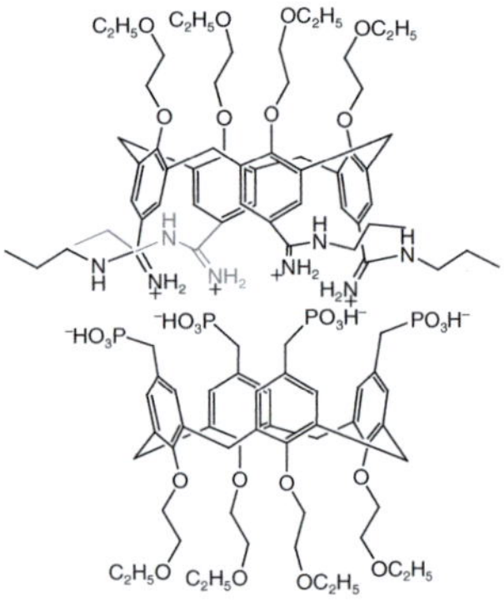

302.

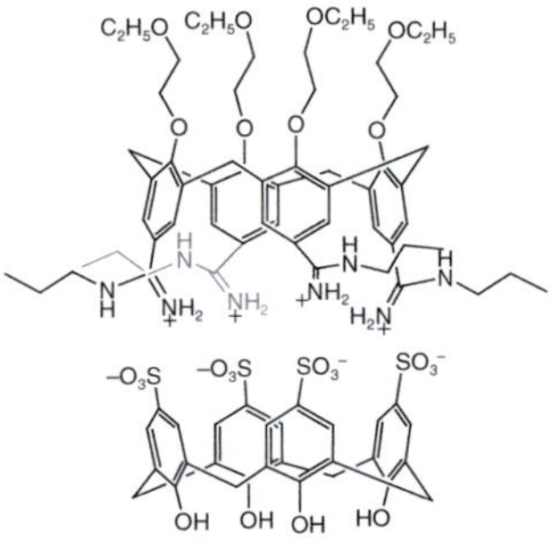

303.

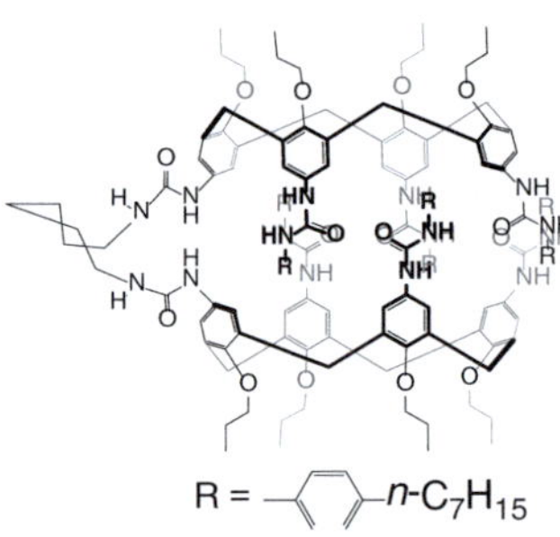

304.

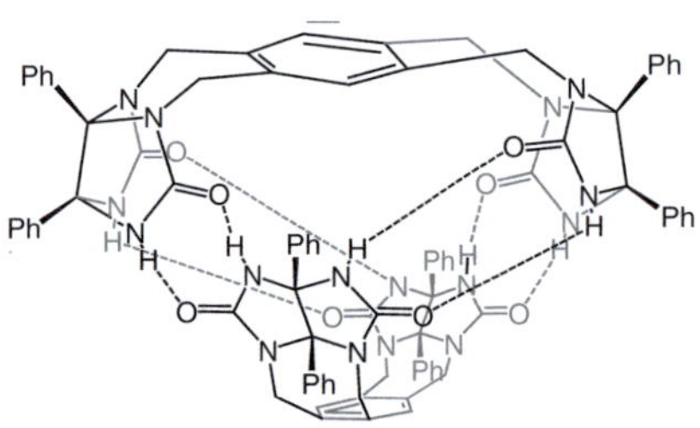

305.

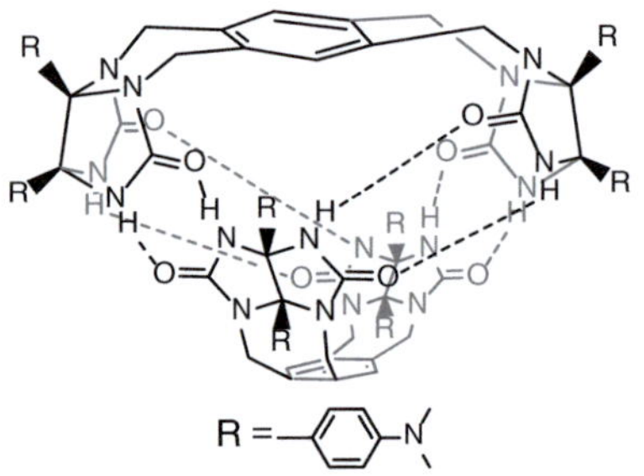

306.

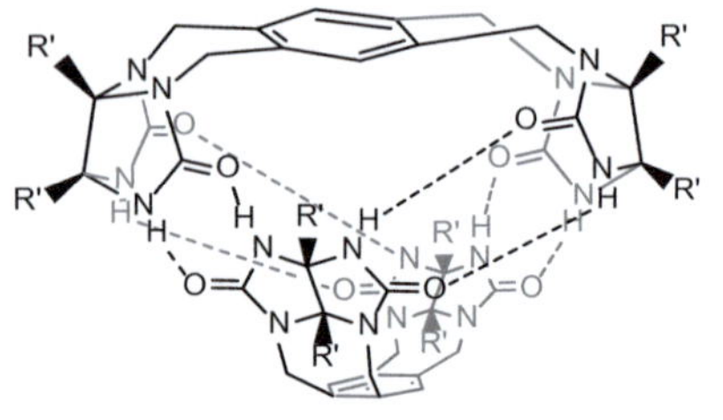

R' = $-COOC_2H_5$

307.

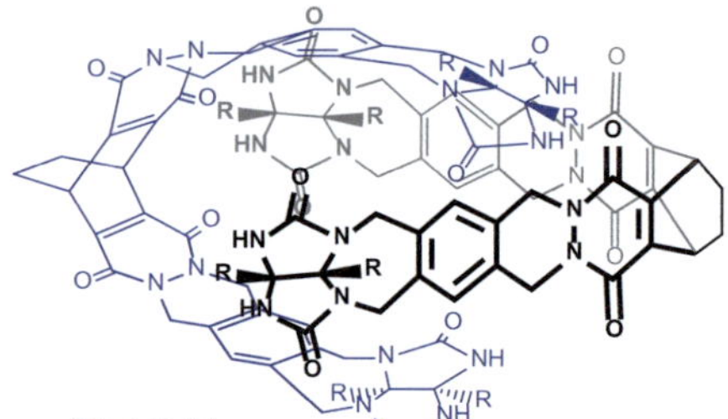

R = C_6H_5

308.

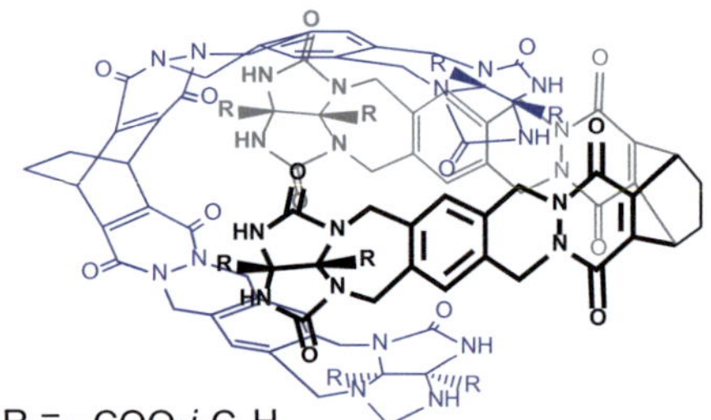

R = $-COO\text{-}i\text{-}C_5H_{11}$

309.

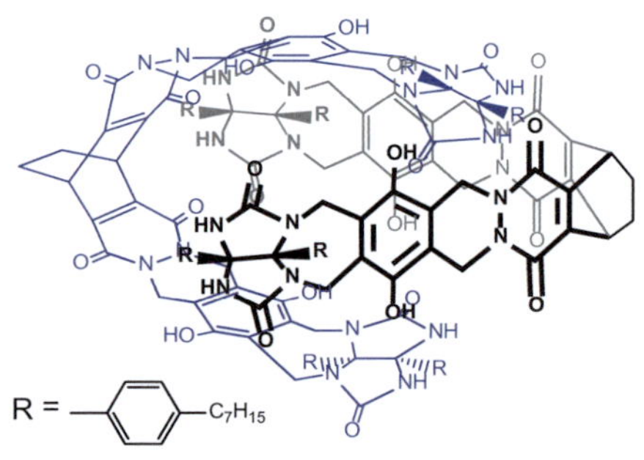

310.

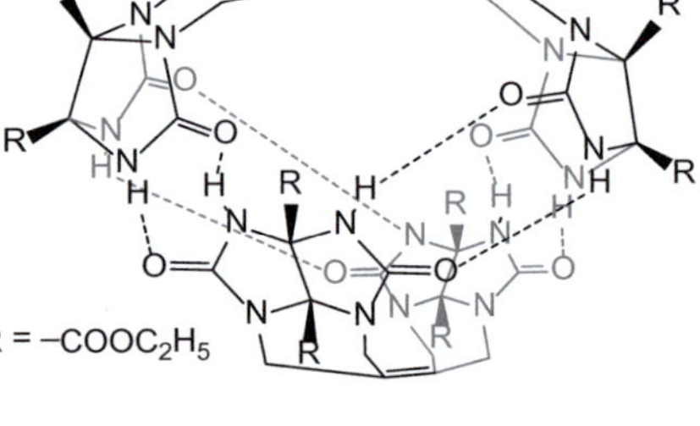

R = $-COOC_2H_5$

311.

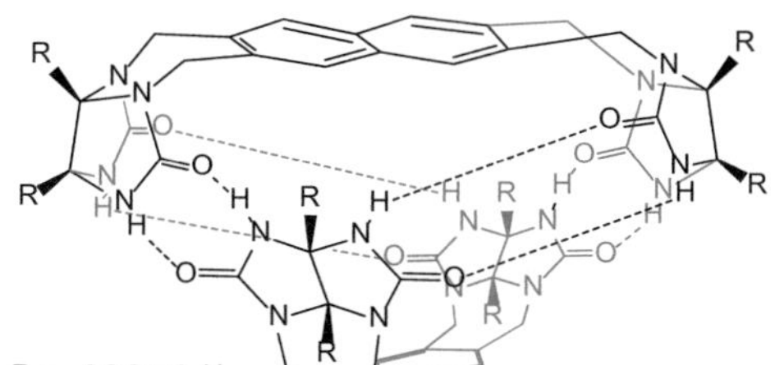

R = –COO*n*-C_4H_9

312.

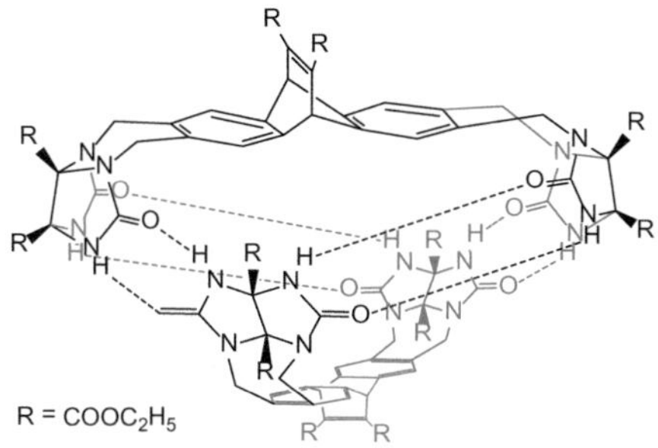

R = $COOC_2H_5$

313.

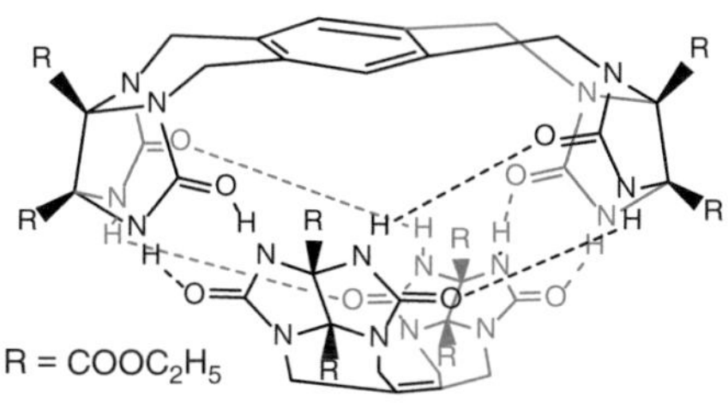

R = $COOC_2H_5$

314.

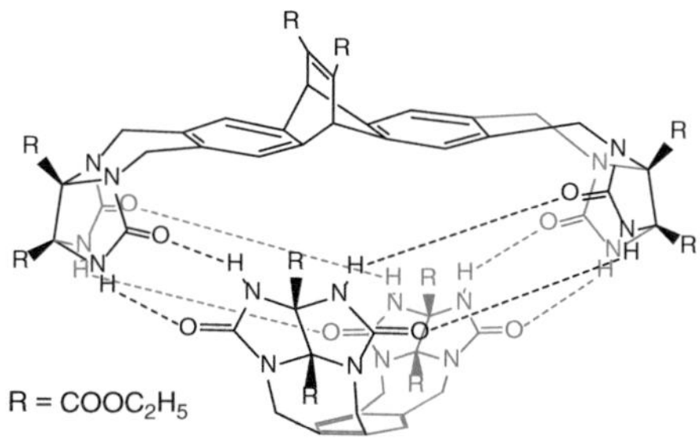

R = $COOC_2H_5$

315.

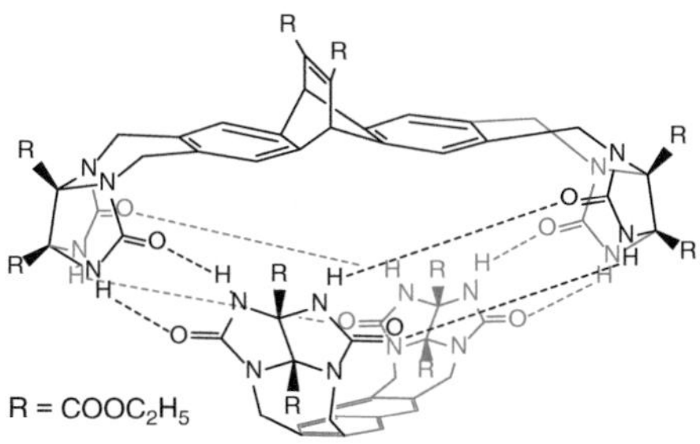

R = $COOC_2H_5$

316.

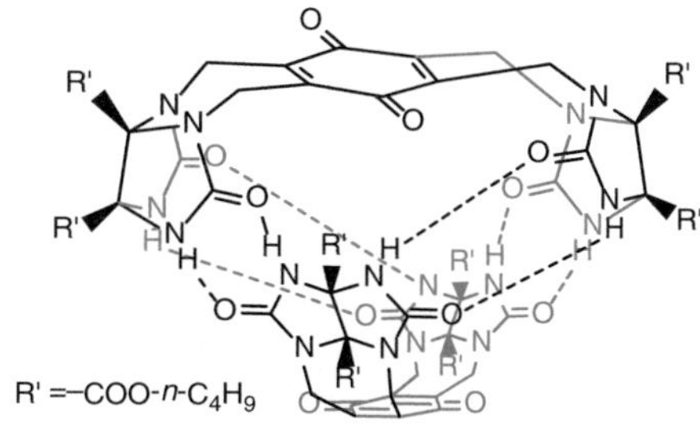

R' =–COO-*n*-C_4H_9

317.

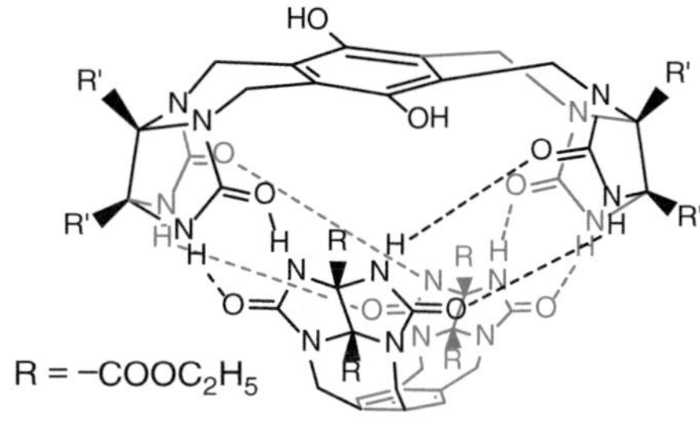

R = –$COOC_2H_5$

318.

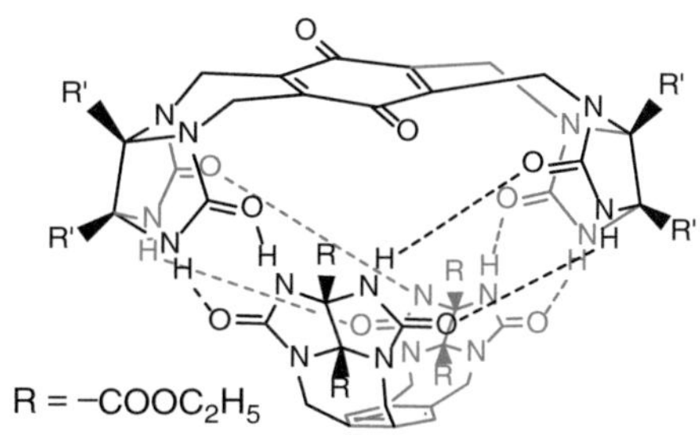

R = –$COOC_2H_5$

319.

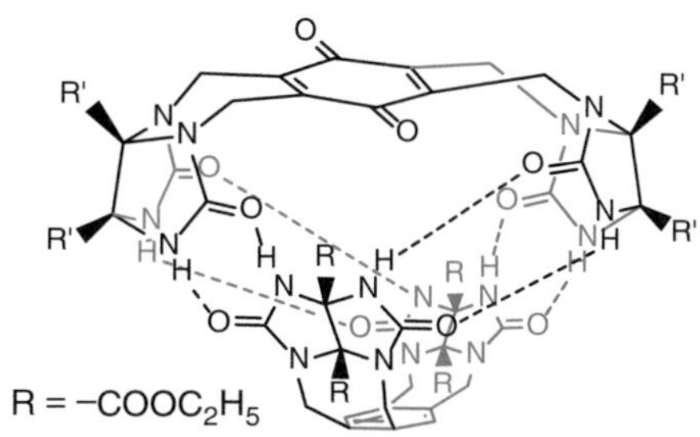

R = –$COOC_2H_5$

320.

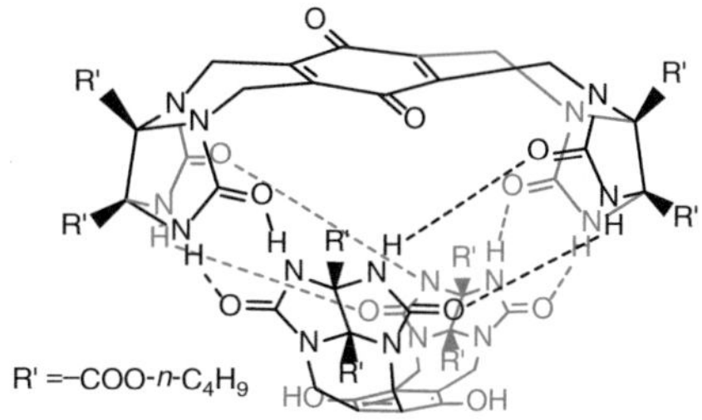

R' =–COO-*n*-C_4H_9

321.

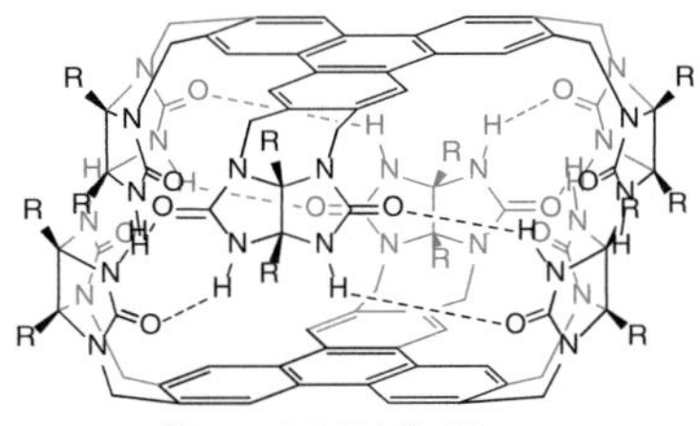

R = −COO*i*-C_5H_{11}

322.

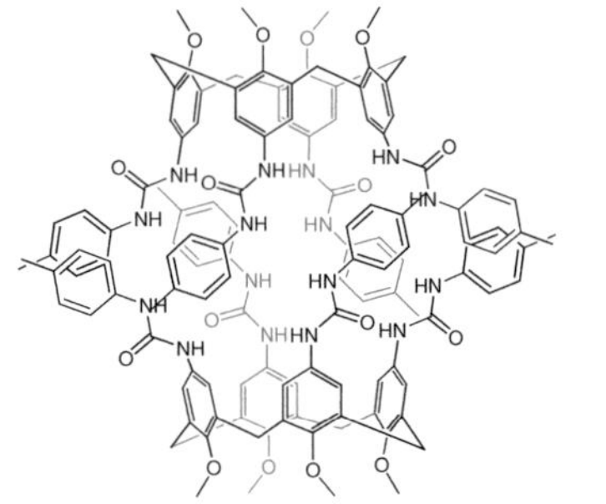

323.

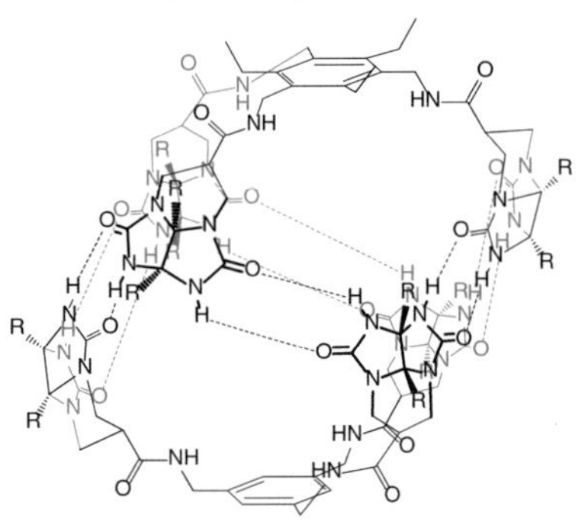

324.

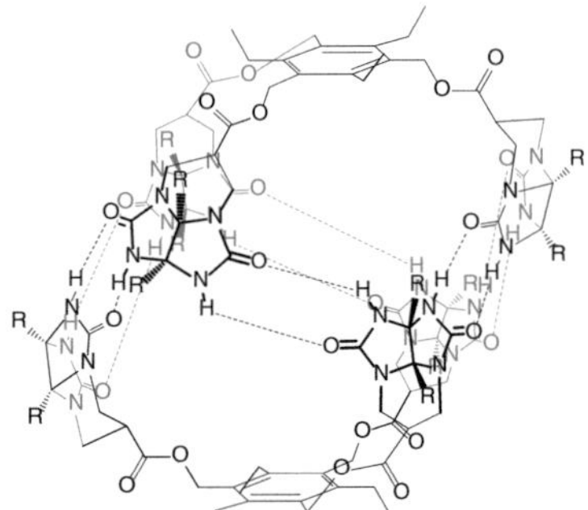

325.

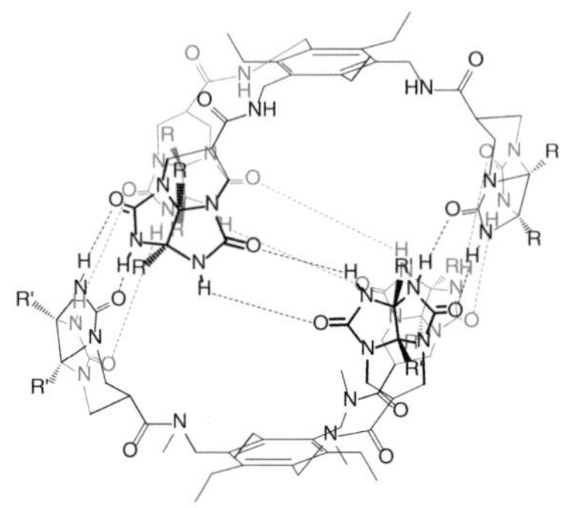

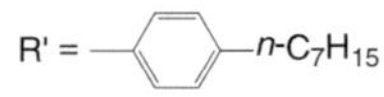

326.

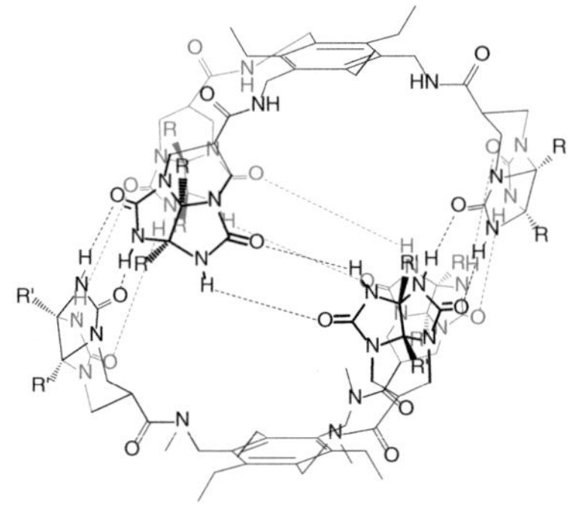

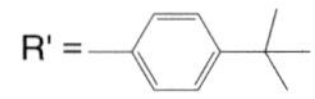

327.

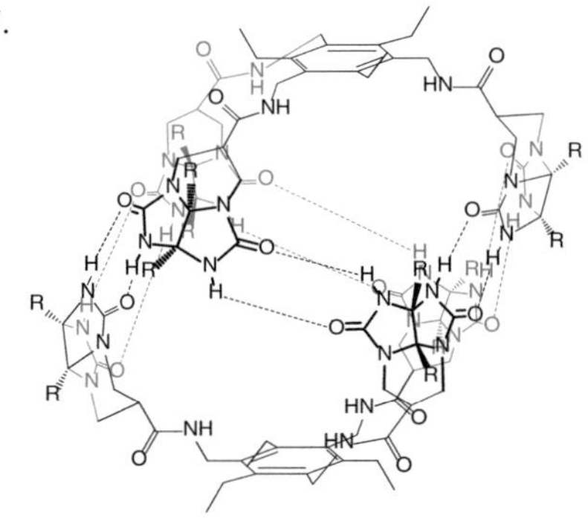

328.

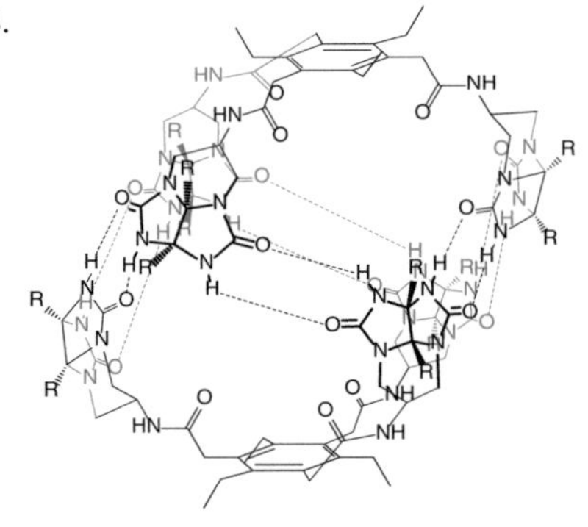

329.

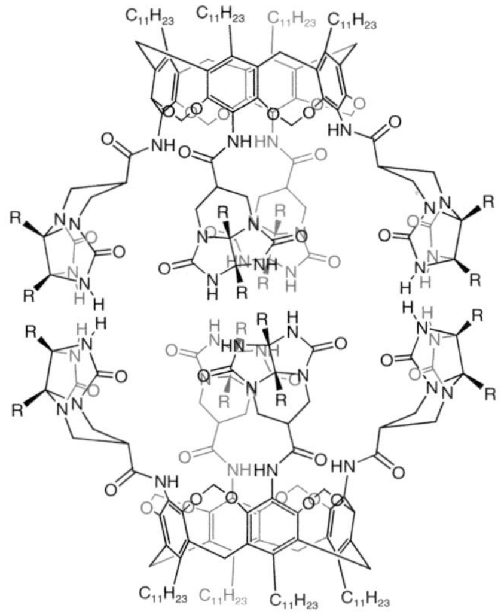

330.

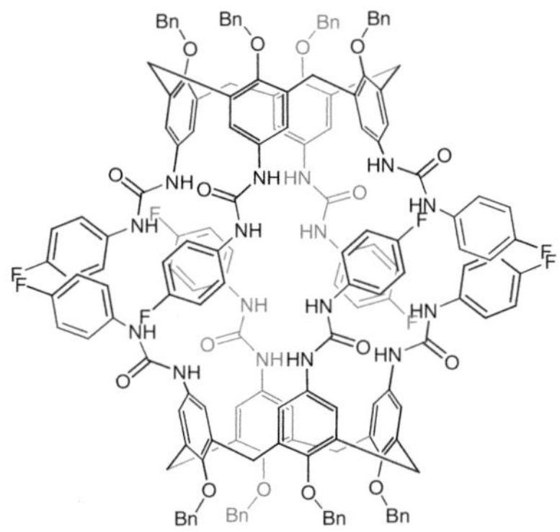

331.

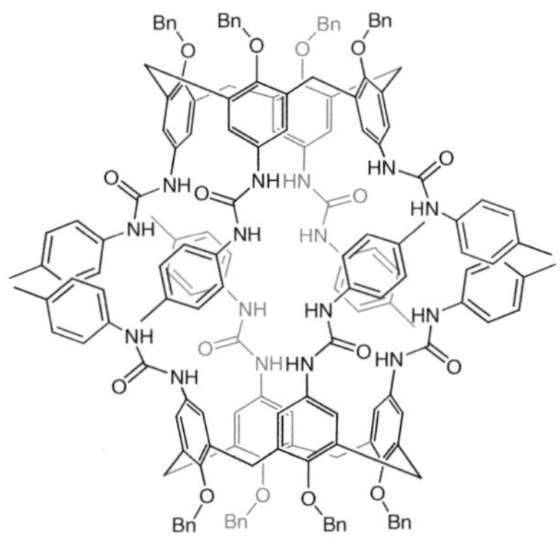

332.

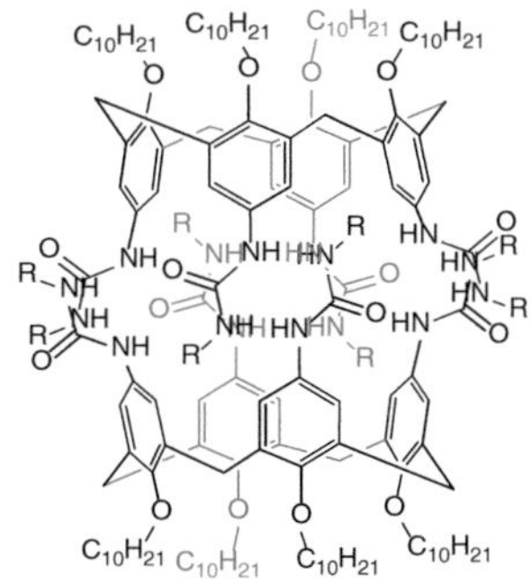

R = C_6H_4-*para*-C_7H_{15}

333.

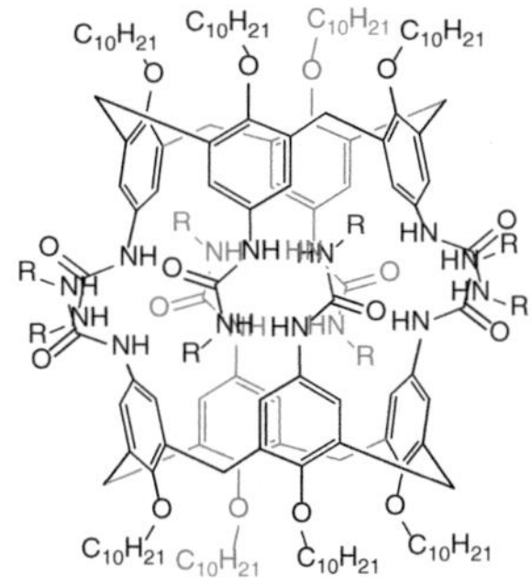

R = $SO_2C_6H_4$-*para*-CH_3

334.

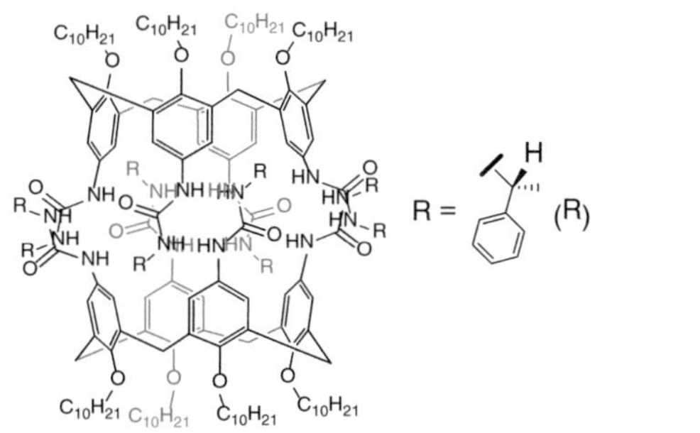

335.

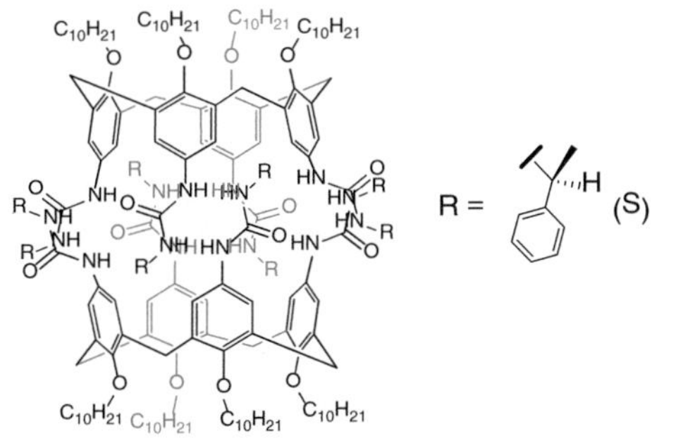

336.

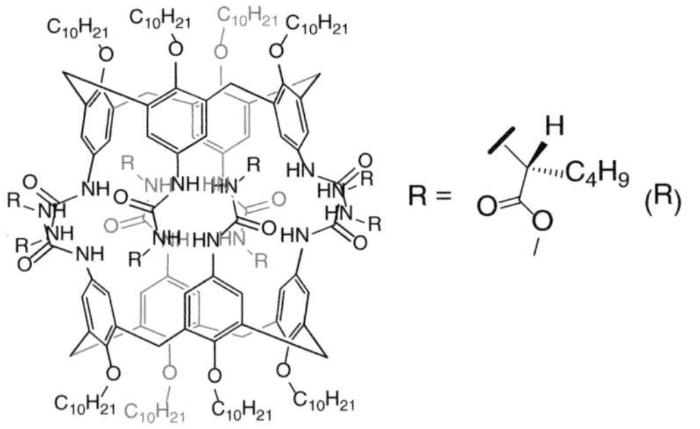

337.

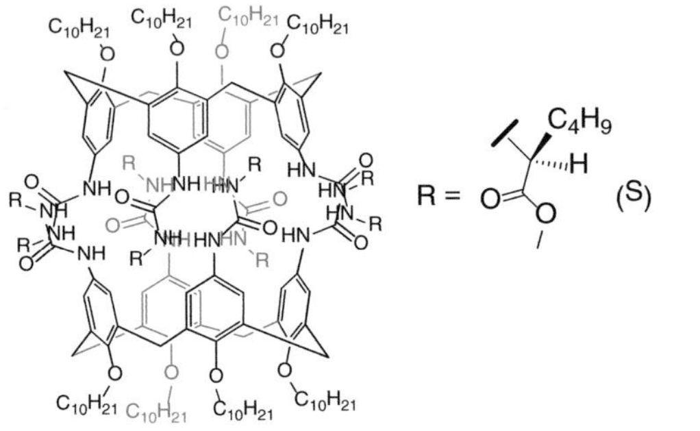

338.

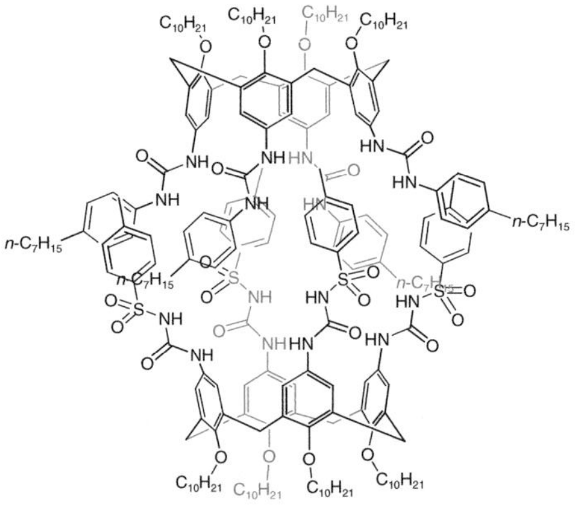

339.

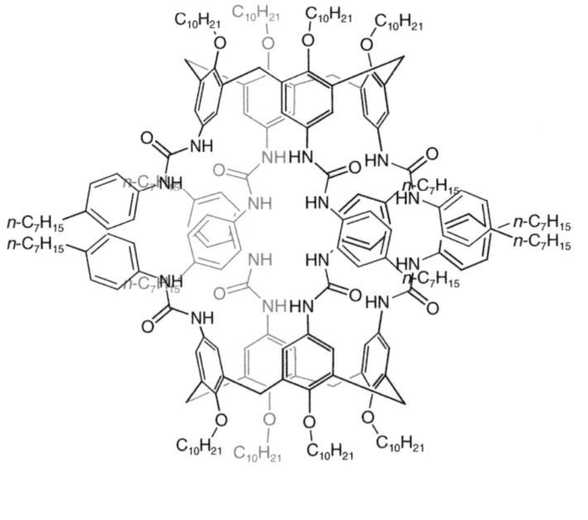

340.

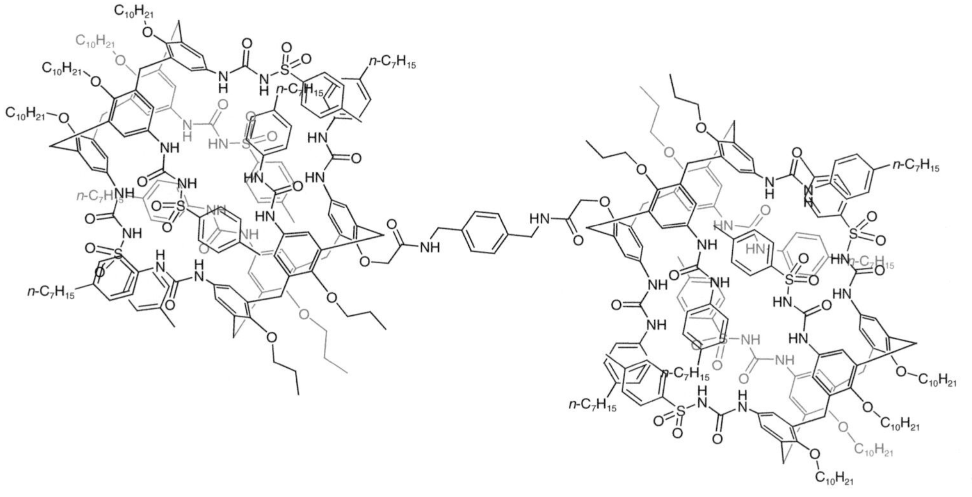

341.

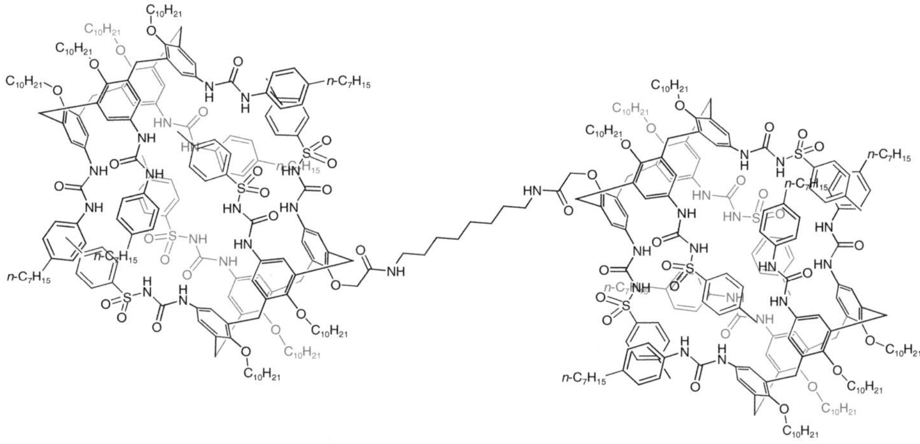

342.

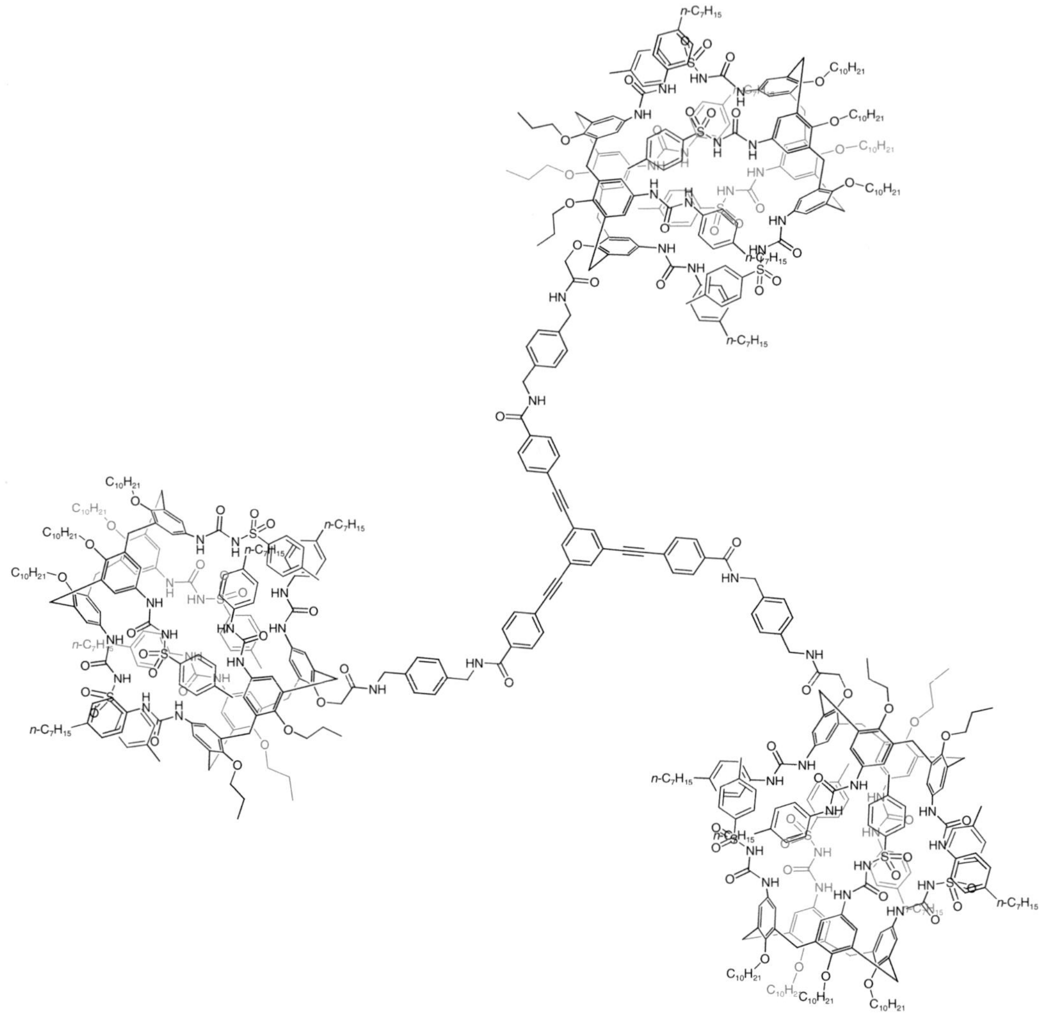

343.

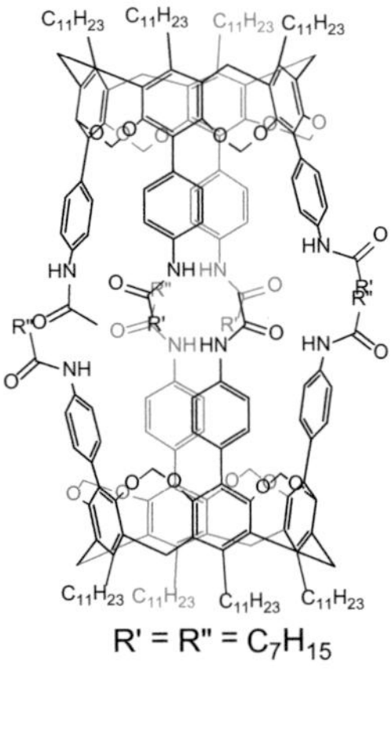

344.

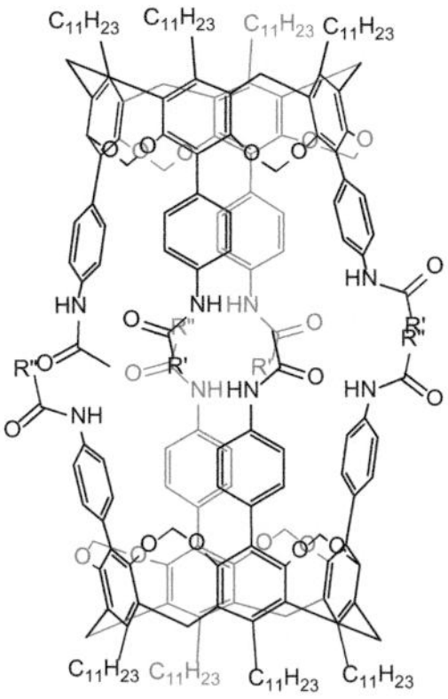

R' = R'' = C_8H_{17}

345.

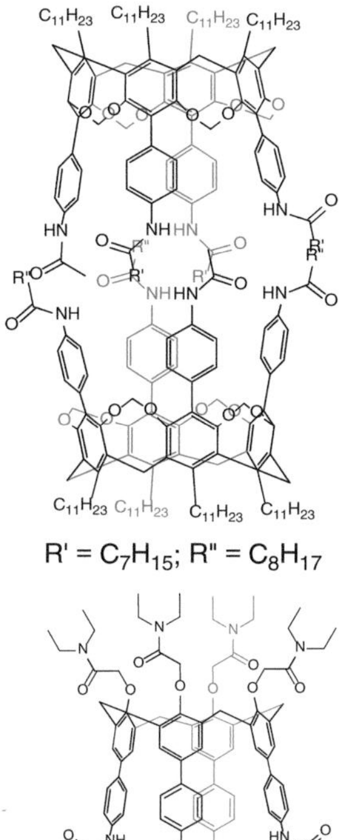

R' = C_7H_{15}; R'' = C_8H_{17}

346.

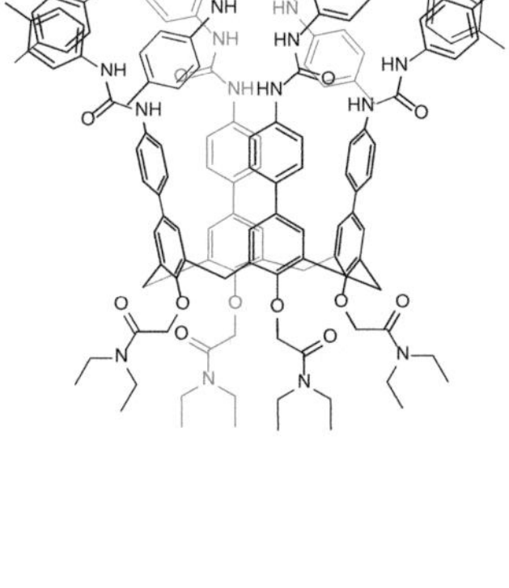

347.

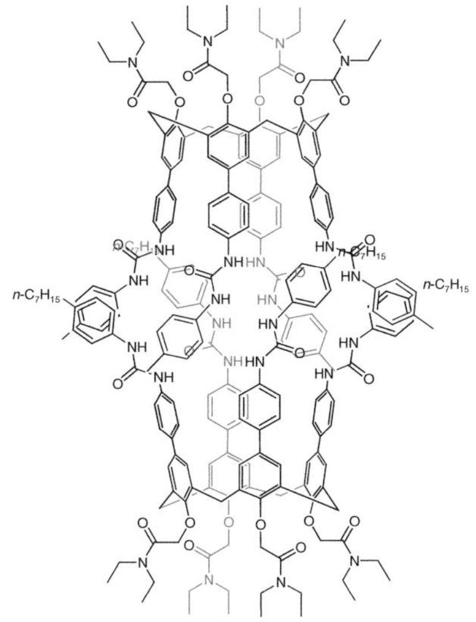

348.

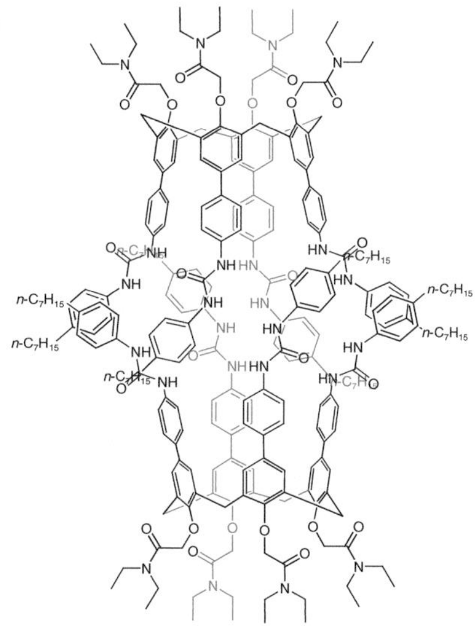

349.

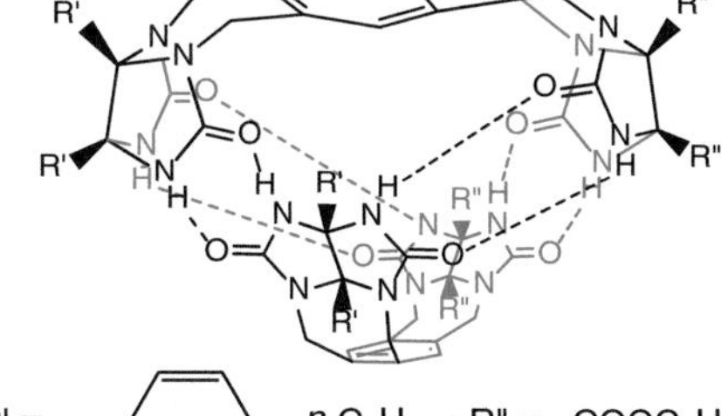

R' = 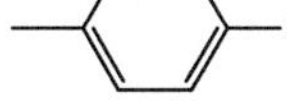 n-C_7H_{15} ; R'' = $-COOC_2H_5$

350.

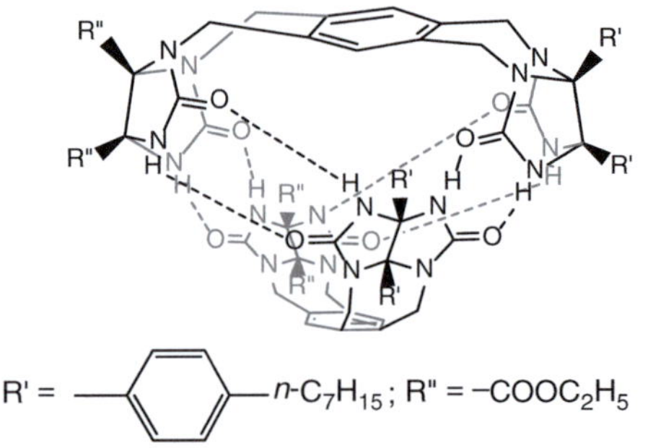

351.

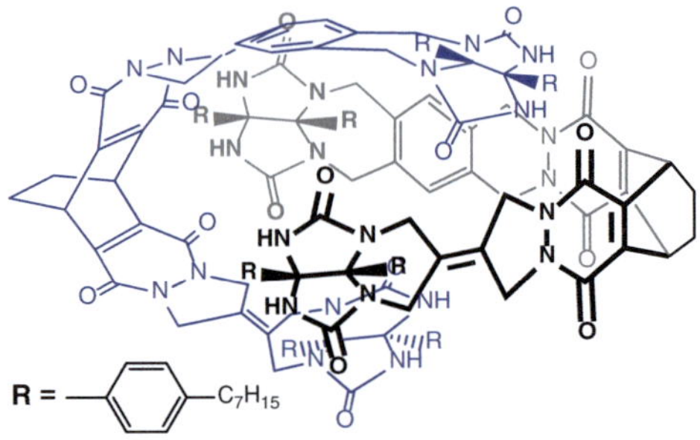

352.

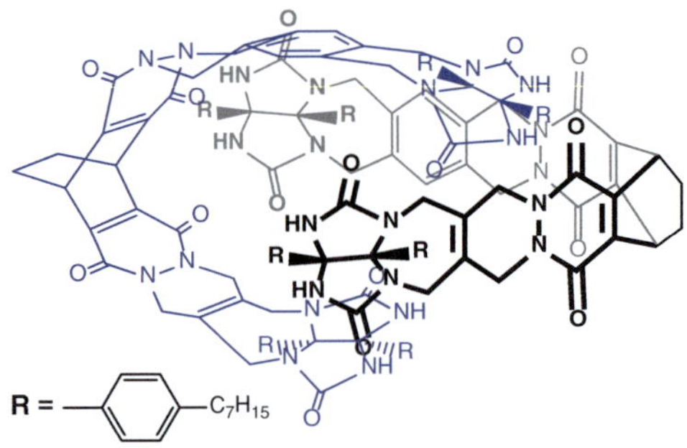

353.

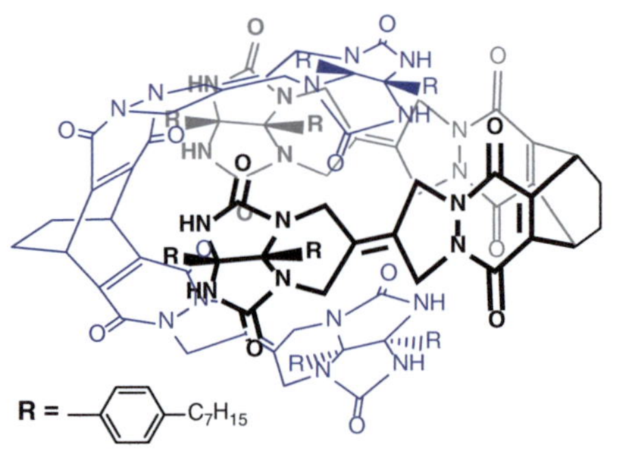

354.

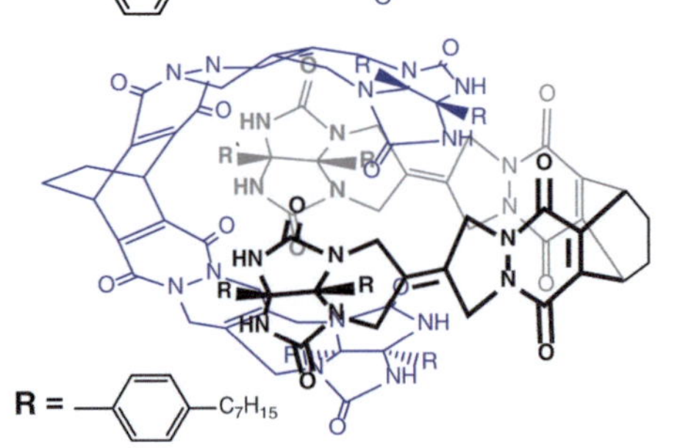

355.

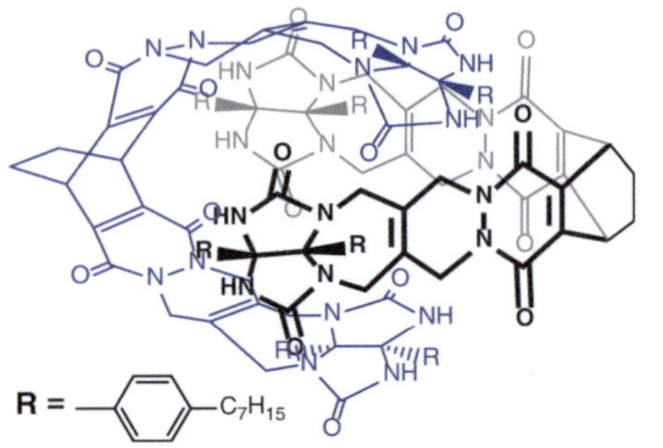

356.

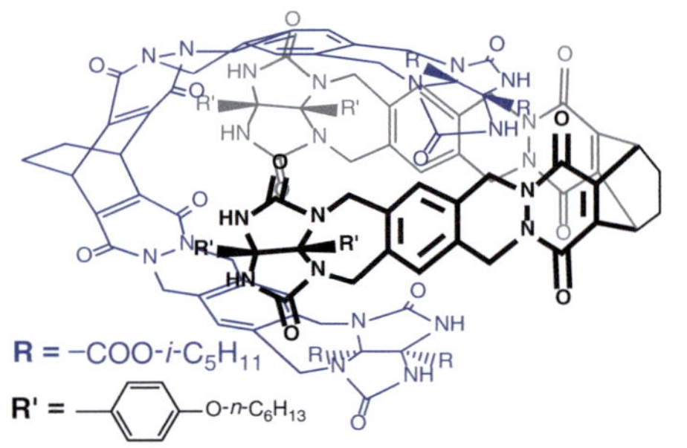

357.

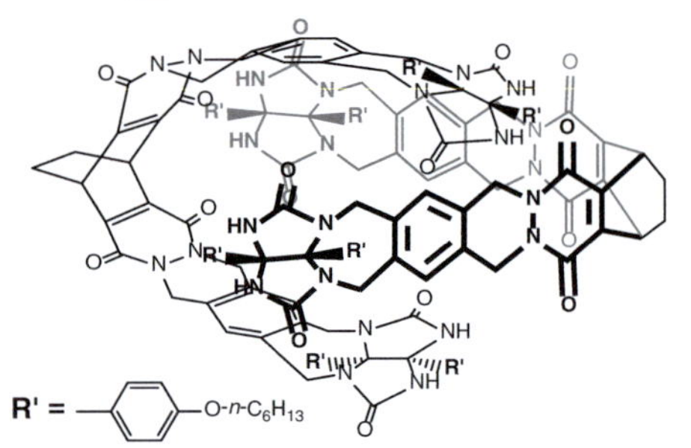

358.

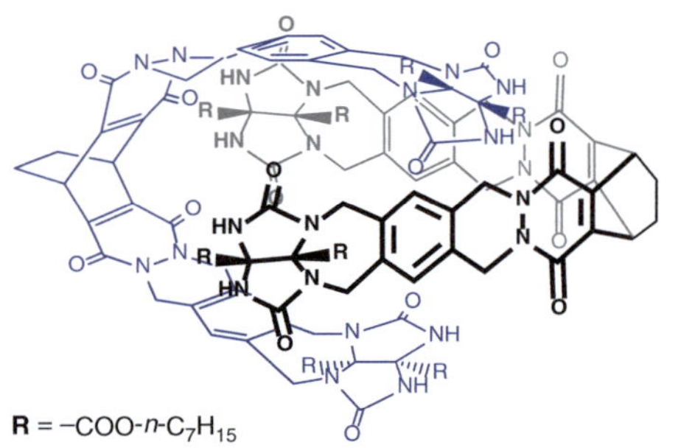

359.

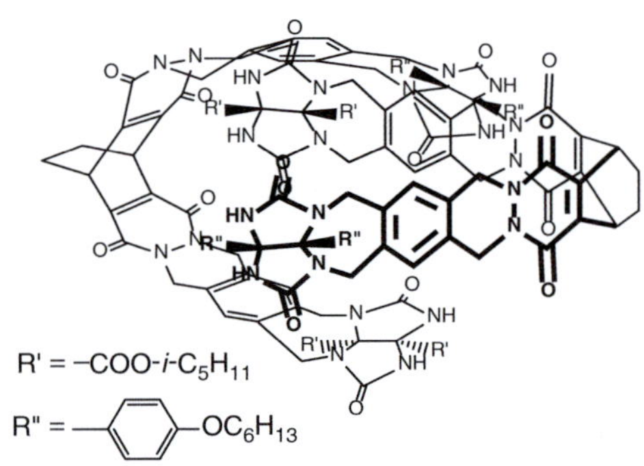

360.

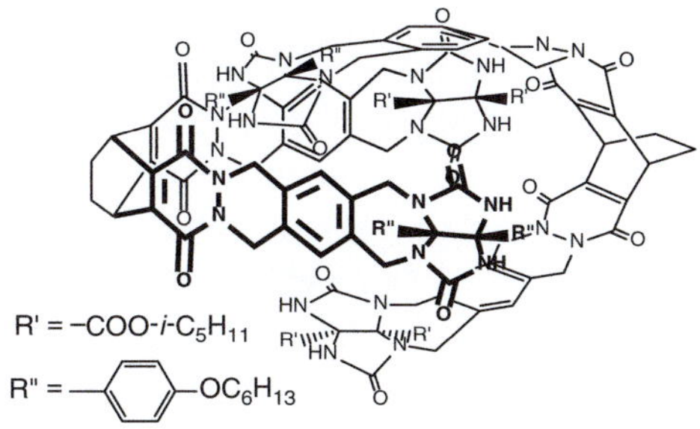

361.

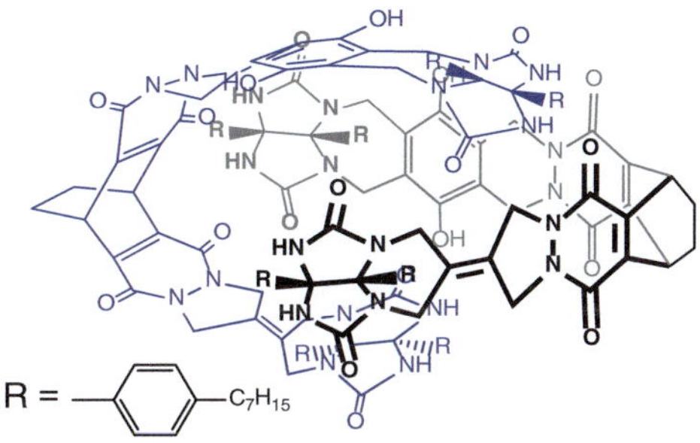

362.

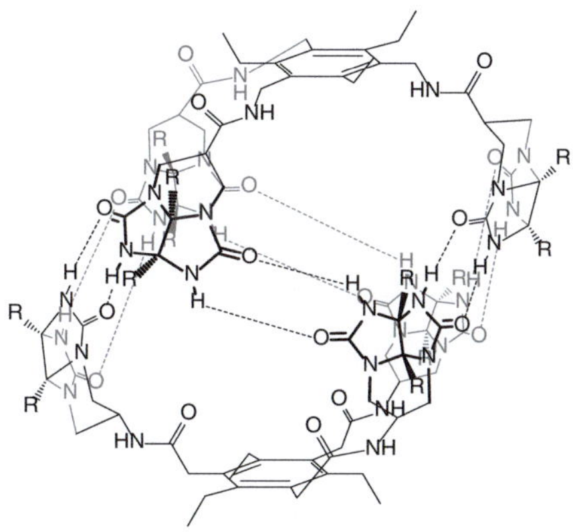

363.

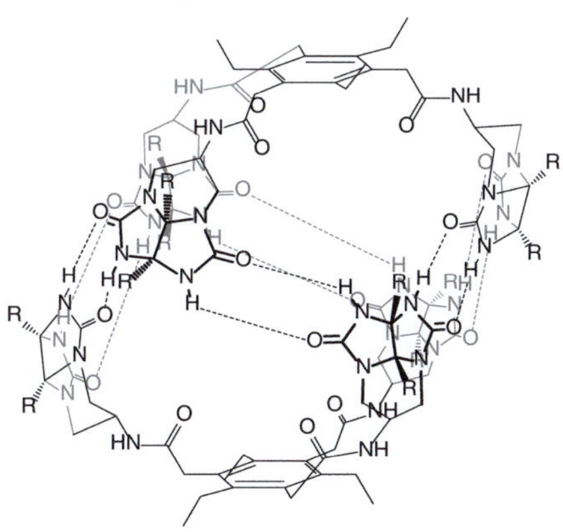

364.

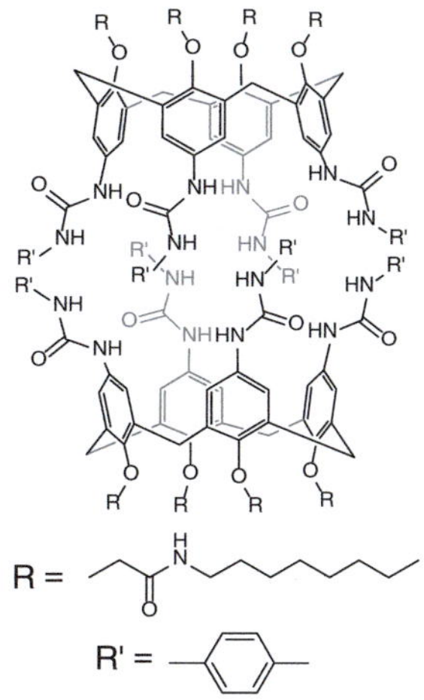

365.

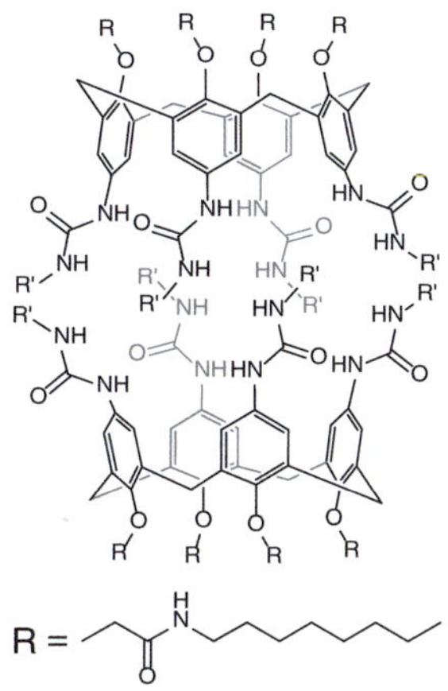

R' = n-C_8H_{15}

366.

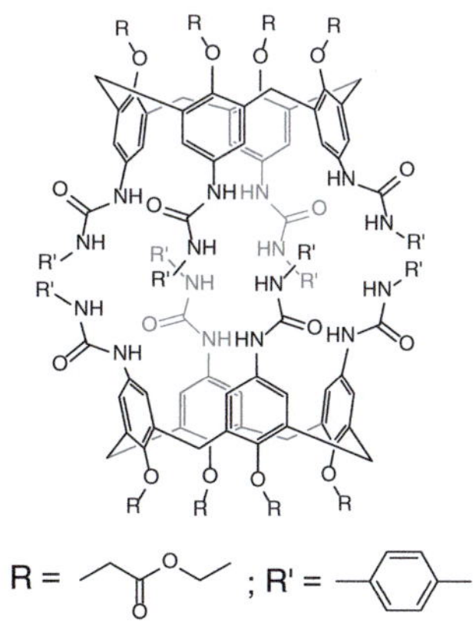

367.

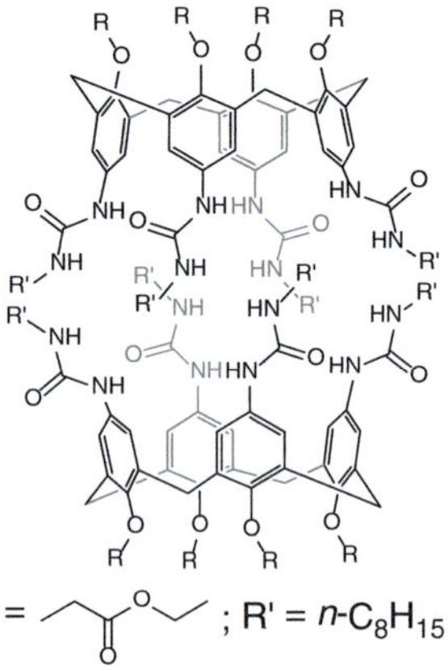

368.

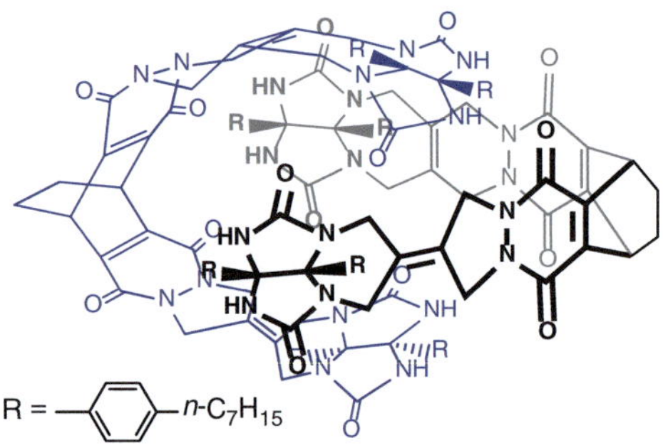

369.

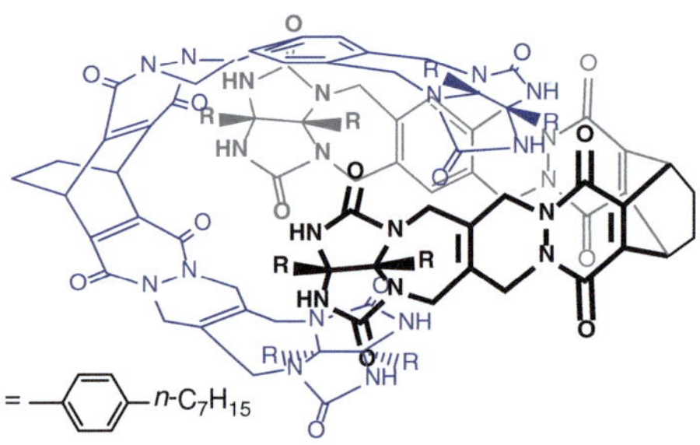

370.

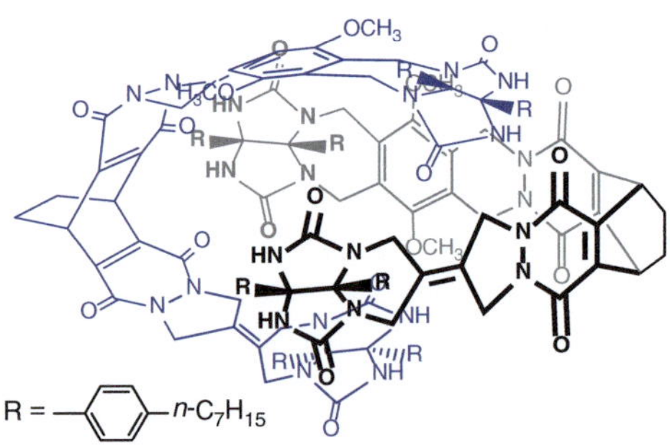

371.

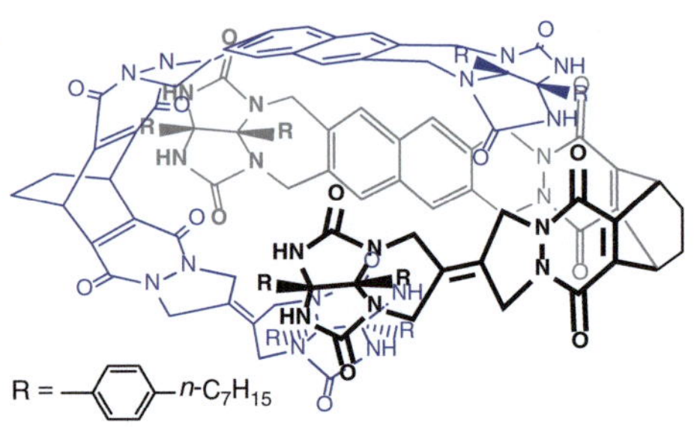

372.

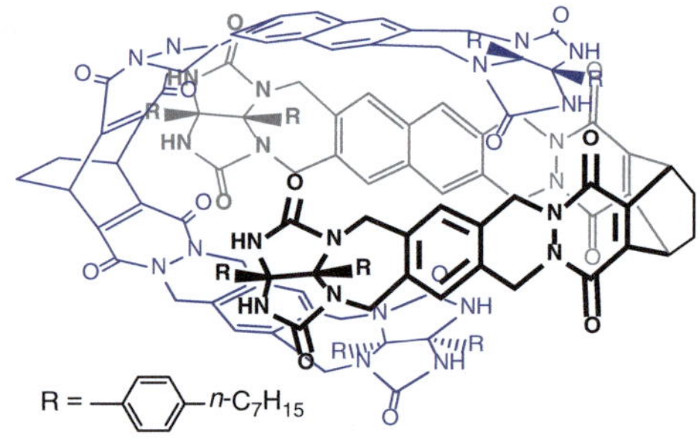

373.

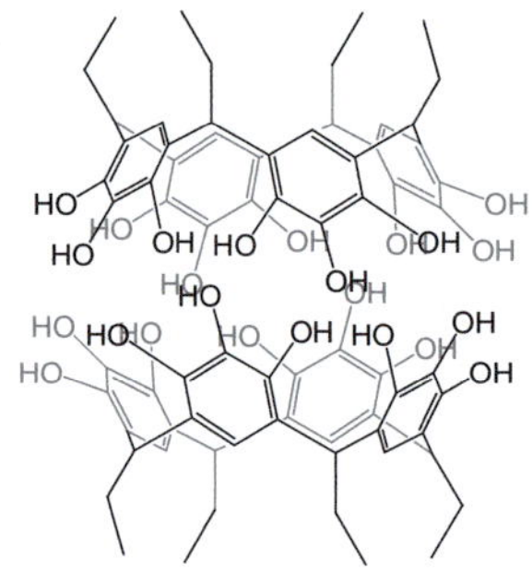

374.

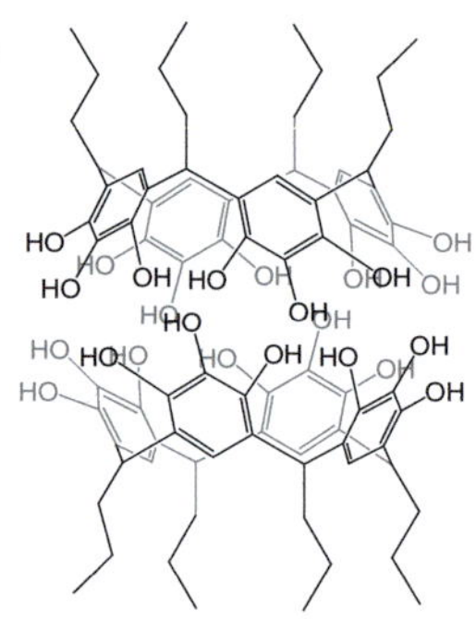

375.

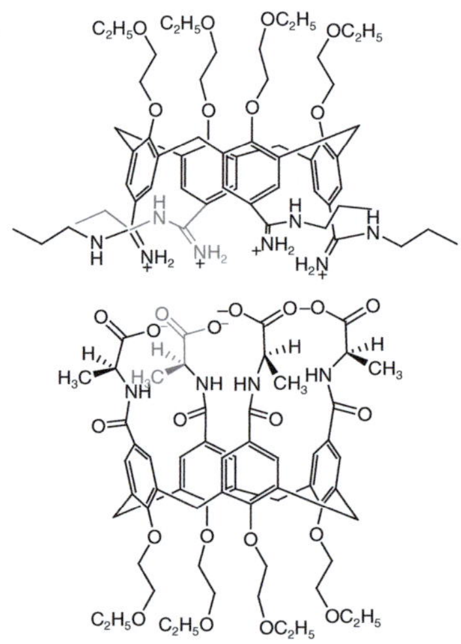

376.

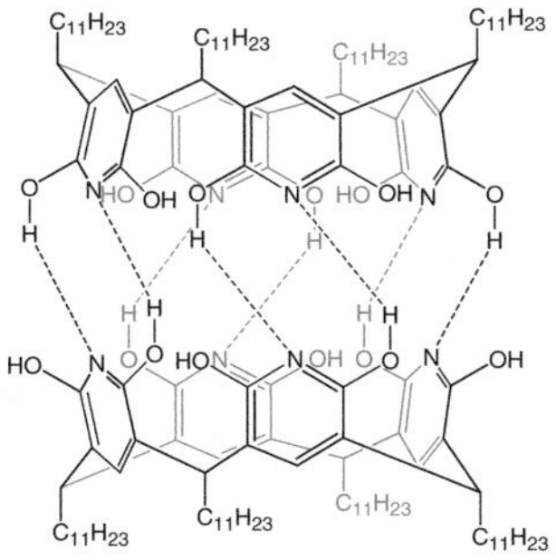

377.

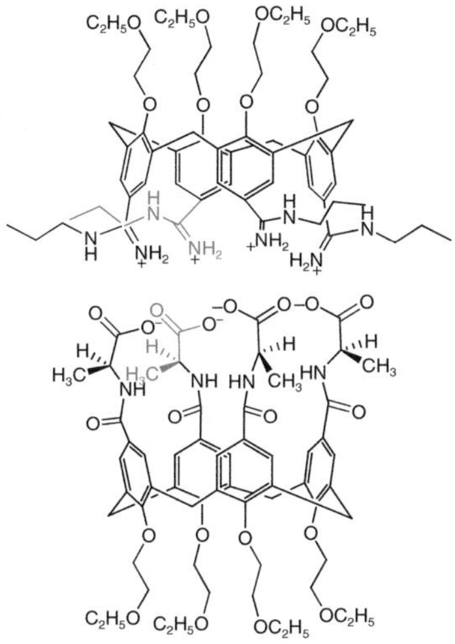

378.

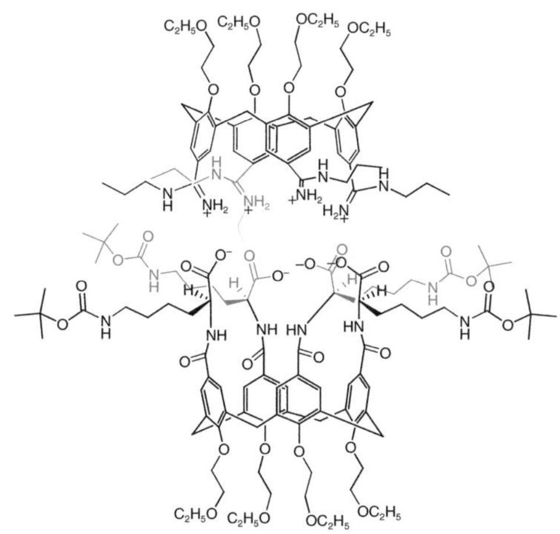

379.

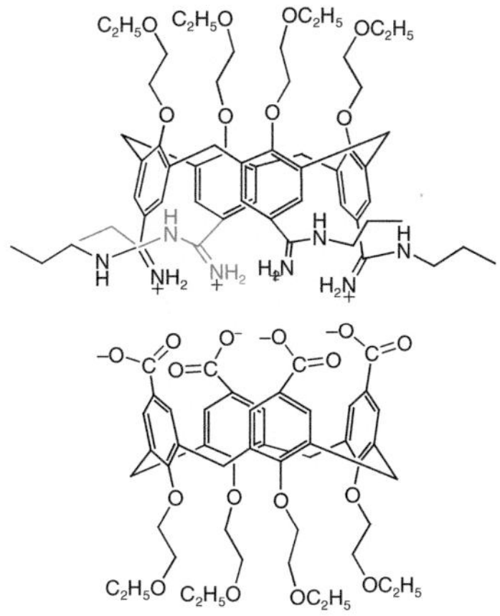

380.

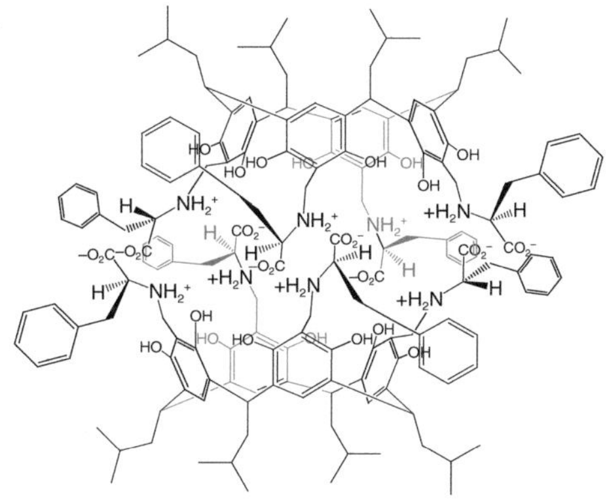

381.

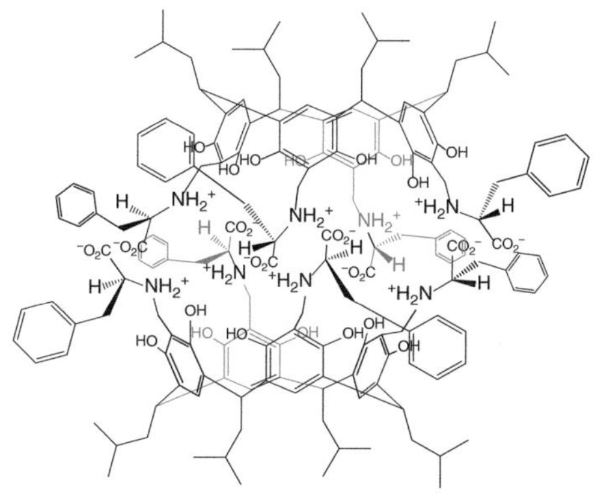

382.

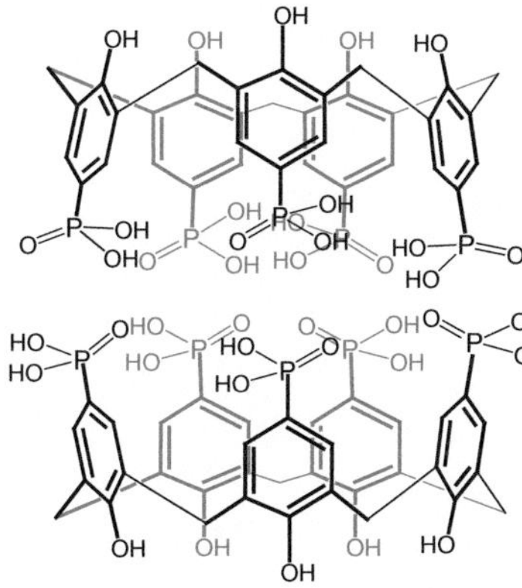

383.

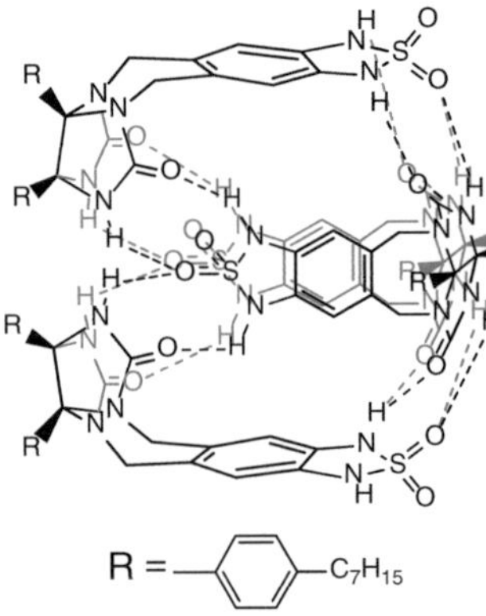

384.

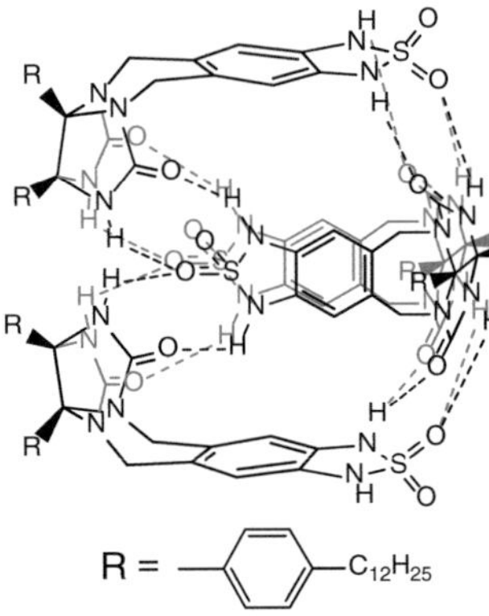

385.

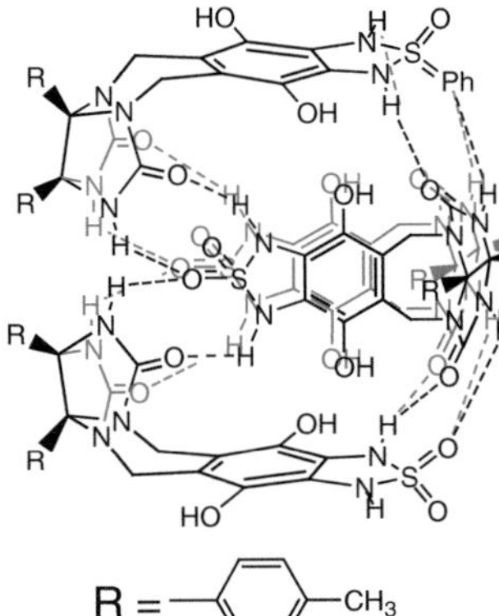

386.

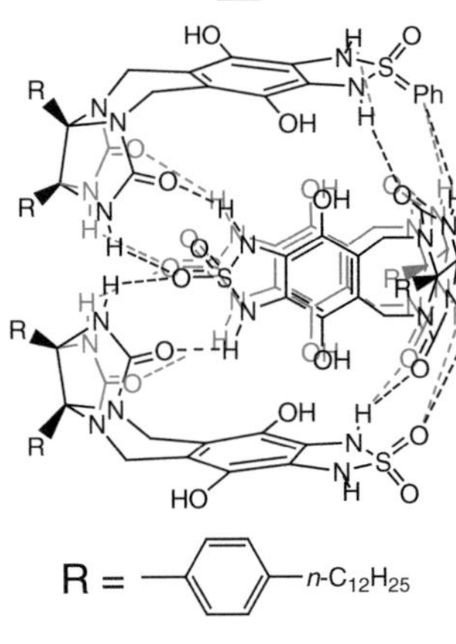

387.

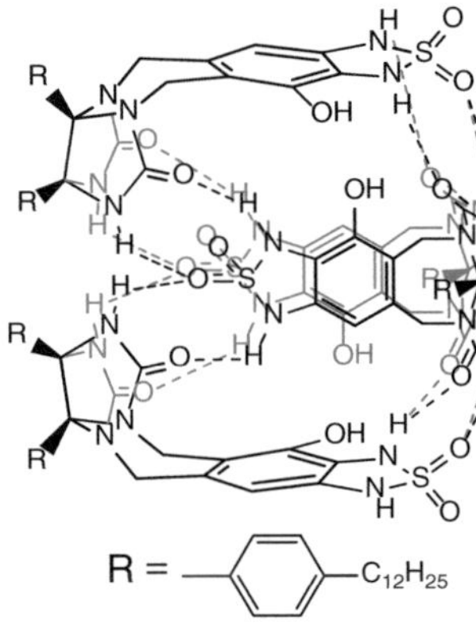

388.

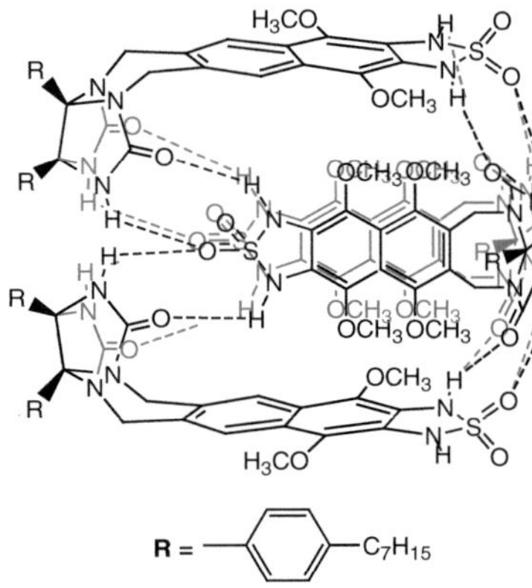

389.

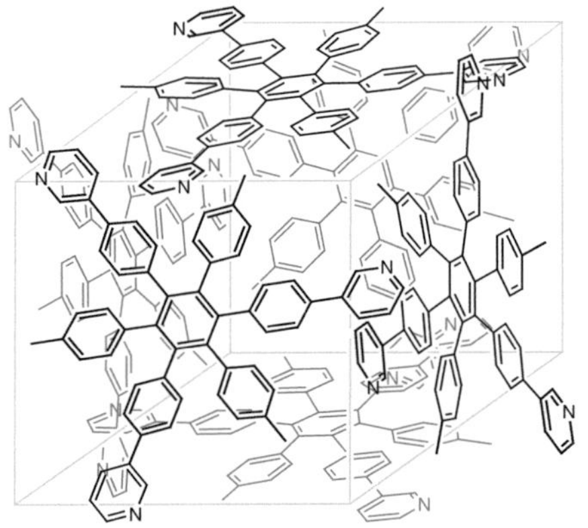

390.

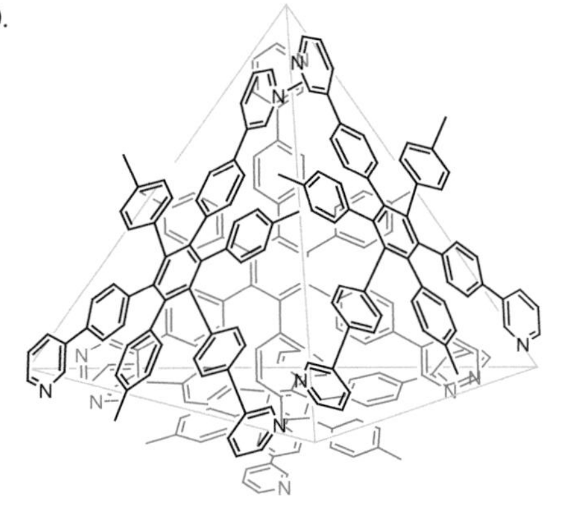

391.

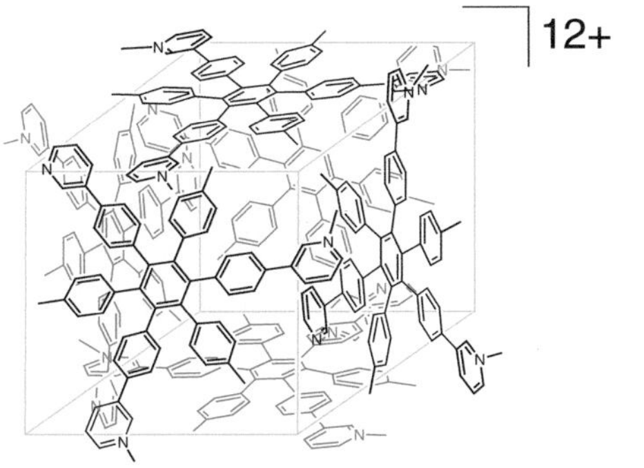

392.

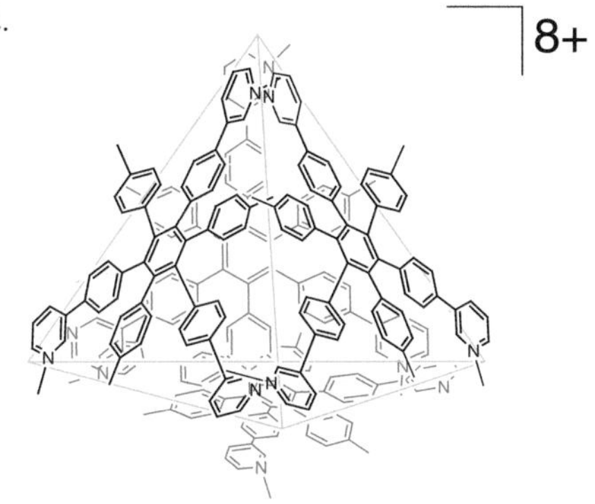

393.

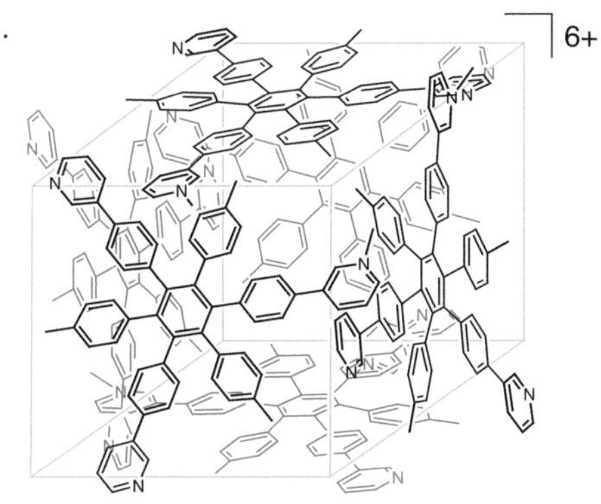

394.

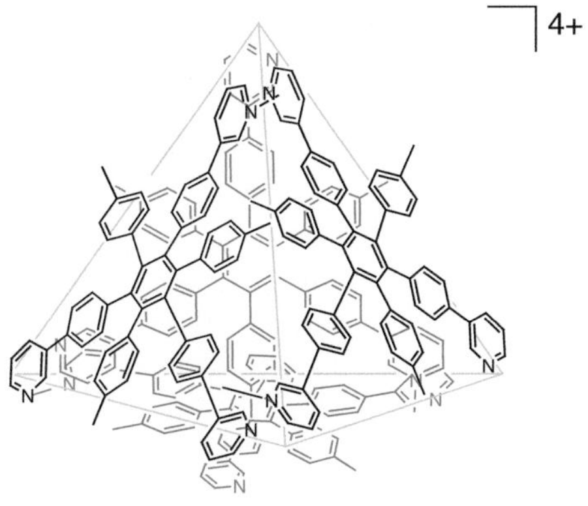

395.

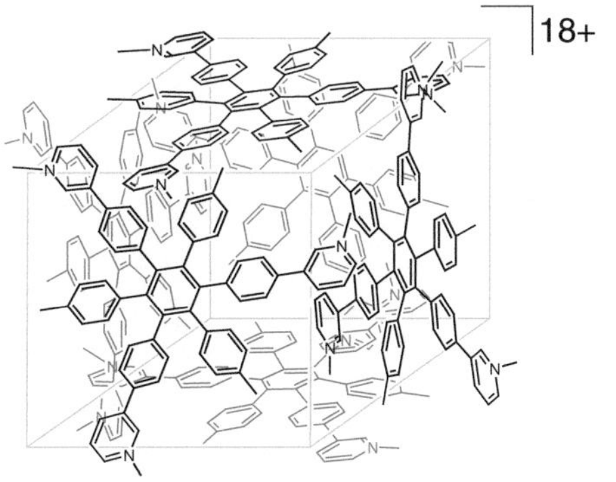

396.

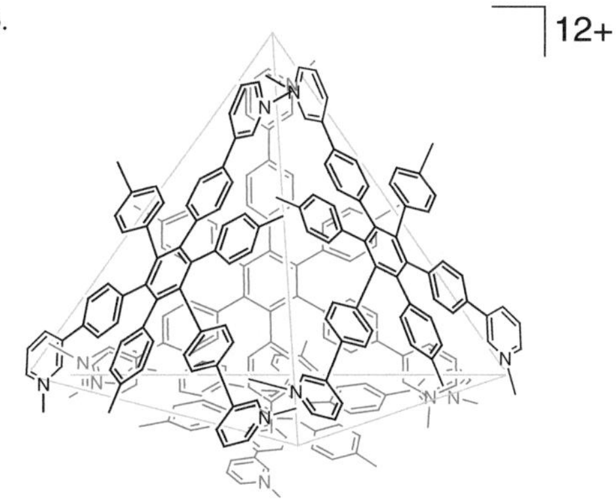

397.

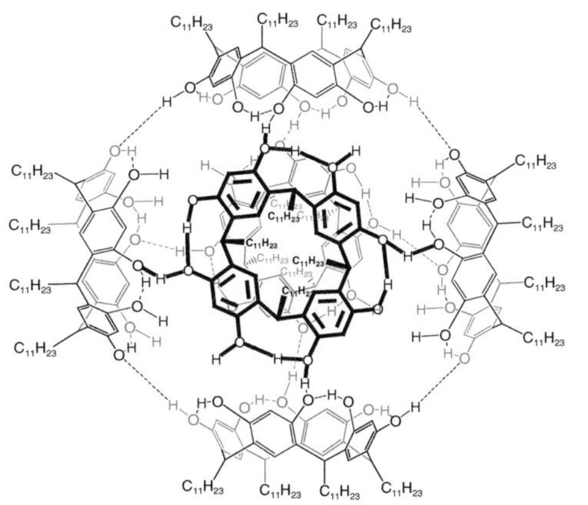

398.

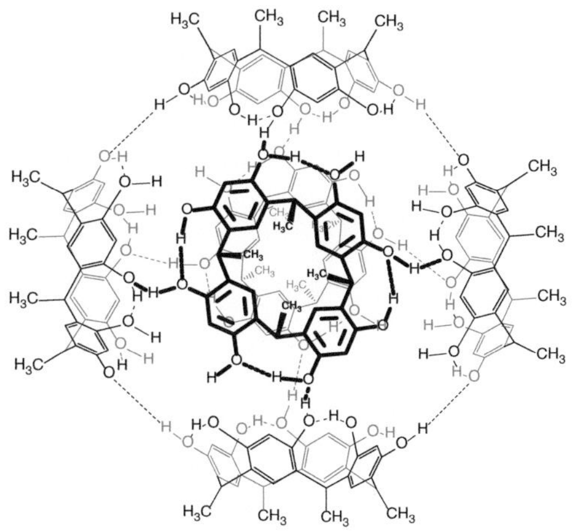

399.

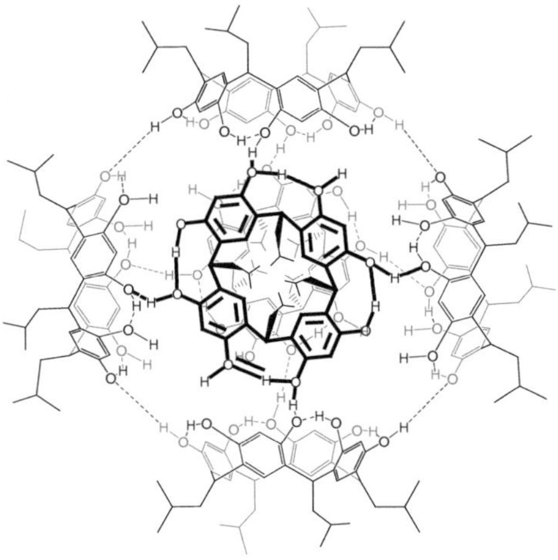

400.

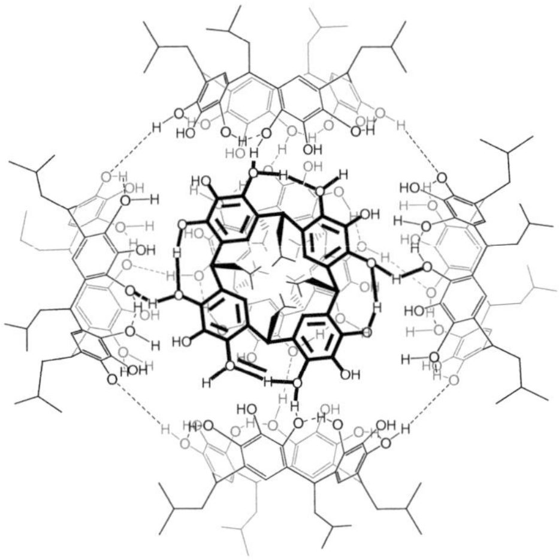

401.

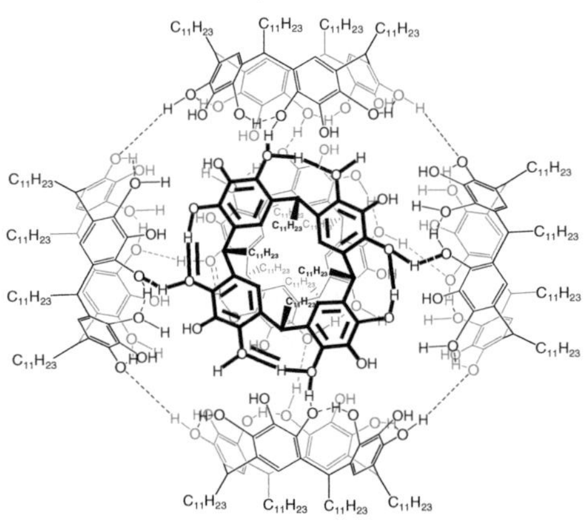

402.

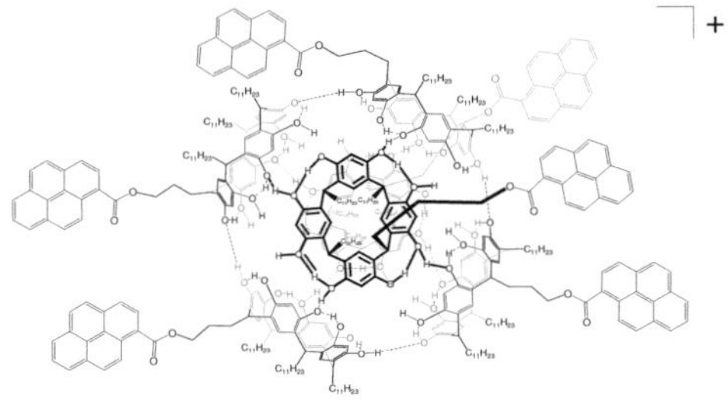

403.

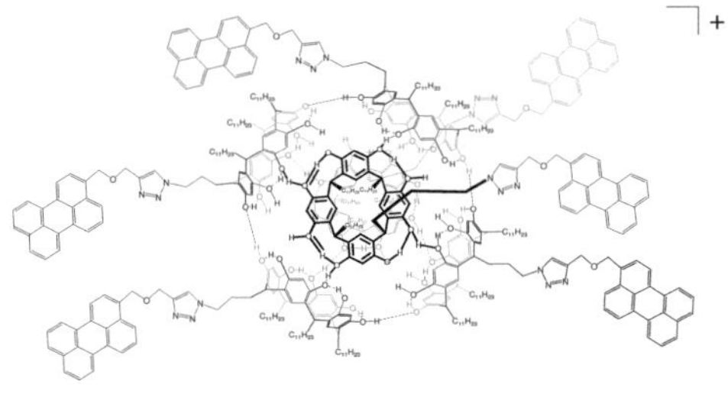

404.

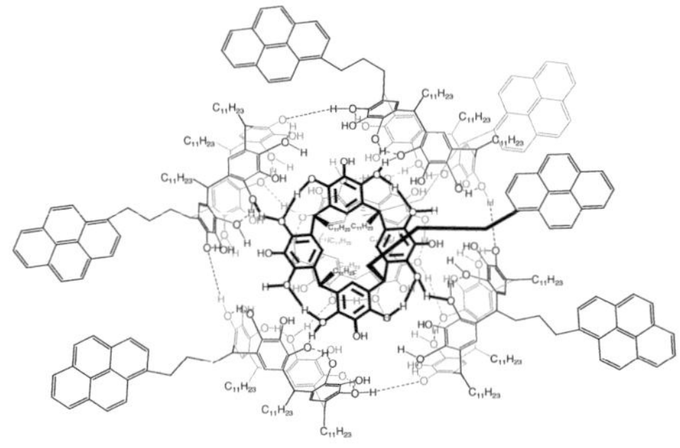

405.

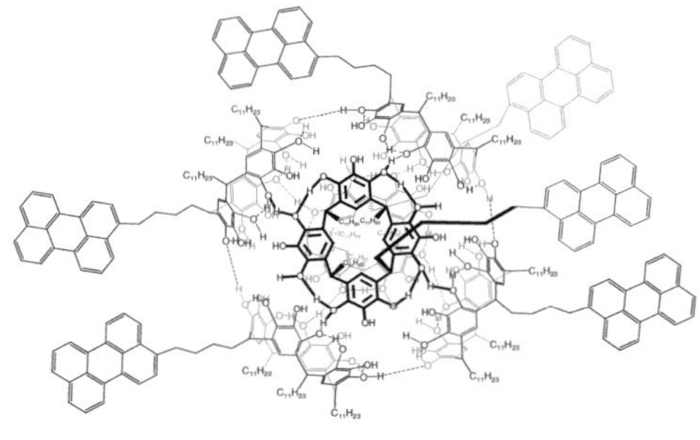

406.

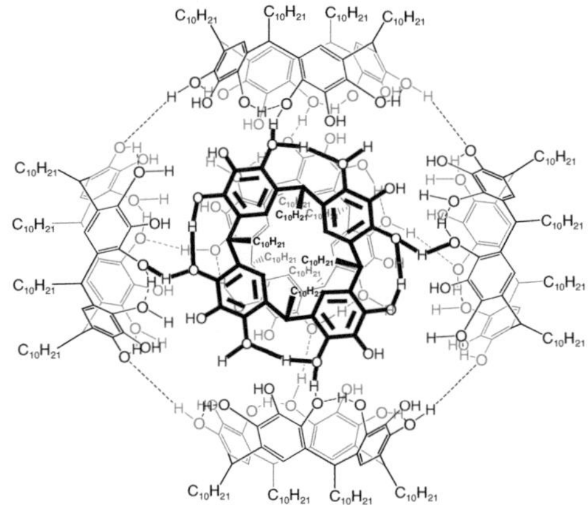

407.

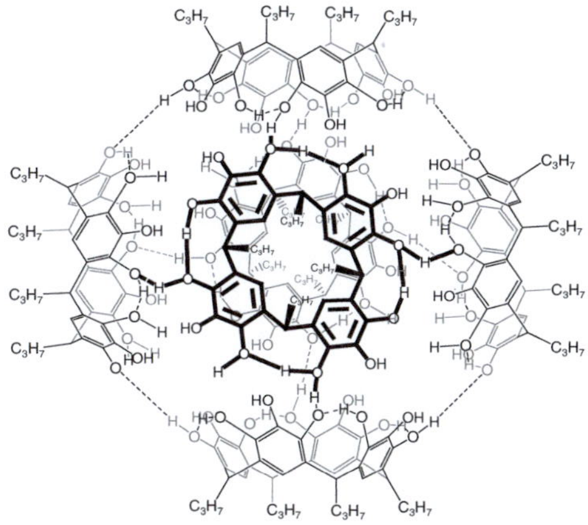

410.

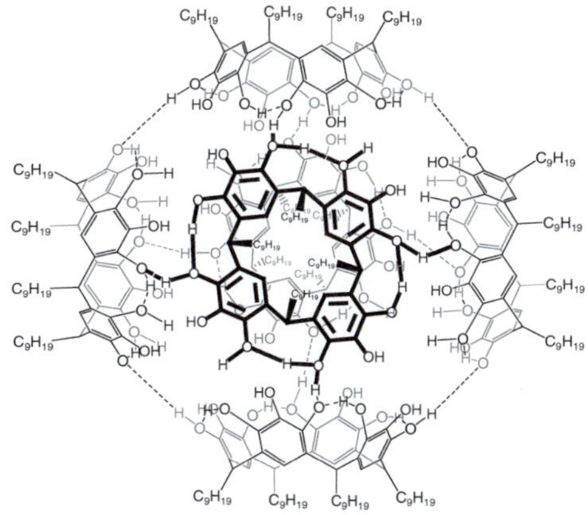

408.

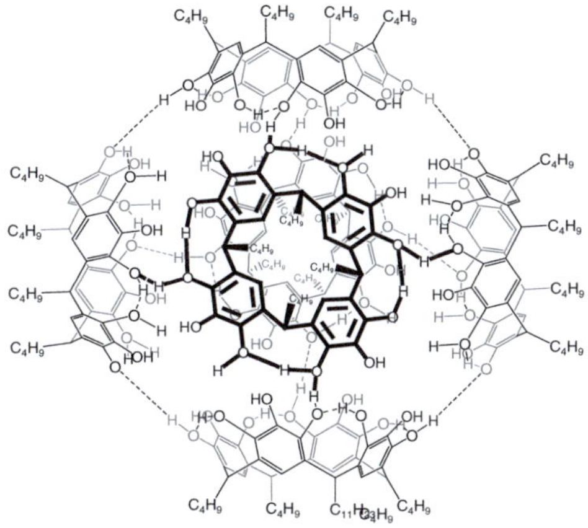

411.

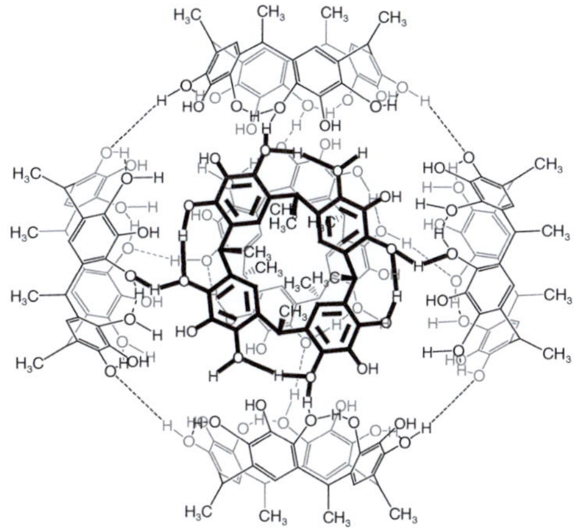

409.

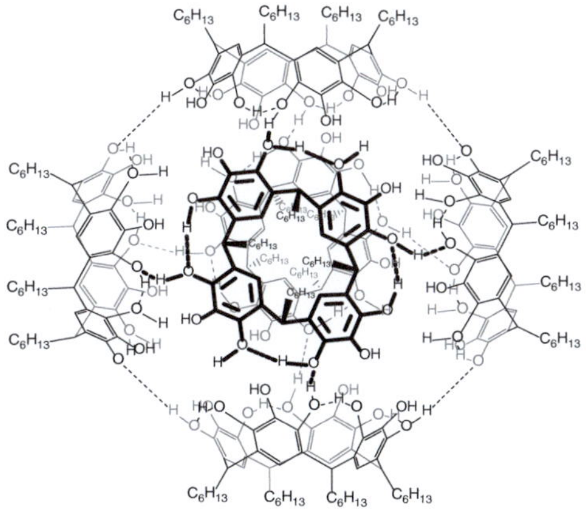

412.

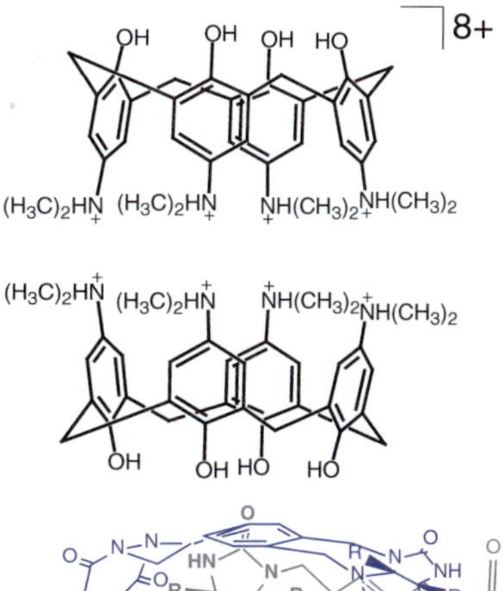

413.

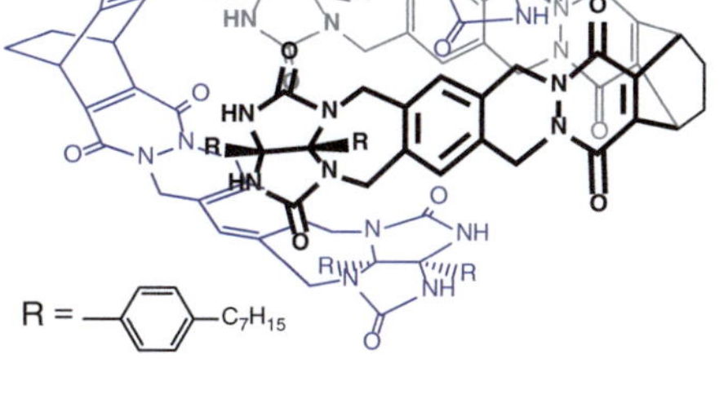

414.

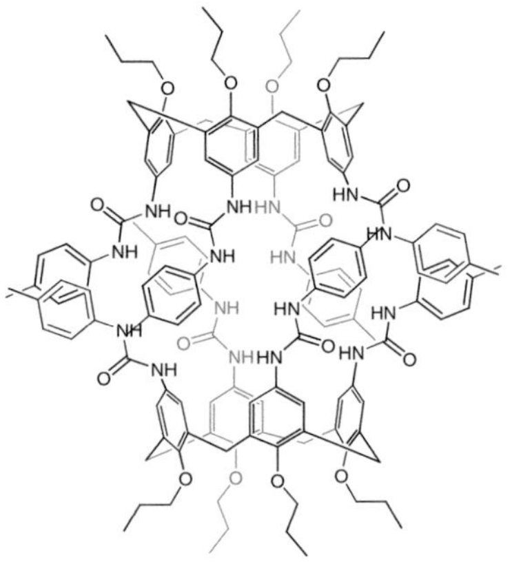

416.

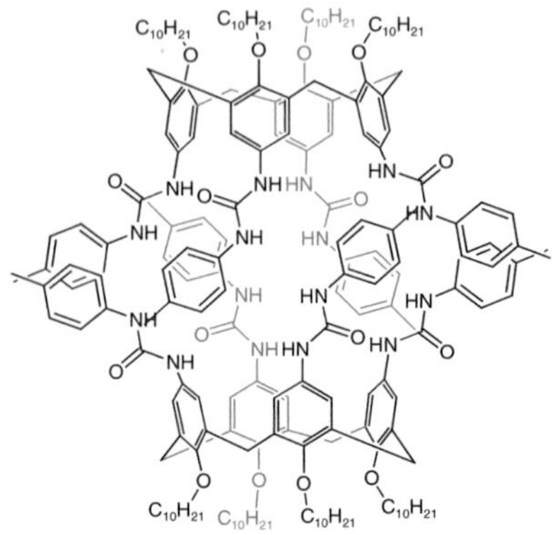

415.

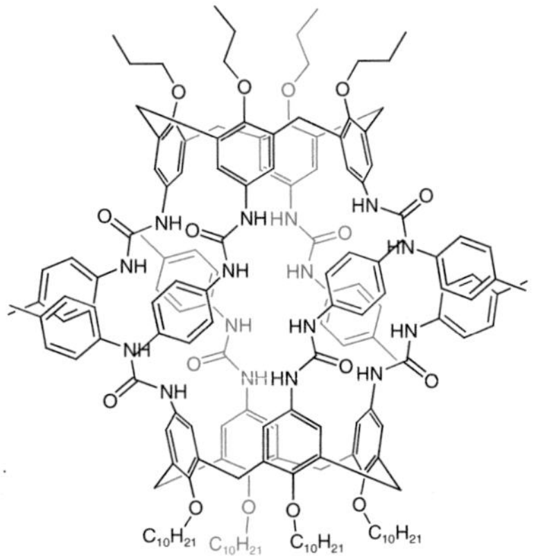

417.

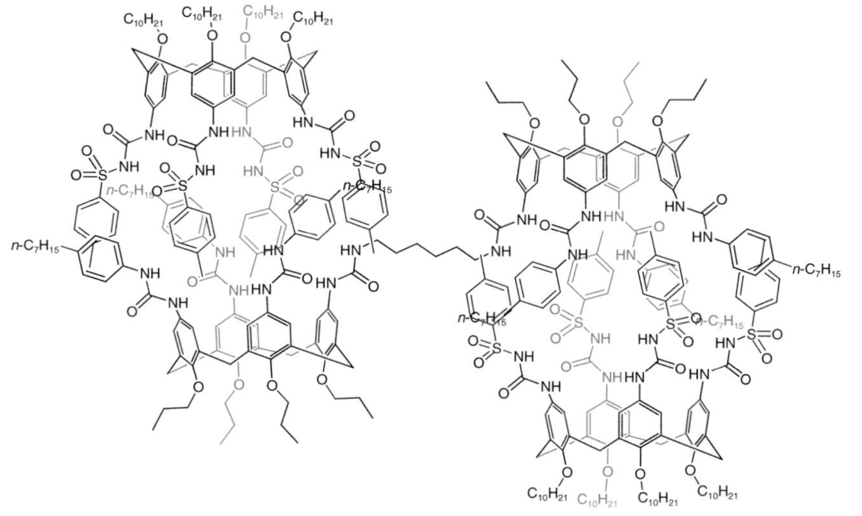

418.

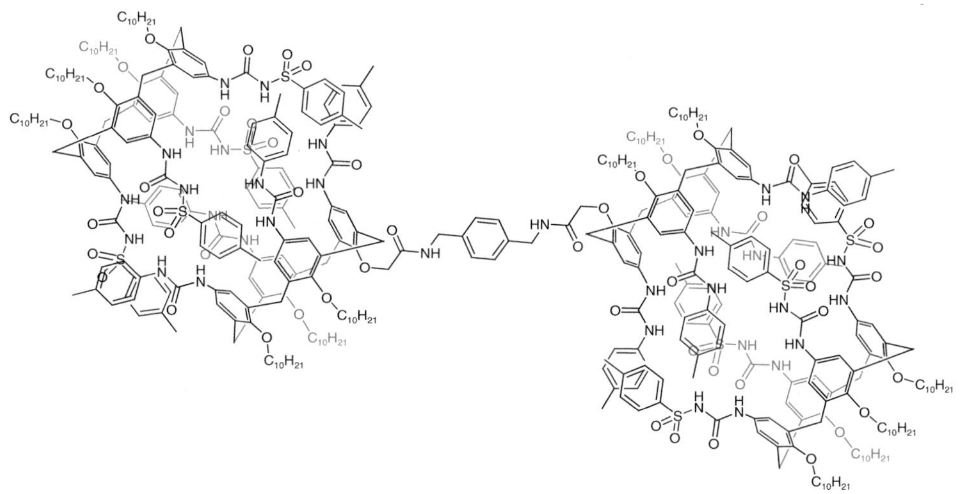

419.

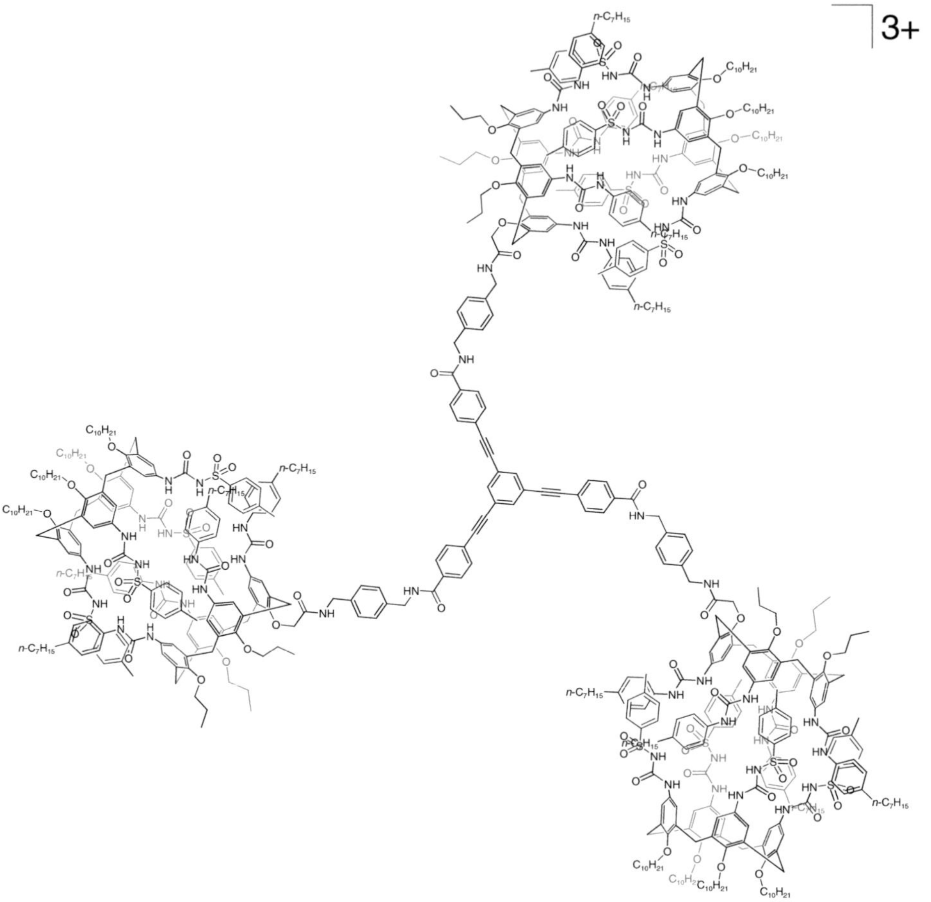

420.

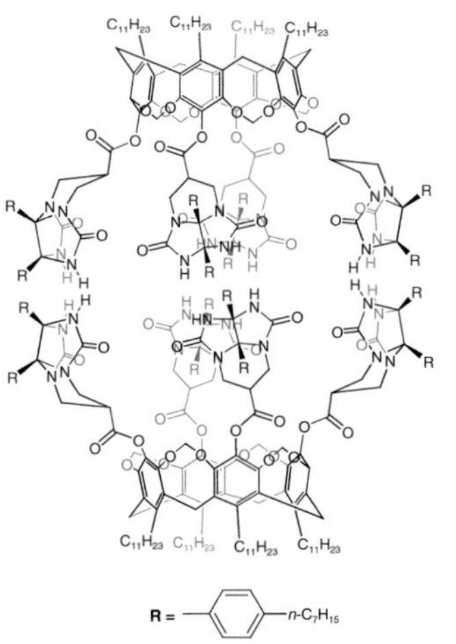

421.

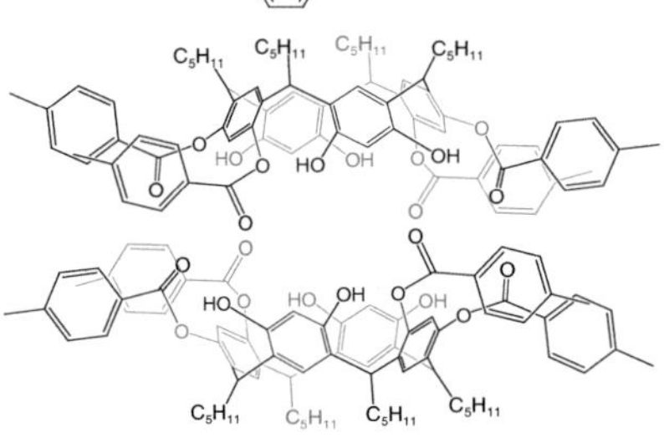

422.

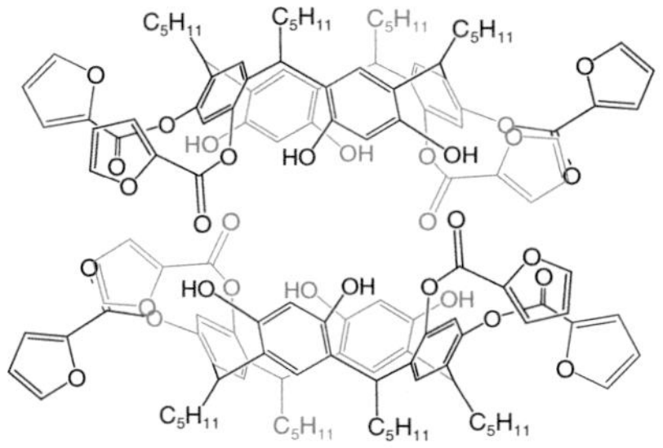

423.

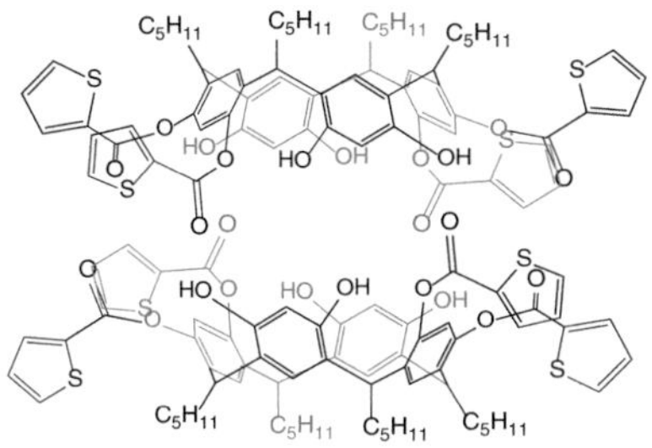

424.

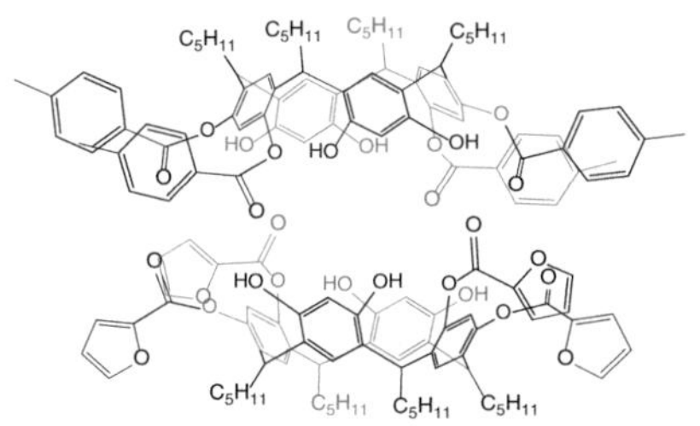

425.

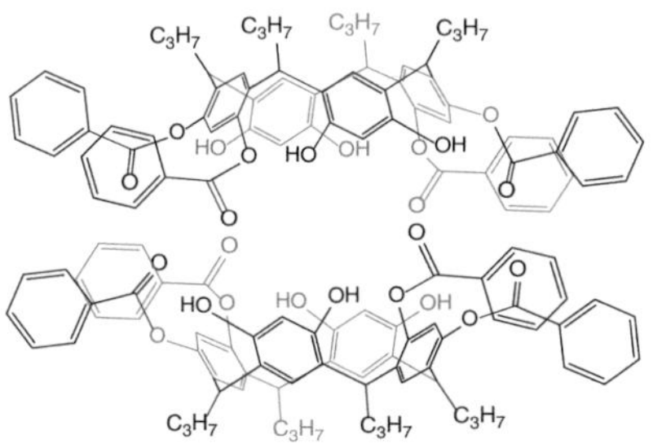

426.

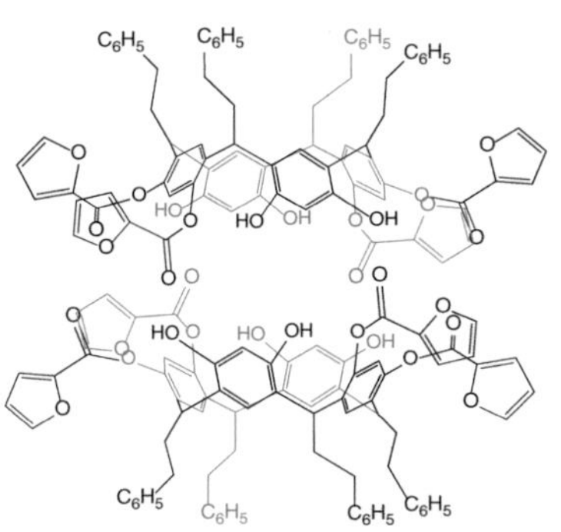

427.

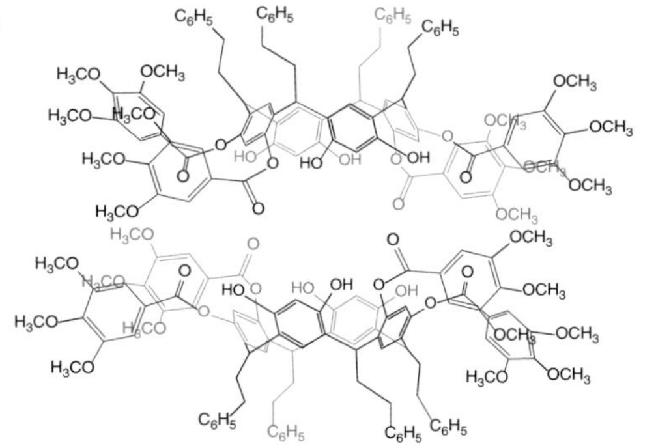

428.

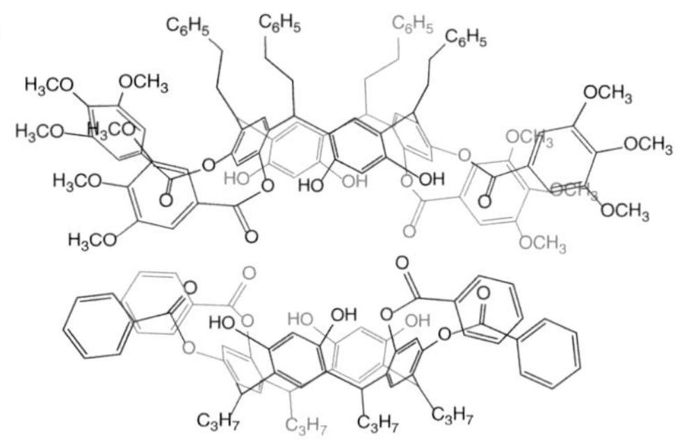

429.

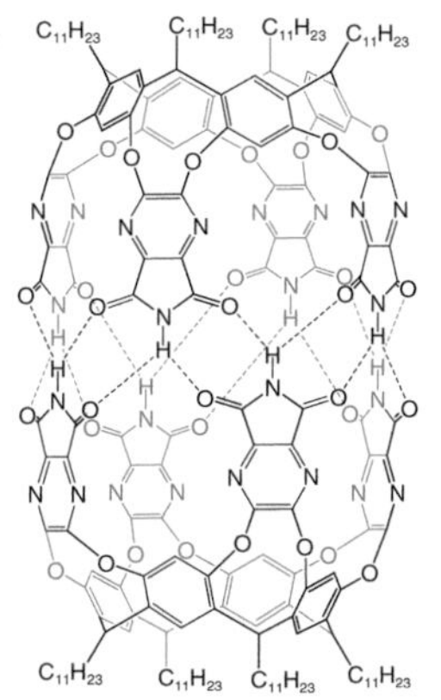

430.

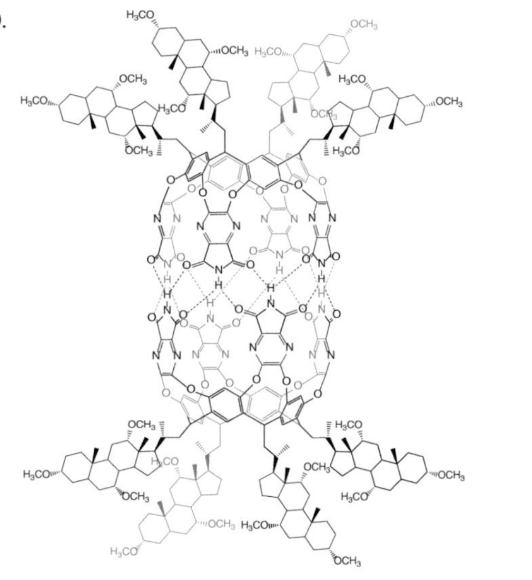

431.

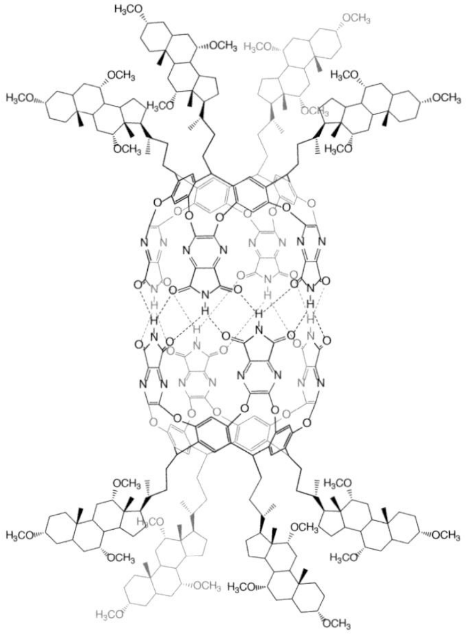

432.

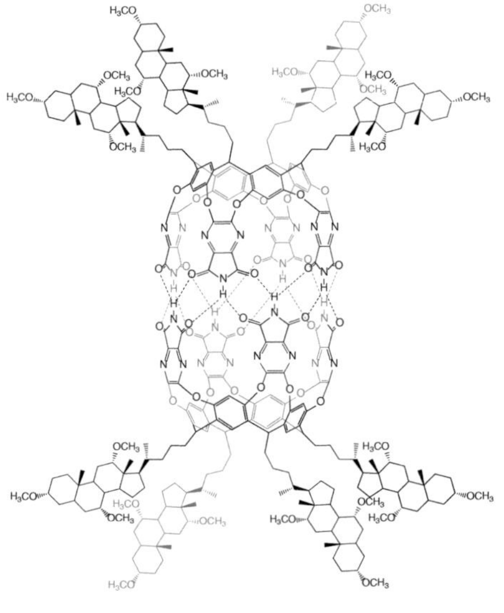

433.

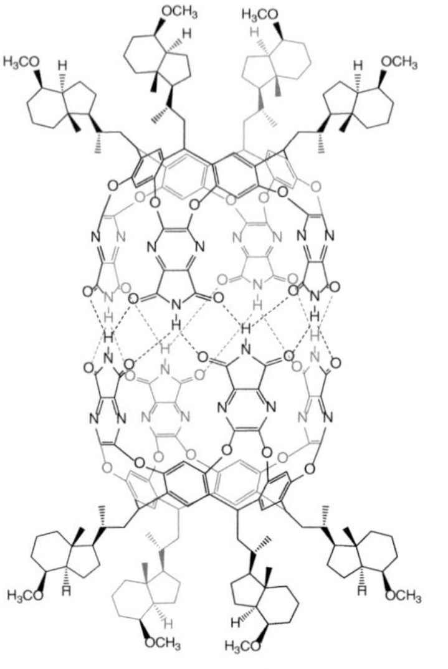

434.

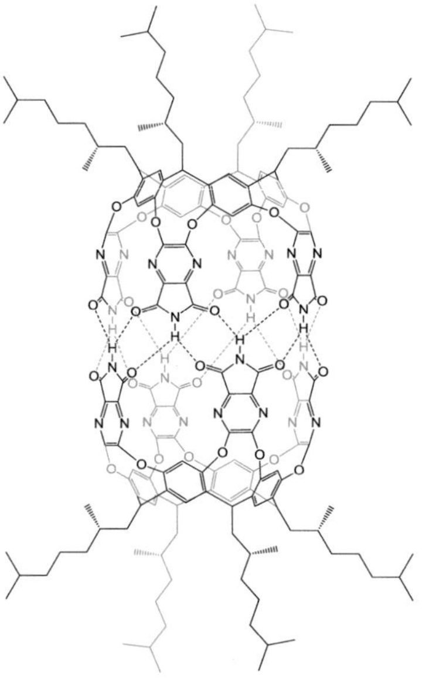

435.

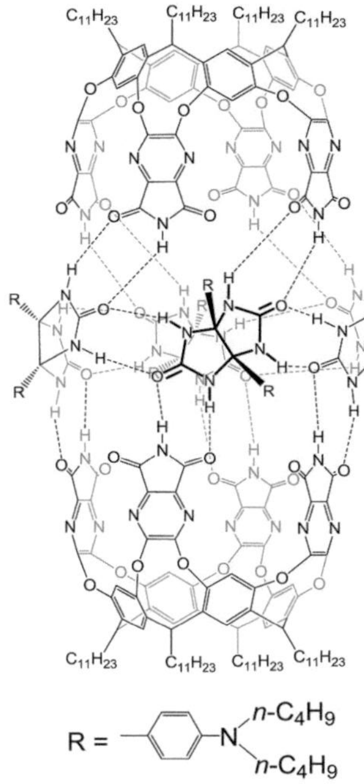

436.

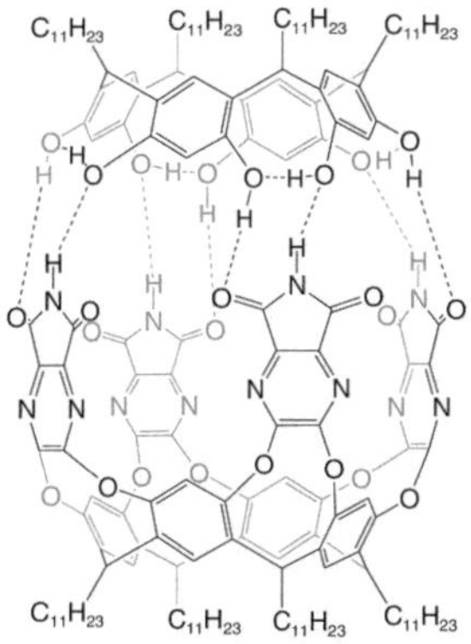

437.

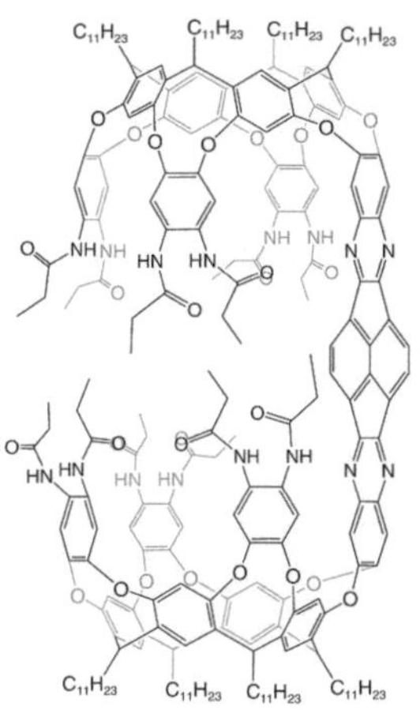

438.

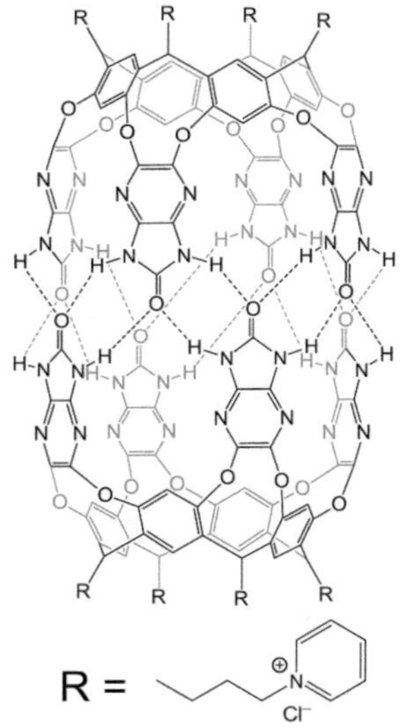

439.

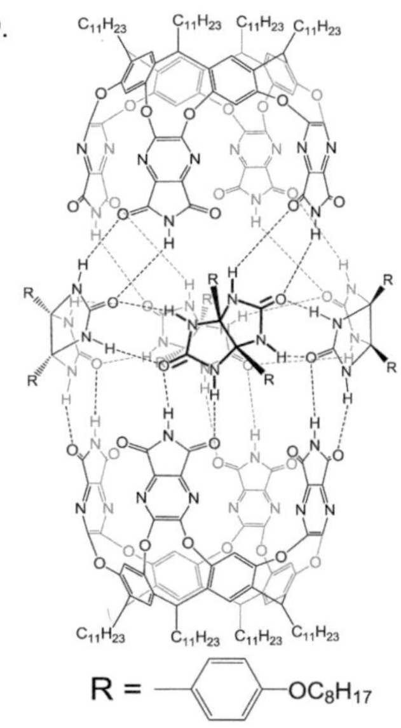

440.

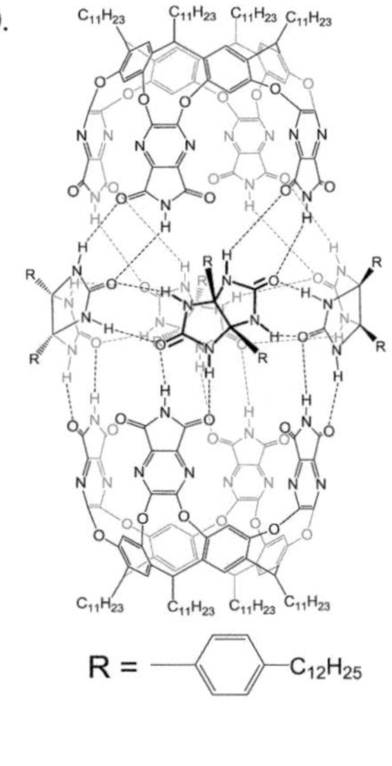

441.

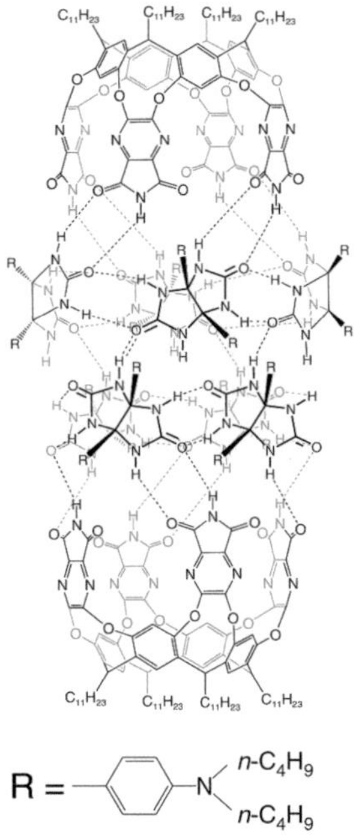

442.

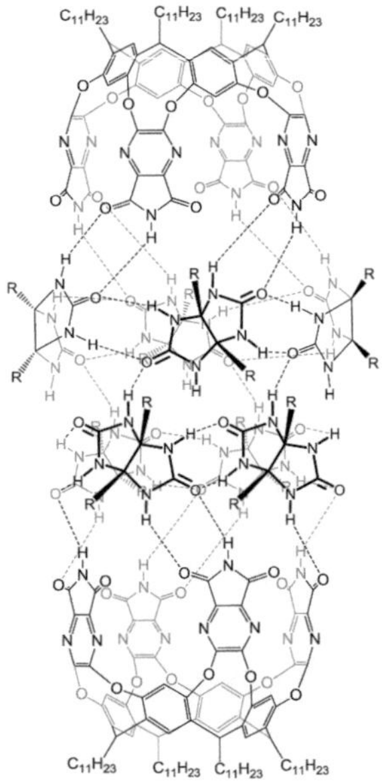

443.

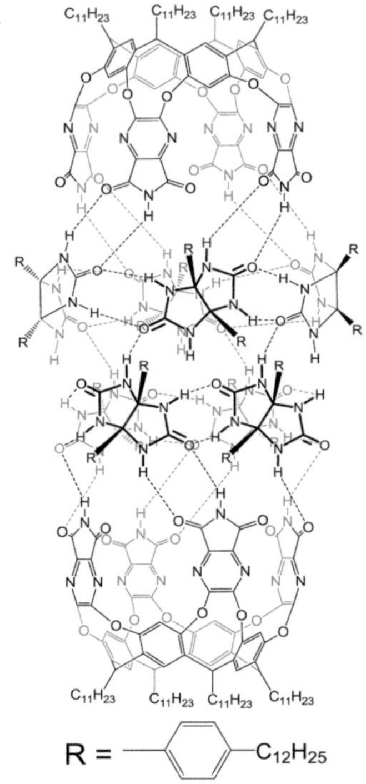

444.

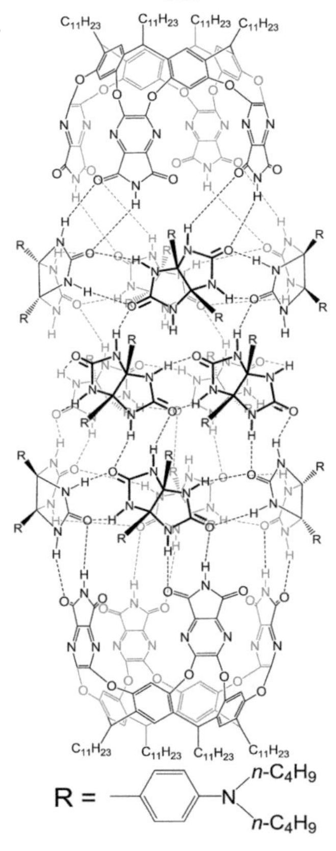

445.

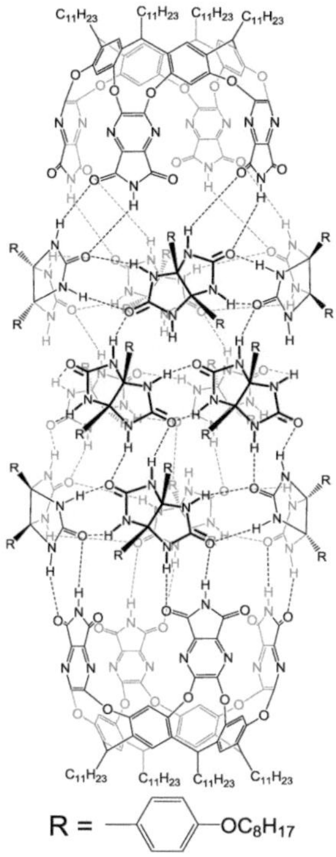

446.

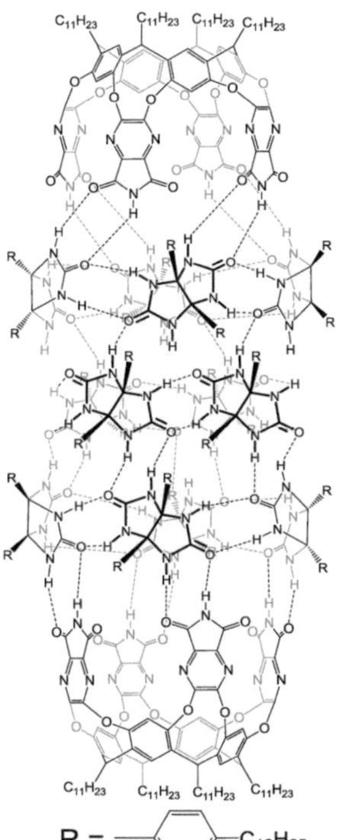

447.

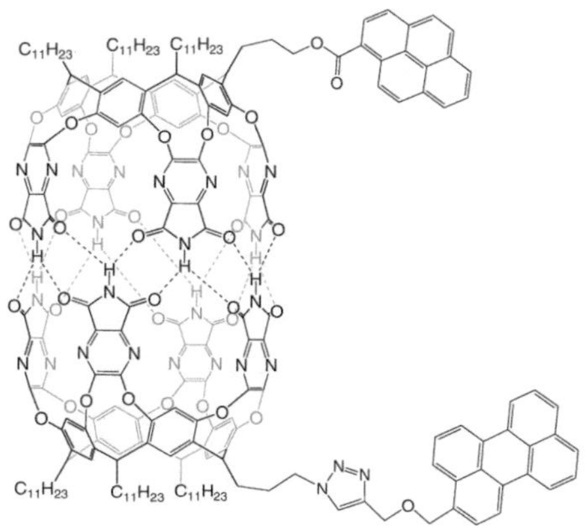

448.

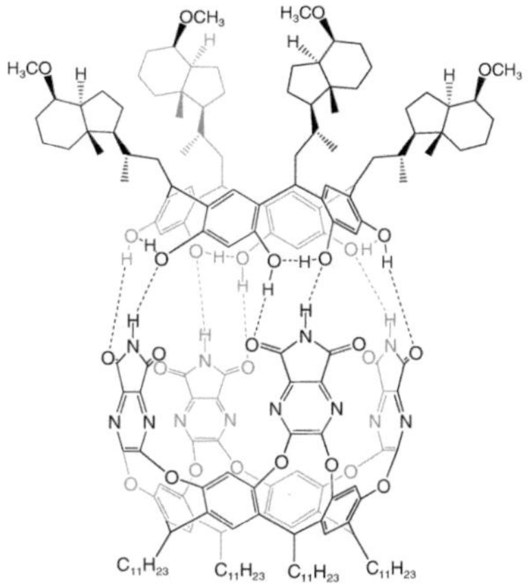

449.

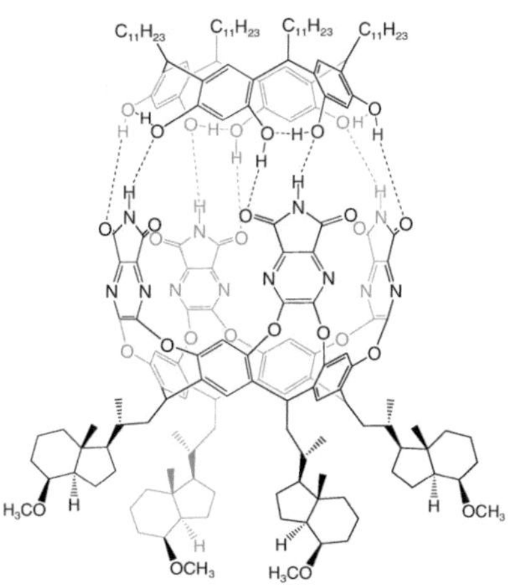

450.

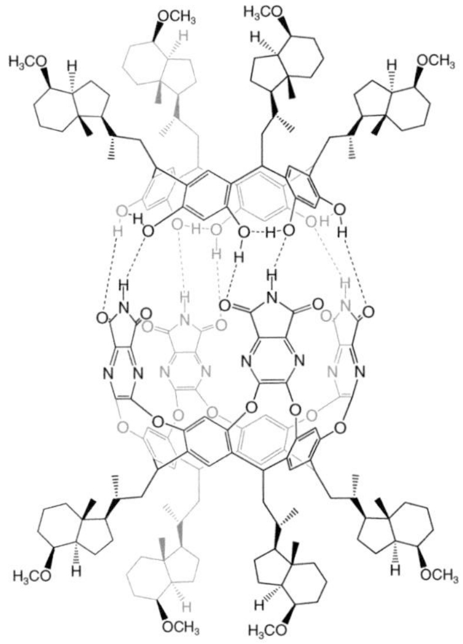

451.

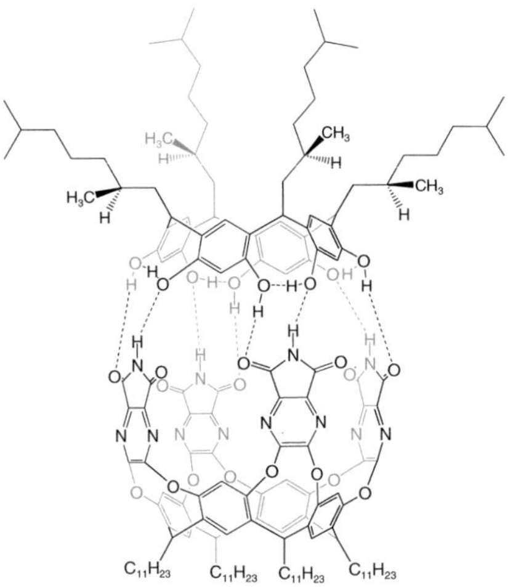

452.

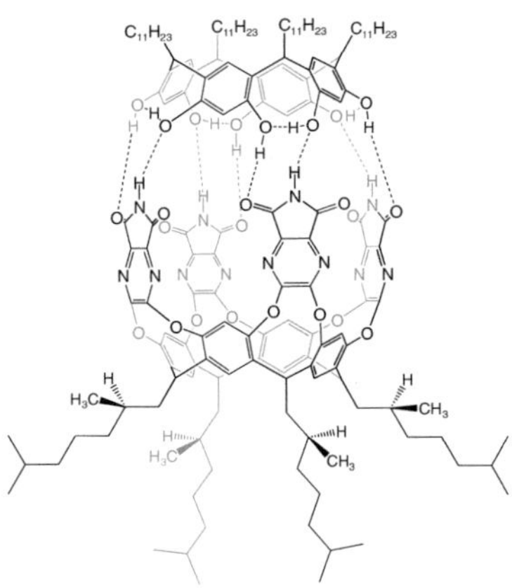

453.

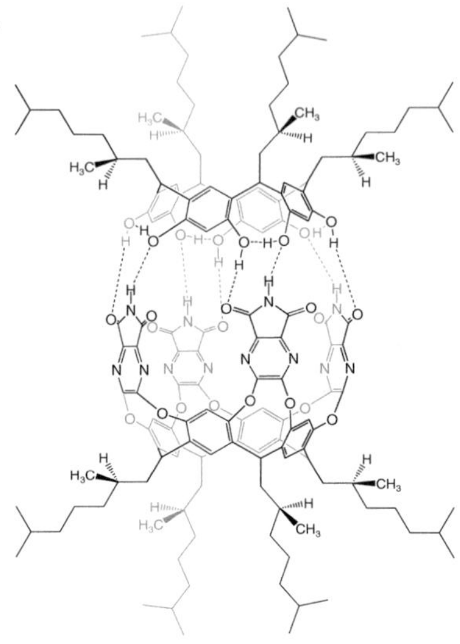

454.

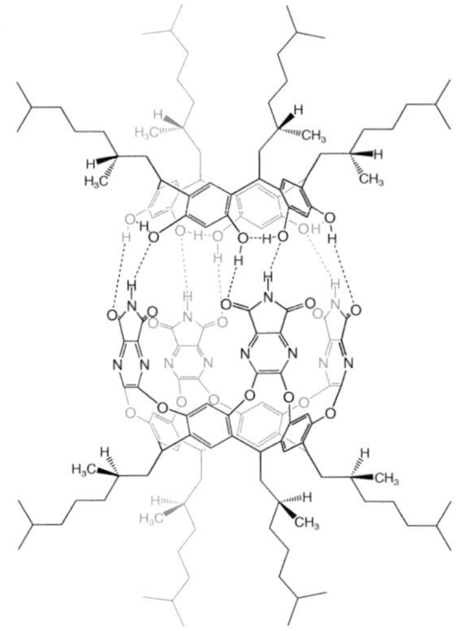

455.

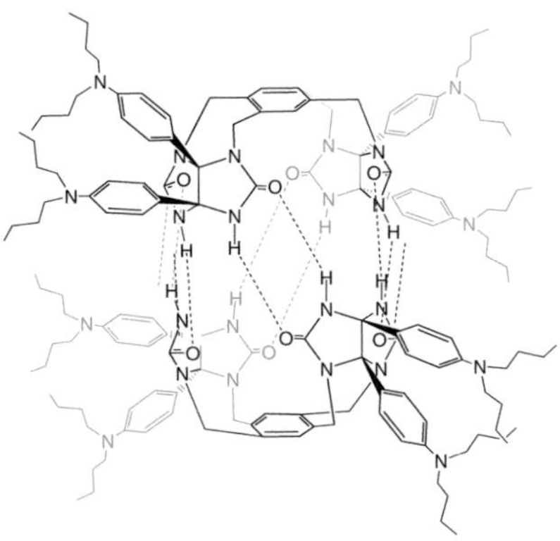

456.

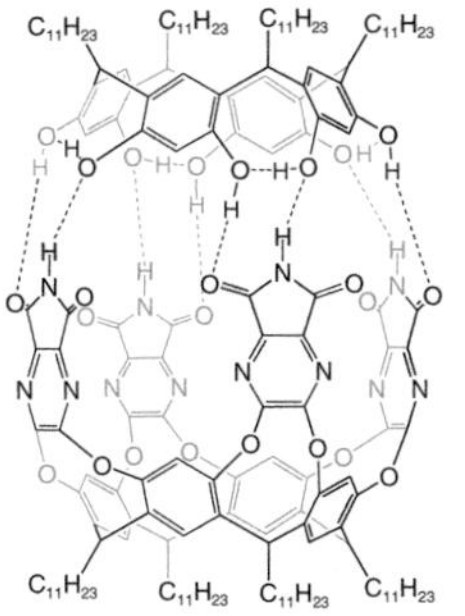

457.

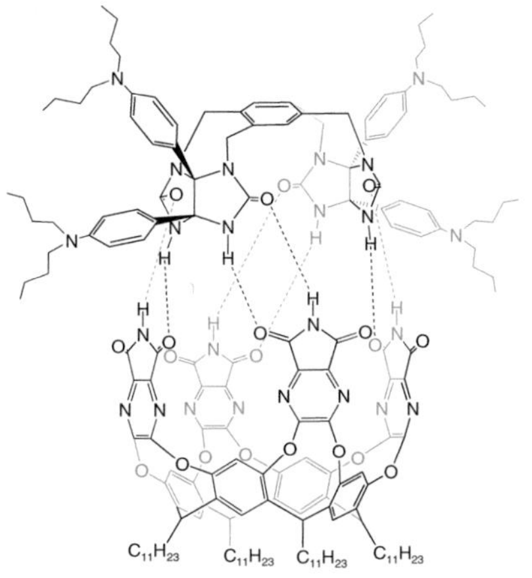

458.

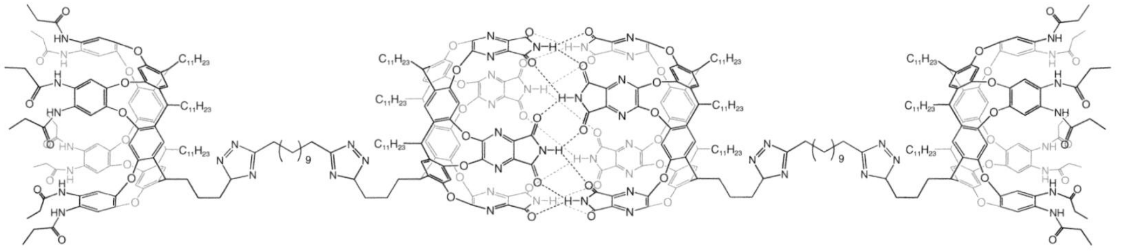

459.

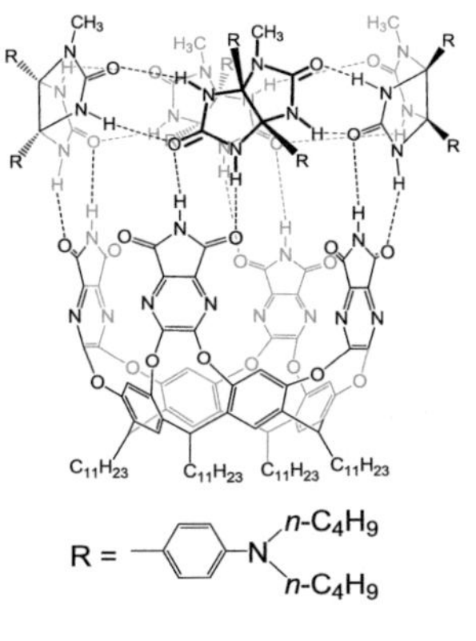

460.

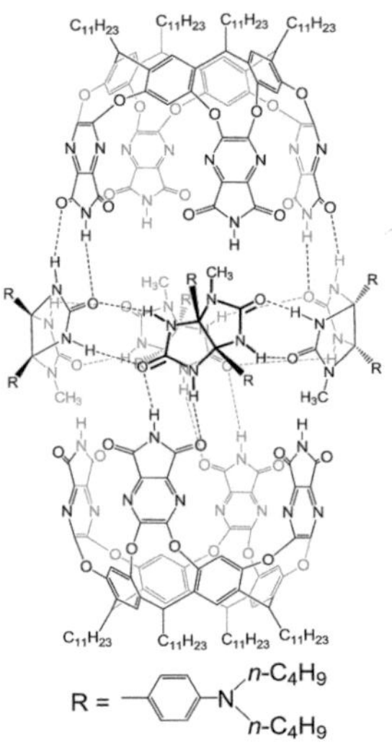

461.

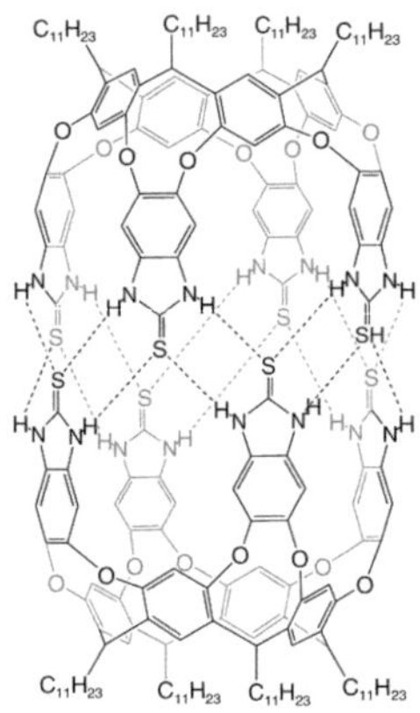

462.

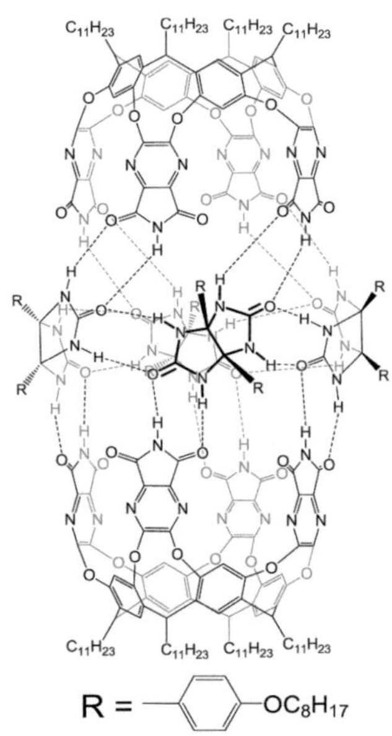

463.

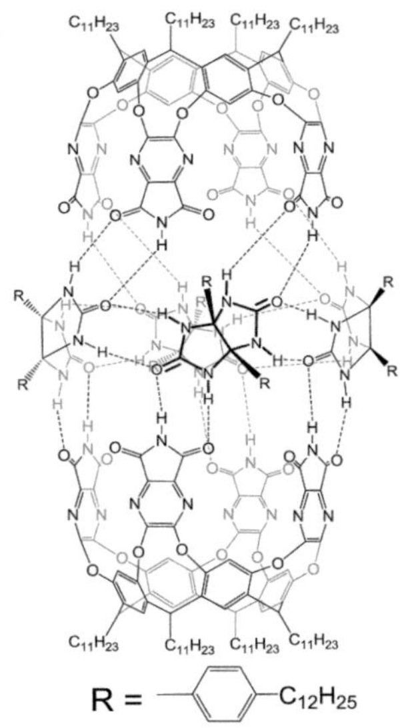

464.

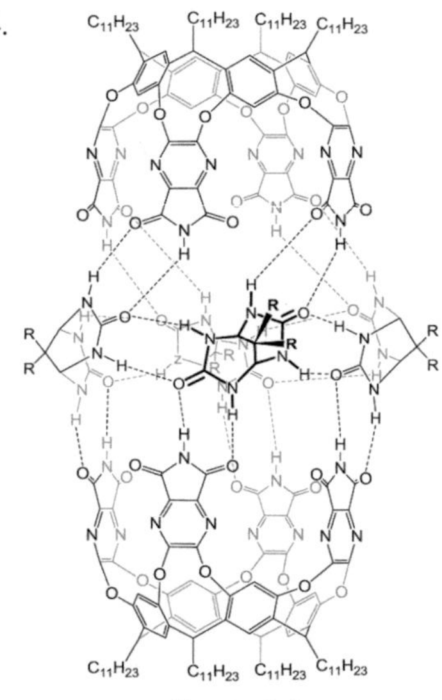

R = *n*-octyl

465.

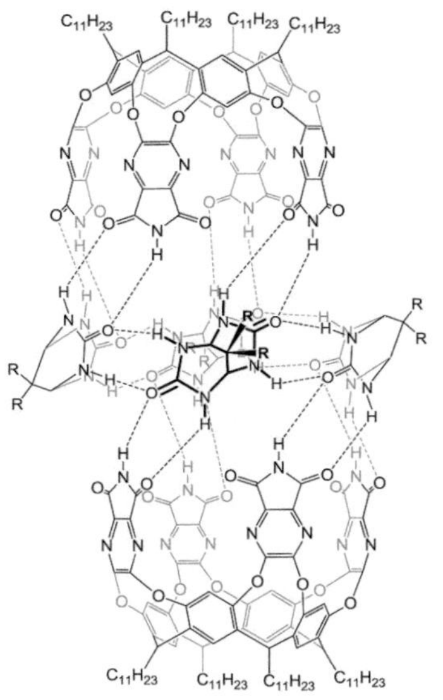

R = *n*-octyl

466.

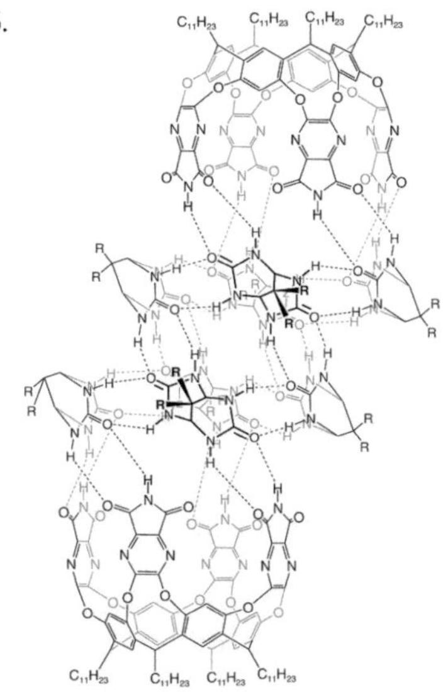

467.

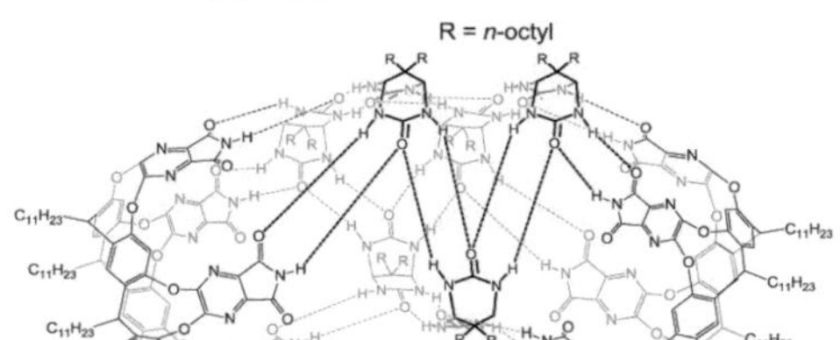

468.

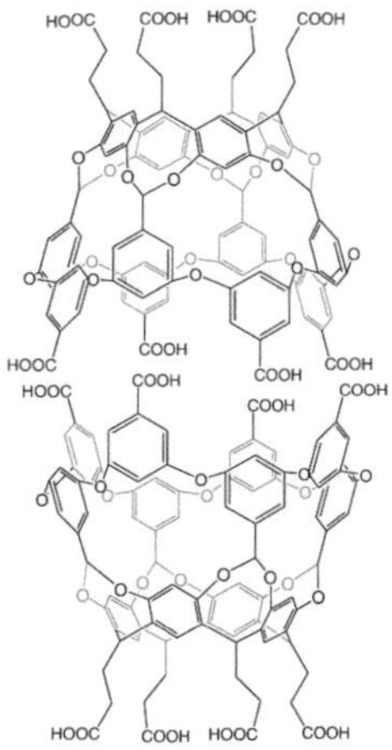

469.

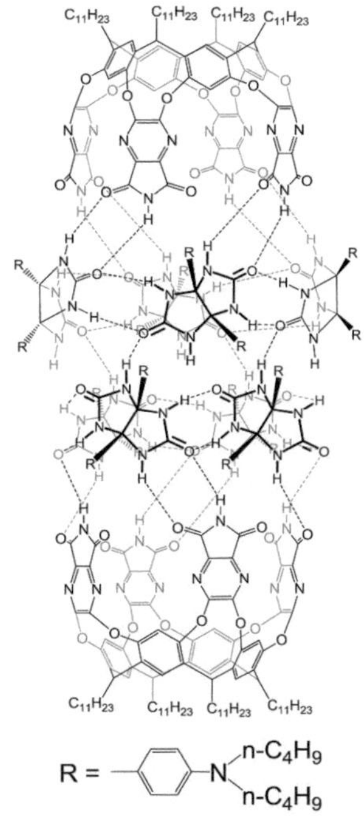

470.

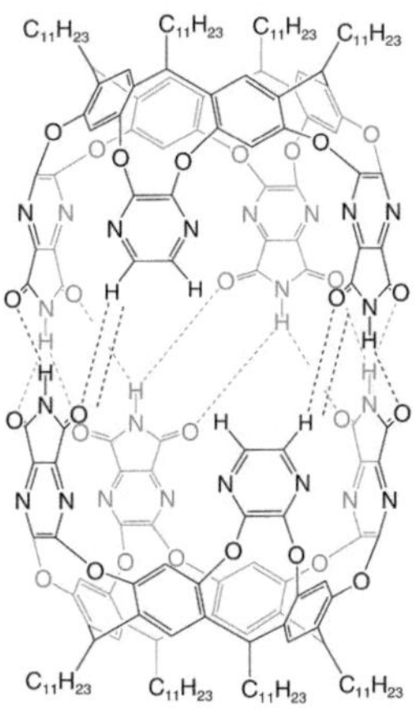

471.

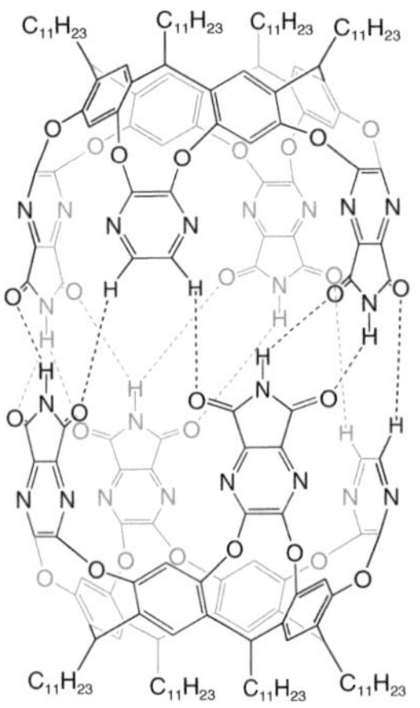

472.

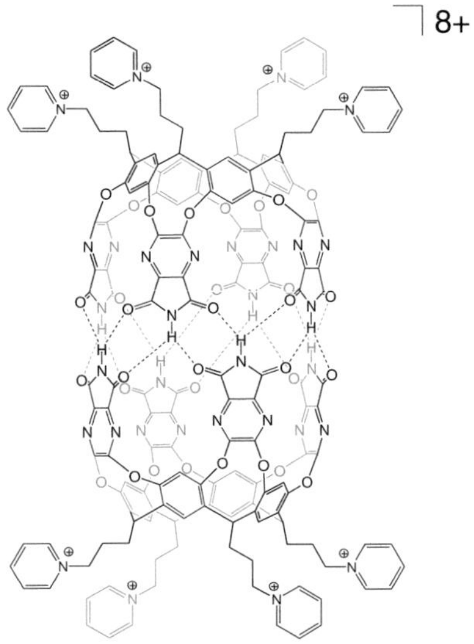

473.

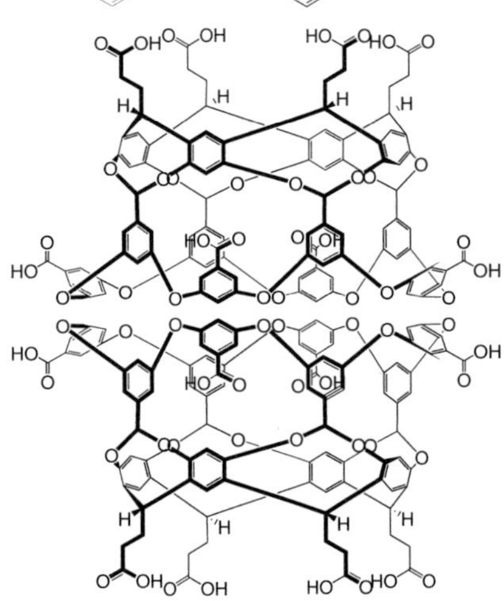

474.

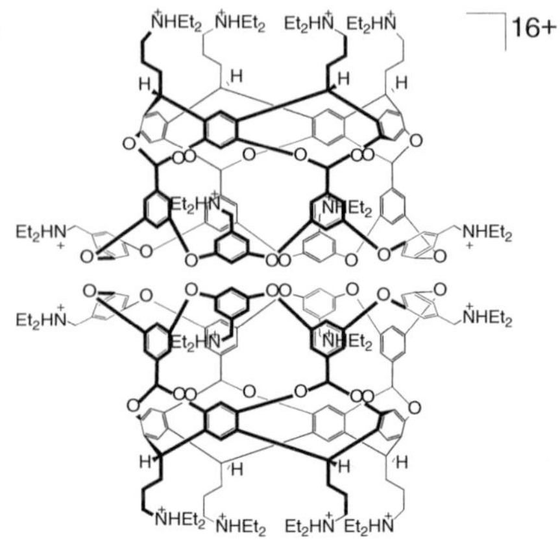

475.

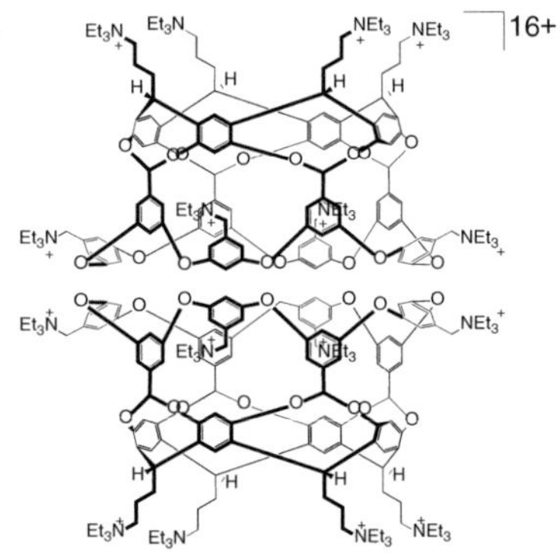

476.

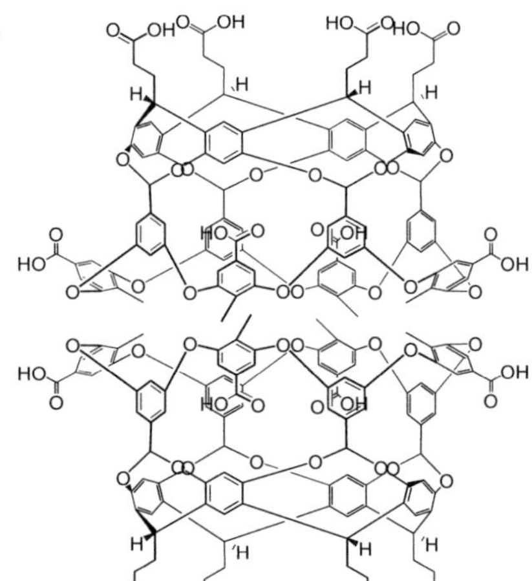

477.

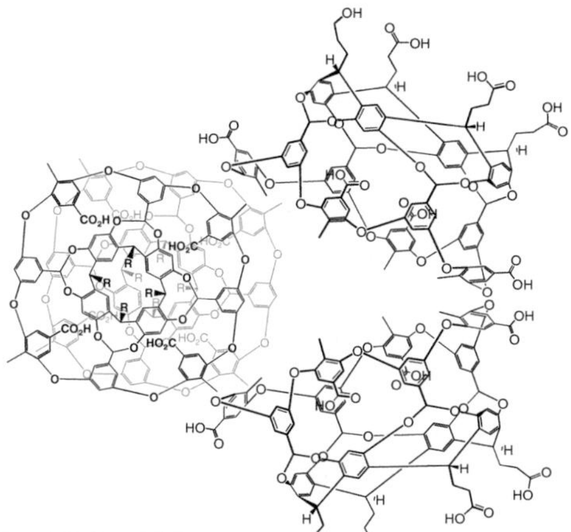

R = CH_2CH_2COOH

478.

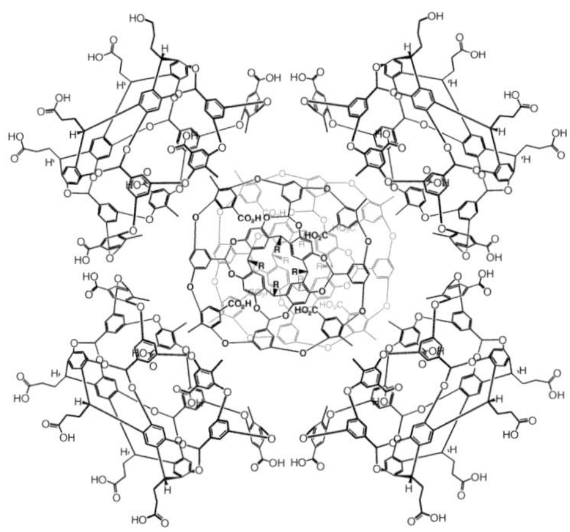

R = CH_2CH_2COOH

479.

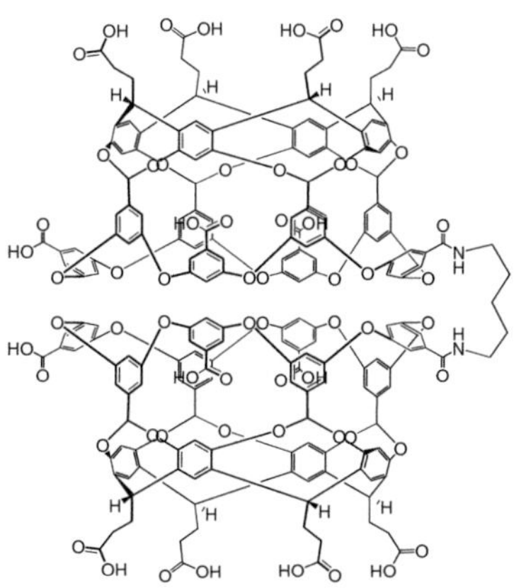

480.

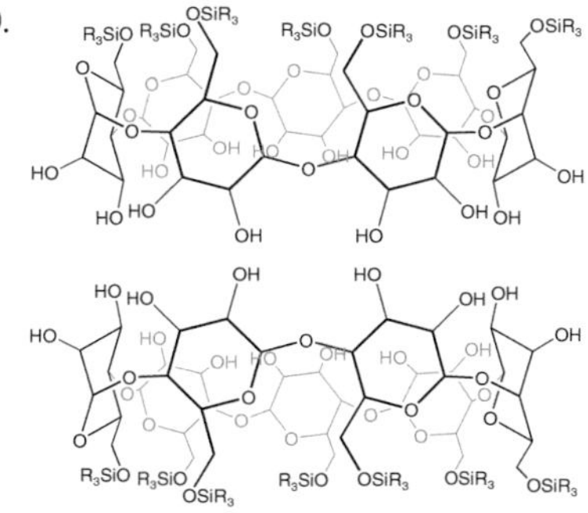

$R_3 = (i\text{-}C_3H_7)_3$

481.

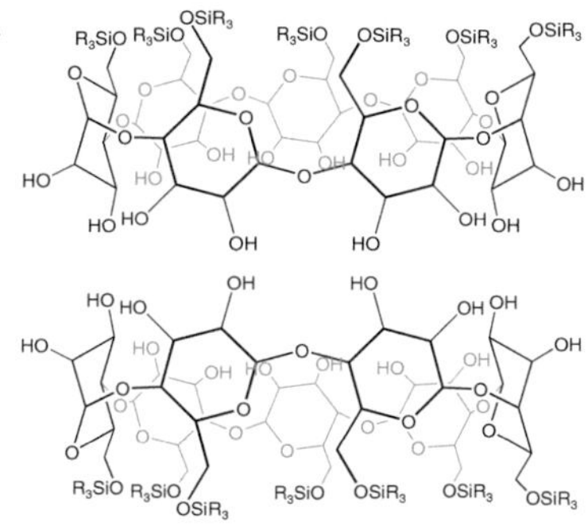

$R_3 = (tert\text{-}C_4H_9)(CH_3)_2$

482.

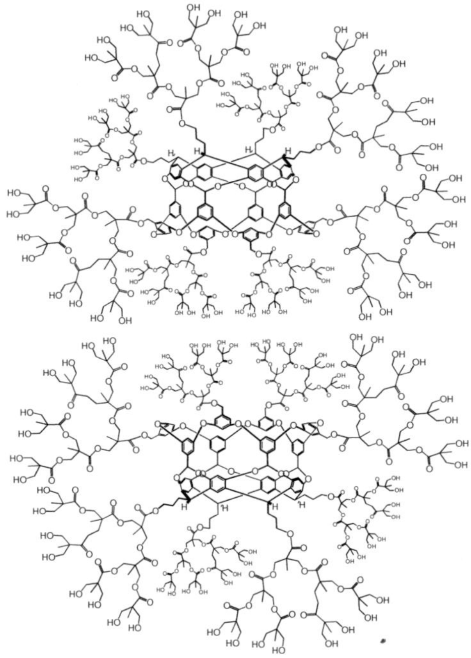

483.

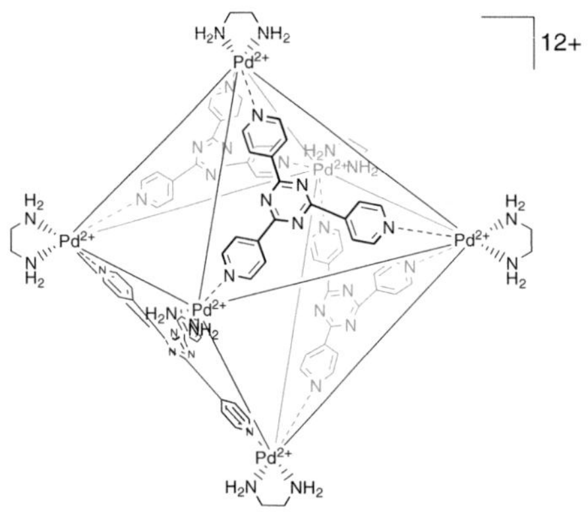

484.

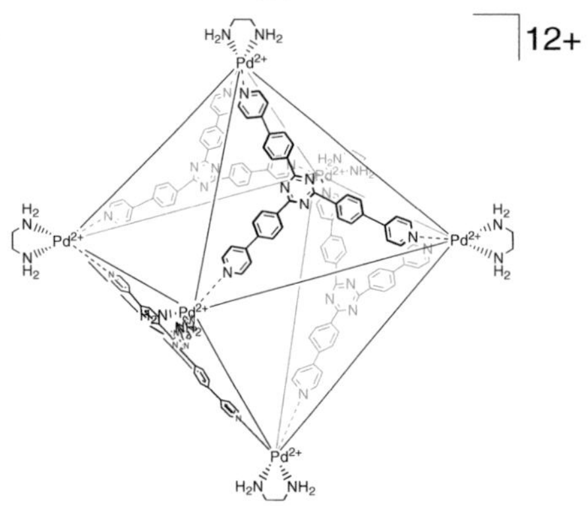

485.

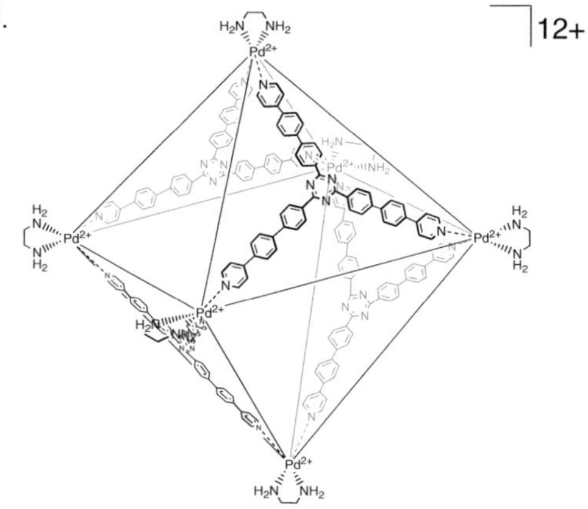

486.

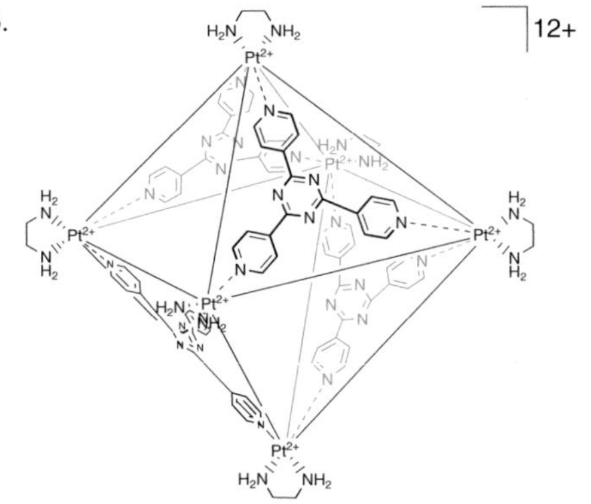

487.

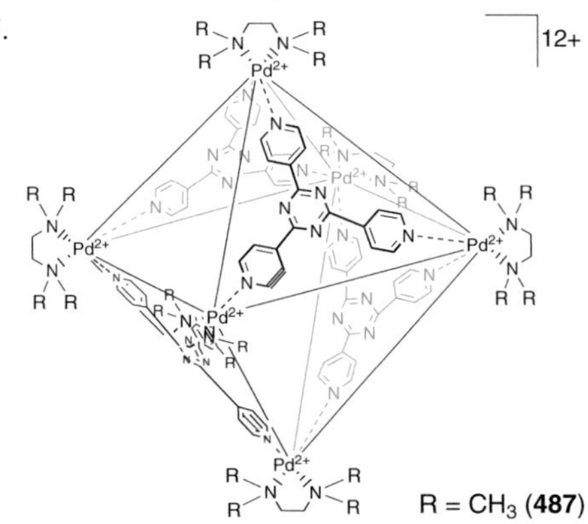

R = CH_3 (**487**)

488.

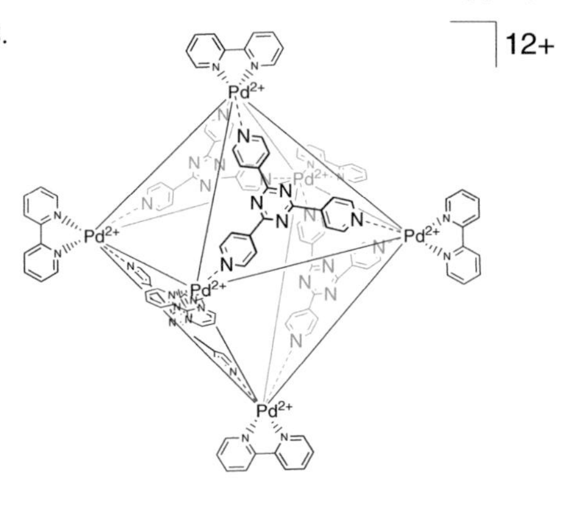

489.

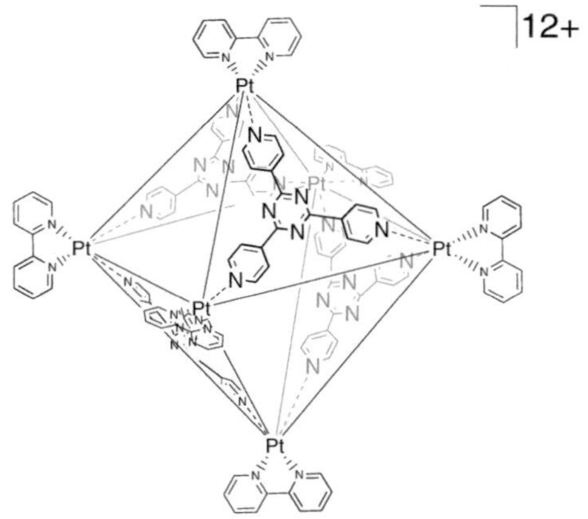

490.

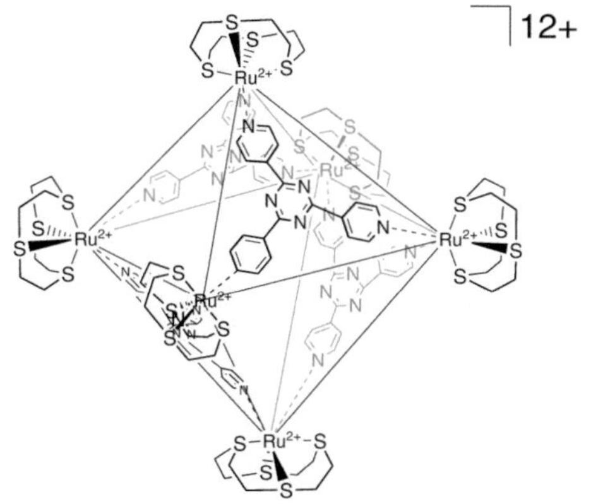

491.

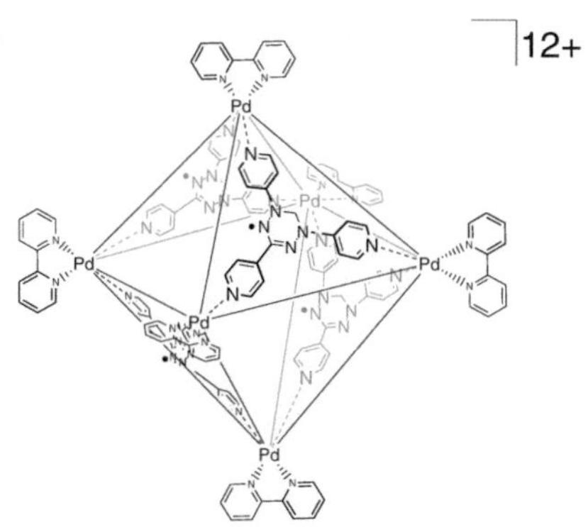

492.

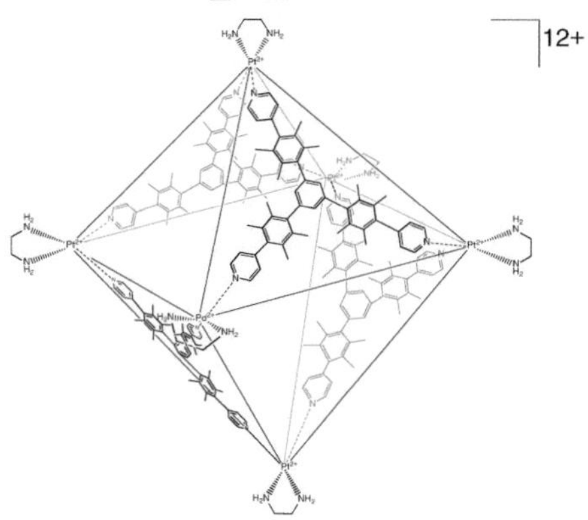

493.

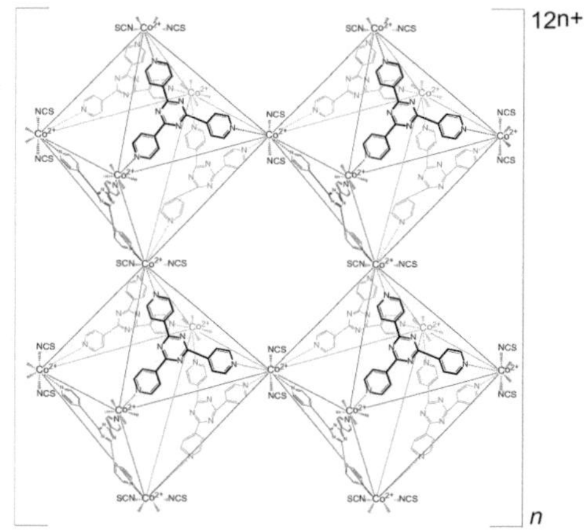

494.

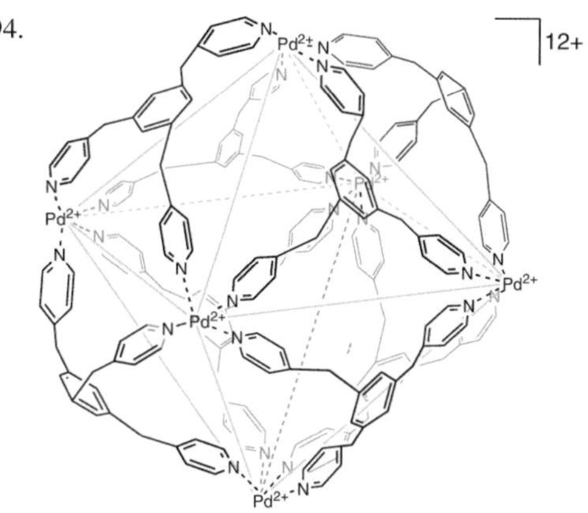

495.

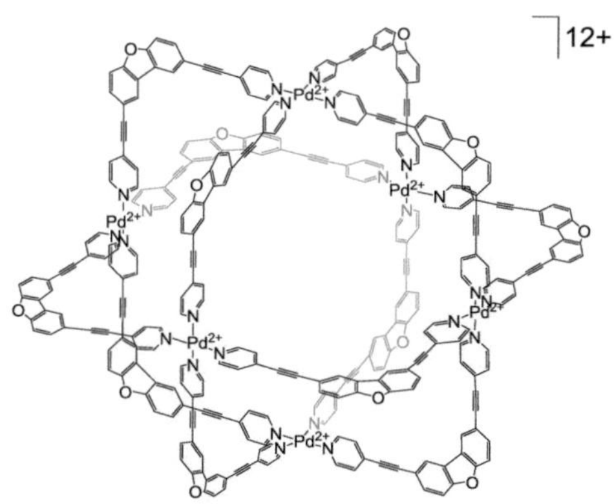

496.

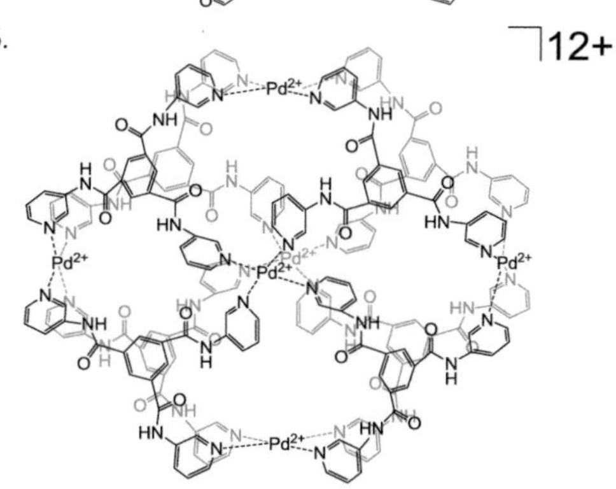

497.

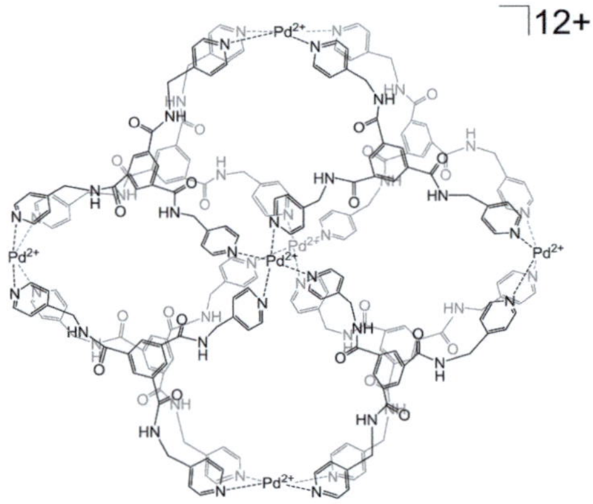

498.

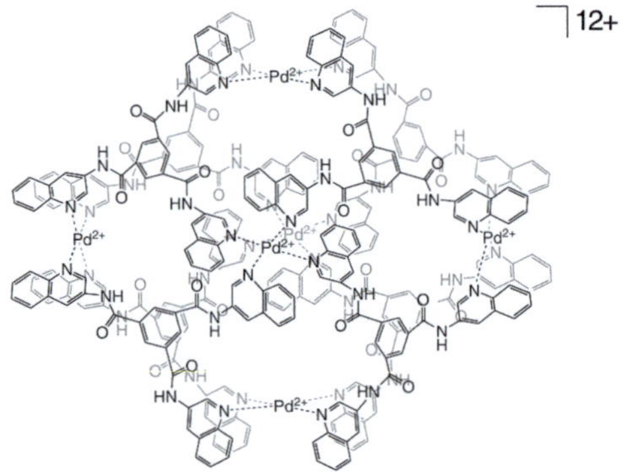

499.

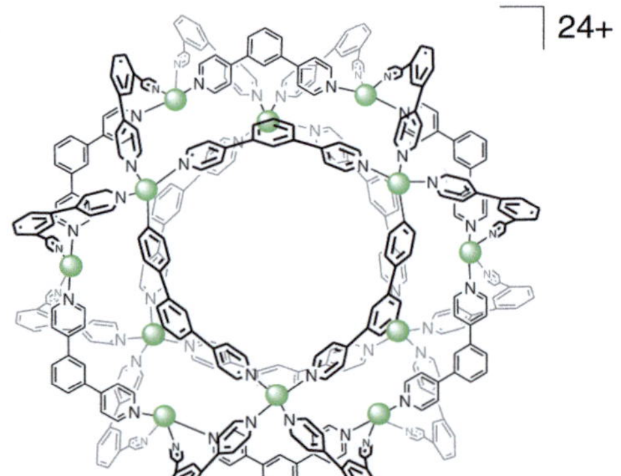

500.

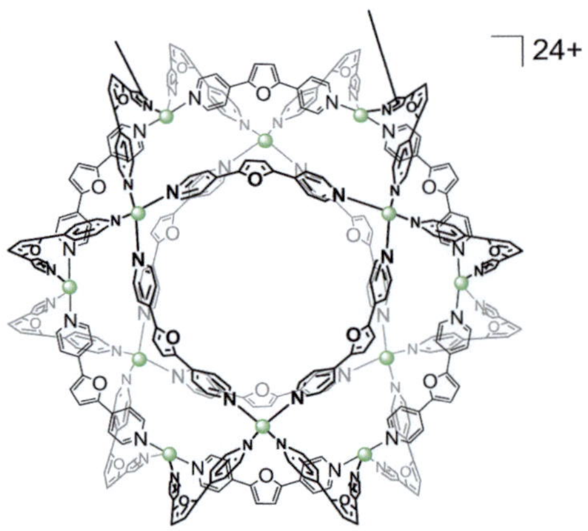

501.

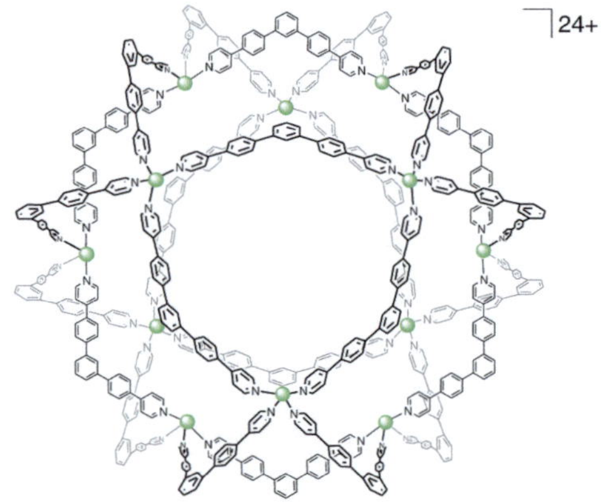

502.

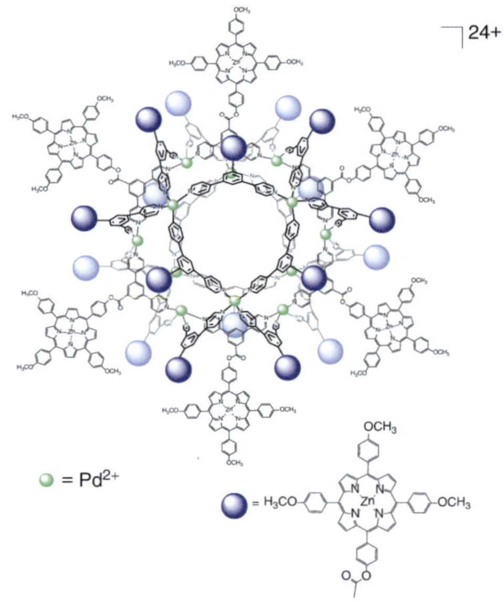

503.

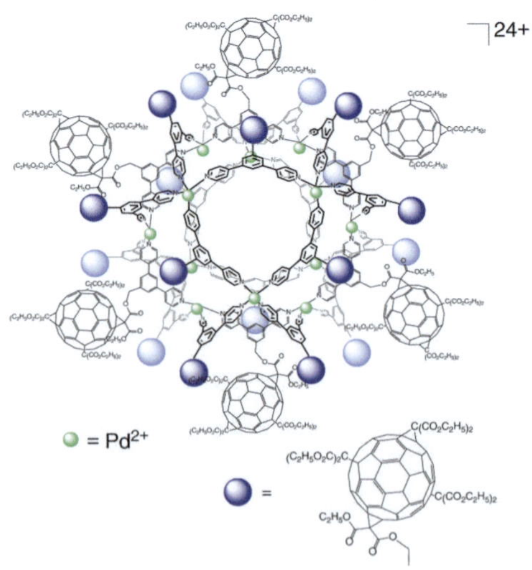

504.

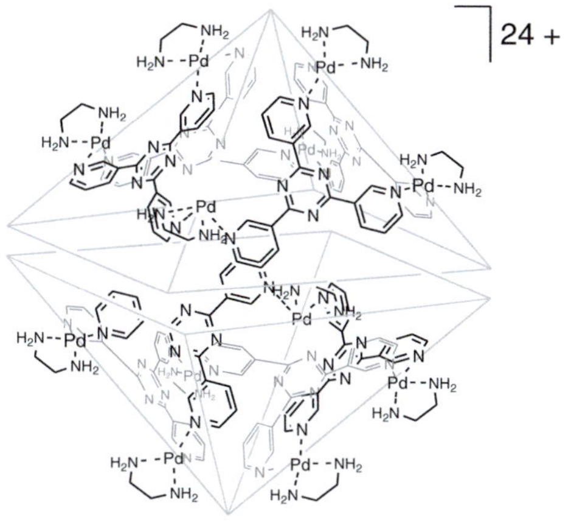

507.

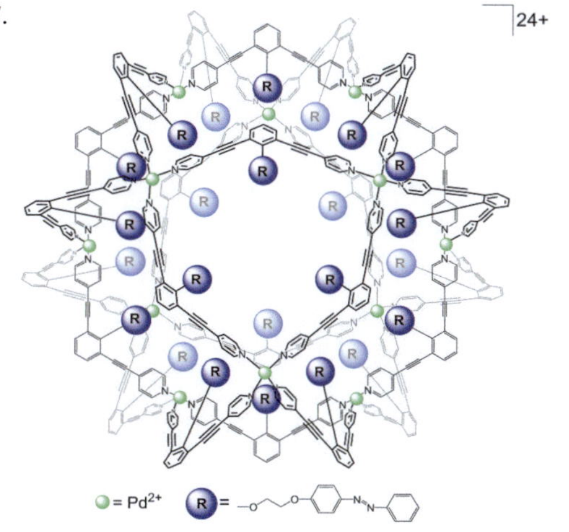

505.

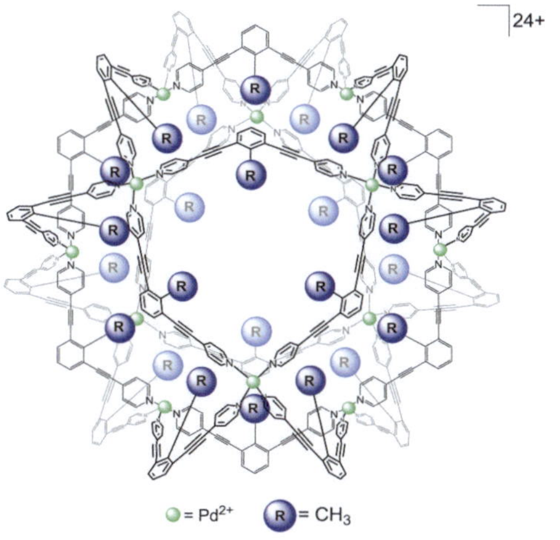

508.

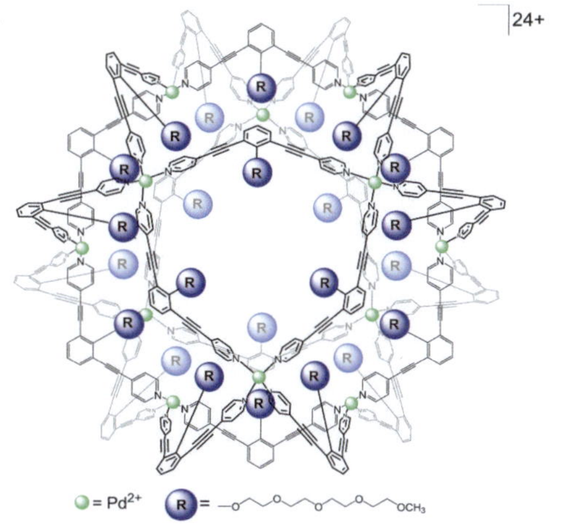

506.

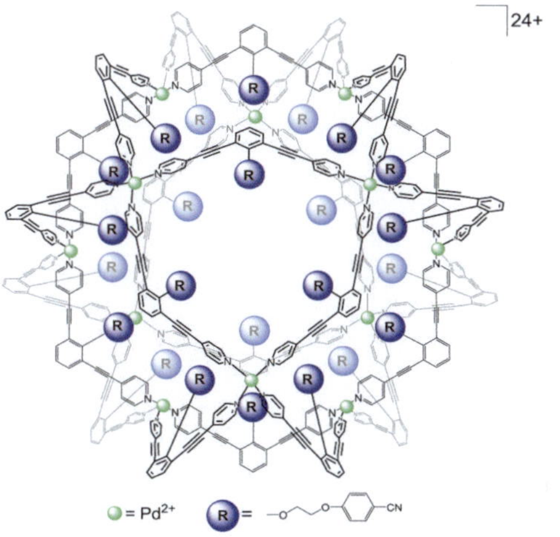

509.

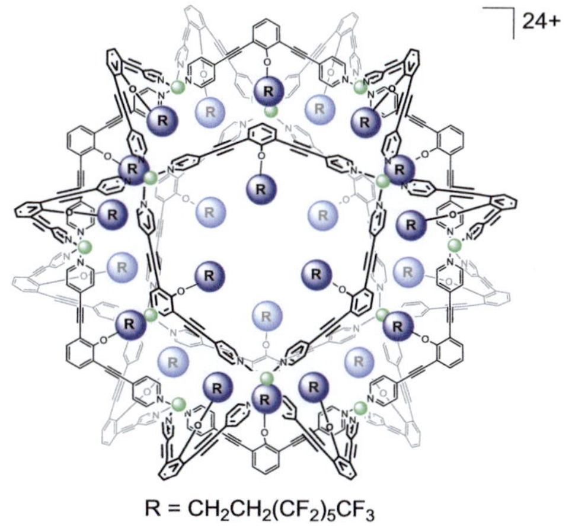

R = $CH_2CH_2(CF_2)_5CF_3$

510.

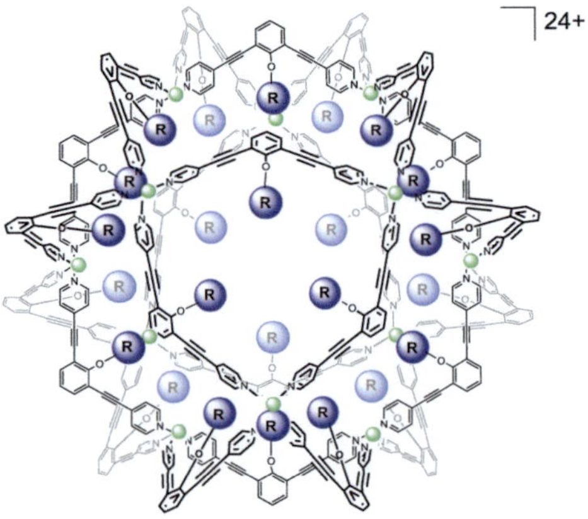

R = $CH_2CH_2(CF_2)_7CF_3$

511.

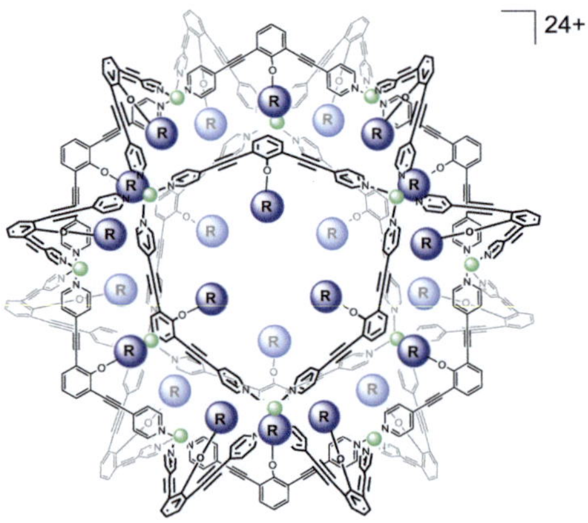

R = $CH_2CH_2(CF_2)_6CF(CF_3)_2$

512

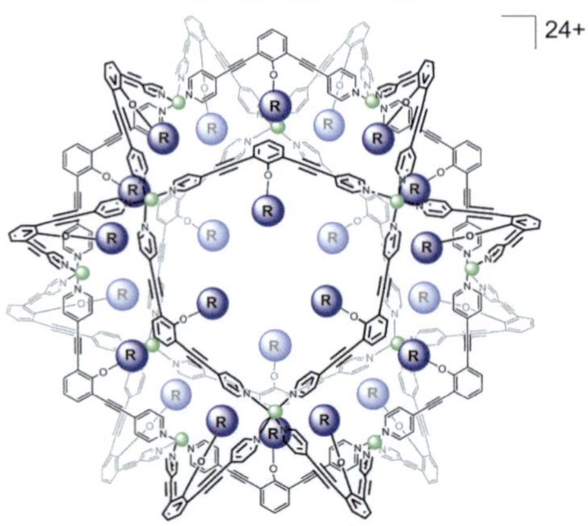

R = $CH_2CH_2(CF_2)_3CF_3$

513.

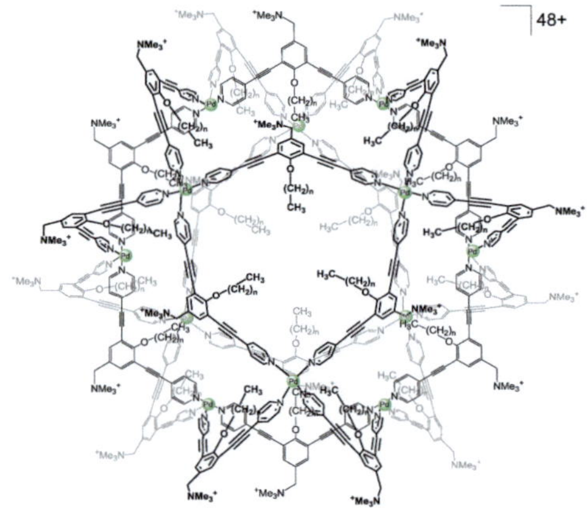

= Pd^{2+} n = 8

514.

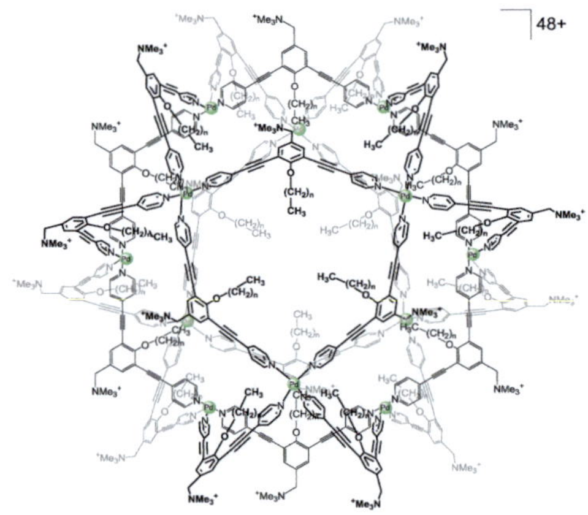

= Pd^{2+} n = 11

515.

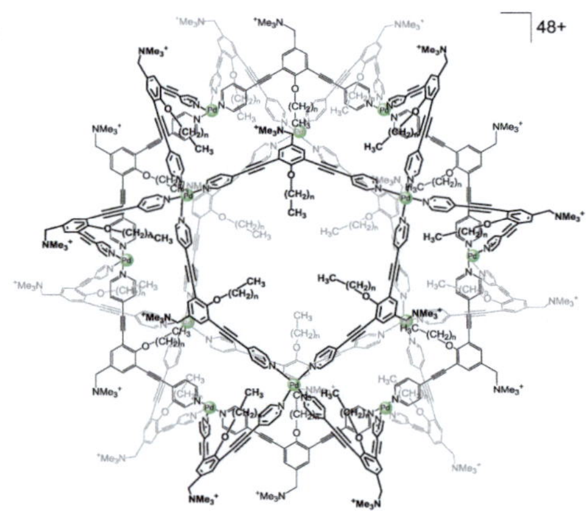

= Pd^{2+} n = 17

516.

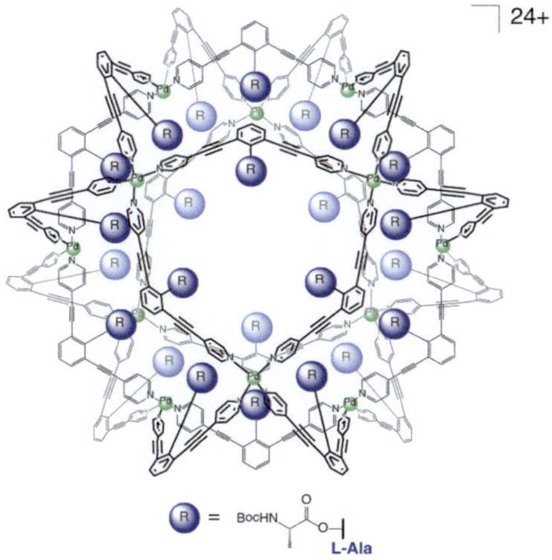

519.

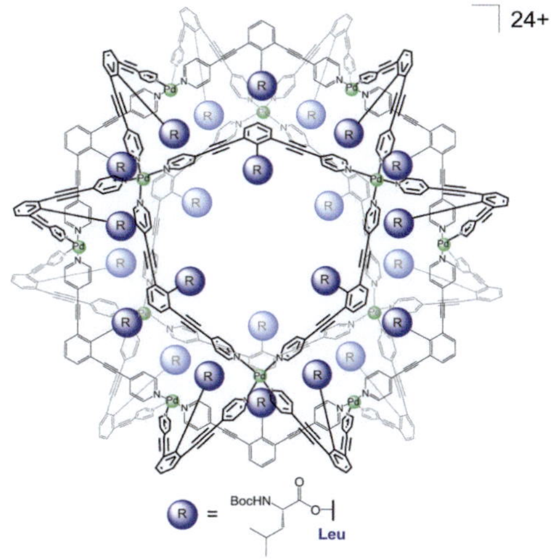

517.

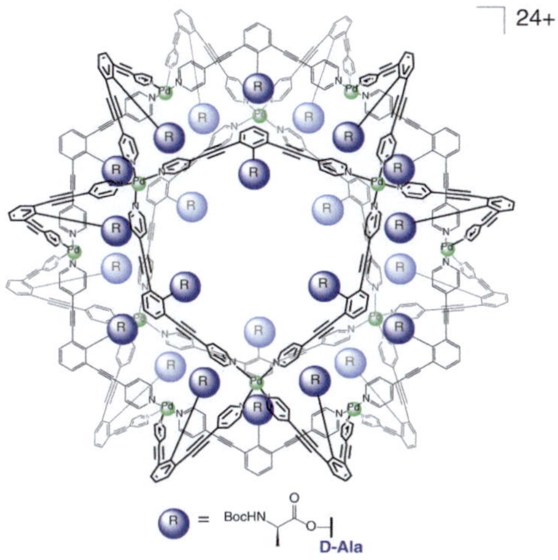

520.

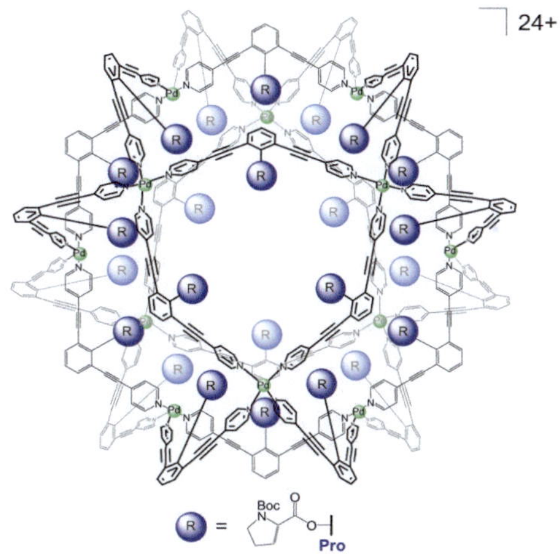

518.

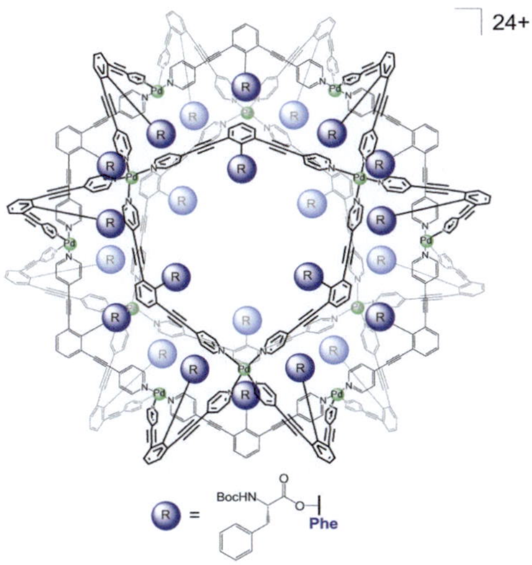

521.

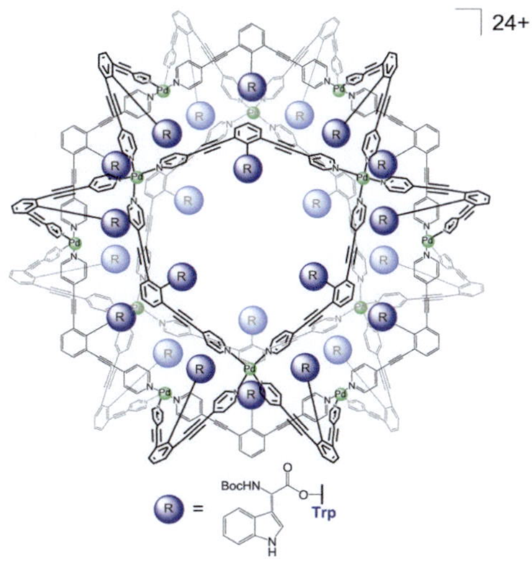

522.

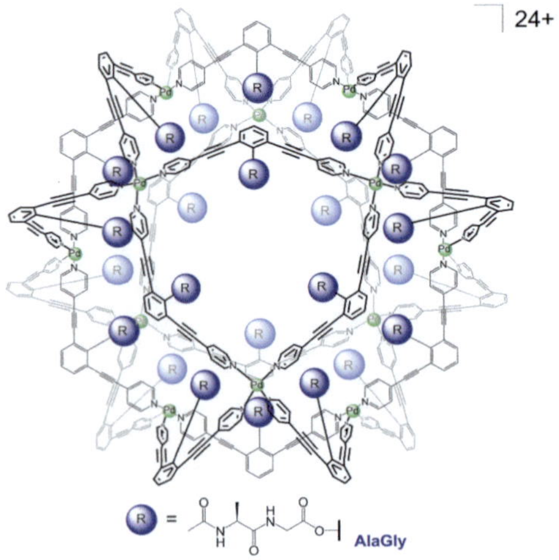

523.

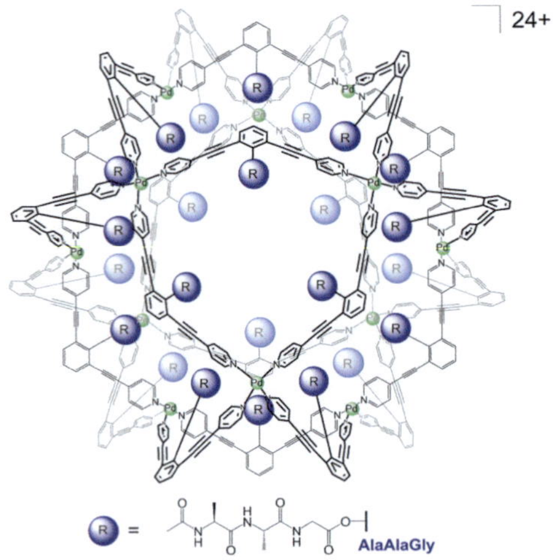

524.

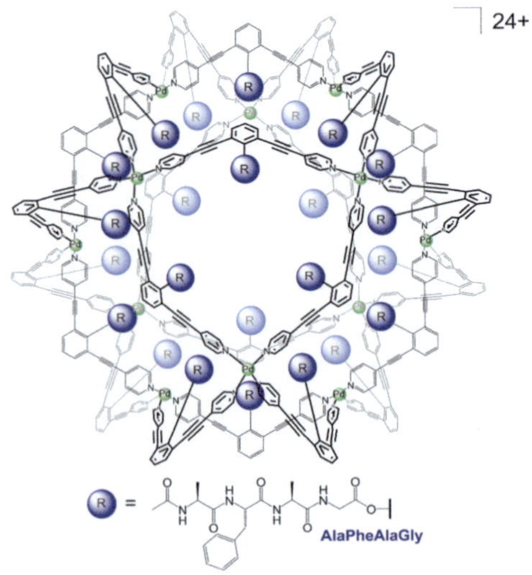

525.

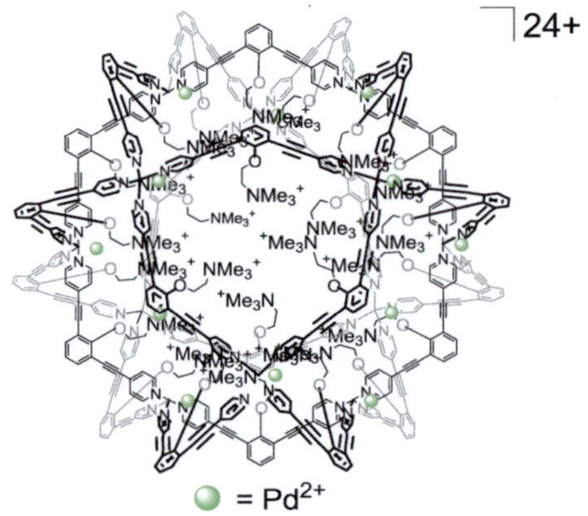

526.

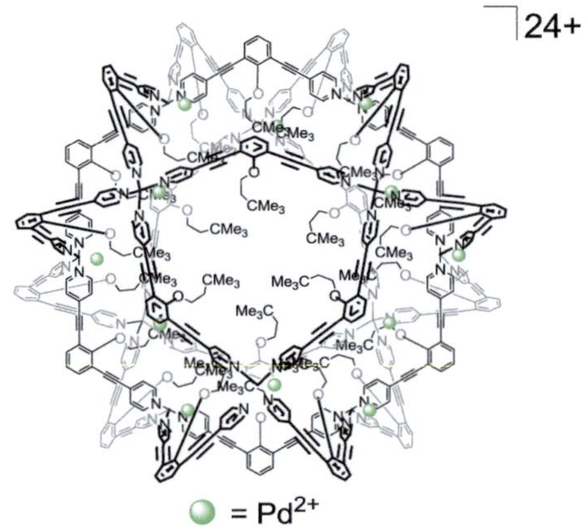

527.

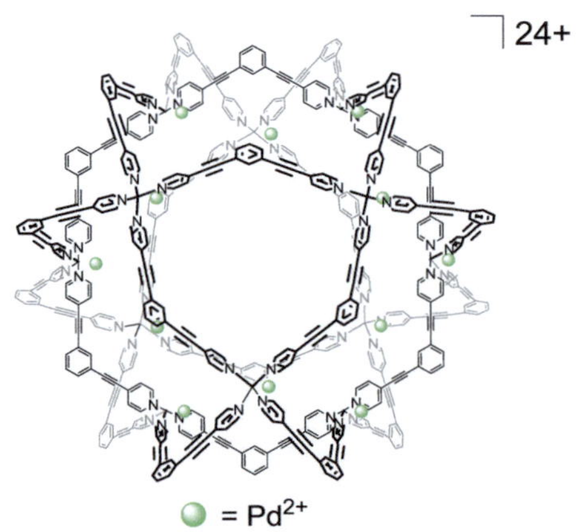

528.

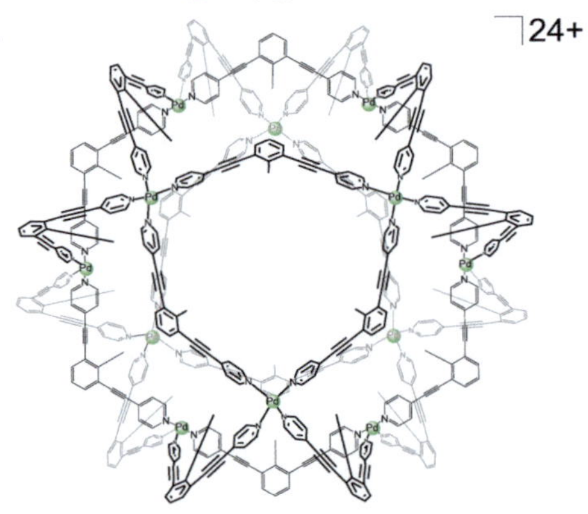

529.

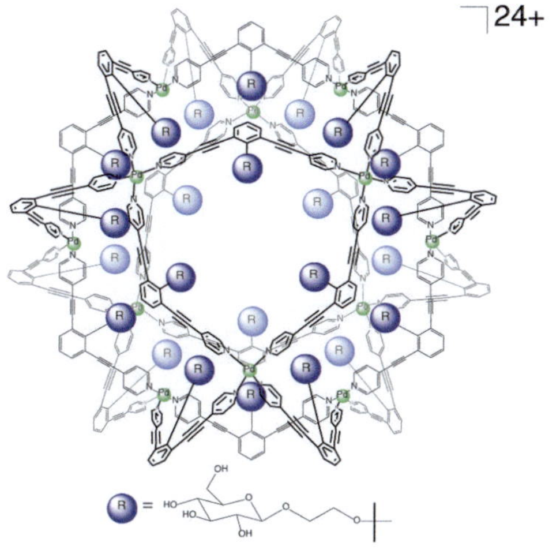

530.

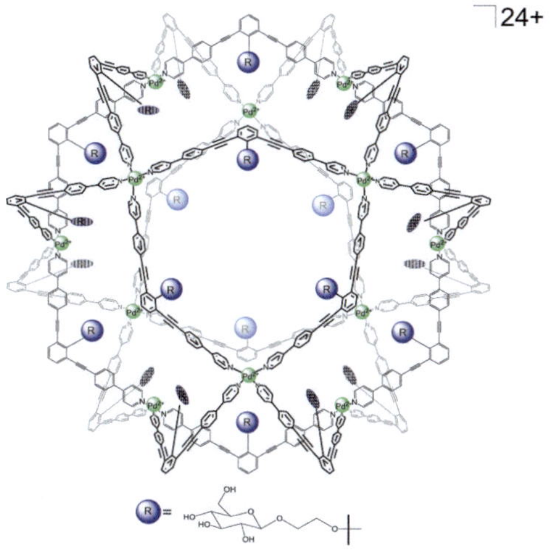

531.

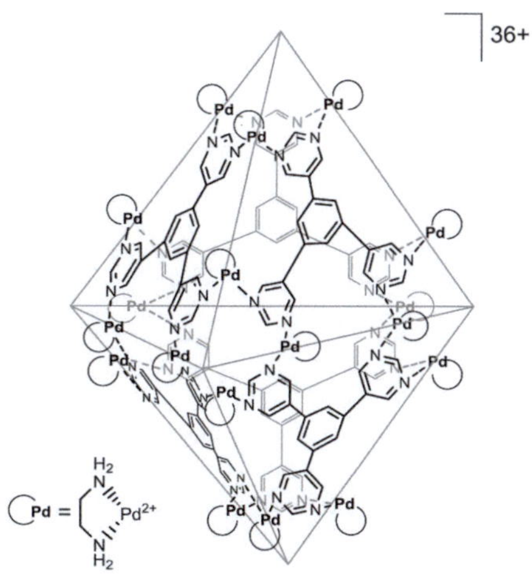

532.

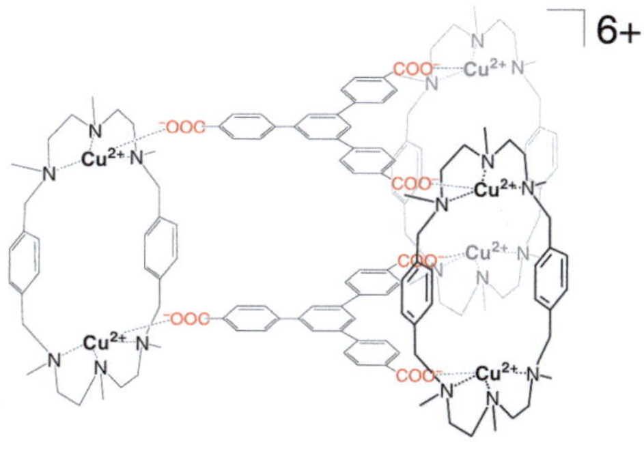

533.

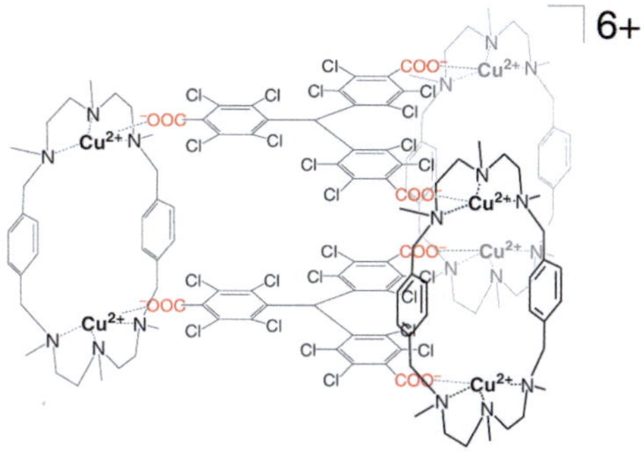

534.

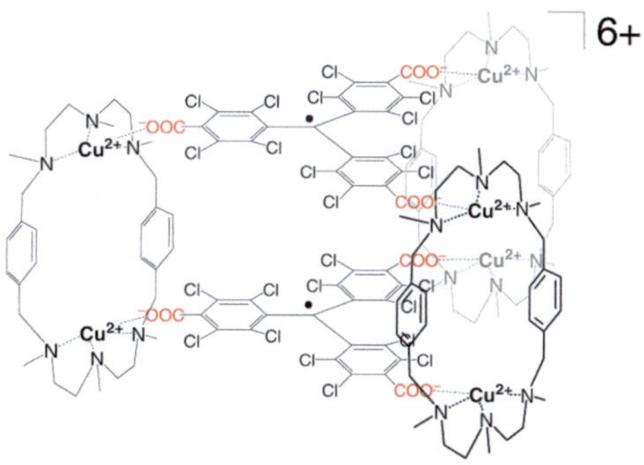

535.

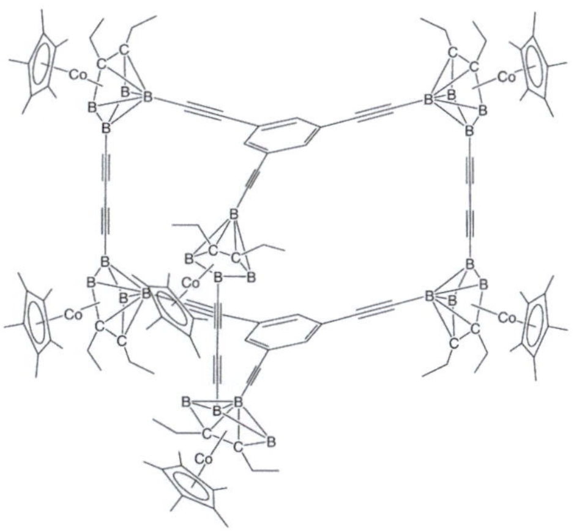

536.

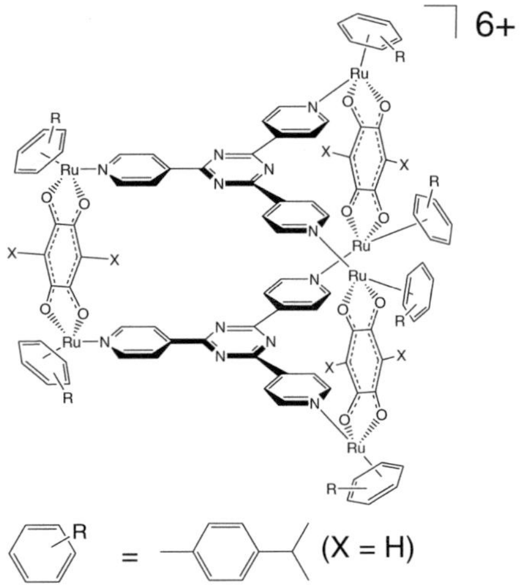

537.

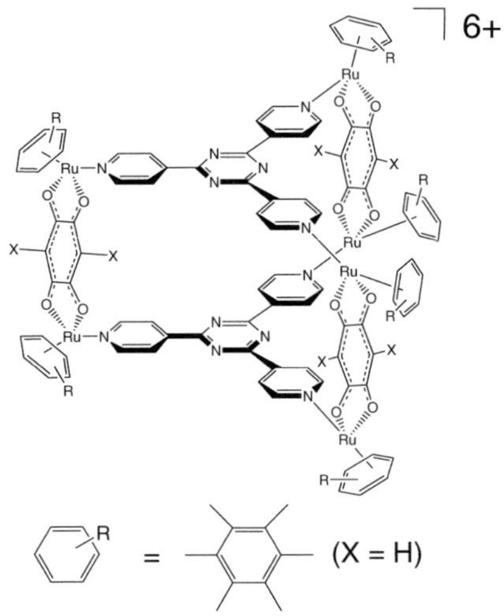

538.

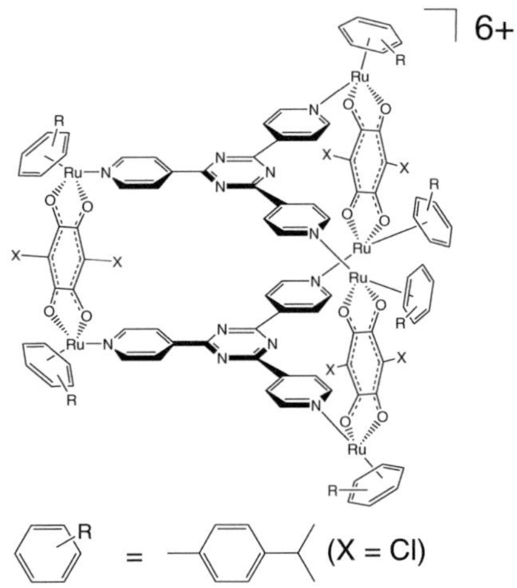

539.

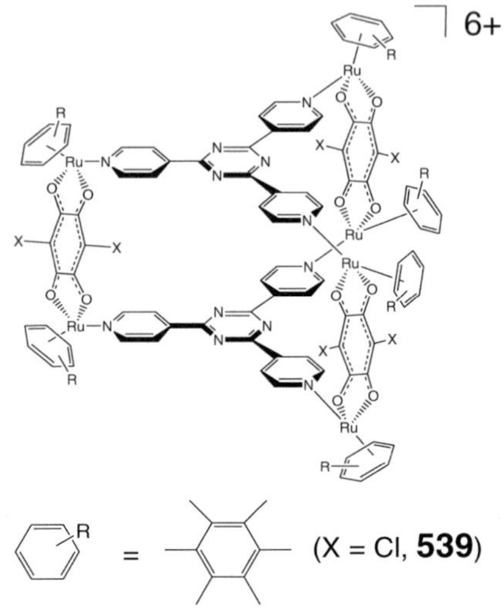

540.

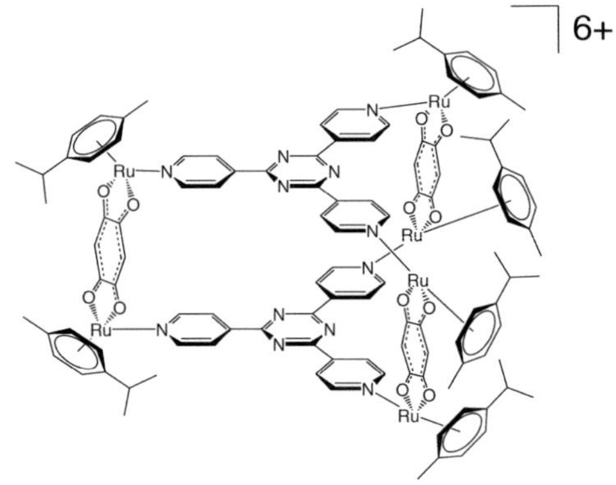

541.

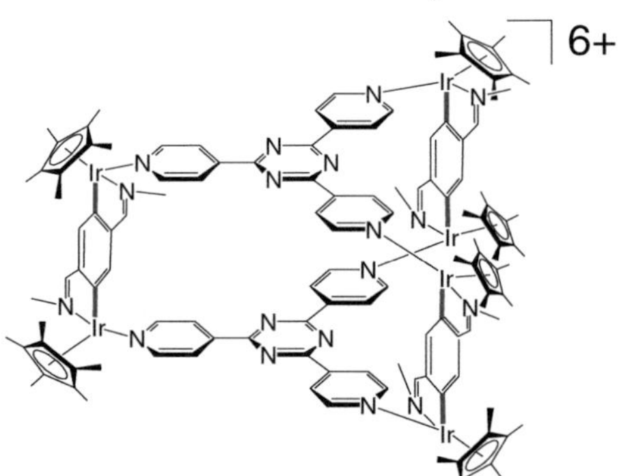

542.

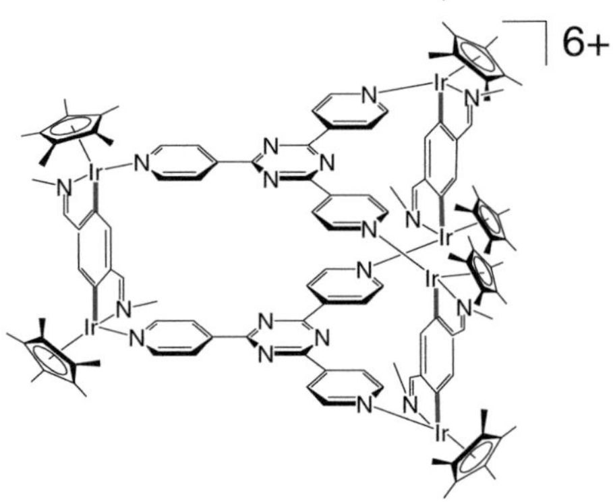

543.

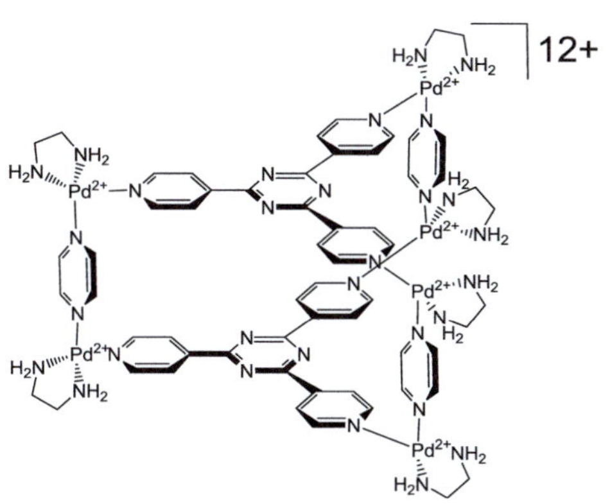

544.

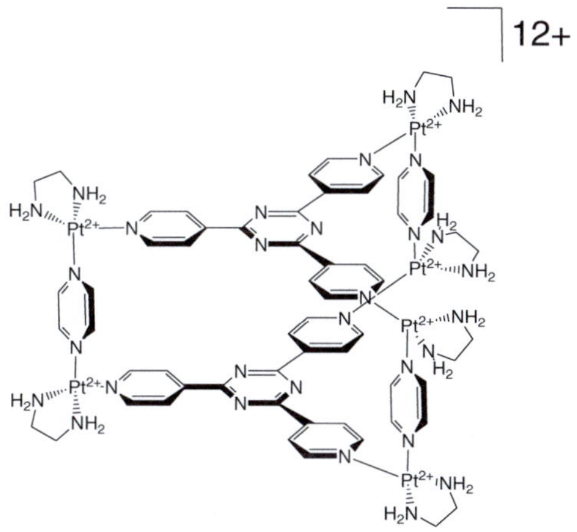

545.

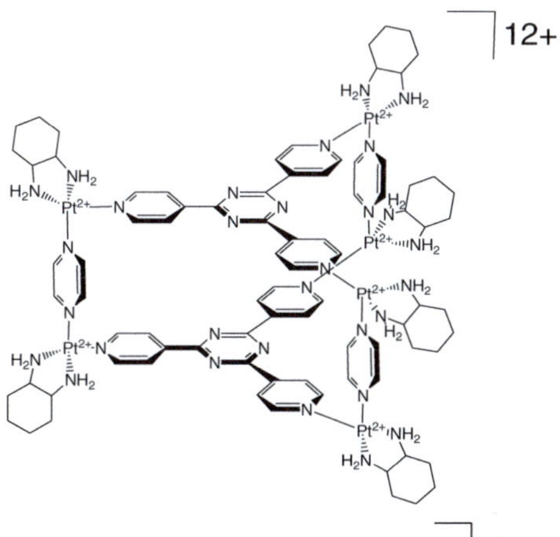

546.

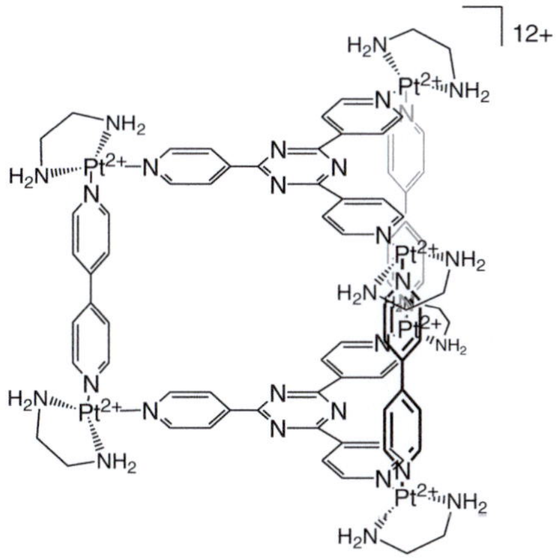

547.

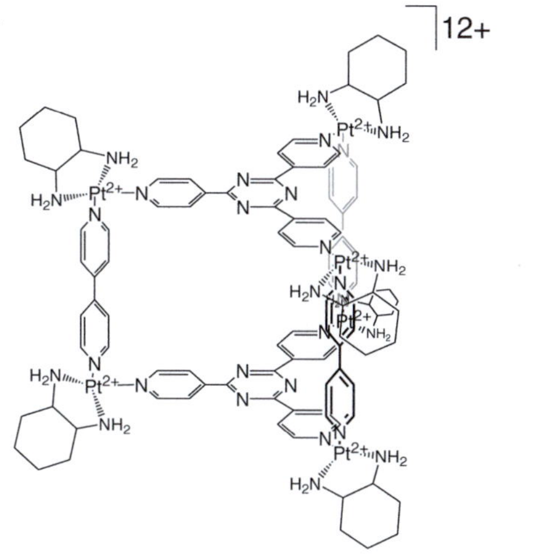

548.

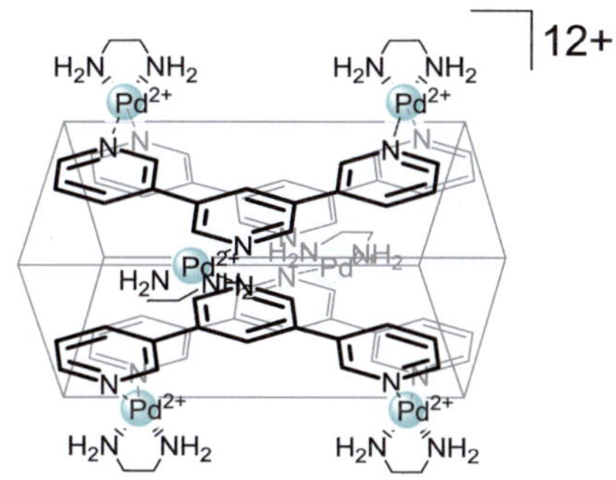

549.

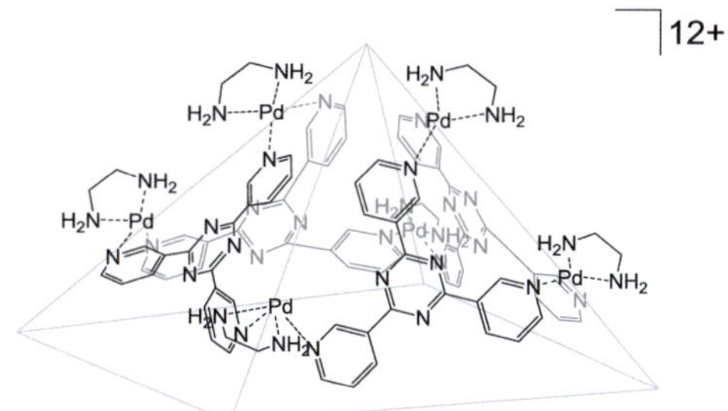

550.

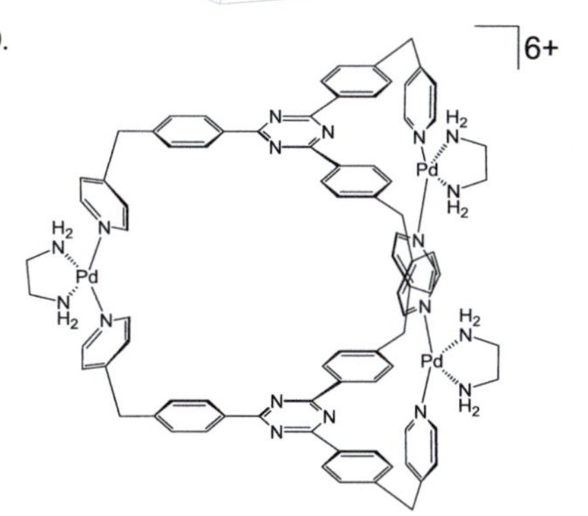

551.

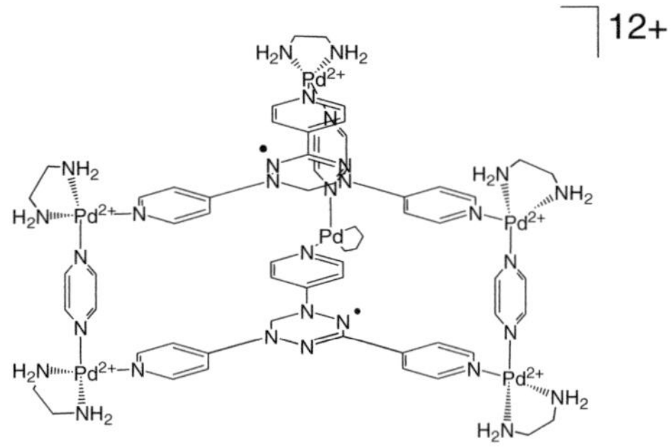

552.

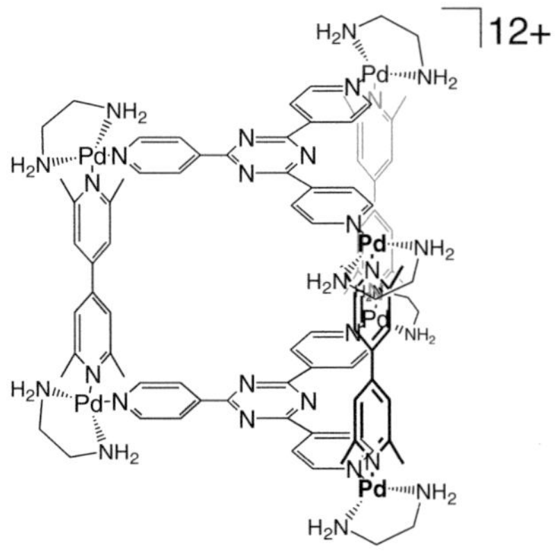

553.

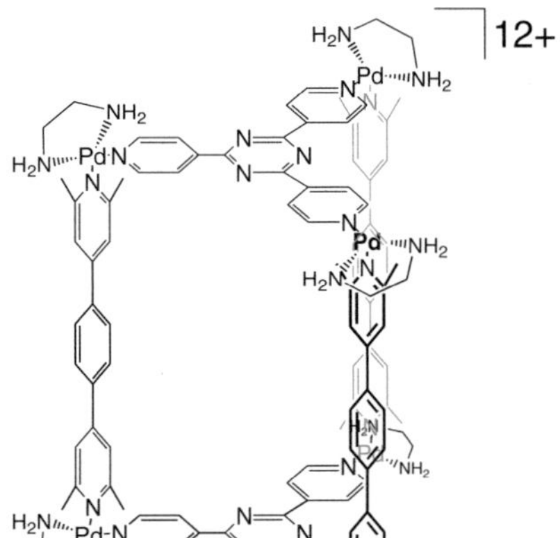

554.

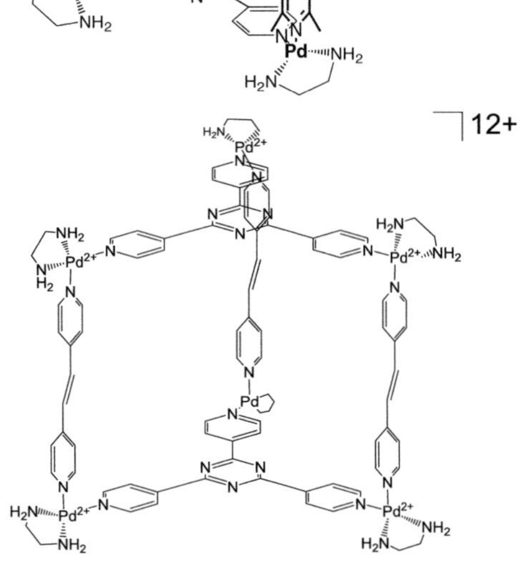

555.

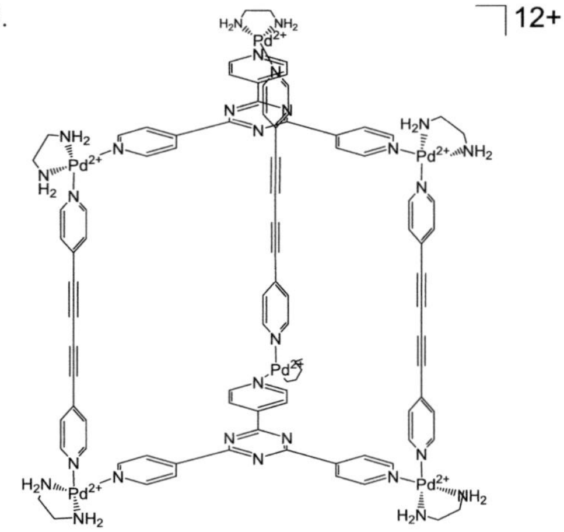

556.

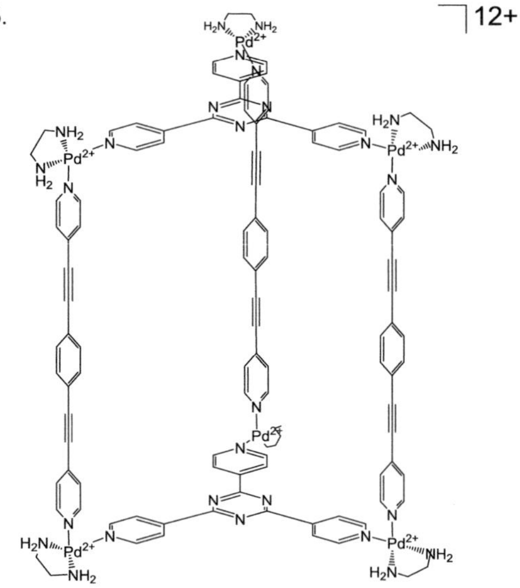

557.

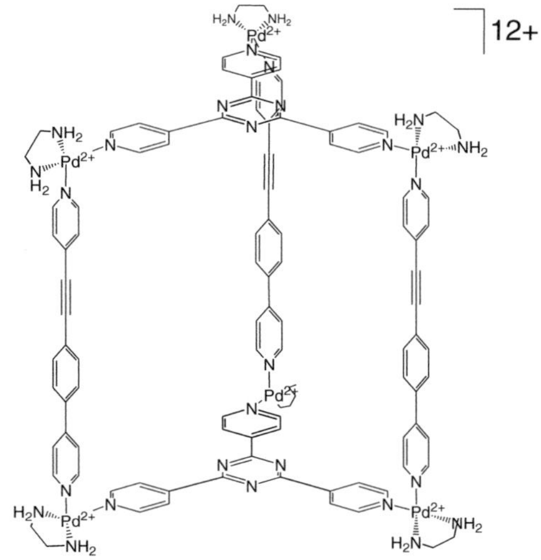

558.

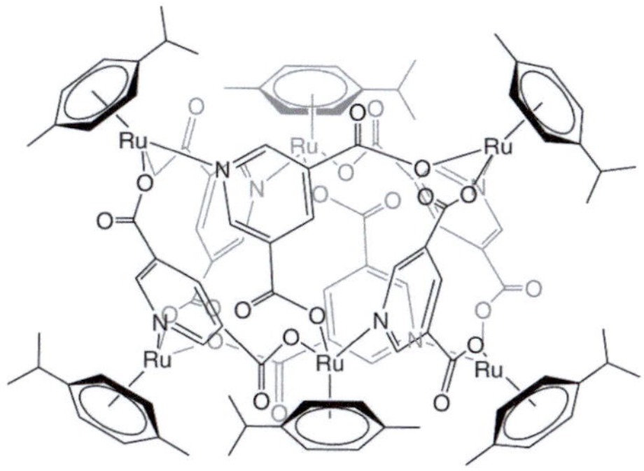

559.

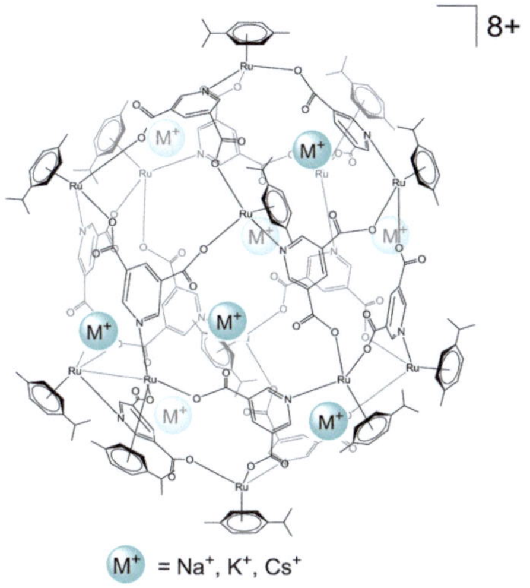

560.

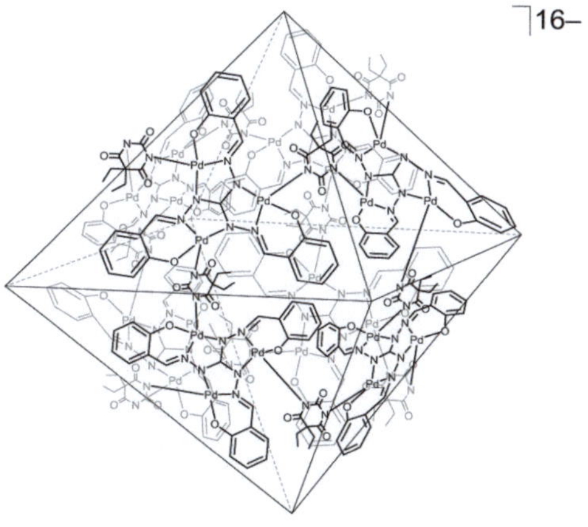

561.

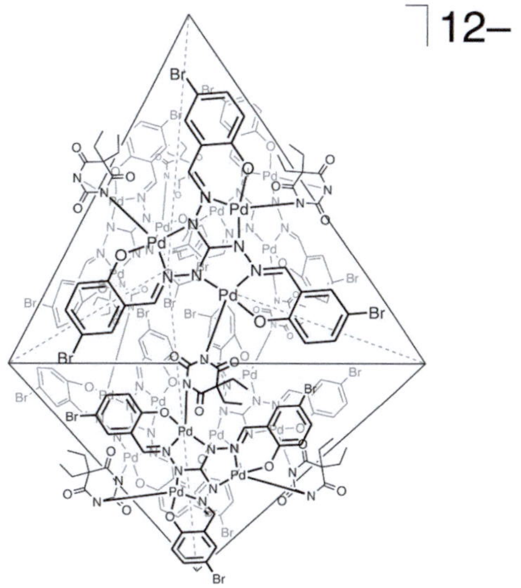

562.

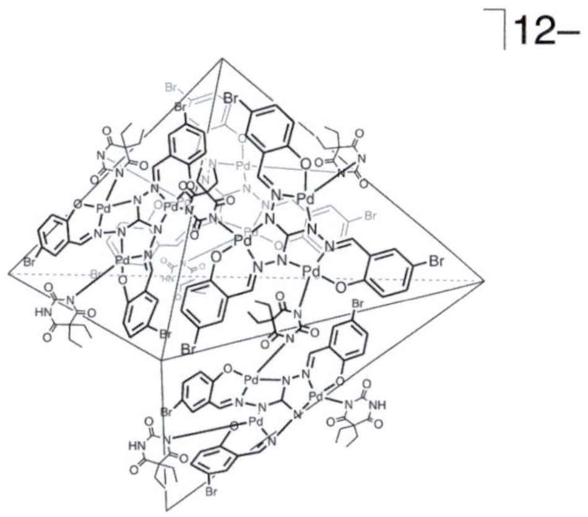

563.

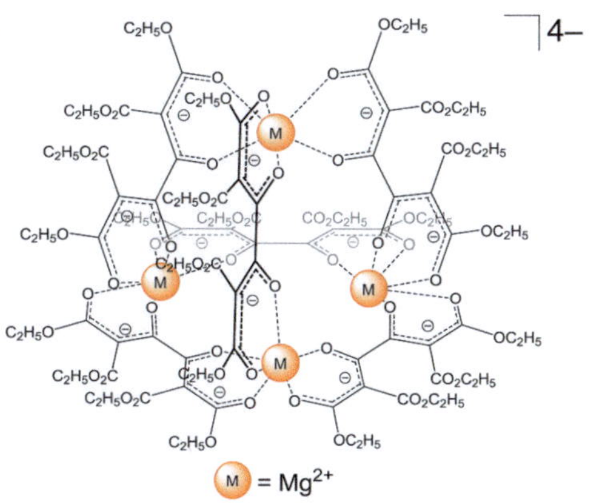

564.

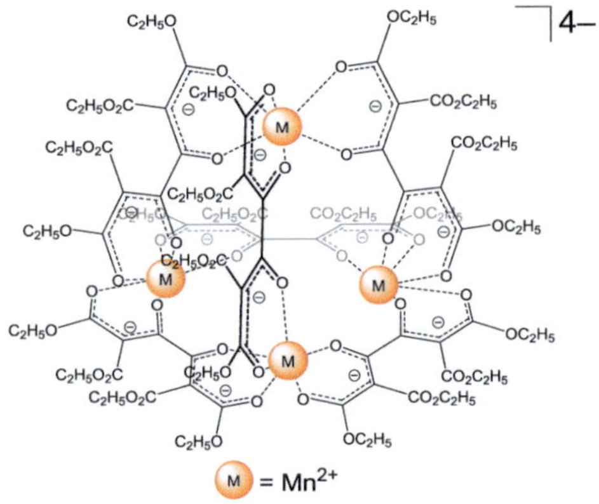

565.

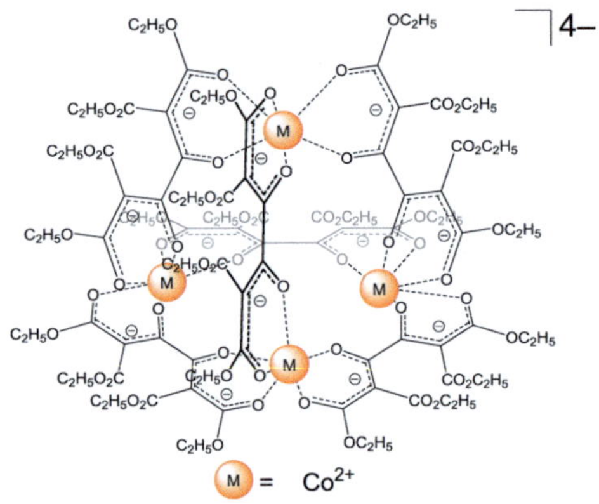

566.

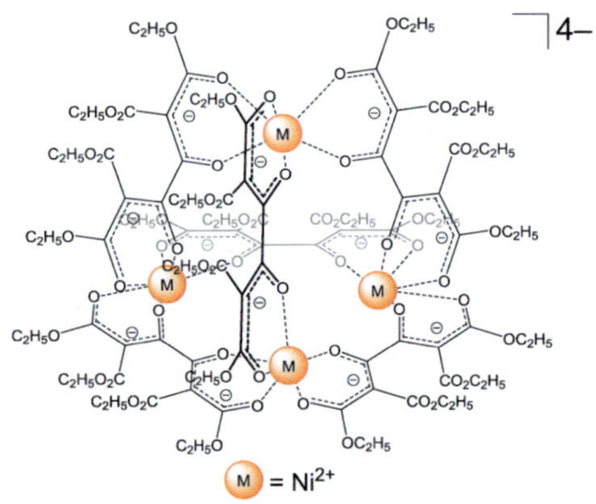

567.

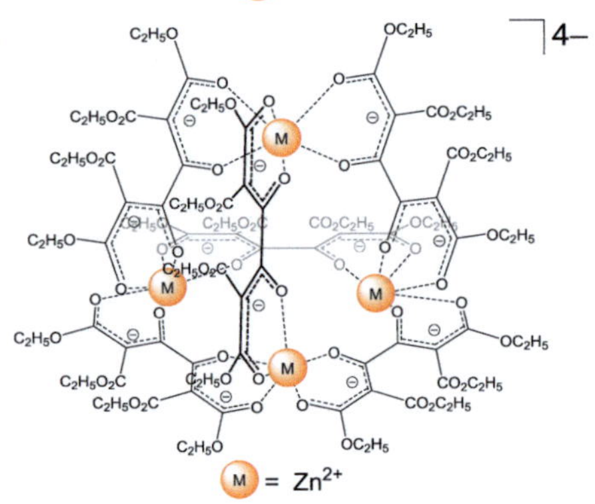

568.

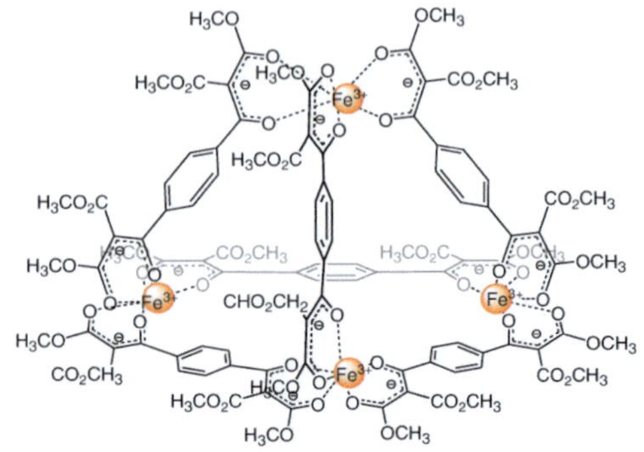

569.

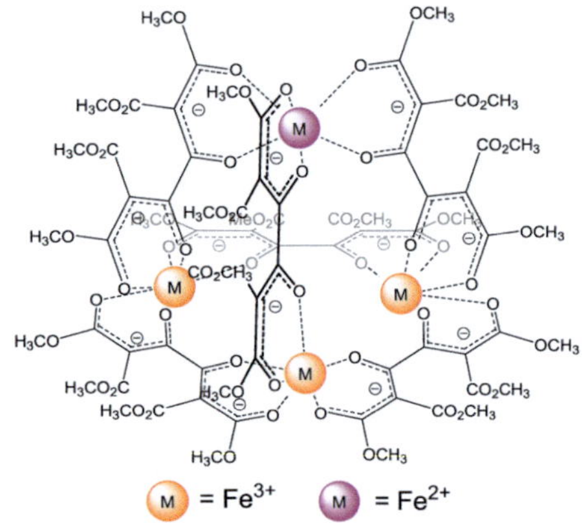

570.

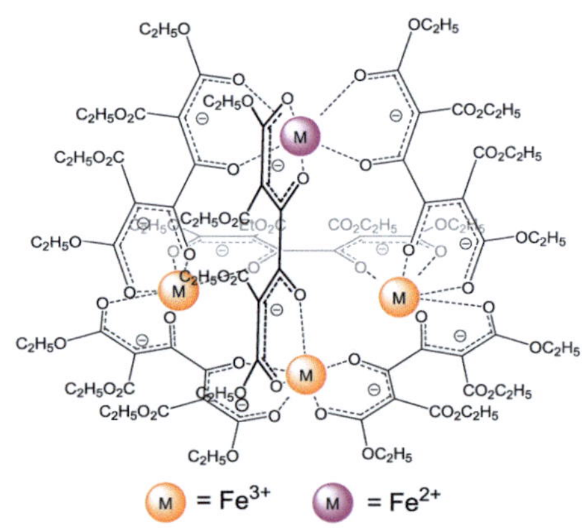

571.

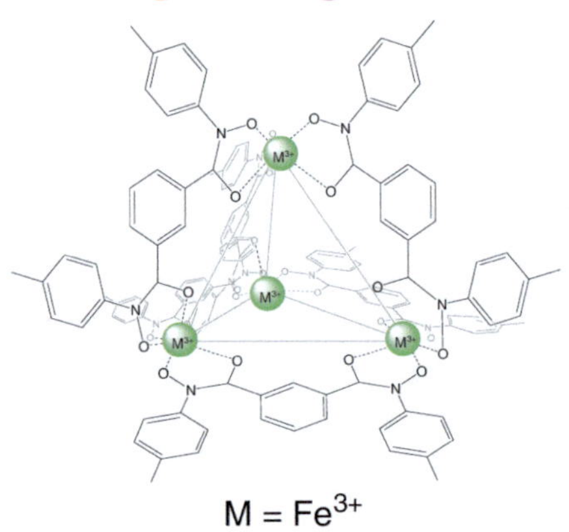

572.

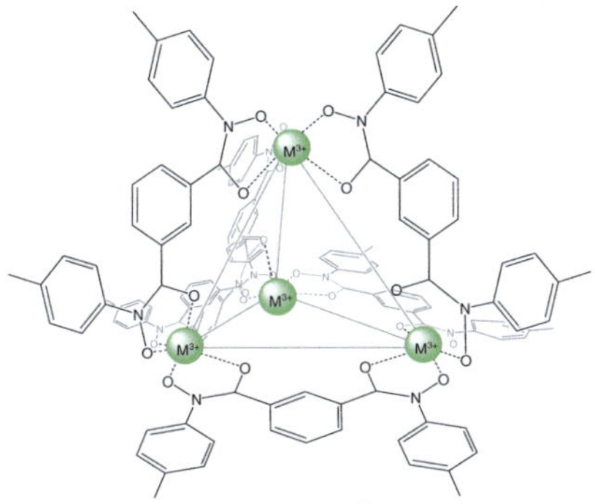

M = Ga^{3+}

573.

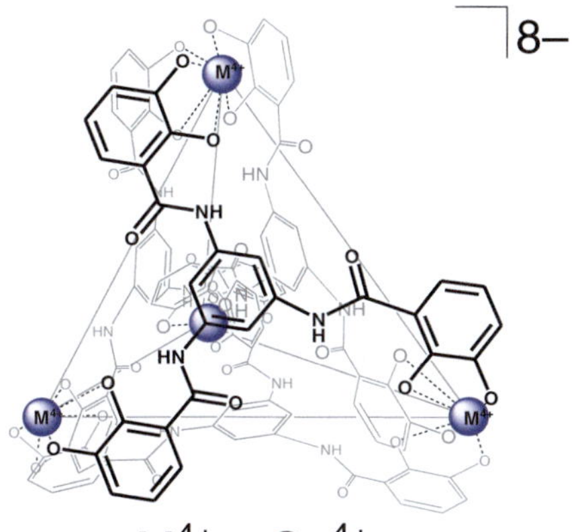

M^{4+} = Sn^{4+}

574.

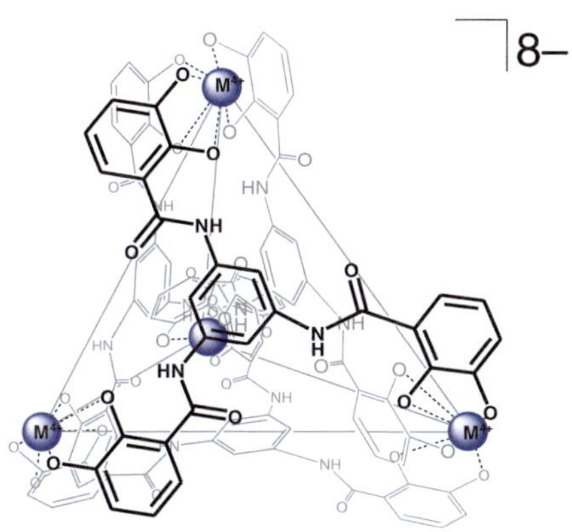

M^{4+} = Ti^{4+}

575.

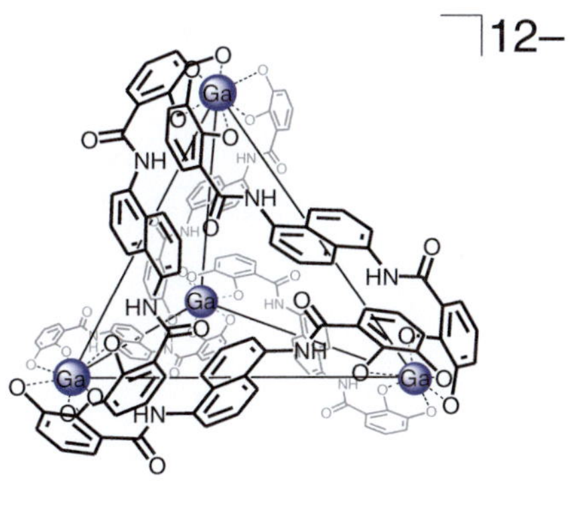

576.

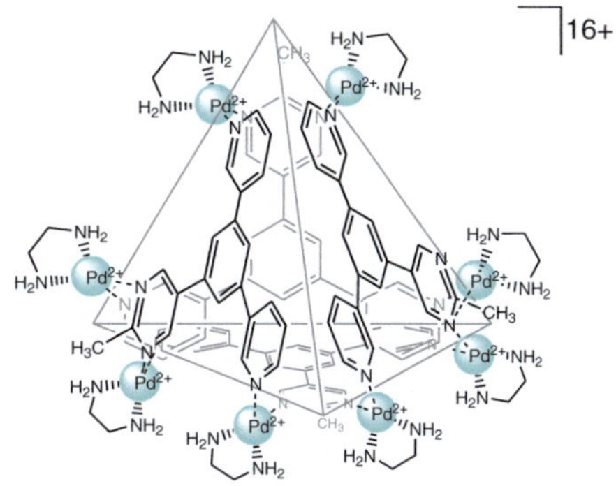

577.

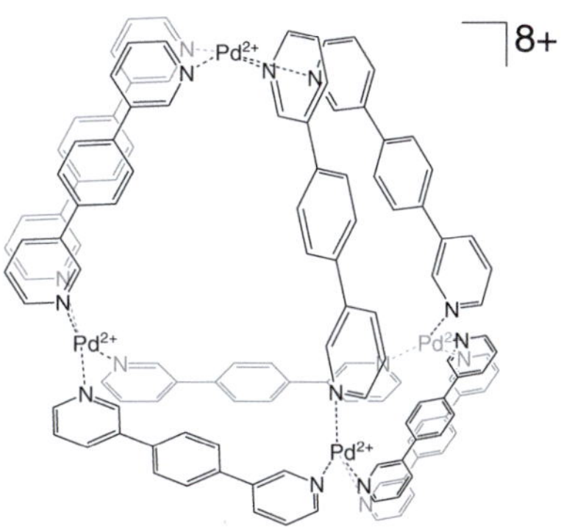

578.

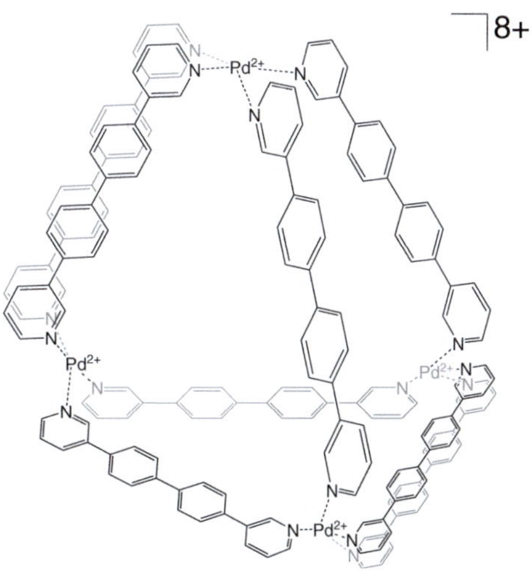

579.

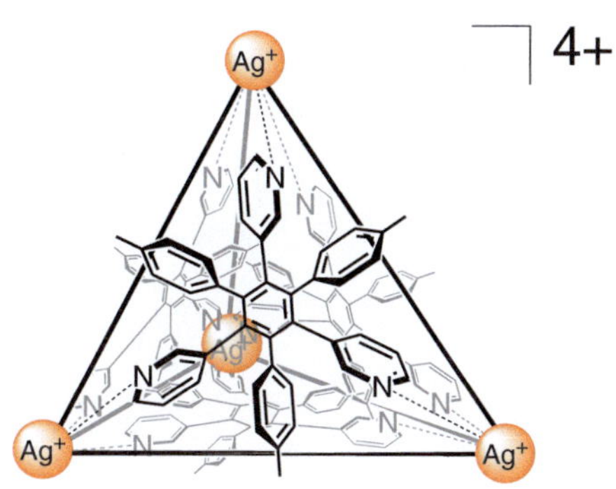

580.

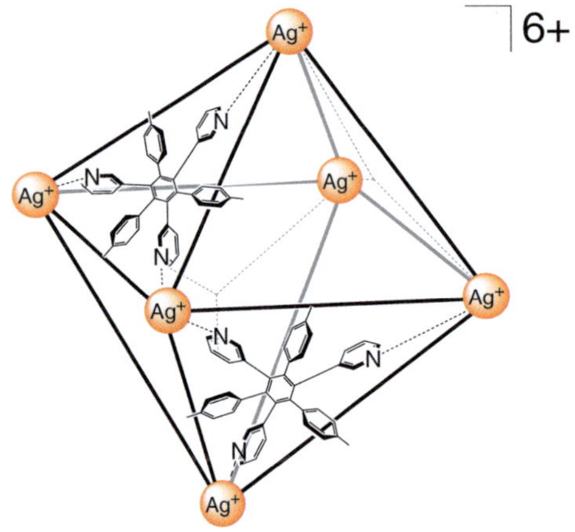

581.

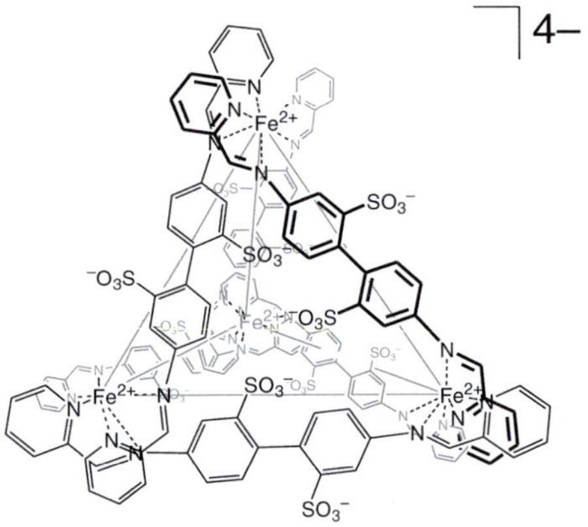

582.

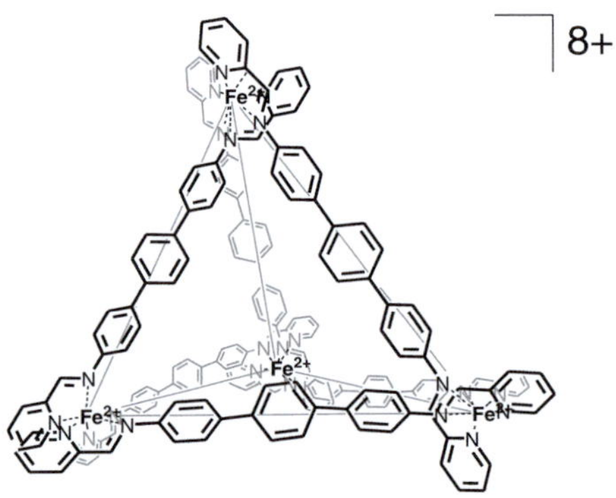

583.

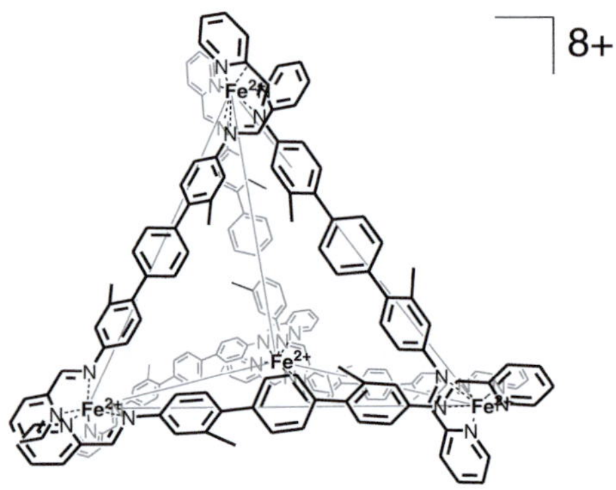

584.

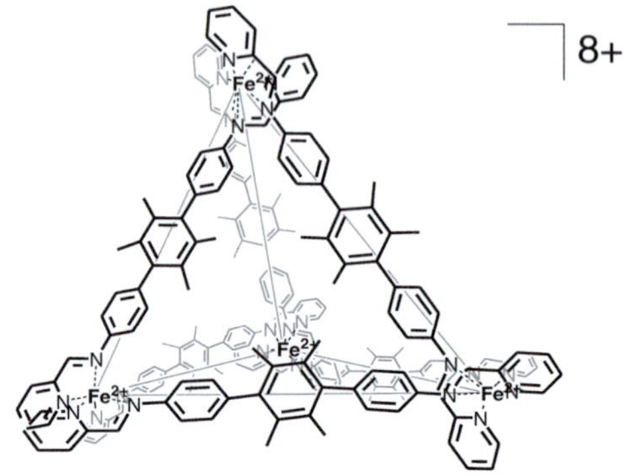

585.

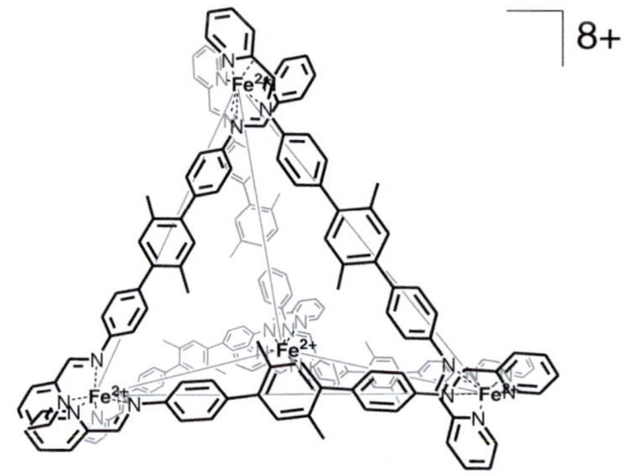

586.

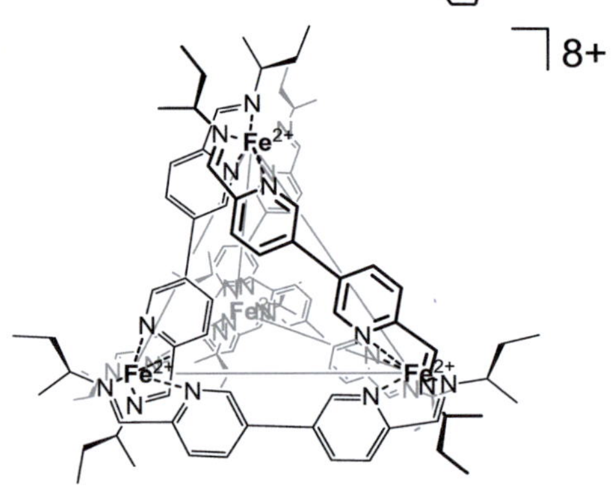

587.

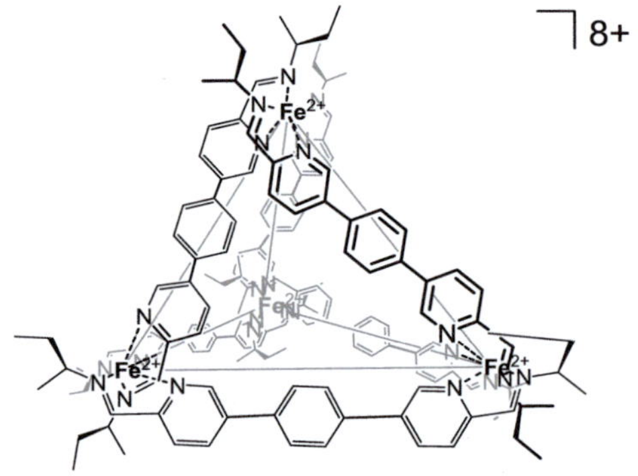

588.

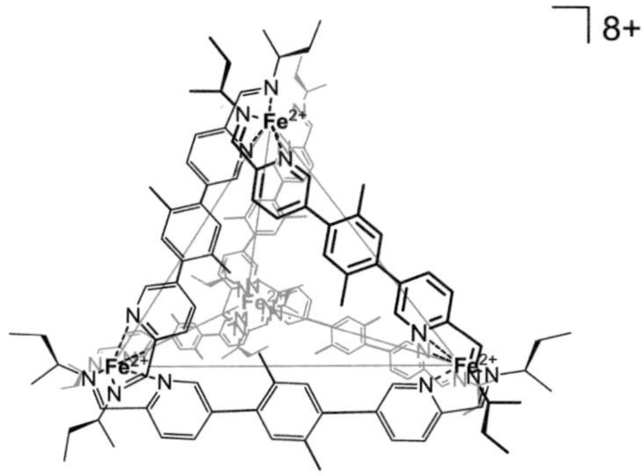

592.

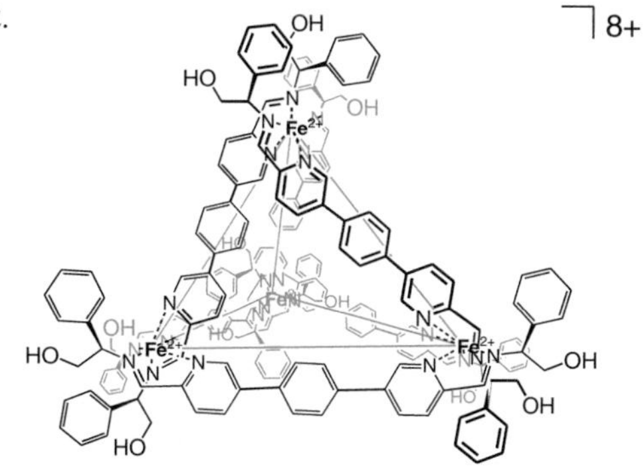

589.

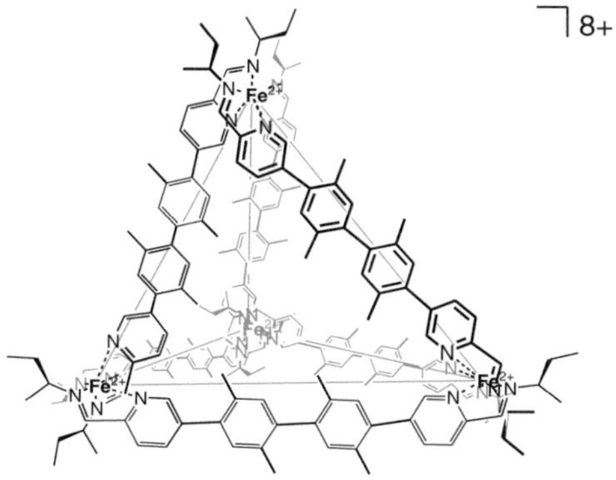

593.

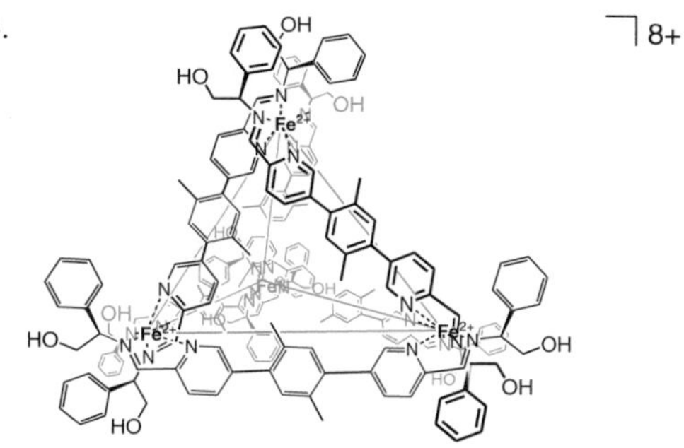

590.

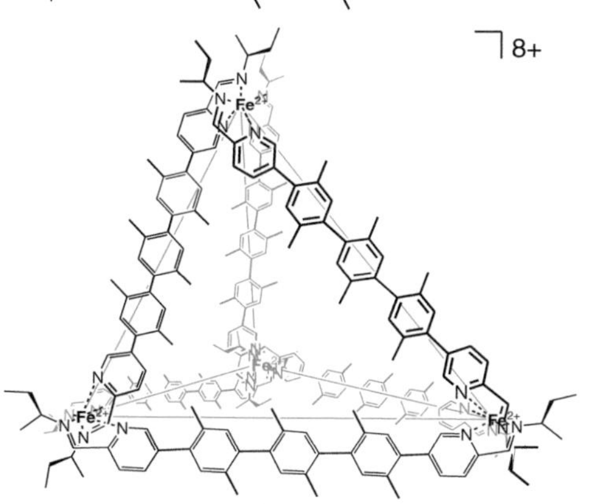

594.

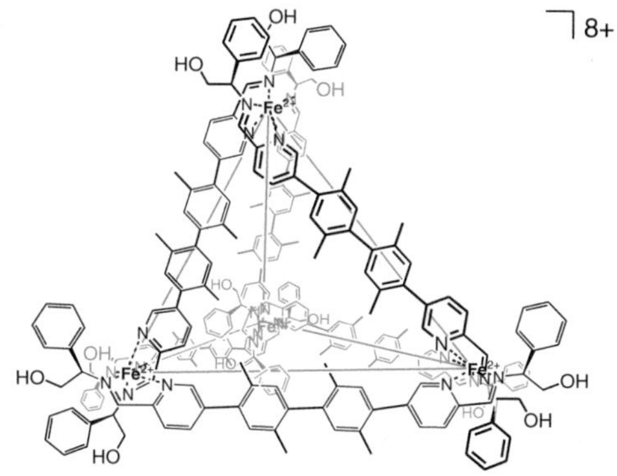

591.

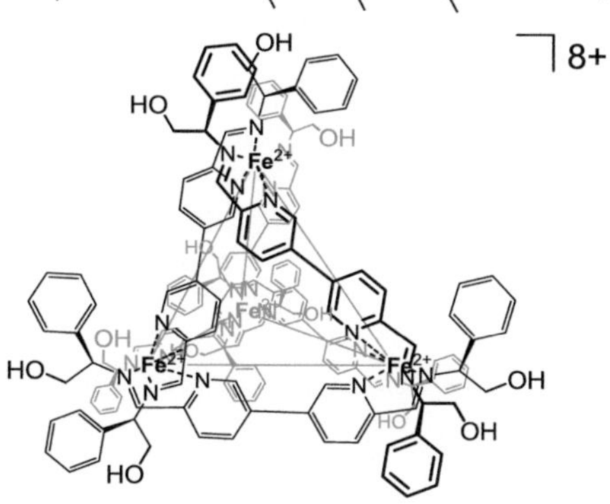

595.

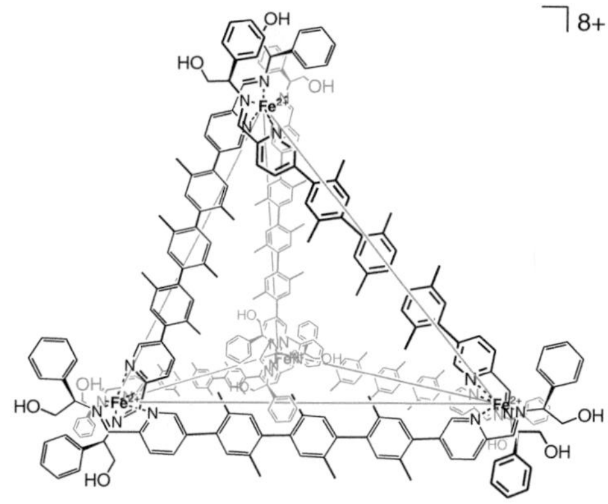

596.

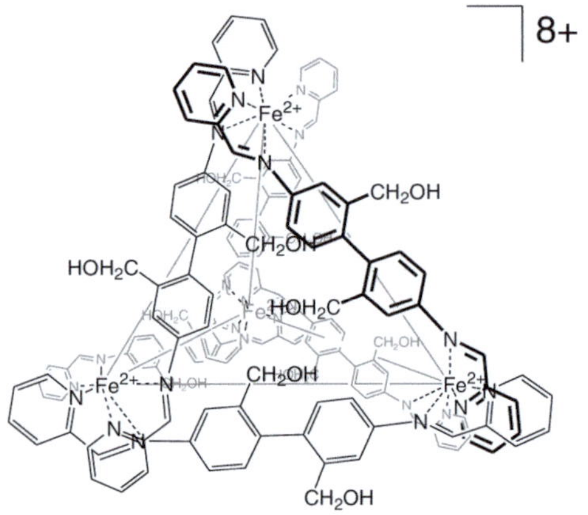

597.

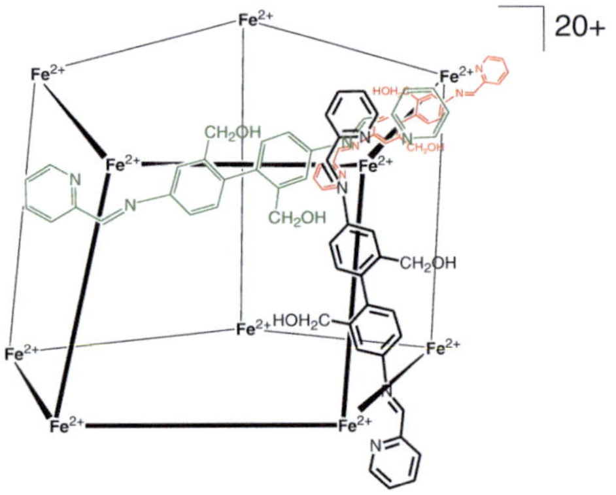

598.

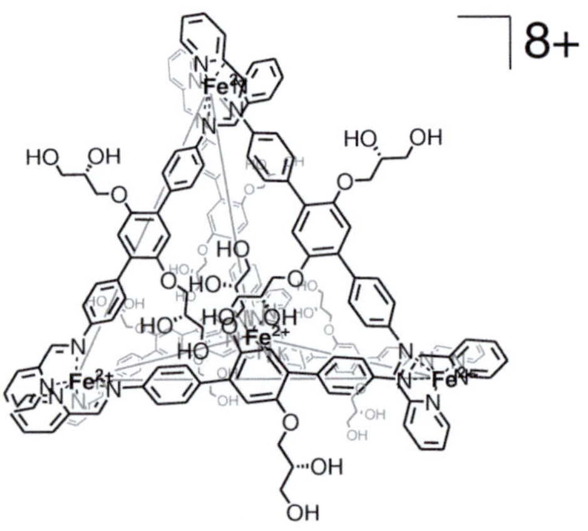

599.

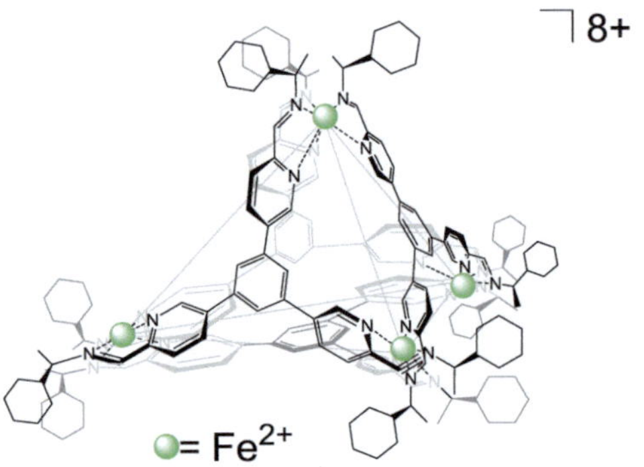

600.

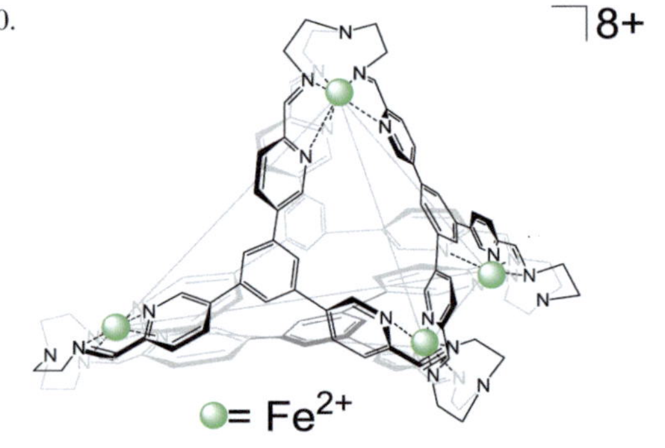

601.

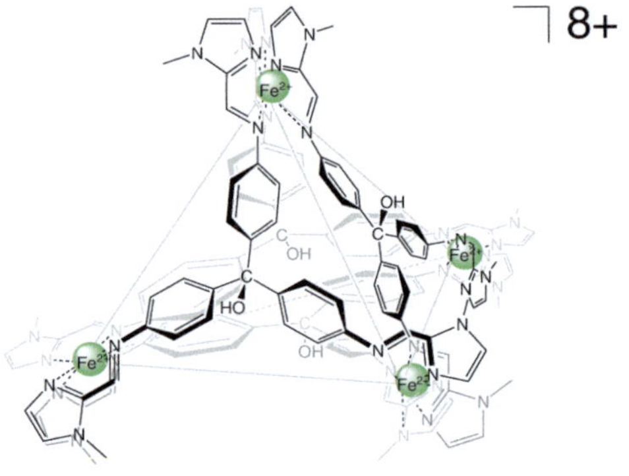

602.

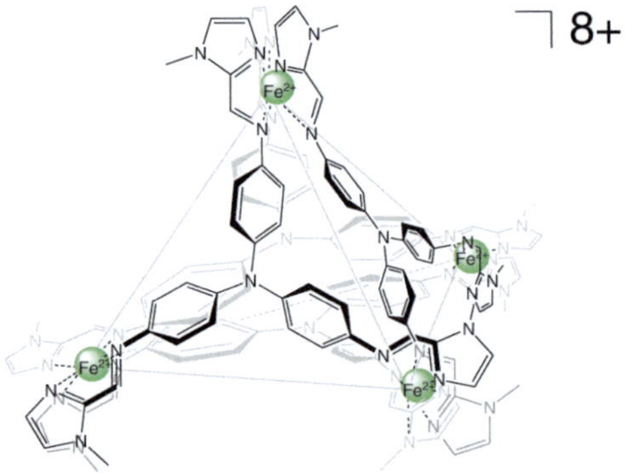

603.

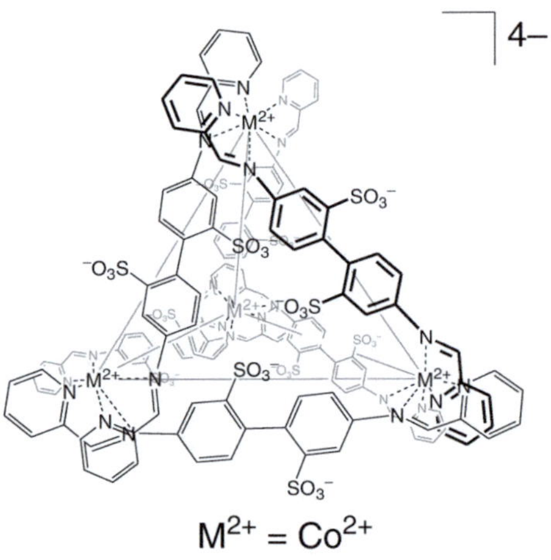

604.

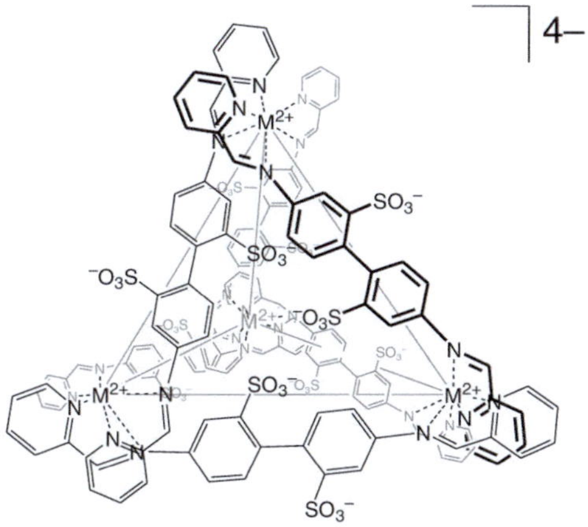

$M^{2+} = Ni^{2+}$

605.

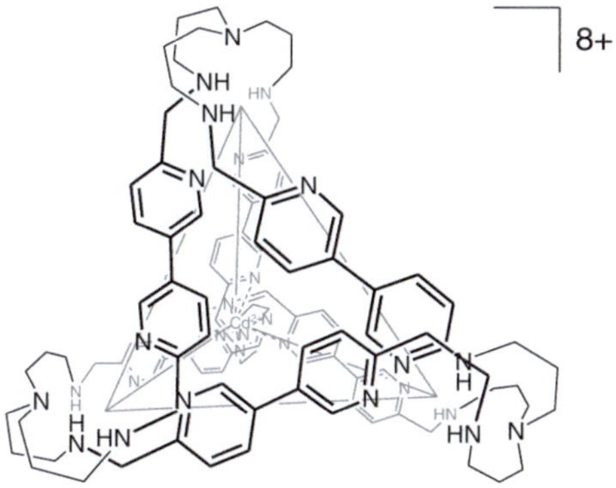

606.

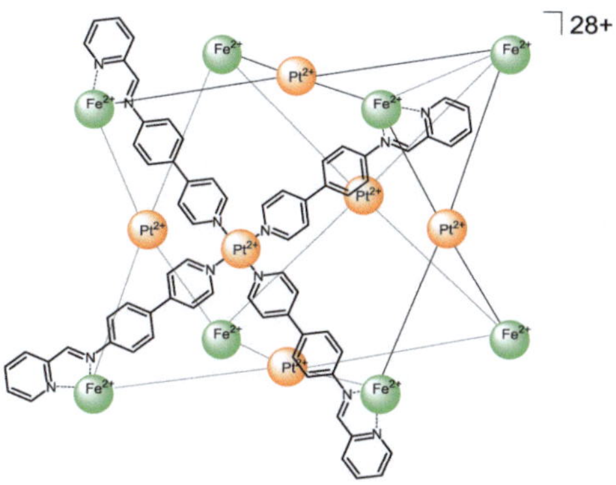

607.

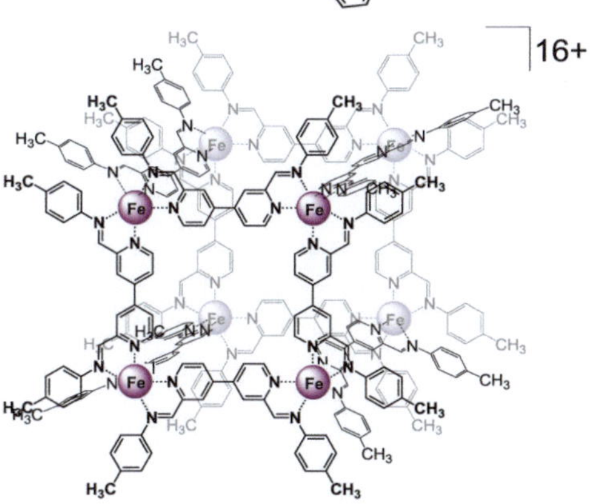

608.

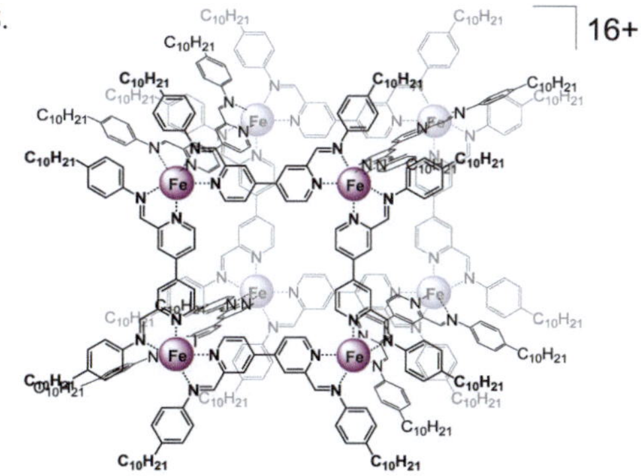

609.

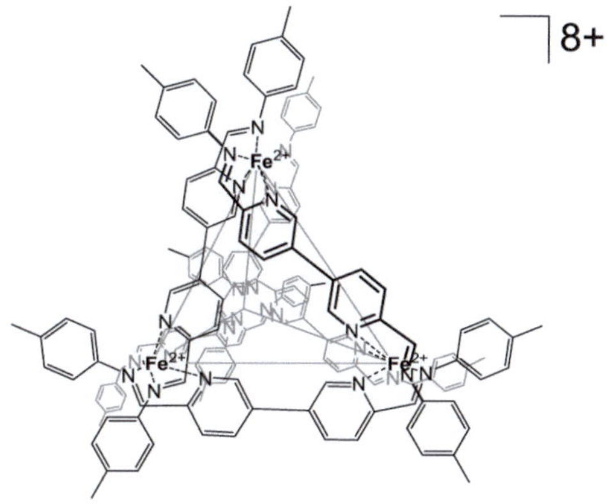

610.

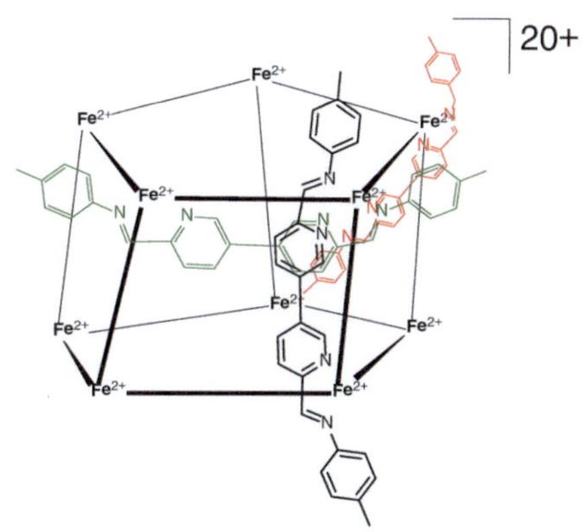

611.

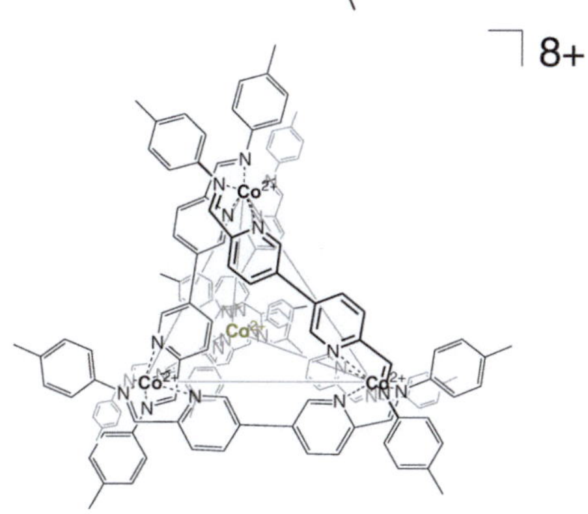

612.

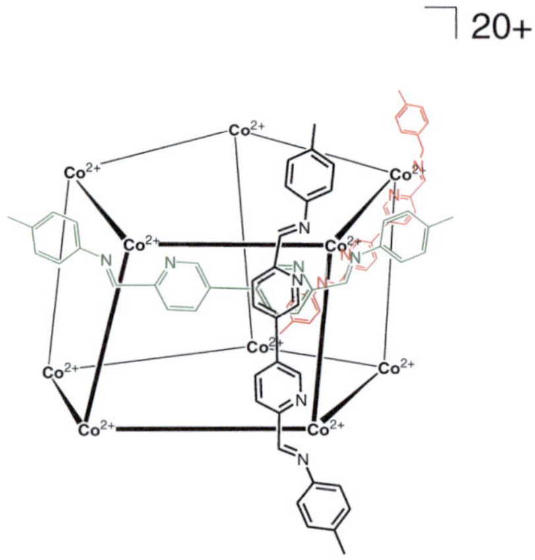

613.

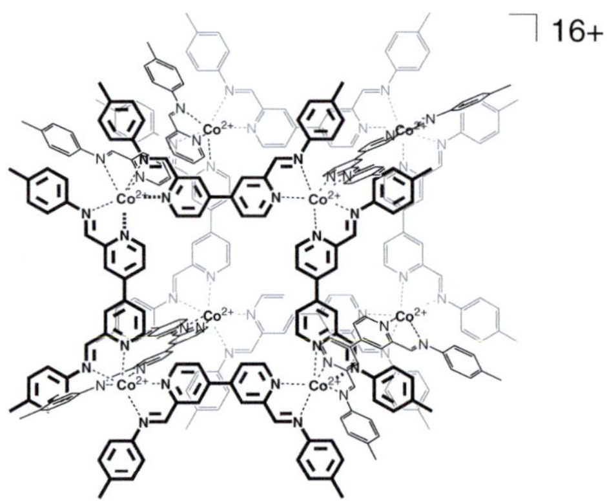

614.

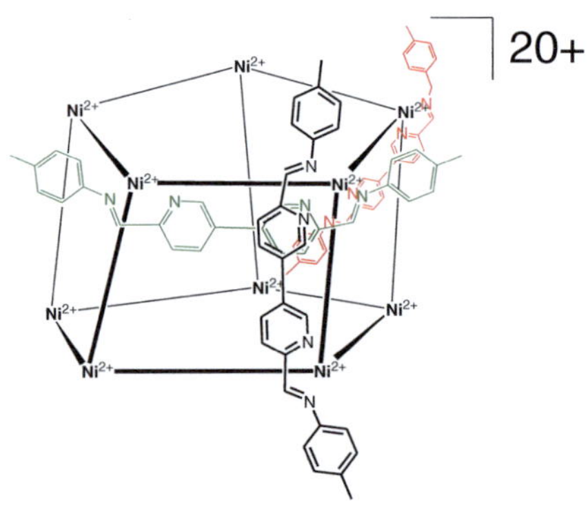

615.

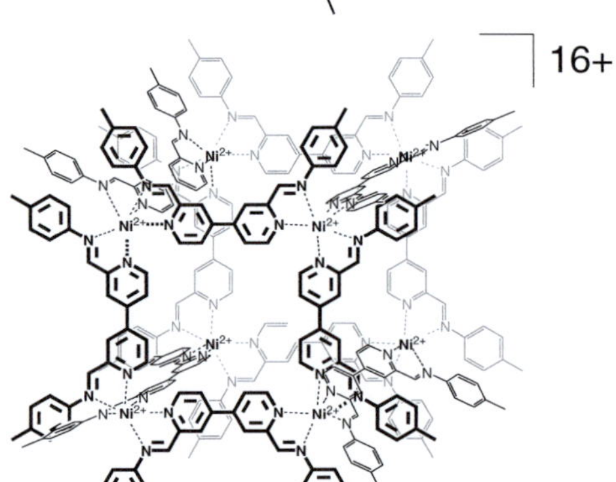

616.

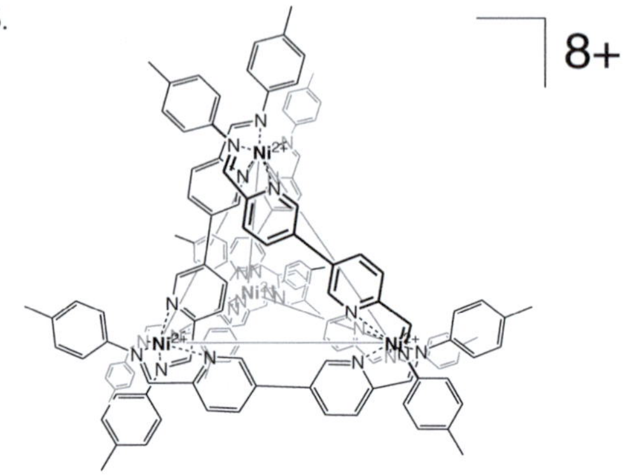

617.

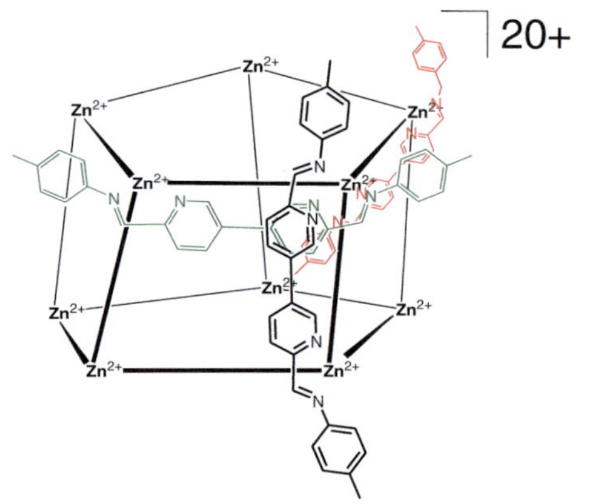

618.

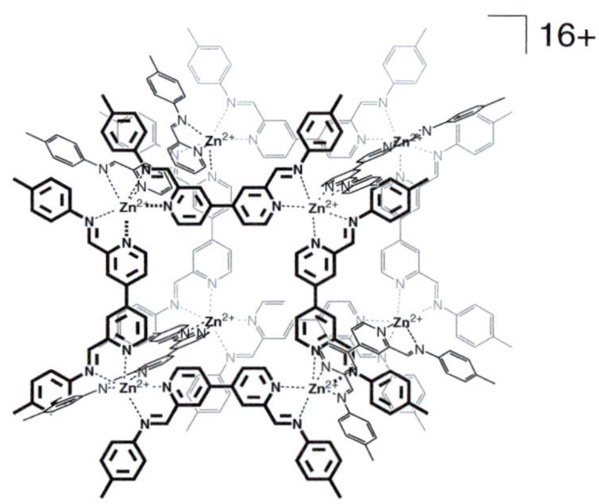

619.

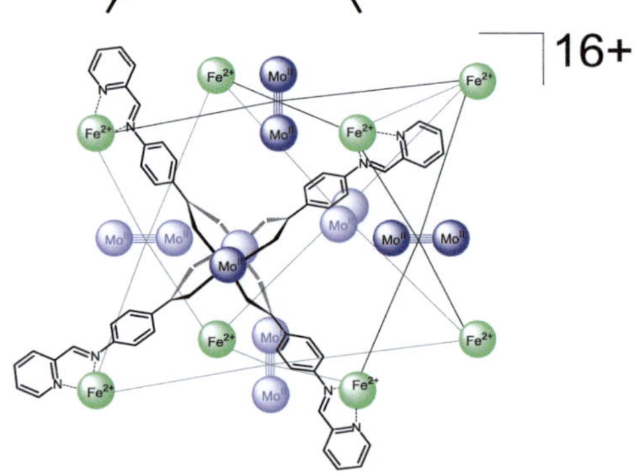

620.

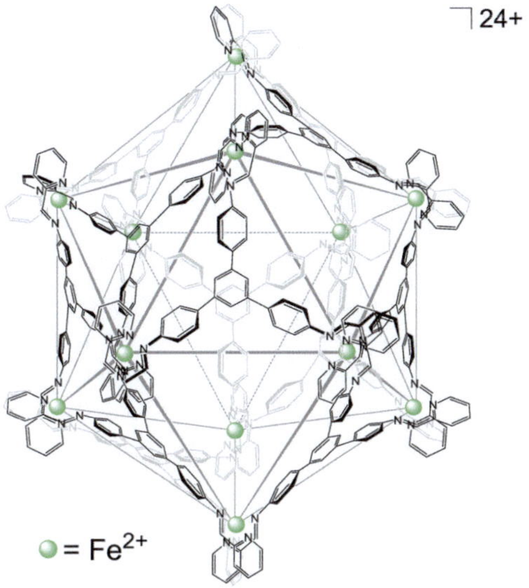

621.

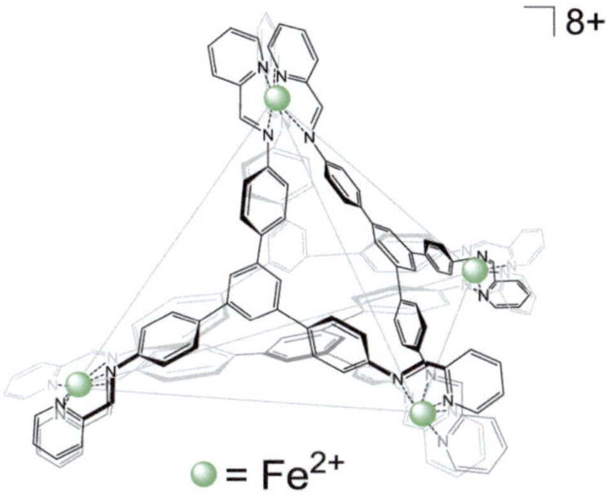

622.

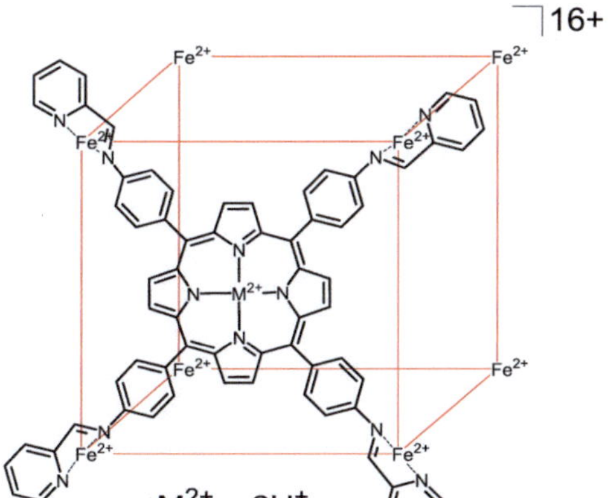

623.

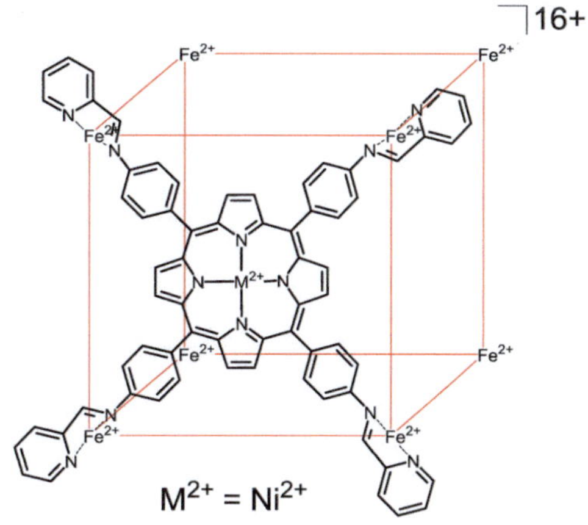

624.

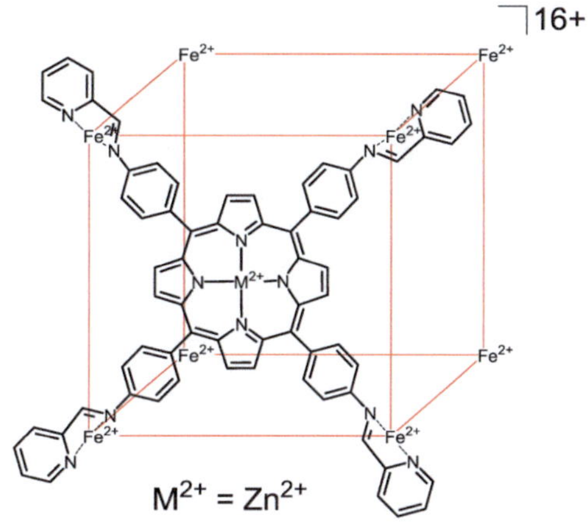

625.

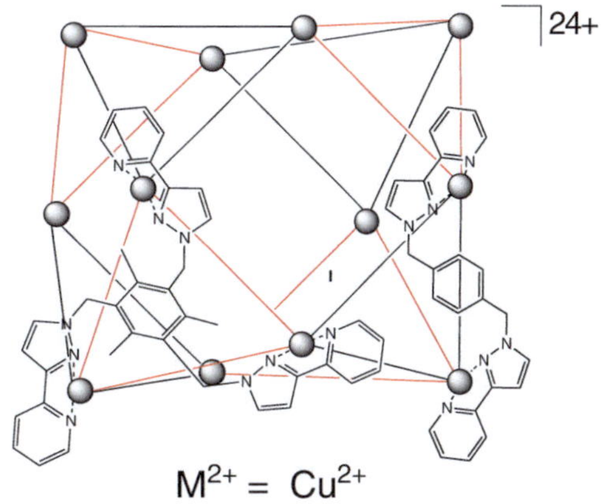

626.

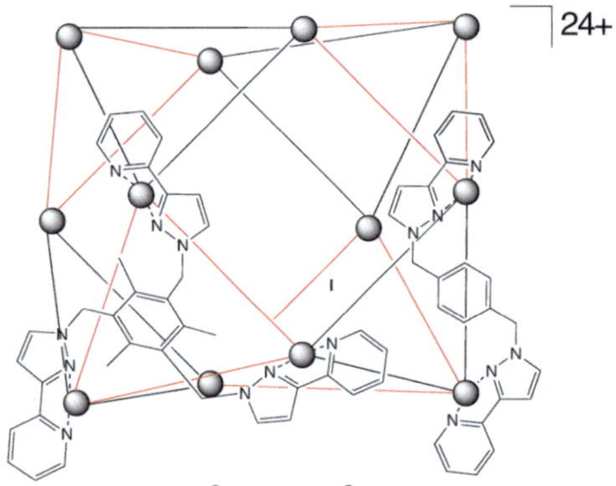

M^{2+} = Co^{2+}

627.

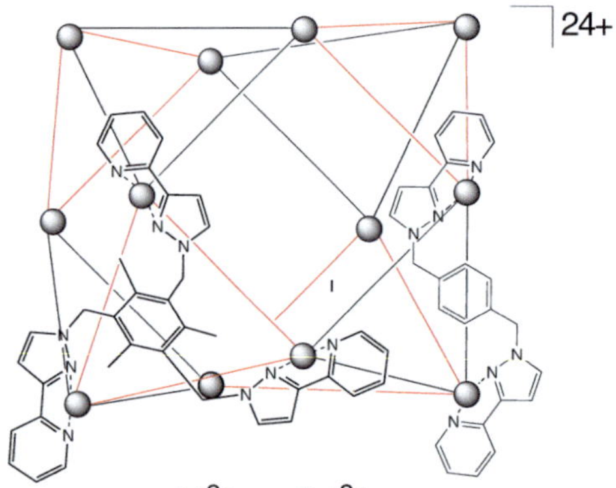

M^{2+} = Cd^{2+}

628.

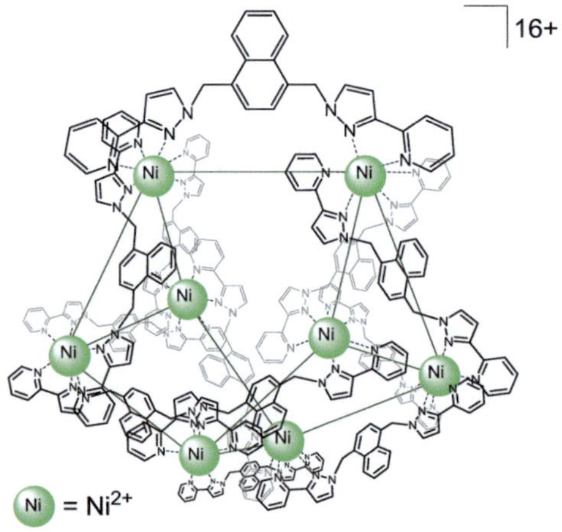

629.

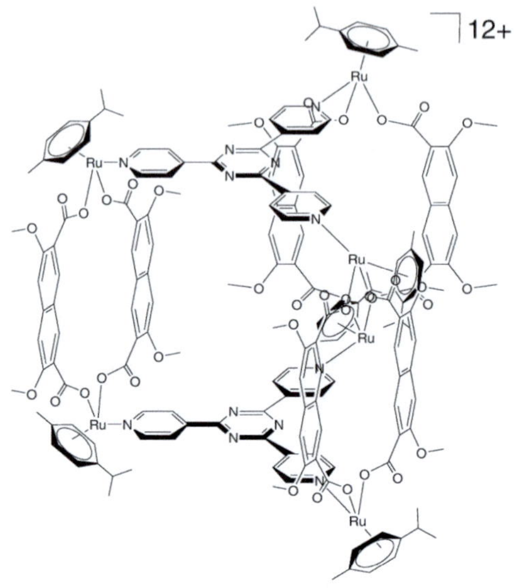

630.

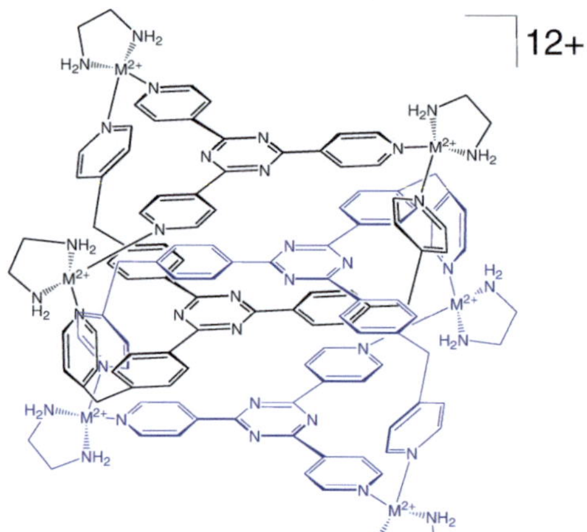

M^{2+} = Pt^{2+}

631.

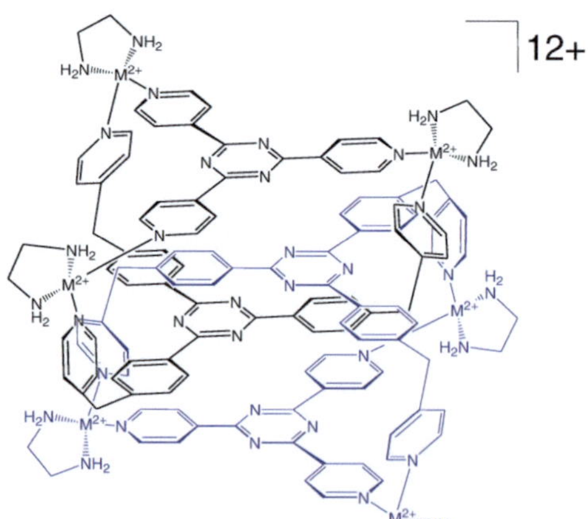

M^{2+} = Pd^{2+}

632.

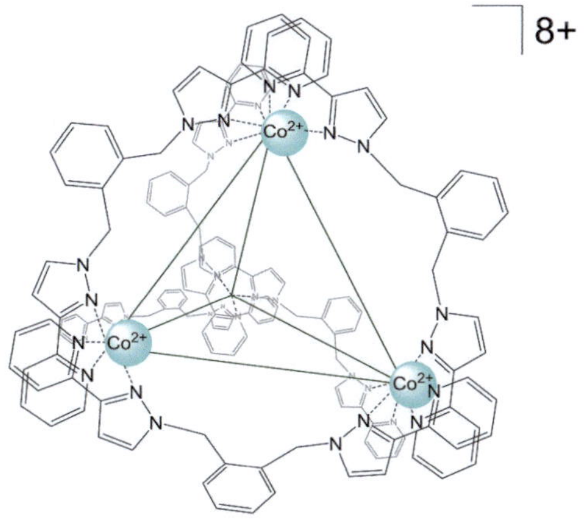

633.

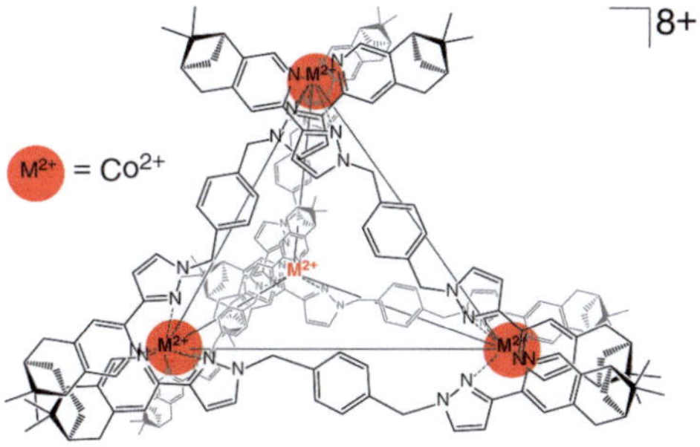

634.

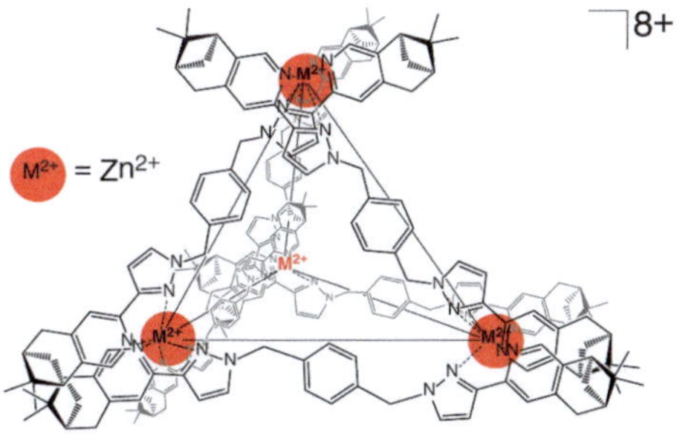

635.

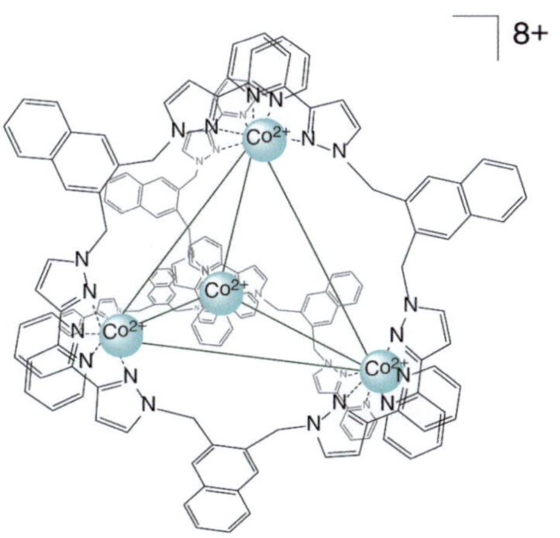

636.

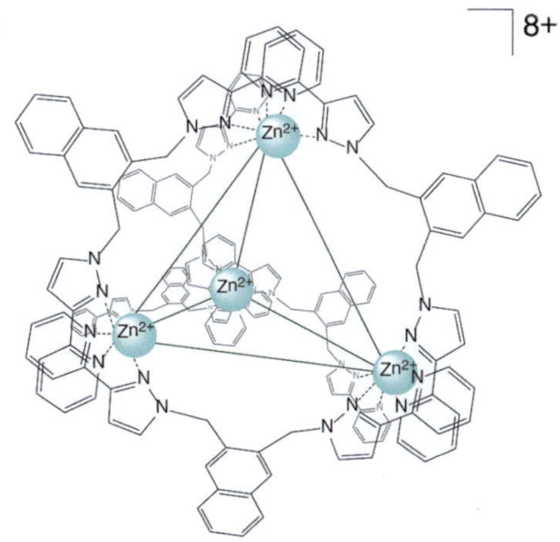

637.

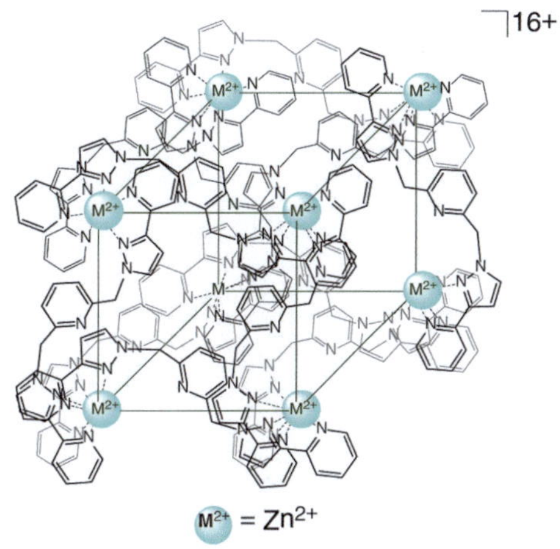

638.

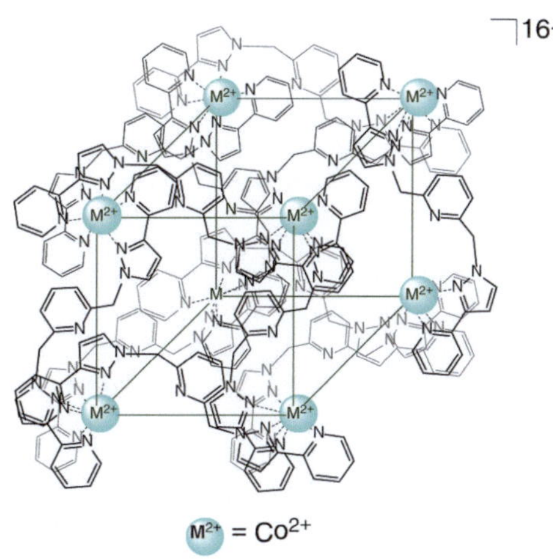

639.

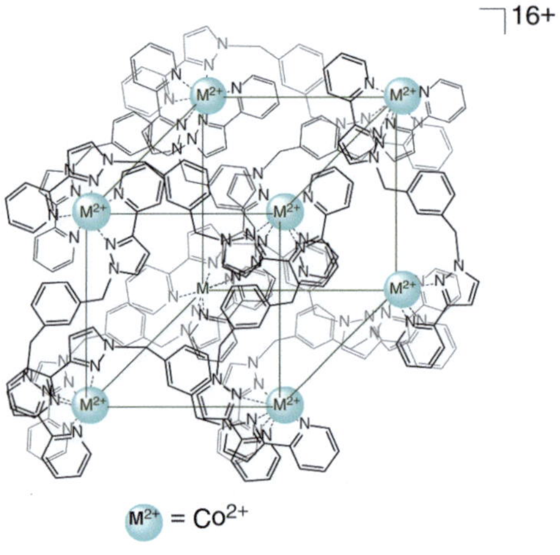

640.

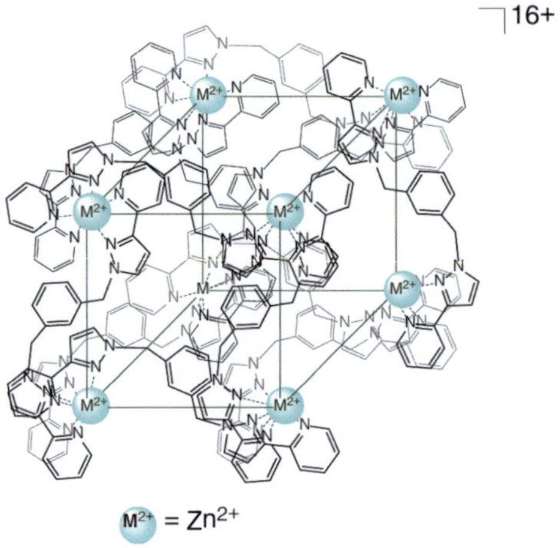

641.

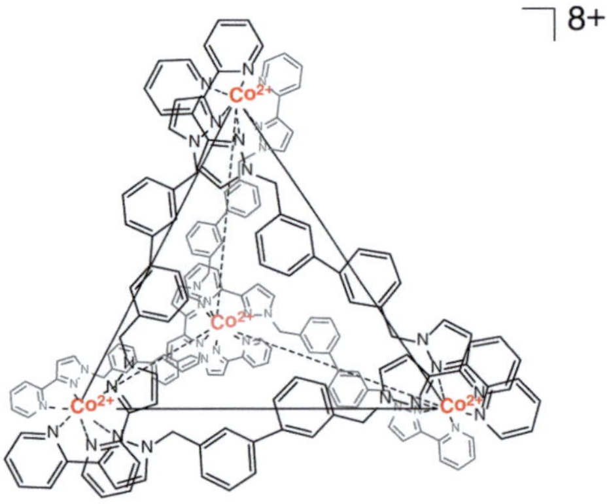

642.

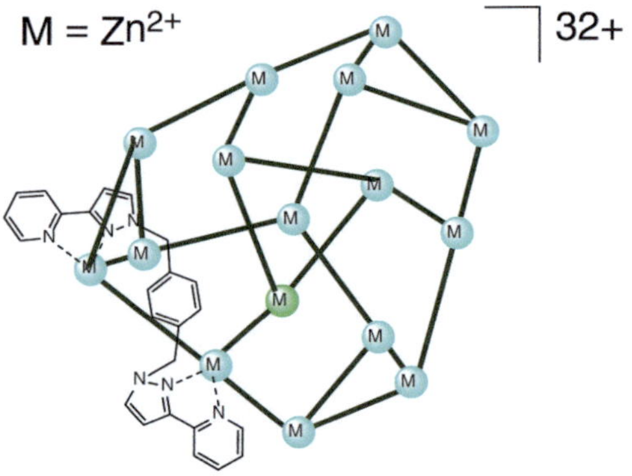

643.

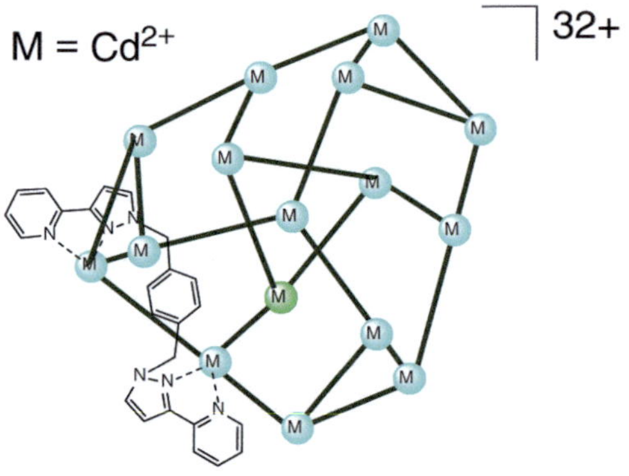

644.

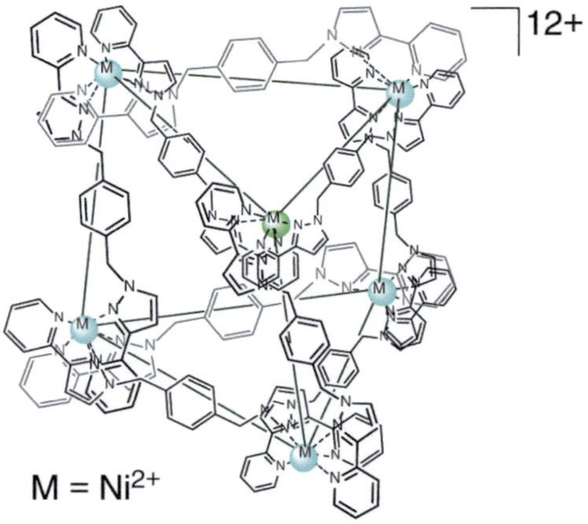

645.

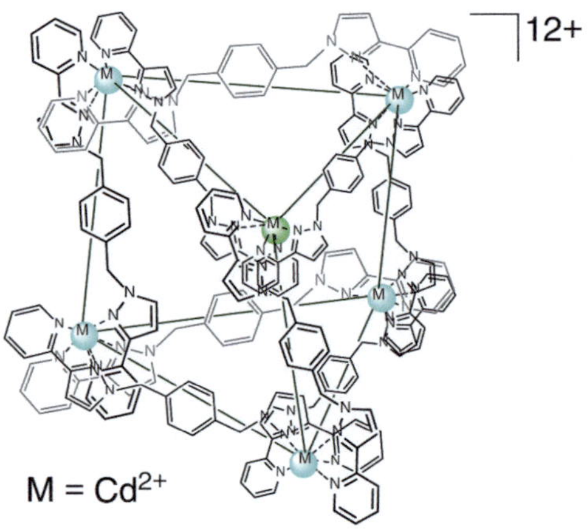

646.

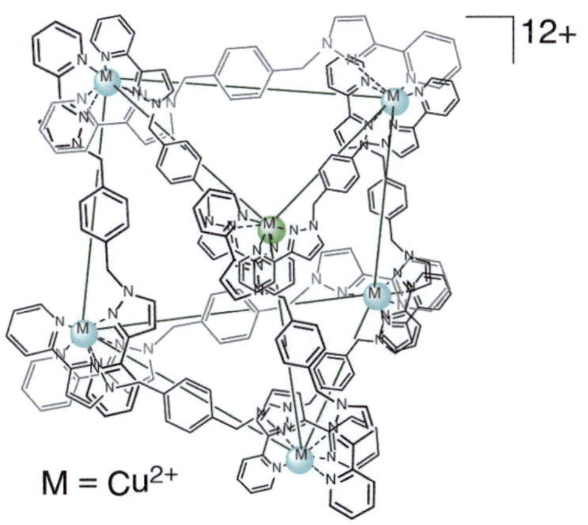

647.

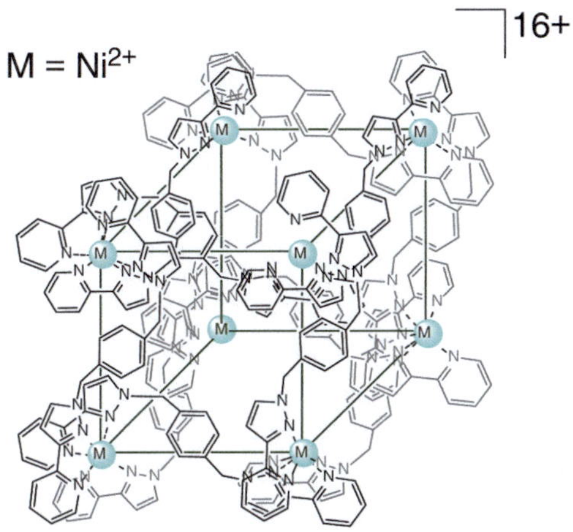

648.

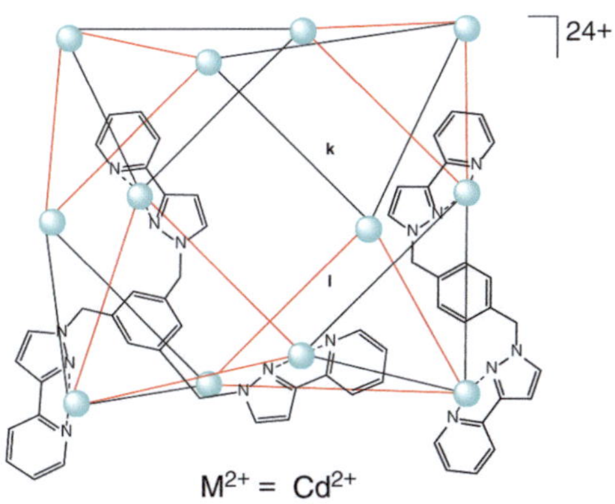

649.

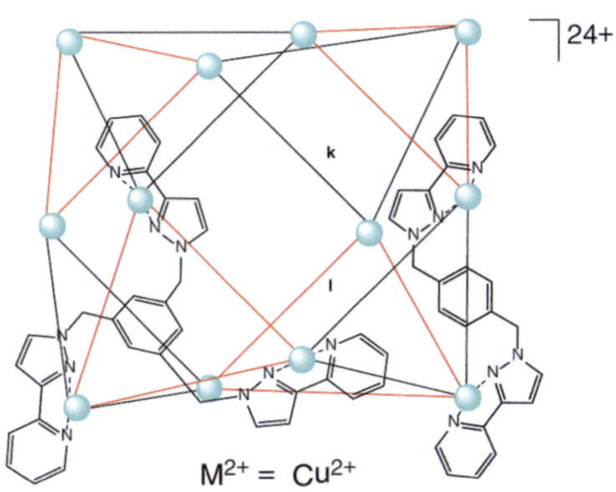

650.

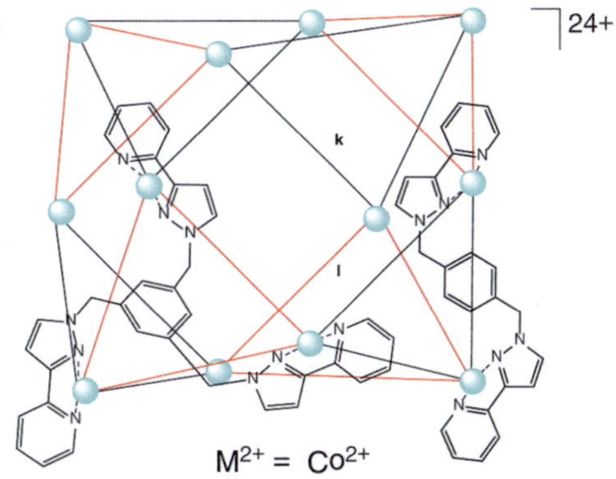

651.

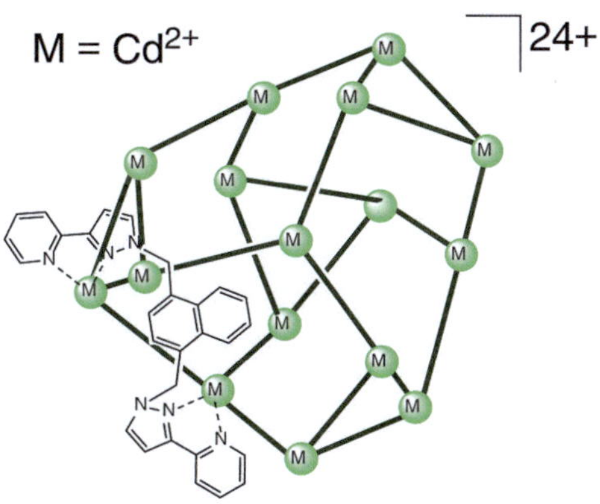

652.

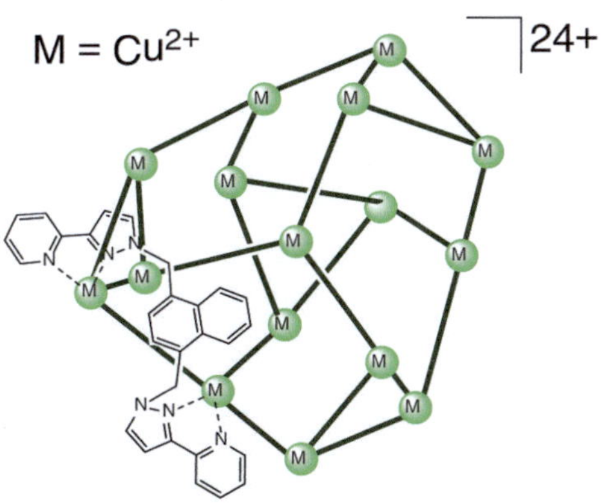

653.

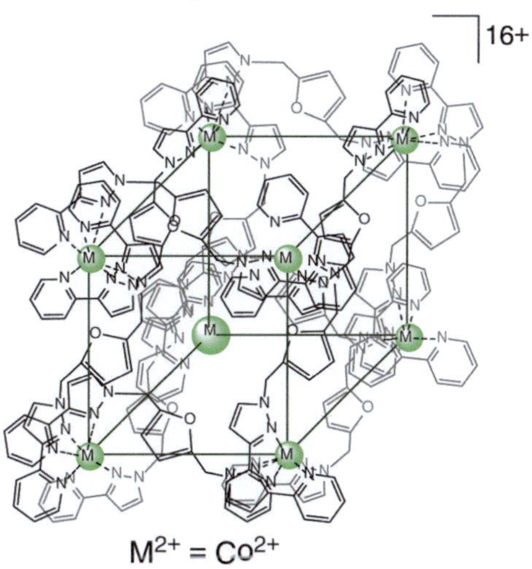

654.

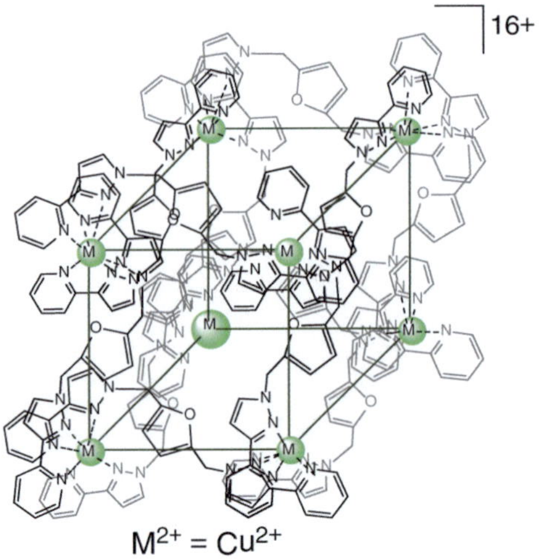

655.

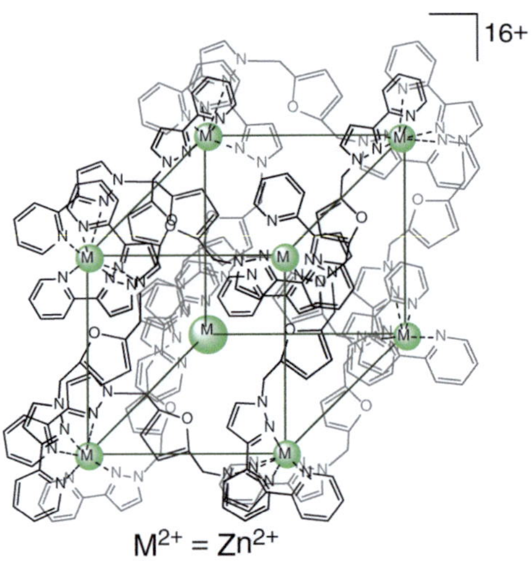

656.

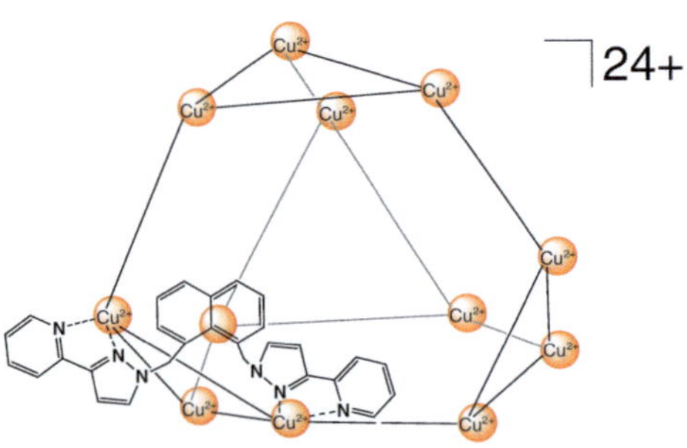

657.

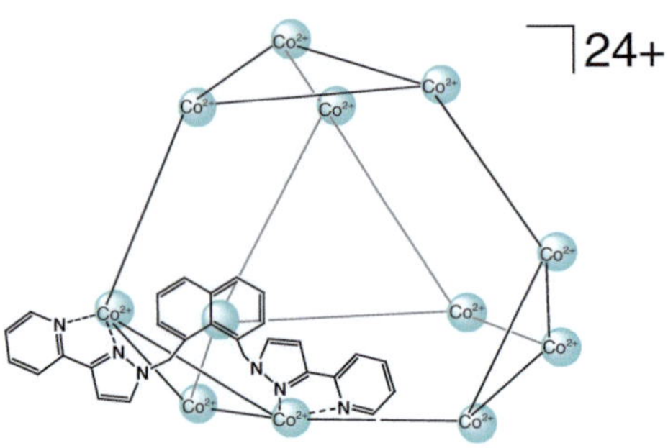

658.

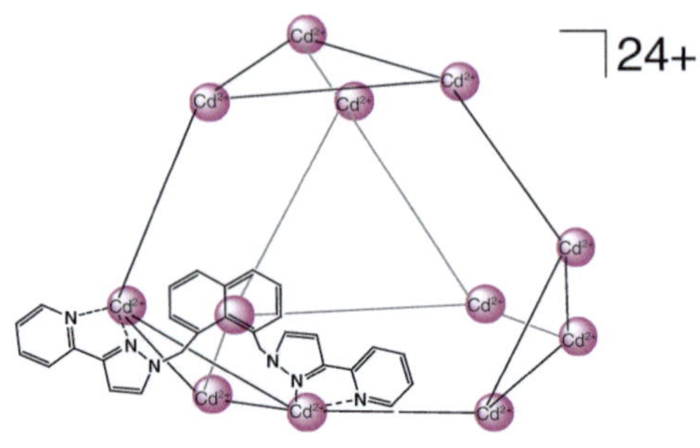

659.

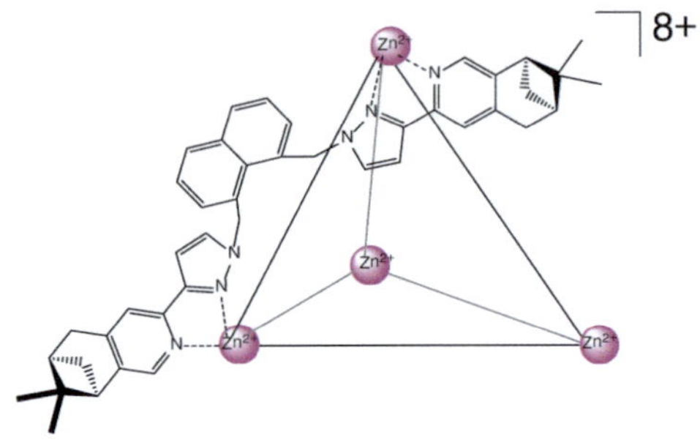

660.

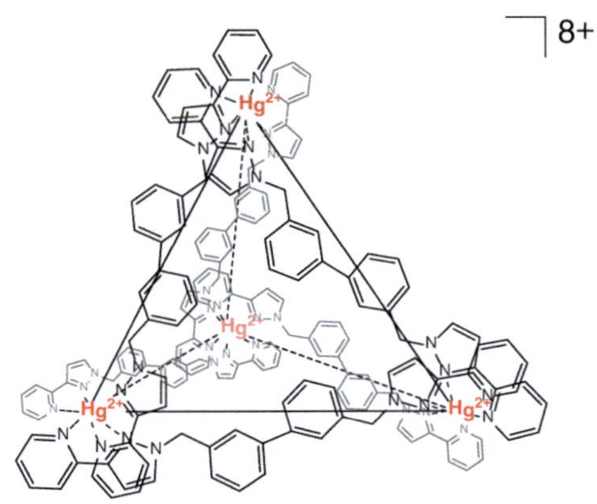

661.

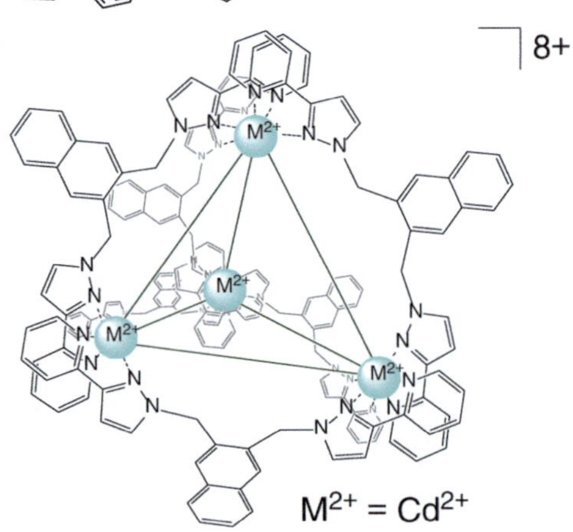

662.

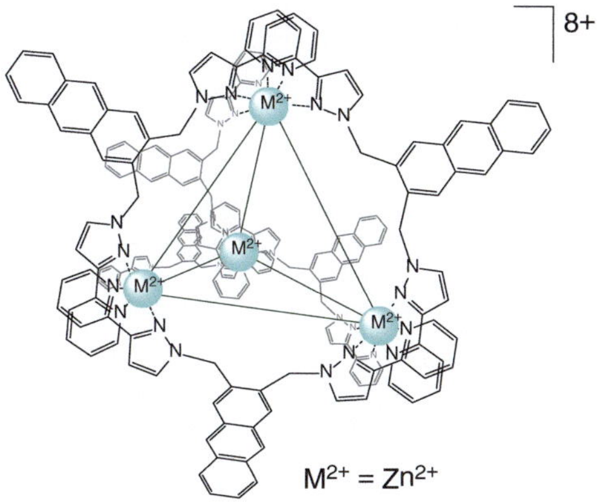

663.

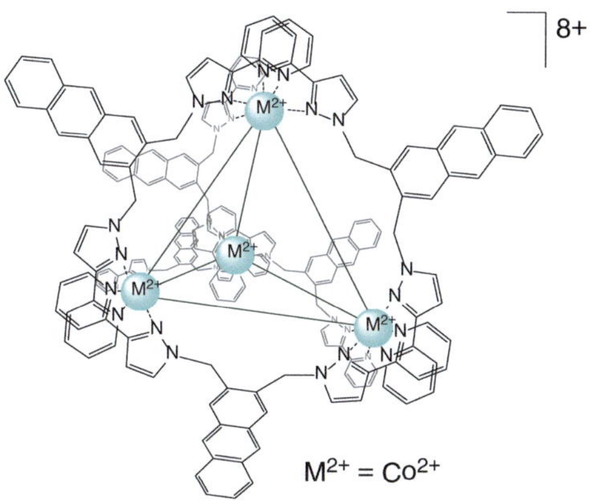

664.

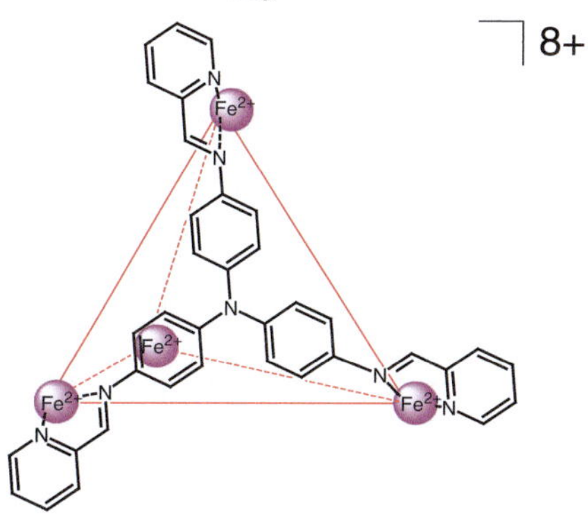

665.

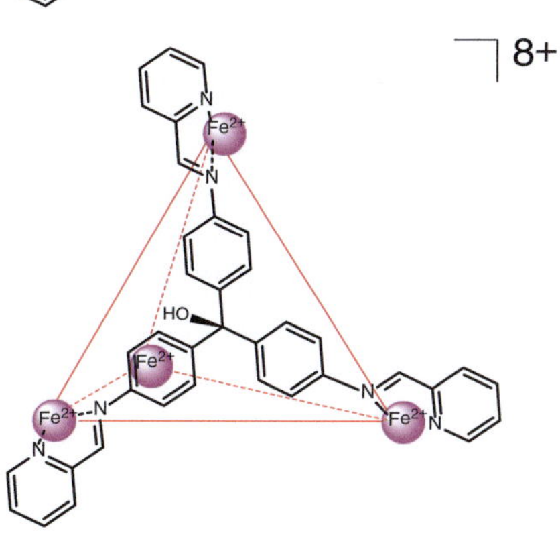

666.

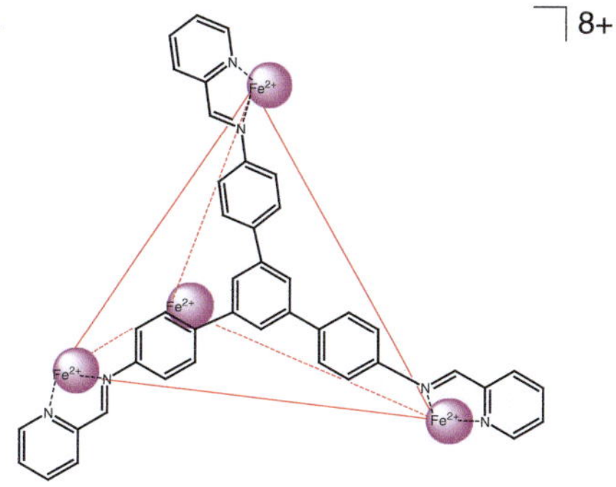

667.

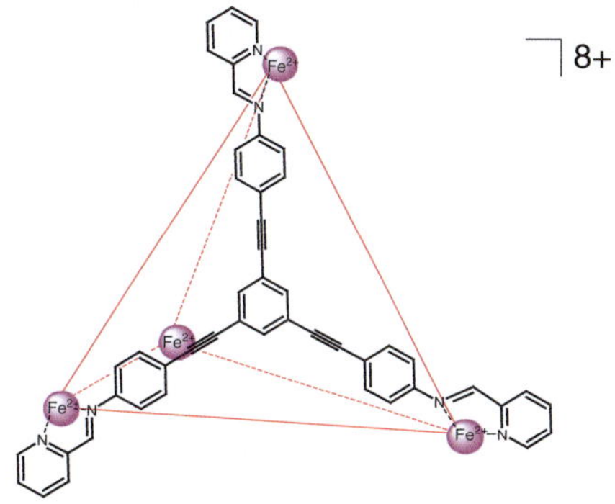

668.

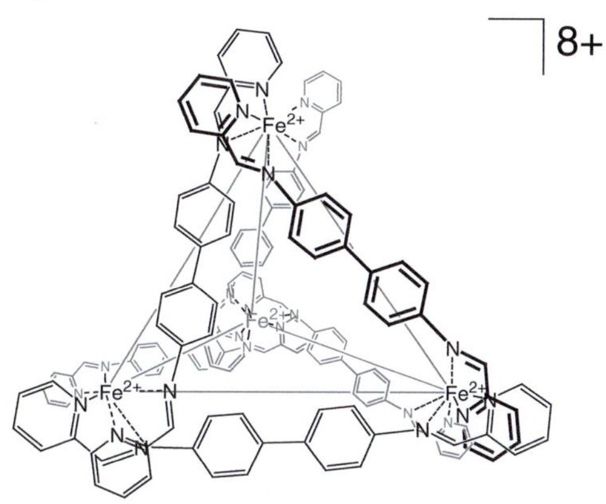

669.

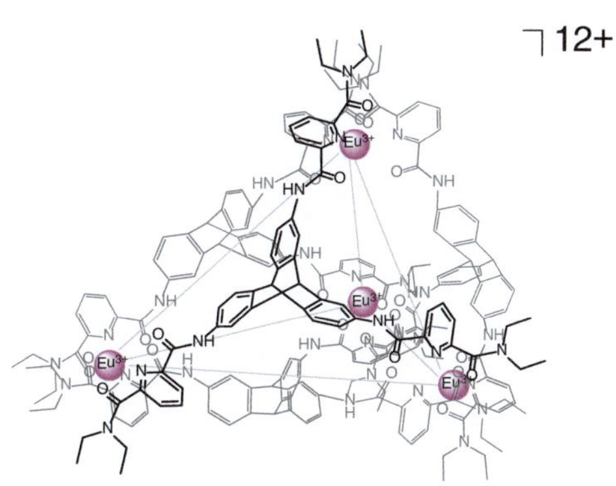

670.

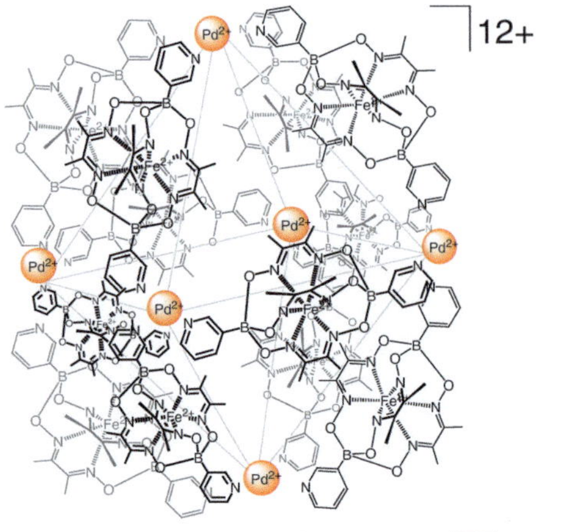

673.

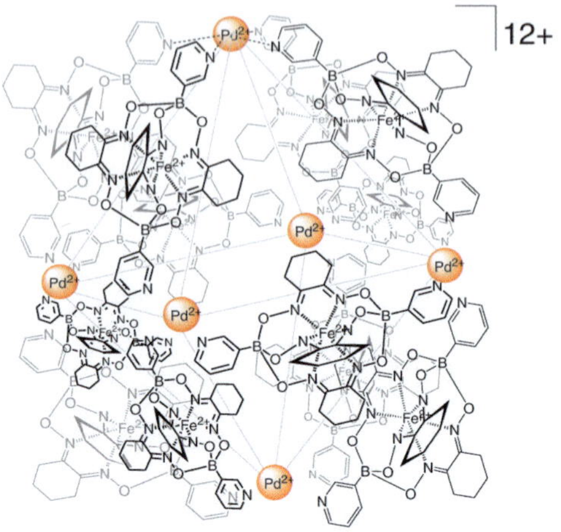

671.

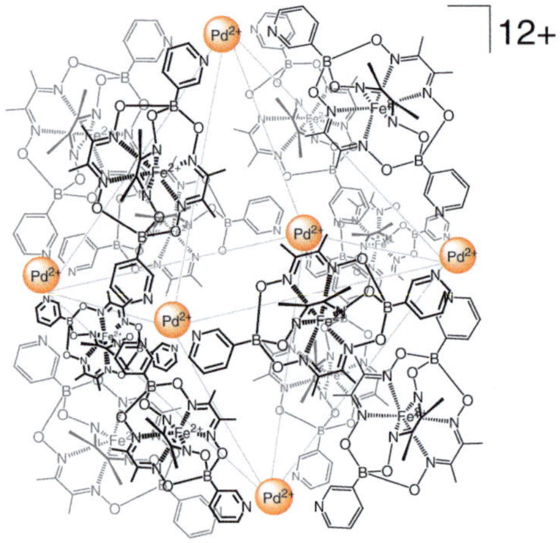

674.

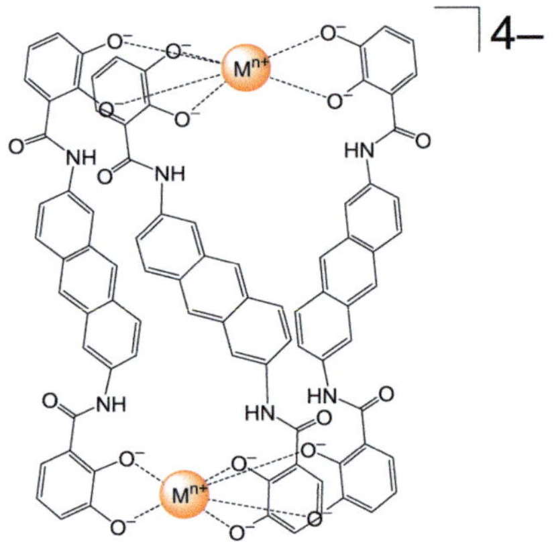

$M^{n+} = Ti^{4+}$

672.

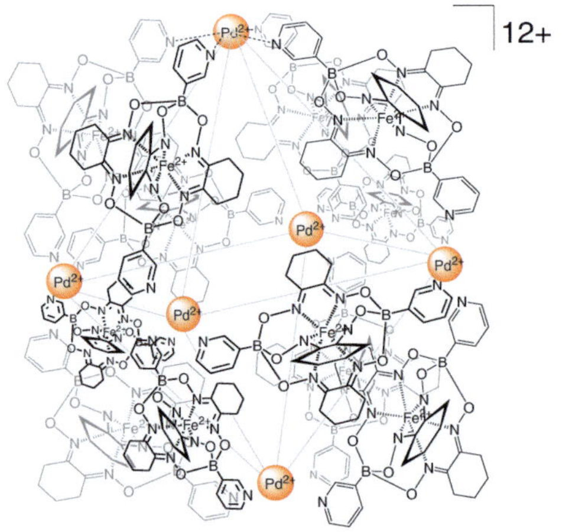

675.

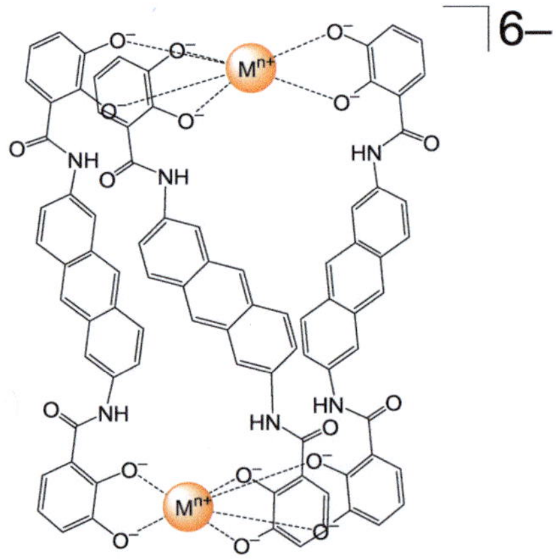

$M^{n+} = Ga^{3+}$

676.

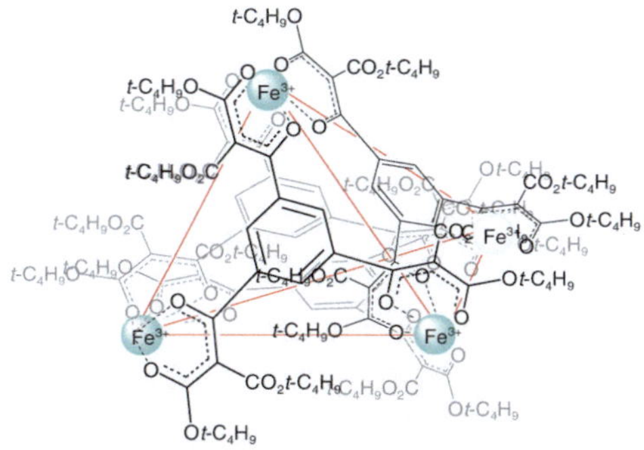

677.

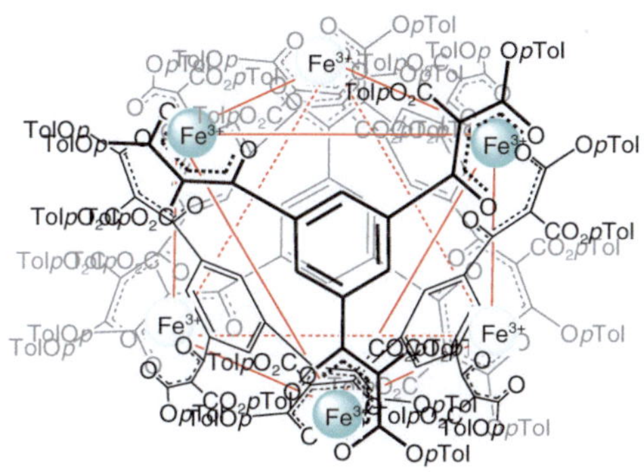

678.

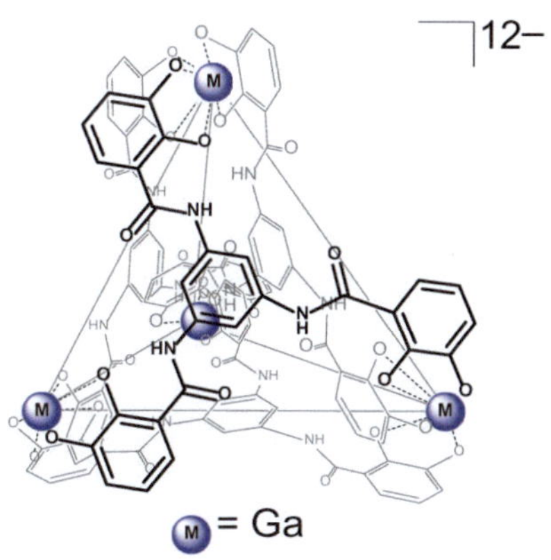

679.

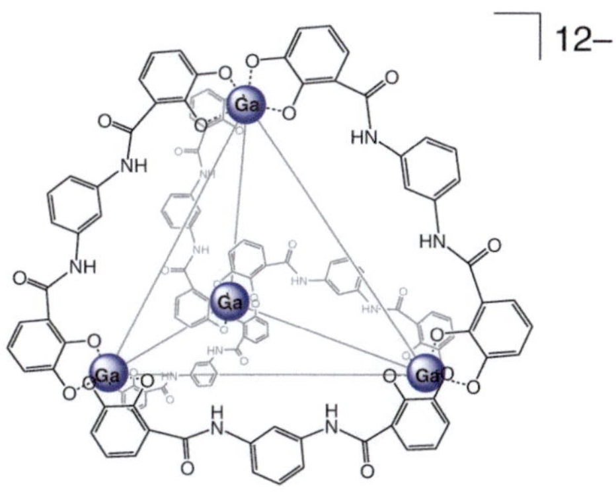

680.

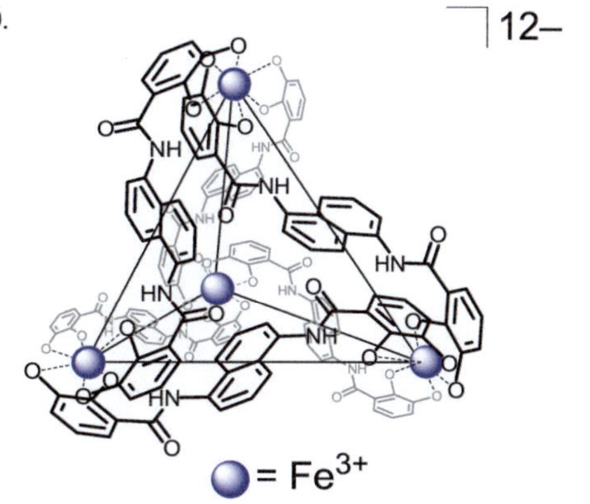

681.

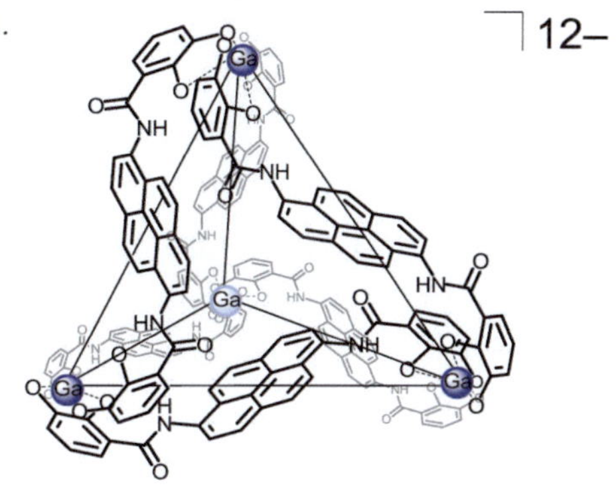

682.

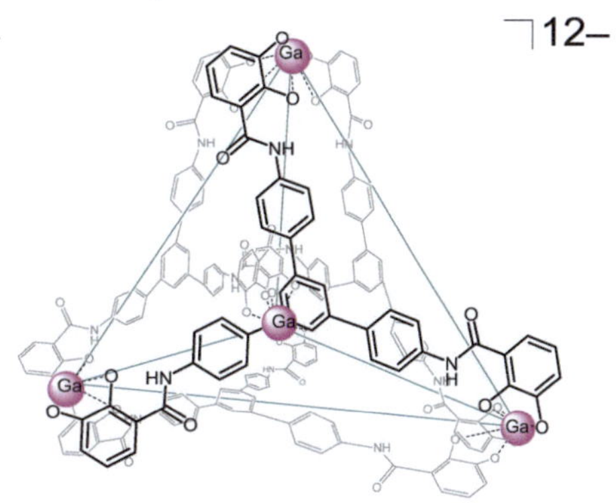

683.

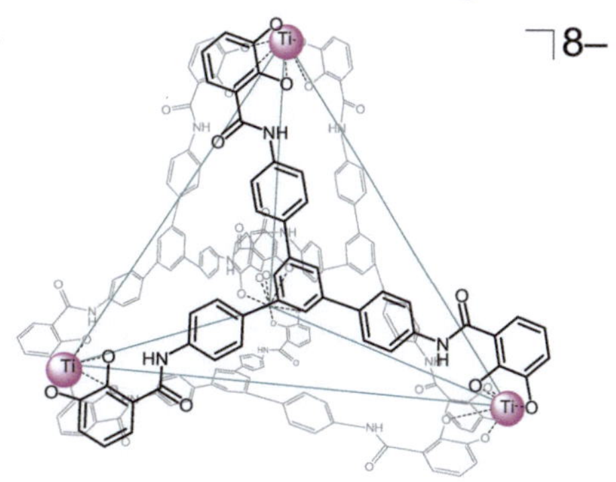

684.

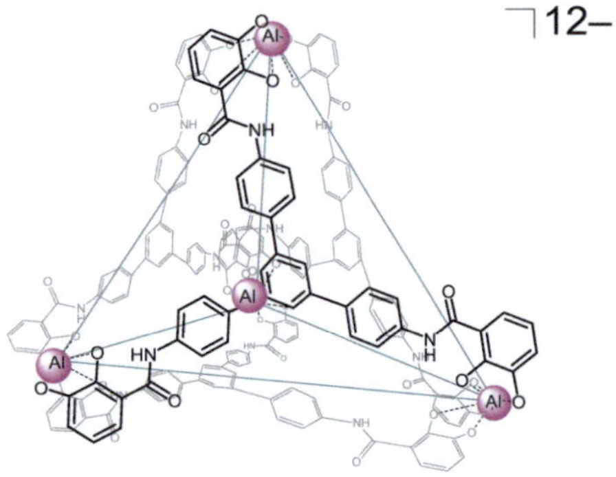

685.

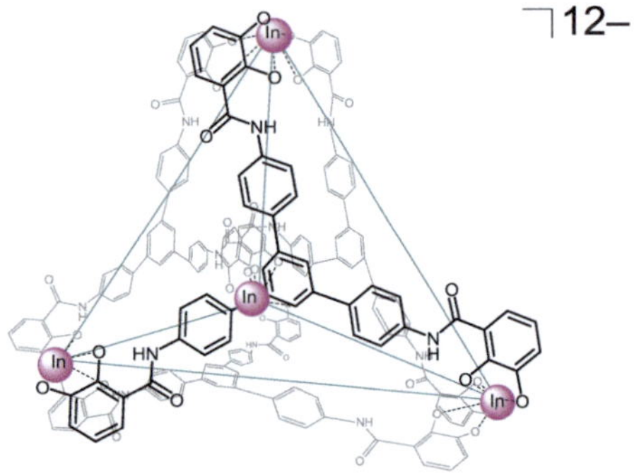

686.

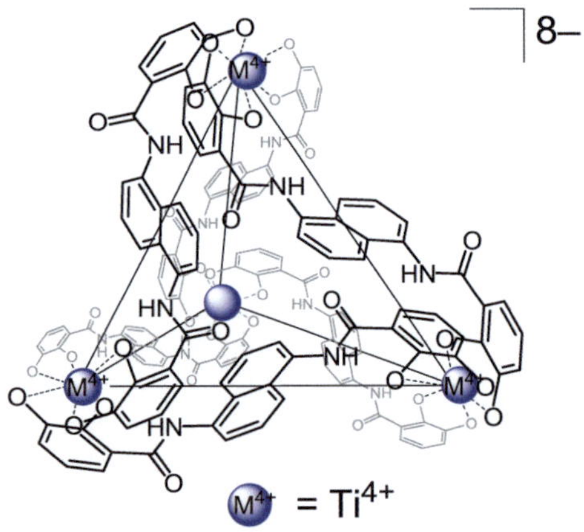

687.

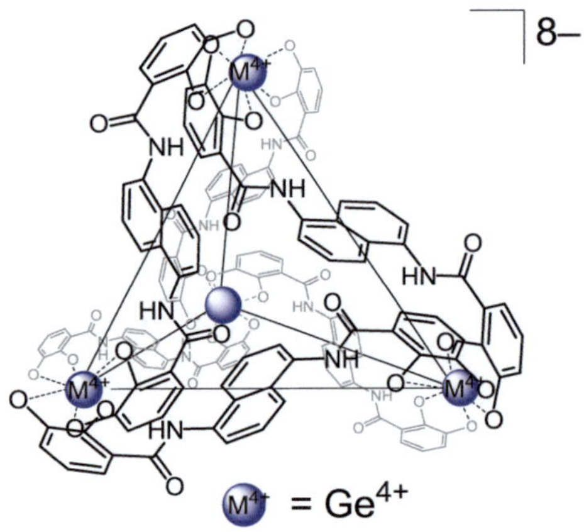

688.

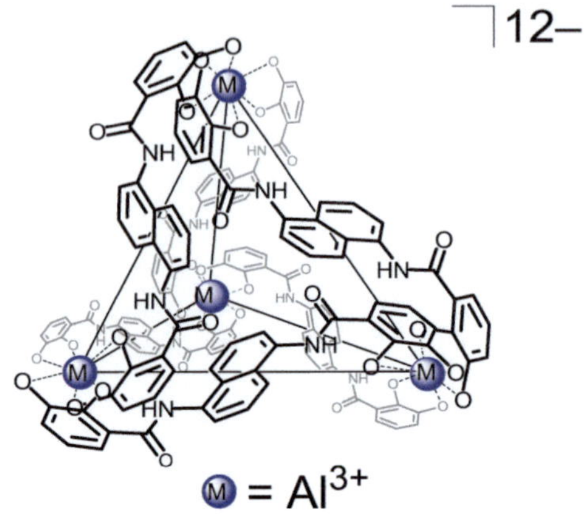

689.

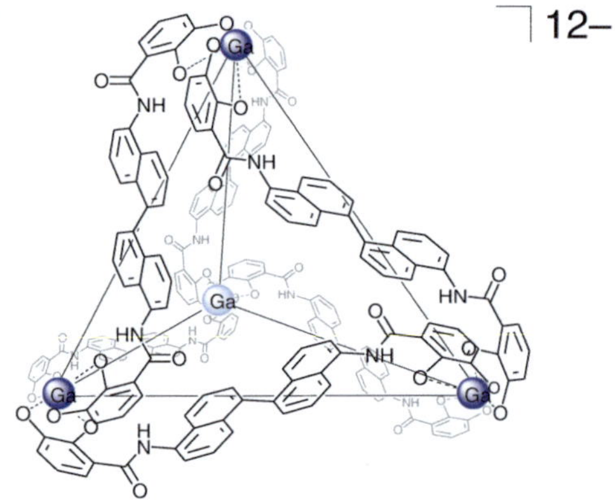

690.

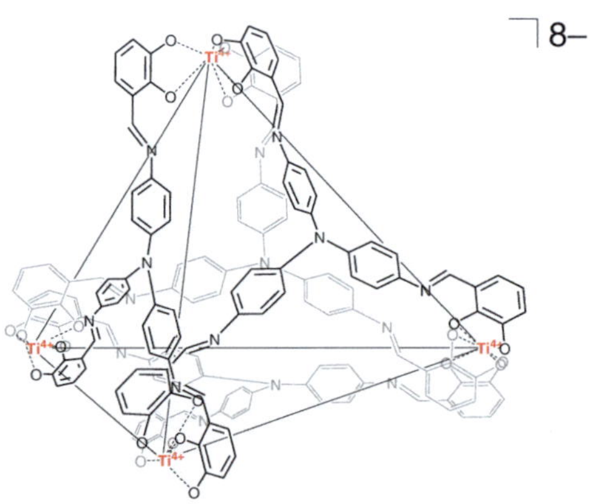

691.

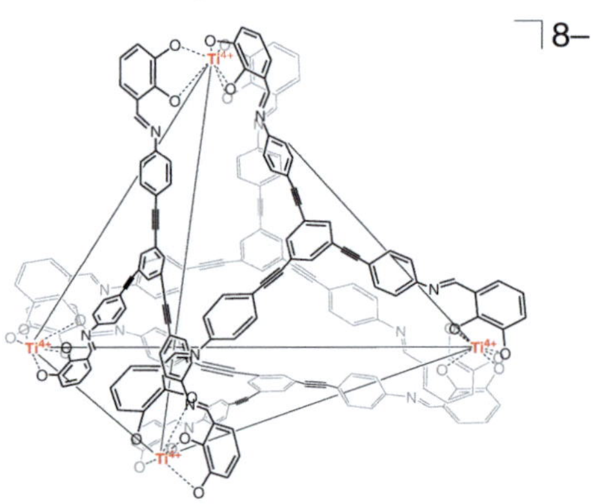

692.

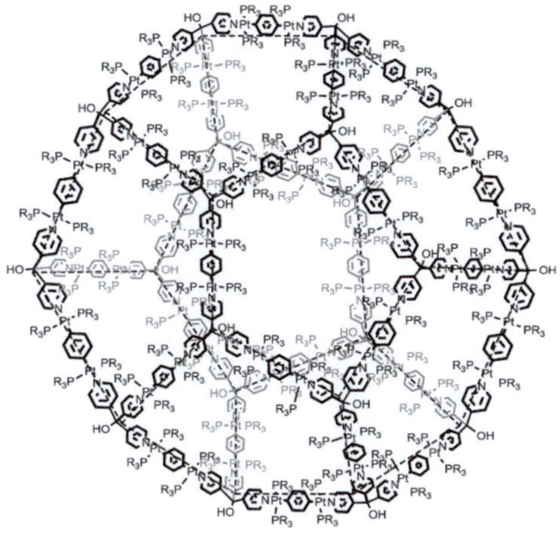

$R = C_2H_5$

693.

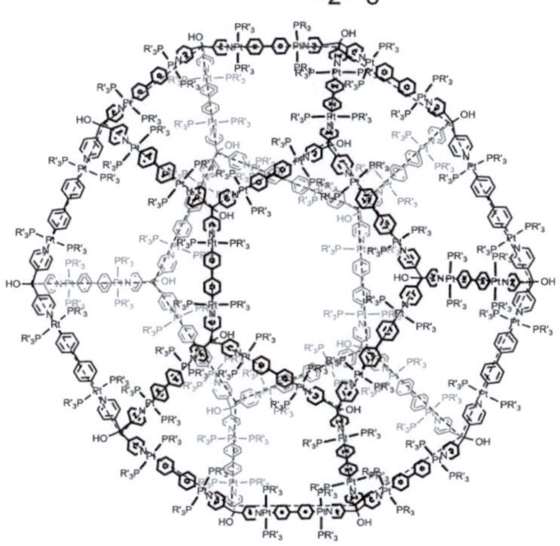

$R' = C_6H_5$

694.

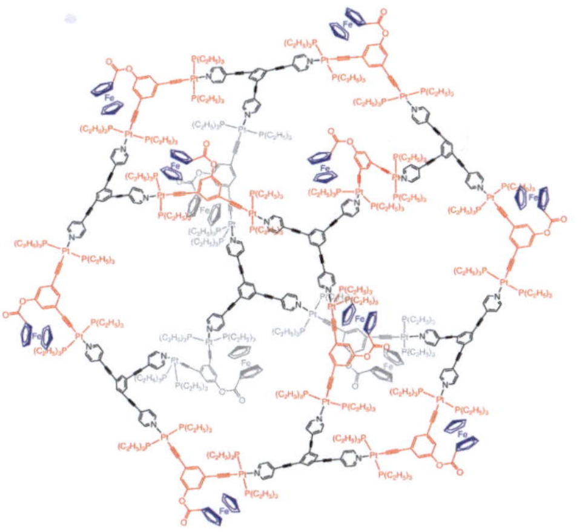

695.

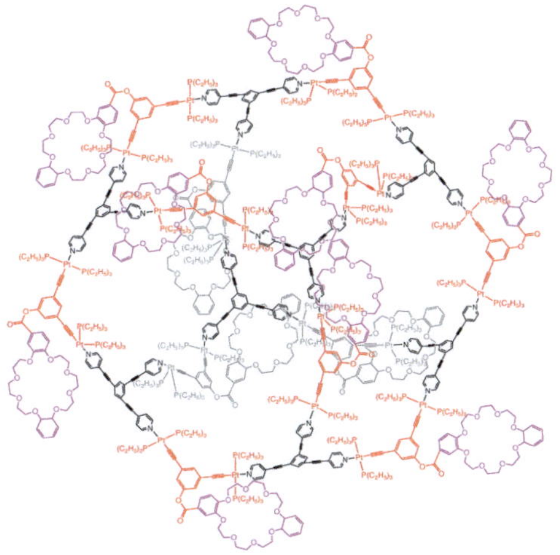

696.

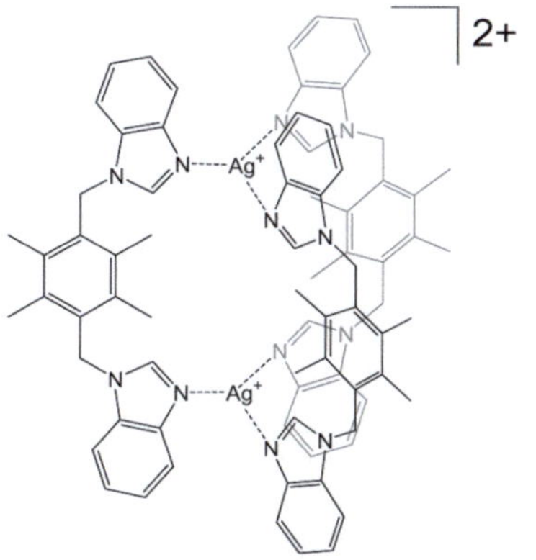

697.

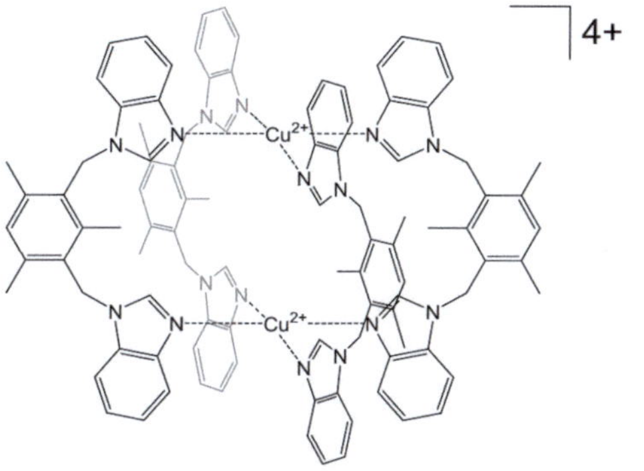

698.

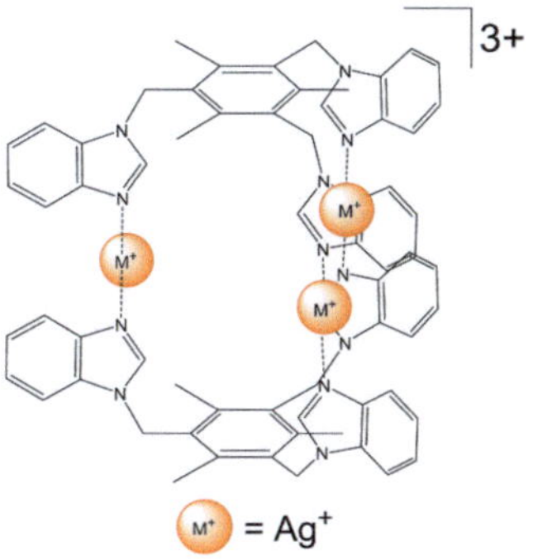

699.

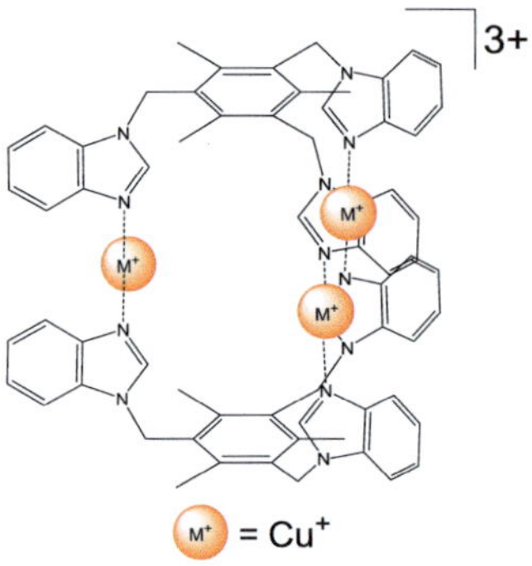

700.

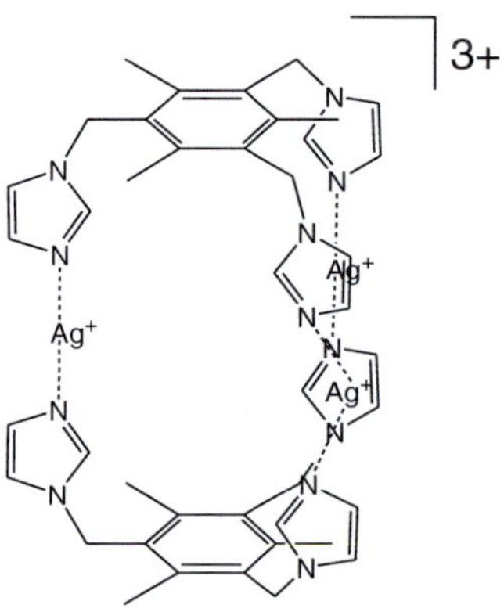

701.

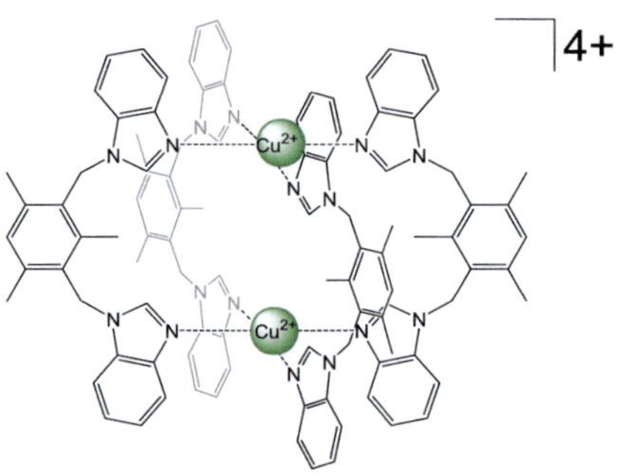

702.

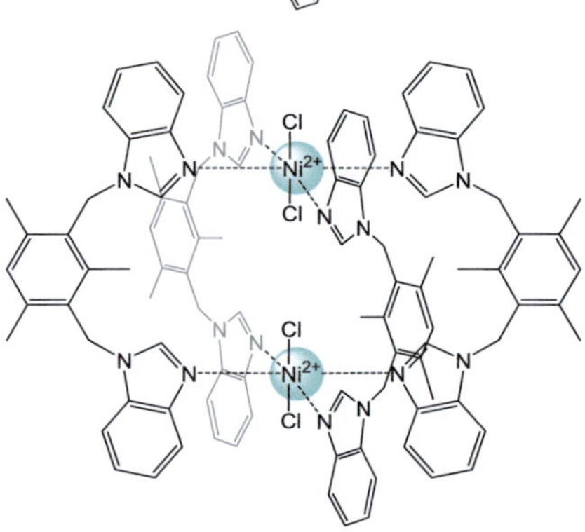

703.

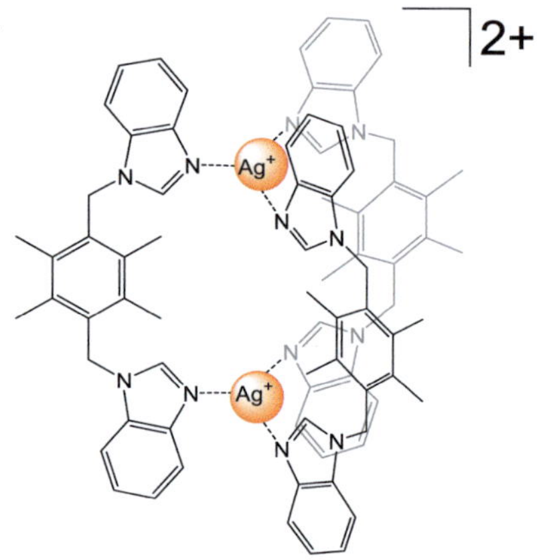

704.

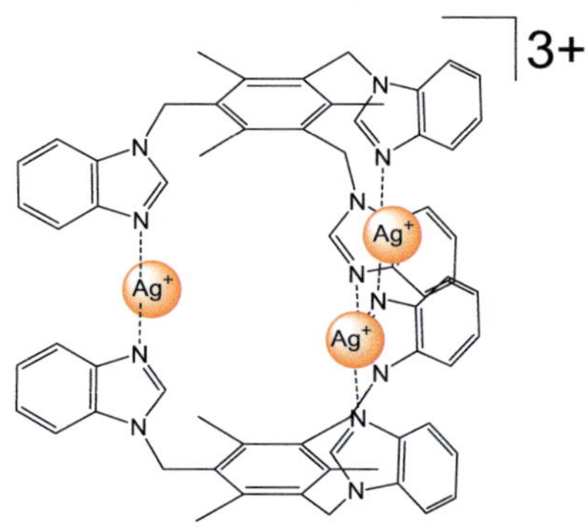

705.

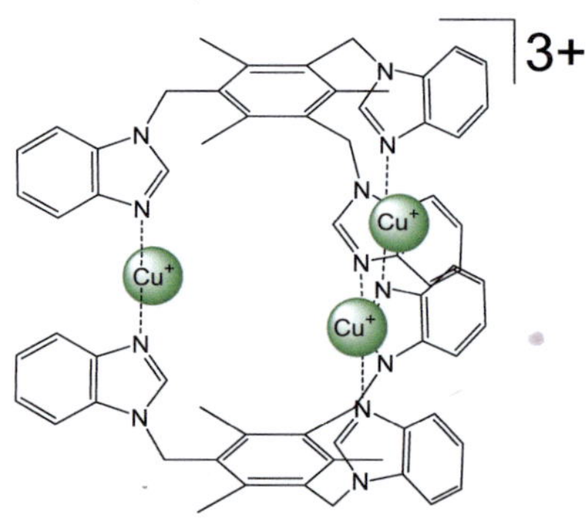

706.

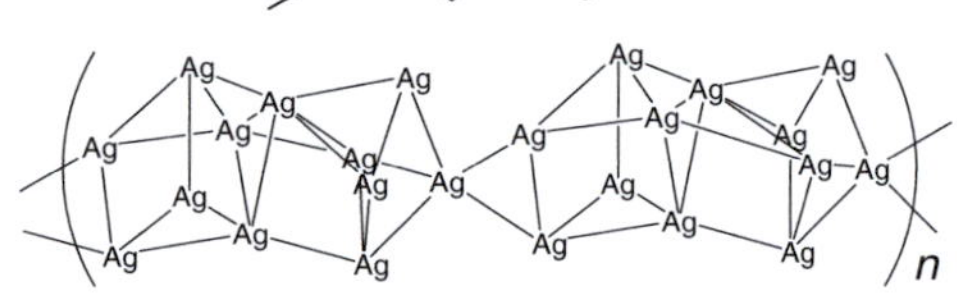

707.

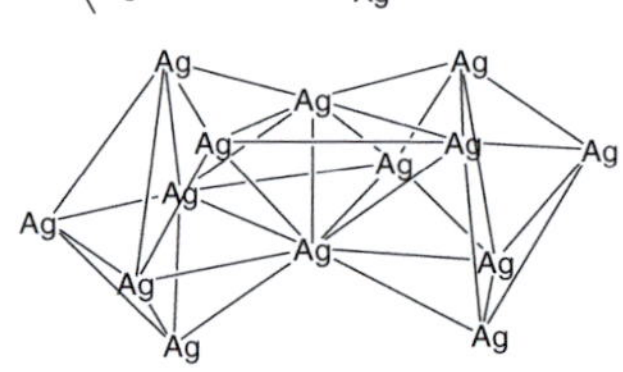

708.

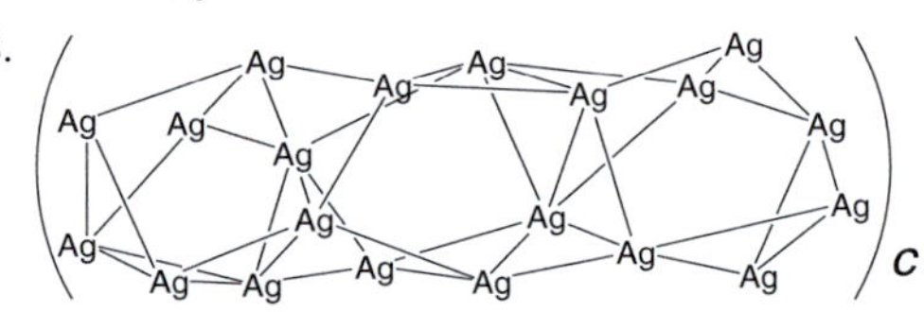

709.

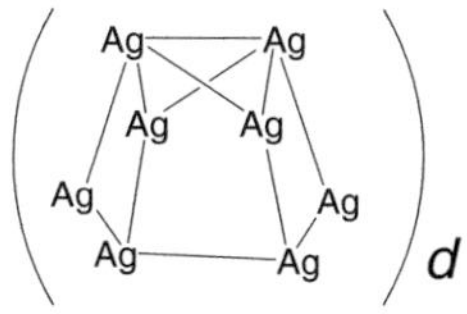

710.

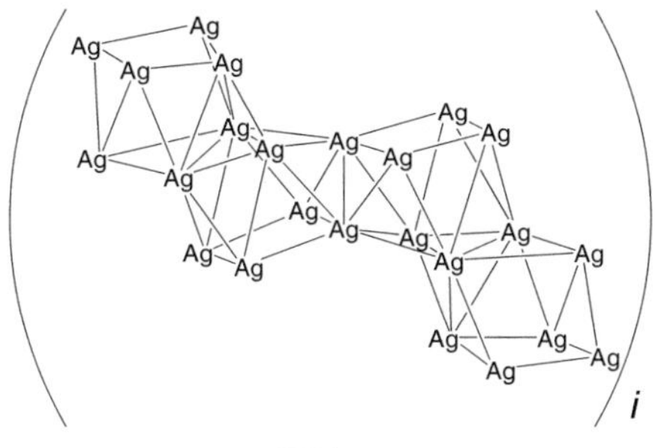

711.

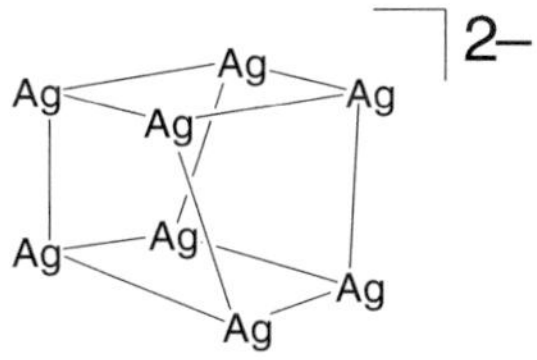

712.

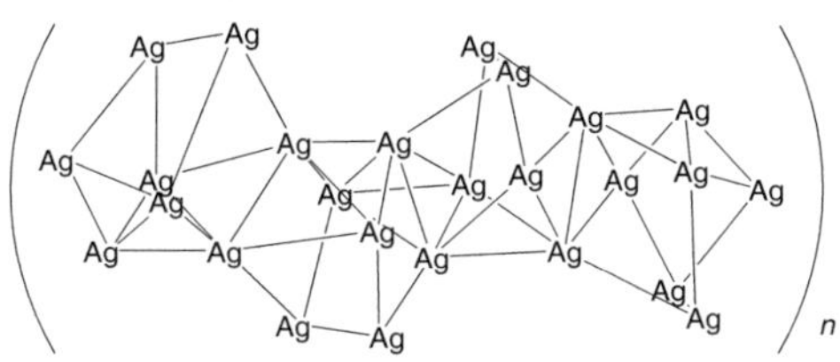

713.

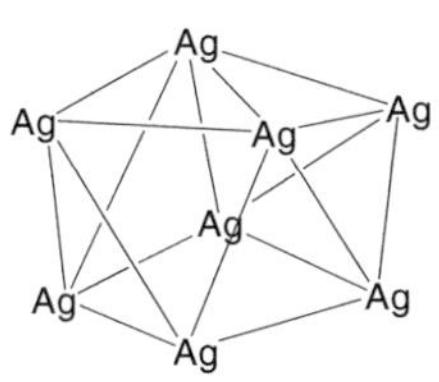

714.

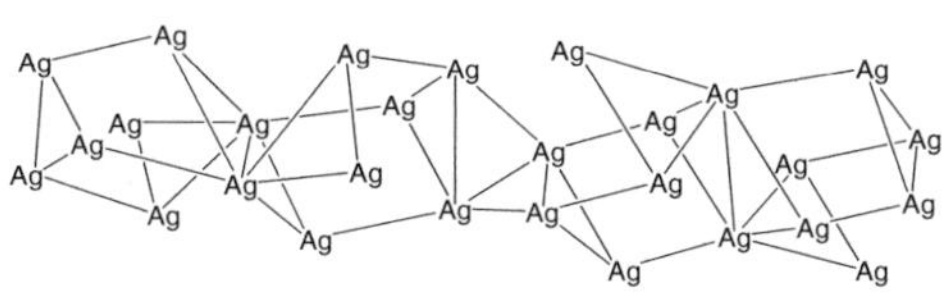

715.

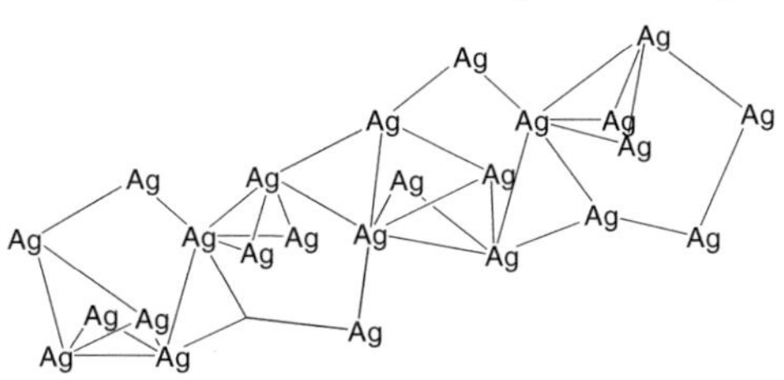

716.

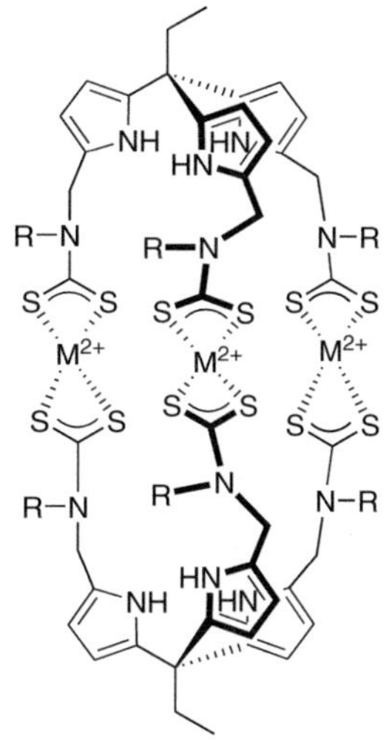

R = benzyl, M = Ni^{2+}

717.

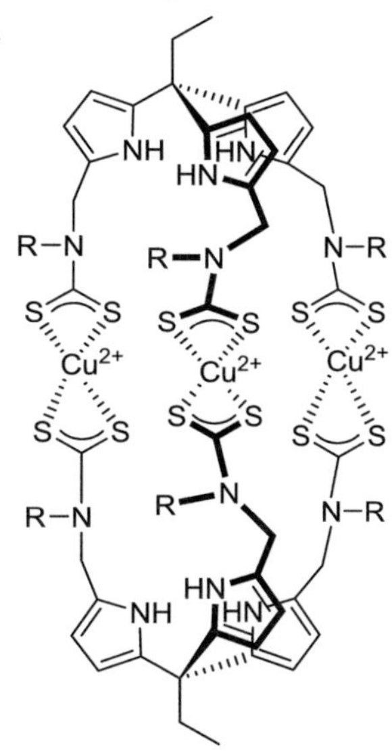

R = benzyl, M = Cu^{2+}

718.

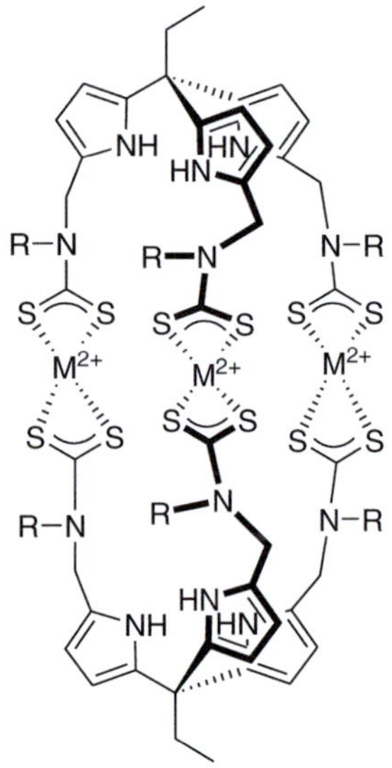

R = benzyl, M = Zn(pyr)$^{2+}$

719.

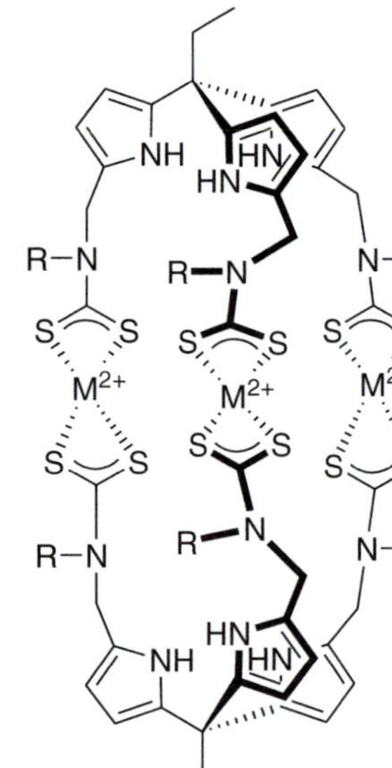

R =cyclohexyl, M = Ni^{2+}

720.

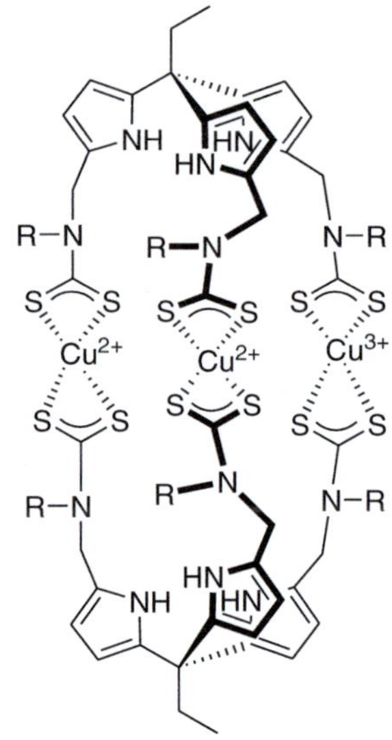

721.

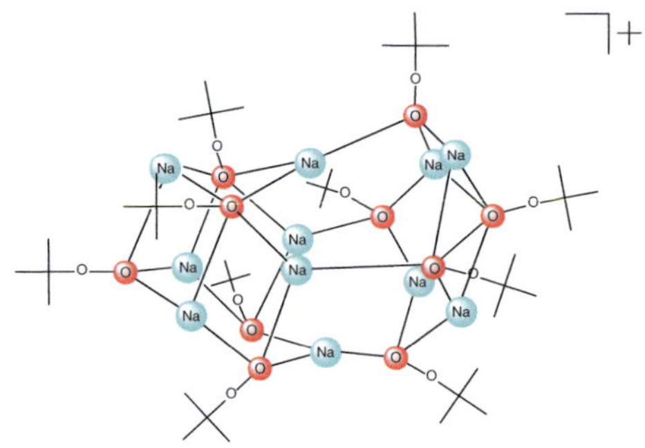

722.

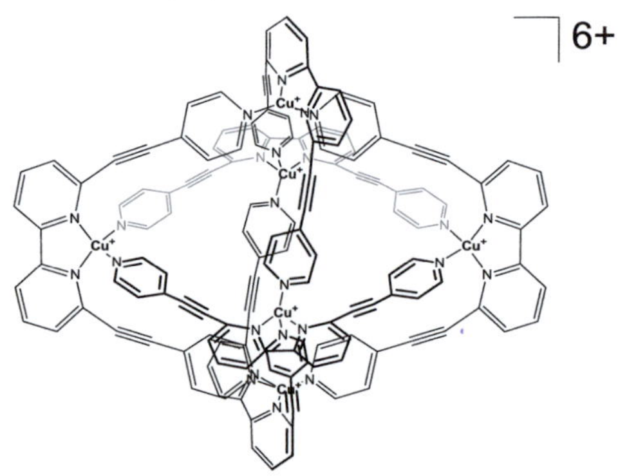

723.

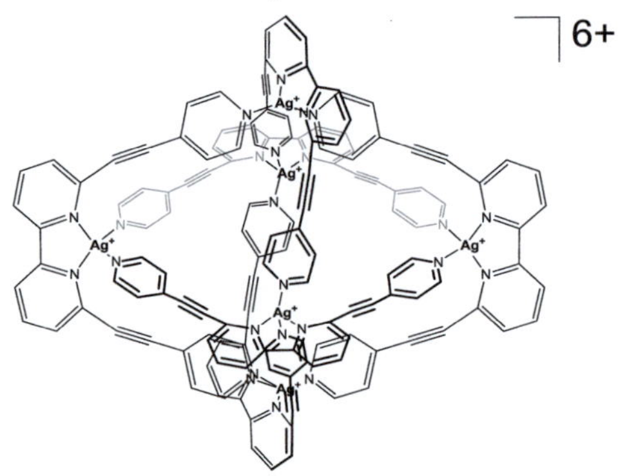

724.

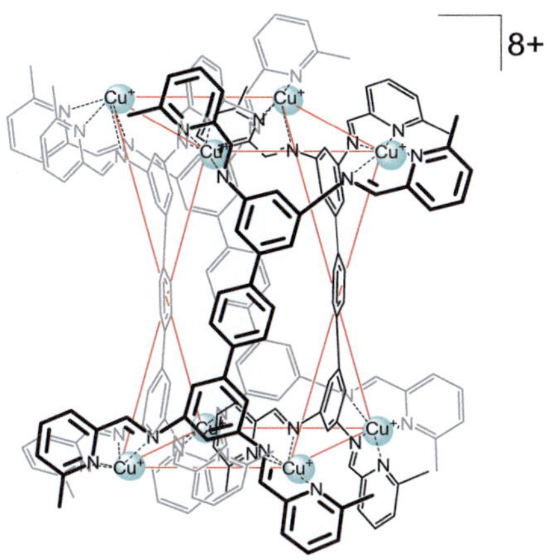

725.

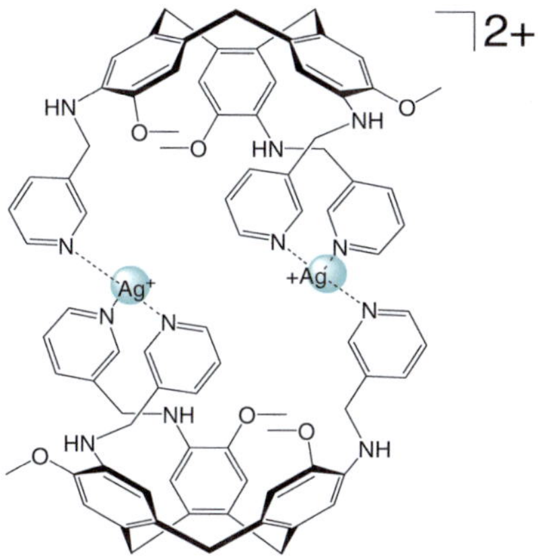

726.

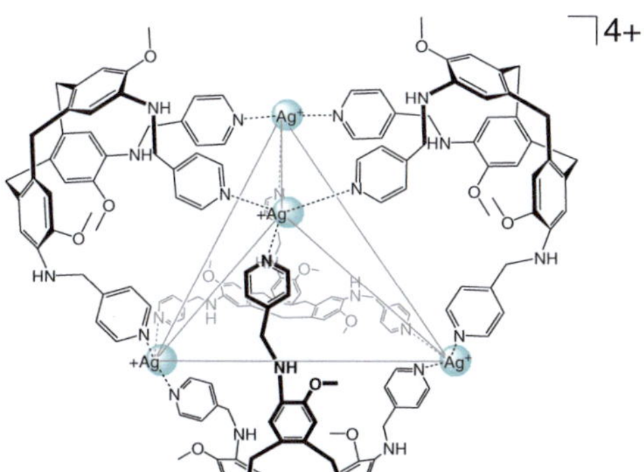

727.

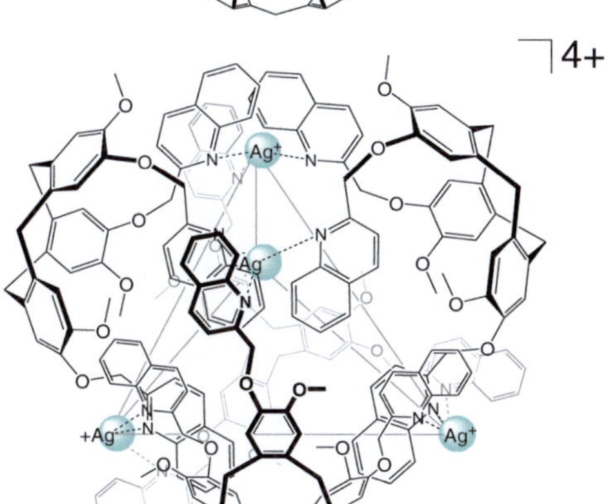

728.

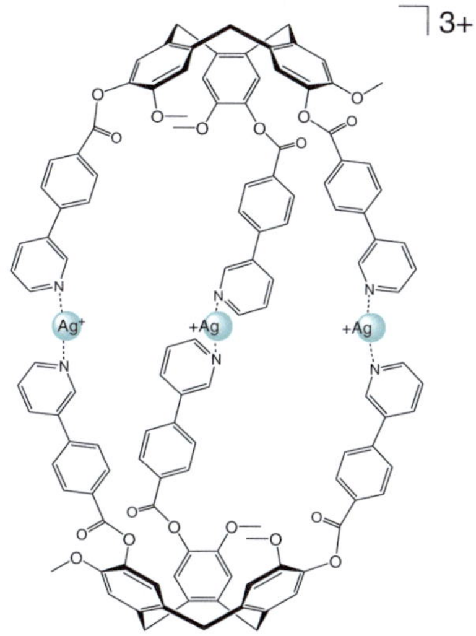

729.

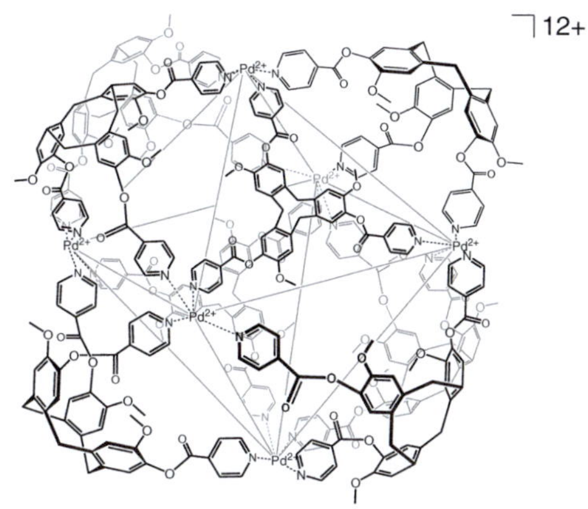

730.

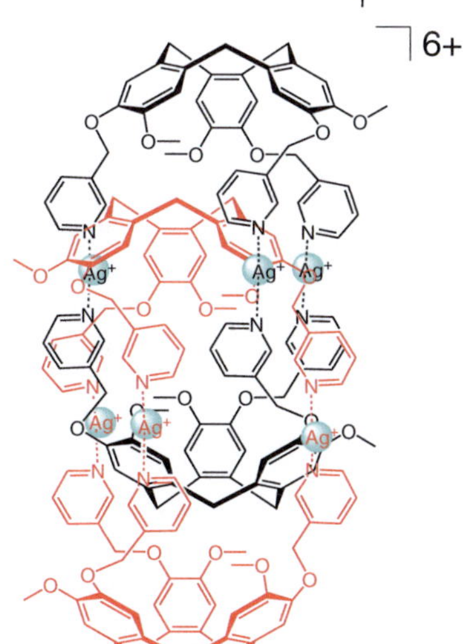

731.

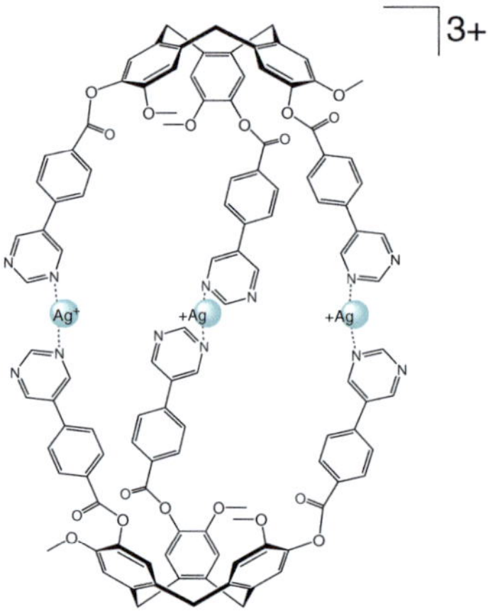

732.

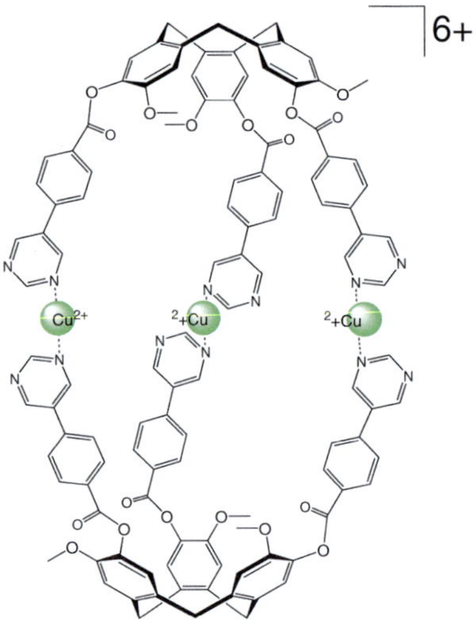

733.

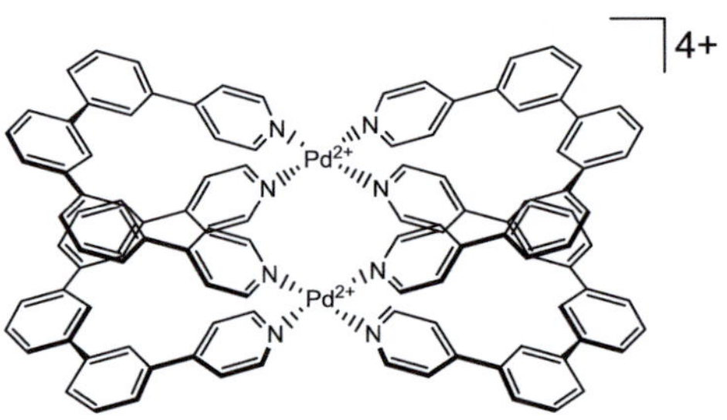

734.

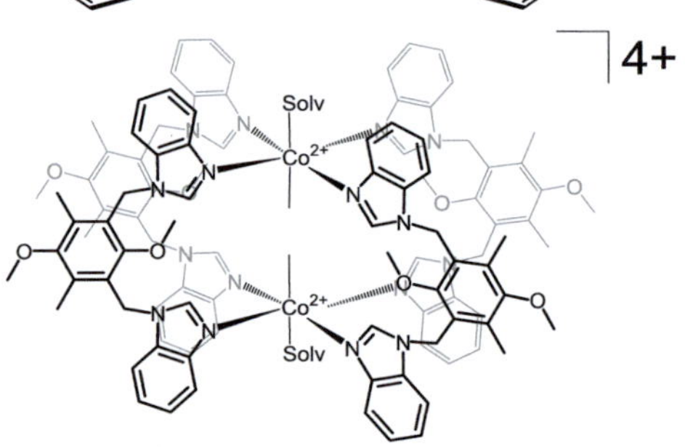

735.

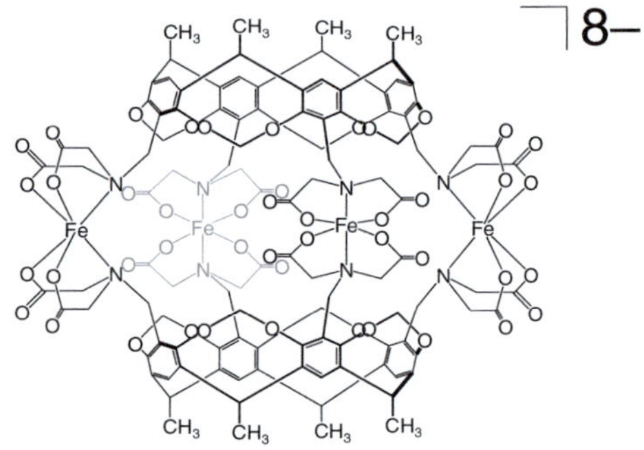

736.

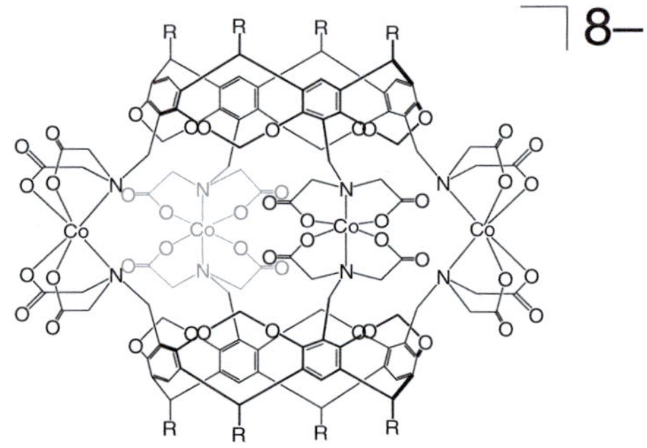

R = CH_3

737.

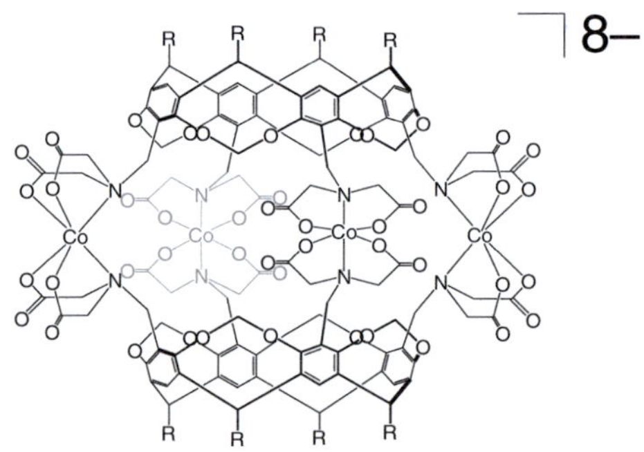

R = CH_2CH_3

738.

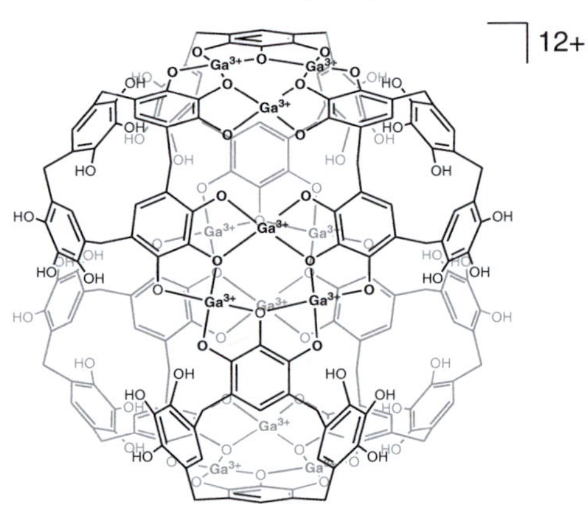

739.

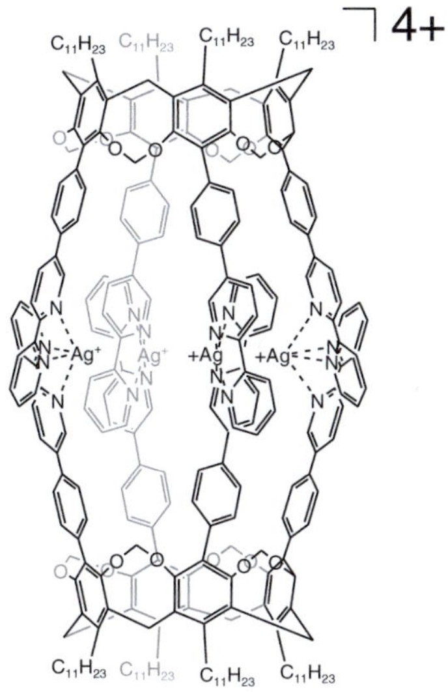

740.

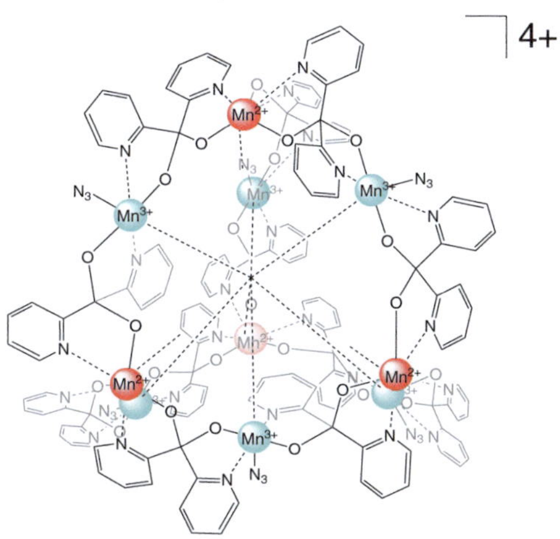

741.

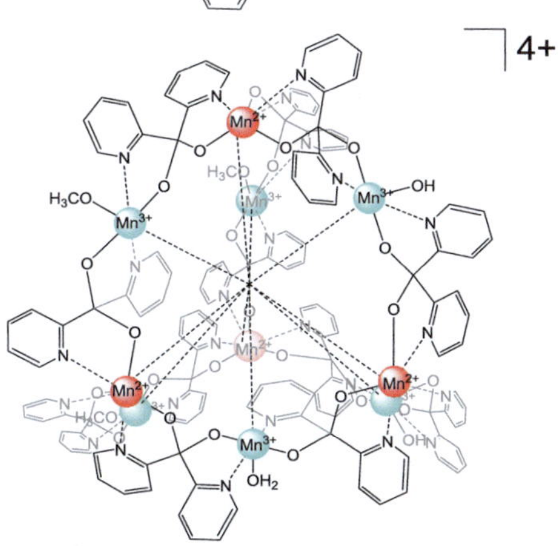

742.

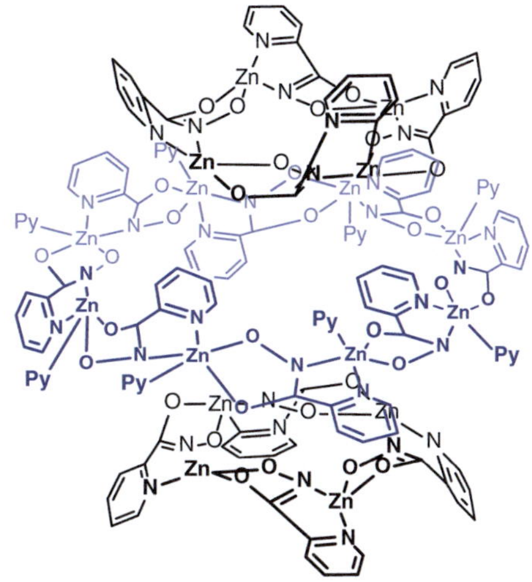

743.

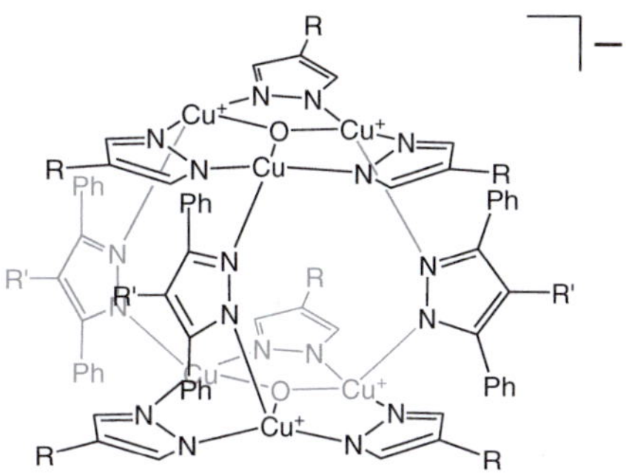

R = R' = H

744.

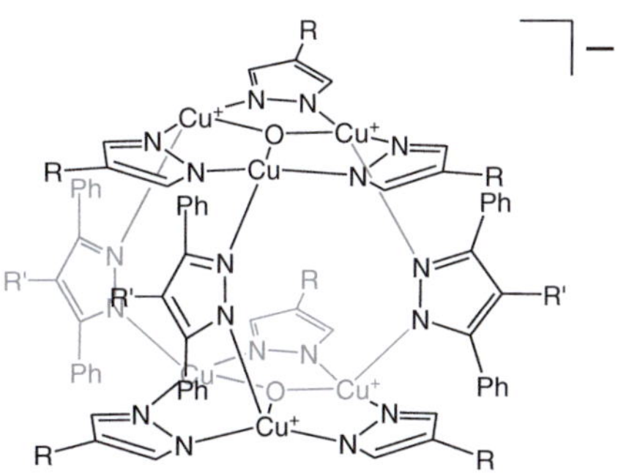

R = Cl, R' = H

745.

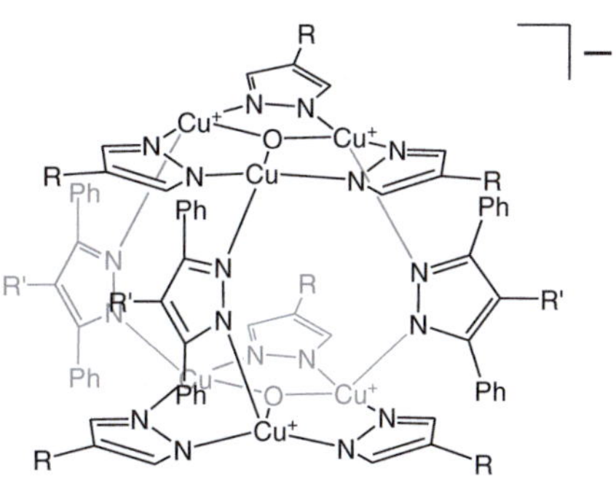

R = H, R' = Cl

746.

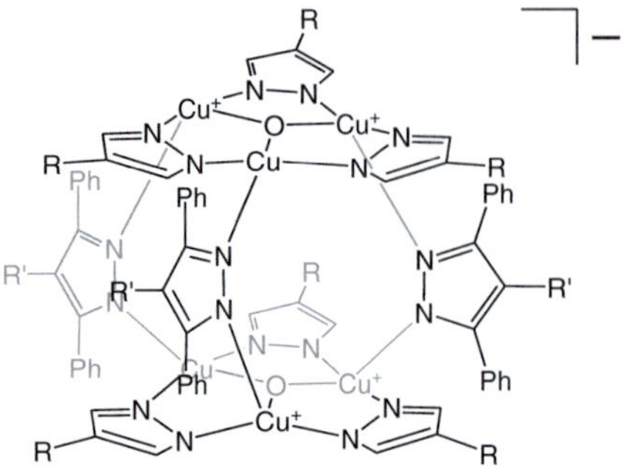

R = R' = Cl

747.

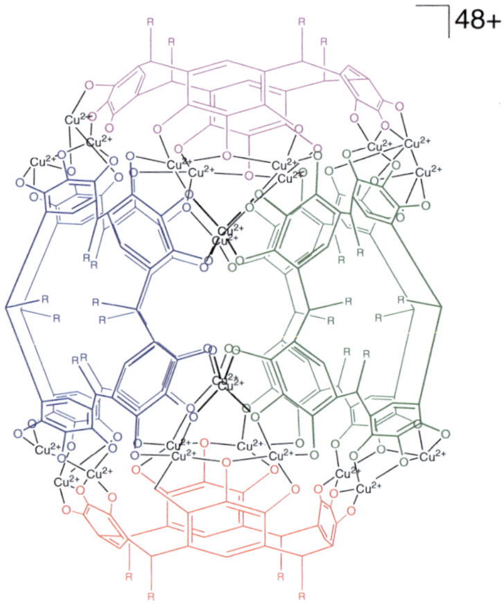

748.

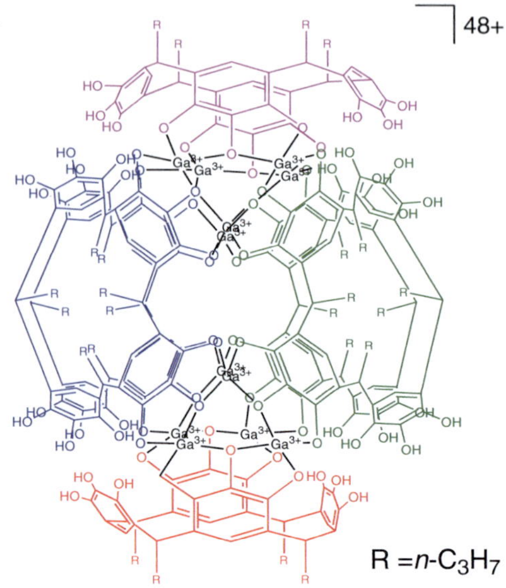

R = n-C_3H_7

749.

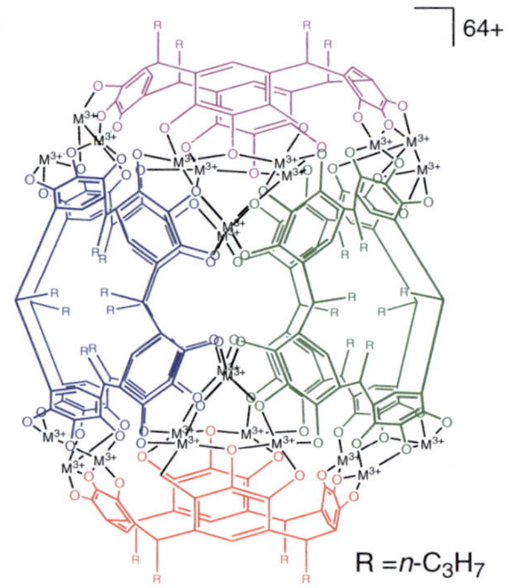

R = n-C_3H_7

750.

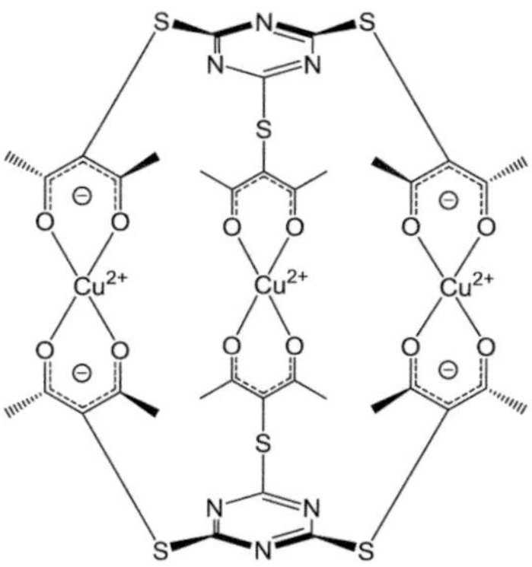

751.

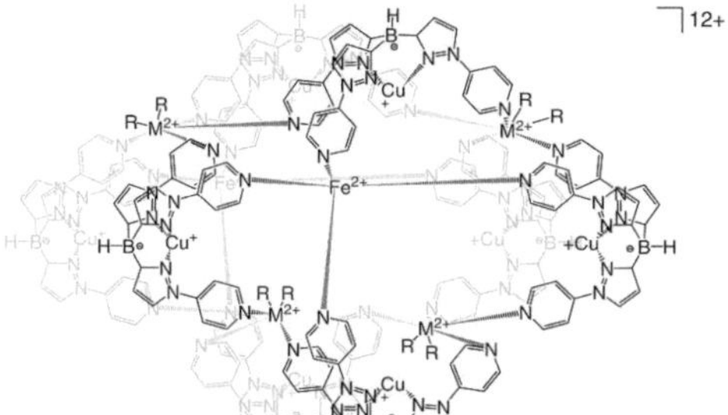

$M^{2+} = Fe^{2+}$
$R = SCN^-$

752.

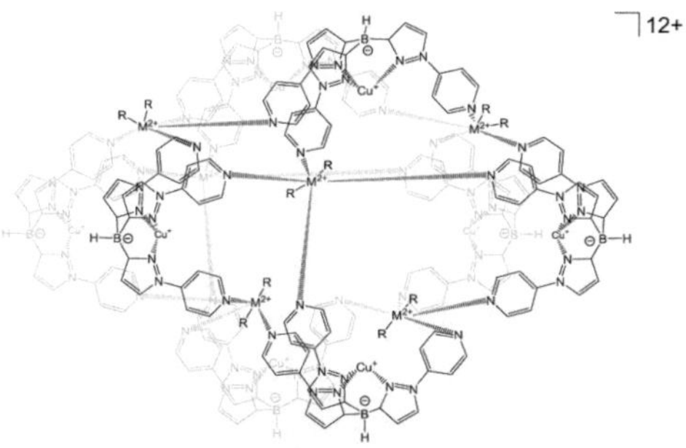

$M^{2+} = Cu^{2+}$
$R = CH_3CN, ClO_4^-, BF_4^-$

753.

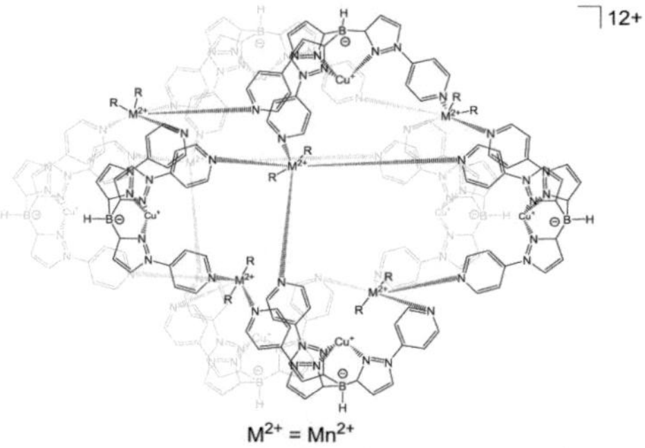

$M^{2+} = Mn^{2+}$
$R = CH_3CN, ClO_4^-, BF_4^-$

754.

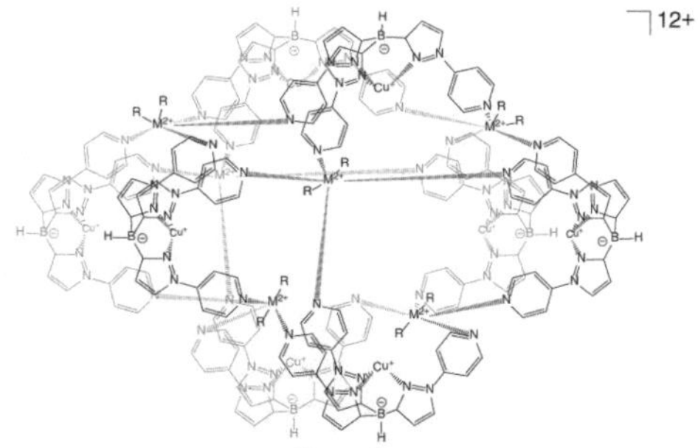

$M^{2+} = Cd^{2+}$
$R = CH_3CN, ClO_4^-, BF_4^-$

755.

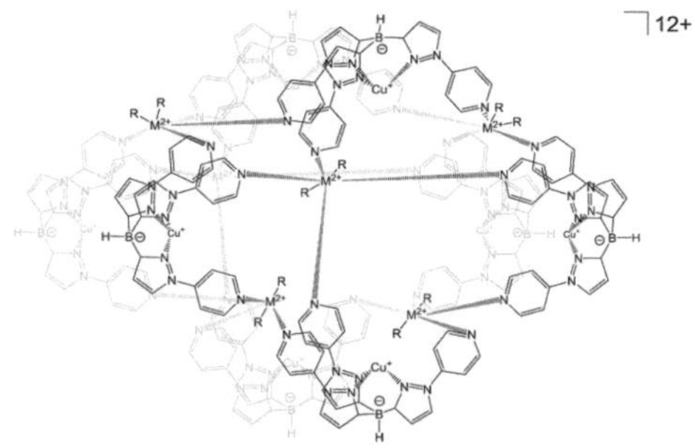

$M^{2+} = Zn^{2+}$
$R = CH_3CN, ClO_4^-, BF_4^-$

756.

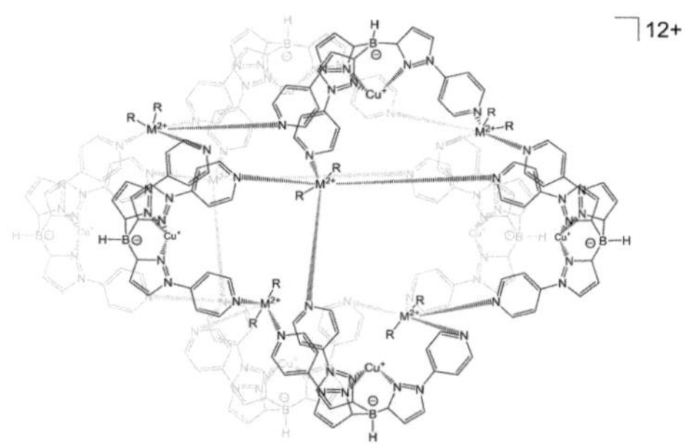

$M^{2+} = Fe^{2+}$
$R = CH_3CN, ClO_4^-, BF_4^-$

757.

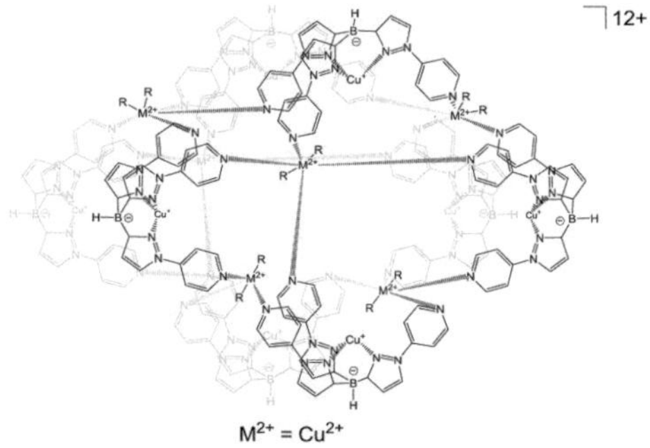

$M^{2+} = Cu^{2+}$
$R = CH_3CN, ClO_4^-, BF_4^-$

758.

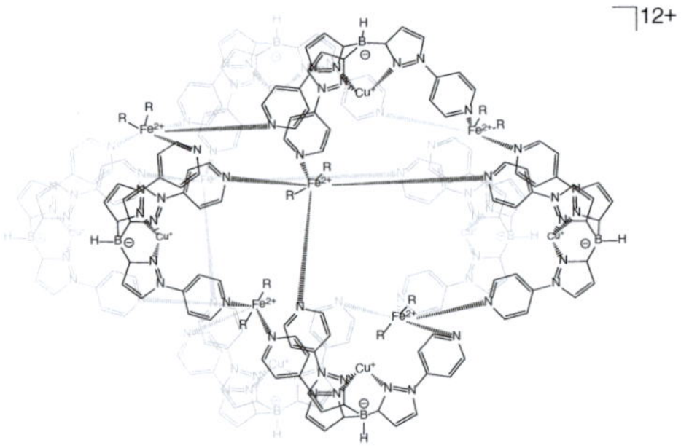

$A^- = NCSe^-$
$R = CH_3CN, NCSe^-, NCBCH_3^-$

759.

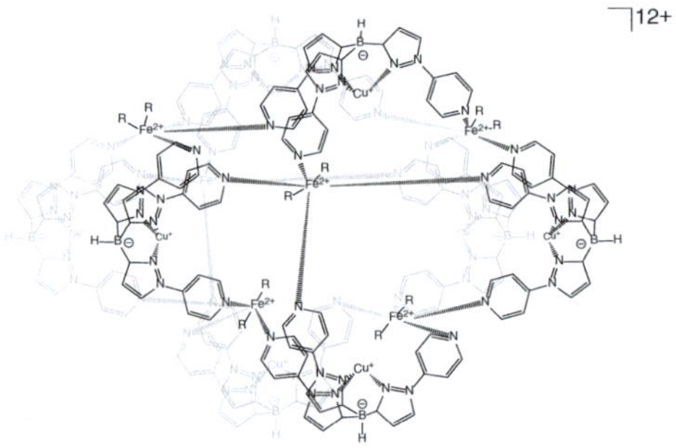

$A^- = NCBCH_3^-$
$R = CH_3CN, NCSe^-, NCBCH_3^-$

760.

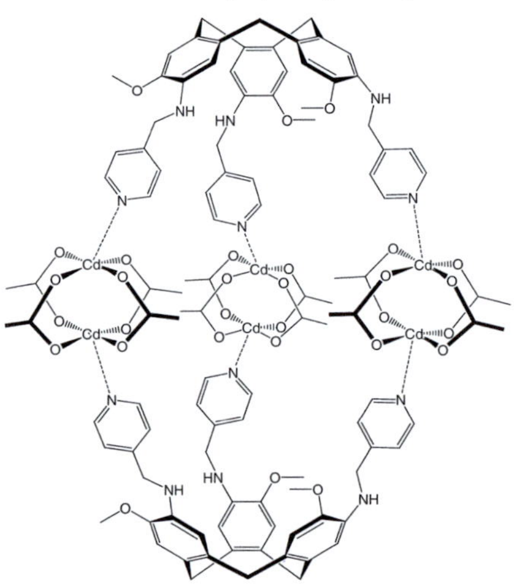

761.

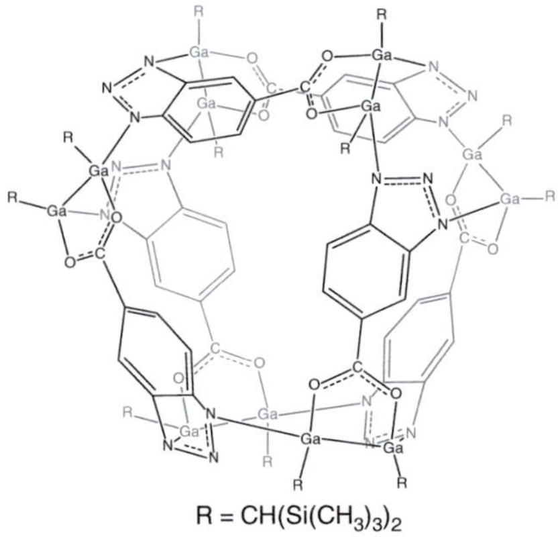

$R = CH(Si(CH_3)_3)_2$

762.

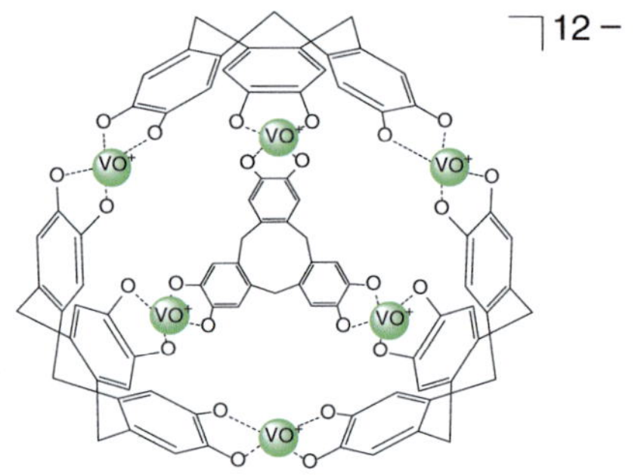

763.

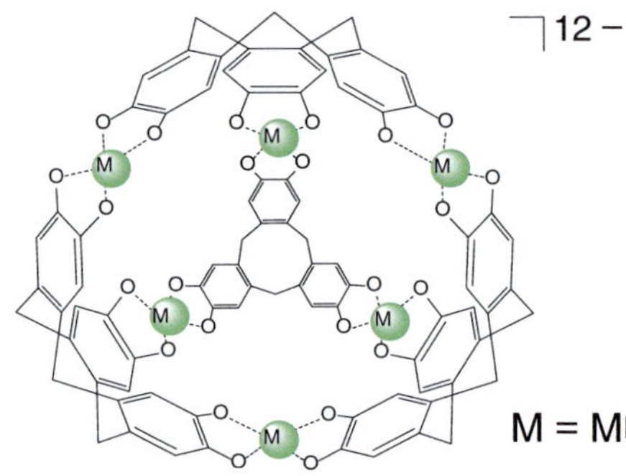

$M = Mn^{2+}$

764.

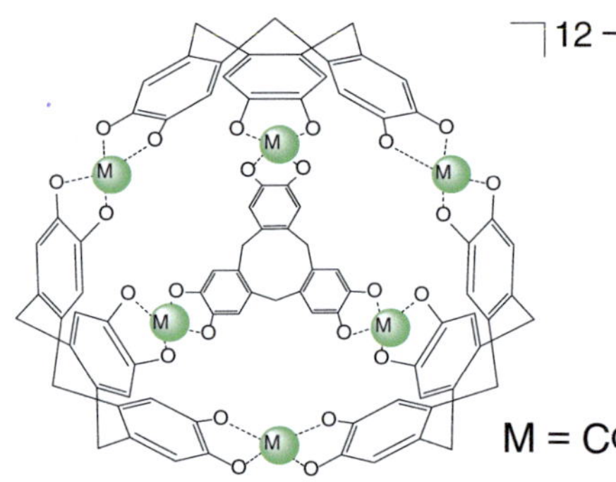

$M = Co^{2+}$

765.

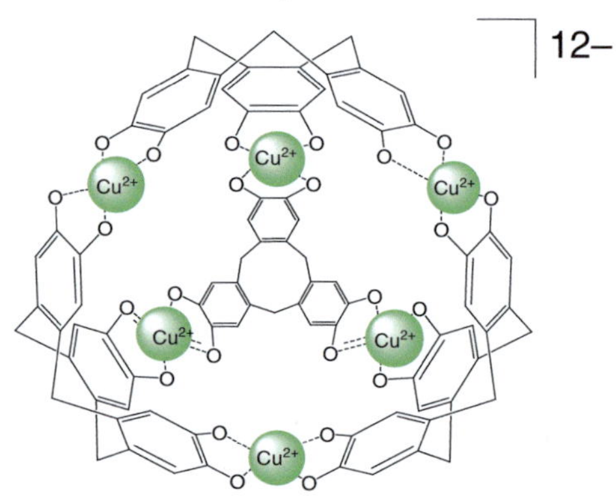

766.

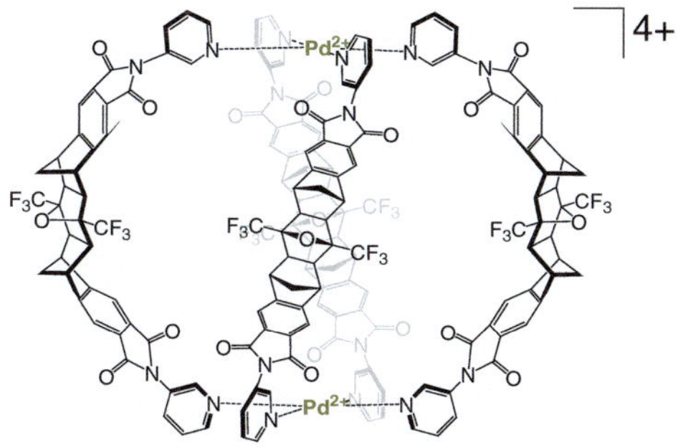

767.

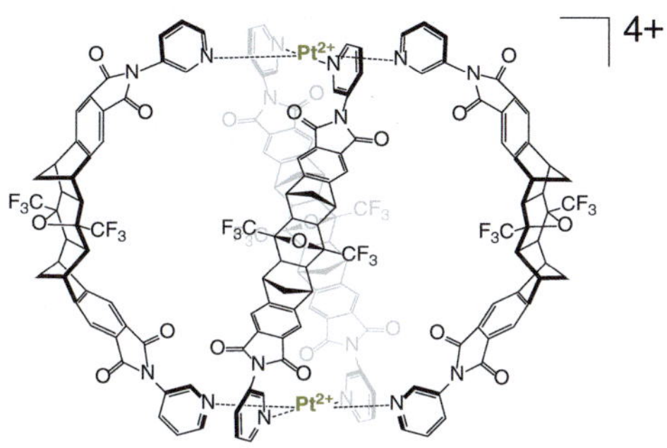

768.

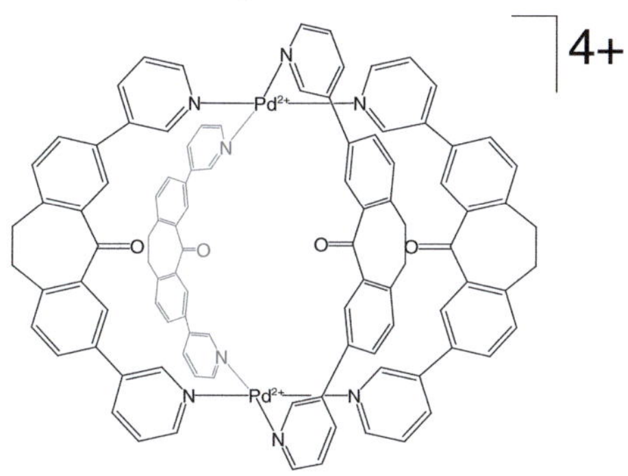

769.

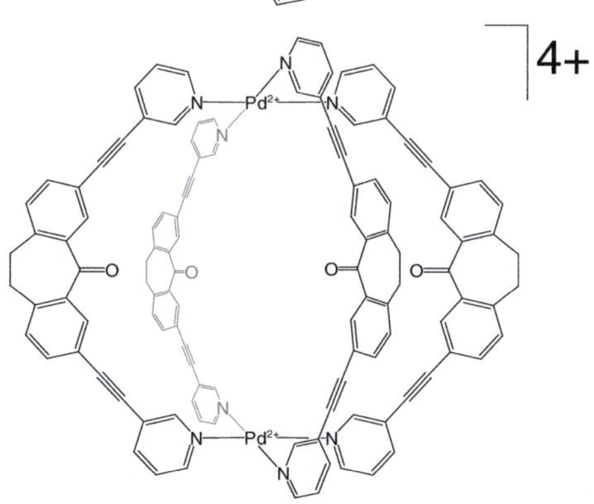

770.

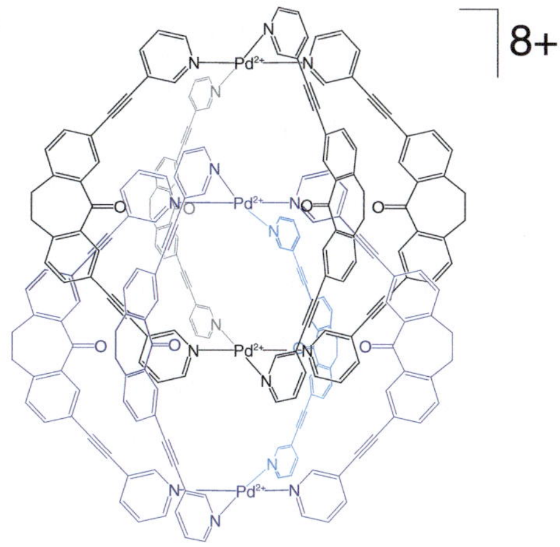

771.

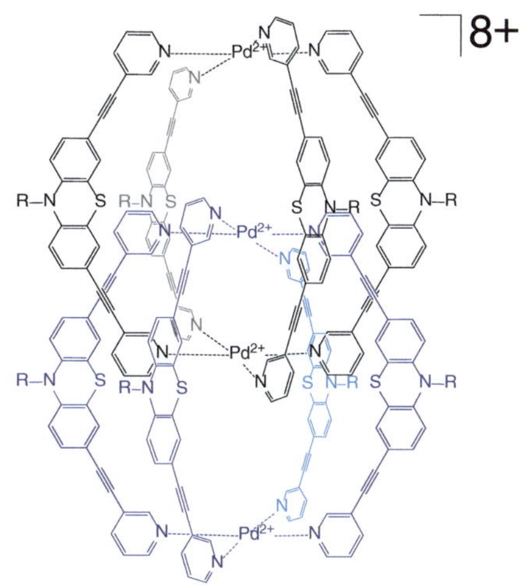

772.

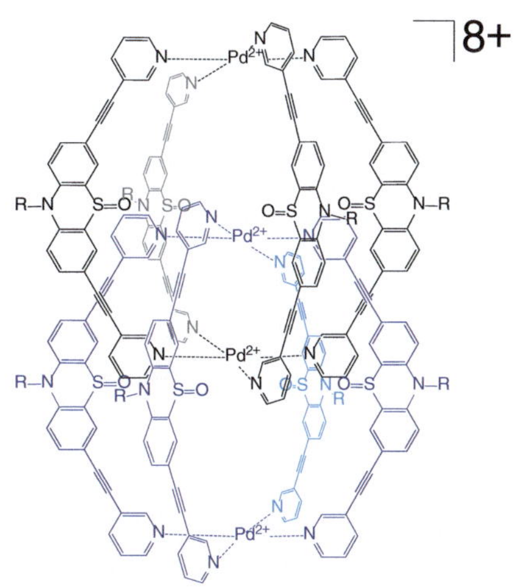

773.

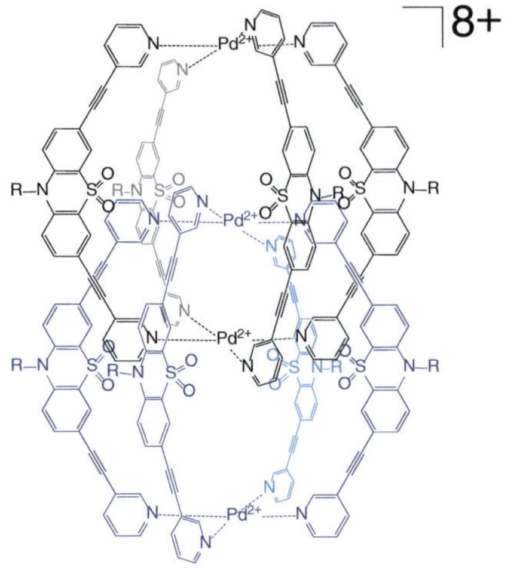

774.

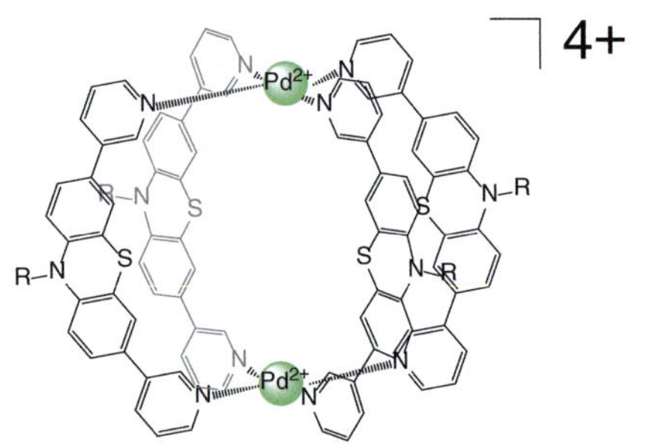

R = $(CH_2)_5CH_3$

775.

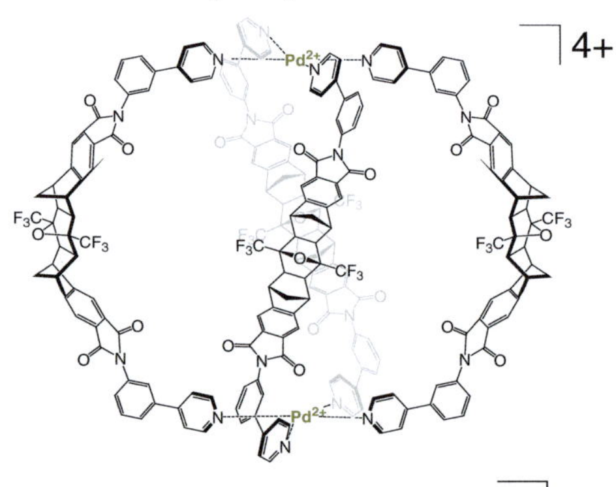

776.

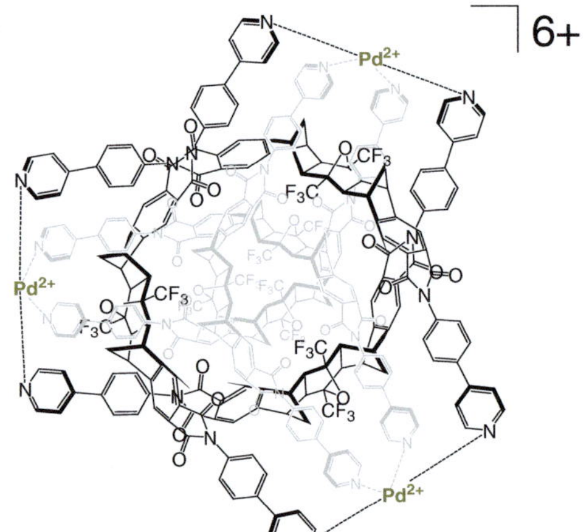

777.

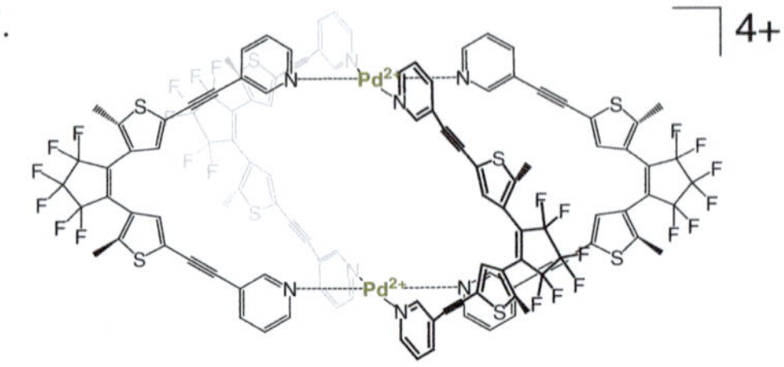

778.

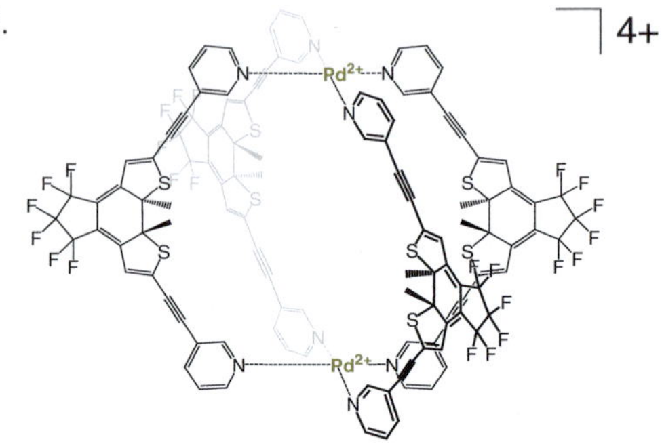

779.

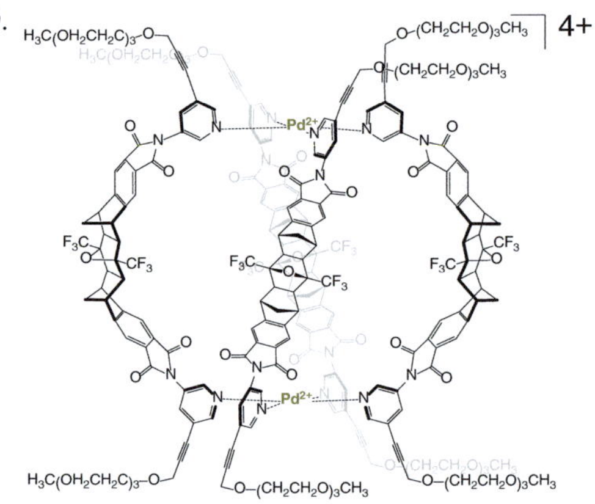

780.

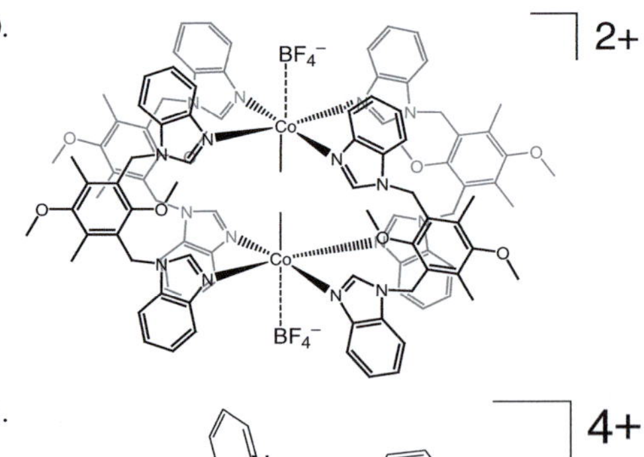

781.

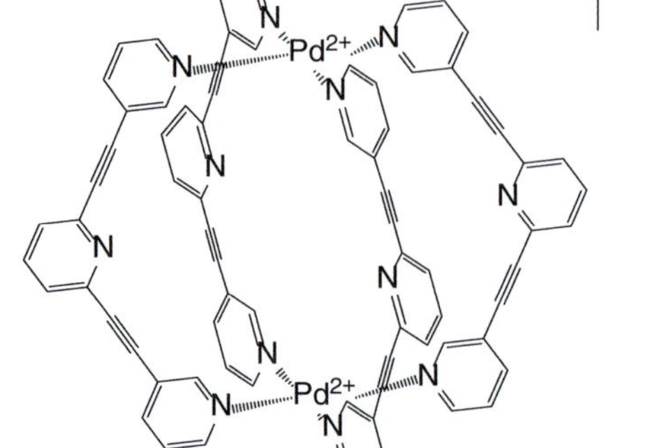

782.

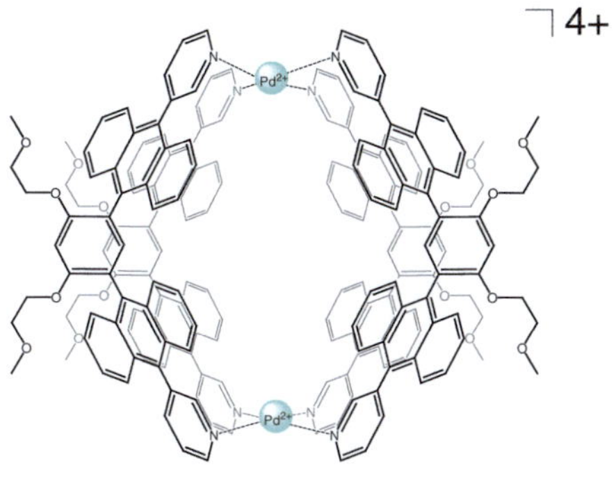

783.

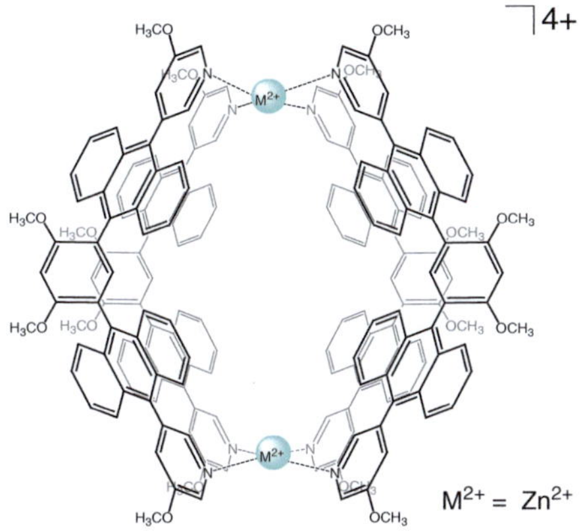

784.

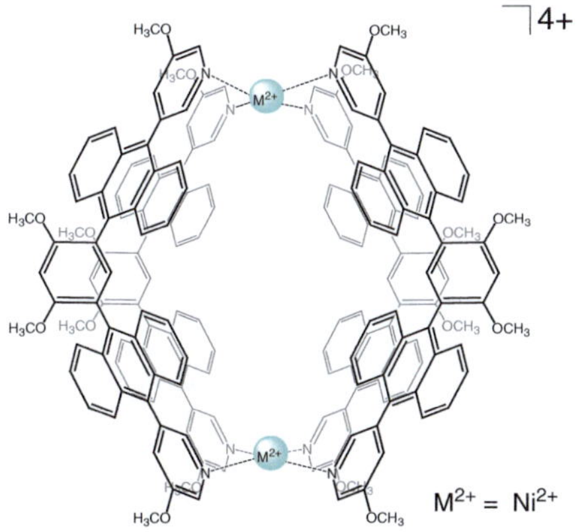

785.

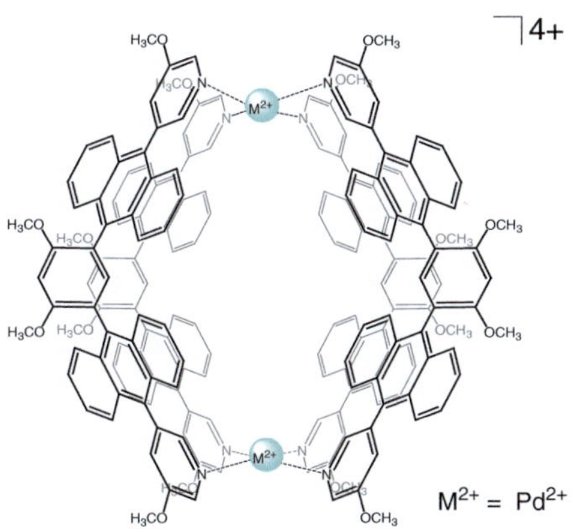

786.

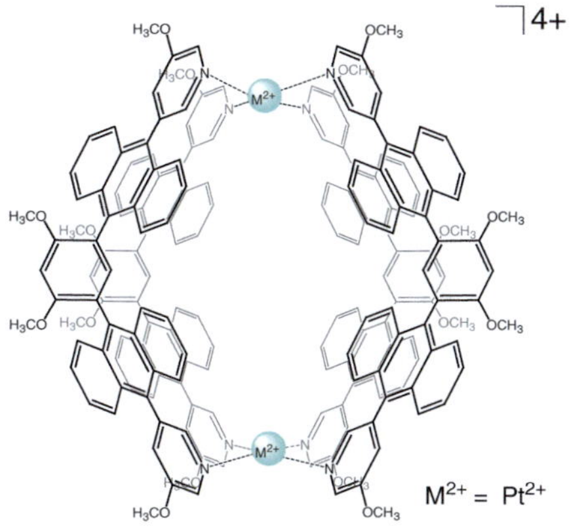

787.

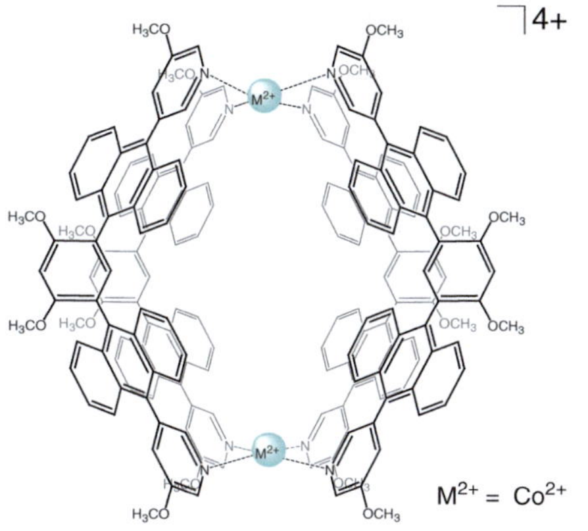

788.

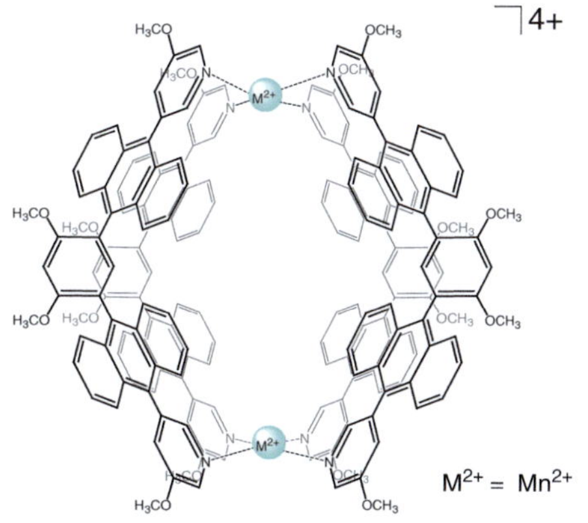

789.

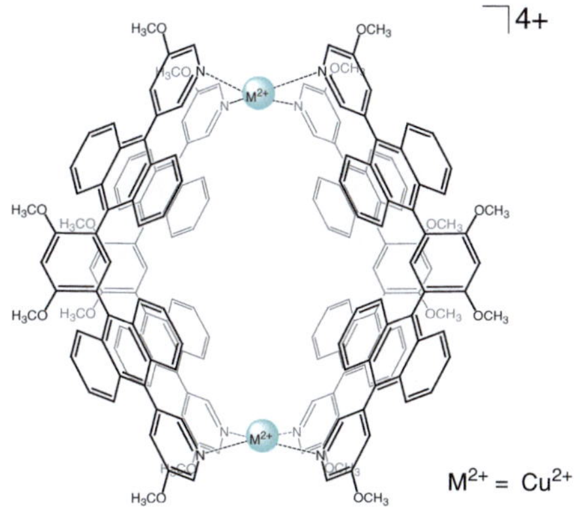

790.

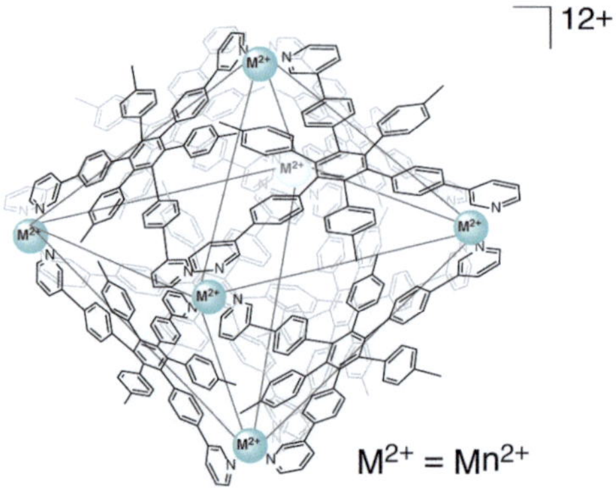

$M^{2+} = Mn^{2+}$

791.

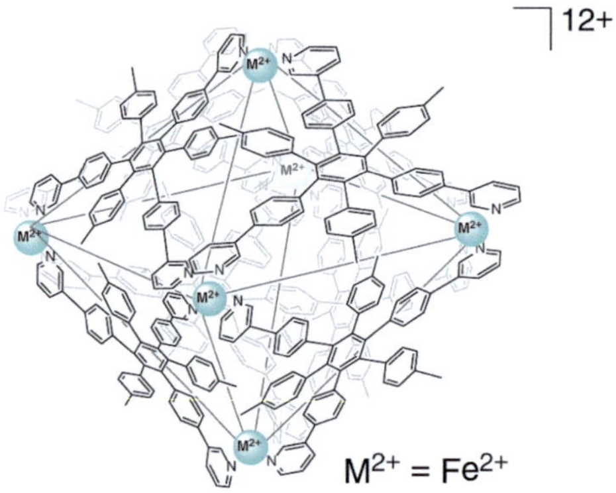

$M^{2+} = Fe^{2+}$

792.

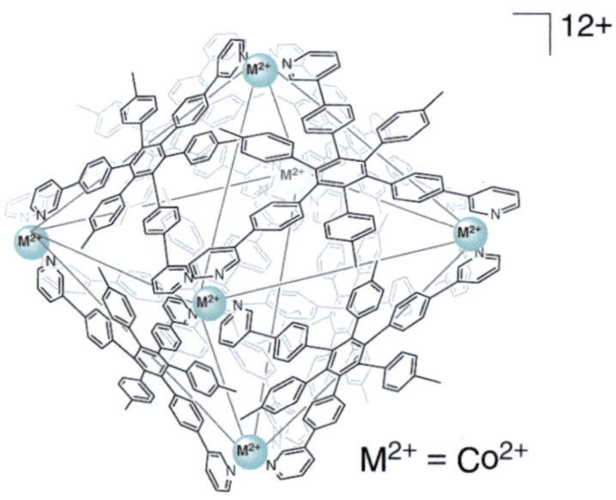

$M^{2+} = Co^{2+}$

793.

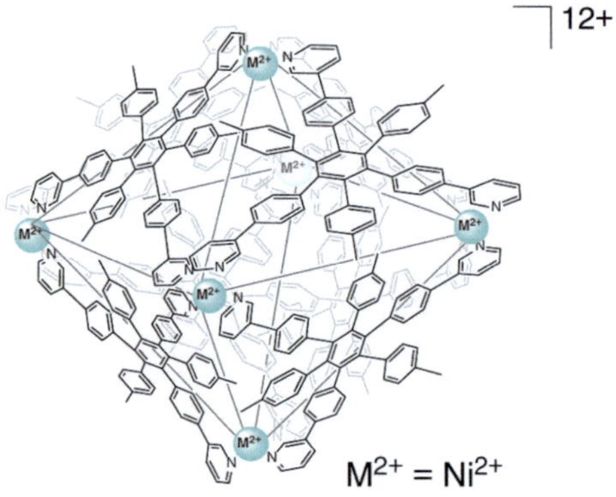

$M^{2+} = Ni^{2+}$

794.

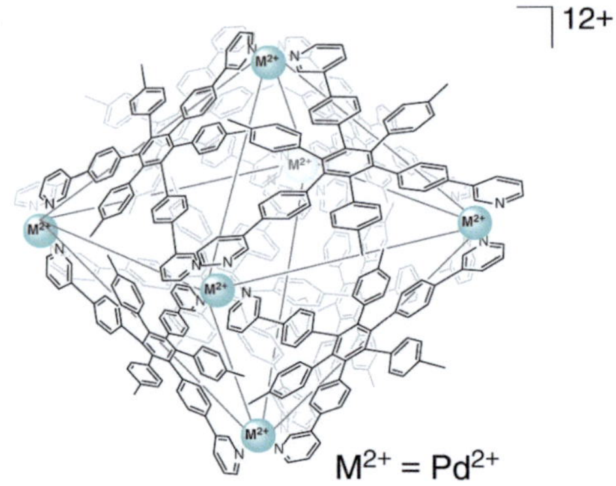

$M^{2+} = Pd^{2+}$

795.

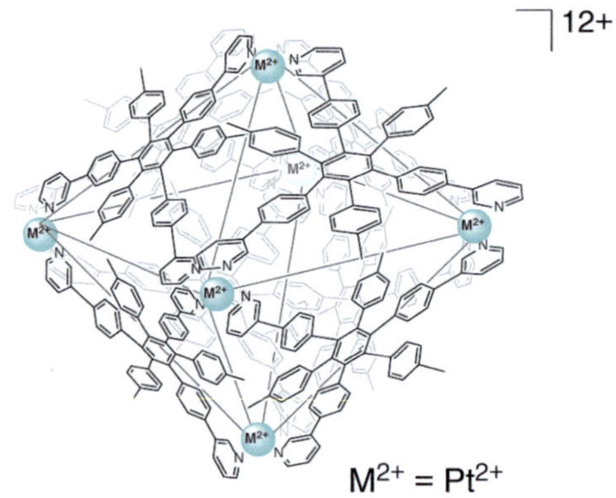

$M^{2+} = Pt^{2+}$

796.

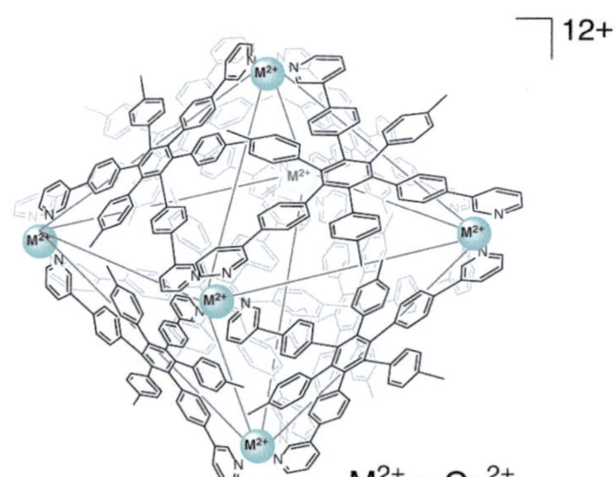

$M^{2+} = Cu^{2+}$

797.

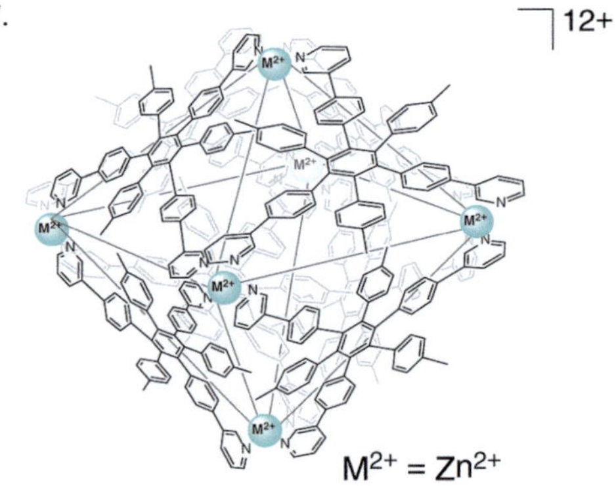

$M^{2+} = Zn^{2+}$

798.

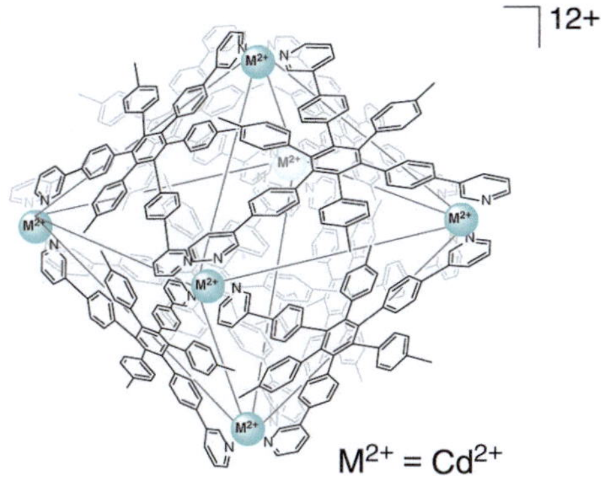

799.

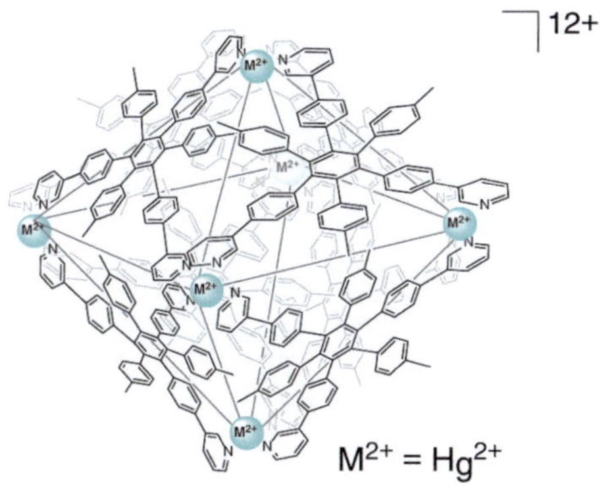

800.

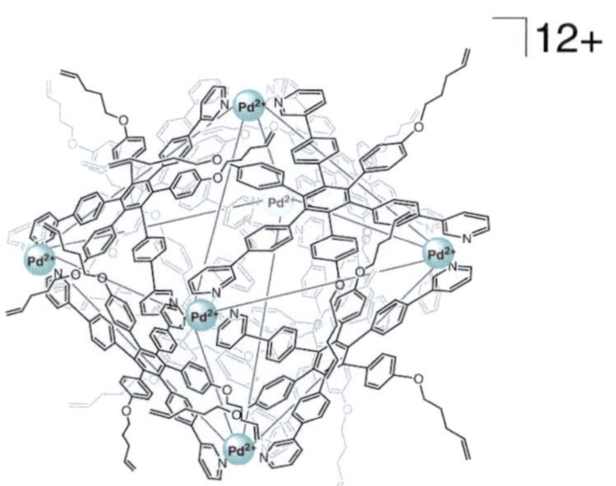

801.

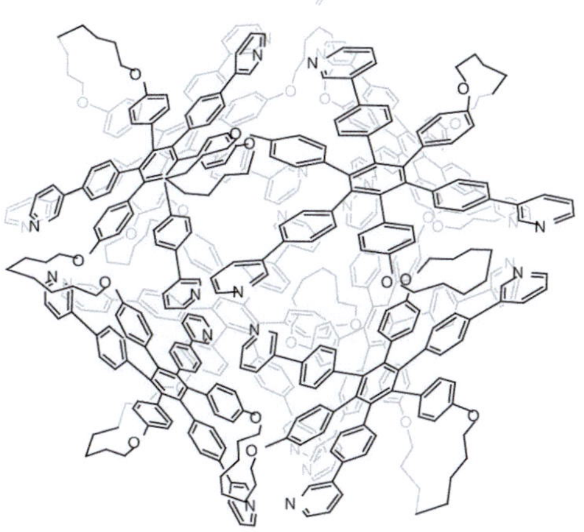

802.

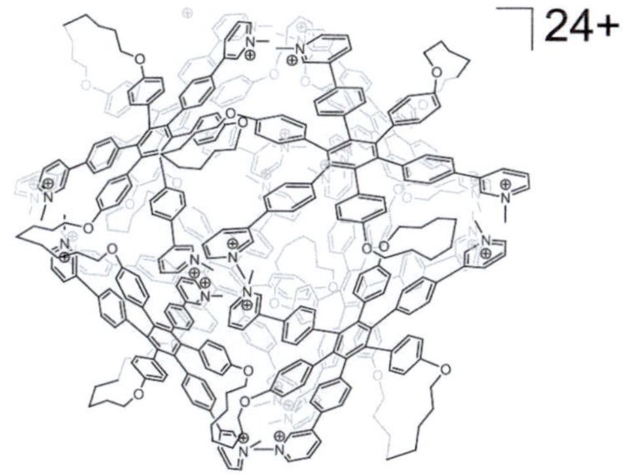

803.

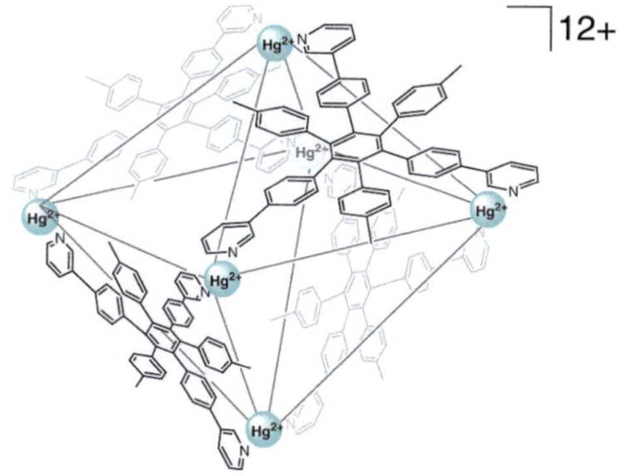

804.

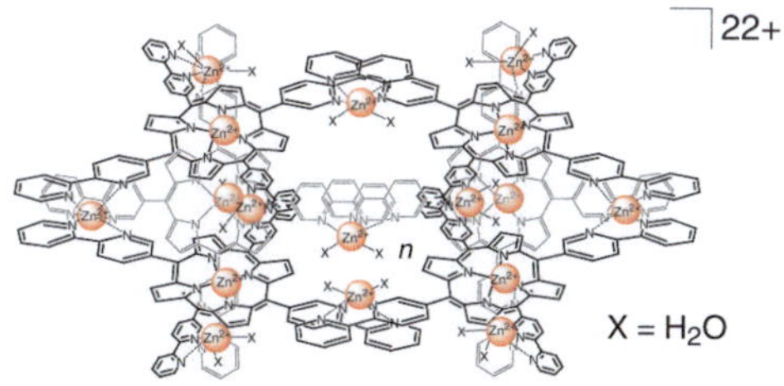

805.

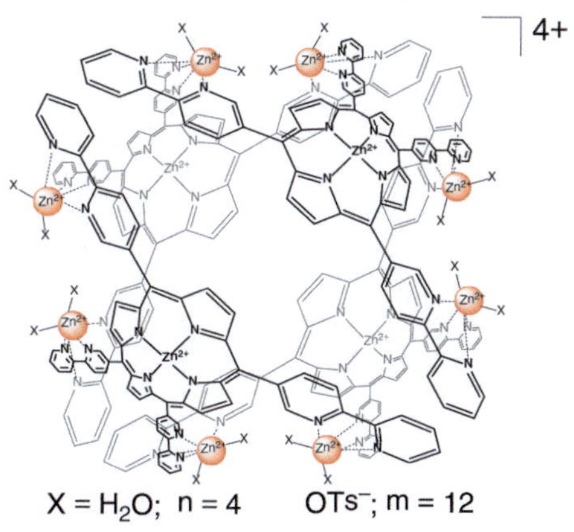

806.

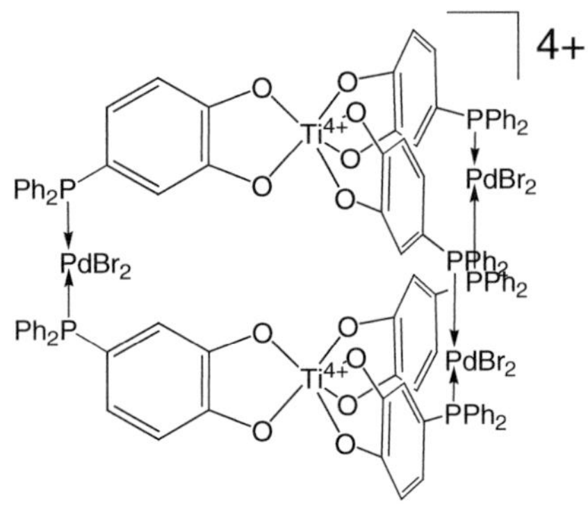

807.

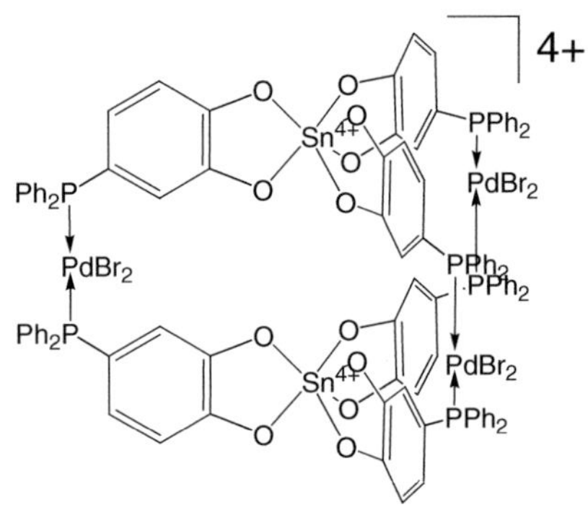

808.

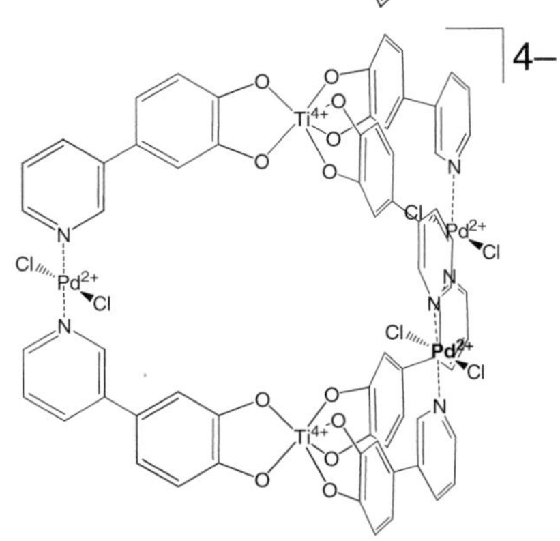

809.

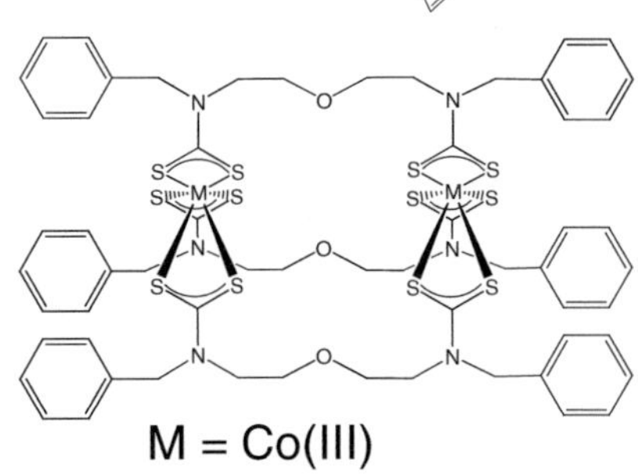

810.

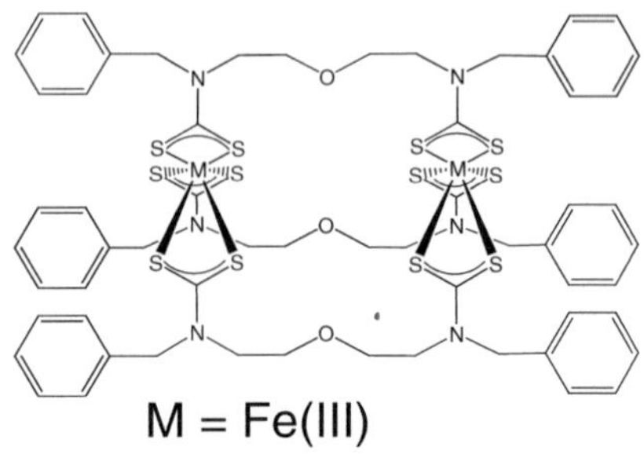

811.

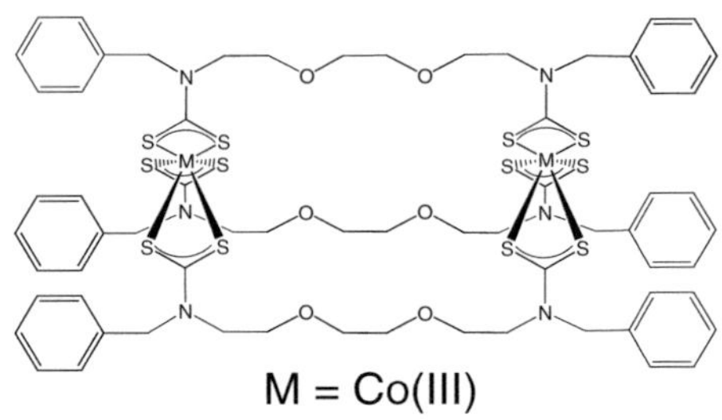

812.

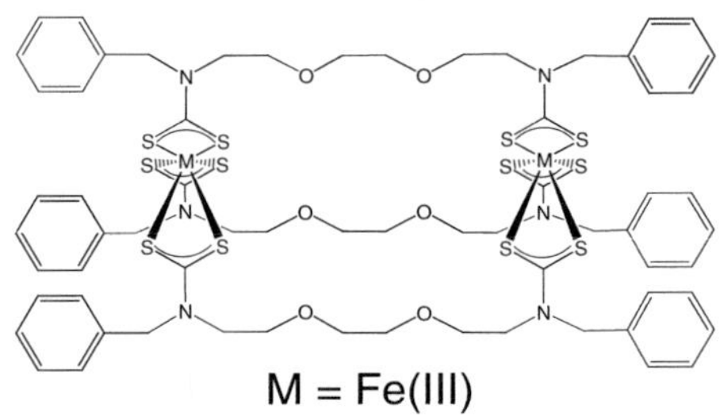

813.

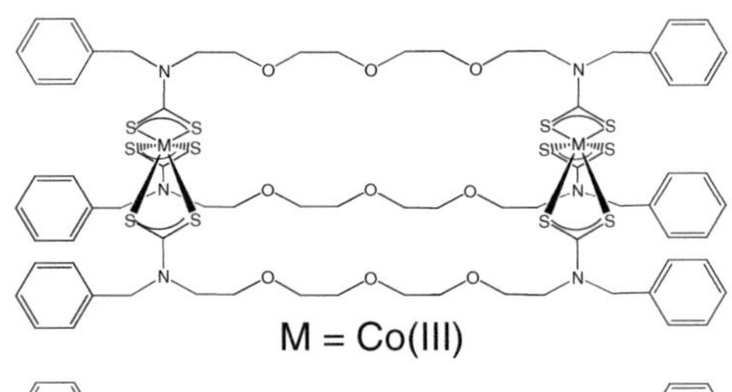

814.

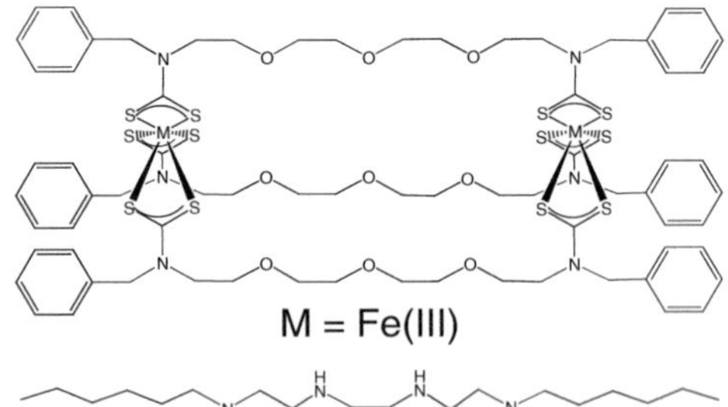

815.

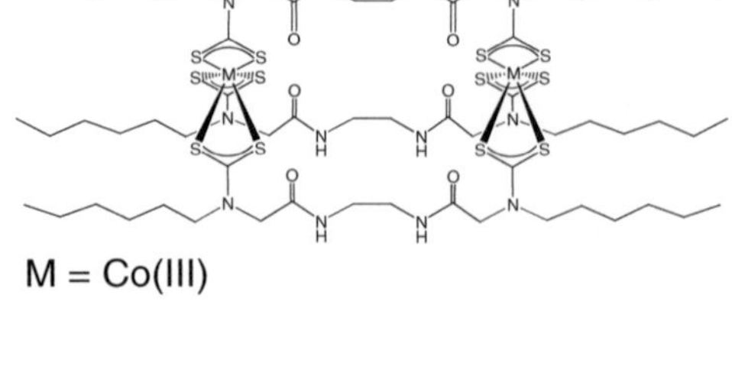

816.

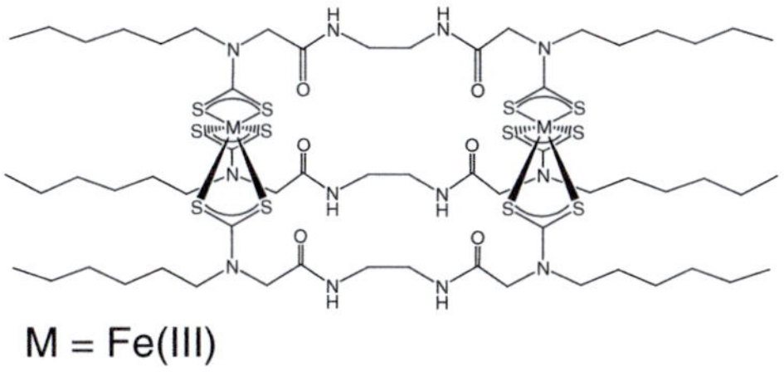

817.

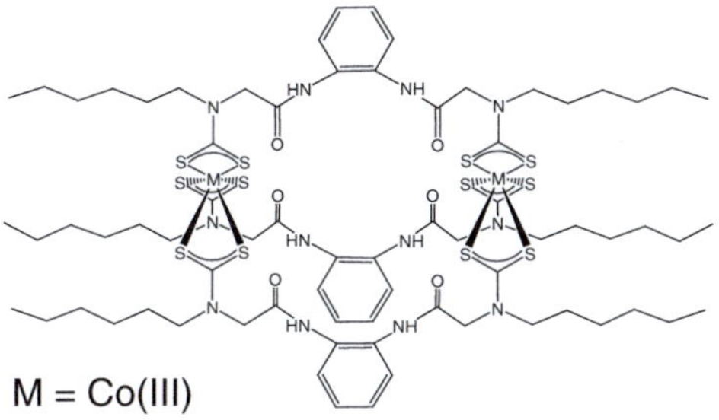

818.

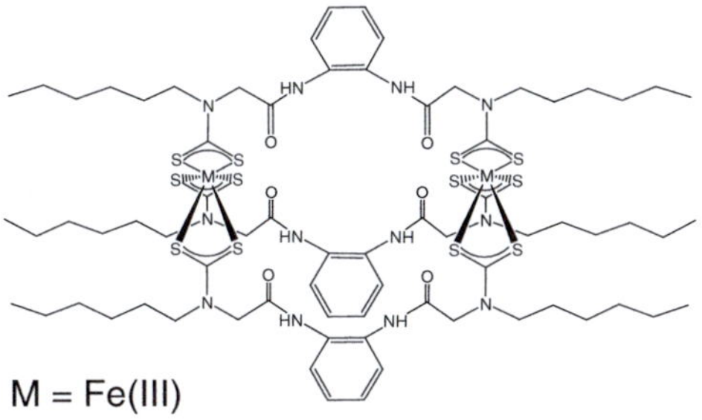

819.

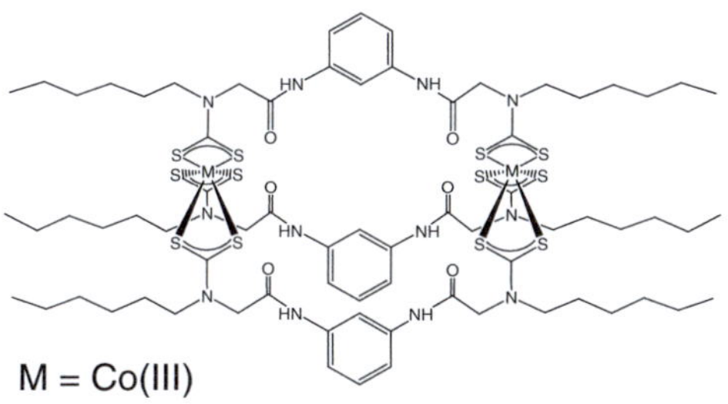

820.

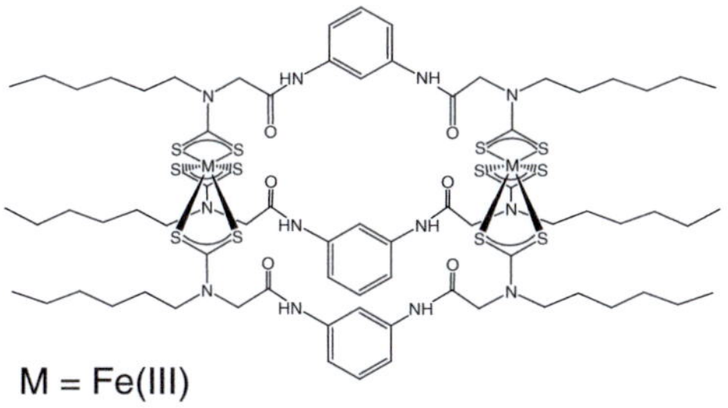

821.

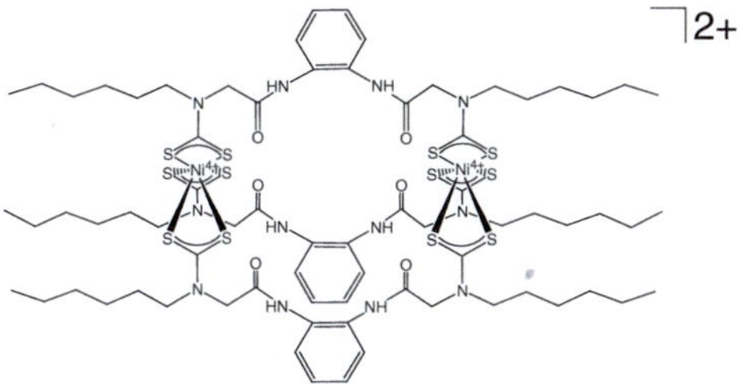

822.

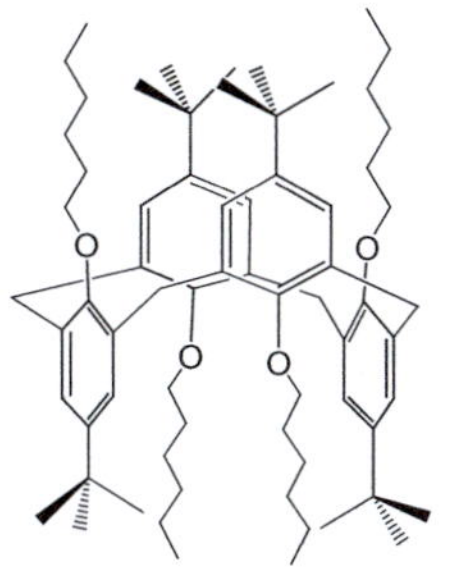

823.

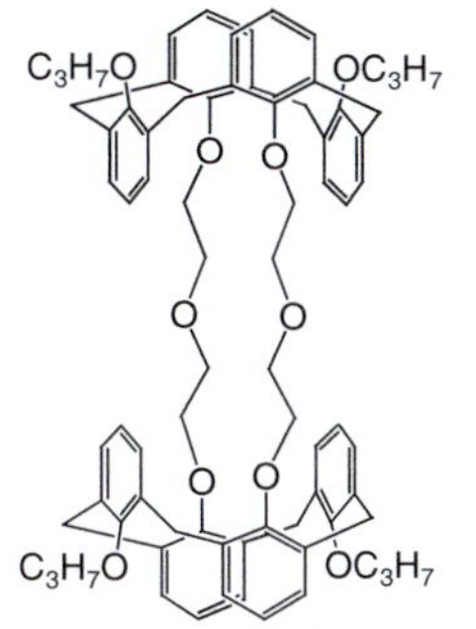

824.

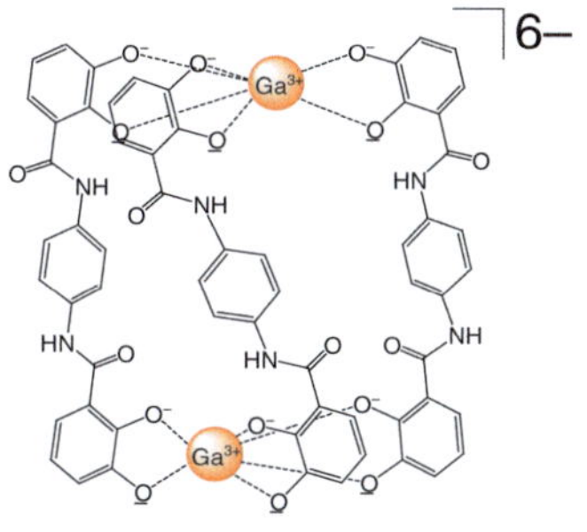

825.

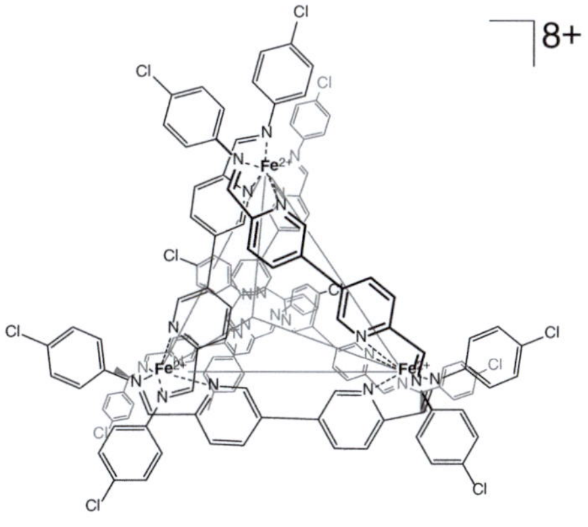

826.

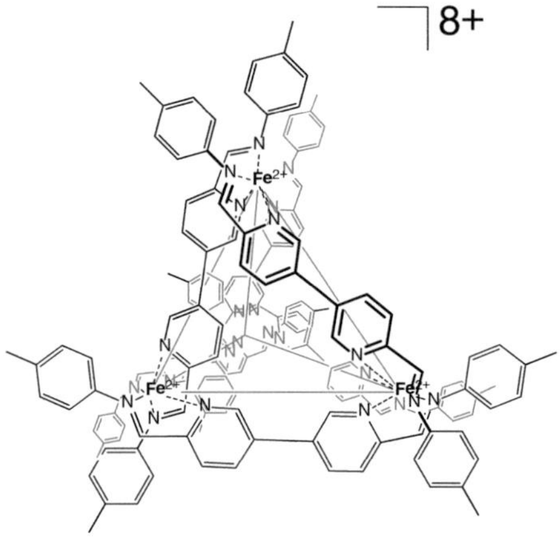

827.

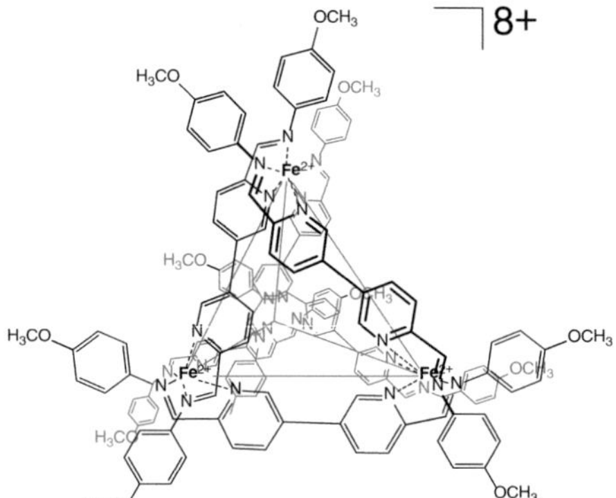

828.

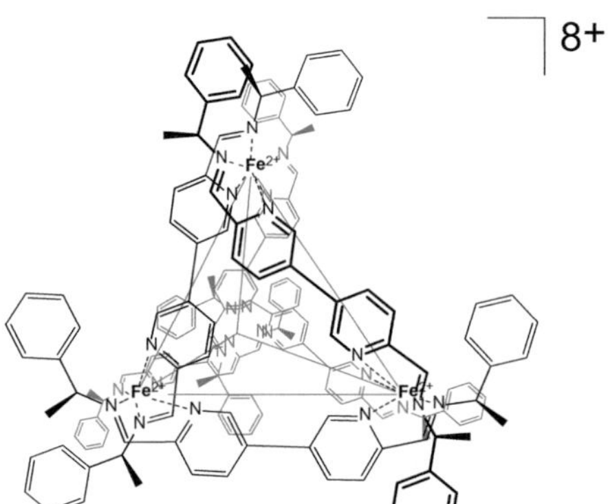

829.

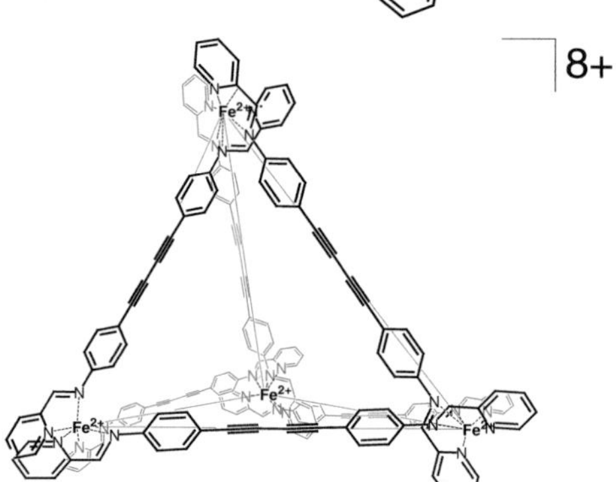

830.

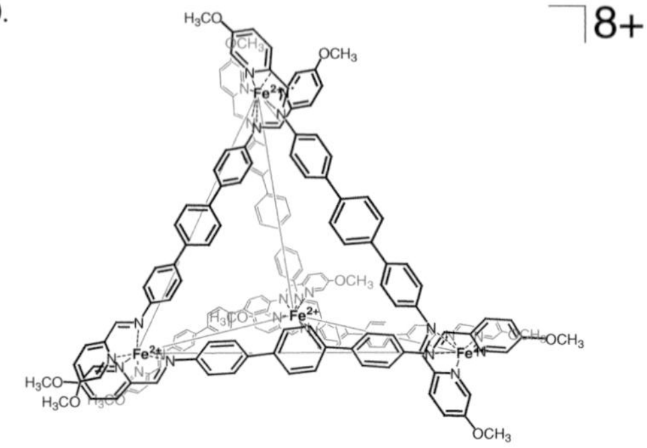

831.

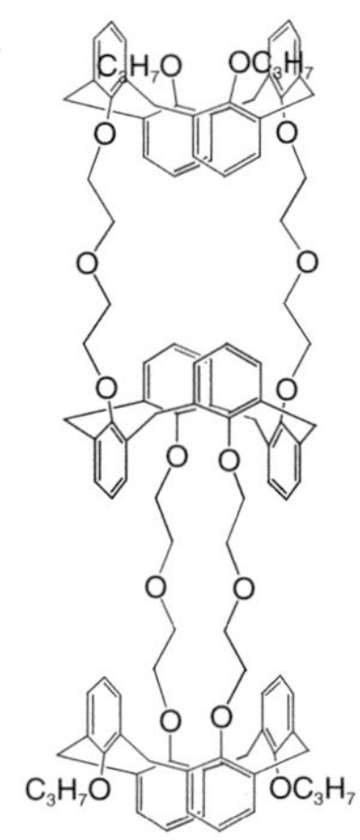

832.

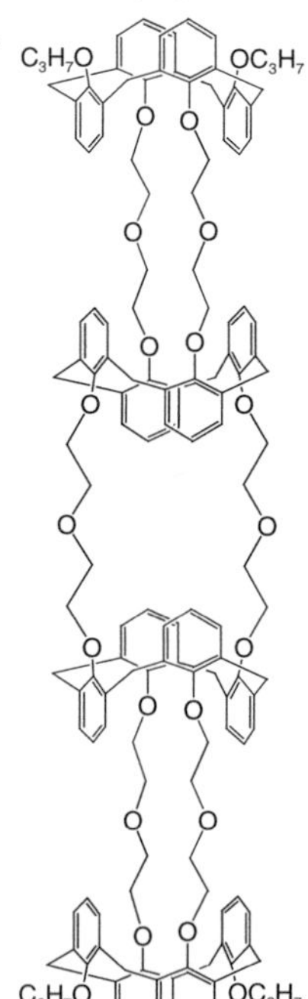

833.

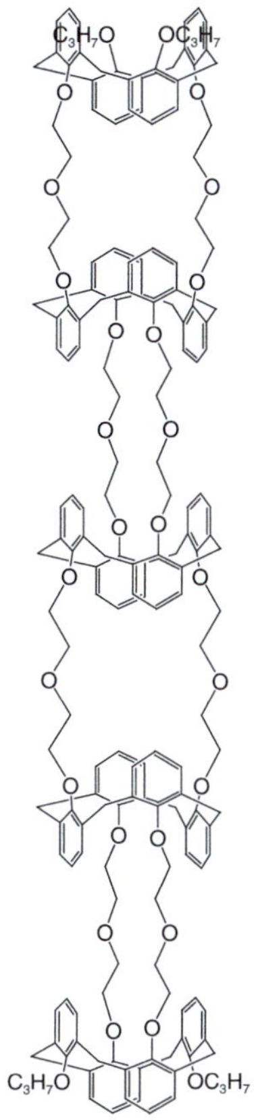

834.

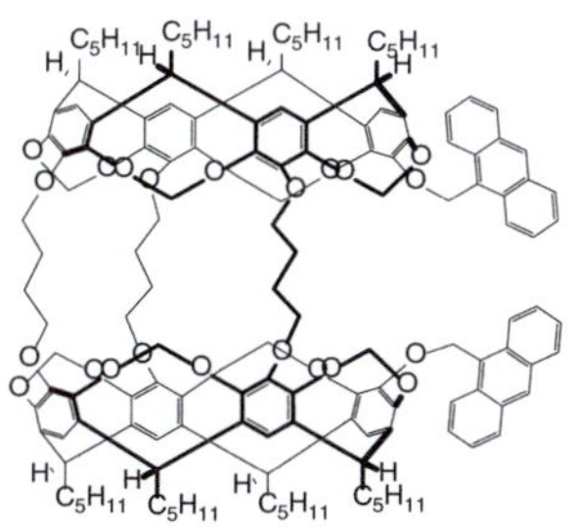

835.

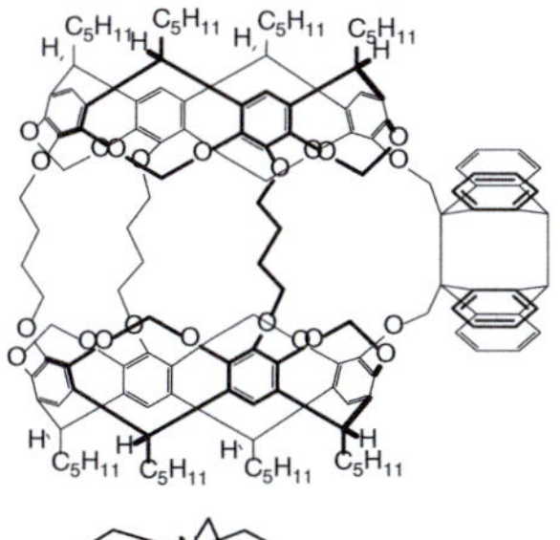

836.

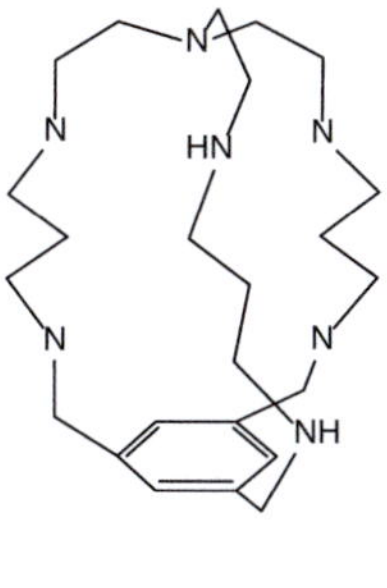

837.

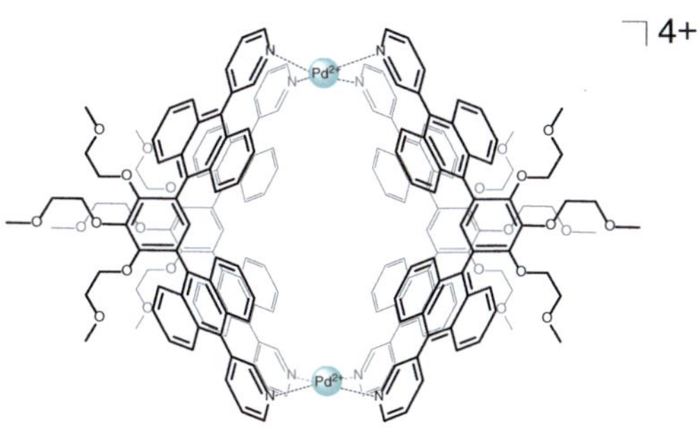

838.

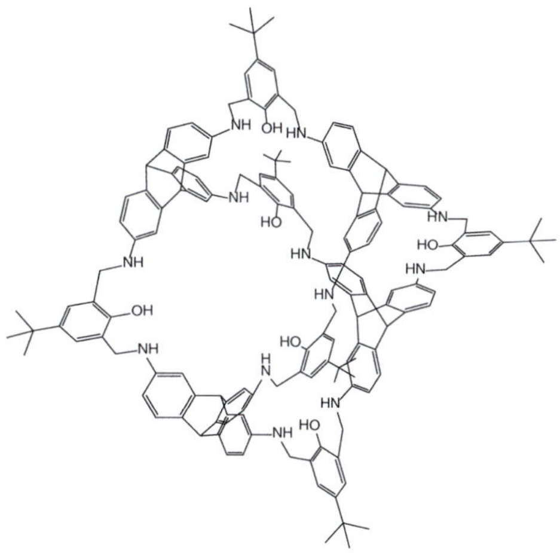

839.

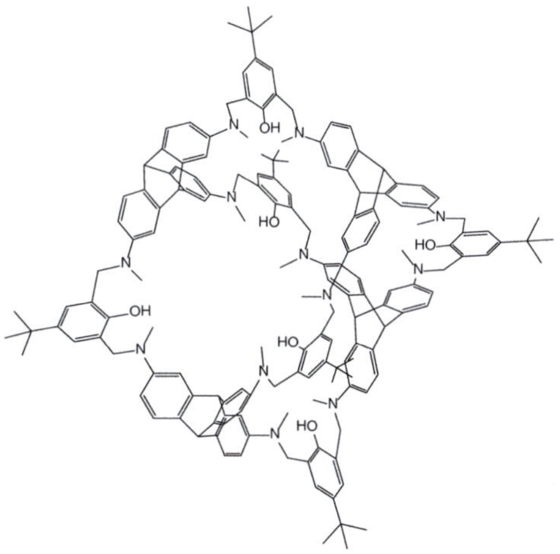

840.

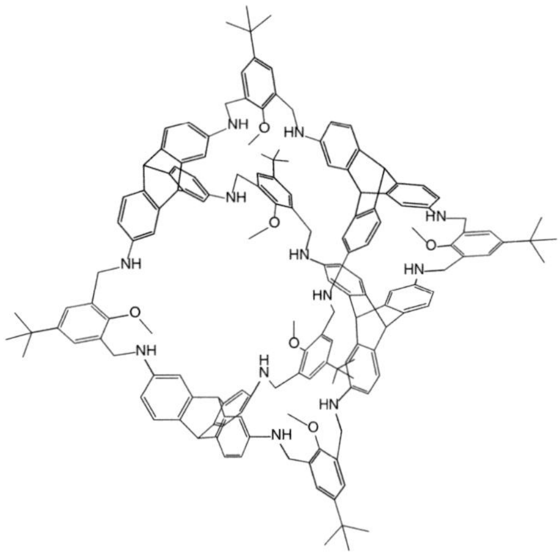

841.

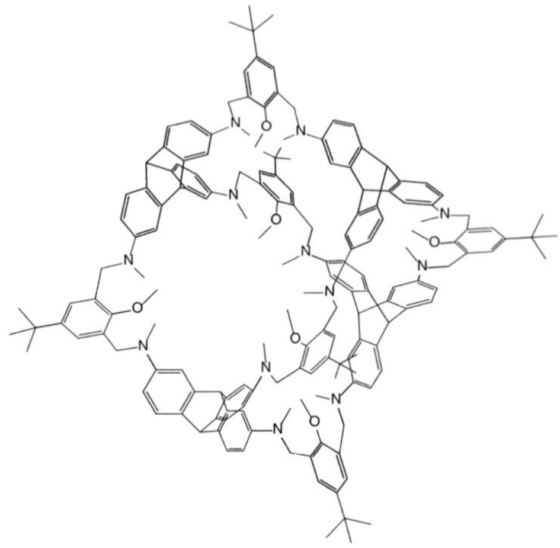

MIX
Papier aus verantwortungsvollen Quellen
Paper from responsible sources
FSC® C105338

If you have any concerns about our products,
you can contact us on
ProductSafety@springernature.com

In case Publisher is established outside the EU,
the EU authorized representative is:
Springer Nature Customer Service Center GmbH
Europaplatz 3, 69115 Heidelberg, Germany

Printed by Libri Plureos GmbH
in Hamburg, Germany